MW00987730

Space Vehicle Dynamics and Control

Second Edition

Space Vehicle Dynamics and Control

Second Edition

Bong Wie
Iowa State University
Ames, Iowa

EDUCATION SERIES
Joseph A. Schetz
Series Editor-in-Chief
Virginia Polytechnic Institute and State University
Blacksburg, Virginia

Published by
American Institute of Aeronautics and Astronautics, Inc.
1801 Alexander Bell Drive, Reston, VA 20191

Cover photo credit: David Murphy, ATK Systems, Goleta, California; Olivier Boisard/ U3P/2006-www.U3P.net.

MATLAB^{TM} is a registered trademark of The MathWorks, Inc.

American Institute of Aeronautics and Astronautics, Inc., Reston, Virginia

1 2 3 4 5

Library of Congress Cataloging-in-Publication Data

Wie, Bong, 1952- -
   Space vehicle dynamics and control / Bong Wie. - - 2nd ed.
     p. cm.
   Includes bibliographical references and index.
   ISBN 978-1-56347-953-3 (alk. paper)
 1. Space vehicles- -Dynamics. 2. Space vehicles- -Control systems. I. Title.

TL1050.W52 2008
629.47$'$1- -dc22

2008028777

Foreword

We are very delighted to present the second edition of *Space Vehicle Dynamics and Control* by Bong Wie of Iowa State University. We are certain that this second edition of his comprehensive and in-depth treatment of this important topic will be very well received by the technical community. The book now has 14 chapters arranged into five parts in over 950 pages, with substantial updated and new material. Part 5, with four chapters, is completely new.

Bong Wie is especially well qualified to write this book because of his broad and deep experience and expertise in the area. His command of the material is excellent, and he is able to organize and present it in a very clear manner.

The AIAA Education Series aims to cover a very broad range of topics in the general aerospace field, including basic theory, applications, and design. A complete list of titles can be found at www.aiaa.org. The philosophy of the series is to develop textbooks that can be used in a university setting, instructional materials for continuing education and professional development courses, and resources that can serve as the basis for independent study or as working references. Suggestions for new topics or authors are always welcome.

Joseph A. Schetz
Editor-in-Chief
AIAA Education Series

Table of Contents

Preface to the Second Edition

The second edition of *Space Vehicle Dynamics and Controls* includes over 260 pages of new material on the recent advances in dynamical modeling and control of advanced spacecraft such as agile imaging satellites equipped with control moment gyros, solar sails, and space solar power satellites. All of the new material contained in the second edition is based on my own or coauthored technical papers published in the *Journal of Guidance, Control, and Dynamics* and/or presented at various technical meetings during the past 10 years. Minor typographical errors of the first edition have been corrected. All of the material of the first edition, except the bibliography, has been retained.

The first edition of this book consisted of four parts. The new Part 5 in the second edition consists of the following four chapters:

Chapter 11   Control Moment Gyros for Agile Imaging Satellites
Chapter 12   Solar-Sail Dynamics and Control
Chapter 13   Solar-Sail Missions for Asteroid Deflection
Chapter 14   Attitude and Orbit Control of Space Solar Power Satellites

Part 5 is a collection of advanced spacecraft control problems and their practical solutions obtained by applying the fundamental principles and techniques emphasized throughout the book.

Chapter 11 presents a comprehensive treatment of the geometric singularity problem inherent to control-moment-gyro (CMG) systems, and it also describes practical CMG steering algorithms and feedback control logic for large-angle, rapid multitarget acquisition and pointing control of agile imaging satellites.

Chapter 12 is concerned with various dynamical modeling and control problems of solar-sail spacecraft, and it also presents the analysis and design of solar-sail attitude control systems for interplanetary solar-sailing missions as well as a solar-sail flight validation mission in a sun-synchronous orbit.

Chapter 13 presents solar-sail mission applications to a complex astrodynamical problem of changing the trajectory of near-Earth objects (NEOs) to mitigate their impact threat to the Earth. This chapter is intended to provide the reader with an overview of such a technically challenging, emerging astrodynamical problem of deflecting NEOs using the nonnuclear alternatives such as the kinetic impactor, gravity tractor, and solar sail.

Chapter 14 presents a preliminary conceptual design of an attitude and orbit control system (AOCS) architecture for a very large, space solar power (SSP) satellite in geostationary orbit.

The solar-sail control research work described in Chapters 12 and 13 was funded by the In-Space Propulsion Technology Program, which was managed by NASA's Science Mission Directorate in Washington, D.C., and implemented by the In-Space Propulsion Technology Office at Marshall Space Flight Center in Huntsville, Alabama. The program objective was to develop in-space propulsion technologies that can enable or benefit near- and mid-term NASA space science missions by significantly reducing cost, mass, or travel times. The research work on the AOCS design for a large geostationary solar-power satellite described in Chapter 14 was funded by the SSP Exploratory Research and Technology Program of NASA.

I would like to thank the following colleagues for their direct and/or indirect contributions to the new material of the second edition: Christopher Heiberg and David Bailey, formerly at Honeywell Space Systems; David Murphy at ATK Space Systems; Vaios Lappas at Surrey Space Centre, University of Surrey; Bernd Dachwald, formerly at German Aerospace Center (DLR); Colin McInnes at University of Strathclyde; Carlos Roithmayr at NASA Langley Research Center; Edward Montgomery, Gregory Garbe, Joan Presson, Andy Heaton, Mark Whorton, and Connie Carrington at NASA Marshall Space Flight Center; and Christopher Moore at NASA Headquarters.

Finally, I wish to thank Tom Shih, Chairperson of the Aerospace Engineering Department at Iowa State University, for providing me with a challenging opportunity to apply space vehicle dynamics and control technology to a complex astrodynamical problem of changing the orbital trajectory of near-Earth objects to mitigate their impact threat to Earth.

Bong Wie
June 2008

Preface to the First Edition

This textbook is intended to provide the reader with a coherent and unified framework for mathematical modeling, analysis, and control of space vehicles. Spacecraft dynamics and control problems of practical and/or theoretical interest are treated from a dynamic systems point of view.

To cover a variety of dynamics and control problems of orbital, rotational, and structural motions of space vehicles, this textbook is organized into four separate parts.

Part 1 (Chapters 1 and 2) provides a comprehensive introduction to dynamic systems modeling, analysis, and control; it provides the necessary background material for the rest of the text. Chapter 1 is intended as a summary of many of the useful results in classical mechanics and dynamic systems theory. Chapter 2, which is somewhat independent of the rest of the text, is concerned with dynamic systems control. In particular, emphasis is placed on both classical and modern robust control of uncertain dynamic systems.

Part 2 (Chapters 3 and 4) is concerned with orbital dynamics and control of space vehicles. Chapter 3 deals with the fundamental problems of orbital mechanics, such as the two-body and restricted three-body problems. Chapter 4 is concerned with orbital maneuvering and control problems. The problems of orbital transfer, rendezvous, and orbit control are treated with special emphasis on a halo orbit determination and control problem.

Part 3 (Chapters 5–7) deals with attitude dynamics and control of rigid spacecraft. Chapters 5 and 6 are concerned with the rotational kinematics and attitude dynamics of rigid spacecraft, respectively. Chapter 7 deals with rotational maneuvering and attitude control problems of rigid spacecraft under the influence of reaction jet firings, internal energy dissipation, or momentum transfer via reaction wheels or control moment gyros.

Part 4 (Chapters 8–10) is concerned with structural dynamics and control of flexible spacecraft. Chapter 8 provides basic physical concepts and mathematical tools necessary for the modeling, analysis, and control of structural dynamic systems in space. Chapter 9 is mainly concerned with the analysis and design of attitude control systems for spacecraft in the presence of structural flexibility and/or propellant sloshing. Active structural vibration control problems are also treated in Chapter 9. Chapter 10 deals with robust fuel- and time-optimal maneuvering control problems of flexible spacecraft and robotic manipulators in the presence of structural modeling uncertainties.

Because of the limited scope of the book, many other important topics, such as orbit and attitude determination, stability of multispin satellites, optimal orbit xvi

transfer, and multiflexible body dynamics, are not treated in detail. However, such interesting topics are discussed briefly at appropriate places throughout the text.

This textbook is intended for use in a variety of undergraduate or graduate courses in dynamics, dynamic systems and control, dynamics and vibration, space-flight dynamics, astrodynamics, spacecraft dynamics and control, and structural vibration control, with proper combinations of the material from the book.

For example, Chapters 1, 3, 5, and 6 are suitable for a one-semester course in dynamics with emphasis on applications of dynamic systems theory. Chapters 1 and 2 contain enough material for a one-semester course in dynamic systems and control at the senior or graduate level. Chapters 3–7 are suitable for a one-semester course in spacecraft dynamics and control at the senior or graduate level; however, control-related topics in these chapters may be omitted for a junior-level course in spacecraft dynamics. Chapters 7–10 (also Secs. 2.6, 2.7, and 4.7) are intended mainly for either a graduate level course or self-study by practicing engineers and researchers who are involved in dynamics and control problems of complex spacecraft.

Prerequisite to reading this text is a thorough knowledge of vector and matrix algebra, calculus, ordinary differential equations, linear system dynamics, and engineering mechanics. Chapter 1 provides a summary of such necessary background material. Some familiarity with structural dynamics and partial differential equations is presumed for studying Chapter 8 on flexible spacecraft dynamics. The reader is also presumed to have some acquaintance with feedback control for studying Chapter 2 on dynamic systems control.

The exercise problems, which appear at appropriate places throughout the text, form an integral part of the text, and they extend the subject matter covered. Hence, the problems should be worked out prior to studying the following sections. Certain problems may require the use of computer software such as MATLAB^{TM} for the analysis, control design, and numerical simulation. This book assumes that the reader has access to computational software such as MATLAB on a personal computer.

References cited in the text are listed at the end of each chapter. A bibliography is provided at the end of the book for those readers who wish to pursue further research in dynamics and control of space vehicles. No attempt was made to provide a complete bibliography nor to provide the original sources of the subject matter. The citation of more of my own technical papers, mostly coauthored with my former graduate students and published in the *Journal of Guidance, Control, and Dynamics*, does not indicate any measure of their relative importance to the subject; they are frequently cited in this book simply because numerous sections of this book are, in fact, based on those papers.

I am indebted to many persons without whose previous work this text would not have been possible. In particular, special thanks go to my former graduate students: Wayne Warren, Tobin Anthony, Marcelo Gonzalez, Evan Wedell, Kuk-Whan Byun, Ravi Sinha, Qiang Liu, David Cielaszyk, and Jianbo Lu for their hardworking, creative research efforts. Their contributions to numerous sections of this text are gratefully acknowledged.

I would like to thank many of my professional colleagues, including Carl Plescia, John Lehner, Nobi Furumoto, Peter Barba, Arun Banerjee, and Peter Chu (at Ford Aerospace and Communications Corporation, Palo Alto, California);

Jason Speyer, David Hull, Roger Broucke, and Heim Weiss (at the University of Texas at Austin); David Schmidt, Karl Bilimoria, and Rafael Livneh (formerly at Arizona State University); Christopher Heiberg and David Bailey (Honeywell Space Systems, Glendale, Arizona); and Srinivas R. Vadali and John Junkins (Texas A & M University) for their direct and indirect contributions to this book. In addition, I would like to thank the many students in my classes at Arizona State University over the past several years. They have patiently read through many drafts of this book and have made numerous corrections.

I am also indebted to John Sunkel, David Geller, Kenneth Cox, Frank Bauer, Harold Frisch, Stanley Carroll, Rudeen Smith-Taylor, and Jerry Newsom at NASA for their support through various research projects. Certainly, this book would not have been possible without such support from NASA. Special thanks also go to the Department of Control Engineering and the Department of Aerospace Engineering at Seoul National University for providing me an opportunity to use my draft text in two graduate courses in dynamics and control during my sabbatical leave in 1995. The effort of writing this book has been in part supported by the Korea Science and Engineering Foundation.

Bong Wie
May 1998

Part 1
Dynamic Systems Modeling, Analysis, and Control

1
Dynamic Systems Modeling and Analysis

This chapter provides mathematical tools and physical concepts that play essential roles in the modeling and analysis of dynamic systems. Fundamental concepts in classical mechanics essential to developing mathematical models of dynamic systems are emphasized. A set of differential or difference equations used to describe a physical system is referred to as the mathematical model of the system; however, no mathematical model of a physical system is exact. This chapter will also acquaint the reader with some of the basic definitions and general results in dynamic systems theory, which are applicable to any dynamic systems that are described by ordinary differential equations. This chapter provides a summary of many of the useful results in classical mechanics and dynamic systems theory.

1.1 Matrix and Vector Analysis

1.1.1 Introduction

It is assumed that the reader has a basic working knowledge of linear algebra as well as vector analysis. This section provides a summary of the basic definitions and fundamental concepts in matrix and vector analysis.*

1.1.1.1 Scalar and Vector A quantity that is characterized by magnitude only is called a scalar; that is, a scalar is simply a single, real number. Typical examples of scalar quantities in dynamic problems are mass, energy, and time. A quantity having both direction and magnitude is called a vector.† Typical vector quantities are velocity, acceleration, force, and moment.

1.1.1.2 Matrix A rectangular array of elements, which may be real numbers, complex numbers, or functions, is called a matrix. A matrix $\mathbf{A}$ of dimension

---

*Matrix algebra, presently of fundamental importance in various fields of engineering and science, was first introduced by Arthur Cayley (1821–1895), whereas three-dimensional vector algebra was developed by John Gibbs (1839–1903).

†The terms "scalar" and "vector" are attributable to William Hamilton (1805–1865).

3

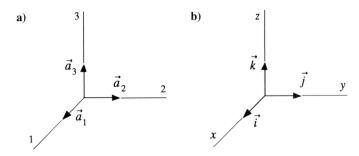

Fig. 1.1 Rectangular Cartesian reference frame with orthogonal basis vectors $\{\vec{a}_1, \vec{a}_2, \vec{a}_3\}$ or $\{\vec{i}, \vec{j}, \vec{k}\}$.

$m \times n$ is denoted by

$$\mathbf{A} = [a_{ij}] = \begin{bmatrix} a_{11} & a_{12} & \cdots & a_{1n} \\ a_{21} & a_{22} & \cdots & a_{2n} \\ \vdots & \vdots & & \vdots \\ a_{m1} & a_{m2} & \cdots & a_{mn} \end{bmatrix}$$

where a_{ij} denotes the ijth element of the matrix $\mathbf{A}$; this matrix has m rows and n columns. If $m = n$, the matrix $\mathbf{A}$ is called a *square matrix* and n is called its *order*. The direction cosine matrix and the inertia matrix are examples of 3×3 square matrices. A rectangular matrix of dimension $n \times 1$ is called a *column matrix*, whereas a rectangular matrix of dimension $1 \times n$ is called a *row matrix*.

1.1.1.3 Three-dimensional Vector Space and Basis Vectors

A reference frame, also called a coordinate system, with a right-hand set of three mutually orthogonal unit vectors, represented in set notation by $\{\vec{a}_1, \vec{a}_2, \vec{a}_3\}$, is illustrated in Fig. 1.1a. Such a set of orthogonal unit vectors forms a *rectangular (orthogonal) Cartesian* coordinate system of a three-dimensional vector space. The three mutually perpendicular lines shown in Fig. 1.1a are called, respectively, the first, second, and third axes of the rectangular Cartesian coordinate system.

A symbol with an overhead arrow indicates a vector in this book, although most books use a boldface symbol to indicate such a vector quantity. (An overbar or underbar is also commonly used to distinguish vector quantities in hand-written material or at the blackboard.) In the literature, a set of unit vectors $\{\vec{i}, \vec{j}, \vec{k}\}$, associated with the x, y, and z axes, is often employed to represent a rectangular Cartesian coordinate system of a three-dimensional vector space, as illustrated in Fig. 1.1b.

In general, sets of vectors $\{\vec{a}_1, \vec{a}_2, \vec{a}_3\}$ or $\{\vec{i}, \vec{j}, \vec{k}\}$, whether the vectors are orthogonal to each other or not, are called *basis vectors* or simply *bases* of a three-dimensional vector space, if they are linearly independent and any vector can be uniquely expressed as a linear combination of $\vec{a}_1$, $\vec{a}_2$, and $\vec{a}_3$. That is, basis vectors are not necessarily of unit magnitude or mutually orthogonal. Consequently, a nonorthogonal Cartesian coordinate system consists of straight-line axes that are not perpendicular to each other. The most familiar examples of non-Cartesian (curvilinear) coordinate systems are the cylindrical and spherical coordinate systems consisting of orthogonal basis vectors. Throughout this text we shall employ a right-hand set of three mutually orthogonal unit vectors as bases of the three-dimensional vector space, unless otherwise stated.

1.1.1.4 Representation of a Vector   Consider an arbitrary vector $\vec{x}$, which is expressed as a linear combination of the basis vectors $\vec{a}_1$, $\vec{a}_2$, and $\vec{a}_3$; i.e., let

$$\vec{x} = x_1\vec{a}_1 + x_2\vec{a}_2 + x_3\vec{a}_3 \equiv \sum_{i=1}^{3} x_i\vec{a}_i \tag{1.1}$$

where the scalar coefficients x_1, x_2, and x_3 are called the *components* of the vector $\vec{x}$. For notational convenience, the summation sign in Eq. (1.1) is sometimes omitted when one adopts the so-called summation convention traditionally used in tensor analysis, in which scalars, vectors, and matrices are often classified as zero-order tensors, first-order tensors, and second-order tensors, respectively. The summation convention and the more general but less familiar tensor notation will not be used in this book, however.

Using matrix notation, we can rewrite Eq. (1.1) as

$$\vec{x} = [\,\vec{a}_1 \quad \vec{a}_2 \quad \vec{a}_3\,] \begin{bmatrix} x_1 \\ x_2 \\ x_3 \end{bmatrix} \tag{1.2}$$

in which we have a somewhat unconventional matrix of the form

$$[\,\vec{a}_1 \quad \vec{a}_2 \quad \vec{a}_3\,]$$

with vectors as its elements. Such a special matrix with vectors as its elements is sometimes called a *vectrix* as suggested in Hughes [1]. Although we may simply call a rectangular array of any elements a matrix, such a vectrix should be distinguished from a standard matrix.

The column matrix in Eq. (1.2) denoted by

$$\mathbf{x} = \begin{bmatrix} x_1 \\ x_2 \\ x_3 \end{bmatrix}$$

is referred to as the *representation* of the vector $\vec{x}$ with respect to the basis vectors $\{\vec{a}_1, \vec{a}_2, \vec{a}_3\}$. The column matrix $\mathbf{x}$ is also frequently referred to as a "column vector," or simply a "vector," in the literature, in which such a column vector is often said to be a tensor of first order. Although a column vector $\mathbf{x}$ should be distinguished from the vector $\vec{x}$ itself, it is also called a vector in this book, though the meaning should be clear from the context and in general we must be clear on what is meant by a vector. Throughout this text, a lowercase boldface symbol indicates a column vector and an uppercase boldface symbol indicates a matrix, unless otherwise stated.

Throughout this text a column vector will also be denoted by

$$\mathbf{x} = (x_1, \ x_2, \ x_3)$$

The transpose of a matrix is denoted by the superscript T, and the transpose of a column vector $\mathbf{x}$ becomes a "row vector" denoted by

$$\mathbf{x}^T = [x_1 \quad x_2 \quad x_3]$$

Similarly, other vectors $\vec{y}$ and $\vec{z}$ can be expressed in terms of basis vectors $\{\vec{a}_1, \vec{a}_2, \vec{a}_3\}$ as follows:

$$\vec{y} = \sum_{i=1}^{3} y_i \vec{a}_i = [y_1 \quad y_2 \quad y_3] \begin{bmatrix} \vec{a}_1 \\ \vec{a}_2 \\ \vec{a}_3 \end{bmatrix} = [\vec{a}_1 \quad \vec{a}_2 \quad \vec{a}_3] \begin{bmatrix} y_1 \\ y_2 \\ y_3 \end{bmatrix}$$

$$\vec{z} = \sum_{i=1}^{3} z_i \vec{a}_i = [z_1 \quad z_2 \quad z_3] \begin{bmatrix} \vec{a}_1 \\ \vec{a}_2 \\ \vec{a}_3 \end{bmatrix} = [\vec{a}_1 \quad \vec{a}_2 \quad \vec{a}_3] \begin{bmatrix} z_1 \\ z_2 \\ z_3 \end{bmatrix}$$

and

$$\mathbf{y} = (y_1, \ y_2, \ y_3) = [y_1 \quad y_2 \quad y_3]^T = \begin{bmatrix} y_1 \\ y_2 \\ y_3 \end{bmatrix}$$

$$\mathbf{z} = (z_1, \ z_2, \ z_3) = [z_1 \quad z_2 \quad z_3]^T = \begin{bmatrix} z_1 \\ z_2 \\ z_3 \end{bmatrix}$$

In the study of dynamics it is often useful to use both vector and matrix notation to represent complex mathematical or physical expressions. Some basic mathematical facts in vector and matrix analysis, which will be used frequently in this text, are summarized as follows.

1.1.2 Matrix Analysis

1.1.2.1 Matrix Addition and Multiplication Let $\mathbf{A} = [a_{ij}]_{m \times n}$ and $\mathbf{B} = [b_{ij}]_{m \times n}$ be $m \times n$ matrices with ijth elements of a_{ij} and b_{ij}, respectively. Then we have

$$\mathbf{A} \pm \mathbf{B} = [a_{ij} \pm b_{ij}]$$

For $\mathbf{A} = [a_{ij}]_{p \times q}$ and $\mathbf{B} = [b_{ij}]_{q \times r}$, we have $\mathbf{AB} = \mathbf{C} = [c_{ij}]_{p \times r}$ where

$$c_{ij} = \sum_{k=1}^{q} a_{ik} b_{kj}$$

If $\mathbf{AB} = \mathbf{BA}$, then $\mathbf{A}$ and $\mathbf{B}$ are said to commute. In general, $\mathbf{AB} \neq \mathbf{BA}$.

1.1.2.2 Transpose of a Matrix The transpose of a matrix $\mathbf{A} = [a_{ij}]$ is denoted by $\mathbf{A}^T$ and $\mathbf{A}^T = [a_{ji}]$. We also have

$$(\mathbf{A} \pm \mathbf{B})^T = \mathbf{A}^T \pm \mathbf{B}^T$$
$$(\mathbf{AB})^T = \mathbf{B}^T \mathbf{A}^T$$

1.1.2.3 Square Matrix A matrix with the same number of columns and rows is called a square matrix. A square matrix is often called a matrix of order n, where n is the number of columns (or rows). A diagonal matrix $\mathbf{A}$ of order n is often denoted by

$$\mathbf{A} = \text{diag}\,(a_{11}, a_{22}, \ldots, a_{nn})$$

An identity or unit matrix is denoted by

$$\mathbf{I} = \text{diag}\,(1, \ldots, 1) = [\delta_{ij}]$$

where δ_{ij} is called the *Kronecker delta* defined as

$$\delta_{ij} = \begin{cases} 1 & \text{if } i = j \\ 0 & \text{if } i \neq j \end{cases}$$

1.1.2.4 Symmetric Matrix A matrix $\mathbf{A}$ is said to be symmetric if $\mathbf{A} = \mathbf{A}^T$. A matrix $\mathbf{A}$ is said to be antisymmetric or skew-symmetric if $\mathbf{A} = -\mathbf{A}^T$. To be precise, we have to state that a matrix $\mathbf{A}$ is said to be symmetric if and only if $\mathbf{A} = \mathbf{A}^T$. That is, mathematical definitions should be understood to be if and only if statements, even though it is customary in mathematics to omit the words "and only if" from definitions. Mathematical theorems are not always if and only if statements, and such brevity is never used for theorems.

1.1.2.5 Determinant of a Matrix Let $\mathbf{A} = [a_{ij}]$ be an $n \times n$ square matrix with the *ij*th element of a_{ij}. Then the determinant of the matrix $\mathbf{A}$, expressed in terms of the cofactors of each element of the *i*th row, is given by

$$|\mathbf{A}| = \sum_{j=1}^{n} a_{ij} c_{ij} = a_{i1} c_{i1} + a_{i2} c_{i2} + \cdots + a_{in} c_{in} \tag{1.3}$$

where c_{ij} is the *ij*th cofactor of $\mathbf{A}$ defined as

$$c_{ij} = (-1)^{i+j} M_{ij}$$

and M_{ij}, called the *ij*th minor, is the determinant of the reduced matrix formed by omitting the *i*th row and the *j*th column of $\mathbf{A}$. For example, we have

$$\begin{vmatrix} x_1 & x_2 & x_3 \\ y_1 & y_2 & y_3 \\ z_1 & z_2 & z_3 \end{vmatrix} = x_1(y_2 z_3 - y_3 z_2) - x_2(y_1 z_3 - y_3 z_1) + x_3(y_1 z_2 - y_2 z_1)$$

Similarly, the determinant of $\mathbf{A}$ may be expressed in terms of the elements of the jth column, as follows:

$$|\mathbf{A}| = \sum_{i=1}^{n} a_{ij}c_{ij} = a_{1j}c_{1j} + a_{2j}c_{2j} + \cdots + a_{nj}c_{nj} \qquad (1.4)$$

A square matrix $\mathbf{A}$ is said to be singular if $|\mathbf{A}| = 0$ and nonsingular if $|\mathbf{A}| \neq 0$. The rank of a matrix is defined as the maximum number of linearly independent columns or rows. An $n \times n$ square matrix with rank less than n is a singular matrix. Note that $|\mathbf{A}| = |\mathbf{A}^T|$ and $|\mathbf{A}\mathbf{B}| = |\mathbf{A}||\mathbf{B}|$ if $\mathbf{A}$ and $\mathbf{B}$ are square matrices. Also note that $|-\mathbf{A}| = (-1)^n |\mathbf{A}|$ where n is the order of the matrix $\mathbf{A}$.

1.1.2.6 Principal Minors of a Square Matrix
The *principal minors* of, for example, a 3×3 matrix $\mathbf{A}$ are

$$a_{11}, a_{22}, a_{33}, \begin{vmatrix} a_{11} & a_{12} \\ a_{21} & a_{22} \end{vmatrix}, \begin{vmatrix} a_{11} & a_{13} \\ a_{31} & a_{33} \end{vmatrix}, \begin{vmatrix} a_{22} & a_{23} \\ a_{32} & a_{33} \end{vmatrix}, \begin{vmatrix} a_{11} & a_{12} & a_{13} \\ a_{21} & a_{22} & a_{23} \\ a_{31} & a_{32} & a_{33} \end{vmatrix}$$

and the *leading principal minors* of a 3×3 matrix $\mathbf{A}$, denoted by $\Delta_i(\mathbf{A})$, are

$$\Delta_1 = a_{11}, \qquad \Delta_2 = \begin{vmatrix} a_{11} & a_{12} \\ a_{21} & a_{22} \end{vmatrix}, \qquad \Delta_3 = \begin{vmatrix} a_{11} & a_{12} & a_{13} \\ a_{21} & a_{22} & a_{23} \\ a_{31} & a_{32} & a_{33} \end{vmatrix}$$

1.1.2.7 Trace of a Matrix
The trace of a square matrix $\mathbf{A} = [a_{ij}]$ of order n is defined to be the sum of the elements of the main diagonal of $\mathbf{A}$; i.e.,

$$\mathrm{tr}\,(\mathbf{A}) = \sum_{i=1}^{n} a_{ii} \qquad (1.5)$$

1.1.2.8 Inverse of a Matrix
The inverse of a nonsingular square matrix $\mathbf{A}$, denoted by $\mathbf{A}^{-1}$, is defined as

$$\mathbf{A}^{-1} = \frac{\mathrm{adj}\,\mathbf{A}}{|\mathbf{A}|} \qquad \text{if } |\mathbf{A}| \neq 0 \qquad (1.6)$$

where the adjoint matrix of $\mathbf{A}$, denoted by adj $\mathbf{A}$, is defined as

$$\mathrm{adj}\,\mathbf{A} = [c_{ij}]^T = \left[(-1)^{i+j}M_{ij}\right]^T$$

where c_{ij} and M_{ij} are, respectively, the ijth cofactor and minor of $\mathbf{A}$.

If $\mathbf{A}$ and $\mathbf{B}$ are nonsingular matrices, then

$$(\mathbf{A}\mathbf{B})^{-1} = \mathbf{B}^{-1}\mathbf{A}^{-1}$$

By definition, we have

$$\mathbf{A}^{-1}\mathbf{A} = \mathbf{A}\mathbf{A}^{-1} = \mathbf{I}$$

where $\mathbf{I}$ is an identity matrix. Also, we have

$$(\mathbf{A}^T)^{-1} = (\mathbf{A}^{-1})^T = \mathbf{A}^{-T}$$

1.1.2.9 Orthonormal Matrix   A square matrix $\mathbf{A}$ is called an *orthogonal* matrix if $\mathbf{A}\mathbf{A}^T$ is a diagonal matrix, and it is called an *orthonormal* matrix if $\mathbf{A}\mathbf{A}^T$ is an identity matrix. For an orthonormal matrix $\mathbf{A}$, we have $\mathbf{A}^{-1} = \mathbf{A}^T$ and $|\mathbf{A}| = \pm 1$. Note that if $\mathbf{A}\mathbf{A}^T$ is a diagonal matrix, then $\mathbf{A}\mathbf{A}^T = \mathbf{A}^T\mathbf{A}$.

1.1.2.10 Cramer's Rule   Given $\mathbf{A}\mathbf{x} = \mathbf{b}$ where, for example,

$$\mathbf{A} = \begin{bmatrix} a_{11} & a_{12} & a_{13} \\ a_{21} & a_{22} & a_{23} \\ a_{31} & a_{32} & a_{33} \end{bmatrix}, \quad \mathbf{x} = \begin{bmatrix} x_1 \\ x_2 \\ x_3 \end{bmatrix}, \quad \mathbf{b} = \begin{bmatrix} b_1 \\ b_2 \\ b_3 \end{bmatrix}$$

and if $|\mathbf{A}| \neq 0$, a unique solution for $\mathbf{x}$ can be obtained as

$$x_1 = \frac{\begin{vmatrix} b_1 & a_{12} & a_{13} \\ b_2 & a_{22} & a_{23} \\ b_3 & a_{32} & a_{33} \end{vmatrix}}{|\mathbf{A}|}, \quad x_2 = \frac{\begin{vmatrix} a_{11} & b_1 & a_{13} \\ a_{21} & b_2 & a_{23} \\ a_{31} & b_3 & a_{33} \end{vmatrix}}{|\mathbf{A}|}, \quad x_3 = \frac{\begin{vmatrix} a_{11} & a_{12} & b_1 \\ a_{21} & a_{22} & b_2 \\ a_{31} & a_{32} & b_3 \end{vmatrix}}{|\mathbf{A}|} \quad (1.7)$$

where

$$|\mathbf{A}| = \begin{vmatrix} a_{11} & a_{12} & a_{13} \\ a_{21} & a_{22} & a_{23} \\ a_{31} & a_{32} & a_{33} \end{vmatrix}$$

1.1.2.11 Inner Product   Given two column vectors $\mathbf{x} = (x_1, x_2, x_3)$ and $\mathbf{y} = (y_1, y_2, y_3)$, the inner or scalar product of $\mathbf{x}$ and $\mathbf{y}$ is defined as

$$\mathbf{x}^T\mathbf{y} = [x_1 \quad x_2 \quad x_3] \begin{bmatrix} y_1 \\ y_2 \\ y_3 \end{bmatrix} = x_1 y_1 + x_2 y_2 + x_3 y_3 = \sum_{i=1}^{3} x_i y_i \quad (1.8)$$

1.1.2.12 Outer Product   The outer product of two column vectors $\mathbf{x}$ and $\mathbf{y}$ is defined as

$$\mathbf{x}\mathbf{y}^T = \begin{bmatrix} x_1 \\ x_2 \\ x_3 \end{bmatrix} [y_1 \quad y_2 \quad y_3] = \begin{bmatrix} x_1 y_1 & x_1 y_2 & x_1 y_3 \\ x_2 y_1 & x_2 y_2 & x_2 y_3 \\ x_3 y_1 & x_3 y_2 & x_3 y_3 \end{bmatrix} \quad (1.9)$$

1.1.2.13 Cross Product   The cross product of two column vectors $\mathbf{x}$ and $\mathbf{y}$ is defined as

$$\mathbf{x} \times \mathbf{y} \equiv \begin{bmatrix} 0 & -x_3 & x_2 \\ x_3 & 0 & -x_1 \\ -x_2 & x_1 & 0 \end{bmatrix} \begin{bmatrix} y_1 \\ y_2 \\ y_3 \end{bmatrix} = \begin{bmatrix} x_2 y_3 - x_3 y_2 \\ x_3 y_1 - x_1 y_3 \\ x_1 y_2 - x_2 y_1 \end{bmatrix} \quad (1.10)$$

The cross product of two column vectors is defined only for the three-dimensional vector space, although the inner or outer product of two column vectors is defined in general for the n-dimensional vector space.

1.1.2.14 Eigenvalues and Eigenvectors   Let $\mathbf{A}$ be an $n \times n$ square matrix. Then a scalar λ is called an *eigenvalue* of $\mathbf{A}$ if there exists a nonzero $n \times 1$ column vector $\mathbf{e}$ such that

$$[\lambda \mathbf{I} - \mathbf{A}]\,\mathbf{e} = 0 \qquad (1.11)$$

where $\mathbf{I}$ is an identity matrix of order n. The eigenvalues are the roots of the nth-order characteristic polynomial equation:

$$|\lambda \mathbf{I} - \mathbf{A}| = \lambda^n + a_1 \lambda^{n-1} + \cdots + a_{n-1}\lambda + a_n = 0 \qquad (1.12)$$

where $a_1, \ldots, a_n$ are scalar coefficients. Any nonzero vector $\mathbf{e}_i$ satisfying

$$[\lambda_i \mathbf{I} - \mathbf{A}]\,\mathbf{e}_i = 0$$

is called the ith *eigenvector* of $\mathbf{A}$ associated with the ith eigenvalue λ_i. Note that $\lambda_i(\mathbf{A}^T) = \lambda_i(\mathbf{A})$ and

$$\lambda_i(\mathbf{A}^{-1}) = \frac{1}{\lambda_i(\mathbf{A})}$$

1.1.2.15 Norm of a Vector   Given an n-dimensional vector $\mathbf{x}$, the Euclidean norm of $\mathbf{x}$ is defined as

$$\|\mathbf{x}\| \equiv \sqrt{\mathbf{x}^T \mathbf{x}} = \sqrt{x_1^2 + \cdots + x_n^2}$$

which is also called the two-norm or the spectral norm of $\mathbf{x}$.

1.1.2.16 Sign-definiteness of a Scalar Function of a Vector   A scalar function of a vector, $f(\mathbf{x})$, is said to be 1) positive definite if $f(\mathbf{x}) > 0$ for all nonzero $\mathbf{x}$ and $f(0) = 0$, 2) positive semidefinite if $f(\mathbf{x}) \geq 0$ for all nonzero $\mathbf{x}$ and $f(0) = 0$, 3) negative definite if $f(\mathbf{x}) < 0$ for all nonzero $\mathbf{x}$ and $f(0) = 0$, and 4) negative semidefinite if $f(\mathbf{x}) \leq 0$ for all nonzero $\mathbf{x}$ and $f(0) = 0$. Otherwise, it is said to be indefinite. The sign-definiteness of $f(\mathbf{x})$ can also be defined, in general, for all $\mathbf{x} \neq \mathbf{x}^*$ such that $f(\mathbf{x}^*) = 0$ where $\mathbf{x}^*$ is a nonzero constant.

1.1.2.17 Sign-definiteness of a Matrix   An $n \times n$ square matrix $\mathbf{A}$ is said to be 1) positive definite if $f(\mathbf{x}) = \mathbf{x}^T \mathbf{A} \mathbf{x} > 0$ for all nonzero $\mathbf{x}$, 2) positive semidefinite if $f(\mathbf{x}) = \mathbf{x}^T \mathbf{A} \mathbf{x} \geq 0$ for all nonzero $\mathbf{x}$, 3) negative definite if $f(\mathbf{x}) = \mathbf{x}^T \mathbf{A} \mathbf{x} < 0$ for all nonzero $\mathbf{x}$, and 4) negative semidefinite if $f(\mathbf{x}) = \mathbf{x}^T \mathbf{A} \mathbf{x} \leq 0$ for all nonzero $\mathbf{x}$. Otherwise, it is said to be indefinite.

The scalar function

$$f(\mathbf{x}) = \mathbf{x}^T \mathbf{A} \mathbf{x} = \sum_{i=1}^{n} \sum_{j=1}^{n} a_{ij} x_i x_j$$

is called a *quadratic form* of $\mathbf{A}$. Because any square matrix $\mathbf{A}$ can be decomposed into

$$\mathbf{A} = \underbrace{\tfrac{1}{2}\left[\mathbf{A} + \mathbf{A}^T\right]}_{\text{symmetric}} + \underbrace{\tfrac{1}{2}\left[\mathbf{A} - \mathbf{A}^T\right]}_{\text{antisymmetric}} \qquad (1.13)$$

and $\mathbf{x}^T\mathbf{B}\mathbf{x} = 0$ for any antisymmetric matrix $\mathbf{B}$, the definiteness of a matrix is the same as that of its symmetric part.

All of the eigenvalues of a symmetric matrix are real, and the definiteness of a symmetric matrix $\mathbf{A}$ can be tested by checking the signs of the eigenvalues $\lambda_i(\mathbf{A})$ or the signs of the leading principal minors $\Delta_i(\mathbf{A})$; i.e., an $n \times n$ symmetric matrix $\mathbf{A}$ is said to be 1) positive definite if all $\lambda_i > 0$ or if all $\Delta_i > 0$; 2) positive semidefinite if all $\lambda_i \geq 0$ or if $\Delta_1 \geq 0$, $\Delta_2 \geq 0, \ldots, \Delta_{n-1} \geq 0$, $\Delta_n = 0$; 3) negative definite if all $\lambda_i < 0$ or if $\Delta_1 < 0$, $\Delta_2 > 0$, $\Delta_3 < 0, \ldots, (-1)^n\Delta_n > 0$; and 4) negative semidefinite if all $\lambda_i \leq 0$ or if $\Delta_1 \leq 0$, $\Delta_2 \geq 0$, $\Delta_3 \leq 0, \ldots,$ $\Delta_n = 0$.

1.1.2.18 Differentiation of Matrices The time derivative of a matrix $\mathbf{A}(t) \equiv [a_{ij}(t)]$ is defined as

$$\frac{\mathrm{d}}{\mathrm{d}t}\mathbf{A}(t) = \left[\frac{\mathrm{d}}{\mathrm{d}t}a_{ij}(t)\right]$$

and we have

$$\frac{\mathrm{d}}{\mathrm{d}t}[\mathbf{A}(t)\mathbf{B}(t)] = \frac{\mathrm{d}\mathbf{A}(t)}{\mathrm{d}t}\mathbf{B}(t) + \mathbf{A}(t)\frac{\mathrm{d}\mathbf{B}(t)}{\mathrm{d}t}$$

1.1.2.19 Differentiation of a Scalar Function of a Vector If $f(\mathbf{x})$ is a scalar function of a vector $\mathbf{x}$, which is itself a function of time t, then we write[†]

$$\frac{\mathrm{d}f}{\mathrm{d}t} = \frac{\partial f}{\partial \mathbf{x}}\frac{\mathrm{d}\mathbf{x}}{\mathrm{d}t} \qquad (1.14)$$

For example, if $\mathbf{x} = (x_1, x_2, x_3)$, then

$$\frac{\mathrm{d}f}{\mathrm{d}t} = \frac{\partial f}{\partial x_1}\frac{\mathrm{d}x_1}{\mathrm{d}t} + \frac{\partial f}{\partial x_2}\frac{\mathrm{d}x_2}{\mathrm{d}t} + \frac{\partial f}{\partial x_3}\frac{\mathrm{d}x_3}{\mathrm{d}t}$$

$$= [\partial f/\partial x_1 \quad \partial f/\partial x_2 \quad \partial f/\partial x_3]\begin{bmatrix} \dot{x}_1 \\ \dot{x}_2 \\ \dot{x}_3 \end{bmatrix}$$

where $\dot{x}_i \equiv \mathrm{d}x_i/\mathrm{d}t$. That is, we have

$$\frac{\partial f}{\partial \mathbf{x}} \equiv [\partial f/\partial x_1 \quad \partial f/\partial x_2 \quad \partial f/\partial x_3]$$

$$\frac{\mathrm{d}\mathbf{x}}{\mathrm{d}t} \equiv \dot{\mathbf{x}} = \begin{bmatrix} \dot{x}_1 \\ \dot{x}_2 \\ \dot{x}_3 \end{bmatrix}$$

[†]Throughout this book, most functions are assumed to be differentiable, unless otherwise stated.

The row vector $\partial f/\partial \mathbf{x}$ is called the *gradient vector* of the scalar function $f(\mathbf{x})$. If the gradient vector is defined to be a column vector, then we write

$$\frac{df}{dt} = \left[\frac{\partial f}{\partial \mathbf{x}}\right]^T \frac{d\mathbf{x}}{dt}$$

where $\partial f/\partial \mathbf{x} \equiv (\partial f/\partial x_1, \partial f/\partial x_2, \partial f/\partial x_3)$.

1.1.2.20 Taylor Series Expansion of a Scalar Function of a Vector   The Taylor series expansion of $f(\mathbf{x})$ about $\mathbf{x}^*$ is given by

$$f(\mathbf{x}) = f(\mathbf{x}^*) + \frac{\partial f}{\partial \mathbf{x}}\bigg|_{\mathbf{x}^*} [\mathbf{x} - \mathbf{x}^*] + \frac{1}{2}[\mathbf{x} - \mathbf{x}^*]^T \frac{\partial^2 f}{\partial \mathbf{x}^2}\bigg|_{\mathbf{x}^*} [\mathbf{x} - \mathbf{x}^*] + \cdots \quad (1.15)$$

where

$$\frac{\partial^2 f}{\partial \mathbf{x}^2} \equiv \frac{\partial}{\partial \mathbf{x}}\left[\frac{\partial f}{\partial \mathbf{x}}\right]^T \equiv \left[\frac{\partial^2 f}{\partial x_i \partial x_j}\right]$$

is called the *Hessian matrix*. All the partial derivatives in Eq. (1.15) are to be evaluated at $\mathbf{x} = \mathbf{x}^*$.

1.1.2.21 Differentiation of a Vector Function   If $\mathbf{f}(\mathbf{x})$ is a vector function of a vector $\mathbf{x}$, which is itself a function of t, then we have

$$\frac{d\mathbf{f}}{dt} = \frac{\partial \mathbf{f}}{\partial \mathbf{x}} \frac{d\mathbf{x}}{dt} \quad (1.16)$$

where $\partial \mathbf{f}/\partial \mathbf{x}$ denotes a matrix with the ijth element of $\partial f_i/\partial x_j$; that is,

$$\frac{\partial \mathbf{f}}{\partial \mathbf{x}} \equiv \left[\frac{\partial f_i}{\partial x_j}\right]$$

which is called the *Jacobian matrix*. For example, if $\mathbf{f} = (f_1, f_2, f_3)$ and $\mathbf{x} = (x_1, x_2, x_3)$, then we have

$$\begin{bmatrix} \dot{f}_1 \\ \dot{f}_2 \\ \dot{f}_3 \end{bmatrix} = \begin{bmatrix} \partial f_1/\partial x_1 & \partial f_1/\partial x_2 & \partial f_1/\partial x_3 \\ \partial f_2/\partial x_1 & \partial f_2/\partial x_2 & \partial f_2/\partial x_3 \\ \partial f_3/\partial x_1 & \partial f_3/\partial x_2 & \partial f_3/\partial x_3 \end{bmatrix} \begin{bmatrix} \dot{x}_1 \\ \dot{x}_2 \\ \dot{x}_3 \end{bmatrix}$$

1.1.2.22 Divergence of a Vector Function   The divergence of $\mathbf{f}(\mathbf{x})$ is defined as

$$\text{div } \mathbf{f}(\mathbf{x}) \equiv \text{tr}\left[\frac{\partial \mathbf{f}}{\partial \mathbf{x}}\right] \quad (1.17)$$

For example, if $\mathbf{f}(\mathbf{x}) = (f_1, f_2, f_3)$ and $\mathbf{x} = (x_1, x_2, x_3)$, then we have

$$\text{div } \mathbf{f}(\mathbf{x}) \equiv \text{tr}\left[\frac{\partial \mathbf{f}}{\partial \mathbf{x}}\right] = \frac{\partial f_1}{\partial x_1} + \frac{\partial f_2}{\partial x_2} + \frac{\partial f_3}{\partial x_3}$$

Consider a vector differential equation of the form

$$\dot{\mathbf{x}} = \mathbf{f}(\mathbf{x}) \tag{1.18}$$

where $\mathbf{x} = (x_1, \ldots, x_n)$ and $\mathbf{f} = (f_1, \ldots, f_n)$. The vectors $\mathbf{x}$ and $\mathbf{f}$ are often called a *state vector* and a *vector field*, respectively. The n-dimensional vector space is also called a *state space*, and the change in a small volume element in the n-dimensional state space is given by

$$
\begin{aligned}
\frac{d}{dt}(\Delta x_1 \Delta x_2 \cdots \Delta x_n) &= \Delta x_1 \cdots \Delta x_n \left\{ \frac{1}{\Delta x_1}\frac{d}{dt}\Delta x_1 + \cdots + \frac{1}{\Delta x_n}\frac{d}{dt}\Delta x_n \right\} \\
&= \Delta x_1 \cdots \Delta x_n \left\{ \frac{\Delta \dot{x}_1}{\Delta x_1} + \cdots + \frac{\Delta \dot{x}_n}{\Delta x_n} \right\} \\
&= \Delta x_1 \cdots \Delta x_n \left\{ \frac{\partial \dot{x}_1}{\partial x_1} + \cdots + \frac{\partial \dot{x}_n}{\partial x_n} \right\} \quad \text{as } \Delta x_i \to 0 \\
&= \Delta x_1 \cdots \Delta x_n \{\text{div } \dot{\mathbf{x}}\} \\
&= \Delta x_1 \cdots \Delta x_n \{\text{div } \mathbf{f}\}
\end{aligned} \tag{1.19}
$$

Thus, the volume of a set of points in the n-dimensional state space is said to be preserved or conserved if the divergence of its vector field is zero, i.e., if div $\mathbf{f} = 0$. In fluid mechanics, a fluid is said to be incompressible if the divergence of the velocity vector field is zero, i.e., if the volume of every element of the fluid is a constant for all times. This statement is often referred to as the *Liouville theorem* in dynamic systems theory.

Problem

1.1 (a) Find the eigenvalues and eigenvectors of the following matrix in so-called companion form:

$$
\mathbf{A} = \begin{bmatrix} 0 & 1 & 0 & 0 \\ 0 & 0 & 1 & 0 \\ 0 & 0 & 0 & 1 \\ 1 & 0 & 0 & 0 \end{bmatrix}
$$

(b) Given a scalar function $f(x_1, x_2) = x_1^2 + 2ax_1x_2 + x_2^2$, show that $f(x_1, x_2)$ is positive semidefinite if $|a| = 1$, positive definite if $|a| < 1$, and indefinite if $|a| > 1$.

(c) Show that the matrix $\mathbf{A}^T\mathbf{A}$ is positive semidefinite for any rectangular matrix $\mathbf{A}$.
Hint: Define a scalar function $f(\mathbf{x})$ of the form: $f(\mathbf{x}) = \mathbf{x}^T\mathbf{A}^T\mathbf{A}\mathbf{x}$.

(d) Given a scalar function $f(\mathbf{x}) = \mathbf{x}^T\mathbf{A}\mathbf{x}$ where $\mathbf{A}$ is a square matrix, show that

$$\frac{\partial f}{\partial \mathbf{x}} = \mathbf{x}^T(\mathbf{A} + \mathbf{A}^T)$$

$$\frac{\partial^2 f}{\partial \mathbf{x}^2} = (\mathbf{A} + \mathbf{A}^T)$$

Hint:

$$\frac{\partial}{\partial \mathbf{x}}(\mathbf{x}^T \mathbf{y}) = \mathbf{y}^T, \qquad \frac{\partial}{\partial \mathbf{y}}(\mathbf{x}^T \mathbf{y}) = \mathbf{x}^T, \qquad \frac{\partial}{\partial \mathbf{x}}(\mathbf{A}\mathbf{x}) = \mathbf{A}$$

1.1.3 Vector Analysis

1.1.3.1 Addition and Subtraction of Vectors Given two vectors expressed as $\vec{x} = x_1 \vec{a}_1 + x_2 \vec{a}_2 + x_3 \vec{a}_3$ and $\vec{y} = y_1 \vec{a}_1 + y_2 \vec{a}_2 + y_3 \vec{a}_3$, we have

$$\vec{x} \pm \vec{y} = (x_1 \pm y_1)\vec{a}_1 + (x_2 \pm y_2)\vec{a}_2 + (x_3 \pm y_3)\vec{a}_3 \qquad (1.20)$$

1.1.3.2 Dot Product The dot product of two vectors $\vec{x}$ and $\vec{y}$, denoted by $\vec{x} \cdot \vec{y}$, is a scalar quantity defined as

$$\vec{x} \cdot \vec{y} = |\vec{x}||\vec{y}| \cos \theta \qquad (1.21)$$

where $|\vec{x}|$ and $|\vec{y}|$ denote the magnitudes of $\vec{x}$ and $\vec{y}$, respectively, and θ is the angle between these two vectors. Consequently, we have

$$\vec{x} \cdot \vec{x} = |\vec{x}|^2, \qquad \vec{y} \cdot \vec{y} = |\vec{y}|^2 \qquad (1.22)$$

The dot product of two vectors $\vec{x}$ and $\vec{y}$ can be considered either as the product of $|\vec{x}|$ and the orthogonal projection of $\vec{y}$ along $\vec{x}$, i.e., $|\vec{y}| \cos \theta$, or as the product of $|\vec{y}|$ and the orthogonal projection of $\vec{x}$ along $\vec{y}$, i.e., $|\vec{x}| \cos \theta$; we have the distributive property for the dot product

$$\vec{x} \cdot (\vec{y} + \vec{z}) = \vec{x} \cdot \vec{y} + \vec{x} \cdot \vec{z} \qquad (1.23)$$

For a right-hand set of orthogonal unit vectors $\{\vec{a}_1, \vec{a}_2, \vec{a}_3\}$, we have

$$\vec{a}_i \cdot \vec{a}_j = \delta_{ij} = \begin{cases} 1 & \text{if } i = j \\ 0 & \text{if } i \neq j \end{cases} \qquad (1.24)$$

and we further define the following vectrix operation:

$$\begin{bmatrix} \vec{a}_1 \\ \vec{a}_2 \\ \vec{a}_3 \end{bmatrix} \cdot \begin{bmatrix} \vec{a}_1 & \vec{a}_2 & \vec{a}_3 \end{bmatrix} \equiv \begin{bmatrix} \vec{a}_1 \cdot \vec{a}_1 & \vec{a}_1 \cdot \vec{a}_2 & \vec{a}_1 \cdot \vec{a}_3 \\ \vec{a}_2 \cdot \vec{a}_1 & \vec{a}_2 \cdot \vec{a}_2 & \vec{a}_2 \cdot \vec{a}_3 \\ \vec{a}_3 \cdot \vec{a}_1 & \vec{a}_3 \cdot \vec{a}_2 & \vec{a}_3 \cdot \vec{a}_3 \end{bmatrix}$$

$$= \begin{bmatrix} 1 & 0 & 0 \\ 0 & 1 & 0 \\ 0 & 0 & 1 \end{bmatrix} = [\delta_{ij}] \qquad (1.25)$$

where δ_{ij} is the Kronecker delta. Furthermore, we have

$$\vec{x} = x_1 \vec{a}_1 + x_2 \vec{a}_2 + x_3 \vec{a}_3$$
$$= (\vec{x} \cdot \vec{a}_1)\vec{a}_1 + (\vec{x} \cdot \vec{a}_2)\vec{a}_2 + (\vec{x} \cdot \vec{a}_3)\vec{a}_3 \qquad (1.26)$$

and $|\vec{x}|^2 = x_1^2 + x_2^2 + x_3^2$.

Given two vectors $\vec{x} = x_1\vec{a}_1 + x_2\vec{a}_2 + x_3\vec{a}_3$ and $\vec{y} = y_1\vec{a}_1 + y_2\vec{a}_2 + y_3\vec{a}_3$, which are expressed in terms of a right-hand set of orthogonal unit vectors $\{\vec{a}_1, \vec{a}_2, \vec{a}_3\}$, we have

$$
\begin{aligned}
\vec{x} \cdot \vec{y} &= (x_1\vec{a}_1 + x_2\vec{a}_2 + x_3\vec{a}_3) \cdot (y_1\vec{a}_1 + y_2\vec{a}_2 + y_3\vec{a}_3) \\
&= x_1y_1(\vec{a}_1 \cdot \vec{a}_1) + x_1y_2(\vec{a}_1 \cdot \vec{a}_2) + x_1y_3(\vec{a}_1 \cdot \vec{a}_3) \\
&\quad + x_2y_1(\vec{a}_2 \cdot \vec{a}_1) + x_2y_2(\vec{a}_2 \cdot \vec{a}_2) + x_2y_3(\vec{a}_2 \cdot \vec{a}_3) \\
&\quad + x_3y_1(\vec{a}_3 \cdot \vec{a}_1) + x_3y_2(\vec{a}_3 \cdot \vec{a}_2) + x_3y_3(\vec{a}_3 \cdot \vec{a}_3) \\
&= x_1y_1 + x_2y_2 + x_3y_3 = [x_1 \quad x_2 \quad x_3]\begin{bmatrix} y_1 \\ y_2 \\ y_3 \end{bmatrix} = \mathbf{x}^T\mathbf{y} \qquad (1.27)
\end{aligned}
$$

because $\vec{a}_i \cdot \vec{a}_j = \delta_{ij}$.

1.1.3.3 Cross Product
The cross product of two vectors $\vec{x}$ and $\vec{y}$, denoted by $\vec{x} \times \vec{y}$, is a vector defined as

$$
\vec{x} \times \vec{y} = |\vec{x}||\vec{y}| \sin\theta\, \vec{n} = -\vec{y} \times \vec{x} \qquad (1.28)
$$

where θ is the smallest angle between these two vectors and $\vec{n}$ is a unit vector perpendicular to both $\vec{x}$ and $\vec{y}$ such that $\{\vec{x}, \vec{y}, \vec{n}\}$ form a right-hand coordinate system. Furthermore, the cross product has the distributive property

$$
\vec{x} \times (\vec{y} + \vec{z}) = \vec{x} \times \vec{y} + \vec{x} \times \vec{z} \qquad (1.29)
$$

For a right-hand set of orthogonal unit vectors $\{\vec{a}_1, \vec{a}_2, \vec{a}_3\}$, we have

$$
\vec{a}_1 \times \vec{a}_2 = \vec{a}_3, \qquad \vec{a}_2 \times \vec{a}_3 = \vec{a}_1, \qquad \vec{a}_3 \times \vec{a}_1 = \vec{a}_2
$$

and $\vec{a}_1 \times \vec{a}_1 = \vec{a}_2 \times \vec{a}_2 = \vec{a}_3 \times \vec{a}_3 = 0$, and we further define the following vectrix operation

$$
\begin{aligned}
\begin{bmatrix} \vec{a}_1 \\ \vec{a}_2 \\ \vec{a}_3 \end{bmatrix} \times [\vec{a}_1 \quad \vec{a}_2 \quad \vec{a}_3] &\equiv \begin{bmatrix} \vec{a}_1 \times \vec{a}_1 & \vec{a}_1 \times \vec{a}_2 & \vec{a}_1 \times \vec{a}_3 \\ \vec{a}_2 \times \vec{a}_1 & \vec{a}_2 \times \vec{a}_2 & \vec{a}_2 \times \vec{a}_3 \\ \vec{a}_3 \times \vec{a}_1 & \vec{a}_3 \times \vec{a}_2 & \vec{a}_3 \times \vec{a}_3 \end{bmatrix} \\
&= \begin{bmatrix} 0 & \vec{a}_3 & -\vec{a}_2 \\ -\vec{a}_3 & 0 & \vec{a}_1 \\ \vec{a}_2 & -\vec{a}_1 & 0 \end{bmatrix} \qquad (1.30)
\end{aligned}
$$

Given two vectors $\vec{x} = x_1\vec{a}_1 + x_2\vec{a}_2 + x_3\vec{a}_3$ and $\vec{y} = y_1\vec{a}_1 + y_2\vec{a}_2 + y_3\vec{a}_3$, which are expressed in terms of a right-hand set of orthogonal unit vectors $\{\vec{a}_1, \vec{a}_2, \vec{a}_3\}$, we obtain

$$
\begin{aligned}
\vec{x} \times \vec{y} &= (x_1\vec{a}_1 + x_2\vec{a}_2 + x_3\vec{a}_3) \times (y_1\vec{a}_1 + y_2\vec{a}_2 + y_3\vec{a}_3) \\
&= (x_2y_3 - x_3y_2)\vec{a}_1 + (x_3y_1 - x_1y_3)\vec{a}_2 + (x_1y_2 - x_2y_1)\vec{a}_3 \qquad (1.31)
\end{aligned}
$$

The cross product of $\vec{x}$ and $\vec{y}$ is often written in determinant form as follows:

$$\vec{x} \times \vec{y} = \begin{vmatrix} \vec{a}_1 & \vec{a}_2 & \vec{a}_3 \\ x_1 & x_2 & x_3 \\ y_1 & y_2 & y_3 \end{vmatrix} \qquad (1.32)$$

which can also be written as

$$\vec{x} \times \vec{y} = [\vec{a}_1 \quad \vec{a}_2 \quad \vec{a}_3] \begin{bmatrix} x_2 y_3 - x_3 y_2 \\ x_3 y_1 - x_1 y_3 \\ x_1 y_2 - x_2 y_1 \end{bmatrix}$$

$$= [\vec{a}_1 \quad \vec{a}_2 \quad \vec{a}_3] \underbrace{\begin{bmatrix} 0 & -x_3 & x_2 \\ x_3 & 0 & -x_1 \\ -x_2 & x_1 & 0 \end{bmatrix} \begin{bmatrix} y_1 \\ y_2 \\ y_3 \end{bmatrix}}_{\text{cross product of } \mathbf{x} \text{ and } \mathbf{y}}$$

1.1.3.4 Scalar Triple Product   Given three vectors $\vec{x}$, $\vec{y}$, and $\vec{z}$, which are expressed in terms of a set of orthogonal unit vectors $\{\vec{a}_1, \vec{a}_2, \vec{a}_3\}$ as $\vec{x} = \sum x_i \vec{a}_i$, $\vec{y} = \sum y_i \vec{a}_i$, and $\vec{z} = \sum z_i \vec{a}_i$, the scalar triple product becomes

$$\vec{x} \cdot (\vec{y} \times \vec{z}) = \begin{vmatrix} x_1 & x_2 & x_3 \\ y_1 & y_2 & y_3 \\ z_1 & z_2 & z_3 \end{vmatrix} \qquad (1.33)$$

The absolute value of the scalar triple product, $|\vec{x} \cdot (\vec{y} \times \vec{z})|$, represents the volume of a parallelepiped having $\vec{x}$, $\vec{y}$, and $\vec{z}$ as edges. We also have

$$\vec{x} \cdot (\vec{y} \times \vec{z}) = \vec{y} \cdot (\vec{z} \times \vec{x}) = \vec{z} \cdot (\vec{x} \times \vec{y}) \qquad (1.34)$$

1.1.3.5 Vector Triple Product   Given three vectors $\vec{x}$, $\vec{y}$, and $\vec{z}$, we have

$$\vec{x} \times (\vec{y} \times \vec{z}) = (\vec{x} \cdot \vec{z})\vec{y} - (\vec{x} \cdot \vec{y})\vec{z} \qquad (1.35)$$

$$(\vec{x} \times \vec{y}) \times \vec{z} = (\vec{z} \cdot \vec{x})\vec{y} - (\vec{z} \cdot \vec{y})\vec{x} \qquad (1.36)$$

It is clear that $\vec{x} \times (\vec{y} \times \vec{z}) \neq (\vec{x} \times \vec{y}) \times \vec{z}$.

1.1.3.6 Differentiation of Vectors   Consider a vector $\vec{x}$ that is a function of time; i.e., $\vec{x} = \vec{x}(t)$. The time derivative of $\vec{x}$, denoted by $d\vec{x}/dt$, is defined as

$$\frac{d\vec{x}}{dt} = \lim_{\Delta t \to 0} \frac{\Delta \vec{x}}{\Delta t} = \lim_{\Delta t \to 0} \frac{\vec{x}(t + \Delta t) - \vec{x}(t)}{\Delta t} \qquad (1.37)$$

and we have

$$\frac{d}{dt}(\vec{x} \pm \vec{y}) = \frac{d\vec{x}}{dt} \pm \frac{d\vec{y}}{dt} \qquad (1.38a)$$

$$\frac{d}{dt}(\vec{x} \cdot \vec{y}) = \frac{d\vec{x}}{dt} \cdot \vec{y} + \vec{x} \cdot \frac{d\vec{y}}{dt} \qquad (1.38b)$$

$$\frac{d}{dt}(\vec{x} \times \vec{y}) = \frac{d\vec{x}}{dt} \times \vec{y} + \vec{x} \times \frac{d\vec{y}}{dt} \qquad (1.38c)$$

and

$$\frac{d}{dt}[\vec{x} \cdot (\vec{y} \times \vec{z})] = \frac{d\vec{x}}{dt} \cdot (\vec{y} \times \vec{z}) + \vec{x} \cdot \left(\frac{d\vec{y}}{dt} \times \vec{z}\right) + \vec{x} \cdot \left(\vec{y} \times \frac{d\vec{z}}{dt}\right) \quad (1.39\text{a})$$

$$\frac{d}{dt}[\vec{x} \times (\vec{y} \times \vec{z})] = \frac{d\vec{x}}{dt} \times (\vec{y} \times \vec{z}) + \vec{x} \times \left(\frac{d\vec{y}}{dt} \times \vec{z}\right)$$
$$+ \vec{x} \times \left(\vec{y} \times \frac{d\vec{z}}{dt}\right) \quad (1.39\text{b})$$

1.1.3.7 Gradient Vector of a Scalar Field

Consider a scalar field (or function) of position vector $\vec{x}$ of the form

$$\phi = \phi(\vec{x}) \quad (1.40)$$

Let the position vector $\vec{x}$ be expressed as

$$\vec{x} = x_1 \vec{a}_1 + x_2 \vec{a}_2 + x_3 \vec{a}_3 \quad (1.41)$$

where $\{\vec{a}_1, \vec{a}_2, \vec{a}_3\}$ is a set of three orthogonal unit vectors of a rectangular Cartesian reference frame. Then the differential position increment $d\vec{x}$ becomes

$$d\vec{x} = dx_1 \vec{a}_1 + dx_2 \vec{a}_2 + dx_3 \vec{a}_3 \quad (1.42)$$

and the total differential of $\phi = \phi(x_1, x_2, x_3)$ can be obtained as

$$d\phi = \frac{\partial \phi}{\partial x_1} dx_1 + \frac{\partial \phi}{\partial x_2} dx_2 + \frac{\partial \phi}{\partial x_3} dx_3 \quad (1.43)$$

where $\partial \phi / \partial x_i$ denotes the partial derivative of ϕ with respect to x_i.

The gradient vector, or simply gradient, of the scalar field ϕ, denoted by grad ϕ or $\nabla \phi$, is in general defined as

$$\nabla \phi \cdot d\vec{x} = d\phi \quad (1.44)$$

and the gradient vector is simply expressed in a rectangular Cartesian reference frame as

$$\text{grad } \phi \equiv \nabla \phi = \frac{\partial \phi}{\partial x_1} \vec{a}_1 + \frac{\partial \phi}{\partial x_2} \vec{a}_2 + \frac{\partial \phi}{\partial x_3} \vec{a}_3 \quad (1.45)$$

Thus, the components of the gradient vector of a scalar function of position are simply the partial derivatives of the function with respect to distances along the three orthogonal basis vectors.

Equation (1.45) may be rewritten as

$$\nabla \phi = \left(\vec{a}_1 \frac{\partial}{\partial x_1} + \vec{a}_2 \frac{\partial}{\partial x_2} + \vec{a}_3 \frac{\partial}{\partial x_3}\right) \phi$$

and in vector analysis we introduce the *nabla* vector, ∇, defined as

$$\nabla \equiv \vec{a}_1 \frac{\partial}{\partial x_1} + \vec{a}_2 \frac{\partial}{\partial x_2} + \vec{a}_3 \frac{\partial}{\partial x_3} \quad (1.46)$$

which is often called the *del* vector.

1.1.3.8 Divergence and Curl of a Vector Field   Consider a vector field $\vec{u}$, which is a function of position vector $\vec{x}$, and let $\vec{u}$ and $\vec{x}$ be expressed as

$$\vec{u} = u_1 \vec{a}_1 + u_2 \vec{a}_2 + u_3 \vec{a}_3$$
$$\vec{x} = x_1 \vec{a}_1 + x_2 \vec{a}_2 + x_3 \vec{a}_3$$

where $\{\vec{a}_1, \vec{a}_2, \vec{a}_3\}$ is a set of three orthogonal unit vectors and $u_i = u_i(x_1, x_2, x_3)$. The divergence and curl of a vector $\vec{u}(x_1, x_2, x_3)$ are then defined as

$$\operatorname{div} \vec{u} \equiv \nabla \cdot \vec{u} = \frac{\partial u_1}{\partial x_1} + \frac{\partial u_2}{\partial x_2} + \frac{\partial u_3}{\partial x_3} \tag{1.47a}$$

$$\operatorname{curl} \vec{u} \equiv \nabla \times \vec{u} = \begin{vmatrix} \vec{a}_1 & \vec{a}_2 & \vec{a}_3 \\ \partial/\partial x_1 & \partial/\partial x_2 & \partial/\partial x_3 \\ u_1 & u_2 & u_3 \end{vmatrix}$$

$$= \left(\frac{\partial u_3}{\partial x_2} - \frac{\partial u_2}{\partial x_3} \right) \vec{a}_1 + \left(\frac{\partial u_1}{\partial x_3} - \frac{\partial u_3}{\partial x_1} \right) \vec{a}_2 + \left(\frac{\partial u_2}{\partial x_1} - \frac{\partial u_1}{\partial x_2} \right) \vec{a}_3 \tag{1.47b}$$

Note that $\nabla \cdot \vec{u}$ becomes a scalar function whereas $\vec{u} \cdot \nabla$ is a scalar operator; i.e., $\nabla \cdot \vec{u} \neq \vec{u} \cdot \nabla$. The curl of a vector vanishes if and only if the vector is the gradient of a scalar function; i.e., $\nabla \times \vec{u} = 0$ if and only if $\vec{u} = \nabla \phi$. Such a vector field $\vec{u}$ is said to be irrotational or conservative, and ϕ is often called the potential function.

The dot product of the del vector with itself becomes a scalar operator, called the *Laplacian* denoted in rectangular Cartesian coordinates x_1, x_2, and x_3 by

$$\nabla \cdot \nabla \equiv \nabla^2 = \frac{\partial^2}{\partial x_1^2} + \frac{\partial^2}{\partial x_2^2} + \frac{\partial^2}{\partial x_3^2}$$

The divergence of the gradient of a scalar function $\phi(x_1, x_2, x_3)$ becomes the Laplacian of $\phi(x_1, x_2, x_3)$; i.e., we have

$$\nabla \cdot \nabla \phi = \nabla^2 \phi = \frac{\partial^2 \phi}{\partial x_1^2} + \frac{\partial^2 \phi}{\partial x_2^2} + \frac{\partial^2 \phi}{\partial x_3^2} \tag{1.48}$$

and $\nabla^2 \phi = 0$ is called Laplace's equation.

The following vector identities can be easily verified:

$$\nabla \cdot \vec{x} = 3$$
$$\nabla \cdot (\vec{x}/x^3) = 0 \qquad \text{where } x = |\vec{x}|$$
$$\nabla \times \vec{x} = 0$$
$$\nabla \cdot (\nabla \times \vec{u}) = 0$$
$$(\nabla \times \nabla) \times \vec{u} = 0$$

Problems

1.2 (a) Given two vectors $\vec{x}$ and $\vec{y}$, show that

$$|\vec{x} \pm \vec{y}|^2 = |\vec{x}|^2 + |\vec{y}|^2 \pm 2|\vec{x}||\vec{y}| \cos \theta$$

where θ denotes the angle between the vectors $\vec{x}$ and $\vec{y}$. This relationship is called the cosine law of trigonometry.
Hint: $|\vec{x} \pm \vec{y}|^2 = (\vec{x} \pm \vec{y}) \cdot (\vec{x} \pm \vec{y})$.

  (b) Given a vector $\vec{\omega}$ expressed as $\vec{\omega} = \omega_1 \vec{b}_1 + \omega_2 \vec{b}_2 + \omega_3 \vec{b}_3$ where $\{\vec{b}_1, \vec{b}_2, \vec{b}_3\}$ is a set of orthogonal unit vectors, verify that

$$\begin{bmatrix} \vec{\omega} \times \vec{b}_1 \\ \vec{\omega} \times \vec{b}_2 \\ \vec{\omega} \times \vec{b}_3 \end{bmatrix} = - \begin{bmatrix} 0 & -\omega_3 & \omega_2 \\ \omega_3 & 0 & -\omega_1 \\ -\omega_2 & \omega_1 & 0 \end{bmatrix} \begin{bmatrix} \vec{b}_1 \\ \vec{b}_2 \\ \vec{b}_3 \end{bmatrix}$$

  (c) Given a vector $\vec{x}$ that is a function of time, show that

$$\vec{x} \cdot \frac{\mathrm{d}\vec{x}}{\mathrm{d}t} = x \frac{\mathrm{d}x}{\mathrm{d}t}$$

where $x \equiv |\vec{x}|$ denotes the magnitude of the vector $\vec{x}$.
Hint: $\vec{x} \cdot \vec{x} = x^2$.

  (d) Given a vector $\vec{x}$ expressed as $\vec{x} = x_1 \vec{a}_1 + x_2 \vec{a}_2 + x_3 \vec{a}_3$ where $\{\vec{a}_1, \vec{a}_2, \vec{a}_3\}$ is a set of any three nonzero, noncoplanar vectors, show that

$$x_1 = \frac{\vec{x} \cdot (\vec{a}_2 \times \vec{a}_3)}{\vec{a}_1 \cdot (\vec{a}_2 \times \vec{a}_3)}, \quad x_2 = \frac{\vec{x} \cdot (\vec{a}_3 \times \vec{a}_1)}{\vec{a}_2 \cdot (\vec{a}_3 \times \vec{a}_1)}, \quad x_3 = \frac{\vec{x} \cdot (\vec{a}_1 \times \vec{a}_2)}{\vec{a}_3 \cdot (\vec{a}_1 \times \vec{a}_2)}$$

Note: $\vec{a}_1 \cdot (\vec{a}_2 \times \vec{a}_3) = \vec{a}_2 \cdot (\vec{a}_3 \times \vec{a}_1) = \vec{a}_3 \cdot (\vec{a}_1 \times \vec{a}_2)$.

1.3 Given a reference frame A with a set of nonorthogonal basis vector $\{\vec{a}_1, \vec{a}_2, \vec{a}_3\}$, a position vector $\vec{x}$ is expressed as

$$\vec{x} = x_1 \vec{a}_1 + x_2 \vec{a}_2 + x_3 \vec{a}_3$$

and the differential position increment $\mathrm{d}\vec{x}$ becomes

$$\mathrm{d}\vec{x} = \mathrm{d}x_1 \vec{a}_1 + \mathrm{d}x_2 \vec{a}_2 + \mathrm{d}x_3 \vec{a}_3$$

The total differential of a scalar function $\phi = \phi(\vec{x}) = \phi(x_1, x_2, x_3)$ is also given by

$$\mathrm{d}\phi = \frac{\partial \phi}{\partial x_1} \mathrm{d}x_1 + \frac{\partial \phi}{\partial x_2} \mathrm{d}x_2 + \frac{\partial \phi}{\partial x_3} \mathrm{d}x_3$$

where $\partial \phi / \partial x_i$ denotes the partial derivative of ϕ with respect to x_i.

Show that the gradient vector of the scalar function ϕ can be defined as

$$\nabla\phi = \frac{\partial\phi}{\partial x_1}\vec{b}_1 + \frac{\partial\phi}{\partial x_2}\vec{b}_2 + \frac{\partial\phi}{\partial x_3}\vec{b}_3$$

such that

$$\nabla\phi \cdot d\vec{x} = d\phi$$

and $\{\vec{b}_1, \vec{b}_2, \vec{b}_3\}$ is a new set of nonorthogonal basis vectors, defined as

$$\vec{b}_1 = \frac{\vec{a}_2 \times \vec{a}_3}{\vec{a}_1 \cdot (\vec{a}_2 \times \vec{a}_3)}, \qquad \vec{b}_2 = \frac{\vec{a}_3 \times \vec{a}_1}{\vec{a}_2 \cdot (\vec{a}_3 \times \vec{a}_1)}, \qquad \vec{b}_3 = \frac{\vec{a}_1 \times \vec{a}_2}{\vec{a}_3 \cdot (\vec{a}_1 \times \vec{a}_2)}$$

Note: The basis vectors $\{\vec{a}_1, \vec{a}_2, \vec{a}_3\}$ and $\{\vec{b}_1, \vec{b}_2, \vec{b}_3\}$ are called the *covariant* and *contravariant* basis vectors, respectively.[*] They are said to be reciprocal to one another, and we have $\vec{a}_i \cdot \vec{b}_j = \delta_{ij}$ and

$$\vec{b}_1 \cdot (\vec{b}_2 \times \vec{b}_3) = \frac{1}{\vec{a}_1 \cdot (\vec{a}_2 \times \vec{a}_3)}$$

1.4 Consider a vector field $\vec{u}$ that depends not only on the position vector $\vec{x}$ but also explicitly on the time t, i.e., $\vec{u} = \vec{u}(\vec{x}, t)$. Let $\vec{u} = u_1\vec{a}_1 + u_2\vec{a}_2 + u_3\vec{a}_3$ and $\vec{x} = x_1\vec{a}_1 + x_2\vec{a}_2 + x_3\vec{a}_3$ where $\{\vec{a}_1, \vec{a}_2, \vec{a}_3\}$ is a set of three orthogonal unit vectors of a rectangular Cartesian coordinate system that is assumed to be inertially fixed; i.e., $\{\vec{a}_i\}$ are constant basis vectors. The time derivative of $\vec{x}$ is simply given as: $d\vec{x}/dt = \vec{u}$.

(a) Show that the time derivative of $\vec{u}$ can be expressed in vector notation as

$$\frac{d\vec{u}}{dt} = \frac{\partial\vec{u}}{\partial t} + (\vec{u} \cdot \nabla)\vec{u}$$

Hint: $u_i = u_i(x_1, x_2, x_3, t)$ and $dx_i/dt = u_i$.

(b) Verify the following vector identities:

$$(\vec{u} \cdot \nabla)\vec{u} = \tfrac{1}{2}\nabla u^2 - \vec{u} \times (\nabla \times \vec{u})$$

where $u = |\vec{u}|$

$$\nabla^2\vec{u} = \nabla(\nabla \cdot \vec{u}) - \nabla \times (\nabla \times \vec{u})$$

where $\nabla^2 = \nabla \cdot \nabla$.

(c) Show that $\nabla \times \vec{u} = 0$ if and only if $\vec{u} = \nabla\phi$ where ϕ is a scalar function. *Note:* The vectors $\vec{u}$ and $\vec{x}$ of this problem, in fact, represent the velocity and position vectors of a fluid particle, respectively. In fluid mechanics, the curl of the velocity vector is defined as the vorticity (or rotation) vector

---

[*]The terms covariant, contravariant, and invariant were introduced by James Sylvester (1814–1897).

$\vec{\Omega}$; i.e., $\vec{\Omega} \equiv \nabla \times \vec{u}$. A fluid is said to be irrotational if $\vec{\Omega} \equiv \nabla \times \vec{u} = 0$ and is said to be incompressible if $\nabla \cdot \vec{u} = 0$. The various vector expressions considered in this problem actually can be seen in the Navier–Stokes equations of a compressible viscous fluid, described by

$$\frac{d\vec{u}}{dt} \equiv \frac{\partial \vec{u}}{\partial t} + (\vec{u} \cdot \nabla)\vec{u} = -\frac{1}{\rho}\nabla p + \nu\nabla^2\vec{u} + \frac{\nu}{3}\nabla(\nabla \cdot \vec{u})$$

where ρ denotes the density, p the pressure, $\nu = \mu/\rho$ the kinematic viscosity, and μ the constant viscosity coefficient of a Newtonian fluid.

1.2 Classical Mechanics

In this section, dynamics of a particle and a system of particles are treated from a dynamic systems point of view. A special emphasis is placed on a kinematic problem in which a vector is expressed in a rotating reference frame but differentiated in an inertial reference frame that is fixed in space or is translating with a constant velocity. This section also introduces the principles of analytical dynamics, including Lagrange's equations of motion and Hamilton's canonical equations of motion. A thorough understanding of the fundamental concepts of analytical dynamics will provide a solid foundation to developing dynamic systems theory. As an introduction to spacecraft dynamics problems of practical concern, a dynamic modeling and computer simulation problem associated with the deployment of solar panel arrays on actual spacecraft is also treated in this section. However, this section is mainly intended as a summary of many of the fundamental concepts in classical mechanics.

We begin the subject of classical mechanics with Kepler's three empirical laws of planetary motion.

1.2.1 Kepler's Laws of Planetary Motion

Using the painstaking observational data of Tycho Brahe,* Kepler was able to discover some regularities in the motions of the planet Mars. The regularities were then summarized into what are known as Kepler's three laws of planetary motion[†]:

1) The orbit of each planet around the sun is an ellipse, with the sun at one focus (the law of orbits).

2) The radius vector from the sun to a planet sweeps out equal areas in equal time intervals (the law of areas).

3) The square of the orbital period of a planet is proportional to the cube of its mean distance from the sun (the law of periods). (The semimajor axis of an ellipse is often called the mean distance, although it is not the average length of the radius vector with respect to time.)

---

*Tycho Brahe (1546–1601) was the greatest astronomical observer before the introduction of the telescope.

[†]Johannes Kepler (1571–1630), who was a contemporary of Galileo Galilei (1564–1642), was the first astronomer to uphold openly the heliocentric theory of Nicholas Copernicus (1473–1543).

These three empirical laws, with the first two laws published in 1609 and the third law in 1619, were used later by Isaac Newton (1642–1727) to deduce his law of gravity. In 1666, Newton began to think of gravity, and later, in 1687, he published his law of gravity and three laws of motion in his *Principia*.

Kepler's three laws of planetary motion describe the motion of planets around the sun, which is considered to be inertially fixed. In Chapter 3, we shall derive these laws of planetary motion by applying Newton's laws of motion and his law of gravity to the general two-body problem in which the primary body is not assumed to be inertially fixed.

1.2.2   Newton's Laws of Motion and Law of Gravity

In *Principia*, Newton published his three laws of motion, which are valid for a particle whose motion is observed in a reference frame that is fixed in space or is translating with a constant velocity. Such a nonrotating and nonaccelerating reference frame as postulated by Newton is frequently referred to as an inertial or Newtonian reference frame. It was, however, Galileo who first introduced the concept of absolute acceleration with respect to an inertial reference frame and first stated the principle of inertia: a body not subject to external forces moves with constant velocity. The principle of inertia is, in fact, Newton's first law, which is a special case of his second law.

1.2.2.1   Newton's Three Laws of Motion   Classical mechanics, which is primarily based on Newton's three laws of motion and his law of gravity, provides the framework for the study of orbital and attitude motions of space vehicles.

Newton's three laws of motion are stated as follows:

1) A particle remains in its state of rest or uniform, straight-line motion unless it is acted upon by forces to change that state; this is the law of inertia. (Note that the term "particle" is a mathematical abstraction of a relatively small body and it is used interchangeably with the term "point mass.")

2) The force acting on a particle equals the mass of the particle times its inertial acceleration.

3) For every applied force, there is an equal and opposite reaction force; this is the law of action and reaction.

In vector notation, the second law takes the standard form

$$\vec{f} = m\vec{a} \tag{1.49}$$

where $\vec{f}$ is the force acting on the particle, m is its constant mass, and $\vec{a}$ is its inertial acceleration. Force, mass, and acceleration are dimensional quantities whose magnitude is defined with respect to a set of the SI units, [Newtons (N), kilograms (kg), meters (m), and seconds (s)], or the U.S. customary units, [pounds (lb), slugs (slug), feet (ft), and seconds (s)]. (See Table 1.1.) The inertial acceleration is related to the inertial velocity $\vec{v}$ and the inertial position $\vec{r}$, as follows:

$$\vec{a} = \frac{d\vec{v}}{dt} = \frac{d^2\vec{r}}{dt^2} \tag{1.50}$$

Table 1.1   Conversion of units of measure

Units	Equivalents
N	kg·m/s^2
lb	slug-ft/s^2
1 ft	0.3048 m
1 slug	14.5939 kg
1 lb	4.4482 N

In formulating the equation of motion of a particle translating in an inertial reference frame, Newton's second law, force equals mass times acceleration, is often rearranged as mass times acceleration equals force, or acceleration equals force divided by mass; i.e., we write the equation of motion of a particle as

$$m\vec{a} = \vec{f} \qquad \text{or} \qquad \vec{a} = \frac{\vec{f}}{m} \qquad (1.51)$$

Furthermore, integrating Eq. (1.51) for $\vec{f} = 0$, we obtain the inertial velocity as

$$\vec{v} = \frac{d\vec{r}}{dt} = \text{constant vector}$$

Thus, it is said that Newton's first law is a simple consequence of his second law. In vector notation, the third law for a system of N particles takes the form

$$\vec{f}_{ij} = -\vec{f}_{ji} \qquad (i,j = 1,\ldots,N) \qquad (1.52)$$

where $\vec{f}_{ij}$ denotes the force acting on the ith particle by the jth particle and, conversely, $\vec{f}_{ji}$ denotes the force acting on the jth particle by the ith particle. Note that the interaction forces between any two particles are assumed to be collinear and $\vec{f}_{ii} = 0$ because there are no interacting forces between a particle and itself. Consequently, the internal action and reaction forces of a system of N particles satisfy the following relationships:

$$\sum_{i=1}^{N}\sum_{j=1}^{N} \vec{f}_{ij} = 0 \qquad (1.53)$$

and

$$\sum_{i=1}^{N}\sum_{j=1}^{N} \vec{r}_i \times \vec{f}_{ij} = \frac{1}{2}\sum_{i=1}^{N}\sum_{j=1}^{N}(\vec{r}_i - \vec{r}_j) \times \vec{f}_{ij} = 0 \qquad (1.54)$$

because $\vec{f}_{ij}$ is collinear with the relative position vector, $\vec{r}_i - \vec{r}_j$, between two particles.

1.2.2.2 Newton's Law of Gravity   Another important physical law in classical mechanics is Newton's law of gravity. According to Newton's law of gravity, two particles of masses m_1 and m_2, separated by a distance r, attract each other with a force of magnitude

$$f = \frac{Gm_1m_2}{r^2} \tag{1.55}$$

where $G = 6.67259 \times 10^{-11} \, \text{N} \cdot \text{m}^2/\text{kg}^2$ is the universal gravitational constant.* Newton's law of gravity is stated in vector form as

$$\vec{f} = -\frac{Gm_1m_2}{r^3}\vec{r} \tag{1.56}$$

where $\vec{f}$ is the gravity force acting on m_2 by m_1, $\vec{r}$ is the position vector of m_2 relative to m_1, and r is the magnitude of $\vec{r}$.

The gravitational acceleration of a point mass near the surface of a nonrotating, spherical model of the Earth, which is commonly denoted by g, is given by

$$g = \frac{GM_\oplus}{R_\oplus^2} \approx 9.8 \, \text{m/s}^2 \qquad (= 32.2 \, \text{ft/s}^2)$$

where $M_\oplus \, (= 5.9737 \times 10^{24} \, \text{kg})$ and $R_\oplus \, (= 6378 \, \text{km})$ are the astronomical symbols for the mass and mean equatorial radius of the Earth, respectively. The product of G and $M_\oplus$, often denoted as $\mu_\oplus$, is called the *gravitational parameter* of the Earth; i.e., we have $\mu_\oplus = GM_\oplus = 398,601 \, \text{km}^3/\text{s}^2$.

Because the gravity force $\vec{f}$ is a function of the position vector $\vec{r}$ only, the gravitational force field is said to be conservative, and we introduce a scalar function ϕ such that

$$\vec{f} \cdot d\vec{r} = d\phi \qquad \text{or} \qquad \vec{f} \cdot \frac{d\vec{r}}{dt} = \frac{d\phi}{dt} \tag{1.57}$$

where $d\vec{r}/dt$ is the time derivative of $\vec{r}$ in an inertial reference frame and ϕ is called the *gravitational potential*. The gravity force is then simply the gradient vector of the gravitational potential ϕ; i.e., we have

$$\vec{f} = \nabla\phi \tag{1.58}$$

Such a conservative force field is also sometimes called irrotational because

$$\text{curl}\,\vec{f} \equiv \nabla \times \vec{f} = \nabla \times \nabla\phi = 0$$

Substituting Eq. (1.56) into Eq. (1.57) and integrating the resulting equation, we obtain

$$\phi = \frac{Gm_1m_2}{r} + C \tag{1.59}$$

---

*In *Principia*, Newton made no attempt to provide a physical explanation of gravitation, and the nature of gravitation still remains mysterious today.

where C is a constant of integration. In orbital mechanics, the constant of integration C is often chosen as zero such that $\phi = 0$ at $r = \infty$, and the negative of the gravitational potential is called the potential energy V; i.e.,

$$V \equiv -\phi = -\frac{Gm_1 m_2}{r} \tag{1.60}$$

We often call V *a* potential energy, rather than *the* potential energy because an arbitrary constant can be added to V.

Newton's three laws of motion and his law of gravity provide the fundamental framework for classical mechanics, and they are of fundamental importance in formulating equations of motion of space vehicles for the study of their orbital and attitude motions.

Problem

1.5 Show that the gravitational force F between a particle of mass m and a solid sphere of radius a and constant mass density ρ is given by

$$F = \frac{GMm}{r^2}$$

where $M = 4\pi \rho a^3/3$ is the total mass of the sphere and r is a distance to the particle from the center of the sphere. That is, a uniformly dense spherical body attracts an external point mass as if all its mass were concentrated at its center.

Note: This problem was not an easy one even for Newton, but eventually it was solved by Newton in 1685 by proving that a homogeneous spherical shell attracts an external particle as if all its mass were concentrated at its center.

1.2.3   Kinematics of a Particle

Kinematics is mainly concerned with the geometry of motion. The subject of kinematics is somewhat mathematical in nature and does not involve any forces associated with the motion. A thorough working knowledge of kinematics is a prerequisite to the successful formulation of the equations of motion of particles and rigid bodies, however. We shall consider here the kinematics of a particle, whereas the rotational kinematics of a rigid body is treated in Chapter 5.

1.2.3.1   Angular Velocity Vector of a Rotating Reference Frame   Consider a reference frame A with a set of three orthogonal unit vectors $\{\vec{a}_1, \vec{a}_2, \vec{a}_3\}$ and a reference frame B with a set of three orthogonal unit vectors $\{\vec{b}_1, \vec{b}_2, \vec{b}_3\}$, as shown in Fig. 1.2.

The angular velocity vector of a reference frame B with respect to a reference frame A is denoted by $\vec{\omega}^{B/A}$. For brevity, the symbol $\vec{\omega}^{B/A}$ is to be read as the angular velocity of B with respect to A. The terms "with respect to" and "relative to" are often used interchangeably in the literature. The angular velocity vector $\vec{\omega}^{B/A}$ lies along the instantaneous axis of rotation of the reference frame B, and it usually

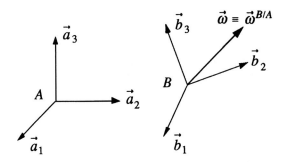

Fig. 1.2   Reference frames A and B.

changes both its direction and magnitude continuously with time. We can also consider the angular velocity of A with respect to B, denoted by $\vec{\omega}^{A/B}$, and we have

$$\vec{\omega}^{A/B} \equiv -\vec{\omega}^{B/A} \qquad (1.61)$$

The angular velocity of B with respect to A can also be expressed as

$$\vec{\omega}^{B/A} = \vec{\omega}^{B/A_1} + \vec{\omega}^{A_1/A_2} + \vec{\omega}^{A_2/A_3} + \cdots + \vec{\omega}^{A_n/A} \qquad (1.62)$$

where $A_1, \ldots, A_n$ are the n auxiliary reference frames.

Because the unit vectors $\vec{b}_i$ fixed in the reference frame B rotate with an angular velocity $\vec{\omega}^{B/A}$ with respect to A, the rate of change of $\vec{b}_i$ is caused only by $\vec{\omega}^{B/A}$ and it must be normal to both $\vec{b}_i$ and $\vec{\omega}^{B/A}$. Thus, the time derivatives of the unit vectors $\vec{b}_i$ measured in A are given by

$$\left\{ \frac{d\vec{b}_1}{dt} \right\}_A = \vec{\omega}^{B/A} \times \vec{b}_1 \qquad (1.63a)$$

$$\left\{ \frac{d\vec{b}_2}{dt} \right\}_A = \vec{\omega}^{B/A} \times \vec{b}_2 \qquad (1.63b)$$

$$\left\{ \frac{d\vec{b}_3}{dt} \right\}_A = \vec{\omega}^{B/A} \times \vec{b}_3 \qquad (1.63c)$$

If a reference frame A is inertially fixed, it is called an inertial or Newtonian reference frame. In that case, a reference frame B is in motion relative to A and becomes a rotating reference frame with an angular velocity of $\vec{\omega}^{B/A}$. For brevity, we write simply $\vec{\omega}$ for $\vec{\omega}^{B/A}$; that is,

$$\vec{\omega} \equiv \vec{\omega}^{B/A}$$

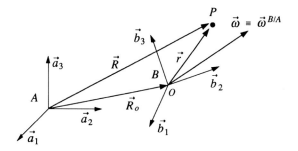

Fig. 1.3   Particle in a rotating and translating reference frame.

Also let an overdot* denote time differentiation in A. Then Eqs. (1.63) are often written as

$$\dot{\vec{b}}_1 = \vec{\omega} \times \vec{b}_1 \tag{1.64a}$$

$$\dot{\vec{b}}_2 = \vec{\omega} \times \vec{b}_2 \tag{1.64b}$$

$$\dot{\vec{b}}_3 = \vec{\omega} \times \vec{b}_3 \tag{1.64c}$$

and the angular velocity vector $\vec{\omega}$ can also be expressed as

$$\begin{aligned}
\vec{\omega} &= \omega_1 \vec{b}_1 + \omega_2 \vec{b}_2 + \omega_3 \vec{b}_3 \\
&= (\dot{\vec{b}}_2 \cdot \vec{b}_3) \vec{b}_1 + (\dot{\vec{b}}_3 \cdot \vec{b}_1) \vec{b}_2 + (\dot{\vec{b}}_1 \cdot \vec{b}_2) \vec{b}_3
\end{aligned} \tag{1.65}$$

Furthermore, Eqs. (1.64) can also be combined as

$$\begin{bmatrix} \dot{\vec{b}}_1 \\ \dot{\vec{b}}_2 \\ \dot{\vec{b}}_3 \end{bmatrix} = - \begin{bmatrix} 0 & -\omega_3 & \omega_2 \\ \omega_3 & 0 & -\omega_1 \\ -\omega_2 & \omega_1 & 0 \end{bmatrix} \begin{bmatrix} \vec{b}_1 \\ \vec{b}_2 \\ \vec{b}_3 \end{bmatrix}$$

1.2.3.2   *Particle Kinematics*   Consider now a particle P that is in motion relative to a reference frame B, which is itself in motion relative to an inertial reference frame A, as shown in Fig. 1.3. The angular velocity of B relative to A is denoted by $\vec{\omega} \equiv \vec{\omega}^{B/A}$. The position vector of the particle from the origin of A is denoted by $\vec{R}$ and the position vector of P from point O of B is denoted by $\vec{r}$.

Suppose that the position vector $\vec{r}$ is expressed in terms of basis vectors of B as follows:

$$\vec{r} = r_1 \vec{b}_1 + r_2 \vec{b}_2 + r_3 \vec{b}_3$$

---

*An overdot often denotes the time derivative measured in an inertial or Newtonian reference frame. It is called the Newtonian "dot" notation.

Then the time derivative of $\vec{r}$ measured in A, or as seen from A, becomes

$$\left\{\frac{d\vec{r}}{dt}\right\}_A \equiv \dot{\vec{r}} = \dot{r}_1\vec{b}_1 + \dot{r}_2\vec{b}_2 + \dot{r}_3\vec{b}_3 + r_1\dot{\vec{b}}_1 + r_2\dot{\vec{b}}_2 + r_3\dot{\vec{b}}_3 \qquad (1.66)$$

Defining the time derivative of $\vec{r}$ measured in B as

$$\left\{\frac{d\vec{r}}{dt}\right\}_B \equiv \dot{r}_1\vec{b}_1 + \dot{r}_2\vec{b}_2 + \dot{r}_3\vec{b}_3 \qquad (1.67)$$

and using Eqs. (1.64) and (1.67), we obtain

$$\left\{\frac{d\vec{r}}{dt}\right\}_A = \left\{\frac{d\vec{r}}{dt}\right\}_B + r_1(\vec{\omega} \times \vec{b}_1) + r_2(\vec{\omega} \times \vec{b}_2) + r_3(\vec{\omega} \times \vec{b}_3)$$

$$= \left\{\frac{d\vec{r}}{dt}\right\}_B + \vec{\omega} \times \vec{r}$$

Note that $\{d\vec{r}/dt\}_A$ is, in fact, the velocity of the particle P with respect to the origin of B measured in A (or as seen from A) and that $\{d\vec{r}/dt\}_B$ is the velocity of the particle P with respect to the origin of B measured in B.

We now summarize this important result for an arbitrary vector $\vec{r}$ as follows:

$$\left\{\frac{d\vec{r}}{dt}\right\}_A = \left\{\frac{d\vec{r}}{dt}\right\}_B + \vec{\omega}^{B/A} \times \vec{r} \qquad (1.68)$$

Because $\vec{\omega}^{B/A} = -\vec{\omega}^{A/B}$, we also have

$$\left\{\frac{d\vec{r}}{dt}\right\}_B = \left\{\frac{d\vec{r}}{dt}\right\}_A + \vec{\omega}^{A/B} \times \vec{r} \qquad (1.69)$$

Equations (1.68) and (1.69) apply to any vector quantity and are of fundamental importance to dynamic problems in which a rotating reference frame is involved.

Problem

1.6 Consider a particle P that is in motion relative to a reference frame B, which is itself in motion relative to an inertial reference frame A, as shown in Fig. 1.3. The angular velocity of B relative to A is denoted by $\vec{\omega} \equiv \vec{\omega}^{B/A}$. The position vector of P from the origin of A is given as

$$\vec{R} = \vec{R}_o + \vec{r}$$

where $\vec{R}_o$ is the position vector of point O of B from the origin of A, and $\vec{r}$ is the position vector of P from point O of B.

Suppose that A is inertially fixed and that $\vec{r}$ is expressed in terms of basis vectors of B as follows:

$$\vec{r} = r_1\vec{b}_1 + r_2\vec{b}_2 + r_3\vec{b}_3$$

The velocity and acceleration of the particle P with respect to the origin of B measured in B are defined, respectively, as

$$\left\{\frac{d\vec{r}}{dt}\right\}_B \equiv \dot{r}_1\vec{b}_1 + \dot{r}_2\vec{b}_2 + \dot{r}_3\vec{b}_3$$

$$\left\{\frac{d^2\vec{r}}{dt^2}\right\}_B \equiv \ddot{r}_1\vec{b}_1 + \ddot{r}_2\vec{b}_2 + \ddot{r}_3\vec{b}_3$$

(a) Show that the angular acceleration of B with respect to A measured in A is the same as the angular acceleration of B with respect to A measured in B; i.e.,

$$\left\{\frac{d\vec{\omega}}{dt}\right\}_A = \left\{\frac{d\vec{\omega}}{dt}\right\}_B$$

(b) Show that the velocity and acceleration of the particle P with respect to the origin of A measured in A can be obtained as

$$\left\{\frac{d\vec{R}}{dt}\right\}_A = \left\{\frac{d\vec{R}_o}{dt}\right\}_A + \left\{\frac{d\vec{r}}{dt}\right\}_A$$

$$= \left\{\frac{d\vec{R}_o}{dt}\right\}_A + \left\{\frac{d\vec{r}}{dt}\right\}_B + \vec{\omega}\times\vec{r}$$

$$\left\{\frac{d^2\vec{R}}{dt^2}\right\}_A = \left\{\frac{d^2\vec{R}_o}{dt^2}\right\}_A + \left\{\frac{d^2\vec{r}}{dt^2}\right\}_A$$

$$= \left\{\frac{d^2\vec{R}_o}{dt^2}\right\}_A + \left\{\frac{d^2\vec{r}}{dt^2}\right\}_B + \dot{\vec{\omega}}\times\vec{r} + \vec{\omega}\times(\vec{\omega}\times\vec{r}) + 2\vec{\omega}\times\left\{\frac{d\vec{r}}{dt}\right\}_B$$

where $\dot{\vec{\omega}} \equiv \{d\vec{\omega}/dt\}_A = \{d\vec{\omega}/dt\}_B$ is the angular acceleration of B with respect to A measured in A or B.

Note: The term $\{d\vec{R}/dt\}_A$ is called the inertial velocity, $\{d^2\vec{R}/dt^2\}_A$ the inertial acceleration, $\vec{\omega}\times(\vec{\omega}\times\vec{r})$ the centripetal acceleration, $-\vec{\omega}\times(\vec{\omega}\times\vec{r})$ the centrifugal acceleration, and $2\vec{\omega}\times\{d\vec{r}/dt\}_B$ the Coriolis acceleration.* The centripetal acceleration vector $\vec{\omega}\times(\vec{\omega}\times\vec{r})$ is orthogonal to both $\vec{\omega}$ and $\vec{\omega}\times\vec{r}$ and it is directed toward the instantaneous axis of rotation.

1.2.4 Dynamics of a Particle

In formulating the equations of motion of a particle (or of a system of particles), it is extremely important to correctly take into account all forces, both reactive and externally applied, that act on the particle (or on the system of particles). This vital step in formulating the equations of motion is called isolating the system

---

*The term "Coriolis acceleration" was named after the French military engineer G. Coriolis (1792–1843), who first disclosed its existence.

or sketching its free-body diagram. It is assumed that the reader is familiar with the concept of drawing free-body diagrams through previous study of statics or mechanics.

1.2.4.1 Formulation of Equations of Motion

Consider a particle P of constant mass m moving in an inertial reference frame A, as shown in Fig. 1.3. Let $\vec{F}$ be the resultant of all forces acting on P, including constraint or reaction forces. Then the equation of motion of the particle P of mass m is simply written as

$$m\ddot{\vec{R}} = \vec{F} \tag{1.70}$$

where $\vec{R}$ is the inertial position vector of P and $\ddot{\vec{R}}$ is the inertial acceleration of P; i.e., $\ddot{\vec{R}} \equiv \{d^2\vec{R}/dt^2\}_A$. Note that $\vec{R} = \vec{R}_o + \vec{r}$ and $\ddot{\vec{R}} = \ddot{\vec{R}}_o + \ddot{\vec{r}}$.

If the inertial position vector $\vec{R}$ of P and the resultant of all forces $\vec{F}$ acting on P are expressed in terms of a set of basis vectors $\{\vec{a}_1, \vec{a}_2, \vec{a}_3\}$ fixed in an inertial reference frame A, as

$$\vec{R} = R_1\vec{a}_1 + R_2\vec{a}_2 + R_3\vec{a}_3$$
$$\vec{F} = F_1\vec{a}_1 + F_2\vec{a}_2 + F_3\vec{a}_3$$

then the vector differential equation (1.70) can be resolved into its three component equations as

$$\vec{a}_1: \quad m\ddot{R}_1 = F_1 \tag{1.71a}$$
$$\vec{a}_2: \quad m\ddot{R}_2 = F_2 \tag{1.71b}$$
$$\vec{a}_3: \quad m\ddot{R}_3 = F_3 \tag{1.71c}$$

These three equations of motion become independent only if the coordinates R_1, R_2, and R_3 are independent.

If the motion of a particle is subject to possible constraints, then a set of a minimum number of independent coordinates necessary for describing the motion of a particle with constraints is often employed in formulating the equations of motion. Such independent coordinates are called the *generalized coordinates*. The number of independent generalized coordinates corresponds to the degrees of freedom of the system.

The equation of motion in the form of Eq. (1.70) can be rewritten as

$$\vec{F} - m\ddot{\vec{R}} = 0 \tag{1.72}$$

which is known as *D'Alembert's principle*. The term $-m\ddot{\vec{R}}$ is often called an inertial force; however, such a fictitious force must be distinguished from an actual force $\vec{F}$ acting on the particle. In formulating the equation of motion of a particle, it is quite often convenient to apply the dynamic equilibrium condition (1.72) to a free-body diagram of a particle. However, considerable care must be taken in formulating the equation of motion by employing D'Alembert's principle.

The equation of motion of a particle also can be rewritten as

$$\frac{\mathrm{d}}{\mathrm{d}t}(m\dot{\vec{R}}) = \vec{F}$$

where $\dot{\vec{R}}$ is the inertial velocity of P; i.e., $\dot{\vec{R}} \equiv \{\mathrm{d}\vec{R}/\mathrm{d}t\}_A$ and $m\dot{\vec{R}}$ is called the *linear momentum vector* of a particle of constant mass m. If the resultant of all forces acting on the particle P is zero; i.e., $\vec{F} = 0$, then

$$m\dot{\vec{R}} = \text{constant vector}$$

which is known as the *principle of conservation of linear momentum*, and we have

$$\vec{R}(t) = \vec{R}(0) + \dot{\vec{R}}(0)t$$

The constant vectors $\vec{R}(0)$ and $\dot{\vec{R}}(0)$, which denote the inertial position and velocity vectors of the particle at $t = 0$, respectively, represent six constant integrals of the equation of motion.

1.2.4.2 Angular Momentum We now consider the angular momentum (or the moment of momentum) of a particle about an arbitrary point O, which is itself in motion relative to an inertial reference frame A, as shown in Fig. 1.3.

The moment about an arbitrary point O of the momentum $m\dot{\vec{R}}$ of a particle P of mass m is

$$\vec{H}_o = \vec{r} \times m\dot{\vec{R}} \tag{1.73}$$

where $\vec{r}$ is the position vector of P from point O of B and $\dot{\vec{R}}$ is the inertial velocity of P. The moment of momentum, $\vec{H}_o$, is often called the *absolute angular momentum* due to the absolute momentum $m\dot{\vec{R}}$ used in defining $\vec{H}_o$.

Differentiating $\vec{H}_o$ with respect to time, we obtain the angular momentum equation as

$$\dot{\vec{H}}_o = \dot{\vec{r}} \times m\dot{\vec{R}} + \vec{r} \times m\ddot{\vec{R}}$$
$$= \dot{\vec{r}} \times m(\dot{\vec{R}}_o + \dot{\vec{r}}) + \vec{r} \times \vec{F}$$
$$= m\dot{\vec{r}} \times \dot{\vec{R}}_o + \vec{r} \times \vec{F}$$

which can be rewritten as

$$\dot{\vec{H}}_o + \dot{\vec{R}}_o \times m\dot{\vec{r}} = \vec{M}_o \tag{1.74}$$

where $\vec{M}_o = \vec{r} \times \vec{F}$ denotes the moment of the resultant force $\vec{F}$ about point O of B.

Similarly, for the *relative* angular momentum of a particle about an arbitrary point O, which is defined as

$$\vec{h}_o = \vec{r} \times m\dot{\vec{r}} \tag{1.75}$$

we also obtain the angular momentum equation of the form

$$\dot{\vec{h}}_o + \vec{r} \times m\ddot{\vec{R}}_o = \vec{M}_o \tag{1.76}$$

The term "relative angular momentum" for $\vec{h}_o$ here is simply due to the relative momentum $m\dot{\vec{r}}$ used in defining $\vec{h}_o$.

If point O is inertially fixed, the distinction between $\vec{H}_o$ and $\vec{h}_o$ disappears, and the angular momentum equation becomes

$$\dot{\vec{H}}_o = \vec{M}_o \qquad \text{or} \qquad \dot{\vec{h}}_o = \vec{M}_o \tag{1.77}$$

Furthermore, if the external moment $\vec{M}_o$ is zero, then the angular momentum vector becomes a constant vector; that is, the angular momentum of the particle is conserved. This is known as the *principle of conservation of angular momentum*.

1.2.4.3 Kinetic and Potential Energy
The kinetic energy T of the particle P of mass m is defined as

$$T = \frac{1}{2} m \dot{\vec{R}} \cdot \dot{\vec{R}} \tag{1.78}$$

where $\dot{\vec{R}}$ is the inertial velocity of P.

If a given force $\vec{F}$ depends only on the position vector $\vec{R}$, then a scalar function V, which is called the potential energy, is defined such that

$$\vec{F} \cdot d\vec{R} = -dV \qquad \text{or} \qquad \vec{F} \cdot \dot{\vec{R}} = -\dot{V} \tag{1.79}$$

Such a force field is said to be conservative, and a conservative force $\vec{F}$ is often expressed as the gradient vector of the potential energy V, as follows:

$$\vec{F} = -\nabla V$$

and $\nabla \times \vec{F} = 0$. For a particle acted upon only by a conservative force, we have the *principle of conservation of energy* described by

$$\frac{d}{dt}(T + V) = 0 \qquad \text{or} \qquad T + V = E = \text{const} \tag{1.80}$$

where the constant of integration E is called the total energy.

Problems

1.7 Consider various mass–spring–damper systems shown in Fig. 1.4, assuming a frictionless horizontal line.

(a) By sketching a free-body diagram and applying Newton's second law of motion, verify that the equation of motion of the single-degree-of-freedom system shown in Fig. 1.4a is simply given as

$$m\ddot{x} + c\dot{x} + kx = 0$$

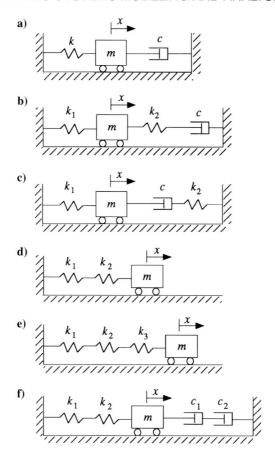

Fig. 1.4   Mass–spring–damper systems.

where x represents the position of the cart from the unstretched spring position, m the mass of the cart, k the linear spring stiffness, and c the viscous damping coefficient of a dashpot.*

Note: This system is referred to as a second-order dynamic system described by a second-order ordinary differential equation with constant coefficients.

(b) Derive the equation of motion of the system shown in Fig. 1.4b as

$$\frac{cm}{k_2}\frac{d^3x}{dt^3} + m\frac{d^2x}{dt^2} + c\left(1 + \frac{k_1}{k_2}\right)\frac{dx}{dt} + k_1x = 0$$

---

*A dashpot, also called a viscous damper, is a device that provides viscous friction or damping. It consists of a piston and oil-filled cylinder. Any relative motion between the cylinder and the piston is resisted by the oil, and the resulting damping force is proportional to the relative velocity between the cylinder and the piston. Consequently, the energy is dissipated as heat.

Note: The junction between the spring k_2 and the dashpot c is massless. Consequently, this system is referred to as a third-order dynamic system with one and one-half degrees of freedom [2].

(c) Derive the third-order differential equation for the system shown in Fig. 1.4c.

(d) Derive the equation of motion of the system, with two springs connected in series, shown in Fig. 1.4d as follows:

$$m\ddot{x} + \frac{k_1 k_2}{k_1 + k_2} x = 0$$

(e) Derive the equation of motion of the mass–spring system, with three springs connected in series, shown in Fig. 1.4e as follows:

$$m\ddot{x} + \frac{k_1 k_2 k_3}{k_1 k_2 + k_2 k_3 + k_1 k_3} x = 0$$

(f) Obtain the equation of motion of the mass–spring–damper system shown in Fig. 1.4f using the result in (d).

1.8   Consider a particle P of mass m that is in motion relative to a reference frame B, which is itself in motion relative to an inertial reference frame A (as was shown in Fig. 1.3), and let $\vec{F}$ be the resultant of all forces acting on P.

Assume that the reference frame B has a known translational motion $\vec{R}_o(t)$ and a known rotational motion $\vec{\omega}(t)$. Also assume that the position of the particle with respect to the origin of B is expressed as

$$\vec{r} = r_1 \vec{b}_1 + r_2 \vec{b}_2 + r_3 \vec{b}_3$$

and

$$\left\{ \frac{d\vec{r}}{dt} \right\}_B \equiv \dot{r}_1 \vec{b}_1 + \dot{r}_2 \vec{b}_2 + \dot{r}_3 \vec{b}_3$$

$$\left\{ \frac{d^2\vec{r}}{dt^2} \right\}_B \equiv \ddot{r}_1 \vec{b}_1 + \ddot{r}_2 \vec{b}_2 + \ddot{r}_3 \vec{b}_3$$

(a) Show that the equation of motion of P can be written as

$$m \left\{ \frac{d^2\vec{r}}{dt^2} \right\}_B = -m \left(\ddot{\vec{R}}_o + \dot{\vec{\omega}} \times \vec{r} + \vec{\omega} \times (\vec{\omega} \times \vec{r}) + 2\vec{\omega} \times \left\{ \frac{d\vec{r}}{dt} \right\}_B \right) + \vec{F}$$

The term $-m\vec{\omega} \times (\vec{\omega} \times \vec{r})$ is known as the centrifugal force and $2m\vec{\omega} \times \{d\vec{r}/dt\}_B$ as the Coriolis force. However, such fictitious forces must be distinguished from an actual force $\vec{F}$ acting on the particle.

(b) Show that the kinetic energy defined as $T = \frac{1}{2}m\dot{\vec{R}} \cdot \dot{\vec{R}}$, where $\dot{\vec{R}}$ is the inertial velocity of P, can be written as

$$T = T_2 + T_1 + T_0$$

where

$$T_2 = \frac{1}{2}m\left|\left\{\frac{d\vec{r}}{dt}\right\}_B\right|^2$$

$$T_1 = m(\dot{\vec{R}}_o + \vec{\omega} \times \vec{r}) \cdot \left\{\frac{d\vec{r}}{dt}\right\}_B$$

$$T_0 = \frac{1}{2}m|\dot{\vec{R}}_o + \vec{\omega} \times \vec{r}|^2$$

Notice that T_2 is a quadratic function of $(\dot{r}_1, \dot{r}_2, \dot{r}_3)$, T_1 is a linear function of $(\dot{r}_1, \dot{r}_2, \dot{r}_3)$, and T_0 is a nonnegative function of only (r_1, r_2, r_3) and time.

1.9 Consider a simple pendulum of mass m and constant length ℓ, connected by a massless rod to the inertially fixed, hinged support point O, as shown in Fig. 1.5a in which θ denotes the angular position of the pendulum from the vertical line. The system is constrained to move in a vertical plane. This well-known pendulum problem with one degree of freedom can be formulated and solved without employing vector notation. However, as an exercise problem of vectorial mechanics, a vectorial approach, which is often referred to as the Newton–Euler formulation of the equations of motion, is to be employed here.

As illustrated in Fig. 1.5a, let $\{\vec{a}_1, \vec{a}_2, \vec{a}_3\}$ and $\{\vec{b}_1, \vec{b}_2, \vec{b}_3\}$ be two sets of right-hand, orthogonal unit vectors of two reference frames A and B fixed at the support point O and the pendulum, respectively. Then the position vector of the pendulum mass from the point O can be expressed as: $\vec{r} = x_1 \vec{a}_1 + x_2 \vec{a}_2$ or $\vec{r} = \ell \vec{b}_1$. Because the rectangular coordinates x_1 and x_2 are related to θ as: $x_1 = \ell \cos \theta$ and $x_2 = \ell \sin \theta$, x_1 and x_2 are not independent but constrained by: $x_1^2 + x_2^2 = \ell^2$, and θ is the generalized coordinate of the system. The angular velocity of B with respect to A is $\vec{\omega}^{B/A} = \dot{\theta} \vec{b}_3$.

(a) Find the absolute acceleration of the pendulum mass as

$$\ddot{\vec{r}} = -\ell \dot{\theta}^2 \vec{b}_1 + \ell \ddot{\theta} \vec{b}_2$$

(b) Derive the equation of motion of the simple pendulum as

$$m\ell\ddot{\theta} = -mg \sin \theta + u$$

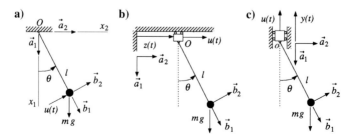

Fig. 1.5   Simple pendulum problems.

where g is the gravitational acceleration and u is the applied force along the $\vec{b}_2$ direction. Also determine the tension, T, in the rod as

$$T = m\ell\dot{\theta}^2 + mg\cos\theta$$

Hint: Sketch a free-body diagram of the pendulum mass, and then apply Newton's second law: $m\ddot{\vec{r}} = \vec{F}$ where $\vec{F} = mg\vec{a}_1 + u\vec{b}_2 - T\vec{b}_1$.

(c) Let the angular momentum vector be defined as $\vec{h}_o = \vec{r} \times m\dot{\vec{r}}$. Then, using the angular momentum equation $\dot{\vec{h}}_o = \vec{M}_o$, derive the equation of motion as

$$m\ell^2\ddot{\theta} = -mg\ell\sin\theta + u\ell$$

which is, in fact, the same as the equation of motion derived in (b).

(d) Obtain the linearized equation of motion for small angle of θ and solve for $\theta(t)$ for a constant u and initial conditions of $\theta(0)$ and $\dot{\theta}(0)$.
Hint: Assume that the solution is of the form

$$\theta(t) = A\cos\sqrt{\frac{g}{\ell}}t + B\sin\sqrt{\frac{g}{\ell}}t + C$$

and then determine constants A, B, and C in terms of constants u, $\theta(0)$, and $\dot{\theta}(0)$.

(e) For this conservative dynamic system without the applied force, i.e., $u = 0$, verify that

$$\frac{d}{dt}(T + V) = 0$$

where $T = \frac{1}{2}m\dot{\vec{r}}\cdot\dot{\vec{r}} = \frac{1}{2}m(\ell\dot{\theta})^2$ and $V = -mg\ell\cos\theta$.

1.10 Consider the pendulum shown in Fig. 1.5b in which the hinged support point is allowed to move with a horizontal displacement $z(t)$ along a frictionless horizontal line and is acted upon by an external force u. The system is constrained to move in a vertical plane. As illustrated in Fig. 1.5b, $\{\vec{a}_1, \vec{a}_2, \vec{a}_3\}$ is a set of basis vectors fixed in an inertial reference frame A and $\{\vec{b}_1, \vec{b}_2, \vec{b}_3\}$ is a set of basis vectors of a rotating reference frame B fixed at the pendulum. The absolute position vector of the pendulum mass is then expressed as $\vec{R} = \vec{R}_o + \vec{r}$ where $\vec{R}_o = z\vec{a}_2$ and $\vec{r} = \ell\vec{b}_1$. For this single particle system with two degrees of freedom, the coordinates θ and z are selected as the generalized coordinates.

(a) Derive the equations of motion of the system as

$$\vec{a}_2: \quad m\ddot{z} + m\ell\ddot{\theta}\cos\theta - m\ell\dot{\theta}^2\sin\theta = u$$
$$\vec{b}_2: \quad m\ell\ddot{\theta} + m\ddot{z}\cos\theta + mg\sin\theta = 0$$

Note that the constraint forces, such as the tension in the rod and the normal force acting on the hinged support point, do not appear in this set of two independent equations.

(b) Obtain the second equation derived in (a) using the relative angular momentum equation: $\dot{\vec{h}}_o + \vec{r} \times m\ddot{\vec{R}}_o = \vec{M}_o$ where $\vec{h}_o = \vec{r} \times m\dot{\vec{r}}$.

(c) Also obtain the second equation derived in (a) using the absolute angular momentum equation $\dot{\vec{H}}_o + \dot{\vec{R}}_o \times m\dot{\vec{r}}1 = \vec{M}_o$, where $\vec{H}_o = \vec{r} \times m\dot{\vec{R}}$.

1.11 Consider the pendulum shown in Fig. 1.5c in which the hinged support point is allowed to move with a vertical displacement $y(t)$ along a frictionless vertical line and is acted upon by an external force u. As illustrated in Fig. 1.5c, $\{\vec{a}_1, \vec{a}_2, \vec{a}_3\}$ is a set of basis vectors fixed in an inertial reference frame A and $\{\vec{b}_1, \vec{b}_2, \vec{b}_3\}$ is a set of basis vectors of a rotating reference frame B fixed at the pendulum. The absolute position vector of the pendulum mass is then expressed as $\vec{R} = \vec{R}_o + \vec{r}$ where $\vec{R}_o = -y\vec{a}_1$ and $\vec{r} = \ell\vec{b}_1$.

(a) Derive the equations of motion as

$$\vec{a}_1: \quad m\ddot{y} + m\ell\ddot{\theta}\sin\theta + m\ell\dot{\theta}^2\cos\theta = u - mg$$

$$\vec{b}_2: \quad m\ell\ddot{\theta} + m\ddot{y}\sin\theta + mg\sin\theta = 0$$

(b) Obtain the second equation derived in (a) using the relative angular momentum equation $\dot{\vec{h}}_o + \vec{r} \times m\ddot{\vec{R}}_o = \vec{M}_o$ where $\vec{h}_o = \vec{r} \times m\dot{\vec{r}}$.

(c) Also obtain the second equation derived in (a) using the absolute angular momentum equation $\dot{\vec{H}}_o + \dot{\vec{R}}_o \times m\dot{\vec{r}} = \vec{M}_o$ where $\vec{H}_o = \vec{r} \times m\dot{\vec{R}}$.

Note: If the vertical motion of the hinged support point is prescribed as $y(t) = A\cos\omega t$, then for the small angular motion of the pendulum, we have the so-called Mathieu equation:

$$\ddot{\theta} + \left(\frac{g}{\ell} - \frac{A\omega^2}{\ell}\cos\omega t \right)\theta = 0$$

1.12 A particle of mass m is constrained to move on a smooth circular hoop of radius ℓ under the action of gravity, as illustrated in Fig. 1.6. It is assumed that the circular hoop rotates with constant angular velocity of Ω about the vertical line. Like the simple pendulum problem, this dynamic problem can also be formulated and solved without employing vector notation. However, as an exercise problem of three-dimensional vectorial mechanics, a vectorial approach to the formulation of the equations of motion is to be employed.

As illustrated in Fig. 1.6, $\{\vec{a}_1, \vec{a}_2, \vec{a}_3\}$ is a set of basis vectors fixed in an inertial reference frame A with the origin O and $\{\vec{b}_1, \vec{b}_2, \vec{b}_3\}$ is a set of basis vectors of a rotating reference frame B fixed at the particle. The position vector of the particle from the inertially fixed point O is then expressed as: $\vec{r} = \ell\vec{b}_1$. The angular velocity of B with respect to A is $\vec{\omega}^{B/A} = -\Omega\vec{a}_1 + \dot{\theta}\vec{b}_3$.

(a) Derive the equation of motion of the particle as

$$m\ell\ddot{\theta} + mg\sin\theta = m\ell\Omega^2\sin\theta\cos\theta$$

where θ denotes the angular position of the particle from the vertical line and g is the gravitational acceleration.

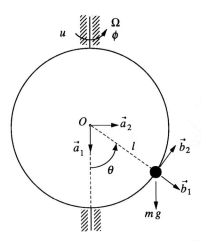

Fig. 1.6   Particle on a rotating circular hoop.

(b) In fact, an external torque is required to maintain the constant angular velocity Ω about the vertical line unless $\dot{\theta} = 0$. Determine such external torque required to be acting upon the massless hoop to maintain constant Ω in the presence of the $\dot{\theta}$ motion of the particle.

Hint: For the same system, but with two degrees of freedom, choosing θ and ϕ ($\dot{\phi} \equiv \Omega \neq$ const) as the generalized coordinates, derive the additional equation of motion as

$$m(\ell \sin \theta)^2 \ddot{\phi} + 2m\ell^2 \dot{\phi}\dot{\theta} \sin \theta \cos \theta = u$$

or

$$\frac{\mathrm{d}}{\mathrm{d}t}(m\ell^2 \sin^2 \theta \dot{\phi}) = u$$

where u is the external torque acting upon the massless hoop along the vertical line.

(c) For this conservative system without the applied torque u, verify that

$$\frac{\mathrm{d}}{\mathrm{d}t}(T + V) = 0$$

where $T = \frac{1}{2}m\dot{\vec{r}} \cdot \dot{\vec{r}} = \frac{1}{2}m(\ell^2\dot{\theta}^2 + \ell^2\dot{\phi}^2\sin^2\theta)$ and $V = -mg\ell \cos \theta$.

1.2.5 Dynamics of a System of Particles

The Newton–Euler formulation of the equations of motion for a system of particles is treated here, with emphasis on the generalized form of Newton's second law of motion and various angular momentum equations of a system of particles.

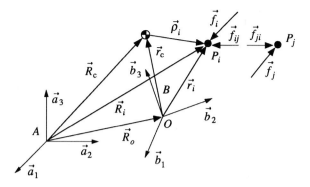

Fig. 1.7   System of particles.

1.2.5.1   Generalized form of Newton's Second Law of Motion   Consider a system of N particles $P_1, \ldots, P_N$ of masses $m_1, \ldots, m_N$ that are in motion relative to a reference frame B with the origin O, which is itself in motion relative to an inertial reference frame A, as shown in Fig. 1.7.

The equation of motion of the ith particle can be written as

$$m_i \ddot{\vec{R}}_i = \vec{f}_i + \sum_{j=1}^{N} \vec{f}_{ij}, \qquad i = 1, 2, \ldots, N \qquad (1.81)$$

where $\vec{f}_i$ is the external force acting on P_i, $\vec{f}_{ij}$ is the internal force acting on P_i due to P_j ($\vec{f}_{ii} = 0$ by the definition of an internal force), $\vec{R}_i$ is the position vector of P_i from the inertial origin of A, and $\ddot{\vec{R}}_i$ is the inertial acceleration of the ith particle P_i.

By summing these N equations, we obtain

$$\sum_{i=1}^{N} m_i \ddot{\vec{R}}_i = \sum_{i=1}^{N} \vec{f}_i \qquad (1.82)$$

since

$$\sum_{i=1}^{N} \sum_{j=1}^{N} \vec{f}_{ij} = 0 \qquad (1.83)$$

by Newton's third law $\vec{f}_{ij} = -\vec{f}_{ji}$.

Introducing the center of mass of the system, as illustrated in Fig. 1.7, such that

$$\vec{R}_i = \vec{R}_c + \vec{\rho}_i, \qquad \sum_{i=1}^{N} m_i \vec{r}_i = m\vec{r}_c, \qquad \sum_{i=1}^{N} m_i \vec{\rho}_i = 0 \qquad (1.84)$$

where $m = \sum_{i=1}^{N} m_i$ denotes the total mass and $\vec{r}_c$ denotes the position vector of the center of mass from the origin O of B, we rewrite Eq. (1.82) as

$$m\ddot{\vec{R}}_c = \vec{F} \tag{1.85}$$

where

$$\vec{F} = \sum_{i=1}^{N} \vec{f}_i$$

denotes the resultant of all external forces acting on the system and $\ddot{\vec{R}}_c$ is the inertial acceleration of the center of mass.

Equation (1.85) is often referred to as the generalized form of Newton's second law of motion for a system of particles of total mass m. This translational equation of motion of the system indicates that the center of mass of the system moves as if the total mass were concentrated at the center of mass and the resultant of all the external forces on the system acted at that point.

Furthermore, if the resultant of all forces acting on the system is zero; i.e., $\vec{F} = 0$, then

$$m\dot{\vec{R}}_c = \text{constant vector}$$

which is known as the principle of conservation of linear momentum for a system of particles.

1.2.5.2 Angular Momentum Equations The angular momentum (also called the moment of momentum), which is one of the fundamental concepts in classical mechanics, is now considered for a system of particles as follows.

The absolute angular momentum of a system of particles about an arbitrary point O, which is itself in motion relative to an inertial reference frame, is defined as

$$\vec{H}_o = \sum_{i=1}^{N} \vec{r}_i \times m_i \dot{\vec{R}}_i \tag{1.86}$$

Differentiating $\vec{H}_o$ with respect to time, we obtain

$$\dot{\vec{H}}_o = \sum_{i=1}^{N} \dot{\vec{r}}_i \times m_i \dot{\vec{R}}_i + \sum_{i=1}^{N} \vec{r}_i \times m_i \ddot{\vec{R}}_i$$

$$= \sum_{i=1}^{N} \dot{\vec{r}}_i \times m_i (\dot{\vec{R}}_o + \dot{\vec{r}}_i) + \sum_{i=1}^{N} \vec{r}_i \times \left(\vec{f}_i + \sum_{j=1}^{N} \vec{f}_{ij} \right)$$

$$= \sum_{i=1}^{N} m_i \dot{\vec{r}}_i \times \dot{\vec{R}}_o + \sum_{i=1}^{N} \vec{r}_i \times \vec{f}_i + \sum_{i=1}^{N} \sum_{j=1}^{N} \vec{r}_i \times \vec{f}_{ij}$$

Noting that $\sum m_i \vec{r}_i = m\vec{r}_c$ and

$$\sum_{i=1}^{N}\sum_{j=1}^{N} \vec{r}_i \times \vec{f}_{ij} = \frac{1}{2}\sum_{i=1}^{N}\sum_{j=1}^{N}(\vec{r}_i - \vec{r}_j) \times \vec{f}_{ij}$$

$$= 0 \quad (\text{as } \vec{f}_{ij} \text{ is collinear with } \vec{r}_i - \vec{r}_j)$$

we obtain the angular momentum equation of the form

$$\dot{\vec{H}}_o + \dot{\vec{R}}_o \times m\dot{\vec{r}}_c = \vec{M}_o \qquad (1.87)$$

where

$$\vec{M}_o = \sum_{i=1}^{N} \vec{r}_i \times \vec{f}_i \qquad (1.88)$$

is the moment of the external forces about the point O. The fact that the internal forces do not appear in $\vec{M}_o$ is a reason that angular momentum is a useful concept for a system of particles.

Similarly, defining the relative angular momentum about an arbitrary point O as

$$\vec{h}_o = \sum_{i=1}^{N} \vec{r}_i \times m_i \dot{\vec{r}}_i \qquad (1.89)$$

we also obtain the angular momentum equation of the form

$$\dot{\vec{h}}_o + \vec{r}_c \times m\ddot{\vec{R}}_o = \vec{M}_o \qquad (1.90)$$

Similar to $\vec{H}_o$ and $\vec{h}_o$, the absolute angular momentum $\vec{H}_c$ and the relative angular momentum $\vec{h}_c$ about the center of mass can be defined as

$$\vec{H}_c = \sum_{i=1}^{N} \vec{\rho}_i \times m_i \dot{\vec{R}}_i \qquad (1.91a)$$

$$\vec{h}_c = \sum_{i=1}^{N} \vec{\rho}_i \times m_i \dot{\vec{\rho}}_i \qquad (1.91b)$$

Because $\sum m_i \vec{\rho}_i = 0$ by the definition of the center of mass, the absolute and relative angular momenta about the center of mass are in fact identical; i.e.,

$$\vec{H}_c \equiv \vec{h}_c$$

and also we have the following relationships:

$$\vec{H}_o = \vec{H}_c + \vec{r}_c \times m\dot{\vec{R}}_c \qquad (1.92a)$$

$$\vec{h}_o = \vec{h}_c + \vec{r}_c \times m\dot{\vec{r}}_c \qquad (1.92b)$$

Consequently, the angular momentum equation (1.87) can be rewritten as

$$\dot{\vec{H}}_c + \vec{r}_c \times m\ddot{\vec{R}}_c = \vec{M}_o \qquad (1.93)$$

If the reference point O is either inertially fixed or at the moving center of mass of the system, the distinction between $\vec{H}_o$ and $\vec{h}_o$ disappears and the angular momentum equation (1.87) simply becomes

$$\dot{\vec{H}}_o = \vec{M}_o \qquad \text{or} \qquad \dot{\vec{H}}_c = \vec{M}_c \qquad (1.94)$$

where

$$\vec{M}_o = \sum_{i=1}^{N} \vec{r}_i \times \vec{f}_i \qquad \text{and} \qquad \vec{M}_c = \sum_{i=1}^{N} \vec{\rho}_i \times \vec{f}_i$$

For this reason, an inertially fixed point or the moving center of mass is often selected as a reference point O. Furthermore, if the external moment is zero, then the angular momentum about an inertially fixed point or the center of mass becomes a constant vector; that is, the angular momentum of the system is conserved. This is known as the principle of conservation of angular momentum.

However, there are many cases in which a reference point is selected to be neither inertially fixed nor the center of mass of the system.

A system of particles in which the distance between any two particles is constant is called a *rigid body*. Because a rigid body is a special case of a system of particles, all of the preceding results developed for a system of particles are also valid for rigid bodies. Rotational kinematics and attitude dynamics of more general rigid bodies will be studied further in Chapters 5 and 6. For a more detailed treatment of classical dynamics, the reader is referred to other standard textbooks on dynamics; e.g., see Greenwood [3].

Problems

1.13 Consider a system of N particles $P_1, \ldots, P_N$ of masses $m_1, \ldots, m_N$, as shown in Fig. 1.7. Suppose that there are no external forces and that the internal forces $\vec{f}_{ij}$ are only due to the gravity forces among the particles. The equation of motion for the ith particle can then be written as

$$m_i \ddot{\vec{R}}_i = -\sum_{j=1}^{N} \frac{G m_i m_j}{R_{ij}^3} \vec{R}_{ij} \qquad (i \neq j)$$

where $\ddot{\vec{R}}_i$ is the inertial acceleration of the ith particle P_i and $R_{ij} \equiv |\vec{R}_{ij}|$ where $\vec{R}_{ij} = \vec{R}_i - \vec{R}_j$ is the position vector of P_i from P_j.

(a) Verify that for such a conservative dynamic system, we have the principle of conservation of energy:

$$T + V = \text{const} \qquad \text{or} \qquad \frac{d}{dt}(T + V) = 0$$

where

$$T = \text{kinetic energy} = \frac{1}{2} \sum_{i=1}^{N} m_i (\dot{\vec{R}}_i \cdot \dot{\vec{R}}_i)$$

$$V = \text{potential energy} = -\frac{1}{2} \sum_{i=1}^{N} \sum_{j=1}^{N} \frac{Gm_i m_j}{R_{ij}} \qquad (i \neq j)$$

Hint: Take the dot product of Eq. (1.81) with $\dot{\vec{R}}_i$ and sum the resulting N equations of all particles. Then show that whether the resulting equation is the same as $\dot{T} + \dot{V} = 0$. Also note that

$$\sum_{i=1}^{N} \sum_{j=1}^{N} \vec{R}_{ij} \cdot \dot{\vec{R}}_i \equiv \sum_{i=1}^{N} \sum_{j=1}^{N} \vec{R}_{ji} \cdot \dot{\vec{R}}_j \equiv - \sum_{i=1}^{N} \sum_{j=1}^{N} \vec{R}_{ij} \cdot \dot{\vec{R}}_j$$

$$\sum_{i=1}^{N} \sum_{j=1}^{N} \vec{R}_{ij} \cdot \dot{\vec{R}}_i = \frac{1}{2} \sum_{i=1}^{N} \sum_{j=1}^{N} \vec{R}_{ij} \cdot (\dot{\vec{R}}_i - \dot{\vec{R}}_j)$$

$$= \frac{1}{2} \sum_{i=1}^{N} \sum_{j=1}^{N} \vec{R}_{ij} \cdot \dot{\vec{R}}_{ij} = \frac{1}{2} \sum_{i=1}^{N} \sum_{j=1}^{N} R_{ij} \dot{R}_{ij}$$

(b) Show that the kinetic energy T can also be expressed as

$$T = \frac{1}{2} m |\dot{\vec{R}}_c|^2 + \frac{1}{2} \sum_{i=1}^{N} m_i |\dot{\vec{\rho}}_i|^2$$

where m is the total mass of the system, $\vec{R}_c$ is the position vector of the composite center of mass from the origin of an inertial reference frame A, and $\vec{\rho}_i$ is the position vector of the ith particle from the composite center of mass.
Hint:

$$\vec{R}_i = \vec{R}_c + \vec{\rho}_i, \qquad \sum_{i=1}^{N} m_i \vec{\rho}_i = 0, \qquad m = \sum_{i=1}^{N} m_i$$

$$\sum_{i=1}^{N} m_i \vec{r}_i = m \vec{r}_c, \qquad \dot{\vec{\rho}}_i \cdot \dot{\vec{\rho}}_i = |\dot{\vec{\rho}}_i|^2$$

and $\vec{r}_c$ is the position vector of the composite center of mass from point O.

(c) Show that this dynamic system of $3N$ degrees of freedom (or $6N$ states) has 10 constants of integration (or integrals of the motion).
Note: In 1846, Urbain J. LeVerrier in France and John C. Adams in England were able to discover the planet Neptune by considering Eq. (1.81)

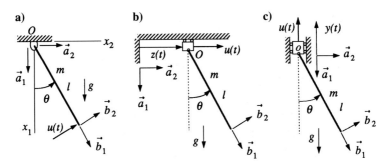

Fig. 1.8   Compound pendulum problems of a thin uniform bar.

for the motion of Uranus and comparing the computational results to the observed motion of Uranus.

1.14 Consider a thin uniform bar of mass m and length ℓ hinged at an inertially fixed point O, as shown in Fig. 1.8a, in which θ denotes the angular position of the bar from the vertical line, u the external force, and g the gravitational acceleration. The system is constrained to move in a vertical plane.

(a) Considering the thin uniform bar as a system of particles of an infinitesimal mass element dm, show that the angular momentum vector about the hinge point O can be expressed as

$$\vec{h}_o = \tfrac{1}{3}m\ell^2\dot\theta\vec{b}_3$$

   Hint: Use the definition of the angular momentum vector of a rigid body about a fixed point O, $\vec{h}_o = \int \vec{r} \times \dot{\vec{r}}\,dm$, and let $\vec{r} = r\vec{b}_1$ be the position vector of an infinitesimal mass element dm from the point O and $dm = (m/\ell)\,dr$.

(b) Using the angular momentum equation, $\dot{\vec{h}}_o = \vec{M}_o$, derive the equation of motion of the compound pendulum shown in Fig. 1.8a as

$$\tfrac{1}{3}m\ell^2\ddot\theta + \tfrac{1}{2}mg\ell\sin\theta = u\ell$$

   Note: The angular momenta, h_o and h_c, of the system can then be expressed as $h_o = I_o\dot\theta$ and $h_c = I_c\dot\theta$, where $I_o = m\ell^2/3$ denotes the moment of inertia of the thin uniform bar about the hinge point O and $I_c = m\ell^2/12$ denotes the moment of inertia of the thin uniform bar about its center of mass. Note that I_o and I_c are related to each other by the parallel-axis theorem: $I_o = I_c + md^2$ where d is the distance between the hinge point O and the center of mass. The thin uniform bar is a simple example of a rigid body. The three-dimensional, rotational kinematics and dynamics of more general rigid bodies will be treated in detail in Chapters 5 and 6.

(c) Also derive the equation of motion derived in (b) using the angular momentum equation about the center of mass and the translational equation of motion of the center of mass.

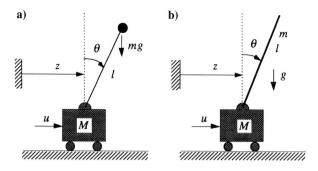

Fig. 1.9   Inverted pendulum problems.

1.15 Derive the equations of motion of the pendulum shown in Fig. 1.8b.
Hint: $m\ddot{\vec{R}}_c = \vec{F}$, $\dot{\vec{h}}_c = \vec{M}_c$, $\vec{h}_c = h_c \vec{b}_3$, $h_c = I_c \dot{\theta}$, $I_c = m\ell^2/12$, $\dot{\vec{h}}_o +$
$\vec{r}_c \times m\ddot{\vec{R}}_o = \vec{M}_o$, $\vec{h}_o = h_o \vec{b}_3$, $h_o = I_o \dot{\theta}$, $I_o = m\ell^2/3$.

1.16 Derive the equations of motion of the pendulum shown in Fig. 1.8c.

1.17 Consider a cart of mass M with a) an inverted, point-mass pendulum or b)
an inverted, thin uniform bar, on a frictionless horizontal line, as illustrated
in Fig. 1.9. Let $z(t)$ be the horizontal position of the cart, $\theta(t)$ the angle of
the pendulum or the thin bar measured from the vertical position, $u(t)$ the
control input force acting on the cart, and g the gravitational acceleration.
The system is constrained to move in a vertical plane.
(a) Derive the equations of motion of the system shown in Fig. 1.9a as

$$(M + m)\ddot{z} + m\ell\ddot{\theta}\cos\theta - m\ell\dot{\theta}^2 \sin\theta = u$$
$$m\ell\ddot{\theta} + m\ddot{z}\cos\theta - mg\sin\theta = 0$$

(b) Derive the equations of motion of the system shown in Fig. 1.9b as

$$(M + m)\ddot{z} + \frac{m\ell}{2}\ddot{\theta}\cos\theta - \frac{m\ell}{2}\dot{\theta}^2 \sin\theta = u$$
$$\frac{2m\ell}{3}\ddot{\theta} + m\ddot{z}\cos\theta - mg\sin\theta = 0$$

(c) Also derive the equations of motion of the system shown in Fig. 1.9b
with an additional tip mass m_o.

1.18 Consider a two-link manipulator that is modeled as a double pendulum
consisting of two massless rods of lengths ℓ_1 and ℓ_2 and point masses
m_1 and m_2, as illustrated in Figs. 1.10a and 1.10b. The first link is pinned
to a fixed point O and the second link is also pinned to the first link. Let
u_1 and u_2 be the shoulder and elbow joint control torques, respectively,
and let g denote the gravitational acceleration. As illustrated in Fig. 1.10a,

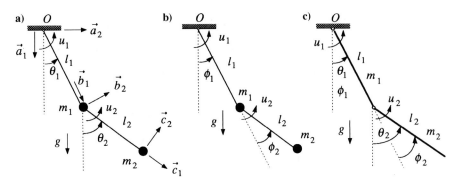

Fig. 1.10 Double pendulum problems.

$\{\vec{a}_1, \vec{a}_2, \vec{a}_3\}$, $\{\vec{b}_1, \vec{b}_2, \vec{b}_3\}$ and $\{\vec{c}_1, \vec{c}_2, \vec{c}_3\}$ are three sets of orthogonal unit vectors fixed at the point O, the first link, and the second link, respectively.

(a) Derive the equations of motion of the double pendulum shown in Fig. 1.10a in which $\{\theta_1, \theta_2\}$ are chosen as a set of the generalized coordinates of the system as follows:

$$(m_1 + m_2)\ell_1^2\ddot{\theta}_1 + m_2\ell_1\ell_2\cos(\theta_2 - \theta_1)\ddot{\theta}_2$$
$$= u_1 - u_2 + m_2\ell_1\ell_2\dot{\theta}_2^2\sin(\theta_2 - \theta_1) - (m_1 + m_2)g\ell_1\sin\theta_1$$
$$m_2\ell_1\ell_2\cos(\theta_2 - \theta_1)\ddot{\theta}_1 + m_2\ell_2^2\ddot{\theta}_2$$
$$= u_2 - m_2\ell_1\ell_2\dot{\theta}_1^2\sin(\theta_2 - \theta_1) - m_2g\ell_2\sin\theta_2$$

(b) Derive the equations of motion for the double pendulum shown in Fig. 1.10b in which $\{\phi_1, \phi_2\}$ are chosen as a set of the generalized coordinates of the system.

1.19   For a double pendulum consisting of two uniform slender bars as shown in Fig. 1.10c, derive the equations of motion in terms of $\{\theta_1, \theta_2\}$ as

$$\left(\tfrac{1}{3}m_1\ell_1^2 + m_2\ell_1^2\right)\ddot{\theta}_1 + \tfrac{1}{2}m_2\ell_1\ell_2\cos(\theta_2 - \theta_1)\ddot{\theta}_2$$
$$= u_1 - u_2 + \tfrac{1}{2}m_2\ell_1\ell_2\dot{\theta}_2^2\sin(\theta_2 - \theta_1) - \tfrac{1}{2}m_1g\ell_1\sin\theta_1 - m_2g\ell_1\sin\theta_1$$
$$\tfrac{1}{2}m_2\ell_1\ell_2\cos(\theta_2 - \theta_1)\ddot{\theta}_1 + \tfrac{1}{3}m_2\ell_2^2\ddot{\theta}_2$$
$$= u_2 - \tfrac{1}{2}m_2\ell_1\ell_2\dot{\theta}_1^2\sin(\theta_2 - \theta_1) - \tfrac{1}{2}m_2g\ell_2\sin\theta_2$$

Note: In terms of $\{\phi_1, \phi_2\}$, these equations can be rewritten as

$$\left(\tfrac{1}{3}m_1\ell_1^2 + m_2\ell_1^2 + \tfrac{1}{2}m_2\ell_1\ell_2\cos\phi_2\right)\ddot{\phi}_1 + \tfrac{1}{2}m_2\ell_1\ell_2\cos\phi_2\ddot{\phi}_2$$
$$= u_1 - u_2 + \tfrac{1}{2}m_2\ell_1\ell_2(\dot{\phi}_1 + \dot{\phi}_2)^2\sin\phi_2$$
$$- \tfrac{1}{2}m_1 g\ell_1\sin\phi_1 - m_2 g\ell_1\sin\phi_1$$
$$\left(\tfrac{1}{3}m_2\ell_2^2 + \tfrac{1}{2}m_2\ell_1\ell_2\cos\phi_2\right)\ddot{\phi}_1 + \tfrac{1}{3}m_2\ell_2^2\ddot{\phi}_2$$
$$= u_2 - \tfrac{1}{2}m_2\ell_1\ell_2\dot{\phi}_1^2\sin\phi_2 - \tfrac{1}{2}m_2 g\ell_2\sin(\phi_1 + \phi_2)$$

And we can also obtain the following equivalent set of equations of motion:

$$\left(\tfrac{1}{3}m_1\ell_1^2 + \tfrac{1}{3}m_2\ell_2^2 + m_2\ell_1^2 + m_2\ell_1\ell_2\cos\phi_2\right)\ddot{\phi}_1$$
$$+ \left(\tfrac{1}{3}m_2\ell_2^2 + \tfrac{1}{2}m_2\ell_1\ell_2\cos\phi_2\right)\ddot{\phi}_2$$
$$= u_1 + \tfrac{1}{2}m_2\ell_1\ell_2\left(2\dot{\phi}_1\dot{\phi}_2 + \dot{\phi}_2^2\right)\sin\phi_2 - \tfrac{1}{2}m_1 g\ell_1\sin\phi_1$$
$$- m_2 g\ell_1\sin\phi_1 - \tfrac{1}{2}m_2 g\ell_2\sin(\phi_1 + \phi_2)$$
$$\left(\tfrac{1}{3}m_2\ell_2^2 + \tfrac{1}{2}m_2\ell_1\ell_2\cos\phi_2\right)\ddot{\phi}_1 + \tfrac{1}{3}m_2\ell_2^2\ddot{\phi}_2$$
$$= u_2 - \tfrac{1}{2}m_2\ell_1\ell_2\dot{\phi}_1^2\sin\phi_2 - \tfrac{1}{2}m_2 g\ell_2\sin(\phi_1 + \phi_2)$$

This set of equations of motion can also be obtained somewhat directly using the so-called Lagrange method, which will be treated in the next section.

1.2.6 Lagrange's Equations of Motion

Classical mechanics is primarily based on Newton's three laws of motion and his law of gravity. However, more theoretical developments of analytical dynamics are possible by introducing the concepts of virtual displacements and virtual work. We shall also introduce some fundamental concepts of analytical dynamics,* including the principle of virtual work, Hamilton's principle, Lagrange's equations of motion, and Hamilton's canonical equations of motion. However, no attempt is made here to cover the details of advanced analytical dynamics. For a more detailed treatment of analytical dynamics, the reader is referred to Greenwood [3] or Meirovitch [4].

1.2.6.1 Principle of Virtual Work
Consider a system of N particles subject to possible constraints. A system is said to be *holonomic* if the constraints can be expressed as functions of coordinates and/or time; otherwise, it is said to be *nonholonomic*. A set of a minimum number of independent coordinates necessary for describing the motion of a dynamic system with constraints is called

---

*Leonhard Euler (1707–1783), who introduced such mathematical symbols as π, e, $\sum$, e^x, $\log x$, $\sin x$, $\cos x$, and $f(x)$, established the foundations for mathematical physics and analytical dynamics. Later, Jean D'Alembert (1717–1783), Joseph Lagrange (1736–1813), and William Hamilton (1805–1865) further developed the theory of analytical dynamics, and Henri Poincaré (1854–1912) further established the foundations for a theory of dynamic systems.

the generalized coordinates, often denoted by $\{q_1, q_2, \ldots, q_n\}$. The number n of independent generalized coordinates corresponds to the degrees of freedom of the system. However, many different sets of generalized coordinates are possible in a given problem.

As an example of a single-degree-of-freedom system with a holonomic constraint, consider a simple pendulum of mass m and length ℓ, connected by a massless rod to the hinged support point O, as was shown in Fig. 1.5a. The position vector of the pendulum mass can be expressed as $\vec{r} = x_1 \vec{a}_1 + x_2 \vec{a}_2$ or $\vec{r} = \ell \vec{b}_1$. The rectangular coordinates x_1 and x_2 are related to the generalized coordinate θ as $x_1 = \ell \cos \theta$ and $x_2 = \ell \sin \theta$. The holonomic constraint is then simply given by $x_1^2 + x_2^2 = \ell^2$. The equation of motion in vector form is

$$m\ddot{\vec{r}} = mg\vec{a}_1 + u\vec{b}_2 - T\vec{b}_1$$

where mg is the gravitational force, u is the applied force, and T is the tension in the massless rod known as a nonworking constraint force.

In general, the equation of motion of the ith particle of mass m_i for a system of N particles is written as

$$m_i \ddot{\vec{r}}_i = \vec{f}_i + \vec{f}'_i, \qquad i = 1, \ldots, N \qquad (1.95)$$

where $\vec{f}_i$ is an applied force acting on the ith particle, $\vec{f}'_i$ is a constraint force acting on the ith particle, and $\vec{r}_i$ is the inertial position vector of the ith particle.

According to the *principle of virtual work*, the virtual work done by all constraint forces is zero; i.e., we have

$$\sum_{i=1}^{N} \vec{f}'_i \cdot \delta \vec{r}_i = 0 \qquad (1.96)$$

where $\delta \vec{r}_i$ is called the virtual displacement associated with the ith particle. The virtual displacement $\delta \vec{r}_i$ is different from the actual displacement $d\vec{r}_i$, but it must be compatible with the constraints irrespective of time.

1.2.6.2 Hamilton's Principle
The principle of virtual work and D'Alembert's principle provide the following equation of motion for a system of N particles:

$$\sum_{i=1}^{N} (\vec{f}_i - m_i \ddot{\vec{r}}_i) \cdot \delta \vec{r}_i = 0 \qquad (1.97)$$

which is rewritten as

$$\sum_{i=1}^{N} m_i \ddot{\vec{r}}_i \cdot \delta \vec{r}_i = \delta W \qquad (1.98)$$

where the virtual work done by all of the applied forces, denoted by δW, is defined as

$$\delta W = \sum_{i=1}^{N} \vec{f}_i \cdot \delta \vec{r}_i \qquad (1.99)$$

Equation (1.98) can also be written as

$$\sum_{i=1}^{N} m_i \frac{\mathrm{d}}{\mathrm{d}t} (\dot{\vec{r}}_i \cdot \delta \vec{r}_i) = \delta T + \delta W \qquad (1.100)$$

where

$$\delta T = \sum_{i=1}^{N} m_i \dot{\vec{r}}_i \cdot \delta \dot{\vec{r}}_i$$

and T, called the kinetic energy of the system, is defined as

$$T = \frac{1}{2} \sum_{i=1}^{N} m_i \dot{\vec{r}}_i \cdot \dot{\vec{r}}_i$$

If we consider times t_1 and t_2 at which $\delta \vec{r}_i = 0$, then Eq. (1.100) can be integrated as

$$\int_{t_1}^{t_2} \sum_{i=1}^{N} m_i \frac{\mathrm{d}}{\mathrm{d}t} (\dot{\vec{r}}_i \cdot \delta \vec{r}_i) \, \mathrm{d}t = \int_{t_1}^{t_2} (\delta T + \delta W) \, \mathrm{d}t \qquad (1.101)$$

Because $\delta \vec{r}_i = 0$ at t_1 and t_2, we have

$$0 = \int_{t_1}^{t_2} (\delta T + \delta W) \, \mathrm{d}t \qquad (1.102)$$

which, in fact, represents Hamilton's principle.

In many dynamic systems, if a given force $\vec{f}$ depends only on the position vector $\vec{r}$, then we introduce a scalar function $V(\vec{r})$ such that

$$\vec{f} \cdot \mathrm{d}\vec{r} = -\mathrm{d}V \qquad \text{or} \qquad \vec{f} = -\nabla V \qquad (1.103)$$

where V is called the potential energy. If the force $\vec{f}$ depends not only on $\vec{r}$ but also explicitly on the time t, then $\vec{f}$ can also be expressed as the gradient vector of a scalar function $V(\vec{r}, t)$. Such a force field is said to be irrotational or *lamellar*, and it becomes a conservative field if $\vec{f}$ does not depend explicitly on the time t.

If all the applied forces in Eq. (1.99) can be expressed as the gradient of a scalar function V, then we have $\delta W = -\delta V$. For such a dynamic system, Hamilton's principle becomes

$$0 = \int_{t_1}^{t_2} (\delta T - \delta V) \, \mathrm{d}t = \int_{t_1}^{t_2} \delta(T - V) \, \mathrm{d}t \qquad (1.104)$$

and it may be written as

$$\delta \int_{t_1}^{t_2} (T - V) \, dt = 0$$

when the system is holonomic.

1.2.6.3 Lagrange's form of D'Alembert's Principle

For a system of N particles with holonomic constraints, the inertial position vector of the ith particle, denoted by $\vec{r}_i$, is expressed in terms of the n independent generalized coordinates, as follows:

$$\vec{r}_i = \vec{r}_i(q_1, \ldots, q_n, t) , \qquad i = 1, \ldots, N \qquad (1.105)$$

Given $\vec{r}_i$ as expressed in Eq. (1.105), we obtain the virtual displacement $\delta \vec{r}_i$ as

$$\delta \vec{r}_i = \sum_{j=1}^{n} \frac{\partial \vec{r}_i}{\partial q_j} \delta q_j \qquad (1.106)$$

Substituting Eq. (1.106) into Eq. (1.97), we obtain

$$\sum_{j=1}^{n} \left\{ \sum_{i=1}^{N} (\vec{f}_i - m_i \ddot{\vec{r}}_i) \cdot \frac{\partial \vec{r}_i}{\partial q_j} \right\} \delta q_j = 0 \qquad (1.107)$$

Because the virtual generalized displacements $\{\delta q_1, \ldots, \delta q_n\}$ are independent, the coefficients of δq_j in Eq. (1.107) must be zero; thus, we have

$$\sum_{i=1}^{N} (\vec{f}_i - m_i \ddot{\vec{r}}_i) \cdot \frac{\partial \vec{r}_i}{\partial q_j} = 0, \qquad j = 1, \ldots, n \qquad (1.108)$$

Note that nonworking constraint forces do not appear in these n equations. The n equations of motion of the form of Eq. (1.108) are known as Lagrange's form of D'Alembert's principle.

1.2.6.4 Lagrange's Equations of Motion for Holonomic Systems

Consider again a system of N particles with holonomic constraints. The inertial position vector of the ith particle is expressed in terms of the n independent generalized coordinates as

$$\vec{r}_i = \vec{r}_i(q_1, \ldots, q_n, t), \qquad i = 1, \ldots, N$$

The velocity vector of the ith particle becomes

$$\dot{\vec{r}}_i = \sum_{j=1}^{n} \frac{\partial \vec{r}_i}{\partial q_j} \dot{q}_j + \frac{\partial \vec{r}_i}{\partial t} \qquad (1.109)$$

and the kinetic energy T can be expressed as

$$T = \frac{1}{2} \sum_{i=1}^{N} m_i \dot{\vec{r}}_i \cdot \dot{\vec{r}}_i = T_2 + T_1 + T_0 \tag{1.110}$$

where

$$T_2 = \frac{1}{2} \sum_{i=1}^{N} m_i \sum_{j=1}^{n} \sum_{k=1}^{n} \frac{\partial \vec{r}_i}{\partial q_j} \cdot \frac{\partial \vec{r}_i}{\partial q_k} \dot{q}_j \dot{q}_k$$

$$T_1 = \sum_{i=1}^{N} m_i \sum_{j=1}^{n} \frac{\partial \vec{r}_i}{\partial q_j} \cdot \frac{\partial \vec{r}_i}{\partial t} \dot{q}_j$$

$$T_0 = \frac{1}{2} \sum_{i=1}^{N} m_i \frac{\partial \vec{r}_i}{\partial t} \cdot \frac{\partial \vec{r}_i}{\partial t}$$

Note that T_2 is a quadratic function in the generalized velocities $\{\dot{q}_1, \ldots, \dot{q}_n\}$, T_1 is a linear function in the generalized velocities, and T_0 is a nonnegative function of only the generalized coordinates $\{q_1, \ldots, q_n\}$ and time.

Consequently, the kinetic energy T becomes a function of q_j, $\dot{q}_j$, and t, and it is expressed as

$$T = T(q_1, \ldots, q_n, \dot{q}_1, \ldots, \dot{q}_n, t) \tag{1.111}$$

The variation δT with t fixed can then be obtained as

$$\delta T = \sum_{j=1}^{n} \frac{\partial T}{\partial q_j} \delta q_j + \sum_{j=1}^{n} \frac{\partial T}{\partial \dot{q}_j} \delta \dot{q}_j \tag{1.112}$$

and

$$\int_{t_1}^{t_2} \delta T \, \mathrm{d}t = \sum_{j=1}^{n} \int_{t_1}^{t_2} \frac{\partial T}{\partial q_j} \delta q_j \, \mathrm{d}t + \sum_{j=1}^{n} \int_{t_1}^{t_2} \frac{\partial T}{\partial \dot{q}_j} \delta \dot{q}_j \, \mathrm{d}t$$

$$= \sum_{j=1}^{n} \int_{t_1}^{t_2} \frac{\partial T}{\partial q_j} \delta q_j \, \mathrm{d}t - \sum_{j=1}^{n} \int_{t_1}^{t_2} \frac{\mathrm{d}}{\mathrm{d}t} \frac{\partial T}{\partial \dot{q}_j} \delta q_j \, \mathrm{d}t$$

$$= - \sum_{j=1}^{n} \int_{t_1}^{t_2} \left(\frac{\mathrm{d}}{\mathrm{d}t} \frac{\partial T}{\partial \dot{q}_j} - \frac{\partial T}{\partial q_j} \right) \delta q_j \, \mathrm{d}t$$

as it is assumed that $\delta q_j = 0$ at t_1 and t_2.

Because the virtual displacement $\delta \vec{r}_i$ can be expressed as

$$\delta \vec{r}_i = \sum_{j=1}^{n} \frac{\partial \vec{r}_i}{\partial q_j} \delta q_j$$

the virtual work δW as defined in Eq. (1.99) becomes

$$\delta W = \sum_{i=1}^{N} \vec{f_i} \cdot \delta \vec{r_i} = \sum_{i=1}^{N} \vec{f_i} \cdot \sum_{j=1}^{n} \frac{\partial \vec{r_i}}{\partial q_j} \delta q_j = \sum_{j=1}^{n} \left\{ \sum_{i=1}^{N} \vec{f_i} \cdot \frac{\partial \vec{r_i}}{\partial q_j} \right\} \delta q_j \qquad (1.113)$$

Furthermore, we express the virtual work as

$$\delta W = \sum_{j=1}^{n} Q_j \delta q_j \qquad (1.114)$$

where Q_j, called the generalized force associated with q_j, is defined as

$$Q_j = \sum_{i=1}^{N} \vec{f_i} \cdot \frac{\partial \vec{r_i}}{\partial q_j} \qquad (1.115)$$

Note that $Q_j \delta q_j$ is the work done by all of the applied forces when q_j alone is changed by a virtual displacement δq_j (time and all other coordinates held fixed).

Finally, Hamilton's equation (1.102) becomes

$$0 = \int_{t_1}^{t_2} (\delta T + \delta W) \, dt = -\sum_{j=1}^{n} \int_{t_1}^{t_2} \left\{ \frac{d}{dt} \frac{\partial T}{\partial \dot{q}_j} - \frac{\partial T}{\partial q_j} - Q_j \right\} \delta q_j \, dt \qquad (1.116)$$

Because $\{q_1, \ldots, q_n\}$ are independent, the coefficients of δq_j in Eq. (1.116) must be zero; thus, Lagrange's equation for the holonomic system is given by

$$\frac{d}{dt} \left(\frac{\partial T}{\partial \dot{q}_j} \right) - \frac{\partial T}{\partial q_j} = Q_j, \qquad j = 1, \ldots, n \qquad (1.117)$$

If some of the applied forces acting on a system of particles are derivable from a potential energy $V(q_1, \ldots, q_n, t)$ whereas other forces are nonpotential forces, then Lagrange's equations of motion of the system are of the form

$$\frac{d}{dt} \left(\frac{\partial L}{\partial \dot{q}_j} \right) - \frac{\partial L}{\partial q_j} = Q_j, \qquad j = 1, \ldots, n \qquad (1.118)$$

where

$$L = T - V = \text{Lagrangian}$$
$$T = T(q_1, \ldots, q_n, \dot{q}_1, \ldots, \dot{q}_n, t) = \text{kinetic energy}$$
$$V = V(q_1, \ldots, q_n, t) = \text{potential energy}$$

and Q_j is the jth generalized force associated with the nonpotential forces in the system. Lagrange's equations of motion of the form of Eq. (1.118) represent a set of n second-order differential equations in terms of the independent generalized coordinates $\{q_1, \ldots, q_n\}$. If the potential energy V does not depend explicitly on time, then Q_j represent the nonconservative forces, including nonholonomic constraint forces to be discussed next.

1.2.6.5 Lagrange's Equations of Motion for Nonholonomic Systems

Consider a system of particles subject to m nonholonomic constraints of the form

$$\sum_{j=1}^{n} a_{ij}\, dq_j + a_i\, dt = 0, \qquad i = 1, \ldots, m \qquad (1.119)$$

or

$$\sum_{j=1}^{n} a_{ij}\dot{q}_j + a_i = 0, \qquad i = 1, \ldots, m \qquad (1.120)$$

where a_{ij} and a_i are, in general, functions of n coordinates q_j and time. Because the nonholonomic constraints are not integrable, the system cannot be described in terms of $(n - m)$ independent generalized coordinates; i.e., more coordinates than there are degrees of freedom are needed. Note that this system has $(n - m)$ degrees of freedom.

Let C_j be the jth generalized constraint force associated with q_j, then, according to the principle of virtual work, the virtual work done by all constraint forces is zero; i.e., we have

$$\sum_{j=1}^{n} C_j \delta q_j = 0 \qquad (1.121)$$

where δq_j is the virtual displacement, which is different from the actual displacement dq_j, but which must be compatible with the constraints irrespective of time; i.e.,

$$\sum_{j=1}^{n} a_{ij}\delta q_j = 0, \qquad i = 1, \ldots, m \qquad (1.122)$$

Multiplying this equation by an arbitrary factor λ_i, known as a *Lagrange multiplier*, and adding the resulting m equations, we obtain

$$\sum_{i=1}^{m}\left(\lambda_i \sum_{j=1}^{n} a_{ij}\delta q_j\right) = 0 \qquad (1.123)$$

Combining Eqs. (1.121) and (1.123), we obtain

$$\sum_{j=1}^{n}\left(C_j - \sum_{i=1}^{m} \lambda_i a_{ij}\right)\delta q_j = 0 \qquad (1.124)$$

If we select the Lagrange multipliers λ_i such that

$$C_j = \sum_{i=1}^{m} \lambda_i a_{ij} \qquad (1.125)$$

then q_j can be considered independent. Consequently, we have Lagrange's equations of motion of the form

$$\frac{d}{dt}\left(\frac{\partial L}{\partial \dot{q}_j}\right) - \frac{\partial L}{\partial q_j} = \sum_{i=1}^{m} \lambda_i a_{ij} + Q_j, \qquad j = 1, \ldots, n \qquad (1.126)$$

where $L = T - V$. Note that the right-hand side simply consists of the non-holonomic constraint force C_j and other generalized force Q_j. Equations (1.120) and (1.126) constitute a set of $(m + n)$ simultaneous differential equations in the unknown functions λ_j and q_j.

1.2.6.6 Hamilton's Canonical Equations of Motion

Let the jth *generalized momentum* p_j be defined as

$$p_j = \frac{\partial L}{\partial \dot{q}_j} = \frac{\partial T}{\partial \dot{q}_j} \qquad (1.127)$$

where $L = T - V$, then Lagrange's equations of motion of the form of Eq. (1.118) can be rewritten as

$$\dot{p}_j = \frac{\partial L}{\partial q_j} + Q_j \qquad (1.128)$$

The generalized momenta p_j are, in general, linear functions of the generalized velocities $\dot{q}_j$. For example, if $T = T_2$ with $T_1 = T_0 = 0$, then T can be expressed as

$$T = \frac{1}{2}\sum_{i=1}^{n}\sum_{j=1}^{n} a_{ij}\dot{q}_i\dot{q}_j \qquad (1.129)$$

where $a_{ij} = a_{ji}$ are constants or functions of $\{q_1, \ldots, q_n\}$, and we obtain

$$\begin{bmatrix} p_1 \\ \vdots \\ p_n \end{bmatrix} = \begin{bmatrix} a_{11} & \cdots & a_{1n} \\ \vdots & & \vdots \\ a_{n1} & \cdots & a_{nn} \end{bmatrix} \begin{bmatrix} \dot{q}_1 \\ \vdots \\ \dot{q}_n \end{bmatrix} \qquad (1.130)$$

In the formulation of Hamilton's canonical equations of motion, the generalized velocities $\dot{q}_j$ need to be expressed in terms of p_j, as follows:

$$\begin{bmatrix} \dot{q}_1 \\ \vdots \\ \dot{q}_n \end{bmatrix} = \begin{bmatrix} a_{11} & \cdots & a_{1n} \\ \vdots & & \vdots \\ a_{n1} & \cdots & a_{nn} \end{bmatrix}^{-1} \begin{bmatrix} p_1 \\ \vdots \\ p_n \end{bmatrix} \qquad (1.131)$$

The *Hamiltonian function*, or simply the *Hamiltonian*, is defined as

$$H = \sum_{j=1}^{n} \frac{\partial L}{\partial \dot{q}_j}\dot{q}_j - L = \sum_{j=1}^{n} p_j\dot{q}_j - L \qquad (1.132)$$

Taking the variation of H, we obtain

$$\delta H = \sum_{j=1}^{n} \left\{ \delta p_j \dot{q}_j + p_j \delta \dot{q}_j - \frac{\partial L}{\partial q_j} \delta q_j - \frac{\partial L}{\partial \dot{q}_j} \delta \dot{q}_j \right\}$$

which becomes

$$\delta H = \sum_{j=1}^{n} \left\{ \delta p_j \dot{q}_j - \frac{\partial L}{\partial q_j} \delta q_j \right\} \qquad (1.133)$$

since $p_j \delta \dot{q}_j = (\partial L / \partial \dot{q}_j) \delta \dot{q}_j$.

Substituting Eq. (1.131) for the generalized velocities $\dot{q}_j$ in Eq. (1.132), we obtain the Hamiltonian of the form

$$H = H(q_1, \ldots, q_n, p_1, \ldots p_n, t)$$

Taking the variation of H of this form, we obtain

$$\delta H = \sum_{j=1}^{n} \left\{ \frac{\partial H}{\partial q_j} \delta q_j + \frac{\partial H}{\partial p_j} \delta p_j \right\} \qquad (1.134)$$

Comparing Eqs. (1.133) and (1.134) and using Lagrange's equations of motion of the form (1.128), we obtain a set of $2n$ first-order differential equations of the form

$$\dot{q}_j = \frac{\partial H}{\partial p_j}, \qquad\qquad j = 1, \ldots, n \qquad (1.135a)$$

$$\dot{p}_j = -\frac{\partial H}{\partial q_j} + Q_j, \qquad j = 1, \ldots, n \qquad (1.135b)$$

where Q_j is the jth generalized force associated with the nonpotential forces in the system, and Eqs. (1.135) are known as Hamilton's canonical equations of motion.

Differentiating the Hamiltonian $H = H(q_1, \ldots, q_n, p_1, \ldots, p_n, t)$ of a holonomic system in an irrotational field, we obtain

$$\frac{dH}{dt} = \sum_{j=1}^{n} \frac{\partial H}{\partial q_j} \dot{q}_j + \sum_{j=1}^{n} \frac{\partial H}{\partial p_j} \dot{p}_j + \frac{\partial H}{\partial t}$$

$$= \sum_{j=1}^{n} \frac{\partial H}{\partial q_j} \frac{\partial H}{\partial p_j} - \sum_{j=1}^{n} \frac{\partial H}{\partial p_j} \frac{\partial H}{\partial q_j} + \frac{\partial H}{\partial t}$$

$$= \frac{\partial H}{\partial t}$$

$$= -\frac{\partial L}{\partial t}$$

Consequently, for a conservative holonomic system in which the Hamiltonian does not depend explicitly on time, we have

$$H = \text{const}$$

When the kinetic energy T is expressed as Eq. (1.110), the Hamiltonian function defined as Eq. (1.132) becomes

$$H = \sum_{j=1}^{n} \frac{\partial L}{\partial \dot{q}_j} \dot{q}_j - L$$
$$= 2T_2 + T_1 - (T_2 + T_1 + T_0 - V)$$
$$= T_2 - T_0 + V \tag{1.136}$$

Furthermore, if $T = T_2$ with $T_1 = T_0 = 0$, then the Hamiltonian simply becomes

$$H = T + V \tag{1.137}$$

which indicates that the Hamiltonian is simply the total energy of the system. Finally, the principle of conservation of energy for a conservative holonomic system can be described as

$$H = T + V = \text{const}$$

Using Hamilton's canonical equations of motion, we describe the motion or trajectory of a dynamic system in the $2n$-dimensional state space or phase space by the set of $2n$ state variables $\{q_1, \ldots, q_n\}$ and $\{p_1, \ldots, p_n\}$. In Lagrange's equations of motion, a set of $2n$ coordinates $\{q_1, \ldots, q_n\}$ and $\{\dot{q}_1, \ldots, \dot{q}_n\}$ can be regarded as the state variables of the system. The concepts of state variables and state space will be studied further in the next section.

Example 1.1

Consider again a particle of mass m that is constrained to move on a massless, circular hoop of radius ℓ under the action of gravity, as was illustrated in Fig. 1.6. The circular hoop is also allowed to rotate freely about the vertical line. This dynamic system with two degrees of freedom is equivalent to a spherical pendulum of mass m and length ℓ with the generalized coordinates θ and ϕ.

As illustrated in Fig. 1.6, let $\{\vec{a}_1, \vec{a}_2, \vec{a}_3\}$ and $\{\vec{b}_1, \vec{b}_2, \vec{b}_3\}$ be two sets of orthogonal unit vectors of two reference frames A and B, fixed at point O and the pendulum mass, respectively. The angular velocity of B with respect to A is: $\vec{\omega}^{B/A} = -\dot{\phi}\vec{a}_1 + \dot{\theta}\vec{b}_3$ and $\vec{a}_1 = \cos\theta\,\vec{b}_1 - \sin\theta\,\vec{b}_2$. The position vector of the pendulum mass from point O is given by $\vec{r} = \ell\,\vec{b}_1$, and the velocity vector can be obtained as $\dot{\vec{r}} = \vec{\omega}^{B/A} \times \vec{r} = \ell\dot{\theta}\,\vec{b}_2 - \ell\dot{\phi}\sin\theta\,\vec{b}_3$.

The Lagrangian then becomes

$$L = T - V = \tfrac{1}{2}m(\ell^2\dot{\theta}^2 + \ell^2\dot{\phi}^2\sin^2\theta) - mg\ell(1 - \cos\theta)$$

and Lagrange's equations of motion of a spherical pendulum can be found as

$$\frac{d}{dt}\left(\frac{\partial L}{\partial \dot\theta}\right) - \frac{\partial L}{\partial \theta} = 0 \Rightarrow m\ell^2\ddot\theta - m\ell^2\dot\phi^2 \sin\theta\cos\theta + mg\ell\sin\theta = 0$$

$$\frac{d}{dt}\left(\frac{\partial L}{\partial \dot\phi}\right) - \frac{\partial L}{\partial \phi} = 0 \Rightarrow m\ell^2\sin^2\theta\ddot\phi + 2m\ell^2\dot\phi\dot\theta\sin\theta\cos\theta = 0$$

which can be rewritten as

$$\begin{bmatrix} m\ell^2 & 0 \\ 0 & m\ell^2\sin^2\theta \end{bmatrix}\begin{bmatrix} \ddot\theta \\ \ddot\phi \end{bmatrix} = \begin{bmatrix} m\ell^2\dot\phi^2\sin\theta\cos\theta - mg\ell\sin\theta \\ -2m\ell^2\dot\theta\dot\phi\sin\theta\cos\theta \end{bmatrix}$$

or as a set of first-order nonlinear differential equations of the form

$$\dot{x}_1 = x_3$$
$$\dot{x}_2 = x_4$$
$$\dot{x}_3 = x_4^2 \sin x_1 \cos x_1 - (g/\ell)\sin x_1$$
$$\dot{x}_4 = -2x_3 x_4 \cos x_1 / \sin x_1$$

where $(x_1, x_2, x_3, x_4) = (\theta, \phi, \dot\theta, \dot\phi)$.

Furthermore, the generalized momenta associated with the generalized coordinates θ and ϕ are related to $\dot\theta$ and $\dot\phi$ as

$$p_\theta = \frac{\partial L}{\partial \dot\theta} = m\ell^2\dot\theta \Rightarrow \dot\theta = \frac{p_\theta}{m\ell^2}$$

$$p_\phi = \frac{\partial L}{\partial \dot\phi} = m\ell^2\dot\phi\sin^2\theta \Rightarrow \dot\phi = \frac{p_\phi}{m\ell^2\sin^2\theta}$$

and we obtain Hamiltonian as

$$H = p_\theta\dot\theta + p_\phi\dot\phi - L = \frac{1}{2m\ell^2}p_\theta^2 + \frac{1}{2m\ell^2\sin^2\theta}p_\phi^2 + mg\ell(1 - \cos\theta)$$

Finally, we obtain Hamilton's canonical equations of motion as

$$\dot\theta = p_\theta/(m\ell^2)$$
$$\dot\phi = p_\phi/(m\ell^2\sin^2\theta)$$
$$\dot{p}_\theta = p_\phi^2 \cos\theta/(m\ell^2\sin^3\theta) - mgl\sin\theta$$
$$\dot{p}_\phi = 0$$

Problems

1.20 Consider a system of N particles of tree-topology connected by mass-less rods, such as a double pendulum. Lagrange's form of D'Alembert's principle for such a system is given by

$$\sum_{i=1}^{N}(\vec{f_i} - m_i\ddot{\vec{r}_i}) \cdot \frac{\partial \vec{r}_i}{\partial q_j} = 0, \qquad j = 1,\ldots,n$$

where $\vec{f_i}$ is an applied force acting on the ith particle (excluding nonworking constraint forces), $\vec{r}_i$ is the inertial position vector of the ith particle represented as

$$\vec{r}_i = \vec{r}_i(q_1,\ldots,q_n,t), \qquad i = 1,\ldots,N$$

and $\{q_1,\ldots,q_n\}$ are the n independent generalized coordinates.

(a) Show that

$$\frac{\partial \vec{r}_i}{\partial q_j} \equiv \frac{\partial \dot{\vec{r}}_i}{\partial \dot{q}_j}$$

Hint: Eq. (1.109).

(b) Let the generalized force Q_j and the kinetic energy T be defined as

$$Q_j = \sum_{i=1}^{N}\vec{f_i} \cdot \frac{\partial \vec{r}_i}{\partial q_j} \equiv \sum_{i=1}^{N}\vec{f_i} \cdot \frac{\partial \dot{\vec{r}}_i}{\partial \dot{q}_j}$$

$$T = \frac{1}{2}\sum_{i=1}^{N}m_i\dot{\vec{r}}_i \cdot \dot{\vec{r}}_i$$

then derive Lagrange's equation of motion of the form

$$\frac{\mathrm{d}}{\mathrm{d}t}\left(\frac{\partial T}{\partial \dot{q}_j}\right) - \frac{\partial T}{\partial q_j} = Q_j, \qquad j = 1,\ldots,n$$

(c) Also show that the generalized force Q_j can also be expressed as

$$Q_j = \sum_{i=1}^{N}\vec{f_i} \cdot \frac{\partial \dot{\vec{r}}_{i-1}}{\partial \dot{q}_j} + \sum_{i=1}^{N}\underbrace{\vec{\rho_i} \times \vec{f_i}}_{\text{moment}} \cdot \frac{\partial \vec{\omega}_i}{\partial \dot{q}_j}$$

where $\vec{\rho_i} \equiv \vec{r}_i - \vec{r}_{i-1}$ is the position vector of the ith particle with respect to the $(i-1)$th particle, and $\vec{\omega}_i$ is the angular velocity vector of the ith particle with respect to the $(i-1)$th particle.

1.21 For the simple pendulum problems illustrated earlier in Fig. 1.5, derive both Lagrange's and Hamilton's equations of motion.

1.22 For the compound pendulum problems illustrated earlier in Fig. 1.8, derive both Lagrange's and Hamilton's equations of motion.

1.23 For the inverted pendulum problems illustrated earlier in Fig. 1.9, derive both Lagrange's and Hamilton's equations of motion.

1.24 Consider a two-link manipulator that is modeled as the double pendulum consisting of two massless rods of lengths ℓ_1 and ℓ_2 and point masses m_1 and m_2, as illustrated earlier in Fig. 1.10a.

(a) Find the kinetic energy T as

$$T = \tfrac{1}{2}\left(a_{11}\dot{\theta}_1^2 + 2a_{12}\dot{\theta}_1\dot{\theta}_2 + a_{22}\dot{\theta}_2^2\right)$$

where $a_{11} = (m_1 + m_2)\ell_1^2$, $a_{12} = a_{21} = m_2\ell_1\ell_2\cos(\theta_2 - \theta_1)$, and $a_{22} = m_2\ell_2^2$.

(b) Derive Lagrange's equations of motion for the double pendulum in terms of the generalized coordinates θ_1 and θ_2, as follows:

$$\begin{bmatrix} a_{11} & a_{12} \\ a_{21} & a_{22} \end{bmatrix}\begin{bmatrix} \ddot{\theta}_1 \\ \ddot{\theta}_2 \end{bmatrix}$$

$$= \begin{bmatrix} m_2\ell_1\ell_2\dot{\theta}_2^2\sin(\theta_2 - \theta_1) - (m_1 + m_2)g\ell_1\sin\theta_1 + u_1 - u_2 \\ -m_2\ell_1\ell_2\dot{\theta}_1^2\sin(\theta_2 - \theta_1) - m_2g\ell_2\sin\theta_2 + u_2 \end{bmatrix}$$

Note: The double pendulum with the shoulder and elbow joint control torques u_1 and u_2 is equivalent to a double pendulum with

$$\vec{f}_1 = \frac{u_1}{\ell_1}\vec{b}_2 - \frac{u_2}{\ell_1}\vec{b}_2 - \frac{u_2}{\ell_2}\vec{c}_2$$

$$\vec{f}_2 = \frac{u_2}{\ell_2}\vec{c}_2$$

(c) Express $\dot{\theta}_1$ and $\dot{\theta}_2$ in terms of the generalized momenta $p_1 = \partial T/\partial\dot{\theta}_1$ and $p_2 = \partial T/\partial\dot{\theta}_2$, as follows:

$$\dot{\theta}_1 = \frac{a_{22}p_1 - a_{12}p_2}{a_{11}a_{22} - a_{12}^2}$$

$$\dot{\theta}_2 = \frac{a_{11}p_2 - a_{12}p_1}{a_{11}a_{22} - a_{12}^2}$$

and then obtain the kinetic energy as

$$T = \frac{a_{22}p_1^2 - 2a_{12}p_1p_2 + a_{11}p_2^2}{2\left(a_{11}a_{22} - a_{12}^2\right)}$$

(d) Derive Hamilton's canonical equations of motion in terms of θ_1, θ_2, p_1, and p_2.

1.25   Consider again the double pendulum shown earlier in Fig. 1.10b with $\{\phi_1, \phi_2\}$ as a set of the generalized coordinates of the system.

(a) Derive Lagrange's equations of motion in terms of the generalized coordinates ϕ_1 and ϕ_2, as follows:

$$
\begin{bmatrix}
(m_1 + m_2)\ell_1^2 + m_2\ell_2^2 & \\
\quad + 2m_2\ell_1\ell_2\cos\phi_2 & m_2\ell_2^2 + m_2\ell_1\ell_2\cos\phi_2 \\
m_2\ell_2^2 + m_2\ell_1\ell_2\cos\phi_2 & m_2\ell_2^2
\end{bmatrix}
\begin{bmatrix}
\ddot{\phi}_1 \\
\ddot{\phi}_2
\end{bmatrix}
$$

$$
=
\begin{bmatrix}
m_2\ell_1\ell_2\sin\phi_2(2\dot{\phi}_1\dot{\phi}_2 + \dot{\phi}_2^2) - (m_1 + m_2)g\ell_1\sin\phi_1 \\
-m_2 g\ell_2\sin(\phi_1 + \phi_2) + u_1 \\
-m_2\ell_1\ell_2\dot{\phi}_1^2\sin\phi_2 - m_2 g\ell_2\sin(\phi_1 + \phi_2) + u_2
\end{bmatrix}
$$

(b) Also derive Hamilton's canonical equations of motion in terms of ϕ_1, ϕ_2, p_1, and p_2 where $p_i = \partial T / \partial \dot{\phi}_i$.

1.26   For the double pendulum consisting of two uniform slender bars shown earlier in Fig. 1.10c, derive Lagrange's equations of motion in terms of the generalized coordinates ϕ_1 and ϕ_2, as follows:

$$
\begin{bmatrix}
\frac{1}{3}m_1\ell_1^2 + \frac{1}{3}m_2\ell_2^2 + m_2\ell_1^2 & \\
\quad + m_2\ell_1\ell_2\cos\phi_2 & \frac{1}{3}m_2\ell_2^2 + \frac{1}{2}m_2\ell_1\ell_2\cos\phi_2 \\
\frac{1}{3}m_2\ell_2^2 + \frac{1}{2}m_2\ell_1\ell_2\cos\phi_2 & \frac{1}{3}m_2\ell_2^2
\end{bmatrix}
\begin{bmatrix}
\ddot{\phi}_1 \\
\ddot{\phi}_2
\end{bmatrix}
$$

$$
=
\begin{bmatrix}
\frac{1}{2}m_2\ell_1\ell_2(2\dot{\phi}_1\dot{\phi}_2 + \dot{\phi}_2^2)S_2 - \frac{1}{2}m_1 g\ell_1 S_1 - m_2 g\ell_1 S_1 \\
-\frac{1}{2}m_2 g\ell_2 S_{12} + u_1 \\
-\frac{1}{2}m_2\ell_1\ell_2\dot{\phi}_1^2 S_2 - \frac{1}{2}m_2 g\ell_2 S_{12} + u_2
\end{bmatrix}
$$

where $S_1 \equiv \sin\phi_1$, $S_2 \equiv \sin\phi_2$, and $S_{12} \equiv \sin(\phi_1 + \phi_2)$.

1.2.7 *Applications to Spacecraft Deployment Dynamics*

Thus far in this section we have presented the fundamentals of classical mechanics and analytical dynamics. This is a textbook about spacecraft dynamics and control. Hence, as an introduction to dynamic problems of practical concern, we consider in this section, based on Wie et al. [5], the dynamic modeling and computer simulation problems associated with the deployment of solar panel arrays on INTELSAT V and INSAT spacecraft. Each spacecraft is characterized by very distinct mechanisms for its solar array deployment.

As spacecraft become larger and more complex, the need to stow the spacecraft within the dimensions of launch vehicle fairing becomes a serious design constraint. For this reason, spacecraft are being built that are stowed in one configuration, then deployed into another configuration once in orbit. INTELSAT V and INSAT spacecraft, shown in Figs. 1.11 and 1.12, respectively, are examples of such spacecraft, with solar panel arrays that must be deployed for the normal on-orbit configuration.

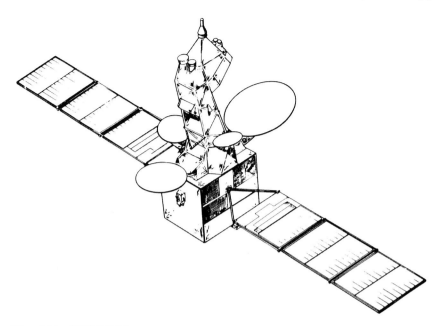

Fig. 1.11 INTELSAT V on-orbit deployed configuration. (Courtesy of Ford Aerospace and Communications Corp., Palo Alto, CA.)

The arrays on INTELSAT V are in a topological tree configuration, whereas the INSAT solar panel array is in a closed-loop configuration because of the four-bar linkage deployment mechanism. The closed-loop multibody configuration poses a unique dynamic problem in formulating the equations of motion. However, the kinematic control rod on the INSAT spacecraft has a synchronizing function very similar to that of the closed cable loop on the INTELSAT V spacecraft.

Such complex deployment mechanisms, to be discussed hereafter, present two major issues that often arise in the modeling and computer simulation of complex dynamic systems: how one can formulate analytically the complex equations of motion, and to what extent one can utilize a multibody computer code for the computerized generation of symbolic equations of motion.

1.2.7.1 INTELSAT V solar array deployment INTELSAT V, shown in Fig. 1.11, has two symmetric solar panel arrays that deploy simultaneously in an accordion manner. Each array consists of a yoke and three panels. A schematic of the deployment mechanism consisting of torsion springs and closed cable loops is shown in Fig. 1.13. The solar panel itself is assumed to be rigid. It can also be assumed that the solar arrays are attached to the fixed base for the deployment analysis, because the two symmetric arrays deploy simultaneously. To prevent large disturbance torques to the mechanism, the attitude control system is disabled during the approximately 20-s deployment period.

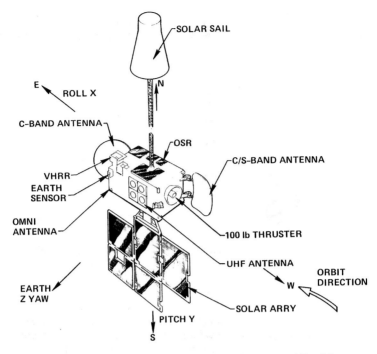

Fig. 1.12   INSAT on-orbit deployed configuration. (Courtesy of Ford Aerospace and Communications Corp., Palo Alto, CA.)

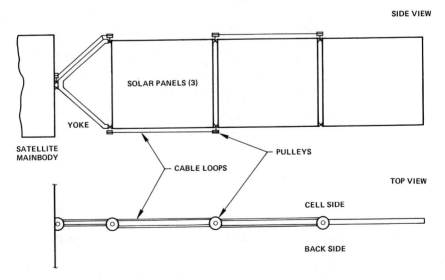

Fig. 1.13   INTELSAT V solar array showing closed cable loops.

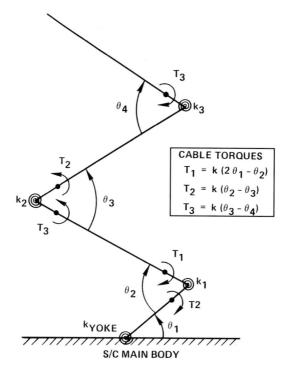

Fig. 1.14   INTELSAT V solar array deployment analysis model.

The preloaded torsion springs located at each hinge provide the energy to deploy the panels, whereas the closed cable loops restrain the deployment by synchronizing the deployment angles. The first cable connects the first panel and the spacecraft main body, the second cable connects the second panel and the yoke, and the third cable connects the third and first panels. These cables synchronize the deployment angles of each panel by applying a passive control torque that is proportional to the angle difference. The cables also serve to slow down the deployment rate to a point below the structural failure rate. Figure 1.14 illustrates a simplified planar model with equivalent cable torques on each body, which, during the deployment, can be simply modeled as

$$T_1 = k(2\theta_1 - \theta_2) \tag{1.138a}$$

$$T_2 = k(\theta_2 - \theta_3) \tag{1.138b}$$

$$T_3 = k(\theta_3 - \theta_4) \tag{1.138c}$$

where θ_i is the ith hinge angle in rad, T_i is the ith cable torque in N·m, $k = 4500r^2$ is the equivalent cable torsional stiffness in N·m/rad, and $r = 0.01835$ m is the cable-pulley radius at the panel hinges. The pulley radius at yoke hinge is $2r$, because the fully deployed yoke hinge angle θ_1 is only 90 deg whereas the fully deployed first hinge angle θ_2 becomes 180 deg.

Table 1.2 INTELSAT V solar array deployment parameters

	Length, m	Mass, kg	Inertia, kg·m^2	Friction, N·m	Hinge spring Stiffness, N·m/rad	Hinge spring Pretorque, N·m
Yoke	1.090	2.222	0.2798	0.66	0.1273	1.70
Panel 1	1.965	12.637	4.2009	0.64	0.1305	0.85
Panel 2	1.965	9.982	3.8773	0.69	0.1305	0.85
Panel 3	1.936	8.920	2.9316	0.70	0.1305	0.85

The most critical parameter in the deployment is the hinge friction level, which is temperature dependent and can be a source of premature partial deployment. The nominal friction levels are given in Table 1.2 with other parameter values. In Table 1.2, the inertia value of each body is the moment-of-inertia value of each body about its center of mass. The moment of inertia of a thin uniform bar of mass m and length ℓ about its center of mass is $I_c = m\ell^2/12$, and its moment of inertia about one end is: $I_o = I_c + m(\ell/2)^2 = m\ell^2/3$. As in the case of a double-pendulum model, one can derive analytically the equations of motion of the planar, four-body model shown in Fig. 1.14 using the Lagrangian approach. Figure 1.15 shows the computer simulation results for such a deployment model of INTELSAT V. As can be seen in this figure, after the yoke initially deploys the deployment occurs fairly evenly due to deployment synchronization by the closed cable loops. At the end of the deployment, the hinges latch into the full deployment position. As latch-up occurs, the locking lever slides into a slot. In the computer simulation, it is adequate to model the latch-up by hinges restrained by very stiff hinges; however, this requires a smaller integration time step near the end of the deployment in the computer simulation.

In practice, extensive computer simulations and ground experiments are often needed to investigate, in particular, the effects of the hinge friction and pretorque level on the overall deployment dynamics. As a result of such efforts the following changes were made in the INTELSAT V case to avoid any premature partial deployment: 1) special application of a dry lubricant to all moving parts in the hinges to reduce the hinge friction level, 2) increase of torsion spring pretorque level, and 3) increase of bearing tolerance to allow greater variation in temperature.

1.2.7.2 INSAT Solar Array Deployment   The INSAT spacecraft shown in Fig. 1.12 has a solar array deployment sequence and mechanism quite different from that of the INTELSAT V spacecraft. Figure 1.16 illustrates the deployment sequence for the solar array and other appendages on the INSAT spacecraft, which occurs in five separate steps, each controlled by pyrotechnic devices. As in the INTELSAT V case, the attitude control system is disabled during each deployment, except the solar-sail deployment. In the fully deployed configuration, the solar array is extended from the south side of the satellite, where it is oriented to the sun by the solar array drive assembly. When stowed for launch, panel 1 and the yoke

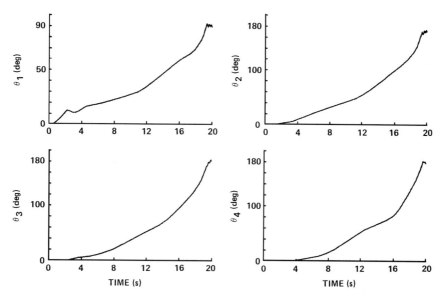

Fig. 1.15   Simulation results for INTELSAT V solar array deployment.

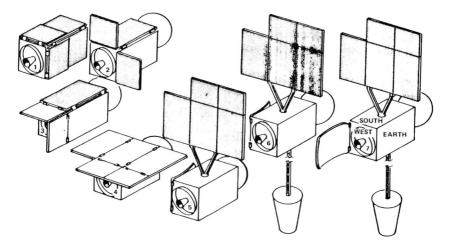

Fig. 1.16   INSAT spacecraft deployment sequence.

are held parallel with the south wall of the satellite. Panels 2 and 4 are also folded on the south face, whereas panels 3 and 5 are folded against the Earth face. Two hinge assemblies are used at each of the four lines to join the panels together.

The final stage solar array deployment is discussed here in detail because of the unique dynamic characteristics of the four-bar linkage deployment mechanism. A schematic of this four-bar linkage for the final stage deployment is shown in Fig. 1.17. To prevent possible interference between the array and the spacecraft

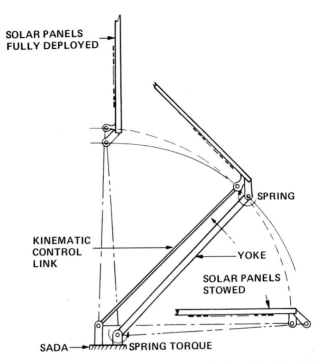

Fig. 1.17 INSAT solar array deployment mechanism (four-bar linkage).

main body, the control link synchronizes the deployment of the solar panels with the motion of the yoke. In fact, the kinematic control rod on the INSAT has a synchronizing function very similar to that of the closed cable loops on the INTELSAT V.

The analytical modeling of the INSAT final stage deployment is not as straightforward as the INTELSAT V case. The four-bar linkage results in a closed-loop configuration, which requires a special consideration in formulating the equations of motion. A schematic of a planar model of the INSAT final-stage deployment is shown in Fig. 1.18. We consider here a simple case with a fixed base. The spacecraft main body is assumed to be fixed in space, and the rotational motion of the linkage and solar panel is assumed to occur about a single axis.

For the Lagrangian formulation of the equations of motion of the model in Fig. 1.18, but with a fixed base, the following kinetic and potential energy terms can be defined as

$$T = \frac{1}{2}(I_1 + m_1 r_1^2 + m_2 d_1^2)\dot{\theta}_1^2 + \frac{1}{2}(I_2 + m_2 r_2^2)(\dot{\theta}_1 + \dot{\theta}_2)^2$$
$$+ m_2 d_1 r_2 \dot{\theta}_1(\dot{\theta}_1 + \dot{\theta}_2)\cos\theta_2 \tag{1.139a}$$

$$V = \frac{1}{2}k_1(\theta_{1F} - \theta_1)^2 + \frac{1}{2}k_2(\theta_{2F} - \theta_2)^2 \tag{1.139b}$$

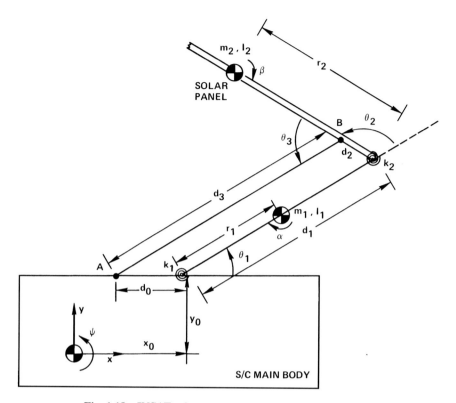

Fig. 1.18   INSAT solar array deployment analysis model.

where θ_{1F} and θ_{2F} are the hinge angles when the hinge springs are unstressed. The hinge angles θ_i and the geometric parameters used in the preceding equations are defined in Fig. 1.18. The INSAT solar-array deployment parameters are summarized in Table 1.3.

Note that the two hinge angles, θ_1 and θ_2, used in Eq. (1.139) are not independent variables. Because the four-bar linkage system has only one degree of freedom, a single coordinate, such as the angle θ_2, suffices for the description of all admissible configurations. The kinetic energy cannot be expressed readily in terms of only θ_2 and $\dot{\theta}_2$, however, because of the transcendental character of the relationship among the three hinge angles. Thus, the third angle θ_3 is introduced, and the angles θ_1 and θ_3 are selected as extraneous coordinates, which must satisfy the following geometrical constraints:

$$d_0 + d_1 \cos\theta_1 + d_2 \cos(\theta_1 + \theta_2) + d_3 \cos(\theta_1 + \theta_2 + \theta_3) = 0 \quad (1.140a)$$

$$d_1 \sin\theta_1 + d_2 \sin(\theta_1 + \theta_2) + d_3 \sin(\theta_1 + \theta_2 + \theta_3) = 0 \quad (1.140b)$$

The selection of θ_2 as an independent coordinate here is, however, arbitrary.

Because it is impractical to express θ_1 and θ_3 in terms of the independent coordinate θ_2, these holonomic-constraint equations are differentiated with respect

Table 1.3   INSAT solar array deployment parameters

Parameters	Values
Spacecraft main body yaw inertia I_z, kg·m^2	360.3
Yoke hinge location with respect to main body center of mass, m	$(x_o, y_o) = (-0.2171, 0.7409)$
Distance d_o from yoke hinge to control rod, m	0.15225
Yoke length d_1, m	1.32588
Distance d_2 from yoke/panel hinge to control rod, m	0.08024
Control rod length d_3, m	(min, max) = (1.3939, 1.4165)
Distance r_1 from yoke hinge to yoke c.m., m	0.8915
Distance r_2 from yoke/panel hinge to panel c.m., m	1.8288
Yoke mass m_1, kg	8.899
Panel mass m_2, kg	36.891
Yoke inertia I_1 about its c.m., kg·m^2	0.9369
Panel inertia I_2 about its c.m., kg·m^2	43.527
Yoke spring constant k_1, N·m	0.6114
Yoke spring zero-torque angle θ_{1F}, deg	240
Simulation data	
$\quad k_2 = -0.776$ N·m/rad for 5 deg $< \theta_2 <$ 172.8 deg	
$\quad k_2 = -40.1$ N·m/rad for -3 deg $< \theta_2 <$ 5 deg	
$\quad \theta_{2F} = 298$ deg for 5 deg $< \theta_2 <$ 172.8 deg	
$\quad \theta_{2F} = 12.7$ deg for -3 deg $< \theta_2 <$ 5 deg	
Initial angles $\theta_1, \theta_2, \theta_3$, deg	0.294, 172.817, 7.283
Deployed yoke/body spring constant, N·m/rad	1582.3
Deployed panel/yoke spring constant, N·m/rad	28523.5
Deployed modal damping ratio	0.02
Fully deployed fixed-base frequencies, Hz	0.5, 2.74

to time and treated as nonholonomic constraints, and the dependent variables are eliminated by the method of Lagrange multipliers. Differentiating Eqs. (1.140), we obtain the following nonholonomic constraints:

$$a_{11}\dot{\theta}_1 + a_{12}\dot{\theta}_2 + a_{13}\dot{\theta}_3 = 0 \tag{1.141a}$$
$$a_{21}\dot{\theta}_1 + a_{22}\dot{\theta}_2 + a_{23}\dot{\theta}_3 = 0 \tag{1.141b}$$

where the coefficients a_{ij} are functions of $\theta_1, \theta_2,$ and θ_3. These equations are linear in $\dot{\theta}_1$ and $\dot{\theta}_3$, and they can be easily solved for these quantities in terms of $\dot{\theta}_2$, as follows:

$$\dot{\theta}_1 = f_1(\dot{\theta}_2, a_{ij}) \tag{1.142a}$$
$$\dot{\theta}_3 = f_3(\dot{\theta}_2, a_{ij}) \tag{1.142b}$$

The equations of motion are then derived using the Lagrange multiplier method, as follows:

$$\frac{d}{dt}\frac{\partial L}{\partial \dot{\theta}_1} - \frac{\partial L}{\partial \theta_1} = \lambda_1 a_{11} + \lambda_2 a_{21} \tag{1.143a}$$

$$\frac{d}{dt}\frac{\partial L}{\partial \dot{\theta}_2} - \frac{\partial L}{\partial \theta_2} = \lambda_1 a_{12} + \lambda_2 a_{22} \tag{1.143b}$$

$$\frac{d}{dt}\frac{\partial L}{\partial \dot{\theta}_3} - \frac{\partial L}{\partial \theta_3} = \lambda_1 a_{13} + \lambda_2 a_{23} \tag{1.143c}$$

where $L = T - V$ is the Lagrangian function and λ_1 and λ_2 are the Lagrange multipliers.

Differentiating Eq. (1.142), substituting the resulting $\ddot{\theta}_1$ and $\ddot{\theta}_3$ into Eq. (1.43), and then eliminating the Lagrange multipliers, one can obtain a differential equation of the form

$$\ddot{\theta}_2 = f_2(\theta_i, \dot{\theta}_i) \tag{1.144}$$

where the right-hand side is somewhat complicated. Equations (1.142) and (1.144) constitute a set of three simultaneous differential equations in the unknown functions θ_1, θ_2, and θ_3.

The results of the computer simulation using these analytically derived equations with parameter values given in Table 1.3 are shown in Fig. 1.19. The complete deployment takes about 10.5 s, which closely matches the actual flight-observed deployment of 11 s.

Figure 1.20 shows the computer simulation results using a more complex model with slotted link and coupled to the free motion of the spacecraft main body. The array vibration after lock-up is modeled by two springs, at the yoke/body and array/yoke hinges. Spring and damping constants are tuned to give the first two fixed-base bending mode frequencies and a modal damping ratio of 0.02. The overall response characteristics between Figs. 1.19 and 1.20 are in good agreement, with the exception of the initial yoke response. The yoke angle plot clearly shows the impact of the slotted link. The yoke also locks up slightly before the array does. The spacecraft yaw offset is about 16 deg, with peak yaw rates up to 10 deg/s. After the yoke initially deployed, the overall deployment occurs fairly evenly. By comparing this with the INTELSAT V yoke response in Fig. 1.15, it can be seen that the kinematic control rod on the INSAT has a synchronizing function very similar to that of the closed cable loop on the INTELSAT V. A peak compressive force of about 220 N (50 lb) is carried by the link. This is well within the designed capacity of the link. The link effectively pushes up the panel to expedite its clearance from the spacecraft main body.

It is possible to derive analytically the equations of motion of a more complex model of a three-axis coupled system; see, e.g., Kane and Levinson [6]. However, a multibody computer code needs to be employed for the automatic generation of symbolic equations of motion for such a complex model. Such computer programs will eliminate the time-consuming part of the analytic formulation and

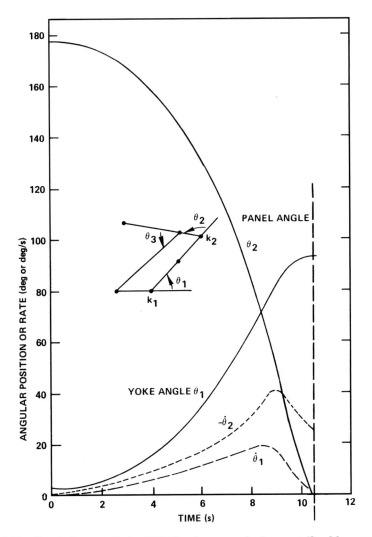

Fig. 1.19 Simulation results for INSAT solar array deployment (fixed base and no slot).

will also automatically generate symbolic equations for more efficient computer simulations, as is discussed in Schaecther and Levinson [7].

Problems

1.27   Derive the equations of motion for the planar, four-body model of the solar array deployment dynamics of the INTELSAT V spacecraft, shown in Fig. 1.14, and verify the computer simulation results shown in Fig. 1.15. (Assume that the center of mass of each body is located at the middle of

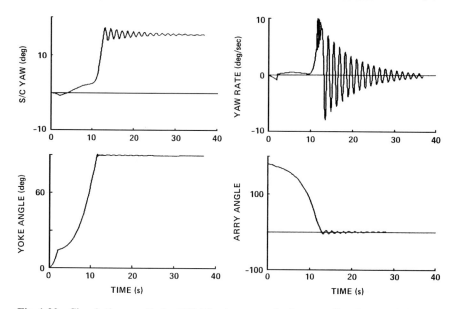

Fig. 1.20   Simulation results for INSAT solar array deployment (free base and slotted link).

each body.) Also perform computer simulations to investigate the effects of the hinge friction and pretorque level on the overall deployment dynamics, with particular emphasis on premature partial deployment.

1.28   Complete the derivation of the equations of motion (1.142) and (1.144), in detail for a planar model of the INSAT solar array deployment dynamics, shown in Fig. 1.18 (but with fixed base and no slot), and verify the computer simulation results shown in Fig. 1.19. Using a multibody computer code (see, e.g., [7] and [8]), verify the computer simulation results shown in Fig. 1.20 for the planar model with free base and slotted link.

1.3   Dynamic Systems Analysis

This section introduces some fundamental concepts in dynamic systems theory, and it is mainly concerned with the stability analysis of dynamic systems. A solid background in dynamic systems theory is essential to understanding the stability and control problems of space vehicles treated in this book. In particular, the subject of linear system dynamics is important for a wide range of dynamics and control problems of practical interest; however, this section is only intended to serve as a brief summary of the subject matter.

We begin the subject of dynamic systems analysis with a brief introduction of the theory of dynamic systems, including Lyapunov's stability theorems.*

1.3.1 Introduction to Dynamic Systems Theory

Consider a dynamic system described by a set of first-order nonlinear differential equations of the following form:

$$\dot{x}_1 = f_1(x_1, \ldots, x_n; u_1, \ldots, u_p; t)$$
$$\dot{x}_2 = f_2(x_1, \ldots, x_n; u_1, \ldots, u_p; t)$$

$$\vdots$$

$$\dot{x}_n = f_n(x_1, \ldots, x_n; u_1, \ldots, u_p; t)$$

where $x_i(t)$ are state variables and $u_i(t)$ are control input variables or control parameters. The state of a system at time t_0 is the amount of information at t_0 that, together with the input, uniquely determines the behavior of the system for all $t \geq t_0$.

In vector notation, the state-variable description of nonlinear dynamic systems becomes

$$\dot{\mathbf{x}} = \mathbf{f}(\mathbf{x}, \mathbf{u}, t) \qquad (1.145)$$

where $\mathbf{x}$ is the state vector and $\mathbf{u}$ the control input vector defined as

$$\mathbf{x} = \begin{bmatrix} x_1 \\ \vdots \\ x_n \end{bmatrix}, \qquad \mathbf{u} = \begin{bmatrix} u_1 \\ \vdots \\ u_p \end{bmatrix}, \qquad \mathbf{f} = \begin{bmatrix} f_1 \\ \vdots \\ f_n \end{bmatrix}$$

If the equations of motion do not explicitly depend on the time, then the system is said to be autonomous; otherwise, it is said to be nonautonomous.

An equilibrium point, or a critical point (also called fixed point, stationary point, singular point, and steady-state point in the literature), of a nonlinear dynamic system corresponding to a known constant input or parameter vector $\mathbf{u}^*$ can be found by solving the steady-state equation $\mathbf{f}(\mathbf{x}, \mathbf{u}, t) = 0$. That is, we have the equilibrium condition

$$\mathbf{f}(\mathbf{x}^*, \mathbf{u}^*, t) = 0 \qquad \text{for all } t \geq t_0 \qquad (1.146)$$

where $\mathbf{x}^*$ denotes the state vector at an equilibrium point for a known constant input or parameter vector $\mathbf{u}^*$.

---

*The Russian mathematician Aleksandr Lyapunov (1857–1918) established the foundations for the stability theory of dynamic systems. Lyapunov was a contemporary of the French mathematician Henri Poincaré (1854–1912), who established the foundations for a theory of dynamic systems. Later, George David Birkhoff (1884–1944) further developed the theory of dynamic systems and published a monograph entitled *Dynamical Systems* in 1926.

A sudden change in the dynamic character of a system when a constant parameter is changed from below to above some critical value is called a *bifurcation* in dynamic systems theory. In particular, bifurcation theory is concerned with how the number and character of equilibrium points depends on parameters.

Example 1.2

Consider the simple pendulum of mass m and length ℓ, connected by a massless rod to the hinged support point O, as shown in Fig. 1.5a. The nonlinear differential equation of motion of this system is given by

$$m\ell\ddot{\theta} = -mg\sin\theta + u$$

where θ is the generalized coordinate of the pendulum from the vertical line, g is the gravitational acceleration, and $u(t)$ is the input force. The natural equilibrium points of this system with $u = 0$ can be directly found by solving the equilibrium condition $\sin\theta = 0$. Thus, the equilibrium points are $\theta = 0, \pm\pi, \ldots$. Note that the equilibrium points $\theta = 0, \pm 2\pi, \ldots$ and $\theta = \pm\pi, \pm 3\pi, \ldots$ correspond to the vertical down and up positions, respectively.

If we choose $x_1 = \theta$ and $x_2 = \dot{\theta}$ as the state variables, then the state-variable description of this nonlinear system becomes

$$\dot{x}_1 = x_2$$

$$\dot{x}_2 = -\frac{g}{\ell}\sin x_1 + \frac{1}{m\ell}u$$

Example 1.3

Consider Hamilton's canonical equations of motion described by

$$\dot{q}_i = \frac{\partial H}{\partial p_i}, \qquad i = 1, \ldots, n \qquad (1.147\text{a})$$

$$\dot{p}_i = -\frac{\partial H}{\partial q_i} + u_i, \qquad i = 1, \ldots, n \qquad (1.147\text{b})$$

where $H = H(q_1, \ldots, q_n, p_1, \ldots, p_n, t)$ and u_i denotes the ith generalized force associated with the nonpotential forces in the system. Defining $\mathbf{q} = (q_1, \ldots, q_n)$, $\mathbf{p} = (p_1, \ldots, p_n)$, and $\mathbf{u} = (u_1, \ldots, u_n)$, we obtain Hamilton's equations of motion in matrix form, as follows:

$$\frac{d\mathbf{q}}{dt} = \left[\frac{\partial H}{\partial \mathbf{p}}\right]^T$$

$$\frac{d\mathbf{p}}{dt} = -\left[\frac{\partial H}{\partial \mathbf{q}}\right]^T + \mathbf{u}$$

where

$$\frac{\partial H}{\partial \mathbf{p}} \equiv \left[\begin{array}{ccc} \dfrac{\partial H}{\partial p_1} & \cdots & \dfrac{\partial H}{\partial p_n} \end{array} \right]$$

$$\frac{\partial H}{\partial \mathbf{q}} \equiv \left[\begin{array}{ccc} \dfrac{\partial H}{\partial q_1} & \cdots & \dfrac{\partial H}{\partial q_n} \end{array} \right]$$

Finally, Hamilton's equations of motion are rewritten in state-vector form as

$$\dot{\mathbf{x}} = \mathbf{f}(\mathbf{x}, \mathbf{u}, t) \tag{1.148}$$

where

$$\mathbf{x} = \begin{bmatrix} \mathbf{q} \\ \mathbf{p} \end{bmatrix}, \qquad \mathbf{f} = \begin{bmatrix} (\partial H/\partial \mathbf{p})^T \\ -(\partial H/\partial \mathbf{q})^T + \mathbf{u} \end{bmatrix}$$

A dynamic system described by Hamilton's canonical equations of the form (1.147) or (1.148) with $\mathbf{u} = 0$ is often called a *Hamiltonian dynamic system*. For such a Hamiltonian system, we have

$$\operatorname{div} \dot{\mathbf{x}} = \operatorname{div} \mathbf{f} = \operatorname{tr}\left[\frac{\partial \mathbf{f}}{\partial \mathbf{x}} \right] = \sum_{i=1}^{n} \frac{\partial}{\partial q_i}\left(\frac{\partial H}{\partial p_i} \right) - \sum_{i=1}^{n} \frac{\partial}{\partial p_i}\left(\frac{\partial H}{\partial q_i} \right) = 0$$

Consequently, it is often said that a Hamiltonian system preserves the volume of a set of points in the $2n$-dimensional state or phase space of $\mathbf{x} = (q_1, \ldots, q_n, p_1, \ldots, p_n)$, whether or not H depends explicitly on t. This statement is often referred to as the Liouville theorem in dynamic systems theory.

Given $H = H(\mathbf{q}, \mathbf{p}, t)$, we obtain

$$\begin{aligned}
\frac{dH}{dt} &= \frac{\partial H}{\partial \mathbf{q}}\frac{d\mathbf{q}}{dt} + \frac{\partial H}{\partial \mathbf{p}}\frac{d\mathbf{p}}{dt} + \frac{\partial H}{\partial t} \\
&= -\left[\frac{d\mathbf{p}}{dt} \right]^T \frac{d\mathbf{q}}{dt} + \left[\frac{d\mathbf{q}}{dt} \right]^T \frac{d\mathbf{p}}{dt} + \frac{\partial H}{\partial t} \\
&= \frac{\partial H}{\partial t}
\end{aligned}$$

Thus, for an autonomous Hamiltonian system with $H = H(\mathbf{q}, \mathbf{p})$, we have $\dot{H} = 0$; i.e., H is constant.

In dynamic systems theory, a nonlinear system described by $\dot{\mathbf{x}} = \mathbf{f}(\mathbf{x}, t)$ with $\operatorname{div} \mathbf{f} = 0$ is called a volume-preserving, nondissipative or conservative system; hence, the Hamiltonian system is said to be a conservative system. A system with $\operatorname{div} \mathbf{f} < 0$ is called a dissipative system, and stable, persistent motion in the $2n$-dimensional dissipative system remains on an attractor that has dimension less than $2n$. It should be emphasized that the dissipativeness or conservativeness here is concerned with whether the volume of a set of points in state space of a particular set of state variables is preserved (conserved) or not.

Furthermore, if div $\mathbf{f}$ of a dynamic system described by $\dot{\mathbf{x}} = \mathbf{f}(\mathbf{x}, t)$ is not identically zero or does not change sign, then periodic solutions (or closed orbits) are not possible for the system. This statement is called *Bendixon's criterion* in dynamic systems theory.

Problem

1.29 Consider a cart of mass M with an inverted pendulum on a frictionless horizontal surface, as was shown in Fig. 1.9a. The inverted pendulum, consisting of a massless rod of length ℓ and a point mass m, is hinged on top of the cart. Let $z(t)$ be the horizontal position of the cart, $\theta(t)$ the angle of the pendulum measured from the vertical position, $u(t)$ the control input force acting on the cart, and g the gravitational acceleration.
  The nonlinear equations of motion for this system can be derived as

$$(M + m)\ddot{z} + m\ell\ddot{\theta}\cos\theta - m\ell\dot{\theta}^2\sin\theta = u$$
$$m\ell^2\ddot{\theta} + m\ell\ddot{z}\cos\theta - mg\ell\sin\theta = 0$$

  (a) Defining $\mathbf{x} = (x_1, x_2, x_3, x_4) = (z, \theta, \dot{z}, \dot{\theta})$ as a state vector, obtain the state-variable description of the form $\dot{\mathbf{x}} = \mathbf{f}(\mathbf{x}, \mathbf{u})$. Then show that div $\dot{\mathbf{x}}$ of this system with $u = 0$ is neither identically zero nor sign definite.
  (b) Defining $\mathbf{x} = (x_1, x_2, x_3, x_4) = (z, \theta, p_z, p_\theta)$ as a state vector, where p_z and p_θ are the generalized momenta associated with the generalized coordinates z and θ, obtain Hamilton's canonical equations of the form $\dot{\mathbf{x}} = \mathbf{f}(\mathbf{x}, \mathbf{u})$, and then show that div $\dot{\mathbf{x}}$ of this system with $u = 0$ is identically zero.
  Note: It is important to point out that the divergence of the vector field of a dynamic system of the form $\dot{\mathbf{x}} = \mathbf{f}(\mathbf{x}, t)$ depends on the selection of a particular state vector $\mathbf{x}$.

1.3.2 Nonlinear System Stability

Consider a nonlinear dynamic system described by

$$\dot{\mathbf{x}} = \mathbf{f}(\mathbf{x}, t) \tag{1.149}$$

where $\mathbf{x} = (x_1, \ldots, x_n)$ is the state vector. An equilibrium point of this dynamic system is a point $\mathbf{x}^*$ such that

$$\mathbf{f}(\mathbf{x}^*, t) = 0 \qquad \text{for all } t \tag{1.150}$$

Definition 1.1   Lyapunov Stability

An isolated equilibrium point $\mathbf{x}^*$ of the system described by Eq. (1.149) is said to be *Lyapunov stable*, or simply called stable, if for any $\epsilon > 0$ there exists a real positive number $\delta(\epsilon, t_0)$ such that

$$\|\mathbf{x}(t_0) - \mathbf{x}^*\| \le \delta \Rightarrow \|\mathbf{x}(t) - \mathbf{x}^*\| \le \epsilon \qquad \text{for all } t \ge t_0$$

where $\|\mathbf{x}\|$ denotes the Euclidean norm of a vector $\mathbf{x}$; i.e.,

$$\|\mathbf{x}\| \equiv \sqrt{\mathbf{x}^T \mathbf{x}}$$

Note that the real number δ depends on ϵ and, in general, also depends on t_0. If δ does not depend on t_0, then it is said to be uniformly Lyapunov stable.

Definition 1.2   Local Asymptotic Stability

An isolated equilibrium point $\mathbf{x}^*$ is said to be *locally asymptotically stable*, or simply asymptotically stable, if it is Lyapunov stable and

$$\|\mathbf{x}(t_0) - \mathbf{x}^*\| \leq \delta \Rightarrow \mathbf{x}(t) \to \mathbf{x}^* \qquad \text{as } t \to \infty$$

Definition 1.3   Global Asymptotic Stability

An equilibrium point $\mathbf{x}^*$ is said to be *globally asymptotically stable* or asymptotically stable in the large if it is Lyapunov stable and $\mathbf{x}(t) \to \mathbf{x}^*$ as $t \to \infty$ for any initial conditions $\mathbf{x}(t_0)$.

Definition 1.4   Instability

An equilibrium point is simply said to be unstable if it is not stable; i.e., if it is neither Lyapunov stable nor asymptotically stable.

A necessary condition for an equilibrium point to be locally asymptotically stable is that it be isolated. A necessary condition for an equilibrium point to be globally asymptotically stable is that it be the only equilibrium point.

In the stability theory of dynamic systems described by

$$\dot{\mathbf{x}} = \mathbf{f}(\mathbf{x}, t), \qquad \mathbf{f}(\mathbf{x}^*, t) = 0 \qquad \text{for all } t \qquad (1.151)$$

where $\mathbf{x}^*$ is an equilibrium point, we often introduce the following transformation

$$\mathbf{x} = \mathbf{x}^* + \mathbf{z}$$

where the perturbation $\mathbf{z}$ from an equilibrium point $\mathbf{x}^*$ is, in general, not necessarily small. A new set of nonlinear differential equations is then obtained as

$$\dot{\mathbf{z}} = \mathbf{f}(\mathbf{x}^* + \mathbf{z}, t) \qquad (1.152)$$

with the origin $\mathbf{z} = 0$ as an equilibrium point.

Assuming that the perturbation $\mathbf{z}$ is small, we further obtain a set of linearized equations about an equilibrium point $\mathbf{x}^*$, as follows:

$$\dot{\mathbf{z}} = \mathbf{A}\mathbf{z} \qquad (1.153)$$

where

$$\mathbf{A} = \left[\frac{\partial \mathbf{f}}{\partial \mathbf{x}}\right]_{\mathbf{x}^*} \equiv \left[\frac{\partial f_i}{\partial x_j}\right]_{\mathbf{x}^*}$$

is the Jacobian matrix to be evaluated at the equilibrium point $\mathbf{x}^*$.

Definition 1.5   Linear System Stability

The origin $z = 0$ of a linearized system described by Eq. (1.153) is said to be *infinitesimally* asymptotically stable, or simply called asymptotically stable if all the eigenvalues of A have negative real parts. A linear system is said to be Lyapunov stable if none of its eigenvalues has a positive real part and if it has no repeated eigenvalues on the imaginary axis. It is said to be unstable if any one of its eigenvalues has a positive real part or if it has repeated eigenvalues on the imaginary axis. (The stability of a dynamic system linearized about an equilibrium point is defined here although linear system dynamics will be studied in detail later in this chapter.)

Note that

$$A = \begin{bmatrix} 0 & 0 \\ 0 & 0 \end{bmatrix} \quad \text{does not have repeated eigenvalues}$$

$$A = \begin{bmatrix} 0 & 1 \\ 0 & 0 \end{bmatrix} \quad \text{has repeated eigenvalues}$$

The following very important stability theorem for determining the stability of an equilibrium point of a nonlinear system from the stability of its linearized system was first established by Lyapunov [9].

Theorem 1.1   Lyapunov's First Stability Theorem

This theorem is often referred to as the indirect (or first) method of Lyapunov in the literature, and can be described as follows:

1) If the origin $z = 0$ of a linearized system described by Eq. (1.153) is asymptotically stable, then the equilibrium point x^* of the nonlinear system described by either Eq. (1.151) or Eq. (1.152) is also asymptotically stable.

2) If the origin $z = 0$ of a linearized system is unstable, then the equilibrium point x^* of the nonlinear system is also unstable.

It is important to note, however, that the Lyapunov stability of the equilibrium point at the origin $z = 0$ of a linearized system does not guarantee the Lyapunov stability of the equilibrium point x^* of the nonlinear system. As an example, consider a system described by

$$\dot{x}_1 = x_2 + bx_1\left(x_1^2 + x_2^2\right) \tag{1.154a}$$

$$\dot{x}_2 = -x_1 + bx_2\left(x_1^2 + x_2^2\right) \tag{1.154b}$$

where b is a constant parameter. It can be easily shown that the origin of a linearized system is Lyapunov stable for any value of b. However, using Lyapunov's second stability theorem (to be presented next), one can show that the origin of the nonlinear system described by Eq. (1.54) is unstable for any $b > 0$ and asymptotically stable for any $b < 0$.

Theorem 1.2   Lyapunov's Second Stability Theorem

This theorem is often referred to as the direct (or second) method of Lyapunov in the literature; however, it is not of the original form of Lyapunov's theorem. In fact, the theorem presented here is an extended version of Lyapunov's second stability theorem by many other researchers. Consider an autonomous nonlinear dynamic system described by

$$\dot{x} = \mathbf{f}(\mathbf{x}), \qquad \mathbf{f}(\mathbf{x}^*) = 0 \qquad\qquad (1.155)$$

where $\mathbf{x}^*$ is an isolated equilibrium point.

If there exists in some finite neighborhood $\mathcal{D}$ of the equilibrium point $\mathbf{x}^*$ a positive-definite scalar function $E(\mathbf{x})$ with continuous first partial derivatives with respect to $\mathbf{x}$ and t such that the following conditions exist:

1) $E(\mathbf{x}) > 0$ for all $\mathbf{x} \neq \mathbf{x}^*$ in $\mathcal{D}$, $E(\mathbf{x}^*) = 0$ for all t
2) $\dot{E}(\mathbf{x}) \leq 0$ for all $\mathbf{x} \neq \mathbf{x}^*$ in $\mathcal{D}$ and t

then the equilibrium point $\mathbf{x}^*$ is Lyapunov stable.

If, in addition,

3) $\dot{E}(\mathbf{x})$ is not identically zero along any solution $\mathbf{x}$ of Eq. (1.155) other than $\mathbf{x}^*$

then the equilibrium point $\mathbf{x}^*$ is locally asymptotically stable.

If, in addition,

4) there exists in the entire state space a positive-definite function $E(\mathbf{x})$ which is radially unbounded; i.e., $E(\mathbf{x}) \to \infty$ as $\|\mathbf{x}\| \to \infty$

then the equilibrium point $\mathbf{x}^*$ is globally asymptotically stable; i.e., $\mathbf{x}(t) \to \mathbf{x}^*$ as $t \to \infty$ for any initial conditions $\mathbf{x}(t_0)$.

If instead condition 2 is

2′) $\dot{E}(\mathbf{x}) > 0$ for all $\mathbf{x} \neq \mathbf{x}^*$ and t, and $\dot{E}(\mathbf{x}^*) = 0$ for all t

then the equilibrium point $\mathbf{x}^*$ is unstable. Such a positive-definite function $E(\mathbf{x})$ is called a Lyapunov function.

This theorem provides only sufficient conditions for checking the stability of an equilibrium point of a nonlinear dynamic system, and does not provide a method for determining a positive-definite Lyapunov function $E(\mathbf{x})$ for a given nonlinear system. A Lyapunov function may be considered as a generalized energy function; however, the total energy or the Hamiltonian of the system may be used as a Lyapunov function.

For example, consider a case in which the total energy of a given dynamic system is chosen as a Lyapunov function, as follows:

$$E = T + V$$

where the kinetic energy T is of the general form

$$T = T_2 + T_1 + T_0$$

and T_2 is a quadratic function in the generalized velocities, T_1 is a linear function in the generalized velocities, and T_0 is a nonnegative function of only the generalized coordinates and time. In this case, it may become difficult to determine the positive definiteness of the chosen Lyapunov function $E = T + V$. In a case where $T = T_2$ with $T_1 = T_0 = 0$, and where V is a quadratic function of the generalized coordinates, however, the total energy E becomes positive definite and it can be used as a Lyapunov function.

Because the Hamiltonian is defined as

$$H = \sum_{i=1}^{n} \frac{\partial L}{\partial \dot{q}_i} \dot{q}_i - L = T_2 - T_0 + V$$

it is also difficult to determine the positive definiteness of the Hamiltonian function unless $T_0 = 0$. Thus, only for dynamic systems with $T = T_2$ and $T_1 = T_0 = 0$, we often choose the total energy or the Hamiltonian as the Lyapunov function, and then determine the sign definiteness of

$$\dot{E} = \sum_{i=1}^{n} \dot{q}_i u_i$$

where u_i is the ith generalized nonconservative force associated with the ith generalized coordinate q_i of the system.

Given a dynamic system of the form

$$\dot{\mathbf{x}} = \mathbf{f}(\mathbf{x}) + \mathbf{u}$$

the system is called *norm-invariant* if the solution $\mathbf{x}(t)$ of the homogeneous system ($\mathbf{u} = 0$) satisfies the following condition:

$$\frac{\mathrm{d}}{\mathrm{d}t} \|\mathbf{x}\| = 0$$

where $\|\mathbf{x}\| \equiv \sqrt{\mathbf{x}^T \mathbf{x}}$; i.e., if $\|\mathbf{x}\|$ is constant for all t. Note that

$$\frac{\mathrm{d}}{\mathrm{d}t} \|\mathbf{x}\| = \frac{\mathbf{x}^T \dot{\mathbf{x}}}{\|\mathbf{x}\|}$$

Given a control problem of dynamic systems described by

$$\dot{\mathbf{x}} = \mathbf{f}(\mathbf{x}, \mathbf{u}) \quad \text{and} \quad \mathbf{f}(\mathbf{x}^*, 0) = 0 \qquad \text{for all } t$$

where $\mathbf{u}$ is the control input vector to be determined, we may select a suitable Lyapunov function $E(\mathbf{x})$, and obtain $\dot{E}$ as

$$\dot{E}(\mathbf{x}) = \frac{\partial E}{\partial \mathbf{x}} \frac{\mathrm{d}\mathbf{x}}{\mathrm{d}t}$$

$$= \frac{\partial E}{\partial \mathbf{x}} \mathbf{f}(\mathbf{x}, \mathbf{u})$$

Then it may be possible to find a globally asymptotically stabilizing control input $\mathbf{u} = \mathbf{u}(\mathbf{x})$, which guarantees $\dot{E} \leq 0$ for all $\mathbf{x} \neq \mathbf{x}^*$ and t and $E(\mathbf{x}^*) = 0$.

Example 1.4

Consider again the simple pendulum described by

$$\dot{x}_1 = x_2$$

$$\dot{x}_2 = -\frac{g}{\ell}\sin x_1 + \frac{1}{m\ell}u$$

where u is the applied input force of the form $u = -cx_2$, where c is a positive constant. The stability of an equilibrium point at the origin $\mathbf{x}^* = (0,\ 0)$ is to be investigated using Lyapunov's second stability theorem.

The total mechanical energy of the system is chosen as a Lyapunov function:

$$E(\mathbf{x}) = T + V = \tfrac{1}{2}m\ell^2 x_2^2 + mg\ell(1 - \cos x_1)$$

since $E(\mathbf{x}^*) = 0$ and $E(\mathbf{x}) > 0$ for all $\mathbf{x} \neq \mathbf{x}^*$. We then, obtain

$$\dot{E} = m\ell^2 x_2 \dot{x}_2 + mg\ell(\sin x_1)\dot{x}_1$$

$$= m\ell^2 x_2 \left(-\frac{g}{\ell}\sin x_1 - \frac{c}{m\ell}x_2\right) + mg\ell(\sin x_1)x_2$$

$$= -c\ell x_2^2 \le 0$$

Because E satisfies condition 3, but not condition 4, of Lyapunov's second stability theorem (Theorem 1.2), the equilibrium point $\mathbf{x}^* = (0,\ 0)$ is said to be asymptotically stable. Intuitively, we also know that the vertical down position of the pendulum with rate damping is asymptotically stable and that the vertical up position is unstable.

Problems

1.30 Consider a system described by

$$\dot{x}_1 = ax_1 + x_2 + bx_1\left(x_1^2 + x_2^2\right)$$
$$\dot{x}_2 = -x_1 - ax_2 + bx_2\left(x_1^2 + x_2^2\right)$$

where a and b are constant parameters.
(a) Show that the origin of a linearized system is Lyapunov stable for $|a| < 1$ and unstable for $|a| > 1$.
(b) Show that the origin of the nonlinear system is asymptotically stable for $|a| < 1$ and $b < 0$, and unstable for $|a| < 1$ and $b > 0$.
  Hint: Use $E = x_1^2 + 2ax_1x_2 + x_2^2$, which is positive definite if $|a| < 1$.

1.31 Consider the rotational motion of a rigid spacecraft described by Euler's equations of motion

$$J_1\dot{\omega}_1 = (J_2 - J_3)\omega_2\omega_3 + u_1$$
$$J_2\dot{\omega}_2 = (J_3 - J_1)\omega_3\omega_1 + u_2$$
$$J_3\dot{\omega}_3 = (J_1 - J_2)\omega_1\omega_2 + u_3$$

where J_i are principal moments of inertia of the spacecraft with $J_1 > J_2 > J_3$, ω_i are the angular velocity components along principal axes, and u_i are the control torque inputs about principal axes.

(a) For a torque-free ($u = 0$), rigid body, show that there exist two constants of integration such as

$$J_1\omega_1^2 + J_2\omega_2^2 + J_3\omega_3^2 = \text{const}$$
$$J_1^2\omega_1^2 + J_2^2\omega_2^2 + J_3^2\omega_3^2 = \text{const}$$

and that, consequently, we have

$$J_1(J_1 - J_3)\omega_1^2 + J_2(J_2 - J_3)\omega_2^2 = \text{const}$$
$$J_2(J_1 - J_2)\omega_2^2 + J_3(J_1 - J_3)\omega_3^2 = \text{const}$$
$$J_1(J_1 - J_2)\omega_1^2 - J_3(J_2 - J_3)\omega_3^2 = \text{const}$$

(b) For a torque-free, rigid body, show that the equilibrium points $(\Omega, 0, 0)$ and $(0, 0, \Omega)$, where Ω is an arbitrary constant spin rate, are both Lyapunov stable; i.e., a pure spinning motion about the major or minor axis is Lyapunov stable. Also show that the equilibrium point $(0, \Omega, 0)$ is unstable; i.e., a pure spinning motion about the intermediate axis is unstable.

Hint: One can show that $(0, \Omega, 0)$ is an unstable equilibrium point by showing that the linearized system about $(0, \Omega, 0)$ is unstable. But note that the Lyapunov stability of the linearized system about $(\Omega, 0, 0)$ or $(0, 0, \Omega)$ does not guarantee the Lyapunov stability of those equilibrium points of the nonlinear system.

(c) For a case of three-axis rate damping control with $u_i = -c_i\omega_i$, where c_i are all positive constants, show that the origin $(0, 0, 0)$ is globally asymptotically stable for any positive constants c_i.

(d) Show that the origin $(0, 0, 0)$ is not asymptotically stable (but Lyapunov stable) for a spacecraft with only two control inputs of the form $u_1 = -c_1\omega_1, u_2 = -c_2\omega_2$, and $u_3 = 0$ where c_1 and c_2 are positive constants.

(e) For a spin-up maneuver control with $u_1 = -c_1(\omega_1 - \Omega), u_2 = -c_2\omega_2$, and $u_3 = -c_3\omega_3$, where Ω is the desired spin rate about the first principal axis, show that the equilibrium point $(\Omega, 0, 0)$ is globally asymptotically stable for any positive constants c_i.

1.32 Defining $x_i = J_i\omega_i$, we rewrite Euler's equations of motion described in the preceding problem as

$$\dot{x}_1 = \frac{J_2 - J_3}{J_2 J_3}x_2 x_3 + u_1$$
$$\dot{x}_2 = \frac{J_3 - J_1}{J_3 J_1}x_3 x_1 + u_2$$
$$\dot{x}_3 = \frac{J_1 - J_2}{J_1 J_2}x_1 x_2 + u_3$$

(a) Show that this dynamic system is norm-invariant.
(b) If this dynamic system is subject to a control input constraint of the form

$$\|\mathbf{u}\| = \sqrt{u_1^2 + u_2^2 + u_3^2} \leq 1$$

the time-optimal control inputs are given by

$$u_i = -\frac{x_i}{\|\mathbf{x}\|}, \qquad i = 1, 2, 3$$

Show that the origin $(0, 0, 0)$ of the system with such time-optimal control inputs is globally asymptotically stable.

1.33 Euler's equations of motion for an axisymmetric body with $J_1 = 2$ and $J_2 = J_3 = 1$ are

$$2\dot{\omega}_1 = u_1$$
$$\dot{\omega}_2 = -\omega_3\omega_1 + u_2$$
$$\dot{\omega}_3 = \omega_1\omega_2 + u_3$$

For a given linear feedback control logic of the form

$$u_1 = -2\sigma\omega_1 + 2\sigma\omega_2$$
$$u_2 = -\omega_2 + r\omega_1$$
$$u_3 = -b\omega_3$$

where σ, r, and b are real positive constant parameters, the closed-loop system is described by

$$\dot{\omega}_1 = -\sigma\omega_1 + \sigma\omega_2$$
$$\dot{\omega}_2 = -\omega_3\omega_1 - \omega_2 + r\omega_1$$
$$\dot{\omega}_3 = \omega_1\omega_2 - b\omega_3$$

These equations are, in fact, the so-called Lorenz equations [10] that are commonly described in x, y, and z coordinates, as follows:

$$\dot{x} = -\sigma x + \sigma y$$
$$\dot{y} = -zx - y + rx$$
$$\dot{z} = xy - bz$$

A dynamic system of the preceding mathematical form is often referred to as the Lorenz system.
(a) Find the equilibrium points of the Lorenz system as follows:

For $r \leq 1$, $(0, 0, 0)$

For $r > 1$, $(0, 0, 0)$ and $C^{\pm} = (\pm\sqrt{b(r-1)}, \pm\sqrt{b(r-1)}, r-1)$

where r is the bifurcation parameter of the system.

(b) Show that the origin of the Lorenz system is globally asymptotically stable when $r < 1$, stable when $r = 1$, and unstable when $r > 1$.
Hint: $E = \frac{1}{2}(x^2 + \sigma y^2 + \sigma z^2)$.

(c) Show that the equilibrium points $C^\pm = (\pm 7.5542, \pm 7.5542, 21.4)$ of the Lorenz system with $\sigma = 10, b = 8/3$, and $r = 22.4$ are both asymptotically stable. Performing a computer simulation of this case, verify that the trajectory starting from $(-10, -10, 30)$ ends up at C^+ and that the trajectory starting from $(-12, -12, 30)$ ends up at C^-.

(d) Show that, for the Lorenz system with $\sigma = 10, b = 8/3$, and $r = 28$, the equilibrium points $C^\pm = (\pm 8.4852, \pm 8.4852, 27)$ are both unstable. Also perform a computer simulation of this case with an initial condition $(0, 1, 0)$ to observe the turbulent or chaotic dynamic behavior of the system, known as a strange attractor.

(e) Show that the Lorenz system is dissipative and that there exists a solid ellipsoid

$$\mathcal{E} = \{(x, y, z) : rx^2 + \sigma y^2 + \sigma(z - 2r)^2 \leq C < \infty\}$$

or

$$\mathcal{E} = \{(x, y, z) : x^2 + y^2 + \sigma(z - r - \sigma)^2 \leq C < \infty\}$$

such that all solutions of the Lorenz system enter $\mathcal{E}$ within finite time and thereafter remain in $\mathcal{E}$.
Hint: Show that $\dot{\mathcal{E}} < 0$ as $(x^2 + y^2 + z^2)^{1/2} \to \infty$.

1.34 Consider a rigid body with a spherical, dissipative fuel slug. Such a simplified model of a rigid body with internal energy dissipation is described by

$$(J_1 - J)\dot{\omega}_1 = (J_2 - J_3)\omega_2\omega_3 + \mu\sigma_1$$

$$(J_2 - J)\dot{\omega}_2 = (J_3 - J_1)\omega_3\omega_1 + \mu\sigma_2$$

$$(J_3 - J)\dot{\omega}_3 = (J_1 - J_2)\omega_1\omega_2 + \mu\sigma_3$$

$$\dot{\sigma}_1 = -\dot{\omega}_1 - \frac{\mu}{J}\sigma_1 - \omega_2\sigma_3 + \omega_3\sigma_2$$

$$\dot{\sigma}_2 = -\dot{\omega}_2 - \frac{\mu}{J}\sigma_2 - \omega_3\sigma_1 + \omega_1\sigma_3$$

$$\dot{\sigma}_3 = -\dot{\omega}_3 - \frac{\mu}{J}\sigma_3 - \omega_1\sigma_2 + \omega_2\sigma_1$$

where (J_1, J_2, J_3) are the principal moments of inertia of the spacecraft including the spherical, dissipative fuel slug of inertia J; $(\omega_1, \omega_2, \omega_3)$ are the body rates about the principal axes; $(\sigma_1, \sigma_2, \sigma_3)$ are the relative rates between the rigid body and the fuel slug about the principal axes; and μ is the viscous damping coefficient of the fuel slug. It is assumed that $J_1 > J_2 > J_3$ without loss of generality.

(a) Show that a necessary condition for the equilibrium points is $\sigma_1^2 + \sigma_2^2 + \sigma_3^2 = 0$; i.e., $\sigma_1 = \sigma_2 = \sigma_3 = 0$.

(b) Show that an equilibrium point $(\omega_1, \omega_2, \omega_3, \sigma_1, \sigma_2, \sigma_3) = (\Omega, 0, 0, 0, 0, 0)$ is stable; i.e., a pure spinning motion about the major axis is stable.

(c) Show that an equilibrium point $(0, 0, \Omega, 0, 0, 0)$ is unstable; i.e., a pure spinning motion about the minor axis is unstable whereas it is Lyapunov stable for a rigid body without energy dissipation.

(d) Show that an equilibrium point $(0, \Omega, 0, 0, 0, 0)$ is also unstable.

(e) Consider a spacecraft with the following numerical values: $(J_1, J_2, J_3, J) = (2000, 1500, 1000, 18)$ kg·m^2 and $\mu = 30$ N·m·s. Performing computer simulation, verify that the trajectory starting from an initial condition $(0.1224, 0, 2.99, 0, 0, 0)$ rad/s ends up at $(-1.5, 0, 0, 0, 0, 0)$ rad/s.

Note: The kinetic energy T and the angular momentum H of the system are defined as

$$H^2 = (J_1\omega_1 + J\sigma_1)^2 + (J_2\omega_2 + J\sigma_2)^2 + (J_3\omega_3 + J\sigma_3)^2$$
$$2T = (J_1 - J)\omega_1^2 + (J_2 - J)\omega_2^2 + (J_3 - J)\omega_3^2$$
$$+ J\{(\omega_1 + \sigma_1)^2 + (\omega_2 + \sigma_2)^2 + (\omega_3 + \sigma_3)^2\}$$

During computer simulation of this case, the angular momentum H needs to be checked whether or not it is maintained at a constant value of 3000 N·m·s.

(f) Also perform a computer simulation with a slightly different initial condition $(0.125, 0, 2.99, 0, 0, 0)$ and verify that the trajectory ends up at $(+1.5, 0, 0, 0, 0, 0)$.

Note: For such a spinning spacecraft with energy dissipation, a small change in initial conditions can lead to a change in the final spin polarity for ω_1. Such sensitive dependence on initial conditions is the property characterizing a chaotic dynamic system.

1.35 Consider the rotational equations of motion of a rigid spacecraft described by

$$2\dot{q}_1 = \omega_1 q_4 - \omega_2 q_3 + \omega_3 q_2$$
$$2\dot{q}_2 = \omega_1 q_3 + \omega_2 q_4 - \omega_3 q_1$$
$$2\dot{q}_3 = -\omega_1 q_2 + \omega_2 q_1 + \omega_3 q_4$$
$$2\dot{q}_4 = -\omega_1 q_1 - \omega_2 q_2 - \omega_3 q_3$$
$$J_1\dot{\omega}_1 = (J_2 - J_3)\omega_2\omega_3 + u_1$$
$$J_2\dot{\omega}_2 = (J_3 - J_1)\omega_3\omega_1 + u_2$$
$$J_3\dot{\omega}_3 = (J_1 - J_2)\omega_1\omega_2 + u_3$$

where q_i are the attitude quaternions constrained by

$$q_1^2 + q_2^2 + q_3^2 + q_4^2 = 1$$

and where ω_i are the body angular rates, J_i are the principal moments of inertia of the spacecraft, and u_i are the control torque inputs.

The stability of the equilibrium point

$$\mathbf{x}^* = (q_1, q_2, q_3, q_4, \omega_1, \omega_2, \omega_3)$$
$$= (0, 0, 0, +1, 0, 0, 0)$$

is to be determined for different control inputs. [In fact, there exists another equili- brium point $(0, 0, 0, -1, 0, 0, 0)$; however, both equilibrium points correspond to the physically identical orientation.]

(a) For a case of rate damping control with $u_i = -c_i\omega_i$ where c_i are positive constants, determine the stability of the equilibrium point $\mathbf{x}^*$.

(b) Given the control torque inputs u_i of the form

$$u_1 = -kq_1 - c_1\omega_1$$
$$u_2 = -kq_2 - c_2\omega_2$$
$$u_3 = -kq_3 - c_3\omega_3$$

where k and c_i are positive constants, show that the equilibrium point $\mathbf{x}^*$ is globally asymptotically stable for any positive values of k and c_i. *Hint:* Select $E = (J_1\omega_1^2 + J_2\omega_2^2 + J_3\omega_3^2)/2 + k[q_1^2 + q_2^2 + q_3^2 + (q_4 - 1)^2]$ as a Lyapunov function.

(c) Given the control torque inputs u_i of the form

$$u_1 = -k_1q_1 - c_1\omega_1$$
$$u_2 = -k_2q_2 - c_2\omega_2$$
$$u_3 = -k_3q_3 - c_3\omega_3$$

where k_i and c_i are positive constants, determine whether or not the equilibrium point $\mathbf{x}^*$ is globally asymptotically stable for any positive values of k_i and c_i. (This is a much harder unsolved problem!)

1.3.3   Linearization and State-Space Equations

Consider a nonlinear dynamic system described by

$$\dot{x}_i = f_i(x_1, \ldots, x_n; u_1, \ldots, u_p; t), \qquad i = 1, \ldots, n \qquad (1.156a)$$
$$y_j = g_j(x_1, \ldots, x_n; u_1, \ldots, u_p; t), \qquad j = 1, \ldots, q \qquad (1.156b)$$

where $x_i(t)$ are called state variables, $u_i(t)$ control input variables, and $y_i(t)$ output variables.

In vector notation, Eqs. (1.156) become

$$\dot{\mathbf{x}} = \mathbf{f}(\mathbf{x}, \mathbf{u}, t) \qquad (1.157a)$$
$$\mathbf{y} = \mathbf{g}(\mathbf{x}, \mathbf{u}, t) \qquad (1.157b)$$

where $\mathbf{x}$ is the state vector, $\mathbf{u}$ the control input or parameter vector, and $\mathbf{y}$ the output vector of the dynamic system, defined as $\mathbf{x} = (x_1, \ldots, x_n)$, $\mathbf{f} = (f_1, \ldots, f_n)$, $\mathbf{u} = (u_1, \ldots, u_p)$, $\mathbf{y} = (y_1, \ldots, y_q)$, and $\mathbf{g} = (g_1, \ldots, g_q)$.

An equilibrium point $\mathbf{x}^*$ is determined from

$$\mathbf{f}(\mathbf{x}^*, \mathbf{u}^*, t) = 0 \qquad \text{for all } t \geq t_0 \qquad (1.158)$$

for a known constant input or parameter vector $\mathbf{u}^*$.

Expanding the right-hand side of Eqs. (1.157) in a first-order Taylor series expansion in terms of small perturbations

$$\delta \mathbf{x} = \mathbf{x} - \mathbf{x}^*$$
$$\delta \mathbf{u} = \mathbf{u} - \mathbf{u}^*$$

we obtain the linearized equations about the equilibrium point as follows:

$$\delta \dot{\mathbf{x}} = \mathbf{A}\,\delta \mathbf{x} + \mathbf{B}\,\delta \mathbf{u} \qquad (1.159\text{a})$$
$$\delta \mathbf{y} = \mathbf{C}\,\delta \mathbf{x} + \mathbf{D}\,\delta \mathbf{u} \qquad (1.159\text{b})$$

where $\delta \mathbf{y} = \mathbf{y} - \mathbf{y}^*$ and $\mathbf{A}$, $\mathbf{B}$, $\mathbf{C}$, and $\mathbf{D}$ are constant matrices defined as

$$\mathbf{A} = \left.\frac{\partial \mathbf{f}}{\partial \mathbf{x}}\right|_*, \qquad \mathbf{B} = \left.\frac{\partial \mathbf{f}}{\partial \mathbf{u}}\right|_* \qquad (1.160\text{a})$$

$$\mathbf{C} = \left.\frac{\partial \mathbf{g}}{\partial \mathbf{x}}\right|_*, \qquad \mathbf{D} = \left.\frac{\partial \mathbf{g}}{\partial \mathbf{u}}\right|_* \qquad (1.160\text{b})$$

The partial derivatives in Eqs. (1.160) are evaluated at the equilibrium point $\mathbf{x} = \mathbf{x}^*$ and $\mathbf{u} = \mathbf{u}^*$. Note that $\partial \mathbf{f}/\partial \mathbf{x}$, $\partial \mathbf{f}/\partial \mathbf{u}$, etc., in Eqs. (1.160) denote the matrices with the ijth elements of $\partial f_i/\partial x_j$, $\partial f_i/\partial u_j$, etc., respectively.

For notational simplicity, the linear time-invariant systems are often described by

$$\dot{\mathbf{x}} = \mathbf{A}\mathbf{x} + \mathbf{B}\mathbf{u} \qquad (1.161)$$
$$\mathbf{y} = \mathbf{C}\mathbf{x} + \mathbf{D}\mathbf{u} \qquad (1.162)$$

where $\mathbf{x}$, $\mathbf{u}$, and $\mathbf{y}$ denote the small perturbed state, input, and output vectors, respectively. Equations (1.161) and (1.162) are called the state-space equation and output equation of the linear time-invariant system, respectively.

Example 1.5

Consider the simple pendulum of mass m and length ℓ, connected by a massless rod to the hinged support point O, as shown earlier in Fig. 1.5a. Choosing $x_1 = \theta$ and $x_2 = \dot{\theta}$ as the state variables, the state-space equation of this system linearized about an equilibrium point $\mathbf{x}^* = (0, 0)$ can be obtained as

$$\begin{bmatrix} \dot{x}_1 \\ \dot{x}_2 \end{bmatrix} = \begin{bmatrix} 0 & 1 \\ -g/\ell & 0 \end{bmatrix} \begin{bmatrix} x_1 \\ x_2 \end{bmatrix} + \begin{bmatrix} 0 \\ 1/m\ell \end{bmatrix} u$$

If we consider $x_1 = \theta$ as an output variable y, then the output equation becomes

$$y = [1 \quad 0] \begin{bmatrix} x_1 \\ x_2 \end{bmatrix}$$

Problem

1.36   For a cart with an inverted pendulum on a frictionless horizontal surface, which was shown previously in Fig. 1.9a, the nonlinear equations of motion can be obtained as

$$(M + m)\ddot{z} + m\ell\ddot{\theta}\cos\theta - m\ell\dot{\theta}^2\sin\theta = u$$
$$m\ell^2\ddot{\theta} + m\ell\ddot{z}\cos\theta - mg\ell\sin\theta = 0$$

(a) Directly linearize these equations of motion for the small angular motion of the pendulum about $\theta = 0$. Defining $\mathbf{x} = (x_1, x_2, x_3, x_4) = (z, \theta, \dot{z}, \dot{\theta})$ as a state vector, obtain the linear state-space equation of the form (1.161).

(b) Also obtain the same linear state-space equation derived in (a) by linearizing the state-variable description of the nonlinear system of the form $\dot{\mathbf{x}} = \mathbf{f}(\mathbf{x}, \mathbf{u})$ where $\mathbf{x} = (x_1, x_2, x_3, x_4) = (z, \theta, \dot{z}, \dot{\theta})$.

1.3.4   Laplace Transformation

The method of Laplace transformation is a mathematical tool particularly useful for characterizing and analyzing linear dynamic systems. Thus, the basic definitions and results in the theory of Laplace transformation are briefly introduced here.

A complex variable s has a real component σ and an imaginary component $j\omega$; i.e.,

$$s = \sigma + j\omega$$

where $j = \sqrt{-1}$. A complex function $G(s)$ has a real part, $\text{Re}[G(s)]$, and an imaginary part, $\text{Im}[G(s)]$; i.e.,

$$G(s) = \text{Re}[G(s)] + j\,\text{Im}[G(s)]$$

and the complex conjugate of $G(s)$ is defined as

$$G^*(s) = \text{Re}[G(s)] - j\,\text{Im}[G(s)]$$

The Laplace transform of a function $f(t)$ is defined as

$$\mathcal{L}\{f(t)\} = F(s) = \int_0^\infty f(t)e^{-st}\,dt \qquad (1.163)$$

where the symbol $\mathcal{L}$ denotes the Laplace transform of a function and s is called the Laplace transform variable, which is in fact a complex variable. The inverse Laplace transform of $F(s)$ is denoted by

$$f(t) = \mathcal{L}^{-1}\{F(s)\}$$

For example, the Laplace transform of a unit impulse function, denoted by $\delta(t)$, is given as

$$\mathcal{L}\{\delta(t)\} = 1 \tag{1.164}$$

where $\delta(t)$, often called the *Dirac delta function*, is defined such that

$$\delta(t) = \begin{cases} 0 & \text{if } t \neq 0 \\ \infty & \text{if } t = 0 \end{cases}$$

$$\int_{-\infty}^{\infty} \delta(t)\, dt = 1$$

$$\int_{-\infty}^{\infty} f(t)\delta(t - t_0)\, dt = f(t_0)$$

Similarly, the Laplace transform of a unit step function, denoted by $u_s(t)$, can be found as

$$\mathcal{L}\{u_s(t)\} = \frac{1}{s} \tag{1.165}$$

where the unit step function is defined as

$$u_s(t) = \begin{cases} 0 & \text{if } t < 0 \\ 1 & \text{if } t > 0 \end{cases} \tag{1.166}$$

The unit impulse function is the derivative of the unit step function; i.e., we have

$$\delta(t) = \frac{d}{dt} u_s(t)$$

The following Laplace transform pairs also exist:

$$\mathcal{L}\{e^{-at}\} = 1/(s + a)$$
$$\mathcal{L}\{1 - e^{-at}\} = a/s(s + a)$$
$$\mathcal{L}\{te^{-at}\} = 1/(s + a)^2$$
$$\mathcal{L}\{t\} = 1/s^2$$
$$\mathcal{L}\{t^2\} = 2/s^3$$
$$\mathcal{L}\{\sin \omega t\} = \omega/(s^2 + \omega^2)$$
$$\mathcal{L}\{\cos \omega t\} = s/(s^2 + \omega^2)$$
$$\mathcal{L}\{e^{-at} \sin \omega t\} = \omega/\left[(s + a)^2 + \omega^2\right]$$
$$\mathcal{L}\{e^{-at} \cos \omega t\} = (s + a)/\left[(s + a)^2 + \omega^2\right]$$

Some useful properties of the Laplace transformation are summarized as follows.

1.3.4.1 Differentiation Theorem

$$\mathcal{L}\{\dot{f}(t)\} = sF(s) - f(0) \tag{1.167a}$$

$$\mathcal{L}\{\ddot{f}(t)\} = s^2 F(s) - sf(0) - \dot{f}(0) \tag{1.167b}$$

$$\vdots$$

$$\mathcal{L}\{f^{(n)}(t)\} = s^n F(s) - s^{n-1} f(0)$$
$$- s^{n-2} f^{(1)}(0) - \cdots - f^{(n-1)}(0) \tag{1.167c}$$

where $F(s) = \mathcal{L}\{f(t)\}$ and $f^{(n)}(t)$ denotes the nth time derivative of $f(t)$.

1.3.4.2 Final Value Theorem

If all poles of $sF(s)$ lie in the left-half s plane,

$$f(\infty) = \lim_{t \to \infty} f(t) = \lim_{s \to 0} sF(s) \tag{1.168}$$

But if $sF(s)$ has poles on the imaginary axis or in the right-half s plane, $f(\infty)$ does not exist.

1.3.4.3 Convolution Integral

The convolution integral of two functions $f_1(t)$ and $f_2(t)$, denoted by $f_1(t) * f_2(t)$, are defined as

$$f_1(t) * f_2(t) = \int_0^t f_1(t - \tau) f_2(\tau) \, d\tau$$
$$= \int_0^t f_1(\tau) f_2(t - \tau) \, d\tau \tag{1.169}$$

The Laplace transform changes the convolution integral in the time domain into an algebraic equation in the s domain; i.e.,

$$\mathcal{L}\{f_1(t) * f_2(t)\} = F_1(s) F_2(s) \tag{1.170}$$

where $F_1(s) = \mathcal{L}\{f_1(t)\}$ and $F_2(s) = \mathcal{L}\{f_2(t)\}$.

1.3.4.4 Time Delay

A delayed function of $f(t)$ by a time T is denoted by $f(t - T)$, and its Laplace transform becomes

$$\mathcal{L}\{f(t - T)\} = F(s) e^{-Ts} \tag{1.171}$$

where $F(s)$ is the Laplace transform of $f(t)$. For example, we have

$$\mathcal{L}\{\delta(t - T)\} = e^{-Ts}$$
$$\mathcal{L}\{u_s(t - T)\} = (1/s) e^{-Ts}$$

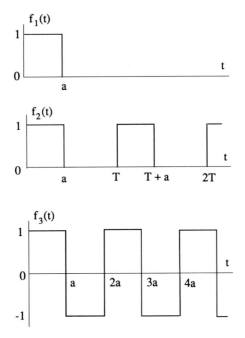

Fig. 1.21   Pulse functions.

1.3.4.5 Periodic Functions If a function $f(t)$ is periodic with period T, then $f(t) = f(t + T)$ and its Laplace transform is

$$F(s) = \bar{F}(s)[1/(1 - e^{-Ts})] \qquad (1.172)$$

where $\bar{F}(s)$ is the Laplace transform of

$$\bar{f}(t) = \begin{cases} f(t) & \text{for } 0 < t \leq T \\ 0 & \text{for } t > T \end{cases}$$

Problem

1.37 Show that the Laplace transforms of various pulse functions shown in Fig. 1.21 are

$$F_1(s) \quad = (1 - e^{-as})/s$$
$$F_2(s) = (1 - e^{-as})/s(1 - e^{-Ts})$$
$$F_3(s) = (1 - e^{-as})/s(1 + e^{-as})$$

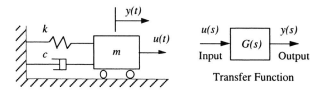

Fig. 1.22   Mass–spring–damper system.

1.3.5   Transfer Function

1.3.5.1   Single-input Single-output System
As an example of the single-input single-output (SISO) system, consider the mass–spring–damper system of mass m, linear spring stiffness k, and viscous damping coefficient c, as shown in Fig. 1.22.

The mathematical model of this system is described by the ordinary differential equation

$$m\ddot{y}(t) + c\dot{y}(t) + ky(t) = u(t) \tag{1.173}$$

where y is the displacement output of the mass and u is the input force acting on the mass. Taking the Laplace transform of this differential equation, we obtain

$$m[s^2 Y(s) - sy(0) - \dot{y}(0)] + c[sY(s) - y(0)] + kY(s) = U(s) \tag{1.174}$$

where $Y(s) = \mathcal{L}\{y(t)\}$ and $U(s) = \mathcal{L}\{u(t)\}$. Rearranging Eq. (1.174) yields

$$Y(s) = \underbrace{\frac{1}{ms^2 + cs + k} U(s)}_{\text{zero-state response}} + \underbrace{\frac{my(0)s + m\dot{y}(0) + cy(0)}{ms^2 + cs + k}}_{\text{zero-input response}} \tag{1.175}$$

The transfer function of a linear dynamic system is, in general, defined as the ratio of the Laplace transform of the output to the Laplace transform of the input, assuming all initial conditions are zero. Customarily, we use $u(s)$ and $y(s)$, instead of $U(s)$ and $Y(s)$, to denote the Laplace transforms of $u(t)$ and $y(t)$, respectively, though the meaning should be clear from the context. Consequently, the mass–spring–damper system can be described in transfer function form as

$$y(s) = G(s)u(s) \tag{1.176}$$

where $u(s)$ and $y(s)$ are called the input and output of the system, respectively, and $G(s)$ is the transfer function given by

$$G(s) = \frac{1}{ms^2 + cs + k} \tag{1.177}$$

A block diagram representation of this SISO system is also given in Fig. 1.22.

If we choose $x_1 = y$ and $x_2 = \dot{y}$ as the state variables, this system can also be represented in state-space form, as follows:

$$\dot{\mathbf{x}} = \mathbf{A}\mathbf{x} + \mathbf{B}u$$

$$y = \mathbf{C}\mathbf{x}$$

where

$$\mathbf{x} = \begin{bmatrix} x_1 \\ x_2 \end{bmatrix}, \qquad \mathbf{A} = \begin{bmatrix} 0 & 1 \\ -k/m & -c/m \end{bmatrix}, \qquad \mathbf{B} = \begin{bmatrix} 0 \\ 1/m \end{bmatrix}, \qquad \mathbf{C} = \begin{bmatrix} 1 & 0 \end{bmatrix}$$

It can be easily verified that

$$\frac{y(s)}{u(s)} = G(s) = \mathbf{C}[s\mathbf{I} - \mathbf{A}]^{-1}\mathbf{B} = \frac{1}{ms^2 + cs + k}$$

1.3.5.2 Poles and Zeros   In general, the transfer function $G(s)$ is given as a ratio of polynomials in s, as follows:

$$G(s) = \frac{N(s)}{D(s)} = \frac{b_0 s^m + b_1 s^{m-1} + \cdots + b_{m-1} s + b_m}{a_0 s^n + a_1 s^{n-1} + \cdots + a_{n-1} s + a_n} \qquad (1.178)$$

where $N(s)$ is called the numerator polynomial and $D(s)$ the denominator polynomial. The roots of $N(s)$ and $D(s)$ are called, respectively, the zeros and poles of the transfer function, and the transfer function in pole–zero form is often represented as

$$G(s) = \frac{k \prod(s/z_i - 1)}{s^\ell \prod(s/p_i - 1)} \qquad (1.179)$$

where k is called the steady-state (or D.C.) gain of the system. A system with ℓ poles at the origin is called the type ℓ system. If all of the poles and zeros of a transfer function lie in the left-half s plane, then it is called the *minimum-phase* transfer function. If a transfer function has at least one pole or zero in the right-half s plane, then it is called the *nonminimum-phase* transfer function. (The term nonminimum-phase comes from the phase shift characteristics of such a system when subjected to sinusoidal inputs.)

A ratio of polynomials is, in general, called a rational function. Consider a rational function $G(s) = N(s)/D(s)$, where $N(s)$ is the numerator polynomial, $D(s)$ the denominator polynomial, and s the complex variable. A rational function $G(s)$ is said to be strictly proper if

$$\lim_{s \to \infty} G(s) = 0$$

and it is said to be proper if $\lim_{s \to \infty} G(s)$ is a finite (zero or nonzero) constant.

A number λ (real or complex) is said to be a pole of a proper rational function $G(s)$ if $|G(\lambda)| = \infty$. It is said to be a zero of $G(s)$ if $|G(\lambda)| = 0$. If a proper rational function is irreducible (relatively prime or coprime; that is, there is no common factor between its numerator polynomial and denominator polynomial), then every root of the denominator of $G(s)$ is a pole of $G(s)$ and every root of the numerator of $G(s)$ is a zero of $G(s)$.

A rational function $G(s)$ is said to be positive real if $G(s)$ is real when s is real and $\text{Re}[G(s)] \geq 0$ when $\text{Re}[s] > 0$. For example, the following transfer functions are positive real:

$$G(s) = \frac{s^2 + z^2}{s(s^2 + p^2)}, \qquad 0 < z \leq p$$

$$G(s) = \frac{s + z}{s(s + p)}, \qquad 0 < z \leq p$$

$$G(s) = \frac{(s + z)^2}{s(s + p_1)(s + p_2)}, \qquad 0 < z \leq p_1 \leq p_2$$

1.3.5.3 Transfer Function Matrix   Consider a linear, time-invariant system described as

$$\dot{\mathbf{x}} = \mathbf{A}\mathbf{x} + \mathbf{B}\mathbf{u} \qquad (1.180a)$$

$$\mathbf{y} = \mathbf{C}\mathbf{x} + \mathbf{D}\mathbf{u} \qquad (1.180b)$$

where $\mathbf{x}$, $\mathbf{u}$, and $\mathbf{y}$ denote the state, input, and output vectors, respectively. This multi-input multi-output (MIMO) system can also be represented as

$$\mathbf{y}(s) = \mathbf{G}(s)\mathbf{u}(s) \qquad (1.181)$$

where $\mathbf{y}(s) = \mathcal{L}\{\mathbf{y}(t)\}$, $\mathbf{u}(s) = \mathcal{L}\{\mathbf{u}(t)\}$, and $\mathbf{G}(s)$, called the transfer function matrix, is defined as

$$\mathbf{G}(s) = \mathbf{C}[s\mathbf{I} - \mathbf{A}]^{-1}\mathbf{B} + \mathbf{D} \qquad (1.182)$$

and $\mathbf{I}$ is an identity matrix. The matrix $\mathbf{D}$ is zero for most dynamic systems.

The characteristic polynomial of a proper rational matrix $\mathbf{G}(s)$, denoted by $D(s)$, is defined to be the least common denominator of all minors of $\mathbf{G}(s)$. The degree of $\mathbf{G}(s)$ is defined to be the degree of the characteristic polynomial of $\mathbf{G}(s)$. The characteristic polynomial of $\mathbf{G}(s)$, is, in general, different from the denominator of the determinant of $\mathbf{G}(s)$ if $\mathbf{G}(s)$ is a square matrix. It is also different from the least common denominator of all the entries of $\mathbf{G}(s)$.

Example 1.6

$$\mathbf{G}(s) = \begin{bmatrix} 1/(s+1) & 1/(s+1) \\ 1/(s+1) & 1/(s+1) \end{bmatrix} = \frac{1}{s+1}\begin{bmatrix} 1 & 1 \\ 1 & 1 \end{bmatrix} \implies D(s) = s+1$$

$$\mathbf{G}(s) = \begin{bmatrix} 2/(s+1) & 1/(s+1) \\ 1/(s+1) & 1/(s+1) \end{bmatrix}$$

$$= \frac{1}{(s+1)^2}\begin{bmatrix} 2(s+1) & s+1 \\ s+1 & s+1 \end{bmatrix} \implies D(s) = (s+1)^2$$

For a square MIMO system with n inputs and n outputs described by

$$\mathbf{y}(s) = \mathbf{G}(s)\mathbf{u}(s) = \frac{1}{D(s)}\mathbf{N}(s)\mathbf{u}(s) \qquad (1.183)$$

we have

$$\det \mathbf{G}(s) = \det\left[\frac{\mathbf{N}(s)}{D(s)}\right] = \frac{\det[\mathbf{N}(s)]}{D^n(s)} = \frac{D^{n-1}(s)N(s)}{D^n(s)} = \frac{N(s)}{D(s)} \qquad (1.184)$$

where $N(s)$ is called the coupling numerator polynomial as described in [11]. The roots of $N(s) = 0$ are defined to be the transmission zeros of the MIMO system described by Eq. (1.183).

Problems

1.38 Given the linearized equations of motion of a cart with an inverted pendulum about an equilibrium point of $\theta = 0$ (see Problem 1.36), obtain the transfer functions from u to z and θ, as follows:

$$\frac{z(s)}{u(s)} = \frac{s^2 - g/\ell}{s^2\{Ms^2 - (M+m)g/\ell\}}$$

$$\frac{\theta(s)}{u(s)} = \frac{-s^2/\ell}{s^2\{Ms^2 - (M+m)g/\ell\}}$$

Note: There are pole–zero cancellations of s^2 in the transfer function θ/u, which indicate a lack of controllability by u and/or observability by θ. (The concept of controllability and observability will be introduced in Chapter 2. The system is, in fact, controllable by u and observable by z because there are no pole–zero cancellations in the transfer function $z(s)/u(s)$. Consequently, the pole–zero cancellations in the transfer function $\theta(s)/u(s)$ indicate that the system is unobservable by θ.)

1.39 (a) Consider a transfer function of the form

$$\frac{y(s)}{u(s)} = G(s) = \frac{N(s)}{(s-p_1)^3(s-p_2)(s-p_3)}$$

$$= \frac{K_{11}}{(s-p_1)^3} + \frac{K_{12}}{(s-p_1)^2} + \frac{K_{13}}{(s-p_1)} + \frac{K_2}{(s-p_2)} + \frac{K_3}{(s-p_3)}$$

where p_i are the poles, $N(s)$ the numerator polynomial, and K_{ij} and K_i are called the residues of the transfer function.

Show that the residues of this transfer function can be determined as

$$K_{11} = \left\{ (s - p_1)^3 G(s) \right\}_{s=p_1}$$

$$K_{12} = \left\{ \frac{d}{ds}(s - p_1)^3 G(s) \right\}_{s=p_1}$$

$$K_{13} = \frac{1}{2!} \left\{ \frac{d^2}{ds^2}(s - p_1)^3 G(s) \right\}_{s=p_1}$$

$$K_2 = \left\{ (s - p_2)G(s) \right\}_{s=p_2}$$

$$K_3 = \left\{ (s - p_3)G(s) \right\}_{s=p_3}$$

Hint: For the determination of K_{11}, K_{12}, and K_{13} of the preceding form, notice that

$$(s - p_1)^3 G(s) = K_{11} + K_{12}(s - p_1) + K_{13}(s - p_1)^2 + \frac{K_2(s - p_1)^3}{(s - p_2)}$$

$$+ \frac{K_3(s - p_1)^3}{(s - p_3)}$$

Also show that this transfer function can be realized into a so-called Jordan canonical form, as follows:

$$\dot{x} = \begin{bmatrix} p_1 & 1 & 0 & 0 & 0 \\ 0 & p_1 & 1 & 0 & 0 \\ 0 & 0 & p_1 & 0 & 0 \\ 0 & 0 & 0 & p_2 & 0 \\ 0 & 0 & 0 & 0 & p_3 \end{bmatrix} x + \begin{bmatrix} 0 \\ 0 \\ 1 \\ 1 \\ 1 \end{bmatrix} u$$

$$y = \begin{bmatrix} K_{11} & K_{12} & K_{13} & K_2 & K_3 \end{bmatrix} x$$

(b) Find the state-space representation of the following transfer function using its partial fraction expansion:

$$\frac{y(s)}{u(s)} = \frac{s^2 + 1}{s^3 + 6s^2 + 12s + 8}$$

1.40 Consider a transfer function of the form

$$\frac{y(s)}{u(s)} = G(s) = \frac{b_1 s^2 + b_2 s + b_3}{s^3 + a_1 s^2 + a_2 s + a_3}$$

where a_i and b_i are real constants, or the equivalent differential equation

$$\frac{d^3 y}{dt^3} + a_1 \frac{d^2 y}{dt^2} + a_2 \frac{dy}{dt} + a_3 y = b_1 \frac{d^2 u}{dt^2} + b_2 \frac{du}{dt} + b_3 u$$

Verify that this system with the input u and the output y can be represented by the following equivalent state-space equations [12]:

(a) Observer canonical form

$$\begin{bmatrix} \dot{x}_1 \\ \dot{x}_2 \\ \dot{x}_3 \end{bmatrix} = \begin{bmatrix} 0 & 0 & -a_3 \\ 1 & 0 & -a_2 \\ 0 & 1 & -a_1 \end{bmatrix} \begin{bmatrix} x_1 \\ x_2 \\ x_3 \end{bmatrix} + \begin{bmatrix} b_3 \\ b_2 \\ b_1 \end{bmatrix} u$$

$$y = [0 \quad 0 \quad 1] \begin{bmatrix} x_1 \\ x_2 \\ x_3 \end{bmatrix}$$

or

$$\begin{bmatrix} \dot{x}_1 \\ \dot{x}_2 \\ \dot{x}_3 \end{bmatrix} = \begin{bmatrix} -a_1 & 1 & 0 \\ -a_2 & 0 & 1 \\ -a_3 & 0 & 0 \end{bmatrix} \begin{bmatrix} x_1 \\ x_2 \\ x_3 \end{bmatrix} + \begin{bmatrix} b_1 \\ b_2 \\ b_3 \end{bmatrix} u$$

$$y = [1 \quad 0 \quad 0] \begin{bmatrix} x_1 \\ x_2 \\ x_3 \end{bmatrix}$$

(b) Controller canonical form

$$\begin{bmatrix} \dot{x}_1 \\ \dot{x}_2 \\ \dot{x}_3 \end{bmatrix} = \begin{bmatrix} 0 & 1 & 0 \\ 0 & 0 & 1 \\ -a_3 & -a_2 & -a_1 \end{bmatrix} \begin{bmatrix} x_1 \\ x_2 \\ x_3 \end{bmatrix} + \begin{bmatrix} 0 \\ 0 \\ 1 \end{bmatrix} u$$

$$y = [b_3 \quad b_2 \quad b_1] \begin{bmatrix} x_1 \\ x_2 \\ x_3 \end{bmatrix}$$

or

$$\begin{bmatrix} \dot{x}_1 \\ \dot{x}_2 \\ \dot{x}_3 \end{bmatrix} = \begin{bmatrix} -a_1 & -a_2 & -a_3 \\ 1 & 0 & 0 \\ 0 & 1 & 0 \end{bmatrix} \begin{bmatrix} x_1 \\ x_2 \\ x_3 \end{bmatrix} + \begin{bmatrix} 1 \\ 0 \\ 0 \end{bmatrix} u$$

$$y = [b_1 \quad b_2 \quad b_3] \begin{bmatrix} x_1 \\ x_2 \\ x_3 \end{bmatrix}$$

(c) Observability canonical form

$$\begin{bmatrix} \dot{x}_1 \\ \dot{x}_2 \\ \dot{x}_3 \end{bmatrix} = \begin{bmatrix} 0 & 1 & 0 \\ 0 & 0 & 1 \\ -a_3 & -a_2 & -a_1 \end{bmatrix} \begin{bmatrix} x_1 \\ x_2 \\ x_3 \end{bmatrix} + \begin{bmatrix} h_1 \\ h_2 \\ h_3 \end{bmatrix} u$$

$$y = [1 \quad 0 \quad 0] \begin{bmatrix} x_1 \\ x_2 \\ x_3 \end{bmatrix}$$

where

$$
\begin{bmatrix} h_1 \\ h_2 \\ h_3 \end{bmatrix} = \begin{bmatrix} 1 & 0 & 0 \\ a_1 & 1 & 0 \\ a_2 & a_1 & 1 \end{bmatrix}^{-1} \begin{bmatrix} b_1 \\ b_2 \\ b_3 \end{bmatrix}
$$

(d) Controllability canonical form

$$
\begin{bmatrix} \dot{x}_1 \\ \dot{x}_2 \\ \dot{x}_3 \end{bmatrix} = \begin{bmatrix} 0 & 0 & -a_3 \\ 1 & 0 & -a_2 \\ 0 & 1 & -a_1 \end{bmatrix} \begin{bmatrix} x_1 \\ x_2 \\ x_3 \end{bmatrix} + \begin{bmatrix} 1 \\ 0 \\ 0 \end{bmatrix} u
$$

$$
y = [h_1 \quad h_2 \quad h_3] \begin{bmatrix} x_1 \\ x_2 \\ x_3 \end{bmatrix}
$$

where

$$
[h_1 \quad h_2 \quad h_3] = [b_1 \quad b_2 \quad b_3] \begin{bmatrix} 1 & a_1 & a_2 \\ 0 & 1 & a_1 \\ 0 & 0 & 1 \end{bmatrix}^{-1}
$$

1.41   Consider a proper rational function of the form

$$
G(s) = \frac{b_0 s^n + b_1 s^{n-1} + \cdots + b_n}{s^n + a_1 s^{n-1} + \cdots + a_n}
$$

where a_i and b_i are real constants.

Show that $G(s)$ can be expanded into an infinite power series of descending power of s, as follows:

$$
G(s) = h_0 + h_1 s^{-1} + h_2 s^{-2} + h_3 s^{-3} + \cdots
$$

where the coefficients h_i, called the Markov parameters, can be obtained recursively, as

$$
h_0 = b_0
$$
$$
h_1 = -a_1 h_0 + b_1
$$
$$
h_2 = -a_1 h_1 - a_2 h_0 + b_2
$$

$$
\vdots
$$

$$
h_n = -a_1 h_{n-1} - a_2 h_{n-2} - \cdots - a_n h_0 + b_n
$$

Hint: Let $b_0 s^n + b_1 s^{n-1} + \cdots + b_n = (s^n + a_1 s^{n-1} + \cdots + a_n)(h_0 + h_1 s^{-1} + h_2 s^{-2} + h_3 s^{-2} + \cdots)$.

1.42   Given a SISO system described by

$$
\dot{x} = Ax + Bu
$$
$$
y = Cx
$$

show that its transfer function

$$\frac{y(s)}{u(s)} = G(s) = C[sI - A]^{-1}B$$

can be expanded as

$$G(s) = h_1 s^{-1} + h_2 s^{-2} + h_3 s^{-3} + \cdots$$

where the Markov parameters h_i are expressed as

$$h_i = CA^{i-1}B, \qquad i = 1, 2, \ldots$$

Hint: $[I - s^{-1}A]^{-1} = \sum_{k=0}^{\infty}(s^{-1}A)^k.$

1.3.6 Linear System Stability

1.3.6.1 Characteristic Equation   Consider a linear dynamic system described by the state-space equation

$$\dot{x} = Ax + Bu$$

Taking the Laplace transform of this equation and ignoring the initial conditions and u, we obtain

$$[sI - A]x = 0 \qquad (1.185)$$

where I is an identity matrix and s is the Laplace transform variable. This set of equations has a solution for x other than the trivial one, $x = 0$, only if the matrix $[sI - A]$ is singular. Consequently, the characteristic equation of a linear dynamic system is defined as

$$|sI - A| = s^n + a_1 s^{n-1} + \cdots + a_{n-1}s + a_n = 0 \qquad (1.186)$$

where a_i are scalar constants. The polynomial in Eq. (1.186) is called the characteristic polynomial. The roots of the characteristic equation are then called the characteristic roots or the eigenvalues of the system. The concepts of characteristic equation and eigenvalues play a very important role in the study of linear dynamic systems.

Let $\lambda_1, \ldots, \lambda_n$ be the eigenvalues of A. Then the ith eigenvector associated with the ith eigenvalue λ_i is defined to be a nonzero e_i such that

$$[\lambda_i I - A]e_i = 0 \qquad (1.187)$$

Furthermore, we have

$$|s\mathbf{I} - \mathbf{A}| = (s - \lambda_1)(s - \lambda_2) \cdots (s - \lambda_n) = \prod_{i=1}^{n} (s - \lambda_i)$$

$$|\mathbf{A}| = \lambda_1 \lambda_2 \cdots \lambda_n = \prod_{i=1}^{n} \lambda_i = (-1)^n a_n$$

$$\mathrm{tr}\,(\mathbf{A}) = \sum_{i=1}^{n} \lambda_i = -a_1$$

Linear dynamic systems are also often described by the second-order matrix differential equations of the form

$$\mathbf{M}\ddot{\mathbf{x}} + (\mathbf{D} + \mathbf{G})\dot{\mathbf{x}} + \mathbf{K}\mathbf{x} = \mathbf{u} \tag{1.188}$$

where $\mathbf{x}$ and $\mathbf{u}$ are the generalized coordinate and input vectors, respectively. $\mathbf{M} = \mathbf{M}^T$, $\mathbf{D} = \mathbf{D}^T$, $\mathbf{G} = -\mathbf{G}^T$, and $\mathbf{K} = \mathbf{K}^T$ are called, respectively, the mass matrix, damping matrix, gyroscopic coupling matrix, and stiffness matrix. For this case, the characteristic equation simply becomes

$$|\mathbf{M}s^2 + (\mathbf{D} + \mathbf{G})s + \mathbf{K}| = 0 \tag{1.189}$$

Because Eq. (1.188) can be transformed into a state-space equation of the form

$$\frac{d}{dt}\begin{bmatrix} \mathbf{x} \\ \dot{\mathbf{x}} \end{bmatrix} = \begin{bmatrix} 0 & \mathbf{I} \\ -\mathbf{M}^{-1}\mathbf{K} & -\mathbf{M}^{-1}(\mathbf{D} + \mathbf{G}) \end{bmatrix} \begin{bmatrix} \mathbf{x} \\ \dot{\mathbf{x}} \end{bmatrix} + \begin{bmatrix} 0 \\ \mathbf{M}^{-1} \end{bmatrix} \mathbf{u} \tag{1.190}$$

we also have the characteristic equation of the form

$$|s\mathbf{I} - \mathbf{A}| = 0$$

where

$$\mathbf{A} = \begin{bmatrix} 0 & \mathbf{I} \\ -\mathbf{M}^{-1}\mathbf{K} & -\mathbf{M}^{-1}(\mathbf{D} + \mathbf{G}) \end{bmatrix}$$

Example 1.7

Given a matrix

$$\mathbf{A} = \begin{bmatrix} 0 & 1 & 0 & 0 \\ 0 & 0 & 1 & 0 \\ 0 & 0 & 0 & 1 \\ 1 & 0 & 0 & 0 \end{bmatrix}$$

the characteristic equation can be found as

$$|s\mathbf{I} - \mathbf{A}| = s^4 - 1 = (s^2 - 1)(s^2 + 1) = 0$$

Thus, the characteristic roots (or eigenvalues) are ± 1 and $\pm j$. The ith eigenvector of $\mathbf{A}$ associated with the ith eigenvalue λ_i can be found as $\mathbf{e}_i = (1, \lambda_i, \lambda_i^2, \lambda_i^3)$.

1.3.6.2 Linear Stability   A linear time-invariant system or its equilibrium state is said to be asymptotically stable if all of its eigenvalues have negative real parts. A linear system is said to be stable in the sense of Lyapunov (or Lyapunov stable) if none of its eigenvalues has a positive real part and if it has no repeated eigenvalues on the imaginary axis. It is said to be unstable if any one of its eigenvalues has a positive real part or if it has repeated eigenvalues on the imaginary axis. In this book, a linear system that is Lyapunov stable is simply said to be stable and a linear system that is not Lyapunov stable is said to be unstable.

A system is also said to be bounded-input bounded-output (BIBO) stable, if, for every bounded input, the output remains bounded for all time. In other words, a linear time-invariant system is BIBO stable provided all of its eigenvalues lie in the left-half of the s plane (not including the imaginary axis). The BIBO stability concept is applicable to nonlinear time-varying dynamic systems, whereas the asymptotic stability concept is applicable only to linear time-invariant systems.

Linear systems are characterized by the principle of superposition, and a sinusoidal input to an asymptotically stable linear system results in a sinusoidal output of the same frequency.

Example 1.8

Consider the mass–spring–damper system described by

$$m\ddot{y} + c\dot{y} + ky = u$$

where y is the output displacement of the mass and u is the input force acting on the mass. The characteristic equation is given by

$$ms^2 + cs + k = 0 \tag{1.191}$$

or

$$s^2 + 2\zeta\omega_n s + \omega_n^2 = 0 \tag{1.192}$$

where $\omega_n = \sqrt{k/m}$ and $\zeta = c/(2\sqrt{mk})$ are called the natural frequency and the damping ratio of the system, respectively. For nonzero positive values of c and k, this system is asymptotically stable, BIBO stable, and also Lyapunov stable.

A mass–spring system with $c = 0$, often called a simple harmonic oscillator, is Lyapunov stable; however, it is neither asymptotically stable nor BIBO stable. A simple cart system with $k = c = 0$, often called a double integrator plant, is not Lyapunov stable because of its double pole at the origin, and it is simply said to be unstable.

Problem

1.43   (a) Consider a set of differential equations, which describes the in-plane motion of a spacecraft near the equilateral equilibrium point of the Earth–moon system, given by

$$\ddot{x} - 2\dot{y} - \tfrac{3}{4}x - \tfrac{3\sqrt{3}}{2}\left(\rho - \tfrac{1}{2}\right)y = 0$$

$$\ddot{y} + 2\dot{x} - \tfrac{3\sqrt{3}}{2}\left(\rho - \tfrac{1}{2}\right)x - \tfrac{9}{4}y = 0$$

where ρ is a constant parameter. Obtain the characteristic equation as

$$s^4 + s^2 + \tfrac{27}{4}\rho(1 - \rho) = 0$$

and discuss the stability of this system with $\rho = 0.01215$.

(b) Consider a set of differential equations, which describes the in-plane motion of a spacecraft near the collinear equilibrium point of the Earth–moon system, given by

$$\ddot{x} - 2\dot{y} - (2\sigma + 1)x = 0$$
$$\ddot{y} + 2\dot{x} + (\sigma - 1)y = 0$$

where σ is a constant parameter. Obtain the characteristic equation as

$$s^4 - (\sigma - 2)s^2 - (2\sigma + 1)(\sigma - 1) = 0$$

and discuss the stability of this system with $\sigma = 3.19043$.

1.3.7 Linear Stability Criteria

1.3.7.1 Historical Background In 1868 the British physicist J. C. Maxwell proposed the mathematical problem of determining the number of roots of a real polynomial in the right-half plane, and in 1875 Routh discovered a solution to the problem. In 1893 Stodola, unaware of Routh's solution, again proposed the same problem, and in 1895 Hurwitz, on the basis of Hermite's paper published in 1856, gave another solution to the same problem (independent of Routh's). The equivalence of the solutions of Routh and Hurwitz was shown by Bompiani in 1911. The determinantal inequalities obtained by Hurwitz are nowadays called the inequalities of Routh–Hurwitz. The Routh–Hurwitz criterion provides an analytical way of determining if all roots of a polynomial have negative real parts. Thus, it provides an analytical way of checking whether or not a given linear time-invariant system is asymptotically stable. For a more detailed, historical, as well as technical treatment of this subject of linear stability criteria, the reader is referred to [9] and [13].

It is interesting to note, however, that in 1892, before Hurwitz's solution, the Russian mathematician Lyapunov had found that a linear time-invariant system described by $\dot{x} = Ax$ is asymptotically stable if and only if for the matrix equation

$$A^T P + PA = -Q \tag{1.193}$$

there exists a positive definite matrix P for any positive definite matrix Q.

1.3.7.2 Hermite–Bieler Theorem Consider a real polynomial of the form

$$f(s) = s^n + a_1 s^{n-1} + \cdots + a_{n-1}s + a_n$$
$$= h(s^2) + sg(s^2)$$
$$= h(\lambda) + sg(\lambda), \qquad \lambda = s^2$$

where $h(s^2)$ and $sg(s^2)$ are the even and odd parts of $f(s)$, respectively. The polynomial $f(s)$ has all its roots in the left-half s plane if and only if $h(\lambda)$ and $g(\lambda)$ have simple real negative alternating roots and $a_1 > 0$. The first root next to zero is of $h(\lambda)$. Such polynomials $h(\lambda)$ and $g(\lambda)$ are called a positive pair, and for even n we have

$$\frac{sg(s^2)}{h(s^2)} = \frac{sa_1(s^2 + \omega_2^2)(s^2 + \omega_4^2)\cdots(s^2 + \omega_{n-2}^2)}{(s^2 + \omega_1^2)(s^2 + \omega_3^2)\cdots(s^2 + \omega_{n-1}^2)} = sa_1 \sum_{i=1,3} \frac{K_i}{s^2 + \omega_i^2} \quad (1.194)$$

where $\omega_1 < \omega_2 < \omega_3 < \cdots < \omega_{n-1}$ and $K_i > 0$ for all i. Such a function described by Eq. (1.194) is called a reactance function.

1.3.7.3 Routh–Hurwitz Criterion   All of the roots of the real polynomial

$$f(s) = s^n + a_1 s^{n-1} + \cdots + a_{n-1}s + a_n \quad (1.195)$$

have negative real parts if and only if the following n inequalities are satisfied:

$$\Delta_1 > 0, \qquad \Delta_2 > 0, \qquad \ldots, \qquad \Delta_n > 0 \quad (1.196)$$

where Δ_i, called the Hurwitz's determinants, are the leading principal minors of the Hurwitz matrix $\mathbf{H}$ defined as

$$\mathbf{H} = \begin{bmatrix} a_1 & a_3 & a_5 & a_7 & \cdots & 0 \\ 1 & a_2 & a_4 & a_6 & \cdots & 0 \\ 0 & a_1 & a_3 & a_5 & \cdots & 0 \\ 0 & 1 & a_2 & a_4 & \cdots & 0 \\ \vdots & \vdots & \vdots & \vdots & \cdots & 0 \\ 0 & 0 & 0 & 0 & \cdots & a_n \end{bmatrix}$$

That is, we have

$$\Delta_1 = a_1$$

$$\Delta_2 = \begin{vmatrix} a_1 & a_3 \\ 1 & a_2 \end{vmatrix}$$

$$\Delta_3 = \begin{vmatrix} a_1 & a_3 & a_5 \\ 1 & a_2 & a_4 \\ 0 & a_1 & a_3 \end{vmatrix}$$

$$\Delta_4 = \begin{vmatrix} a_1 & a_3 & a_5 & a_7 \\ 1 & a_2 & a_4 & a_6 \\ 0 & a_1 & a_3 & a_5 \\ 0 & 1 & a_2 & a_4 \end{vmatrix}$$

$$\vdots$$

$$\Delta_n = \begin{vmatrix} a_1 & a_3 & a_5 & a_7 & \cdots & 0 \\ 1 & a_2 & a_4 & a_6 & \cdots & 0 \\ 0 & a_1 & a_3 & a_5 & \cdots & 0 \\ 0 & 1 & a_2 & a_4 & \cdots & 0 \\ \vdots & \vdots & \vdots & \vdots & \cdots & 0 \\ 0 & 0 & 0 & 0 & \cdots & a_n \end{vmatrix} = a_n \Delta_{n-1}$$

A real polynomial whose coefficients satisfy the Routh–Hurwitz criterion is often called a *Hurwitz polynomial*.

Because $f(s)$ can be expressed as

$$f(s) = (s - \lambda_1)(s - \lambda_2) \cdots (s - \lambda_n) = \prod_{i=1}^{n}(s - \lambda_i)$$

where λ_i are the roots of $f(s)$, we have

$$a_1 = -(\lambda_1 + \lambda_2 + \cdots + \lambda_n) = -\sum_{i=1}^{n} \lambda_i$$

$$a_n = (-1)^n \lambda_1 \lambda_2 \cdots \lambda_n = (-1)^n \prod_{i=1}^{n} \lambda_i$$

The necessary conditions for a polynomial of the form of Eq. (1.195) to be a Hurwitz polynomial are that all of the coefficients are positive; that is,

$$a_1 > 0, \qquad a_2 > 0, \qquad \ldots, \qquad a_n > 0 \qquad (1.197)$$

When these conditions hold, the Routh–Hurwitz inequalities are not independent. This fact was investigated by Liénard and Chipart in 1914 and enabled them to develop a stability criterion different from the Routh–Hurwitz criterion. The Liénard and Chipart criterion has a definite advantage over the Routh–Hurwitz criterion, because it involves only about half the number of determinantal inequalities.

1.3.7.4 Liénard and Chipart Criterion

The necessary and sufficient conditions for all of the roots of the polynomial $f(s) = s^n + a_1 s^{n-1} + \cdots + a_{n-1}s +$

a_n to have negative real parts can be given in any one of the following four forms:

$$a_n > 0, \quad a_{n-2} > 0, \quad \ldots, \quad \Delta_1 > 0, \quad \Delta_3 > 0, \quad \ldots$$

$$a_n > 0, \quad a_{n-2} > 0, \quad \ldots, \quad \Delta_2 > 0, \quad \Delta_4 > 0, \quad \ldots$$

$$a_n > 0, \quad a_{n-1} > 0, \quad a_{n-3} > 0, \quad \ldots, \quad \Delta_1 > 0, \quad \Delta_3 > 0, \quad \ldots$$

$$a_n > 0, \quad a_{n-1} > 0, \quad a_{n-3} > 0, \quad \ldots, \quad \Delta_2 > 0, \quad \Delta_4 > 0, \quad \ldots$$

Example 1.9

Given a characteristic polynomial

$$f(s) = s^4 + a_1 s^3 + a_2 s^2 + a_3 s + a_4$$

where all of the coefficients are positive numbers, the necessary and sufficient condition for stable roots is

$$\Delta_3 = \begin{vmatrix} a_1 & a_3 & a_5 \\ 1 & a_2 & a_4 \\ 0 & a_1 & a_3 \end{vmatrix} = a_1(a_2 a_3 - a_1 a_4) - a_3^2 > 0 \qquad (1.198)$$

1.3.7.5 Kharitonov's Theorem The following stability criterion by Kharitonov [14], which is often called Kharitonov's theorem in the literature, is concerned with the stability problem of a so-called interval polynomial. Consider an interval polynomial

$$f(s) = s^n + a_1 s^{n-1} + \cdots + a_{n-1} s + a_n$$

with the coefficients a_i in prescribed intervals

$$\underline{a}_i \leq a_i \leq \bar{a}_i, \qquad i = 1, \ldots, n$$

The interval polynomial $f(s)$ is a Hurwitz polynomial if and only if the four extreme polynomials (often called Kharitonov's polynomials):

$$f_1(s) = \bar{a}_n + \underline{a}_{n-1}s + \underline{a}_{n-2}s^2 + \bar{a}_{n-3}s^3 + \cdots \qquad (1.199a)$$

$$f_2(s) = \bar{a}_n + \bar{a}_{n-1}s + \underline{a}_{n-2}s^2 + \underline{a}_{n-3}s^3 + \cdots \qquad (1.199b)$$

$$f_3(s) = \underline{a}_n + \bar{a}_{n-1}s + \bar{a}_{n-2}s^2 + \underline{a}_{n-3}s^3 + \cdots \qquad (1.199c)$$

$$f_4(s) = \underline{a}_n + \underline{a}_{n-1}s + \bar{a}_{n-2}s^2 + \bar{a}_{n-3}s^3 + \cdots \qquad (1.199d)$$

are Hurwitz polynomials.

A proof of this theorem will be given in Chapter 2. According to Kharitonov's theorem, stability of an interval polynomial can be determined by examining only four members of the set of all of the polynomials obtained by restricting coefficients to the extreme corner (or vertex) points of their range of variations. This important result is limited, however, by an assumption of independent coefficient perturbations in the interval polynomial, which introduces conservative stability bounds if the coefficients are, in fact, functionally dependent. The subject of uncertainty modeling and stability robustness analysis will be further studied in Chapter 2. For more details of Kharitonov's theorem and the subject of uncertainty modeling and stability robustness analysis, the reader is referred to [15] and [16].

Problems

1.44 Consider a set of differential equations, which describes the linearized attitude motion of a spacecraft in a circular orbit, given by

$$\ddot{\theta}_1 + (k_1 - 1)n\dot{\theta}_3 + 4n^2 k_1 \theta_1 = 0$$
$$\ddot{\theta}_2 + 3n^2 k_2 \theta_2 = 0$$
$$\ddot{\theta}_3 + (1 - k_3)n\dot{\theta}_1 + n^2 k_3 \theta_3 = 0$$

where n is a constant orbital rate and k_i are constant parameters. Find the necessary and sufficient conditions for the Lyapunov stability of the system.

1.45 Consider a cubic polynomial

$$f(s) = s^3 + a_1 s^2 + a_2 s + a_3$$

with the coefficients a_i in prescribed intervals $0 < \underline{a}_i \le a_i \le \bar{a}_i$ for all i. Show that $f(s)$ is a Hurwitz polynomial if and only if Kharitonov's polynomial $f_1(s)$ is a Hurwitz polynomial; i.e.,

$$\underline{a}_1 \underline{a}_2 > \bar{a}_3$$

1.46 Consider a quartic polynomial

$$f(s) = s^4 + a_1 s^3 + a_2 s^2 + a_3 s + a_4$$

with the coefficients a_i in prescribed intervals: $0 < \underline{a}_i \le a_i \le \bar{a}_i$ for all i. Show that $f(s)$ is a Hurwitz polynomial if and only if Kharitonov's polynomials $f_1(s)$ and $f_2(s)$ are Hurwitz polynomials; i.e.,

$$\bar{a}_1 \underline{a}_2 \underline{a}_3 > \bar{a}_1^2 \bar{a}_4 + \underline{a}_3^2$$
$$\underline{a}_1 \underline{a}_2 \bar{a}_3 > \underline{a}_1^2 \bar{a}_4 + \bar{a}_3^2$$

1.47 Consider a fifth-order polynomial of the form $f(s) = s^5 + a_1 s^4 + a_2 s^3 + a_3 s^2 + a_4 s + a_5$ with the coefficients a_i in prescribed intervals: $0 < \underline{a}_i \le a_i \le \bar{a}_i$ for all i. Show that $f(s)$ is a Hurwitz polynomial if and only if Kharitonov's polynomials $f_1(s), f_2(s)$, and $f_3(s)$ are Hurwitz polynomials.

1.3.8 Linear System Dynamics

1.3.8.1 First-order System   Consider a first-order system represented in transfer function form as

$$y(s) = \frac{1}{Ts + 1} u(s) \qquad (1.200)$$

where $u(s)$ is called the input, $y(s)$ the output, T the time constant, and $1/T$ the bandwidth of the system. The bandwidth of a linear system is often defined as the frequency at which the magnitude of its frequency-response function drops by a factor of 0.707 (3 dB) from its low-frequency gain.

For the unit-impulse input $u(s) = 1$, the unit impulse response of this first-order system can be obtained as

$$y(t) = \mathcal{L}^{-1} \left\{ \frac{1}{Ts + 1} \right\} = \frac{1}{T} e^{-t/T} \qquad (1.201)$$

Similarly, the unit step response for $u(s) = 1/s$ can be found as

$$y(t) = \mathcal{L}^{-1} \left\{ \frac{1}{s(Ts + 1)} \right\} = \mathcal{L}^{-1} \left\{ \frac{1}{s} - \frac{T}{Ts + 1} \right\} = 1 - e^{-t/T} \qquad (1.202)$$

and its steady-state value becomes $y(\infty) = 1$. The value of the unit step response $y(t)$ at $t = T$ becomes

$$y(T) = 1 - e^{-1} = 0.632$$

That is, the unit step response at $t = T$ reaches 63.2% of its steady-state value, and, for $t \geq 4T$, the unit-step response remains within 2% of its steady-state value.

1.3.8.2 Second-order System   A second-order system is often represented in transfer function form as

$$y(s) = \frac{\omega_n^2}{s^2 + 2\zeta \omega_n s + \omega_n^2} u(s) \qquad (1.203)$$

where ζ and ω_n are, respectively, the damping ratio and the natural frequency of this second-order system. The poles are

$$-\zeta \omega_n \pm j \omega_n \sqrt{1 - \zeta^2}$$

The unit impulse response of this system with $0 \leq \zeta < 1$ is given as

$$y(t) = \mathcal{L}^{-1} \left\{ \frac{\omega_n^2}{s^2 + 2\zeta \omega_n s + \omega_n^2} \right\} = \frac{\omega_n}{\sqrt{1 - \zeta^2}} e^{-\zeta \omega_n t} \sin \omega_n \sqrt{1 - \zeta^2} t \qquad (1.204)$$

DYNAMIC SYSTEMS MODELING AND ANALYSIS   107

and the unit step response can be found as

$$
y(t) = \mathcal{L}^{-1} \left\{ \frac{\omega_n^2}{s\left(s^2 + 2\zeta\omega_n s + \omega_n^2\right)} \right\} = \mathcal{L}^{-1} \left\{ \frac{1}{s} - \frac{s + 2\zeta\omega_n}{s^2 + 2\zeta\omega_n s + \omega_n^2} \right\}
$$

$$
= \mathcal{L}^{-1} \left\{ \frac{1}{s} - \frac{s + \zeta\omega_n}{(s + \zeta\omega_n)^2 + \omega_d^2} - \frac{\zeta\omega_n}{(s + \zeta\omega_n)^2 + \omega_d^2} \right\}
$$

$$
= 1 - e^{-\zeta\omega_n t} \cos \omega_d t - \frac{\zeta\omega_n}{\omega_d} e^{-\zeta\omega_n t} \sin \omega_d t \tag{1.205}
$$

where $\omega_d = \omega_n \sqrt{1 - \zeta^2}$ is called the damped natural frequency, and the steady-state value becomes $y(\infty) = 1$.

The time constant of a second-order system is defined as

$$
T = 1/\zeta\omega_n
$$

and ω_n is often considered the bandwidth of a second-order system with $\zeta \approx 0.707$. For $t \geq 3T$, the unit-step response remains within 5% of its steady-state value, whereas it remains within 2% of its steady-state value for $t \geq 4T$. Consequently, either $t_s = 3T$ or $t_s = 4T$ is called the settling time of an underdamped second-order system, corresponding to the 5% or 2% criterion, respectively.

1.3.8.3 Impulse Response Function

We now consider a general linear time-invariant dynamic system represented in transfer function form, as follows:

$$
y(s) = G(s)u(s) \tag{1.206}
$$

where $u(s)$ and $y(s)$ are called the input and output of the system, respectively, and $G(s)$ is the transfer function of the system.

The input–output description in the time domain is then given by

$$
y(t) = \int_0^t g(t - \tau)u(\tau)\,d\tau = \int_0^t g(\tau)u(t - \tau)\,d\tau \tag{1.207}
$$

where $g(t)$, called the *impulse response function* at time t due to an impulse input applied at time 0, is defined as

$$
g(t) = \mathcal{L}^{-1}\{G(s)\} \quad \text{or} \quad G(s) = \mathcal{L}\{g(t)\} \tag{1.208}
$$

That is, the Laplace transform of the impulse response function is, in fact, the transfer function of the system.

Similarly, for a system described by a transfer function matrix of the form

$$
\mathbf{y}(s) = \mathbf{G}(s)\mathbf{u}(s) \tag{1.209}
$$

we have

$$
\mathbf{y}(t) = \int_0^t \mathbf{G}(t - \tau)\mathbf{u}(\tau)\,d\tau = \int_0^t \mathbf{G}(\tau)\mathbf{u}(t - \tau)\,d\tau \tag{1.210}
$$

where the impulse response matrix of the system, denoted as $\mathbf{G}(t)$, is defined such that

$$\mathbf{G}(s) = \mathcal{L}\{\mathbf{G}(t)\} = \int_0^\infty \mathbf{G}(t)e^{-st}\,dt \qquad (1.211)$$

That is, the transfer function matrix is simply the Laplace transform of the impulse response matrix.

1.3.8.4 State Transition Matrix
Consider a linear time-invariant dynamic system described by

$$\dot{\mathbf{x}} = \mathbf{A}\mathbf{x} + \mathbf{B}\mathbf{u} \qquad (1.212)$$

$$\mathbf{y} = \mathbf{C}\mathbf{x} \qquad (1.213)$$

where $\mathbf{x}$ is the state vector, $\mathbf{u}$ the control input vector, and $\mathbf{y}$ the output vector. Equations (1.212) and (1.213) are called the state-space equation and output equation of the linear time-invariant system, respectively.

Taking the Laplace transform of Eq. (1.212), we obtain

$$s\mathbf{x}(s) - \mathbf{x}(0) = \mathbf{A}\mathbf{x}(s) + \mathbf{B}\mathbf{u}(s)$$

or

$$[s\mathbf{I} - \mathbf{A}]\mathbf{x}(s) = \mathbf{x}(0) + \mathbf{B}\mathbf{u}(s)$$

where $\mathbf{x}(0)$ denotes the state vector $\mathbf{x}(t)$ at $t = 0$. This equation can be rewritten as

$$\mathbf{x}(s) = [s\mathbf{I} - \mathbf{A}]^{-1}\mathbf{x}(0) + [s\mathbf{I} - \mathbf{A}]^{-1}\mathbf{B}\mathbf{u}(s) \qquad (1.214)$$

Taking the inverse Laplace transform of Eq. (1.214), we obtain the solution of Eq. (1.212) as

$$\mathbf{x}(t) = \mathcal{L}^{-1}\{[s\mathbf{I} - \mathbf{A}]^{-1}\}\mathbf{x}(0) + \mathcal{L}^{-1}\{[s\mathbf{I} - \mathbf{A}]^{-1}\mathbf{B}\mathbf{u}(s)\} \qquad (1.215)$$

which is rewritten as

$$\mathbf{x}(t) = e^{\mathbf{A}t}\mathbf{x}(0) + \int_0^t e^{\mathbf{A}(t-\tau)}\mathbf{B}\mathbf{u}(\tau)\,d\tau \qquad (1.216)$$

where

$$e^{\mathbf{A}t} \equiv \mathcal{L}^{-1}\{[s\mathbf{I} - \mathbf{A}]^{-1}\} \qquad (1.217)$$

and the matrix $e^{\mathbf{A}t}$ is called the state transition matrix.

Using the infinite series expression of $e^{\mathbf{A}t}$, the state transition matrix $e^{\mathbf{A}t}$ can also be expressed as

$$e^{\mathbf{A}t} = \sum_{k=0}^{\infty} \frac{\mathbf{A}^k t^k}{k!} = \mathbf{I} + \mathbf{A}t + \frac{1}{2}\mathbf{A}^2 t^2 + \cdots \qquad (1.218)$$

Some interesting properties of e^{At} are

$$e^{A(t_1+t_2)} = e^{At_1}e^{At_2}$$

$$[e^{At}]^{-1} = e^{-At}$$

$$\frac{d}{dt}e^{At} = Ae^{At} = e^{At}A$$

and we also have

$$e^{(A+B)t} = e^{At}e^{Bt} \text{ if and only if } AB = BA$$

Example 1.10

$$A = \begin{bmatrix} 0 & 1 \\ 0 & 0 \end{bmatrix} \Rightarrow [sI - A]^{-1} = \begin{bmatrix} s^{-1} & s^{-2} \\ 0 & s^{-1} \end{bmatrix} \Rightarrow e^{At} = \begin{bmatrix} 1 & t \\ 0 & 1 \end{bmatrix}$$

$$A = \begin{bmatrix} 0 & 1 \\ -1 & 0 \end{bmatrix} \Rightarrow [sI - A]^{-1} = \frac{1}{s^2+1}\begin{bmatrix} s & 1 \\ -1 & s \end{bmatrix} \Rightarrow e^{At} = \begin{bmatrix} \cos t & \sin t \\ -\sin t & \cos t \end{bmatrix}$$

$$A = \begin{bmatrix} 0 & 1 \\ 1 & 0 \end{bmatrix} \Rightarrow [sI - A]^{-1} = \frac{1}{s^2-1}\begin{bmatrix} s & -1 \\ -1 & s \end{bmatrix} \Rightarrow e^{At} = \begin{bmatrix} \cosh t & \sinh t \\ \sinh t & \cosh t \end{bmatrix}$$

Problems

1.48   (a) Find the state transition matrix of the following matrix:

$$A = \begin{bmatrix} 0 & 0 & -2 \\ 0 & 1 & 0 \\ 1 & 0 & 3 \end{bmatrix}$$

(b) Given a matrix

$$A = \begin{bmatrix} \lambda_1 & 1 & 0 & 0 & 0 \\ 0 & \lambda_1 & 1 & 0 & 0 \\ 0 & 0 & \lambda_1 & 0 & 0 \\ 0 & 0 & 0 & \lambda_2 & 1 \\ 0 & 0 & 0 & 0 & \lambda_2 \end{bmatrix}$$

where λ_1 and λ_2 are real numbers, show that

$$[sI - A]^{-1} = \begin{bmatrix} (s-\lambda_1)^{-1} & (s-\lambda_1)^{-2} & (s-\lambda_1)^{-3} & 0 & 0 \\ 0 & (s-\lambda_1)^{-1} & (s-\lambda_1)^{-2} & 0 & 0 \\ 0 & 0 & (s-\lambda_1)^{-1} & 0 & 0 \\ 0 & 0 & 0 & (s-\lambda_2)^{-1} & (s-\lambda_2)^{-2} \\ 0 & 0 & 0 & 0 & (s-\lambda_2)^{-1} \end{bmatrix}$$

and

$$e^{At} = \mathcal{L}^{-1}\{[s\mathbf{I} - \mathbf{A}]^{-1}\} = \begin{bmatrix} e^{\lambda_1 t} & te^{\lambda_1 t} & t^2 e^{\lambda_1 t}/2 & 0 & 0 \\ 0 & e^{\lambda_1 t} & te^{\lambda_1 t} & 0 & 0 \\ 0 & 0 & e^{\lambda_1 t} & 0 & 0 \\ 0 & 0 & 0 & e^{\lambda_2 t} & te^{\lambda_2 t} \\ 0 & 0 & 0 & 0 & e^{\lambda_2 t} \end{bmatrix}$$

1.49 Consider a set of differential equations, which describes the nutational motion of a spinning rocket caused by a thrust vector misalignment, of the form

$$\dot{\omega}_1 - \lambda\omega_2 = \mu$$
$$\dot{\omega}_2 + \lambda\omega_1 = 0$$

where ω_1 and ω_2 are the angular velocity components and λ and μ are constants. Obtain the solution as

$$\omega_1(t) = \omega_1(0)\cos\lambda t + \omega_2(0)\sin\lambda t + (\mu/\lambda)\sin\lambda t$$
$$\omega_2(t) = \omega_2(0)\cos\lambda t - \omega_1(0)\sin\lambda t - (\mu/\lambda)(1 - \cos\lambda t)$$

1.50 Consider the following set of differential equations, often called the Clohessy–Wiltshire equations in orbital mechanics, which describe the small relative motion of a spacecraft with respect a circular target orbit:

$$\ddot{x} - 2n\dot{y} - 3n^2 x = 0$$
$$\ddot{y} + 2n\dot{x} = 0$$
$$\ddot{z} + n^2 z = 0$$

where n is a constant orbital rate.
Obtain the solution of this set of differential equations as

$$\begin{bmatrix} x(t) \\ y(t) \\ \dot{x}(t) \\ \dot{y}(t) \end{bmatrix} = \begin{bmatrix} 4 - 3\cos nt & 0 & \sin nt/n & 2(1 - \cos nt)/n \\ 6\sin nt - 6nt & 1 & 2(-1 + \cos nt)/n & 4\sin nt/n - 3t \\ 3n\sin nt & 0 & \cos nt & 2\sin nt \\ 6n(-1 + \cos nt) & 0 & -2\sin nt & -3 + 4\cos nt \end{bmatrix} \begin{bmatrix} x_0 \\ y_0 \\ \dot{x}_0 \\ \dot{y}_0 \end{bmatrix}$$

$$\begin{bmatrix} z(t) \\ \dot{z}(t) \end{bmatrix} = \begin{bmatrix} \cos nt & \sin nt/n \\ -n\sin nt & \cos nt \end{bmatrix} \begin{bmatrix} z_0 \\ \dot{z}_0 \end{bmatrix}$$

where x_0, y_0, z_0, $\dot{x}_0$, $\dot{y}_0$, and $\dot{z}_0$ are initial conditions at $t = 0$.

1.3.9 Linear Oscillatory Systems

In the analysis of linear dynamic systems, the equations of motion in physical coordinates are often transformed to decoupled *modal equations* by means of a linear coordinate transformation known as the *modal transformation* or *similarity transformation*. The transformation matrix is often called the modal matrix, whose columns are the eigenvectors of the system.

In linear system theory, these decoupled first-order equations are called the Jordan canonical equations and, in general, involve complex numbers. In linear vibration analysis of mechanical systems, the modal equations are decoupled second-order equations and involve only real numbers. Whether the system to be studied is lumped parameter or distributed parameter, its modal form gives much physical insight into the system dynamics.

Consider a conservative, nongyroscopic, lumped-parameter dynamic system described by

$$\mathbf{M}\ddot{\mathbf{x}} + \mathbf{K}\mathbf{x} = \mathbf{u} \tag{1.219}$$

where $\mathbf{M}$ is an $n \times n$ symmetric mass matrix, $\mathbf{K}$ an $n \times n$ symmetric stiffness matrix, $\mathbf{x}$ an n-dimensional generalized coordinate vector, and $\mathbf{u}$ an n-dimensional generalized input vector associated with $\mathbf{x}$.

The characteristic equation of the system is simply given as

$$|\mathbf{M}s^2 + \mathbf{K}| = 0$$

where s is the Laplace transform variable. The characteristic roots of such an undamped nongyroscopic system are pure imaginary numbers including zeros at the origin. Thus, letting $s = \pm j\omega$ and $\mathbf{u} = 0$, we formulate the modal analysis problem of undamped structural dynamic systems, as follows:

$$\left| -\mathbf{M}\omega_i^2 + \mathbf{K} \right| = 0 \tag{1.220}$$

$$\left[-\mathbf{M}\omega_i^2 + \mathbf{K} \right]\phi_i = 0 \tag{1.221}$$

where ω_i is the ith modal frequency, ϕ_i is the ith modal vector (or eigenvector), and $i = 1, \ldots, n$.

Equation (1.219) is then transformed into a set of decoupled equations, called modal equations, by the modal transformation

$$\mathbf{x} = \mathbf{\Phi}\mathbf{q} \tag{1.222}$$

where $\mathbf{q}$ is the n-dimensional modal coordinate vector and $\mathbf{\Phi}$ is the $n \times n$ modal matrix defined as

$$\mathbf{\Phi} = [\phi_1 \quad \cdots \quad \phi_n] \tag{1.223}$$

Note that the modal matrix $\mathbf{\Phi}$ is an orthogonal matrix. Substituting Eq. (1.222) into Eq. (1.219) and premultiplying it by $\mathbf{\Phi}^T$, we obtain

$$\mathbf{\Phi}^T \mathbf{M} \mathbf{\Phi} \ddot{\mathbf{q}} + \mathbf{\Phi}^T \mathbf{K} \mathbf{\Phi} \mathbf{q} = \mathbf{\Phi}^T \mathbf{u} \tag{1.224}$$

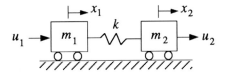

Fig. 1.23   Two-mass–spring system.

The modal matrix $\mathbf{\Phi}$ is often normalized such that

$$\mathbf{\Phi}^T \mathbf{M} \mathbf{\Phi} = \mathbf{I}$$
$$\mathbf{\Phi}^T \mathbf{K} \mathbf{\Phi} = \mathbf{\Omega}^2$$

where $\mathbf{I}$ is an identity matrix and $\mathbf{\Omega}^2 = \text{diag}\{\omega_i^2\}$.

Finally, we obtain the modal equations in matrix form as

$$\ddot{\mathbf{q}} + \mathbf{\Omega}^2 \mathbf{q} = \mathbf{\Phi}^T \mathbf{u} \tag{1.225}$$

and the decoupled equation for the ith mode as

$$\ddot{q}_i + \omega_i^2 q_i = \phi_i^T \mathbf{u} \tag{1.226}$$

Spacecraft dynamics and control problems can often be characterized or analyzed using a simple two- or three-mass–spring system, which provides much physical insights into the various spacecraft control problems. Therefore, we shall treat such simple dynamic systems here in preparation for real spacecraft control problems that are to be studied later in this book.

1.3.9.1 Two-mass–spring System

Consider two carts of masses m_1 and m_2 on a frictionless horizontal surface, which are connected by a massless, linear spring of stiffness k, as illustrated in Fig. 1.23. External forces u_1 and u_2 act on m_1 and m_2, respectively.

The equations of motion can be derived as

$$\begin{bmatrix} m_1 & 0 \\ 0 & m_2 \end{bmatrix} \begin{bmatrix} \ddot{x}_1 \\ \ddot{x}_2 \end{bmatrix} + \begin{bmatrix} k & -k \\ -k & k \end{bmatrix} \begin{bmatrix} x_1 \\ x_2 \end{bmatrix} = \begin{bmatrix} u_1 \\ u_2 \end{bmatrix} \tag{1.227}$$

where x_1 and x_2 are the generalized coordinates of m_1 and m_2, respectively. It is assumed that the spring is unstressed when $x_1 = x_2$.

Taking the Laplace transform of Eq. (1.227) and ignoring the initial conditions, we obtain the Laplace transformed equations of motion as

$$\begin{bmatrix} m_1 s^2 + k & -k \\ -k & m_2 s^2 + k \end{bmatrix} \begin{bmatrix} x_1(s) \\ x_2(s) \end{bmatrix} = \begin{bmatrix} u_1(s) \\ u_2(s) \end{bmatrix} \tag{1.228}$$

The transfer function matrix from u_1 and u_2 to x_1 and x_2 can then be obtained as

$$\begin{bmatrix} x_1(s) \\ x_2(s) \end{bmatrix} = \frac{1}{D(s)} \begin{bmatrix} m_2 s^2 + k & k \\ k & m_1 s^2 + k \end{bmatrix} \begin{bmatrix} u_1(s) \\ u_2(s) \end{bmatrix} \tag{1.229}$$

where

$$D(s) = s^2 \left[m_1 m_2 s^2 + (m_1 + m_2)k \right]$$

denotes the characteristic polynomial of the system. The characteristic roots of the system are

$$0, \; 0, \; \pm j \sqrt{\frac{(m_1 + m_2)k}{m_1 m_2}}$$

where $j = \sqrt{-1}$. Note that this system is said to be unstable because it has repeated characteristic roots on the imaginary axis, i.e., the double pole at the origin.

Letting $s = \pm j\omega$ and $u_1 = u_2 = 0$, we have

$$\begin{bmatrix} -m_1 \omega_i^2 + k & -k \\ -k & -m_2 \omega_i^2 + k \end{bmatrix} \begin{bmatrix} \phi_{1i} \\ \phi_{2i} \end{bmatrix} = \begin{bmatrix} 0 \\ 0 \end{bmatrix}, \qquad i = 1, 2 \qquad (1.230)$$

where $\omega_1 = 0$, $\omega_2 = \sqrt{(m_1 + m_2)k/(m_1 m_2)}$ and (ϕ_{1i}, ϕ_{2i}) is the ith modal vector.

For example, when $m_1 = m_2 = k = 1$ with appropriate units, we have the modal equations in matrix form as

$$\ddot{\mathbf{q}} + \mathbf{\Omega}^2 \mathbf{q} = \mathbf{\Phi}^T \mathbf{u} \qquad (1.231)$$

where $\mathbf{q} = (q_1, q_2)$, $\mathbf{u} = (u_1, u_2)$, $\mathbf{\Omega}^2 \equiv \mathrm{diag}\{\omega_1^2, \omega_2^2\}$, $\omega_1 = 0$, $\omega_2 = \sqrt{2}$, $\mathbf{x} = \mathbf{\Phi}\mathbf{q}$, and

$$\mathbf{\Phi} = \begin{bmatrix} \phi_{11} & \phi_{12} \\ \phi_{21} & \phi_{22} \end{bmatrix} = \begin{bmatrix} 1/\sqrt{2} & 1/\sqrt{2} \\ 1/\sqrt{2} & -1/\sqrt{2} \end{bmatrix}$$

This system has two modes: a rigid-body mode q_1 with zero natural frequency, and a symmetric flexible mode q_2 with the natural frequency of $\omega_2 = \sqrt{2}$.

1.3.9.2 Three-Mass–Spring System

Consider three carts of masses m_1, m_2, and m_3 on a frictionless horizontal surface, which are connected by massless, linear springs, as illustrated in Fig. 1.24. External forces u_1, u_2, and u_3 act on m_1, m_2, and m_3, respectively.

The equations of motion can be derived as

$$\begin{bmatrix} m_1 & 0 & 0 \\ 0 & m_2 & 0 \\ 0 & 0 & m_3 \end{bmatrix} \begin{bmatrix} \ddot{x}_1 \\ \ddot{x}_2 \\ \ddot{x}_3 \end{bmatrix} + \begin{bmatrix} k_1 & -k_1 & 0 \\ -k_1 & k_1 + k_2 & -k_2 \\ 0 & -k_2 & k_2 \end{bmatrix} \begin{bmatrix} x_1 \\ x_2 \\ x_3 \end{bmatrix} = \begin{bmatrix} u_1 \\ u_2 \\ u_3 \end{bmatrix} \qquad (1.232)$$

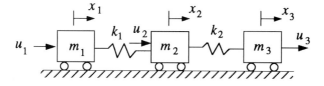

Fig. 1.24   Three-mass–spring system.

where x_1, x_2, and x_3 are the generalized coordinates of m_1, m_2, and m_3, respectively.

The Laplace transformed equations of motion become

$$\begin{bmatrix} m_1 s^2 + k_1 & -k_1 & 0 \\ -k_1 & m_2 s^2 + (k_1 + k_2) & -k_2 \\ 0 & -k_2 & m_3 s^2 + k_2 \end{bmatrix} \begin{bmatrix} x_1(s) \\ x_2(s) \\ x_3(s) \end{bmatrix} = \begin{bmatrix} u_1(s) \\ u_2(s) \\ u_3(s) \end{bmatrix} \quad (1.233)$$

The characteristic equation is

$$D(s) = s^2\big[m_1 m_2 m_3 s^4 + \{(m_1 + m_2)m_3 k_1 + (m_2 + m_3)m_1 k_2\}s^2$$
$$+ k_1 k_2(m_1 + m_2 + m_3)\big] = 0$$

Like the two-mass–spring system, consider a case in which $m_1 = m_2 = m_3 = k_1 = k_2 = 1$ with appropriate units. The transfer function matrix from u_1, u_2, and u_3 to x_1, x_2, and x_3 can then be obtained as

$$\begin{bmatrix} x_1(s) \\ x_2(s) \\ x_3(s) \end{bmatrix} = \frac{1}{D(s)} \begin{bmatrix} s^4 + 3s^2 + 1 & s^2 + 1 & 1 \\ s^2 + 1 & (s^2 + 1)^2 & s^2 + 1 \\ 1 & s^2 + 1 & s^4 + 3s^2 + 1 \end{bmatrix} \begin{bmatrix} u_1(s) \\ u_2(s) \\ u_3(s) \end{bmatrix}$$
$$(1.234)$$

where

$$D(s) = s^2(s^2 + 1)(s^2 + 3)$$

and the modal equations become

$$\ddot{q} + \Omega^2 q = \Phi^T u \quad (1.235)$$

where $q = (q_1, q_2, q_3)$, $u = (u_1, u_2, u_3)$, $\Omega^2 \equiv \text{diag}\{\omega_1^2, \omega_2^2, \omega_3^2\}$, $\omega_1 = 0$, $\omega_2 = 1$, $\omega_3 = \sqrt{3}$, $x = \Phi q$, and

$$\Phi = \begin{bmatrix} 0.3333 & 0.5 & 0.1667 \\ 0.3333 & 0 & -0.3333 \\ 0.3333 & -0.5 & 0.1667 \end{bmatrix}$$

This system has a rigid-body mode and two flexible modes.

Problem

1.51 Consider two carts of masses m_1 and m_2 on a frictionless horizontal surface, which are connected by a massless, linear spring of stiffness k, as was illustrated in Fig. 1.23. External forces u_1 and u_2 act on m_1 and m_2, respectively.

Let new coordinates q_1 and q_2 be defined in terms of the physical coordinates x_1 and x_2 as

$$q_1 = \frac{m_1 x_1 + m_2 x_2}{m_1 + m_2}$$
$$q_2 = x_2 - x_1$$

where q_1 is the position of the system center of mass and q_2 is the relative position of m_2 with respect to m_1. It is assumed that the spring is unstressed when $x_1 = x_2$.

(a) Derive the decoupled equations of motion in terms of the modal coordinates q_1 and q_2 as

$$(m_1 + m_2)\ddot{q}_1 = u_1 + u_2$$

$$\ddot{q}_2 + k\frac{m_1 + m_2}{m_1 m_2}q_2 = -\frac{u_1}{m_1} + \frac{u_2}{m_2}$$

(b) Derive the equations of motion in terms of the hybrid coordinates x_1 and q_2 as

$$(m_1 + m_2)\ddot{x}_1 + m_2\ddot{q}_2 = u_1 + u_2$$

$$m_2\ddot{q}_2 + kq_2 + m_2\ddot{x}_1 = u_2$$

which can be rewritten as

$$(m_1 + m_2)\ddot{x}_1 + \sqrt{m_2}\ddot{\eta}_2 = u_1 + u_2$$

$$\ddot{\eta}_2 + \omega_z^2\eta_2 + \sqrt{m_2}\ddot{x}_1 = \frac{u_2}{\sqrt{m_2}}$$

where $\eta_2 \equiv \sqrt{m_2}q_2$ is called the normalized cantilever modal coordinate, $\omega_z \equiv \sqrt{k/m_2}$ the cantilever frequency, and $\sqrt{m_2}$ the coupling scalar of this two-mass–spring system. Also show that the transfer function from u_1 to x_1 can be expressed as

$$\frac{x_1(s)}{u_1(s)} = \frac{s^2 + \omega_z^2}{Ms^2\{(1 - m_2/M)s^2 + \omega_z^2\}}$$

where $M \equiv m_1 + m_2$ is the total mass of the system.

(c) Assume that the system is initially at rest with $x_1 = x_2 = 0$ and that $m_1 = m_2 = m$. For a case in which u_1 is a unit impulse force at $t = 0$ and $u_2 = 0$, solve for $x_1(t)$ and $x_2(t)$ as

$$x_1 = \frac{1}{2m}\left(t + \frac{1}{\omega_n}\sin\omega_n t\right)$$

$$x_2 = \frac{1}{2m}\left(t - \frac{1}{\omega_n}\sin\omega_n t\right)$$

where

$$\omega_n = \sqrt{2k/m}$$

1.3.10 Phase-Plane Analysis of Nonlinear Systems

Both the linear and nonlinear stability concepts are further examined here via the phase-plane analysis of nonlinear dynamic systems.

Consider a nonlinear system described by

$$\dot{\mathbf{x}} = \mathbf{f}(\mathbf{x}, \mathbf{u})$$

where $\mathbf{x}$ is the state vector and $\mathbf{u}$ the control input vector. An equilibrium point of a nonlinear dynamic system can be found by solving the equilibrium condition

$$\mathbf{f}(\mathbf{x}^*, \mathbf{u}^*) = 0 \qquad \text{for all } t \geq t_0$$

where $\mathbf{x}^*$ denotes the state vector at an equilibrium point for a known constant input or parameter vector $\mathbf{u}^*$.

The linearized equations about an equilibrium point can then be obtained as

$$\delta\dot{\mathbf{x}} = \mathbf{A}\,\delta\mathbf{x} + \mathbf{B}\,\delta\mathbf{u} \qquad (1.236)$$

where $\delta\mathbf{x} = \mathbf{x} - \mathbf{x}^*$, $\delta\mathbf{u} = \mathbf{u} - \mathbf{u}^*$, and $\mathbf{A}$ and $\mathbf{B}$ are constant matrices defined as

$$\mathbf{A} = \left.\frac{\partial\mathbf{f}}{\partial\mathbf{x}}\right|_*, \qquad \mathbf{B} = \left.\frac{\partial\mathbf{f}}{\partial\mathbf{u}}\right|_* \qquad (1.237)$$

The partial derivatives in Eq. (1.237) are to be evaluated at the equilibrium point.

If none of the eigenvalues of $\mathbf{A}$ have zero real parts, then the equilibrium point is said to be a *hyperbolic* point. In the phase-plane analysis of dynamic systems, a hyperbolic equilibrium point is called a saddle if some of the eigenvalues have positive real parts and the rest of the eigenvalues have negative real parts. If all of the eigenvalues have negative real parts, then the hyperbolic point is called a stable node or a stable focus. If all of the eigenvalues have positive real parts, then the hyperbolic point is called an unstable node or an unstable focus. If the eigenvalues are pure imaginary and nonzero, then the equilibrium point is said to be an elliptic point, and the elliptic equilibrium point is called a center.

A stable node and a stable focus are attractors, and an unstable node and an unstable focus are repellers. A saddle point is neither an attractor nor a repeller. A center is not asymptotically stable and, therefore, is not an attractor. In dynamic systems theory, attractors and repellers are called invariant sets, and they occur only in dissipative dynamic systems. In a conservative system, all equilibrium points are either centers or saddle points, and attractors and repellers are not possible.

As was discussed earlier, if an equilibrium point of a linearized system is asymptotically stable, then the equilibrium point of the nonlinear system is asymptotically stable. However, the Lyapunov stability of the equilibrium point of a linearized system does not guarantee the Lyapunov stability of the equilibrium point of the nonlinear system.

For further details of dynamic systems theory and its applications, the reader is referred to [17–21].

Problems

1.52 Consider a simple pendulum of mass m and length ℓ, as described by

$$m\ell\ddot{\theta} + mg \sin\theta = 0$$

where θ is the angular displacement of the pendulum from its vertical down position. The equilibrium points are $\theta = 0, \pm\pi, \ldots$.
(a) Find the eigenvalues for each equilibrium point and discuss its stability.
(b) Sketch the phase-plane trajectories in the $(\theta, \dot{\theta})$ plane and identify *center manifolds* and a *separatrix*. (A manifold is a generalization of a point, curve, or volume in the n-dimensional state space. The closed trajectory through saddle points is called a separatrix or a homoclinic orbit.)

1.53 Consider an inverted pendulum of mass m and length ℓ. The equation of motion is simply given by

$$m\ell^2\ddot{\theta} - mg\ell \sin\theta = u$$

where θ is the angular displacement of the pendulum from its vertical up position and u is a constant torque acting on the pendulum. This equation can be rewritten as

$$\ddot{\theta} = \mu + \omega_n^2 \sin\theta$$

where $\mu = u/m\ell^2$ and $\omega_n^2 = g/\ell$.
  For the following three cases: 1) $|\mu| > \omega_n^2$, 2) $|\mu| = \omega_n^2$, and 3) $|\mu| < \omega_n^2$,
(a) Find equilibrium points and discuss the stability of each equilibrium point.
(b) Sketch the phase-plane trajectories.

1.54 Consider a particle of mass m that is constrained to move on a smooth circular hoop of radius ℓ under the action of gravity, as was illustrated in Fig. 1.6. The equation of motion of the particle is given by

$$m\ell^2\ddot{\theta} + mg\ell \sin\theta = m\ell^2\Omega^2 \sin\theta \cos\theta$$

where θ denotes the angular position of the particle from the vertical line, g is the gravitational acceleration, and Ω is the constant angular velocity of the circular hoop.
(a) Determine the equilibrium points as the dimensionless bifurcation parameter $\mu = \ell\Omega^2/g$ varies.
(b) Discuss the stability of each equilibrium point.
(c) Sketch the phase-plane trajectories.

1.55 Consider an undamped Duffing oscillator described by

$$\ddot{x} - x + x^3 = 0$$

The equilibrium points are $(0, 0)$ and $(\pm 1, 0)$.
(a) Find the eigenvalues for each equilibrium point and discuss its stability.
(b) Sketch the phase-plane trajectories and identify center manifolds and a separatrix.

1.56 Consider a damped Duffing oscillator described by

$$\ddot{x} + \mu \dot{x} - x + x^3 = 0, \qquad \mu > 0$$

The equilibrium points are $(0, 0)$ and $(\pm 1, 0)$.
(a) Find the eigenvalues for each equilibrium point and discuss its stability.
(b) Plot the phase-plane trajectories that illustrate stable manifolds as well as two domains or basins of attraction.

1.57 Consider the van der Pol oscillator described by

$$\ddot{x} + \mu(x^2 - 1)\dot{x} + x = 0$$

where μ is a constant parameter.
(a) Show that for $\mu > 0$, the origin $(x, \dot{x}) = (0,0)$ is an unstable equilibrium point. Plot the phase-plane trajectories of the van der Pol oscillator for $\mu = 1.0$ and verify the existence of a periodic solution, i.e., a closed curve in the phase plane, called a stable limit cycle.
(b) Show that for $\mu < 0$, the origin $(x, \dot{x}) = (0,0)$ is a stable equilibrium point. Plot the phase-plane trajectories of the van der Pol oscillator for $\mu = -1.0$ and verify the existence of an unstable limit cycle.
(c) Predict analytically the existence of such limit cycles for the van der Pol oscillator.
Note: In 1926, van der Pol derived this equation of an electrical circuit with a vacuum tube, now called the van der Pol oscillator.

1.58 Consider again the so-called Lorenz system described by

$$\dot{x} = -\sigma x + \sigma y$$
$$\dot{y} = -zx - y + rx$$
$$\dot{z} = xy - bz$$

where σ, r, and b are positive constant parameters. The equilibrium points of the Lorenz system are given by the following: for $r \leq 1, (0,0,0)$ and for $r > 1, (0,0,0)$ and $C^{\pm} = (\pm\sqrt{b(r-1)}, \pm\sqrt{b(r-1)}, r-1)$.
(a) For the Lorenz system with $\sigma = 10$ and $b = 8/3$, find the eigenvalues for each equilibrium point and discuss its stability as the bifurcation parameter r varies.
(b) For the Lorenz system with $\sigma = 10, b = 8/3$, and $r = 22.4$, the equilibrium points $C^{\pm} = (\pm 7.5542, \pm 7.5542, 21.4)$ are both asymptotically stable. Perform computer simulation of this case with various initial

conditions and identify two basins of attraction in the (x, y, z) state space.

(c) For the Lorenz system with $\sigma = 10$, $b = 8/3$, and $r = 28$, the equilibrium points $C^{\pm} = (\pm 8.4852, \pm 8.4852, 27)$ are both unstable. Perform computer simulation of this case with various initial conditions and investigate the nonperiodic, turbulent (or chaotic) dynamic behavior of the Lorenz system, known as a strange attractor.

References

[1] Hughes, P. C., *Spacecraft Attitude Dynamics*, Wiley, New York, 1986, Appendix B.

[2] Meirovitch, L., *Elements of Vibration Analysis*, McGraw–Hill, New York, 1975, pp. 28–32.

[3] Greenwood, D. T., *Principles of Dynamics*, 2nd ed., Prentice–Hall, Englewood Cliffs, NJ, 1988, Chaps. 2–4.

[4] Meirovitch, L., *Methods of Analytical Dynamics*, McGraw–Hill, New York, 1970, Chaps. 2 and 9.

[5] Wie, B., Furumoto, N., Banerjee, A. K., and Barba, P. M., "Modeling and Simulation of Spacecraft Solar Array Deployment," *Journal of Guidance, Control, and Dynamics*, Vol. 9, No. 5, 1986, pp. 593–598.

[6] Kane, T. R., and Levinson, D. A., "Formulation of Equations of Motion for Complex Spacecraft," *Journal of Guidance, Control, and Dynamics*, Vol. 3, No. 2, 1980, pp. 99–112.

[7] Schaecther, D. B., and Levinson, D. A., "Interactive Computerized Symbolic Dynamics for the Dynamicist," *Journal of the Astronautical Sciences*, Vol. 36, No. 4, 1988, pp. 365–388.

[8] Bodley, C., Devers, D., and Park, C., "A Digital Computer Program for the Dynamic Interaction Simulation of Controls and Structure (DISCOS)," Vols. I and II, NASA TP-1219, May 1978.

[9] Gantmacher, F. R., *Theory of Matrices*, Vol. II, Chelsea, New York, 1959, Chap. 15.

[10] Lorenz, E. N., "Deterministic Nonperiodic Flow," *Journal of the Atmospheric Sciences*, Vol. 20, 1963, pp. 130–141; also Lorenz, E. N., *The Essence of Chaos*, Univ. of Washington Press, Seattle, WA, 1993, Chaps. 3 and 4.

[11] McRuer, D., Ashkenas, I., and Graham, D., *Aircraft Dynamics and Automatic Control*, Princeton Univ. Press, Princeton, NJ, 1973, Chap. 3.

[12] Kailath, T., *Linear Systems*, Prentice–Hall, Englewood Cliffs, NJ, 1980, Chap. 2.

[13] Bellman, R., and Kalaba, R. (eds.), *Mathematical Trends in Control Theory*, Dover, New York, 1964, pp. 72–82 (translation of Hurwitz, "On the Conditions under which an Equation has only Roots with Negative Real Parts," *Mathematishe Annalen*, Vol. 46, 1895, pp. 273–284).

[14] Kharitonov, V. L., "Asymptotic Stability of an Equilibrium Position of a Family of Systems of Linear Differential Equations," *Differential'nye Uraveniya*, Vol. 14, No. 11, 1978, pp. 1483–1485.

[15] Weinmann, A., *Uncertain Models and Robust Control*, Springer–Verlag, New York, 1991, Chap. 4.

[16] Bhattacharyya, S. P., Chapellat, H., and Keel, L. H., *Robust Control: The Parametric Approach*, Prentice–Hall PTR, Upper Saddle River, NJ, 1995, Chap. 5.

[17] Arrowsmith, D. K., and Place, C. M., *An Introduction to Dynamic Systems*, Cambridge Univ. Press, Cambridge, England, UK, 1990.

[18] Wiggins, S., *Introduction to Applied Nonlinear Dynamic Systems and Chaos*, Springer–Verlag, New York, 1990.

[19] Moon, F. C., *Chaotic and Fractal Dynamics*, Wiley, New York, 1992.

[20] Drazin, P. G., *Nonlinear Systems*, Cambridge Univ. Press, Cambridge, England, UK, 1992.

[21] Nayfeh, A. H., and Balachandran, B., *Applied Nonlinear Dynamics*, Wiley, New York, 1995.

2
Dynamic Systems Control

This chapter is concerned with the feedback control of linear dynamic systems, with particular emphasis on robust control of uncertain dynamic systems. Classical control techniques, digital control, and modern state-space methods are introduced. Furthermore, the fundamental concepts of classical gain-phase stabilization, nonminimum-phase compensation, disturbance accommodating control, and uncertainty modeling are emphasized. The problems of designing robust H_∞ compensators and of computing parameter margins for uncertain dynamic systems are also introduced. This chapter is mainly intended to summarize the fundamental concepts and techniques in dynamic systems control, however, which will be essential in the analysis and design of spacecraft control systems. The reader is presumed to have access to computational software such as MATLAB^{TM}, which will be essential in solving various control problems treated in this chapter and in the remainder of this textbook. For an introduction to feedback control analysis, design, and simulation using MATLAB, the reader is referred to Refs. 1 and 2.

2.1   Feedback Control Systems

Block diagram representations of a feedback control system are shown in Fig. 2.1. Figure 2.1a is called a functional block diagram representation. Physical systems to be controlled are often referred to as plants. A set of differential or difference equations used to describe a physical system is called a mathematical model of the system. In the analysis and design of a feedback control system, we often deal with a mathematical model of the plant, not with the actual physical plant. Consequently, special care must be taken regarding uncertainties in the mathematical model because no mathematical model of a physical system is exact.

A closed-loop feedback control system maintains a specified relationship between the actual output and the desired output (or the reference input) by using the difference of these outputs, called the *error signal*. A control system in which the output has no effect on the control decision is called an open-loop control system. In a feedback control system, a *controller*, also called a compensator or control logic, is designed to manipulate or process the error signal so that certain specifications are satisfied in the presence of plant disturbances and sensor noise. In the analysis of control systems, we analyze the dynamic behavior or characteristics of the system under consideration. In the design or synthesis, we are concerned with designing a feedback control system that will achieve the desired

a)

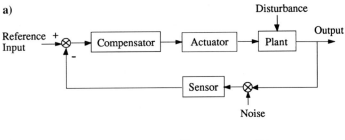

b)

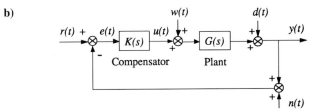

Fig. 2.1   Block diagram representations of a feedback control system.

system characteristics. A feedback control system can also be represented as it is in Fig. 2.1b, using transfer functions. In this figure, for simplicity, the actuator and sensor dynamics are neglected, and $r(t)$ denotes the reference input, $y(t)$ the plant output, $G(s)$ the plant transfer function, $K(s)$ the compensator, $u(t)$ the control input, $e(t)$ the error signal, $w(t)$ the disturbance, $d(t)$ the output disturbance, and $n(t)$ a sensor noise.

The output of this closed-loop system, neglecting the sensor noise $n(t)$, can then be represented as

$$y(s) = \frac{K(s)G(s)}{1 + K(s)G(s)} r(s) + \frac{G(s)}{1 + K(s)G(s)} w(s) + \frac{1}{1 + K(s)G(s)} d(s) \quad (2.1)$$

where $y(s) = \mathcal{L}[y(t)]$, $r(s) = \mathcal{L}[r(t)]$, $w(s) = \mathcal{L}[w(t)]$, and $d(s) = \mathcal{L}[d(t)]$. In particular, the closed-loop transfer functions from $d(s)$ and $r(s)$ to $y(s)$ are

$$\frac{y(s)}{d(s)} = \frac{1}{1 + K(s)G(s)} = S(s) \quad (2.2)$$

$$\frac{y(s)}{r(s)} = \frac{K(s)G(s)}{1 + K(s)G(s)} = T(s) \quad (2.3)$$

and $S(s)$ and $T(s)$ are called the *sensitivity function* and the *complementary sensitivity function*, respectively. Furthermore, we have the following relationship:

$$S(s) + T(s) = 1 \quad (2.4)$$

The closed-loop characteristic equation is defined as

$$1 + K(s)G(s) = 0 \quad (2.5)$$

and $K(s)G(s)$ is called the loop (or open-loop) *transfer function*. The importance of the loop transfer function cannot be overemphasized because it is used extensively in the analysis and design of closed-loop systems. The roots of the closed-loop characteristic equation are called the closed-loop poles.

The error signal, ignoring the sensor noise $n(t)$, is defined as

$$e(t) = r(t) - y(t) \tag{2.6}$$

and the steady-state error can be found as

$$e_{ss} = \lim_{t \to \infty} e(t) = \lim_{s \to 0} se(s) \tag{2.7}$$

where $e(s) = \mathcal{L}[e(t)]$, provided that $e(t)$ has a final value. For the system shown in Fig. 2.1, ignoring $w(s)$ and $d(s)$, we have

$$e(s) = \frac{1}{1 + K(s)G(s)} r(s) \tag{2.8}$$

and

$$e_{ss} = \lim_{s \to 0} \frac{sr(s)}{1 + K(s)G(s)} \tag{2.9}$$

Thus, it is required that

$$\lim_{s \to 0} K(s)G(s) = \infty \tag{2.10}$$

to have zero steady-state tracking error for a constant reference input command.

A feedback control system is often characterized by its system type. The system type is defined as the number of poles of the loop transfer function $K(s)G(s)$ at the origin. Therefore, a type 1 system has zero steady-state error for a constant reference input, a type 2 system has zero steady-state error for a constant or ramp reference input, and so forth.

To reduce the effects of the disturbance, the magnitude of the loop transfer function $K(s)G(s)$ must be large over the frequency band of the disturbance $d(t)$. For good command following at any frequency, the steady-state or D.C. gain must be large. In general, a fast transient response, good tracking accuracy, good disturbance rejection, and good sensitivity require a high loop gain over a wide band of frequencies. Because the high loop gain may degrade the overall system stability margins, proper tradeoffs between performance and stability are always necessary in practical control designs.

2.2 Classical Frequency-Domain Methods

2.2.1 Root Locus Method

One of the classical control analysis and design techniques is the root locus method developed by Evans [3] in 1950 (see also Ref. 4). In Evans's root locus method, the closed-loop characteristic equation is described by

$$1 + KG(s) = 0 \tag{2.11}$$

where $KG(s)$ denotes the loop transfer function, $G(s)$ includes both the compensator transfer function and the plant transfer function, and K is called the overall loop gain. Note that the roots of the closed-loop characteristic equation are called the closed-loop poles.

In Evans's root locus plot, the poles and zeros of the loop transfer function $KG(s)$ are shown, where the poles are represented as crosses, ×, and zeros as circles, o. A root locus is then simply a plot of the closed-loop poles as the overall loop gain K is usually varied from zero to infinity.

Using a root locus plot, one can easily determine a gain margin, which is one of the most important measures of the relative stability of a feedback control system. A gain margin indicates how much the loop gain K can be increased or decreased from its chosen nominal value until the closed-loop system becomes unstable. For example, if the loop gain K can be increased by a factor of two until a root locus crosses the imaginary axis toward the right-half s plane, then the gain margin becomes $20 \log 2 \approx +6$ dB. In some cases of an open-loop unstable system, the closed-loop system may become unstable if the loop gain is decreased from its chosen nominal value. For example, if the gain can be decreased by a factor of 0.707 until the closed-loop system becomes unstable, then the (negative) gain margin is $20 \log 0.707 \approx -3$ dB.

The root locus method also allows the designer to properly select at least some of the closed-loop pole locations and thus control the transient response characteristics.

Example 2.1

Consider a simple example of root locus plot vs overall loop gain K as illustrated in Fig. 2.2. The closed-loop characteristic equation of this system is

$$1 + K(s+1)/s^2 = 0 \qquad \text{or} \qquad s^2 + Ks + K = 0$$

For this simple case, the closed-loop poles can be analytically determined as

$$s = \tfrac{1}{2}(-K \pm \sqrt{K^2 - 4K})$$

and they can be computed and plotted for different values of K, as shown in Fig. 2.2. Because this closed-loop system is asymptotically stable for any nonzero positive value of K, it is often said that this system has $\pm\infty$ dB gain omargins.

Many useful properties of the root locus will aid in quick, rough sketching of root loci. Presently, however, computer programs are usually employed for plotting accurate root loci in practical design of complex control systems. As mentioned earlier, it is assumed that the reader has access to, and some acquaintance with, computer-aided control design software.

The root locus method can also be employed for the analysis and design of multiloop feedback control systems [5]. Consider a plant with two inputs and two outputs described by

$$\begin{bmatrix} y_1 \\ y_2 \end{bmatrix} = \frac{1}{D(s)} \begin{bmatrix} N_{11}(s) & N_{12}(s) \\ N_{21}(s) & N_{22}(s) \end{bmatrix} \begin{bmatrix} u_1 \\ u_2 \end{bmatrix} \qquad (2.12)$$

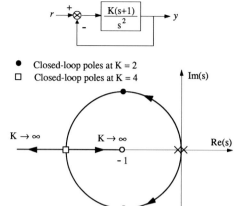

Fig. 2.2 Root locus vs overall loop gain K.

where y_1 and y_2 are the outputs, u_1 and u_2 are the control inputs, $D(s)$ is the characteristic polynomial, and $N_{ij}(s)$ are the numerator polynomials. A diagonal feedback control logic is assumed as

$$u_1 = -K_1(s)y_1 \tag{2.13a}$$

$$u_2 = -K_2(s)y_2 \tag{2.13b}$$

where $K_1(s)$ and $K_2(s)$ are the diagonal feedback compensators.

The closed-loop characteristic equation can be obtained as

$$D^2 + (K_1N_{11} + K_2N_{22})D + K_1K_2(N_{11}N_{22} - N_{12}N_{21}) = 0 \tag{2.14}$$

which becomes

$$D + K_1N_{11} + K_2N_{22} + K_1K_2N = 0 \tag{2.15}$$

because

$$N_{11}N_{22} - N_{12}N_{21} = D(s)N(s) \tag{2.16}$$

where $N(s)$ is called the coupling numerator polynomial.

The first-loop analysis or synthesis can be performed by considering a characteristic equation of the form

$$1 + (K_1N_{11}/D) = 0 \tag{2.17}$$

After synthesizing $K_1(s)$ in the first-loop closure, the characteristic equation of the second loop can be found as

$$1 + \frac{K_2(N_{22} + K_1N)}{D + K_1N_{11}} = 0 \tag{2.18}$$

The second-loop compensator $K_2(s)$ can then be analyzed or synthesized in a manner similar to the method for the first-loop design; however, the zeros and the poles of the second-loop transfer function are changed by the first-loop closure. The new zeros are related to the coupling numerator. This property is useful in finding the new zeros for the second-loop closure in the root locus analysis.

Classical stability analysis of multiloop feedback control systems, by breaking loops one at a time, is known to be an unreliable way of testing robustness or sensitivity to simultaneous perturbations of all of the loops. As can be learned from the following two problems [6], however, the successive-loop-closure approach described earlier can be effectively used to predict a lack of robustness in the nominal designs and, furthermore, provides insights into how the design can be changed so as to be more robust.

Problems

2.1 Consider a system described by

$$\begin{bmatrix} y_1 \\ y_2 \end{bmatrix} = \frac{1}{s^2 + 100} \begin{bmatrix} s - 100 & 10(s+1) \\ -10(s+1) & s - 100 \end{bmatrix} \begin{bmatrix} u_1 \\ u_2 \end{bmatrix}$$

where (y_1, y_2) are the outputs and (u_1, u_2) are the control inputs. A constant-gain diagonal feedback control logic is given as

$$u_1 = -K_1 y_1$$
$$u_2 = -K_2 y_2$$

with the nominal gains of $K_1 = K_2 = 1$.
(a) Sketch root locus vs K_1 of the first-loop closure and indicate the closed-loop poles at $K_1 = 0.9, 1.0$, and 1.1.
(b) Show that, after the first-loop closure, the closed-loop characteristic equation for the second-loop closure becomes

$$1 + K_2 \left\{ \frac{s + (101K_1 - 100)}{s^2 + K_1 s + 100(1 - K_1)} \right\} = 0$$

and that the coupling numerator polynomial $N(s) = 101$.
(c) After selecting the nominal gain $(K_1 = 1)$ for the first-loop closure, sketch root locus vs K_2 of the second-loop closure and indicate the closed-loop poles at $K_2 = 0.5, 1.0$, and 10.
(d) According to the result in (c), one may conclude that the second-loop gain K_2 can be increased to infinity (but keeping $K_1 = 1$), without destabilizing the overall closed-loop system. Even so, discuss the effect of independent perturbations of loop gains K_1 and K_2 on the closed-loop stability by sketching root locus plots vs K_2 of the second-loop closure with two different values of the nominal gains: $K_1 = 0.9$ and 1.1.

(e) Although this problem may not represent any particular physical system, show that a controller should be of the form

$$u_1 = +K_1 y_2$$
$$u_2 = -K_2 y_1$$

where K_1 and K_2 are positive gains.

2.2 Consider a system described by

$$\begin{bmatrix} y_1 \\ y_2 \end{bmatrix} = \frac{1}{(s+1)(s+2)} \begin{bmatrix} -47s+2 & 56s \\ -42s & 50s+2 \end{bmatrix} \begin{bmatrix} u_1 \\ u_2 \end{bmatrix}$$

where (y_1, y_2) are the outputs and (u_1, u_2) are the control inputs. A constant-gain diagonal feedback control logic is given as

$$u_1 = -K_1 y_1$$
$$u_2 = -K_2 y_2$$

with the nominal gains of $K_1 = K_2 = 1$.

(a) Sketch root locus vs K_1 of the first-loop closure and indicate the closed-loop poles at $K_1 = 0.064$ and 1.

(b) Show that, after the first-loop closure, the closed-loop characteristic equation for the second-loop closure becomes

$$1 + K_2 \left\{ \frac{50s + 2(1 + K_1)}{s^2 + (3 - 47K_1)s + 2(1 + K_1)} \right\} = 0$$

and that the coupling numerator polynomial $N(s) = 2$.

(c) After selecting the nominal gain $(K_1 = 1)$ for the first-loop closure, sketch root locus vs K_2 of the second-loop closure and indicate the closed-loop poles at $K_2 = 0.88$ and 1.

(d) According to the result in (c), one may conclude that the second loop has $+\infty$- and only -1.1-dB gain margins. Similarly, by closing the K_2 loop first, show that the K_1 loop has $-\infty$- and only $+1.0$-dB gain margins.

(e) Also show that the system has $\pm\infty$-dB gain margins for simultaneous gain perturbations of $K_1 \equiv K_2$.

2.2.2 Frequency-Response Methods

Frequency-response analysis and synthesis methods are among the most commonly used techniques for feedback control system analysis and design, and they are based on the concept of *frequency-response function*.

The frequency-response function is defined by the transfer function evaluated at $s = j\omega$; that is, the frequency response function of a transfer function $G(s)$ is given by

$$G(s)|_{s=j\omega} = G(j\omega) = \text{Re}[G(j\omega)] + j\,\text{Im}[G(j\omega)] = |G(j\omega)|e^{j\phi(\omega)} \qquad (2.19)$$

where $|G(j\omega)|$ and $\phi(\omega)$ denote, respectively, the magnitude and phase of $G(j\omega)$ defined as

$$|G(j\omega)| = \sqrt{\{\text{Re}[G(j\omega)]\}^2 + \{\text{Im}[G(j\omega)]\}^2}$$

$$\phi(\omega) = \tan^{-1}\frac{\text{Im}[G(j\omega)]}{\text{Re}[G(j\omega)]}$$

For a given value of ω, $G(j\omega)$ is a complex number. Thus, the frequency-response function $G(j\omega)$ is a complex function of ω. Mathematically, the frequency-response function is a mapping from the s plane to the $G(j\omega)$ plane. The upper-half of the $j\omega$ axis, which is a straight line, is mapped into the complex plane via mapping $G(j\omega)$.

One common method of displaying the frequency-response function is a *polar plot* (also called a *Nyquist plot*) where the magnitude and phase angle of $G(j\omega)$, or its real and imaginary parts, are plotted in a plane as the frequency ω is varied. Another form of displaying $G(j\omega)$ is to plot the magnitude of $G(j\omega)$ vs ω and to plot the phase angle of $G(j\omega)$ vs ω. In a *Bode plot*, the magnitude and phase angle are plotted with frequency on a logarithmic scale. Also, we often plot the magnitude of the frequency-response function in decibels; that is, we plot $20 \log |G(j\omega)|$. A plot of the logarithmic magnitude in decibels vs the phase angle for a frequency range of interest is called a *Nichols plot*.

For a feedback control system, as shown in Fig. 2.1, the loop transfer function $K(s)G(s)$ evaluated at $s = j\omega$ is used extensively in the analysis and design of the system using frequency-response methods. The closed-loop frequency response functions defined as

$$\frac{y(j\omega)}{d(j\omega)} = S(j\omega) = \frac{1}{1 + K(j\omega)G(j\omega)} \qquad (2.20)$$

$$\frac{y(j\omega)}{r(j\omega)} = T(j\omega) = \frac{K(j\omega)G(j\omega)}{1 + K(j\omega)G(j\omega)} \qquad (2.21)$$

are also used in classical frequency-domain control systems design.

Among the most important measures of the relative stability of a feedback control system are the gain and phase margins as defined as follows.

2.2.2.1 Gain Margin   Given the loop transfer function $K(s)G(s)$ of a feedback control system, the gain margin is defined as the reciprocal of the magnitude $|K(j\omega)G(j\omega)|$ at the phase-crossover frequency at which the phase angle $\phi(\omega)$ is -180 deg; that is, the gain margin, denoted by g_m, is defined as

$$g_m = \frac{1}{|K(j\omega_c)G(j\omega_c)|} \qquad (2.22)$$

or

$$g_m = -20 \log |K(j\omega_c)G(j\omega_c)| \quad \text{dB} \qquad (2.23)$$

where ω_c is the phase-crossover frequency. For a stable minimum-phase system, the gain margin indicates how much the gain can be increased before the closed-loop system becomes unstable.

2.2.2.2 *Phase Margin*   The phase margin is the amount of additional phase lag at the gain-crossover frequency ω_c at which $|K(j\omega_c)G(j\omega_c)| = 1$ required to make the system unstable; that is,

$$\phi_m = \phi[K(j\omega_c)G(j\omega_c)] + 180 \text{ deg} \qquad (2.24)$$

Although the gain and phase margins may be obtained directly from a Nyquist plot, they can also be determined from a Bode plot or a Nichols plot of the loop transfer function $K(j\omega)G(j\omega)$.

Problems

2.3   Consider a feedback control system shown in Fig. 2.2 with the closed-loop characteristic equation

$$1 + K(s+1)/s^2 = 0$$

with the nominal loop gain of $K = 2$.
   (a) Sketch the Bode, Nyquist, and Nichols plots of the loop transfer function with $K = 2$. Using these plots, determine the gain and phase margins of this feedback control system with the nominal loop gain of $K = 2$.
   (b) Sketch the Bode plot of the closed-loop frequency response function with $K = 2$ from r to y, and determine the bandwidth of this closed-loop system, i.e., the frequency at which the magnitude of its frequency response function drops by 3 dB from its low frequency gain.
   (c) Sketch the closed-loop transient response $y(t)$ when $r(t)$ is a unit step function, for three different values of $K = 1, 2,$ and 4.

2.4   Consider a feedback control system illustrated in Fig. 2.3 with the nominal loop gain of $K = 1$.
   (a) Plot root locus vs the overall loop gain K from 0 to ∞. Indicate the nominal closed-loop poles at $K = 1$. (Note that the root locus plot must be symmetric about the real axis.)
   (b) Determine the gain and phase margins of this feedback control system.
   (c) Discuss the inherent sensitivity of this closed-loop system.

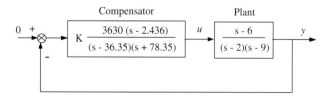

Fig. 2.3   Stabilization of an unstable plant.

2.2.3 Classical Proportional-Integral-Derivative Control Design

The proportional-integral-derivative (PID) control logic is commonly used in most feedback controllers. To illustrate the basic concept of the PID control, consider a cart of mass m on a frictionless horizontal surface, as shown in Fig. 2.4a. This so-called double integrator plant is described by

$$m\ddot{y}(t) = u(t) + w(t) \tag{2.25}$$

where y is the output displacement of the cart, u is the input force acting on the cart, and w is a disturbance force. This system with a rigid-body mode is unstable; thus the system needs to be stabilized and the desired output is assumed to be zero.

Assuming that the position and velocity of the system can be directly measured, consider a direct velocity and position feedback control logic expressed as

$$u(t) = -ky(t) - c\dot{y}(t) \tag{2.26}$$

or

$$u = -(k + cs)y$$

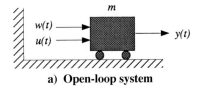

a) **Open-loop system**

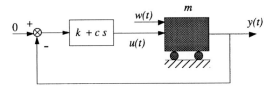

b) **Closed-loop system with position and velocity feedback**

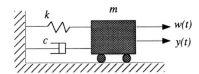

c) **Equivalent closed-loop system representation**

Fig. 2.4   Control of a double integrator plant by direct velocity and position feedback.

where k and c are controller gains to be determined. The closed-loop system illustrated by Fig. 2.4b is then described by

$$m\ddot{y}(t) + c\dot{y}(t) + ky(t) = w(t)$$

which is, in fact, a mathematical representation of a mass–spring–damper system forced by an external disturbance $w(t)$, as illustrated in Fig. 2.4c.

The closed-loop characteristic equation of the system shown in Fig. 2.4 is

$$ms^2 + cs + k = 0$$

The control design task is to tune the active damper and active spring to meet given performance/stability specifications of the closed-loop system. Let ω_n and ζ be the desired natural frequency and damping ratio of the closed-loop poles. Then the desired closed-loop characteristic equation becomes

$$s^2 + 2\zeta\omega_n s + \omega_n^2 = 0$$

and the controller gains c and k can be determined as

$$c = 2m\zeta\omega_n \tag{2.27a}$$

$$k = m\omega_n^2 \tag{2.27b}$$

The damping ratio ζ is often selected as $0.5 \le \zeta \le 0.707$, and the natural frequency ω_n is then considered the bandwidth of the proportional-derivative (PD) controller of a system with a rigid-body mode. For a unit-step disturbance, this closed-loop system with the PD controller results in a nonzero steady-state output $y(\infty) = 1/k$. The steady-state output error $y(\infty)$ can be made small, however, by designing a high-bandwidth control system.

To keep the cart at the desired position $y = 0$ at steady state in the presence of a constant disturbance, consider a PID controller of the form

$$u(t) = -K_P y(t) - K_I \int y(t)\,\mathrm{d}t - K_D \dot{y}(t) \tag{2.28}$$

or

$$u(s) = -\left[K_P + \frac{K_I}{s} + K_D s\right] y(s)$$

In practical analog circuit implementation of a PID controller when $\dot{y}$ is not directly measured, differentiation is always preceded by a low-pass filter to reduce noise effects. It can be shown that for a constant disturbance, the closed-loop system with the PID controller, in fact, results in a zero steady-state output $y(\infty) = 0$.

The closed-loop characteristic equation of the cart with the PID controller can be found as

$$ms^3 + K_D s^2 + K_P s + K_I = 0$$

and let the desired closed-loop characteristic equation be expressed as

$$\left(s^2 + 2\zeta\omega_n s + \omega_n^2\right)(s + 1/T) = 0$$

where ω_n and ζ denote, respectively, the natural frequency and damping ratio of the complex poles associated with the rigid-body mode, and T is the time constant of the real pole associated with integral control.

The PID controller gains can then be determined as

$$K_P = m\left(\omega_n^2 + 2\zeta\omega_n/T\right) \qquad (2.29a)$$

$$K_I = m\left(\omega_n^2/T\right) \qquad (2.29b)$$

$$K_D = m(2\zeta\omega_n + 1/T) \qquad (2.29c)$$

The time constant T of integral control is often selected as

$$T \approx 10/\zeta\omega_n$$

A more detailed treatment of a classical approach to control logic design for high-order dynamic systems will be presented in the next section, followed by modern state-space methods.

Problems

2.5 Consider the control problem of a double integrator plant with measurement of position only. A common method of stabilizing the double integrator plant with noisy position measurement is to employ a phase-lead compensator of the form

$$u(s) = -K\frac{T_1 s + 1}{T_2 s + 1}y(s)$$

as illustrated in Fig. 2.5a.

Show that an equivalent closed-loop system can be represented using two springs and a damper as in Fig. 2.5b and that

$$K = k_1, \qquad T_1 = \frac{c(k_1 + k_2)}{k_1 k_2}, \qquad T_2 = \frac{c}{k_2}$$

Note: For further details of designing a passive three-parameter isolator known as the D-Strut^{TM} that can be modeled as Fig. 2.5b, see Davis et al. [7]. Also see Ogata (Ref. 1, pp. 439–440) for other types of a mechanical lead network.

2.6 Consider again the control problem of a double integrator plant employing a phase-lead compensator of the form

$$u(s) = -K\frac{T_1 s + 1}{T_2 s + 1}y(s)$$

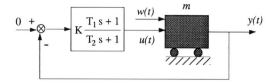

a) Closed-loop system with a phase-lead compensator

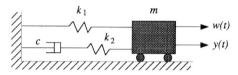

b) Equivalent closed-loop system representation using two springs and a damper

Fig. 2.5 Control of a double integrator plant using a phase-lead compensator.

(a) Determine T_1, T_2, and K such that the closed-loop plant poles be located near $s = -2 \pm j2$ for the nominal plant with $m = 1$.
   Hint: One possible design is $T_1 = 1, T_2 = 1/6$, and $K = 16/6$.
(b) Sketch root locus vs the overall loop gain K and indicate the closed-loop poles at the selected nominal gain of K.
(c) Sketch the Bode, Nyquist, and Nichols plots of the loop transfer function with the selected nominal design. Using these plots, determine the gain and phase margins of the nominal closed-loop system.
(d) Sketch the Bode plot of the closed-loop frequency-response function from w to y.
(e) Sketch (or determine analytically/numerically) the closed-loop time response of the position output $y(t)$ to a unit-step disturbance $w(t) = 1$ for $t \geq 0$.

2.7 Consider the problem of designing a PID controller for the double integrator plant shown in Fig. 2.4a with $m = 1$. The PID control logic is represented as

$$u(t) = -K_P y(t) - K_I \int y(t)\,dt - K_D \dot{y}(t)$$

in which the desired output is assumed to be zero.
(a) Determine the PID controller gains such that the closed-loop poles are located at $s = -0.1$ and $s = -1 \pm j$ for the nominal plant with $m = 1$. (Answers: $K_P = 2.2, K_I = 0.2$, and $K_D = 2.1$.)
(b) Sketch root locus vs an overall loop gain K of the selected PID controller of the form

$$u(s) = -K\,\{2.2 + 0.2/s + 2.1s\}\,y(s)$$

and indicate the closed-loop poles at the selected nominal loop gain of $K = 1$.

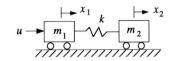

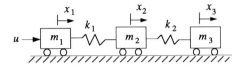

Fig. 2.6   Control of mass–spring systems.

(c) Find the gain and phase margins of the closed-loop system.
(d) Sketch (or determine analytically/numerically) the closed-loop time response of the position output $y(t)$ to a unit-step disturbance.

2.8   Consider the two-mass spring system shown in Fig. 2.6. Two carts of masses m_1 and m_2 on a frictionless horizontal surface are connected by a massless, linear spring of stiffness k. This unstable system is to be stabilized using a single control input force u acting on m_1. The equations of motion of the system are

$$m_1\ddot{x}_1 + k(x_1 - x_2) = u$$
$$m_2\ddot{x}_2 + k(x_2 - x_1) = 0$$

(a) Assume that only the position and velocity of m_1 are directly measured and are fed back to u, as follows:

$$u = -K_1 x_1 - C_1 \dot{x}_1$$

where K_1 and C_1 are positive gains. For this so-called collocated actuator and sensor control problem, show that the closed-loop system is asymptotically stable for any positive values of m_1, m_2, k, K_1, and C_1. *Hint:* See Fig. 2.7.

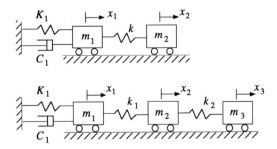

Fig. 2.7   Equivalent active spring and active damper representations of the collocated control logic $u = -K_1 x_1 - C_1 \dot{x}_1$.

(b) Assume that only the position and velocity of m_2 are directly measured and are fed back to u, as follows:

$$u = -K_2 x_2 - C_2 \dot{x}_2$$

where K_2 and C_2 are positive gains. For this so-called noncollocated actuator and sensor control problem, show that the closed-loop system is unstable for any positive values of m_1, m_2, k, K_2, and C_2.
Hint: For a characteristic equation of the form $a_0 s^4 + a_1 s^3 + a_2 s^2 + a_3 s + a_4 = 0$, we have $a_1 = -\sum \lambda_i$ where λ_i are the characteristic roots.

(c) Assume that all of the states of the system are directly measured and are fed back to u, as follows:

$$u = -K_1 x_1 - C_1 \dot{x}_1 - K_2 x_2 - C_2 \dot{x}_2$$

where K_1, C_1, K_2, and C_2 are all real scalars. For the nominal system with $m_1 = m_2 = k = 1$ with appropriate units, determine the full-state feedback controller gains such that the closed-loop poles be located at $s = -0.1 \pm 0.1j$ and $s = -0.1 \pm 1.4j$. (Answers: $K_1 = 0.03, C_1 = 0.4, K_2 = 0.0094$, and $C_2 = -0.002$.)

2.9 Consider the three-mass–spring system shown in Fig. 2.6. This unstable system is to be stabilized using a single control input force u acting on m_1. The equations of motion of this system are

$$m_1 \ddot{x}_1 + k_1 (x_1 - x_2) = u$$
$$m_2 \ddot{x}_2 + k_1 (x_2 - x_1) + k_2 (x_2 - x_3) = 0$$
$$m_3 \ddot{x}_3 + k_2 (x_3 - x_2) = 0$$

(a) Assume that only the position and velocity of m_1 are directly measured and are fed back to u, as follows:

$$u = -K_1 x_1 - C_1 \dot{x}_1$$

where K_1 and C_1 are positive gains. Show that the closed-loop system is asymptotically stable for any positive values of m_i, k_i, K_1, and C_1.
Hint: See Fig. 2.7.

(b) Assume that only the position and velocity of m_2 are directly measured and are fed back to u, as follows:

$$u = -K_2 x_2 - C_2 \dot{x}_2$$

where K_2 and C_2 are positive scalars. Show that the closed-loop system is unstable for any positive values of m_i, k_i, K_2, and C_2.

(c) Assume that only the position and velocity of m_3 are directly measured and are fed back to u, as follows:

$$u = -K_3 x_3 - C_3 \dot{x}_3$$

where K_3 and C_3 are positive scalars. Show that the closed-loop system is unstable for any positive values of m_i, k_i, K_3, and C_3.

2.3   Classical Gain-Phase Stabilization

In the preceding sections we introduced the fundamentals of classical control. In this section, based on Refs. 8–10, we present a classical gain-phase stabilization approach to compensator design, in particular, for a structural dynamic system that has a rigid-body mode and lightly damped, oscillatory flexible modes. The approach allows the control designer to properly gain-phase stabilize each mode, one by one, resulting in a meaningful control design with physical insight. The use of nonminimum-phase compensation for a certain class of dynamic systems is emphasized. The classical gain-phase stabilization method is, however, primarily restricted to the single-input single-output (SISO) control problems.

2.3.1   Introduction

The classical concepts of gain-phase stabilization of a rigid body and flexible modes can be summarized briefly as follows.

1) Gain stabilization of a flexible mode provides attenuation of the control loop gain at the desired frequency, to ensure stability regardless of the control loop phase uncertainty. A lightly damped, flexible mode is said to be gain stabilized if it is closed-loop stable for the selected loop gain, but it becomes unstable if the loop gain is raised or its passive damping reduced. Hence, a gain stabilized mode has a finite gain margin, but is closed-loop stable regardless of the phase uncertainty.

2) Phase stabilization of a flexible mode provides the proper phase characteristics at the desired frequency to obtain a closed-loop damping that is greater than the passive damping of the mode. A lightly damped, flexible mode is said to be phase stabilized if it is closed-loop stable for arbitrarily small passive damping. Hence, a phase stabilized mode has a finite phase margin, but is closed-loop stable regardless of the loop gain uncertainty.

3) A rigid body or flexible mode is said to be gain-phase stabilized if it is closed-loop stable with finite gain and phase margins.

When an actuator and a sensor are collocated on flexible structures in space, the rigid-body mode and all of the flexible modes are said to be stably interacting with each other. For such a collocated case, position feedback with a phase-lead compensator or direct rate and position feedback can be used to stabilize all of the flexible and rigid-body modes. Because all of the modes are phase stabilized in this case, special care must be taken regarding the phase uncertainty from the control loop time delay and actuator/sensor dynamics. As frequency increases, the phase lag due to a time delay will eventually exceed the maximum phase lead of 90 deg from the direct rate feedback. Thus, rolloff filtering, i.e., gain stabilization, of high-frequency modes is often needed to attenuate the control loop gain at frequencies above the control bandwidth. The selection of rolloff filter corner frequency depends on many factors. When a collocated actuator/sensor pair is used, the corner frequency is often selected between the primary flexible modes and the secondary flexible modes. An attempt to gain stabilize all of the flexible modes should be avoided, unless the spacecraft or structures are nearly rigid. In practice, the actual phase uncertainty of the control loop must be taken into account for the proper tradeoff between phase stabilization and gain stabilization.

When an actuator and a sensor are not collocated, the rigid-body mode and some of the flexible modes are said to be unstably interacting with each other. Unless gain stabilization of all of the flexible modes is possible for a low-bandwidth control, a proper combination of gain-phase stabilization is unavoidable. Gain stabilization of an unstably interacting flexible mode can be achieved only if that mode has a certain amount of passive damping. The larger the passive damping at a particular mode, the more conveniently it can be gain stabilized. Usually, gain stabilization is applied to stabilize high-frequency modes that have no significant effects on the overall performance. In practice, a structure has always a certain amount of passive damping, which allows for the convenient gain stabilization of such flexible modes.

Notch filtering is a conventional way of suppressing an unwanted oscillatory signal in the control loop, resulting in gain stabilization of a particular flexible mode. The use of notch filtering ensures that the specific mode is not destabilized by feedback control; however, it does not introduce any active damping, which often results in too much ringing that may not be acceptable in certain cases. In general, rolloff of the control loop gain at frequencies above the control bandwidth is always needed to avoid destabilizing unmodeled high-frequency modes and to attenuate high-frequency noise, and it is often simply achieved by using a double-pole low-pass filter. To sharply attenuate a signal at high frequencies while affecting the magnitude and phase of the signal at low frequencies as little as possible, various high-order low-pass filters such as Bessel, Butterworth, Chevyshev, or elliptical filters, are also used in feedback control systems, but mostly in open-loop signal processing [11]. The common characteristic of these conventional filters is that they are minimum-phase filters.

2.3.2 Generalized Second-Order Filters

Although the last several decades have brought major developments in advanced control theory, the most usual approach to the design of practical control systems has been repetitive, trial-and-error synthesis using the root locus method by Evans or the frequency-domain methods by Bode, Nyquist, and Nichols. Classical control designs employ primarily a PID-type controller with notch or rolloff filtering. Such classical control designs for a certain class of dynamic systems become difficult, however, especially if a high control bandwidth is required in the presence of many closely spaced, unstably interacting, lightly damped modes with a wide range of parameter variations.

In this section we introduce the concept of a generalized second-order filter so that the classical s-domain or frequency-domain methods can still be employed to solve such a difficult structural control problem. The concept is a natural extension of the classical notch and phase lead/lag filtering, and it is based on various pole–zero patterns that can be realized from a second-order filter of the form

$$\frac{s^2/\omega_z^2 + 2\zeta_z s/\omega_z + 1}{s^2/\omega_p^2 + 2\zeta_p s/\omega_p + 1} \tag{2.30}$$

where $\omega_z, \zeta_z, \omega_p,$ and ζ_p are filter parameters to be properly selected.

For different choices of the coefficients of this second-order filter, several well-known filters such as notch, bandpass, low-pass, high-pass, phase-lead, and phase-lag filters can be realized. In addition to these minimum-phase filters, various nonminimum-phase filters can also be realized from this second-order filter; however, we only consider here stable filters with poles in the left-half s plane.

Pole–zero patterns of conventional filters, which are commonly employed in classical control systems design, are shown in Fig. 2.8. Some of the typical pole–zero patterns and the gain-phase characteristics of various filters that can be realized from the generalized second-order filter are summarized in Figs. 2.9–2.13. These basic pole–zero patterns of a second-order filter, especially the nonminimum-phase filters, are the essence of the generalized second-order filtering concept. Any other compensator pole–zero patterns are basically a combination of those basic filters shown in Figs. 2.9–2.13.

2.3.2.1   Minimum-phase Lead or Lag Filter

A phase-lead or phase-lag filter can be realized from the second-order filter of the form of Eq. (2.30), as illustrated in Figs. 2.9a and 2.9b, in which $\zeta_c \equiv \zeta_p = \zeta_z > 0$. The maximum phase lead or lag, denoted by ϕ_{max}, is obtained at $\omega_c = \sqrt{\omega_z \omega_p}$, as follows:

$$\phi_{max} = \cos^{-1}\left[\frac{(2\zeta_c\sqrt{\omega_p/\omega_z})^2 - (\omega_p/\omega_z - 1)^2}{(2\zeta_c\sqrt{\omega_p/\omega_z})^2 + (\omega_p/\omega_z - 1)^2}\right] \qquad (2.31)$$

The gain increase or decrease at high frequencies becomes

$$K_\infty = 40\log_{10}(\omega_p/\omega_z) \quad \text{dB} \qquad (2.32)$$

For a small ζ_c, i.e., a case with the filter poles and zeros near the imaginary axis, the effective phase-lead or phase-lag region lies between ω_z and ω_p and the maximum phase shift approaches ± 180 deg. For $\zeta_c = 1$, a conventional double-lead or double-lag filter with poles and zeros on the real axis can be realized. At $\omega_c = \sqrt{\omega_z \omega_p}$, one-half of K_∞ is increased (lead filter) or attenuated (lag filter). In practice, ζ_p is often selected to be greater than ζ_z, i.e., the filter poles are placed sufficiently far to the left from the imaginary axis. For example, the control system of the OSO-8 spacecraft [12] employed a phase-lead filter with $\zeta_p = 0.6$, $\zeta_z = 0.3$, and $\omega_p/\omega_z = 2$. The use of a phase-lead filter with a large ω_p/ω_z ratio greater than two should be avoided from a practical viewpoint.

2.3.2.2   Minimum-phase Notch or Bandpass Filter

For $\omega_p = \omega_z$, a notch (band-reject) or bandpass filter is obtained as shown in Figs. 2.10a and 2.10b. The minimum or maximum gain of the filter is obtained at $\omega_c \equiv \omega_p = \omega_z$ as

$$K_{max} = 20\log_{10}(\zeta_z/\zeta_p) \quad \text{dB} \qquad (2.33)$$

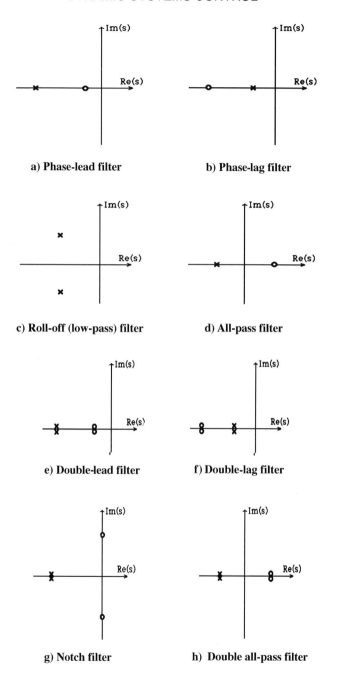

Fig. 2.8   Conventional phase-lead, phase-lag, low-pass, all-pass, and notch filters.

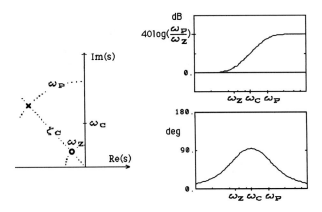

Fig. 2.9a   Minimum-phase lead filter.

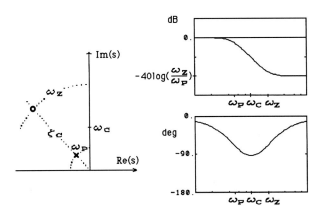

Fig. 2.9b   Minimum-phase lag filter.

Both phase lead and lag appear near ω_c. For the notch filter, the maximum-phase lag and lead occur at ω_1 and ω_2, respectively, where

$$\omega_1/\omega_c = \sqrt{2\zeta_z\zeta_p + 1 - \sqrt{(2\zeta_z\zeta_p + 1)^2 - 1}}$$

$$\omega_2/\omega_c = \sqrt{2\zeta_z\zeta_p + 1 + \sqrt{(2\zeta_z\zeta_p + 1)^2 - 1}}$$

(2.34)

Because ω_1/ω_c and ω_2/ω_c depend only on $\zeta_z\zeta_p$, the filter damping ratios determine the effective notch region. Typical values for the damping ratios of a notch filter are $\zeta_p = 1$ and $\zeta_z = 0$.

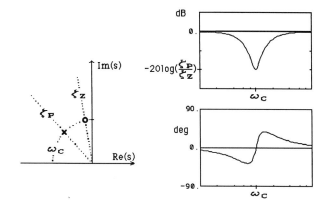

Fig. 2.10a   Minimum-phase notch filter.

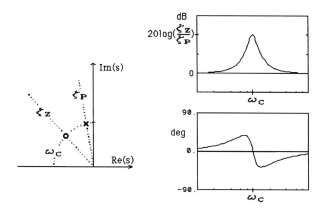

Fig. 2.10b   Minimum-phase bandpass filter.

2.3.2.3   Nonminimum-phase all-pass Filter   For $\omega_c \equiv \omega_p = \omega_z$ and $\zeta_c \equiv \zeta_p = |\zeta_z|$ ($\zeta_z < 0$), a large phase lag can be obtained from the second-order filter, while the gain is being held constant, as shown in Fig. 2.11 as follows:

$$\phi(\omega) = \cos^{-1}\left\{\frac{\left[1 - (\omega/\omega_c)^2\right]^2 - [2\zeta_c(\omega/\omega_c)]^2}{\left[1 - (\omega/\omega_c)^2\right]^2 + [2\zeta_c(\omega/\omega_c)]^2}\right\} \qquad (2.35)$$

The phase varies from 0 deg to −360 deg. The slope of the phase change depends on ζ_c; a smaller ζ_c results in a steeper slope. A typical value for ζ_c might be between 0.3 and 0.7 depending on the specific application. This nonminimum-phase filter is useful for stabilizing the unstably interacting, flexible modes that may need a 180-deg phase change (lead or lag). This filter, when used for flexible-mode

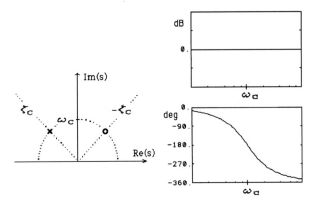

Fig. 2.11 Nonminimum-phase all-pass filter.

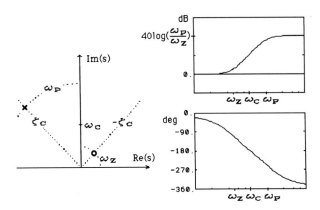

Fig. 2.12a Nonminimum-phase high-pass filter.

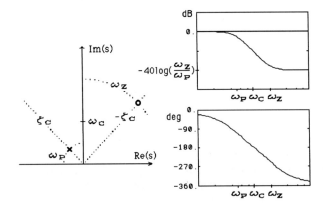

Fig. 2.12b Nonminimum-phase low-pass filter.

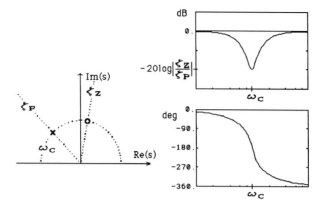

Fig. 2.13a   Nonminimum-phase notch filter.

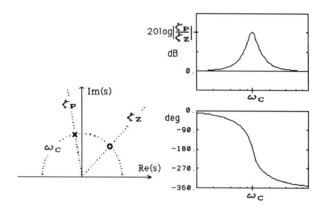

Fig. 2.13b   Nonminimum-phase bandpass filter.

stabilization, maintains the control loop gain at all frequency ranges and provides the proper phasing of the particular signals necessary to increase the closed-loop damping ratio of the flexible modes.

2.3.2.4   Nonminimum-phase High-pass Filter   For $\omega_p > \omega_z$ and $\zeta_c \equiv \zeta_p = |\zeta_z|(\zeta_z < 0)$, a nonminimum-phase filter with a high-pass characteristic (a gain increase) is realized, as shown in Fig. 2.12a. The phase lag is continuous from 0 to -360 deg, but the phase curve shows a flat region near -180 deg that is accompanied by a gain increase. By adjusting the ω_p/ω_z ratio, a broad regional -180-deg phase change can be achieved. This filter, as opposed to the previous nonminimum-phase all-pass filter, provides a more robust phase shift to a flexible mode that needs a 180-deg phase change. Special care must be taken, however, regarding the gain increase at a higher frequency region.

2.3.2.5   Nonminimum-phase Low-pass Filter   As shown in Fig. 2.12b, the nonminimum-phase low-pass filter has a different gain curve compared with the nonminimum-phase high-pass filter, even though the phase curve is very similar to that of the nonminimum-phase high-pass filter. If a less active damping is allowed, this filter can be used for stabilizing the unstably interacting, flexible mode because it provides gain attenuation with a proper phase shift.

2.3.2.6 Nonminimum-phase Notch Filter

For $\omega_p = \omega_z$ and $\zeta_p > |\zeta_z|$ ($\zeta_z < 0$), a nonminimum-phase notch filter is realized, as shown in Fig. 2.13a. If passive damping (in an ideal case), does not exist, the conventional notch filter shown in Fig. 2.10a cannot be used for the stabilization of unstably interacting, flexible modes. On the other hand, this nonminimum-phase notch filter provides the desired phase shift along with sharp gain attenuation at a particular frequency. As the pole–zero pair of this nonminimum-phase notch filter is placed farther from the imaginary axis, the robustness of the closed-loop system is enhanced, while the nominal stability margins become smaller.

2.3.2.7 Nonminimum-phase Bandpass Filter

As shown in Fig. 2.13b, this filter has a gain characteristic that is the opposite of that of the nonminimum-phase notch filter. This filter can be used for a regional gain increase with a proper phasing to increase the active damping, but the overall stability margin may be decreased if the filter poles are placed too close to the imaginary axis.

2.3.3 Classical Compensator Synthesis

An intuitively meaningful procedure for designing a compensator, using the concept of generalized structural filtering, is now presented.

The basic idea of this approach is to synthesize a compensator for each mode, one by one. This successive-mode-stabilization method of synthesizing a compensator provides physical insight into how the design can become more robust and practical. This trial-and-error approach is difficult to use, especially for high-order systems, but the proper use of the basic filter patterns shown in Figs. 2.9–2.13 will result in a straightforward design with a few iterations. In particular, the use of the nonminimum-phase filtering concept significantly enhances the classical approach; however, the usefulness of this design procedure depends on the familiarity of the designer with the classical control techniques.

The design of a SISO feedback control system for a flexible structure can be carried out starting with the stabilization of the rigid-body mode and subsequent analysis and stabilization of unstably interacting flexible modes. Feedback control with a noncollocated actuator and sensor pair generally results in the presence of unstably interacting flexible modes. After the unstably interacting modes have been identified, proper filtering to phase or gain stabilize those modes is then introduced. Aided by the root locus method and/or Bode plots, as well as a certain amount of trial and error, a robust compensator design can be obtained.

A procedure for SISO compensator synthesis for a system with a rigid-body mode and many flexible modes is now summarized as follows.

2.3.3.1 Step 1: Control Bandwidth Selection

The control loop bandwidth is one of the key parameters in control design. Selection of the control bandwidth depends on many factors including performance, noise sensitivity, limited control authority, etc. The control bandwidth is closely related to the settling time, which is determined primarily by the closed-loop poles of the rigid-body mode. In many cases, it is specified a priori.

2.3.3.2 Step 2: Rigid-Body Mode Compensation
Control of the rigid-body mode is simply achieved with proportional plus derivative (PD) type feedback. In practice, a phase-lead filter or PD control with a rolloff filter is employed. That is, neglecting the flexible modes, determine the position and rate gains of an ideal PD control logic represented as $u = -K(1 + Ts)y$ or $u = -(K_P + K_D s)y$, to achieve the desired control bandwidth. If direct rate feedback is not possible, synthesize a first-order phase-lead compensator, $u = -K(T_1 s + 1)/(T_2 s + 1)y$, with a lead ratio not much greater than 10. The selection of this ratio depends on the sensor noise, dominant flexible-mode frequency, and unmodeled high-frequency modes. The rigid-body mode should have reasonable damping for a satisfactory settling time, because the overall transient response is often dominated by the rigid-body mode.

2.3.3.3 Step 3: Flexible Mode Compensation
Examine the closed-loop stability of each flexible mode after closing the loop with the controller designed in step 2. The types and degrees of the closed-loop behavior of each flexible mode depends on the actuator/sensor location and the relative spectral separation of each mode. Examine whether gain stabilization of all of the flexible modes is possible or not. If not, then determine the necessary phase lead or lag angles for each destabilized mode. Synthesize an appropriate structural filter for each destabilized mode, one by one, using the various second-order filters shown in Figs. 2.9–2.13. In this step, some skill and intuition in the classical direct frequency-shaping approaches are needed, which may be the most significant shortcoming of the successive-mode-stabilization approach.

2.3.3.4 Step 4: Design Iteration
Repeat the design process to compromise some interactions between each compensation. A few iterations using computer software packages will result in a quick and straightforward design with physical insight. When a trial design is completed, then perform the closed-loop stability analysis to ensure that the design has adequate robustness to a specified or assumed range of parameter variations. Checking the closed-loop stability for all possible situations is not a trivial problem. The closed-loop stability may be checked by a uniform increase or decrease of each flexible mode frequency. This provides a simple verification of the effect of stiffening or softening the structure on the overall closed-loop stability.

Next, the classical gain-phase stabilization approach enhanced by the concept of generalized second-order structural filtering is applied to a flexible structure control problem.

Example 2.2

Consider the problem of controlling a two-mass–spring system, as illustrated in Fig. 2.14. A control force u acts on m_1, and the position and velocity of m_2 are directly measured for feedback control. It is assumed that for the nominal system,

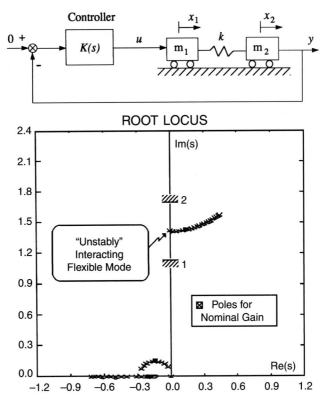

Fig. 2.14   **Two-mass–spring system and root locus vs overall loop gain of a PD controller** $u = -0.086\,(1 + s/0.15)\,y$ **where** $y = x_2$.

$m_1 = m_2 = 1$ and $k = 1$ with appropriate units and time measured in seconds. For simplicity, only the spring constant k is assumed to be uncertain. The stability robustness of the control loop to the mass uncertainty is to be measured by the control-loop gain margin. A transfer function description of the system is given by

$$\frac{y}{u} = \frac{k}{s^2(s^2 + 2k)} \tag{2.36}$$

The closed-loop poles of the rigid-body mode, with the natural frequency (bandwidth) of $\omega_n = 0.2$ rad/s and the closed-loop damping ratio of $\zeta = 0.7$, are assumed to be specified. As a first attempt, a PD controller can be selected as $u = -0.086(1 + s/0.15)y$. It can be seen from Fig. 2.14 that the flexible mode becomes unstable due to the unstable interaction between the flexible mode and the rigid-body control logic.

To properly stabilize the destabilized flexible mode, an approximately ±180-deg phase shift is needed at the flexible mode frequency. The first approach to solving this problem is to provide a 180-deg phase lead at the flexible mode frequency by using the minimum-phase lead filter shown in Fig. 2.9a. In this case, the filter zeros with a frequency lower than the flexible mode frequency are placed near the imaginary axis. The filter poles associated with the zeros are then usually placed sufficiently far to the left of the imaginary axis. A similar design based on this approach can be found in various places where such phase-lead filtering is misleadingly called notch filtering [13]. This approach with phase-lead compensation may not be acceptable if the ω_p/ω_z ratio is chosen to be too large, which would amplify any measurement noise intolerably. A typical value for this ratio would be about two and some compromise between performance and noise sensitivity should be made in selecting this ratio.

The second approach is to employ the conventional notch filter shown in Fig. 2.10a or the nonminimum-phase notch filter shown in Fig. 2.13a. The conventional notch filtering gain stabilizes the flexible mode without adding any active damping to the system. If more active structural damping is required, or if there is no natural passive damping in the flexible mode, the conventional notch filtering is not an appropriate solution. The passive vibration suppression for a case with no natural damping requires the use of nonminimum-phase notch filtering, which provides the proper gain and phase adjustments at the flexible mode frequency.

The third approach is to employ the nonminimum-phase all-pass filter, shown in Fig. 2.11, which maintains the control loop gain and provides the proper phase lag of the flexible mode signals, resulting in an increased closed-loop damping ratio of the flexible mode, i.e., active damping. The filter poles and zeros are selected as $\omega_p = \omega_z = \sqrt{2k}$ ($k = 1$ for nominal case) and $\zeta_p = -\zeta_z = 0.5$. Fig. 2.15 shows the root locus vs overall loop gain of the PD controller with this nonminimum-phase all-pass filter.

For this particular design, the gain margin is 5 dB, and the rigid-body and flexible modes have phase margins of 37 and 64 deg, respectively. The flexible mode has a closed-loop damping ratio of 0.1 and the rigid-body mode has a 0.7 damping ratio.

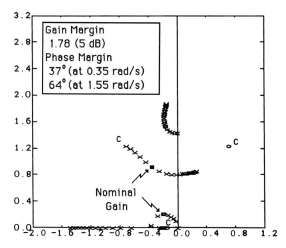

Fig. 2.15   Root locus vs overall loop gain of a controller $u = -0.086(1 + s/0.15)y$ with nonminimum-phase all-pass filtering.

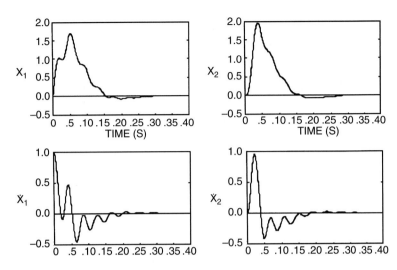

Fig. 2.16   Closed-loop time responses to an impulse disturbance at m_1.

It can be verified that the closed-loop system is stable for $0.5 < k < 2.1$. Responses of the nominal system to an impulse disturbance at m_1 are shown in Fig. 2.16; it can be seen that the closed-loop system has a settling time of about 20 s.

Problems

2.10   The preceding example design has a control bandwidth of $\omega_n \approx 0.2$ rad/s. Assuming that only the position of m_2 is measured, design a higher-bandwidth controller with $\omega_n \approx 0.5$ rad/s. The closed-loop system must have a gain margin >3 dB and a phase margin >25 deg, and it must be stable for $\pm 25\%$ variations of k.

2.11   Consider the problem of stabilizing a high-order system with many closely spaced modal frequencies using a paired noncollocated actuator and sensor, as illustrated in Fig. 2.17. It is assumed that for the nominal system, $m_1 = m_2 = m_3 = m_4 = 1$ and $k = 4$. For simplicity, only the spring constant k is assumed to be uncertain. The natural frequencies of the flexible modes are $\omega_1^2 = (2 - \sqrt{2})k$, $\omega_2^2 = 2k$, and $\omega_3^2 = (2 + \sqrt{2})k$. It is also assumed that each dashpot has a damping coefficient of 0.004. The flexible modes then have nominal passive damping ratios of $\zeta_1 = 0.0008$, $\zeta_2 = 0.0014$, and $\zeta_3 = 0.0018$.

For this eighth-order flexible structure model, the locus of the closed-loop poles vs the overall loop gain of a noncollocated PD controller is also shown in Fig. 2.17. As can be seen in this figure, the first and third flexible modes are unstably interacting with the rigid-body mode control, whereas the second mode is stably interacting with the rigid-body mode control. Assuming that only position of m_4 is measured, design a stabilizing feedback compensator with a control bandwidth of $\omega_n = 0.2$ rad/s. The

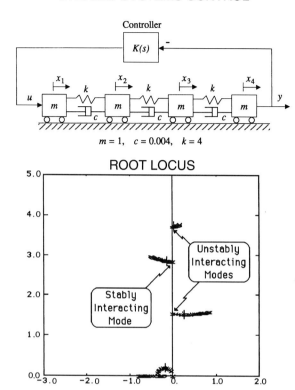

Fig. 2.17   Four-mass–spring system and root locus vs overall loop gain of a PD controller $u = -K(1 + 5s)y$ **where** $y = x_4$.

closed-loop system must have a gain margin >3 dB and a phase margin >25 deg, and it must be stable for $\pm 25\%$ variations of the spring stiffness k from its nominal value of 4.

2.3.4 Persistent Disturbance Rejection

A classical approach to disturbance accommodating control of dynamic systems in the presence of persistent or quasiperiodic disturbances is presented here. The method exploits the so-called *internal model principle* for asymptotic disturbance rejection. The concept of a disturbance rejection dipole is introduced from a classical control viewpoint. The method invariably makes use of disturbance rejection dipoles and nonminimum-phase compensation for a class of noncollocated control problems in the presence of persistent disturbances.

After successful stabilization of the rigid-body mode, as well as any other unstably interacting flexible modes, active disturbance rejection is simply achieved by introducing a model of the disturbance into the feedback loop. A block diagram representation of a persistent disturbance rejection control system is shown in Fig. 2.18a.

a)

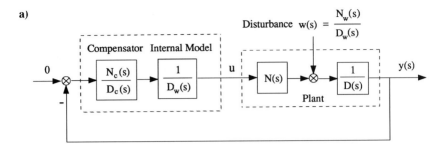

b)

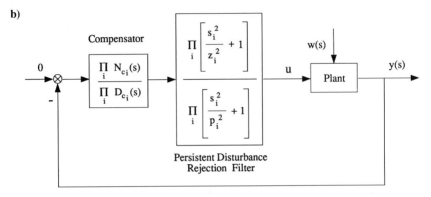

Fig. 2.18   Persistent disturbance rejection control system.

It is assumed that a persistent (or quasi-periodic) disturbance is represented as

$$w(t) = \sum_{i=1}^{n} A_i \sin(2\pi f_i t + \phi_i)$$

with unknown magnitudes A_i and phases ϕ_i, but known frequencies f_i. Note that if, for example, $f_1 = 2f_2 = \cdots = nf_n$, then $w(t)$ becomes a periodic disturbance.

In general, the disturbance $w(t)$ can be described by a Laplace transformation

$$w(s) = \frac{N_w(s)}{D_w(s)}$$

where $N_w(s)$ is arbitrary as long as $w(s)$ remains proper. The roots of $D_w(s)$ correspond to the frequencies at which the persistent excitation takes place. The inclusion of the disturbance model $1/D_w$ inside the control loop is often referred to as the internal modeling of the disturbance. In classical design, the internal disturbance model is regarded as being part of the compensator as shown in Fig. 2.18a. The presence of $1/D_w$ in the control loop results in the effective cancellation of the poles of $w(s)$, provided that no root of $D_w(s)$ is a zero of the plant transfer function. This is shown in the following closed-loop transfer function:

$$y(s) = \frac{1/D(s)}{1 + N_c(s)N(s)/D_c(s)D_w(s)D(s)} w(s)$$

$$= \frac{D_c(s)D_w(s)}{D_w(s)D_c(s)D(s) + N_c(s)N(s)} \frac{N_w(s)}{D_w(s)} \tag{2.37}$$

where we can see the cancellation of $D_w(s)$.

The compensator can be viewed as a series of individual first-order or second-order filters as follows:

$$\frac{N_c(s)}{D_c(s)} = \prod_i \frac{N_{c_i}(s)}{D_{c_i}(s)}$$

Each filter is designed to perform a specific task, like the stabilization of a particular mode. In the same manner, a disturbance rejection filter can be designed that has a proper transfer function and uses the internal disturbance model $1/D_w$. Thus, a proper numerator is chosen in the compensator to go with the disturbance model as shown in Fig. 2.18b. The numerator is chosen to be of the same order as D_w so that there is a zero for each pole of the disturbance model $1/D_w$.

Although the asymptotic disturbance rejection based on the internal model principle has been well known, an interesting interpretation of the concept from a classical control viewpoint is presented here. Each pole–zero combination of the disturbance rejection filter

$$\prod_i \frac{s^2/\omega_{z_i}^2 + 2\zeta_{z_i}s/\omega_{z_i} + 1}{s^2/\omega_{p_i}^2 + 1}$$

can be called a *dipole*, where ζ_{z_i} is included for generality. The filter thus consists of as many dipoles as there are frequency components in the persistent disturbance. The separation between the zero and the pole is generally referred to as the strength of the dipole. The strength of the dipole affects the settling time of the closed-loop system; in general, the larger the separation between the pole and zero of the filter the shorter the settling time is. This is caused by the position of the closed-loop eigenvalue corresponding to the filter dipole. As the strength of the dipole is increased, this eigenvalue is pushed farther to the left, speeding up the response time of the disturbance rejection. This separation influences the gain-phase characteristics of the system, however, because the dipole causes a certain amount of gain-phase changes in its neighborhood. Moreover, at frequencies higher than the dipole there is a net gain increase or reduction. The magnitude of this gain increases with the separation between pole and zero. Therefore, as the strength of the dipole is changed to meet a chosen settling time the compensation must be readjusted. A compromise has to be reached often between the settling time and the stability of the compensated system.

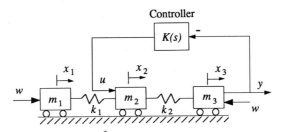

Fig. 2.19 Noncollocated control of a three-mass–spring system.

Problem

2.12 Consider the two-mass–spring system discussed earlier. The nominal values of the system parameters are $m_1 = m_2 = 1$ and $k = 1$ with appropriate units. A control force u acts on m_1 and the position of m_2 is measured as y, resulting in a noncollocated control problem. Assume that a sinusoidal disturbance of $w(t) = \sin 0.5t$, with unknown magnitude and phase, is exerted on m_1 and/or m_2 and that asymptotic disturbance rejection for the position of m_2 with a settling time of about 20 s is to be achieved. Design a controller for this problem using a classical approach and perform computer simulations of the closed-loop system.
Note: Later in this chapter, a controller of this problem will be designed using modern state-space approaches.

2.13 Consider the three-mass–spring system illustrated in Fig. 2.19. The nominal values of the system parameters are $m_1 = m_2 = m_3 = 1$ and $k_1 = k_2 = 1$ with appropriate units. A control force u acts on m_2 and the position of m_3 is measured as y, resulting in a noncollocated control problem. Assume that a sinusoidal disturbance of $w(t) = \sin 0.5t$, with unknown magnitude and phase, is acting on m_1 and m_3 simultaneously, as illustrated in Fig. 2.19 and that asymptotic disturbance rejection for the position of m_3 with a settling time of about 20 s is to be achieved. Design a controller $K(s)$ for this problem.
Note: This control problem of a three-mass–spring system represents the essential nature of the control redesign problem of the Hubble Space Telescope for reducing the effects of solar array vibrations on the telescope pointing jitter, which is to be studied in Chapter 9.

2.4 Digital Control

2.4.1 Discrete-Time Systems

A system described by a difference equation is called a *discrete-time system* or *sampled-data system*. A signal within a discrete-time system is described by a number sequence. In digital control systems in which a digital computer is used as a controller, these number sequences are obtained by sampling a continuous-time or analog signal. An analog-to-digital (A/D) converter converts the analog signal into a binary form that a digital computer can process. A digital-to-analog (D/A) converter then changes the binary signal out of the computer to an analog signal to drive the actuator or the plant.

The signal into the computer with the sampling period of T is described by the number sequence

$$y(0), \ y(T), \ y(2T), \ldots$$

which can be expressed by the notation $\{y(kT)\}$, or simply by $\{y(k)\}$. A digital control logic, also called a digital compensator or filter, can then be represented by a difference equation of the form

$$\begin{aligned}
u(k) &= a_0 y(k) + a_1 y(k-1) + \cdots + a_n y(k-n) \\
&\quad - b_1 u(k-1) - b_2 u(k-2) - \cdots - b_n u(k-n)
\end{aligned} \tag{2.38}$$

where $y(k)$ denotes the signal into the computer at the kth sampling and $u(k)$ denotes the signal from the computer at the kth sampling.

2.4.2   The z Transformation

The z domain is used for a sampled-data or discrete-time system in much the same way that the s domain is used for a continuous-time system. The z and s domains are simply related as follows:

$$z = e^{Ts} \qquad \text{or} \qquad s = (1/T)\ell n \, z \tag{2.39}$$

where T is the sampling period. The z transform of a number sequence $\{y(k)\}$ is defined as a power series in z^{-k} with coefficients equal to the values of $y(k)$; i.e., we have

$$y(z) \equiv \mathcal{Z}[\{y(k)\}] = \sum_{k=0}^{\infty} y(k) \, z^{-k} = y(0) + y(1) \, z^{-1} + y(2) \, z^{-2} + \cdots \tag{2.40}$$

Some interesting properties of the z transform are summarized as follows.

2.4.2.1   Translation Theorem

$$\mathcal{Z}[\{y(k-n)\}] = \sum_{k=0}^{\infty} y(k-n) \, z^{-k} = z^{-n} y(z) \tag{2.41}$$

because $y(k) = 0$ for $k < 0$. We also have

$$\mathcal{Z}[\{y(k+n)\}] = \sum_{k=0}^{\infty} y(k+n) \, z^{-k} = z^n \left[y(z) - \sum_{k=0}^{n-1} y(k) z^{-k} \right] \tag{2.42}$$

2.4.2.2 Initial Value Theorem If $y(k)$ has the z transform $y(z)$ and $\lim y(z)$ exists as $z \to \infty$, then the initial value $y(0)$ can be found as

$$y(0) = \lim_{z \to \infty} y(z) \tag{2.43}$$

2.4.2.3 Final Value Theorem If $y(k)$ has the z transform $y(z)$, then we have

$$\lim_{k \to \infty} y(k) = \lim_{z \to 1} [(z-1)y(z)] \tag{2.44}$$

provided that the left-side limit exists.

Using Eq. (2.41), we obtain the z-domain transfer function representation of the difference equation (2.38), as follows:

$$\frac{u(z)}{y(z)} = \frac{a_0 + a_1 z^{-1} + \cdots + a_n z^{-n}}{1 + b_1 z^{-1} + b_2 z^{-2} + \cdots + b_n z^{-n}} \tag{2.45}$$

In digital signal processing, digital filters with both poles and zeros are called infinite impulse response (IIR) filters, whereas digital filters with only zeros are called finite impulse response (FIR) filters. The output of a FIR filter becomes exactly zero in a finite amount of time after the input is removed. The FIR filters are also called moving average filters, whereas the IIR filters are often called auto regressive moving average (ARMA) filters.

2.4.3 Sampling and Zero-Order Hold

In digital control systems, the number sequences $\{y(kT)\}$ are obtained by sampling a continuous-time signal. As shown in Fig. 2.20, the output of an ideal sampler with a sampling period T is a train of impulses. One of the most commonly used data holds is the zero-order hold (ZOH). The output of the ZOH is a staircase approximation of the input signal $y(t)$. In Fig. 2.20, the sampled signal is denoted as $y^*(t)$ and the output signal of the ZOH is denoted as $\hat{y}(t)$. The corresponding Laplace transformed variables are denoted as $y^*(s)$ and $\hat{y}(s)$, respectively.

The Laplace transform of $\hat{y}(t)$ can be found as

$$\hat{y}(s) = y(0) \left\{ \frac{1}{s} - \frac{e^{-Ts}}{s} \right\} + y(T) \left\{ \frac{e^{-Ts}}{s} - \frac{e^{-2Ts}}{s} \right\} + \cdots$$

$$= \left\{ \frac{1 - e^{-Ts}}{s} \right\} \sum_{k=0}^{\infty} y(kT) e^{-kTs} \tag{2.46}$$

Defining

$$y^*(s) = \sum_{k=0}^{\infty} y(kT) e^{-kTs} \tag{2.47}$$

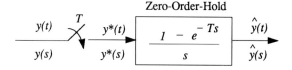

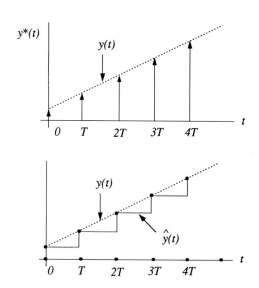

Fig. 2.20   Sampling and ZOH.

we have

$$\hat{y}(s) = \frac{1 - e^{-Ts}}{s} y^*(s) \qquad (2.48)$$

Because the number sequence, $\{y(kT)\} = y^*(t)$, is obtained by sampling the signal $y(t)$ every T s, we have

$$y(z) \equiv \mathcal{Z}[\{y(k)\}] = \sum_{k=0}^{\infty} y(kT) z^{-k}$$

and

$$y^*(s) = y(z)|_{z=e^{Ts}} \qquad (2.49)$$

The sampled signal $y^*(t)$ is a train of impulse functions whose weights are equal to the values of the signal at the instants of sampling; i.e., we have

$$y^*(t) = \sum_{k=0}^{\infty} y(kT)\delta(t - kT) \qquad (2.50)$$

where $\delta(t)$ denotes the unit impulse function occurring at $t = 0$. The sampled signal $y^*(t)$ can also be expressed as

$$y^*(t) = y(t) \sum_{k=-\infty}^{\infty} \delta(t - kT) = y(t) \sum_{n=-\infty}^{\infty} C_n \exp(j2\pi nt/T) \qquad (2.51)$$

where C_n is the Fourier coefficient given by

$$C_n = \frac{1}{T} \int_0^T \sum_{k=0}^{\infty} \delta(t - kT) \exp(-j2\pi nt/T)\, dt = \frac{1}{T}$$

Consequently, we have

$$y^*(t) = \frac{1}{T} \sum_{n=-\infty}^{\infty} y(t) \exp(j\omega_s nt) \qquad (2.52a)$$

$$y^*(s) = \mathcal{L}[y^*(t)] = \frac{1}{T} \sum_{n=-\infty}^{\infty} y(s - j\omega_s n) \qquad (2.52b)$$

$$\approx \frac{1}{T} y(s) \qquad \text{for} \qquad 0 \le \omega < \frac{\omega_s}{2} \qquad (2.52c)$$

where $y(s) = \mathcal{L}[y(t)]$, $\omega_s = 2\pi/T$ is called the sampling frequency and $\omega_s/2$ is called the Nyquist frequency. Note that an ideal sampler has an effective scale factor of $1/T$.

The ideal sampler and the ZOH can then be approximated as

$$\hat{y}(s) \approx \frac{1 - e^{-Ts}}{Ts} y(s) \qquad (2.53)$$

For the classical s-domain analysis and synthesis of a digital control system, the sampler and ZOH can be further approximated as

$$\frac{1 - e^{-Ts}}{Ts} \approx \frac{1}{(Ts)^2/12 + Ts/2 + 1} \qquad (2.54)$$

Similarly, a pure computational time delay $e^{-T_d s}$ inherent to any digital control systems can be approximated as

$$e^{-T_d s} \approx \frac{(T_d s)^2/12 - T_d s/2 + 1}{(T_d s)^2/12 + T_d s/2 + 1} \qquad (2.55)$$

2.4.4 Digital Proportional-Integral-Derivative Controller

Consider a continuous-time PID controller represented as

$$u(t) = -K_P y(t) - K_I \int y(t)\, dt - K_D \dot{y}(t)$$

Using Euler's approximation of differentiation

$$s \approx \frac{1 - z^{-1}}{T} = \frac{z - 1}{Tz} \qquad (2.56)$$

we obtain an equivalent digital PID controller represented in z-domain transfer function form as

$$u = -\left\{ K_P + K_I \frac{T}{1 - z^{-1}} + K_D \frac{1 - z^{-1}}{T} \right\} y \qquad (2.57)$$

This digital PID control logic can be implemented in a computer as follows:

$$u(k) = -K_P y(k) - K_I \hat{u}(k) - K_D \frac{y(k) - y(k-1)}{T} \qquad (2.58)$$

where

$$\hat{u}(k) = \hat{u}(k-1) + Ty(k)$$

A single-axis block diagram representation of a digital control system of the Hubble Space Telescope is shown in Fig. 2.21. As can be seen in this figure, the baseline digital control system of the Hubble Space Telescope, with a sampling period $T = 0.025$ s and a computational delay of $T_d = 0.008$ s, is in fact a digital PID controller with an FIR filter in the rate loop. A control redesign problem of the Hubble Space Telescope will be treated in Chapter 9.

2.4.5 Discretization of Continuous-Time Systems

2.4.5.1 Bilinear or Tustin Transformation
For the digital PID control logic, represented as Eq. (2.57), we simply used Euler's method to approximate the continuous-time integration or differentiation. Euler's approximation described by Eq. (2.56) is a simple method for converting functions of s to functions of z. If we need a more accurate digital integration or differentiation, however, we may use the bilinear or Tustin transformation

$$s \approx \frac{2(1 - z^{-1})}{T(1 + z^{-1})} = \frac{2(z - 1)}{T(z + 1)} \qquad (2.59)$$

which is based on the Taylor series approximation for e^{Ts} given as

$$z = e^{Ts} = \frac{e^{Ts/2}}{e^{-Ts/2}} \approx \frac{1 + Ts/2}{1 - Ts/2}$$

If the gain and phase of both functions in s domain and z domain are desired to be identical at a specified frequency, then the following bilinear transformation with prewarping can also be used

$$s \approx \frac{\omega_0}{\tan(\omega_0 T / 2)} \frac{1 - z^{-1}}{1 + z^{-1}} \qquad (2.60)$$

where ω_0 is the prewarping frequency.

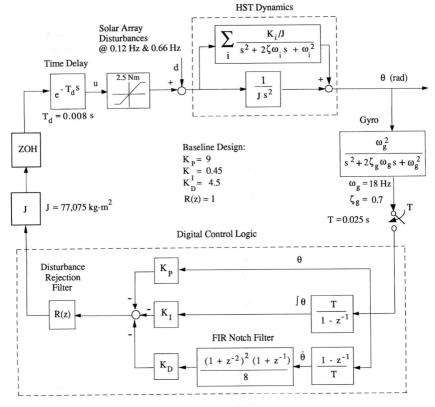

Fig. 2.21 Simplified block diagram of the pitch-axis pointing control system of the Hubble Space Telescope [14].

2.4.5.2 Pole–zero Mapping In addition to the bilinear transformation, the pole–zero mapping method can be employed to convert functions of s to functions of z. Given an s-domain function that is a ratio of two polynomials, $G(s) = N(s)/D(s)$, this method substitutes $z = e^{Ts}$ in the zeros of $N(s)$ and $D(s)$ to obtain $N(z)$ and $D(z)$. The gain of the digital filter is selected such that

$$G(s)|_{s=0} = G(z)|_{z=1}$$

and the zeros of $G(s)$ at $s = \infty$ are mapped to the point $z = -1$ of the digital filter $G(z)$.

For example, consider a transfer function of the form

$$G(s) = \frac{a}{s+a} \frac{\omega_n^2}{s^2 + 2\zeta\omega_n s + \omega_n^2}$$

Using the pole–zero mapping method, we obtain the equivalent z-domain transfer function, as follows:

$$G(z) = \frac{(1 - e^{-aT})(z + 1)}{2(z - e^{-aT})} \frac{(1 - 2e^{\zeta \omega_n T} \cos \omega_d + e^{-2\zeta \omega_n T})z^2}{z^2 - (2e^{\zeta \omega_n T} \cos \omega_d)z + e^{-2\zeta \omega_n T}}$$

where $\omega_d = \omega_n \sqrt{1 - \zeta^2}$. A transfer function $G(s)$ with zeros can also be similarly transformed to the z-domain transfer function using the pole–zero mapping method as was illustrated.

2.4.5.3 State Transition Matrix Consider a linear time-invariant, continuous-time system described by

$$\dot{x} = Ax + Bu \tag{2.61}$$

where x is the state vector and u the control input vector. This continuous-time equation is discretized to a discrete-time equation as follows. Assume that the input $u(t)$ is constant over the sampling interval between any two consecutive sampling instants; i.e., we assume that $u(t) = u(kT)$ for the kth sampling period.

The solution of Eq. (2.61) is, in general, given as

$$x(t) = e^{At}x(0) + \int_0^t e^{A(t-\tau)}Bu(\tau)\, d\tau \tag{2.62}$$

where the matrix e^{At} is called the *state transition matrix* and $x(0)$ denotes the state vector at $t = 0$. Therefore, we have the following discrete-time equation for Eq. (2.61):

$$x(k + 1) = e^{AT}x(k) + \left[\int_0^T e^{At}\, dt\right]Bu(k) \tag{2.63}$$

Problems

2.14 Given the continuous-time state equation

$$\begin{bmatrix} \dot{x}_1 \\ \dot{x}_2 \end{bmatrix} = \begin{bmatrix} 0 & 1 \\ 0 & 0 \end{bmatrix} \begin{bmatrix} x_1 \\ x_2 \end{bmatrix} + \begin{bmatrix} 0 \\ 1 \end{bmatrix} u^-$$

show that the equivalent discrete-time equation for $u(t) = u(kT)$ for the kth sampling period can be found as

$$\begin{bmatrix} x_1(k + 1) \\ x_2(k + 1) \end{bmatrix} = \begin{bmatrix} 1 & T \\ 0 & 1 \end{bmatrix} \begin{bmatrix} x_1(k) \\ x_2(k) \end{bmatrix} + \begin{bmatrix} T^2/2 \\ T \end{bmatrix} u(k)$$

2.15 Given the continuous-time state equation

$$\begin{bmatrix} \dot{x}_1 \\ \dot{x}_2 \end{bmatrix} = \begin{bmatrix} 0 & 1 \\ -1 & 0 \end{bmatrix} \begin{bmatrix} x_1 \\ x_2 \end{bmatrix} + \begin{bmatrix} 0 \\ 1 \end{bmatrix} u$$

show that the equivalent discrete-time equation for $u(t) = u(kT)$ for the kth sampling period can be found as

$$\begin{bmatrix} x_1(k+1) \\ x_2(k+1) \end{bmatrix} = \begin{bmatrix} \cos T & \sin T \\ -\sin T & \cos T \end{bmatrix} \begin{bmatrix} x_1(k) \\ x_2(k) \end{bmatrix} + \begin{bmatrix} 1 - \cos T \\ \sin T \end{bmatrix} u(k)$$

2.16 Consider a preliminary control design problem of the X-ray Timing Explorer (XTE) spacecraft, in which a slow sampling rate and an additional computational delay of a digital control system cause an unstable control-structure interaction for the spacecraft even with collocated actuator and sensor. A single-axis block diagram representation of a digital control system of the XTE is shown in Fig. 2.22. The nominal value of the spacecraft inertia is $J = 2690$ kg·m^2, and the nominal values of the modal frequencies and modal gains are given in Table 2.1.

The control design specifications are $\omega_n = 0.5$ rad/s (0.08 Hz), $\zeta = 0.707$, and $\tau = 10$ s where ω_n and ζ are the natural frequency (bandwidth) and damping ratio of the closed-loop poles associated with the rigid-body mode, respectively, and τ is the time constant of the integral control. The stability margin requirements are gain margin >6 dB and phase margin >30 deg. The control system is also required to maintain a pointing accuracy of 15 arcsec with respect to a disturbance $d(t)$ induced by periodic angular motions of payload instruments with a period of 32 s.

Design a digital PID controller of the form shown in Fig. 2.22. Also perform computer simulation of the closed-loop system, in particular, with the disturbance shown in Fig. 2.22.

Hint: One of the following continuous-time filters may be utilized for designing a digital filter $D(z)$:

$$C_1(s) = \frac{1}{0.25s^2 + 0.4s + 1}, \qquad C_2(s) = \frac{0.0625s^2 + 0.1s + 1}{0.0256s^2 + 0.08s + 1}$$

$$C_3(s) = \frac{0.04s^2 + 0.002s + 1}{0.04s^2 + 0.4s + 1}, \qquad C_4(s) = \frac{0.0494s^2 - 0.2222s + 1}{0.0494s^2 + 0.2222s + 1}$$

Table 2.1   Modal frequencies and gains of XTE spacecraft[a]

Modes	ω_i, rad/s	K_i
1	5.0	0.4168
2	21.1	0.0901
3	26.3	0.1684
4	32.3	0.0831
5	93.3	0.0991

[a]It is assumed that $\zeta_i = 0.001$ for all of the flexible modes.

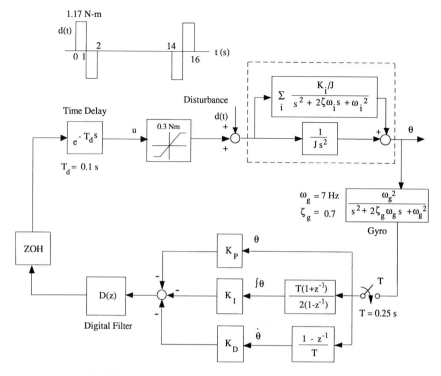

Fig. 2.22   Digital control design for the XTE spacecraft.

Other direct digital design techniques [15] may also be employed for this problem. Instead of using one of the preceding structural filters, an additional time delay may also be utilized to stabilize the closed-loop system.

2.5   Modern State-Space Methods

In this section we introduce modern state-space approaches to linear systems control. This somewhat mathematical subject is also of practical importance for a wide range of dynamics and control problems of space vehicles.

2.5.1   Mathematical Preliminaries

Some mathematical facts from advanced linear algebra are briefly summarized here. Familiarity with these concepts will be useful in studying advanced control techniques for complex dynamic systems.

2.5.1.1   Set   If x is a member of the set X, we write $x \in X$. A set X is a subset of the set Y if every element of X is also an element of Y. In this case, we write $X \subset Y$. The set of all $m \times n$ matrices with coefficients in the field of real numbers

$t_0 \le t \le t_1$ suffices to determine the state $\mathbf{x}(t_0)$. Otherwise, the system is said to be unobservable at time t_0.

A dynamic system described by Eqs. (2.99) is said to be observable by the following equivalent conditions is satisfied:

1) All columns of a matrix Ce^{At} are linearly independent or all columns of $C[s\mathbf{I} - \mathbf{A}]^{-1}$ are linearly independent.

2) The observability grammian

$$\int_0^t e^{A^T \tau} \mathbf{C}^T \mathbf{C} e^{A\tau} \, d\tau$$

is nonsingular for any $t > 0$.

3) The $nq \times n$ observability matrix

$$\begin{bmatrix} \mathbf{C} \\ \mathbf{CA} \\ \vdots \\ \mathbf{CA}^{n-1} \end{bmatrix}$$

has rank n.

4) For every eigenvalue λ of $\mathbf{A}$, the $(n+p) \times n$ complex matrix

$$\begin{bmatrix} \lambda \mathbf{I} - \mathbf{A} \\ \mathbf{C} \end{bmatrix}$$

has rank n. (This condition implies that $[\lambda \mathbf{I} - \mathbf{A}]$ and $\mathbf{C}$ are right coprime.)

2.5.2.3 Pole–Zero Cancellation

If a dynamic system is controllable, all of the modes of the system can be excited by the control input $\mathbf{u}(t)$; if a dynamic system is observable, all of the modes can be observed by the output $\mathbf{y}(t)$. Consequently, a pole–zero cancellation in a transfer function $y(s)/u(s)$ indicates, in general, a lack of controllability by u and/or observability by y.

Example 2.3

Consider a cart with an inverted pendulum, as was illustrated in Fig. 1.9a. The linearized equations of motion are

$$M\ddot{z} + mg\theta = u$$
$$m\ddot{z} + m\ell\ddot{\theta} - mg\theta = 0$$

Let z and θ be the two outputs of the system; then the transfer functions from u to z and θ can be found, respectively, as

$$\frac{z(s)}{u(s)} = \frac{s^2 - g/\ell}{s^2[Ms^2 - (M+m)g/\ell]}$$

$$\frac{\theta(s)}{u(s)} = \frac{-s^2/\ell}{s^2[Ms^2 - (M+m)g/\ell]}$$

There are pole–zero cancellations of s^2 in the transfer function from u to θ, which indicate, in general, a lack of controllability by u and/or observability by θ. The system is controllable by u and observable by z, however, because there are no pole–zero cancellations in $z(s)/u(s)$. Consequently, the pole–zero cancellations in $\theta(s)/u(s)$ indicate that the system is unobservable by θ. Also we notice that the rigid-body mode of the system cannot be observed by measuring the pendulum deflection.

Problems

2.19 Consider a cart of mass M with two inverted pendulums of lengths ℓ_1, ℓ_2 and tip masses m_1, m_2, as illustrated in Fig. 2.23. Let z be the horizontal distance of the cart, θ_1 and θ_2 be the angles of the pendulums measured from the vertical position, u the control input force acting on the cart, and g the gravitational acceleration.

(a) Derive the linearized equations of motion for small angles of θ_1 and θ_2 as

$$M\ddot{z} + m_1 g\theta_1 + m_2 g\theta_2 = u$$
$$m_1\ddot{z} + m_1\ell_1\ddot{\theta}_1 - m_1 g\theta_1 = 0$$
$$m_2\ddot{z} + m_2\ell_2\ddot{\theta}_2 - m_2 g\theta_2 = 0$$

(b) Show that the system with $\ell_1 \neq \ell_2$ is controllable by u and observable by z.

(c) Show that the system with $\ell_1 = \ell_2$ is uncontrollable by u and unobservable by z, even when $m_1 \neq m_2$.

(d) Show that the system is unobservable by θ_1 and/or θ_2 for any values of m_i and ℓ_i.

2.20 Consider a three-mass–spring system ($m_1 = m_2 = m_3 = 1$; $\ k_1 = k_2 = 1$)

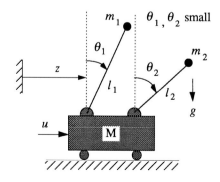

Fig. 2.23   Cart with two inverted pendulums.

described by

$$\ddot{y}_1 + y_1 - y_2 = u_1$$
$$\ddot{y}_2 + 2y_2 - y_1 - y_3 = u_2$$
$$\ddot{y}_3 + y_3 - y_2 = u_3$$

(a) Is the system controllable by u_1?
(b) Is the system controllable by u_2?
(c) Is the system controllable by u_3?
(d) Is the system observable by y_1?
(e) Is the system observable by y_2?
(f) Is the system observable by y_3?

2.21 Consider the translational motion of a satellite in circular orbit described by

$$\ddot{y}_1 - 2n\dot{y}_2 - 3n^2 y_1 = u_1$$
$$\ddot{y}_2 + 2n\dot{y}_1 = u_2$$
$$\ddot{y}_3 + n^2 y_3 = u_3$$

where (y_1, y_2, y_3) are small perturbations from a nominal circular orbit with an orbital rate of n, and (u_1, u_2, u_3) are the control inputs.
(a) Is the system controllable by u_1 and u_3?
(b) Is the system controllable by u_2 and u_3?
(c) Is the system observable by y_1 and y_3?
(d) Is the system observable by y_2 and y_3?

2.5.3 State Feedback and State Estimation

2.5.3.1 State Feedback Control Consider a linear time-invariant dynamic system described by

$$\dot{\mathbf{x}} = \mathbf{A}\mathbf{x} + \mathbf{B}\mathbf{u} \qquad (2.100)$$

where $\mathbf{x} \in R^n$ and $\mathbf{u} \in R^p$. In state feedback control, the state vector is multiplied by a gain matrix $\mathbf{K}$ and fed back into the control input; that is,

$$\mathbf{u} = -\mathbf{K}\mathbf{x} \qquad (2.101)$$

where $\mathbf{K} \in R^{p \times n}$. The closed-loop system is then described by

$$\dot{\mathbf{x}} = (\mathbf{A} - \mathbf{B}\mathbf{K})\mathbf{x} \qquad (2.102)$$

and the closed-loop characteristic equation becomes

$$|s\mathbf{I} - \mathbf{A} + \mathbf{B}\mathbf{K}| = 0$$

If the system is controllable, the eigenvalues of the closed-loop system can be arbitrarily assigned, provided that the complex conjugate eigenvalues appear in pairs.

2.5.3.2 State Estimator Consider a linear time-invariant dynamic system described by

$$\dot{\mathbf{x}} = \mathbf{A}\mathbf{x} + \mathbf{B}\mathbf{u} \tag{2.103a}$$

$$\mathbf{y} = \mathbf{C}\mathbf{x} \tag{2.103b}$$

where $\mathbf{x} \in R^n$, $\mathbf{u} \in R^p$, and $\mathbf{y} \in R^q$. Because all state variables are not directly measured by the output $\mathbf{y}$, we consider an *asymptotic state estimator* (or observer) of the form:

$$\dot{\hat{\mathbf{x}}} = \mathbf{A}\hat{\mathbf{x}} + \mathbf{B}\mathbf{u} + \mathbf{L}(\mathbf{y} - \mathbf{C}\hat{\mathbf{x}})$$

$$= (\mathbf{A} - \mathbf{L}\mathbf{C})\hat{\mathbf{x}} + \mathbf{B}\mathbf{u} + \mathbf{L}\mathbf{y} \tag{2.104}$$

where $\hat{\mathbf{x}}$ denotes an estimate of $\mathbf{x}$, and $\mathbf{L}$ denotes an $n \times q$ gain matrix. Note that the estimator is driven by the input $\mathbf{u}(t)$ as well as the output $\mathbf{y}(t)$ of the system. The difference of $\mathbf{y}$ and $\hat{\mathbf{y}} = \mathbf{C}\hat{\mathbf{x}}$ is used as a correction term in Eq. (2.104).

Let $\mathbf{e}$ be the error between the actual state and the estimated state; that is,

$$\mathbf{e} = \mathbf{x} - \hat{\mathbf{x}} \tag{2.105}$$

Then, we have the following estimate-error equation:

$$\dot{\mathbf{e}} = (\mathbf{A} - \mathbf{L}\mathbf{C})\mathbf{e} \tag{2.106}$$

and the estimate-error characteristic equation is obtained as

$$|s\mathbf{I} - \mathbf{A} + \mathbf{L}\mathbf{C}| = 0$$

If the system is state observable by $\mathbf{y}$, then all the eigenvalues of $(\mathbf{A} - \mathbf{L}\mathbf{C})$ can be arbitrarily assigned and the behavior of the estimate error $\mathbf{e}$ can be arbitrarily controlled.

2.5.3.3 Estimated-State Feedback Controller If the actual state is not available for state feedback control, we use the estimated state $\hat{\mathbf{x}}$ for feedback as follows:

$$\mathbf{u} = -\mathbf{K}\hat{\mathbf{x}} \tag{2.107}$$

Then the overall closed-loop system with the estimated-state feedback controller is described by the state equation

$$\dot{\mathbf{x}} = \mathbf{A}\mathbf{x} + \mathbf{B}\mathbf{u}$$

the output equation

$$\mathbf{y} = \mathbf{C}\mathbf{x}$$

the regulator

$$\mathbf{u} = -\mathbf{K}\hat{\mathbf{x}}$$

and the estimator

$$\dot{\hat{\mathbf{x}}} = \mathbf{A}\hat{\mathbf{x}} + \mathbf{B}\mathbf{u} + \mathbf{L}(\mathbf{y} - \mathbf{C}\hat{\mathbf{x}})$$

where $\mathbf{K}$ is called the regulator gain matrix and $\mathbf{L}$ the estimator gain matrix.
The composite dynamic system can then be described as

$$\begin{bmatrix} \dot{\mathbf{x}} \\ \dot{\hat{\mathbf{x}}} \end{bmatrix} = \begin{bmatrix} \mathbf{A} & -\mathbf{BK} \\ \mathbf{LC} & \mathbf{A} - \mathbf{BK} - \mathbf{LC} \end{bmatrix} \begin{bmatrix} \mathbf{x} \\ \hat{\mathbf{x}} \end{bmatrix} \qquad (2.108)$$

which can be rewritten in terms of $\mathbf{x}$ and $\mathbf{e}$, as follows:

$$\begin{bmatrix} \dot{\mathbf{x}} \\ \dot{\mathbf{e}} \end{bmatrix} = \begin{bmatrix} \mathbf{A} - \mathbf{BK} & \mathbf{BK} \\ 0 & \mathbf{A} - \mathbf{LC} \end{bmatrix} \begin{bmatrix} \mathbf{x} \\ \mathbf{e} \end{bmatrix} \qquad (2.109)$$

The closed-loop characteristic equation can then be found as

$$\begin{vmatrix} s\mathbf{I} - \mathbf{A} + \mathbf{BK} & -\mathbf{BK} \\ 0 & s\mathbf{I} - \mathbf{A} + \mathbf{LC} \end{vmatrix} = 0 \qquad (2.110)$$

which becomes

$$|s\mathbf{I} - \mathbf{A} + \mathbf{BK}| \; |s\mathbf{I} - \mathbf{A} + \mathbf{LC}| = 0 \qquad (2.111)$$

Therefore, the closed-loop characteristic equation can be decomposed as

$$|s\mathbf{I} - \mathbf{A} + \mathbf{BK}| = 0 \quad \text{and} \quad |s\mathbf{I} - \mathbf{A} + \mathbf{LC}| = 0 \qquad (2.112)$$

This result shows that there is no difference in estimated-state or the actual-state feedback, as far as the closed-loop eigenvalues are concerned. Consequently, the design of state feedback controller (or regulator) and the design of state estimator (or observer) can be carried out independently. This property is often called the *separation property*. We often choose the eigenvalues of the estimator to be two or three times faster than the eigenvalues of the state feedback controller.

The state feedback controller and the estimator can be combined to yield a *compensator*, as follows:

$$\mathbf{u}(s) = \underbrace{-\mathbf{K}[s\mathbf{I} - \mathbf{A} + \mathbf{BK} + \mathbf{LC}]^{-1}\mathbf{L}}_{\text{Compensator}} \mathbf{y}(s) \qquad (2.113)$$

It is emphasized that the compensator poles are different from the regulator and estimate-error eigenvalues.

2.5.3.4 Pole-Placement Method for Selecting K and L   The pole-placement met- hod basically allows the designer to directly choose the closed-loop regulator and estimator eigenvalues to meet desired criteria.

For simplicity, consider a SISO system described by the state-space equation

$$\dot{\mathbf{x}} = \mathbf{A}\mathbf{x} + \mathbf{B}u$$

$$y = \mathbf{C}\mathbf{x}$$

The desired closed-loop regulator characteristic equation is assumed to be given by

$$s^n + \hat{a}_1 s^{n-1} + \cdots + \hat{a}_{n-1}s + \hat{a}_n = 0 \qquad (2.114)$$

and the closed-loop regulator characteristic equation is given by

$$|s\mathbf{I} - \mathbf{A} + \mathbf{BK}| = 0 \qquad (2.115)$$

Thus, matching the coefficients of the characteristic polynomials in Eq. (2.114) and Eq. (2.115), we can determine the state feedback (regulator) gain matrix $\mathbf{K}$.

Similarly, the estimator gain matrix $\mathbf{L}$ can also be determined by matching the coefficients of the characteristic polynomial $|s\mathbf{I} - \mathbf{A} + \mathbf{LC}|$ with the coefficients of the desired estimate-error characteristic polynomial.

In general, for a single-input system, we can employ the Bass–Gura method that computes the gain matrix $\mathbf{K}$ as

$$\mathbf{K} = [\hat{\mathbf{a}} - \mathbf{a}][\mathcal{CT}]^{-1} \qquad (2.116)$$

where $\mathbf{a} = [a_1, \ldots, a_n]$ is a row vector containing the coefficients of the characteristic polynomial of $\mathbf{A}$: $s^n + a_1 s^{n-1} + \cdots + a_{n-1}s + a_n$; $\hat{\mathbf{a}} = [\hat{a}_1, \ldots, \hat{a}_n]$ is a row vector containing the coefficients of the desired characteristic polynomial of $\hat{\mathbf{A}} = \mathbf{A} - \mathbf{BK}$; $\mathcal{T}$ is an upper-triangular Toeplitz matrix defined as

$$\mathcal{T} = \begin{bmatrix} 1 & a_1 & a_2 & \cdots & a_{n-1} \\ 0 & 1 & a_1 & \cdots & a_{n-2} \\ 0 & 0 & 1 & \cdots & a_{n-3} \\ \vdots & \vdots & \vdots & \cdots & \vdots \\ 0 & 0 & 0 & \cdots & 1 \end{bmatrix}$$

and $\mathcal{C}$ is the controllability matrix defined as

$$\mathcal{C} = [\mathbf{B} \quad \mathbf{AB} \quad \mathbf{A}^2\mathbf{B} \quad \cdots \quad \mathbf{A}^{n-1}\mathbf{B}]$$

The estimator gain matrix $\mathbf{L}$ can also be determined using duality, as follows:

$$\mathbf{A} \rightarrow \mathbf{A}^T, \qquad \mathbf{B} \rightarrow \mathbf{C}^T, \qquad \mathbf{K} \rightarrow \mathbf{L}^T$$

and

$$\mathbf{L} = [\mathcal{T}^T \mathcal{O}]^{-1}[\breve{\mathbf{a}} - \mathbf{a}]^T \qquad (2.117)$$

where $\mathbf{a}$ and $\breve{\mathbf{a}}$ are row vectors containing the coefficients of the characteristic polynomials of $\mathbf{A}$ and $\breve{\mathbf{A}} = \mathbf{A} - \mathbf{LC}$, respectively; and $\mathcal{O}$ is the observability matrix defined as

$$\mathcal{O} = \begin{bmatrix} \mathbf{C} \\ \mathbf{CA} \\ \vdots \\ \mathbf{CA}^{n-1} \end{bmatrix}$$

Although the closed-loop eigenvalues can be arbitrarily chosen, not all selections result in good designs. Although the settling time of the compensated system depends on the real part of the regulator closed-loop eigenvalues, the real part of these eigenvalues cannot be arbitrarily large because a faster decay means a larger input signal. Also, this influence increases with the frequency of the eigenvalue, so that faster decay of high-frequency modes means even more control input effort.

For a more general, multi-input multi-output (MIMO) system, other pole placement methods with robust eigenstructure assignment can be employed for selecting the gain matrices $\mathbf{K}$ and $\mathbf{L}$; see, e.g., Junkins and Kim [16].

2.5.3.5 Linear Quadratic Regulator/Linear Quadratic Estimator Method for Selecting K and L
The linear-quadratic-regulator (LQR) and linear-quadratic-estimator (LQE) methods are now briefly introduced for the selection of $\mathbf{K}$ and $\mathbf{L}$. The combined LQR/LQE method is also referred to as the linear-quadratic-Gaussian (LQG) design method in the literature, and the resulting controller is called the LQG compensator. This LQR/LQE method is directly applicable to MIMO systems.

The gain matrix $\mathbf{K}$ of the state feedback control logic $\mathbf{u} = -\mathbf{Kx}$ can be determined by minimizing the linear quadratic performance index

$$ J = \frac{1}{2} \int_0^\infty (\mathbf{x}^T\mathbf{Qx} + \mathbf{u}^T\mathbf{Ru})\, dt \qquad (2.118) $$

where $\mathbf{Q}$ is the state weighting matrix and $\mathbf{R}$ is the control input weighting matrix. The gain matrix $\mathbf{K}$ is then obtained as

$$ \mathbf{K} = \mathbf{R}^{-1}\mathbf{B}^T\mathbf{X} \qquad (2.119) $$

by solving the algebraic Riccati equation

$$ 0 = \mathbf{A}^T\mathbf{X} + \mathbf{XA} - \mathbf{XBR}^{-1}\mathbf{B}^T\mathbf{X} + \mathbf{Q} \qquad (2.120) $$

Certain conditions must be met for a unique positive-definite solution to the above Riccati equation to exist: 1) $\mathbf{Q}$ must be symmetric and positive semidefinite, i.e., $\mathbf{Q} = \mathbf{Q}^T \geq 0$; 2) $\mathbf{R}$ must be symmetric positive definite, i.e., $\mathbf{R} = \mathbf{R}^T > 0$; 3) the $(\mathbf{A}, \mathbf{B})$ pair must be controllable (stabilizable); and 4) the $(\mathbf{A}, \mathbf{H})$ pair must be observable (detectable) where $\mathbf{H}^T\mathbf{H} = \mathbf{Q}$ and rank $\mathbf{H}$ = rank $\mathbf{Q}$. For further details of LQR control theory and applications, the reader is referred to Refs. 16–18.

To determine the estimator gain matrix $\mathbf{L}$ using the LQE method, we consider a plant described by the following state-space equation:

$$ \dot{\mathbf{x}} = \mathbf{Ax} + \mathbf{Bu} + \mathbf{Gw} \qquad (2.121a) $$

$$ \mathbf{y} = \mathbf{Cx} + \mathbf{v} \qquad (2.121b) $$

where $\mathbf{w}$ is the process noise and $\mathbf{v}$ is the measurement noise. Both $\mathbf{w}$ and $\mathbf{v}$ are assumed to be white noise processes with

$$ E[\mathbf{w}(t)\mathbf{w}^T(\tau)] = \mathbf{W}\delta(t - \tau) $$
$$ E[\mathbf{v}(t)\mathbf{v}^T(\tau)] = \mathbf{V}\delta(t - \tau) $$

where $\mathbf{W}$ and $\mathbf{V}$ are the corresponding spectral density matrices [17].

The gain matrix $\mathbf{L}$ of the LQE is then selected such that the observation error

$$\mathbf{e} = \mathbf{x} - \hat{\mathbf{x}}$$

is minimized in the presence of noise, by solving the algebraic Riccati equation

$$0 = \mathbf{AY} + \mathbf{YA}^T - \mathbf{YC}^T\mathbf{V}^{-1}\mathbf{CY} + \mathbf{GWG}^T \tag{2.122}$$

where $\mathbf{Y}$ is the estimate-error covariance matrix, and $\mathbf{L}$ is computed as

$$\mathbf{L} = \mathbf{YC}^T\mathbf{V}^{-1} \tag{2.123}$$

A more detailed treatment of LQG control theory and applications to aerospace dynamic systems can be found in Bryson and Ho [17], and Bryson [18].

Problems

2.22 Consider the LQR problem for a linear time-invariant system

$$\dot{\mathbf{x}} = \mathbf{Ax} + \mathbf{Bu}$$

with a quadratic performance index

$$J = \frac{1}{2}\int_0^\infty (\mathbf{x}^T\mathbf{Qx} + \mathbf{u}^T\mathbf{Ru})\, dt$$

The optimal LQR solution is given as

$$\mathbf{u}(t) = -\mathbf{R}^{-1}\mathbf{B}^T\boldsymbol{\lambda}(t)$$

where $\boldsymbol{\lambda}$ satisfies the Euler–Lagrange equations

$$\begin{bmatrix} \dot{\mathbf{x}} \\ \dot{\boldsymbol{\lambda}} \end{bmatrix} = \begin{bmatrix} \mathbf{A} & -\mathbf{BR}^{-1}\mathbf{B}^T \\ -\mathbf{Q} & -\mathbf{A}^T \end{bmatrix}\begin{bmatrix} \mathbf{x} \\ \boldsymbol{\lambda} \end{bmatrix}$$

given $\mathbf{x}(0)$ and $\boldsymbol{\lambda}(\infty) = 0$.

(a) By letting $\boldsymbol{\lambda}(t) = \mathbf{Xx}(t)$, derive the following matrix Riccati equation:

$$0 = \mathbf{A}^T\mathbf{X} + \mathbf{XA} - \mathbf{XBR}^{-1}\mathbf{B}^T\mathbf{X} + \mathbf{Q}$$

(b) Defining the so-called Hamiltonian matrix $\mathbf{H}$ as

$$\mathbf{H} = \begin{bmatrix} \mathbf{A} & -\mathbf{BR}^{-1}\mathbf{B}^T \\ -\mathbf{Q} & -\mathbf{A}^T \end{bmatrix}$$

and also defining

$$\mathbf{E} = \begin{bmatrix} \mathbf{I} & 0 \\ \mathbf{X} & \mathbf{I} \end{bmatrix}$$

where $\mathbf{I}$ is the identity matrix, show that

$$\mathbf{E}^{-1}\mathbf{H}\mathbf{E} = \begin{bmatrix} \mathbf{A} - \mathbf{B}\mathbf{K} & -\mathbf{B}\mathbf{R}^{-1}\mathbf{B}^T \\ 0 & -(\mathbf{A} - \mathbf{B}\mathbf{K})^T \end{bmatrix}$$

where $\mathbf{K}$ is the LQR gain matrix defined as $\mathbf{K} = \mathbf{R}^{-1}\mathbf{B}^T\mathbf{X}$ such that $\mathbf{u} = -\mathbf{K}\mathbf{x}$.

(c) Show that $\det[s\mathbf{I} - \mathbf{E}^{-1}\mathbf{H}\mathbf{E}] = \det[s\mathbf{I} - \mathbf{H}]$ where s is the Laplace transform variable.
Hint: $|\mathbf{A}| = |\mathbf{A}^T|$ and $|\mathbf{A}^{-1}| = 1/|\mathbf{A}|$.

(d) Using the result from (c), show that if s_i is an eigenvalue of $\mathbf{H}$, $-s_i$ is also an eigenvalue of $\mathbf{H}$.
Hint: The eigenvalues of $\mathbf{A}$ are the same as the eigenvalues of $\mathbf{A}^T$.

(e) Show that the Hamiltonian matrix $\mathbf{H}$ satisfies $\mathbf{J}^{-1}\mathbf{H}^T\mathbf{J} = -\mathbf{H}$ where

$$\mathbf{J} = \begin{bmatrix} 0 & \mathbf{I} \\ -\mathbf{I} & 0 \end{bmatrix}$$

Hint: $\mathbf{J}^{-1} = \mathbf{J}^T = -\mathbf{J}$.

(f) Using the result from (e), show that if s_i is an eigenvalue of $\mathbf{H}$, $-s_i$ is also an eigenvalue of $\mathbf{H}$.

(g) Show that the state transition matrix, $\boldsymbol{\Phi}$, of the Euler–Lagrange equations satisfies the *sympletic property:* $\mathbf{J}^{-1}\boldsymbol{\Phi}^T\mathbf{J} = \boldsymbol{\Phi}^{-1}$.
Hint: Differentiate $\mathbf{J}^{-1}\boldsymbol{\Phi}^T\mathbf{J} = \boldsymbol{\Phi}^{-1}$ and use $\dot{\boldsymbol{\Phi}} = \mathbf{H}\boldsymbol{\Phi}$ and $\mathbf{J}^{-1}\mathbf{H}^T\mathbf{J} = -\mathbf{H}$.

(h) Using the result from (g), also show that if s_i is an eigenvalue of $\boldsymbol{\Phi}$, $1/s_i$ is also an eigenvalue of $\boldsymbol{\Phi}$.

(i) Show that $\underline{\sigma}\,[\mathbf{I} + \mathbf{G}(j\omega)] \geq 1$ for all ω where $\mathbf{G}(j\omega) = \mathbf{K}(j\omega\mathbf{I} - \mathbf{A})^{-1}\mathbf{B}$ with $\mathbf{R} = \mathbf{I}$.

(j) Show that the result from (i) implies that

$$\underline{\sigma}\,[\mathbf{I} + \mathbf{G}^{-1}(j\omega)] \geq \tfrac{1}{2}$$

for all ω.

(k) For the single-input LQR design, show that a Nyquist plot of $G(j\omega)$, as ω varies from $-\infty$ to ∞, remains outside a circle of center $-1 + j0$ and radius 1. Also show that an inverse Nyquist plot for $G^{-1}(j\omega)$, as ω varies from $-\infty$ to ∞, always remains outside a circle of center $-1 + j0$ and radius 1/2.
Hint: Use the results from (i) and (j).

(l) Show that the results in (k) imply that the single-input LQR design has -6 dB and $+\infty$ gain margins and a phase margin of at least 60 deg.

2.23 Consider a double integrator plant described by

$$\begin{bmatrix} \dot{x}_1 \\ \dot{x}_2 \end{bmatrix} = \begin{bmatrix} 0 & 1 \\ 0 & 0 \end{bmatrix} \begin{bmatrix} x_1 \\ x_2 \end{bmatrix} + \begin{bmatrix} 0 \\ 1 \end{bmatrix} u$$

$$y = \begin{bmatrix} 1 & 0 \end{bmatrix} \begin{bmatrix} x_1 \\ x_2 \end{bmatrix}$$

(a) Determine the state-feedback gain matrix for $u = -\mathbf{K}x$ such that the closed-loop regulator poles be located at $s = -1 \pm j$.
(b) Determine the estimator gain matrix $\mathbf{L}$ such that the closed-loop estimator poles be located at $s = -2 \pm 2j$.
(c) Find the equivalent compensator from y to u.
(d) Sketch root locus of the closed-loop system vs overall loop gain K_o of the preceding compensator with the nominal gain of $K_o = 1$, and find the gain and phase margins of the closed-loop system.
(e) Also synthesize a compensator for this same problem ($\omega_n \approx 1.4$ rad/s) using the LQR/LQE method by trying a few different sets of weighting matrices: $\mathbf{Q}$, $\mathbf{R}$, $\mathbf{W}$, and $\mathbf{V}$.
Note: Access to computer software such as MATLAB is needed to solve the algebraic Riccati equation.

2.24 Consider a dynamic system described in state-space form as

$$\begin{bmatrix} \dot{x}_1 \\ \dot{x}_2 \end{bmatrix} = \begin{bmatrix} 5 & -3 \\ -4 & 6 \end{bmatrix} \begin{bmatrix} x_1 \\ x_2 \end{bmatrix} + \begin{bmatrix} 1 \\ 0 \end{bmatrix} u$$

$$y = \begin{bmatrix} 1 & 0 \end{bmatrix} \begin{bmatrix} x_1 \\ x_2 \end{bmatrix}$$

or in transfer function as

$$\frac{y}{u} = \frac{s-6}{(s-2)(s-9)}$$

(a) Determine the state-feedback gain matrix for $u = -\mathbf{K}x$ such that the closed-loop regulator poles be located at $s = -2$ and -9.
(b) Determine the estimator gain matrix $\mathbf{L}$ such that the closed-loop estimator poles be located at $s = -10$ and -10.
(c) Obtain the equivalent compensator from y to u as

$$u(s) = -\frac{3630(s-2.436)}{(s-36.35)(s+78.35)} y(s)$$

(d) Plot root locus of the closed-loop system vs overall loop gain K_o of the preceding compensator with the nominal gain of $K_o = 1$. Determine the gain and phase margins of the closed-loop system.
(e) After trying a few different sets of closed-loop eigenvalues, discuss the inherent difficulty of controlling the system.
(f) Also synthesize a compensator for this problem using the LQR/LQE method by trying a few different sets of weighting matrices: $\mathbf{Q}$, $\mathbf{R}$, $\mathbf{W}$, and $\mathbf{V}$. Any concluding remarks?

2.25 Repeat Problem 2.10 using both the pole-placement and LQR/LQE methods. In particular, determine the stability robustness of each controller with respect to the three uncertain parameters m_1, m_2, and k.

2.26 Repeat Problem 2.11 using both the pole-placement and LQR/LQE methods. In particular, determine the stability robustness of each controller with
respect to the three uncertain parameters m, c, and k where $m_i = m$ for all i.

2.5.4 Persistent Disturbance Rejection

The internal model principle for persistent disturbance rejection, which was
considered in Sec. 2.3.4, is now incorporated with the standard state-space control design problem. Active disturbance rejection for the measured output $\mathbf{y}$ is to
be achieved by introducing a model of the disturbance inside the control loop,
therefore using again the concept of internal modeling, as illustrated in Fig. 2.24.

For example, consider a scalar disturbance $d(t)$ with one or more frequency
components represented as

$$d(t) = \sum_i A_i \sin(\omega_i t + \phi_i)$$

with unknown magnitudes A_i and phases ϕ_i but known frequencies ω_i. The
disturbance rejection filter is then described by

$$\dot{\mathbf{x}}_d = \mathbf{A}_d \mathbf{x}_d + \mathbf{B}_d \mathbf{y} \tag{2.124}$$

where $\mathbf{x}_d$ is the state vector introduced by the disturbance model and, for example,

$$\mathbf{A}_d = \begin{bmatrix} 0 & 1 & 0 & 0 \\ -\omega_1^2 & 0 & 0 & 0 \\ 0 & 0 & 0 & 1 \\ 0 & 0 & -\omega_2^2 & 0 \end{bmatrix}, \qquad \mathbf{B}_d = \begin{bmatrix} 0 \\ 1 \\ 0 \\ 1 \end{bmatrix}$$

for a scalar output $y(t)$ with $d(t)$ of two frequency components. The disturbance
rejection filter can include as many frequency components as the given disturbance,

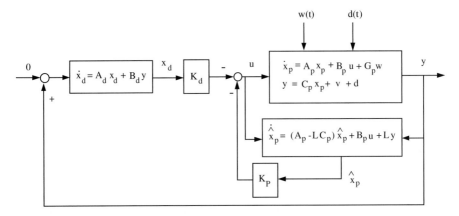

Fig. 2.24   Persistent disturbance rejection control system.

and is driven by the measured output $\mathbf{y}$ of the plant. This procedure is equivalent to the one used in the classical approach with the disturbance model now consisting of a state-space model.

We now consider a plant described by the state-space equation

$$\dot{\mathbf{x}}_p = \mathbf{A}_p\mathbf{x}_p + \mathbf{B}_p\mathbf{u} + \mathbf{G}_p\mathbf{w} \tag{2.125a}$$

$$\mathbf{y} = \mathbf{C}_p\mathbf{x}_p + \mathbf{v} + \mathbf{d} \tag{2.125b}$$

where $\mathbf{x}_p$ denotes the plant's state vector, $\mathbf{u}$ the control input vector, $\mathbf{w}$ the process noise, $\mathbf{v}$ the measurement noise, and $\mathbf{d}$ the output equivalent persistent disturbance. Both $\mathbf{w}$ and $\mathbf{v}$ are assumed to be white noise processes with

$$E[\mathbf{w}(t)\mathbf{w}^T(\tau)] = \mathbf{W}\delta(t-\tau)$$

$$E[\mathbf{v}(t)\mathbf{v}^T(\tau)] = \mathbf{V}\delta(t-\tau)$$

where $\mathbf{W}$ and $\mathbf{V}$ are the corresponding spectral density matrices.

In general, a compensator designed for this plant will consist of a regulator and an estimator that will approximate the states $\mathbf{x}_p$ with estimated states $\hat{\mathbf{x}}_p$ using the information from the measured output $\mathbf{y}$. The estimator that attempts to asymptotically reduce the error term $\mathbf{e} = \mathbf{x}_p - \hat{\mathbf{x}}_p$ is given by

$$\dot{\hat{\mathbf{x}}}_p = \mathbf{A}_p\hat{\mathbf{x}}_p + \mathbf{B}_p\mathbf{u} + \mathbf{L}(\mathbf{y} - \mathbf{C}_p\hat{\mathbf{x}}_p)$$
$$= (\mathbf{A}_p - \mathbf{L}\mathbf{C}_p)\hat{\mathbf{x}}_p + \mathbf{B}_p\mathbf{u} + \mathbf{L}\mathbf{y} \tag{2.126}$$

where the term $(\mathbf{y} - \mathbf{C}_p\hat{\mathbf{x}}_p)$ represents the error between the output of the plant and the estimated output and $\mathbf{L}$ is the estimator gain matrix to be determined.

The disturbance filter model described by Eq. (2.124) is then augmented to a plant described by Eqs. (2.125) as follows:

$$\dot{\mathbf{x}} = \mathbf{A}\mathbf{x} + \mathbf{B}\mathbf{u} + \mathbf{G}\mathbf{w} \tag{2.127a}$$

$$\mathbf{y} = \mathbf{C}\mathbf{x} + \mathbf{v} + \mathbf{d} \tag{2.127b}$$

where

$$\mathbf{x} = \begin{bmatrix} \mathbf{x}_p \\ \mathbf{x}_d \end{bmatrix}, \qquad \mathbf{A} = \begin{bmatrix} \mathbf{A}_p & 0 \\ \mathbf{B}_d\mathbf{C}_p & \mathbf{A}_d \end{bmatrix}, \qquad \mathbf{B} = \begin{bmatrix} \mathbf{B}_p \\ 0 \end{bmatrix}$$

$$\mathbf{C} = [\mathbf{C}_p \quad 0], \qquad \mathbf{G} = \begin{bmatrix} \mathbf{G}_p \\ 0 \end{bmatrix}$$

An estimated state feedback controller is then given as

$$\mathbf{u} = -\mathbf{K}\hat{\mathbf{x}}$$

where $\hat{\mathbf{x}} = \begin{bmatrix} \hat{\mathbf{x}}_p^T & \hat{\mathbf{x}}_d^T \end{bmatrix}^T$ and the gain matrix $\mathbf{K} = \begin{bmatrix} \mathbf{K}_p & \mathbf{K}_d \end{bmatrix}$ is to be determined for the augmented system described by Eqs. (2.127).

As shown in Fig. 2.24, however, $\mathbf{x}_d$ can be directly fed back as

$$\mathbf{u} = -[\mathbf{K}_p \quad \mathbf{K}_d]\begin{bmatrix} \hat{\mathbf{x}}_p \\ \mathbf{x}_d \end{bmatrix} \qquad (2.128)$$

because $\mathbf{x}_d$ is directly available from Eq. (2.124).

An active disturbance rejection controller in state-space form is then given by

$$\begin{bmatrix} \dot{\hat{\mathbf{x}}}_p \\ \dot{\mathbf{x}}_d \end{bmatrix} = \begin{bmatrix} \mathbf{A}_p - \mathbf{B}_p\mathbf{K}_p - \mathbf{L}\mathbf{C}_p & -\mathbf{B}_p\mathbf{K}_d \\ 0 & \mathbf{A}_d \end{bmatrix}\begin{bmatrix} \hat{\mathbf{x}}_p \\ \mathbf{x}_d \end{bmatrix} + \begin{bmatrix} \mathbf{L} \\ \mathbf{B}_d \end{bmatrix}\mathbf{y} \qquad (2.129a)$$

$$\mathbf{u} = -[\mathbf{K}_p \quad \mathbf{K}_d]\begin{bmatrix} \hat{\mathbf{x}}_p \\ \mathbf{x}_d \end{bmatrix} \qquad (2.129b)$$

And the closed-loop system with $\mathbf{w} = \mathbf{v} = \mathbf{d} = 0$ is described as

$$\begin{bmatrix} \dot{\mathbf{x}}_p \\ \dot{\hat{\mathbf{x}}}_p \\ \dot{\mathbf{x}}_d \end{bmatrix} = \begin{bmatrix} \mathbf{A}_p & -\mathbf{B}_p\mathbf{K}_p & -\mathbf{B}_p\mathbf{K}_d \\ \mathbf{L}\mathbf{C}_p & \mathbf{A}_p - \mathbf{B}_p\mathbf{K}_p - \mathbf{L}\mathbf{C}_p & -\mathbf{B}_p\mathbf{K}_d \\ \mathbf{B}_d\mathbf{C}_p & 0 & \mathbf{A}_d \end{bmatrix}\begin{bmatrix} \mathbf{x}_p \\ \hat{\mathbf{x}}_p \\ \mathbf{x}_d \end{bmatrix}$$

which can be modified using the error term $\mathbf{e} = \mathbf{x}_p - \hat{\mathbf{x}}_p$, resulting in a partially decoupled system of equations, as follows:

$$\begin{bmatrix} \dot{\mathbf{x}}_p \\ \dot{\mathbf{x}}_d \\ \dot{\mathbf{e}} \end{bmatrix} = \begin{bmatrix} \mathbf{A}_p - \mathbf{B}_p\mathbf{K}_p & -\mathbf{B}_p\mathbf{K}_d & \mathbf{B}_p\mathbf{K}_p \\ \mathbf{B}_d\mathbf{C}_p & \mathbf{A}_d & 0 \\ 0 & 0 & \mathbf{A}_p - \mathbf{L}\mathbf{C}_p \end{bmatrix}\begin{bmatrix} \mathbf{x}_p \\ \mathbf{x}_d \\ \mathbf{e} \end{bmatrix}$$

The closed-loop characteristic equation can then be written as

$$\begin{vmatrix} s\mathbf{I} - \mathbf{A}_p + \mathbf{B}_p\mathbf{K}_p & \mathbf{B}_p\mathbf{K}_d & -\mathbf{B}_p\mathbf{K}_p \\ -\mathbf{B}_d\mathbf{C}_p & s\mathbf{I} - \mathbf{A}_d & 0 \\ 0 & 0 & s\mathbf{I} - \mathbf{A}_p + \mathbf{L}\mathbf{C}_p \end{vmatrix} = 0 \qquad (2.130)$$

The determinant in Eq. (2.130) is equal to the determinants of the diagonal submatrices multiplied together, one giving the regulator eigenvalues for the augmented system including the internal model and the other giving the estimator eigenvalues for only the plant. Hence, we have shown that the separation principle for regulator and estimator holds for a closed-loop system even with an internal model for asymptotic disturbance rejection.

Example 2.4

Consider again the two-mass–spring problem considered in Problem 2.12 for the illustration of the state-space approaches to disturbance rejection control of a flexible spacecraft [10]. First, the LQR/LQE method is applied to the problem, followed by the pole-placement method.

After trial and error using an LQR/LQE design code, an LQG compensator can be found as

$$u(s) = \frac{-0.355[(s/0.2375) + 1][(s/0.496)^2 + 2(0.204)(s/0.496) + 1]}{[(s/2.545)^2 + 2(0.204)(s/2.545) + 1][(s/0.5)^2 + 1]}$$

$$\times \frac{[(s/1.41)^2 - 2(0.0546)(s/1.41) + 1]}{[(s/2.873)^2 + 2(0.859)(s/2.873) + 1]} y(s) \qquad (2.131)$$

It can be seen that for persistent disturbance rejection, the compensator has poles at $\pm 0.5j$ with the associated zeros near $\pm 0.5j$. Such a pole–zero pair is called a disturbance rejection filter dipole.

The closed-loop system with this compensator has a relatively small gain margin of 1.8 dB, and is stable only for $0.9 < k < 1.16$. The standard LQR/LQE control design is necessarily tuned closely to the plant model for high performance; hence it is not robust to plant parameter uncertainty. The responses to $w_2 = \sin 0.5t$ show that the transient peak is, however, very small compared to the responses of other classical control designs with nonminimum-phase zeros. It is clear that for the LQR/LQE design, high performance (small transient peak and fast settling time) has been achieved at the expense of a small stability robustness margin with respect to parameter uncertainty. Hence, some tradeoffs between performance and parameter robustness must be considered in practical control design.

The regulator gain matrix $\mathbf{K}$ can also be determined for the augmented system, including the internal model, for given desired closed-loop eigenvalues. The estimator gain matrix $\mathbf{L}$ is computed using only the plant system matrix, because only the plant states are to be estimated. The regulator eigenvalues are tentatively chosen to be similar to the closed-loop eigenvalues resulting from a classical design as follows:

$$-0.2 \pm 0.2j, \quad -0.1 \pm 0.4j, \quad -0.5 \pm 1.45j$$

The estimator eigenvalues (only for the plant) are then chosen to be twice as far as the regulator eigenvalues. The resulting compensator is

$$u(s) = \frac{-0.0354[(s/0.0942) + 1][-(s/0.544) + 1][(s/4.617) + 1]}{[(s/0.5)^2 + 1][(s/1.672)^2 + 2(0.815)(s/1.672) + 1]}$$

$$\times \frac{[(s/0.467)^2 - 2(0.073)(s/0.467) + 1]}{[(s/2.849)^2 + 2(0.29)(s/2.849) + 1]} y(s)$$

It can be seen that for asymptotic disturbance rejection, the compensator has poles at $\pm 0.5j$ with the associated zeros near them.

Again, the rigid-body mode is stabilized by a PD-type compensator with a second-order rolloff filter and the flexible mode is stabilized with a nonminimum-phase filter. Also, note that the pole-placement design also introduces a dipole for disturbance rejection. It also introduces a real, positive zero to go with the complex pair of poles for the stabilization of the unstably interacting flexible mode. The placement of the eigenvalue corresponding to the disturbance rejection filter determines the location of this zero. The real part of the eigenvalue influences the settling time whereas the imaginary part affects the magnitude of the response. The smaller the imaginary component of the eigenvalue the larger the magnitude of the overshoot in the response of the system, as well as in the control input signal. The settling time determined by the real component of the eigenvalue is not altered by the overshoot.

The closed-loop system is stable for $0.57 \leq k \leq 3.55$, and has a 2.48-dB gain margin. The controller designed here using the pole-placement technique has a parameter robustness margin larger than that of the LQR/LQE design, but it has very large transient peak. It is again evident that some tradeoffs between performance and robustness are needed even for this simple example problem.

Problem

2.27   For the preceding example problem, perform a standard LQR/LQE control design so that the closed-loop system has a gain margin >3 dB and a phase margin >30 deg. Perform computer simulation of the closed-loop system to verify whether or not asymptotic disturbance rejection for the position of m_2 with a settling time of about 20 has been achieved. Also determine the stability robustness of the closed-loop system with respect to the three uncertain parameters m_1, m_2, and k.

2.5.5   Classical vs Modern Control Issues

State-space approaches to control design are currently emphasized in the literature and more widely explored than classical methods. This arises from the convenience of obtaining a compensator for the whole system given one set of design parameters, e.g., given $\mathbf{Q}$, $\mathbf{R}$, $\mathbf{W}$, and $\mathbf{V}$, or desired closed-loop eigenvalues. In classical design, on the other hand, a compensator must be constructed piece by piece, or mode by mode. However, both classical and state-space methods have their drawbacks as well as advantages. All these methods require, nevertheless, a certain amount of trial and error.

As discussed in this section, both state-space techniques (pole-placement and LQR/LQE) introduce nonminimum-phase filtering of the unstably interacting flexible mode of the two-mass–spring problem. The LQR/LQE technique offers an optimal compensator design in the presence of random disturbances given certain weighting parameters for the states and the control inputs, and certain parameters describing the random disturbances. The question remains of how to choose these parameters and what choice provides the best optimal design. The designer must find an acceptable set of parameters for a good optimal design. The use of state-space methods for control design usually results in a compensator of the same order as the system to be controlled. This means that for systems having several flexible modes, the compensator adds compensation even to modes that are stable and need no compensation. This may result in a complicated compensator design.

The classical design is particularly convenient for the control of dynamic systems with well-separated modes. The concept of nonminimum-phase compensation also provides an extremely convenient way of stabilizing unstably interacting flexible modes. The resulting compensator is usually of a lower order than the system to be controlled because not all flexible modes in a structure tend to be destabilized by a reduced-order controller. A helpful characteristic of most flexible space structures is their inherent passive damping. This gives the designer the opportunity of phase stabilizing significant modes and to gain stabilize all other higher frequency modes that have less influence on the structure. On the other hand, successive-mode stabilization presents problems of its own, and a retuning of the compensated system becomes necessary. It is also noticed that reducing the damping in a frequency shaping filter reduces its influence on neighboring frequencies, and it also reduces the phase lag at lower frequencies; however, reducing the damping of the filters increases the sensitivity of the phase stabilized modes to plant parameter uncertainties.

Active disturbance rejection is achieved in both the classical methods and state-space methods, with the introduction of an internal model of the disturbance into the feedback loop. The concept of internal modeling of the disturbance works as well with a classical transfer function description as with a state-space description. In the classical design, the internal modeling of the disturbance leads to the introduction of a disturbance rejection dipole, or filter, for each frequency component of the disturbance. In the state-space design the introduction of the internal model results in the addition of two states for each frequency component of the disturbance.

2.6 Stability Robustness Analysis

This section, based on Refs. 19 and 20, is concerned with the problem of computing the structured singular values μ for uncertain dynamic systems. In particular, this section deals with the problem of computing ∞-norm real parameter margins or real μ for structural dynamic systems with masses, stiffness constants, and damping constants as uncertain parameters. The real μ problem is essentially the same as the problem of determining the largest stable hypercube in the uncertain parameter space. In this section, the concepts of the critical gains, critical frequencies, and critical parameters are introduced for stabilized conservative plants. A concept of two real critical constraints is also introduced to solve the problem of determining the largest stable hypercube in parameter space that touches the stability boundary on one of its corners. The concept is simply based on the idea of separating the real and imaginary parts of a characteristic polynomial equation.

2.6.1 Stabilized Conservative Plants

Conservative plants have special properties that aid in real parameter margin computations; i.e., the plant transfer function is real valued for all frequencies. This allows for the identification of compensator-dependent frequencies where the loop transfer function becomes real valued. If uncertain parameters appear multilinearly in the plant transfer function, real parameter margin computations can be further reduced to checking system stability at only a few frequencies and at only the corners of a parameter space hypercube.

2.6.1.1 Critical Frequency and Gain Consider a single-input single-output (SISO) feedback control system with the closed-loop characteristic equation

$$1 + G(s, \mathbf{p})K(s) = 0 \qquad (2.132)$$

where $G(s, \mathbf{p})$ is the transfer function of a conservative plant with an uncertain parameter vector $\mathbf{p} = (p_1, \ldots, p_\ell)$, $K(s)$ the compensator transfer function, and s the Laplace transform variable. Because the plant is conservative, $G(s, \mathbf{p})$ is a function of even powers of s and is a real number for every $s = j\omega$, where $j = \sqrt{-1}$; that is, $G(j\omega, \mathbf{p}) = G(\omega^2, \mathbf{p})$. Thus, for a conservative plant, we have

$$1 + G(\omega^2, \mathbf{p})K(j\omega) = 0 \qquad (2.133)$$

where $G(\omega^2, \mathbf{p})$ is real.

Let the compensator be expressed as

$$K(j\omega) = \text{Re}[K(j\omega)] + j\,\text{Im}[K(j\omega)] \qquad (2.134)$$

Then Eq. (2.133) becomes

$$\{1 + G(\omega^2, \mathbf{p})\,\text{Re}[K(j\omega)]\} + jG(\omega^2, \mathbf{p})\,\text{Im}[K(j\omega)] = 0 \qquad (2.135)$$

Because the real and imaginary parts of Eq. (2.135) must be zero, we have the following two critical instability constraints:

$$G(\omega^2, \mathbf{p})\,\text{Im}[K(j\omega)] = 0 \qquad (2.136)$$

$$1 + G(\omega^2, \mathbf{p})\,\text{Re}[K(j\omega)] = 0 \qquad (2.137)$$

Note that a solution ω of $G(\omega^2, \mathbf{p}) = 0$ cannot be a solution of Eqs. (2.136) and (2.137). Consequently, Eq. (2.136) simplifies to

$$\text{Im}[K(j\omega)] = 0$$

and it can be said that the closed-loop system with uncertain conservative plant becomes unstable only at frequencies that depend only on the compensator parameters. Such a frequency, denoted by ω_c, is called the *critical frequency* and the corresponding parameter vector $\mathbf{p}_c$ is called the *critical parameter vector*.

Solving for $\mathbf{p}_c$ from Eq. (2.137) for each ω_c is quite numerically complicated. However, it is interesting to notice that $G(\omega_c^2, \mathbf{p}_c)$ can be expressed as

$$G(\omega_c^2, \mathbf{p}_c) = \kappa_c G(\omega_c^2, \bar{\mathbf{p}}) \qquad (2.138)$$

where $\bar{\mathbf{p}}$ is the nominal parameter vector and κ_c is a real scalar. Thus, Eq. (2.137) becomes

$$1 + \kappa_c G(\omega_c^2, \bar{\mathbf{p}})\,\text{Re}[K(j\omega_c)] = 0 \qquad (2.139)$$

where κ_c is referred to as the critical gain, which represents an overall gain change due to parameter variations.

A conventional root locus plot of the nominal closed-loop system vs overall loop gain may be used to identify the critical gains and frequencies where root loci cross the imaginary axis. The closed-loop system becomes unstable only at these critical frequencies, including $\omega = 0$, for all possible parameter variations. Thus, the classical gain margin concept may still be used as a measure of the overall parameter robustness for a system whose uncertain parameters do not necessarily appear multilinearly. The smallest critical gain corresponds to the conventional gain margin of a SISO closed-loop system.

2.6.1.2 Critical Polynomial Equations   Because a SISO system composed of a fixed compensation and an uncertain conservative plant becomes unstable only at critical frequencies that depend on the compensator parameters alone, the computation of the ∞-norm parameter margin can be performed as follows.

Let

$$K(j\omega) = \frac{\tilde{N}(j\omega)}{\tilde{D}(j\omega)} = \frac{\tilde{N}_r(\omega^2) + j\omega\tilde{N}_i(\omega^2)}{\tilde{D}_r(\omega^2) + j\omega\tilde{D}_i(\omega^2)} \tag{2.140}$$

and

$$G(j\omega, \mathbf{p}) = G(\omega^2, \mathbf{p}) = \frac{N(\omega^2, \mathbf{p})}{D(\omega^2, \mathbf{p})} \tag{2.141}$$

Substituting Eqs. (2.140) and (2.141) into Eq. (2.133), we obtain the *critical polynomial equations*

$$\omega\left[\tilde{N}_i(\omega^2)\tilde{D}_r(\omega^2) - \tilde{N}_r(\omega^2)\tilde{D}_i(\omega^2)\right] = 0 \tag{2.142}$$

$$D(\omega^2, \mathbf{p})\tilde{D}_r(\omega^2) + N(\omega^2, \mathbf{p})\tilde{N}_r(\omega^2) = 0 \tag{2.143}$$

For each critical frequency obtained by solving Eq. (2.142), we need to find the largest stable hypercube centered about $\bar{\mathbf{p}}$ in parameter space. That is, the problem is to find $\mathbf{p}_c$ to minimize $\|\boldsymbol{\delta}\|_\infty$, where $\boldsymbol{\delta} = (\delta_1, \ldots, \delta_\ell)$ may be actual perturbations as in $p_i = \bar{p}_i + \delta_i$, or percentage variations as in $p_i = \bar{p}_i(1 + \delta_i)$, subject to Eq. (2.143) for each critical frequency. Then the solution with the smallest magnitude becomes the ∞-norm parameter margin. The computation of such a parameter margin for general cases is quite numerically complicated. The ∞-norm parameter margin computation can be greatly simplified, however, by making use of the multilinear property of conservative dynamic systems, as is to be discussed next.

2.6.1.3 Corner Directions in Parameter Space   As first shown by Ghaoui and Bryson [21], for a stabilized conservative system with multilinearly uncertain parameters, one needs only to check for instability in the corner directions of the parameter space hypercube, at a finite number of critical frequencies. An alternative geometric proof of such an elegant corner (or vertex) property using the mapping theorem is provided here as follows.

For a SISO conservative system, the loop transfer function $G(s, \mathbf{p})K(s)$ becomes real valued at critical frequencies. When the mapping theorem [22, 23] is applied at critical frequencies to a SISO conservative system with multilinear parameters, having fixed compensation and independent parameter perturbations, the convex hull of the $G(j\omega, \mathbf{p})K(j\omega)$-plane image of the parameter space hypercube collapses to a line segment on the real axis. Because the extreme points of the convex hull are defined by the vertices of the parameter space hypercube, the endpoints of this line segment correspond to one or more of these vertices. Therefore, it is sufficient to check system stability only in the corner directions of the parameter space hypercube for each critical frequency, and the largest stable hypercube touches the stability boundary on one of its corners.

The parameter margin is then defined by the parameter changes that cause an endpoint of the line segment to touch the critical stability point. In addition, the overall change in gain for the plant transfer function due to parameter changes is equal to the associated critical gain. When parameter margin computations for each corner of a parameter space hypercube and each critical frequency are completed, the parameter margin of smallest magnitude becomes the overall parameter margin for the system.

For a stabilized conservative plant with ℓ independently uncertain parameters that appear multilinearly, there are 2^ℓ hypercube corner directions that need to be checked in the ∞-norm real parameter margin computation. These corners correspond to the 2^ℓ possible combinations of parameter values, where the uncertain value may be an increase or a decrease in any particular parameter.

2.6.1.4 Real Parameter Margin

Consider a closed-loop system described as in Fig. 2.25, where $\mathbf{G}(s)$ is the nominal plant, $\mathbf{K}(s)$ a stabilizing controller, $\mathbf{\Delta}$ the structured uncertainty matrix, and $\mathbf{M}(s) \in \mathcal{C}^{\ell \times \ell}$ the stable, nominal transfer function matrix from perturbation inputs $\mathbf{d}$ to perturbation outputs $\mathbf{z}$. For this uncertain system, the real uncertain parameter vector $\mathbf{p} = (p_1, \ldots, p_\ell)$ and the perturbation vector $\boldsymbol{\delta} = (\delta_1, \ldots, \delta_\ell)$ are related as $p_i = \bar{p}_i(1 + \delta_i)$.

The closed-loop characteristic equation is then obtained as

$$\det[\mathbf{I} - \mathbf{M}(s)\mathbf{\Delta}] = 0 \qquad (2.144)$$

where $\mathbf{M}(s)$ may contain input and output scaling factors, $\mathbf{I}$ is an identity matrix, and

$$\mathbf{\Delta} = \operatorname{diag}(\delta_1, \ldots, \delta_\ell) \qquad (2.145)$$

is the diagonal uncertainty matrix of independent parameter perturbations $\delta_i \in \mathcal{R}$.

Note that $\det[\mathbf{I} - \mathbf{M}(s)\mathbf{\Delta}]$ is a polynomial of δ_i, and is affine with respect to each δ_i, and that the coefficients of the characteristic polynomial of the perturbed system are multilinear functions of δ_i.

A characteristic polynomial, which has coefficients affine with respect to each uncertain parameter δ_i, is called a *multilinearly uncertain polynomial*. A dynamic system with such characteristic polynomial is called a multilinearly uncertain system or a system with multilinearly uncertain parameters.

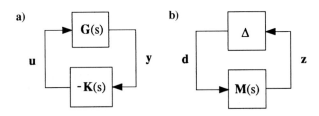

Fig. 2.25 Block diagram representations of a closed-loop control system with uncertain plant parameters.

If a system is described by a closed-loop characteristic equation (2.144) and Eq. (2.145) with nonrepeated entries, then the system is simply called a multilinearly uncertain system; however, not all multilinearly uncertain polynomials can be expressed in the form of Eqs. (2.144) and (2.145) with nonrepeated entries. In many cases, we have $\mathbf{\Delta}$ with repeated entries.

Because only the corners of a parameter space hypercube are to be checked in ∞-norm parameter margin computations for a conservative plant with multilinear parameters, $\mathbf{\Delta}$ can be expressed as

$$\mathbf{\Delta} = \kappa \mathbf{E} \tag{2.146}$$

where $\kappa \in [0, \infty)$ and

$$\mathcal{E} = \{\mathbf{E} : \mathbf{E} = \mathrm{diag}(e_i), \ e_i = \pm 1 \ \forall i\}$$

In this case, κ represents the size of the parameter space hypercube whereas the 2^ℓ possible sets of ± 1 in $\mathcal{E}$ define the 2^ℓ corner directions. The task is to find κ and the particular $\mathbf{E}$ matrix corresponding to the stable, parameter space hypercube and its particular corner that touches the stability boundary.

For a conservative plant with real multilinear parameter variations, we may introduce the *real parameter robustness measure* $\kappa(\omega)$ and the *real parameter margin* κ^*, as follows:

$$\kappa(\omega) = \inf_{\mathbf{E} \in \mathcal{E}} \{\kappa : \det[\mathbf{I} - \kappa \mathbf{M}(j\omega)\mathbf{E}] = 0\} \tag{2.147}$$

$$\kappa^* = \inf_{\omega} \kappa(\omega) = \inf_{\omega_c} \kappa(\omega) \tag{2.148}$$

where ω_c denotes the critical frequencies defined earlier. Note that the real parameter robustness measure $\kappa(\omega)$ of a conservative plant is discontinuous at each critical frequency.

2.6.2 Stabilized Nonconservative Plants

Some of the simplifications that are possible in real parameter margin computations for conservative plants are no longer valid for nonconservative plants. Consequently, a more general algorithm is needed for computing real parameter margins (or real μ).

In practice, the damping constant is often the most uncertain parameter for structural dynamic systems. In that case, we may consider 1) a worst case with no damping, i.e., a conservative plant; 2) a case with fixed, nominal values of passive damping; or 3) a case in which the damping constant is considered as one of the uncertain parameters.

A direct approach to the cases 2 and 3 is to define a parameter space hypercube, of dimension equal to the number of uncertain parameters and centered about the nominal parameter values, and then increase the size of the hypercube, always checking closed-loop stability for parameter values corresponding to points on the surface of the hypercube, until system instability occurs. Then the computation of the ∞-norm parameter margin corresponds to finding the largest stable hypercube in the parameter space. An obvious advantage of this method is that it is applicable for systems whose uncertain parameters do not necessarily appear multilinearly; however, the amount of computation increases dramatically as the number of parameters increases.

2.6.3 Two-Mass–Spring–Damper Example

2.6.3.1 Structured Parameter Uncertainty Modeling
Consider the two-mass–spring–damper system shown in Fig. 2.26, which is a generic model of an uncertain dynamic system with a rigid-body mode and one vibration mode. A control force acts on m_1 and the position of m_2 is measured, resulting in a noncollocated control problem.

This system can be described as

$$m_1\ddot{x}_1 + c(\dot{x}_1 - \dot{x}_2) + k(x_1 - x_2) = u$$
$$m_2\ddot{x}_2 + c(\dot{x}_2 - \dot{x}_1) + k(x_2 - x_1) = 0$$
$$y = x_2$$

where x_1 and x_2 are the positions of m_1 and m_2, respectively; u is the control input acting on m_1; y is the measured output; k is the spring stiffness coefficient; c is the damping constant; and all parameters have the appropriate units and time is in seconds.

The transfer function from the control input u to the measured output y is

$$\frac{y(s)}{u(s)} = \frac{cs + k}{s^2\{m_1 m_2 s^2 + c(m_1 + m_2)s + k(m_1 + m_2)\}}$$

It can be seen that the uncertain parameters, m_1, m_2, k, and c, appear multilinearly in the numerator and denominator of the plant transfer function [24].

The uncertain parameters are then modeled as

$$m_1 = \bar{m}_1(1 + \delta_1)$$
$$m_2 = \bar{m}_2(1 + \delta_2)$$
$$k = \bar{k}(1 + \delta_3)$$
$$c = \bar{c}(1 + \delta_4)$$

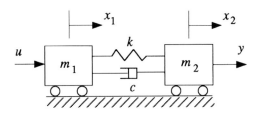

Fig. 2.26 Two-mass–spring–damper system with multilinearly uncertain parameters.

where $\bar{m}_1, \bar{m}_2, \bar{k}$, and $\bar{c}$ are the nominal parameters, and $\delta_1, \delta_2, \delta_3$, and δ_4 represent percentage variations in each parameter.

The plant equations may now be written as

$$\bar{m}_1(1 + \delta_1)\ddot{x}_1 = \bar{c}(1 + \delta_4)(\dot{x}_2 - \dot{x}_1) + \bar{k}(1 + \delta_3)(x_2 - x_1) + u$$
$$\bar{m}_2(1 + \delta_2)\ddot{x}_2 = -\bar{c}(1 + \delta_4)(\dot{x}_2 - \dot{x}_1) - \bar{k}(1 + \delta_3)(x_2 - x_1)$$

After rearranging terms, we obtain

$$\bar{m}_1\ddot{x}_1 = \bar{c}(\dot{x}_2 - \dot{x}_1) + \bar{k}(x_2 - x_1) + u - d_1 + d_3 + d_4$$
$$\bar{m}_2\ddot{x}_2 = -\bar{c}(\dot{x}_2 - \dot{x}_1) - \bar{k}(x_2 - x_1) - d_2 - d_3 - d_4$$

where

$$
\begin{aligned}
d_1 &= \delta_1 z_1, & z_1 &= \bar{m}_1\ddot{x}_1 \\
d_2 &= \delta_2 z_2, & z_2 &= \bar{m}_2\ddot{x}_2 \\
d_3 &= \delta_3 z_3, & z_3 &= \bar{k}(x_2 - x_1) \\
d_4 &= \delta_4 z_4, & z_4 &= \bar{c}(\dot{x}_2 - \dot{x}_1)
\end{aligned}
$$

and d_i and z_i are referred to as the fictitious inputs and outputs, respectively.

This system is then described in state-space form with $x_3 = \dot{x}_1$ and $x_4 = \dot{x}_2$ as

$$\dot{\mathbf{x}} = \mathbf{A}\mathbf{x} + \mathbf{B}_1\mathbf{d} + \mathbf{B}_2 u \qquad (2.149a)$$
$$\mathbf{z} = \mathbf{C}_1\mathbf{x} + \mathbf{D}_{11}\mathbf{d} + \mathbf{D}_{12}u \qquad (2.149b)$$
$$y = \mathbf{C}_2\mathbf{x} + \mathbf{D}_{21}\mathbf{d} + \mathbf{D}_{22}u \qquad (2.149c)$$
$$\mathbf{d} = \boldsymbol{\Delta}\mathbf{z} \qquad (2.149d)$$

where

$$\mathbf{x} = (x_1, \ x_2, \ x_3, \ x_4)$$
$$\mathbf{d} = (d_1, \ d_2, \ d_3, \ d_4)$$
$$\mathbf{z} = (z_1, \ z_2, \ z_3, \ z_4)$$
$$\boldsymbol{\Delta} = \text{diag}(\delta_1, \ \delta_2, \ \delta_3, \ \delta_4)$$

and

$$\mathbf{A} = \begin{bmatrix} 0 & 0 & 1 & 0 \\ 0 & 0 & 0 & 1 \\ -\bar{k}/\bar{m}_1 & \bar{k}/\bar{m}_1 & -\bar{c}/\bar{m}_1 & \bar{c}/\bar{m}_1 \\ \bar{k}/\bar{m}_2 & -\bar{k}/\bar{m}_2 & \bar{c}/\bar{m}_2 & -\bar{c}/\bar{m}_2 \end{bmatrix}$$

$$\mathbf{B}_1 = \begin{bmatrix} 0 & 0 & 0 & 0 \\ 0 & 0 & 0 & 0 \\ -1/\bar{m}_1 & 0 & 1/\bar{m}_1 & 1/\bar{m}_1 \\ 0 & -1/\bar{m}_2 & -1/\bar{m}_2 & -1/\bar{m}_2 \end{bmatrix}$$

$$\mathbf{B}_2 = [0 \quad 0 \quad 1/\bar{m}_1 \quad 0]^T$$

$$
\mathbf{C}_1 = \begin{bmatrix} -\bar{k} & \bar{k} & -\bar{c} & \bar{c} \\ \bar{k} & -\bar{k} & \bar{c} & -\bar{c} \\ -\bar{k} & \bar{k} & 0 & 0 \\ 0 & 0 & -\bar{c} & \bar{c} \end{bmatrix}
$$

$$
\mathbf{C}_2 = \begin{bmatrix} 0 & 1 & 0 & 0 \end{bmatrix}
$$

$$
\mathbf{D}_{11} = \begin{bmatrix} -1 & 0 & 1 & 1 \\ 0 & -1 & -1 & -1 \\ 0 & 0 & 0 & 0 \\ 0 & 0 & 0 & 0 \end{bmatrix}
$$

$$
\mathbf{D}_{12} = \begin{bmatrix} 1 & 0 & 0 & 0 \end{bmatrix}^T
$$

$$
\mathbf{D}_{21} = \begin{bmatrix} 0 & 0 & 0 & 0 \end{bmatrix}
$$

$$
\mathbf{D}_{22} = 0
$$

Given a controller in state-space form as

$$
\dot{\mathbf{x}}_c = \mathbf{A}_c \mathbf{x}_c + \mathbf{B}_c y
$$
$$
u = \mathbf{C}_c \mathbf{x}_c
$$

the overall closed-loop system can be represented in state-space form as

$$
\begin{bmatrix} \dot{\mathbf{x}} \\ \dot{\mathbf{x}}_c \end{bmatrix} = \begin{bmatrix} \mathbf{A} & \mathbf{B}_2 \mathbf{C}_c \\ \mathbf{B}_c \mathbf{C}_2 & \mathbf{A}_c \end{bmatrix} \begin{bmatrix} \mathbf{x} \\ \mathbf{x}_c \end{bmatrix} + \begin{bmatrix} \mathbf{B}_1 \\ \mathbf{B}_c \mathbf{D}_{21} \end{bmatrix} \mathbf{d}
$$

$$
\mathbf{z} = \begin{bmatrix} \mathbf{C}_1 & \mathbf{D}_{12} \mathbf{C}_c \end{bmatrix} \begin{bmatrix} \mathbf{x} \\ \mathbf{x}_c \end{bmatrix} + \mathbf{D}_{11} \mathbf{d}
$$

or in transfer function form as

$$
\mathbf{z} = \mathbf{M}(s)\, \mathbf{d}
$$

and

$$
\mathbf{d} = \mathbf{\Delta} \mathbf{z}
$$

2.6.3.2 *Conservative Case*   Consider a case with zero damping. The two mass elements m_1 and m_2 and the spring constant k are assumed to be uncertain. The parameter space hypercube is three dimensional, which has $2^3 = 8$ corners of the cube centered about the assumed nominal parameter values $(\bar{m}_1, \bar{m}_2, \bar{k}) = (1, 1, 1)$.

The nominal system has open-loop eigenvalues of $s = 0, 0, \pm\sqrt{2}j$ on the imaginary axis. The closed-loop stability robustness is to be analyzed for a particular controller given in state-space form as

$$
\dot{\mathbf{x}}_c = \mathbf{A}_c \mathbf{x}_c + \mathbf{B}_c y
$$
$$
u = \mathbf{C}_c \mathbf{x}_c
$$

where

$$\mathbf{A}_c = \begin{bmatrix} 0.1250 & -0.2879 & 1.0587 & 0.0076 \\ -0.1116 & -0.5530 & -0.0524 & 0.9932 \\ -2.5747 & 1.8080 & -2.1485 & -0.2774 \\ 1.0069 & -1.1966 & 0.0042 & 0.0005 \end{bmatrix}$$

$$\mathbf{B}_c = [0.1946 \quad 0.6791 \quad -0.0359 \quad 0.2013]^T$$

$$\mathbf{C}_c = [-1.5717 \quad 0.7722 \quad -2.1450 \quad -0.2769]$$

The nominal closed-loop system has eigenvalues at $s = -0.2322 \pm 0.1919j$, $-0.4591 \pm 0.3936j$, $-0.4251 \pm 1.3177j$, and $-0.1717 \pm 1.4312j$. The nominal closed-loop system has a 6.1-dB gain margin and a 34-deg phase margin. Root loci vs overall loop gain cross the imaginary axis at $\omega = 0.747801$ rad/s ($\kappa_c = 2.0198$) and $\omega = 2.811543$ rad/s ($\kappa_c = 247.8$). As was discussed earlier, this system becomes unstable at these two critical frequencies for all possible variations of m_1, m_2, and k.

The critical frequencies can also be identified by solving Eq. (2.136) or Eq. (2.142), and $\kappa(\omega_c)$ and the corresponding $\mathbf{E}$ can be found as

$$\omega_c = 0.747801, \qquad \kappa(\omega_c) = 0.459848, \qquad \mathbf{E} = \mathrm{diag}(1, 1, -1)$$

$$\omega_c = 2.811534, \qquad \kappa(\omega_c) = 0.586918, \qquad \mathbf{E} = \mathrm{diag}(-1, -1, 1)$$

Note that $\omega_c = 0$ is always a critical frequency, and the corresponding $\kappa(0)$ can be found as $\kappa(0) = 1$ for this example. Thus, the ∞-norm real parameter margin for the system becomes

$$\kappa^* = \inf_{\omega_c} \kappa(\omega_c) = 0.459848$$

and the instability occurs at the critical corner at

$$(\delta_1, \delta_2, \delta_3) = (0.459848, \ 0.459848, \ -0.459848)$$

The corresponding critical parameter values are

$$(m_1, m_2, k) = (1.459848, \ 1.459848, \ 0.540152)$$

The real parameter robustness measure $\kappa(\omega)$, which is discontinuous in frequency, is shown Fig. 2.27. Because the system becomes unstable only at the two nonzero critical frequencies for all possible real parameter variations, $\kappa(\omega) = 1$ at all other frequencies. This corresponds to the trivial case where some or all of the parameter values become zero.

2.6.3.3 Nonconservative Case

To assess any practical significance of including passive damping, nonconservative cases have been studied in Ref. 19. The study results show that the critical instability occurs at one of the corners of the parameter space hypercube for cases in which the damping constant c is assumed to be a known nonzero constant. For cases in which the damping coefficient c is actually modeled as one of four uncertain parameters ($\ell = 4$), the

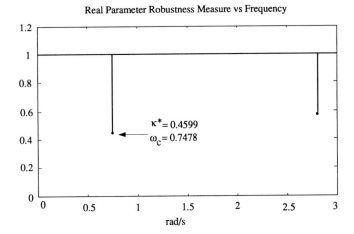

Fig. 2.27 Real parameter robustness measure $\kappa(\omega)$ for a conservative case $(c = \bar{c} = 0)$.

study results of Ref. 19 also show that the critical instability occurs at one of the corners of the parameter space hypercube. For all these cases, real parameter margins are very close to the zero-damping case of 0.459848. Therefore, for a certain class of structural dynamic systems with small passive damping, the computational complexity of the real μ problem may be avoided by modeling the system as a conservative plant, without loss of any practical significance.

Problem

2.28 For the example problem considered in this section, generate $1/\mu$ plots using the ssv.m file of MATLAB Robust Control Toolbox for the following two cases:
(a) A conservative plant with $c = 0$.
(b) A nonconservative plant with $\bar{c} = 0.02$.
    Discuss the results by comparing the $1/\mu$ plots with the real parameter robustness measure $\kappa(\omega)$ shown in Fig. 2.27.

2.6.4 Two Real Critical Constraints

A concept of two real critical constraints is introduced here to solve the problem of determining the largest stable hypercube in parameter space that touches the stability boundary on one of its corners. The concept is based on sufficient conditions for checking for critical instability only in the corner directions of the parameter space hypercube [20].

2.6.4.1 Two Real Critical Constraints
Consider a characteristic polynomial $\phi(s; \delta)$ with the real perturbation vector $\delta = (\delta_1, \ldots, \delta_\ell)$. The nominal system

with $\delta = 0$ is assumed to be asymptotically stable. Then, there exists an $n \times n$ rational matrix $\mathbf{M}(s)$ and a diagonal matrix $\mathbf{\Delta} \in \mathcal{X}$ such that

$$\phi(s; \delta) = \phi(s; 0) \det[\mathbf{I} - \mathbf{M}(s)\mathbf{\Delta}] \tag{2.150}$$

and

$$\mathcal{X} = \{\mathbf{\Delta} \ : \ \mathbf{\Delta} = \mathrm{diag}(\delta_i \mathbf{I}_i), \quad i = 1, \ldots, \ell\} \tag{2.151}$$

where $\mathbf{I}_i$ denotes an $m_i \times m_i$ identity matrix and

$$\sum_{i=1}^{\ell} m_i = n$$

Consequently, the critical stability constraint equation

$$\phi(j\omega; \delta) = 0 \tag{2.152}$$

can be rewritten as

$$\det[\mathbf{I} - \mathbf{M}(j\omega)\mathbf{\Delta}] = 0 \tag{2.153}$$

because $\phi(j\omega; 0) \neq 0$ for all ω.

The real parameter robustness measure $\kappa(\omega)$ and the *real structured singular value measure* $\mu(\omega)$ associated with the critical constraint equation (2.153) are then defined as

$$
\begin{aligned}
\kappa(\omega) &\equiv 1/\mu(\omega) \\
&= \inf_{\mathbf{\Delta} \in \mathcal{X}} \{\kappa \ : \ \det[\mathbf{I} - \mathbf{M}(j\omega)\mathbf{\Delta}] = 0, \ \bar{\sigma}(\mathbf{\Delta}) \leq \kappa\} \\
&= \sup_{\mathbf{\Delta} \in \mathcal{X}} \{\kappa \ : \ \det[\mathbf{I} - \mathbf{M}(j\omega)\mathbf{\Delta}] \neq 0, \ \bar{\sigma}(\mathbf{\Delta}) \leq \kappa\}
\end{aligned}
$$

where $\mathcal{X}$ is the set of all repeated blocks defined as Eq. (2.151) and $\bar{\sigma}(\mathbf{\Delta})$ denotes the largest singular value of $\mathbf{\Delta}$. The real parameter margin κ^* and the associated real structured singular value μ^* are defined as

$$\kappa^* \equiv 1/\mu^* = \inf_{\omega} \kappa(\omega) \tag{2.154}$$

and the corresponding uncertain parameter vector is called the critical parameter vector and is denoted by δ^*.

The critical stability constraint equation (2.152) or (2.153) is a complex constraint. We now exploit the idea of separating the real and imaginary parts of the constraint equation (2.152), as follows:

$$\mathrm{Re}[\phi(j\omega; \delta)] = f_1(\omega)\phi_1(\omega; \delta) = 0 \tag{2.155a}$$

$$\mathrm{Im}[\phi(j\omega; \delta)] = f_2(\omega)\phi_2(\omega; \delta) = 0 \tag{2.155b}$$

where $f_1(\omega)$ and $f_2(\omega)$ are polynomials that are independent of δ and $\phi_1(\omega; 0) \neq 0$ and $\phi_2(\omega; 0) \neq 0$ for all $\omega \geq 0$.

Because there exist real rational matrices $\mathbf{M}_1(\omega)$ and $\mathbf{M}_2(\omega)$ such that

$$\phi_i(\omega; \boldsymbol{\delta}) = \phi_i(\omega; 0) \det[\mathbf{I} - \mathbf{M}_i(\omega)\boldsymbol{\Delta}_i], \qquad i = 1, 2 \tag{2.156}$$

we have the following two real critical constraints

$$f_1(\omega) \det[\mathbf{I} - \mathbf{M}_1(\omega)\boldsymbol{\Delta}_1] = 0 \tag{2.157a}$$

$$f_2(\omega) \det[\mathbf{I} - \mathbf{M}_2(\omega)\boldsymbol{\Delta}_2] = 0 \tag{2.157b}$$

with

$$\mathcal{X}_i := \{\boldsymbol{\Delta}_i : \boldsymbol{\Delta}_i = \mathrm{diag}(\delta_{ij}\mathbf{I}_{ij}), \delta_{ij} \in \mathcal{R}, j = 1, \ldots, \ell_i\} \tag{2.158}$$

where $\{\delta_{1j}, j = 1, \ldots, \ell_1\}$ and $\{\delta_{2j}, j = 1, \ldots, \ell_2\}$ are two subsets of $\{\delta_i, i = 1, \ldots, \ell\}$, and $\mathbf{I}_{ij}$ is an $m_{ij} \times m_{ij}$ identity matrix with

$$\sum_{j=1}^{\ell_i} m_{ij} = n_i, \qquad i = 1, 2$$

Because polynomials with coefficients linearly dependent on uncertain parameters δ_i can be expressed in a form with rank-one matrices $\mathbf{M}_1(\omega)$ and $\mathbf{M}_2(\omega)$, the critical stability constraint (2.152) of an interval polynomial or a polytopic polynomial can be expressed as two real critical constraints of the form of Eqs. (2.157) with rank-one matrices $\mathbf{M}_1(\omega)$ and $\mathbf{M}_2(\omega)$.

A frequency at which the two real critical constraints (2.157) reduce to a single constraint is called the *degenerate frequency*, and the real nonnegative roots of the polynomials $f_1(\omega)$ and $f_2(\omega)$ of Eqs. (2.157) are the degenerate frequencies that cause isolated discontinuities in $\mu(\omega)$.

The two-constraint real μ measure, associated with the two constraints (2.157), is defined as

$$\frac{1}{\mu_{12}(\omega)} = \inf_{\boldsymbol{\Delta}\in\mathcal{X}} \{\bar{\sigma}[\mathrm{diag}(\boldsymbol{\Delta}_1, \boldsymbol{\Delta}_2)] : \det[\mathbf{I} - \mathbf{M}_1\boldsymbol{\Delta}_1] = 0$$

$$\det[\mathbf{I} - \mathbf{M}_2\boldsymbol{\Delta}_2] = 0\} \tag{2.159}$$

The single-constraint real μ measures, $\mu_1(\omega)$ and $\mu_2(\omega)$, associated with each constraint in Eqs. (2.157), are defined as

$$\frac{1}{\mu_1(\omega)} = \inf_{\boldsymbol{\Delta}_1\in\mathcal{X}_1} \{\bar{\sigma}(\boldsymbol{\Delta}_1) : \det[\mathbf{I} - \mathbf{M}_1\boldsymbol{\Delta}_1] = 0\} \tag{2.160}$$

$$\frac{1}{\mu_2(\omega)} = \inf_{\boldsymbol{\Delta}_2\in\mathcal{X}_2} \{\bar{\sigma}(\boldsymbol{\Delta}_2) : \det[\mathbf{I} - \mathbf{M}_2\boldsymbol{\Delta}_2] = 0\} \tag{2.161}$$

The real μ measure is related to the two-constraint real μ measure and the single-constraint real μ measures at each frequency ω, as follows:

$$\mu(\omega) = \begin{cases} \mu_{12}(\omega) & \text{if } f_1(\omega) \neq 0, f_2(\omega) \neq 0 \\ \mu_1(\omega) & \text{if } f_1(\omega) \neq 0, f_2(\omega) = 0 \\ \mu_2(\omega) & \text{if } f_1(\omega) = 0, f_2(\omega) \neq 0 \end{cases} \tag{2.162}$$

where $f_1(\omega)$ and $f_2(\omega)$ are the two polynomials defined as in Eqs. (2.155).

Let $\mathcal{S} = \{\delta_i, \ i = 1, \ldots, \ell\}$. Also let $\mathcal{S}_1$ and $\mathcal{S}_2$ be two subsets of $\mathcal{S}$, and $\mathcal{S}_1 \cup \mathcal{S}_2 = \mathcal{S}$. If $\mathcal{S}_1 \cap \mathcal{S}_2 \neq \emptyset$ we define the *restricted parameter vector* $\mathbf{p}$ in $\mathcal{S}_1 \cap \mathcal{S}_2$, as follows:

$$\mathbf{p} = (p_1, \ldots, p_m), \qquad p_i \in \mathcal{S}_1 \cap \mathcal{S}_2, \qquad i = 1, \ldots, m \qquad (2.163)$$

The restricted parameter vectors associated with μ_1 and μ_2 are denoted by $\mathbf{p}_{s_1}$ and $\mathbf{p}_{s_2}$, respectively.

If $\mathcal{S}_1 \cap \mathcal{S}_2 = \emptyset$, the real μ^* can be found as

$$\mu^* = \sup_{\omega} \mu_{12}(\omega) = \max \left\{ \sup_{\omega} \mu_1(\omega), \ \sup_{\omega} \mu_2(\omega) \right\} \qquad (2.164)$$

If $\mu_1(\omega_c) = \mu_2(\omega_c)$ at some critical frequencies ω_c, and if the restricted parameter vectors in $\mathcal{S}_1 \cap \mathcal{S}_2$, associated with $\mu_1(\omega_c)$ and $\mu_2(\omega_c)$, become $\mathbf{p}_{s_1} = \mathbf{p}_{s_2}$, then the real μ^* is

$$\mu^* = \sup_{\omega} \mu_{12}(\omega) = \max_{\omega_c} \mu_1(\omega_c) = \max_{\omega_c} \mu_2(\omega_c) \qquad (2.165)$$

2.6.4.2 Sufficient Conditions for Corner Property
If the critical instability of the constraint (2.153) with possible repeated entries in $\mathbf{\Delta}$ occurs at one of the corners of the parameter space hypercube, then

$$\kappa(\omega) = \left\{ \max_{\mathbf{E} \in \mathcal{E}} \rho[\mathbf{M}(j\omega)\mathbf{E}] \right\}^{-1} \qquad (2.166)$$

where $\rho(\mathbf{ME})$ denotes the maximum real eigenvalue of $\mathbf{ME}$ and it is defined to be zero if $\mathbf{ME}$ does not have real eigenvalues. Also, the corner matrix, denoted by $\mathbf{E}$, is defined as

$$\mathcal{E} = \{\mathbf{E} \ : \ \mathbf{E} = \text{diag}(e_i \mathbf{I}_i), \ e_i = +1 \ or \ -1, \ i = 1, \ldots, \ell\}$$

The real parameter margin κ^*, or real μ^*, is then determined as

$$\kappa^* \equiv 1/\mu^* = \inf_{\omega} \kappa(\omega)$$

The corresponding critical corner matrix $\mathbf{E}^*$ and critical corner vector $\mathbf{e}^*$ are, respectively, given by

$$\mathbf{E}^* = \text{diag}(e_i^* \mathbf{I}_i) \qquad (2.167)$$
$$\mathbf{e}^* = (e_1^*, \ e_2^*, \ldots, \ e_\ell^*) \qquad (2.168)$$

Furthermore, the critical parameter vector $\boldsymbol{\delta}^*$ can be determined as

$$\boldsymbol{\delta}^* = \kappa^* \mathbf{e}^*$$

Note that, if $\mathbf{E} \in \mathcal{E}$, then $-\mathbf{E} \in \mathcal{E}$ and $\lambda(\mathbf{ME}) = -\lambda(-\mathbf{ME})$, where $\lambda(\mathbf{ME})$ denotes the eigenvalues of $\mathbf{ME}$. Thus, $\kappa(\omega)$ defined as Eq. (2.166) is always positive real.

We now summarize the sufficient condition for the corner property of a multilinearly uncertain system, as follows.

Theorem 2.1

If a multilinearly uncertain polynomial $\phi(j\omega; \delta)$ is always valued as real at some frequency ω, then the critical instability at that frequency occurs at one of the corners of the parameter space hypercube $\kappa^* \mathcal{D}$, where $\delta \in \kappa^* \mathcal{D}$, or $-\kappa^* \leq \delta_i \leq \kappa^*$ ($i = 1, 2, \ldots, \ell$).

For a multilinearly uncertain system, the single-constraint real μ must attain their values at one of the corner of the parameter space hypercube $\kappa^* \mathcal{D}$. Also, at the degenerate frequencies, the critical instability of a multilinearly uncertain system occurs at one of the corners of the parameter space hypercube. At $\omega = 0$, the critical instability occurs at one of the corners of the parameter space hypercube.

Theorem 2.2

Consider the two real critical constraints (2.157) with multilinearly uncertain parameters.

1) Case 1 ($\mathcal{S}_1 \cap \mathcal{S}_2 = \emptyset$): The critical instability occurs at one of the corners of the parameter space hypercube.

2) Case 2 ($\mathcal{S}_1 \cap \mathcal{S}_2 \neq \emptyset$): If $\mu_1(\omega)$ and $\mu_2(\omega)$ plots intersect at some frequencies ω_c, and if the restricted parameters vectors subject to $\mathcal{S}_1 \cap \mathcal{S}_2$, associated with $\mu_1(\omega_c)$ and $\mu_2(\omega_c)$, become $\mathbf{p}_{s_1} = \mathbf{p}_{s_2}$, then the critical instability occurs at one of the corners of the parameter space hypercube.

2.6.4.3 Interval Polynomial As an application of the concept of the two real critical constraints, consider a family of real polynomials

$$\phi(s, \mathbf{a}) = s^n + a_1 s^{n-1} + \cdots + a_{n-1} s + a_n \qquad (2.169)$$

where $\mathbf{a}$ denotes the uncertain parameter vector and each uncertain coefficient a_i has a prescribed interval as

$$a_i^- \leq a_i \leq a_i^+, \qquad i = 1, \ldots, n$$

The nominal values of a_i are assumed as

$$\bar{a}_i = \tfrac{1}{2}(a_i^+ + a_i^-), \qquad i = 1, \ldots, n$$

An interval polynomial with the normalized uncertain parameters δ_i will be used in the subsequent discussion.

A polynomial of the form of Eq. (2.169) whose zeros lie on the open left-half s plane is called a Hurwitz polynomial. Kharitonov's theorem provides a simple way of checking whether a given polynomial whose coefficients have prescribed intervals is a Hurwitz polynomial; however, it is not directly applicable to determining the size of the largest stable hypercube in the coefficient parameter space. Now we employ $\mu_1(\omega)$ and $\mu_2(\omega)$ to determine the real parameter margin κ^* of an uncertain dynamic system described by an interval characteristic polynomial.

An interval polynomial can be transformed into the two real constraints with the rank-one matrices $\mathbf{M}_1$ and $\mathbf{M}_2$. If n is an even integer, we have

$$\mathbf{M}_i(\omega) = \frac{\boldsymbol{\alpha}_i \boldsymbol{\beta}_i^T(\omega)}{g_i(\omega)}, \qquad\qquad i = 1, 2$$

$$\boldsymbol{\Delta}_1 = \text{diag}(\delta_2, \delta_4, \dots, \delta_n), \qquad \boldsymbol{\Delta}_2 = \text{diag}(\delta_1, \delta_3, \dots, \delta_{n-1})$$

$$\boldsymbol{\alpha}_1 = \boldsymbol{\alpha}_2 = [1, 1, \dots, 1]^T$$

$$\boldsymbol{\beta}_1 = \boldsymbol{\beta}_2 = \left[-\omega^{n-2}, \omega^{n-4}, \dots, -\omega^2, (-1)^{n/2} \right]^T$$

$$g_1(\omega) = -\omega^n + \bar{a}_2 \omega^{n-2} - \dots - (-1)^{n/2} \bar{a}_n$$

$$g_2(\omega) = \bar{a}_1 \omega^{n-2} - \bar{a}_3 \omega^{n-4} + \dots - (-1)^{n/2} \bar{a}_{(n-1)}$$

Similar results can be obtained for the case of odd n.

Because $\mathbf{M}_1(\omega)$ and $\mathbf{M}_2(\omega)$ are rank-one matrices, we obtain the following theorem.

Theorem 2.3

The $\mu_1(\omega)$ and $\mu_2(\omega)$ of an interval polynomial of the form of Eq. (2.169) attain their values at one of the corners of the parameter space hypercube, and then can be expressed as

$$\mu_i(\omega) = \frac{\boldsymbol{\alpha}_i^T \mathbf{E}_i \boldsymbol{\beta}_i(\omega)}{g_i(\omega)}, \qquad\qquad i = 1, 2 \qquad (2.170)$$

where

$$\mathbf{E}_i = \text{sgn}(g_i)\, \text{diag}\{\text{sgn}(\beta_{i1}), \text{sgn}(\beta_{i2}), \dots, 1\}, \qquad i = 1, 2 \qquad (2.171)$$

β_{ij} is the jth element of the column vector $\boldsymbol{\beta}_i$, and $\text{sgn}(\cdot)$ denotes the signum function. The real parameter margin is then obtained as

$$1/\kappa^* \equiv \mu^* \equiv \sup_{\omega} \mu_{12}(\omega)$$

$$= \sup_{\omega} \left\{ \frac{\boldsymbol{\alpha}_1^T \mathbf{E}_1 \boldsymbol{\beta}_1(\omega)}{g_1(\omega)}, \frac{\boldsymbol{\alpha}_2^T \mathbf{E}_2 \boldsymbol{\beta}_2(\omega)}{g_2(\omega)} \right\} \qquad (2.172)$$

Proof: Because $\mu_1(\omega)$ and $\mu_2(\omega)$ attain their values at one of the corners of the parameter space hypercube, we have

$$\mu_i(\omega) = \max_{E_i \in \mathcal{E}_i} \rho[\mathbf{M}_i(j\omega)\mathbf{E}_i], \qquad\qquad i = 1, 2$$

Because

$$\det\left[\lambda\mathbf{I} - \frac{\boldsymbol{\alpha}_i(\omega)\boldsymbol{\beta}_i^T(\omega)\mathbf{E}_i}{g_i(\omega)}\right] = \lambda - \frac{\boldsymbol{\beta}_i^T(\omega)\mathbf{E}_i\boldsymbol{\alpha}_i(\omega)}{g_i(\omega)}$$

$$= \lambda - \frac{\boldsymbol{\alpha}_i^T(\omega)\mathbf{E}_i\boldsymbol{\beta}_i(\omega)}{g_i(\omega)}$$

we have

$$\mu_i(\omega) = \max_{\mathbf{E}_i\in\mathcal{E}_i}\left\{\frac{\boldsymbol{\alpha}_i^T(\omega)\mathbf{E}_i\boldsymbol{\beta}_i(\omega)}{g_i(\omega)}\right\}$$

Let

$$\mathbf{E}_i = \text{sgn}(g_i)\,\text{diag}\{\text{sgn}(\beta_{i1}), \text{sgn}(\beta_{i2}), \dots, 1\}$$

then $\mu_i(\omega)$ will attain their maximum values.

2.6.4.4 Kharitonov's Theorem
An alternative proof of Kharitonov's theorem using the concept of $\mu_1(\omega)$ and $\mu_2(\omega)$ and Theorem 2.3 is given here. Without loss of generality, we consider the case of even n.

The uncertain parameter sets corresponding to μ_1 and μ_2 of an interval polynomial are disjoint; i.e., $S_1 \cap S_2 = \emptyset$, and these disjoint parameter sets should have the same bound κ^*. Consequently, from Theorem 2.3, we have either

$$\mu^* = \frac{\boldsymbol{\alpha}_1^T\mathbf{E}_1\boldsymbol{\beta}_1(\omega_c)}{g_1(\omega_c)} = \left|\frac{\boldsymbol{\alpha}_2^T\mathbf{E}_2\boldsymbol{\beta}_2(\omega_c)}{g_2(\omega_c)}\right| \tag{2.173}$$

or

$$\mu^* = \frac{\boldsymbol{\alpha}_2^T\mathbf{E}_2\boldsymbol{\beta}_2(\omega_c)}{g_2(\omega_c)} = \left|\frac{\boldsymbol{\alpha}_1^T\mathbf{E}_1\boldsymbol{\beta}_1(\omega_c)}{g_1(\omega_c)}\right| \tag{2.174}$$

where $\mathbf{E}_i$ are defined as Eq. (2.171).

For the case of Eq. (2.173), the possible critical parameters are

$$(\delta_2, \delta_4, \dots, \delta_n) = \kappa^*[\text{sgn}(\beta_{11}), \text{sgn}(\beta_{12}), \dots, 1]^T\,\text{sgn}(g_1)$$

$$(\delta_1, \delta_3, \dots, \delta_{n-1}) = \pm\kappa^*[\text{sgn}(\beta_{21}), \text{sgn}(\beta_{22}), \dots, 1]^T\,\text{sgn}(g_2)$$

which become

$$(\delta_2, \delta_4, \dots, \delta_n) = \kappa^*\left[1, -1, 1, \dots, (-1)^{n/2}\right]^T\,\text{sgn}(g_1)$$

$$(\delta_1, \delta_3, \dots, \delta_{n-1}) = \pm\kappa^*\left[1, -1, 1, \dots, (-1)^{n/2}\right]^T\,\text{sgn}(g_2)$$

For the case of Eq. (2.174), the possible critical parameters are

$$(\delta_1, \delta_3, \ldots, \delta_{n-1}) = \kappa^*[\text{sgn}(\beta_{21}), \text{sgn}(\beta_{22}), \ldots, 1]^T \, \text{sgn}(g_2)$$

$$(\delta_2, \delta_4, \ldots, \delta_n) = \pm\kappa^*[\text{sgn}(\beta_{11}), \text{sgn}(\beta_{12}), \ldots, 1]^T \, \text{sgn}(g_1)$$

which become

$$(\delta_1, \delta_3, \ldots, \delta_{n-1}) = \kappa^*\left[1, -1, 1, \ldots, (-1)^{n/2}\right]^T \, \text{sgn}(g_2)$$

$$(\delta_2, \delta_4, \ldots, \delta_n) = \pm\kappa^*\left[1, -1, 1, \ldots, (-1)^{n/2}\right]^T \, \text{sgn}(g_1)$$

There are a total of 16 combinations of possible critical parameters, but only 4 of them are different from each other. The four corner vectors of the parameter space hypercube for possible critical instability at a corner are then obtained as

$$\delta^{(1)} = \left[1, -1, -1, 1, \ldots, (-1)^{n/2+1}, (-1)^{n/2}\right]^T$$

$$\delta^{(2)} = \left[1, 1, -1, -1, \ldots, (-1)^{n/2}, (-1)^{n/2}\right]^T$$

$$\delta^{(3)} = \left[-1, 1, 1, -1, \ldots, (-1)^{n/2}, (-1)^{n/2+1}\right]^T$$

$$\delta^{(4)} = \left[-1, -1, 1, 1, \ldots, (-1)^{n/2+1}, (-1)^{n/2+1}\right]^T$$

where $\delta = (\delta_1, \delta_2, \delta_3, \delta_4, \ldots, \delta_{n-1}, \delta_n)$. These are, in fact, Kharitonov's four corners for the case of even n, corresponding to the four extreme polynomials of Kharitonov's theorem discussed in Section 1.3.7.

2.6.4.5 Polytopic Polynomial

Consider a polynomial whose coefficients depend linearly on the normalized, perturbation parameter vector $\delta \in \kappa^*\mathcal{D}$:

$$\phi(s; \delta) = s^n + \sum_{i=1}^{n} a_i(\delta)s^{n-i} \tag{2.175}$$

where

$$a_i(\delta) = \bar{a}_i + \sum_{j=1}^{\ell} a_{ij}\delta_j$$

and where a_{ij} are constants. The critical constraints can be written as Eq. (2.157) with the rank-one matrices $\mathbf{M}_1(\omega)$ and $\mathbf{M}_2(\omega)$ such that

$$\mathbf{M}_i(\omega) = \frac{\alpha_i(\omega)\beta_i^T(\omega)}{g_i(\omega)}, \qquad i = 1, 2$$

where $g_i(\omega)$ is a scalar function of ω, and α_i and β_i are column vectors that are functions of ω.

Because of the rank-one property of $\mathbf{M}_1$ and $\mathbf{M}_2$, we obtain the following result.

Theorem 2.4

The two single-constraint real μ measures of a polytopic polynomial of the form of Eq. (2.175) will attain their values at one of the corners of the parameter space hypercube and can be expressed as

$$\mu_i(\omega) = \frac{\boldsymbol{\alpha}_i^T(\omega)\mathbf{E}_i\boldsymbol{\beta}_i(\omega)}{g_i(\omega)}, \qquad i = 1, 2 \qquad (2.176)$$

where

$$\mathbf{E}_i = \mathrm{diag}\{\mathrm{sgn}(\alpha_{i1}\beta_{i1}), \ldots, \mathrm{sgn}(\alpha_{i\ell_i}\beta_{i\ell_i})\}\mathrm{sgn}(g_i)$$

for $i = 1, 2$, and α_{ij} and β_{ij} are, respectively, the jth elements of the column vectors $\boldsymbol{\alpha}_i$ and $\boldsymbol{\beta}_i$. If μ_1 and μ_2 intersect at some frequencies ω_c and if $\mathbf{p}_{s_1} = \mathbf{p}_{s_2}$ at these frequencies, then the critical instability occurs at one of the corners of the parameter space hypercube.

2.6.4.6 Multilinearly Uncertain Polynomial For a general case of multilinearly uncertain polynomial, we know $\mu_1(\omega)$ and $\mu_2(\omega)$ will attain their values at one of the corners of the parameter space hypercube $\mathcal{D}$ at any frequency ω. Consequently, we have the following theorem.

Theorem 2.5

The two single-constraint real μ measures will attain their values at one of the corners of the parameter space hypercube and can be expressed as

$$\mu_i(\omega) = \max_{\mathbf{E}_i \in \mathcal{E}_i} \rho[\mathbf{M}_i(\omega)\mathbf{E}_i], \qquad i = 1, 2$$

where

$$\mathcal{E}_i = \{\mathbf{E}_i \ : \ \mathbf{E}_i = \mathrm{diag}(e_j\mathbf{I}_j), \ e_j = +1 \text{ or } -1 \ \forall j\}$$

and $\rho(\cdot)$ denotes the maximum real eigenvalue of a matrix. If μ_1 and μ_2 intersect at some frequencies ω_c and if $\mathbf{p}_{s_1} = \mathbf{p}_{s_2}$ at these frequencies, then, the critical instability occurs at one of the corners of the parameter space hypercube.

For a more detailed treatment of robustness analysis and robust control, see Refs. 25 and 26.

Problems

2.29 Consider a polynomial whose coefficients linearly dependent on uncertain parameters δ_1 and δ_2 as follows:

$$\phi(s; \delta_1, \delta_2) = s^4 + (\delta_2 + 3)s^3 + (\delta_1 + 5.5)s^2$$
$$+ (\delta_1 + \delta_2 + 4.5)s + 3\delta_1 - \delta_2 + 5.5$$

(a) Show that the two real critical constraints can be found as

$$\text{Re}[\phi(j\omega; \delta)] = 0 \rightarrow f_1(\omega) \det[\mathbf{I}_1 - \mathbf{M}_1(\omega)\mathbf{\Delta}_1] = 0$$
$$\text{Im}[\phi(j\omega; \delta)] = 0 \rightarrow f_2(\omega) \det[\mathbf{I}_2 - \mathbf{M}_2(\omega)\mathbf{\Delta}_2] = 0$$

where

$$f_1(\omega) = 1, \qquad f_2(\omega) = \omega, \qquad \mathbf{\Delta}_1 = \mathbf{\Delta}_2 = \text{diag}(\delta_1, \delta_2)$$

and

$$\mathbf{M}_i(\omega) = \frac{\alpha_i(\omega)\beta_i^T(\omega)}{g_i(\omega)}$$

with

$$\alpha_1 = \alpha_2 = (1, \ 1)$$
$$\beta_1 = (3 - \omega^2, \ -1)$$
$$\beta_2 = (1, \ 1 - \omega^2)$$
$$g_1(\omega) = -\omega^4 + 5.5\omega^2 - 5.5$$
$$g_2(\omega) = 3\omega^2 - 4.5$$

(b) Verify that μ_1 and μ_2 are

$$\mu_1(\omega) = \frac{|3 - \omega^2| + 1}{|-\omega^4 + 5.5\omega^2 - 5.5|}$$

$$\mu_2(\omega) = \frac{|1 - \omega^2| + 1}{|3\omega^2 - 4.5|}$$

and that the critical corner matrices of $\mu_1(\omega)$ and $\mu_2(\omega)$ are

$$\mathbf{E}_1 = \text{diag}\{\text{sgn}(3 - \omega^2), \ -1\}\text{sgn}(-\omega^4 + 5.5\omega^2 - 5.5)$$
$$\mathbf{E}_2 = \text{diag}\{1, \ \text{sgn}(1 - \omega^2)\}\text{sgn}(3\omega^2 - 4.5)$$

(c) Determine the real parameter margin κ^* as

$$\kappa^* = 1/\mu_{12} = 1/\mu_1(\omega_c) = 1/\mu_2(\omega_c) = 0.75$$

where $\omega_c = 1.4142$ and also the critical parameter values as

$$(\delta_1^*, \delta_2^*) = \kappa^* \mathbf{e}^* = \kappa^*(1, -1) = (0.75, -0.75)$$

2.30 Consider a feedback control system consisting of a plant transfer function $G(s)$ and a compensator $K(s)$ given by

$$G(s) = \frac{p_1}{s(s + p_2)(s + p_3)}, \qquad K(s) = \frac{s + 2}{s + 10}$$

and where the uncertain parameters are described by [22]

$$p_1 = 800(1 + \epsilon_1), \qquad |\epsilon_1| \le 0.1$$
$$p_2 = 4 + \epsilon_2, \qquad\qquad |\epsilon_2| \le 0.2$$
$$p_3 = 6 + \epsilon_3, \qquad\qquad |\epsilon_3| \le 0.3$$

The closed-loop characteristic polynomial is

$$\phi(s; \epsilon_1, \epsilon_2, \epsilon_3) = s^4 + a_1 s^3 + a_2 s^2 + a_3 s + a_4$$

where

$$a_1 = 20 + \epsilon_2 + \epsilon_3$$
$$a_2 = 124 + 16\epsilon_2 + 14\epsilon_3 + \epsilon_2\epsilon_3$$
$$a_3 = 1040 + 800\epsilon_1 + 60\epsilon_2 + 40\epsilon_3 + 10\epsilon_2\epsilon_3$$
$$a_4 = 1600(1 + \epsilon_1)$$

The uncertain parameters are normalized as

$$\delta_1 = \epsilon_1/0.1, \qquad \delta_2 = \epsilon_2/0.2, \qquad \delta_3 = \epsilon_3/0.3$$

(a) Show that the two real critical constraints can be obtained as

$$\text{Re}[\phi(j\omega; \delta)] = 0 \rightarrow f_1(\omega) \det[\mathbf{I}_1 - \mathbf{M}_1(\omega)\mathbf{\Delta}_1] = 0$$
$$\text{Im}[\phi(j\omega; \delta)] = 0 \rightarrow f_2(\omega) \det[\mathbf{I}_2 - \mathbf{M}_2(\omega)\mathbf{\Delta}_2] = 0$$

where

$$f_1(\omega) = 1, \qquad f_2(\omega) = \omega$$
$$\mathbf{\Delta}_1 = \mathbf{\Delta}_2 = \mathbf{\Delta} = \text{diag}(\delta_1, \delta_2, \delta_3)$$
$$\mathbf{M}_i(\omega) = -\mathbf{R}_i(\omega)\mathbf{A}_i^{-1}(\omega)\mathbf{L}_i(\omega), \qquad i = 1, 2$$

and

$$\mathbf{R}_1(\omega) = \begin{bmatrix} 1 & 0 \\ -3.2\omega^2 & 1 \\ 1 & 0 \end{bmatrix}$$

$$\mathbf{R}_2(\omega) = \begin{bmatrix} 1 & 0 \\ 12 - 0.2\omega^2 & -1 \\ 1 & 0 \end{bmatrix}$$

$$\mathbf{L}_1(\omega) = \begin{bmatrix} 160 & 1 & -4.2\omega^2 \\ 0 & 0 & 0.06\omega^2 \end{bmatrix}$$

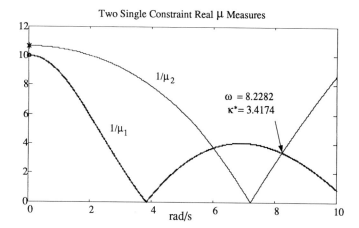

Fig. 2.28   Plots of $1/\mu_1\ (\omega)$ and $1/\mu_2\ (\omega)$.

$$\mathbf{L}_2(\omega) = \begin{bmatrix} 80 & 1 & 12 - 0.3\omega^2 \\ 0 & 0 & 0.6 \end{bmatrix}$$

$$\mathbf{A}_1^{-1}(\omega) = \begin{bmatrix} (1600 - 124\omega^2 + \omega^4)^{-1} & 0 \\ 0 & 1 \end{bmatrix}$$

$$\mathbf{A}_2^{-1}(\omega) = \begin{bmatrix} (1040 - 20\omega^2)^{-1} & 0 \\ 0 & 1 \end{bmatrix}$$

(b) Verify the plots of $1/\mu_1(\omega)$ and $1/\mu_2(\omega)$ as shown in Fig. 2.28. Also show that the critical corner matrices are

$$\mathbf{E}_1^* = \mathbf{E}_2^* = \operatorname{diag}(1, -1, -1)$$

and the real parameter margin is

$$\kappa^* = 1/\mu^* = 1/\mu_1(\omega_c) = 1/\mu_2(\omega_c) = 3.4174$$

where $\omega_c = 8.2282$.

2.7   Robust Control of Uncertain Dynamic Systems

In this section, which is based on Refs. 27–29, a design method is introduced that is particularly useful for uncertain structural dynamic systems with uncertainties in inertia, damping, and stiffness matrices. The method is based on the recent advances in the state-space solution to a standard H_∞ control problem. By defining the structured parameter variations in terms of fictitious inputs and outputs for an internal feedback loop, the state-space solution to a standard H_∞ control problem is easily utilized. In this section, the concept of the internal feedback loop, originally developed for uncertain systems described by first-order state-space equations, is extended to uncertain dynamic systems described by second-order matrix differential equations.

2.7.1 Standard H_∞ Control Problem Formulation

Consider a linear, time-invariant system described by

$$\dot{x}(t) = Ax(t) + B_1 d(t) + B_2 u(t)$$
$$z(t) = C_1 x(t) + D_{11} d(t) + D_{12} u(t) \qquad (2.177)$$
$$y(t) = C_2 x(t) + D_{21} d(t) + D_{22} u(t)$$

where $x \in \mathcal{R}^n$, $d \in \mathcal{R}^{m_1}$, $u \in \mathcal{R}^{m_2}$, $z \in \mathcal{R}^{p_1}$, and $y \in \mathcal{R}^{p_2}$ are, respectively, the state, disturbance input, control input, controlled output, and measured output vectors.

The transfer function representation of this system is given by

$$\begin{bmatrix} z(s) \\ y(s) \end{bmatrix} = \begin{bmatrix} P_{11}(s) & P_{12}(s) \\ P_{21}(s) & P_{22}(s) \end{bmatrix} \begin{bmatrix} d(s) \\ u(s) \end{bmatrix}$$

$$= \left\{ \begin{bmatrix} C_1 \\ C_2 \end{bmatrix} [sI - A]^{-1} \begin{bmatrix} B_1 & B_2 \end{bmatrix} + \begin{bmatrix} D_{11} & D_{12} \\ D_{21} & D_{22} \end{bmatrix} \right\} \begin{bmatrix} d(s) \\ u(s) \end{bmatrix} \qquad (2.178)$$

where $P_{ij}(s)$ are real-rational transfer function matrices and $P(s) \equiv [P_{ij}(s)]$ is called the generalized plant, which may include the internal feedback loop model and frequency-dependent weightings, as will be discussed later in this section.

For a linear system described by Eq. (2.177) or Eq. (2.178) with a feedback control logic of the form

$$u(s) = -K(s)y(s) \qquad (2.179)$$

the closed-loop transfer function from d to z can be derived as

$$T_{zd}(P, K) = P_{11} + P_{12}K[I - P_{22}K]^{-1}P_{21} \qquad (2.180)$$

where $K(s)$ is a compensator transfer function matrix to be synthesized.

The H_∞-norm of a real-rational transfer function matrix $T(s)$ is defined as

$$\|T(s)\|_\infty = \sup_\omega \bar{\sigma}[T(j\omega)] \qquad (2.181)$$

where $\bar{\sigma}[T(j\omega)]$ denotes the largest singular value of $T(j\omega)$ for a given ω. The H_∞ space consists of functions that are stable and bounded.

The design objective of standard H_∞ control problem is then to find $K(s)$ such that the closed-loop system is stable and $\|T_{zd}(P, K)\|_\infty$ is minimized. An H_∞ suboptimal control problem is to find a stabilizing controller $K(s)$ such that $\|T_{zd}(P, K)\|_\infty < \gamma$ for some prespecified $\gamma \in \mathcal{R}$.

Using the input–output decomposition of structured plant parameter variations in terms of the fictitious inputs and outputs of an internal uncertainty loop to be discussed next, the state-space solution to the standard H_∞ control problem will be utilized to design parameter-insensitive controllers.

2.7.2 Modeling of Structured Parameter Uncertainty

2.7.2.1 Uncertainty Modeling of Linear Systems Consider an uncertain linear dynamic system described by

$$\mathbf{E}\dot{\mathbf{x}} = \mathbf{F}\mathbf{x} + \mathbf{G}_d\mathbf{d} + \mathbf{G}_u\mathbf{u} \tag{2.182}$$

where $\mathbf{x}$, $\mathbf{d}$, and $\mathbf{u}$ are the state, external disturbance, and control input vectors, respectively; $\mathbf{G}_d$ is the disturbance distribution matrix; $\mathbf{G}_u$ is the control input distribution matrix; and the matrices $\mathbf{E}$ and $\mathbf{F}$ are subject to structured parameter variations.

Suppose that there are ℓ independent, uncertain parameter variables δ_i and assume that the perturbed matrices $\mathbf{E}$ and $\mathbf{F}$ in Eq. (2.182) can be linearly decomposed as follows:

$$\mathbf{E} = \mathbf{E}_0 + \Delta\mathbf{E} \tag{2.183a}$$

$$\mathbf{F} = \mathbf{F}_0 + \Delta\mathbf{F} \tag{2.183b}$$

where $\mathbf{E}_0$ and $\mathbf{F}_0$ are the nominal matrices and $\Delta\mathbf{E}$ and $\Delta\mathbf{F}$ are the perturbation matrices defined as

$$\Delta\mathbf{E} = \sum_{i=1}^{\ell} \Delta\mathbf{E}_i\delta_i = \sum_{i=1}^{\ell} \mathbf{M}_E^{(i)}\delta_i\mathbf{I}_{\kappa_i}\mathbf{N}_E^{(i)} = \mathbf{M}_E\mathcal{E}_E\mathbf{N}_E \tag{2.184a}$$

$$\Delta\mathbf{F} = \sum_{i=1}^{\ell} \Delta\mathbf{F}_i\delta_i = \sum_{i=1}^{\ell} \mathbf{M}_F^{(i)}\delta_i\mathbf{I}_{\nu_i}\mathbf{N}_F^{(i)} = \mathbf{M}_F\mathcal{E}_F\mathbf{N}_F \tag{2.184b}$$

where κ_i is the rank of $\Delta\mathbf{E}_i$, ν_i the rank of $\Delta\mathbf{F}_i$, and $\mathcal{E}_E$ and $\mathcal{E}_F$ are diagonal matrices with δ_i as their diagonal elements. If $\kappa_i = \nu_i = 1$ for $i = 1\ldots\ell$, i.e., a special case of rank-one dependency, $\mathbf{M}_E^{(i)}$ and $\mathbf{M}_F^{(i)}$ become column vectors and $\mathbf{N}_E^{(i)}$ and $\mathbf{N}_F^{(i)}$ become row vectors. In this case, there are no repeated elements δ_i in $\mathcal{E}_E$ and $\mathcal{E}_F$.

Let

$$\mathcal{E} \stackrel{\triangle}{=} \text{diag}\{\mathcal{E}_E, \mathcal{E}_F\} \tag{2.185a}$$

$$\tilde{\mathbf{z}} \stackrel{\triangle}{=} \begin{bmatrix} \tilde{\mathbf{z}}_E \\ \tilde{\mathbf{z}}_F \end{bmatrix} = \begin{bmatrix} \mathbf{N}_E\dot{\mathbf{x}} \\ \mathbf{N}_F\mathbf{x} \end{bmatrix} \tag{2.185b}$$

$$\tilde{\mathbf{d}} \stackrel{\triangle}{=} -\mathcal{E}\tilde{\mathbf{z}} \tag{2.185c}$$

where $\tilde{\mathbf{d}}$ is called the fictitious disturbance input, $\tilde{\mathbf{z}}$ the fictitious output, and $\mathcal{E}$ the gain matrix of a fictitious internal uncertainty loop, which is caused by uncertainty in the matrices $\mathbf{E}$ and $\mathbf{F}$. Then, substituting Eqs. (2.183) into Eq. (2.182), we obtain

$$\mathbf{E}_0\dot{\mathbf{x}} = \mathbf{F}_0\mathbf{x} + \mathbf{G}_{\tilde{d}}\tilde{\mathbf{d}} + \mathbf{G}_d\mathbf{d} + \mathbf{G}_u\mathbf{u} \tag{2.186}$$

where $\mathbf{G}_{\tilde{d}}$, the fictitious disturbance distribution matrix, is defined as

$$\mathbf{G}_{\tilde{d}} = [\mathbf{M}_E \quad -\mathbf{M}_F]$$

Defining the controlled output vector as

$$\mathbf{z} = \begin{bmatrix} \mathbf{C}_{11} \\ 0 \end{bmatrix} \mathbf{x} + \begin{bmatrix} 0 \\ \mathbf{I} \end{bmatrix} \mathbf{u}$$

and introducing new variables

$$\hat{\mathbf{d}} \triangleq \begin{bmatrix} \tilde{\mathbf{d}} \\ \mathbf{d} \end{bmatrix}, \qquad \hat{\mathbf{z}} \triangleq \begin{bmatrix} \tilde{\mathbf{z}} \\ \mathbf{z} \end{bmatrix}$$

we obtain a modified state-space representation of the system as follows:

$$\dot{\mathbf{x}} = \mathbf{A}\mathbf{x} + \mathbf{B}_1\hat{\mathbf{d}} + \mathbf{B}_2\mathbf{u} \qquad (2.187a)$$

$$\hat{\mathbf{z}} = \mathbf{C}_1\mathbf{x} + \mathbf{D}_{11}\hat{\mathbf{d}} + \mathbf{D}_{12}\mathbf{u} \qquad (2.187b)$$

where

$$\mathbf{A} = \mathbf{E}_0^{-1}\mathbf{F}_0, \qquad \mathbf{B}_1 = \mathbf{E}_0^{-1}[\mathbf{G}_{\tilde{d}} \quad \mathbf{G}_d], \qquad \mathbf{B}_2 = \mathbf{E}_0^{-1}\mathbf{G}_u$$

$$\mathbf{C}_1 = \begin{bmatrix} \mathbf{N}_E\mathbf{E}_0^{-1}\mathbf{F}_0 \\ \mathbf{N}_F \\ \mathbf{C}_{11} \\ 0 \end{bmatrix}, \qquad \mathbf{D}_{11} = \mathbf{N}_E\mathbf{E}_0^{-1}\begin{bmatrix} \mathbf{G}_{\tilde{d}} & \mathbf{G}_d \\ 0 & 0 \\ 0 & 0 \\ 0 & 0 \end{bmatrix}$$

$$\mathbf{D}_{12} = \begin{bmatrix} \mathbf{N}_E\mathbf{E}_0^{-1}\mathbf{G}_u \\ 0 \\ 0 \\ \mathbf{I} \end{bmatrix}$$

Note that $\mathbf{D}_{11} = 0$ if there is no uncertainty in $\mathbf{E}$.

2.7.2.2 Uncertainty Modeling of Structural Dynamic Systems Consider an uncertain structural dynamic system described by

$$\mathbf{M}\ddot{\mathbf{q}} + \mathbf{D}\dot{\mathbf{q}} + \mathbf{K}\mathbf{q} = \mathbf{G}_d\mathbf{d} + \mathbf{G}_u\mathbf{u} \qquad (2.188)$$

where $\mathbf{q}$ is the generalized displacement vector, $\mathbf{M}$ the mass matrix, $\mathbf{D}$ the damping matrix, $\mathbf{K}$ the stiffness matrix, $\mathbf{G}_d$ the disturbance distribution matrix, $\mathbf{G}_u$ the control input distribution matrix, $\mathbf{d}$ the external disturbance vector, and $\mathbf{u}$ the control input vector.

Suppose that matrices $\mathbf{M}$, $\mathbf{D}$, and $\mathbf{K}$ are subject to ℓ independent parameter variations represented by δ_i, $i = 1, \ldots, \ell$. Then the perturbed matrices in Eq. (2.188) can be linearly decomposed as

$$\mathbf{M} = \mathbf{M}_0 + \mathbf{L}_M\boldsymbol{\Delta}_M\mathbf{R}_M$$
$$\mathbf{D} = \mathbf{D}_0 + \mathbf{L}_D\boldsymbol{\Delta}_D\mathbf{R}_D \qquad (2.189)$$
$$\mathbf{K} = \mathbf{K}_0 + \mathbf{L}_K\boldsymbol{\Delta}_K\mathbf{R}_K$$

where $\mathbf{\Delta}_M$, $\mathbf{\Delta}_D$, and $\mathbf{\Delta}_K$ are diagonal matrices with independent parameter variations δ_i as diagonal elements. The first matrices in the right-hand side of Eq. (2.189) are the nominal matrices and the second ones are the perturbation matrices, which are linearly decomposed.

Define

$$\mathbf{\Delta} = \text{diag}\{\mathbf{\Delta}_M, \mathbf{\Delta}_D, \mathbf{\Delta}_K\} \tag{2.190a}$$

$$\tilde{\mathbf{z}} = \begin{bmatrix} \tilde{\mathbf{z}}_M \\ \tilde{\mathbf{z}}_D \\ \tilde{\mathbf{z}}_K \end{bmatrix} = \begin{bmatrix} \mathbf{R}_M \ddot{\mathbf{q}} \\ \mathbf{R}_D \ddot{\mathbf{q}} \\ \mathbf{R}_K \mathbf{q} \end{bmatrix} \tag{2.190b}$$

$$\tilde{\mathbf{d}} = \mathbf{\Delta} \tilde{\mathbf{z}} \tag{2.190c}$$

where $\tilde{\mathbf{d}}$ is the fictitious input, $\tilde{\mathbf{z}}$ the fictitious output, and $\mathbf{\Delta}$ the gain matrix of internal uncertainty loop. Substituting Eq. (2.189) into Eq. (2.188), we obtain

$$\mathbf{M}_0 \ddot{\mathbf{q}} + \mathbf{D}_0 \dot{\mathbf{q}} + \mathbf{K}_0 \mathbf{q} = \mathbf{G}_{\tilde{d}} \tilde{\mathbf{d}} + \mathbf{G}_d \mathbf{d} + \mathbf{G}_u \mathbf{u} \tag{2.191}$$

where $\mathbf{G}_{\tilde{d}}$, called the fictitious disturbance distribution matrix, is defined as

$$\mathbf{G}_{\tilde{d}} = [-\mathbf{L}_M, \quad -\mathbf{L}_D, \quad -\mathbf{L}_K] \tag{2.192}$$

Let the state vector and controlled output vector be defined as

$$\mathbf{x} = \begin{bmatrix} \mathbf{q} \\ \dot{\mathbf{q}} \end{bmatrix} \tag{2.193a}$$

$$\mathbf{z} = \begin{bmatrix} \mathbf{C}_{11} & \mathbf{C}_{12} \\ 0 & 0 \end{bmatrix} \begin{bmatrix} \mathbf{q} \\ \dot{\mathbf{q}} \end{bmatrix} + \begin{bmatrix} 0 \\ \mathbf{I} \end{bmatrix} \mathbf{u} \tag{2.193b}$$

Then the parametric variations are incorporated into the standard state-space formulation by introducing new variables

$$\hat{\mathbf{d}} = \begin{bmatrix} \tilde{\mathbf{d}} \\ \mathbf{d} \end{bmatrix}, \qquad \hat{\mathbf{z}} = \begin{bmatrix} \tilde{\mathbf{z}} \\ \mathbf{z} \end{bmatrix} \tag{2.194}$$

The resulting, modified state-space representation is then given by

$$\dot{\mathbf{x}}(t) = \mathbf{A}\mathbf{x}(t) + \mathbf{B}_1 \hat{\mathbf{d}}(t) + \mathbf{B}_2 \mathbf{u}(t)$$
$$\hat{\mathbf{z}}(t) = \mathbf{C}_1 \mathbf{x}(t) + \mathbf{D}_{11} \hat{\mathbf{d}}(t) + \mathbf{D}_{12} \mathbf{u}(t) \tag{2.195}$$
$$\mathbf{y}(t) = \mathbf{C}_2 \mathbf{x}(t) + \mathbf{D}_{21} \hat{\mathbf{d}}(t) + \mathbf{D}_{22} \mathbf{u}(t)$$

where

$$A = \begin{bmatrix} 0 & I \\ -M_0^{-1}K_0 & -M_0^{-1}D_0 \end{bmatrix}$$

$$B_1 = \begin{bmatrix} 0 & 0 \\ M_0^{-1}G_{\tilde{d}} & M_0^{-1}G_d \end{bmatrix}, \qquad B_2 = \begin{bmatrix} 0 \\ M_0^{-1}G_u \end{bmatrix}$$

$$C_1 = \begin{bmatrix} -R_M M_0^{-1}K_0 & -R_M M_0^{-1}D_0 \\ 0 & R_D \\ R_K & 0 \\ C_{11} & C_{12} \\ 0 & 0 \end{bmatrix}$$

$$D_{11} = \begin{bmatrix} R_M M_0^{-1}G_{\tilde{d}} & R_M M_0^{-1}G_d \\ 0 & 0 \\ 0 & 0 \\ 0 & 0 \\ 0 & 0 \end{bmatrix}$$

$$D_{12} = \begin{bmatrix} R_M M_0^{-1}G_u \\ 0 \\ 0 \\ 0 \\ I \end{bmatrix}$$

If there is no uncertainty in the mass matrix M, then $D_{11} = 0$.

2.7.3 Robust H_∞ Compensator Design

Consider an uncertain linear system described by

$$\begin{bmatrix} \hat{z}(s) \\ y(s) \end{bmatrix} = \begin{bmatrix} P_{11}(s) & P_{12}(s) \\ P_{21}(s) & P_{22}(s) \end{bmatrix} \begin{bmatrix} \hat{d}(s) \\ u(s) \end{bmatrix}$$

$$= \left\{ \begin{bmatrix} C_1 \\ C_2 \end{bmatrix} [sI - A]^{-1} [B_1 \quad B_2] + \begin{bmatrix} D_{11} & D_{12} \\ D_{21} & D_{22} \end{bmatrix} \right\} \begin{bmatrix} \hat{d}(s) \\ u(s) \end{bmatrix} \quad (2.196)$$

where

$$\hat{z} = \begin{bmatrix} \tilde{z} \\ z \end{bmatrix}, \qquad \hat{d} = \begin{bmatrix} \tilde{d} \\ d \end{bmatrix}$$

and $x \in \mathcal{R}^n$, $\hat{d} \in \mathcal{R}^{m_1}$, $u \in \mathcal{R}^{m_2}$, $\hat{z} \in \mathcal{R}^{p_1}$, and $y \in \mathcal{R}^{p_2}$ are, respectively, the state, augmented disturbance input, control input, augmented controlled output, and measured output vectors. Furthermore, the fictitious disturbance input $\tilde{d}$ and the fictitious output $\tilde{z}$ are defined as

$$\tilde{d} = \Delta\tilde{z}$$

where Δ is the gain matrix of the internal uncertainty loop.

This uncertain system to be controlled can also be described by

$$\begin{bmatrix} \tilde{z} \\ z \\ y \end{bmatrix} = \begin{bmatrix} G_{11} & G_{12} & G_{13} \\ G_{21} & G_{22} & G_{23} \\ G_{31} & G_{32} & G_{33} \end{bmatrix} \begin{bmatrix} \tilde{d} \\ d \\ u \end{bmatrix} \qquad (2.197a)$$

$$\tilde{d} = \Delta \tilde{z} \qquad (2.197b)$$

$$u = -K(s)y \qquad (2.197c)$$

where $K(s)$ is a feedback compensator to be designed. The overall system described by Eqs. (2.197) is visualized in Fig. 2.29.

After closing the control loop with a stabilizing controller $K(s)$, we obtain the following representation of the closed-loop system (but with the internal uncertainty loop broken):

$$\hat{z} = T_{\hat{z}\hat{d}} \hat{d} \qquad (2.198)$$

where

$$T_{\hat{z}\hat{d}} = \begin{bmatrix} T_{\tilde{z}\tilde{d}} & T_{\tilde{z}d} \\ T_{z\tilde{d}} & T_{zd} \end{bmatrix} \equiv \begin{bmatrix} T_{11} & T_{12} \\ T_{21} & T_{22} \end{bmatrix} \qquad (2.199a)$$

$$T_{11} = G_{11} - G_{13}K(I + G_{33}K)^{-1}G_{31} \qquad (2.199b)$$

$$T_{12} = G_{12} - G_{13}K(I + G_{33}K)^{-1}G_{32} \qquad (2.199c)$$

$$T_{21} = G_{21} - G_{23}K(I + G_{33}K)^{-1}G_{31} \qquad (2.199d)$$

$$T_{22} = G_{22} - G_{23}K(I + G_{33}K)^{-1}G_{32} \qquad (2.199e)$$

The actual closed-loop transfer function matrix from d to z under plant perturbations becomes

$$T_{zd} = T_{22} + T_{21}\Delta(I - T_{11}\Delta)^{-1}T_{12} \qquad (2.200)$$

The following two theorems provide sufficient conditions for stability/performance robustness [29, 30].

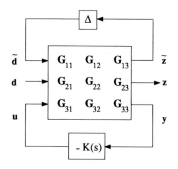

Fig. 2.29   Overall closed-loop system with internal uncertainty loop.

Theorem 2.6 Stability Robustness

If $\|\mathbf{T}_{11}(s)\|_\infty < \gamma$, then $\mathbf{T}_{zd}(s, \alpha\mathbf{\Delta})\ \forall \alpha \in [0, 1]$ is stable for $\|\mathbf{\Delta}\| \le \gamma^{-1}$.

Theorem 2.7 Performance Robustness

If $\|\mathbf{T}_{\hat{z}\hat{d}}\|_\infty < \gamma$, then $\mathbf{T}_{zd}(s, \alpha\mathbf{\Delta})\ \forall \alpha \in [0, 1]$ is stable and $\|\mathbf{T}_{zd}(s, \alpha\mathbf{\Delta})\|_\infty < \gamma$ $\forall \alpha \in [0, 1]$ with $\|\mathbf{\Delta}\| \le \gamma^{-1}$.

Because $\mathbf{T}_{11}$ can be represented as

$$\mathbf{T}_{11} = [\mathbf{I} \quad 0] \begin{bmatrix} \mathbf{T}_{11} & \mathbf{T}_{12} \\ \mathbf{T}_{21} & \mathbf{T}_{22} \end{bmatrix} \begin{bmatrix} \mathbf{I} \\ 0 \end{bmatrix} \tag{2.201}$$

we have $\|\mathbf{T}_{11}\|_\infty \le \|\mathbf{T}_{\hat{z}\hat{d}}\|_\infty$. Consequently, if the condition in Theorem 2.7 is satisfied, i.e., $\|\mathbf{T}_{\hat{z}\hat{d}}\|_\infty < \gamma$, both stability and performance robustness will be achieved with respect to bounded uncertainty $\|\mathbf{\Delta}\| \le \gamma^{-1}$.

Theorem 2.8 H_∞-Suboptimal Controller

This theorem from Ref. 29 provides a robust H_∞-suboptimal controller, which satisfies the condition in Theorem 2.7.

Consider a linear system described by Eq. (2.196). Assume the following:
1) $(\mathbf{A}, \mathbf{B}_2)$ is stabilizable and $(\mathbf{C}_2, \mathbf{A})$ is detectable.
2) $\mathbf{D}_{12}^T[\mathbf{C}_1 \quad \mathbf{D}_{12}] = [0 \quad \mathbf{I}]$.
3)

$$\begin{bmatrix} \mathbf{B}_1 \\ \mathbf{D}_{21} \end{bmatrix} \mathbf{D}_{21}^T = \begin{bmatrix} 0 \\ \mathbf{I} \end{bmatrix}$$

4) The rank of $\mathbf{P}_{12}(j\omega)$ and $\mathbf{P}_{21}(j\omega)$ is m_2 and p_2, respectively, for all ω.
5) $\mathbf{D}_{11} = 0$ and $\mathbf{D}_{22} = 0$.

There exists an internally stabilizing controller such that $\|\mathbf{T}_{\hat{z}\hat{d}}\|_\infty < \gamma$, if and only if the following Riccati equations

$$0 = \mathbf{A}^T\mathbf{X} + \mathbf{X}\mathbf{A} - \mathbf{X}(\mathbf{B}_2\mathbf{B}_2^T - \gamma^{-2}\mathbf{B}_1\mathbf{B}_1^T)\mathbf{X} + \mathbf{C}_1^T\mathbf{C}_1 \tag{2.202}$$

$$0 = \mathbf{A}\mathbf{Y} + \mathbf{Y}\mathbf{A}^T - \mathbf{Y}(\mathbf{C}_2^T\mathbf{C}_2 - \gamma^{-2}\mathbf{C}_1^T\mathbf{C}_1)\mathbf{Y} + \mathbf{B}_1\mathbf{B}_1^T \tag{2.203}$$

have solutions $\mathbf{X}$ and $\mathbf{Y}$. An H_∞-suboptimal controller that satisfies $\|\mathbf{T}_{\hat{z}\hat{d}}\|_\infty < \gamma$, where γ is a design tradeoff variable specifying an upper bound of the perturbed closed-loop transfer matrix $\mathbf{T}_{\hat{z}\hat{d}}$, is then obtained as

$$\dot{\hat{\mathbf{x}}} = \mathbf{A}_c\hat{\mathbf{x}} + \mathbf{L}\mathbf{y} \tag{2.204a}$$

$$\mathbf{u} = -\mathbf{K}\hat{\mathbf{x}} \tag{2.204b}$$

where

$$\mathbf{K} = \mathbf{B}_2^T\mathbf{X} \tag{2.205}$$

$$\mathbf{L} = (\mathbf{I} - \gamma^{-2}\mathbf{Y}\mathbf{X})^{-1}\mathbf{Y}\mathbf{C}_2^T \tag{2.206}$$

$$\mathbf{A}_c = \mathbf{A} + \gamma^{-2}\mathbf{B}_1\mathbf{B}_1^T\mathbf{X} - \mathbf{B}_2\mathbf{K} - \mathbf{L}\mathbf{C}_2 \qquad (2.207)$$

and $\hat{\mathbf{x}}$ represents the controller state vector.

The closed-loop system (neglecting all of the external inputs) is then described as

$$\begin{bmatrix} \dot{\mathbf{x}} \\ \dot{\hat{\mathbf{x}}} \end{bmatrix} = \begin{bmatrix} \mathbf{A} & -\mathbf{B}_2\mathbf{K} \\ \mathbf{L}\mathbf{C}_2 & \mathbf{A}_c \end{bmatrix} \begin{bmatrix} \mathbf{x} \\ \hat{\mathbf{x}} \end{bmatrix} \qquad (2.208)$$

Note that the H_∞ controller has a structure similar to a conventional state-space controller, consisting of an estimator and a regulator, but is designed for a plant system matrix

$$\mathbf{A} + \gamma^{-2}\mathbf{B}_1\mathbf{B}_1^T\mathbf{X}$$

Consequently, the separation principle of the conventional LQG technique does not hold here.

Relaxing the assumptions in Theorem 2.8, especially $\mathbf{D}_{11} = 0$, will significantly complicate the formulas [30].

2.7.4 Benchmark Problems for Robust Control Design

Simple, yet meaningful, control problems used to highlight issues in robust control design and to provide a forum for the application of a variety of robust control design methodologies were formulated by Wie and Bernstein [31] in 1990. These problems were then refined and addressed as benchmark problems for robust control design, and various solutions to these problems were presented at the American Control Conferences in 1990–1992. The original three problems are concerned with a disturbance rejection control problem in the presence of parametric uncertainty. These problems were later augmented with a command tracking control problem in the presence of plant modeling uncertainty [31]. For this fast tracking problem, the control input saturation limit is specified explicitly.

A special section of the *Journal of Guidance, Control, and Dynamics* (Vol. 15, No. 5, 1992) was devoted to the 11 different control designs for the original three benchmark problems [31]. See Thompson [32] for further discussion of these various solutions to benchmark problems.

It is emphasized that in the statement of problems, certain aspects, such as parameter uncertainty with given nominal parameter values and nominal desired performance, are specified concretely, whereas other aspects, such as the sensor noise model, definition of settling time, measure of control effort, controller complexity, bandwidth, etc., are deliberately left vague. Each designer is thus given the opportunity to emphasize additional design tradeoffs for a realistic control design as desired.

Problem

2.31 Consider the two-mass–spring system shown in Fig. 2.30. It is assumed that for the nominal system, $m_1 = m_2 = 1$ and $k = 1$, with appropriate units and time in seconds. A control force acts on body 1 and the position of body 2 is measured, resulting in a noncollocated control problem.

This system can be represented in state-space form as

$$
\begin{bmatrix} \dot{x}_1 \\ \dot{x}_2 \\ \dot{x}_3 \\ \dot{x}_4 \end{bmatrix} = \begin{bmatrix} 0 & 0 & 1 & 0 \\ 0 & 0 & 0 & 1 \\ -k/m_1 & k/m_1 & 0 & 0 \\ k/m_2 & -k/m_2 & 0 & 0 \end{bmatrix} \begin{bmatrix} x_1 \\ x_2 \\ x_3 \\ x_4 \end{bmatrix}
$$
$$
+ \begin{bmatrix} 0 \\ 0 \\ 1/m_1 \\ 0 \end{bmatrix} (u + d_1) + \begin{bmatrix} 0 \\ 0 \\ 0 \\ 1/m_2 \end{bmatrix} d_2
$$

$$
y = x_2 + \text{noise}
$$

$$
z = x_2
$$

where x_1 and x_2 are the positions of body 1 and body 2, respectively; x_3 and x_4 are the velocities of body 1 and body 2, respectively; u is the control input acting on body 1; y is the measured output; d_1 and d_2 are the plant disturbances acting on body 1 and body 2, respectively; and z is the output to be controlled.

The transfer function description of the plant is

$$
y(s) = \frac{k}{s^2[m_1 m_2 s^2 + k(m_1 + m_2)]} u(s)
$$

Constant-gain linear feedback controllers are to be designed for the following four different problems:

(a) For a unit impulse disturbance exerted on body 1 and/or body 2, the controlled output ($z = x_2$) must have a settling time of about 15 s for the nominal system with $m_1 = m_2 = k = 1$. The closed-loop system should be stable for $0.5 \leq k \leq 2.0$ and $m_1 = m_2 = 1$. The sensor noise, actuator saturation, and high-frequency rolloff must be considered to reflect practical control design tradeoffs.

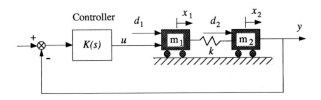

Fig. 2.30   Noncollocated control of the two-mass–spring system.

(b) Maximize a stability robustness measure with respect to the three uncertain parameters m_1, m_2, and k whose nominal values are $m_1 = m_2 = k = 1$. For a unit impulse disturbance exerted on body 1 and/or body 2, the controlled output must have a settling time of about 15 s for the nominal system with $m_1 = m_2 = k = 1$.

(c) There is a sinusoidal disturbance with known frequency of 0.5 rad/s acting on body 1 and/or body 2, but whose amplitude and phase, although constant, are not available to the designer. The closed-loop system must achieve asymptotic disturbance rejection for the controlled output with a 20-s settling time for $m_1 = m_2 = 1$ and $0.5 \leq k \leq 2.0$.

(d) Design a feedback/feedforward controller for a unit-step output command tracking problem for the controlled output, z, with the following properties: 1) the control input is limited as $|u| \leq 1$; 2) settling time and overshoot are both to be minimized; 3) performance robustness and stability robustness with respect to the three uncertain parameters m_1, m_2, and k (with the nominal values of $m_1 = m_2 = k = 1$) are both to be maximized; and 4) if there are conflicts between properties 2 and 3, then performance vs robustness tradeoffs must be considered.

References

[1] Ogata, K., *Modern Control Engineering*, 3rd ed., Prentice–Hall, Upper Saddle River, NJ, 1997.

[2] Shahian, B., and Hassul, M., *Control System Design Using MATLAB*, Prentice–Hall, Englewood Cliffs, NJ, 1993.

[3] Evans, W. R., "Control System Synthesis by Root Locus Method," *AIEE Transactions*, Vol. 69, 1950, pp. 66–69.

[4] MacFarlane, A. G. J. (ed.), *Frequency-Response Methods in Control Systems*, IEEE Press, New York, 1979, pp. 57–60.

[5] McRuer, D., Ashkenas, I., and Graham, D., *Aircraft Dynamics and Automatic Control*, Princeton Univ. Press, Princeton, NJ, 1973, Chap. 3.

[6] Wie, B., and Bryson, A. E., Jr., "On Multivariable Robustness Examples: A Classical Approach," *Journal of Guidance, Control, and Dynamics*, Vol. 10, No. 1, 1987, pp. 118–120.

[7] Davis, L. P., Cunningham, D., and Harrell, J., "Advanced 1.5 Hz Passive Viscous Isolation System," *Proceedings of AIAA Structures, Structural Dynamics, and Materials Conference*, AIAA, Washington, DC, 1994.

[8] Wie, B., and Byun, K.-W., "New Generalized Structural Filtering Concept for Active Vibration Control Synthesis," *Journal of Guidance, Control, and Dynamics*, Vol. 12, No. 2, 1989, pp. 147–154.

[9] Byun, K.-W., Wie, B., and Sunkel, J., "Robust Nonminimum-Phase Compensation for a Class of Uncertain Dynamical Systems," *Journal of Guidance, Control, and Dynamics*, Vol. 14, No. 6, 1991, pp. 1191–1199.

[10] Wie, B., and Gonzalez, M., "Control Synthesis for Flexible Space Structures Excited by Persistent Disturbances," *Journal of Guidance, Control, and Dynamics*, Vol. 15, No. 1, 1992, pp. 73–80.

[11] Van Valkenburg, M. E., *Analog Filter Design*, Holt, Rinehart, and Winston, Orlando, FL, 1982.

[12] Yokum, J. F., and Slafer, L. I., "Control System Design in the Presence of Severe Structural Dynamic Interactions," *Journal of Guidance, Control, and Dynamics*, Vol. 1, No. 2, 1978, pp. 109–116.

[13] Franklin, G. F., Powell, J. D., and Emami-Naeni, A., *Feedback Control of Dynamic Systems*, 3rd ed., Addison–Wesley, Menlo Park, CA, 1994, Chap. 9, pp. 669–683.

[14] Wie, B., Liu, Q., and Bauer, F., "Classical and Robust H_∞ Control Redesign for the Hubble Space Telescope," *Journal of Guidance, Control, and Dynamics*, Vol. 16, No. 6, 1993, pp. 1069–1077.

[15] Franklin, G., Powell, J. D., and Workman, M. L., *Digital Control of Dynamic Systems*, 2nd ed., Addison–Wesley, Menlo Park, CA, 1990.

[16] Junkins, J., and Kim, Y., *Introduction to Dynamics and Control of Flexible Structures*, AIAA Education Series, AIAA, Washington, DC, 1993.

[17] Bryson, A. E., Jr., and Ho, Y.-C., *Applied Optimal Control*, Hemisphere, Washington, DC, 1975.

[18] Bryson, A. E., Jr., *Control of Spacecraft and Aircraft*, Princeton Univ. Press, Princeton, NJ, 1994.

[19] Wie, B., Lu, J., and Warren, W., "Real Parameter Margin Computation for Uncertain Structural Dynamic Systems," *Journal of Guidance, Control, and Dynamics*, Vol. 16, No. 1, 1993, pp. 26–33.

[20] Wie, B., and Lu, J., "Two Real Critical Constraints for Real Parameter Margin Computation," *Journal of Guidance, Control, and Dynamics*, Vol. 17, No. 3, 1994, pp. 561–569.

[21] Ghaoui, E. L., and Bryson, A. E., Jr., "Worst Case Parameter Changes for Stabilized Conservative SISO Systems," *Proceedings of AIAA Guidance, Navigation, and Control Conference*, AIAA, Washington, DC, 1991, pp. 1490–1495.

[22] De Gaston, R .E., and Safonov, M. G., "Exact Calculation of the Multiloop Stability Margin," *IEEE Transactions on Automatic Control*, Vol. AC-33, No. 2, 1988, pp. 156–171.

[23] Zadeh, L. A., and Desoer, C. A., *Linear System Theory*, McGraw–Hill, New York, 1963.

[24] Wedell, E., Chuang, C.-H., and Wie, B., "Parameter Margin Computation for Structured Real-Parameter Perturbations," *Journal of Guidance, Control, and Dynamics*, Vol. 14, No. 3, 1991, pp. 607–614.

[25] Weinmann, A., *Uncertain Models and Robust Control*, Springer–Verlag, New York, 1991, Chap. 4.

[26] Bhattacharyya, S. P., Chapellat, H., and Keel, L. H., *Robust Control: The Parametric Approach*, Prentice-Hall PTR, Upper Saddle River, NJ, 1995, Chap. 5.

[27] Wie, B., Liu, Q., and Byun, K.-W., "Robust H_∞ Control Synthesis Method and Its Application to Benchmark Problems," *Journal of Guidance, Control, and Dynamics*, Vol. 15, No. 5, 1992, pp. 1140–1148.

[28] Wie, B., Liu, Q., and Sunkel, J., "Robust Stabilization of the Space Station in the Presence of Inertia Matrix Uncertainty," *Journal of Guidance, Control, and Dynamics*, Vol. 18, No. 3, 1995, pp. 611–617.

[29] Doyle, J., Glover, K., Khargonekar, P., and Francis, B., "State-Space Solutions to Standard H_2 and H_∞ Control Problems," *IEEE Transactions on Automatic Control*, Vol. AC-34, No. 8, 1989, pp. 831–847.

[30] Glover, K., and Doyle, J., "State-Space Formulae for All Stabilizing Controllers that Satisfy an H_∞-norm Bound and Relations to Risk Sensitivity," *System and Control Letters*, Vol. 11, 1988, pp. 167–172.

[31] Wie, B., and Bernstein, D., "Benchmark Problems for Robust Control Design," *Journal of Guidance, Control, and Dynamics*, Vol. 15, No. 5, 1992, pp. 1057–1059.

[32] Thompson, P. M., "Classical/H_2 Solution for a Robust Control Design Benchmark Problem," *Journal of Guidance, Control, and Dynamics*, Vol. 18, No. 1, 1995, pp. 160–169.

Part 2
Orbital Dynamics and Control

Orbital Dynamics

Orbital dynamics* is concerned with the orbital motion of space vehicles under the influence of gravitational and other external forces. It is founded on the basic physical principles of celestial mechanics,[†] which is concerned with the natural motion of planets or celestial bodies. Newton's laws of motion and law of gravity, as well as Kepler's three laws of planetary motion, provide the fundamental framework of both disciplines and were discussed in Chapter 1.

This chapter is primarily concerned with the orbital motion of a particle in a gravitational field. Fundamental problems of orbital mechanics, such as the *two-body* and *restricted three-body* problems, are treated. This chapter also provides the foundation for the further study of orbital maneuvering problems, interplanetary mission analysis, and halo orbit control problems to be treated in the next chapter.

3.1 Two-Body Problem

In this section, we consider the simplest problem in orbital mechanics, that is, the dynamic problem of two point masses under the influence of their mutual gravitational attraction. Such a problem is called the two-body problem in celestial mechanics. Using Newton's laws of motion and law of gravity, we will formulate the two-body problem and verify Kepler's three empirical laws of planetary motion, which are stated as follows:

1) The orbit of each planet around the sun is an ellipse, with the sun at one focus (the law of orbits).

2) The radius vector from the sun to a planet sweeps out equal areas in equal time intervals (the law of areas).

3) The square of the orbital period of a planet is proportional to the cube of its mean distance from the sun (the law of periods).

These three empirical laws, with the first two laws published in 1609 and the third law in 1619, were used later by Newton to deduce his law of gravity.

---

*The terms "orbital dynamics," "astrodynamics," "orbital mechanics," "spaceflight dynamics," and "astronautics" are often used interchangeably in the literature.

[†]The term "celestial mechanics" was introduced by Pierre-Simon de Laplace (1749–1827). He showed that the secular variation of Jupiter and Saturn, which was observed by Edmond Halley (1656–1742) in 1695, was actually periodic, with Jupiter and Saturn returning to their initial positions every 929 years.

Kepler's three laws of planetary motion describe the motion of planets around the sun, which is considered to be inertially fixed. In this chapter, we shall derive these laws of planetary motion by applying both Newton's laws of motion and law of gravity to the general two-body problem in which the primary body is not assumed to be inertially fixed.

3.1.1 Equation of Relative Motion

Consider two particles of masses m_1 and m_2 with position vectors $\vec{R}_1$ and $\vec{R}_2$, respectively, in an inertial reference frame, as shown in Fig. 3.1.

Applying Newton's second law and his law of gravity to each mass, we write the equations of motion as

$$m_1\ddot{\vec{R}}_1 = +\frac{Gm_1m_2}{r^3}\vec{r} \qquad (3.1)$$

$$m_2\ddot{\vec{R}}_2 = -\frac{Gm_1m_2}{r^3}\vec{r} \qquad (3.2)$$

where $\vec{r} = \vec{R}_2 - \vec{R}_1$ is the position vector of m_2 from m_1, $r = |\vec{r}|$, $\ddot{\vec{R}}_i = \mathrm{d}^2\vec{R}_i/\mathrm{d}t^2$ is the inertial acceleration of the ith body, and $G = 6.6695 \times 10^{-11}$ N $\cdot$ m^2/kg^2 is the universal gravitational constant.

This nonlinear dynamic system with six degrees of freedom can also be described by two decoupled sets of equations of motion as follows. By adding Eqs. (3.1) and (3.2), we obtain

$$m_1\ddot{\vec{R}}_1 + m_2\ddot{\vec{R}}_2 = 0 \qquad (3.3)$$

Defining the position vector $\vec{R}_c$ to the system center of mass as

$$\vec{R}_c = \frac{m_1\vec{R}_1 + m_2\vec{R}_2}{m_1 + m_2} \qquad (3.4)$$

we obtain

$$\ddot{\vec{R}}_c = 0 \qquad (3.5)$$

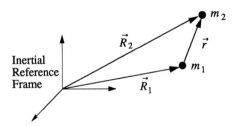

Fig. 3.1 Two-body problem.

which can be integrated as

$$\dot{\vec{R}}_c(t) = \dot{\vec{R}}_c(0) = \text{constant vector} \tag{3.6}$$

$$\vec{R}_c(t) = \vec{R}_c(0) + \dot{\vec{R}}_c(0)t \tag{3.7}$$

where $\vec{R}_c(0)$ and $\dot{\vec{R}}_c(0)$ are the position and inertial velocity of the system center of mass at $t = 0$. Thus, one-half of the 12 states of this dynamic system have been solved. The remaining 6 states of the system can be solved by considering the relative motion of the two masses.

Subtracting Eq. (3.1) times m_2 from Eq. (3.2) times m_1, we obtain

$$\ddot{\vec{r}} + \mu \frac{\vec{r}}{r^3} = 0 \tag{3.8}$$

where $\ddot{\vec{r}} = d^2\vec{r}/dt^2$ is the inertial acceleration of m_2 with respect to m_1, $r = |\vec{r}|$, and $\mu = G(m_1 + m_2)$ is called the *gravitational parameter* of the two-body system under consideration. Equation (3.8) describes the motion of m_2 relative to m_1 in an inertial reference frame, and it is the fundamental equation in the two-body problem.

In most practical cases of interest in orbital mechanics, the mass of the primary body is much greater than that of the secondary body, i.e., $m_1 \gg m_2$, which results in $\mu \approx Gm_1$. For example, for a sun–planet system, we have $\mu \approx \mu_\odot \equiv GM_\odot$, where $\mu_\odot$ denotes the gravitational parameter of the sun and $M_\odot$ denotes the mass of the sun. Also, for an Earth–satellite system, we have $\mu \approx \mu_\oplus \equiv GM_\oplus$, where $\mu_\oplus$ denotes the gravitational parameter of the Earth and $M_\oplus$ denotes the mass of the Earth. It is worth emphasizing that the primary body is not inertially fixed in the two-body problem. The two-body problem must be distinguished from a so-called restricted two-body problem in which the primary body of mass m_1 is assumed to be inertially fixed. Such a restricted two-body problem is often described by *central force motion* of a particle of mass m_2 around the inertially fixed primary body of mass m_1.

3.1.2 Conservation of Energy

From our basic knowledge of dynamics, we know that the two-body system is a conservative system. That is, the mechanical energy of the system, which is the sum of kinetic and potential energy, remains constant. Such a principle of conservation of energy for the two-body system can be derived as follows.

Taking the dot product of Eq. (3.8) with $\dot{\vec{r}}$ yields

$$\ddot{\vec{r}} \cdot \dot{\vec{r}} + (\mu/r^3)\vec{r} \cdot \dot{\vec{r}} = 0 \tag{3.9}$$

which can be written as

$$\frac{1}{2}\frac{d}{dt}(\dot{\vec{r}} \cdot \dot{\vec{r}}) + \frac{\mu}{r^3}r\dot{r} = 0$$

This equation can be rewritten as

$$\frac{d}{dt}\left(\frac{v^2}{2} - \frac{\mu}{r}\right) = 0 \quad \text{or} \quad \frac{v^2}{2} - \frac{\mu}{r} = \text{const} \tag{3.10}$$

where $v \equiv |\vec{v}| = |\dot{\vec{r}}| = \sqrt{\dot{\vec{r}} \cdot \dot{\vec{r}}}$. Finally we obtain the *energy equation*

$$\frac{v^2}{2} - \frac{\mu}{r} = \mathcal{E} \tag{3.11}$$

where the constant $\mathcal{E}$ is called the *total mechanical energy per unit mass* or the *specific mechanical energy*, $v^2/2$ is the *kinetic energy per unit mass*, and $-\mu/r$ is a *potential energy per unit mass*. This equation represents the law of conservation of energy for the two-body system.

Given the specific (mechanical) energy $\mathcal{E}$, the orbital velocity v can be expressed as

$$v = \sqrt{2\mu/r + 2\mathcal{E}} \tag{3.12}$$

The magnitude of a velocity vector is often called the "speed." The term "velocity" is, however, used loosely in the same sense throughout this book.

3.1.3 Conservation of Angular Momentum

From our basic knowledge of dynamics, we also know that the angular momentum of the two-body system must be conserved because the gravitational force is an internal force and there is no external force acting on the two-body system. Such a principle of conservation of angular momentum for the two-body system can be derived as follows.

Taking the cross product of Eq. (3.8) with $\vec{r}$, we have

$$\vec{r} \times \ddot{\vec{r}} + \vec{r} \times \mu \frac{\vec{r}}{r^3} = 0 \tag{3.13}$$

which can be written as

$$\frac{d}{dt}(\vec{r} \times \dot{\vec{r}}) = 0 \tag{3.14}$$

Defining the *angular momentum per unit mass* or the *specific angular momentum* as

$$\vec{h} = \vec{r} \times \dot{\vec{r}} \equiv \vec{r} \times \vec{v} \tag{3.15}$$

we obtain

$$\frac{d\vec{h}}{dt} = 0 \quad \text{or} \quad \vec{h} = \text{constant vector} \tag{3.16}$$

Thus we have the law of conservation of angular momentum for the two-body system. Because $\vec{h}$ is the vector cross product of $\vec{r}$ and $\vec{v}$, it is always perpendicular to the plane containing $\vec{r}$ and $\vec{v}$. Furthermore, because $\vec{h}$ is a constant vector, $\vec{r}$ and $\vec{v}$ always remain in the same plane, called an *orbital plane*. Therefore, we conclude that the orbital plane is fixed in space and the angular momentum vector $\vec{h}$ is perpendicular to the orbital plane.

3.1.4 Orbit Equation

Thus far we have found two constants of the two-body system: the specific energy constant $\mathcal{E}$ and the specific angular momentum constant h. Kepler's first law can now be verified by deriving the equation of a conic section as a partial solution to the equation of relative motion, as follows.

Taking the post–cross product of Eq. (3.8) with $\vec{h}$, we have

$$\ddot{\vec{r}} \times \vec{h} + \mu \frac{\vec{r}}{r^3} \times \vec{h} = 0 \tag{3.17}$$

which can be written as

$$\frac{d}{dt}[\dot{\vec{r}} \times \vec{h} - (\mu/r)\vec{r}] = 0 \tag{3.18}$$

because

$$\vec{r} \times \vec{h} = \vec{r} \times (\vec{r} \times \dot{\vec{r}}) = (\vec{r} \cdot \dot{\vec{r}})\vec{r} - (\vec{r} \cdot \vec{r})\dot{\vec{r}} \tag{3.19}$$

Integrating Eq. (3.18) gives

$$\dot{\vec{r}} \times \vec{h} - (\mu/r)\vec{r} = \text{constant vector} = \mu\vec{e} \tag{3.20}$$

where a constant vector $\mu\vec{e}$ is introduced and $\vec{e}$ is called the *eccentricity vector*. Note that the constant vector $\mu\vec{e}$ can also be written as

$$\mu\vec{e} = \vec{v} \times \vec{h} - \frac{\mu}{r}\vec{r} = \vec{v} \times (\vec{r} \times \vec{v}) - \frac{\mu}{r}\vec{r}$$
$$= [v^2 - (\mu/r)]\vec{r} - (\vec{r} \cdot \vec{v})\vec{v}$$

Taking the dot product of Eq. (3.20) with $\vec{r}$ gives

$$\vec{r} \cdot \dot{\vec{r}} \times \vec{h} - \vec{r} \cdot (\mu/r)\vec{r} = \vec{r} \cdot \mu\vec{e} \tag{3.21}$$

Because $\vec{r} \cdot \dot{\vec{r}} \times \vec{h} = (\vec{r} \times \dot{\vec{r}}) \cdot \vec{h} = h^2$, Eq. (3.21) becomes

$$h^2 - \mu r = \mu r e \cos\theta \tag{3.22}$$

where $h \equiv |\vec{h}|$, $e \equiv |\vec{e}|$, and θ is the angle between $\vec{r}$ and $\vec{e}$. The angle θ is called the *true anomaly* (the classical term "anomaly" is interchangeable with the term "angle"), and e is called the eccentricity of the orbit.

3.1.5 Kepler's First Law

Equation (3.22) can be further transformed into the *orbit equation* of the form

$$r = \frac{h^2/\mu}{1 + e\cos\theta} \tag{3.23}$$

which can be rewritten as

$$r = \frac{p}{1 + e\cos\theta} \tag{3.24}$$

where p, called the parameter, is defined as

$$p = h^2/\mu \tag{3.25}$$

Equation (3.24) is the equation of a conic section, written in terms of polar coordinates r and θ with the origin located at a focus, whereas θ is measured from the point on the conic nearest the focus. This equation is, in fact, a statement of Kepler's first law: the orbit of each planet around the sun is an ellipse, with the sun at one focus. The size and shape of the orbit depends on the parameter p and the eccentricity e, respectively.

Problems

3.1   Consider the total kinetic energy of the two-body system shown in Fig. 3.1, defined by

$$T = \tfrac{1}{2}m_1\dot{\vec{R}}_1 \cdot \dot{\vec{R}}_1 + \tfrac{1}{2}m_2\dot{\vec{R}}_2 \cdot \dot{\vec{R}}_2$$

Show that this total kinetic energy can be written as

$$T = \frac{1}{2}(m_1 + m_2)\dot{\vec{R}}_c \cdot \dot{\vec{R}}_c + \frac{1}{2}\left(\frac{m_1 m_2}{m_1 + m_2}\right)\dot{\vec{r}} \cdot \dot{\vec{r}}$$

where $\vec{R}_c$ is the inertial position vector of the system center of mass and $\vec{r} \equiv \vec{R}_2 - \vec{R}_1$.

Note: The first term on the right-hand side of the preceding equation is the translational kinetic energy of the center of mass of the two-body system. The second term is the rotational kinetic energy of the two-body system about the system center of mass. Because a potential energy of the two-body system is given by $-Gm_1 m_2/r$, we write the energy equation of the two-body system (without including the translational kinetic energy of the center of mass of the two-body system), as follows:

$$\frac{1}{2}\left(\frac{m_1 m_2}{m_1 + m_2}\right)v^2 - \frac{Gm_1 m_2}{r} = \text{const}$$

where $v^2 \equiv |\dot{\vec{r}}|^2 = \dot{\vec{r}} \cdot \dot{\vec{r}}$. We can also rewrite the equation of relative motion, given by Eq. (3.8), as

$$\frac{m_1 m_2}{m_1 + m_2} \ddot{\vec{r}} + \frac{Gm_1 m_2}{r^3} \vec{r} = 0$$

where $m_1 m_2 / (m_1 + m_2)$ is called the reduced mass of the system.

3.2 Consider the two-body system described by the equation of relative motion, $\ddot{\vec{r}} + (\mu/r^3)\vec{r} = 0$; the energy equation, $v^2/2 - \mu/r = \mathcal{E} = \text{const}$; and the angular momentum equation, $\vec{h} = \vec{r} \times \vec{v} = \text{constant vector}$, where $\mu = G(m_1 + m_2)$, $r = |\vec{r}|$, and $v = |\vec{v}| = |\dot{\vec{r}}|$.

(a) Using the fact that the orbital motion is confined to the plane containing $\vec{r}$ and $\vec{v}$, show that the equation of relative motion can be written in terms of polar coordinates (r, θ), as follows:

$$\vec{e}_r: \quad \ddot{r} - r\dot{\theta}^2 = -\mu/r^2$$
$$\vec{e}_\theta: \quad r\ddot{\theta} + 2\dot{r}\dot{\theta} = 0$$

Hint: As illustrated in Fig. 3.2, let $\vec{e}_r$ and $\vec{e}_\theta$ be unit vectors along the radial vector direction and the transverse orbit direction, respectively, such that $\vec{r} = r\vec{e}_r$, $\dot{\vec{e}}_r = \dot{\theta}\vec{e}_\theta$, and $\dot{\vec{e}}_\theta = -\dot{\theta}\vec{e}_r$.

(b) Show that the specific angular momentum, denoted by $h = |\vec{h}|$, can be found as

$$h = r^2\dot{\theta}$$

Also verify that h is, in fact, constant.

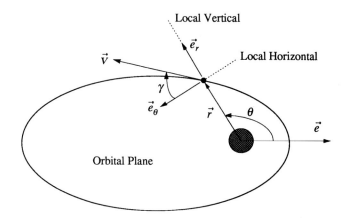

Fig. 3.2   Orbital motion.

(c) Show that the specific angular momentum can also be expressed as

$$h = rv \cos \gamma$$

where γ denotes the *flight-path angle* as shown in Fig. 3.2.

(d) By eliminating the independent variable t of the equation of motion using the result in (b), obtain

$$-\frac{1}{r^2}\frac{d^2 r}{d\theta^2} + \frac{2}{r^3}\left(\frac{dr}{d\theta}\right)^2 + \frac{1}{r} = \frac{\mu}{h^2}$$

which can be rewritten as

$$\frac{d^2}{d\theta^2}\left(\frac{1}{r}\right) + \frac{1}{r} = \frac{\mu}{h^2}$$

Hint: From the equation in (b), we have

$$\frac{d}{dt}(\ ) \equiv \frac{h}{r^2}\frac{d}{d\theta}(\ )$$

(e) The solution of the equation in (d) is, in general, given as

$$\frac{1}{r} = \frac{\mu}{h^2} + C \cos(\theta + \theta_0)$$

By measuring θ from a point of minimum distance from the origin of $\vec{r}$, the constant θ_0 can be made to be zero. Determine the constant C as

$$C = \frac{\mu}{h^2}\sqrt{1 + (2\mathcal{E}h^2/\mu^2)}$$

Hint: To determine the constant C in (e), use the energy equation for a special case of $\theta = 0$ and $h = r_p v_p$.
   Also note that

$$v^2 = \dot{r}^2 + (r\dot{\theta})^2 = h^2\left[\left\{\frac{d}{d\theta}\left(\frac{1}{r}\right)\right\}^2 + \left(\frac{1}{r}\right)^2\right]$$

(f) Let $p = h^2/\mu$ and also let e be defined as

$$e = \sqrt{1 + (2\mathcal{E}h^2/\mu^2)} = \sqrt{1 + (2\mathcal{E}p/\mu)}$$

then obtain the orbit equation of the form

$$r = p/(1 + e \cos \theta)$$

which is, in fact, the orbit equation derived in Sec. (3.1.4), whereas e is the eccentricity and p is the parameter introduced in Sec. (3.1.5).

(g) Show that

$$\dot{\theta} = \sqrt{\mu/p^3}\,(1 + e\cos\theta)^2$$

$$\dot{r} = \sqrt{\mu/p}\,e\sin\theta$$

$$\tan\gamma = e\sin\theta/(1 + e\cos\theta)$$

3.2 Geometry of Conic Sections

This section is concerned with the geometric characteristics of conic sections, established by Apollonius,* that are important in orbital mechanics. The geometrical characteristics of the conic sections are illustrated in Fig. 3.3.

The equation of a general conic section in terms of polar coordinates (r, θ) is given by

$$r = p/(1 + e\cos\theta) \qquad (3.26)$$

where p is the parameter or *semilatus rectum* and e is the eccentricity. The size of the conic section is determined by p and its shape is determined by the eccentricity e. Note that p is the radial distance r to a point on a conic section when $\theta = \pm 90$ deg.

3.2.1 Ellipse

For any point on the ellipse, the sum of its distance to two fixed points, called *foci*, is defined to be a constant $2a$. One of the foci is occupied by the primary attracting body and is called the true focus, whereas the other is called the empty focus. An ellipse can also be defined as the locus of points whose distance from a focus is equal to its distance, multiplied by e, from a straight line known as the *directrix*.

The eccentricity e of the ellipse is defined as

$$e = c/a \qquad (3.27)$$

where a is the *semimajor axis* and c is the distance from the center of the ellipse to either focus, as shown in Fig. 3.3a. The ellipse is a conic section with $0 \le e < 1$. The circle is considered as a special case of an ellipse with $e = 0$.

The *semiminor axis* b of the ellipse is expressed in terms of a and e as follows:

$$b = a\sqrt{1 - e^2} \qquad (3.28)$$

because a, b, and c of the ellipse are related as follows:

$$a^2 = b^2 + c^2$$

---

*Apollonius, known as "the Great Geometer," was born in about 262 B.C. In his masterpiece, *Conics*, he showed that conic sections, such as ellipses, parabolas, and hyperbolas could be generated by taking sections of a circular cone by varying the inclination of the cutting plane.

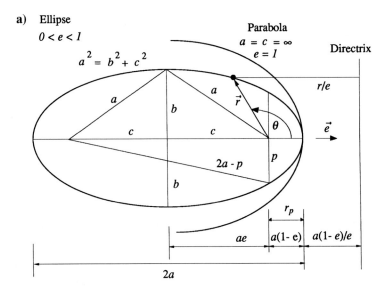

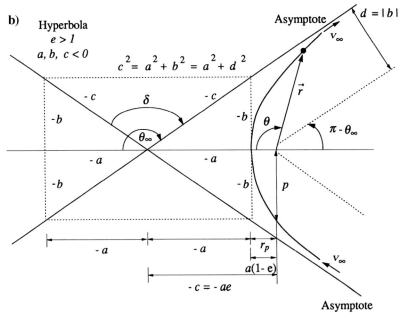

Fig. 3.3   Geometric characteristics of the conic sections.

If a line is drawn perpendicular to the major axis at the true focus, it will intersect any conic section at two points. The distance between these two points, denoted by $2p$, is called *latus rectum*, and p is called semilatus rectum or parameter. The parameter p of the ellipse can be expressed in terms of a and e as

$$p = a(1 - e^2) \tag{3.29}$$

because, for any point on the ellipse, the sum of its distance to two foci is defined as a constant $2a$.

Finally, the equation of the ellipse can be expressed as

$$r = \frac{a(1 - e^2)}{1 + e \cos \theta} \tag{3.30}$$

which is in terms of the *geometric constants* a and e. (The specific energy $\mathcal{E}$ and the specific angular momentum h are often referred to as the *dynamic constants*, whereas a and e are called the geometric constants.)

The points in an orbit closest to and farthest from the focus of gravitational attraction are called *apsides*. The point of the orbit nearest to the primary attracting body is called the *periapsis* and the point farthest from the primary body is called the *apoapsis*. The closest point of the orbit to the Earth (sun) is called the *perigee (perihelion)*, whereas the farthest point from the Earth (sun) is called the *apogee (aphelion)*. The periapsis distance r_p and the apoapsis distance r_a of the ellipse can be found as

$$r_p = a(1 - e) \tag{3.31a}$$
$$r_a = a(1 + e) \tag{3.31b}$$

which are combined to give

$$a = (r_p + r_a)/2 \tag{3.32a}$$
$$e = (r_a - r_p)/(r_a + r_p) \tag{3.32b}$$

3.2.2 Parabola

A parabola is the locus of points whose distance from a focus is equal to the distance from the directrix. The parabola is a conic section with $a = c = \infty$ and $e = 1$; i.e.,

$$r = p/(1 + \cos \theta)$$

The parameter p of a parabola is equal to twice the periapsis distance; i.e., $p = 2r_p$.

3.2.3 Hyperbola

The hyperbola is defined as the locus of points such that the difference of their distances from two fixed foci is a constant length $-2a$ ($a < 0$). For the hyperbola it is convenient to consider the semimajor axis a (also b and c) to be negative, as shown in Fig. 3.3b. The hyperbola has two branches that are separated by a distance $-2c$ ($c < 0$); only one branch is shown in Fig. 3.3b.

As in the case of ellipse, the eccentricity e of the hyperbola is also defined as

$$e = c/a$$

and the hyperbola is a conic section with $e > 1$.

The semiminor axis b of the hyperbola can also be expressed in terms of the semimajor axis a and the eccentricity e as

$$b = a\sqrt{e^2 - 1} \tag{3.33}$$

because a, b, and c of the hyperbola, taken as negative, are related as

$$c^2 = a^2 + b^2$$

Note that the asymptote distance, denoted as d in Fig. 3.3b, is, in fact, identical to the absolute value of the semiminor axis b of the hyperbola. The hyperbola is also characterized by the asymptote angle θ_∞ and the deflection angle δ as illustrated in Fig. 3.3b. Also shown in this figure is the hyperbolic excess velocity, denoted by v_∞, which will be discussed later in this section.

The parameter p and the periapsis distance r_p of the hyperbola can be expressed in terms of a and e as follows:

$$p = a(1 - e^2) \tag{3.34}$$

$$r_p = a(1 - e) \tag{3.35}$$

Consequently, the equation of the hyperbola can also be written as

$$r = \frac{a(1 - e^2)}{1 + e \cos \theta} = \frac{p}{1 + e \cos \theta}$$

where $a < 0$, $e > 1$, and $p > 0$.

3.2.4 Kepler's Second and Third Laws

The orbital area ΔA, swept out by the radius vector $\vec{r}$ as it moves through a small angle $\Delta\theta$ in a time interval Δt, is given as

$$\Delta A = \tfrac{1}{2} r(r\Delta\theta)$$

Then the *areal velocity* of the orbit, denoted by dA/dt, can be shown to be constant, as follows:

$$\frac{dA}{dt} = \lim_{\Delta t \to 0} \frac{\Delta A}{\Delta t} = \lim_{\Delta t \to 0} \frac{1}{2} r^2 \frac{\Delta\theta}{\Delta t} = \frac{1}{2} r^2 \dot{\theta} = \frac{1}{2} h = \text{const} \tag{3.36}$$

which is a statement of Kepler's second law: the radius vector from the sun to a planet sweeps out equal areas in equal time intervals.

The period of an elliptical orbit can be found by dividing the total orbital area by the areal velocity, as follows:

$$P = \frac{A}{dA/dt} = \frac{\pi ab}{h/2} = \frac{\pi a^2 \sqrt{1 - e^2}}{\sqrt{\mu a(1 - e^2)}/2} = 2\pi \sqrt{\frac{a^3}{\mu}} \qquad (3.37)$$

This can be rewritten as

$$P^2 = (4\pi^2/\mu)a^3$$

which is, in fact, a statement of Kepler's third law: the square of the orbital period of a planet is proportional to the cube of the semimajor axis of the ellipse. Note that the ratio P^2/a^3 is not constant for all planets because $\mu = G(M_\odot + m_2)$, where $M_\odot$ is the mass of the sun and m_2 is the mass of the planet. Therefore, the ratio differs slightly for each planet.

3.3 Vis-Viva Equation

The energy equation at periapsis and apoapsis is expressed as

$$\mathcal{E} = \frac{v_p^2}{2} - \frac{\mu}{r_p} = \frac{v_a^2}{2} - \frac{\mu}{r_a} \qquad (3.38)$$

Because the velocity vector at periapsis or apoapsis of any orbit is directed horizontally and the flight-path angle γ is zero, we have the angular momentum equation simply expressed as

$$h = r_p v_p = r_a v_a \qquad (3.39)$$

where $h = \sqrt{\mu p}$, $p = a(1 - e^2)$, $r_p = a(1 - e)$, and $r_a = a(1 + e)$.

Combining Eqs. (3.38) and (3.39) for either periapsis or apoapsis yields

$$\mathcal{E} = -\frac{\mu}{2a} \qquad (3.40)$$

That is, the total specific energy depends only on the semimajor axis and is independent of eccentricity. Note that an elliptical orbit (with $a > 0$) has $\mathcal{E} < 0$, a parabolic orbit (with $a = \infty$) has $\mathcal{E} = 0$, and a hyperbolic orbit (with $a < 0$) has $\mathcal{E} > 0$.

Consequently, we have the energy equation expressed as

$$\mathcal{E} = \frac{v^2}{2} - \frac{\mu}{r} = -\frac{\mu}{2a} \qquad (3.41)$$

and the velocity equation

$$v = \sqrt{2\mu/r + 2\mathcal{E}}$$

can be rewritten as

$$v = \sqrt{\frac{2\mu}{r} - \frac{\mu}{a}}$$

(3.42)

which is known as the *vis-viva equation*. In classical mechanics, *vis viva* and *vis mortua* mean living and dead forces, respectively.

The orbital velocity, denoted as v_c, of a circular orbit with the constant radius $r = a$ is simply given as

$$v_c = \sqrt{\mu/r}$$

(3.43)

which is often called the *circular velocity*.

For a parabolic orbit with $a = \infty$ and $\mathcal{E} = 0$, we have

$$v = \sqrt{2\mu/r}$$

Consequently, we define the *escape velocity* v_e as

$$v_e = \sqrt{2\mu/r} = \sqrt{2}v_c$$

(3.44)

for any given orbital position r. As an example, the escape velocity of an object from the surface of the Earth ($\mu_\oplus = GM_\oplus = 398{,}601$ km^3/s^2 and $R_\oplus = 6{,}378$ km) can be estimated as

$$v_e = \sqrt{2\mu_\oplus/R_\oplus} = 11.18 \text{ km/s}$$

(3.45)

which is about 36,679 ft/s or 25,000 mph.*

The energy equation for a hyperbolic orbit as $r \to \infty$, becomes

$$\mathcal{E} = v_\infty^2/2 = -\mu/2a$$

(3.46)

where v_∞, called the *hyperbolic excess velocity*, (This will also be called the hyperbolic escape or approach velocity in Chapter 4.) is the velocity v as $r \to \infty$, and

$$v_\infty = \sqrt{\frac{\mu}{-a}}$$

Problem

3.3   Consider a hyperbolic orbit illustrated in Fig. 3.3b.

(a) Show that the asymptote angle θ_∞ and the deflection angle δ are related to the eccentricity e, as follows:

$$\cos\theta_\infty = -\frac{1}{e} \quad \text{and} \quad \sin\frac{\delta}{2} = \frac{1}{e}$$

Hint: $\theta_\infty = \pi/2 + \delta/2$.

---

*1 ft/s $= 0.3048$ m/s and 1 mph $= 0.44704$ m/s.

(b) Show that the specific angular momentum h is simply given by

$$h = v_\infty d$$

where v_∞ is the hyperbolic excess velocity and d is the asymptote distance, as defined in Fig. 3.3b.

Hint: $h = |\vec{h}| = |\vec{r} \times \vec{v}|$.

(c) Show that the eccentricity of the hyperbolic orbit can be expressed as

$$e = \sqrt{1 + \left(v_\infty^2 d/\mu\right)^2} \qquad \text{or} \qquad e = 1 + \left(r_p v_\infty^2/\mu\right)$$

where r_p is the periapsis distance.

Hint: $\mathcal{E} \equiv v^2/2 - \mu/r = v_\infty^2/2 = -\mu/(2a), h \equiv \sqrt{\mu p}, p = a(1 - e^2),$ and $r_p = a(1 - e)$.

3.4   Kepler's Time Equation

In this section we introduce a geometric parameter known as the *eccentric anomaly* to find the position in an orbit as a function of time or vice versa.

Consider an auxiliary circle, which was first introduced by Kepler, as shown in Fig. 3.4. From Fig. 3.4, we have

$$a \cos E + r \cos(\pi - \theta) = ae \qquad (3.47)$$

where E is the *eccentric anomaly* and θ is the *true anomaly*. Using the orbit equation

$$r = \frac{p}{1 + e \cos \theta} = \frac{a(1 - e^2)}{1 + e \cos \theta} \qquad (3.48)$$

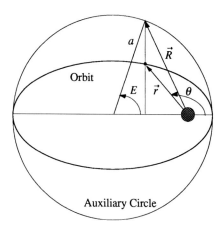

Fig. 3.4   Eccentric anomaly E of an elliptic orbit.

we rewrite Eq. (3.47) as

$$\cos E = \frac{e + \cos \theta}{1 + e \cos \theta} \tag{3.49}$$

Using the fact that all lines parallel to the minor axis of an ellipse have a foreshortening factor of b/a with respect to a circle with a radius of a, we obtain

$$r \sin \theta = (b/a)(a \sin E) = a\sqrt{1 - e^2} \sin E \tag{3.50}$$

Combining this with the orbit equation, we obtain

$$\sin E = \frac{\sqrt{1 - e^2} \sin \theta}{1 + e \cos \theta} \tag{3.51}$$

Furthermore, we have

$$\tan \frac{E}{2} = \frac{\sin E}{1 + \cos E} = \sqrt{\frac{1 - e}{1 + e}} \tan \frac{\theta}{2} \tag{3.52}$$

from which E or θ can be determined without quadrant ambiguity.

Equation (3.47) can be rewritten as

$$r \cos \theta = a(\cos E - e) \tag{3.53}$$

Thus, squaring Eqs. (3.50) and (3.53) and adding them, we obtain

$$r = a(1 - e \cos E) \tag{3.54}$$

which is the orbit equation in terms of the eccentric anomaly E and its geometric constants a and e.

The area swept out by the position vector $\vec{r}$ is

$$(t - t_p)\dot{A} = (t - t_p)\frac{ab}{2}\sqrt{\frac{\mu}{a^3}} \tag{3.55}$$

where t_p is the perigee passage time, $(t - t_p)$ is the elapsed time since perigee passage, and $\dot{A}$ is the constant areal velocity given by Kepler's third law,

$$\dot{A} = \frac{\pi ab}{P} = \frac{\pi ab}{2\pi \sqrt{a^3/\mu}} = \frac{ab}{2}\sqrt{\frac{\mu}{a^3}} \tag{3.56}$$

This area of the ellipse is the same as the area of the auxiliary circle swept out by the vector $\vec{R}$, multiplied by the factor b/a. Thus, we have

$$\frac{ab}{2}\sqrt{\frac{\mu}{a^3}}(t - t_p) = \frac{b}{a}\left(\frac{1}{2}a^2 E - \frac{ae}{2}a \sin E\right)$$

$$= \frac{ab}{2}(E - e \sin E) \tag{3.57}$$

which becomes

$$\sqrt{\mu/a^3}\,(t - t_p) = E - e\sin E \qquad (3.58)$$

where E is in radians.

Defining the *mean anomaly M* and the orbital *mean motion n*, as follows:

$$M = n(t - t_p) \qquad (3.59a)$$

$$n = \sqrt{\mu/a^3} \qquad (3.59b)$$

we obtain

$$M = E - e\sin E \qquad (3.60)$$

which is known as *Kepler's time equation* for relating time to position in orbit.

The time required to travel between any two points in an elliptical orbit can be computed simply by first determining the eccentric anomaly E corresponding to a given true anomaly θ, and then using Kepler's time equation.

However, Kepler's time equation (3.60) does not provide time values $(t - t_p)$ greater than one-half of the orbit period, but it gives the elapsed time since perigee passage in the shortest direction. Thus, for $\theta > \pi$, the result obtained from Eq. (3.60) must be subtracted from the orbit period to obtain the correct time since perigee passage.

Problems

3.4 Show that the area in the ellipse swept out by the position vector $\vec{r}$ is

$$\frac{1}{2}\int_0^\theta r^2\,d\theta = \frac{a^2}{2}\int_0^\theta (1 - e\cos E)^2\,d\theta$$

$$= \frac{ab}{2}\int_0^E (1 - e\cos E)\,dE$$

and that

$$\frac{ab}{2}\sqrt{\frac{\mu}{a^3}}\,(t - t_p) = \frac{ab}{2}(E - e\sin E)$$

which becomes

$$M = E - e\sin E$$

Hint:

$$\sin\theta = \frac{\sqrt{1 - e^2}\,\sin E}{1 - e\cos E}, \qquad \cos\theta = \frac{\cos E - e}{1 - e\cos E}$$

3.5   The auxiliary circle of an elliptic orbit is often described by

$$(x^2/a^2) + (y^2/a^2) = 1$$

where $x = a \cos E$ and $y = a \sin E$ (since $\cos^2 E + \sin^2 E = 1$), and E is the eccentric anomaly. Similarly, an *equilateral hyperbola* with a 45-deg branch angle, i.e., $e = \sqrt{2}$, $a = b$, $p = -a$, can be introduced for a hyperbola such that

$$(x^2/a^2) - (y^2/a^2) = 1$$

where $x = -a \cosh H$ and $y = -a \sinh H$ (since $\cosh^2 H - \sinh^2 H = 1$), and H is called the *hyperbolic eccentric anomaly*.

Derive the following relationships for the hyperbolic form of Kepler's time equation:

$$\sinh H = \frac{\sqrt{e^2 - 1}\, \sin \theta}{1 + e \cos \theta}$$

$$\cosh H = \frac{e + \cos \theta}{1 + e \cos \theta}$$

$$\tanh \frac{H}{2} = \sqrt{\frac{e-1}{1+e}}\, \tan \frac{\theta}{2}$$

$$r = a(1 - e \cosh H)$$

$$\frac{1}{2} \int_0^\theta r^2 \, d\theta = \frac{a^2}{2} \int_0^\theta (1 - e \cosh H)^2 \, d\theta$$

$$= \frac{ab}{2} \int_0^H (e \cosh H - 1) \, dH$$

$$\frac{ab}{2} \sqrt{\frac{\mu}{(-a)^3}} (t - t_p) = \frac{ab}{2} (e \sinh H - H)$$

$$N = e \sinh H - H$$

where N corresponds to the mean anomaly M of an elliptic orbit.
Note: The hyperbolic form of Kepler's time equation can also be obtained from Kepler's time equation by substituting $E = -jH$ and $M = jN$ where $j = \sqrt{-1}$. The following hyperbolic relations are also useful in solving for time,

$$H = \ln \left(\cosh H + \sqrt{\cosh^2 H - 1} \right)$$

$$\sinh H = \tfrac{1}{2}[\exp(H) - \exp(-H)]$$

3.5   Orbital Position and Velocity

In this section we consider the motion of a satellite that is revolving (or orbiting) around the Earth.

3.5.1   Orbital Elements

In general, the two-body dynamic system characterized by the equation of relative motion of the form

$$\ddot{\vec{r}} + (\mu/r^3)\vec{r} = 0 \tag{3.61}$$

has three degrees of freedom, and the orbit is uniquely determined if the six initial conditions $\vec{r}$ and $\vec{v} \equiv \dot{\vec{r}}$ are specified. In orbital mechanics, the constants of integration or integrals of the motion are also referred to as orbital elements and such initial conditions can be considered as six possible orbital elements.

To describe a satellite orbit about the Earth, we often employ six other scalars, called the six classical orbital elements. Three of these scalars specify the orientation of the orbit plane with respect to the *geocentric-equatorial* reference frame, which has its origin at the center of the Earth. This geocentric-equatorial reference frame has an inclination of 23.45 deg with respect to the *heliocentric-ecliptic* reference frame that has its origin at the center of the sun. A set of orthogonal unit vectors $\{\vec{I}, \vec{J}, \vec{K}\}$ is selected as basis vectors of the geocentric-equatorial reference frame with (X, Y, Z) coordinates, as shown in Fig. 3.5.

Note that this reference frame is not fixed to the Earth and is not rotating with it; rather the Earth rotates around it. The (X, Y) plane of the geocentric-equatorial

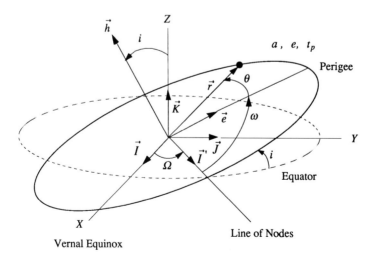

Fig. 3.5   Orbit orientation with respect to the geocentric-equatorial reference frame. (A near circular orbit is shown in this figure.)

reference frame is the Earth's equatorial plane, simply called the equator. The Z axis is along the Earth's polar axis of rotation. The X axis is pointing toward the vernal equinox, the point in the sky where the sun crosses the equator from south to north on the first day of spring. The vernal equinox direction is often denoted by the symbol Υ.

The six classical orbital elements consist of five independent quantities, which are sufficient to completely describe the size, shape, and orientation of an orbit, and one quantity required to pinpoint the position of a satellite along the orbit at any particular time. The six classical orbital elements are:

a = semimajor axis
e = eccentricity
t_p = time of perigee passage
Ω = right ascension longitude of the ascending node
i = inclination of the orbit plane
ω = argument of the perigee

The elements a and e determine the size and shape of the elliptic orbit, respectively, and t_p relates position in orbit to time. The angles Ω and i specify the orientation of the orbit plane with respect to the geocentric-equatorial reference frame. The angle ω specifies the orientation of the orbit in its plane. Orbits with $i < 90$ deg are called *prograde orbits*, whereas orbits with $i > 90$ deg are called *retrograde orbits*. The term prograde means easterly direction in which the sun, Earth, and most of the planets and their moons rotate on their axes. The term retrograde means westerly direction, which is simply the opposite of prograde. An orbit whose inclination is near 90 deg is called a polar orbit. An equatorial orbit has zero inclination.

3.5.2 Orbit Determination

Consider the problem of determining the six classical orbital elements of a satellite using its known position and velocity vectors at a specified time t_0. We assume that the position and velocity vectors $\vec{r}$ and $\vec{v} \equiv \dot{\vec{r}}$ at t_0 are expressed in terms of the basis vectors $\{\vec{I}, \vec{J}, \vec{K}\}$ of the geocentric-equatorial reference frame; i.e.,

$$\vec{r} = X\vec{I} + Y\vec{J} + Z\vec{K} \qquad (3.62a)$$
$$\vec{v} = \dot{X}\vec{I} + \dot{Y}\vec{J} + \dot{Z}\vec{K} \qquad (3.62b)$$

whose six components $(X, Y, Z, \dot{X}, \dot{Y}, \dot{Z})$ are known constants at t_0.

Because $\vec{r}$ and $\vec{v}$ are known at t_0, the specific angular momentum vector is first determined as

$$\vec{h} = \vec{r} \times \vec{v} \qquad (3.63)$$

and the eccentricity $e \equiv |\vec{e}|$ is determined from the eccentricity vector given as

$$\mu\vec{e} = \vec{v} \times \vec{h} - (\mu/r)\vec{r} = [v^2 - (\mu/r)]\vec{r} - (\vec{r} \cdot \vec{v})\vec{v} \qquad (3.64)$$

Since $r \equiv |\vec{r}|$ and $v \equiv |\vec{v}|$ are known at t_0, the specific energy can be found as

$$\mathcal{E} = (v^2/2) - (\mu/r) \qquad (3.65)$$

and the semimajor axis a is determined as

$$a = -\mu/2\mathcal{E} \tag{3.66}$$

The right ascension of the ascending node, Ω, can be determined from a unit vector $\vec{I}'$ toward the ascending node, given by

$$\vec{I}' = \cos\Omega\,\vec{I} + \sin\Omega\,\vec{J} \tag{3.67}$$

which is perpendicular to both $\vec{h}$ and $\vec{K}$; i.e.,

$$\vec{I}' = \vec{K} \times (\vec{h}/h) \tag{3.68}$$

The inclination angle i is obtained from

$$\cos i = \vec{K} \cdot (\vec{h}/h) \tag{3.69}$$

and the argument of the perigee, ω, is also obtained as

$$\cos\omega = \vec{I}' \cdot (\vec{e}/e) \tag{3.70}$$

where a proper quadrant correction must be made for $\vec{e} \cdot \vec{K} < 0$.

The time of perigee passage, t_p, is determined by first computing the true anomaly θ at t_0, from

$$\cos\theta = \frac{\vec{r} \cdot \vec{e}}{re} \tag{3.71}$$

where a proper quadrant correction must be made when $\vec{r} \cdot \vec{v} > 0$. The eccentric anomaly E at t_0 is then obtained from

$$\tan\frac{E}{2} = \sqrt{\frac{1-e}{1+e}}\,\tan\frac{\theta}{2} \tag{3.72}$$

and then Kepler's time equation

$$M = \sqrt{\mu/a^3}\,(t_0 - t_p) = E - e\sin E \tag{3.73}$$

is used to obtain the perigee passage time t_p. The mean anomaly M or the true anomaly θ at a particular time t_0, called the *epoch*, often replaces t_p in a set of modified classical orbital elements.

The inverse problem of determining the position and velocity vectors at any other time t, given the six classical orbital elements, begins with solving Kepler's time equation

$$M = \sqrt{\mu/a^3}\,(t - t_p) = E - e\sin E \tag{3.74}$$

for the eccentric anomaly E.

After solving such a transcendental equation by using a numerical method, we obtain the true anomaly θ from

$$\tan \frac{\theta}{2} = \sqrt{\frac{1+e}{1-e}} \tan \frac{E}{2} \tag{3.75}$$

and the radial position is then obtained using the orbit equation

$$r = \frac{a(1-e^2)}{1+e\cos\theta} \tag{3.76}$$

or directly using the orbit equation expressed in terms of the eccentric anomaly

$$r = a(1 - e\cos E) \tag{3.77}$$

To determine the position and velocity vectors for given r and θ at a specified time t, we introduce a so-called *perifocal* reference frame with (x, y, z) coordinates and with a set of basis vectors $\{\vec{i}, \vec{j}, \vec{k}\}$, as shown in Fig. 3.6. This reference frame is fixed to the orbit plane. The x axis points toward the perigee of the orbit, the y axis is in the orbit plane, and the z axis is out of the orbit plane and completes a right-handed reference frame. Note that $\vec{e} = e\vec{i}$ and $\vec{h} = h\vec{k}$. Because the perifocal reference frame does not have a relative motion with respect to the geocentric-equatorial reference frame, which is considered to be inertially fixed in space, the perifocal reference frame is also an inertial reference frame.

In terms of basis vectors of the perifocal reference frame, the position and velocity vectors are expressed as

$$\vec{r} = x\vec{i} + y\vec{j} + z\vec{k} \tag{3.78a}$$

$$\vec{v} = \dot{x}\vec{i} + \dot{y}\vec{j} + \dot{z}\vec{k} \tag{3.78b}$$

whose six components $(x, y, z, \dot{x}, \dot{y}, \dot{z})$ are to be determined from the radial position r and the true anomaly θ at a given time t.

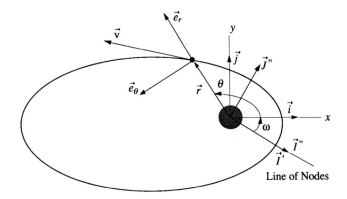

Fig. 3.6   Perifocal reference frame.

As shown in Fig. 3.6, let $\vec{e}_r$ and $\vec{e}_\theta$ be unit vectors along the radial vector direction and the transverse orbit direction, respectively. Then the position vector $\vec{r}$ and the velocity vector $\vec{v} \equiv \dot{\vec{r}}$ can be expressed as

$$\vec{r} = r\vec{e}_r = r(\cos\theta\,\vec{i} + \sin\theta\,\vec{j}) \tag{3.79}$$

$$\vec{v} = \dot{r}\vec{e}_r + r\dot{\theta}\vec{e}_\theta = \dot{r}(\cos\theta\,\vec{i} + \sin\theta\,\vec{j}) + r\dot{\theta}(-\sin\theta\,\vec{i} + \cos\theta\,\vec{j}) \tag{3.80}$$

Using the relationship $h = \sqrt{\mu p} = r^2\dot{\theta}$ and the orbit equation (3.76), we represent the transverse velocity component, denoted by v_θ, as follows:

$$v_\theta \equiv r\dot{\theta} = \sqrt{\mu/p}\,(1 + e\cos\theta) \tag{3.81}$$

Also the radial velocity component, denoted by v_r, can be obtained by differentiating Eq. (3.76), as follows:

$$v_r \equiv \dot{r} = \frac{a(1-e^2)e\sin\theta\,\dot{\theta}}{(1 + e\cos\theta)^2} = \sqrt{\frac{\mu}{p}}\,e\sin\theta \tag{3.82}$$

The velocity vector expressed as Eq. (3.80) then becomes

$$\vec{v} = \sqrt{\mu/p}\{-\sin\theta\,\vec{i} + (e + \cos\theta)\vec{j}\} \tag{3.83}$$

Finally, we have the results in matrix notation, as follows:

$$\begin{bmatrix} x \\ y \\ z \end{bmatrix} = \begin{bmatrix} r\cos\theta \\ r\sin\theta \\ 0 \end{bmatrix} \quad \text{and} \quad \begin{bmatrix} \dot{x} \\ \dot{y} \\ \dot{z} \end{bmatrix} = \begin{bmatrix} -\sqrt{\mu/p}\sin\theta \\ \sqrt{\mu/p}(e + \cos\theta) \\ 0 \end{bmatrix} \tag{3.84}$$

Problem

3.6 (a) Given Kepler's time equation $\sqrt{\mu/a^3}\,(t - t_p) = E - e\sin E$, show that we simply obtain $\dot{E}$ as

$$\dot{E} = \frac{\sqrt{\mu/a^3}}{1 - e\cos E}$$

(b) Given the position vector $\vec{r}$ expressed as

$$\vec{r} = a(\cos E - e)\vec{i} + b\sin E\,\vec{j}$$

where $\vec{i}$ and $\vec{j}$ are unit vectors of the perifocal reference frame, show that

$$\vec{v} \equiv \dot{\vec{r}} = \sqrt{\frac{\mu}{a^3}}\left\{\frac{-a\sin E}{1 - e\cos E}\vec{i} + \frac{b\cos E}{1 - e\cos E}\vec{j}\right\}$$

where $b = a\sqrt{1 - e^2}$.

3.5.3 Coordinate Transformation

Given the geocentric-equatorial (X, Y, Z) reference frame with basis vectors $\{\vec{I}, \vec{J}, \vec{K}\}$ and a perifocal (x, y, z) reference frame with basis vectors $\{\vec{i}, \vec{j}, \vec{k}\}$, the position vector is represented as

$$\vec{r} = X\vec{I} + Y\vec{J} + Z\vec{K} = x\vec{i} + y\vec{j} + z\vec{k} \tag{3.85}$$

The position vector $\vec{r}$ can also be expressed as

$$\vec{r} = X'\vec{I'} + Y'\vec{J'} + Z'\vec{K'} = X''\vec{I''} + Y''\vec{J''} + Z''\vec{K''} \tag{3.86}$$

where (X', Y', Z') and (X'', Y'', Z'') are the components of the position vector $\vec{r}$ in two intermediate reference frames with basis vectors $\{\vec{I'}, \vec{J'}, \vec{K'}\}$ and $\{\vec{I''}, \vec{J''}, \vec{K''}\}$, respectively.

The perifocal reference frame is then related to the geocentric-equatorial reference frame through three successive rotations as follows:

$$\begin{bmatrix} \vec{I'} \\ \vec{J'} \\ \vec{K'} \end{bmatrix} = \begin{bmatrix} \cos\Omega & \sin\Omega & 0 \\ -\sin\Omega & \cos\Omega & 0 \\ 0 & 0 & 1 \end{bmatrix} \begin{bmatrix} \vec{I} \\ \vec{J} \\ \vec{K} \end{bmatrix} \tag{3.87a}$$

$$\begin{bmatrix} \vec{I''} \\ \vec{J''} \\ \vec{K''} \end{bmatrix} = \begin{bmatrix} 1 & 0 & 0 \\ 0 & \cos i & \sin i \\ 0 & -\sin i & \cos i \end{bmatrix} \begin{bmatrix} \vec{I'} \\ \vec{J'} \\ \vec{K'} \end{bmatrix} \tag{3.87b}$$

$$\begin{bmatrix} \vec{i} \\ \vec{j} \\ \vec{k} \end{bmatrix} = \begin{bmatrix} \cos\omega & \sin\omega & 0 \\ -\sin\omega & \cos\omega & 0 \\ 0 & 0 & 1 \end{bmatrix} \begin{bmatrix} \vec{I''} \\ \vec{J''} \\ \vec{K''} \end{bmatrix} \tag{3.87c}$$

The orbital elements Ω, i, and ω are, in fact, the Euler angles of the so-called $C_3(\omega) \leftarrow C_1(i) \leftarrow C_3(\Omega)$ rotational sequence to be treated in Chapter 5.

By combining the preceding sequence of rotations, we obtain

$$\begin{bmatrix} \vec{i} \\ \vec{j} \\ \vec{k} \end{bmatrix} = \begin{bmatrix} \cos\omega & \sin\omega & 0 \\ -\sin\omega & \cos\omega & 0 \\ 0 & 0 & 1 \end{bmatrix} \begin{bmatrix} 1 & 0 & 0 \\ 0 & \cos i & \sin i \\ 0 & -\sin i & \cos i \end{bmatrix}$$

$$\times \begin{bmatrix} \cos\Omega & \sin\Omega & 0 \\ -\sin\Omega & \cos\Omega & 0 \\ 0 & 0 & 1 \end{bmatrix} \begin{bmatrix} \vec{I} \\ \vec{J} \\ \vec{K} \end{bmatrix}$$

which becomes

$$\begin{bmatrix} \vec{i} \\ \vec{j} \\ \vec{k} \end{bmatrix} = \begin{bmatrix} C_{11} & C_{12} & C_{13} \\ C_{21} & C_{22} & C_{23} \\ C_{31} & C_{32} & C_{33} \end{bmatrix} \begin{bmatrix} \vec{I} \\ \vec{J} \\ \vec{K} \end{bmatrix} \tag{3.88}$$

where

$$C_{11} = \cos \Omega \cos \omega - \sin \Omega \sin \omega \cos i$$
$$C_{12} = \sin \Omega \cos \omega + \cos \Omega \sin \omega \cos i$$
$$C_{13} = \sin \omega \sin i$$
$$C_{21} = -\cos \Omega \sin \omega - \sin \Omega \cos \omega \cos i$$
$$C_{22} = -\sin \Omega \sin \omega + \cos \Omega \cos \omega \cos i$$
$$C_{23} = \cos \omega \sin i$$
$$C_{31} = \sin \Omega \sin i$$
$$C_{32} = -\cos \Omega \sin i$$
$$C_{33} = \cos i$$

The matrix $[C_{ij}]$ is called the *direction cosine matrix*, which describes the orientation of the perifocal reference frame with respect to the geocentric-equatorial reference frame.

The components (x, y, z) of the position vector in the perifocal reference frame are then related to the components (X, Y, Z) of the position vector in the geocentric-equatorial reference frame via the same direction cosine matrix $[C_{ij}]$ as

$$\begin{bmatrix} x \\ y \\ z \end{bmatrix} = \begin{bmatrix} C_{11} & C_{12} & C_{13} \\ C_{21} & C_{22} & C_{23} \\ C_{31} & C_{32} & C_{33} \end{bmatrix} \begin{bmatrix} X \\ Y \\ Z \end{bmatrix} \tag{3.89}$$

Because the direction cosine matrix $[C_{ij}]$ is an orthonormal matrix, i.e., $[C_{ij}]^{-1} = [C_{ij}]^T$, we also have

$$\begin{bmatrix} X \\ Y \\ Z \end{bmatrix} = \begin{bmatrix} C_{11} & C_{21} & C_{31} \\ C_{12} & C_{22} & C_{32} \\ C_{13} & C_{23} & C_{33} \end{bmatrix} \begin{bmatrix} x \\ y \\ z \end{bmatrix} \tag{3.90}$$

The components of the velocity vector represented as

$$\vec{v} = \dot{X}\vec{I} + \dot{Y}\vec{J} + \dot{Z}\vec{K} = \dot{x}\vec{i} + \dot{y}\vec{j} + \dot{z}\vec{k} \tag{3.91}$$

are also related as

$$\begin{bmatrix} \dot{x} \\ \dot{y} \\ \dot{z} \end{bmatrix} = \begin{bmatrix} C_{11} & C_{12} & C_{13} \\ C_{21} & C_{22} & C_{23} \\ C_{31} & C_{32} & C_{33} \end{bmatrix} \begin{bmatrix} \dot{X} \\ \dot{Y} \\ \dot{Z} \end{bmatrix} \tag{3.92}$$

or

$$\begin{bmatrix} \dot{X} \\ \dot{Y} \\ \dot{Z} \end{bmatrix} = \begin{bmatrix} C_{11} & C_{21} & C_{31} \\ C_{12} & C_{22} & C_{32} \\ C_{13} & C_{23} & C_{33} \end{bmatrix} \begin{bmatrix} \dot{x} \\ \dot{y} \\ \dot{z} \end{bmatrix} \tag{3.93}$$

3.5.4   Earth Satellite Applications

3.5.4.1   Orbit Inclination   The orbit inclination of a satellite launched from a launch site at latitude ϕ, and with a launch azimuth A_z measured clockwise from north, can be determined from spherical trigonometry as

$$\cos i = \sin A_z \cos \phi \qquad (3.94)$$

assuming a nonrotating Earth.

For launch sites in the Northern Hemisphere with $0 < \phi < 90$ deg, a prograde orbit with $0 < i < 90$ deg requires a launch azimuth with an easterly component and a retrograde orbit with $90 < i < 180$ deg requires a launch azimuth with a westerly component. Equation (3.94) implies that the minimum inclination that can be achieved from a launch site at latitude ϕ is equal to the latitude of the launch site. Therefore, a satellite cannot be injected directly into an equatorial orbit ($i = 0$ deg) from a launch site that is not on the equator. (The launch site of the European Space Agency is on the equator and the Cape Canaveral launch site is located at $\phi = 28.5$ deg.)

3.5.4.2   Ground Track   The orbit of an Earth satellite always lies in a plane passing through the Earth center. The track of this plane, called a ground track (or trace), on the surface of a nonrotating Earth is a great circle. The maximum latitude north or south of the equator over which the satellite passes is equal to the orbit inclination i (deg) for a prograde orbit ($180 - i$ for a retrograde orbit) and the ground track often looks nearly sinusoidal for a low-altitude circular orbit. Because the Earth rotates about the polar axis easterly at a rate of 360 deg per one *siderial* day the ground track is displaced westward on each successive revolution of the satellite by the number of degrees the Earth turns during one orbital period. (A siderial day is defined as the period required for one rotation of the Earth about its polar axis relative to an inertially fixed reference frame. One siderial day is 23 h 56 min 4 s, whereas one solar day is 24 h.)

3.5.4.3   Geosynchronous Orbits   If the period of a satellite in a circular prograde equatorial orbit is exactly one siderial day, it will appear to hover motionlessly over a point on the equator. Such a satellite, located at 42,164 km ($\approx 6.6R_\oplus$) from the Earth center (or at an altitude of 35,786 km), is called a *geostationary* satellite. A satellite with the orbital period of one siderial day but with a nonzero inclination is called a *geosynchronous* satellite. Its ground track is often characterized by a figure-eight curve. Note that regardless of the satellite's orbital inclination, geosynchronous satellites still take 23 h 56 min 4 s to make one complete revolution around the Earth.

3.6   Orbital Perturbations

Thus far in this chapter, we have considered an ideal Keplerian orbit of the two-body problem in which the primary body has a spherically symmetric mass distribution and its orbital plane is fixed in space. In general, however, we should consider a non-Keplerian orbit whose orbital plane is not fixed in space due to the *asphericity* of the primary body. The small deviations from the ideal Keplerian orbital motion are called *orbital perturbations*. In this section, we consider the effects of the Earth's oblateness on the orbital motions of near-Earth satellites.

3.6.1 Earth's Oblateness Effects

The Earth is not a perfect sphere but it is an oblate spheroid of revolution; that is, the Earth is flattened at the poles to produce geoid or ellipsoid of revolution. There are also minor harmonics of the Earth's shape that produce a pear shape. The pear shape is not significant for most cases and so we will focus on perturbations due to the polar flattening. The equatorial bulge caused by the polar flattening is only about 21 km. However, this bulge distorts the path of a satellite each time it passes either the ascending node or descending node. The attractive force from the bulge shifts the satellite path northward as the satellite approaches the equatorial plane from the south. As the satellite leaves the equatorial plane, it is shifted southward. The net result is the ascending node having shifted or regressed opposite the direction of satellite motion. The Earth's oblateness causes motion of the orbital plane and also affects the position of satellites within the orbital plane. In this section, we analyze the effects of the Earth's oblateness on the precession of the node line and the regression of the apsidal line of near-Earth satellites' orbits. When the major axis is used as a reference line for measuring the true anomaly, it is called the line of apsides or the apsidal line.

Consider the equation of motion of a satellite about the Earth described by

$$\ddot{\vec{r}} + (\mu/r^3)\vec{r} = \vec{f} \tag{3.95}$$

where $\vec{r}$ is the position vector of the satellite from the center of the Earth, $\mu \approx \mu_\oplus$, and $\vec{f}$, which is the sum of all of the perturbing forces per unit mass, is called the *perturbing acceleration* acting on the satellite. The position of a satellite acted upon by the perturbing acceleration is often referred to as a plane containing $\vec{r}$ and $\dot{\vec{r}}$, called the *osculating orbital* plane.

Taking the dot product of Eq. (3.95) with $\dot{\vec{r}}$ yields

$$\ddot{\vec{r}} \cdot \dot{\vec{r}} + (\mu/r^3)\vec{r} \cdot \dot{\vec{r}} = \vec{f} \cdot \dot{\vec{r}} \tag{3.96}$$

which is rewritten as

$$\frac{d}{dt}\left(\frac{v^2}{2} - \frac{\mu}{r}\right) = \vec{f} \cdot \dot{\vec{r}} \tag{3.97}$$

Substituting the specific energy $\mathcal{E}$ defined as

$$\mathcal{E} = \frac{v^2}{2} - \frac{\mu}{r} = -\frac{\mu}{2a}$$

into Eq. (3.97), we obtain

$$\dot{a} = (2a^2/\mu)\vec{f} \cdot \dot{\vec{r}}$$

(3.98)

Note that $\mathcal{E}$ is not a constant unless $\vec{f} = 0$ or $\vec{f} \cdot \dot{\vec{r}} = 0$.

Taking the cross product of Eq. (3.95) with $\vec{r}$, we have

$$\vec{r} \times \ddot{\vec{r}} = \vec{r} \times \vec{f}$$

(3.99)

Differentiating the specific angular momentum defined as

$$\vec{h} = \vec{r} \times \dot{\vec{r}}$$

(3.100)

we obtain

$$\dot{\vec{h}} = \vec{r} \times \ddot{\vec{r}} = \vec{r} \times \vec{f}$$

(3.101)

Note that $\vec{h}$ is not a constant vector unless $\vec{f} = 0$ or $\vec{r} \times \vec{f} = 0$.

Taking the post–cross product of Eq. (3.95) with $\vec{h}$, we have

$$\ddot{\vec{r}} \times \vec{h} + (\mu/r^3)\vec{r} \times \vec{h} = \vec{f} \times \vec{h}$$

(3.102)

which is rewritten as

$$\frac{\mathrm{d}}{\mathrm{d}t}\left(\dot{\vec{r}} \times \vec{h} - \frac{\mu}{r}\vec{r}\right) = \dot{\vec{r}} \times \dot{\vec{h}} + \vec{f} \times \vec{h}$$

(3.103)

Substituting the eccentricity vector $\vec{e}$ defined as

$$\mu\vec{e} = \dot{\vec{r}} \times \vec{h} - (\mu/r)\vec{r}$$

(3.104)

and Eq. (3.101) into Eq. (3.103), we obtain

$$\mu\dot{\vec{e}} = \dot{\vec{r}} \times (\vec{r} \times \vec{f}) + \vec{f} \times \vec{h}$$

(3.105)

Here, $\vec{e}$ is not a constant vector unless the right-hand side of Eq. (3.105) is zero.

Let $\vec{e}_r$, $\vec{e}_\theta$, and $\vec{e}_z$ be unit vectors along the radial vector direction, the transverse orbit direction, and the direction normal to the orbit plane, respectively, such that $\vec{e}_r \times \vec{e}_\theta = \vec{e}_z$. Then the perturbing acceleration $\vec{f}$ and the velocity vector $\vec{v} \equiv \dot{\vec{r}}$ are represented in terms of the unit vectors $\{\vec{e}_r, \vec{e}_\theta, \vec{e}_z\}$, as follows:

$$\vec{f} = f_r \vec{e}_r + f_\theta \vec{e}_\theta + f_z \vec{e}_z$$

(3.106)

$$\vec{v} = v_r \vec{e}_r + v_\theta \vec{e}_\theta + v_z \vec{e}_z$$

(3.107)

From Sec. (3.5), we have

$$v_r \equiv \dot{r} = \sqrt{\mu/p}\, e \sin\theta$$

(3.108a)

$$v_\theta \equiv r\dot{\theta} = \sqrt{\mu/p}\,(1 + e\cos\theta)$$

(3.108b)

and $v_z = 0$ due to the assumptions of the osculating orbit. Consequently, the term $\vec{f} \cdot \dot{\vec{r}}$ in Eq. (3.98) becomes

$$\vec{f} \cdot \dot{\vec{r}} = f_r v_r + f_\theta v_\theta \qquad (3.109)$$

and we obtain

$$\dot{a} = \frac{2a^2}{\sqrt{\mu p}} \{f_r e \sin \theta + f_\theta (1 + e \cos \theta)\} \qquad (3.110)$$

Differentiating the specific angular momentum vector expressed as

$$\vec{h} = \sqrt{\mu p} \, \vec{k} \qquad (3.111)$$

where $\vec{k} \, (= \vec{e}_z)$ is a unit vector normal to the orbit plane, we obtain

$$\dot{\vec{h}} = \frac{1}{2} \sqrt{\mu/p} \, \dot{p} \vec{k} + \sqrt{\mu p} \, \dot{\vec{k}} \qquad (3.112)$$

Furthermore, we have

$$\dot{\vec{k}} = (\dot{\Omega} \vec{K} + \dot{i} \vec{I}'' + \dot{\omega} \vec{k}) \times \vec{k}$$
$$= \dot{\Omega} \sin i \vec{I}'' - \dot{i} \vec{J}'' \qquad (3.113)$$

where $\vec{I}''$ is a unit vector toward the ascending node and $\vec{J}''$ is orthogonal to $\vec{I}''$ (see Figs. 3.5 and 3.6). Thus, we have

$$\dot{\vec{h}} = \frac{1}{2} \sqrt{\mu/p} \, \dot{p} \vec{k} + \sqrt{\mu p} (\dot{\Omega} \sin i \vec{I}'' - \dot{i} \vec{J}'') \qquad (3.114)$$

The term $\vec{r} \times \vec{f}$ in Eq. (3.101) is also written as

$$\vec{r} \times \vec{f} = r f_\theta \vec{k} - r f_z \vec{e}_\theta \qquad (3.115)$$

In terms of unit vectors $\vec{I}''$, $\vec{J}''$, and $\vec{k}$, this equation becomes

$$\vec{r} \times \vec{f} = r f_\theta \vec{k} - r f_z [-\sin (\omega + \theta) \vec{I}'' + \cos (\omega + \theta) \vec{J}''] \qquad (3.116)$$

Because $\dot{\vec{h}} = \vec{r} \times \vec{f}$, equating the coefficients of Eqs. (3.114) and (3.116) gives

$$\dot{p} = 2 \sqrt{p/\mu} \, r f_\theta \qquad (3.117)$$

$$\dot{\Omega} \sin i = \frac{r f_z}{\sqrt{\mu p}} \sin (\omega + \theta) \qquad (3.118)$$

$$\dot{i} = \frac{r f_z}{\sqrt{\mu p}} \cos (\omega + \theta) \qquad (3.119)$$

Differentiating the relation, $p = a(1 - e^2)$, gives

$$\dot{e} = \frac{\dot{a}(1 - e^2) - \dot{p}}{2ea} \qquad (3.120)$$

Combining this equation with Eqs. (3.110) and (3.117) and using the following relationships:

$$p = a(1 - e^2)$$
$$r = a(1 - e \cos E)$$

we obtain

$$\dot{e} = \sqrt{p/\mu} \{ f_r \sin \theta + f_\theta (\cos \theta + \cos E) \} \qquad (3.121)$$

The effects of the Earth's oblateness on the precession of the node line and the regression of the apsidal line of near-Earth satellites' orbits can now be analyzed considering the gravitational potential of the oblate Earth given by

$$U_\oplus(r, \phi) = \frac{\mu}{r} \left\{ 1 + \frac{J_2 R_\oplus^2}{2r^2} (1 - 3 \sin^2\phi) - \frac{J_3 R_\oplus^3}{2r^3} (5 \sin^3\phi - 3 \sin \phi) - \cdots \right\}$$

$$(3.122)$$

where r is the geocentric distance, ϕ is the geocentric latitude, $R_\oplus$ (= 6,378 km) is the mean equatorial radius of the Earth, $\mu \approx \mu_\oplus = 398,601$ km^3/s^2, and $J_2 = 1082.64 \times 10^{-6}$ and $J_3 = -2.56 \times 10^{-6}$ are the harmonic constants of the Earth due to its oblateness. As illustrated in Fig. 3.7, the angle between the equatorial plane and the radius from the geocenter is called geocentric latitude, whereas the angle between the equatorial plane and the normal to the surface of the ellipsoid is called geodetic latitude. The commonly used geodetic altitude is also illustrated in Fig. 3.7.

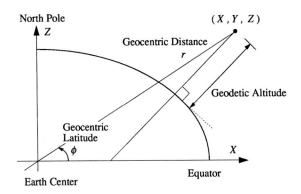

Fig. 3.7   Two-dimensional view of the oblate Earth.

Ignoring higher-order terms, the perturbing gravitational potential due to J_2, denoted by U, is then defined as

$$U = U_\oplus - \frac{\mu}{r} = \frac{\mu J_2 R_\oplus^2}{2r^3}(1 - 3\sin^2\phi) \tag{3.123}$$

Because the geocentric latitude ϕ is related to the orbital elements as

$$\sin\phi = \frac{Z}{r} = \frac{r\sin(\omega + \theta)\sin i}{r} = \sin(\omega + \theta)\sin i \tag{3.124}$$

Eq. (3.123) is rewritten as

$$U = \frac{\mu J_2 R_\oplus^2}{2}\left\{\frac{1}{r^3} - \frac{3\sin^2 i \sin^2(\omega + \theta)}{r^3}\right\} \tag{3.125}$$

Since $\vec{r} = r\vec{e}_r$, and $dz = r\sin(\omega + \theta)\,di$, the perturbing acceleration due to J_2 is described by

$$\vec{f} = \nabla U = \frac{\partial U}{\partial r}\vec{e}_r + \frac{1}{r}\frac{\partial U}{\partial\theta}\vec{e}_\theta + \frac{1}{r\sin(\omega + \theta)}\frac{\partial U}{\partial i}\vec{e}_z \tag{3.126}$$

Taking the partial derivatives of U with respect to r, θ, and i, and substituting them into Eq. (3.126), we obtain the radial, transverse, and normal components of $\vec{f}$, as follows:

$$f_r = -\frac{3\mu J_2 R_\oplus^2}{2r^4}\{1 - 3\sin^2 i \sin^2(\omega + \theta)\} \tag{3.127}$$

$$f_\theta = -\frac{3\mu J_2 R_\oplus^2}{2r^4}\sin^2 i \sin 2(\omega + \theta) \tag{3.128}$$

$$f_z = -\frac{3\mu J_2 R_\oplus^2}{2r^4}\sin 2i \sin(\omega + \theta) \tag{3.129}$$

where $\sin 2i = 2\sin i \cos i$.

Substituting Eq. (3.129) into Eq. (3.118), we obtain the precession of the node line as

$$\dot{\Omega} = -\frac{3\mu J_2 R_\oplus^2}{r^3\sqrt{\mu p}}\cos i \sin^2(\omega + \theta) \tag{3.130}$$

Integrating this equation over an entire orbit of period P yields

$$\Delta\Omega = -\frac{3\mu J_2 R_\oplus^2}{\sqrt{\mu p}}\cos i \int_0^P \frac{\sin^2(\omega + \theta)}{r^3}\,dt \tag{3.131}$$

where $\Delta\Omega$ denotes the change of Ω over an entire orbit, assuming that changes in other orbital elements are second-order terms. (Note that the average rate of change of i over the orbital period is zero.)

Because the angular momentum $h \equiv |\vec{h}|$ can be expressed as

$$h = \sqrt{\mu p} \approx r^2(\dot{\Omega}\cos i + \dot{\omega} + \dot{\theta}) \tag{3.132}$$

we have

$$\dot{\omega} + \dot{\theta} \approx \sqrt{\mu p}/r^2 \tag{3.133}$$

in which the second-order term $\dot{\Omega}\cos i$ is further neglected. This equation is used to change the independent variable t into $(\omega + \theta)$, as follows:

$$d(\omega + \theta) = \frac{\sqrt{\mu p}}{r^2}\, dt$$

Thus, Eq. (3.131) can be rewritten as

$$\Delta\Omega = -\frac{3J_2 R_\oplus^2}{p}\cos i \int_0^{2\pi} \frac{\sin^2(\omega + \theta)}{r}\, d(\omega + \theta)$$

$$= -\frac{3J_2 R_\oplus^2}{p}\cos i \int_0^{2\pi} \frac{1 - \cos 2(\omega + \theta)}{2r}\, d(\omega + \theta)$$

Performing the integration after a substitution of $r = p/(1 + e\cos\theta)$ yields

$$\Delta\Omega = -\frac{3\pi J_2 R_\oplus^2}{p^2}\cos i + \text{ higher-order terms} \tag{3.134}$$

Dividing this by the average orbital period $P = 2\pi/n$, where $n = \sqrt{\mu/a^3}$ is the orbital mean motion, we obtain the average rate of change of Ω, as follows:

$$\dot{\Omega} \approx -\frac{3J_2 R_\oplus^2}{2p^2} n \cos i \tag{3.135}$$

Similarly, assuming that the eccentricity and the semimajor axis of the orbit remain unperturbed by the oblateness of the Earth to a first-order approximation, we can obtain the average rate of change of ω, as follows:

$$\dot{\omega} \approx -\frac{3J_2 R_\oplus^2}{2p^2} n \left(\frac{5\sin^2 i}{2} - 2\right) \tag{3.136}$$

Problem

3.7 Consider the perturbing gravitational potential U of the Earth due to its J_2 term expressed as

$$U = \frac{\mu J_2 R_\oplus^2}{2r^3}(1 - 3\sin^2\phi) = \frac{\mu J_2 R_\oplus^2}{2}\left(\frac{1}{r^3} - \frac{3Z^2}{r^5}\right)$$

where $r = \sqrt{X^2 + Y^2 + Z^2}$ and (X, Y, Z) are the coordinates of the radial position vector in the geocentric-equatorial reference frame.

(a) Show that the perturbing acceleration due to J_2 can be expressed as

$$\vec{f} = \nabla U = \left.\frac{\partial U}{\partial r}\right|_Z \vec{e}_r + \left.\frac{\partial U}{\partial Z}\right|_r \vec{K}$$

where $\vec{e}_r$ is a unit vector along the radial vector direction and $\vec{K}$ is a unit vector perpendicular to the Earth's equator. (Note that $\vec{e}_r$ and $\vec{K}$ are not orthogonal to each other.)

Hint: Let $U = U(r, Z)$ where $r = r(X, Y, Z)$, then

$$\nabla U = \frac{\partial U}{\partial X}\vec{I} + \frac{\partial U}{\partial Y}\vec{J} + \frac{\partial U}{\partial Z}\vec{K} = \left.\frac{\partial U}{\partial r}\right|_Z \frac{\partial r}{\partial X}\vec{I} + \left.\frac{\partial U}{\partial r}\right|_Z \frac{\partial r}{\partial Y}\vec{J} + \cdots$$

(b) Show that $\vec{f}$ then becomes

$$\vec{f} = -\frac{\mu J_2 R_\oplus^2}{2}\left(\frac{3}{r^4} - \frac{15Z^2}{r^6}\right)\vec{e}_r - \frac{\mu J_2 R_\oplus^2}{2}\left(\frac{6Z}{r^5}\right)\vec{K}$$

(c) Verify that

$$Z = r\sin(\omega + \theta)\sin i$$

$$\vec{K} = \sin(\omega + \theta)\sin i\vec{e}_r + \cos(\omega + \theta)\sin i\vec{e}_\theta + \cos i\vec{e}_z$$

(d) Finally, show that the radial, transverse, and normal components of $\vec{f}$ can then be obtained as Eqs. (3.127), (3.128), and (3.129), respectively.

3.6.2 Earth Satellite Applications

The *nodal regression rate* $\dot{\Omega}$ and the *apsidal line rotation rate* $\dot{\omega}$ are of primary importance for near-Earth satellites. Some practical applications of the effects of the Earth's oblateness are briefly described here.

3.6.2.1 Sun-Synchronous Orbits
The Earth revolves around the sun in a nearly circular orbit ($e = 0.016726$) with a period of 365.24 solar days. The Earth also rotates about its own axis with a period of one siderial day (23 h 56 min 4.09 s). The sun-synchronized (or sun-synchronous) orbit has a nodal regression rate equal to the Earth's mean rate of revolution around the sun, i.e., 360 deg in 365.24 solar days or 0.985 deg/day. This regression must be in the direction of the Earth's rotation because the Earth rotates about its axis in the same direction that it revolves around the sun. Therefore, a sun-synchronous satellite must have a retrograde orbit so that its nodal regression can be prograde. Also, the satellite must have a combination of altitude and inclination that produces 0.985 deg/day regression.

The sun-synchronous orbits maintain their initial orientation relative to sun. These orbits are retrograde and lie between inclination angles of 95.7 deg and

180 deg at altitudes up to 5970 km. For certain mission requirements, a noon–midnight sun-synchronous orbit (with a 12:00 crossing time) can be selected, which provides good photography for about one-half of every revolution. Conversely, twilight or sunrise-sunset orbits (with a 6:00 crossing time) could be established in which the satellite is never in shadow, relieving the need for power storage if solar power is used. In practice, a sun-synchronized orbit with a 9:30 crossing time is often selected; however, this requires a solar-array cant angle of 37.5 deg.

3.6.2.2 Critical Inclination From Eq. (3.136), two critical inclinations (63.4 deg and 116.6 deg) can be found that cause the apsidal line rotational rate to be zero. Such critical inclination angles are very important for some missions where the position of apogee must remain fixed in space. The former Soviet Union's Molniya communications satellites with an eccentricity of 0.73 and a period of 12 h have an inclination of 63.4 deg so that their apogee remains fixed in space.

3.6.2.3 Triaxiality As discussed in the preceding section, a polar cross section of the Earth is not circular; it has an approximately 21-km bulge along the equator, which produces two perturbations into the ideal Keplerian orbit. In fact, an equatorial cross section of the Earth is also not circular and it is nearly elliptical, with a 65-m deviation from circular. This ellipse introduces two more axes into the Earth's shape, called the *triaxiality*. The second tesseral harmonic J_{22} of the Earth gravity harmonics is related to the ellipticity of the Earth's equatorial plane.

There are four equilibrium points separated by approximately 90 deg along the equator: two stable points and two unstable points. The effect of the triaxiality is to cause geosynchronous satellites to oscillate about the nearest stable point on the minor axis. These two stable points, at 75° E longitude and 255° E longitude, are called gravitational valleys. A geosynchronous satellite at the bottom of a gravitational valley is in stable equilibrium. Satellites placed at other longitudes will drift with a five-year period of oscillation; thus, they require east–west stationkeeping maneuvers to maintain their orbital positions. The stable equilibrium points are used among other things as a junkyard for deactivated geosynchronous satellites.

3.6.2.4 Orbital Decay Another very important perturbation for low-altitude satellites is atmospheric drag. Lifetimes of satellites in low Earth orbits are affected by atmospheric drag, although the Earth's atmosphere is only 81 km thick. The exact effect of atmospheric drag is difficult to predict due to uncertainties in the dynamics of the upper atmosphere. Air density is constantly changing in this region; there are diurnal variations because the sun heats up the air on the daylight side of the Earth, causing the air to expand. This heating increases the number of air molecules encountered by near-Earth satellites. There is a similar seasonal variation between summer and winter. There is also a 27-day cycle in atmospheric density, as well as an 11-year cycle. Magnetic storms can heat the atmosphere as can solar flares. Major solar events emit charged particles that heat the outer atmosphere and produce significant changes in satellite orbits.

Because the atmosphere drops off so rapidly with altitude, most drag is experienced at perigee. The less time the satellite spends at near-perigee altitudes, the less total mechanical energy the satellite dissipates by air drag. A reduction in total energy produces a corresponding reduction in the length of the semimajor axis. Also, air friction causes the eccentricity of the orbit to diminish toward zero, making orbit more circular. Thus, the apogee drops faster than perigee in elliptical orbits. Air density drops off so rapidly with increasing height that high-altitude satellites can essentially ignore air drag.

For more details of orbital mechanics and orbital applications, the reader is referred to other textbooks [1–10].

3.7 Circular Restricted Three-Body Problem

3.7.1 Introduction

The classic restricted three-body problem is concerned with the motion of an infinitesimal body in the gravitational field of two massive primary bodies. In this section, we consider the *circular restricted three-body problem* in which the motion of the two primary bodies is constrained to circular orbits about their barycenter. If the motion of the two primary bodies is constrained to elliptic orbits about their barycenter, we have the *elliptic restricted three-body problem* in which the distance between the two primary bodies varies periodically. This case will be studied in Sec. 3.8.

Both the circular and elliptic restricted three-body problems are treated in this text for the purpose of introducing future space missions that involve the stationing spacecraft around the collinear equilibrium points of the Earth–moon system or the sun–Mars system. The same face of the moon always faces the Earth; therefore, communications with the far side of the moon is impossible without a relay network. One method of providing this communications network involves the use of two or more relay satellites in a lunar polar orbit. Another method, introduced by Farquhar [11] in 1968, would be to position one communications satellite in a halo orbit about the translunar L_2 libration point, as illustrated in Fig. 3.8. In addition, if a communications satellite was located at the cislunar L_1 libration point, there could be continuous communications coverage between the Earth and most of the lunar surface.

In fact, a spacecraft called International Sun–Earth Explorer (ISEE-3) was placed in a halo orbit around the interior sun–Earth equilibrium point in November 1978, and it remained in the halo orbit until June 1982. A mission objective of the ISEE-3 spacecraft was to continuously monitor the characteristics of the solar wind and other solar induced phenomena, such as solar flares, about an hour before they could disturb the space environment near the Earth [12].

3.7.2 Problem Formulation

The circular restricted three-body problem was originally formulated by Euler in 1772 for the sun–Earth–moon system to study the motion of the moon about the Earth, but perturbed by the sun. In this section, we consider the Earth–moon–spacecraft system, illustrated in Figs. 3.8 and 3.9, as an example of the circular restricted three-body problem without loss of generality. It is assumed that the spacecraft mass is insignificant compared to the masses of the two primary bodies. Hence, the orbital motion of the two primary bodies is not

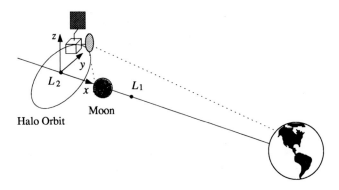

Fig. 3.8   Lunar far-side communications.

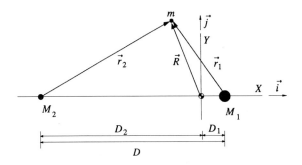

Fig. 3.9   Circular restricted three-body problem.

affected by the spacecraft; that is, it is simply described by the two-body problem in which the two primary bodies rotate about their composite center of mass (barycenter).

It can be further assumed that the two primary bodies rotate about their barycenter in circular orbits. That is, it is assumed that the Earth–moon system rotates with a constant angular velocity

$$n = \sqrt{G(M_1 + M_2)/D^3} \tag{3.137}$$

where M_1 and M_2 are the masses of the Earth and moon, respectively, and D is the constant distance between them. For the Earth–moon system, we have

$$\mu_1 = GM_1 = 398,601 \text{ km}^3/\text{s}^2$$

$$\mu_2 = GM_2 = 4887 \text{ km}^3/\text{s}^2$$

$$M_1 = 81.3045 M_2$$

$$D = 384,748 \text{ km}$$

$$D_1 = 0.01215D = 4674 \text{ km}$$
$$D_2 = 0.98785D = 380{,}073 \text{ km}$$
$$n = 2.661699 \times 10^{-6} \text{ rad/s}$$

where D_1 is the distance between M_1 and the barycenter, D_2 is the distance between M_2 and the barycenter, and $D = D_1 + D_2$. Note that the period of the moon's orbit is 27.3 days.

The position vector of the spacecraft relative to the barycenter is expressed in terms of basis vectors $\{\vec{i}, \vec{j}, \vec{k}\}$ of a rotating reference frame with an angular velocity of $n\vec{k}$ and with its origin at the barycenter, as follows:

$$\vec{R} = X\vec{i} + Y\vec{j} + Z\vec{k} \tag{3.138}$$

Noting that $\dot{\vec{i}} = n\vec{j}, \dot{\vec{j}} = -n\vec{i}$, and $\dot{\vec{k}} = 0$, the inertial acceleration of the spacecraft can be found as

$$\ddot{\vec{R}} = (\ddot{X} - 2n\dot{Y} - n^2 X)\vec{i} + (\ddot{Y} + 2n\dot{X} - n^2 Y)\vec{j} + \ddot{Z}\vec{k} \tag{3.139}$$

The equation of motion of the spacecraft is then simply given by

$$m\ddot{\vec{R}} = -\frac{GM_1 m}{r_1^3}\vec{r}_1 - \frac{GM_2 m}{r_2^3}\vec{r}_2 \tag{3.140}$$

where m is the mass of the spacecraft, $r_1 = |\vec{r}_1|$, $r_2 = |\vec{r}_2|$, and

$$\vec{r}_1 = -D_1\vec{i} + \vec{R} = (X - D_1)\vec{i} + Y\vec{j} + Z\vec{k}$$
$$\vec{r}_2 = D_2\vec{i} + \vec{R} = (X + D_2)\vec{i} + Y\vec{j} + Z\vec{k}$$

Equation (3.140) is rewritten as

$$\ddot{\vec{R}} = -\left(\mu_1/r_1^3\right)\vec{r}_1 - \left(\mu_2/r_2^3\right)\vec{r}_2 \tag{3.141}$$

where $\mu_1 = GM_1$ and $\mu_2 = GM_2$. Equating the components of the inertial acceleration and the gravitational acceleration in Eq. (3.141), we obtain the equations of motion, as follows:

$$\ddot{X} - 2n\dot{Y} - n^2 X = -\frac{\mu_1(X - D_1)}{r_1^3} - \frac{\mu_2(X + D_2)}{r_2^3} \tag{3.142}$$

$$\ddot{Y} + 2n\dot{X} - n^2 Y = -\frac{\mu_1 Y}{r_1^3} - \frac{\mu_2 Y}{r_2^3} \tag{3.143}$$

$$\ddot{Z} = -\frac{\mu_1 Z}{r_1^3} - \frac{\mu_2 Z}{r_2^3} \tag{3.144}$$

The terms $2n\dot{Y}$ and $2n\dot{X}$ are the Coriolis accelerations, and $n^2 X$ and $n^2 Y$ are centrifugal acceleration terms.

Adding Eqs. (3.142), (3.143), and (3.144) after multiplying them by $\dot{X}$, $\dot{Y}$, and $\dot{Z}$, respectively, we obtain

$$\ddot{X}\dot{X} + \ddot{Y}\dot{Y} + \ddot{Z}\dot{Z} - n^2 X\dot{X} - n^2 Y\dot{Y} = -(\mu_1/r_1^3)[(X - D_1)\dot{X} + Y\dot{Y} + Z\dot{Z}]$$
$$- (\mu_2/r_2^3)[(X + D_2)\dot{X} + Y\dot{Y} + Z\dot{Z}] \qquad (3.145)$$

which, after integration, becomes

$$\frac{1}{2}(\dot{X}^2 + \dot{Y}^2 + \dot{Z}^2) - \frac{1}{2}n^2(X^2 + Y^2) - (\mu_1/r_1) - (\mu_2/r_2) = C \qquad (3.146)$$

where C is a constant called *Jacobi's integral*.*

The equations of motion, given by Eqs. (3.142), (3.143), and (3.144), for the restricted three-body problem can also be expressed in terms of a pseudopotential $U = U(X, Y, Z)$ as follows:

$$\ddot{X} - 2n\dot{Y} = \frac{\partial U}{\partial X} \qquad (3.147)$$

$$\ddot{Y} + 2n\dot{X} = \frac{\partial U}{\partial Y} \qquad (3.148)$$

$$\ddot{Z} = \frac{\partial U}{\partial Z} \qquad (3.149)$$

where the pseudopotential U, which is, in fact, the centrifugal plus gravitational force potential, is defined as

$$U = \frac{1}{2}n^2(X^2 + Y^2) + (\mu_1/r_1) + (\mu_2/r_2) \qquad (3.150)$$

where $\mu_1 = GM_1$, $\mu_2 = GM_2$, $r_1 = \sqrt{(X - D_1)^2 + Y^2 + Z^2}$, $r_2 = \sqrt{(X + D_2)^2 + Y^2 + Z^2}$, and Jacobi's integral is simply given by

$$C = \frac{1}{2}(\dot{X}^2 + \dot{Y}^2 + \dot{Z}^2) - U \qquad (3.151)$$

3.7.3 Lagrangian Points

3.7.3.1 Equilibrium Points Introducing the mass ratio ρ of the Earth–moon system as

$$M_2/(M_1 + M_2) = \rho = 0.01215$$
$$M_1/(M_1 + M_2) = 1 - \rho = 0.98785$$

---

*Carl Gustav Jacob Jacobi (1804–1851), who was a great admirer of Euler and a contemporary of Gauss and Bessel, is remembered for his contributions to mathematics and mechanics.

we rewrite the equations of motion in nondimensional form, as follows:

$$\ddot{X} - 2\dot{Y} - X = -\frac{(1 - \rho)(X - \rho)}{r_1^3} - \frac{\rho(X + 1 - \rho)}{r_2^3} \tag{3.152}$$

$$\ddot{Y} + 2\dot{X} - Y = -\frac{(1 - \rho)Y}{r_1^3} - \frac{\rho Y}{r_2^3} \tag{3.153}$$

$$\ddot{Z} = -\frac{(1 - \rho)Z}{r_1^3} - \frac{\rho Z}{r_2^3} \tag{3.154}$$

where

$$r_1 = \sqrt{(X - \rho)^2 + Y^2 + Z^2}$$
$$r_2 = \sqrt{(X + 1 - \rho)^2 + Y^2 + Z^2}$$

Time is in units of $1/n$, and X, Y, Z, r_1, and r_2 are in units of D.

Setting all derivatives in Eqs. (3.152–3.154) to zero, we can find equilibrium points of the Earth–moon system. At such equilibrium points, the gravitational forces and the centrifugal force acting on the spacecraft are balanced. Five equilibrium points of the restricted three-body problem exist, as illustrated in Fig. 3.10 for the Earth–moon system. Euler discovered three equilibrium points, called the collinear equilibrium points, on the X axis with $Y = Z = 0$. Lagrange found two other equilibrium points, called the equilateral equilibrium points with $r_1 = r_2$. These five equilibrium points, denoted by L_1 through L_5, are called the Lagrangian or *libration points* of the restricted three-body problem. The locations of these five libration points are summarized in Table 3.1.

For the Earth–moon system, L_1 at $X = -0.83692$ is called the *cislunar* point, L_2 at $X = -1.15568$ called the *translunar* point, and L_3 at $X = 1.00506$ called the

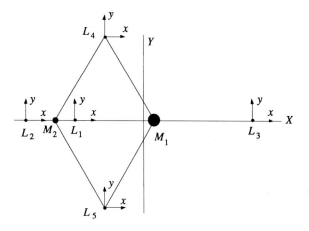

Fig. 3.10   Lagrangian points.

Table 3.1   Earth–moon libration point locations

Libration points	X	Y	Z
L_1	−0.83692	0	0
L_2	−1.15568	0	0
L_3	1.00506	0	0
L_4	−0.48785	$\sqrt{3}/2$	0
L_5	−0.48785	$-\sqrt{3}/2$	0

trans-Earth point. It will be shown that the three collinear points, L_1, L_2, and L_3, are unstable, whereas the two equilateral points, L_4 and L_5, are stable.

The libration points do appear in the natural solar system. The Trojan asteroids are situated at Jupiter's orbit on the vertices of equilateral triangles with the sun and Jupiter at the two other vertices. These positions are the triangular libration points in the circular restricted three-body problem of the sun–Jupiter–asteroid system.

3.7.3.2 Linearized Equations of Motion   The linearized equations of motion about an equilibrium point are now derived for the linear stability analysis of the Lagrangian points. Let

$$X = X_o + x, \qquad Y = Y_o + y, \qquad Z = Z_o + z \qquad (3.155)$$

where (X_o, Y_o, Z_o) are the coordinates of the equilibrium point with $Z_o = 0$ and (x, y, z) are the components of the position vector of the spacecraft relative to the equilibrium point.

In 1665 Newton established the general binomial theorem

$$(1 + x)^n = 1 + nx + \frac{n(n-1)}{2!}x^2 + \frac{n(n-1)(n-2)}{3!}x^3 + \cdots$$

where n is any real number. If $x \ll 1$, then $(1 + x)^n \approx 1 + nx$. Assuming that x, y, and z are small, i.e., $x^2 = y^2 = z^2 \approx 0$, we approximate the terms $1/r_1^3$ and $1/r_2^3$ appearing in the nondimensionalized equations (3.152–3.154), using the general binomial theorem, as follows:

$$r_1^{-3} = \left[(X_o + x - \rho)^2 + (Y_o + y)^2 + z^2\right]^{-\frac{3}{2}}$$
$$\approx R_1^{-3}\left\{1 - 3R_1^{-2}[(X_o - \rho)x + Y_o y]\right\} \qquad (3.156)$$

and

$$r_2^{-3} = \left[(X_o + x + 1 - \rho)^2 + (Y_o + y)^2 + z^2\right]^{-\frac{3}{2}}$$
$$\approx R_2^{-3}\left\{1 - 3R_2^{-2}[(X_o + 1 - \rho)x + Y_o y]\right\} \qquad (3.157)$$

where R_1 and R_2, in units of D, are the distances to the equilibrium point from the Earth and moon, respectively; that is,

$$R_1 = \sqrt{(X_o - \rho)^2 + Y_o^2}$$

$$R_2 = \sqrt{(X_o + 1 - \rho)^2 + Y_o^2}$$

The nondimensional equations of motion given by Eqs. (3.152–3.154) then become

$$\ddot{x} - 2\dot{y} - x = -\left\{ (1 - \rho) \left[\frac{1}{R_1^3} - 3\frac{(X_o - \rho)^2}{R_1^5} \right] \right.$$

$$+ \rho \left[\frac{1}{R_2^3} - 3\frac{(X_o + 1 - \rho)^2}{R_2^5} \right] \right\} x$$

$$+ \left\{ 3(1 - \rho)\frac{(X_o - \rho)Y_o}{R_1^5} + 3\rho\frac{(X_o + 1 - \rho)Y_o}{R_2^5} \right\} y \quad (3.158a)$$

$$\ddot{y} + 2\dot{x} - y = \left\{ 3(1 - \rho)\frac{(X_o - \rho)Y_o}{R_1^5} + 3\rho\frac{(X_o + 1 - \rho)Y_o}{R_2^5} \right\} x$$

$$- \left\{ (1 - \rho) \left[\frac{1}{R_1^3} - 3\frac{Y_o^2}{R_1^5} \right] + \rho \left[\frac{1}{R_2^3} - 3\frac{Y_o^2}{R_2^5} \right] \right\} y \quad (3.158b)$$

$$\ddot{z} = -\left\{ \frac{(1 - \rho)}{R_1^3} + \frac{\rho}{R_2^3} \right\} z \quad (3.158c)$$

These are the linearized equations of motion of the spacecraft with respect to the equilibrium point. Similar to the original nonlinear equations of motion, the Z axis out-of-plane equation is decoupled from the in-plane equations of motion. The out-of-plane motion is described by a simple harmonic oscillator and is said to be (Lyapunov) stable.

3.7.3.3 Stability of the Equilateral Equilibrium Points

Consider the equilateral equilibrium point L_4, with $R_1 = R_2 = 1$, at $X_o = \rho - 0.5 = -0.48785$ and $Y_o = \sqrt{3}/2 = 0.86602$. Substituting these values into Eqs. (3.158a) and (3.158b), we obtain the in-plane equations of motion, as follows:

$$\ddot{x} - 2\dot{y} - \frac{3}{4}x - \frac{3\sqrt{3}}{2}\left(\rho - \frac{1}{2} \right)y = 0 \quad (3.159)$$

$$\ddot{y} + 2\dot{x} - \frac{3\sqrt{3}}{2}\left(\rho - \frac{1}{2} \right)x - \frac{9}{4}y = 0 \quad (3.160)$$

The characteristic equation can then be found as

$$\lambda^4 + \lambda^2 + (27/4)\rho(1 - \rho) = 0 \tag{3.161}$$

The characteristic roots are

$$\lambda = \pm\sqrt{\frac{-1 \pm \sqrt{1 - 27\rho(1 - \rho)}}{2}}$$

For $\rho \le 0.03852$ or $0.96148 \le \rho$, the four eigenvalues become pure imaginary numbers, and the in-plane motion is said to be stable. The L_4 point of the Earth–moon system with $\rho = 0.01215$ is thus a stable equilibrium point. Because of the symmetry of the system, the L_5 point of the Earth–moon system is also a stable equilibrium point; however, the L_4 and L_5 points of the Earth–moon system can be found to be unstable if the gravitational effect of the sun is also included. The L_4 and L_5 points in the sun–Jupiter system with $\rho = 9.5387 \times 10^{-4}$ are also stable points, and the Trojan asteriods are located at these stable libration points.

3.7.3.4 Stability of the Collinear Equilibrium Points

The linearized equations of motion about the collinear equilibrium point can be described by

$$\ddot{x} - 2\dot{y} - (2\sigma + 1)x = 0 \tag{3.162a}$$

$$\ddot{y} + 2\dot{x} + (\sigma - 1)y = 0 \tag{3.162b}$$

$$\ddot{z} + \sigma z = 0 \tag{3.162c}$$

where

$$\sigma = \frac{(1 - \rho)}{|X_o - \rho|^3} + \frac{\rho}{|X_o + 1 - \rho|^3}$$

The characteristic equation of the in-plane motion can be found as

$$\lambda^4 - (\sigma - 2)\lambda^2 - (2\sigma + 1)(\sigma - 1) = 0$$

For the L_2 point with $\sigma = 3.19043$, for example, the in-plane characteristic equation has two real and two imaginary roots: ± 2.15868 and $\pm 1.86265j$. Thus, the in-plane motion has a divergent mode as well as an oscillatory mode, and the L_2 point is said to be an unstable equilibrium point. The out-of-plane motion is simple harmonic with a nondimensional frequency of $\sqrt{\sigma} = 1.78618$.

Similarly, it can be shown that the other collinear points, L_1 and L_3, are unstable equilibrium points for the Earth–moon system. Although such collinear points are unstable in nature, the cislunar L_1 and translunar L_2 points are of practical importance for future space missions involving the stationing of a communication platform or a lunar space station. The cislunar L_1 point could serve as a transportation node for lunar transfer trajectories. The translunar L_2 point would be an excellent orbital location to station a satellite to provide a communications link to the far side of the moon.

Problem

3.8 Consider the nonlinear equations of motion of the restricted three-body problem in nondimensional form, given by

$$\ddot{X} - 2\dot{Y} = \frac{\partial U}{\partial X}$$

$$\ddot{Y} + 2\dot{X} = \frac{\partial U}{\partial Y}$$

$$\ddot{Z} = \frac{\partial U}{\partial Z}$$

where the pseudopotential $U = U(X, Y, Z)$ in nondimensional form is defined as

$$U = \frac{1}{2}(X^2 + Y^2) + \frac{1 - \rho}{r_1} + \frac{\rho}{r_2}$$

and

$$\rho = M_2/(M_1 + M_2)$$

$$r_1 = \sqrt{(X - \rho)^2 + Y^2 + Z^2}$$

$$r_2 = \sqrt{(X + 1 - \rho)^2 + Y^2 + Z^2}$$

By the definition of the equilibrium point, the partial derivatives evaluated at the equilibrium point (X_o, Y_o, Z_o) become zero; i.e., we have

$$\left.\frac{\partial U}{\partial X}\right|_o = \left.\frac{\partial U}{\partial Y}\right|_o = \left.\frac{\partial U}{\partial Z}\right|_o = 0$$

(a) Show that the linearized equations of motion about an equilibrium point (X_o, Y_o, Z_o) can be written as

$$\ddot{x} - 2\dot{y} = \left.\frac{\partial^2 U}{\partial X^2}\right|_o x + \left.\frac{\partial^2 U}{\partial Y \partial X}\right|_o y + \left.\frac{\partial^2 U}{\partial Z \partial X}\right|_o z$$

$$\ddot{y} + 2\dot{x} = \left.\frac{\partial^2 U}{\partial X \partial Y}\right|_o x + \left.\frac{\partial^2 U}{\partial Y^2}\right|_o y + \left.\frac{\partial^2 U}{\partial Z \partial Y}\right|_o z$$

$$\ddot{z} = \left.\frac{\partial^2 U}{\partial X \partial Z}\right|_o x + \left.\frac{\partial^2 U}{\partial Y \partial Z}\right|_o y + \left.\frac{\partial^2 U}{\partial Z^2}\right|_o z$$

where $x = X - X_o$, $y = Y - Y_o$, $z = Z - Z_o$, and the partial derivatives are to be evaluated at the equilibrium point (X_o, Y_o, Z_o).
Hint: Use a first-order Taylor series expansion of $\partial U/\partial X$, $\partial U/\partial Y$, and $\partial U/\partial Z$ about the equilibrium point (X_o, Y_o, Z_o).

(b) Also show that the preceding equations become

$$\ddot{x} - 2\dot{y} = U_{XX}\, x + U_{XY}\, y$$

$$\ddot{y} + 2\dot{x} = U_{XY}\, x + U_{YY}\, y$$

$$\ddot{z} = U_{ZZ}\, z$$

where the partial derivatives evaluated at the equilibrium point (X_o, Y_o, Z_o) can be found as

$$U_{XX} \equiv \left.\frac{\partial^2 U}{\partial X^2}\right|_o = 1 - \left\{(1-\rho)\left[\frac{1}{R_1^3} - 3\frac{(X_o - \rho)^2}{R_1^5}\right]\right.$$

$$\left. + \rho\left[\frac{1}{R_2^3} - 3\frac{(X_o + 1 - \rho)^2}{R_2^5}\right]\right\}$$

$$U_{YY} \equiv \left.\frac{\partial^2 U}{\partial Y^2}\right|_o = 1 - \left\{(1-\rho)\left[\frac{1}{R_1^3} - 3\frac{Y_o^2}{R_1^5}\right] + \rho\left[\frac{1}{R_2^3} - 3\frac{Y_o^2}{R_2^5}\right]\right\}$$

$$U_{XY} \equiv \left.\frac{\partial^2 U}{\partial X \partial Y}\right|_o = 3(1-\rho)\frac{(X_o - \rho)Y_o}{R_1^5} + 3\rho\frac{(X_o + 1 - \rho)Y_o}{R_2^5}$$

$$U_{ZZ} \equiv \left.\frac{\partial^2 U}{\partial Z^2}\right|_o = -\left\{\frac{(1-\rho)}{R_1^3} + \frac{\rho}{R_2^3}\right\}$$

where $R_1 = \sqrt{(X_o - \rho)^2 + Y_o^2}$ and $R_2 = \sqrt{(X_o + 1 - \rho)^2 + Y_o^2}$.

(c) For the collinear libration points of the restricted three-body system, show that

$$U_{XX} = 2\sigma + 1$$
$$U_{YY} = -\sigma + 1$$
$$U_{ZZ} = -\sigma$$

where

$$\sigma = \frac{(1-\rho)}{|X_o - \rho|^3} + \frac{\rho}{|X_o + 1 - \rho|^3}$$

(d) For the three collinear libration points of the restricted three-body system with $0 < \rho < 0.5$, show that

$$U_{XX} > 0, \qquad U_{YY} < 0, \qquad U_{ZZ} < 0, \qquad U_{XY} = 0$$

In particular, find the numerical values of U_{XX}, U_{YY}, and U_{ZZ} for the translunar L_2 point in the Earth–moon system.

3.7.4 Quasi-periodic Orbits

The nondimensional linearized equations of motion of a spacecraft near the collinear libration point are given by

$$\ddot{x} - 2\dot{y} - U_{XX} x = 0 \qquad (3.163a)$$

$$\ddot{y} + 2\dot{x} - U_{YY} y = 0 \qquad (3.163b)$$

$$\ddot{z} - U_{ZZ} z = 0 \qquad (3.163c)$$

where the partial derivatives are evaluated at the collinear libration point.

The in-plane characteristic equation can be obtained as

$$\lambda^4 + (4 - U_{XX} - U_{YY})\lambda^2 + U_{XX} U_{YY} = 0 \qquad (3.164)$$

and the in-plane eigenvalues can be expressed as [4]

$$\lambda_{1,2} = \pm\sqrt{-\beta_1 + \sqrt{\beta_1^2 + \beta_2^2}}$$

$$\lambda_{3,4} = \pm j\sqrt{\beta_1 + \sqrt{\beta_1^2 + \beta_2^2}} = \pm j\omega_{xy}$$

where

$$\beta_1 = 2 - (U_{XX} + U_{YY})/2$$

$$\beta_2^2 = -U_{XX} U_{YY} > 0$$

and ω_{xy} is called the nondimensional frequency of the in-plane oscillatory mode. The out-of-plane characteristic equation is

$$\lambda^2 - U_{ZZ} = 0 \qquad (3.165)$$

and the out-of-plane eigenvalues are

$$\lambda_{5,6} = \pm j\sqrt{|U_{ZZ}|} = \pm j\omega_z \qquad (3.166)$$

where ω_z is called the nondimensional frequency of the out-of-plane oscillatory mode.

For the translunar L_2 point in the Earth–moon system, we have

$$U_{XX} = 7.3809, \qquad U_{YY} = -2.1904, \qquad U_{ZZ} = -3.1904$$

and the eigenvalues are

$$\lambda_{1,2} = \pm 2.15868$$

$$\lambda_{3,4} = \pm j1.86265$$

$$\lambda_{5,6} = \pm j1.78618$$

Thus for the translunar L_2 point, the in-plane motion has a divergent mode as well as an oscillatory mode with a nondimensional frequency $\omega_{xy} = 1.86265$. The out-of-plane motion is simple harmonic with a nondimensional frequency $\omega_z = 1.78618$. Note that the period of the in-plane oscillatory mode is 14.7 days and the period of the out-of-plane oscillatory mode is 15.3 days, compared to the moon's orbital period of 27.3 days.

The in-plane equations of motion can be written in state-space form as

$$\dot{\mathbf{x}} = \mathbf{A}\mathbf{x} \tag{3.167}$$

where

$$\mathbf{x} = \begin{bmatrix} x \\ y \\ \dot{x} \\ \dot{y} \end{bmatrix}, \qquad \mathbf{A} = \begin{bmatrix} 0 & 0 & 1 & 0 \\ 0 & 0 & 0 & 1 \\ U_{XX} & 0 & 0 & 2 \\ 0 & U_{YY} & -2 & 0 \end{bmatrix}$$

We now consider the modal decomposition of the state matrix $\mathbf{A}$ into a diagonal matrix $\mathbf{\Lambda}$. (The modal decomposition of a square matrix was discussed in Chapter 2; however, we shall briefly review the concept again here.)

Let $\mathbf{q}_j$ be a right eigenvector of $\mathbf{A}$ associated with the jth eigenvalue λ_j; i.e.,

$$\mathbf{A}\mathbf{q}_j = \lambda_j \mathbf{q}_j$$

Then, we have

$$\mathbf{A} = \mathbf{Q}\mathbf{\Lambda}\mathbf{Q}^{-1}$$

where

$$\mathbf{Q} = [\mathbf{q}_1 \quad \cdots \quad \mathbf{q}_4]$$
$$\mathbf{\Lambda} = \mathrm{diag}(\lambda_1, \ldots, \lambda_4)$$

Also let $\mathbf{p}_i$ be a left eigenvector of $\mathbf{A}$ associated with λ_i; i.e.,

$$\mathbf{p}_i^T \mathbf{A} = \lambda_i \mathbf{p}_i^T$$

Then, we have

$$\mathbf{A} = \mathbf{P}^{-T}\mathbf{\Lambda}\mathbf{P}^T$$

where

$$\mathbf{P} = [\mathbf{p}_1 \quad \cdots \quad \mathbf{p}_4]$$

Consequently, we have the orthogonality conditions

$$\mathbf{p}_i^T \mathbf{q}_j = 0 \qquad (\text{because } \lambda_i \neq \lambda_j \quad \text{when } i \neq j)$$

Furthermore, if $\mathbf{q}_i$ and $\mathbf{p}_i$ are normalized such that $\mathbf{p}_i^T \mathbf{q}_i = 1$ for all i, then we have

$$\mathbf{p}_i^T \mathbf{q}_j = \delta_{ij} \qquad \text{or} \qquad \mathbf{P}^T \mathbf{Q} = \mathbf{I}$$

Finally, the matrix $\mathbf{A}$ with distinct eigenvalues can then be decomposed into the form

$$\mathbf{A} = \mathbf{Q}\mathbf{\Lambda}\mathbf{Q}^{-1} = \mathbf{Q}\mathbf{\Lambda}\mathbf{P}^T$$

which becomes

$$\mathbf{A} = [\mathbf{q}_1 \cdots \mathbf{q}_4] \begin{bmatrix} \lambda_1 & & \\ & \ddots & \\ & & \lambda_4 \end{bmatrix} \begin{bmatrix} \mathbf{p}_1^T \\ \vdots \\ \mathbf{p}_4^T \end{bmatrix} = \sum_{i=1}^{4} \lambda_i \mathbf{q}_i \mathbf{p}_i^T$$

Consequently, the solution of Eq. (3.167) can be obtained as

$$\mathbf{x}(t) = e^{\mathbf{A}t}\mathbf{x}(0) = \sum_{i=1}^{4} e^{\lambda_i t} \mathbf{q}_i \underbrace{\mathbf{p}_i^T \mathbf{x}(0)}_{\text{scalar}} \qquad (3.168)$$

This solution can be made to contain only the oscillatory modes with the proper choice of initial conditions. Because $\lambda_{3,4} = \pm j\omega_{xy}$, the conditions to eliminate the exponential terms are

$$\mathbf{p}_1^T \mathbf{x}(0) = 0 \qquad (3.169a)$$

$$\mathbf{p}_2^T \mathbf{x}(0) = 0 \qquad (3.169b)$$

These conditions can be met if we choose

$$\dot{x}(0) = (\omega_{xy}/k)\, y(0) \qquad (3.170a)$$

$$\dot{y}(0) = -k\omega_{xy}\, x(0) \qquad (3.170b)$$

where

$$k = \frac{\omega_{xy}^2 + U_{XX}}{2\omega_{xy}}$$

For the L_2 point of the Earth–moon system, $k = 2.91261$.

Substituting these conditions into Eq. (3.168), we obtain

$$x(t) = x(0) \cos \omega_{xy} t + (1/k)y(0) \sin \omega_{xy} t \qquad (3.171a)$$

$$y(t) = y(0) \cos \omega_{xy} t - kx(0) \sin \omega_{xy} t \qquad (3.171b)$$

$$z(t) = z(0) \cos \omega_z t + [\dot{z}(0)/\omega_z] \sin \omega_z t \qquad (3.171c)$$

In addition, if $x(0) = z(0) = 0$ and $\dot{z}(0) = -y(0)\omega_z$, the solution further reduces to the following form:

$$x(t) = (1/k)y(0) \sin \omega_{xy}t \tag{3.172a}$$

$$y(t) = y(0) \cos \omega_{xy}t \tag{3.172b}$$

$$z(t) = -y(0) \sin \omega_z t \tag{3.172c}$$

The difference between the in-plane and out-of-plane frequencies results in a quasi-periodic *Lissajous* trajectory, shown in Fig. 3.11, for an example case of $y(0) = -0.00911$, which corresponds to a maximum amplitude of 3500 km for the Earth–moon system. The y–z projection is a view from the Earth toward the moon and the x–y and x–z projections are top and side views, respectively. The moon and Earth are in the positive x direction. The preceding initial conditions correspond to zero in the x direction, maximum amplitude in the negative y direction, and zero in the z direction. The motion is clockwise in the x–y projection. Such a properly selected reference trajectory can also be described by

$$x_r(t) = -A_x \sin \omega_{xy}t$$

$$y_r(t) = -A_y \cos \omega_{xy}t$$

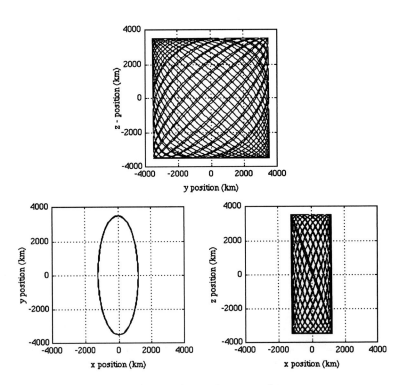

Fig. 3.11   Lissajous reference trajectory.

$$[1pt]z_r(t) = A_z \sin \omega_z t$$
$$\dot{x}_r(t) = -A_x \omega_{xy} \cos \omega_{xy} t$$
$$[1pt]\dot{y}_r(t) = A_y \omega_{xy} \sin \omega_{xy} t$$
$$[1pt]\dot{z}_r(t) = A_z \omega_z \cos \omega_z t$$

where A_x, $A_y = kA_x$, and $A_z = A_y$ denote the amplitudes of the reference trajectory for x, y, and z axis, respectively.

Unless the frequency ratio is a rational number, the Lissajous trajectory does not close. For the case of rational frequency ratios, the trajectory becomes periodic and is called a *halo orbit"*. For most cases, the solution of the linearized equations of motion is not periodic and some control effort is needed to achieve equal in-plane and out-of-plane frequencies. This is often called period or frequency control in the literature. The resulting periodic orbit due to frequency control will also be called a halo orbit in this book, although the term "halo orbit" in celestial mechanics usually means a larger, periodic orbit that is a solution of the nonlinear differential equations of motion. A halo orbit with such period control is shown in Fig. 3.12, which also shows the disk of the moon, which has a radius of 1738 km.

The Lissajous trajectory and halo orbit are the results of the linear analysis of the restricted three-body problem. The actual motions of a spacecraft are governed by

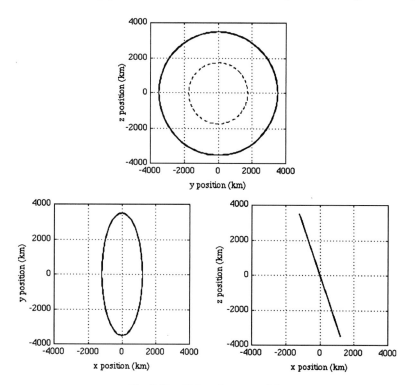

Fig. 3.12   Halo reference orbit.

the nonlinear equations of motion, and the spacecraft will not follow the Lissajous trajectory or halo orbit naturally. The Lissajous trajectory and halo orbit can be used as reference trajectories for a spacecraft control problem described briefly as follows.

Consider the problem of stationing a communications satellite at the translunar libration point, and assume that the communications satellite is placed into a quasi-periodic Lissajous trajectory, shown in Fig. 3.11, around the L_2 libration point. Because the moon, with a radius of 1738 km, is centered at the origin of Fig. 3.11, the satellite will spend some fraction of time obscured by the moon. Consequently, halo orbit control is needed to adjust the frequency of the z-axis motion to keep the communications satellite visible from the Earth at all times, as illustrated in Fig. 3.12. Such halo orbit control is also possible if the amplitudes of the in-plane and out-of-plane motions are of sufficient magnitude that the nonlinear contributions to the system result in an identical z-axis and y-axis frequency. Such halo orbit determination and control problems will be treated in detail in Chapter 4.

It should be noted that when a spacecraft is actively controlled to follow a periodic halo orbit, the orbit will, in general, not close due to tracking error. Therefore, when discussing orbits about the libration points, both halo orbits and Lissajous trajectories will be called quasi-periodic orbits. When particular orbits are being discussed, they will be referred to as halo orbits or Lissajous trajectories. The basic shape of the motion will determine the classification of either halo orbit or Lissajous trajectory.

Problem

3.9   Given the in-plane equations of motion written in state-space form as

$$\dot{\mathbf{x}} = \mathbf{A}\mathbf{x}$$

where

$$\mathbf{x} = \begin{bmatrix} x \\ y \\ \dot{x} \\ \dot{y} \end{bmatrix}, \qquad \mathbf{A} = \begin{bmatrix} 0 & 0 & 1 & 0 \\ 0 & 0 & 0 & 1 \\ U_{XX} & 0 & 0 & 2 \\ 0 & U_{YY} & -2 & 0 \end{bmatrix}$$

(a) Find the right eigenvector $\mathbf{q}_i$ and the left eigenvector $\mathbf{p}_i$ associated with the ith eigenvalue λ_i, as follows:

$$\mathbf{q}_i = \begin{bmatrix} 2\lambda_i \\ \lambda_i^2 - U_{XX} \\ 2\lambda_i^2 \\ \lambda_i(\lambda_i^2 - U_{XX}) \end{bmatrix}, \qquad \mathbf{p}_i = \begin{bmatrix} 2\lambda_i U_{XX} \\ (U_{XX} - \lambda_i^2)U_{YY} \\ 2\lambda_i^2 \\ (U_{XX} - \lambda_i^2)\lambda_i \end{bmatrix}$$

(b) Verify that if we choose

$$\dot{x}(0) = (\omega_{xy}/k)\, y(0)$$
$$\dot{y}(0) = -k\omega_{xy}\, x(0)$$

where

$$k = \frac{\omega_{xy}^2 + U_{XX}}{2\omega_{xy}}$$

then we have

$$\mathbf{p}_1^T \mathbf{x}(0) = 0$$
$$\mathbf{p}_2^T \mathbf{x}(0) = 0$$

3.8 Elliptic Restricted Three-Body Problem

The effects of the eccentricity of the moon's orbit ($e = 0.05490$) are in fact, larger than the nonlinear effects for a spacecraft in a typical 3500-km quasi-periodic orbit about the translunar libration point. For this reason, the circular restricted three-body model treated in the preceding section is not an accurate model for the analysis of quasi-periodic orbits in the Earth–moon system. The effects of the sun's gravitational field and radiation pressure are smaller than the effects of eccentricity and nonlinearities for a spacecraft in a typical 3500-km quasi-periodic orbit. In this section, the elliptic restricted three-body problem is considered as the next step in complexity for the halo orbit determination and control problem.

Consider the system of three bodies shown in Fig. 3.13. The masses of the two primary bodies are denoted by M_1 and M_2, respectively, whereas the infinitesimal mass of the third body (a spacecraft) is denoted by m. The point O in Fig. 3.13 is assumed to be inertially fixed.

The equation of motion of the spacecraft is

$$m\ddot{\vec{R}}_3 = -\left(GM_1 m / r_1^3\right)\vec{r}_1 - \left(GM_2 m / r_2^3\right)\vec{r}_2 \qquad (3.173)$$

and the inertial acceleration of the spacecraft, $\ddot{\vec{R}}_3$, simply becomes

$$\ddot{\vec{R}}_3 = -\left(GM_1 / r_1^3\right)\vec{r}_1 - \left(GM_2 / r_2^3\right)\vec{r}_2 \qquad (3.174)$$

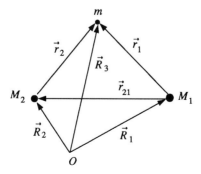

Fig. 3.13   Three-body system.

Defining $\vec{r}_{21}$ as the position vector from M_1 to M_2 such that

$$\vec{r}_{21} = \vec{R}_2 - \vec{R}_1 \tag{3.175}$$

we also obtain the inertial acceleration of mass M_1 as

$$\ddot{\vec{R}}_1 = \left(GM_2/r_{21}^3\right)\vec{r}_{21} + \left(Gm/r_1^3\right)\vec{r}_1 \tag{3.176}$$

The relative motion of m with respect to M_1 can then be described by

$$\ddot{\vec{r}}_1 = -G(M_1 + m)\frac{\vec{r}_1}{r_1^3} - GM_2\left\{\frac{\vec{r}_2}{r_2^3} + \frac{\vec{r}_{21}}{r_{21}^3}\right\} \tag{3.177}$$

since $\ddot{\vec{r}}_1 = \ddot{\vec{R}}_3 - \ddot{\vec{R}}_1$.

Similarly, the relative motion of m with respect to M_2 can also be described by

$$\ddot{\vec{r}}_2 = -G(M_2 + m)\frac{\vec{r}_2}{r_2^3} - GM_1\left\{\frac{\vec{r}_1}{r_1^3} - \frac{\vec{r}_{21}}{r_{21}^3}\right\} \tag{3.178}$$

Consider now the elliptic restricted three-body problem as illustrated in Fig. 3.14. The basis vectors $\{\vec{i}, \vec{j}, \vec{k}\}$ define a rotating reference frame with its origin at the barycenter. The angular velocity of this rotating reference frame is not constant because M_2 rotates about M_1 in an elliptic orbit. In Fig. 3.14, ℓ_2 denotes the distance between M_2 and L_2, and ℓ_1 denotes the distance between M_2 and L_1.

Because the distance D between M_1 and M_2 is not constant, ℓ_1 and ℓ_2 are not constant, and L_1 and L_2 denote the instantaneous libration points; however, the ratio of ℓ_2 and the instantaneous distance D is a constant and is denoted as

$$\ell_2/D = \gamma = \text{const} \tag{3.179}$$

There is a similar constant corresponding to ℓ_1 and the instantaneous distance D.

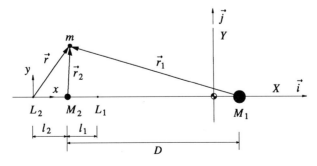

Fig. 3.14   Elliptic restricted three-body problem.

Equation (3.177) can be used to describe the motion of the spacecraft about one of the collinear libration points by defining the position of the spacecraft with respect to a collinear libration point as

$$\vec{r} = x\vec{i} + y\vec{j} + z\vec{k} \tag{3.180}$$

Referring to Fig. 3.14, we obtain

$$\vec{r}_{21} = -D\vec{i} \tag{3.181}$$

$$\vec{r}_1 = [-(1+\gamma)D + x]\vec{i} + y\vec{j} + z\vec{k} \tag{3.182}$$

$$\vec{r}_2 = (-\gamma D + x)\vec{i} + y\vec{j} + z\vec{k} \tag{3.183}$$

Noting that the distance, D, between M_1 and M_2 and the angular velocity $\dot{\theta}$ of the rotating reference frame are not constant, we obtain the second derivative of $\vec{r}_1$, as follows:

$$\ddot{\vec{r}}_1 = \{-(1+\gamma)\ddot{D} + \ddot{x} - \ddot{\theta}y - 2\dot{\theta}\dot{y} - \dot{\theta}^2[-(1+\gamma)D + x]\}\vec{i}$$
$$+ \{\ddot{y} - \ddot{\theta}(1+\gamma)D - 2\dot{\theta}(1+\gamma)\dot{D} + \ddot{\theta}x + 2\dot{\theta}\dot{x} - \dot{\theta}^2y\}\vec{j} + \ddot{z}\vec{k} \tag{3.184}$$

Introducing the mass ratio ρ of the three-body system as

$$\rho = M_2/(M_1 + M_2)$$
$$1 - \rho = M_1/(M_1 + M_2)$$

and by equating components of Eqs. (3.177) and (3.184), we obtain the nondimensional equations of motion for the spacecraft as follows:

$$\ddot{x} - 2\dot{\theta}\dot{y} - \ddot{\theta}y - \dot{\theta}^2[-(1+\gamma)D + x] = (1+\gamma)\ddot{D} - \frac{(1-\rho)[-(1+\gamma)D + x]}{r_1^3}$$
$$- \frac{\rho(-\gamma D + x)}{r_2^3} + \frac{\rho}{D^2} \tag{3.185}$$

$$\ddot{y} + \ddot{\theta}[-(1+\gamma)D + x] - \dot{\theta}^2y = -2\dot{\theta}[-(1+\gamma)\dot{D} + \dot{x}]$$
$$- \frac{(1-\rho)y}{r_1^3} - \frac{\rho y}{r_2^3} \tag{3.186}$$

$$\ddot{z} = -\frac{(1-\rho)z}{r_1^3} - \frac{\rho z}{r_2^3} \tag{3.187}$$

where

$$r_1 = \sqrt{[-(1+\gamma)D + x]^2 + y^2 + z^2}$$
$$r_2 = \sqrt{(-\gamma D + x)^2 + y^2 + z^2}$$

In these nondimensionalized equations, the distances x, y, z, r_1, r_2, and D are in units of the semimajor axis a; $\dot{\theta}$ is in units of the mean angular rate n; and time is in units of $1/n$.

The motion of the two primary bodies is a solution of the two-body problem. The distance between the two primaries, D, and the angular rate $\dot{\theta}$ are written in nondimensional form as a series in eccentricity as [13]

$$
\begin{aligned}
D = 1 + \frac{1}{2}e^2 &+ \left(-e + \frac{3}{8}e^3 - \frac{5}{192}e^5 + \frac{7}{9216}e^7\right)\cos M \\
&+ \left(-\frac{1}{2}e^2 + \frac{1}{3}e^4 - \frac{1}{16}e^6\right)\cos 2M \\
&+ \left(-\frac{3}{8}e^3 + \frac{45}{128}e^5 - \frac{567}{5120}e^7\right)\cos 3M \\
&+ \left(-\frac{1}{3}e^4 + \frac{2}{5}e^6\right)\cos 4M + \left(-\frac{125}{384}e^5 + \frac{4375}{9216}e^7\right)\cos 5M \\
&- \frac{27}{80}e^6 \cos 6M - \frac{16807}{46080}e^7 \cos 7M + \cdots
\end{aligned}
\tag{3.188}
$$

$$
\begin{aligned}
\dot{\theta} = 1 &+ \left(2e - \frac{1}{4}e^3 + \frac{5}{96}e^5 + \frac{107}{4608}e^7\right)\cos M \\
&+ 2\left(\frac{5}{4}e^2 - \frac{11}{24}e^4 + \frac{17}{192}e^6\right)\cos 2M \\
&+ 3\left(\frac{13}{12}e^3 - \frac{43}{64}e^5 + \frac{95}{512}e^7\right)\cos 3M \\
&+ 4\left(\frac{103}{96}e^4 - \frac{451}{480}e^6\right)\cos 4M + 5\left(\frac{1097}{960}e^5 - \frac{5957}{4608}e^7\right)\cos 5M \\
&+ 6\frac{1223}{960}e^6 \cos 6M + 7\frac{47273}{32256}e^7 \cos 7M + \cdots
\end{aligned}
\tag{3.189}
$$

where M is the nondimensional mean anomaly defined as $M = t - t_p$ and t_p is the time of perigee passage. It can be simply assumed that $t_p = 0$. This means that the spacecraft begins its quasi-periodic orbit when M_2 is at its closest position to M_1.

It is noted that these series converge for small eccentricity and are divergent when $e > 0.6627$. Most two-body systems in the solar system follow near circular orbits, therefore, the preceding series describe most of the two-body systems encountered in the solar system. As a measure of the accuracy of the series, it is noted that the last term in the series for D represents about 21 cm for the Earth–moon system and the last term in the series for $\dot{\theta}$ represents about 5.86×10^{-15} rad/s.

It is also possible to obtain the equations of motion without the use of a series expansion. This derivation makes use of the conic equation for motion between two bodies by changing the independent variable to the true anomaly.

Although the equations of motion have been derived with respect to the L_2 point, they can also be used to describe the motion relative to the L_1 point. For the Earth–moon system the constants in the equations of motion are

$$\rho = 0.01215$$

$$a = 384{,}748 \text{ km}$$

$$e = 0.05490$$

$$n = 2.661699 \times 10^{-6} \text{ rad/s}$$

$$\gamma = -0.150935 \quad \text{for } L_1$$

$$\gamma = 0.167833 \quad \text{for } L_2$$

The elliptic restricted three-body model derived in this section will be further utilized in Chapter 4 when we study the halo orbit determination and control problem [14, 15].

References

[1] Ehricke, K. A., *Space Flight*, Vol. 1: Environment and Celestial Mechanics, and Vol. 2: Dynamics, Van Nostrand, Princeton, NJ, 1960.

[2] Thompson, W. T., *Introduction to Space Dynamics*, Wiley, New York, 1961.

[3] Danby, J. M., *Fundamentals of Celestial Mechanics*, Macmillan, New York, 1962.

[4] Szebehely, V., *Theory of Orbits*, Academic, San Diego, CA, 1967.

[5] Kaplan, M. H., *Modern Spacecraft Dynamics and Control*, Wiley, New York, 1976.

[6] Battin, R. H., *An Introduction to the Mathematics and Methods of Astrodynamics*, AIAA Education Series, AIAA, Washington, DC, 1987.

[7] Chobotov, V. A. (ed.), *Orbital Mechanics*, AIAA Education Series, AIAA, Washington, DC, 1991.

[8] Prussing, J. E., and Conway, B. A., *Orbital Mechanics*, Oxford Univ. Press, London, 1993.

[9] Hale, F. J., *Introduction to Space Flight*, Prentice–Hall, Englewood Cliffs, NJ, 1994.

[10] Wiesel, W. E., *Spaceflight Dynamics*, 2nd ed., McGraw–Hill, New York, 1996.

[11] Farquhar, R. W., "The Control and Use of Libration-Point Satellites," Stanford Univ., SUDAAR-350, Stanford, CA, July 1968; also NASA TR R-346, Sept. 1970.

[12] Farquhar, R. W., Muhonen, D. P., Newman, C. R., and Heuberger, H. S., "Trajectories and Orbital Maneuvers for the First Libration-Point Satellite," *Journal of Guidance, Control, and Dynamics*, Vol. 3, No. 6, 1980, pp. 549–554.

[13] Taff, L. G., *Celestial Mechanics: A Computational Guide for the Practitioner*, Wiley, New York, 1985, p. 61.

[14] Cielaszyk, D., and Wie, B., "A New Approach to Halo Orbit Determination and Control," *Proceedings of the AIAA/AAS Astrodynamics Conference*, AIAA, Washington, DC, 1994, pp. 166–175; also, *Journal of Guidance, Control, and Dynamics*, Vol. 19, No. 2, 1996, pp. 266–273.

[15] Cielaszyk, D., and Wie, B., "Halo Orbit Determination and Control for the Eccentric-Restricted Three-Body Problem," *Proceedings of the AIAA/AAS Astrodynamics Conference*, AIAA, Washington, DC, 1994, pp. 176–185.

4

Orbital Maneuvers and Control

Fundamental problems in orbital mechanics, such as the two-body and restricted three-body problems, were treated in Chapter 3. This chapter is concerned with orbital maneuvering and control problems of spacecraft under the influence of rocket firings. This chapter also briefly introduces the practical problems of analyzing launch vehicle trajectories and placing a satellite into an orbit. The problems of orbital transfer, rendezvous, and orbit control, which are of fundamental importance to space missions, are covered. In particular, the halo orbit determination and control problem is introduced in this chapter.

4.1 Launch Vehicle Trajectories

In this section, we briefly introduce basic physical concepts and terminologies that are often encountered when we deal with rocket propulsion and launch vehicles, and then we discuss the basic trajectory equations during various launch phases: 1) vertical ascent, 2) turn-over flight, and 3) gravity turn maneuver.

The subject of launch vehicle guidance and control is beyond the scope of this book; however, the material briefly discussed in this section will provide the foundations for studying the orbit injection and orbital maneuvering problems of spacecraft under the influence of rocket firings.

4.1.1 *Rocket Propulsion*

4.1.1.1 Rocket Thrust At a specified position in space, a launch vehicle achieves the velocity required to place a payload spacecraft into a desired orbit. Most launch vehicles are propelled by liquid-propellant rockets and/or solid-propellant rockets. Customarily, a liquid-propellant rocket is called a *liquid rocket engine* and a solid-propellant rocket is called a *solid rocket motor*. Propellants are working substances used in rockets to produce thrust. These substances can be liquids, solids, or gases. Thrust that is developed by a rocket is a direct result of a chemical reaction between the fuel and the oxidizer in a propellant. This reaction takes place in the rocket's combustion chamber and produces gas at a higher temperature. This gas is then exhausted out the rocket nozzle at a high velocity, thereby imparting a pushing force on the rocket. The nozzle is a part of the rocket thrust chamber assembly in which the gases are accelerated to high velocities.

The rocket thrust T is expressed as [1, 2]

$$T = -V_e \frac{dm}{dt} + (p_e - p_o)A_e \qquad (4.1)$$

where V_e is called the *exit velocity* of a rocket, p_e is the nozzle exit pressure, p_o is the local atmospheric pressure, and A_e is the nozzle exit area. The rocket thrust consists of the momentum thrust, $-V_e dm/dt$, and the pressure thrust, $(p_e - p_o)A_e$. Note that the mass flow rate dm/dt is negative because the rocket is losing mass.

The pressure thrust term in Eq. (4.1) normally accounts for only a small part of the total thrust and vanishes when the exit pressure is exactly matched to the local atmospheric pressure. The exit (or exhaust) velocity V_e is often expressed as

$$V_e = g_o I_{sp} \qquad (4.2)$$

where $g_o = 9.8$ m/s^2 is the gravitational acceleration at sea level and I_{sp} is called the *specific impulse*. The specific impulse, with units of seconds, is a measure of propellant quality; that is, it is a measure of rocket thrust obtained per unit of propellant weight flow.

4.1.1.2 Propellants

Liquid propellants are commonly classified as *monopropellants* and *bipropellants*. A monopropellant contains a fuel and oxidizer combined into one substance. A bipropellant is a combination of two propellants, a fuel and an oxidizer. The chemicals are not mixed until after they have been injected into the combustion chamber. Liquid propellants are also commonly classified as being either *cryogenic* or *storable* propellants. A cryogenic propellant is an oxidizer or fuel that has a very low boiling point and must be kept at a very low temperature. A storable propellant is an oxidizer and fuel that is a liquid at normal temperatures and pressures.

Solid propellants burn on their exposed surfaces producing hot gases that provide the propulsive force for a solid rocket motor. A solid propellant, often called a *grain*, contains all of the substances needed to sustain chemical combustion; it consists of a fuel and an oxidizer that do not react chemically below some minimum temperature. Solid propellants are divided into two common classes: the composite (or heterogeneous) propellants and the homogeneous propellants. Composite propellants are a heterogeneous mixture of an oxidizer and an organic fuel binder. Small particles of oxidizer are dispersed throughout the fuel. The fuel is called a binder because the oxidizer has no mechanical strength. Homogeneous propellants contain chemical compounds that have the oxidizer and the fuel in a single molecule.

4.1.1.3 Multistage Rockets

A propulsion unit of a multistage launch vehicle is called a *stage*, and it is generally designated as either first, second, or third stage. The first stage contains the main engines, and the second and third stages provide additional thrust capability as required. An upper stage is often needed to inject a payload into a geosynchronous transfer orbit or into a heliocentric transfer orbit. Some of the currently used upper stages are the inertial upper stage (IUS), payload assist module (PAM), and Centaur upper stage. Some of the

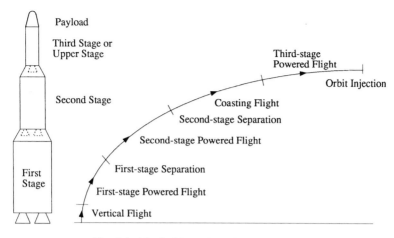

Fig. 4.1 Typical launch vehicle trajectory.

launch vehicles that have been used successfully on many space missions are the Jupiter, Atlas, Saturn, Scout, Titan, Delta, Space Shuttle, and Ariane. The launch systems that are currently available include: Delta II, Titan, Ariane, Proton, Long March, Space Shuttle, Pegasus, and Taurus [3].

A typical launch vehicle trajectory is illustrated in Fig. 4.1. The trajectory begins with a brief initial period of vertical ascent, followed by first-stage powered flight, second-stage powered flight, coasting phase, third-stage powered flight, and orbit injection. The determination of the trajectory involves the solution of sets of differential equations describing the motion of the launch vehicle.

4.1.2 Vertical Ascent

Most launch vehicles usually ascend vertically (or near vertically) through the denser regions of the Earth's atmosphere before performing any turn-over maneuvers. Because of aerodynamic heating considerations a turn-over maneuver is usually not initiated until after the vehicle has ascended through the denser regions of the Earth's atmosphere.

Consider a launch vehicle flying vertically during the early portion of its flight, immediately after liftoff from the launch pad. The equation of motion of the vertically ascending vehicle is simply given by

$$m\frac{\mathrm{d}v}{\mathrm{d}t} = T - D - mg \qquad (4.3)$$

where m is the vehicle mass, v is the vertical velocity, T is the rocket thrust, D is the aerodynamic drag, and g is the local gravitational acceleration.

The rocket thrust T, neglecting the pressure thrust term in Eq. (4.1), is given by

$$T = -V_e\frac{\mathrm{d}m}{\mathrm{d}t} = -g_0 I_{\mathrm{sp}}\frac{\mathrm{d}m}{\mathrm{d}t} \qquad (4.4)$$

The aerodynamic drag D acting on a launch vehicle is, in general, expressed as

$$D = \tfrac{1}{2}\rho v^2 C_D S \tag{4.5}$$

where C_D is the coefficient of drag, ρ is the aerodynamic density, v is the velocity of the vehicle, and S is the reference area. The coefficient of drag, C_D, varies both with altitude and Mach number, which is defined as the ratio of the velocity of the vehicle to the local speed of sound.

The solution of Eq. (4.3) can be obtained by numerical simulation on a digital computer; however, an approximate solution to Eq. (4.3) easily can be obtained by ignoring the aerodynamic drag term and assuming a constant thrust. For most rockets, the propellant flow rate $\dot{m}$ is nearly constant. The inclusion of the aerodynamic drag in the calculation of a trajectory causes a decrease in rocket performance of about 10%.

Neglecting the aerodynamic drag term and assuming the thrust remains constant, Eq. (4.3) becomes

$$m\frac{dv}{dt} = -g_o I_{sp}\frac{dm}{dt} - mg \tag{4.6}$$

which is rewritten as

$$dv = -g_o I_{sp}\frac{dm}{m} - g\,dt \tag{4.7}$$

Integrating this equation yields

$$\int_{v_0}^{v} dv = -g_o I_{sp}\int_{m_0}^{m}\frac{dm}{m} - g\int_{0}^{t} dt \tag{4.8}$$

which becomes

$$v(t) = -g_o I_{sp}\ln\frac{m}{m_0} - gt \tag{4.9}$$

where m_0 is the initial mass, $m = m_0 + \dot{m}t$, and the initial velocity v_0 is assumed to be zero. (The mass flow rate $\dot{m}$ is negative because the rocket is losing mass.)

Since $v = dy/dt$ and

$$\int_{0}^{t}\ell n\frac{m}{m_0}\,dt = \int_{0}^{t}\ell n\,m\,dt - \int_{0}^{t}\ell n\,m_0\,dt = \frac{1}{\dot{m}}\int_{m_0}^{m}\ell n\,m\,dm - t\ell n\,m_0$$

$$= \frac{1}{\dot{m}}[m(\ell n\,m - 1)]_{m_0}^{m} - t\ell n\,m_0$$

Eq. (4.9) can be integrated further with respect to time to yield the altitude y at a specified time t, as follows:

$$y(t) = g_o I_{sp}t\left\{1 - \frac{1}{(m_0/m) - 1}\ell n\frac{m_0}{m}\right\} - \frac{1}{2}gt^2 \tag{4.10}$$

where $m = m_0 + \dot{m}t$. Given the burning time of the rocket and its initial mass, the burnout velocity and the altitude at burnout can be estimated using Eqs. (4.9) and (4.10), respectively.

4.1.3 Turn-Over Trajectory

After a launch vehicle has ascended through the denser regions of the Earth's atmosphere, a turn-over maneuver is initiated to achieve a predetermined horizontal velocity component. This can be obtained by launching the vehicle at a fixed inclined angle to the horizontal from the launch pad or by a turn-over maneuver during its ascent.

Consider a launch vehicle in a planar, two-dimensional motion, as illustrated in Fig. 4.2. The translational equations of motion of the vehicle's center of mass are

$$\vec{i}: \quad m\ddot{x} = T\cos(\theta + \delta) - N\sin\theta - D\cos\theta \tag{4.11}$$

$$\vec{j}: \quad m\ddot{y} = T\sin(\theta + \delta) + N\cos\theta - D\sin\theta - mg \tag{4.12}$$

where m is the vehicle mass; x and y are, respectively, the horizontal distance and altitude; T is the rocket thrust; N is the normal force acting through the center of pressure of the vehicle; D is the aerodynamic drag; g is the local gravitational acceleration; θ is the pitch angle; and δ is the thrust gimbal angle.

The pitch angle is defined as

$$\theta = \gamma + \alpha \tag{4.13}$$

where γ is called the flight-path angle, which measures the angle between the local horizontal and the flight path, and α is called the angle of attack, which measures the angle between the flight path and the vehicle's roll axis, as illustrated

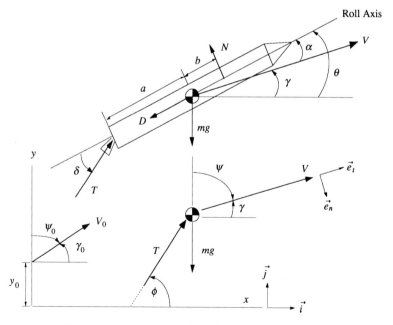

Fig. 4.2 Two-dimensional motion of a launch vehicle.

in Fig. 4.2. Also shown in this figure are the thrust vector angle ϕ, measured from the horizontal, and the velocity vector angle ψ, measured from the vertical. These angles, instead of γ or θ, will be used later in this section.

The aerodynamic forces N and D are often expressed as

$$N = \tfrac{1}{2}\rho V^2 C_N S \tag{4.14a}$$

$$D = \tfrac{1}{2}\rho V^2 C_D S \tag{4.14b}$$

where $V^2 = \dot{x}^2 + \dot{y}^2$. The aerodynamic coefficients C_D and C_N vary both with altitude and Mach number and are functions of the angle of attack.

The rotational equation of motion of the vehicle about its center of mass is

$$J\ddot{\theta} = -Ta\sin\delta + Nb \tag{4.15}$$

where J is the pitch moment of inertia of the vehicle about its center of mass, a is the distance between the center of mass and the rocket nozzle, and b is the distance between the center of mass and the center of pressure.

Now consider a launch vehicle performing a constant turn-over maneuver in which a constant negative pitch rate, i.e., $\dot{\theta} =$ a negative constant and $\ddot{\theta} = 0$, is maintained by having an offset gimbal angle δ. The offset gimbal angle needed for a constant turn-over maneuver can be determined from

$$\sin\delta = Nb/Ta \tag{4.16}$$

whereas the aerodynamic normal force N and the center of pressure distance b must be known as functions of time.

Because of the curvature of the Earth, the local horizontal along the flight path does not coincide with the horizontal at launch, except at liftoff. The angle between the flight path and the local horizontal can be obtained as

$$\gamma = \gamma_0 + \dot{\theta}t + \int_0^x \frac{dx}{R+y} \tag{4.17}$$

where γ_0 is the initial flight-path angle at $t = 0$ and R is the Earth's radius.

The vehicle's desired angle of attack for a constant turn-over maneuver can then be determined from

$$\sin\alpha = \frac{\dot{x}\sin(\gamma+\alpha) - \dot{y}\cos(\gamma+\alpha)}{\sqrt{\dot{x}^2 + \dot{y}^2}} \tag{4.18}$$

No analytical solution to the preceding equations exists, but they may be solved using an iterative procedure.

4.1.4 Gravity Turn Trajectory

One of many trajectory optimization problems of launch vehicles is concerned with determining the thrust vector angle $\phi(t)$, shown in Fig. 4.2, necessary to achieve maximum horizontal velocity at a specified altitude. The solution to this classical optimization problem, assuming no aerodynamic effects, can be found as

$$\tan \phi = \left(1 - \frac{t}{t_f}\right) \tan \gamma_0 \tag{4.19}$$

where γ_0 is the initial flight-path angle, t_f is the unspecified time at the specified altitude, and ϕ varies from γ_0 to zero.

This optimum thrust vector angle may demand large angle-of-attack changes, however, which is not practical due to aerodynamic heating considerations. In practice, a so-called *gravity turn trajectory*, in which the thrust vector is always kept parallel to the velocity vector, is employed, and the angle of attack is kept near zero in a gravity turn maneuver; i.e., $\delta = -\alpha = 0$.

Consider again a simplified two-dimensional model of a launch vehicle as shown in Fig. 4.2. It is sometimes convenient to use the velocity vector angle ψ in writing the equations of motion in terms of the tangential and normal components to the trajectory. The equations of motion in terms of the tangential acceleration $\dot{V}$ and the normal acceleration $V\dot{\psi}$ are written as

$$\vec{e}_t: \quad m\dot{V} = T\cos(\delta + \alpha) - mg\cos\psi - D\cos\alpha - N\sin\alpha \tag{4.20a}$$

$$\vec{e}_n: \quad mV\dot{\psi} = -T\sin(\delta + \alpha) + mg\sin\psi + D\sin\alpha - N\cos\alpha \tag{4.20b}$$

where $\vec{e}_t$ and $\vec{e}_n$ are unit vectors along the tangential and normal directions of the trajectory, as shown in Fig. 4.2.

For a gravity turn maneuver ($\delta = -\alpha = 0$, and thus $N = D = 0$), these equations of motion can be simplified as

$$\dot{V} = (T/m) - g\cos\psi \tag{4.21a}$$

$$\dot{\psi} = (g/V)\sin\psi \tag{4.21b}$$

and we have

$$\dot{x} = V\sin\psi \tag{4.22a}$$

$$\dot{y} = V\cos\psi \tag{4.22b}$$

For even the most practical cases, in which T is constant and $m = m_0 + \dot{m}t$, these equations are still nonlinear and no analytical solution exists. But these equations can be numerically integrated to determine the gravity turn trajectory for given initial conditions: $x(0) = 0$, $y(0) = y_0$, $V(0) = V_0$, and $\psi(0) = \psi_0$. A case in which the thrust-to-weight ratio T/mg is a constant is also of practical interest and some analytical results can be found in Ref. 4.

In general, the final burnout velocities required for most space flight missions cannot be easily achieved by a single-stage launch vehicle, because it has to carry its entire structural weight up to the final burnout point. Thus, a multistage launch vehicle is needed to inject a spacecraft into even a low Earth orbit. In a multistage launch vehicle, the first stage contains the main engines, and the second and third stages provide additional thrust capability as required. More details of this important, classical subject of performance analysis and configuration optimization of multistage launch vehicles can be found in the literature [4, 5] and we shall not pursue this subject further in this text.

In the next section, we shall study an orbit determination problem of a spacecraft given its orbital injection conditions at the final burnout point.

4.2 Orbit Injection

Consider a problem of determining an orbit of a spacecraft when only its orbital injection conditions are known. Assume that a spacecraft is injected into an orbit with the following conditions at final burnout:

$$r = r_0, \qquad v = v_0, \qquad \gamma = \gamma_0$$

where γ is the flight-path angle measured from the local horizontal to the velocity vector as illustrated in Fig. 4.3.

First we will determine the eccentricity e and then the true anomaly θ_0 at the burnout point to locate the perigee of the resulting orbit.

Because we know the values of v and r at the burnout point, the specific energy $\mathcal{E}$ can be found as

$$\mathcal{E} = -(\mu/2a) = (v^2/2) - (\mu/r) = \left(v_0^2/2\right) - (\mu/r_0) \qquad (4.23)$$

The specific angular momentum h of the resulting orbit can also be determined as

$$h = r^2\dot{\theta} = r_0^2\dot{\theta}_0 = r_0(r_0\dot{\theta}_0) = r_0 v_0 \cos\gamma_0 \qquad (4.24)$$

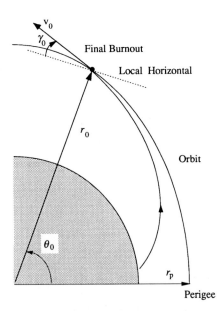

Fig. 4.3   Orbit injection.

Once we know $\mathcal{E}$ and h, we can determine the eccentricity e as

$$e = \sqrt{1 + (2\mathcal{E}h^2/\mu^2)} \qquad (4.25)$$

Furthermore, the semimajor axis a and the perigee distance r_p can be determined as follows:

$$a = -(\mu/2\mathcal{E}) \qquad (4.26a)$$

$$r_p = a(1 - e) \qquad (4.26b)$$

At the burnout point, we have

$$r_0 = \frac{h^2/\mu}{1 + e\cos\theta_0} \qquad (4.27)$$

and the true anomaly θ_0 can be determined from

$$\cos\theta_0 = \frac{1}{e}\left(\frac{h^2}{\mu r_0} - 1\right) \qquad (4.28)$$

Using Eq. (3.82), we express the radial velocity component, $\dot{r} \equiv v\sin\gamma$, at the burnout point, as follows:

$$v_0\sin\gamma_0 = (\mu/h)e\sin\theta_0 \qquad (4.29)$$

which can be rewritten as

$$\sin\theta_0 = (hv_0/\mu e)\sin\gamma_0 \qquad (4.30)$$

Substituting Eqs. (4.23) and (4.24) into Eqs. (4.25), (4.28), and (4.30), and defining a new variable λ as

$$\lambda = \frac{v_0^2}{\mu/r_0} = \text{twice the ratio of kinetic to potential energy at burnout} \qquad (4.31)$$

we obtain

$$e^2 = (\lambda - 1)^2\cos^2\gamma_0 + \sin^2\gamma_0 \qquad (4.32)$$

and

$$\tan\theta_0 = \frac{\lambda\sin\gamma_0\cos\gamma_0}{\lambda\cos^2\gamma_0 - 1} \qquad (4.33)$$

These equations indicate that an orbit for any burnout condition is uniquely established by two parameters, λ and γ_0.

If $\lambda = 2$, then $e = 1$ for any values of γ_0, which corresponds to an escape parabola as long as the orbital path does not intercept the Earth. If $\lambda = 1$, then

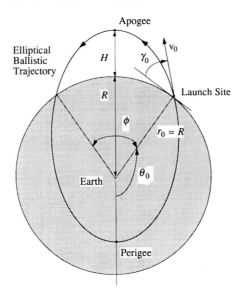

Fig. 4.4   Suborbital flight.

γ_0 must be zero to have a circular orbit, although the condition of $\lambda = 1$ itself corresponds to the case of the circular velocity: $v_0 = \sqrt{\mu/r_0}$.

As an application of Eqs. (4.32) and (4.33), consider an elliptical ballistic trajectory illustrated in Fig. 4.4. The point of maximum height corresponds to the apogee, and the perigee is inside the Earth.

A problem of practical interest is the determination of the range $R\phi$ for given initial conditions: $r_0 = R$, v_0, and γ_0 at launch site. First, the eccentricity is determined from Eq. (4.32) as

$$e^2 = (\lambda - 1)^2 \cos^2 \gamma_0 + \sin^2 \gamma_0$$

where $\lambda = Rv_0^2/\mu$. Because $\phi/2 = \pi - \theta_0$, the flight range angle ϕ can be determined from

$$\tan \frac{\phi}{2} = -\frac{\lambda \sin \gamma_0 \cos \gamma_0}{\lambda \cos^2 \gamma_0 - 1}$$

Note that the maximum height H can also be simply determined from

$$r_a = a(1 + e) = H + R$$

where the semimajor axis a is determined from Eq. (4.23).

4.3   Single-Impulse Maneuvers

In orbital dynamics, the term Δv maneuver refers to an impulsive maneuver that is characterized by an instantaneous change in orbital velocity of a space vehicle.

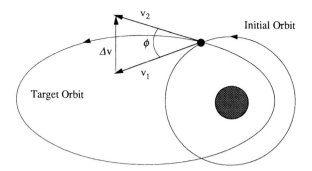

Fig. 4.5 Single-impulse orbital maneuver.

Rocket burn times (a few minutes at most) are short compared to typical orbital periods of 100 min or longer; thus, typical rocket firings for orbital maneuvers are considered to be impulsive. As a result, in orbital maneuvering problems, we frequently speak of Δv as a measure of the velocity impulse needed for a space vehicle to enter a new orbit.

As illustrated in Fig. 4.5 for an impulsive maneuver between intersecting coplanar orbits, most basic single-impulse maneuvering problems can be analyzed by simply using the velocity vector equation

$$\vec{v}_2 = \vec{v}_1 + \Delta\vec{v} \tag{4.34}$$

where $\vec{v}_1$ and $\vec{v}_2$ are the velocity vectors of the initial and final target orbits, at the maneuvering point, respectively, and $\Delta\vec{v}$ is the required velocity change needed to enter the target orbit.

Using the law of cosines, we compute the magnitude of $\Delta\vec{v}$, as follows:

$$\Delta v \equiv |\Delta\vec{v}| = \sqrt{v_1^2 + v_2^2 - 2v_1 v_2 \cos\phi} \tag{4.35}$$

where $v_1 = |\vec{v}_1|$, $v_2 = |\vec{v}_2|$, and ϕ is the angle between $\vec{v}_1$ and $\vec{v}_2$.

Similarly, the velocity change required for an inclination change maneuver between two circular orbits with the same period is given by

$$\Delta v = 2v_c \sin(\Delta i/2) \tag{4.36}$$

where v_c is the circular speed of both orbits and Δi is the inclination angle between them. The maneuver should occur at either the ascending or the descending node.

The propellant mass needed for a given value of Δv can be determined from the following relationship:

$$\Delta v = -g_o I_{\text{sp}} \ln (m/m_0) \tag{4.37}$$

where g_o is the gravitational acceleration at sea level, I_{sp} is the propellant specific impulse, and m_0 and m are the space vehicle mass before and after a Δv maneuver, respectively. The mass of the propellant burned during the Δv maneuver, denoted as $|\Delta m|$, is then expressed as

$$|\Delta m| = m_0 \left\{ 1 - \exp\left(-\frac{\Delta v}{g_0 I_{sp}} \right) \right\} \tag{4.38}$$

because $m = m_0 + \Delta m$ and $\Delta m < 0$. (Note that the mass flow rate $\dot{m}$ is defined to be negative because the rocket is losing mass.)

4.4   Hohmann Transfer

This section is concerned with an orbital transfer problem between two coplanar circular orbits, as illustrated in Fig. 4.6. The elliptic transfer orbit is tangent to both the inner and outer circular orbits at the perigee and apogee of the transfer orbit, respectively. This problem was first solved by Walter Hohmann [6] in 1925, and the associated orbital maneuver is called the *Hohmann transfer*.

4.4.1   Two-Impulse Elliptic Transfer

Consider an outward Hohmann transfer as illustrated in Fig. 4.6 without loss of generality. The velocity of a space vehicle orbiting along the inner circular orbit of radius r_1 is given by

$$v_{c_1} = \sqrt{\mu/r_1} \tag{4.39}$$

The velocity required to leave the inner circular orbit and travel along the elliptic transfer orbit to reach the outer circular target orbit of radius r_2 is

$$v_1 = \sqrt{(2\mu/r_1) - (\mu/a)} \tag{4.40}$$

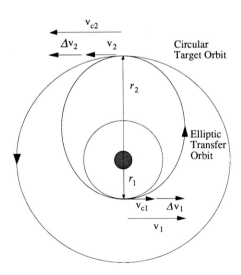

Fig. 4.6   Hohmann transfer.

where

$$a = \tfrac{1}{2}(r_1 + r_2)$$

is the semimajor axis of the elliptic transfer orbit whose eccentricity is given by

$$e = (r_2 - r_1)/(r_2 + r_1) \qquad (4.41)$$

The velocity change required at the perigee of the transfer orbit to transfer from the inner circular orbit is then given by

$$\Delta v_1 = v_1 - v_{c_1} = \sqrt{(2\mu/r_1) - (\mu/a)} - \sqrt{\mu/r_1} \qquad (4.42)$$

Similarly, the velocity change required at the apogee of the transfer orbit to enter the outer circular orbit of radius r_2 is given by

$$\Delta v_2 = v_{c_2} - v_2 = \sqrt{\mu/r_2} - \sqrt{(2\mu/r_2) - (\mu/a)} \qquad (4.43)$$

The total velocity change required in the Hohmann transfer between two coplanar circular orbits is

$$\Delta v = \Delta v_1 + \Delta v_2 \qquad (4.44)$$

The transfer time, T, is half the period P of the elliptic transfer orbit; that is,

$$T = P/2 = \pi\sqrt{a^3/\mu} \qquad (4.45)$$

where $a = (r_1 + r_2)/2$.

Problem

4.1 A spacecraft initially in a 200-km circular parking orbit above the Earth ($\mu_\oplus = 398{,}601$ km^3/s^2 and $R_\oplus = 6{,}378$ km) is injected into a Hohmann transfer orbit to reach a geosynchronous orbit with a period of 23 h 56 min 4 s. Find the radius of the target orbit, r_2, the eccentricity of the transfer orbit, e, the required velocity changes, Δv_1 and Δv_2, and the transfer time T. *Answers:* $r_2 = 42{,}164$ km, $e = 0.73$, $\Delta v_1 = 2.45$ km/s, $\Delta v_2 = 1.48$ km/s, and $T = 5.26$ h.

4.4.2 Optimality of the Hohmann Transfer

The fuel-optimality of the Hohmann transfer among two-impulse maneuvers between two coplanar circular orbits can be verified as follows [7, 8].

Consider an elliptic transfer orbit with semimajor axis a, semilatus rectum p, and eccentricity e. It is assumed that this elliptic transfer orbit intersects the inner and outer circular orbits of radii r_1 and r_2; that is,

$$r_p = p/(1 + e) \le r_1 \qquad (4.46a)$$
$$r_a = p/(1 - e) \ge r_2 \qquad (4.46b)$$

Note that the minimum value of e corresponds to an elliptic orbit with $r_p = r_1$ and $r_a = r_2$.

The velocity changes Δv_1 and Δv_2 are given by the law of cosines, as follows:

$$(\Delta v_1)^2 = v_1^2 + v_{c_1}^2 - 2v_{c_1} v_1 \cos \phi_1 = v_1^2 + v_{c_1}^2 - 2v_{c_1} \sqrt{\mu p}/r_1 \quad (4.47a)$$

$$(\Delta v_2)^2 = v_2^2 + v_{c_2}^2 - 2v_{c_2} v_2 \cos \phi_2 = v_2^2 + v_{c_2}^2 - 2v_{c_2} \sqrt{\mu p}/r_2 \quad (4.47b)$$

where

$$v_1^2 = \mu \left(\frac{2}{r_1} - \frac{1}{a} \right) = \mu \left(\frac{2}{r_1} + \frac{e^2 - 1}{p} \right) \quad (4.48a)$$

$$v_2^2 = \mu \left(\frac{2}{r_2} - \frac{1}{a} \right) = \mu \left(\frac{2}{r_2} + \frac{e^2 - 1}{p} \right) \quad (4.48b)$$

and

$$v_{c_1}^2 = \mu/r_1 \quad (4.49a)$$

$$v_{c_2}^2 = \mu/r_2 \quad (4.49b)$$

The total velocity change, which represents the total fuel cost, is then

$$\Delta v = \Delta v_1 + \Delta v_2 \quad (4.50)$$

and the partial derivative of Δv with respect to e, while keeping p constant, can be found as

$$\frac{\partial \Delta v}{\partial e} = \frac{e\mu}{p} \left[(\Delta v_1)^{-1} + (\Delta v_2)^{-1} \right] > 0 \quad (4.51)$$

Consequently, the minimum value of Δv occurs at the minimum value of e, which corresponds to the Hohmann transfer; that is, the first impulse at $r_p = r_1$ and the second impulse at $r_a = r_2$. This proves that among two-impulse maneuvers, the Hohmann transfer is the minimum-fuel transfer between coplanar circular orbits; however, for values of r_2/r_1 greater than approximately 11.94, a three-impulse bielliptic transfer can be shown to have a lower fuel cost than the two-impulse Hohmann transfer. For a more formal proof of the optimality of the Hohmann transfer, the reader is referred to Refs. 7 and 8.

4.5 Interplanetary Flight

A typical interplanetary flight consists of three phases: Earth escape, heliocentric orbital transfer, and planet encounter. A simple method developed by Hohmann, which is very useful in preliminary mission design of such interplanetary flights, is presented here. The method is based on the concept of a sphere of influence and it is often referred to as the *patched conic method*.

4.5.1 Sphere of Influence

Consider the system of three bodies as shown in Fig. 3.13 of Sec. 3.8. The masses of the two primary bodies are denoted by M_1 and M_2, respectively, whereas the infinitesimal mass of the third body (a spacecraft) is denoted by m.

As was derived in Sec. 3.8, the relative motion of m with respect to M_1 can be described by

$$\ddot{\vec{r}}_1 + G(M_1 + m)\frac{\vec{r}_1}{r_1^3} = -GM_2 \left\{ \frac{\vec{r}_2}{r_2^3} + \frac{\vec{r}_{21}}{r_{21}^3} \right\} \qquad (4.52)$$

where $\vec{r}_1$ and $\vec{r}_2$ are the position vectors of m from M_1 and M_2, respectively, and $\vec{r}_{21}$ is the position vector of M_2 from M_1. The right-hand side of this equation can be considered as the perturbing acceleration to the two-body problem of M_1 and m due to the presence of M_2.

Similarly, the relative motion of m with respect to M_2 can be described by

$$\ddot{\vec{r}}_2 + G(M_2 + m)\frac{\vec{r}_2}{r_2^3} = -GM_1 \left\{ \frac{\vec{r}_1}{r_1^3} - \frac{\vec{r}_{21}}{r_{21}^3} \right\} \qquad (4.53)$$

The right-hand side of this equation can also be considered as the perturbing acceleration to the two-body problem of M_2 and m due to the presence of M_1.

Now consider a spacecraft in orbit between two primary attracting bodies of masses M_1 and M_2, which are separated by a distance D.

The radius of a sphere of influence, simply called the *activity radius r* of M_2 relative to M_1 can be defined as (see e.g., [9])

$$r/D \approx (M_2/M_1)^{\frac{2}{5}} \qquad (4.54)$$

When the space vehicle is outside the activity sphere of M_2, then its orbital motion is influenced mainly by M_1 and is described by

$$\ddot{\vec{r}}_1 + G(M_1 + m)\left(\vec{r}_1/r_1^3\right) \approx 0$$

If a space vehicle is within the sphere of influence of M_2, the orbital motion of the vehicle is influenced mainly by M_2 and is described by

$$\ddot{\vec{r}}_2 + G(M_2 + m)\left(\vec{r}_2/r_2^3\right) \approx 0$$

For the sun–Earth system, the Earth's activity radius r is given as

$$\frac{r}{D} \approx \left(\frac{M_\oplus}{M_\odot}\right)^{\frac{2}{5}} = \left(\frac{1}{328,900}\right)^{\frac{2}{5}} = 0.00621 \qquad (4.55)$$

where $D = 1$ AU $= 1.4959789 \times 10^8$ km. That is, the Earth's activity radius relative to sun is only 0.00621 AU or $145R_\oplus$. (See Table 4.1 for activity radii of

Table 4.1   Activity radii of nine planets in our solar system

Planet	Activity radii, AU
Mercury	0.00075
Venus	0.00411
Earth	0.00621
Mars	0.00385
Jupiter	0.32220
Saturn	0.36400
Uranus	0.34600
Neptune	0.58000
Pluto	0.00056

all planets in our solar system.) Note that the moon located at a distance of $60R_\oplus$ from the Earth is certainly inside the Earth's activity sphere.

For the Earth–moon system, the moon's activity radius r is given as

$$\frac{r}{D} \approx \left(\frac{M_{\text{moon}}}{M_\oplus}\right)^{\frac{2}{5}} = \left(\frac{1}{81.56}\right)^{\frac{2}{5}} = 0.17 \tag{4.56}$$

where $D = 384{,}748$ km. Therefore, the moon's activity radius relative to Earth is approximately the same as the distance from the moon to the L_1 point.

4.5.2 Patched Conic Method

Consider a spacecraft departing from its low-altitude circular parking orbit above the Earth for an interplanetary trip to an outer planet. A minimum Δv Hohmann heliocentric transfer orbit is selected here for such an outer-planet mission, although it is not the minimum-time path. It is assumed that the motion of the planet and the spacecraft lie in the ecliptic plane. Such a simplified interplanetary flight mission is analyzed here using the method of patched conics.

As illustrated in Fig. 4.7, the spacecraft departs the Earth at a proper time, called the launch window, so that it can encounter the target planet at aphelion of the heliocentric transfer orbit. In this figure, $r_\oplus$ denotes the radius of the Earth's orbit about the sun and $r_\otimes$ denotes the orbital radius of the target planet about the sun.*

The transfer time is simply half the period P of the transfer orbit; that is,

$$T = P/2 = \pi\sqrt{a^3/\mu_\odot} \tag{4.57}$$

where $a = (r_\oplus + r_\otimes)/2$ is the semimajor axis of the elliptic transfer orbit and $\mu_\odot$ is the gravitational parameter of the sun.

___

*In this chapter, the symbol $\otimes$ is used to denote a target planet, although such a symbol is not one of the astronomical symbols such as $\odot$ and $\oplus$.

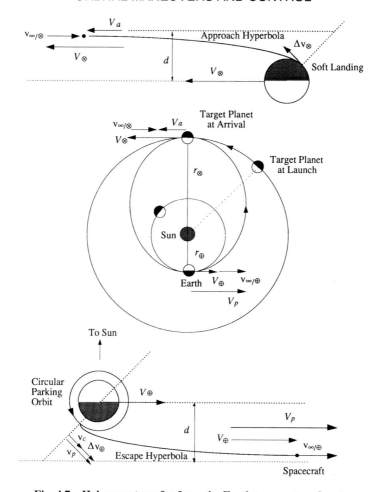

Fig. 4.7   Hohmann transfer from the Earth to an outer planet.

The velocity required to leave the Earth's orbit and travel along a heliocentric transfer orbit to reach the target planet's orbit is

$$V_p = \sqrt{\frac{2\mu_\odot}{r_\oplus} - \frac{2\mu_\odot}{r_\oplus + r_\otimes}} \qquad (4.58)$$

which is called the perihelion velocity of the Hohmann transfer orbit.

Because the Earth's activity sphere is very small compared to the distance from the Earth to the sun, the velocity of the space vehicle relative to the Earth when it leaves the Earth's activity sphere can be determined as

$$v_{\infty/\oplus} = V_p - V_\oplus = \sqrt{\frac{2\mu_\odot}{r_\oplus} - \frac{2\mu_\odot}{r_\oplus + r_\otimes}} - \sqrt{\frac{\mu_\odot}{r_\oplus}} \qquad (4.59)$$

where $V_{\oplus}$ is the circular velocity of the Earth about the sun and $V_p > V_{\oplus}$. The symbol $v_{\infty/\oplus}$ denotes the hyperbolic escape velocity of the spacecraft with respect to the Earth. Note that the velocity relationship (4.59) is obtained from the corresponding vector equation

$$\vec{V}_p = \vec{v}_{\infty/\oplus} + \vec{V}_{\oplus} \tag{4.60}$$

where $\vec{V}_p$ is, in fact, the inertial velocity vector of the spacecraft with respect to the sun when it leaves the Earth's activity sphere.

Assume that an actual Δv maneuver occurs at perigee of the escape (or departure) hyperbola. The energy equation for a hyperbolic orbit about the Earth is given by

$$\mathcal{E} = -\frac{\mu_{\oplus}}{2a} = \frac{v_p^2}{2} - \frac{\mu_{\oplus}}{r_p} = \frac{v_{\infty/\oplus}^2}{2} \tag{4.61}$$

where r_p is the perigee distance, v_p is the perigee velocity, and a is the semimajor axis of the escape hyperbola. The perigee velocity required for the escape hyperbola is then determined as

$$v_p = \sqrt{v_{\infty/\oplus}^2 + (2\mu_{\oplus}/r_p)} \tag{4.62}$$

This perigee velocity is achieved by an impulsive Δv maneuver, as follows:

$$v_p = v_c + \Delta v_{\oplus} \tag{4.63}$$

where $v_c = \sqrt{\mu_{\oplus}/r_p}$ is the circular velocity of a parking orbit about the Earth. Finally, the velocity change $\Delta v_{\oplus}$ required at perigee is computed as

$$\Delta v_{\oplus} = \sqrt{v_{\infty/\oplus}^2 + (2\mu_{\oplus}/r_p)} - \sqrt{\mu_{\oplus}/r_p} \tag{4.64}$$

The semimajor axis a and the eccentricity e of the escape hyperbola can be determined as

$$a = -(\mu_{\oplus}/2\mathcal{E}) = -\left(\mu_{\oplus}/v_{\infty/\oplus}^2\right) \tag{4.65a}$$

$$e = 1 + \left(r_p v_{\infty/\oplus}^2/\mu_{\oplus}\right) \tag{4.65b}$$

The parameter p and the specific angular momentum h of the escape hyperbola can be determined as

$$p = a(1 - e^2) \tag{4.66a}$$

$$h = \sqrt{\mu_{\oplus} p} \tag{4.66b}$$

Furthermore, the distance d between the Earth and the asymptote of the escape hyperbola can be found from the relationship $h = v_{\infty} d$. Note that the outgoing asymptote of the escape hyperbola with the asymptote angle

$$\theta_{\infty} = \cos^{-1}(-1/e) \tag{4.67}$$

is aligned with the Earth's velocity vector, as a Hohmann transfer orbit is considered.

Similarly, when the spacecraft approaches the activity sphere of the target planet along a heliocentric Hohmann transfer orbit, we have

$$\vec{V}_a = \vec{v}_{\infty/\otimes} + \vec{V}_\otimes \qquad (4.68)$$

where $\vec{V}_a$ is the inertial velocity vector of the spacecraft with respect to the sun, called the aphelion velocity of the Hohmann transfer orbit, $\vec{V}_\otimes$ is the circular velocity vector of the target planet with respect to the sun, and $\vec{v}_{\infty/\otimes}$ is the hyperbolic approach velocity vector of the space vehicle relative to the target planet.

Because $V_\otimes > V_a$, we have

$$v_{\infty/\otimes} = V_\otimes - V_a = \sqrt{\frac{\mu_\odot}{r_\otimes}} - \sqrt{\frac{2\mu_\odot}{r_\otimes} - \frac{2\mu_\odot}{r_\oplus + r_\otimes}} \qquad (4.69)$$

Consequently, the direction of $\vec{v}_{\infty/\otimes}$ is the opposite of the direction of $\vec{V}_\otimes$, as illustrated in Fig. 4.7. That is, during the encounter with the outer planet, the spacecraft approaches the target planet from a point ahead of the planet, relative to the view of an observer on the target planet. Also note that the spacecraft could approach the target planet on either the sunlit side or the dark side with the same $v_{\infty/\otimes}$.

The energy equation of an approach hyperbola is

$$\mathcal{E} = -(\mu_\otimes/2a) = (v_p^2/2) - (\mu_\otimes/r_p) = v_{\infty/\otimes}^2/2 \qquad (4.70)$$

where a, r_p, and v_p are, respectively, the semimajor axis, the periapsis distance, and the periapsis velocity of the approach hyperbola about the target planet, and where $\mu_\otimes$ is the gravitational parameter of the target planet.

If the spacecraft approaches the target planet with the hyperbolic approach velocity $v_{\infty/\otimes}$ and the asymptote distance d, the eccentricity of the hyperbolic orbit can be found as

$$e = \sqrt{1 + \left(v_{\infty/\otimes}^2 d/\mu_\otimes\right)^2} \qquad (4.71)$$

and the periapsis distance of the hyperbolic orbit about the target planet can be determined from

$$e = 1 + \left(r_p v_{\infty/\otimes}^2/\mu_\otimes\right) \qquad (4.72)$$

The periapsis velocity is then found as

$$v_p = \sqrt{v_{\infty/\otimes}^2 + (2\mu_\otimes/r_p)} \qquad (4.73)$$

Neglecting the atmospheric drag effect and the rotational motion of the target planet, we can estimate the retro velocity required for a "soft landing" on the target planet's surface, i.e., $r_p = R_\otimes$, as follows:

$$\Delta v_\otimes = \sqrt{v_{\infty/\otimes}^2 + (2\mu_\otimes/R_\otimes)} \qquad (4.74)$$

where $R_\otimes$ is the equatorial radius of the target planet.

Problems

4.2 A spacecraft initially in a 200-km circular parking orbit above the Earth
($R_\oplus = 6378$ km) is injected into a Hohmann heliocentric transfer orbit to
the outer planet Mars. It is assumed that the motion of Mars and the spacecraft lie in the ecliptic plane. The radius of the Earth's orbit about the sun
is $r_\oplus = 1.496 \times 10^8$ km $(= 1$ AU$)$ and the radius of the orbit of Mars about
the sun is $r_\otimes = 1.523691$ AU. The gravitational parameters of the sun and
Earth are: $\mu_\odot = 1.32715 \times 10^{11}$ km^3/s^2 and $\mu_\oplus = 3.98601 \times 10^5$ km^3/s^2.

  (a) Estimate the trip time T from the Earth to Mars along the Hohmann
transfer orbit.

  (b) Compute the perihelion velocity V_p of the Hohmann heliocentric transfer
orbit, the Earth's velocity $V_\oplus$, the hyperbolic escape velocity $v_{\infty/\oplus}$, and
$\Delta v_\oplus$ required at perigee. Also sketch the escape hyperbola by showing
$\theta_\infty, d, V_p, V_\oplus, v_{\infty/\oplus}, \Delta v_\oplus, v_p, v_c,$ and r_p.

  (c) Compute the aphelion velocity V_a of the Hohmann heliocentric transfer
orbit, the velocity of the target planet Mars, denoted as $V_\otimes$, and the
hyperbolic approach velocity $v_{\infty/\otimes}$.

  (d) Compute $\Delta v_\otimes$ required for a soft landing on the target planet Mars.
The gravitational parameter of Mars is $\mu_\otimes = 43{,}058$ km^3/s^2 and the
equatorial radius of Mars is $R_\otimes = 3{,}379$ km.

  (e) Assuming that the spacecraft is required to approach the target planet on
the sunlit side, sketch the approach hyperbola for a soft landing mission
by showing $\theta_\infty, d, V_a, V_\otimes, v_{\infty/\otimes}, \Delta v_\otimes,$ and $R_\otimes$.

    Answers: T $= 259$ days, $V_p = 32.729$ km/s, $V_\oplus = 29.784$ km/s,
$v_{\infty/\oplus} = 2.945$ km/s, $v_p = 11.395$ km/s, $v_c = 7.784$ km/s, $\Delta v_\oplus = 3.612$
km/s, $V_\otimes = 24.13$ km/s, $V_a = 21.48$ km/s, $v_{\infty/\otimes} = 2.648$ km/s, and
$\Delta v_\otimes = 5.70$ km/s. The total Δv requirement for a soft landing mission
to Mars is about 9.3 km/s.

4.3 A spacecraft initially in a 200-km circular parking orbit above the Earth
($R_\oplus = 6378$ km) is injected into a Hohmann heliocentric transfer orbit to
the inner planet Venus ($R_\otimes = 6200$ km and $\mu_\otimes = 3.257 \times 10^5$ km^3/s^2). It is
assumed that the motion of Venus and the spacecraft lie in the ecliptic plane.
The radius of the Earth's orbit about the sun is $r_\oplus = 1.496 \times 10^8$ km $(= 1$
AU$)$ and the radius of the orbit of Venus about the sun is $r_\otimes = 0.723332$
AU. The gravitational parameters of the sun and Earth are: $\mu_\odot = 1.32715 \times
10^{11}$ km^3/s^2 and $\mu_\oplus = 3.98601 \times 10^5$ km^3/s^2.

  (a) Estimate the trip time T from the Earth to Venus along the Hohmann
transfer orbit.

  (b) Compute the aphelion velocity V_a of the Hohmann heliocentric transfer
orbit, the Earth's velocity $V_\oplus$, the hyperbolic escape velocity $v_{\infty/\oplus}$, and
$\Delta v_\oplus$ required at perigee. Also sketch the escape hyperbola by showing
$\theta_\infty, d, V_a, V_\oplus, v_{\infty/\oplus}, \Delta v_\oplus, v_p, v_c,$ and r_p.

  (c) Compute the perihelion velocity V_p of the Hohmann heliocentric transfer
orbit, the velocity of the target planet Venus $V_\otimes$, and the hyperbolic
approach velocity $v_{\infty/\otimes}$.

  (d) Compute $\Delta v_\otimes$ required for a soft landing on the target planet Venus.

(e) Assuming that the spacecraft is required to approach the target planet on the dark side, sketch the approach hyperbola for a soft landing mission by showing θ_∞, d, V_p, $V_\otimes$, $v_{\infty/\otimes}$, $\Delta v_\otimes$, and $R_\otimes$.
(f) Assuming that the spacecraft is required to approach the target planet on the sunlit side, sketch the approach hyperbola for a soft landing mission by showing θ_∞, d, V_p, $V_\otimes$, $v_{\infty/\otimes}$, $\Delta v_\otimes$, and $R_\otimes$.
Note: For a Hohmann heliocentric transfer orbit to the inner planet Venus, we have

$$\vec{V}_a = \vec{v}_{\infty/\oplus} + \vec{V}_\oplus$$

where $\vec{V}_a$ is the inertial velocity vector of the spacecraft at aphelion of the heliocentric transfer orbit and $V_\oplus > V_a$.
Answers: $T = 146$ days, $V_a = 27.30$ km/s, $V_\oplus = 29.78$ km/s, $v_{\infty/\oplus} = 2.48$ km/s, $v_p = 11.28$ km/s, $\Delta v_\oplus = 3.49$ km/s, the escape hyperbola ($e = 1.10$, $\theta_\infty = 155.38$ deg, $\delta = 130.4$ deg), $V_p = 37.71$ km/s, $V_\otimes = 35.0$ km/s, $v_{\infty/\otimes} = 2.71$ km/s, $\Delta v_\otimes = 10.6$ km/s.

4.5.3 Planetary Flyby

Thus far in this section, a Hohmann transfer orbit, which is tangential both to the Earth's orbit and the target planet's orbit, was considered for an interplanetary flight mission. The trip time to Mars along a Hohmann heliocentric transfer orbit, for example, can be estimated as 259 days and the total Δv required for a soft landing mission on Mars from a 200-km parking orbit can be estimated as 9.3 km/s. In certain missions, a non-Hohmann transfer orbit is used to shorten the trip time at the expense of the increased Δv requirements. Some interplanetary missions also utilize a planetary flyby to increase or decrease the energy of the spacecraft with respect to the sun so that other planets can also be encountered by the same spacecraft.

A spacecraft after its planetary flyby (or swingby) may gain or lose its energy depending on whether it passes behind or ahead of the planet. The gain or loss of energy is caused by the rotation of the spacecraft's velocity vector with respect to the planet, as illustrated in Fig. 4.8. In this figure, v_∞^- and v_∞^+ denote the hyperbolic excess velocity relative to the target planet before and after the planetary flyby, respectively; V^- and V^+ denote the inertial velocity of the spacecraft with respect to the sun before and after the planetary flyby, respectively; and $V_\otimes$ is the inertial velocity of the target planet. In a trailing-side flyby of a planet as shown in Fig. 4.8a, the spacecraft's inertial velocity is increased. On the other hand, the spacecraft's inertial velocity is decreased in a leading-side flyby, as shown in Fig. 4.8b. Note that the magnitude of relative velocity before and after the planetary flyby is the same; only its direction is changed by the flyby.

Such a planetary flyby maneuver is often called a *gravity-assist* or *gravitational slingshot* maneuver. One of the most recent space exploration missions that utilized such a gravity-assist maneuver to reach a target planet is the Galileo spacecraft that was sent to Jupiter. In October 1989, after a three-year delay, the Galileo spacecraft was launched from a Space Shuttle parking orbit toward Jupiter, using an inertial

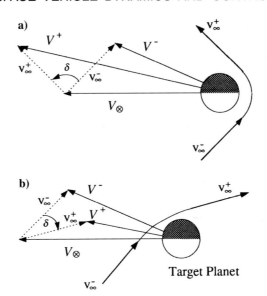

Fig. 4.8   Planetary flybys: a) trailing-side flyby and b) leading-side flyby.

upper stage (IUS). Changing from the originally planned use of a liquid Centaur upper stage to a less powerful solid IUS (because of a safety concern after the Shuttle Challenger accident) has resulted in a new trajectory. The Galileo spacecraft took about six years of travel time to reach Jupiter after performing one gravity-assist maneuver at Venus and two such maneuvers at Earth.

4.5.4 Planetary Capture

Some interplanetary missions require a circularization about the target planet, instead of a soft landing or a flyby. If such a planetary capture maneuver is performed at periapsis, then the required retro velocity is simply determined as

$$\Delta v = \sqrt{v_\infty^2 + (2\mu_\otimes/r_p)} - \sqrt{\mu_\otimes/r_p} \qquad (4.75)$$

where $\mu_\otimes$ is the gravitational parameter of the target planet, r_p is the periapsis distance, and v_∞ is the hyperbolic approach velocity of a spacecraft relative to the target planet.

If the final circular orbit radius is not specified and the required Δv, i.e., amount of propellant, is to be minimized, then the minimum value of Δv and the associated r_p can be found as follows. Taking the partial derivative of Eq. (4.75) with respect to r_p and setting it equal to zero, we obtain

$$r_p = 2\mu_\otimes/v_\infty^2 \qquad (4.76)$$

and the minimum value of Δv

$$\Delta v_{\min} = v_\infty/\sqrt{2} \qquad (4.77)$$

Note that a soft landing requires the maximum value of Δv.

For further details of space mission design and orbital applications, the reader is referred to Refs. 10–12.

4.6 Orbital Rendezvous

Consider a problem of describing the relative motion of a chase vehicle with respect to a target vehicle that is in a circular orbit. Such a rendezvous or proximity operation problem of two space vehicles is of primary importance to many current and planned space missions. The linearized equations of motion of a chase vehicle that is in close proximity to a target vehicle are derived in this section. Such equations are sometimes called the *Clohessy–Wiltshire equations* or *Hill's equations*. These relative motion equations were first studied by Hill in 1878 for a somewhat different purpose, and further investigated by Clohessy and Wiltshire [13] in 1960 for practical purposes.

As illustrated in Fig. 4.9, a target vehicle is in a circular orbit of radius R_o with an orbital rate of $n = \sqrt{\mu_\oplus / R_o^3}$. A reference frame with (x, y, z) coordinates and with basis vectors $\{\vec{i}, \vec{j}, \vec{k}\}$ is fixed at the center of mass of the target vehicle and rotates with an angular velocity of $\vec{\omega} = n\vec{k}$ with respect to an inertial reference frame. The x axis is along the radial direction, the y axis is along the flight direction of the target orbit, and the z axis is out of the orbit plane and completes a right-handed reference frame.

The position vector of the target vehicle from the center of the Earth is expressed as

$$\vec{R}_o = R_o \vec{i} \tag{4.78}$$

and the relative position vector of the chase vehicle from the target vehicle is also expressed as

$$\vec{r} = x\vec{i} + y\vec{j} + z\vec{k} \tag{4.79}$$

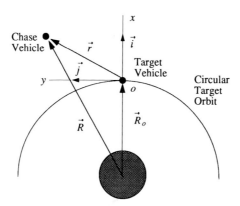

Fig. 4.9 Relative motion between chase and target vehicles.

The position vector of the chase vehicle from the center of the Earth is then given by

$$\vec{R} = \vec{R}_o + \vec{r} = (R_o + x)\vec{i} + y\vec{j} + z\vec{k} \tag{4.80}$$

and the inertial acceleration of the chase vehicle can be found as

$$\ddot{\vec{R}} = [\ddot{x} - 2n\dot{y} - n^2(R_o + x)]\vec{i} + (\ddot{y} + 2n\dot{x} - n^2 y)\vec{j} + \ddot{z}\vec{k} \tag{4.81}$$

The gravitational acceleration of the chase vehicle is

$$\vec{g} = -\frac{\mu_\oplus}{R^3}\vec{R} = -\frac{\mu_\oplus[(R_o + x)\vec{i} + y\vec{j} + z\vec{k}]}{[(R_o + x)^2 + y^2 + z^2]^{\frac{3}{2}}} \tag{4.82}$$

The equation of motion of the chase vehicle is then simply given by Newton's second law

$$m\ddot{\vec{R}} = m\vec{g} \quad \text{or} \quad \ddot{\vec{R}} = \vec{g} \tag{4.83}$$

where m is the mass of the chase vehicle.

For small relative motion with $x^2 + y^2 + z^2 \approx 0$, the denominator of Eq. (4.82) becomes

$$[(R_o + x)^2 + y^2 + z^2]^{\frac{3}{2}} \approx [R_o^2 + 2R_o x]^{\frac{3}{2}} = R_o^3[1 + (2x/R_o)]^{\frac{3}{2}} \tag{4.84}$$

As a result, the gravitational acceleration given by Eq. (4.82) becomes

$$\vec{g} \approx -(\mu_\oplus/R_o^3)(1 + 2x/R_o)^{-\frac{3}{2}}[(R_o + x)\vec{i} + y\vec{j} + z\vec{k}] \tag{4.85}$$

which can be further approximated, using the generalized binomial theorem, as follows:

$$\vec{g} \approx -(\mu_\oplus/R_o^3)[1 - (3x/R_o)][(R_o + x)\vec{i} + y\vec{j} + z\vec{k}] \tag{4.86}$$

Again neglecting the second-order terms such as x^2, xy, and xz for small relative motion, we have

$$\vec{g} \approx -n^2[(R_o - 2x)\vec{i} + y\vec{j} + z\vec{k}] \tag{4.87}$$

where $n = \sqrt{\mu_\oplus/R_o^3}$.

Finally, a set of linear differential equations, often called the Clohessy–Wiltshire equations, which describe the small relative motion of the chase vehicle with respect the target vehicle, can be found as

$$\ddot{x} - 2n\dot{y} - 3n^2 x = 0 \tag{4.88}$$

$$\ddot{y} + 2n\dot{x} = 0 \tag{4.89}$$

$$\ddot{z} + n^2 z = 0 \tag{4.90}$$

Note that for these equations, the y-axis displacement does not have to be small if it is measured circumferentially, as follows:

$$y = R_o\theta \tag{4.91}$$

The out-of-plane equation (4.90) is decoupled from the in-plane equations (4.88) and (4.89), and its solution is given by

$$\begin{bmatrix} z(t) \\ \dot{z}(t) \end{bmatrix} = \begin{bmatrix} \cos nt & \sin nt/n \\ -n\sin nt & \cos nt \end{bmatrix} \begin{bmatrix} z_0 \\ \dot{z}_0 \end{bmatrix} \tag{4.92}$$

where z_0 and $\dot{z}_0$ are initial conditions at $t = 0$.

Equation (4.89) is integrated to yield

$$\dot{y}(t) = -2nx(t) + 2nx_0 + \dot{y}_0 \tag{4.93}$$

where x_0 and $\dot{y}_0$ are initial conditions at $t = 0$. Substituting this equation into Eq. (4.88), we obtain

$$\ddot{x} + n^2x = 4n^2x_0 + 2n\dot{y}_0 \tag{4.94}$$

and its solution can be found as

$$x(t) = -\left(\frac{2\dot{y}_0}{n} + 3x_0\right)\cos nt + \frac{\dot{x}_0}{n}\sin nt + \left(4x_0 + \frac{2\dot{y}_0}{n}\right) \tag{4.95}$$

The complete solution of the in-plane motion is then represented in matrix form, as follows:

$$\begin{bmatrix} x(t) \\ y(t) \\ \dot{x}(t) \\ \dot{y}(t) \end{bmatrix} = \begin{bmatrix} 4 - 3\cos nt & 0 & \sin nt/n & 2(1 - \cos nt)/n \\ 6\sin nt - 6nt & 1 & 2(-1 + \cos nt)/n & 4\sin nt/n - 3t \\ 3n\sin nt & 0 & \cos nt & 2\sin nt \\ 6n(-1 + \cos nt) & 0 & -2\sin nt & -3 + 4\cos nt \end{bmatrix} \begin{bmatrix} x_0 \\ y_0 \\ \dot{x}_0 \\ \dot{y}_0 \end{bmatrix} \tag{4.96}$$

These solutions, as well as the solution of the out-of-plane motion, can be used for many practical applications, such as orbital rendezvous and proximity operations of two space vehicles. Given the initial position and velocity components of the chase vehicle, one can easily determine the position and velocity components of the chase vehicle relative to the target vehicle at a specified time t from these equations.

If the chase vehicle with initial position components x_0^-, y_0^-, and z_0^- relative to the target vehicle is required to rendezvous with the target vehicle at a specified time t, then Eqs. (4.92) and (4.96) can be used to determine a proper set of initial velocity components $\dot{x}_0^+$, $\dot{y}_0^+$, and $\dot{z}_0^+$ for a Δv maneuver at $t = 0$. (The superscripts $-$ and $+$ denote just before and after an impulsive maneuver at $t = 0$, respectively.)

For example, consider a two-impulse rendezvous maneuver where the first impulse maneuver provides a proper initial velocity vector to rendezvous with the target vehicle at a specified time t and the second impulse maneuver is needed to stop the chase vehicle at the target. Given the initial position components x_0^-, y_0^-, z_0^- and the desired final position $x = y = z = 0$ at a specified time t, the required velocity components at $t = 0^+$ can be found as

$$\dot{x}_0^+ = \frac{x_0^- n(4 \sin nt - 3nt \cos nt) - 2y_0^- n(1 - \cos nt)}{3nt \sin nt - 8(1 - \cos nt)} \tag{4.97a}$$

$$\dot{y}_0^+ = \frac{-x_0^- n[6nt \sin nt - 14(1 - \cos nt)] + y_0^- n \sin nt}{3nt \sin nt - 8(1 - \cos nt)} \tag{4.97b}$$

$$\dot{z}_0^+ = \frac{-z_0^- n}{\tan nt} \tag{4.97c}$$

The velocity change $\Delta \vec{v}_1$ required for the first impulse maneuver is then given by

$$\Delta \vec{v}_1 = \left(\dot{x}_0^+ - \dot{x}_0^-\right)\vec{i} + \left(\dot{y}_0^+ - \dot{y}_0^-\right)\vec{j} + \left(\dot{z}_0^+ - \dot{z}_0^-\right)\vec{k} \tag{4.98}$$

where $\dot{x}_0^-$, $\dot{y}_0^-$, and $\dot{z}_0^-$ are the actual initial velocity components of the chase vehicle relative to the target vehicle at $t = 0^-$, i.e., just before the impulsive maneuver, and we have

$$\Delta v_1 \equiv |\Delta \vec{v}_1| = \sqrt{\left(\dot{x}_0^+ - \dot{x}_0^-\right)^2 + \left(\dot{y}_0^+ - \dot{y}_0^-\right)^2 + \left(\dot{z}_0^+ - \dot{z}_0^-\right)^2} \tag{4.99}$$

The second velocity change $\Delta \vec{v}_2$ needed to stop the chase vehicle at a specified time t is also given by

$$\Delta \vec{v}_2 = -\dot{x}(t)\vec{i} - \dot{y}(t)\vec{j} - \dot{z}(t)\vec{k} \tag{4.100}$$

and

$$\Delta v_2 \equiv |\Delta \vec{v}_2| = \sqrt{\dot{x}^2(t) + \dot{y}^2(t) + \dot{z}^2(t)} \tag{4.101}$$

where $\dot{x}(t)$, $\dot{y}(t)$, and $\dot{z}(t)$ are the velocity components of the target vehicle at a specified time t resulting from the initial position and velocity components: $x_0^+ = x_0^-$, $y_0^+ = y_0^-$, $z_0^+ = z_0^-$, $\dot{x}_0^+$, $\dot{y}_0^+$, and $\dot{z}_0^+$, just after the impulsive maneuver Δv_1 at $t = 0$.

There exists a *standoff position* that does not require continuous thrusting to maintain the relative position. Such a situation is possible if the chase vehicle has zero radial and cross-track (out-of-plane) position components and zero relative velocity, in other words, if the chase vehicle is in the target orbit, but with a nonzero value of the in-track position $y(t)$.

A chase vehicle located at such a standoff position in a target orbit with a nonzero value of the in-track position y_0^- will move along a trajectory about the target vehicle if a radial velocity impulse $\dot{x}_0^+$ is imparted to the chase vehicle. The resulting *elliptical flyaround trajectory* is described as follows:

$$x(t) = \left(\dot{x}_0^+/n\right) \sin nt \tag{4.102a}$$

$$y(t) = y_0^- - \left(2\dot{x}_0^+/n\right)(1 - \cos nt) \tag{4.102b}$$

which can be combined to yield

$$x^2 + \frac{1}{4}\left(y - y_0^- + \frac{2\dot{x}_0^+}{n}\right)^2 = \left(\frac{\dot{x}_0^+}{n}\right)^2 \tag{4.103}$$

Note that the target vehicle is located at the origin: $x = y = z = 0$.

4.7 Halo Orbit Determination and Control

The circular and elliptic restricted three-body problems were studied in Secs. 3.7 and 3.8, respectively. An interesting feature of the circular restricted three-body problem is the existence of equilibrium points, called libration or Lagrangian points. These are points where the gravitational and centrifugal forces acting on the third body cancel each other.

The existence of periodic orbits about the libration points has been known for many years. In the late 1960s and early 1970s, Farquhar [14,15] proposed the use of lunar libration points and orbits about libration points for lunar far-side communications. As illustrated in Fig. 4.10, it may be desirable to maintain a 3500-km halo orbit about the translunar L_2 point. A spacecraft following this trajectory will always lie within the 11.1-deg beamwidth of a fixed lunar-surface antenna even when the latitudinal and longitudinal oscillations of the moon are taken into account. If the halo orbit amplitude is increased to 3700 km, the spacecraft would also be visible from the cislunar L_1 point. Therefore, if another spacecraft were stationed at the L_1 point there could be continuous communications between the Earth and most of the lunar surface.

In this section, which is based on Refs. 16 and 17, such a halo orbit determination and control problem is considered first for the circular restricted Earth–moon system. Later in this section, the effects of lunar orbital eccentricity are also considered. Solar gravitation and radiation effects could be included in a more accurate model; however, these effects are smaller than the elliptic and nonlinear effects for a spacecraft in a 3500-km orbit about the translunar libration point.

For the purpose of preliminary mission analyses, it is assumed that all of the states can be measured for feedback control and that proportional control accelerations can be produced by pulse-modulated thrusters. The stationkeeping control

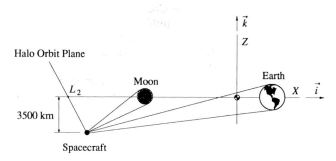

Fig. 4.10   Halo orbit control problem of the Earth–moon–spacecraft system.

problem of practical concern, including the effects of tracking and propulsion errors, is not considered here. Other practical issues regarding the use of constant low-thrust jets are also not considered. We are mainly concerned here with the computation of a fuel-efficient trajectory about a libration point using a disturbance accommodating control approach.

4.7.1 Linear State-Feedback Control

Defining u_x, u_y, and u_z as the control acceleration components along the X, Y, and Z axes, respectively, we can obtain nondimensionalized, nonlinear equations of motion of the spacecraft as

$$\ddot{X} - 2\dot{Y} - X = -\frac{(1-\rho)(X-\rho)}{r_1^3} - \frac{\rho(X+1-\rho)}{r_2^3} + u_x \qquad (4.104a)$$

$$\ddot{Y} + 2\dot{X} - Y = -\frac{(1-\rho)Y}{r_1^3} - \frac{\rho Y}{r_2^3} + u_y \qquad (4.104b)$$

$$\ddot{Z} = -\frac{(1-\rho)Z}{r_1^3} - \frac{\rho Z}{r_2^3} + u_z \qquad (4.104c)$$

where

$$X = X_o + x, \qquad Y = Y_o + y; \qquad Z = Z_o + z$$

$$r_1 = \sqrt{(X-\rho)^2 + Y^2 + Z^2}$$

$$r_2 = \sqrt{(X+1-\rho)^2 + Y^2 + Z^2}$$

(X_o, Y_o, Z_o) are the coordinates of the libration point, and (x, y, z) are the components of the position vector of the spacecraft relative to the libration point. (See Sec. 3.7 for a detailed description of the circular restricted three-body model.)

The nondimensional linearized equations of motion in terms of x, y, and z can be derived in state space form as

$$\dot{x} = Ax + Bu \qquad (4.105)$$

where $x = (x,\ y,\ z,\ \dot{x},\ \dot{y},\ \dot{z})$, $u = (u_x, u_y, u_z)$, and

$$A = \begin{bmatrix} 0 & 0 & 0 & 1 & 0 & 0 \\ 0 & 0 & 0 & 0 & 1 & 0 \\ 0 & 0 & 0 & 0 & 0 & 1 \\ U_{XX} & U_{XY} & 0 & 0 & 2 & 0 \\ U_{YX} & U_{YY} & 0 & -2 & 0 & 0 \\ 0 & 0 & U_{ZZ} & 0 & 0 & 0 \end{bmatrix}, \qquad B = \begin{bmatrix} 0 & 0 & 0 \\ 0 & 0 & 0 \\ 0 & 0 & 0 \\ 1 & 0 & 0 \\ 0 & 1 & 0 \\ 0 & 0 & 1 \end{bmatrix} \qquad (4.106)$$

For the translunar L_2 libration point in the Earth–moon system with

$$U_{XX} = 7.3809, \qquad U_{YY} = -2.1904, \qquad U_{ZZ} = -3.1904$$

the eigenvalues can be found as

$$\lambda_{1,2} = \pm 2.15868$$
$$\lambda_{3,4} = \pm j\omega_{xy} = \pm j1.86265$$
$$\lambda_{5,6} = \pm j\omega_z = \pm j1.78618$$

As discussed in Sec. 3.7, a Lissajous reference trajectory is a force-free solution to the linearized equations of motion with a proper set of initial conditions and is described by

$$x_r(t) = -A_x \sin \omega_{xy}t$$
$$y_r(t) = -A_y \cos \omega_{xy}t$$
$$z_r(t) = A_z \sin \omega_z t$$
$$\dot{x}_r(t) = -A_x \omega_{xy} \cos \omega_{xy}t$$
$$\dot{y}_r(t) = A_y \omega_{xy} \sin \omega_{xy}t$$
$$\dot{z}_r(t) = A_z \omega_z \cos \omega_z t$$

where A_x, A_y, and A_z denote the amplitudes of the reference trajectory for the x, y, and z axes, respectively.

Given a linear model of the system described by Eq. (4.105), we can consider a linear state-feedback controller of the form

$$\mathbf{u} = -\mathbf{K}(\mathbf{x} - \mathbf{x}_r) \tag{4.107}$$

where $\mathbf{K}$ is the gain matrix to be properly determined; i.e., we have

$$\begin{bmatrix} u_x \\ u_y \\ u_z \end{bmatrix} = - \begin{bmatrix} K_{xx} & K_{xy} & K_{xz} & K_{x\dot{x}} & K_{x\dot{y}} & K_{x\dot{z}} \\ K_{yx} & K_{yy} & K_{yz} & K_{y\dot{x}} & K_{y\dot{y}} & K_{y\dot{z}} \\ K_{zx} & K_{zy} & K_{zz} & K_{z\dot{x}} & K_{z\dot{y}} & K_{z\dot{z}} \end{bmatrix} \begin{bmatrix} x - x_r \\ y - y_r \\ z - z_r \\ \dot{x} - \dot{x}_r \\ \dot{y} - \dot{y}_r \\ \dot{z} - \dot{z}_r \end{bmatrix} \tag{4.108}$$

Using the LQR method described in Chapter 2, one may find the following gain matrix:

$$\mathbf{K} = \begin{bmatrix} 39.016 & -7.154 & 0 & 9.876 & 0.122 & 0 \\ 7.329 & 28.69 & 0 & 0.122 & 8.823 & 0 \\ 0 & 0 & 28.593 & 0 & 0 & 8.786 \end{bmatrix}$$

Notice the decoupled nature of the z-axis control from the coupled x- and y-axes control.

Nonlinear simulation results for the control acceleration components needed to maintain a 3500-km Lissajous reference trajectory are given in Fig. 4.11, in which the circular restricted three-body system with a linear state-feedback controller, along with the preceding gain matrix, is used for closed-loop nonlinear simulation.

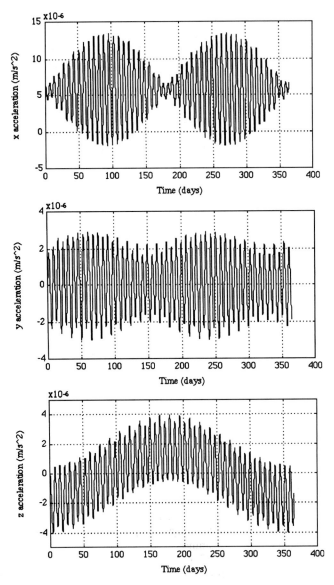

Fig. 4.11   Control acceleration inputs needed for maintaining a 3500-km Lissajous reference trajectory.

It is important to note that such a cyclic nature of the control inputs is a result of the nonlinear dynamic effects neglected in the derivation of the Lissajous reference trajectory. The actual trajectory is very close to the commanded reference trajectory shown in Fig. 3.11 of Sec. 3.7, and thus not included here.

In orbital mechanics, ΔV is often used as a criteria to evaluate the performance of orbit control. The fuel needed for orbit control cannot be determined unless the mass of a spacecraft is known. Therefore, it is common practice to use the total velocity change ΔV as a measure of performance. The ΔV can be found by integrating the control acceleration inputs with respect to time. The units of ΔV to be used here are meters per second (m/s). A spacecraft would require approximately 285 (m/s)/year to maintain the 3500-km Lissajous trajectory, and it would require approximately 375 (m/s)/year to maintain a 3500-km halo reference orbit shown in Fig. 3.12 of Sec. 3.7. This increase in ΔV for the halo orbit is attributed to period control, which is required to produce equal in-plane and out-of-plane frequencies. The average ΔV for a geosynchronous satellite is about 50 (m/s)/year.

4.7.2   Disturbance Accommodating Control

The accuracy of the reference trajectory is critical in reducing the ΔV needed to maintain a quasiperiodic orbit in the restricted three-body problem. An important area of research pertaining to the restricted three-body problem involves the determination of reference trajectories that are closer to a periodic solution of the nonlinear equations of motion. For example, a third-order analytic solution for a quasiperiodic Lissajous trajectory has been derived by Farquhar and Kamel [18]. This solution takes into account nonlinearities, lunar orbital eccentricity, and effects from solar gravitation and radiation.

The persistent disturbance-rejection (or disturbance-accommodation) technique, described in Chapter 2, is another method that can be employed to reduce the ΔV for the halo orbit control problem. As first demonstrated by Hoffman [19], disturbance accommodation can be used to eliminate the control acceleration that results from forcing the spacecraft to follow the reference trajectory derived with the linearized equations of motion. The elimination of this control acceleration allows the spacecraft to follow a trajectory that is closer to a periodic solution of the nonlinear equations of motion.

The ΔV required to maintain the reference quasiperiodic orbit is large as a result of the neglected nonlinear terms in the equations of motion. Such nonlinear effects can be considered constant and periodic persistent disturbances, which are functions of the position, velocity, and acceleration of the spacecraft to be controlled. They are therefore termed *trajectory-dependent disturbances*. An iterative method for designing a disturbance-accommodating controller is briefly introduced here. The method allows the spacecraft to deviate from the reference trajectory and follow a trajectory that requires substantially less ΔV. The resulting deviation from the reference trajectory is, however, relatively small compared to the size of the orbit itself.

The analytic series solutions of Farquhar and Kamel [18], contain bias, sine, and cosine terms with frequencies similar to those included in the disturbance-accommodating controller. This should be expected because the disturbance-accommodating controller allows the spacecraft to follow a trajectory that is closer to a periodic solution of the nonlinear equations of motion. Therefore, the iterative design of a disturbance-accommodating controller and an analytic solution should produce similar results.

The libration points are at fixed positions in the rotating reference frame because it is assumed that the primaries move in circular orbits. Therefore, the position of the spacecraft following a reference trajectory can be written as

$$X_r = X_o + x_r, \qquad Y_r = Y_o + y_r, \qquad Z_r = Z_o + z_r \qquad (4.109)$$

where X_o, Y_o, and Z_o are the coordinates of a libration point in the rotating reference frame. The control acceleration needed to maintain a reference trajectory is determined by solving the nondimensional equations of motion for the control acceleration, as follows:

$$u_x = \ddot{X}_r - 2\dot{Y}_r - X_r + \frac{(1 - \rho)(X_r - \rho)}{r_1^3} + \frac{\rho(X_r + 1 - \rho)}{r_2^3} \qquad (4.110a)$$

$$u_y = \ddot{Y}_r + 2\dot{X}_r - Y_r + \frac{(1 - \rho)Y_r}{r_1^3} + \frac{\rho Y_r}{r_2^3} \qquad (4.110b)$$

$$u_z = \ddot{Z}_r + \frac{(1 - \rho)Z_r}{r_1^3} + \frac{\rho Z_r}{r_2^3} \qquad (4.110c)$$

where

$$r_1 = \sqrt{(X_r - \rho)^2 + Y_r^2 + Z_r^2}$$

$$r_2 = \sqrt{(X_r + 1 - \rho)^2 + Y_r^2 + Z_r^2}$$

This is an exact expression for the control acceleration needed to maintain a reference trajectory about one of the libration points. If the reference trajectory is a periodic solution of the nonlinear equations of motion with a proper set of initial conditions, then the control acceleration components (u_x, u_y, u_z) will be equal to zero.

For the collinear libration points with $Y_o = Z_o = 0$, Eqs. (4.110) become

$$u_x = \ddot{x}_r - 2\dot{y}_r - (X_o + x_r) + (1 - \rho)(X_o + x_r - \rho)r_1^{-3}$$
$$+ \rho(X_o + x_r + 1 - \rho)r_2^{-3} \qquad (4.111a)$$

$$u_y = \ddot{y}_r + 2\dot{x}_r - y_r + (1 - \rho)y_r r_1^{-3} + \rho y_r r_2^{-3} \qquad (4.111b)$$

$$u_z = \ddot{z}_r + (1 - \rho)z_r r_1^{-3} + \rho z_r r_2^{-3} \qquad (4.111c)$$

where

$$r_1^{-3} = \left[(X_o + x_r - \rho)^2 + y_r^2 + z_r^2\right]^{-\frac{3}{2}}$$

$$r_2^{-3} = \left[(X_o + x_r + 1 - \rho)^2 + y_r^2 + z_r^2\right]^{-\frac{3}{2}}$$

and the reference trajectory is described by

$$x_r(t) = -A_x \sin \omega_{xy} t \tag{4.111d}$$

$$y_r(t) = -A_y \cos \omega_{xy} t \tag{4.111e}$$

$$z_r(t) = A_z \sin \omega_z t \tag{4.111f}$$

$$\dot{x}_r(t) = -A_x \omega_{xy} \cos \omega_{xy} t \tag{4.111g}$$

$$\dot{y}_r(t) = A_y \omega_{xy} \sin \omega_{xy} t \tag{4.111h}$$

$$\dot{z}_r(t) = A_z \omega_z \cos \omega_z t \tag{4.111i}$$

Using Eqs. (4.111), one can determine numerically the spectral components of the control acceleration needed to maintain a given reference trajectory. The spectral components represent the nonlinear effects neglected in the derivation of the reference trajectory.

After determining the spectral components of the control acceleration, one can design the periodic-disturbance accommodation filters of the form

$$\ddot{\alpha}_i + (\omega_{xi})^2 \alpha_i = u_x$$

$$\ddot{\beta}_i + (\omega_{yi})^2 \beta_i = u_y$$

$$\ddot{\gamma}_i + (\omega_{zi})^2 \gamma_i = u_z$$

where ω_{xi}, ω_{yi}, and ω_{zi} are the ith frequency components in each axis. The constant-disturbance accommodation filters can also be included as

$$\dot{\tau}_x = u_x$$

$$\dot{\tau}_y = u_y$$

$$\dot{\tau}_z = u_z$$

where τ_x, τ_y, and τ_z are the filter states necessary to eliminate any bias components of the control acceleration inputs (u_x, u_y, u_z).

The disturbance filter can include as many frequencies as the given persistent disturbance model and is driven by the control inputs. The disturbance-accommodation filter is then described in state-space form as

$$\dot{\mathbf{x}}_d = \mathbf{A}_d \mathbf{x}_d + \mathbf{B}_d \mathbf{u} \tag{4.112}$$

where $\mathbf{x}_d$ is the disturbance filter state vector. The disturbance filter described by Eq. (4.112) can then be augmented to the plant described by Eq. (4.105), as follows:

$$\begin{bmatrix} \dot{\mathbf{x}} \\ \dot{\mathbf{x}}_d \end{bmatrix} = \begin{bmatrix} \mathbf{A} & 0 \\ 0 & \mathbf{A}_d \end{bmatrix} \begin{bmatrix} \mathbf{x} \\ \mathbf{x}_d \end{bmatrix} + \begin{bmatrix} \mathbf{B} \\ \mathbf{B}_d \end{bmatrix} \mathbf{u} \tag{4.113}$$

The standard linear quadratic regulator (LQR) design technique can be similarly applied to the augmented system described by Eq. (4.113) for the design of a linear state-feedback controller of the form

$$\mathbf{u} = -\mathbf{K} \begin{bmatrix} \mathbf{x} - \mathbf{x}_r \\ \mathbf{x}_d \end{bmatrix}$$

One can obtain a disturbance-accommodating controller that gives a considerable reduction in the control acceleration components, which results in a ΔV of about 10 (m/s)/year, as demonstrated by Cielaszyk and Wie [16] and Hoffman [19].

4.7.3 Large Lissajous Trajectory

As discussed in Sec. 3.7, period control may be needed to guarantee no periods of lunar occultation, at the expense of a large ΔV needed to maintain a halo reference orbit. If this constraint is relaxed, and small periods of lunar occultation are allowed, a Lissajous trajectory may be used to provide a lunar far-side communications. With a Lissajous trajectory, the percentage of time the spacecraft is visible from Earth increases with an increase in orbit size. As the size of the orbit is increased, however, the amount of lunar surface visible to the spacecraft decreases. There is, therefore, a tradeoff between visibility of the lunar surface and visibility to an Earth observer.

A Lissajous reference trajectory with an amplitude of 30,000 km, studied by Howell and Pernicka [20], is considered here to illustrate the use of disturbance accommodation as a method of determining a trajectory that is a solution of the nonlinear equations of motion.

The large orbit size results in the spacecraft being far away from the linear domain about the libration point; however, the solution of the linearized equations of motion can be still used as a starting point for the determination of a trajectory that is closer to a solution of the nonlinear equations of motion. As a result of the spacecraft being far away from the linear domain, there is a large number of disturbance frequencies required to accommodate the nonlinear effects. This application of disturbance accommodation is not meant to provide an implementable control logic; it demonstrates that disturbance accommodation can be used as a simple method of numerically determining a quasi-periodic solution of the nonlinear equations of motion.

The results of an iterative application of disturbance accommodation are presented in Table 4.2. The resulting 30,000-km Lissajous trajectory is also shown in Fig. 4.12.

A trajectory similar to that shown in Fig. 4.12 was found by Howell and Pernicka [20], and they used a third-order analytic solution to identify target positions at specified intervals along the trajectory. A path with velocity discontinuities is located passing through the target points. The velocity discontinuities are simultaneously reduced in an iterative process by allowing the continuous path to deviate from the target positions. The ΔV after this iterative process is essentially zero. The ΔV associated with the disturbance-accommodation trajectories is much larger than the ΔV quoted by Howell and Pernicka [20]. One obvious reason is that with disturbance-accommodation the control is continuous; therefore, control acceleration is being applied constantly as opposed to only at selected target points. Despite the difference in ΔV, the iterative design method is supported by the fact that linear state-feedback with a disturbance-accommodation filter can be used to determine a trajectory with such a unique shape and similarity to trajectories determined by Howell and Pernicka [20].

Table 4.2   Design iterations for fuel-efficient, large Lissajous trajectory

Iteration	x axis	y axis	z axis	ΔV, (m/s)/year
0				18,517
1	0	0	0	14,652
2	$2\omega_z$			9,010
	$2\omega_{xy}$			
3	$2\omega_z + \omega_{xy}$	$2\omega_z$	$\omega_{xy} + \omega_z$	3,036
	$3\omega_{xy}$			
4	ω_{xy}	$2\omega_z - \omega_{xy}$	$2\omega_{xy} - \omega_z$	2,233
	$2\omega_{xy} + 2\omega_z$	$2\omega_z + \omega_{xy}$	$2\omega_{xy} + \omega_z$	
	$4\omega_{xy}$			
5	$2\omega_{xy} - 2\omega_z$	$2\omega_{xy} - 2\omega_z$	$\omega_{xy} - \omega_z$	1,427
	$2\omega_z - \omega_{xy}$	$2\omega_{xy} + 2\omega_z$	$3\omega_z + \omega_{xy}$	
6	$4\omega_z + \omega_{xy}$			1,267
	$2\omega_z + 3\omega_{xy}$			
7	$4\omega_z - 2\omega_{xy}$	$2\omega_{xy}$	ω_z	466
	$4\omega_z - \omega_{xy}$	$4\omega_z + \omega_{xy}$	$3\omega_z - \omega_{xy}$	
	$4\omega_z$	$2\omega_z + 3\omega_{xy}$	$3\omega_{xy} - \omega_z$	
			$3\omega_z$	
8		$3\omega_{xy}$		362
9	$2\omega_z + 4\omega_{xy}$	$4\omega_{xy} - 2\omega_z$	$3\omega_z + 2\omega_{xy}$	293
10	$3\omega_{xy} - 2\omega_z$	ω_{xy}	$3\omega_z - 2\omega_{xy}$	162
		$3\omega_{xy} - 2\omega_z$		
		$4\omega_{xy}$		
		$2\omega_z + 4\omega_{xy}$		
11	$4\omega_{xy} - 2\omega_z$			152
12		$4\omega_z - \omega_{xy}$		143
		$4\omega_z$		

4.7.4   Elliptic Restricted Three-Body Problem

It is important to note that the effects from eccentricity are much larger than the effects from nonlinearities for a typical 3500-km halo orbit. There are considerably more periodic disturbances affecting the spacecraft when the elliptic restricted three-body model is used. The accommodation of nonlinear and eccentric effects results in a large number of frequencies being required in the disturbance-accommodation filter.

Table 4.3 contains the results of a halo orbit design and the ΔV associated with each iteration step. Therefore, the halo orbit requires considerably more ΔV than the Lissajous trajectory even after disturbance accommodation has been applied. A halo orbit after three iterations is shown in Fig. 4.13. The control acceleration components needed to maintain such a halo orbit are also shown in Fig. 4.14. It can be seen that the z-axis control acceleration u_z has a frequency component of ω_{xy} for period control needed to maintain a halo orbit, whereas the

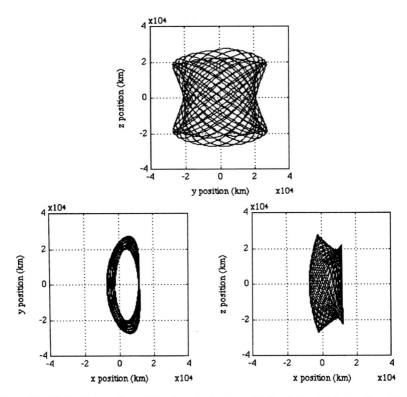

Fig. 4.12   Fuel-efficient, large Lissajous trajectory obtained after 12 design iterations.

x- and y-axes control accelerations become near zero at steady state as a result of disturbance-accommodating control.

4.7.5   International Sun–Earth Explorer-3 Spacecraft

We now apply the disturbance-accommodation approach to the halo orbit control problem of the International Sun–Earth Explorer-3 (ISEE-3) spacecraft discussed in Refs. 21 and 22. This problem is of interest for a couple of reasons.

Table 4.3   Design iterations for a fuel-efficient halo orbit

Iteration	x axis	y axis	z axis	ΔV, (m/s)/year
0				548
1	0	0	0	401.4
2	$\omega_{xy} - 1$	$\omega_{xy} - 1$	$\omega_{xy} - 1$	196.7
	$\omega_{xy} + 1$	$\omega_{xy} + 1$	$\omega_{xy} + 1$	
3	$2\omega_{xy}$	$2\omega_{xy}$	$2\omega_{xy}$	160

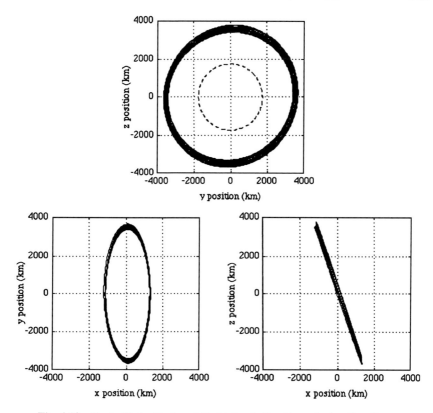

Fig. 4.13   Fuel-efficient halo orbit obtained after three design iterations.

First, this mission was actually flown in the late 1970s and early 1980s, and there is actual flight information available for comparison. Second, this represents what is known as a large halo orbit in the literature. This type of halo orbit utilizes the nonlinear effects to naturally create equal in-plane and out-of-plane frequencies. Thus, the large cost for period control found for the small halo orbit about the translunar libration point is absent. Such a halo orbit control problem about the L_1 point of the sun–Earth–moon elliptic restricted three-body system is illustrated in Fig. 4.15.

To use the sun-Earth restricted three-body system as a model for the ISEE-3 mission, the moon's gravitational effect would need to be neglected. It has been known that it is more accurate to combine the mass of the moon with that of the Earth. This results in the collinear libration points being located on a line connecting the sun and the barycenter of the Earth and moon. The gravitational perturbations due to the orbit of the moon about the Earth are neglected in this model. In addition, solar radiation pressure and planetary perturbations are of small significance and are neglected.

The equations of motion derived in Sec. 3.8 can also be used to represent motion relative to the L_1 point of the sun–Earth–moon elliptic restricted three-body system.

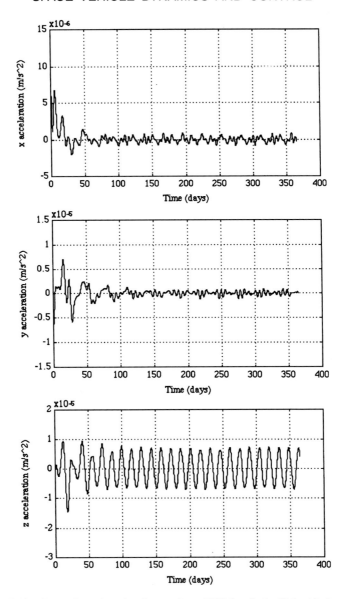

Fig. 4.14   Control acceleration inputs for a 3500-km fuel-efficient halo orbit.

As in Sec. 3.8, the time of perihelion passage of the Earth–moon barycenter is assumed to be zero. The math model describing the motion of the ISEE-3 spacecraft about the L_1 point in the sun–Earth–moon system has the following characteristics:

$$\rho = 3.040357143 \times 10^{-6}$$

$$a = 1 \text{ AU} = 1.495978714 \times 10^8 \text{ km}$$

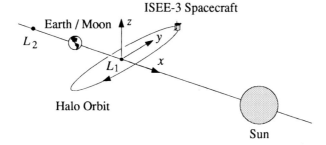

Fig. 4.15   ISEE-3 mission.

$$e = 0.01673$$

$$n = 1.990986606 \times 10^{-7} \text{ rad/s}$$

$$\gamma = -0.0100109$$

The reference trajectories derived using the linearized equations of motion are no longer accurate for an orbit with the amplitude of the ISEE-3 spacecraft. There were actually two reference orbits derived for this mission including an analytical solution and a differentially corrected solution. A standard Lissajous trajectory is unacceptable because the spacecraft will pass close to the L_1 point and the downlink telemetry will be overwhelmed by the intense solar noise background. In the planning for the ISEE-3 mission, it was found that a slight fuel savings could be made by following the numerically generated path. It should be noted that both reference orbits were derived with the use of the circular restricted three-body model, whereas closed-loop simulations need to be performed with the elliptic restricted three-body model.

The third-order analytic reference trajectory derived for the ISEE-3 mission by Richardson [22] is provided here,

$$x_r = a_{21}A_x^2 + a_{22}A_z^2 - A_x\cos(\lambda\tau + \phi) + (a_{23}A_x^2 - a_{24}A_z^2)\cos 2(\lambda\tau + \phi)$$
$$+ (a_{31}A_x^3 - a_{32}A_xA_z^2)\cos 3(\lambda\tau + \phi) \tag{4.114a}$$
$$y_r = A_y\sin(\lambda\tau + \phi) + (b_{21}A_x^2 - b_{22}A_z^2)\sin 2(\lambda\tau + \phi)$$
$$+ (b_{31}A_x^3 - b_{32}A_xA_z^2)\sin 3(\lambda\tau + \phi) \tag{4.114b}$$
$$z_r = -3d_{21}A_xA_z + A_z\cos(\lambda\tau + \phi) + d_{21}A_xA_z\cos 2(\lambda\tau + \phi)$$
$$+ (d_{32}A_zA_x^2 - d_{31}A_z^3)\cos 3(\lambda\tau + \phi) \tag{4.114c}$$

where λ denotes the nondimensional in-plane frequency ω_{xy}; A_x, $A_y \equiv kA_x$, and A_z are the amplitudes of the orbit in units of the distance between the Earth–moon barycenter and the L_1 libration point ($|\gamma a| = 1.49761 \times 10^6$ km); the phase angle ϕ determines the initial position in the orbit; τ is a new independent variable with a frequency correction ω such that

$$\tau = \omega t$$
$$\omega = 1 - 0.8246605235A_x^2 + 0.1210986087A_z^2$$

time t is in units of $1/n$; and the remaining constants in this third-order reference orbit are given as

$$\lambda = 2.086453455, \qquad\qquad k = 3.2292680962$$

$$a_{21} = 2.092695581, \qquad\qquad b_{21} = -4.924458751 \times 10^{-1}$$

$$a_{22} = 2.482976703 \times 10^{-1}, \qquad b_{22} = 6.074646717 \times 10^{-2}$$

$$a_{23} = -9.059647954 \times 10^{-1}, \quad b_{31} = 8.857007762 \times 10^{-1}$$

$$a_{24} = -1.044641164 \times 10^{-1}, \quad b_{32} = 3.980954252 \times 10^{-1}$$

$$a_{31} = 7.938201951 \times 10^{-1}, \quad d_{21} = -3.468654605 \times 10^{-1}$$

$$a_{32} = 8.268538529 \times 10^{-2}, \quad d_{31} = 1.904387005 \times 10^{-2}$$

$$d_{32} = 3.980954252 \times 10^{-1}$$

Figure 4.16 shows a third-order halo reference orbit of a period of 177.73 days with the following amplitudes:

$$A_x = 206,000 \text{ km}$$
$$A_y = 665,000 \text{ km}$$
$$A_z = 110,000 \text{ km}$$

The y–z projection is a view from the sun toward the Earth–moon barycenter. The x–y and x–z projections are top and side views, respectively. The Earth is in the negative x direction nearly 1,500,000 km from the origin of these plots. Initial conditions approximating the halo orbit injection conditions can be found by using $\phi = \pi$ and time equal to zero. The initial conditions correspond to maximum amplitude in the negative x direction, zero in the y direction, and maximum amplitude in the negative z direction. The direction of the orbit is counterclockwise in the y–z projection and clockwise in the x–y projection.

The linearized equations of motion about the collinear libration points contain a positive real eigenvalue and are unstable. Therefore, orbit control is required to maintain the orbit about the L_1 libration point in the sun–Earth–moon system. To illustrate the application of the iterative disturbance accommodating method to the ISEE-3 mission, it is assumed that the motion of the Earth–moon barycenter around the sun is circular. The location of the L_1 libration point is given as $X_o = -0.9900266$. This position is nearly 1,500,000 km away from the Earth. For

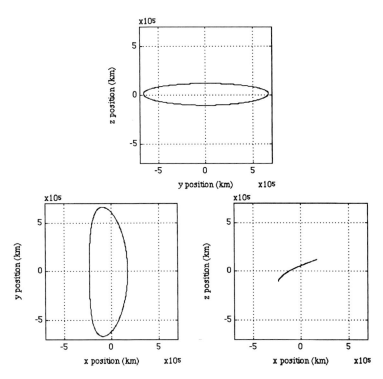

Fig. 4.16   ISEE-3 halo reference orbit.

Table 4.4   Design iterations for a fuel-efficient halo orbit of ISEE-3

Iteration	x axis	y axis	z axis	ΔV, (m/s)/year
0				146.6
1	0	0	0	158.4
2	ω_{xy} $4\omega_{xy}$	$\omega_{xy} - 1$ $\omega_{xy} + 1$		148.6
3	$\omega_{xy} - 1$ $2\omega_{xy} + 1$	$4\omega_{xy}$	$\omega_{xy} - 1$ $\omega_{xy} + 1$	94.3
4	$\omega_{xy} + 1$	$2\omega_{xy} + 1$	$4\omega_{xy}$	70.8
5	$2\omega_{xy}$	$2\omega_{xy}$		58.0
6	$3\omega_{xy} + 1$ $5\omega_{xy}$	$3\omega_{xy} + 1$ $5\omega_{xy}$		30.8
7	$3\omega_{xy}$	$3\omega_{xy}$		22.2
8	$2\omega_{xy} - 1$ $4\omega_{xy} + 1$	$4\omega_{xy} + 1$	$2\omega_{xy} + 1$	12.8
9	$6\omega_{xy}$	$6\omega_{xy}$		9.8
10	$5\omega_{xy} + 1$		$2\omega_{xy}$	8.3

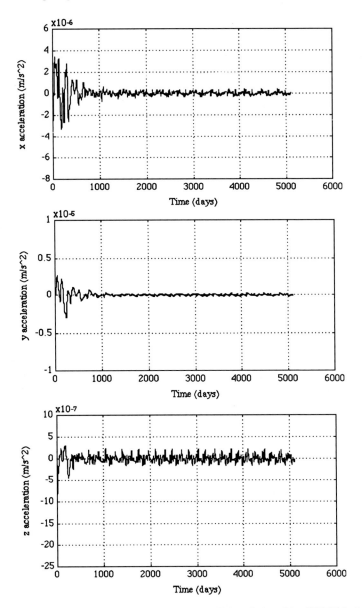

Fig. 4.17   Control acceleration inputs for a fuel-efficient halo orbit of ISEE-3 obtained after 10 iterations.

the L_1 point in the sun–Earth–moon system, the elements in the state matrix have the following values:

$$U_{XX} = 9.12214, \qquad U_{YY} = -3.06107,$$

$$U_{ZZ} = -4.06107, \qquad U_{XY} = U_{YX} = 0$$

Furthermore, the control acceleration is assumed to be continuous to allow the design of a linear state-feedback controller. The nondimensionalized natural frequencies of the sun–Earth–moon system are: $\omega_{xy} = 2.08645$ (period $= 174.3$ days) and $\omega_z = 2.01521$ (period $= 180.4$ days).

A linear state-feedback controller without disturbance accommodation for the halo reference orbit of Fig. 4.16 results in ΔV of approximately 146 (m/s)/year. It was estimated that the ISEE-3 mission would require approximately 10–15 (m/s)/year by loosely controlling the spacecraft about the reference halo orbit. Consequently, the iterative method for disturbance accommodation can be used to determine a trajectory that is close to a solution of the nonlinear equations of motion, regardless of the assumptions required for the formulation of a spacecraft control problem.

The large size of the reference orbit results in many disturbances being included in the disturbance accommodation filter. Table 4.4 contains the results of a disturbance-accommodating control design for the ISEE-3 halo orbit. Notice that when constant disturbance accommodation is included in iteration 1, the ΔV actually increases. This is attributed to the dependence of the disturbances on the trajectory. In other words, the trajectory that the spacecraft maintained after disturbance accommodation required an increase in ΔV because magnitudes of some disturbances increased.

The closed-loop trajectory is very close to the reference trajectory shown in Fig. 4.16, and thus is not shown in this text. Figure 4.17 shows the control acceleration needed to maintain such a fuel-efficient halo orbit. Note that the z-axis control acceleration u_z does not contain a frequency component of ω_{xy} for period control. The ΔV per year can be estimated to be approximately 8.3 (m/s)/year.

These results demonstrate the application of disturbance accommodation as a method of determining trajectories that are closer to solutions of the nonlinear equations of motion. They also show that a solution of a simplified model, such as the circular restricted three-body model, along with disturbance accommodation, can be used to determine a halo orbit for a more complex model such as the elliptic restricted three-body model. The results also demonstrate that disturbance accommodation can be used to determine a trajectory for the interesting case of naturally equal in-plane and out-of-plane frequencies.

References

[1] Seifert, H. S. (ed.), *Space Technology*, Wiley, New York, 1959.

[2] Sutton, G. P., *Rocket Propulsion Elements,* Wiley, New York, 1963.

[3] Isakowitz, S. J., and Samella, J. (eds.), *International Reference Guide to Space Launch Systems*, 2nd ed., AIAA, Washington, DC, 1995.

[4] Thompson, W. T., *Introduction to Space Dynamics*, Wiley, New York, 1961, pp. 257–259.

[5] Ball, K. J., and Osborne, G. F., *Space Vehicle Dynamics*, Oxford Univ. Press, Oxford, England, UK, 1967, pp. 18–31.

[6] Hohmann, W., "Die Ereichbarkeit der Himmelskorper (The Attainability of Heavenly Bodies)," Oldenbourg, Munich, Germany, 1925; also, NASA Technical Translation F-44, 1960.

[7] Palmore, J., "An Elementary Proof of the Optimality of Hohmann Transfers," *Journal of Guidance, Control, and Dynamics,* Vol. 7, No. 5, 1984, pp. 629–630.

[8] Prussing, J. E., "A Simple Proof of the Global Optimality of the Hohmann Transfer," *Journal of Guidance, Control, and Dynamics,* Vol. 15, No. 4, 1992, pp. 1037–1038.

[9] Kaplan, M. H., *Modern Spacecraft Dynamics & Control,* Wiley, New York, 1976, pp. 287–289.

[10] Wertz, J. R., and Larson, W. J. (eds.), *Space Mission Analysis and Design,* Kluwer Academic, Dordrecht, The Netherlands, 1991.

[11] Chobotov, V. A. (ed.), *Orbital Mechanics,* AIAA Education Series, AIAA, Washington, DC, 1991.

[12] Brown, C. D., *Spacecraft Mission Design,* AIAA Education Series, AIAA, Washington, DC, 1992.

[13] Clohessy, W. H., and Wiltshire, R. S., "Terminal Guidance System for Satellite Rendezvous," *Journal of the Aerospace Sciences,* Vol. 27, Sept. 1960, pp. 653–658.

[14] Farquhar, R. W., "The Control and Use of Libration-Point Satellites," Stanford Univ. SUDAAR-350, July 1968; also, NASA TR R-346, Sept. 1970.

[15] Farquhar, R. W., "The Utilization of Halo Orbits in Advanced Lunar Operations," NASA TN D-6365, July 1971.

[16] Cielaszyk, D., and Wie, B., "A New Approach to Halo Orbit Determination and Control," *Proceedings of the AIAA/AAS Astrodynamics Conference,* AIAA, Washington, DC, 1994, pp. 166–175; also, *Journal of Guidance, Control, and Dynamics,* Vol. 19, No. 2, 1996, pp. 266–273.

[17] Cielaszyk, D., and Wie, B., "Halo Orbit Determination and Control for the Eccentric-Restricted Three-Body Problem," *Proceedings of the AIAA/AAS Astrodynamics Conference,* AIAA, Washington, DC, 1994, pp. 176–185.

[18] Farquhar, R. W., and Kamel, A. A., "Quasi-Periodic Orbits About the Translunar Libration Point," *Celestial Mechanics,* Vol. 7, 1973, pp. 458–473.

[19] Hoffman, D., "Station-keeping at the Collinear Equilibrium Points of the Earth–Moon System," Systems Engineering Div., NASA Johnson Space Center, JSC-261898, Houston, TX, Sept. 1993.

[20] Howell, K. C., and Pernicka, H. J., "Numerical Determination of Lissajous Trajectories in the Restricted Three-Body Problem," *Celestial Mechanics,* Vol. 41, 1988, pp. 107–124.

[21] Farquhar, R. W., Muhonen, D. P., Newman, C. R., and Heuberger, H. S., "Trajectories and Orbital Maneuvers for the First Libration-Point Satellite," *Journal of Guidance, Control, and Dynamics,* Vol. 3, No. 6, 1980, pp. 549–554.

[22] Richardson, D. L., "Halo Orbit Formulation for the ISEE-3 Mission," *Journal of Guidance, Control, and Dynamics,* Vol. 3, No. 6, 1980, pp. 543–548.

Part 3
Attitude Dynamics and Control

The formulation of spacecraft attitude dynamics and control problems involves considerations of kinematics. This chapter is concerned with rotational kinematics of a rigid body. In kinematics, we are primarily interested in describing the orientation of a body that is in rotational motion. The subject of rotational kinematics is somewhat mathematical in nature because it does not involve any forces associated with motion. Throughout this chapter, we will speak of the orientation of a reference frame fixed in a body to describe the orientation of the body itself.

5.1 Direction Cosine Matrix

Consider a reference frame A with a right-hand set of three orthogonal unit vectors $\{\vec{a}_1, \vec{a}_2, \vec{a}_3\}$ and a reference frame B with another right-hand set of three orthogonal unit vectors $\{\vec{b}_1, \vec{b}_2, \vec{b}_3\}$, as shown in Fig. 5.1. Basis vectors $\{\vec{b}_1, \vec{b}_2, \vec{b}_3\}$ of B are expressed in terms of basis vectors $\{\vec{a}_1, \vec{a}_2, \vec{a}_3\}$ of A as follows:

$$\vec{b}_1 = C_{11}\vec{a}_1 + C_{12}\vec{a}_2 + C_{13}\vec{a}_3 \tag{5.1a}$$

$$\vec{b}_2 = C_{21}\vec{a}_1 + C_{22}\vec{a}_2 + C_{23}\vec{a}_3 \tag{5.1b}$$

$$\vec{b}_3 = C_{31}\vec{a}_1 + C_{32}\vec{a}_2 + C_{33}\vec{a}_3 \tag{5.1c}$$

where $C_{ij} \equiv \vec{b}_i \cdot \vec{a}_j$ is the cosine of the angle between $\vec{b}_i$ and $\vec{a}_j$, and C_{ij} is simply called the *direction cosine*.

For convenience, we write Eqs. (5.1) in matrix (or vectrix) notation, as follows:

$$\begin{bmatrix} \vec{b}_1 \\ \vec{b}_2 \\ \vec{b}_3 \end{bmatrix} = \begin{bmatrix} C_{11} & C_{12} & C_{13} \\ C_{21} & C_{22} & C_{23} \\ C_{31} & C_{32} & C_{33} \end{bmatrix} \begin{bmatrix} \vec{a}_1 \\ \vec{a}_2 \\ \vec{a}_3 \end{bmatrix} = \mathbf{C}^{B/A} \begin{bmatrix} \vec{a}_1 \\ \vec{a}_2 \\ \vec{a}_3 \end{bmatrix} \tag{5.2}$$

where $\mathbf{C}^{B/A} \equiv [C_{ij}]$ is called the direction cosine matrix, which describes the orientation of B relative to A and which can be written as

$$\mathbf{C}^{B/A} = \begin{bmatrix} \vec{b}_1 \cdot \vec{a}_1 & \vec{b}_1 \cdot \vec{a}_2 & \vec{b}_1 \cdot \vec{a}_3 \\ \vec{b}_2 \cdot \vec{a}_1 & \vec{b}_2 \cdot \vec{a}_2 & \vec{b}_2 \cdot \vec{a}_3 \\ \vec{b}_3 \cdot \vec{a}_1 & \vec{b}_3 \cdot \vec{a}_2 & \vec{b}_3 \cdot \vec{a}_3 \end{bmatrix} \equiv \begin{bmatrix} \vec{b}_1 \\ \vec{b}_2 \\ \vec{b}_3 \end{bmatrix} \cdot \begin{bmatrix} \vec{a}_1 & \vec{a}_2 & \vec{a}_3 \end{bmatrix} \tag{5.3}$$

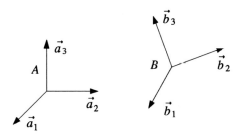

Fig. 5.1   Two reference frames A and B.

The direction cosine matrix $\mathbf{C}^{B/A}$ is also called the *rotation matrix* or *coordinate transformation matrix* to B from A. Such a coordinate transformation is symbolically represented as

$$\mathbf{C}^{B/A} : \quad B \leftarrow A$$

For brevity, we often use $\mathbf{C}$ for $\mathbf{C}^{B/A}$. Because each set of basis vectors of A and B consists of orthogonal unit vectors, the direction cosine matrix $\mathbf{C}$ is an orthonormal matrix; thus, we have

$$\mathbf{C}^{-1} = \mathbf{C}^{T} \tag{5.4}$$

which is equivalent to

$$\mathbf{C}\mathbf{C}^{T} = \mathbf{I} = \mathbf{C}^{T}\mathbf{C} \tag{5.5}$$

In general, a square matrix $\mathbf{A}$ is called an orthogonal matrix if $\mathbf{A}\mathbf{A}^{T}$ is a diagonal matrix, and it is called an orthonormal matrix if $\mathbf{A}\mathbf{A}^{T}$ is an identity matrix. For an orthonormal matrix $\mathbf{A}$, we have $\mathbf{A}^{-1} = \mathbf{A}^{T}$ and $|\mathbf{A}| = \pm 1$.

We also use $\mathbf{C}^{A/B}$ to denote a coordinate transformation matrix to A from B or a direction cosine matrix of A relative to B; i.e., we have

$$\mathbf{C}^{A/B} = \begin{bmatrix} \vec{a}_1 \cdot \vec{b}_1 & \vec{a}_1 \cdot \vec{b}_2 & \vec{a}_1 \cdot \vec{b}_3 \\ \vec{a}_2 \cdot \vec{b}_1 & \vec{a}_2 \cdot \vec{b}_2 & \vec{a}_2 \cdot \vec{b}_3 \\ \vec{a}_3 \cdot \vec{b}_1 & \vec{a}_3 \cdot \vec{b}_2 & \vec{a}_3 \cdot \vec{b}_3 \end{bmatrix} \equiv \begin{bmatrix} \vec{a}_1 \\ \vec{a}_2 \\ \vec{a}_3 \end{bmatrix} \cdot \begin{bmatrix} \vec{b}_1 & \vec{b}_2 & \vec{b}_3 \end{bmatrix}$$

Consequently, we have the following intimate relationships between $\mathbf{C}^{A/B}$ and $\mathbf{C}^{B/A}$:

$$\left[\mathbf{C}^{A/B}\right]_{ij} = \vec{a}_i \cdot \vec{b}_j$$

$$\left[\mathbf{C}^{B/A}\right]_{ij} = \vec{b}_i \cdot \vec{a}_j$$

$$[\mathbf{C}^{A/B}]^{-1} = [\mathbf{C}^{A/B}]^T = \mathbf{C}^{B/A}$$

$$[\mathbf{C}^{B/A}]^{-1} = [\mathbf{C}^{B/A}]^T = \mathbf{C}^{A/B}$$

Given the two sets of reference frames A and B, an arbitrary vector $\vec{H}$ can be expressed in terms of basis vectors of A and B, as follows:

$$\begin{aligned}
\vec{H} &= H_1\vec{a}_1 + H_2\vec{a}_2 + H_3\vec{a}_3 \\
&= H_1'\vec{b}_1 + H_2'\vec{b}_2 + H_3'\vec{b}_3
\end{aligned} \tag{5.6}$$

and we have

$$H_1' \equiv \vec{b}_1 \cdot \vec{H} = \vec{b}_1 \cdot (H_1\vec{a}_1 + H_2\vec{a}_2 + H_3\vec{a}_3) \tag{5.7a}$$

$$H_2' \equiv \vec{b}_2 \cdot \vec{H} = \vec{b}_2 \cdot (H_1\vec{a}_1 + H_2\vec{a}_2 + H_3\vec{a}_3) \tag{5.7b}$$

$$H_3' \equiv \vec{b}_3 \cdot \vec{H} = \vec{b}_3 \cdot (H_1\vec{a}_1 + H_2\vec{a}_2 + H_3\vec{a}_3) \tag{5.7c}$$

which can be written in matrix form, as follows:

$$\begin{bmatrix} H_1' \\ H_2' \\ H_3' \end{bmatrix} = \begin{bmatrix} \vec{b}_1 \cdot \vec{a}_1 & \vec{b}_1 \cdot \vec{a}_2 & \vec{b}_1 \cdot \vec{a}_3 \\ \vec{b}_2 \cdot \vec{a}_1 & \vec{b}_2 \cdot \vec{a}_2 & \vec{b}_2 \cdot \vec{a}_3 \\ \vec{b}_3 \cdot \vec{a}_1 & \vec{b}_3 \cdot \vec{a}_2 & \vec{b}_3 \cdot \vec{a}_3 \end{bmatrix} \begin{bmatrix} H_1 \\ H_2 \\ H_3 \end{bmatrix} = \mathbf{C}^{B/A} \begin{bmatrix} H_1 \\ H_2 \\ H_3 \end{bmatrix} \tag{5.8}$$

Thus, the components of a vector $\vec{H}$ are also transformed to B from A using the direction cosine matrix $\mathbf{C}^{B/A}$, which was defined in Eq. (5.2) for the transformation of orthogonal basis vectors.

Three elementary rotations respectively about the first, second, and third axes of the reference frame A are described by the following rotation matrices:

$$\mathbf{C}_1(\theta_1) = \begin{bmatrix} 1 & 0 & 0 \\ 0 & \cos\theta_1 & \sin\theta_1 \\ 0 & -\sin\theta_1 & \cos\theta_1 \end{bmatrix} \tag{5.9a}$$

$$\mathbf{C}_2(\theta_2) = \begin{bmatrix} \cos\theta_2 & 0 & -\sin\theta_2 \\ 0 & 1 & 0 \\ \sin\theta_2 & 0 & \cos\theta_2 \end{bmatrix} \tag{5.9b}$$

$$\mathbf{C}_3(\theta_3) = \begin{bmatrix} \cos\theta_3 & \sin\theta_3 & 0 \\ -\sin\theta_3 & \cos\theta_3 & 0 \\ 0 & 0 & 1 \end{bmatrix} \tag{5.9c}$$

where $\mathbf{C}_i(\theta_i)$ denotes the direction cosine matrix $\mathbf{C}$ of an elementary rotation about the ith axis of A with an angle θ_i.

Problem

5.1 Consider the direction cosine matrix, $\mathbf{C} \equiv [C_{ij}]$, between two sets of right-hand orthogonal unit vectors $\{\vec{a}_1, \vec{a}_2, \vec{a}_3\}$ and $\{\vec{b}_1, \vec{b}_2, \vec{b}_3\}$, defined as

$$\begin{bmatrix} \vec{b}_1 \\ \vec{b}_2 \\ \vec{b}_3 \end{bmatrix} = \begin{bmatrix} C_{11} & C_{12} & C_{13} \\ C_{21} & C_{22} & C_{23} \\ C_{31} & C_{32} & C_{33} \end{bmatrix} \begin{bmatrix} \vec{a}_1 \\ \vec{a}_2 \\ \vec{a}_3 \end{bmatrix}$$

(a) Show that the direction cosine matrix $\mathbf{C}$ is an orthonormal matrix; i.e., $\mathbf{CC}^T = \mathbf{I} = \mathbf{C}^T\mathbf{C}$.
Hint: $\begin{bmatrix} \vec{b}_1 & \vec{b}_2 & \vec{b}_3 \end{bmatrix} = \begin{bmatrix} \vec{a}_1 & \vec{a}_2 & \vec{a}_3 \end{bmatrix}\mathbf{C}^T$ and

$$\begin{bmatrix} \vec{b}_1 \\ \vec{b}_2 \\ \vec{b}_3 \end{bmatrix} \cdot \begin{bmatrix} \vec{b}_1 & \vec{b}_2 & \vec{b}_3 \end{bmatrix} \equiv \begin{bmatrix} \vec{b}_1 \cdot \vec{b}_1 & \vec{b}_1 \cdot \vec{b}_2 & \vec{b}_1 \cdot \vec{b}_3 \\ \vec{b}_2 \cdot \vec{b}_1 & \vec{b}_2 \cdot \vec{b}_2 & \vec{b}_2 \cdot \vec{b}_3 \\ \vec{b}_3 \cdot \vec{b}_1 & \vec{b}_3 \cdot \vec{b}_2 & \vec{b}_3 \cdot \vec{b}_3 \end{bmatrix} = \begin{bmatrix} 1 & 0 & 0 \\ 0 & 1 & 0 \\ 0 & 0 & 1 \end{bmatrix}$$

(b) Show that the ijth element of the direction cosine matrix $\mathbf{C}$ is equal to the ijth cofactor of $\mathbf{C}$; i.e., $C_{ij} = (-1)^{i+j}M_{ij}$ or adj $\mathbf{C} \equiv [(-1)^{i+j}M_{ij}]^T = [C_{ij}]^T = \mathbf{C}^T$.
Hint: Using $\vec{b}_1 = \vec{b}_2 \times \vec{b}_3$, show that

$$C_{11} = C_{22}C_{33} - C_{23}C_{32}$$
$$C_{12} = C_{23}C_{31} - C_{21}C_{33}$$
$$C_{13} = C_{21}C_{32} - C_{22}C_{31}$$

(c) Show that $|\mathbf{C}| = 1$.
Hint: $\mathbf{C}^{-1} \equiv$ adj $\mathbf{C}/|\mathbf{C}|$.
(d) Find six independent equations for C_{ij} using the row-orthonormality condition, $\mathbf{CC}^T = \mathbf{I}$.
(e) Find six independent equations for C_{ij} using the column-orthonormality condition, $\mathbf{C}^T\mathbf{C} = \mathbf{I}$. Are these six equations independent of those of part (d)?
(f) Finally, show that only three of the nine direction cosines are independent. Note: Three direction cosines, however, do not uniquely define the orientation of two reference frames.

5.2 Euler Angles

One scheme for orienting a rigid body to a desired attitude is called a *body-axis rotation*; it involves successively rotating three times about the axes of the rotated, body-fixed reference frame. The first rotation is about any axis. The second rotation is about either of the two axes not used for the first rotation. The third rotation is then about either of the two axes not used for the second rotation. There are 12 sets of Euler angles for such successive rotations about the axes fixed in the body.*

It is also possible to bring a rigid body into an arbitrary orientation by performing three successive rotations that involve the axes fixed in an inertial reference frame. This scheme will then provide another 12 sets of Euler angles for the so-called *space-axis rotations* [1]. Because the coordinate transformation matrices for the body-axis rotation and the space-axis rotation are intimately related to each other, we often only consider the 12 sets of body-axis rotations.

Consider three successive body-axis rotations that describe the orientation of a reference frame B relative to a reference frame A. A particular sequence chosen here is symbolically represented as

$$\mathbf{C}_3(\theta_3): \quad A' \leftarrow A \quad (5.10a)$$

$$\mathbf{C}_2(\theta_2): \quad A'' \leftarrow A' \quad (5.10b)$$

$$\mathbf{C}_1(\theta_1): \quad B \leftarrow A'' \quad (5.10c)$$

where each rotation is described as

$$\begin{bmatrix} \vec{a}_1' \\ \vec{a}_2' \\ \vec{a}_3' \end{bmatrix} = \begin{bmatrix} \cos\theta_3 & \sin\theta_3 & 0 \\ -\sin\theta_3 & \cos\theta_3 & 0 \\ 0 & 0 & 1 \end{bmatrix} \begin{bmatrix} \vec{a}_1 \\ \vec{a}_2 \\ \vec{a}_3 \end{bmatrix} = \mathbf{C}_3(\theta_3) \begin{bmatrix} \vec{a}_1 \\ \vec{a}_2 \\ \vec{a}_3 \end{bmatrix} \quad (5.11a)$$

$$\begin{bmatrix} \vec{a}_1'' \\ \vec{a}_2'' \\ \vec{a}_3'' \end{bmatrix} = \begin{bmatrix} \cos\theta_2 & 0 & -\sin\theta_2 \\ 0 & 1 & 0 \\ \sin\theta_2 & 0 & \cos\theta_2 \end{bmatrix} \begin{bmatrix} \vec{a}_1' \\ \vec{a}_2' \\ \vec{a}_3' \end{bmatrix} = \mathbf{C}_2(\theta_2) \begin{bmatrix} \vec{a}_1' \\ \vec{a}_2' \\ \vec{a}_3' \end{bmatrix} \quad (5.11b)$$

$$\begin{bmatrix} \vec{b}_1 \\ \vec{b}_2 \\ \vec{b}_3 \end{bmatrix} = \begin{bmatrix} 1 & 0 & 0 \\ 0 & \cos\theta_1 & \sin\theta_1 \\ 0 & -\sin\theta_1 & \cos\theta_1 \end{bmatrix} \begin{bmatrix} \vec{a}_1'' \\ \vec{a}_2'' \\ \vec{a}_3'' \end{bmatrix} = \mathbf{C}_1(\theta_1) \begin{bmatrix} \vec{a}_1'' \\ \vec{a}_2'' \\ \vec{a}_3'' \end{bmatrix} \quad (5.11c)$$

and A' and A'' are two intermediate reference frames with basis vectors $\{\vec{a}_1', \vec{a}_2', \vec{a}_3'\}$ and $\{\vec{a}_1'', \vec{a}_2'', \vec{a}_3''\}$, respectively. The three angles θ_1, θ_2, and θ_3 are called *Euler angles*.

By combining the preceding sequence of rotations, we obtain

$$\begin{bmatrix} \vec{b}_1 \\ \vec{b}_2 \\ \vec{b}_3 \end{bmatrix} = \mathbf{C}_1(\theta_1) \begin{bmatrix} \vec{a}_1'' \\ \vec{a}_2'' \\ \vec{a}_3'' \end{bmatrix} = \mathbf{C}_1(\theta_1)\mathbf{C}_2(\theta_2) \begin{bmatrix} \vec{a}_1' \\ \vec{a}_2' \\ \vec{a}_3' \end{bmatrix} = \mathbf{C}_1(\theta_1)\mathbf{C}_2(\theta_2)\mathbf{C}_3(\theta_3) \begin{bmatrix} \vec{a}_1 \\ \vec{a}_2 \\ \vec{a}_3 \end{bmatrix} \quad (5.12)$$

---

*Leonard Euler (1707–1783), who was the leading mathematician and theoretical physicist of the 18th century, first introduced the concept of three successive rotations to describe the orientation of an orbit plane using the three angles Ω, i, and ω.

The rotation matrix to B from A, or the direction cosine matrix of B relative to A, is then defined as

$$\mathbf{C}^{B/A} \equiv \mathbf{C}_1(\theta_1)\mathbf{C}_2(\theta_2)\mathbf{C}_3(\theta_3)$$

$$= \begin{bmatrix} c_2\,c_3 & c_2\,s_3 & -s_2 \\ s_1\,s_2\,c_3 - c_1\,s_3 & s_1\,s_2\,s_3 + c_1\,c_3 & s_1\,c_2 \\ c_1\,s_2\,c_3 + s_1\,s_3 & c_1\,s_2\,s_3 - s_1\,c_3 & c_1\,c_2 \end{bmatrix} \qquad (5.13)$$

where $c_i \equiv \cos\theta_i$ and $s_i \equiv \sin\theta_i$.

The preceding sequence of rotations to B from A is also symbolically denoted by*

$$\mathbf{C}_1(\theta_1) \leftarrow \mathbf{C}_2(\theta_2) \leftarrow \mathbf{C}_3(\theta_3)$$

where $\mathbf{C}_i(\theta_i)$ indicates a rotation about the ith axis of the body-fixed frame with an angle θ_i, or by

$$\theta_1\,\vec{a}_1'' \leftarrow \theta_2\,\vec{a}_2' \leftarrow \theta_3\,\vec{a}_3$$

in which, for example, $\theta_3\,\vec{a}_3$ denotes a rotation about the $\vec{a}_3$ axis with an angle θ_3.

In general, there are 12 sets of Euler angles, each resulting in a different form for the rotation matrix $\mathbf{C}^{B/A}$. For example, we may consider the sequence of $\mathbf{C}_1(\theta_1) \leftarrow \mathbf{C}_3(\theta_3) \leftarrow \mathbf{C}_2(\theta_2)$ to B from A. For this case, the rotation matrix becomes

$$\mathbf{C}^{B/A} \equiv \mathbf{C}_1(\theta_1)\mathbf{C}_3(\theta_3)\mathbf{C}_2(\theta_2)$$

$$= \begin{bmatrix} c_2\,c_3 & s_3 & -s_2\,c_3 \\ -c_1\,c_2\,s_3 + s_1\,s_2 & c_1\,c_3 & c_1\,s_2\,s_3 + s_1\,c_2 \\ s_1\,c_2\,s_3 + c_1\,s_2 & -s_1\,c_3 & -s_1\,s_2\,s_3 + c_1\,c_2 \end{bmatrix} \qquad (5.14)$$

Note that for small (infinitesimal) Euler angles of θ_1, θ_2, and θ_3, the direction cosine matrices in Eqs. (5.13) and (5.14) become

$$\mathbf{C} \approx \begin{bmatrix} 1 & \theta_3 & -\theta_2 \\ -\theta_3 & 1 & \theta_1 \\ \theta_2 & -\theta_1 & 1 \end{bmatrix} \qquad (5.15)$$

That is, the rotation sequence of Euler angles becomes unimportant for infinitesimal rotations, whereas rotation sequence is important for finite rotations.

For other sequences, such as a classical "$3 \leftarrow 1 \leftarrow 3$" rotational sequence in which the third axis is used twice, we use the following notational convention: $\mathbf{C}_3(\psi) \leftarrow \mathbf{C}_1(\theta) \leftarrow \mathbf{C}_3(\phi)$ to B from A, in which, for example, $\mathbf{C}_3(\phi)$ indicates a

---

*The notation $\mathbf{C}_1(\theta_1) \leftarrow \mathbf{C}_2(\theta_2) \leftarrow \mathbf{C}_3(\theta_3)$, which, in fact, denotes the same rotational sequence as denoted by the notation $\mathbf{C}_3(\theta_3) \rightarrow \mathbf{C}_2(\theta_2) \rightarrow \mathbf{C}_1(\theta_1)$, is introduced in this book to emphasize the resulting structure of the total rotation matrix $\mathbf{C}^{B/A} = \mathbf{C}_1(\theta_1)\mathbf{C}_2(\theta_2)\mathbf{C}_3(\theta_3)$.

rotation about the third axis with an angle ϕ. Thus, for such a classical $\mathbf{C}_3(\psi) \leftarrow \mathbf{C}_1(\theta) \leftarrow \mathbf{C}_3(\phi)$ rotational sequence, the rotation matrix to B from A becomes

$$\mathbf{C}^{B/A} \equiv \mathbf{C}_3(\psi)\mathbf{C}_1(\theta)\mathbf{C}_3(\phi)$$

$$= \begin{bmatrix} c\,\phi\,c\,\psi - s\,\phi\,c\,\theta\,s\,\psi & s\,\phi\,c\,\psi + c\,\phi\,c\,\theta\,s\,\psi & s\,\theta\,s\,\psi \\ -c\,\phi\,s\,\psi - s\,\phi\,c\,\theta\,c\,\psi & -s\,\phi\,s\,\psi + c\,\phi\,c\,\theta\,c\,\psi & s\,\theta\,c\,\psi \\ s\,\phi\,s\,\theta & -c\,\phi\,s\,\theta & c\,\theta \end{bmatrix}$$

where $c\,\phi \equiv \cos\phi$, $s\,\phi \equiv \sin\phi$, etc.

In general, Euler angles have an advantage over direction cosines in that three Euler angles determine a unique orientation, although there is no unique set of Euler angles for a given orientation.

5.3 Euler's Eigenaxis Rotation

In this section, we consider rotation of a rigid body (or a reference frame) about an arbitrary axis that is fixed to the body and stationary in an inertial reference frame. An intimate relationship between the body-axis and space-axis rotations is derived. Such a relationship provides insights into the understanding of Euler's eigenaxis rotation and the space-axis rotation.

5.3.1 Euler's Eigenaxis Rotation Theorem

Euler's eigenaxis rotation theorem states that by rotating a rigid body about an axis that is fixed to the body and stationary in an inertial reference frame, the rigid-body attitude can be changed from any given orientation to any other orientation. Such an axis of rotation, whose orientation relative to both an inertial reference frame and the body remains unchanged throughout the motion, is called the *Euler axis* or *eigenaxis*.

Various different approaches can be used to develop several different parameterizations of the direction cosine matrix of the Euler axis rotation. Almost every formula can be derived in a variety of ways; however, the approach to be taken here is simple and it will provide insights into the understanding of the intimate relationship between the body-axis and space-axis rotations.

Suppose unit vectors $\vec{a}_i$ and $\vec{b}_i$ ($i = 1, 2, 3$) are fixed in reference frames A and B, respectively. The orientation of B with respect to A is characterized by a unit vector $\vec{e}$ along the Euler axis and the rotation angle θ about that axis, as follows:

$$\vec{e} = e_1\vec{a}_1 + e_2\vec{a}_2 + e_3\vec{a}_3$$
$$= e_1\vec{b}_1 + e_2\vec{b}_2 + e_3\vec{b}_3 \tag{5.16}$$

where e_i are the direction cosines of the Euler axis relative to both A and B and $e_1^2 + e_2^2 + e_3^2 = 1$. Let $\mathbf{C}^{B/A} = \mathbf{C} = [C_{ij}]$ be the direction cosine matrix of B relative to A, then Euler's eigenaxis rotation is also characterized by

$$\begin{bmatrix} e_1 \\ e_2 \\ e_3 \end{bmatrix} = \begin{bmatrix} C_{11} & C_{12} & C_{13} \\ C_{21} & C_{22} & C_{23} \\ C_{31} & C_{32} & C_{33} \end{bmatrix} \begin{bmatrix} e_1 \\ e_2 \\ e_3 \end{bmatrix} \tag{5.17}$$

To parameterize the direction cosine matrix $\mathbf{C}$ in terms of e_i and θ, a sequence of Euler's successive rotations is used as follows:

1) Rotate the reference frame A, using a rotation matrix $\mathbf{R}$, to align the $\vec{a}_1$ axis of A with the chosen direction $\vec{e}$. Let A' be the new reference frame after this rotation and also let A remain the original frame with basis vectors $\{\vec{a}_1, \vec{a}_2, \vec{a}_3\}$ before this rotation; i.e., we have

$$\mathbf{C}^{A'/A} = \mathbf{R} = \begin{bmatrix} e_1 & e_2 & e_3 \\ R_{21} & R_{22} & R_{23} \\ R_{31} & R_{32} & R_{33} \end{bmatrix}. \tag{5.18}$$

2) Rotate both frames A and A' as a rigid body around direction $\vec{e}$ through an angle θ. After this eigenaxis rotation, the frame A will be aligned with the reference frame B with basis vectors $\{\vec{b}_1, \vec{b}_2, \vec{b}_3\}$, and A' will become another reference frame A'' via the rotation matrix

$$\mathbf{C}^{A''/A'} = \mathbf{C}_1(\theta) = \begin{bmatrix} 1 & 0 & 0 \\ 0 & \cos\theta & \sin\theta \\ 0 & -\sin\theta & \cos\theta \end{bmatrix} \tag{5.19}$$

The orientation of B relative to A is described by the direction cosine matrix $\mathbf{C}^{B/A}$. It is important to notice that the relative orientation of A'' and B is the same as that of A' and A; i.e., $\mathbf{C}^{A''/B} = \mathbf{C}^{A'/A} = \mathbf{R}$.

3) Rotate A'' through an inverse matrix $\mathbf{R}^{-1} = \mathbf{R}^T$, then the frame A'' will be aligned with B since $\mathbf{C}^{B/A''} = \mathbf{R}^{-1}$.

These three successive rotations can be combined as

$$\begin{bmatrix} \vec{b}_1 \\ \vec{b}_2 \\ \vec{b}_3 \end{bmatrix} = \mathbf{C}^{B/A} \begin{bmatrix} \vec{a}_1 \\ \vec{a}_2 \\ \vec{a}_3 \end{bmatrix} \tag{5.20}$$

where

$$\mathbf{C}^{B/A} = \mathbf{C}^{B/A''}\,\mathbf{C}^{A''/A'}\,\mathbf{C}^{A'/A} = \mathbf{C}^{A/A'}\mathbf{C}_1(\theta)\mathbf{C}^{A'/A} = \mathbf{R}^T\mathbf{C}_1(\theta)\mathbf{R} \tag{5.21}$$

If the $\vec{a}_2$ or $\vec{a}_3$ axis, instead of $\vec{a}_1$ axis, is aligned with the chosen direction $\vec{e}$ for the first rotation, then the rotation matrix $\mathbf{C}_1(\theta)$ in Eq. (5.21) is replaced by $\mathbf{C}_2(\theta)$ or $\mathbf{C}_3(\theta)$, respectively, and the rotation matrix $\mathbf{R}$ is replaced, respectively, by

$$\begin{bmatrix} R_{11} & R_{12} & R_{13} \\ e_1 & e_2 & e_3 \\ R_{31} & R_{32} & R_{33} \end{bmatrix} \quad \text{or} \quad \begin{bmatrix} R_{11} & R_{12} & R_{13} \\ R_{21} & R_{22} & R_{23} \\ e_1 & e_2 & e_3 \end{bmatrix}$$

Substituting Eqs. (5.18) and (5.19) into Eq. (5.21) and defining $\mathbf{C} = [C_{ij}] = \mathbf{C}^{B/A}$, we obtain

$$C_{11} = e_1^2 + \left(R_{21}^2 + R_{31}^2\right)\cos\theta$$
$$C_{12} = e_1 e_2 + (R_{21}R_{22} + R_{31}R_{32})\cos\theta + (R_{21}R_{32} - R_{22}R_{31})\sin\theta$$
$$C_{13} = e_1 e_3 + (R_{21}R_{23} + R_{31}R_{33})\cos\theta + (R_{21}R_{33} - R_{23}R_{31})\sin\theta$$

$$\vdots$$

$$C_{33} = e_3^2 + \left(R_{23}^2 + R_{33}^2\right)\cos\theta$$

The column-orthonormality condition of the rotation matrix gives

$$e_1^2 + R_{21}^2 + R_{31}^2 = 1$$
$$e_2^2 + R_{22}^2 + R_{32}^2 = 1$$
$$e_3^2 + R_{23}^2 + R_{33}^2 = 1$$
$$e_1 e_2 + R_{21}R_{22} + R_{31}R_{32} = 0$$
$$e_2 e_3 + R_{22}R_{23} + R_{32}R_{33} = 0$$
$$e_1 e_3 + R_{21}R_{23} + R_{31}R_{33} = 0$$

Because each element of the rotation matrix $\mathbf{R}$ of Eq. (5.18) is equal to its cofactor, we also have

$$e_1 = R_{22}R_{33} - R_{23}R_{32}$$
$$e_2 = R_{23}R_{31} - R_{21}R_{33}$$
$$e_3 = R_{21}R_{32} - R_{22}R_{31}$$

Using these relationships, we obtain $\mathbf{C} = \mathbf{R}^T \mathbf{C}_1(\theta)\mathbf{R}$ as

$$\mathbf{C} = \begin{bmatrix} c\theta + e_1^2(1 - c\theta) & e_1 e_2(1 - c\theta) + e_3\,s\theta & e_1 e_3(1 - c\theta) - e_2\,s\theta \\ e_2 e_1(1 - c\theta) - e_3\,s\theta & c\theta + e_2^2(1 - c\theta) & e_2 e_3(1 - c\theta) + e_1\,s\theta \\ e_3 e_1(1 - c\theta) + e_2\,s\theta & e_3 e_2(1 - c\theta) - e_1\,s\theta & c\theta + e_3^2(1 - c\theta) \end{bmatrix}$$

$$(5.22)$$

where $c\theta \equiv \cos\theta$ and $s\theta \equiv \sin\theta$. This is the parameterization of the direction cosine matrix $\mathbf{C}$ in terms of e_i and θ. Note that e_1, e_2, and e_3 are not independent of each other, but constrained by the relationship $e_1^2 + e_2^2 + e_3^2 = 1$.

By defining

$$\mathbf{e} = \begin{bmatrix} e_1 \\ e_2 \\ e_3 \end{bmatrix} \quad \text{and} \quad \mathbf{E} = \begin{bmatrix} 0 & -e_3 & e_2 \\ e_3 & 0 & -e_1 \\ -e_2 & e_1 & 0 \end{bmatrix} \tag{5.23}$$

we can also express the direction cosine matrix $\mathbf{C}$ in Eq. (5.22) as

$$\mathbf{C} = \cos\theta\mathbf{I} + (1 - \cos\theta)\mathbf{e}\mathbf{e}^T - \sin\theta\mathbf{E} \qquad (5.24)$$

where $\mathbf{I}$ is the identity matrix; i.e., $C_{ij} = \delta_{ij}\cos\theta + (1 - \cos\theta)e_i e_j - \sin\theta E_{ij}$. Given a direction cosine matrix $\mathbf{C} = [C_{ij}]$, θ can be found from

$$\cos\theta = \tfrac{1}{2}(C_{11} + C_{22} + C_{33} - 1) \qquad (5.25)$$

From Eq. (5.24), we obtain

$$\mathbf{E} = \frac{1}{2\sin\theta}(\mathbf{C}^T - \mathbf{C}) \qquad \text{if } \theta \neq 0, \pm\pi, \pm2\pi, \ldots \qquad (5.26)$$

from which the eigenaxis $\mathbf{e}$ can be found as

$$\mathbf{e} = \begin{bmatrix} e_1 \\ e_2 \\ e_3 \end{bmatrix} = \frac{1}{2\sin\theta} \begin{bmatrix} C_{23} - C_{32} \\ C_{31} - C_{13} \\ C_{12} - C_{21} \end{bmatrix} \qquad (5.27)$$

Problems

5.2 Verify that $\vec{b}_i$ can be expressed as

$$\begin{aligned} \vec{b}_i &= \vec{a}_i \cos\theta + \vec{e}(\vec{a}_i \cdot \vec{e})(1 - \cos\theta) - \vec{a}_i \times \vec{e}\sin\theta \\ &= \vec{a}_i + \vec{e} \times (\vec{e} \times \vec{a}_i)(1 - \cos\theta) - \vec{a}_i \times \vec{e}\sin\theta, \qquad i = 1, 2, 3 \end{aligned}$$

Hint: $\vec{e} = e_1\vec{a}_1 + e_2\vec{a}_2 + e_3\vec{a}_3$ and $\vec{b}_i = C_{i1}\vec{a}_1 + C_{i2}\vec{a}_2 + C_{i3}\vec{a}_3$ where C_{ij} are given by Eq. (5.22).

5.3 Consider two successive eigenaxis rotations to A'' from A represented by

$$\begin{aligned} \mathbf{C}(\mathbf{e}_1, \theta_1): & \quad A' \leftarrow A \\ \mathbf{C}(\mathbf{e}_2, \theta_2): & \quad A'' \leftarrow A' \end{aligned}$$

where

$$\begin{aligned} \mathbf{C}(\mathbf{e}_1, \theta_1) &= \left[\cos\theta_1\mathbf{I} + (1 - \cos\theta_1)\mathbf{e}_1\mathbf{e}_1^T - \sin\theta_1\mathbf{E}_1\right] \\ \mathbf{C}(\mathbf{e}_2, \theta_2) &= \left[\cos\theta_2\mathbf{I} + (1 - \cos\theta_2)\mathbf{e}_2\mathbf{e}_2^T - \sin\theta_2\mathbf{E}_2\right] \end{aligned}$$

and $\mathbf{e}_1$ and $\mathbf{e}_2$ are the eigenaxes associated with the first and second eigenaxis rotations, respectively.

These successive rotations are also represented by an equivalent single eigenaxis rotation to A'' directly from A, as follows:

$$\mathbf{C}(\mathbf{e}, \theta): \quad A'' \leftarrow A$$

where

$$C(e, \theta) = [\cos \theta I + (1 - \cos \theta)ee^T - \sin \theta E]$$

and we have $C(e, \theta) = C(e_2, \theta_2)C(e_1, \theta_1)$.

(a) Show that the equivalent eigenangle θ can be determined as

$$\cos \theta = \tfrac{1}{2}(\text{tr } C - 1)$$

and

$$\begin{aligned} \text{tr } C = \ & \cos \theta_1 + \cos \theta_2 + \cos \theta_1 \cos \theta_2 \\ & + (1 - \cos \theta_1)(1 - \cos \theta_2)\cos^2 \gamma - 2 \sin \theta_1 \sin \theta_2 \cos \gamma \end{aligned}$$

and γ is the angle between the two eigenaxes e_1 and e_2; i.e., $\cos \gamma = e_1^T e_2$.

(b) Show that $\cos \theta$ obtained in (a) can be expressed in terms of half-angles:

$$\cos \frac{\theta}{2} = \cos \frac{\theta_1}{2} \cos \frac{\theta_2}{2} - \sin \frac{\theta_1}{2} \sin \frac{\theta_2}{2} \cos \gamma$$

Hint: $\sin^2(\theta/2) = (1 - \cos \theta)/2$ and $\cos^2(\theta/2) = (1 + \cos \theta)/2$.

(c) Show that the equivalent eigenaxis of rotation e can be found as

$$\begin{aligned} 2 \sin \theta \; e = \ & e_1\{\sin \theta_1(1 + \cos \theta_2) - \sin \theta_2(1 - \cos \theta_1) \cos \gamma\} \\ & + e_2\{\sin \theta_2(1 + \cos \theta_1) - \sin \theta_1(1 - \cos \theta_2) \cos \gamma\} \\ & + (e_1 \times e_2)\{\sin \theta_1 \sin \theta_2 - (1 - \cos \theta_1)(1 - \cos \theta_2) \cos \gamma\} \end{aligned}$$

which can be rewritten as

$$e \sin \frac{\theta}{2} = e_1 \sin \frac{\theta_1}{2} \cos \frac{\theta_2}{2} + e_2 \sin \frac{\theta_2}{2} \cos \frac{\theta_1}{2} + (e_1 \times e_2) \sin \frac{\theta_1}{2} \sin \frac{\theta_2}{2}$$

where $e_1 \times e_2 \equiv E_1 e_2$.

Note: See Ref. 2 for additional information pertaining to Problem 5.3.

5.3.2 Space-Axis Rotation

The space-axis and body-axis rotations are defined as a successive rotation about the space-fixed axes and body-fixed axes, respectively. An interesting relationship between these different schemes of successive rotations exists [1,3,4].

Consider a space-axis rotation in the sequence of $\theta_3 \vec{a}_3 \leftarrow \theta_2 \vec{a}_2 \leftarrow \theta_1 \vec{a}_1$, in which $\theta_i \vec{a}_i$ means an $\vec{a}_i$ axis rotation through an angle θ_i. Its total rotation matrix is defined as

$$C^{B/A} = C^{B/A''} C^{A''/A'} C^{A'/A} \tag{5.28}$$

in which the first rotation matrix is simply

$$\mathbf{C}^{A'/A} = \mathbf{C}_1(\theta_1) \tag{5.29}$$

Next, to construct a matrix $\mathbf{C}^{A''/A'}$ that characterizes the $\vec{a}_2$ axis rotation with an angle θ_2, we use the approach discussed in the preceding section, as follows:

$$\begin{aligned}
\mathbf{C}^{A''/A'} &= \mathbf{C}^{A''/A'''} \mathbf{C}^{A'''/A} \mathbf{C}^{A/A'} \\
&= \mathbf{C}^{A'/A} \mathbf{C}_2(\theta_2) \mathbf{C}^{A/A'}
\end{aligned} \tag{5.30}$$

Combining Eqs. (5.29) and (5.30), we obtain

$$\begin{aligned}
\mathbf{C}^{A''/A} &= \mathbf{C}^{A''/A'} \mathbf{C}^{A'/A} \\
&= \mathbf{C}^{A'/A} \mathbf{C}_2(\theta_2) \mathbf{C}^{A/A'} \mathbf{C}^{A'/A} \\
&= \mathbf{C}_1(\theta_1) \mathbf{C}_2(\theta_2)
\end{aligned}$$

Similarly, for the $\vec{a}_3$ axis rotation through an angle θ_3, we have

$$\mathbf{C}^{B/A''} = \mathbf{C}^{A''/A} \mathbf{C}_3(\theta_3) \mathbf{C}^{A/A''} \tag{5.31}$$

Finally, the total rotation matrix becomes

$$\begin{aligned}
\mathbf{C}^{B/A} &= \mathbf{C}^{B/A''} \mathbf{C}^{A''/A} \\
&= \mathbf{C}^{A''/A} \mathbf{C}_3(\theta_3) \mathbf{C}^{A/A''} \mathbf{C}^{A''/A} \\
&= \mathbf{C}_1(\theta_1) \mathbf{C}_2(\theta_2) \mathbf{C}_3(\theta_3)
\end{aligned} \tag{5.32}$$

Thus, the total rotation matrix for the space-axis rotation of $\theta_3 \vec{a}_3 \leftarrow \theta_2 \vec{a}_2 \leftarrow \theta_1 \vec{a}_1$ is identical to the total rotation matrix for the body-axis rotation of $\theta_1 \vec{a}_1'' \leftarrow \theta_2 \vec{a}_2' \leftarrow \theta_3 \vec{a}_3$, which has been denoted by $\mathbf{C}_1(\theta_1) \leftarrow \mathbf{C}_2(\theta_2) \leftarrow \mathbf{C}_3(\theta_3)$ in preceding sections.

Although the total rotation matrix for the space-axis rotation has a simple form as Eq. (5.32), each intermediate rotation matrix is rather complicated as can be seen from Eqs. (5.30) and (5.31); however, the approach used here does not require an explicit determination of these intermediate matrices to find the total rotation matrix. Indeed, an intimate relationship between the two different rotation schemes has been obtained directly [4].

5.4 Quaternions

5.4.1 Euler Parameters or Quaternions

Consider again Euler's eigenaxis rotation about an arbitrary axis fixed both in a body-fixed reference frame B and in an inertial reference frame A. In Sec. 5.3, a unit vector $\vec{e}$ along the Euler axis was defined as

$$\begin{aligned}
\vec{e} &= e_1 \vec{a}_1 + e_2 \vec{a}_2 + e_3 \vec{a}_3 \\
&= e_1 \vec{b}_1 + e_2 \vec{b}_2 + e_3 \vec{b}_3
\end{aligned}$$

where e_i are the direction cosines of the Euler axis relative to both A and B, and $e_1^2 + e_2^2 + e_3^2 = 1$.

Then we define the four *Euler parameters* as follows:

$$q_1 = e_1 \sin(\theta/2) \tag{5.33a}$$

$$q_2 = e_2 \sin(\theta/2) \tag{5.33b}$$

$$q_3 = e_3 \sin(\theta/2) \tag{5.33c}$$

$$q_4 = \cos(\theta/2) \tag{5.33d}$$

where θ is the rotation angle about the Euler axis. Like the eigenaxis vector $\mathbf{e} = (e_1, e_2, e_3)$, we define a vector $\mathbf{q} = (q_1, q_2, q_3)$ such that

$$\mathbf{q} = \mathbf{e} \sin(\theta/2) \tag{5.34}$$

Note that the Euler parameters are not independent of each other, but constrained by the relationship

$$\mathbf{q}^T \mathbf{q} + q_4^2 = q_1^2 + q_2^2 + q_3^2 + q_4^2 = 1 \tag{5.35}$$

because $e_1^2 + e_2^2 + e_3^2 = 1$.

The Euler parameters are also called *quaternions*. Hamilton invented quaternions as a result of searching for hypercomplex numbers that could be represented by points in three-dimensional space.* Although the historical importance of quaternions is significant, we will not discuss quaternion algebra here. Instead, we simply use the terms *quaternions* and *Euler parameters* interchangeably.

The direction cosine matrix parameterized as Eq. (5.22) can also be parameterized in terms of quaternions, as follows:

$$\mathbf{C}^{B/A} = \mathbf{C}(\mathbf{q}, q_4) = \begin{bmatrix} 1 - 2(q_2^2 + q_3^2) & 2(q_1 q_2 + q_3 q_4) & 2(q_1 q_3 - q_2 q_4) \\ 2(q_2 q_1 - q_3 q_4) & 1 - 2(q_1^2 + q_3^2) & 2(q_2 q_3 + q_1 q_4) \\ 2(q_3 q_1 + q_2 q_4) & 2(q_3 q_2 - q_1 q_4) & 1 - 2(q_1^2 + q_2^2) \end{bmatrix} \tag{5.36}$$

where (q_1, q_2, q_3, q_4) is the quaternion associated with the direction cosine matrix $\mathbf{C}^{B/A}$. Note that $\sin \theta = 2 \sin(\theta/2) \cos(\theta/2)$ and $\cos \theta = \cos^2(\theta/2) - \sin^2(\theta/2) = 2\cos^2(\theta/2) - 1 = 1 - 2\sin^2(\theta/2)$.

In terms of the quaternion vector $\mathbf{q}$ and a skew-symmetric matrix $\mathbf{Q}$ defined, respectively, as

$$\mathbf{q} = \begin{bmatrix} q_1 \\ q_2 \\ q_3 \end{bmatrix}, \qquad \mathbf{Q} = \begin{bmatrix} 0 & -q_3 & q_2 \\ q_3 & 0 & -q_1 \\ -q_2 & q_1 & 0 \end{bmatrix} \tag{5.37}$$

---

*William Hamilton (1805–1865) regarded his discovery of quaternions as his greatest achievement, whereas we may consider his contributions to analytical dynamics as his greatest achievement.

the direction cosine matrix (5.36) becomes

$$\mathbf{C} = (q_4^2 - \mathbf{q}^T\mathbf{q})\mathbf{I} + 2\mathbf{q}\mathbf{q}^T - 2q_4\mathbf{Q} \tag{5.38}$$

Given a direction cosine matrix $\mathbf{C}$, we can determine q_4 and $\mathbf{q}$ as follows:

$$q_4 = \frac{1}{2}(1 + C_{11} + C_{22} + C_{33})^{\frac{1}{2}} \qquad \text{for } 0 \le \theta \le \pi \tag{5.39}$$

$$\mathbf{q} = \frac{1}{4q_4}\begin{bmatrix} C_{23} - C_{32} \\ C_{31} - C_{13} \\ C_{12} - C_{21} \end{bmatrix} \qquad \text{if } q_4 \ne 0 \tag{5.40}$$

Consider two successive rotations to A'' from A represented by

$$\mathbf{C}(\mathbf{q}', q_4') : \quad A' \leftarrow A \tag{5.41a}$$
$$\mathbf{C}(\mathbf{q}'', q_4'') : \quad A'' \leftarrow A' \tag{5.41b}$$

where $(\mathbf{q}', q_4')$ is the quaternion associated with the coordinate transformation $A' \leftarrow A$, and $(\mathbf{q}'', q_4'')$ is the quaternion associated with the coordinate transformation $A'' \leftarrow A'$. These successive rotations are also represented by a single rotation to A'' directly from A, as follows:

$$\mathbf{C}(\mathbf{q}, q_4) : \quad A'' \leftarrow A \tag{5.42}$$

where $(\mathbf{q}, q_4)$ is the quaternion associated with the coordinate transformation $A'' \leftarrow A$, and we have

$$\mathbf{C}(\mathbf{q}, q_4) = \mathbf{C}(\mathbf{q}'', q_4'')\mathbf{C}(\mathbf{q}', q_4') \tag{5.43}$$

Note that Eq. (5.43) can also be represented as

$$\mathbf{C}(\mathbf{e}, \theta) = \mathbf{C}(\mathbf{e}_2, \theta_2)\mathbf{C}(\mathbf{e}_1, \theta_1)$$

where $(\mathbf{e}_1, \theta_1)$ and $(\mathbf{e}_2, \theta_2)$ are the eigenaxes and angles associated with the first and second eigenaxis rotations, respectively, and $(\mathbf{e}, \theta)$ are the eigenaxis and angle associated with the equivalent single eigenaxis rotation (see Problem 5.3). Using the result of Problem 5.3(c), and defining

$$q_4' = \cos\frac{\theta_1}{2}, \qquad q_4'' = \cos\frac{\theta_2}{2}$$

and

$$\mathbf{q}' = \begin{bmatrix} q_1' \\ q_2' \\ q_3' \end{bmatrix} = \mathbf{e}_1 \sin\frac{\theta_1}{2}, \qquad \mathbf{q}'' = \begin{bmatrix} q_1'' \\ q_2'' \\ q_3'' \end{bmatrix} = \mathbf{e}_2 \sin\frac{\theta_2}{2}$$

we obtain

$$\mathbf{q} = q_4'' \mathbf{q}' + q_4' \mathbf{q}'' + \mathbf{q}' \times \mathbf{q}'' \tag{5.44}$$

$$q_4 = q_4' q_4'' - (\mathbf{q}')^T \mathbf{q}'' \tag{5.45}$$

These equations can be combined as

$$\begin{bmatrix} q_1 \\ q_2 \\ q_3 \\ q_4 \end{bmatrix} = \begin{bmatrix} q_4'' & q_3'' & -q_2'' & q_1'' \\ -q_3'' & q_4'' & q_1'' & q_2'' \\ q_2'' & -q_1'' & q_4'' & q_3'' \\ -q_1'' & -q_2'' & -q_3'' & q_4'' \end{bmatrix} \begin{bmatrix} q_1' \\ q_2' \\ q_3' \\ q_4' \end{bmatrix} \tag{5.46}$$

which is known as the quaternion multiplication rule in matrix form. The 4×4 orthonormal matrix in Eq. (5.46) is called the quaternion matrix. Equation (5.46) can also be written as

$$\begin{bmatrix} q_1 \\ q_2 \\ q_3 \\ q_4 \end{bmatrix} = \begin{bmatrix} q_4' & -q_3' & q_2' & q_1' \\ q_3' & q_4' & -q_1' & q_2' \\ -q_2' & q_1' & q_4' & q_3' \\ -q_1' & -q_2' & -q_3' & q_4' \end{bmatrix} \begin{bmatrix} q_1'' \\ q_2'' \\ q_3'' \\ q_4'' \end{bmatrix} \tag{5.47}$$

The 4×4 matrix in Eq. (5.47) is also orthonormal and is called the quaternion transmuted matrix [5].

5.4.2 Gibbs Parameters

The direction cosine matrix can also be parameterized in terms of the *Gibbs vector*, which is defined as

$$\mathbf{g} = \begin{bmatrix} g_1 \\ g_2 \\ g_3 \end{bmatrix} = \begin{bmatrix} q_1/q_4 \\ q_2/q_4 \\ q_3/q_4 \end{bmatrix} = \mathbf{e} \tan \frac{\theta}{2} \tag{5.48}$$

The components of the Gibbs vector, called the *Gibbs parameters*, are also referred to as the *Rodrigues parameters* in the literature, and the direction cosine matrix can be parameterized in terms of them as follows:

$$C = \frac{1}{1 + g_1^2 + g_2^2 + g_3^2}$$
$$\times \begin{bmatrix} 1 + g_1^2 - g_2^2 - g_3^2 & 2(g_1 g_2 + g_3) & 2(g_1 g_3 - g_2) \\ 2(g_2 g_1 - g_3) & 1 - g_1^2 + g_2^2 - g_3^2 & 2(g_2 g_3 + g_1) \\ 2(g_3 g_1 + g_2) & 2(g_3 g_2 - g_1) & 1 - g_1^2 - g_2^2 + g_3^2 \end{bmatrix} \tag{5.49}$$

which can be rewritten as

$$C = \frac{(1 - \mathbf{g}^T \mathbf{g})\mathbf{I} + 2\mathbf{g}\mathbf{g}^T - 2\mathbf{G}}{1 + \mathbf{g}^T \mathbf{g}} \equiv [\mathbf{I} - \mathbf{G}][\mathbf{I} + \mathbf{G}]^{-1} \tag{5.50}$$

where

$$
\mathbf{G} = \begin{bmatrix} 0 & -g_3 & g_2 \\ g_3 & 0 & -g_1 \\ -g_2 & g_1 & 0 \end{bmatrix}
$$

For a given direction cosine matrix $\mathbf{C}$, the Gibbs vector can be determined as

$$
\mathbf{g} = \begin{bmatrix} g_1 \\ g_2 \\ g_3 \end{bmatrix} = \frac{1}{1 + C_{11} + C_{22} + C_{33}} \begin{bmatrix} C_{23} - C_{32} \\ C_{31} - C_{13} \\ C_{12} - C_{21} \end{bmatrix} \tag{5.51}
$$

Problems

5.4 Show that

$$
\vec{b}_i = \vec{a}_i + 2\{q_4\vec{q} \times \vec{a}_i + \vec{q} \times (\vec{q} \times \vec{a}_i)\}, \qquad i = 1, 2, 3
$$

where $\vec{q} = \vec{e}\sin(\theta/2)$.

5.5 Consider the body-fixed rotational sequence to B from A: $\mathbf{C}_1(\theta_1) \leftarrow \mathbf{C}_2(\theta_2) \leftarrow \mathbf{C}_3(\theta_3)$.

(a) Show that the three Euler angles of this rotational sequence are related to quaternions, as follows:

$$
\begin{bmatrix} q_1 \\ q_2 \\ q_3 \\ q_4 \end{bmatrix} = \begin{bmatrix} s_1\, c_2\, c_3 - c_1\, s_2\, s_3 \\ c_1\, s_2\, c_3 + s_1\, c_2\, s_3 \\ c_1\, c_2\, s_3 - s_1\, s_2\, c_3 \\ c_1\, c_2\, c_3 + s_1\, s_2\, s_3 \end{bmatrix}
$$

where $s_i = \sin(\theta_i/2)$, $c_i = \cos(\theta_i/2)$, and (q_1, q_2, q_3, q_4) is the quaternion associated with the coordinate transformation $B \leftarrow A$.

Hint: Use Eq. (5.46). The quaternions associated with $\mathbf{C}_1(\theta_1) \leftarrow \mathbf{C}_2(\theta_2) \leftarrow \mathbf{C}_3(\theta_3)$ are represented as

$$
\begin{bmatrix} \sin(\theta_1/2) \\ 0 \\ 0 \\ \cos(\theta_1/2) \end{bmatrix} \leftarrow \begin{bmatrix} 0 \\ \sin(\theta_2/2) \\ 0 \\ \cos(\theta_2/2) \end{bmatrix} \leftarrow \begin{bmatrix} 0 \\ 0 \\ \sin(\theta_3/2) \\ \cos(\theta_3/2) \end{bmatrix}
$$

(b) Also verify that, for small (infinitesimal) rotational angles of θ_1, θ_2, and θ_3, we simply have

$$
q_1 \approx \theta_1/2
$$
$$
q_2 \approx \theta_2/2
$$
$$
q_3 \approx \theta_3/2
$$
$$
q_4 \approx 1
$$

5.5   Kinematic Differential Equations

In preceding sections, we have studied the problem of describing the orientation of a reference frame (or a rigid body) in terms of the direction cosine matrix, Euler angles, and quaternions. In this section, we treat *kinematics* in which the relative orientation between two reference frames is time dependent. The time-dependent relationship between two reference frames is described by the so-called *kinematic differential equations*. In this section, we derive the kinematic differential equations for the direction cosine matrix, Euler angles, and quaternions.

5.5.1   *Direction Cosine Matrix*

Consider two reference frames A and B, shown in Fig. 5.1, which are moving relative to each other. The angular velocity vector of a reference frame B with respect to a reference frame A is denoted by $\vec{\omega} \equiv \vec{\omega}^{B/A}$, and it is expressed in terms of basis vectors of B as follows:

$$\vec{\omega} = \omega_1 \vec{b}_1 + \omega_2 \vec{b}_2 + \omega_3 \vec{b}_3 \tag{5.52}$$

where the angular velocity vector $\vec{\omega}$ is time dependent.

In Sec. 5.1, we have defined the direction cosine matrix $\mathbf{C} \equiv \mathbf{C}^{B/A}$ such that

$$\begin{bmatrix} \vec{b}_1 \\ \vec{b}_2 \\ \vec{b}_3 \end{bmatrix} = \mathbf{C} \begin{bmatrix} \vec{a}_1 \\ \vec{a}_2 \\ \vec{a}_3 \end{bmatrix} \tag{5.53}$$

which can be rewritten as

$$\begin{bmatrix} \vec{a}_1 \\ \vec{a}_2 \\ \vec{a}_3 \end{bmatrix} = \mathbf{C}^{-1} \begin{bmatrix} \vec{b}_1 \\ \vec{b}_2 \\ \vec{b}_3 \end{bmatrix} = \mathbf{C}^T \begin{bmatrix} \vec{b}_1 \\ \vec{b}_2 \\ \vec{b}_3 \end{bmatrix} \tag{5.54}$$

Because the two reference frames are rotating relative to each other, the direction cosine matrix and its elements C_{ij} are functions of time. Taking the time derivative of Eq. (5.54) in A and denoting it by an overdot, we obtain

$$\begin{bmatrix} 0 \\ 0 \\ 0 \end{bmatrix} = \dot{\mathbf{C}}^T \begin{bmatrix} \vec{b}_1 \\ \vec{b}_2 \\ \vec{b}_3 \end{bmatrix} + \mathbf{C}^T \begin{bmatrix} \dot{\vec{b}}_1 \\ \dot{\vec{b}}_2 \\ \dot{\vec{b}}_3 \end{bmatrix}$$

$$= \dot{\mathbf{C}}^T \begin{bmatrix} \vec{b}_1 \\ \vec{b}_2 \\ \vec{b}_3 \end{bmatrix} + \mathbf{C}^T \begin{bmatrix} \vec{\omega} \times \vec{b}_1 \\ \vec{\omega} \times \vec{b}_2 \\ \vec{\omega} \times \vec{b}_3 \end{bmatrix}$$

$$= \dot{\mathbf{C}}^T \begin{bmatrix} \vec{b}_1 \\ \vec{b}_2 \\ \vec{b}_3 \end{bmatrix} - \mathbf{C}^T \begin{bmatrix} 0 & -\omega_3 & \omega_2 \\ \omega_3 & 0 & -\omega_1 \\ -\omega_2 & \omega_1 & 0 \end{bmatrix} \begin{bmatrix} \vec{b}_1 \\ \vec{b}_2 \\ \vec{b}_3 \end{bmatrix} \qquad (5.55)$$

where

$$\dot{\mathbf{C}} \equiv \begin{bmatrix} \dot{C}_{11} & \dot{C}_{12} & \dot{C}_{13} \\ \dot{C}_{21} & \dot{C}_{22} & \dot{C}_{23} \\ \dot{C}_{31} & \dot{C}_{32} & \dot{C}_{33} \end{bmatrix} \qquad (5.56)$$

By defining the skew-symmetric matrix in Eq. (5.55) as

$$\mathbf{\Omega} \equiv \begin{bmatrix} 0 & -\omega_3 & \omega_2 \\ \omega_3 & 0 & -\omega_1 \\ -\omega_2 & \omega_1 & 0 \end{bmatrix} \qquad (5.57)$$

we obtain

$$[\dot{\mathbf{C}}^T - \mathbf{C}^T \mathbf{\Omega}] \begin{bmatrix} \vec{b}_1 \\ \vec{b}_2 \\ \vec{b}_3 \end{bmatrix} = \begin{bmatrix} 0 \\ 0 \\ 0 \end{bmatrix}$$

from which we obtain

$$\dot{\mathbf{C}}^T - \mathbf{C}^T \mathbf{\Omega} = 0 \qquad (5.58)$$

Taking the transpose of Eq. (5.58) and using the relationship $\mathbf{\Omega}^T = -\mathbf{\Omega}$, we obtain

$$\dot{\mathbf{C}} + \mathbf{\Omega}\mathbf{C} = 0 \qquad (5.59)$$

which is called the kinematic differential equation for the direction cosine matrix $\mathbf{C}$. Differential equations for each element of $\mathbf{C}$ can be written as

$$\dot{C}_{11} = \omega_3 C_{21} - \omega_2 C_{31}$$

$$\dot{C}_{12} = \omega_3 C_{22} - \omega_2 C_{32}$$

$$\dot{C}_{13} = \omega_3 C_{23} - \omega_2 C_{33}$$

$$\dot{C}_{21} = \omega_1 C_{31} - \omega_3 C_{11}$$

$$\dot{C}_{22} = \omega_1 C_{32} - \omega_3 C_{12}$$

$$\dot{C}_{23} = \omega_1 C_{33} - \omega_3 C_{13}$$

$$\dot{C}_{31} = \omega_2 C_{11} - \omega_1 C_{21}$$

$$\dot{C}_{32} = \omega_2 C_{12} - \omega_1 C_{22}$$

$$\dot{C}_{33} = \omega_2 C_{13} - \omega_1 C_{23}$$

If ω_1, ω_2, and ω_3 are known as functions of time, then the orientation of B relative to A as a function of time can be determined by solving Eq. (5.59). In general, it is difficult to solve Eq. (5.59) analytically in closed form except in special cases; hence, in most cases, Eq. (5.59) is integrated numerically using a digital computer. It can be shown that the orthonormality condition $\mathbf{CC}^T = \mathbf{I} = \mathbf{C}^T\mathbf{C}$ is a constant integral of Eq. (5.59); that is, if the orthonormality condition is satisfied at $t = 0$, then any (exact) solution of Eq. (5.59) automatically satisfies the orthonormality condition of $\mathbf{C}$ for all $t > 0$. However, the orthonormality condition is often used to check the accuracy of numerical integration on a digital computer.

Problem

5.6 Given the kinematic differential equation (5.59), show that

$$\omega_1 = \dot{C}_{21} C_{31} + \dot{C}_{22} C_{32} + \dot{C}_{23} C_{33}$$

$$\omega_2 = \dot{C}_{31} C_{11} + \dot{C}_{32} C_{12} + \dot{C}_{33} C_{13}$$

$$\omega_3 = \dot{C}_{11} C_{21} + \dot{C}_{12} C_{22} + \dot{C}_{13} C_{23}$$

5.5.2 Euler Angles

Like the kinematic differential equation for the direction cosine matrix $\mathbf{C}$, the orientation of a reference frame B relative to a reference frame A can also be described by introducing the time dependence of Euler angles.

Consider the rotational sequence of $\mathbf{C}_1(\theta_1) \leftarrow \mathbf{C}_2(\theta_2) \leftarrow \mathbf{C}_3(\theta_3)$ to B from A, which is symbolically represented as

$$\mathbf{C}_3(\theta_3): \quad A' \leftarrow A \qquad (5.60a)$$

$$\mathbf{C}_2(\theta_2): \quad A'' \leftarrow A' \qquad (5.60b)$$

$$\mathbf{C}_1(\theta_1): \quad B \leftarrow A'' \qquad (5.60c)$$

The time derivatives of Euler angles, called Euler rates, are denoted by $\dot{\theta}_3$, $\dot{\theta}_2$, and $\dot{\theta}_1$. These successive rotations are also represented as

$$\vec{\omega}^{A'/A}: \quad A' \leftarrow A \qquad (5.61a)$$

$$\vec{\omega}^{A''/A'}: \quad A'' \leftarrow A' \qquad (5.61b)$$

$$\vec{\omega}^{B/A''}: \quad B \leftarrow A'' \qquad (5.61c)$$

and the angular velocity vectors $\vec{\omega}^{A'/A}$, $\vec{\omega}^{A''/A'}$, and $\vec{\omega}^{B/A''}$ are expressed as

$$\vec{\omega}^{A'/A} = \dot{\theta}_3\,\vec{a}_3 = \dot{\theta}_3\,\vec{a}_3' \tag{5.62a}$$

$$\vec{\omega}^{A''/A'} = \dot{\theta}_2\,\vec{a}_2' = \dot{\theta}_2\,\vec{a}_2'' \tag{5.62b}$$

$$\vec{\omega}^{B/A''} = \dot{\theta}_1\,\vec{a}_1'' = \dot{\theta}_1\,\vec{b}_1 \tag{5.62c}$$

The angular velocity vector $\vec{\omega}^{B/A}$ then becomes

$$\vec{\omega}^{B/A} = \vec{\omega}^{B/A''} + \vec{\omega}^{A''/A'} + \vec{\omega}^{A'/A} = \dot{\theta}_1\,\vec{b}_1 + \dot{\theta}_2\,\vec{a}_2'' + \dot{\theta}_3\,\vec{a}_3' \tag{5.63}$$

which can be rewritten as

$$\vec{\omega}^{B/A} = [\vec{b}_1 \quad \vec{b}_2 \quad \vec{b}_3]\begin{bmatrix}\dot{\theta}_1\\0\\0\end{bmatrix} + [\vec{a}_1'' \quad \vec{a}_2'' \quad \vec{a}_3'']\begin{bmatrix}0\\\dot{\theta}_2\\0\end{bmatrix} + [\vec{a}_1' \quad \vec{a}_2' \quad \vec{a}_3']\begin{bmatrix}0\\0\\\dot{\theta}_3\end{bmatrix}$$

$$\tag{5.64}$$

and we have

$$[\vec{a}_1'' \quad \vec{a}_2'' \quad \vec{a}_3''] = [\vec{b}_1 \quad \vec{b}_2 \quad \vec{b}_3]\mathbf{C}_1(\theta_1)$$

$$[\vec{a}_1' \quad \vec{a}_2' \quad \vec{a}_3'] = [\vec{b}_1 \quad \vec{b}_2 \quad \vec{b}_3]\,\mathbf{C}_1(\theta_1)\mathbf{C}_2(\theta_2)$$

Because the angular velocity vector $\vec{\omega} \equiv \vec{\omega}^{B/A}$ can also be represented as

$$\vec{\omega} = \omega_1\,\vec{b}_1 + \omega_2\,\vec{b}_2 + \omega_3\,\vec{b}_3 = [\vec{b}_1 \quad \vec{b}_2 \quad \vec{b}_3]\begin{bmatrix}\omega_1\\\omega_2\\\omega_3\end{bmatrix} \tag{5.65}$$

we obtain

$$\begin{bmatrix}\omega_1\\\omega_2\\\omega_3\end{bmatrix} = \begin{bmatrix}\dot{\theta}_1\\0\\0\end{bmatrix} + \mathbf{C}_1(\theta_1)\begin{bmatrix}0\\\dot{\theta}_2\\0\end{bmatrix} + \mathbf{C}_1(\theta_1)\mathbf{C}_2(\theta_2)\begin{bmatrix}0\\0\\\dot{\theta}_3\end{bmatrix}$$

$$= \begin{bmatrix}1 & 0 & -\sin\theta_2\\0 & \cos\theta_1 & \sin\theta_1\cos\theta_2\\0 & -\sin\theta_1 & \cos\theta_1\cos\theta_2\end{bmatrix}\begin{bmatrix}\dot{\theta}_1\\\dot{\theta}_2\\\dot{\theta}_3\end{bmatrix} \tag{5.66}$$

Note that the 3×3 matrix in Eq. (5.66) is not an orthogonal matrix because $\vec{b}_1$, $\vec{a}_2''$, and $\vec{a}_3'$ do not constitute a set of orthogonal unit vectors. The inverse relationship can be found by inverting the 3×3 nonorthogonal matrix in Eq. (5.66), as follows:

$$
\begin{bmatrix} \dot{\theta}_1 \\ \dot{\theta}_2 \\ \dot{\theta}_3 \end{bmatrix} = \frac{1}{\cos \theta_2} \begin{bmatrix} \cos \theta_2 & \sin \theta_1 \sin \theta_2 & \cos \theta_1 \sin \theta_2 \\ 0 & \cos \theta_1 \cos \theta_2 & -\sin \theta_1 \cos \theta_2 \\ 0 & \sin \theta_1 & \cos \theta_1 \end{bmatrix} \begin{bmatrix} \omega_1 \\ \omega_2 \\ \omega_3 \end{bmatrix} \quad (5.67)
$$

which is the kinematic differential equation for the sequence of $C_1(\theta_1) \leftarrow C_2(\theta_2) \leftarrow C_3(\theta_3)$.

If ω_1, ω_2, and ω_3 are known as functions of time, then the orientation of B relative to A as a function of time can be determined by solving Eq. (5.67). Numerical integration of Eq. (5.67), however, involves the computation of trigonometric functions of the angles. Also note that Eq. (5.67) becomes singular when $\theta_2 = \pi/2$. Such a mathematical singularity problem for a certain orientation angle can be avoided by selecting a different set of Euler angles, but it is an inherent property of all different sets of Euler angles.

Similarly, for the sequence of $C_3(\psi) \leftarrow C_1(\theta) \leftarrow C_3(\phi)$, we have

$$
\begin{aligned}
\vec{\omega} \equiv \vec{\omega}^{B/A} &= \omega_1 \vec{b}_1 + \omega_2 \vec{b}_2 + \omega_3 \vec{b}_3 \\
&= \dot{\psi} \vec{b}_3 + \dot{\theta} \vec{a}_1'' + \dot{\phi} \vec{a}_3'
\end{aligned} \quad (5.68)
$$

$$
\begin{aligned}
\begin{bmatrix} \omega_1 \\ \omega_2 \\ \omega_3 \end{bmatrix} &= \begin{bmatrix} 0 \\ 0 \\ \dot{\psi} \end{bmatrix} + C_3(\psi) \begin{bmatrix} \dot{\theta} \\ 0 \\ 0 \end{bmatrix} + C_3(\psi)C_1(\theta) \begin{bmatrix} 0 \\ 0 \\ \dot{\phi} \end{bmatrix} \\
&= \begin{bmatrix} \sin \theta \sin \psi & \cos \psi & 0 \\ \sin \theta \cos \psi & -\sin \psi & 0 \\ \cos \theta & 0 & 1 \end{bmatrix} \begin{bmatrix} \dot{\phi} \\ \dot{\theta} \\ \dot{\psi} \end{bmatrix}
\end{aligned} \quad (5.69)
$$

and

$$
\begin{bmatrix} \dot{\phi} \\ \dot{\theta} \\ \dot{\psi} \end{bmatrix} = \frac{1}{\sin \theta} \begin{bmatrix} \sin \psi & \cos \psi & 0 \\ \cos \psi \sin \theta & -\sin \psi \sin \theta & 0 \\ -\sin \psi \cos \theta & -\cos \psi \cos \theta & \sin \theta \end{bmatrix} \begin{bmatrix} \omega_1 \\ \omega_2 \\ \omega_3 \end{bmatrix} \quad (5.70)
$$

which is the kinematic differential equation for the $C_3(\psi) \leftarrow C_1(\theta) \leftarrow C_3(\phi)$ sequence.

Problems

5.7 For the sequence of $C_1(\theta_1) \leftarrow C_3(\theta_3) \leftarrow C_2(\theta_2)$, derive the following kinematic differential equation:

$$
\begin{bmatrix} \dot{\theta}_1 \\ \dot{\theta}_2 \\ \dot{\theta}_3 \end{bmatrix} = \frac{1}{\cos \theta_3} \begin{bmatrix} \cos \theta_3 & -\cos \theta_1 \sin \theta_3 & \sin \theta_1 \sin \theta_3 \\ 0 & \cos \theta_1 & -\sin \theta_1 \\ 0 & \sin \theta_1 \cos \theta_3 & \cos \theta_1 \cos \theta_3 \end{bmatrix} \begin{bmatrix} \omega_1 \\ \omega_2 \\ \omega_3 \end{bmatrix}
$$

5.8 For the sequence of $C_3(\theta_3) \leftarrow C_2(\theta_2) \leftarrow C_1(\theta_1)$, derive the following kinematic differential equation:

$$
\begin{bmatrix} \dot{\theta}_1 \\ \dot{\theta}_2 \\ \dot{\theta}_3 \end{bmatrix} = \frac{1}{\cos\theta_2}
\begin{bmatrix} \cos\theta_3 & -\sin\theta_3 & 0 \\ \cos\theta_2\sin\theta_3 & \cos\theta_2\cos\theta_3 & 0 \\ -\sin\theta_2\cos\theta_3 & \sin\theta_2\sin\theta_3 & \cos\theta_2 \end{bmatrix}
\begin{bmatrix} \omega_1 \\ \omega_2 \\ \omega_3 \end{bmatrix}
$$

5.5.3 Quaternions

Substituting C_{ij} of Eq. (5.36) into the equations derived in Problem 5.6, we obtain

$$\omega_1 = 2(\dot{q}_1 q_4 + \dot{q}_2 q_3 - \dot{q}_3 q_2 - \dot{q}_4 q_1) \tag{5.71a}$$

$$\omega_2 = 2(\dot{q}_2 q_4 + \dot{q}_3 q_1 - \dot{q}_1 q_3 - \dot{q}_4 q_2) \tag{5.71b}$$

$$\omega_3 = 2(\dot{q}_3 q_4 + \dot{q}_1 q_2 - \dot{q}_2 q_1 - \dot{q}_4 q_3) \tag{5.71c}$$

Differentiating Eq. (5.35) gives

$$0 = 2(\dot{q}_1 q_1 + \dot{q}_2 q_2 + \dot{q}_3 q_3 + \dot{q}_4 q_4) \tag{5.72}$$

These four equations can be combined into matrix form, as follows:

$$
\begin{bmatrix} \omega_1 \\ \omega_2 \\ \omega_3 \\ 0 \end{bmatrix} = 2
\begin{bmatrix} q_4 & q_3 & -q_2 & -q_1 \\ -q_3 & q_4 & q_1 & -q_2 \\ q_2 & -q_1 & q_4 & -q_3 \\ q_1 & q_2 & q_3 & q_4 \end{bmatrix}
\begin{bmatrix} \dot{q}_1 \\ \dot{q}_2 \\ \dot{q}_3 \\ \dot{q}_4 \end{bmatrix} \tag{5.73}
$$

Because the 4×4 matrix in this equation is orthonormal, we simply obtain the kinematic differential equation for quaternions, as follows:

$$
\begin{bmatrix} \dot{q}_1 \\ \dot{q}_2 \\ \dot{q}_3 \\ \dot{q}_4 \end{bmatrix} = \frac{1}{2}
\begin{bmatrix} q_4 & -q_3 & q_2 & q_1 \\ q_3 & q_4 & -q_1 & q_2 \\ -q_2 & q_1 & q_4 & q_3 \\ -q_1 & -q_2 & -q_3 & q_4 \end{bmatrix}
\begin{bmatrix} \omega_1 \\ \omega_2 \\ \omega_3 \\ 0 \end{bmatrix} \tag{5.74}
$$

which can be rewritten as

$$
\begin{bmatrix} \dot{q}_1 \\ \dot{q}_2 \\ \dot{q}_3 \\ \dot{q}_4 \end{bmatrix} = \frac{1}{2}
\begin{bmatrix} 0 & \omega_3 & -\omega_2 & \omega_1 \\ -\omega_3 & 0 & \omega_1 & \omega_2 \\ \omega_2 & -\omega_1 & 0 & \omega_3 \\ -\omega_1 & -\omega_2 & -\omega_3 & 0 \end{bmatrix}
\begin{bmatrix} q_1 \\ q_2 \\ q_3 \\ q_4 \end{bmatrix} \tag{5.75}
$$

In terms of $\mathbf{q}$ and $\boldsymbol{\omega}$ defined as

$$
\mathbf{q} = \begin{bmatrix} q_1 \\ q_2 \\ q_3 \end{bmatrix}, \qquad \boldsymbol{\omega} = \begin{bmatrix} \omega_1 \\ \omega_2 \\ \omega_3 \end{bmatrix}
$$

we can rewrite the kinematic differential equation (5.75) as follows:

$$\dot{q} = \tfrac{1}{2}(q_4\boldsymbol{\omega} - \boldsymbol{\omega} \times \mathbf{q}) \tag{5.76a}$$

$$\dot{q}_4 = -\tfrac{1}{2}\boldsymbol{\omega}^T\mathbf{q} \tag{5.76b}$$

where

$$\boldsymbol{\omega} \times \mathbf{q} \equiv \begin{bmatrix} 0 & -\omega_3 & \omega_2 \\ \omega_3 & 0 & -\omega_1 \\ -\omega_2 & \omega_1 & 0 \end{bmatrix}\begin{bmatrix} q_1 \\ q_2 \\ q_3 \end{bmatrix}$$

It is historically interesting to note that Eq. (5.75) was first published by Robinson [6] in 1958 and derived independently by Harding [7], Mortenson [8], and Margulies (see [9]) in the mid-1960s.

In *strapdown inertial reference systems* of aerospace vehicles, the body rates, ω_1, ω_2, and ω_3 are measured by rate gyros that are "strapped down" to the vehicles. The kinematic differential Eq. (5.75) is then integrated numerically using an onboard flight computer to determine the orientation of the vehicles in terms of quaternions. Quaternions have no inherent geometric singularity as do Euler angles. Moreover, quaternions are well suited for onboard real-time computation because only products and no trigonometric relations exist in the quaternion kinematic differential equations. Thus, spacecraft orientation is now commonly described in terms of quaternions.

There are a number of numerical methods available for solving Eq. (5.75). Methods that can be applied to the strapdown attitude algorithms include Taylor series expansion, the rotation vector concept, Runge–Kutta algorithms, and the state transition matrix. Of these methods, the Taylor series expansion lends itself well to the use of an incremental angle output from the digital rate integrating gyros. A tradeoff between algorithm complexity vs algorithm truncation and roundoff errors is generally required (see, e.g., [10]).

For further details of rotational kinematics and spacecraft attitude determination, the reader is referred to Refs. 11–15.

Currently, spacecraft attitude determination using the Global Positioning System (GPS) is also being considered for near-Earth satellites. The GPS, consisting of a constellation of 24 satellites and a ground monitoring and control network, is widely used for positioning vehicles near the surface of the Earth and for orbit determination of near-Earth satellites [16]. The GPS is also capable of providing vehicle attitude using L-band carrier phase interferometry between multiple antennas. Consequently, GPS-based attitude determination is of current research interest for near-Earth satellites because of the potential for reducing the number of onboard navigation and attitude sensors [17].

Problems

5.9 Show that the constraint $q_1^2 + q_2^2 + q_3^2 + q_4^2 = 1$ is a constant integral of Eq. (5.75). If this constraint is satisfied at $t = 0$, then any exact solution of Eq. (5.75) automatically satisfies the constraint for all $t > 0$; however, this constraint is often used to check the accuracy of numerical integration on a digital computer.

5.10   Consider the Gibbs vector defined as

$$\mathbf{g} = \mathbf{e} \tan \frac{\theta}{2}$$

where $\mathbf{g} = (g_1, g_2, g_3)$, $\mathbf{e} = (e_1, e_2, e_3)$ is Euler's eigenaxis vector, and θ is the angle associated with Euler's eigenaxis rotation.

(a) Show that the kinematic differential equation for the Gibbs vector can be found as

$$
\begin{bmatrix} \dot{g}_1 \\ \dot{g}_2 \\ \dot{g}_3 \end{bmatrix} = \frac{1}{2}
\begin{bmatrix}
1 + g_1^2 & g_1 g_2 - g_3 & g_1 g_3 + g_2 \\
g_2 g_1 + g_3 & 1 + g_2^2 & g_2 g_3 - g_1 \\
g_3 g_1 - g_2 & g_3 g_2 + g_1 & 1 + g_3^2
\end{bmatrix}
\begin{bmatrix} \omega_1 \\ \omega_2 \\ \omega_3 \end{bmatrix}
$$

or

$$\dot{\mathbf{g}} = \tfrac{1}{2}[\boldsymbol{\omega} - \boldsymbol{\omega} \times \mathbf{g} + (\boldsymbol{\omega}^T \mathbf{g})\mathbf{g}] = \tfrac{1}{2}[\mathbf{I} + \mathbf{G} + \mathbf{g}\mathbf{g}^T]\boldsymbol{\omega}$$

where

$$
\mathbf{G} = \begin{bmatrix}
0 & -g_3 & g_2 \\
g_3 & 0 & -g_1 \\
-g_2 & g_1 & 0
\end{bmatrix}
$$

(b) Also show that for infinitesimal rotations, we have

$$\dot{\mathbf{g}} \approx \tfrac{1}{2}\boldsymbol{\omega}$$

Detailed discussions of the recent progresses in spacecraft attitude estimation methods can be found in Refs. 18–20.

References

[1] Kane, T. R., Likins, P. W., and Levinson, D. A., *Spacecraft Dynamics*, McGraw–Hill, New York, 1983.

[2] Hughes, P. C., *Spacecraft Attitude Dynamics*, Wiley, New York, 1986, pp. 16, 17.

[3] Fu, K. S., Gonzalez, R. C., and Lee, C. S. G., *Robotics: Control, Sensing, Vision, and Intelligence*, McGraw–Hill, New York, 1987, pp. 14–25.

[4] Wie, B., "A New Approach to the Space-Axis Rotation," *Journal of Guidance, Control, and Dynamics*, Vol. 10, No. 4, 1987, pp. 411–412.

[5] Ickes, B. F., "A New Method for Performing Control System Attitude Computations Using Quaternions," *AIAA Journal*, Vol. 8, No. 1, 1970, pp. 13–17.

[6] Robinson, A. C., "On the Use of Quaternions in Simulation of Rigid-Body Motion," WADC Technical Rept. 58-17, Dec. 1958.

[7] Harding, C. F., "Solution of Euler Gyrodynamics," *Journal of Applied Mechanics*, Vol. 31, Ser. E, No. 2, 1964, pp. 325–328.

[8] Mortenson, R. E., "Comment on Solution of Euler Gyrodynamics," *Journal of Applied Mechanics*, Vol. 32, Ser. E, No. 1, 1965, pp. 228–230.

[9] Wie, B., Weiss, H., and Arapostathis, A., "Quaternion Feedback Regulator for Spacecraft Eigenaxis Rotations," *Journal of Guidance, Control, and Dynamics*, Vol. 12, No. 3, 1989, pp. 375–380.

[10] Wie, B., and Barba, P. M., "Quaternion Feedback for Spacecraft Large Angle Maneuvers," *Journal of Guidance, Control, and Dynamics*, Vol. 8, No. 3, 1985, pp. 360–365.

[11] Garg, S. C., Morrow, L. D., and Mamen, R., "Strapdown Navigation Technology: A Literature Survey," *Journal of Guidance and Control*, Vol. 1, No. 3, 1978, pp. 161–172.

[12] Wertz, J. R. (ed.), *Spacecraft Attitude Determination and Control*, Kluwer Academic, Dordrecht, The Netherlands, 1978.

[13] Lefferts, E. G., Markley, F. L., and Shuster, M. D., "Kalman Filtering for Spacecraft Attitude Estimation," *Journal of Guidance, Control, and Dynamics*, Vol. 5, No. 5, 1982, pp. 417–429.

[14] Junkins, J. L., and Turner, J. D., *Optimal Spacecraft Rotational Maneuvers*, Elsevier, New York, 1985.

[15] Shuster, M. D., "A Survey of Attitude Representations," *Journal of the Astronautical Sciences*, Vol. 41, No. 4, 1993, pp. 439–518.

[16] Parkinson, B. W., and Spilker, J. J. (eds.), *Global Positioning System: Theory and Applications*, Vol. I and II, Progress in Astronautics and Aeronautics, Vol. 164, AIAA, Washington, DC, 1996.

[17] Axelrad, P., and Ward, L. M., "Spacecraft Attitude Estimation Using the Global Positioning System: Methodology and Results for RADCAL," *Journal of Guidance, Control, and Dynamics*, Vol. 19, No. 6, 1996, pp. 1201–1209.

[18] Markley, F. L., "Attitude Estimation or Quaternion Estimation?," *Journal of the Astronautical Sciences*, Vol. 52, Nos. 1/2, 2004, pp. 221–238.

[19] Crassidis, J. L., and Junkins, J. L., *Optimal Estimation of Dynamic Systems*, Chapman & Hall/CRC, Boca Raton, FL, 2004.

[20] Crassidis, J. L., Markley, F. L., and Cheng, Y., "Survey of Nonlinear Attitude Estimation Methods," *Journal of Guidance, Control, and Dynamics*, Vol. 30, No. 1, 2007, pp. 12–28.

6
Rigid-Body Dynamics

The motion of a rigid body in space consists of the translational motion of its center of mass and the rotational motion of the body about its center of mass; thus, a rigid body in space is a dynamic system with six degrees of freedom. The translational motion of a rigid body in space was treated in Part II. This chapter is concerned with the rotational motion of a rigid vehicle with or without the influence of gravitational and other external forces. Rotational maneuvering and attitude control problems of rigid space vehicles will be covered in Chapter 7.

6.1 Angular Momentum of a Rigid Body

Consider a rigid body that is in motion relative to a Newtonian inertial reference frame N, as shown in Fig. 6.1. The rotational equation of motion of the rigid body about an arbitrary point O is given as

$$\int \vec{r} \times \ddot{\vec{R}} \, dm = \vec{M}_o \tag{6.1}$$

where $\vec{r}$ is the position vector of a small (infinitesimal) mass element dm relative to point O, $\vec{R}$ is the position vector of dm from an inertial origin of N, $\ddot{\vec{R}}$ is the inertial acceleration of dm, and $\vec{M}_o$ is the total external moment (or torque) about point O.

Let $\vec{r}_c$ be the position vector of the center of mass relative to point O and also let $\vec{\rho}$ be the position vector of dm relative to the center of mass. Then we have

$$\int \vec{r} \, dm = m\vec{r}_c \tag{6.2a}$$

$$\int \vec{\rho} \, dm = 0 \tag{6.2b}$$

where m denotes the mass of the rigid body.

Because $\vec{R} = \vec{R}_o + \vec{r}$, we can rewrite Eq. (6.1) as

$$\dot{\vec{h}}_o + m\vec{r}_c \times \ddot{\vec{R}}_o = \vec{M}_o \tag{6.3}$$

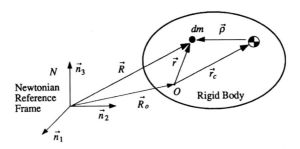

Fig. 6.1   Rigid body in motion relative to a Newtonian reference frame.

where $\vec{h}_o$, called the *relative angular momentum* about point O, is defined as

$$\vec{h}_o = \int \vec{r} \times \dot{\vec{r}} \, dm \tag{6.4}$$

Note that the time derivative is taken with respect to an inertial reference frame.

Like the relative angular momentum defined as Eq. (6.4), the *absolute angular momentum* about point O is defined as

$$\vec{H}_o = \int \vec{r} \times \dot{\vec{R}} \, dm \tag{6.5}$$

Combining Eqs. (6.1) and (6.5), we obtain

$$\dot{\vec{H}}_o + m\dot{\vec{R}}_o \times \dot{\vec{r}}_c = \vec{M}_o \tag{6.6}$$

If the reference point O is either inertially fixed or at the center of mass of the rigid body, the distinction between $\vec{H}_o$ and $\vec{h}_o$ disappears and the angular momentum equation simply becomes

$$\dot{\vec{H}}_o = \vec{M}_o \qquad \text{or} \qquad \dot{\vec{h}}_o = \vec{M}_o \tag{6.7}$$

Furthermore, if the moment of forces $\vec{M}_o$ is zero, then the angular momentum vector becomes a constant vector, that is, the angular momentum of the rigid body is conserved. This is known as the *principle of conservation of angular momentum*. For this reason, the center of mass is often selected as a reference point O of the rigid body.

6.2   Inertia Matrix and Inertia Dyadic

Consider a rigid body with a body-fixed reference frame B with its origin at the center of mass of the rigid body, as shown in Fig. 6.2. In this figure, $\vec{\rho}$ denotes the position vector of a small mass element dm from the center of mass, $\vec{R}_c$ is the position vector of the center of mass from an inertial origin of N, and $\vec{R}$ is the position vector of dm from an inertial origin of N.

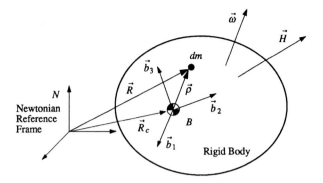

Fig. 6.2   Rigid body with a body-fixed reference frame B with its origin at the center of mass.

Let $\vec{\omega} \equiv \vec{\omega}^{B/N}$ be the angular velocity vector of the rigid body in an inertial reference frame N. The angular momentum vector $\vec{H}$ of a rigid body about its center of mass is then defined as

$$\vec{H} = \int \vec{\rho} \times \dot{\vec{R}} \, dm = \int \vec{\rho} \times \dot{\vec{\rho}} \, dm = \int \vec{\rho} \times (\vec{\omega} \times \vec{\rho}) \, dm \qquad (6.8)$$

as $\vec{R} = \vec{R}_c + \vec{\rho}$, $\int \vec{\rho} \, dm = 0$, $\dot{\vec{R}} \equiv \{d\vec{R}/dt\}_N$, and

$$\dot{\vec{\rho}} \equiv \left\{ \frac{d\vec{\rho}}{dt} \right\}_N = \left\{ \frac{d\vec{\rho}}{dt} \right\}_B + \vec{\omega}^{B/N} \times \vec{\rho} \qquad (6.9)$$

Note that $\{d\vec{\rho}/dt\}_B = 0$ for a rigid body.

Let $\vec{\rho}$ and $\vec{\omega}$ be expressed as

$$\vec{\rho} = \rho_1 \vec{b}_1 + \rho_2 \vec{b}_2 + \rho_3 \vec{b}_3 \qquad (6.10a)$$

$$\vec{\omega} = \omega_1 \vec{b}_1 + \omega_2 \vec{b}_2 + \omega_3 \vec{b}_3 \qquad (6.10b)$$

where $\{\vec{b}_1, \vec{b}_2, \vec{b}_3\}$ is a set of three orthogonal unit vectors, called basis vectors, of a body-fixed reference frame B. The angular momentum vector described by Eq. (6.8) can then be written as

$$\vec{H} = (J_{11}\omega_1 + J_{12}\omega_2 + J_{13}\omega_3)\vec{b}_1 + (J_{21}\omega_1 + J_{22}\omega_2 + J_{23}\omega_3)\vec{b}_2$$
$$+ (J_{31}\omega_1 + J_{32}\omega_2 + J_{33}\omega_3)\vec{b}_3 \qquad (6.11)$$

where J_{11}, J_{22}, and J_{33} are the *moments of inertia* defined as*

$$J_{11} = \int \left(\rho_2^2 + \rho_3^2\right) dm \qquad (6.12a)$$

$$J_{22} = \int \left(\rho_1^2 + \rho_3^2\right) dm \qquad (6.12b)$$

$$J_{33} = \int \left(\rho_1^2 + \rho_2^2\right) dm \qquad (6.12c)$$

and J_{ij} $(i \neq j)$ are the *products of inertia* defined as

$$J_{12} = J_{21} = -\int \rho_1 \rho_2 \, dm \qquad (6.13a)$$

$$J_{13} = J_{31} = -\int \rho_1 \rho_3 \, dm \qquad (6.13b)$$

$$J_{23} = J_{32} = -\int \rho_2 \rho_3 \, dm \qquad (6.13c)$$

Expressing the angular momentum vector as

$$\vec{H} = H_1 \vec{b}_1 + H_2 \vec{b}_2 + H_3 \vec{b}_3 \qquad (6.14)$$

we obtain

$$H_1 = J_{11}\omega_1 + J_{12}\omega_2 + J_{13}\omega_3 \qquad (6.15a)$$
$$H_2 = J_{21}\omega_1 + J_{22}\omega_2 + J_{23}\omega_3 \qquad (6.15b)$$
$$H_3 = J_{31}\omega_1 + J_{32}\omega_2 + J_{33}\omega_3 \qquad (6.15c)$$

which can be rewritten in matrix form, as follows:

$$\begin{bmatrix} H_1 \\ H_2 \\ H_3 \end{bmatrix} = \begin{bmatrix} J_{11} & J_{12} & J_{13} \\ J_{21} & J_{22} & J_{23} \\ J_{31} & J_{32} & J_{33} \end{bmatrix} \begin{bmatrix} \omega_1 \\ \omega_2 \\ \omega_3 \end{bmatrix} \qquad (6.16)$$

or

$$\mathbf{H} = \mathbf{J}\boldsymbol{\omega} \qquad (6.17)$$

where

$$\mathbf{H} = \begin{bmatrix} H_1 \\ H_2 \\ H_3 \end{bmatrix}, \qquad \mathbf{J} = \begin{bmatrix} J_{11} & J_{12} & J_{13} \\ J_{21} & J_{22} & J_{23} \\ J_{31} & J_{32} & J_{33} \end{bmatrix}, \qquad \boldsymbol{\omega} = \begin{bmatrix} \omega_1 \\ \omega_2 \\ \omega_3 \end{bmatrix}$$

---

*The term "moments of inertia" is attributed to Euler.

and $\mathbf{J}$ is called the *inertia matrix* of a rigid body about a body-fixed reference frame B with its origin at the center of mass.* Note that the inertia matrix is a symmetric matrix; i.e., $\mathbf{J} = \mathbf{J}^T$.

To introduce the *inertia dyadic* of a rigid body, we consider a pair of vectors with neither a dot nor a cross between them, such as $\vec{b}_i \vec{b}_j$. Such a pair of vectors is called a dyadic with the following properties:

$$(\vec{b}_i \vec{b}_j) \cdot \vec{b}_k = \vec{b}_i(\vec{b}_j \cdot \vec{b}_k) \tag{6.18a}$$

$$\vec{b}_i \cdot (\vec{b}_j \vec{b}_k) = (\vec{b}_i \cdot \vec{b}_j)\vec{b}_k \tag{6.18b}$$

The *unit dyadic*, denoted as $\hat{I}$, is defined as

$$\hat{I} = \vec{b}_1 \vec{b}_1 + \vec{b}_2 \vec{b}_2 + \vec{b}_3 \vec{b}_3 = [\vec{b}_1 \quad \vec{b}_2 \quad \vec{b}_3] \begin{bmatrix} \vec{b}_1 \\ \vec{b}_2 \\ \vec{b}_3 \end{bmatrix} \tag{6.19}$$

and we have $\hat{I} \cdot \vec{\omega} = \vec{\omega}$.

Using the vector identity

$$\vec{x} \times (\vec{y} \times \vec{z}) = (\vec{x} \cdot \vec{z})\vec{y} - \vec{z}(\vec{x} \cdot \vec{y})$$

and employing dyadic notation, we express the angular momentum vector $\vec{H}$ as

$$\vec{H} = \int \vec{\rho} \times (\vec{\omega} \times \vec{\rho})\, dm$$

$$= \int [\rho^2 \vec{\omega} - \vec{\rho}(\vec{\rho} \cdot \vec{\omega})]\, dm, \qquad \rho = |\vec{\rho}|$$

$$= \int [(\rho^2 \hat{I} \cdot \vec{\omega} - (\vec{\rho}\vec{\rho}) \cdot \vec{\omega}]\, dm$$

$$= \left[\int (\rho^2 \hat{I} - \vec{\rho}\vec{\rho})\, dm \right] \cdot \vec{\omega}$$

$$= \hat{J} \cdot \vec{\omega} \tag{6.20}$$

where $\hat{J}$ is the inertia dyadic defined as

$$\hat{J} = \int (\rho^2 \hat{I} - \vec{\rho}\vec{\rho})\, dm \tag{6.21}$$

The inertia dyadic is related to the inertia matrix by

$$\hat{J} = \sum_{i=1}^{3} \sum_{j=1}^{3} J_{ij} \vec{b}_i \vec{b}_j = [\vec{b}_1 \quad \vec{b}_2 \quad \vec{b}_3] \begin{bmatrix} J_{11} & J_{12} & J_{13} \\ J_{21} & J_{22} & J_{23} \\ J_{31} & J_{32} & J_{33} \end{bmatrix} \begin{bmatrix} \vec{b}_1 \\ \vec{b}_2 \\ \vec{b}_3 \end{bmatrix} \tag{6.22}$$

---

*In most textbooks on dynamics, the symbol $\mathbf{I}$ is commonly used to denote an inertia matrix; however, throughout this book, the symbol $\mathbf{J}$ indicates an inertia matrix and the symbol $\mathbf{I}$ indicates an identity matrix.

Consequently, the ijth element of the inertia matrix is related to the inertia dyadic by

$$J_{ij} = \vec{b}_i \cdot \hat{J} \cdot \vec{b}_j$$

Like the inertia dyadic described as Eq. (6.22), the unit dyadic that is related to the unit (or identity) matrix can also be expressed as follows:

$$\hat{I} = \sum_{i=1}^{3} \sum_{j=1}^{3} \delta_{ij} \vec{b}_i \vec{b}_j = [\vec{b}_1 \quad \vec{b}_2 \quad \vec{b}_3] \begin{bmatrix} 1 & 0 & 0 \\ 0 & 1 & 0 \\ 0 & 0 & 1 \end{bmatrix} \begin{bmatrix} \vec{b}_1 \\ \vec{b}_2 \\ \vec{b}_3 \end{bmatrix} \qquad (6.23)$$

where δ_{ij} is the Kronecker delta.

Note that Eq. (6.20) is a concise mathematical expression of the relationship between the angular velocity vector and angular momentum vector using the inertia dyadic $\hat{J}$. The equivalent relationship in matrix notation is given as Eq. (6.17), and the angular momentum vector of Eq. (6.20) can also be expressed as

$$\vec{H} = [\vec{b}_1 \quad \vec{b}_2 \quad \vec{b}_3] \begin{bmatrix} H_1 \\ H_2 \\ H_3 \end{bmatrix} = [\vec{b}_1 \quad \vec{b}_2 \quad \vec{b}_3] \begin{bmatrix} J_{11} & J_{12} & J_{13} \\ J_{21} & J_{22} & J_{23} \\ J_{31} & J_{32} & J_{33} \end{bmatrix} \begin{bmatrix} \omega_1 \\ \omega_2 \\ \omega_3 \end{bmatrix} \qquad (6.24)$$

The rotational kinetic energy of a rigid body is defined as

$$T = \frac{1}{2} \int \dot{\vec{\rho}} \cdot \dot{\vec{\rho}} \, dm$$

which can be written as

$$\begin{aligned} T &= \frac{1}{2} \int \dot{\vec{\rho}} \cdot (\vec{\omega} \times \vec{\rho}) \, dm \\ &= \frac{1}{2} \vec{\omega} \cdot \int \vec{\rho} \times \dot{\vec{\rho}} \, dm \\ &= \frac{1}{2} \vec{\omega} \cdot \vec{H} \\ &= \frac{1}{2} \vec{\omega} \cdot \hat{J} \cdot \vec{\omega} \end{aligned} \qquad (6.25)$$

In matrix notation, the rotational kinetic energy of a rigid body is written as

$$T = \frac{1}{2} \omega^T \mathbf{H} = \frac{1}{2} \omega^T \mathbf{J} \omega = \frac{1}{2} \sum_{i=1}^{3} \sum_{j=1}^{3} J_{ij} \omega_i \omega_j \qquad (6.26)$$

Problem

6.1 Consider a rigid body as illustrated in Fig. 6.1. Let $\hat{J}$ be the inertia dyadic of
the body relative to its center of mass; i.e.,

$$\hat{J} = \int (\rho^2 \hat{I} - \vec{\rho}\vec{\rho}) \, dm$$

where $\vec{\rho}$ is the position vector of dm relative to the center of mass and $\rho = |\vec{\rho}|$.
Also let $\hat{J}'$ be the inertia dyadic of the body relative to an arbitrary point O;
i.e.,

$$\hat{J}' = \int (r^2 \hat{I} - \vec{r}\vec{r}) \, dm$$

where $\vec{r}$ is the position vector of dm relative to point O and $r = |\vec{r}|$. The
position vector of the center of mass from point O is denoted by $\vec{r}_c$ and
$\vec{r} = \vec{r}_c + \vec{\rho}$.
(a) Verify the parallel-axis theorem,

$$\hat{J}' = \hat{J} + m\left[r_c^2 \hat{I} - \vec{r}_c \vec{r}_c \right]$$

where m is the mass of the body and $r_c \equiv |\vec{r}_c|$.
(b) Let $\{\vec{b}_1, \vec{b}_2, \vec{b}_3\}$ be a set of basis vectors of a body-fixed reference frame
with its origin at the center of mass and let $\vec{r}_c = \ell_1 \vec{b}_1 + \ell_2 \vec{b}_2 + \ell_3 \vec{b}_3$.
Also let $J_{ij} = \vec{b}_i \cdot \hat{J} \cdot \vec{b}_j$ and $J'_{ij} = \vec{b}_i \cdot \hat{J}' \cdot \vec{b}_j$. Then show that the parallel-
axis theorem can also be represented as

$$J'_{ij} = J_{ij} + m\left[(\ell_1^2 + \ell_2^2 + \ell_3^2)\delta_{ij} - \ell_i \ell_j \right]$$

where δ_{ij} is the Kronecker delta.

6.3 Principal Axes

Consider two body-fixed reference frames B and B' with basis vectors
$\{\vec{b}_1, \vec{b}_2, \vec{b}_3\}$ and $\{\vec{b}'_1, \vec{b}'_2, \vec{b}'_3\}$, respectively, as shown in Fig. 6.3. The angular
momentum vector and angular velocity vector of a rigid body are expressed
in terms of these basis vectors, as follows:

$$\vec{H} = \begin{bmatrix} \vec{b}_1 & \vec{b}_2 & \vec{b}_3 \end{bmatrix} \begin{bmatrix} H_1 \\ H_2 \\ H_3 \end{bmatrix} = \begin{bmatrix} \vec{b}'_1 & \vec{b}'_2 & \vec{b}'_3 \end{bmatrix} \begin{bmatrix} H'_1 \\ H'_2 \\ H'_3 \end{bmatrix} \qquad (6.27\text{a})$$

$$\vec{\omega} = \begin{bmatrix} \vec{b}_1 & \vec{b}_2 & \vec{b}_3 \end{bmatrix} \begin{bmatrix} \omega_1 \\ \omega_2 \\ \omega_3 \end{bmatrix} = \begin{bmatrix} \vec{b}'_1 & \vec{b}'_2 & \vec{b}'_3 \end{bmatrix} \begin{bmatrix} \omega'_1 \\ \omega'_2 \\ \omega'_3 \end{bmatrix} \qquad (6.27\text{b})$$

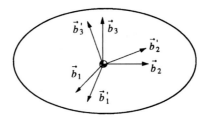

Fig. 6.3   Rigid body with two body-fixed reference frames B and B'.

The components of the angular momentum and angular velocity vectors are related to each other by

$$\mathbf{H} = \mathbf{J}\omega \tag{6.28a}$$

$$\mathbf{H}' = \mathbf{J}'\omega' \tag{6.28b}$$

where $\mathbf{H} = (H_1, H_2, H_3)$, $\mathbf{H}' = (H_1', H_2', H_3')$, $\omega = (\omega_1, \omega_2, \omega_3)$, $\omega' = (\omega_1', \omega_2', \omega_3')$, $\mathbf{J}$ is the inertia matrix about the reference frame B, and $\mathbf{J}'$ is the inertia matrix about the reference frame B'.

Given a direction cosine matrix $\mathbf{C} \equiv \mathbf{C}^{B'/B}$ of B' relative to B, we have the following transformation relationships:

$$\mathbf{H}' = \mathbf{C}\mathbf{H}, \qquad \omega' = \mathbf{C}\omega \tag{6.29}$$

Combining Eqs. (6.28) and (6.29), we obtain

$$\mathbf{J}' = \mathbf{C}\mathbf{J}\mathbf{C}^T \quad \text{or} \quad \mathbf{J} = \mathbf{C}^T\mathbf{J}'\mathbf{C} \tag{6.30}$$

because $\mathbf{C}^{-1} = \mathbf{C}^T$.

For any shape of a rigid body, there exists a set of axes for which the products of inertia are all zero. Such axes are called *principal axes* and the corresponding moments of inertia are called the *principal moments of inertia.** The principal axes with the smallest, intermediate, and largest inertia are often called the minor, intermediate, and major axes, respectively.

Given an inertia matrix $\mathbf{J}$ about a reference frame B with basis vectors $\{\vec{b}_1, \vec{b}_2, \vec{b}_3\}$, the principal moments of inertia and the corresponding set of principal axes B' with basis vectors $\{\vec{b}_1', \vec{b}_2', \vec{b}_3'\}$ can be found by solving the following eigenvalue/eigenvector problem:

$$[\lambda \mathbf{I} - \mathbf{J}]\mathbf{e} = 0 \tag{6.31}$$

where $\mathbf{I}$ is a 3×3 identity matrix, and λ and $\mathbf{e}$ are the eigenvalue and eigenvector, respectively, to be determined. The eigenvalues become the principal moments

---

*In 1750 Euler discovered the existence of principal axes and principal moments of inertia of a rigid body.

of inertia, whereas the eigenvectors provide the orientation of the corresponding principal axes relative to a reference frame B.

The eigenvalues are simply the roots of the characteristic equation

$$|\lambda \mathbf{I} - \mathbf{J}| = 0 \tag{6.32}$$

After determining the three eigenvalues λ_1, λ_2, and λ_3, which are the principal moments of inertia, we also find the corresponding eigenvectors $\mathbf{e}_1$, $\mathbf{e}_2$, and $\mathbf{e}_3$ by solving the following equations:

$$[\lambda_i \mathbf{I} - \mathbf{J}]\mathbf{e}_i = 0, \qquad i = 1, 2, 3 \tag{6.33}$$

where $\mathbf{e}_i = (e_{i1}, e_{i2}, e_{i3})$, or

$$\begin{bmatrix} \lambda_i - J_{11} & -J_{12} & -J_{13} \\ -J_{21} & \lambda_i - J_{22} & -J_{23} \\ -J_{31} & -J_{32} & \lambda_i - J_{33} \end{bmatrix} \begin{bmatrix} e_{i1} \\ e_{i2} \\ e_{i3} \end{bmatrix} = \begin{bmatrix} 0 \\ 0 \\ 0 \end{bmatrix}, \qquad i = 1, 2, 3$$

If each eigenvector $\mathbf{e}_i$ is normalized such that

$$e_{i1}^2 + e_{i2}^2 + e_{i3}^2 = 1, \qquad i = 1, 2, 3$$

then $\{\vec{b}_1', \vec{b}_2', \vec{b}_3'\}$ of the principal-axis reference frame B' are written as

$$\vec{b}_1' \equiv \vec{e}_1 = e_{11}\vec{b}_1 + e_{12}\vec{b}_2 + e_{13}\vec{b}_3 \tag{6.34a}$$

$$\vec{b}_2' \equiv \vec{e}_2 = e_{21}\vec{b}_1 + e_{22}\vec{b}_2 + e_{23}\vec{b}_3 \tag{6.34b}$$

$$\vec{b}_3' \equiv \vec{e}_3 = e_{31}\vec{b}_1 + e_{32}\vec{b}_2 + e_{33}\vec{b}_3 \tag{6.34c}$$

Because e_{ij} is, in fact, the direction cosine between $\vec{b}_i'$ and $\vec{b}_j$, the direction cosine matrix of B' relative to B is defined as

$$\mathbf{C} = \mathbf{C}^{B'/B} = \begin{bmatrix} e_{11} & e_{12} & e_{13} \\ e_{21} & e_{22} & e_{23} \\ e_{31} & e_{32} & e_{33} \end{bmatrix} \tag{6.35}$$

Furthermore, we have

$$\mathbf{C}\mathbf{J}\mathbf{C}^T = \mathbf{J}' = \mathrm{diag}(J_1, J_2, J_3) \tag{6.36}$$

where $J_i \equiv \lambda_i$ denote the principal moments of inertia. The corresponding inertia dyadic about the principal axes is

$$\hat{J}' = J_1 \vec{b}_1' \vec{b}_1' + J_2 \vec{b}_2' \vec{b}_2' + J_3 \vec{b}_3' \vec{b}_3' \tag{6.37}$$

As an example, consider a rigid body with the following inertia matrix

$$\mathbf{J} = \begin{bmatrix} 1500 & 0 & -1000 \\ 0 & 2700 & 0 \\ -1000 & 0 & 3000 \end{bmatrix} \text{ kg·m}^2$$

about a body-fixed reference frame B with its origin at the center of mass.

The three eigenvalues of the inertia matrix $\mathbf{J}$ can be obtained as 1000, 2700, and 3500. Letting $(\lambda_1, \lambda_2, \lambda_3) = (1000, 2700, 3500)$, we find the inertia matrix about the principal axes as $\mathbf{J}' = \text{diag}(1000, 2700, 3500)$. Furthermore, the direction cosine matrix of the corresponding principal axes B' relative to B can be obtained as

$$\mathbf{C} \equiv \mathbf{C}^{B'/B} = \begin{bmatrix} e_{11} & e_{12} & e_{13} \\ e_{21} & e_{22} & e_{23} \\ e_{31} & e_{32} & e_{33} \end{bmatrix} = \begin{bmatrix} 2/\sqrt{5} & 0 & 1/\sqrt{5} \\ 0 & 1 & 0 \\ -1/\sqrt{5} & 0 & 2/\sqrt{5} \end{bmatrix}$$

One can also verify that $\mathbf{CJC}^T = \mathbf{J}' = \text{diag}(1000, 2700, 3500)$.

Problem

6.2   Consider an inertia matrix of the form

$$\mathbf{J} = \begin{bmatrix} J_{11} & J_{12} & J_{13} \\ J_{21} & J_{22} & J_{23} \\ J_{31} & J_{32} & J_{33} \end{bmatrix}$$

which is symmetric and positive definite. (All of the eigenvalues of a symmetric matrix are real. A symmetric matrix is said to be positive definite if all its eigenvalues or all its leading principal minors are positive.)

(a) Show that the moments of inertia are interrelated by

$$J_{11} + J_{22} > J_{33}, \qquad J_{22} + J_{33} > J_{11}, \qquad J_{33} + J_{11} > J_{22}$$

which are called the *triangle inequalities*.
Hint: Use Eqs. (6.12).

(b) Show that

$$J_{11} > 0, \qquad J_{22} > 0, \qquad J_{33} > 0$$

$$J_{11}J_{22} - J_{12}^2 > 0, \qquad J_{22}J_{33} - J_{23}^2 > 0, \qquad J_{33}J_{11} - J_{31}^2 > 0$$

$$J_{11}J_{22}J_{33} + 2J_{12}J_{23}J_{31} - J_{11}J_{23}^2 - J_{22}J_{31}^2 - J_{33}J_{12}^2 > 0$$

$$J_{11} > 2|J_{23}|, \qquad J_{22} > 2|J_{31}|, \qquad J_{33} > 2|J_{12}|$$

Hint: Use the definition as well as the positive definiteness of the inertia matrix.

(c) Let J_1, J_2, and J_3 be the principal moments of inertia. Then show that

$$J_{11} + J_{22} + J_{33} = J_1 + J_2 + J_3$$

$$J_{11}J_{22} + J_{22}J_{33} + J_{33}J_{11} - J_{12}^2 - J_{23}^2 - J_{31}^2 = J_1J_2 + J_2J_3 + J_3J_1$$

$$J_{11}J_{22}J_{33} + 2J_{12}J_{23}J_{31} - J_{11}J_{23}^2 - J_{22}J_{31}^2 - J_{33}J_{12}^2 = J_1J_2J_3$$

Hint: $|\lambda\mathbf{I} - \mathbf{J}| = (\lambda - J_1)(\lambda - J_2)(\lambda - J_3)$.

6.4   Euler's Rotational Equations of Motion

As discussed in Sec. 6.1, the angular momentum equation of a rigid body about its center of mass is simply given as

$$\vec{M} = \dot{\vec{H}} \tag{6.38}$$

where $\vec{H}$ is the angular momentum vector of a rigid body about its mass center, $\vec{M}$ is the external moment acting on the body about its mass center, and we have

$$\dot{\vec{H}} \equiv \left\{\frac{d\vec{H}}{dt}\right\}_N = \left\{\frac{d\vec{H}}{dt}\right\}_B + \vec{\omega}^{B/N} \times \vec{H} \tag{6.39}$$

where $\vec{H} = \hat{J} \cdot \vec{\omega}^{B/N}$.

The rotational equation of motion of a rigid body about its center of mass is then written as

$$\vec{M} = \left\{\frac{d\vec{H}}{dt}\right\}_B + \vec{\omega}^{B/N} \times \vec{H} \tag{6.40}$$

For notational convenience, let $\vec{\omega} \equiv \vec{\omega}^{B/N}$, then Eq. (6.40) becomes

$$\vec{M} = \left\{\frac{d}{dt}(\hat{J} \cdot \vec{\omega})\right\}_B + \vec{\omega} \times \hat{J} \cdot \vec{\omega}$$

$$= \left\{\frac{d\hat{J}}{dt}\right\}_B \cdot \vec{\omega} + \hat{J} \cdot \left\{\frac{d\vec{\omega}}{dt}\right\}_B + \vec{\omega} \times \hat{J} \cdot \vec{\omega} \tag{6.41}$$

where $\{d\hat{J}/dt\}_B = 0$ and $\{d\vec{\omega}/dt\}_B = \{d\vec{\omega}/dt\}_N = \dot{\vec{\omega}}$. Finally, we obtain

$$\vec{M} = \hat{J} \cdot \dot{\vec{\omega}} + \vec{\omega} \times \hat{J} \cdot \vec{\omega} \tag{6.42}$$

which is called Euler's rotational equation of motion in vector/dyadic form.

Let $\vec{M}$, $\vec{H}$, and $\vec{\omega}$ be expressed in terms of body-fixed basis vectors $\{\vec{b}_1, \vec{b}_2, \vec{b}_3\}$, as follows:

$$\vec{M} = M_1\vec{b}_1 + M_2\vec{b}_2 + M_3\vec{b}_3$$
$$\vec{H} = H_1\vec{b}_1 + H_2\vec{b}_2 + H_3\vec{b}_3$$
$$\vec{\omega} = \omega_1\vec{b}_1 + \omega_2\vec{b}_2 + \omega_3\vec{b}_3$$

Substituting these into Eq. (6.40), we also obtain the rotational equation of motion in matrix form, as follows:

$$\begin{bmatrix} M_1 \\ M_2 \\ M_3 \end{bmatrix} = \begin{bmatrix} \dot{H}_1 \\ \dot{H}_2 \\ \dot{H}_3 \end{bmatrix} + \begin{bmatrix} 0 & -\omega_3 & \omega_2 \\ \omega_3 & 0 & -\omega_1 \\ -\omega_2 & \omega_1 & 0 \end{bmatrix} \begin{bmatrix} H_1 \\ H_2 \\ H_3 \end{bmatrix} \quad (6.43)$$

Because

$$\begin{bmatrix} H_1 \\ H_2 \\ H_3 \end{bmatrix} = \begin{bmatrix} J_{11} & J_{12} & J_{13} \\ J_{21} & J_{22} & J_{23} \\ J_{31} & J_{32} & J_{33} \end{bmatrix} \begin{bmatrix} \omega_1 \\ \omega_2 \\ \omega_3 \end{bmatrix}$$

we obtain

$$\begin{bmatrix} M_1 \\ M_2 \\ M_3 \end{bmatrix} = \begin{bmatrix} J_{11} & J_{12} & J_{13} \\ J_{21} & J_{22} & J_{23} \\ J_{31} & J_{32} & J_{33} \end{bmatrix} \begin{bmatrix} \dot{\omega}_1 \\ \dot{\omega}_2 \\ \dot{\omega}_3 \end{bmatrix}$$
$$+ \begin{bmatrix} 0 & -\omega_3 & \omega_2 \\ \omega_3 & 0 & -\omega_1 \\ -\omega_2 & \omega_1 & 0 \end{bmatrix} \begin{bmatrix} J_{11} & J_{12} & J_{13} \\ J_{21} & J_{22} & J_{23} \\ J_{31} & J_{32} & J_{33} \end{bmatrix} \begin{bmatrix} \omega_1 \\ \omega_2 \\ \omega_3 \end{bmatrix} \quad (6.44)$$

Defining a skew-symmetric matrix

$$\Omega = \begin{bmatrix} 0 & -\omega_3 & \omega_2 \\ \omega_3 & 0 & -\omega_1 \\ -\omega_2 & \omega_1 & 0 \end{bmatrix} \quad (6.45)$$

we rewrite Eq. (6.44) concisely as

$$\mathbf{J}\dot{\omega} + \Omega \mathbf{J}\omega = \mathbf{M} \quad (6.46)$$

where

$$\mathbf{J} = \begin{bmatrix} J_{11} & J_{12} & J_{13} \\ J_{21} & J_{22} & J_{23} \\ J_{31} & J_{32} & J_{33} \end{bmatrix}, \qquad \omega = \begin{bmatrix} \omega_1 \\ \omega_2 \\ \omega_3 \end{bmatrix}, \qquad \mathbf{M} = \begin{bmatrix} M_1 \\ M_2 \\ M_3 \end{bmatrix}$$

Using cross product notation of two column vectors, ω and $\mathbf{J}\omega$, defined as

$$\omega \times \mathbf{J}\omega \equiv \Omega \mathbf{J}\omega$$

we often write Eq. (6.46) as

$$\mathbf{J}\dot{\boldsymbol{\omega}} + \boldsymbol{\omega} \times \mathbf{J}\boldsymbol{\omega} = \mathbf{M} \qquad (6.47)$$

which must be distinguished from the vector/dyadic form of Euler's rotational equation of motion

$$\hat{J} \cdot \dot{\vec{\omega}} + \vec{\omega} \times \hat{J} \cdot \vec{\omega} = \vec{M}$$

For a principal-axis reference frame with a set of basis vectors $\{\vec{b}_1, \vec{b}_2, \vec{b}_3\}$, Euler's rotational equations of motion of a rigid body become

$$J_1\dot{\omega}_1 - (J_2 - J_3)\omega_2\omega_3 = M_1 \qquad (6.48a)$$
$$J_2\dot{\omega}_2 - (J_3 - J_1)\omega_3\omega_1 = M_2 \qquad (6.48b)$$
$$J_3\dot{\omega}_3 - (J_1 - J_2)\omega_1\omega_2 = M_3 \qquad (6.48c)$$

where J_1, J_2, and J_3 are the principal moments of inertia, $M_i = \vec{M} \cdot \vec{b}_i$, $\omega_i = \vec{\omega} \cdot \vec{b}_i$. These are three coupled, nonlinear ordinary differential equations for state variables ω_1, ω_2, and ω_3 of a rigid body. These dynamical equations and the kinematic differential equations of the preceding chapter completely describe the rotational motions of a rigid body with three rotational degrees of freedom (i.e., six state variables).

Problems

6.3 Consider Euler's equations of motion of a rigid spacecraft given by Eqs. (6.48).
   (a) Show that the rotational kinetic energy T and the angular momentum $H = |\vec{H}|$ are simply given by

$$2T = J_1\omega_1^2 + J_2\omega_2^2 + J_3\omega_3^2$$
$$H^2 = (J_1\omega_1)^2 + (J_2\omega_2)^2 + (J_3\omega_3)^2$$

   (b) Show that

$$\dot{T} = \omega_1 M_1 + \omega_2 M_2 + \omega_3 M_3$$
$$\dot{H} = \frac{J_1\omega_1 M_1 + J_2\omega_2 M_2 + J_3\omega_3 M_3}{H}$$

6.4 Consider Euler's equations of motion at steady state of the form

$$0 = (J_2 - J_3)\omega_2\omega_3 + M_1$$
$$0 = (J_3 - J_1)\omega_3\omega_1 + M_2$$
$$0 = (J_1 - J_2)\omega_1\omega_2 + M_3$$

where M_1, M_2, and M_3 are constant body-fixed torque components. Show that equilibrium points exist if and only if $M_1 M_2 M_3 > 0$ (only if $M_1 M_2 M_3 \geq 0$).

6.5 Consider again Euler's equations of a rigid body described by Eqs. (6.48) which can be rewritten as

$$\dot{\omega}_1 - k_1\omega_2\omega_3 = M_1/J_1$$
$$\dot{\omega}_2 + k_2\omega_3\omega_1 = M_2/J_2$$
$$\dot{\omega}_3 - k_3\omega_1\omega_2 = M_3/J_3$$

where $\dot{\omega}_i \equiv d\omega_i/dt$ and

$$k_1 = \frac{J_2 - J_3}{J_1}, \qquad k_2 = \frac{J_1 - J_3}{J_2}, \qquad k_3 = \frac{J_1 - J_2}{J_3}$$

It is assumed that $J_1 > J_2 > J_3$ and $k_i > 0$, without loss of generality.

(a) Show that Euler's equations of motion can also be written as

$$\frac{dx_1}{d\tau} - x_2x_3 = \mu_1$$
$$\frac{dx_2}{d\tau} + x_3x_1 = \mu_2$$
$$\frac{dx_3}{d\tau} - x_1x_2 = \mu_3$$

where $\tau = t\sqrt{k_1k_2k_3}$, and

$$x_1 = \frac{\omega_1}{\sqrt{k_1}}, \qquad x_2 = \frac{\omega_2}{\sqrt{k_2}}, \qquad x_3 = \frac{\omega_3}{\sqrt{k_3}}$$

and

$$\mu_1 = \frac{M_1}{J_1k_1\sqrt{k_2k_3}}, \qquad \mu_2 = \frac{M_2}{J_2k_2\sqrt{k_3k_1}}, \qquad \mu_3 = \frac{M_3}{J_3k_3\sqrt{k_1k_2}}$$

(b) For a special case of constant $M_1 > 0$ and $M_2 = M_3 = 0$, show that Euler's equations of motion can be written in nondimensional form as

$$\frac{dx_1}{d\tau} - x_2x_3 = 1$$
$$\frac{dx_2}{d\tau} + x_3x_1 = 0$$
$$\frac{dx_3}{d\tau} - x_1x_2 = 0$$

where $\tau = t\sqrt{\mu k_1 k_2 k_3}$, and

$$\mu = \frac{M_1}{J_1k_1\sqrt{k_2k_3}}, \qquad x_1 = \frac{\omega_1}{\sqrt{\mu k_1}}, \qquad x_2 = \frac{\omega_2}{\sqrt{\mu k_2}}, \qquad x_3 = \frac{\omega_3}{\sqrt{\mu k_3}}$$

6.5 Torque-Free Motion of an Axisymmetric Rigid Body

Most spin-stabilized spacecraft are nearly axisymmetric, and they rotate about one of their principal axes. The stability of torque-free rotational motion of such spin-stabilized spacecraft is of practical importance. The term "torque-free motion" commonly employed in spacecraft attitude dynamics refers to the rotational motion of a rigid body in the presence of no external torques.

Consider a torque-free, axisymmetric rigid body with a body-fixed reference frame B, which has basis vectors $\{\vec{b}_1, \vec{b}_2, \vec{b}_3\}$, and which has its origin at the center of mass, as illustrated in Fig. 6.4. The reference frame B coincides with a set of principal axes, and the $\vec{b}_3$ axis is the axis of symmetry; thus, $J_1 = J_2$.

Euler's rotational equations of motion of a torque-free, axisymmetric spacecraft with $J_1 = J_2 = J$ become

$$J\dot{\omega}_1 - (J - J_3)\omega_3\omega_2 = 0 \tag{6.49}$$

$$J\dot{\omega}_2 + (J - J_3)\omega_3\omega_1 = 0 \tag{6.50}$$

$$J_3\dot{\omega}_3 = 0 \tag{6.51}$$

where $\omega_i \equiv \vec{b}_i \cdot \vec{\omega}$ are the body-fixed components of the angular velocity of the spacecraft.

From Eq. (6.51), we have

$$\omega_3 = \text{const} = n \tag{6.52}$$

where the constant n is called the *spin rate* of the spacecraft about its symmetry axis $\vec{b}_3$.

Defining the relative spin rate λ as

$$\lambda = \frac{(J - J_3)n}{J}$$

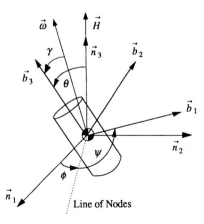

Fig. 6.4   Torque-free motion of an axisymmetric rigid body.

we rewrite Eqs. (6.49) and (6.50) as

$$\dot{\omega}_1 - \lambda\omega_2 = 0 \tag{6.53}$$
$$\dot{\omega}_2 + \lambda\omega_1 = 0 \tag{6.54}$$

Adding Eqs. (6.53) and (6.54) after multiplying by ω_1 and ω_2, respectively, we obtain

$$\omega_1\dot{\omega}_1 + \omega_2\dot{\omega}_2 = 0 \tag{6.55}$$

which, after integration, leads to

$$\omega_1^2 + \omega_2^2 = \text{const} = \omega_{12}^2 \tag{6.56}$$

where ω_{12} is called the constant transverse angular velocity.

Consequently, we find that the magnitude of the angular velocity vector $\vec{\omega}$ is constant; i.e.,

$$|\vec{\omega}| = \sqrt{\omega_1^2 + \omega_2^2 + \omega_3^2} = \sqrt{\omega_{12}^2 + n^2} = \text{const} \tag{6.57}$$

Note that the angular velocity vector $\vec{\omega}$ itself is not a constant vector as ω_1 and ω_2 are not constant and they are, in fact, the solutions of Eqs. (6.53) and (6.54) given by

$$\omega_1(t) = \omega_1(0)\cos\lambda t + \omega_2(0)\sin\lambda t \tag{6.58a}$$
$$\omega_2(t) = \omega_2(0)\cos\lambda t - \omega_1(0)\sin\lambda t \tag{6.58b}$$

where $\omega_1(0) \equiv -\dot{\omega}_2(0)/\lambda$ and $\omega_2(0) \equiv \dot{\omega}_1(0)/\lambda$ are the initial conditions at $t = 0$.

Because no external torque is acting on the spacecraft, the angular momentum vector $\vec{H}$ of the spacecraft is constant and it is inertially fixed in space; i.e.,

$$\dot{\vec{H}} = \vec{M} = 0 \quad\text{or}\quad \vec{H} = \text{constant vector} \tag{6.59}$$

The constant angular momentum vector is expressed in terms of basis vectors $\{\vec{b}_1, \vec{b}_2, \vec{b}_3\}$ of a rotating reference frame B, as follows:

$$\begin{aligned}
\vec{H} &= H_1\vec{b}_1 + H_2\vec{b}_2 + H_3\vec{b}_3 \\
&= J_1\omega_1\vec{b}_1 + J_2\omega_2\vec{b}_2 + J_3\omega_3\vec{b}_3 \\
&= J(\omega_1\vec{b}_1 + \omega_2\vec{b}_2) + J_3n\vec{b}_3
\end{aligned}$$

and

$$H \equiv |\vec{H}| = \sqrt{H_1^2 + H_2^2 + H_3^2} = \sqrt{(J\omega_{12})^2 + (J_3n)^2} = \text{const} \tag{6.60}$$

The angle θ between $\vec{H}$ and the axis of symmetry ($\vec{b}_3$ axis) is called the *nutation angle*, and it can be determined as

$$\cos\theta = \frac{\vec{H} \cdot \vec{b}_3}{H} = \frac{J_3 n}{\sqrt{(J\omega_{12})^2 + (J_3 n)^2}} = \text{const} \qquad (6.61)$$

or

$$\tan\theta = \frac{\sqrt{H_1^2 + H_2^2}}{H_3} = \frac{J\omega_{12}}{J_3 n} \qquad (6.62)$$

The angle γ between $\vec{b}_3$ and $\vec{\omega}$ can also be determined as

$$\tan\gamma = \frac{\sqrt{\omega_1^2 + \omega_2^2}}{\omega_3} = \frac{\omega_{12}}{n} = \text{const} \qquad (6.63)$$

The angles θ and γ are related to each other by

$$\tan\theta = (J/J_3)\tan\gamma \qquad (6.64)$$

For a disk-shaped (oblate) body with $J_3 > J$, we have $\gamma > \theta$. For a rod-shaped (prolate) body with $J_3 < J$, we have $\gamma < \theta$. The three vectors $\vec{H}$, $\vec{\omega}$, and $\vec{b}_3$ are coplanar at all times, and this plane rotates about $\vec{H}$.

To describe the rotational motion of a spinning spacecraft as seen from an inertial reference frame, we consider the body-fixed rotational sequence of $C_3(\psi) \leftarrow C_1(\theta) \leftarrow C_3(\phi)$ treated in Chapter 5. The three Euler angles ϕ, θ, and ψ of this rotational sequence are shown in Fig. 6.4, and we have

$$\omega_1 = \dot{\phi}\sin\theta\sin\psi + \dot{\theta}\cos\psi \qquad (6.65a)$$
$$\omega_2 = \dot{\phi}\sin\theta\cos\psi - \dot{\theta}\sin\psi \qquad (6.65b)$$
$$\omega_3 = \dot{\psi} + \dot{\phi}\cos\theta \qquad (6.65c)$$

Because both the nutation angle θ and ω_3 are constant, we have

$$\omega_1 = \dot{\phi}\sin\theta\sin\psi \qquad (6.66a)$$
$$\omega_2 = \dot{\phi}\sin\theta\cos\psi \qquad (6.66b)$$
$$\omega_3 = n = \dot{\psi} + \dot{\phi}\cos\theta \qquad (6.66c)$$

Furthermore, both $\dot{\phi}$ and $\dot{\psi}$ become constant as $\omega_1^2 + \omega_2^2 = \text{const}$.
Substituting Eqs. (6.66) into Eq. (6.49), we obtain

$$J\dot{\phi}\sin\theta\cos\psi\dot{\psi} - (J - J_3)\dot{\phi}\sin\theta\cos\psi n = 0 \qquad (6.67)$$

which can be rewritten as

$$\dot{\psi} = \frac{(J - J_3)n}{J} = \lambda \qquad (6.68)$$

Consequently, $\dot{\psi}$ is also called the *relative spin rate*, whereas $\dot{\phi}$ is called the *precession rate* or *coning speed*.

Using the relationship $\omega_3 = n = \dot{\psi} + \dot{\phi}\cos\theta$, we can also express the precession rate $\dot{\phi}$ as

$$\dot{\phi} = (n - \dot{\psi})/\cos\theta = J_3 n/J \cos\theta \qquad (6.69)$$

or

$$\dot{\phi} = J_3\dot{\psi}/(J - J_3)\cos\theta \qquad (6.70)$$

If $J > J_3$, the right-hand side of this equation becomes positive and we have a *direct precession*; this is a situation in which the spin and precession are in the same direction. On the other hand, if $J < J_3$, the right-hand side of this equation becomes negative and we have a *retrograde precession*, where the spin and precession are in opposite directions.

Because $H_3 = H\cos\theta = J_3 n$, we also have

$$H = J_3 n/\cos\theta = J\dot{\phi} \qquad (6.71)$$

The coning motion, or precession, of a spinning axisymmetric body described here has an elegant geometrical interpretation using the so-called body and space cones. However, this classic material, which can be found in most textbooks on rigid-body dynamics, is not further pursued here.

6.6 General Torque-Free Motion

Consider the general case of an unsymmetrical, or asymmetrical, rigid body in torque-free motion, described by

$$J_1\dot{\omega}_1 = (J_2 - J_3)\omega_2\omega_3 \qquad (6.72a)$$
$$J_2\dot{\omega}_2 = (J_3 - J_1)\omega_3\omega_1 \qquad (6.72b)$$
$$J_3\dot{\omega}_3 = (J_1 - J_2)\omega_1\omega_2 \qquad (6.72c)$$

where J_1, J_2, and J_3 are the principal moments of inertia. We assume that $J_1 > J_2 > J_3$ without loss of generality.

Let the angular momentum vector and angular velocity vector be represented as

$$\vec{H} = H_1\vec{b}_1 + H_2\vec{b}_2 + H_3\vec{b}_3 \qquad (6.73a)$$
$$\vec{\omega} = \omega_1\vec{b}_1 + \omega_2\vec{b}_2 + \omega_3\vec{b}_3 \qquad (6.73b)$$

where $\{\vec{b}_1, \vec{b}_2, \vec{b}_3\}$ is a set of basis vectors for principal axes. The angular momentum vector of a rigid body is constant for a torque-free motion; i.e.,

$$\vec{H} = \text{constant vector} \qquad (6.74)$$

and

$$H^2 \equiv |\vec{H}|^2 = (J_1\omega_1)^2 + (J_2\omega_2)^2 + (J_3\omega_3)^2 = \text{const} \qquad (6.75)$$

The conservation of rotational kinetic energy implies that

$$2T = \vec{\omega} \cdot \vec{H} = \omega_1 H_1 + \omega_2 H_2 + \omega_3 H_3$$
$$= J_1\omega_1^2 + J_2\omega_2^2 + J_3\omega_3^2 = \text{const} \qquad (6.76)$$

Equations (6.75) and (6.76) can be rewritten, respectively, as

$$\frac{\omega_1^2}{(H/J_1)^2} + \frac{\omega_2^2}{(H/J_2)^2} + \frac{\omega_3^2}{(H/J_3)^2} = 1 \qquad (6.77)$$

$$\frac{\omega_1^2}{(2T/J_1)} + \frac{\omega_2^2}{(2T/J_2)} + \frac{\omega_3^2}{(2T/J_3)} = 1 \qquad (6.78)$$

Geometrically, it can be said that the angular velocity vector must lie on the surface of the *angular momentum ellipsoid*, Eq. (6.77), and at the same time it must lie on the surface of the *kinetic energy ellipsoid*, Eq. (6.78). The curve in which these ellipsoids intersect is the path of the angular velocity vector as seen from the body-fixed reference frame, called a *polhode*.

Let a constant parameter J^* be defined as

$$J^* \equiv \frac{H^2}{2T} = \frac{(J_1\omega_1)^2 + (J_2\omega_2)^2 + (J_3\omega_3)^2}{J_1\omega_1^2 + J_2\omega_2^2 + J_3\omega_3^2} \qquad (6.79)$$

which has the dimensions of a moment of inertia. Then, combining Eqs. (6.77) and (6.78), we obtain the polhode equation

$$J_1(J_1 - J^*)\omega_1^2 + J_2(J_2 - J^*)\omega_2^2 + J_3(J_3 - J^*)\omega_3^2 = 0 \qquad (6.80)$$

Because all of the coefficients of the left-hand side of this equation must not have the same sign and $J_1 > J_2 > J_3$, the constant parameter J^* must be between J_1 and J_3; i.e.,

$$J_1 \geq J^* \geq J_3$$

For an ideal rigid body without internal energy dissipation, H, T, and J^* are constant and the polhode is a closed path.

Eliminating ω_3 from Eqs. (6.77) and (6.78), we obtain the polhode projection onto the (ω_1, ω_2) plane normal to the minor axis, as follows:

$$J_1(J_1 - J_3)\omega_1^2 + J_2(J_2 - J_3)\omega_2^2 = 2T(J^* - J_3) \qquad (6.81)$$

which represents an ellipse because $J_1 > J_2 > J_3$, and the right-hand side of this equation is positive. Similarly, for the polhode projection onto the (ω_2, ω_3) plane normal to the major axis, we have

$$J_2(J_1 - J_2)\omega_2^2 + J_3(J_1 - J_3)\omega_3^2 = 2T(J_1 - J^*) \qquad (6.82)$$

which also represents an ellipse. However, the polhode projection onto the (ω_1, ω_3) plane that is normal to the intermediate axis becomes

$$J_1(J_1 - J_2)\omega_1^2 - J_3(J_2 - J_3)\omega_3^2 = 2T(J^* - J_2) \qquad (6.83)$$

which represents a hyperbola. Note that the right-hand side can be either positive or negative.

Boundaries between the polhode paths about the major and minor axes are called the *separatrices*. If $J^* = J_2$, then Eq. (6.83) becomes

$$J_1(J_1 - J_2)\omega_1^2 - J_3(J_2 - J_3)\omega_3^2 = 0 \qquad (6.84)$$

This is, in fact, the projection of the separatrices onto the (ω_1, ω_3) plane normal to the intermediate axis. The slopes of the separatrices are

$$\frac{\omega_3}{\omega_1} = \pm\sqrt{\frac{J_1(J_1 - J_2)}{J_3(J_2 - J_3)}} \qquad (6.85)$$

which are independent of H and T.

The preceding results regarding the polhode projections are, in fact, closely related to the stability problem of a rigid body spinning about one of its principal axes, to be discussed in the next section.

There exists the analytical closed-form solution to the torque-free motion of an asymmetric rigid body in terms of Jacobi elliptic functions, and the reader is referred to Leimanis [1] or Hughes [2].

Problems

6.6 Consider Euler's equations of an asymmetrical rigid body in torque-free motion described by Eqs. (6.72).

(a) Show that

$$\ddot{\omega}_1 = \frac{J_2 - J_3}{J_1}\left\{\frac{J_1 - J_2}{J_3}\omega_2^2 + \frac{J_3 - J_1}{J_2}\omega_3^2\right\}\omega_1$$

$$\ddot{\omega}_2 = \frac{J_3 - J_1}{J_2}\left\{\frac{J_1 - J_2}{J_3}\omega_1^2 + \frac{J_2 - J_3}{J_1}\omega_3^2\right\}\omega_2$$

$$\ddot{\omega}_3 = \frac{J_1 - J_2}{J_3}\left\{\frac{J_3 - J_1}{J_2}\omega_1^2 + \frac{J_2 - J_3}{J_1}\omega_2^2\right\}\omega_3$$

(b) Show that the equations derived in (a) can be rewritten as

$$\ddot{\omega}_1 + A_1\omega_1 + B_1\omega_1^3 = 0$$

$$\ddot{\omega}_2 + A_2\omega_2 + B_2\omega_2^3 = 0$$

$$\ddot{\omega}_3 + A_3\omega_3 + B_3\omega_3^3 = 0$$

where the constant coefficients A_i and B_i are defined as

$$A_1 = \frac{(J_1 - J_2)(2J_3T - H^2) + (J_3 - J_1)(H^2 - 2J_2T)}{J_1J_2J_3}$$

$$A_2 = \frac{(J_2 - J_3)(2J_1T - H^2) + (J_1 - J_2)(H^2 - 2J_3T)}{J_1J_2J_3}$$

$$A_3 = \frac{(J_3 - J_1)(2J_2T - H^2) + (J_2 - J_3)(H^2 - 2J_1T)}{J_1J_2J_3}$$

$$B_1 = \frac{2(J_1 - J_2)(J_1 - J_3)}{J_2J_3}$$

$$B_2 = \frac{2(J_2 - J_1)(J_2 - J_3)}{J_1J_3}$$

$$B_3 = \frac{2(J_1 - J_3)(J_2 - J_3)}{J_1J_2}$$

Note that each of these uncoupled equations is a homogeneous, undamped Duffing equation with constant coefficients [3].

Hint: Combine Eqs. (6.77) and (6.78) to solve for the squares of any two ω_i as a function of the third, e.g.,

$$\omega_2^2 = \frac{2J_3T - H^2}{J_2(J_3 - J_2)} - \frac{J_1(J_3 - J_1)}{J_2(J_3 - J_2)}\omega_1^2$$

$$\omega_3^2 = \frac{2J_2T - H^2}{J_3(J_2 - J_3)} - \frac{J_1(J_2 - J_1)}{J_3(J_2 - J_3)}\omega_1^2$$

6.7 As shown in Problem 6.5, Euler's rotational equations of an asymmetric rigid body with $J_1 > J_2 > J_3$ in torque-free motion can be written as

$$\dot{x}_1 - x_2x_3 = 0$$
$$\dot{x}_2 + x_3x_1 = 0$$
$$\dot{x}_3 - x_1x_2 = 0$$

where $\dot{x}_i \equiv dx_i/d\tau$, $\tau = t\sqrt{k_1k_2k_3}$, and

$$x_1 = \frac{\omega_1}{\sqrt{k_1}}, \qquad x_2 = \frac{\omega_2}{\sqrt{k_2}}, \qquad x_3 = \frac{\omega_3}{\sqrt{k_3}}$$

(a) Show that there exist two constants of integration such as

$$x_2^2 + x_3^2 = \text{const} = A$$
$$x_1^2 + x_2^2 = \text{const} = B$$

(Consequently, we have $x_1^2 - x_3^2 = B - A$ and $x_1^2 + 2x_2^2 + x_3^2 = A + B$.)

(b) Show that

$$\ddot{x}_1 + (x_3^2 - x_2^2)x_1 = 0$$
$$\ddot{x}_2 + (x_1^2 + x_3^2)x_2 = 0$$
$$\ddot{x}_3 + (x_1^2 - x_2^2)x_3 = 0$$

and then derive the following decoupled equations:

$$\ddot{x}_1 + (A - 2B)x_1 + 2x_1^3 = 0$$
$$\ddot{x}_2 + (A + B)x_2 - 2x_2^3 = 0$$
$$\ddot{x}_3 + (B - 2A)x_3 + 2x_3^3 = 0$$

where A and B are the two constants defined in (a).

6.7 Stability of Torque-Free Motion About Principal Axes

Consider a rigid body that is rotating about one of its principal axes. If there are no external torques acting on the body and if, for example, the rotation was initially about the third axis, then the rigid body will continue to spin about the third axis; i.e., $\omega_3 = \text{const}$ and $\omega_1 = \omega_2 = 0$. On the other hand, if a small impulsive disturbance torque is applied to the body, then the body will no longer rotate purely about the third axis.

For linear stability analysis, we assume that the perturbation terms ω_1 and ω_2 are much smaller in magnitude than ω_3. Neglecting the products of the small perturbation terms, we obtain the linearized rotational equations of motion, as follows:

$$J_1\dot{\omega}_1 - (J_2 - J_3)n\omega_2 = 0 \qquad (6.86a)$$

$$J_2\dot{\omega}_2 - (J_3 - J_1)n\omega_1 = 0 \qquad (6.86b)$$

where n is the constant spin rate about the third axis. Combining these equations, we obtain

$$\ddot{\omega}_1 + n^2\frac{(J_2 - J_3)(J_1 - J_3)}{J_1 J_2}\omega_1 = 0 \qquad (6.87)$$

The characteristic equation is

$$s^2 + n^2\frac{(J_2 - J_3)(J_1 - J_3)}{J_1 J_2} = 0 \qquad (6.88)$$

and the characteristic roots are

$$s = \pm jn \sqrt{\frac{(J_2 - J_3)(J_1 - J_3)}{J_1 J_2}} \qquad (6.89)$$

where $j = \sqrt{-1}$.

If the spin axis (the third axis) is either the major axis ($J_1 < J_3$, $J_2 < J_3$) or minor axis ($J_1 > J_3$, $J_2 > J_3$), then the characteristic roots become pure imaginary numbers and the rotational motion is said to be (Lyapunov) stable, though not asymptotically stable. If the spin axis is the intermediate axis, then one of the characteristic roots is positive real number and the motion is said to be unstable.

Similar conclusions about the stability of a rigid body spinning about one of its principal axes can also be made using the polhode projection equations derived in the preceding section. The hyperbolic nature of the polhode projection onto a plane normal to the intermediate axis indicates the instability of a rigid body spinning about its intermediate axis. The elliptic nature of the polhode projection onto a plane normal to either the major or minor axis indicates the stability of a rigid body spinning about its major or minor axis.

Similar to the term "particle" (or "point mass"), which is a mathematical abstraction of a relatively small body, a rigid body is also a mathematical abstraction of a relatively rigid body. Space vehicles are never perfect rigid bodies; most space vehicles have flexible appendages, e.g., solar panel arrays or antennas, and propellant tanks with sloshing fluid. It has been observed that such semirigid spacecraft with internal energy dissipation caused by fuel slosh or structural vibration is stable only when spinning about its major axis.

For a semirigid body with internal energy dissipation, the kinetic energy decreases and the energy ellipsoid becomes smaller with time. This results in an open polhode path that spirals outward from the minor axis, crosses the separatrix, and approaches the major axis. Consequently, a spacecraft spinning about its minor axis in the presence of energy dissipation is unstable; that is, the spacecraft will eventually reorient and rotate about its major axis with either a positive or negative spin rate. A classic example of the minor-axis instability phenomenon is the first U.S. satellite, Explorer I, launched in 1958 [4].

Problem

6.8 Consider a rigid body with a spherical, dissipative fuel slug. Such a simplified model of a rigid body with internal energy dissipation is described by [5]

$$(J_1 - J)\dot{\omega}_1 = (J_2 - J_3)\omega_2\omega_3 + \mu\sigma_1$$
$$(J_2 - J)\dot{\omega}_2 = (J_3 - J_1)\omega_3\omega_1 + \mu\sigma_2$$
$$(J_3 - J)\dot{\omega}_3 = (J_1 - J_2)\omega_1\omega_2 + \mu\sigma_3$$
$$\dot{\sigma}_1 = -\dot{\omega}_1 - (\mu/J)\sigma_1 - \omega_2\sigma_3 + \omega_3\sigma_2$$
$$\dot{\sigma}_2 = -\dot{\omega}_2 - (\mu/J)\sigma_2 - \omega_3\sigma_1 + \omega_1\sigma_3$$
$$\dot{\sigma}_3 = -\dot{\omega}_3 - (\mu/J)\sigma_3 - \omega_1\sigma_2 + \omega_2\sigma_1$$

where (J_1, J_2, J_3) are the principal moments of inertia of the spacecraft including the spherical, dissipative fuel slug of inertia J; $(\omega_1, \omega_2, \omega_3)$ are the body rates about the principal axes; $(\sigma_1, \sigma_2, \sigma_3)$ are the relative rates between the rigid body and the fuel slug about the principal axes; and μ is the viscous damping coefficient of the fuel slug. It is assumed that $J_1 > J_2 > J_3$ without loss of generality.

(a) Show that a necessary condition for the equilibrium points is $\sigma_1^2 + \sigma_2^2 + \sigma_3^2 = 0$, i.e., $\sigma_1 = \sigma_2 = \sigma_3 = 0$.

(b) Show that an equilibrium point $(\omega_1, \omega_2, \omega_3, \sigma_1, \sigma_2, \sigma_3) = (\Omega, 0, 0, 0, 0, 0)$ is stable; i.e., a pure spinning motion about the major axis is stable.

(c) Show that an equilibrium point $(0, 0, \Omega, 0, 0, 0)$ is unstable; i.e., a pure spinning motion about the minor axis is unstable, whereas it is Lyapunov stable for a rigid body without energy dissipation.

(d) Show that an equilibrium point $(0, \Omega, 0, 0, 0, 0)$ is also unstable.

(e) Consider a spacecraft with the following numerical values: $(J_1, J_2, J_3, J) = (2000, 1500, 1000, 18)$ kg·m^2 and $\mu = 30$ N·m·s. Performing computer simulation, verify that the trajectory starting from an initial condition $(0.1224, 0, 2.99, 0, 0, 0)$ rad/s ends up at $(-1.5, 0, 0, 0, 0, 0)$ rad/s.

Note: The kinetic energy T and the angular momentum H of the system are

$$H^2 = (J_1\omega_1 + J\sigma_1)^2 + (J_2\omega_2 + J\sigma_2)^2 + (J_3\omega_3 + J\sigma_3)^2$$

$$2T = (J_1 - J)\omega_1^2 + (J_2 - J)\omega_2^2 + (J_3 - J)\omega_3^2$$
$$+ J\{(\omega_1 + \sigma_1)^2 + (\omega_2 + \sigma_2)^2 + (\omega_3 + \sigma_3)^2\}$$

During computer simulation of this case, the angular momentum H needs to be checked to see whether or not it is maintained at a constant value of 3000 N·m·s.

(f) Also perform computer simulation with a slightly different initial condition $(0.125, 0, 2.99, 0, 0, 0)$ and verify that the trajectory ends up at $(+1.5, 0, 0, 0, 0, 0)$.

Note: For such a spinning spacecraft with energy dissipation, a small change in initial conditions can lead to a change in the final spin polarity for ω_1. Such sensitive dependence on initial conditions is the property characterizing a chaotic dynamic system.

6.8 Spinning Axisymmetric Body with Constant Body-Fixed Torque

A simple solution to the problem of maintaining a desired orientation of a space vehicle during thrusting maneuvers is to spin the vehicle in the fashion of a football or a spinning rocket about its longitudinal axis. A thrust vector misalignment with the longitudinal axis will cause the vehicle to tumble in the absence of spinning. A spinning axisymmetric body possesses a gyroscopic stiffness to external disturbances, however, and its motion under the influence of disturbances is characterized by the precession and nutation of the longitudinal axis about the desired direction of the longitudinal axis.

Consider an axisymmetric rigid body possessing a body-fixed reference frame B with basis vectors $\{\vec{b}_1, \vec{b}_2, \vec{b}_3\}$ and with its origin at the center of mass. The reference frame B coincides with principal axes. It is assumed that the first and second axes are the transverse axes and that the third axis is the axis of symmetry. A longitudinal thrust vector is nominally aligned along $\vec{b}_3$ through the center of mass of the spacecraft.

Euler's rotational equations of motion of an axisymmetric spacecraft with $J_1 = J_2 = J$ are

$$J\dot{\omega}_1 - (J - J_3)\omega_3\omega_2 = M_1 \tag{6.90}$$

$$J\dot{\omega}_2 + (J - J_3)\omega_3\omega_1 = 0 \tag{6.91}$$

$$J_3\dot{\omega}_3 = 0 \tag{6.92}$$

where $\omega_i \equiv \vec{b}_i \cdot \vec{\omega}$ are the body-fixed components of the angular velocity of the spacecraft in an inertial reference frame and M_1 is the transverse torque component due to misalignment of the thrust vector.

From Eq. (6.92), we have

$$\omega_3 = \text{const} = n \tag{6.93}$$

where the constant n is called the spin rate of the spacecraft about its symmetry axis $\vec{b}_3$. Equations (6.90) and (6.91) then become

$$\dot{\omega}_1 = \lambda\omega_2 + \mu \tag{6.94}$$

$$\dot{\omega}_2 = -\lambda\omega_1 \tag{6.95}$$

where

$$\lambda = \frac{(J - J_3)n}{J}$$

and $\mu \equiv M_1/J$ denotes the constant disturbance acceleration resulting from misalignment of the thrust vector. Note that λ can be either positive or negative depending on whether the third axis is the minor or major axis.

To describe the rotational motion of the spinning spacecraft as seen from an inertial reference frame, we consider the body-fixed rotational sequence of $\mathbf{C}_3(\theta_3) \leftarrow \mathbf{C}_2(\theta_2) \leftarrow \mathbf{C}_1(\theta_1)$. For this rotational sequence, we have the following kinematic differential equations:

$$\dot{\theta}_1 = (\omega_1 \cos\theta_3 - \omega_2 \sin\theta_3)/\cos\theta_2 \tag{6.96a}$$

$$\dot{\theta}_2 = \omega_1 \sin\theta_3 + \omega_2 \cos\theta_3 \tag{6.96b}$$

$$\dot{\theta}_3 = (-\omega_1 \cos\theta_3 + \omega_2 \sin\theta_3)\tan\theta_2 + \omega_3 \tag{6.96c}$$

For small θ_2, the kinematic differential equations become

$$\dot{\theta}_1 = \omega_1 \cos\theta_3 - \omega_2 \sin\theta_3 \tag{6.97a}$$

$$\dot{\theta}_2 = \omega_1 \sin\theta_3 + \omega_2 \cos\theta_3 \tag{6.97b}$$

$$\dot{\theta}_3 = -\theta_2\dot{\theta}_1 + \omega_3 \tag{6.97c}$$

Assuming $\theta_2 \dot{\theta}_1 \ll \omega_3$, we can further approximate $\dot{\theta}_3$ as

$$\dot{\theta}_3 \approx \omega_3 = n = \text{const}$$

and $\theta_3 \approx nt$. Finally, we have a set of linearized equations of motion

$$\dot{\omega}_1 = \lambda \omega_2 + \mu \tag{6.98}$$

$$\dot{\omega}_2 = -\lambda \omega_1 \tag{6.99}$$

$$\dot{\theta}_1 = \omega_1 \cos nt - \omega_2 \sin nt \tag{6.100}$$

$$\dot{\theta}_2 = \omega_1 \sin nt + \omega_2 \cos nt \tag{6.101}$$

The solutions of Eqs. (6.98) and (6.99) for a constant μ can be found as

$$\omega_1(t) = \omega_1(0) \cos \lambda t + \omega_2(0) \sin \lambda t + (\mu/\lambda) \sin \lambda t \tag{6.102a}$$

$$\omega_2(t) = \omega_2(0) \cos \lambda t - \omega_1(0) \sin \lambda t - (\mu/\lambda)(1 - \cos \lambda t) \tag{6.102b}$$

For a case with $\omega_1(0) = \omega_2(0) = 0$, Eqs. (6.100) and (6.101) become

$$\dot{\theta}_1 = \frac{\mu}{\lambda} \left\{ -\sin \frac{J_3}{J} nt + \sin nt \right\} \tag{6.103a}$$

$$\dot{\theta}_2 = \frac{\mu}{\lambda} \left\{ \cos \frac{J_3}{J} nt - \cos nt \right\} \tag{6.103b}$$

because $\sin(A \pm B) = \sin A \cos B \pm \cos A \sin B$ and $\cos(A \pm B) = \cos A \cos B \mp \sin A \sin B$. Integrating these equations with respect to time for the initial conditions $\theta_1(0) = \theta_2(0) = 0$, we obtain

$$\theta_1(t) = -A_p(1 - \cos \omega_p t) + A_n(1 - \cos \omega_n t) \tag{6.104a}$$

$$\theta_2(t) = A_p \sin \omega_p t - A_n \sin \omega_n t \tag{6.104b}$$

where

$$A_p = \frac{\mu J}{\lambda n J_3} = \text{precessional amplitude}$$

$$A_n = \frac{\mu}{\lambda n} = \text{nutational amplitude}$$

$$\omega_p = \frac{J_3 n}{J} = \text{precessional frequency}$$

$$\omega_n = n = \text{nutational frequency}$$

The path of the tip of the axis of symmetry in space is an *epicycloid* formed by a point on a circle of radius A_n rolling on the outside of a circle of radius A_p, centered at $\theta_1 = -A_p$ and $\theta_2 = 0$, when $J > J_3$.

Problem

6.9 Consider a spinning axisymmetric rocket with misaligned longitudinal thrust described by a set of differential equations of the form

$$\dot{\omega}_1 = \lambda \omega_2 + \mu$$

$$\dot{\omega}_2 = -\lambda \omega_1$$

$$\dot{\theta}_1 = (\omega_1 \cos \theta_3 - \omega_2 \sin \theta_3)/\cos \theta_2$$

$$\dot{\theta}_2 = \omega_1 \sin \theta_3 + \omega_2 \cos \theta_3$$

$$\dot{\theta}_3 = (-\omega_1 \cos \theta_3 + \omega_2 \sin \theta_3) \tan \theta_2 + n$$

in which the kinematic differential equations are not linearized yet.
(a) For the following parameter values and initial conditions:

$$J_3/J = 0.05, \qquad n = 15 \text{ rad/s}$$

$$\lambda = n(J - J_3)/J = 14.25 \text{ rad/s}$$

$$\mu = M_1/J = 0.1875 \text{ rad/s}^2$$

$$\omega_1(0) = \omega_2(0) = 0 \text{ rad/s}$$

$$\theta_1(0) = \theta_2(0) = \theta_3(0) = 0$$

perform computer simulations of both the nonlinear and linear models. In particular, plot the paths of the tip of the axis of symmetry in the (θ_1, θ_2) plane. Compare the computer simulation results with the linear analysis results given by Eqs. (6.104).
(b) For the same parameter values and initial conditions as given in (a), but with $\omega_2(0) = 0.025$ rad/s, perform computer simulations of both the nonlinear and linear models, and compare the results in terms of the numerical values of A_p, A_n, ω_p, and ω_n.
Note: In Jarmolow [6] and Kolk [7], ω_n was modified as

$$\omega_n \approx n(J - J_3)/J$$

based on their computer simulation results. The discrepancy was attributed to the linearizing assumption: $\theta_2 \dot{\theta}_1 \ll \omega_3$. On the contrary, however, the nonlinear and linear simulation results agree very well, as is verified in this problem.

6.9 Asymmetric Rigid Body with Constant Body-Fixed Torques

In the preceding section, we studied the problem of a spinning axisymmetric body under the influence of a constant torque along one of the transverse axes. The rotational motion of such an axisymmetric body was characterized by the precession and nutation of the longitudinal axis.

In this section, based on Refs. 8 and 9, we consider the general motion of an asymmetric rigid body under the influence of constant body-fixed torques.

6.9.1   Linear Stability of Equilibrium Points

Euler's rotational equations of motion are given by

$$J_1\dot{\omega}_1 = (J_2 - J_3)\omega_2\omega_3 + M_1$$
$$J_2\dot{\omega}_2 = (J_3 - J_1)\omega_3\omega_1 + M_2$$
$$J_3\dot{\omega}_3 = (J_1 - J_2)\omega_1\omega_2 + M_3$$

where J_i are the principal moments of inertia and M_i are the constant torque components along the body-fixed principal axes. It is assumed that $J_1 > J_2 > J_3$ without loss of generality. The equations of motion at steady state become

$$- (J_2 - J_3)\Omega_2\Omega_3 = M_1$$
$$+ (J_1 - J_3)\Omega_3\Omega_1 = M_2$$
$$- (J_1 - J_2)\Omega_1\Omega_2 = M_3$$

where $(\Omega_1, \Omega_2, \Omega_3)$ is an equilibrium point. Combining these equations, we obtain

$$(J_2 - J_3)(J_1 - J_3)(J_1 - J_2)\Omega_1^2\Omega_2^2\Omega_3^2 = M_1M_2M_3 \qquad (6.105)$$

which indicates that equilibrium points exist if and only if $M_1M_2M_3 > 0$ (only if $M_1M_2M_3 \geq 0$).

Given a constant torque vector (M_1, M_2, M_3) with $M_1M_2M_3 > 0$, eight equilibrium points $(\pm\Omega_1, \pm\Omega_2, \pm\Omega_3)$ exist where

$$\Omega_1 = \sqrt{\frac{J_2 - J_3}{(J_1 - J_2)(J_1 - J_3)}\frac{M_2M_3}{M_1}}$$

$$\Omega_2 = \sqrt{\frac{J_1 - J_3}{(J_1 - J_2)(J_2 - J_3)}\frac{M_3M_1}{M_2}}$$

$$\Omega_3 = \sqrt{\frac{J_1 - J_2}{(J_1 - J_3)(J_2 - J_3)}\frac{M_1M_2}{M_3}}$$

Equilibrium points associated with a torque vector (M_1, M_2, M_3) with $M_1M_2M_3 = 0$ are

$$(M_1, 0, 0) \Rightarrow \{(0, \Omega_2, \Omega_3) : -(J_2 - J_3)\Omega_2\Omega_3 = M_1\}$$
$$(0, M_2, 0) \Rightarrow \{(\Omega_1, 0, \Omega_3) : +(J_1 - J_3)\Omega_3\Omega_1 = M_2\}$$
$$(0, 0, M_3) \Rightarrow \{(\Omega_1, \Omega_2, 0) : -(J_1 - J_2)\Omega_1\Omega_2 = M_3\}$$
$$(0, 0, 0) \Rightarrow \{(0, 0, 0), (\Omega_1, 0, 0), (0, \Omega_2, 0), (0, 0, \Omega_3)\}$$

where M_i and Ω_i are nonzero constants. For other cases of $(0, M_2, M_3)$, $(M_1, 0, M_3)$, and $(M_1, M_2, 0)$, no equilibrium points exist.

Let $(\Omega_1, \Omega_2, \Omega_3)$ be such an equilibrium point of a rigid body with $M_1 M_2 M_3 \geq 0$ and also let

$$\omega_1 = \Omega_1 + \Delta\omega_1$$
$$\omega_2 = \Omega_2 + \Delta\omega_2$$
$$\omega_3 = \Omega_3 + \Delta\omega_3$$

then the linearized equations of motion can be obtained as

$$\begin{bmatrix} \Delta\dot{\omega}_1 \\ \Delta\dot{\omega}_2 \\ \Delta\dot{\omega}_3 \end{bmatrix} = \begin{bmatrix} 0 & k_1\Omega_3 & k_1\Omega_2 \\ -k_2\Omega_3 & 0 & -k_2\Omega_1 \\ k_3\Omega_2 & k_3\Omega_1 & 0 \end{bmatrix} \begin{bmatrix} \Delta\omega_1 \\ \Delta\omega_2 \\ \Delta\omega_3 \end{bmatrix} \tag{6.106}$$

where k_i are all positive constants defined as

$$k_1 = \frac{J_2 - J_3}{J_1}, \qquad k_2 = \frac{J_1 - J_3}{J_2}, \qquad k_3 = \frac{J_1 - J_2}{J_3} \tag{6.107}$$

The characteristic equation is then obtained as

$$s^3 + \left(k_2 k_3 \Omega_1^2 - k_1 k_3 \Omega_2^2 + k_1 k_2 \Omega_3^2\right)s + 2k_1 k_2 k_3 \Omega_1 \Omega_2 \Omega_3 = 0 \tag{6.108}$$

and the linear stability of different types of equilibrium points can be summarized as in Table 6.1, where $\Omega_i \neq 0$.

As discussed in Chapter 1, however, the Lyapunov stability of a dynamic system linearized about an equilibrium point does not guarantee the Lyapunov stability of the equilibrium point of the nonlinear system. Furthermore, the linear stability analysis does not provide any information about the domains of attraction. Consequently, a nonlinear analysis is needed and will be discussed next.

Table 6.1   Linear stability of equilibrium points

Equilibrium points	Characteristic equations	Stability
$(0, 0, 0)$	$s^3 = 0$	Unstable
$(\Omega_1, 0, 0)$	$s\left(s^2 + k_2 k_3 \Omega_1^2\right) = 0$	Stable
$(0, \Omega_2, 0)$	$s\left(s^2 - k_1 k_3 \Omega_2^2\right) = 0$	Unstable
$(0, 0, \Omega_3)$	$s\left(s^2 + k_1 k_2 \Omega_3^2\right) = 0$	Stable
$(0, \Omega_2, \Omega_3)$	$s\left(s^2 - k_1 k_3 \Omega_2^2 + k_1 k_2 \Omega_3^2\right) = 0$	Stable for $k_3 \Omega_2^2 < k_2 \Omega_3^2$
$(\Omega_1, 0, \Omega_3)$	$s\left(s^2 + k_2 k_3 \Omega_1^2 + k_1 k_2 \Omega_3^2\right) = 0$	Stable
$(\Omega_1, \Omega_2, 0)$	$s\left(s^2 + k_2 k_3 \Omega_1^2 - k_1 k_3 \Omega_2^2\right) = 0$	Stable for $k_2 \Omega_1^2 > k_1 \Omega_2^2$
$(\Omega_1, \Omega_2, \Omega_3)$	Eq. (6.108)	Unstable

6.9.2 Constant Torque About the Major or Minor Axis

Consider a case in which a constant body-fixed torque acts along the major axis; i.e., $M_1 \neq 0$ and $M_2 = M_3 = 0$. For such a case, the equations of motion are simply given by

$$\frac{d\omega_1}{dt} - k_1\omega_2\omega_3 = M_1/J_1 \tag{6.109a}$$

$$\frac{d\omega_2}{dt} + k_2\omega_3\omega_1 = 0 \tag{6.109b}$$

$$\frac{d\omega_3}{dt} - k_3\omega_1\omega_2 = 0 \tag{6.109c}$$

For the convenience of mathematical derivations, we will employ the equations of motion in nondimensional form in the subsequent analysis and consider only the positive torque case ($M_1 > 0$) without loss of generality.

The equations of motion in nondimensional form for constant $M_1 > 0$ can be obtained as

$$\frac{dx_1}{d\tau} - x_2x_3 = 1 \tag{6.110a}$$

$$\frac{dx_2}{d\tau} + x_3x_1 = 0 \tag{6.110b}$$

$$\frac{dx_3}{d\tau} - x_1x_2 = 0 \tag{6.110c}$$

where $\tau = t\sqrt{\mu k_1 k_2 k_3}$, and

$$\mu = \frac{M_1}{J_1 k_1\sqrt{k_2 k_3}}, \qquad x_1 = \frac{\omega_1}{\sqrt{\mu k_1}}, \qquad x_2 = \frac{\omega_2}{\sqrt{\mu k_2}}, \qquad x_3 = \frac{\omega_3}{\sqrt{\mu k_3}}$$

Equilibrium curves (or manifolds) are then described by

$$\{(0, x_2, x_3) \ : \ -x_2x_3 = 1\}$$

From linear stability analysis, it has been shown that the equilibrium manifold is Lyapunov stable when $x_2^2 < x_3^2$ and unstable when $x_2^2 \geq x_3^2$. A stability diagram of equilibrium manifolds in the (x_2, x_3) plane is illustrated in Fig. 6.5. Also shown in this figure is a circle that touches the equilibrium manifolds at $(1, -1)$ and $(-1, 1)$.

Introducing a new variable θ_1 such that

$$\frac{d\theta_1}{d\tau} = x_1 \tag{6.111}$$

and $\theta_1(0) = 0$, we rewrite Eqs. (6.110b) and (6.110c) as

$$\frac{dx_2}{d\theta_1} + x_3 = 0 \tag{6.112a}$$

$$\frac{dx_3}{d\theta_1} - x_2 = 0 \tag{6.112b}$$

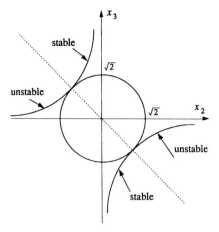

Fig. 6.5 Linear stability diagram of equilibrium manifolds in the (x_2, x_3) plane ($\mu > 0$).

The solution of these equations is simply given by

$$x_2 = x_2(0) \cos\theta_1 - x_3(0) \sin\theta_1 \qquad (6.113a)$$

$$x_3 = x_2(0) \sin\theta_1 + x_3(0) \cos\theta_1 \qquad (6.113b)$$

which can be rewritten as

$$x_2 = A \cos(\theta_1 + \phi) \qquad (6.114a)$$

$$x_3 = A \sin(\theta_1 + \phi) \qquad (6.114b)$$

where

$$A = \sqrt{x_2^2(0) + x_3^2(0)}$$

$$\phi = \tan^{-1}\left\{\frac{x_3(0)}{x_2(0)}\right\}$$

and we have

$$x_2^2 + x_3^2 = A^2 \qquad \text{for all } \tau \geq 0 \qquad (6.115)$$

That is, the projection of the tip of the nondimensional angular velocity vector $\vec{x}$ onto the (x_2, x_3) plane normal to the major axis is a complete circle or a portion of a circle, although the body is acted upon by a constant torque along the major axis. Consequently, the endpoint of the angular velocity vector always lies on the surface of a circular cylinder defined by Eq. (6.115). Note that the end point of the angular velocity vector $\vec{\omega}$ actually lies on the surface of an elliptic cylinder.

Substituting Eqs. (6.114) into Eq. (6.110a), we obtain

$$\frac{d^2\theta_1}{d\tau^2} - \frac{A^2}{2} \sin 2(\theta_1 + \phi) = 1 \qquad (6.116)$$

which can be rewritten as

$$\frac{d^2\theta}{d\tau^2} - A^2 \sin\theta = 2 \qquad (6.117)$$

where $\theta = 2(\theta_1 + \phi)$. Note that Eq. (6.117) is similar to the equation of motion of an inverted pendulum with a constant external torque, but with a specified initial condition of

$$\theta(0) = 2\phi = 2\tan^{-1}\left\{\frac{x_3(0)}{x_2(0)}\right\}$$

and

$$-2\pi < \theta(0) < 2\pi$$

because we define the arctangent function such that

$$-\pi < \tan^{-1}\left\{\frac{x_3(0)}{x_2(0)}\right\} < \pi$$

For the phase-plane analysis, Eq. (6.117) is rewritten as

$$\frac{d\theta}{d\tau} = x \qquad (6.118a)$$

$$\frac{dx}{d\tau} = A^2 \sin\theta + 2 \qquad (6.118b)$$

which can be combined as

$$\frac{dx}{d\theta} = \frac{A^2 \sin\theta + 2}{x} \qquad (6.119)$$

Integrating this equation after separation of variables, we obtain the trajectory equation in the (x, θ) plane for given initial conditions $x_2(0)$ and $x_3(0)$, as follows:

$$\tfrac{1}{2}x^2 + A^2 \cos\theta - 2\theta = E \qquad (6.120)$$

where E is the constant integral of the system.

Because $x = 2x_1$, $\theta = 2(\theta_1 + \phi)$, and $\theta_1(0) = 0$, the trajectory equation (6.120) at $\tau = 0$ becomes

$$2x_1^2(0) + A^2 \cos 2\phi - 4\phi = E \tag{6.121}$$

where

$$A = \sqrt{x_2^2(0) + x_3^2(0)}$$

$$\phi = \tan^{-1}\left\{\frac{x_3(0)}{x_2(0)}\right\}$$

Noting that

$$\cos 2\phi = \frac{1 - \tan^2 \phi}{1 + \tan^2 \phi} = \frac{x_2^2(0) - x_3^2(0)}{x_2^2(0) + x_3^2(0)}$$

one can rewrite the trajectory equation (6.121) at $\tau = 0$ as

$$2x_1^2(0) + x_2^2(0) - x_3^2(0) - 4\tan^{-1}\left\{\frac{x_3(0)}{x_2(0)}\right\} = E \tag{6.122}$$

Finally, we obtain the general trajectory equation in the (x_1, x_2, x_3) space, as follows:

$$2x_1^2 + x_2^2 - x_3^2 - 4\tan^{-1}\left\{\frac{x_3}{x_2}\right\} = E \tag{6.123}$$

The rotational motion of an asymmetric rigid body when a constant body-fixed torque acts along its major axis can now be analyzed using the phase-plane method as follows. The equilibrium points of Eq. (6.117) or Eqs. (6.118) are first determined by the equation

$$A^2 \sin \theta + 2 = 0 \tag{6.124}$$

and we shall consider the following three cases.

1) If $A < \sqrt{2}$, then there exist no equilibrium points and all trajectories in the (x, θ) plane approach infinity. This corresponds to a case in which $x_2(0)$ and $x_3(0)$ lie inside the circle shown in Fig. 6.5.

2) If $A = \sqrt{2}$, then there exists an equilibrium point that is unstable and all trajectories in the (x, θ) plane approach infinity. This corresponds to a case in which $x_2(0)$ and $x_3(0)$ lie on the circle shown in Fig. 6.5.

3) If $A > \sqrt{2}$, then there exists an infinite number of stable and unstable equilibrium points along the θ axis. This corresponds to a case in which $x_2(0)$ and $x_3(0)$ lie outside the circle shown in Fig. 6.5. Furthermore, a stable periodic motion does exist (i.e., x is bounded and does not approach infinity) for certain initial conditions, as illustrated in Fig. 6.6. The separatrix that passes through an unstable equilibrium point θ^* is described by

$$\tfrac{1}{2}x^2 + A^2 \cos \theta - 2\theta = E^* \tag{6.125}$$

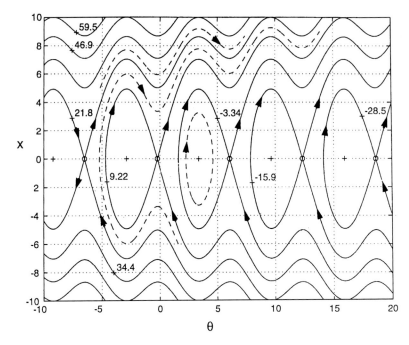

Fig. 6.6   Phase-plane trajectories in the (x, θ) plane.

where

$$E^* = A^2 \cos \theta^* - 2\theta^*$$
$$A^2 = x_2^2(0) + x_3^2(0)$$

$$\theta^* = \sin^{-1}\left(\frac{-2}{A^2}\right) \pm 2n\pi, \qquad n = 0, 1, 2, \ldots$$

and the arcsine function is defined as $-\pi/2 \le \sin^{-1}(\cdot) \le \pi/2$. Such separatrices, which separate the stable and unstable domains in the (x, θ) plane, are indicated by solid lines in Fig. 6.6 for the case of $x_2(0) = 0$, $x_3(0) = 3$, $\theta(0) = 2\phi = \pi$, and $\theta^* = -0.224 \pm 2n\pi$.

Using the general trajectory equation (6.123), we can also find the equation of the *separatrix surface* in the (x_1, x_2, x_3) space of the form:

$$2x_1^2 + x_2^2 - x_3^2 - 4\tan^{-1}\left\{\frac{x_3}{x_2}\right\} = E^* \qquad (6.126)$$

Such separatrix surfaces, which separate the stable and unstable domains in the (x_1, x_2, x_3) space, are shown in Fig. 6.7. Further details of such separatrix surfaces can be found in Refs. 8 and 9.

If a constant body-fixed torque acts along the minor axis, the resulting motion is also characterized as similar to the preceding case of a constant body-fixed torque

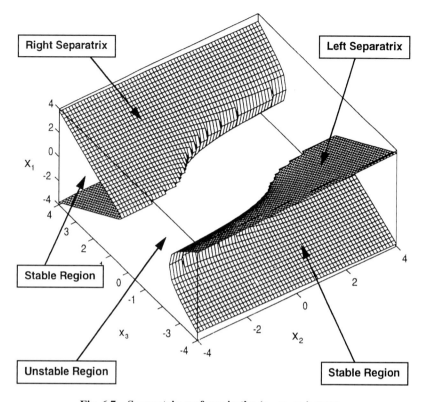

Fig. 6.7   Separatrix surfaces in the (x_1, x_2, x_3) space.

along the major axis. If a constant body-fixed torque acts along the intermediate axis, however, the resulting motion is quite different from the preceding case, as will be discussed next.

6.9.3 Constant Torque About the Intermediate Axis

Consider a case in which a constant body-fixed torque acts along the intermediate axis; i.e., $M_2 \neq 0$ and $M_1 = M_3 = 0$. For such a case, Euler's rotational equations are given by

$$\frac{d\omega_1}{dt} - k_1 \omega_2 \omega_3 = 0$$

$$\frac{d\omega_2}{dt} + k_2 \omega_3 \omega_1 = M_2/J_2$$

$$\frac{d\omega_3}{dt} - k_3 \omega_1 \omega_2 = 0$$

Similar to the preceding case, we will employ the following equations of motion in nondimensional form for $M_2 > 0$ in the subsequent analysis

$$\frac{dx_1}{d\tau} - x_2 x_3 = 0 \qquad (6.127a)$$

$$\frac{dx_2}{d\tau} + x_3 x_1 = 1 \qquad (6.127b)$$

$$\frac{dx_3}{d\tau} - x_1 x_2 = 0 \qquad (6.127c)$$

where $\tau = t\sqrt{\mu k_1 k_2 k_3}$, and

$$\mu = \frac{M_2}{J_2 k_2 \sqrt{k_1 k_3}}, \qquad x_1 = \frac{\omega_1}{\sqrt{\mu k_1}}, \qquad x_2 = \frac{\omega_2}{\sqrt{\mu k_2}}, \qquad x_3 = \frac{\omega_3}{\sqrt{\mu k_3}}$$

Much like the new variable θ_1 introduced earlier, define a new variable θ_2 such that

$$\frac{d\theta_2}{d\tau} = x_2 \qquad (6.128)$$

and $\theta_2(0) = 0$, then we rewrite Eqs. (6.127a) and (6.127c) as

$$\frac{dx_1}{d\theta_2} - x_3 = 0 \qquad (6.129a)$$

$$\frac{dx_3}{d\theta_2} - x_1 = 0 \qquad (6.129b)$$

The solution of these equations is given by

$$x_1 = x_1(0) \cosh \theta_2 + x_3(0) \sinh \theta_2 \qquad (6.130a)$$

$$x_3 = x_1(0) \sinh \theta_2 + x_3(0) \cosh \theta_2 \qquad (6.130b)$$

where $x_1(0)$ and $x_3(0)$ are initial conditions at $\tau = 0$. Note that $\cosh x = (e^x + e^{-x})/2$, $\sinh x = (e^x - e^{-x})/2$, and $\cosh^2 x - \sinh^2 x = 1$. And we have

$$x_1^2 - x_3^2 = \text{const} = x_1^2(0) - x_3^2(0) \qquad \text{for all } \tau \geq 0 \qquad (6.131)$$

That is, the projection of the tip of the nondimensional angular velocity vector $\vec{x}$ onto the (x_1, x_3) plane normal to the intermediate axis is a hyperbola, although the body is acted upon by a constant torque along the intermediate axis. Consequently, the end point of the nondimensional angular velocity vector always lies on the surface of one of the hyperbolic cylinders defined by Eq. (6.131). For particular initial conditions such that $|x_1(0)| = |x_3(0)|$, the resulting motion is along the separatrices described by

$$x_3 = \pm x_1$$

Substituting Eqs. (6.130) into Eq. (6.127b), and using the relationships $\sinh^2 x = (\cosh 2x - 1)/2$, $\cosh^2 x = (\cosh 2x + 1)/2$, and $\sinh 2x = 2 \sinh x \cosh x$, we obtain

$$\frac{d^2\theta_2}{d\tau^2} = 1 - A \sinh 2\theta_2 - B \cosh 2\theta_2 \tag{6.132}$$

where

$$A = \tfrac{1}{2}\{x_1^2(0) + x_3^2(0)\}$$
$$B = x_1(0)x_3(0)$$

Because $A > B$ when $|x_1(0)| \neq |x_3(0)|$, Eq. (6.132) can be rewritten using $\sinh(x \pm y) = \sinh x \cosh y \pm \cosh x \sinh y$, as follows:

$$\frac{d^2\theta_2}{d\tau^2} = 1 - \sqrt{A^2 - B^2}\, \sinh(2\theta_2 + \phi) \tag{6.133}$$

where $\phi = \tanh^{-1}(B/A)$.

Defining $\theta = 2\theta_2 + \phi$, we rewrite Eq. (6.133) as

$$\frac{d\theta}{d\tau} = x \tag{6.134a}$$

$$\frac{dx}{d\tau} = -2\sqrt{A^2 - B^2}\, \sinh \theta + 2 \tag{6.134b}$$

which can be combined as

$$\frac{dx}{d\theta} = \frac{-2\sqrt{A^2 - B^2}\, \sinh \theta + 2}{x} \tag{6.135}$$

Integrating this equation after separation of variables, we obtain the trajectory equation on the (x, θ) plane, as follows:

$$\tfrac{1}{2}x^2 + 2\sqrt{A^2 - B^2}\, \cosh \theta - 2\theta = E \tag{6.136}$$

where E is the integral constant.

It can be shown that all the equilibrium points of Eq. (6.133) or Eqs. (6.134) are stable. Consequently, x is bounded and does not approach infinity for any values of the constant torque along the intermediate axis if $|x_1(0)| \neq |x_3(0)|$. However, for certain particular initial conditions such that $x_1(0) = x_3(0)$, x approaches $-\infty$ if $\mu < 0$ but it is bounded for $\mu \geq 0$. When $x_1(0) = -x_3(0)$, x approaches $+\infty$ if $\mu > 0$ but it is bounded for $\mu \leq 0$.

Like the preceding case of a constant torque about the major axis, the general trajectory equation in the (x_1, x_2, x_3) space can be found as

$$x_1^2 + 2x_2^2 + x_3^2 - 2\tanh^{-1}\left\{\frac{2x_1 x_3}{x_1^2 + x_3^2}\right\} = E \tag{6.137}$$

For a more detailed treatment of this subject, the reader is referred to Leimanis [1] and Livneh and Wie [8, 9].

6.10   Rigid Body in a Circular Orbit

The solution of most spacecraft dynamics and control problems requires a consideration of gravitational forces and moments. When a body is in a uniform gravitational field, its center of mass becomes the center of gravity and the gravitational torque about its center of mass is zero. The gravitational field is not uniform over a body in space, however, and a gravitational torque exists about the body's center of mass. This effect was first considered by D'Alembert and Euler in 1749. Later, in 1780, Lagrange used it to explain why the moon always has the same face toward the Earth. In this section, we derive the equations of motion of a rigid spacecraft in a circular orbit and study its stability.

6.10.1   Equations of Motion

Consider a rigid body in a circular orbit. A local vertical and local horizontal (LVLH) reference frame A with its origin at the center of mass of an orbiting spacecraft has a set of unit vectors $\{\vec{a}_1, \vec{a}_2, \vec{a}_3\}$, with $\vec{a}_1$ along the orbit direction, $\vec{a}_2$ perpendicular to the orbit plane, and $\vec{a}_3$ toward the Earth, as illustrated in Fig. 6.8. The angular velocity of A with respect to N is

$$\vec{\omega}^{A/N} = -n\vec{a}_2 \tag{6.138}$$

where n is the constant orbital rate. The angular velocity of the body-fixed reference frame B with basis vectors $\{\vec{b}_1, \vec{b}_2, \vec{b}_3\}$ is then given by

$$\vec{\omega}^{B/N} = \vec{\omega}^{B/A} + \vec{\omega}^{A/N} = \vec{\omega}^{B/A} - n\vec{a}_2 \tag{6.139}$$

where $\vec{\omega}^{B/A}$ is the angular velocity of B relative to A.

The orientation of the body-fixed reference frame B with respect to the LVLH reference frame A is in general described by the direction cosine matrix $\mathbf{C} = \mathbf{C}^{B/A}$ such that

$$\begin{bmatrix} \vec{b}_1 \\ \vec{b}_2 \\ \vec{b}_3 \end{bmatrix} = \begin{bmatrix} C_{11} & C_{12} & C_{13} \\ C_{21} & C_{22} & C_{23} \\ C_{31} & C_{32} & C_{33} \end{bmatrix} \begin{bmatrix} \vec{a}_1 \\ \vec{a}_2 \\ \vec{a}_3 \end{bmatrix} \tag{6.140}$$

or

$$\begin{bmatrix} \vec{a}_1 \\ \vec{a}_2 \\ \vec{a}_3 \end{bmatrix} = \begin{bmatrix} C_{11} & C_{21} & C_{31} \\ C_{12} & C_{22} & C_{32} \\ C_{13} & C_{23} & C_{33} \end{bmatrix} \begin{bmatrix} \vec{b}_1 \\ \vec{b}_2 \\ \vec{b}_3 \end{bmatrix} \tag{6.141}$$

The gravitational force acting on a small mass element dm is given by

$$\vec{df} = -\frac{\mu \vec{R}\, dm}{|\vec{R}|^3} = -\frac{\mu(\vec{R}_c + \vec{\rho})\, dm}{|\vec{R}_c + \vec{\rho}|^3} \tag{6.142}$$

where μ is the gravitational parameter of the Earth, $\vec{R}$ and $\vec{\rho}$ are the position vectors of dm from the Earth's center and the spacecraft's mass center, respectively, and $\vec{R}_c$ is the position vector of the spacecraft's mass center from the Earth's center.

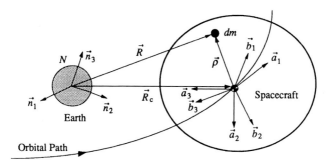

Fig. 6.8   Rigid body in a circular orbit.

The gravity-gradient torque about the spacecraft's mass center is then expressed as

$$\vec{M} = \int \vec{\rho} \times d\vec{f} = -\mu \int \frac{\vec{\rho} \times \vec{R}_c}{|\vec{R}_c + \vec{\rho}|^3}\, dm \qquad (6.143)$$

and we have the following approximation:

$$|\vec{R}_c + \vec{\rho}|^{-3} = R_c^{-3}\left\{1 + \frac{2(\vec{R}_c \cdot \vec{\rho})}{R_c^2} + \frac{\rho^2}{R_c^2}\right\}^{-\frac{3}{2}}$$

$$= R_c^{-3}\left\{1 - \frac{3(\vec{R}_c \cdot \vec{\rho})}{R_c^2} + \text{higher-order terms}\right\} \qquad (6.144)$$

where $R_c = |\vec{R}_c|$ and $\rho = |\vec{\rho}|$. Because $\int \vec{\rho}\, dm = 0$, the gravity-gradient torque neglecting the higher-order terms can be written as

$$\vec{M} = \frac{3\mu}{R_c^5} \int (\vec{R}_c \cdot \vec{\rho})\,(\vec{\rho} \times \vec{R}_c)\, dm \qquad (6.145)$$

This equation is further manipulated as follows:

$$\vec{M} = -\frac{3\mu}{R_c^5}\vec{R}_c \times \int \vec{\rho}(\vec{\rho} \cdot \vec{R}_c)\, dm$$

$$= -\frac{3\mu}{R_c^5}\vec{R}_c \times \int \vec{\rho}\vec{\rho}\, dm \cdot \vec{R}_c$$

$$= -\frac{3\mu}{R_c^5}\vec{R}_c \times \left[\int \rho^2 \hat{I}\, dm - \hat{J}\right] \cdot \vec{R}_c$$

$$= -\frac{3\mu}{R_c^5}\vec{R}_c \times \int \rho^2 \hat{I}\, dm \cdot \vec{R}_c + \frac{3\mu}{R_c^5}\vec{R}_c \times \hat{J} \cdot \vec{R}_c$$

$$= \frac{3\mu}{R_c^5}\vec{R}_c \times \hat{J} \cdot \vec{R}_c$$

because $\hat{J} = \int (\rho^2 \hat{I} - \vec{\rho}\vec{\rho}) \, dm$ and $\vec{R}_c \times \hat{I} \cdot \vec{R}_c = \vec{R}_c \times \vec{R}_c = 0$.

Finally, the gravity-gradient torque is expressed in vector/dyadic form as

$$\vec{M} = 3n^2 \vec{a}_3 \times \hat{J} \cdot \vec{a}_3 \qquad (6.146)$$

where $n = \sqrt{\mu/R_c^3}$ is the orbital rate and $\vec{a}_3 \equiv -\vec{R}_c/R_c$.

The rotational equation of motion of a rigid body with an angular momentum $\vec{H} = \hat{J} \cdot \vec{\omega}^{B/N}$ in a circular orbit is then given by

$$\left\{ \frac{d\vec{H}}{dt} \right\}_N \equiv \left\{ \frac{d\vec{H}}{dt} \right\}_B + \vec{\omega}^{B/N} \times \vec{H} = \vec{M}$$

which can be written as

$$\hat{J} \cdot \dot{\vec{\omega}} + \vec{\omega} \times \hat{J} \cdot \vec{\omega} = 3n^2 \, \vec{a}_3 \times \hat{J} \cdot \vec{a}_3 \qquad (6.147)$$

where $\vec{\omega} \equiv \vec{\omega}^{B/N}$.

Because $\vec{\omega}$ and $\vec{a}_3$ can be expressed in terms of basis vectors of the body-fixed reference frame B as

$$\vec{\omega} = \omega_1 \vec{b}_1 + \omega_2 \vec{b}_2 + \omega_3 \vec{b}_3 \qquad (6.148a)$$

$$\vec{a}_3 = C_{13} \vec{b}_1 + C_{23} \vec{b}_2 + C_{33} \vec{b}_3 \qquad (6.148b)$$

the equation of motion in matrix form becomes

$$\begin{bmatrix} J_{11} & J_{12} & J_{13} \\ J_{21} & J_{22} & J_{23} \\ J_{31} & J_{32} & J_{33} \end{bmatrix} \begin{bmatrix} \dot{\omega}_1 \\ \dot{\omega}_2 \\ \dot{\omega}_3 \end{bmatrix} + \begin{bmatrix} 0 & -\omega_3 & \omega_2 \\ \omega_3 & 0 & -\omega_1 \\ -\omega_2 & \omega_1 & 0 \end{bmatrix} \begin{bmatrix} J_{11} & J_{12} & J_{13} \\ J_{21} & J_{22} & J_{23} \\ J_{31} & J_{32} & J_{33} \end{bmatrix} \begin{bmatrix} \omega_1 \\ \omega_2 \\ \omega_3 \end{bmatrix}$$

$$= 3n^2 \begin{bmatrix} 0 & -C_{33} & C_{23} \\ C_{33} & 0 & -C_{13} \\ -C_{23} & C_{13} & 0 \end{bmatrix} \begin{bmatrix} J_{11} & J_{12} & J_{13} \\ J_{21} & J_{22} & J_{23} \\ J_{31} & J_{32} & J_{33} \end{bmatrix} \begin{bmatrix} C_{13} \\ C_{23} \\ C_{33} \end{bmatrix} \qquad (6.149)$$

To describe the orientation of the body-fixed reference frame B with respect to the LVLH reference frame A in terms of three Euler angles θ_i ($i = 1, 2, 3$), consider the rotational sequence of $\mathbf{C}_1(\theta_1) \leftarrow \mathbf{C}_2(\theta_2) \leftarrow \mathbf{C}_3(\theta_3)$ to B from A. For this sequence, we have

$$\begin{bmatrix} \vec{b}_1 \\ \vec{b}_2 \\ \vec{b}_3 \end{bmatrix} = \begin{bmatrix} C_{11} & C_{12} & C_{13} \\ C_{21} & C_{22} & C_{23} \\ C_{31} & C_{32} & C_{33} \end{bmatrix} \begin{bmatrix} \vec{a}_1 \\ \vec{a}_2 \\ \vec{a}_3 \end{bmatrix}$$

$$= \begin{bmatrix} c\,\theta_2\,c\,\theta_3 & c\,\theta_2\,s\,\theta_3 & -s\,\theta_2 \\ s\,\theta_1\,s\,\theta_2\,c\,\theta_3 - c\,\theta_1\,s\,\theta_3 & s\,\theta_1\,s\,\theta_2\,s\,\theta_3 + c\,\theta_1\,c\,\theta_3 & s\,\theta_1\,c\,\theta_2 \\ c\,\theta_1\,s\,\theta_2\,c\,\theta_3 + s\,\theta_1\,s\,\theta_3 & c\,\theta_1\,s\,\theta_2\,s\,\theta_3 - s\,\theta_1\,c\,\theta_3 & c\,\theta_1\,c\,\theta_2 \end{bmatrix} \begin{bmatrix} \vec{a}_1 \\ \vec{a}_2 \\ \vec{a}_3 \end{bmatrix}$$

where $c\,\theta_i \equiv \cos \theta_i$ and $s\,\theta_i \equiv \sin \theta_i$.

Also, for the sequence of $C_1(\theta_1) \leftarrow C_2(\theta_2) \leftarrow C_3(\theta_3)$, the angular velocity of B relative to A is represented as

$$\vec{\omega}^{B/A} = \omega_1' \vec{b}_1 + \omega_2' \vec{b}_2 + \omega_3' \vec{b}_3$$

where

$$\begin{bmatrix} \omega_1' \\ \omega_2' \\ \omega_3' \end{bmatrix} = \begin{bmatrix} 1 & 0 & -s\,\theta_2 \\ 0 & c\,\theta_1 & s\,\theta_1\,c\,\theta_2 \\ 0 & -s\,\theta_1 & c\,\theta_1\,c\,\theta_2 \end{bmatrix} \begin{bmatrix} \dot{\theta}_1 \\ \dot{\theta}_2 \\ \dot{\theta}_3 \end{bmatrix} \tag{6.150}$$

Because

$$\vec{\omega} \equiv \vec{\omega}^{B/N} = \vec{\omega}^{B/A} + \vec{\omega}^{A/N} = \vec{\omega}^{B/A} - n\vec{a}_2$$

where $\vec{\omega} = \omega_1 \vec{b}_1 + \omega_2 \vec{b}_2 + \omega_3 \vec{b}_3$ and

$$\begin{aligned}
\vec{a}_2 &= C_{12}\vec{b}_1 + C_{22}\vec{b}_2 + C_{32}\vec{b}_3 \\
&= c\,\theta_2\,s\,\theta_3\vec{b}_1 + (s\,\theta_1\,s\,\theta_2\,s\,\theta_3 + c\,\theta_1\,c\,\theta_3)\vec{b}_2 + (c\,\theta_1\,s\,\theta_2\,s\,\theta_3 - s\,\theta_1\,c\,\theta_3)\vec{b}_3
\end{aligned}$$

we have

$$\begin{bmatrix} \omega_1 \\ \omega_2 \\ \omega_3 \end{bmatrix} = \begin{bmatrix} 1 & 0 & -s\,\theta_2 \\ 0 & c\,\theta_1 & s\,\theta_1\,c\,\theta_2 \\ 0 & -s\,\theta_1 & c\,\theta_1\,c\,\theta_2 \end{bmatrix} \begin{bmatrix} \dot{\theta}_1 \\ \dot{\theta}_2 \\ \dot{\theta}_3 \end{bmatrix} - n \begin{bmatrix} c\,\theta_2\,s\,\theta_3 \\ s\,\theta_1\,s\,\theta_2\,s\,\theta_3 + c\,\theta_1\,c\,\theta_3 \\ c\,\theta_1\,s\,\theta_2\,s\,\theta_3 - s\,\theta_1\,c\,\theta_3 \end{bmatrix} \tag{6.151}$$

Finally, the kinematic differential equations of an orbiting rigid body can be found as

$$\begin{bmatrix} \dot{\theta}_1 \\ \dot{\theta}_2 \\ \dot{\theta}_3 \end{bmatrix} = \frac{1}{c\,\theta_2} \begin{bmatrix} c\,\theta_2 & s\,\theta_1\,s\,\theta_2 & c\,\theta_1\,s\,\theta_2 \\ 0 & c\,\theta_1\,c\,\theta_2 & -s\,\theta_1\,c\,\theta_2 \\ 0 & s\,\theta_1 & c\,\theta_1 \end{bmatrix} \begin{bmatrix} \omega_1 \\ \omega_2 \\ \omega_3 \end{bmatrix} + \frac{n}{c\,\theta_2} \begin{bmatrix} s\,\theta_3 \\ c\,\theta_2\,c\,\theta_3 \\ s\,\theta_2\,s\,\theta_3 \end{bmatrix} \tag{6.152}$$

The dynamic equations of motion about body-fixed principal axes become

$$J_1\dot{\omega}_1 - (J_2 - J_3)\omega_2\omega_3 = -3n^2(J_2 - J_3)C_{23}C_{33} \tag{6.153a}$$

$$J_2\dot{\omega}_2 - (J_3 - J_1)\omega_3\omega_1 = -3n^2(J_3 - J_1)C_{33}C_{13} \tag{6.153b}$$

$$J_3\dot{\omega}_3 - (J_1 - J_2)\omega_1\omega_2 = -3n^2(J_1 - J_2)C_{13}C_{23} \tag{6.153c}$$

where $C_{13} = -\sin\theta_2$, $C_{23} = \sin\theta_1\cos\theta_2$, and $C_{33} = \cos\theta_1\cos\theta_2$ for the sequence of $\mathbf{C}_1(\theta_1) \leftarrow \mathbf{C}_2(\theta_2) \leftarrow \mathbf{C}_3(\theta_3)$. Furthermore, for small angles ($\sin\theta_i \approx \theta_i$ and $\cos\theta_i \approx 1$), these dynamic equations become

$$J_1\dot{\omega}_1 - (J_2 - J_3)\omega_2\omega_3 = -3n^2(J_2 - J_3)\theta_1 \tag{6.154a}$$

$$J_2\dot{\omega}_2 - (J_3 - J_1)\omega_3\omega_1 = 3n^2(J_3 - J_1)\theta_2 \tag{6.154b}$$

$$J_3\dot{\omega}_3 - (J_1 - J_2)\omega_1\omega_2 = 0 \tag{6.154c}$$

Also, for small θ_i and $\dot{\theta}_i$, Eq. (6.151) can be linearized as

$$\omega_1 = \dot{\theta}_1 - n\theta_3 \tag{6.155a}$$

$$\omega_2 = \dot{\theta}_2 - n \tag{6.155b}$$

$$\omega_3 = \dot{\theta}_3 + n\theta_1 \tag{6.155c}$$

Substituting Eqs. (6.155) into Eqs. (6.154), we obtain the linearized equations of motion of a rigid body in a circular orbit, as follows, for roll, pitch, and yaw, respectively:

$$J_1\ddot{\theta}_1 - n(J_1 - J_2 + J_3)\dot{\theta}_3 + 4n^2(J_2 - J_3)\theta_1 = 0 \tag{6.156}$$

$$J_2\ddot{\theta}_2 + 3n^2(J_1 - J_3)\theta_2 = 0 \tag{6.157}$$

$$J_3\ddot{\theta}_3 + n(J_1 - J_2 + J_3)\dot{\theta}_1 + n^2(J_2 - J_1)\theta_3 = 0 \tag{6.158}$$

where θ_1, θ_2, and θ_3 are often called, respectively, the roll, pitch, and yaw attitude angles of the spacecraft relative to the LVLH reference frame A.

For these small angles θ_1, θ_2, and θ_3, the body-fixed reference frame B is, in fact, related to the LVLH frame A by

$$\begin{bmatrix} \vec{b}_1 \\ \vec{b}_2 \\ \vec{b}_3 \end{bmatrix} \approx \begin{bmatrix} 1 & \theta_3 & -\theta_2 \\ -\theta_3 & 1 & \theta_1 \\ \theta_2 & -\theta_1 & 1 \end{bmatrix} \begin{bmatrix} \vec{a}_1 \\ \vec{a}_2 \\ \vec{a}_3 \end{bmatrix} \tag{6.159}$$

or

$$\begin{bmatrix} \vec{a}_1 \\ \vec{a}_2 \\ \vec{a}_3 \end{bmatrix} \approx \begin{bmatrix} 1 & -\theta_3 & \theta_2 \\ \theta_3 & 1 & -\theta_1 \\ -\theta_2 & \theta_1 & 1 \end{bmatrix} \begin{bmatrix} \vec{b}_1 \\ \vec{b}_2 \\ \vec{b}_3 \end{bmatrix} \tag{6.160}$$

The angular velocity of B in N for this case of small relative angles of B with respect to A is also given by

$$\begin{aligned} \vec{\omega} \equiv \vec{\omega}^{B/N} &= \omega_1\vec{b}_1 + \omega_2\vec{b}_2 + \omega_3\vec{b}_3 \\ &\approx (\dot{\theta}_1 - n\theta_3)\vec{b}_1 + (\dot{\theta}_2 - n)\vec{b}_2 + (\dot{\theta}_3 + n\theta_1)\vec{b}_3 \end{aligned} \tag{6.161}$$

6.10.2   Linear Stability Analysis

Because the pitch-axis equation (6.157) is decoupled from the roll/yaw equations (6.156) and (6.158), consider first the characteristic equation of the pitch axis given by

$$s^2 + \left[3n^2(J_1 - J_3)/J_2\right] = 0 \qquad (6.162)$$

If $J_1 > J_3$, then the characteristic roots are pure imaginary numbers and it is said to be (Lyapunov) stable. If $J_1 < J_3$, then one of the characteristic roots is a positive real number and it is said to be unstable. Therefore, the necessary and sufficient condition for pitch stability is

$$J_1 > J_3 \qquad (6.163)$$

For the roll/yaw stability analysis, the roll/yaw equations (6.156) and (6.158) are rewritten as

$$\ddot{\theta}_1 + (k_1 - 1)n\dot{\theta}_3 + 4n^2k_1\theta_1 = 0 \qquad (6.164a)$$
$$\ddot{\theta}_3 + (1 - k_3)n\dot{\theta}_1 + n^2k_3\theta_3 = 0 \qquad (6.164b)$$

where

$$k_1 = (J_2 - J_3)/J_1, \qquad k_3 = (J_2 - J_1)/J_3 \qquad (6.165)$$

Because of the physical properties of the moments of inertia ($J_1 + J_2 > J_3$, $J_2 + J_3 > J_1$, and $J_1 + J_3 > J_2$), k_1 and k_3 are, in fact, bounded as

$$|k_1| < 1, \qquad |k_3| < 1 \qquad (6.166)$$

The roll/yaw characteristic equation can then be found as

$$s^4 + (1 + 3k_1 + k_1k_3)n^2s^2 + 4k_1k_3n^4 = 0 \qquad (6.167)$$

The roll/yaw characteristic roots become pure imaginary numbers if and only if

$$k_1k_3 > 0 \qquad (6.168a)$$
$$1 + 3k_1 + k_1k_3 > 0 \qquad (6.168b)$$
$$(1 + 3k_1 + k_1k_3)^2 - 16k_1k_3 > 0 \qquad (6.168c)$$

which are the necessary and sufficient conditions for roll/yaw stability.

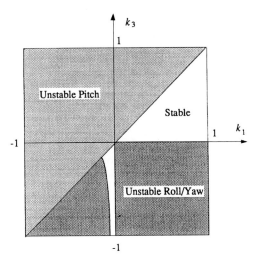

Fig. 6.9　Gravity-gradient stability plot.

The preceding results for linear stability of a rigid body in a circular orbit can be summarized using a stability diagram in the (k_1, k_3) plane, as shown in Fig. 6.9. For a further treatment of this subject, see Hughes [2].

Problems

6.10　Consider the sequence of $C_1(\theta_1) \leftarrow C_3(\theta_3) \leftarrow C_2(\theta_2)$ from the LVLH reference frame A to a body-fixed reference frame B for a rigid spacecraft in a circular orbit.

(a) Verify the following relationship:

$$\begin{bmatrix} \vec{b}_1 \\ \vec{b}_2 \\ \vec{b}_3 \end{bmatrix} = \begin{bmatrix} c\,\theta_2\,c\,\theta_3 & s\,\theta_3 & -s\,\theta_2\,c\,\theta_3 \\ -c\,\theta_1\,c\,\theta_2\,s\,\theta_3 + s\,\theta_1\,s\,\theta_2 & c\,\theta_1\,c\,\theta_3 & c\,\theta_1\,s\,\theta_2\,s\,\theta_3 + s\,\theta_1\,c\,\theta_2 \\ s\,\theta_1\,c\,\theta_2\,s\,\theta_3 + c\,\theta_1\,s\,\theta_2 & -s\,\theta_1\,c\,\theta_3 & -s\,\theta_1\,s\,\theta_2\,s\,\theta_3 + c\,\theta_1\,c\,\theta_2 \end{bmatrix} \begin{bmatrix} \vec{a}_1 \\ \vec{a}_2 \\ \vec{a}_3 \end{bmatrix}$$

where $c\,\theta_i = \cos\theta_i$ and $s\,\theta_i = \sin\theta_i$.

(b) Derive the following kinematic differential equation:

$$\begin{bmatrix} \dot{\theta}_1 \\ \dot{\theta}_2 \\ \dot{\theta}_3 \end{bmatrix} = \frac{1}{\cos\theta_3} \begin{bmatrix} \cos\theta_3 & -\cos\theta_1\sin\theta_3 & \sin\theta_1\sin\theta_3 \\ 0 & \cos\theta_1 & -\sin\theta_1 \\ 0 & \sin\theta_1\cos\theta_3 & \cos\theta_1\cos\theta_3 \end{bmatrix} \begin{bmatrix} \omega_1 \\ \omega_2 \\ \omega_3 \end{bmatrix} + \begin{bmatrix} 0 \\ n \\ 0 \end{bmatrix}$$

(c) For small attitude deviations from LVLH orientation, show that the linearized dynamic equations of motion, including the products of inertia, can be written as

$$
\begin{bmatrix} J_{11} & J_{12} & J_{13} \\ J_{21} & J_{22} & J_{23} \\ J_{31} & J_{32} & J_{33} \end{bmatrix} \begin{bmatrix} \dot{\omega}_1 \\ \dot{\omega}_2 \\ \dot{\omega}_3 \end{bmatrix} = n \begin{bmatrix} J_{31} & 2J_{32} & J_{33} - J_{22} \\ -J_{32} & 0 & J_{12} \\ J_{22} - J_{11} & -2J_{12} & -J_{13} \end{bmatrix} \begin{bmatrix} \omega_1 \\ \omega_2 \\ \omega_3 \end{bmatrix}
$$

$$
+ 3n^2 \begin{bmatrix} J_{33} - J_{22} & J_{21} & 0 \\ J_{12} & J_{33} - J_{11} & 0 \\ -J_{13} & -J_{23} & 0 \end{bmatrix} \begin{bmatrix} \theta_1 \\ \theta_2 \\ \theta_3 \end{bmatrix} + n^2 \begin{bmatrix} -2J_{23} \\ 3J_{13} \\ -J_{12} \end{bmatrix}
$$

(d) Verify that the linearized equations of motion can also be written in terms of Euler angles, as follows:

$$
\begin{bmatrix} J_{11} & J_{12} & J_{13} \\ J_{21} & J_{22} & J_{23} \\ J_{31} & J_{32} & J_{33} \end{bmatrix} \begin{bmatrix} \ddot{\theta}_1 \\ \ddot{\theta}_2 \\ \ddot{\theta}_3 \end{bmatrix}
$$

$$
= n \begin{bmatrix} 0 & 2J_{32} & J_{11} - J_{22} + J_{33} \\ -2J_{32} & 0 & 2J_{12} \\ -J_{11} + J_{22} - J_{33} & -2J_{12} & 0 \end{bmatrix} \begin{bmatrix} \dot{\theta}_1 \\ \dot{\theta}_2 \\ \dot{\theta}_3 \end{bmatrix}
$$

$$
+ n^2 \begin{bmatrix} 4(J_{33} - J_{22}) & 3J_{21} & -J_{31} \\ 4J_{12} & 3(J_{33} - J_{11}) & J_{32} \\ -4J_{13} & -3J_{23} & J_{11} - J_{22} \end{bmatrix} \begin{bmatrix} \theta_1 \\ \theta_2 \\ \theta_3 \end{bmatrix} + n^2 \begin{bmatrix} -4J_{23} \\ 3J_{13} \\ J_{12} \end{bmatrix}
$$

(e) For large, pitch-axis angular motion but with small roll/yaw angles, derive the following equations of motion about principal axes:

$$
J_1\ddot{\theta}_1 + (1 + 3\cos^2\theta_2)n^2(J_2 - J_3)\theta_1 - n(J_1 - J_2 + J_3)\dot{\theta}_3
$$

$$
+ 3(J_2 - J_3)n^2(\sin\theta_2 \cos\theta_2)\theta_3 = 0
$$

$$
J_2\ddot{\theta}_2 + 3n^2(J_1 - J_3)\sin\theta_2 \cos\theta_2 = 0
$$

$$
J_3\ddot{\theta}_3 + (1 + 3\sin^2\theta_2)n^2(J_2 - J_1)\theta_3 + n(J_1 - J_2 + J_3)\dot{\theta}_1
$$

$$
+ 3(J_2 - J_1)n^2(\sin\theta_2 \cos\theta_2)\theta_1 = 0
$$

Note: See Ref. 10 for additional information pertaining to Problem 6.10.

6.11 Consider a rigid body in a circular orbit possessing a body-fixed reference frame B with basis vectors $\{\vec{b}_1, \vec{b}_2, \vec{b}_3\}$. The LVLH reference frame A with its origin at the center of mass of an orbiting spacecraft has a set of unit vectors $\{\vec{a}_1, \vec{a}_2, \vec{a}_3\}$, with $\vec{a}_1$ along the orbit direction, $\vec{a}_2$ perpendicular to the orbit plane, and $\vec{a}_3$ toward the Earth, as illustrated in Fig. 6.8.

where

$$\begin{bmatrix} H_x \\ H_y \\ H_z \end{bmatrix} = \begin{bmatrix} J_1\dot{\theta}_x + n(J_2 - J_1)\theta_z \\ J_2(\dot{\theta}_y - n) \\ J_3\dot{\theta}_z + n(J_3 - J_2)\theta_x \end{bmatrix}$$

and T_x, T_y, and T_z are the components of any other external torque expressed in the LVLH frame.

Note: See Ref. 12 for additional information pertaining to Problem 6.12.

6.11 Gyrostat in a Circular Orbit

There are basically two different types of spacecraft: 1) a three-axis stabilized spacecraft and 2) a dual-spin stabilized spacecraft.

A three-axis stabilized spacecraft with a bias-momentum wheel is often called a bias-momentum stabilized spacecraft. INTELSAT V and INTELSAT VII satellites are typical examples of a bias-momentum stabilized spacecraft. In this kind of spacecraft configuration, a wheel is spun up to maintain a certain level of gyroscopic stiffness and the wheel is aligned along the pitch axis, nominally parallel to orbit normal.

A spacecraft with a large external rotor is called a dual-spinner or dual-spin stabilized spacecraft. INTELSAT IV and INTELSAT VI satellites are typical examples of a dual-spin stabilized spacecraft. The angular momentum, typically 2000 N·m·s, of a dual-spin stabilized spacecraft is much larger than that of a bias-momentum stabilized spacecraft. For example, INTELSAT V, a bias-momentum stabilized satellite, has an angular momentum of 35 N·m·s.

In this section we formulate the equations of motion of an Earth-pointing spacecraft equipped with reaction wheels. A rigid body, consisting of a main platform and spinning wheels, is often referred to as a gyrostat.

Consider a generic model of a gyrostat equipped with two reaction wheels aligned along roll and yaw axes and a pitch momentum wheel, as illustrated in Fig. 6.10. The pitch momentum wheel is nominally spun up along the negative pitch axis. Like Fig. 6.8 of the preceding section, a LVLH reference frame A with its origin at the center of mass of an orbiting gyrostat has a set of unit vectors $\{\vec{a}_1, \vec{a}_2, \vec{a}_3\}$, with $\vec{a}_1$ along the orbit direction, $\vec{a}_2$ perpendicular to the orbit plane, and $\vec{a}_3$ toward the Earth. Let $\{\vec{b}_1, \vec{b}_2, \vec{b}_3\}$ be a set of basis vectors of a body-fixed reference frame B, which is assumed to be aligned with principal axes of the gyrostat.

The total angular momentum vector of the spacecraft is then expressed as

$$\vec{H} = (J_1\omega_1 + h_1)\vec{b}_1 + (J_2\omega_2 - H_0 + h_2)\vec{b}_2 + (J_3\omega_3 + h_3)\vec{b}_3 \quad (6.169)$$

where J_1, J_2, and J_3 are the principal moments of inertia of the gyrostat spacecraft; ω_1, ω_2, and ω_3 are the body-fixed components of the angular velocity of the spacecraft, i.e., $\vec{\omega}^{B/N} \equiv \vec{\omega} = \omega_1\vec{b}_1 + \omega_2\vec{b}_2 + \omega_3\vec{b}_3$; h_1, $-H_0 + h_2$, and h_3 are the body-fixed components of the angular momentum of the three wheels; and H_0 is the nominal pitch bias momentum along the negative pitch axis.

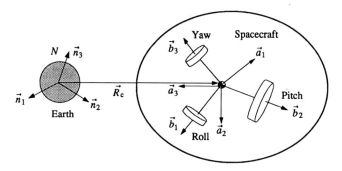

Fig. 6.10   Gyrostat in a circular orbit.

The rotational equation of motion is then simply given by

$$\dot{\vec{H}} = \left\{ \frac{\mathrm{d}\vec{H}}{\mathrm{d}t} \right\}_B + \vec{\omega}^{B/N} \times \vec{H} = \vec{M} \tag{6.170}$$

where $\vec{M}$ is the gravity-gradient torque acting on the vehicle. For the principal-axis frame B, the equations of motion can be written as

$$J_1\dot{\omega}_1 - (J_2 - J_3)\omega_2\omega_3 + \dot{h}_1 + \omega_2 h_3 - \omega_3(-H_0 + h_2) = M_1 \tag{6.171a}$$
$$J_2\dot{\omega}_2 - (J_3 - J_1)\omega_3\omega_1 + \dot{h}_2 + \omega_3 h_1 - \omega_1 h_3 = M_2 \tag{6.171b}$$
$$J_3\dot{\omega}_3 - (J_1 - J_2)\omega_1\omega_2 + \dot{h}_3 + \omega_1(-H_0 + h_2) - \omega_2 h_1 = M_3 \tag{6.171c}$$

where $M_i = \vec{M} \cdot \vec{b}_i$.

For small relative angles between B and A, we have

$$\omega_1 = \dot{\theta}_1 - n\theta_3 \tag{6.172a}$$
$$\omega_2 = \dot{\theta}_2 - n \tag{6.172b}$$
$$\omega_3 = \dot{\theta}_3 + n\theta_1 \tag{6.172c}$$

and

$$M_1 = -3n^2(J_2 - J_3)\theta_1 \tag{6.173a}$$
$$M_2 = 3n^2(J_3 - J_1)\theta_2 \tag{6.173b}$$
$$M_3 = 0 \tag{6.173c}$$

where n is the orbital rate and θ_1, θ_2, and θ_3 are called the roll, pitch, and yaw attitude angles of the spacecraft relative to the LVLH reference frame A.

The linearized equations of motion of a gyrostat spacecraft in a circular orbit in terms of small roll, pitch, and yaw angles can be obtained as

$$J_1\ddot{\theta}_1 + \left[4n^2(J_2 - J_3) + nH_0\right]\theta_1$$
$$+ [-n(J_1 - J_2 + J_3) + H_0]\dot{\theta}_3 + \dot{h}_1 - nh_3 = 0 \qquad (6.174)$$

$$J_2\ddot{\theta}_2 + 3n^2(J_1 - J_3)\theta_2 + \dot{h}_2 = 0 \qquad (6.175)$$

$$J_3\ddot{\theta}_3 + \left[n^2(J_2 - J_1) + nH_0\right]\theta_3$$
$$- [-n(J_1 - J_2 + J_3) + H_0]\dot{\theta}_1 + \dot{h}_3 + nh_1 = 0 \qquad (6.176)$$

These linearized equations of motion will be used in Chapter 7 when we design attitude control systems of a bias-momentum stabilized spacecraft.

6.12   Dual-Spinner with a Platform Damper

A dual-spin spacecraft consisting of an external, axisymmetric rotor and a platform with a mass–spring–damper model is illustrated in Fig. 6.11. The so-called "despun platform" of dual-spin stabilized, geosynchronous communications satellites maintains continuous Earth-pointing, and thus it is actually spinning at orbital rate. The large external rotor with a much higher spin rate provides gyroscopic stiffness for attitude stability.

Most dual spinners such as INTELSAT III, launched in the early 1960s, had their rotor spin axis aligned with the major principal axis of the spacecraft. That is, the early, small dual spinners were disk shaped because of the major-axis stability condition of a spinning body with energy dissipation. In the mid-1960s, however, larger communications satellites had to be designed and dual spinners were no longer limited to the oblate (disk-shaped) configuration because of fairing constraints of launch vehicles. Consequently, some stability criteria for a prolate (rod-shaped) dual spinner were developed independently [13]. It was argued that the addition of energy dissipating devices on the despun platform would offset the destabilizing effect of energy dissipation in the rotor of a prolate dual spinner. Although such arguments were not rigorous at that time, an experimental prolate dual spinner, called the tactical communications satellite (TACSAT) was launched in 1969. Its successful mission led to many theoretical analysis results [14–16] and, furthermore, resulted in many prolate dual spinners such as the INTELSAT IV series starting in 1971.

In this section, we formulate the equations of motion of a dual-spin spacecraft with a despun platform damper, illustrated in Fig. 6.11. The problem of a rigid spacecraft with internal moving mass was first investigated independently by Roberson [17] in 1958 and then by Grubin [18, 19] in the early 1960s. For dynamic problems with internal moving mass, we may choose the composite center of mass of the total system as a reference point for the equations of motion. This formulation leads to a time-varying inertia matrix of the main rigid body, because the

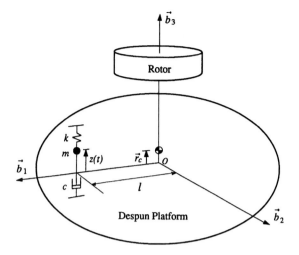

Fig. 6.11   Dual-spinner with a despun platform damper.

reference point is not fixed at the main body as the internal mass moves relative to the main body. On the other hand, we may choose the center of mass of the main body as the reference point, which leads to a constant inertia matrix of the main body relative to the reference point.

As illustrated in Fig. 6.11, a dual-spin spacecraft consists of an external rotor and a despun platform. The despun platform is considered to be the main body of the total system of mass M, and it has an internal moving part of mass m connected by a spring of stiffness k and a dashpot of damping coefficient c. The main body has a body-fixed reference frame B with basis vectors $\{\vec{b}_1, \vec{b}_2, \vec{b}_3\}$. The point mass m is located at a fixed distance ℓ from the reference point O along $\vec{b}_1$ direction and has a relative displacement z along $\vec{b}_3$ direction. The reference point O of B is assumed to be the center of mass of the main body when $z = 0$. The reference frame B is also assumed to be aligned with principal axes of the main body when $z = 0$. The rotor has an angular momentum of $J\Omega$ relative to the main body.

The angular momentum equation of an external torque-free, dual spinner with respect to the reference point O can be written as

$$\dot{\vec{h}}_o + M\vec{r}_c \times \vec{a}_o = 0 \qquad (6.177)$$

where M is the total mass, $\vec{r}_c$ is the position vector of the composite center of mass from the reference point O, and $\vec{a}_o$ is the inertial acceleration of the point O. The relative angular momentum of the total system about point O, denoted by $\vec{h}_o$, is given by

$$\vec{h}_o = \hat{J} \cdot \vec{\omega} + J\Omega\vec{b}_3 - m\ell\dot{z}\vec{b}_2 \qquad (6.178)$$

where $\vec{\omega}$ is the angular velocity vector of the main body and

$$\hat{J} = \begin{bmatrix} \vec{b}_1 & \vec{b}_2 & \vec{b}_3 \end{bmatrix} \begin{bmatrix} J_1 + mz^2 & 0 & -m\ell z \\ 0 & J_2 + mz^2 & 0 \\ -m\ell z & 0 & J_3 \end{bmatrix} \begin{bmatrix} \vec{b}_1 \\ \vec{b}_2 \\ \vec{b}_3 \end{bmatrix}$$

is the inertia dyadic of the main body, including the damper mass, about the reference point O.

From the geometry of the system, we find the position vector of the composite center of mass from the reference point O, as follows:

$$\vec{r}_c = (m/M)\, z\vec{b}_3 \tag{6.179}$$

The inertial acceleration of the reference point O is related to the inertial acceleration of the composite center of mass as

$$\vec{a}_c = \vec{a}_o + \ddot{\vec{r}}_c \tag{6.180}$$

Because $\vec{a}_c = 0$ for dynamic systems with zero external force, we have

$$\vec{a}_o = -\ddot{\vec{r}}_c = -\frac{d^2}{dt^2}\left(\frac{m}{M}z\vec{b}_3\right) \tag{6.181}$$

Expressing the angular velocity vector $\vec{\omega}$ as

$$\vec{\omega} = \omega_1\vec{b}_1 + \omega_2\vec{b}_2 + \omega_3\vec{b}_3 \tag{6.182}$$

we obtain

$$\vec{a}_o = -\frac{m}{M}\big[(2\omega_2\dot{z} + \dot{\omega}_2 z + \omega_2\omega_3 z)\vec{b}_1 + (-2\omega_1\dot{z} - z\dot{\omega}_1 + z\omega_2\omega_3)\vec{b}_2$$
$$+ (\ddot{z} - \omega_1^2 z - \omega_2^2 z)\vec{b}_3\big] \tag{6.183}$$

and

$$\vec{h}_o = \big[(J_1 + mz^2)\omega_1 - m\ell\omega_3 z\big]\vec{b}_1 + \big[(J_2 + mz^2)\omega_2 - m\ell\dot{z}\big]\vec{b}_2$$
$$+ \big[-m\ell\omega_1 z + J_3\omega_3 + J\Omega\big]\vec{b}_3 \tag{6.184}$$

Finally, the rotational equations of motion of the main body about the reference point O can be obtained as

$$J_1\dot{\omega}_1 - (J_2 - J_3)\omega_2\omega_3 + J\Omega\omega_2 + m(1 - m/M)\dot{\omega}_1 z^2 - m(1 - m/M)\omega_2\omega_3 z^2$$
$$+ 2m(1 - m/M)\omega_1 z\dot{z} - m\ell\dot{\omega}_3 z - m\ell\omega_1\omega_2 z = 0 \tag{6.185}$$

$$J_2\dot{\omega}_2 - (J_3 - J_1)\omega_3\omega_1 - J\Omega\omega_1 + m(1 - m/M)\dot{\omega}_2 z^2 + m(1 - m/M)\omega_1\omega_3 z^2$$
$$+ 2m(1 - m/M)\omega_2 z\dot{z} - m\ell\ddot{z} + m\ell\omega_1^2 z - m\ell\omega_3^2 z = 0 \tag{6.186}$$

$$J_3\dot{\omega}_3 - (J_1 - J_2)\omega_1\omega_2 + J\dot{\Omega} + m\ell\omega_2\omega_3 z - 2m\ell\omega_1\dot{z} - m\ell\dot{\omega}_1 z = 0 \tag{6.187}$$

The equation of motion for the rotor is given by

$$J(\dot{\omega}_3 + \dot{\Omega}) = T \tag{6.188}$$

where T is the rotor spin control torque.

The dynamic equation of the internal moving mass itself can also be found by applying Newton's second law, as follows:

$$m\vec{a} = F_1\vec{b}_1 + F_2\vec{b}_2 - (c\dot{z} + kz)\vec{b}_3 \tag{6.189}$$

where F_1 and F_2 are constraint forces, and $\vec{a}$ is the inertial acceleration of the internal moving mass, which can be expressed as

$$\vec{a} = \vec{a}_c + \frac{d^2}{dt^2}\left\{\ell\vec{b}_1 + \left(1 - \frac{m}{M}\right)z\vec{b}_3\right\} \tag{6.190}$$

Because $\vec{a}_c = 0$, we have

$$\vec{a} = \frac{d^2}{dt^2}\left\{\ell\vec{b}_1 + \left(1 - \frac{m}{M}\right)z\vec{b}_3\right\} \tag{6.191}$$

and the equation of motion of the damper mass along the $\vec{b}_3$ direction can be written as

$$m(1 - m/M)\ddot{z} + c\dot{z} + kz - m(1 - m/M)\left(\omega_1^2 + \omega_2^2\right)z + m\ell\omega_1\omega_3 - m\ell\dot{\omega}_2 = 0 \tag{6.192}$$

The rigorous stability analysis of a dual-spin spacecraft with energy dissipation is beyond the scope of this book and will not be pursued further in this text. For a more detailed treatment of this subject, the reader is referred to Hughes [2] and Kaplan [20], and also Refs. 14–16, and 21–26.

References

[1] Leimanis, E., *The General Problem of the Motion of Coupled Rigid Bodies About a Fixed Point,* Springer–Verlag, New York, 1965.

[2] Hughes, P. C., *Spacecraft Attitude Dynamics,* Wiley, New York, 1986.

[3] Junkins, J. L., Jacobson, I. D., and Blanton, J. N., "A Nonlinear Oscillator Analog of Rigid Body Dynamics," *Celestial Mechanics,* Vol. 7, 1973, pp. 398–407.

[4] Bracewell, R. N., and Garriott, O. K., "Rotation of Artificial Earth Satellites," *Nature,* Vol. 182, Sept. 1958, pp. 760–762.

[5] Rahn, C. D., and Barba, P. M., "Reorientation Maneuver for Spinning Spacecraft," *Journal of Guidance, Control, and Dynamics,* Vol. 14, No. 4, 1991, pp. 724–728.

[6] Jarmolow, K., "Dynamics of a Spinning Rocket with Varying Inertia and Applied Moment," *Journal of Applied Physics,* Vol. 28, No. 3, 1957, pp. 308–313.

[7] Kolk, W. R., *Modern Flight Dynamics,* Prentice–Hall, Englewood Cliffs, NJ, 1961, Chap. 12.

[8] Livneh, R., and Wie, B., "New Results for an Asymmetric Body with Constant Body-Fixed Torques," *Proceedings of the AIAA/AAS Astrodynamics Conference*, AIAA, Washington, DC, 1996, pp. 828–838; also *Journal of Guidance, Control, and Dynamics*, Vol. 20, No. 5, 1997, pp. 873–881.

[9] Livneh, R., and Wie, B., "The Effect of Energy Dissipation on a Rigid Body with Constant Body-Fixed Torques," *Proceedings of the AIAA/AAS Astrodynamics Conference*, AIAA, Washington, DC, 1996, pp. 839–847.

[10] Wie, B., Byun, K.-W., Warren, W., Geller, D., Long, D., and Sunkel, J., "New Approach to Momentum/Attitude Control for the Space Station," *Journal of Guidance, Control, and Dynamics*, Vol. 12, No. 5, 1989, pp. 714–722.

[11] Warren, W., Wie, B., and Geller, D., "Periodic Disturbance Accommodating Control of the Space Station for Asymptotic Momentum Management," *Journal of Guidance, Control, and Dynamics*, Vol. 13, No. 6, 1990, pp. 984–992.

[12] Wie, B., Hu, A., and Singh, R., "Multi-Body Interaction Effects on Space Station Attitude Control and Momentum Management," *Journal of Guidance, Control, and Dynamics*, Vol. 13, No. 6, 1990, pp. 993–999.

[13] Landon, V. D., and Stewart, B., "Nutational Stability of an Axisymmetric Body Containing a Rotor," *Journal of Spacecraft and Rockets*, Vol. 1, No. 6, 1964, pp. 682–684.

[14] Likins, P. W., "Attitude Stability Criteria for Dual-Spin Spacecraft," *Journal of Spacecraft and Rockets*, Vol. 4, No. 12, 1967, pp. 1638–1643.

[15] Mingori, D. L., "Effects of Energy Dissipation on the Attitude Stability of Dual-Spin Satellites," *AIAA Journal*, Vol. 7, No. 1, 1969, pp. 20–27.

[16] Likins, P. W., "Spacecraft Attitude Dynamics and Control: A Personal Perspective on Early Developments," *Journal of Guidance, Control, and Dynamics*, Vol. 9, No. 2, 1986, pp. 129–134.

[17] Roberson, R. E., "Torques on a Satellite Vehicle from Internal Moving Parts," *Journal of Applied Mechanics*, Vol. 25, 1958, pp. 196–200.

[18] Grubin, C., "On Generalization of the Angular Momentum Equation," *Journal of Engineering Education*, Vol. 51, 1960, pp. 237–238.

[19] Grubin, C., "Dynamics of a Vehicle Containing Moving Parts," *Journal of Applied Mechanics*, Vol. 29, 1962, pp. 486–488.

[20] Kaplan, M. H., *Modern Spacecraft Dynamics and Control*, Wiley, New York, 1976.

[21] Bainum, P. M., Fuechsel, P. G., and Mackison, D. L., "Motion and Stability of a Dual-Spin Satellite with Nutation Damping," *Journal of Spacecraft and Rockets*, Vol. 7, No. 6, 1970, pp. 690–696.

[22] Alfriend, K. T., and Hubert, C. H., "Stability of a Dual-Spin Satellite with Two Dampers," *Journal of Spacecraft and Rockets*, Vol. 11, No. 7, 1974, pp. 469–474.

[23] Cochran, J. E., and Thompson, J. A., "Nutation Dampers vs. Precession Dampers for Asymmetric Spacecraft," *Journal of Guidance and Control*, Vol. 3, No. 1, 1980, pp. 22–28.

[24] Adams, G. J., "Dual-Spin Spacecraft Dynamics During Platform Spinup," *Journal of Guidance and Control*, Vol. 3, No. 1, 1980, pp. 29–36.

[25] Cochran, J. E., and Shu, P. H., "Effects of Energy Addition and Dissipation on Dual-Spin Spacecraft," *Journal of Guidance, Control, and Dynamics*, Vol. 6, No. 5, 1983, pp. 368–373.

[26] Or, A. C., "Resonances in the Despin Dynamics of Dual-Spin Spacecraft," *Journal of Guidance, Control, and Dynamics*, Vol. 14, No. 2, 1991, pp. 321–329.

7

Rotational Maneuvers and Attitude Control

Chapters 5 and 6 were concerned with the rotational kinematics and attitude dynamics of rigid spacecraft, respectively. This chapter deals with attitude control and stabilization problems of rigid spacecraft under the influence of reaction jet firings, internal energy dissipation, or momentum transfer via reaction wheels or control moment gyros (CMGs). A variety of control problems of spinning as well as three-axis stabilized spacecraft are treated. Emphasis is placed on a large-angle reorientation maneuver in which a spacecraft is required to maneuver about an inertially fixed axis as fast as possible, but within the saturation limits of rate gyros as well as reaction wheels. The attitude control and momentum management problem of a large space vehicle in low Earth orbit, such as the International Space Station, is also treated. Advanced spacecraft control problems of developing CMG steering logic and optimal jet selection logic are also treated. The attitude control problem of nonrigid spacecraft in the presence of propellant sloshing and/or structural flexibility will be covered in Chapter 9.

7.1 Control of Spinning Spacecraft

7.1.1 Introduction

Although a spacecraft spinning about its minor axis is unstable in the presence of internal energy dissipation, spacecraft are often required to spin about their minor axis for several reasons. Fairing constraints of most launch vehicles require that the minor axis of payload spacecraft be aligned with the longitudinal axis of the launch vehicles. Furthermore, most launch vehicles spin about their longitudinal axis prior to payload separation, resulting in a minor axis spin of the spacecraft after separation. Some launch vehicles or upper stages do not have spin-up capability for payload spacecraft and spin up of the spacecraft is achieved after separation. Because of initial angular rates at separation, a typical spin-up maneuver usually results in a residual nutation angle and a spin-axis precession from the separation attitude. The spin rate selection depends on many factors, including the pointing accuracy requirement.

Spacecraft spinning about their minor axis are often stabilized using an active nutation control system consisting of thrusters and accelerometers. Spinning spacecraft are also required to spin down or reorient their spin axis during the various

403

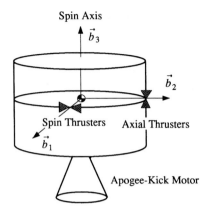

Spin Axis

$\vec{b}_3$

$\vec{b}_2$

Spin Thrusters   Axial Thrusters

$\vec{b}_1$

Apogee-Kick Motor

Fig. 7.1   Axisymmetric, spin-stabilized spacecraft with axial and spin thrusters.

phases of spacecraft operation.* For example, a spin-axis reorientation is required to align the spacecraft spin axis in the proper direction for apogee-kick motor firing. After apogee-kick motor burn, a spin-axis reorientation is also required to orient the spin axis to the orbit normal. During the operational life of spin-stabilized spacecraft, periodic spin-axis reorientation maneuvers are also required to compensate for the effects of external disturbance torques, primarily caused by solar pressures.

In this section we treat such various problems of controlling spin-stabilized spacecraft, including a spin-up maneuver, active nutation control, spin-axis reorientation, flat-spin recovery, and attitude acquisition by momentum transfer.

7.1.2   Spin-Up Maneuver

An axisymmetric, spin-stabilized spacecraft equipped with axial and spin thrusters is illustrated in Fig. 7.1. The axial thrusters are used during spin-axis reorientation maneuvers and the spin thrusters are used during spin-up (or spin-down) maneuvers as well as for spin rate control.

Consider an axisymmetric rigid body possessing a body-fixed reference frame B with basis vectors $\{\vec{b}_1, \vec{b}_2, \vec{b}_3\}$ and with its origin at the center of mass, as illustrated in Fig. 7.1. The reference frame B coincides with principal axes. The first and second axes are called the transverse axes and the third is the axis of symmetry.

Euler's rotational equations of motion of an axisymmetric spacecraft with $J_1 = J_2 = J$ are

$$J\dot{\omega}_1 - (J - J_3)\omega_3\omega_2 = 0 \tag{7.1}$$

$$J\dot{\omega}_2 + (J - J_3)\omega_3\omega_1 = 0 \tag{7.2}$$

$$J_3\dot{\omega}_3 = M_3 \tag{7.3}$$

---

*The term "spin axis" is ambiguous; however, it is used in this chapter to denote the axis of symmetry of a spinning axisymmetric spacecraft.

where $\omega_i \equiv \vec{b}_i \cdot \vec{\omega}$ are the body-fixed components of the angular velocity of the spacecraft and M_3 is the constant spin-up moment.

From Eq. (7.3), we have

$$\omega_3 = \omega_3(0) + (M_3/J_3)t \qquad (7.4)$$

and Eqs. (7.1) and (7.2) are rewritten as

$$\dot{\omega}_1 - \lambda\omega_2 = 0 \qquad (7.5)$$

$$\dot{\omega}_2 + \lambda\omega_1 = 0 \qquad (7.6)$$

where

$$\lambda = \frac{J - J_3}{J}\left\{\omega_3(0) + \frac{M_3}{J_3}t\right\}$$

To describe the rotational motion of the spacecraft during a spin-up maneuver as seen from an inertial reference frame, we consider the following kinematic differential equations:

$$\dot{\theta}_1 = (\omega_1 \cos\theta_3 - \omega_2 \sin\theta_3)/\cos\theta_2 \qquad (7.7a)$$

$$\dot{\theta}_2 = \omega_1 \sin\theta_3 + \omega_2 \cos\theta_3 \qquad (7.7b)$$

$$\dot{\theta}_3 = (-\omega_1 \cos\theta_3 + \omega_2 \sin\theta_3)\tan\theta_2 + \omega_3 \qquad (7.7c)$$

for the rotational sequence of $C_3(\theta_3) \leftarrow C_2(\theta_2) \leftarrow C_1(\theta_1)$.

For small angles of θ_1 and θ_2, which are often the cases of practical interest, these kinematic differential equations can be approximated as

$$\dot{\theta}_1 = \omega_1 \cos\theta_3 - \omega_2 \sin\theta_3 \qquad (7.8a)$$

$$\dot{\theta}_2 = \omega_1 \sin\theta_3 + \omega_2 \cos\theta_3 \qquad (7.8b)$$

$$\dot{\theta}_3 = \omega_3 \qquad (7.8c)$$

Furthermore, if $\omega_3(0) = 0$ and $\theta_3(0) = 0$, then

$$\omega_3(t) = (M_3/J_3)t \qquad \text{and} \qquad \theta_3(t) = (M_3/2J_3)t^2 \qquad (7.9)$$

For a special case of an inertially spherical body with $J_1 = J_2 = J_3$, we have

$$\theta_1(t) = \omega_1(0)\int_0^t \cos\frac{M_3}{2J_3}t^2\,dt - \omega_2(0)\int_0^t \sin\frac{M_3}{2J_3}t^2\,dt \qquad (7.10a)$$

$$\theta_2(t) = \omega_1(0)\int_0^t \sin\frac{M_3}{2J_3}t^2\,dt + \omega_2(0)\int_0^t \cos\frac{M_3}{2J_3}t^2\,dt \qquad (7.10b)$$

In terms of the Fresnel integrals, these equations become

$$\theta_1(t) = \omega_1(0)\sqrt{\frac{J_3}{2M_3}}\int_0^{\theta_3}\frac{\cos\theta}{\sqrt{\theta}}\,d\theta - \omega_2(0)\sqrt{\frac{J_3}{2M_3}}\int_0^{\theta_3}\frac{\sin\theta}{\sqrt{\theta}}\,d\theta \qquad (7.11a)$$

$$\theta_2(t) = \omega_1(0)\sqrt{\frac{J_3}{2M_3}}\int_0^{\theta_3}\frac{\sin\theta}{\sqrt{\theta}}\,d\theta + \omega_2(0)\sqrt{\frac{J_3}{2M_3}}\int_0^{\theta_3}\frac{\cos\theta}{\sqrt{\theta}}\,d\theta \qquad (7.11b)$$

and

$$\theta_1(\infty) = \omega_1(0)\sqrt{\frac{J_3}{2M_3}}\sqrt{\frac{\pi}{2}} - \omega_2(0)\sqrt{\frac{J_3}{2M_3}}\sqrt{\frac{\pi}{2}} \tag{7.12a}$$

$$\theta_2(\infty) = \omega_1(0)\sqrt{\frac{J_3}{2M_3}}\sqrt{\frac{\pi}{2}} + \omega_2(0)\sqrt{\frac{J_3}{2M_3}}\sqrt{\frac{\pi}{2}} \tag{7.12b}$$

For an axisymmetric body with $J_1 = J_2 = J \neq J_3$, we have

$$\frac{d\omega_1}{dt} - \sigma t\omega_2 = 0 \tag{7.13a}$$

$$\frac{d\omega_2}{dt} + \sigma t\omega_1 = 0 \tag{7.13b}$$

where

$$\sigma = \frac{J - J_3}{J}\frac{M_3}{J_3}$$

By changing the independent variable t to τ as $\tau = \sigma t^2/2$, we obtain

$$\frac{d\omega_1}{d\tau} - \omega_2 = 0$$

$$\frac{d\omega_2}{d\tau} + \omega_1 = 0$$

The solutions of Eqs. (7.13) are then simply found as

$$\omega_1(t) = \omega_1(0)\cos\frac{\sigma t^2}{2} + \omega_2(0)\sin\frac{\sigma t^2}{2} \tag{7.14a}$$

$$\omega_2(t) = \omega_2(0)\cos\frac{\sigma t^2}{2} - \omega_1(0)\sin\frac{\sigma t^2}{2} \tag{7.14b}$$

Substituting these analytic solutions into Eq. (7.8a), we obtain

$$\dot{\theta}_1 = \left\{\omega_1(0)\cos\frac{\sigma t^2}{2} + \omega_2(0)\sin\frac{\sigma t^2}{2}\right\}\cos\theta_3$$

$$- \left\{\omega_2(0)\cos\frac{\sigma t^2}{2} - \omega_1(0)\sin\frac{\sigma t^2}{2}\right\}\sin\theta_3$$

$$= \omega_1(0)\left\{\cos\frac{M_3 t^2}{2J_3}\cos\frac{\sigma t^2}{2} + \sin\frac{M_3 t^2}{2J_3}\sin\frac{\sigma t^2}{2}\right\}$$

$$+ \omega_2(0)\left\{\cos\frac{M_3 t^2}{2J_3}\sin\frac{\sigma t^2}{2} - \sin\frac{M_3 t^2}{2J_3}\cos\frac{\sigma t^2}{2}\right\}$$

which, using the trigonometric relationships $\cos A \cos B \pm \sin A \sin B = \cos(A \mp B)$ and $\sin A \cos B \pm \cos A \sin B = \sin(A \pm B)$, can be simplified as

$$\dot{\theta}_1 = \omega_1(0) \cos \frac{M_3}{2J} t^2 - \omega_2(0) \sin \frac{M_3}{2J} t^2 \qquad (7.15)$$

This equation, after integration, is written in terms of the Fresnel integrals, as follows:

$$\theta_1(t) = \omega_1(0) \sqrt{\frac{J}{2M_3}} \int_0^{\theta^*} \frac{\cos \theta}{\sqrt{\theta}} d\theta - \omega_2(0) \sqrt{\frac{J}{2M_3}} \int_0^{\theta^*} \frac{\sin \theta}{\sqrt{\theta}} d\theta \qquad (7.16)$$

where

$$\theta^* = \frac{M^3}{2J} t^2$$

Similarly, we obtain

$$\theta_2(t) = \omega_1(0) \sqrt{\frac{J}{2M_3}} \int_0^{\theta^*} \frac{\sin \theta}{\sqrt{\theta}} d\theta + \omega_2(0) \sqrt{\frac{J}{2M_3}} \int_0^{\theta^*} \frac{\cos \theta}{\sqrt{\theta}} d\theta \qquad (7.17)$$

Furthermore, the steady-state precession angle of the spin axis from an initial attitude can be found as

$$\theta_1(\infty) = \omega_1(0) \sqrt{\frac{J}{2M_3}} \sqrt{\frac{\pi}{2}} - \omega_2(0) \sqrt{\frac{J}{2M_3}} \sqrt{\frac{\pi}{2}} \qquad (7.18a)$$

$$\theta_2(\infty) = \omega_1(0) \sqrt{\frac{J}{2M_3}} \sqrt{\frac{\pi}{2}} + \omega_2(0) \sqrt{\frac{J}{2M_3}} \sqrt{\frac{\pi}{2}} \qquad (7.18b)$$

The Fresnel integrals, when plotted one against the other, describe a curve called a *Cornu spiral* or *clothoid*.

Problems

7.1 Consider a nearly axisymmetric spacecraft with $(J_1, J_2, J_3) = (4223, 4133, 768)$ kg·m^2, $(M_1, M_2, M_3) = (0, 0, 10)$ N·m, and $(\omega_1, \omega_2, \omega_3) = (0.0001, 0, 0)$ rad/s at $t = 0$.
(a) Perform nonlinear simulation for a 100-s spin-up maneuver. In particular, plot θ_2 vs θ_1.
(b) Compare the nonlinear simulation results with the analytical solutions Eqs. (7.16), (7.17), and (7.18) obtained for an axisymmetric body.
(c) Discuss the rotational motion of the spacecraft after the spin-up maneuver in terms of a residual nutation angle and a spin-axis precession from the initial attitude.
Note: The spin-up maneuvering time of this problem is somewhat long compared to the few seconds of burn time of solid-propellant spin motors.

7.2 Repeat Problem 7.1(a) but with $(\omega_1, \omega_2, \omega_3) = (0, 0, 0)$ at $t = 0$ and $(M_1, M_2, M_3) = (1, 0, 10)$ N·m. Discuss the effect of nonzero M_1, which is caused by thruster misalignment.

7.1.3 Flat-Spin Transition Maneuver

One of the simplest rotational maneuvers is the reorientation of the spin axis of a spacecraft using internal energy dissipation. A semirigid spacecraft with internal energy dissipation is stable only when spinning about its major axis. A spacecraft spinning about its minor axis in the presence of energy dissipation is unstable; that is, the spacecraft will eventually reorient to spin about its major axis. Such a passive reorientation maneuver is called a flat-spin transition maneuver. The orientation of the spacecraft relative to the inertially fixed angular momentum vector at the end of the maneuver is, however, unpredictable; i.e., the spacecraft can end up with either a positive or a negative spin about the major axis.

To study such a flat-spin transition maneuver, consider a rigid body with a spherical fuel slug of inertia J, which is surrounded by a viscous fluid layer [1]. The rigid body has body rates of ω_1, ω_2, and ω_3 about the major, intermediate, and minor principal axes, respectively. Let the relative rates between the rigid body and the fuel slug be σ_1, σ_2, and σ_3 about the principal axes. The total angular momentum H and kinetic energy T are then expressed as

$$H^2 = (J_1\omega_1 + J\sigma_1)^2 + (J_2\omega_2 + J\sigma_2)^2 + (J_3\omega_3 + J\sigma_3)^2 \qquad (7.19)$$

$$2T = (J_1 - J)\omega_1^2 + (J_2 - J)\omega_2^2 + (J_3 - J)\omega_3^2$$
$$+ J\left\{(\omega_1 + \sigma_1)^2 + (\omega_2 + \sigma_2)^2 + (\omega_3 + \sigma_3)^2\right\} \qquad (7.20)$$

where (J_1, J_2, J_3) are the principal moments of inertia of the spacecraft including the slug. The rotational equations of motion are written as

$$(J_1 - J)\dot{\omega}_1 = (J_2 - J_3)\omega_2\omega_3 + \mu\sigma_1 + M_1 \qquad (7.21\text{a})$$

$$(J_2 - J)\dot{\omega}_2 = (J_3 - J_1)\omega_3\omega_1 + \mu\sigma_2 + M_2 \qquad (7.21\text{b})$$

$$(J_3 - J)\dot{\omega}_3 = (J_1 - J_2)\omega_1\omega_2 + \mu\sigma_3 + M_3 \qquad (7.21\text{c})$$

$$\dot{\sigma}_1 = -\dot{\omega}_1 - (\mu/J)\sigma_1 - \omega_2\sigma_3 + \omega_3\sigma_2 \qquad (7.22\text{a})$$

$$\dot{\sigma}_2 = -\dot{\omega}_2 - (\mu/J)\sigma_2 - \omega_3\sigma_1 + \omega_1\sigma_3 \qquad (7.22\text{b})$$

$$\dot{\sigma}_3 = -\dot{\omega}_3 - (\mu/J)\sigma_3 - \omega_1\sigma_2 + \omega_2\sigma_1 \qquad (7.22\text{c})$$

where μ is the the viscous damping coefficient of the fuel slug and (M_1, M_2, M_3) are the control torques about the principal axes.

When $(M_1, M_2, M_3) = (0, 0, 0)$, the following can be shown.

1) A necessary condition for the equilibrium points is $\sigma_1^2 + \sigma_2^2 + \sigma_3^2 = 0$; i.e., $\sigma_1 = \sigma_2 = \sigma_3 = 0$.

2) An equilibrium point $(\omega_1, \omega_2, \omega_3, \sigma_1, \sigma_2, \sigma_3) = (\Omega, 0, 0, 0, 0, 0)$ is asymptotically stable; i.e., a pure spinning motion about the major axis is asymptotically stable.

3) An equilibrium point $(0, 0, \Omega, 0, 0, 0)$ is unstable; i.e., a pure spinning motion about the minor axis is unstable whereas it is Lyapunov stable for a rigid body without energy dissipation.

4) An equilibrium point $(0, \Omega, 0, 0, 0, 0)$ is also unstable.

As discussed in Chapter 6, the angular velocity vector $\vec{\omega}$ lies on the surface of the momentum ellipsoid and at the same time it lies on the surface of the energy ellipsoid. The curve in which these ellipsoids intersect is the path of the angular velocity vector as seen from a body-fixed reference frame, and it is called a polhode.

If there is no energy dissipation, H and T are constant and the polhode is a closed path. If there is energy dissipation, T decreases and the energy ellipsoid shrinks with time while H is kept constant. This results in an open polhode path that spirals outward from the minor axis toward the major axis by crossing a separatrix. The exact point at which the polhode crosses the separatrix is very sensitively dependent on the initial conditions and the energy dissipation rate of the spacecraft. Consequently, the orientation of the spacecraft relative to the inertially fixed angular momentum vector at the end of the flat-spin transition maneuver is unpredictable from a practical point of view.

The passive, flat-spin transition maneuver, when augmented by few thruster firings based on rate gyro signals, can, however, provide a predetermined, final spin polarity. For example, a control logic studied by Rahn and Barba [1] utilizes angular rate sign changes to determine when a separatrix has been crossed and when thrusters must be fired. The sizing of the thruster firings is, however, based on estimates of the energy dissipation in the spacecraft.

Problem

7.3 Consider a spacecraft with the following numerical values: $(J_1, J_2, J_3, J) = (2000, 1500, 1000, 18)$ kg·m^2, $\mu = 30$ N·m·s, and $|M_i| = 20$ N·m ($i = 1, 2, 3$).

(a) Perform computer simulation to verify that the trajectory starting from an initial condition $(0.1224, 0, 2.99, 0, 0, 0)$ rad/s ends up at $(-1.5, 0, 0, 0, 0, 0)$ rad/s. In particular, plot ω_3 vs ω_1 to show the separatrix crossing.

   Note: During computer simulation of this case with $M_i = 0$, the angular momentum H needs to be checked regardless of whether or not it is maintained at a constant value of 3000 N·m·s.

(b) Also perform computer simulation with a slightly different initial condition $(0.125, 0, 2.99, 0, 0, 0)$ and $M_i = 0$, to verify that the trajectory ends up at $(+1.5, 0, 0, 0, 0, 0)$. In particular, plot ω_3 vs ω_1 to show the separatrix crossing.

(c) Develop a thruster firing logic that provides a predetermined, final spin polarity using only the sign information of angular rates ω_1, ω_2, and ω_3. The control logic must be robust with respect to system modeling uncertainty, and the total thrust impulse needs to be minimized.

7.1.4 Active Nutation Control

An active nutation control (ANC) system of a spacecraft spinning about its minor axis usually consists of a pair of thrusters and a nutation sensor that measures the spacecraft acceleration associated with the nutational motion. The nutation sensor (accelerometer) is mounted with its sensitive axis aligned parallel to the spacecraft spin axis. The amplitude of sinusoidal acceleration output of the nutation sensor is proportional to the nutational angle. The ANC system generates a thruster pulse command with a proper timing when the nutation angle exceeds a certain threshold level.

Consider an axisymmetric rigid body spinning about its minor axis. As was illustrated in Fig. 7.1, the spacecraft has a body-fixed reference frame B with basis vectors $\{\vec{b}_1, \vec{b}_2, \vec{b}_3\}$ and with its origin at the center of mass. Euler's rotational equations of motion of such an axisymmetric spacecraft with $J_1 = J_2 = J > J_3$ become

$$J\dot{\omega}_1 - (J - J_3)\omega_3\omega_2 = M_1 \qquad (7.23)$$

$$J\dot{\omega}_2 + (J - J_3)\omega_3\omega_1 = M_2 \qquad (7.24)$$

$$J_3 \dot{\omega}_3 = 0 \qquad (7.25)$$

where $\omega_i \equiv \vec{b}_i \cdot \vec{\omega}$ are the body-fixed components of the angular velocity of the spacecraft; and M_1 and M_2 are the control torques along the transverse axes.

From Eq. (7.25), we have

$$\omega_3 = \text{const} = n \qquad (7.26)$$

where the constant n is called the spin rate of the spacecraft about its symmetry axis. Equations (7.23) and (7.24) then become

$$\dot{\omega}_1 = \lambda\omega_2 + M_1/J \qquad (7.27a)$$

$$\dot{\omega}_2 = -\lambda\omega_1 + M_2/J \qquad (7.27b)$$

where

$$\lambda = n(J - J_3)/J > 0$$

For a case in which $M_1 = 0$ and M_2 is a pulse applied at $t = t_1$, Eqs. (7.27) become

$$\dot{\omega}_1 = \lambda\omega_2 \qquad (7.28a)$$

$$\dot{\omega}_2 = -\lambda\omega_1 + (M/J)\{u_s(t - t_1) - u_s(t - t_1 - T)\} \qquad (7.28b)$$

where M is the constant thrust magnitude, $u_s(t)$ denotes the unit-step function, and T denotes the pulse width.

For $t > t_1 + T$, the solution $\omega_1(t)$ can be found as

$$\omega_1(t) = \omega_1(0) \cos \lambda t + \omega_2(0) \sin \lambda t$$
$$+ (M/\lambda J)\{-\cos \lambda(t - t_1) + \cos \lambda(t - t_1 - T)\}$$

which can be rewritten as

$$\omega_1(t) = [\omega_1(0) + (M/\lambda J)\{\cos \lambda t_1 + \cos \lambda(t_1 + T)\}] \cos \lambda t$$
$$+ [\omega_2(0) + (M/\lambda J)\{-\sin \lambda t_1 + \sin \lambda(t_1 + T)\}] \sin \lambda t \qquad (7.29)$$

Without loss of generality, let $t = 0$ be the initial time such that $\omega_1(t)$ is at its positive peak value and $\omega_2(0) = 0$. Then we have

$$
\begin{aligned}
\omega_1(t) = {}& [\omega_1(0) + (M/\lambda J)\{\cos \lambda t_1 + \cos \lambda (t_1 + T)\}] \cos \lambda t \\
& + [(M/\lambda J)\{- \sin \lambda t_1 + \sin \lambda (t_1 + T)\}] \sin \lambda t
\end{aligned}
\tag{7.30}
$$

and $\omega_2(t)$, for $t > t_1 + T$, also becomes

$$
\begin{aligned}
\omega_2(t) = {}& [\omega_1(0) + (M/\lambda J)\{\cos \lambda t_1 + \cos \lambda (t_1 + T)\}] \sin \lambda t \\
& - [(M/\lambda J)\{- \sin \lambda t_1 + \sin \lambda (t_1 + T)\}] \cos \lambda t
\end{aligned}
\tag{7.31}
$$

The transverse angular rate $\omega_{12} = (\omega_1^2 + \omega_2^2)^{1/2}$ after one thrust firing can be obtained as

$$
\begin{aligned}
\omega_{12}(t) = \Big\{ & \omega_1^2(0) + 2\left(\frac{M}{\lambda J}\right)^2 - 2\left(\frac{M}{\lambda J}\right)^2 \cos \lambda T \\
& + \frac{2M}{\lambda J}[\cos \lambda (t_1 + T) - \cos \lambda t_1]\omega_1(0) \Big\}^{\frac{1}{2}}
\end{aligned}
\tag{7.32}
$$

Taking the time derivative of Eq. (7.32) and setting the derivative equal to zero, we obtain the necessary condition for selecting t_1 to minimize the transverse angular rate ω_{12}, as follows:

$$
\lambda t_1 = \frac{\pi}{2} - \frac{\lambda T}{2}
\tag{7.33}
$$

This condition indicates that the thrust pulse should be centered over the negative peak of ω_2, because it is assumed that $\omega_2(0) = 0$.

The transverse angular rate reduced after one thrust pulse is then found as

$$
\Delta \omega_{12} = \frac{2M}{\lambda J} \sin \frac{\lambda T}{2}
\tag{7.34}
$$

Because the nutation angle θ is defined as

$$
\tan \theta = \frac{\sqrt{H_1^2 + H_2^2}}{H_3} = \frac{J\omega_{12}}{J_3 n}
$$

where $\omega_{12} = (\omega_1^2 + \omega_2^2)^{1/2}$, the nutation angle is approximated as

$$
\theta \approx \frac{J\omega_{12}}{J_3 n}
$$

and the nutation angle reduced after one thrust pulse becomes

$$
\Delta \theta \approx \frac{J\Delta \omega_{12}}{J_3 n} = \frac{2M}{\lambda J_3 n} \sin \frac{\lambda T}{2}
\tag{7.35}
$$

This result indicates that $\Delta\theta$ is maximized when the pulse width T is one-half of the nutation period, i.e., when $T = \pi/\lambda$.

The divergent nutation dynamics of a spacecraft spinning about its minor axis due to energy dissipation is often modeled as

$$\theta(t) = \theta_0\left(e^{t/\tau} - 1\right) \tag{7.36}$$

where θ_0 is the initial nutation angle and τ is called the divergent time constant of the spacecraft due to energy dissipation. The nutation angle reduction, Eq. (7.35), must be greater than the nutation angle increased by energy dissipation during one thrust cycle.

The spin-axis precession due to one thruster pulse can also be determined as

$$\Delta\phi = \frac{\int_{-T/2}^{T/2} M \cos nt \, dt}{J_3 n} = \frac{2M}{J_3 n^2} \sin \frac{nT}{2} \tag{7.37}$$

Thus, it should be noted that an active nutation control also results in a spin-axis precession. In practice, the thruster pulse width T is selected such that $\Delta\phi$ per thruster firing does not exceed a specified value of the spin-axis precession angle.

7.1.5 Reorientation of a Gyrostat

Consider the reorientation of the spin axis of an axisymmetric gyrostat or a dual-spin satellite using an axial thruster located on the despun platform as illustrated in Fig. 7.2. The main body (despun platform) has a body-fixed principal-axis frame B with basis vectors $\{\vec{b}_1, \vec{b}_2, \vec{b}_3\}$. Consequently, the $\vec{b}_3$ axis is the spin or symmetry axis of the spacecraft.

It is assumed that the rotor has a constant angular momentum of H_o relative to the main body and that the angular momentum vector of the spacecraft (platform plus rotor) is simply $\vec{H} = H_o \vec{b}_3$ before a reorientation maneuver. It is also assumed that a reorientation of the spin axis about the $\vec{b}_2$ axis is required without loss of generality. From the following momentum vector consideration

$$\Delta\vec{H}/\Delta t = \vec{M}$$

an axial thruster located on the despun platform as shown in Fig. 7.2, which generates a body-fixed torque about the $\vec{b}_1$ axis, is to be used to reorient the spacecraft about the $\vec{b}_2$ axis.

The linearized equations of motion of an axisymmetric gyrostat ($J_1 = J_2 = J$) with a control torque M_1 about the $\vec{b}_1$ axis are

$$\dot{\omega}_1 + \lambda\omega_2 = \mu \tag{7.38a}$$

$$\dot{\omega}_2 - \lambda\omega_1 = 0 \tag{7.38b}$$

$$\dot{\theta}_1 = \omega_1 \tag{7.38c}$$

$$\dot{\theta}_2 = \omega_2 \tag{7.38d}$$

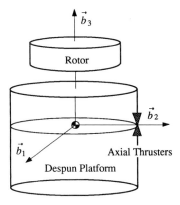

Fig. 7.2   Gyrostat with despun platform thrusters.

where $\lambda = H_o/J$ is called the nutation frequency; $\mu = M_1/J$ is the rotational control acceleration about the $\vec{b}_1$ axis; ω_1 and ω_2 are the body-fixed components of the angular velocity of the spacecraft along the $\vec{b}_1$ and $\vec{b}_2$ axes, respectively; and θ_1 and θ_2 are the small orientation angles of the $\vec{b}_3$ axis relative to inertial space.

For continuous thrusting with constant μ and with zero initial conditions, we have

$$\omega_1(t) = \frac{\mu}{\lambda} \sin \lambda t \qquad (7.39\text{a})$$

$$\omega_2(t) = \frac{\mu}{\lambda}(1 - \cos \lambda t) \qquad (7.39\text{b})$$

$$\theta_1(t) = \frac{\mu}{\lambda^2}(1 - \cos \lambda t) \qquad (7.39\text{c})$$

$$\theta_2(t) = \frac{\mu}{\lambda^2}(\lambda t - \sin \lambda t) \qquad (7.39\text{d})$$

The motion with constant thrusting about the $\vec{b}_1$ axis consists of a *precession* of the angular momentum vector about the $\vec{b}_2$ axis and a *nutation* of the $\vec{b}_3$ axis about the angular momentum vector at the nutation frequency λ. The path of the tip of the symmetry axis is a *cycloid* in an inertial frame; i.e., it is the path traced by a point on a circle of radius μ/λ^2 that is rolling on the bottom side of the θ_2 axis with angular velocity λ.

Another simple maneuvering scheme pulses the thruster at twice the nutation frequency λ for the reorientation of an axisymmetric gyrostat. The solutions to Eqs. (7.38) with zero initial conditions but an impulsive torque about the $\vec{b}_1$ axis are given by

$$\omega_1(t) = \omega_1(0^+) \cos \lambda t \qquad (7.40\text{a})$$

$$\omega_2(t) = \omega_1(0^+) \sin \lambda t \qquad (7.40\text{b})$$

$$\theta_1(t) = \frac{\omega_1(0^+)}{\lambda} \sin \lambda t \qquad (7.40\text{c})$$

$$\theta_2(t) = \frac{\omega_1(0^+)}{\lambda}(1 - \cos \lambda t) \qquad (7.40\text{d})$$

where $\omega_1(0^+)$ is due to an impulsive rotational acceleration μ at $t = 0$. The resulting motion is simply a nutation of the $\vec{b}_3$ axis about the angular momentum vector at frequency λ. The resulting path of the tip of the symmetry axis is a circle centered at $\theta_1 = 0$ and $\theta_2 = 2\omega_1(0^+)/\mu$.

If a second impulse equal in magnitude to the first impulse is applied at the half of the nutation cycle, however, then the resulting path of the tip of the symmetry axis is a semicircle. And if the thrust is pulsed repeatedly at twice the nutation frequency, then the semicircle will be repeated while the $\vec{b}_3$ axis is stationary for half a nutation cycle between semicircles. For a given reorientation maneuver angle, the same thrust impulse is required for both maneuvers as shown in Bryson [2].

7.1.6 Attitude Acquisition by Momentum Transfer

A transition from a spinning mode to a bias-momentum stabilized mode of a three-axis stabilized spacecraft is often achieved by applying a constant torque to a wheel that is initially orthogonal to the spacecraft momentum vector, thus causing angular momentum transfer from the spinning spacecraft to the wheel. Such an attitude acquisition maneuver by momentum transfer simultaneously achieves final despin of a spinning spacecraft, wheel spin up, and a proper reorientation of the spacecraft. A similar procedure is also applicable to the recovery of dual-spin stabilized spacecraft, which is known as the flat-spin recovery maneuver.

More details of this subject can be found in Kaplan [3] and Barba and Aubrun [4].

Problems

7.4 For a dual-spin spacecraft considered in Chapter 6, Sec. 6.12, but with no damper, i.e., $m = 0$, the equations of motion can be simplified as

$$J_1\dot{\omega}_1 = (J_2 - J_3)\omega_2\omega_3 - J\Omega\omega_2$$
$$J_2\dot{\omega}_2 = (J_3 - J_1)\omega_3\omega_1 + J\Omega\omega_1$$
$$J_3\dot{\omega}_3 = (J_1 - J_2)\omega_1\omega_2 - J\dot{\Omega}$$

and the rotor equation can be approximated as $J\dot{\Omega} = T$ because $J \ll J_3$ and $\omega_3 \ll \Omega$. Consider an attitude acquisition and wheel spin-up maneuver of a satellite with $(J_1, J_2, J_3) = (706, 645, 543)$ kg·m^2 and $J = 0.08$ kg·m^2.

(a) Perform computer simulation for the following case: $\omega_1(0) = 0.0581$ rad/s, $\omega_2(0) = \omega_3(0) = \Omega(0) = 0$, and $T = -0.01$ N·m for $0 \le t \le 4000$ s. In particular, obtain the time history of the nutation angle θ between $\vec{b}_3$ and the constant angular momentum vector $\vec{H} = J_1\omega_1\vec{b}_1 + J_2\omega_2\vec{b}_2 + (J_3\omega_3 + J\Omega)\vec{b}_3$.

    Note: The nutation angle θ starts at 90 deg and approaches a steady-state value of approximately 23 deg, but not 0 deg as desired.

(b) Discuss the simulation results noting that the total angular momentum vector $\vec{H}$ is a constant vector.

    Note: See Ref. 3 for additional information pertaining to this problem.

7.5 Consider a satellite attitude acquisition maneuver by momentum transfer. The equations of motion of a spacecraft equipped with a pitch wheel are given by

$$J_1\dot{\omega}_1 = (J_2 - J_3)\omega_2\omega_3 + h\omega_3$$
$$J_2\dot{\omega}_2 = (J_3 - J_1)\omega_3\omega_1 - \dot{h}$$
$$J_3\dot{\omega}_3 = (J_1 - J_2)\omega_1\omega_2 - h\omega_1$$

where $(J_1, J_2, J_3) = (86.24, 85.07, 113.59)$ slug-ft^2 (1 slug-ft$^2 = 1.356$ kg·m^2 and 1 ft-lb $= 1.356$ N·m).

(a) Perform computer simulation for the following case: $\omega_3(0) = 0.1761$ rad/s, $\omega_1(0) = \omega_2(0) = 0$, and a constant motor torque $\dot{h} = T = 0.005$ ft-lb for $0 \le t \le 4000$ s with $h(0) = 0$. In particular, obtain the time history of the nutation angle θ between $\vec{b}_2$ and the total angular momentum vector $\vec{H} = J_1\omega_1\vec{b}_1 + (J_2\omega_2 + h)\vec{b}_2 + J_3\omega_3\vec{b}_3$.

Note: The nutation angle θ starts at 90 deg and approaches a steady-state value of approximately 7.7 deg, but not 0 deg as desired. In fact, as shown by Barba and Aubrun, [4] the residual nutation angle is a function of (increases with) the magnitude of the motor torque T. The attitude acquisition time decreases as T is increased.

(b) Is there any simple explanation of the simulation results?

7.2   Time-Optimal Reorientation Maneuvers

Time-optimal control of dynamic systems is a class of optimization problems of interest in many different research areas. In particular, spacecraft, including robot manipulators and optical pointing systems in space, are sometimes required to reorient or reposition as quickly as possible within the physical limits of actuators. Consequently, the problem of spacecraft time-optimal slew maneuvers has been the subject of extensive research (e.g., see [5]).

In this section, we present the mathematical foundation of a time-optimal control problem and then apply the optimal control theory to the spacecraft time-optimal control problem. In particular, the time-optimal maneuvering control problem of a rigid spacecraft is examined by considering a simple case of an inertially symmetric rigid body. The so-called eigenaxis rotation, which provides the minimum angular path between two orientations, has often been considered as a natural approach to a time-optimal rotational maneuver. An eigenaxis rotation about one of the principal axes of even such a simple spherical body is not time optimal, however, as first shown by Bilimoria and Wie [6, 7].

7.2.1   Introduction to Time-Optimal Control

Consider a nonlinear dynamic system described by

$$\dot{\mathbf{x}} = \mathbf{f}(\mathbf{x}, \mathbf{u}, t) \tag{7.41}$$

where $\mathbf{x} = (x_1, \ldots, x_n)$ is the state vector and $\mathbf{u} = (u_1, \ldots, u_m)$ is the control input vector. The problem is to determine the optimal control input $\mathbf{u}$ to transfer the system from its given initial state $\mathbf{x}(0)$ to a specified final state $\mathbf{x}(t_f)$ in minimum time; i.e., the final time t_f is to be minimized. For such a time-optimal control problem, the control inputs are often constrained by

$$|u_i| \leq 1, \qquad i = 1, \ldots, m$$

which is, in fact, equivalent to the hypercube in m-dimensional space described by

$$\|\mathbf{u}\|_\infty \equiv \lim_{\zeta \to \infty} \left\{ \sum_{i=1}^{m} u_i^{2\zeta} \right\}^{\frac{1}{2\zeta}} \equiv \max_i |u_i| \leq 1$$

Similarly, the control inputs may also be constrained by the hypershere in m-dimensional space

$$\|\mathbf{u}\|_2 \equiv \left\{ \sum_{i=1}^{m} u_i^2 \right\}^{\frac{1}{2}} \leq 1$$

The performance index or cost function to be minimized is, in general, of the form

$$J = \int_0^{t_f} L(\mathbf{x}, \mathbf{u}, t)\, dt \qquad (7.42)$$

and $L = 1$ for the time-optimal control problem. Adjoining Eq. (7.41) to the performance index J with the so-called Lagrange multiplier $\boldsymbol{\lambda}(t)$, we obtain

$$J = \int_0^{t_f} \{L + \boldsymbol{\lambda}^T (\mathbf{f} - \dot{\mathbf{x}})\}\, dt \qquad (7.43)$$

where $\boldsymbol{\lambda} = (\lambda_1, \ldots, \lambda_n)$ is also referred to as the adjoint or costate vector. Defining a scalar function H, called the Hamiltonian, as

$$H = L + \boldsymbol{\lambda}^T \mathbf{f}(\mathbf{x}, \mathbf{u}, t) \qquad (7.44)$$

we rewrite the performance index as

$$J = \int_0^{t_f} \{H - \boldsymbol{\lambda}^T \dot{\mathbf{x}}\}\, dt \qquad (7.45)$$

Taking the differential of J with respect to dt_f, $\delta\mathbf{x}$, $\delta\mathbf{u}$, and $\delta\dot{\mathbf{x}}$, we obtain

$$dJ = L\, dt_f + \int_0^{t_f} \left\{ \frac{\partial H}{\partial \mathbf{x}} \delta\mathbf{x} + \frac{\partial H}{\partial \mathbf{u}} \delta\mathbf{u} - \boldsymbol{\lambda}^T \delta\dot{\mathbf{x}} \right\} dt \qquad (7.46)$$

which, after the integration by parts of $\lambda^T \delta \dot{\mathbf{x}}$, becomes

$$dJ = L\, dt_f - \left[\lambda^T \delta \mathbf{x}\right]_0^{t_f} + \int_0^{t_f} \left\{ \frac{\partial H}{\partial \mathbf{x}} \delta \mathbf{x} + \frac{\partial H}{\partial \mathbf{u}} \delta \mathbf{u} + \dot{\lambda}^T \delta \mathbf{x} \right\} dt \qquad (7.47)$$

Using the relationship $d\mathbf{x}(t_f) = \delta\mathbf{x}(t_f) + \dot{\mathbf{x}}(t_f)\, dt_f$ (Ref. 8, page 72) and $\delta\mathbf{x}(0) = 0$, we obtain

$$dJ = \left\{ L + \lambda^T(t_f)\, \dot{\mathbf{x}}(t_f) \right\} dt_f - \lambda^T(t_f)\, d\mathbf{x}(t_f)$$
$$+ \int_0^{t_f} \left[\left\{ \frac{\partial H}{\partial \mathbf{x}} + \dot{\lambda}^T \right\} \delta\mathbf{x} + \frac{\partial H}{\partial \mathbf{u}} \delta\mathbf{u} \right] dt \qquad (7.48)$$

Note that $d\mathbf{x}(t_f) = 0$ because $\mathbf{x}(t_f)$ is specified. Furthermore, choosing λ to make the coefficients of dt_f and $\delta\mathbf{x}$ vanish, we obtain the transversality condition

$$H(t_f) = 0 \qquad (7.49)$$

and the costate equation

$$\dot{\lambda} = -\left[\frac{\partial H}{\partial \mathbf{x}} \right]^T \qquad (7.50)$$

Consequently, dJ is simplified to

$$dJ = \int_0^{t_f} \frac{\partial H}{\partial \mathbf{u}} \delta\mathbf{u}\, dt$$

For a stationary value of J, dJ must be zero and, thus, we obtain the following necessary condition for optimality:

$$\frac{\partial H}{\partial \mathbf{u}} = 0 \qquad (7.51)$$

In summary, the optimal control input must satisfy the following set of necessary conditions:

$$\dot{\mathbf{x}} = \left[\frac{\partial H}{\partial \lambda} \right]^T = \mathbf{f}(\mathbf{x}, \mathbf{u}, t), \qquad \mathbf{x}(0) \text{ and } \mathbf{x}(t_f) \text{ specified} \qquad (7.52)$$

$$\dot{\lambda} = -\left[\frac{\partial H}{\partial \mathbf{x}} \right]^T = -\left[\frac{\partial \mathbf{f}}{\partial \mathbf{x}} \right]^T \lambda, \qquad \lambda(0) \text{ and } \lambda(t_f) \text{ free} \qquad (7.53)$$

with the optimality condition

$$\frac{\partial H}{\partial \mathbf{u}} = 0 \qquad (7.54)$$

and the transversality condition

$$H(t_f) = 0 \qquad (7.55)$$

Notice that the costate variables are "free," i.e., unspecified, at both the initial and final times because the corresponding state variables of the system are specified. This set of differential equations with specified boundary conditions and with specified control input constraints provides the necessary conditions for the time-optimal control and forms a nonlinear two-point boundary-value problem (TPBVP) which is, in general, a difficult problem to solve. In general, the optimality condition (7.54) is used to determine the optimal control input and the transversality condition (7.55) determines the final time t_f.

A first integral of the TPBVP exists if $\mathbf{f}(\mathbf{x}, \mathbf{u})$ is not an explicit function of time t; i.e., we have

$$
\begin{aligned}
\dot{H} &= \frac{\partial H}{\partial t} + \frac{\partial H}{\partial \mathbf{x}}\dot{\mathbf{x}} + \frac{\partial H}{\partial \mathbf{u}}\dot{\mathbf{u}} + \frac{\partial H}{\partial \boldsymbol{\lambda}}\dot{\boldsymbol{\lambda}} \\
&= \frac{\partial H}{\partial t} + \{-\dot{\boldsymbol{\lambda}}^T\dot{\mathbf{x}} + \dot{\mathbf{x}}^T\dot{\boldsymbol{\lambda}}\} + \frac{\partial H}{\partial \mathbf{u}}\dot{\mathbf{u}} \\
&= 0
\end{aligned}
$$

and thus H is constant on the optimal trajectory. Because $H(t_f) = 0$ from the transversality condition, we have $H = 0$ for $0 \le t \le t_f$.

Consider the time-optimal control problem of a dynamic system described by

$$
\dot{\mathbf{x}} = \mathbf{f}(\mathbf{x}, t) + \mathbf{B}(\mathbf{x}, t)\mathbf{u} \tag{7.56}
$$

or, equivalently, in component form as

$$
\dot{x}_i = f_i(\mathbf{x}, t) + \sum_{j=1}^{m} B_{ij}(\mathbf{x}, t)u_j, \qquad i = 1, \dots, n
$$

where the control inputs are constrained by the hypercube in m-dimensional space

$$
|u_j| \le 1, \qquad j = 1, \dots, m
$$

Because the Hamiltonian is linear in the control input $\mathbf{u}$ for this particular case, the optimality condition (7.54) does not provide any information about the optimal control input. However, according to Pontryagin's minimum principle, the optimal control input must minimize the Hamiltonian

$$
\begin{aligned}
H &= 1 + \boldsymbol{\lambda}^T\mathbf{f}(\mathbf{x}, t) + \boldsymbol{\lambda}^T\mathbf{B}(\mathbf{x}, t)\mathbf{u} \\
&= 1 + \sum_{i}^{n} \lambda_i f_i + \sum_{i=1}^{n} \lambda_i \left\{ \sum_{j=1}^{m} B_{ij} u_j \right\}
\end{aligned}
$$

and minimization of the Hamiltonian function subject to the preceding constraints requires that

$$
u_j = -\operatorname{sgn}\{S_j\} = -\operatorname{sgn}\left\{\frac{\partial H}{\partial u_j}\right\} = -\operatorname{sgn}\left\{\sum_{i=1}^{n} B_{ij}\lambda_i\right\}, \qquad j = 1, \dots, m \tag{7.57}
$$

where S_j is the switching function associated with the jth control input u_j and the signum function is defined as

$$\text{sgn}\{S_j\} = \begin{cases} +1 & \text{if } S_j > 0 \\ -1 & \text{if } S_j < 0 \end{cases} \tag{7.58}$$

and it becomes singular $(-1 < u_j < +1)$ if $S_j = 0$. In vector form, the optimal control input is expressed as

$$\mathbf{u}(t) = -\text{sgn}\{\mathbf{B}^T \boldsymbol{\lambda}(t)\}$$

Problems

7.6 Consider a double-integrator plant described by

$$\dot{x}_1 = x_2$$
$$\dot{x}_2 = u(t)$$

subject to $|u| \leq 1$. The optimal control input that transfers the system from an initial state $\mathbf{x}(0) \neq 0$ to the final state $\mathbf{x}(t_f) = (0,0)$ in minimum time is to be determined.

(a) Show that the optimal control input changes sign, at most, once; i.e.,

$$u(t) = -\text{sgn}\{\lambda_2\} = -\text{sgn}\{\lambda_2(0) - \lambda_1(0)t\}$$

(b) Show that the optimal control input can also be expressed in nonlinear feedback control form as

$$u(t) = -\text{sgn}\left\{x_1 + \tfrac{1}{2}x_2|x_2|\right\}$$

(c) Sketch typical time-optimal trajectories in the (x_1, x_2) phase plane, including the switching curve, which is made up of two parabolas.

Hint: See, e.g., Bryson and Ho ([8], pages 110–113).

7.7 Consider a linear system described by

$$\dot{x}_1 = \omega_n x_2$$
$$\dot{x}_2 = -\omega_n x_1 + u$$

where ω_n is the natural frequency of this undamped oscillator system, and the control input is subject to $|u| \leq 1$. The optimal control input that transfers the system from an initial state $\mathbf{x}(0) \neq 0$ to the final state $\mathbf{x}(t_f) = (0,0)$ in minimum time is to be determined.

(a) Show that the optimal control input $u(t)$ changes sign with period π/ω_n; i.e.,

$$u(t) = -\text{sgn}\{\lambda_2\} = -\text{sgn}\{\lambda_1(0) \sin \omega_n t + \lambda_2(0) \cos \omega_n t\}$$

(b) Find the trajectory equation in the (x_1, x_2) phase plane, as follows:

$$[\omega_n x_1(t) - u]^2 + [\omega_n x_2(t)]^2 = [\omega_n x_1(0) - u]^2 + [\omega_n x_2(0)]^2$$

where $u = \pm 1$.

(c) Sketch typical time-optimal trajectories in the (x_1, x_2) phase plane, including the terminal arcs and switching curve.

Hint: See Chapter 7 of Athans and Falb [9].

7.8 Consider a spinning axisymmetric body described by

$$\dot{x}_1 = \omega_n x_2 + u_1$$
$$\dot{x}_2 = -\omega_n x_1 + u_2$$

where ω_n is the nutation frequency observed in a rotating reference frame, and the control inputs are subject to $|u_i| \leq 1$. The optimal control input vector that transfers the system from an initial state $\mathbf{x}(0) \neq 0$ to the final state $\mathbf{x}(t_f) = (0, 0)$ in minimum time is to be determined.

(a) Show that the time-optimal, rate damping control inputs u_i change sign with period π / ω_n.

(b) Find the trajectory equation in the (x_1, x_2) phase plane, as follows:

$$[\omega_n x_1(t) - u_2]^2 + [\omega_n x_2(t) + u_1]^2 = [\omega_n x_1(0) - u_2]^2 + [\omega_n x_2(0) + u_1]^2$$

where $u_i = \pm 1$.

(c) Sketch typical time-optimal trajectories in the (x_1, x_2) phase plane, including the terminal arcs and switching curve.

7.9 Consider a dynamic system described by

$$\dot{\mathbf{x}} = \mathbf{f}(\mathbf{x}, t) + \mathbf{B}(t)\mathbf{u}$$

where the control input vector is constrained by the hypersphere

$$\|\mathbf{u}\| \equiv \sqrt{u_1^2 + u_2^2 + \cdots + u_m^2} \leq 1$$

Show that the time-optimal control input vector is of the form

$$\mathbf{u}(t) = -\frac{\mathbf{B}^T \boldsymbol{\lambda}(t)}{\|\mathbf{B}^T \boldsymbol{\lambda}\|} \quad \text{if } \|\mathbf{B}^T \boldsymbol{\lambda}\| \neq 0$$

Hint: See Chapter 10 of Athans and Falb [9].

7.10 Defining $x_i = J_i \omega_i$, we rewrite Euler's equations of motion of a rigid spacecraft about principal axes, as follows:

$$\dot{x}_1 = \frac{J_2 - J_3}{J_2 J_3} x_2 x_3 + u_1$$

$$\dot{x}_2 = \frac{J_3 - J_1}{J_3 J_1} x_3 x_1 + u_2$$

$$\dot{x}_3 = \frac{J_1 - J_2}{J_1 J_2} x_1 x_2 + u_3$$

which is said to be a norm-invariant system.

Show that the time-optimal, rate damping control input vector is of the form

$$\mathbf{u}(t) = -\frac{\mathbf{x}(t)}{\|\mathbf{x}\|} \qquad \text{if } \|\mathbf{x}\| \neq 0$$

if the spacecraft is subject to a control input constraint of the form

$$\|\mathbf{u}\| \equiv \sqrt{u_1^2 + u_2^2 + u_3^2} \leq 1$$

7.2.2 Time-Optimal Control of an Inertially Symmetric Body

Consider the time-optimal reorientation problem of an inertially symmetric, e.g., spherical or cubical, rigid body. Although most real spacecraft are not inertially symmetric, the study of this simple case is likely lead to more fundamental understanding of the time-optimal control problem of complex spacecraft.

It is assumed that the control axes are aligned with principal axes with the principal moment of inertia J_o and that the control torque inputs are bounded as $\pm \tau_o$ for each axis.

The nondimensionalized equations of motion of an inertially symmetric body are simply given by

$$\dot{\omega}_1 = u_1 \tag{7.59a}$$

$$\dot{\omega}_2 = u_2 \tag{7.59b}$$

$$\dot{\omega}_3 = u_3 \tag{7.59c}$$

where ω_i are the angular velocity components in units of $1/\sqrt{J_o/\tau_o}$, u_i are the control torque inputs in units of τ_o, and time is in units of $\sqrt{J_o/\tau_o}$. The kinematic differential equations in terms of quaternions are also given by

$$\dot{q}_1 = (+\omega_1 q_4 - \omega_2 q_3 + \omega_3 q_2)/2 \tag{7.60a}$$

$$\dot{q}_2 = (+\omega_1 q_3 + \omega_2 q_4 - \omega_3 q_1)/2 \tag{7.60b}$$

$$\dot{q}_3 = (-\omega_1 q_2 + \omega_2 q_1 + \omega_3 q_4)/2 \tag{7.60c}$$

$$\dot{q}_4 = (-\omega_1 q_1 - \omega_2 q_2 - \omega_3 q_3)/2 \tag{7.60d}$$

Equations (7.59) and (7.60) are, in fact, the state equations of the system with the state variables ω_1, ω_2, ω_3, q_1, q_2, q_3, and q_4.

The time-optimal, rest-to-rest reorientation problem is then to determine the optimal control inputs (u_1, u_2, u_3) that drive the dynamic system described by Eqs. (7.59) and (7.60) from rest at its initial orientation $(q_1, q_2, q_3, q_4)_0$ to rest at its final orientation $(q_1, q_2, q_3, q_4)_f$, while minimizing the cost function

$$J = \int_0^{t_f} dt = t_f \qquad (7.61)$$

subject to the control input constraints

$$-1 \le u_i \le +1, \qquad i = 1, 2, 3 \qquad (7.62)$$

The Hamiltonian is formed by adjoining the state equations (7.59) and (7.60) with the appropriate adjoint variables λ_{ω_1}, λ_{ω_2}, λ_{ω_3}, λ_{q_1}, λ_{q_2}, λ_{q_3}, and λ_{q_4}, and it becomes

$$
\begin{aligned}
H = {} & 1 + \lambda_{\omega_1} u_1 + \lambda_{\omega_2} u_2 + \lambda_{\omega_3} u_3 \\
& + \lambda_{q_1}(+\omega_1 q_4 - \omega_2 q_3 + \omega_3 q_2)/2 \\
& + \lambda_{q_2}(+\omega_1 q_3 + \omega_2 q_4 - \omega_3 q_1)/2 \\
& + \lambda_{q_3}(-\omega_1 q_2 + \omega_2 q_1 + \omega_3 q_4)/2 \\
& + \lambda_{q_4}(-\omega_1 q_1 - \omega_2 q_2 - \omega_3 q_3)/2
\end{aligned}
$$

where H denotes the Hamiltonian, which is constant along an optimal trajectory of the time-optimal control problem under consideration. From the transversality condition its value along an optimal trajectory is given by

$$H \equiv H(t_f) = 0$$

The costate equations are then obtained in the standard fashion by differentiating the negative of the Hamiltonian with respect to the states and are given by

$$\dot{\lambda}_{\omega_1} = -\left(+\lambda_{q_1} q_4 + \lambda_{q_2} q_3 - \lambda_{q_3} q_2 - \lambda_{q_4} q_1\right)/2 \qquad (7.63a)$$

$$\dot{\lambda}_{\omega_2} = -\left(-\lambda_{q_1} q_3 + \lambda_{q_2} q_4 + \lambda_{q_3} q_1 - \lambda_{q_4} q_2\right)/2 \qquad (7.63b)$$

$$\dot{\lambda}_{\omega_3} = -\left(+\lambda_{q_1} q_2 - \lambda_{q_2} q_1 + \lambda_{q_3} q_4 - \lambda_{q_4} q_3\right)/2 \qquad (7.63c)$$

$$\dot{\lambda}_{q_1} = -\left(-\lambda_{q_2} \omega_3 + \lambda_{q_3} \omega_2 - \lambda_{q_4} \omega_1\right)/2 \qquad (7.63d)$$

$$\dot{\lambda}_{q_2} = -\left(+\lambda_{q_1} \omega_3 - \lambda_{q_3} \omega_1 - \lambda_{q_4} \omega_2\right)/2 \qquad (7.63e)$$

$$\dot{\lambda}_{q_3} = -\left(-\lambda_{q_1} \omega_2 + \lambda_{q_2} \omega_1 - \lambda_{q_4} \omega_3\right)/2 \qquad (7.63f)$$

$$\dot{\lambda}_{q_4} = -\left(+\lambda_{q_1} \omega_1 + \lambda_{q_2} \omega_2 + \lambda_{q_3} \omega_3\right)/2 \qquad (7.63g)$$

The adjoint variables are free, i.e., unspecified, at both the initial and final times because the corresponding state variables of the system are specified. The set of quaternion adjoint variables $(\lambda_{q1}, \lambda_{q2}, \lambda_{q3}, \lambda_{q4})$ is not unique, however, because of the following relationship:

$$q_1^2 + q_2^2 + q_3^2 + q_4^2 = 1$$

Hence, any one of the quaternion adjoint variables can be normalized to unity at the initial (or final) time.

The set of differential equations consisting of Eqs. (7.59), (7.60), and (7.63) with specified boundary conditions provides the necessary conditions for the time-optimal control, and forms a nonlinear TPBVP that is, in general, a difficult problem to solve.

According to Pontryagin's minimum principle, the optimal control inputs that minimize the Hamiltonian subject to the constraints (7.62) are given by

$$u_i(t) = -\text{sgn}\left\{\frac{\partial H}{\partial u_i}\right\} = -\text{sgn}\{\lambda_{\omega_i}(t)\}, \qquad i = 1, 2, 3 \qquad (7.64)$$

It can be shown that for this problem, at least one control component must be saturated at any given instant of time. In geometric terms, the control input constraints given by Eq. (7.62) may be visualized as a unit cube in three-dimensional space. Such constraints require that an admissible control vector lie inside or on the surface of this cube. The optimal control input vector cannot lie inside the cube, however, because this would imply that all three components of the control input vector are simultaneously singular. Consequently, the optimal control vector must lie on the cubical constraint surface.

For further details of this subject, the reader is referred to Li and Bainum [10], Bilimoria and Wie [6,7,11], Byers and Vadali [12], and Seywald and Kumar [13].

Problems

7.11 Given the time-optimal reorientation problem of an inertially symmetric rigid body, as formulated by Eqs. (7.59), (7.60), and (7.62), consider the following rest-to-rest maneuver boundary conditions:

$$\omega_1(0) = \omega_2(0) = \omega_3(0) = 0$$
$$q_1(0) = q_2(0) = q_3(0) = 0, \qquad q_4(0) = 1$$
$$\omega_1(t_f) = \omega_2(t_f) = \omega_3(t_f) = 0$$
$$q_1(t_f) = q_2(t_f) = 0, \qquad q_3(t_f) = \sin\frac{\theta}{2}, \qquad q_4(t_f) = \cos\frac{\theta}{2}$$

where $\theta = 180$ deg is the required rotation angle about the eigenaxis. For these boundary conditions, the eigenaxis rotation is simply a bang-bang maneuver about the third principal axis of the body, resulting in $t_f = 3.5449$.

(a) Verify that by performing computer simulation for the control inputs given in Fig. 7.3, an eigenaxis rotation about one of the principal axes of even such a simple body is not time optimal. In particular, plot three

In this section we introduce a feedback control logic for three-axis, large-angle reorientation maneuvers, and we further extend such a simple feedback control logic to a case in which the spacecraft is required to maneuver about an inertially fixed axis as fast as possible within the saturation limits of rate gyros as well as reaction wheels. (This section is based on [15–17].)

7.3.1 Quaternion Feedback Control

Consider the attitude dynamics of a rigid spacecraft described by Euler's rotational equation of motion

$$\mathbf{J}\dot{\boldsymbol{\omega}} + \boldsymbol{\omega} \times \mathbf{J}\boldsymbol{\omega} = \mathbf{u} \tag{7.65}$$

where $\mathbf{J}$ is the inertia matrix, $\boldsymbol{\omega} = (\omega_1, \omega_2, \omega_3)$ the angular velocity vector, and $\mathbf{u} = (u_1, u_2, u_3)$ the control torque input vector. The cross product of two vectors is represented in matrix notation as

$$\boldsymbol{\omega} \times \mathbf{h} \equiv \begin{bmatrix} 0 & -\omega_3 & \omega_2 \\ \omega_3 & 0 & -\omega_1 \\ -\omega_2 & \omega_1 & 0 \end{bmatrix} \begin{bmatrix} h_1 \\ h_2 \\ h_3 \end{bmatrix}$$

where $\mathbf{h} = \mathbf{J}\boldsymbol{\omega}$ is the angular momentum vector. It is assumed that the angular velocity vector components ω_i along the body-fixed control axes are measured by rate gyros.

Euler's rotational theorem states that the rigid-body attitude can be changed from any given orientation to any other orientation by rotating the body about an axis, called the Euler axis, that is fixed to the rigid body and stationary in inertial space. Such a rigid-body rotation about an Euler axis is often called the eigenaxis rotation.

Let a unit vector along the Euler axis be denoted by $\mathbf{e} = (e_1, e_2, e_3)$ where e_1, e_2, and e_3 are the direction cosines of the Euler axis relative to either an inertial reference frame or the body-fixed control axes. The four elements of quaternions are then defined as

$$q_1 = e_1 \sin(\theta/2)$$
$$q_2 = e_2 \sin(\theta/2)$$
$$q_3 = e_3 \sin(\theta/2)$$
$$q_4 = \cos(\theta/2)$$

where θ denotes the rotation angle about the Euler axis, and we have

$$q_1^2 + q_2^2 + q_3^2 + q_4^2 = 1$$

The quaternion kinematic differential equations are given by

$$\begin{bmatrix} \dot{q}_1 \\ \dot{q}_2 \\ \dot{q}_3 \\ \dot{q}_4 \end{bmatrix} = \frac{1}{2} \begin{bmatrix} 0 & \omega_3 & -\omega_2 & \omega_1 \\ -\omega_3 & 0 & \omega_1 & \omega_2 \\ \omega_2 & -\omega_1 & 0 & \omega_3 \\ -\omega_1 & -\omega_2 & -\omega_3 & 0 \end{bmatrix} \begin{bmatrix} q_1 \\ q_2 \\ q_3 \\ q_4 \end{bmatrix} \tag{7.66}$$

Like the Euler-axis vector $\mathbf{e} = (e_1, \ e_2, \ e_3)$, defining a quaternion vector $\mathbf{q} = (q_1, q_2, q_3)$ as

$$\mathbf{q} = \mathbf{e} \sin \frac{\theta}{2}$$

we rewrite Eq. (7.66) as

$$2\dot{\mathbf{q}} = q_4\boldsymbol{\omega} - \boldsymbol{\omega} \times \mathbf{q} \tag{7.67a}$$

$$2\dot{q}_4 = -\boldsymbol{\omega}^T\mathbf{q} \tag{7.67b}$$

where

$$\boldsymbol{\omega} \times \mathbf{q} \equiv \begin{bmatrix} 0 & -\omega_3 & \omega_2 \\ \omega_3 & 0 & -\omega_1 \\ -\omega_2 & \omega_1 & 0 \end{bmatrix} \begin{bmatrix} q_1 \\ q_2 \\ q_3 \end{bmatrix}$$

Because quaternions are well suited for onboard real-time computation, space-craft orientation is nowadays commonly described in terms of the quaternions, and a linear state feedback controller of the following form can be considered for real-time implementation:

$$\mathbf{u} = -\mathbf{K}\mathbf{q}_e - \mathbf{C}\boldsymbol{\omega} \tag{7.68}$$

where $\mathbf{q}_e = (q_{1e}, q_{2e}, q_{3e})$ is the attitude error quaternion vector and $\mathbf{K}$ and $\mathbf{C}$ are controller gain matrices to be properly determined. The attitude error quaternions $(q_{1e}, q_{2e}, q_{3e}, q_{4e})$ are computed using the desired or commanded attitude quaternions $(q_{1c}, q_{2c}, q_{3c}, q_{4c})$ and the current attitude quaternions (q_1, q_2, q_3, q_4), as follows:

$$\begin{bmatrix} q_{1e} \\ q_{2e} \\ q_{3e} \\ q_{4e} \end{bmatrix} = \begin{bmatrix} q_{4c} & q_{3c} & -q_{2c} & -q_{1c} \\ -q_{3c} & q_{4c} & q_{1c} & -q_{2c} \\ q_{2c} & -q_{1c} & q_{4c} & -q_{3c} \\ q_{1c} & q_{2c} & q_{3c} & q_{4c} \end{bmatrix} \begin{bmatrix} q_1 \\ q_2 \\ q_3 \\ q_4 \end{bmatrix} \tag{7.69}$$

If the commanded attitude quaternion vector is simply the origin defined as

$$(q_{1c}, q_{2c}, q_{3c}, q_{4c}) = (0, 0, 0, +1)$$

then the control logic (7.68) becomes

$$\mathbf{u} = -\mathbf{K}\mathbf{q} - \mathbf{C}\boldsymbol{\omega} \tag{7.70}$$

On the other hand, if the origin is chosen as $(0, 0, 0, -1)$, then the control logic (7.68) becomes

$$\mathbf{u} = +\mathbf{K}\mathbf{q} - \mathbf{C}\boldsymbol{\omega} \tag{7.71}$$

Note, however, that both quaternions $(0, 0, 0, +1)$ and $(0, 0, 0, -1)$ correspond to the physically identical orientation.

Without loss of generality, we consider here the control logic of the form (7.70). As shown by Wie and Barba [15] and Wie et al., [16] the origin, either $(0, 0, 0, +1)$ or $(0, 0, 0, -1)$, of the closed-loop nonlinear systems of a rigid spacecraft with such control logic is globally asymptotically stable for the following gain selections.

Controller 1:

$$\mathbf{K} = k\mathbf{I}, \qquad\qquad \mathbf{C} = \mathrm{diag}(c_1, c_2, c_3) \qquad\qquad (7.72\text{a})$$

Controller 2:

$$\mathbf{K} = \frac{k}{q_4^3}\mathbf{I}, \qquad\qquad \mathbf{C} = \mathrm{diag}(c_1, c_2, c_3) \qquad\qquad (7.72\text{b})$$

Controller 3:

$$\mathbf{K} = k\,\mathrm{sgn}(q_4)\mathbf{I}, \qquad \mathbf{C} = \mathrm{diag}(c_1, c_2, c_3) \qquad\qquad (7.72\text{c})$$

Controller 4:

$$\mathbf{K} = [\alpha\mathbf{J} + \beta\mathbf{I}]^{-1}, \qquad \mathbf{K}^{-1}\mathbf{C} > 0 \qquad\qquad (7.72\text{d})$$

where k and c_i are positive scalar constants, $\mathbf{I}$ is a 3×3 identity matrix, $\mathrm{sgn}(\cdot)$ denotes the signum function, and α and β are nonnegative scalars.

Note that controller 1 is a special case of controller 4 with $\alpha = 0$, and that β can also be simply selected as zero when $\alpha \neq 0$. Controllers 2 and 3 approach the origin, either $(0, 0, 0, +1)$ or $(0, 0, 0, -1)$, by taking a shorter angular path.

Problems

7.13 Consider the rotational equations of motion of a rigid spacecraft about principal axes described by

$$2\dot{q}_1 = \omega_1 q_4 - \omega_2 q_3 + \omega_3 q_2$$
$$2\dot{q}_2 = \omega_1 q_3 + \omega_2 q_4 - \omega_3 q_1$$
$$2\dot{q}_3 = -\omega_1 q_2 + \omega_2 q_1 + \omega_3 q_4$$
$$2\dot{q}_4 = -\omega_1 q_1 - \omega_2 q_2 - \omega_3 q_3$$
$$J_1\dot{\omega}_1 = (J_2 - J_3)\omega_2\omega_3 + u_1$$
$$J_2\dot{\omega}_2 = (J_3 - J_1)\omega_3\omega_1 + u_2$$
$$J_3\dot{\omega}_3 = (J_1 - J_2)\omega_1\omega_2 + u_3$$

The stability of the origin defined as

$$\mathbf{x}^* = (q_1, q_2, q_3, q_4, \omega_1, \omega_2, \omega_3)$$
$$= (0, 0, 0, +1, 0, 0, 0)$$

is to be studied for the control torque inputs u_i of the form

$$u_1 = -k_1 q_1 - c_1 \omega_1$$
$$u_2 = -k_2 q_2 - c_2 \omega_2$$
$$u_3 = -k_3 q_3 - c_3 \omega_3$$

where k_i and c_i are positive constants.

(a) Show that for any positive constants c_i, the equilibrium point $\mathbf{x}^*$ is globally asymptotically stable if k_i are selected such that

$$\frac{J_2 - J_3}{k_1} + \frac{J_3 - J_1}{k_2} + \frac{J_1 - J_2}{k_3} = 0$$

Hint: Choose the following positive-definite function

$$E = \frac{J_1 \omega_1^2}{2k_1} + \frac{J_2 \omega_2^2}{2k_2} + \frac{J_3 \omega_3^2}{2k_3} + q_1^2 + q_2^2 + q_3^2 + (q_4 - 1)^2$$

as a Lyapunov function.

(b) Also determine whether or not the equilibrium point $\mathbf{x}^*$ is globally asymptotically stable for any positive values of k_i and c_i. (This is a much harder unsolved problem.)

7.14 Consider a rigid spacecraft with

$$\mathbf{J} = \begin{bmatrix} 1200 & 100 & -200 \\ 100 & 2200 & 300 \\ -200 & 300 & 3100 \end{bmatrix} \text{ kg·m}^2$$

It is assumed that $(q_1, q_2, q_3, q_4) = (0.5, \ 0.5, \ 0.5, \ -0.5)$ and $(\omega_1, \omega_2, \omega_3) = (0, \ 0, \ 0)$ at $t = 0$ and that the spacecraft needs to be reoriented within approximately 500 s to the origin $(q_{1c}, q_{2c}, q_{3c}, q_{4c}) = (0, \ 0, \ 0, \ \pm 1)$. This given initial orientation corresponds to an eigenangle-to-go of 240 deg or 120 deg depending on the direction of reorientation.

Neglecting the products of inertia of the spacecraft, synthesize quaternion feedback control logic of the form

$$\mathbf{u} = -\mathbf{Kq} - \mathbf{C\omega}$$

with the four different types of gain matrices given by Eqs. (7.72)

For each controller, perform computer simulation of the closed-loop system including the products of inertia. In particular, plot the time history of the eigenangle θ and also plot q_i vs q_j ($i, j = 1, 2, 3$).

7.3.2 Eigenaxis Rotational Maneuvers

The gyroscopic term of Euler's rotational equation of motion is not significant for most practical rotational maneuvers. In some cases, however, it may be desirable to directly counteract the term by control torque, as follows:

$$\mathbf{u} = -\mathbf{Kq} - \mathbf{C}\boldsymbol{\omega} + \boldsymbol{\omega} \times \mathbf{J}\boldsymbol{\omega} \qquad (7.73)$$

The origin of the closed-loop system with the controller (7.73) is globally asymptotically stable if the matrix $\mathbf{K}^{-1}\mathbf{C}$ is positive definite [16]. A natural selection of $\mathbf{K}$ and $\mathbf{C}$ for guaranteeing such condition is $\mathbf{K} = k\mathbf{J}$ and $\mathbf{C} = c\mathbf{J}$ where k and c are positive scalar constants to be properly selected. Furthermore, a rigid spacecraft with a controller of the form

$$\mathbf{u} = -k\mathbf{Jq} - c\mathbf{J}\boldsymbol{\omega} + \boldsymbol{\omega} \times \mathbf{J}\boldsymbol{\omega} \qquad (7.74)$$

performs a rest-to-rest reorientation maneuver about an eigenaxis along the initial quaternion vector, $\mathbf{q}(0)$.

Euler's rotational theorem is only concerned with the kinematics of the eigenaxis rotation, and it does not deal with the dynamics of the eigenaxis rotation; however, the following eigenaxis rotation theorems describe the complete closed-loop rotational dynamics of a rigid spacecraft [17].

Theorem 7.1

The closed-loop rotational motion of a rigid spacecraft with the quaternion feedback control logic (7.74) is described by

$$\dot{\boldsymbol{\omega}} = -k\mathbf{q} - c\boldsymbol{\omega} \qquad (7.75a)$$

$$2\,\dot{\mathbf{q}} = q_4\boldsymbol{\omega} - \boldsymbol{\omega} \times \mathbf{q} \qquad (7.75b)$$

$$2\dot{q}_4 = -\boldsymbol{\omega}^T\mathbf{q} \qquad (7.75c)$$

If $\boldsymbol{\omega}(0)$ and $\mathbf{q}(0)$ are collinear at $t = 0$, then the resulting rotational motion is an eigenaxis rotation about $\mathbf{q}(0)$, and the solution $\boldsymbol{\omega}(t)$ and $\mathbf{q}(t)$ of the closed-loop system dynamics described by Eqs. (7.75) will become collinear with $\mathbf{q}(0)$ for all $t \geq 0$, i.e., $\boldsymbol{\omega} \times \mathbf{q} = 0$ for all $t \geq 0$.

Proof: If $\boldsymbol{\omega}(0)$ and $\mathbf{q}(0)$ are collinear at $t = 0$, then Eq. (7.75a) indicates that $\boldsymbol{\omega}(t)$ and $\mathbf{q}(t)$ are collinear for all $t \geq 0$; i.e., $\boldsymbol{\omega} \times \mathbf{q} = 0$ for all $t \geq 0$. Also, Eq. (7.75b), with $\boldsymbol{\omega} \times \mathbf{q} = 0$, indicates that $\boldsymbol{\omega}(t)$ and $\mathbf{q}(t)$ are collinear for all $t \geq 0$. Consequently, the resulting rotational motion is an eigenaxis rotation about $\mathbf{q}(0)$.

Theorem 7.2

If the angular velocity vector $\boldsymbol{\omega}(t)$ lies along the direction of $\mathbf{q}(0)$, i.e.,

$$\boldsymbol{\omega}(t) = a(t)\mathbf{q}(0)$$

where $a(t)$ is a scalar function with $a(0) = 0$, then $\mathbf{q}(t)$ of Eqs. (7.75) will remain along the same direction of $\mathbf{q}(0)$, i.e., the resulting motion is an eigenaxis rotation about $\mathbf{q}(0)$.

Proof: This is a special case of Theorem 7.1 with $\boldsymbol{\omega}(0) = 0$.

Problems

7.15 Show that the closed-loop equation for a rest-to-rest reorientation maneuver about an eigenaxis using a control logic of the form

$$\mathbf{u} = -k\mathbf{Jq} - c\mathbf{J\omega} + \mathbf{\omega} \times \mathbf{J\omega}$$

simply becomes

$$\ddot{\theta} + c\dot{\theta} + k \sin \frac{\theta}{2} = 0$$

Note: For a specified maneuver time ($\approx 4/\zeta\omega_n$), the controller gain constants k and c can be determined approximately from the following relationship:

$$\ddot{\theta} + c\dot{\theta} + \frac{k}{2}\theta = \ddot{\theta} + 2\zeta\omega_n\dot{\theta} + \omega_n^2\theta = 0$$

7.16. Repeat Problem 7.14 using a control logic of the form

$$\mathbf{u} = -k\mathbf{Jq} - c\mathbf{J\omega} + \mathbf{\omega} \times \mathbf{J\omega}$$

Note: A rest-to-rest eigenaxis rotation is characterized by a straight line in q_i vs q_j plots ($i, j = 1, 2, 3$).

7.3.3   *Cascade-Saturation Control Logic*

Consider the rotational equations of motion of a rigid spacecraft described by

$$2\,\dot{\mathbf{q}} = \mathbf{f}(\mathbf{q}, \mathbf{\omega}) = \pm\sqrt{1 - \|\mathbf{q}\|^2}\,\mathbf{\omega} - \mathbf{\omega} \times \mathbf{q} \qquad (7.76a)$$

$$\dot{\mathbf{\omega}} = \mathbf{g}(\mathbf{\omega}, \mathbf{u}) = \mathbf{J}^{-1}(-\mathbf{\omega} \times \mathbf{J\omega} + \mathbf{u}) \qquad (7.76b)$$

where $\mathbf{J}$ is the inertia matrix, $\mathbf{q} = (q_1, q_2, q_3)$ is the quaternion vector, $\mathbf{\omega} = (\omega_1, \omega_2, \omega_3)$ is the angular velocity vector, $\mathbf{u} = (u_1, u_2, u_3)$ is the control input vector, and

$$\|\mathbf{q}\|^2 \equiv \mathbf{q}^T\mathbf{q} = q_1^2 + q_2^2 + q_3^2$$

The state vector of the system, denoted by $\mathbf{x}$, is then defined as

$$\mathbf{x} = \begin{bmatrix} \mathbf{q} \\ \mathbf{\omega} \end{bmatrix}$$

A dynamic system described by a set of differential equations of the form of Eqs. (7.76) is called a *cascaded system* because $\mathbf{q}$ does not appear in Eq. (7.76b). A cascade-saturation controller is introduced here for such a cascaded system, based on Ref. 17.

Saturation functions to be employed for the cascade-saturation controller are first defined as follows.

Definition 7.1

A *saturation function* of an n-dimensional vector $\mathbf{x} = (x_1, \ldots, x_n)$ is defined as

$$\text{sat}(\mathbf{x}) = \begin{bmatrix} \text{sat}_1(x_1) \\ \text{sat}_2(x_2) \\ \vdots \\ \text{sat}_n(x_n) \end{bmatrix}$$

where

$$\text{sat}_i(x_i) = \begin{cases} x_i^+ & \text{if } x_i > x_i^+ \\ x_i & \text{if } x_i^- \leq x_i \leq x_i^+ \\ x_i^- & \text{if } x_i < x_i^- \end{cases} \tag{7.77}$$

The normalized lower and upper bounds can be assumed as ± 1 for all i, without loss of generality.

Similarly, a *signum function* of an n-dimensional vector $\mathbf{x}$ is defined as

$$\text{sgn}(\mathbf{x}) = \begin{bmatrix} \text{sgn}(x_1) \\ \text{sgn}(x_2) \\ \vdots \\ \text{sgn}(x_n) \end{bmatrix}$$

where

$$\text{sgn}(x_i) = \begin{cases} +1 & \text{if } x_i > 0 \\ 0 & \text{if } x_i = 0 \\ -1 & \text{if } x_i < 0 \end{cases} \tag{7.78}$$

Definition 7.2

A *normalized saturation* function of an n-dimensional vector $\mathbf{x}$ is defined as

$$\text{sat}_\sigma(\mathbf{x}) = \begin{cases} \mathbf{x} & \text{if } \sigma(\mathbf{x}) < 1 \\ \mathbf{x}/\sigma(\mathbf{x}) & \text{if } \sigma(\mathbf{x}) \geq 1 \end{cases} \tag{7.79}$$

where $\sigma(\mathbf{x})$ is a positive scalar function of $\mathbf{x}$, which characterizes the largeness of the vector $\mathbf{x}$.

Because the largeness of a vector $\mathbf{x}$ is often characterized by its norms, we may choose $\sigma(\mathbf{x}) = \|\mathbf{x}\|_2 = \sqrt{\mathbf{x}^T \mathbf{x}}$ or $\sigma(\mathbf{x}) = \|\mathbf{x}\|_\infty = \max_i |x_i|$. Note that the normalized saturation of a vector $\mathbf{x}$ as defined here has the same direction of the vector $\mathbf{x}$ itself before saturation; i.e., it maintains the direction of the vector.

Definition 7.3

A state feedback controller of the following form is called the *m-layer cascade-saturation controller:*

$$\mathbf{u} = \mathbf{Q}_m \, \text{sat}(\mathbf{P}_m\mathbf{x} + \cdots + \mathbf{Q}_2 \, \text{sat}[\mathbf{P}_2\mathbf{x} + \mathbf{Q}_1 \, \text{sat}(\mathbf{P}_1\mathbf{x})]) \tag{7.80}$$

where $\mathbf{P}_i$ and $\mathbf{Q}_i$ are the controller gain matrices to be properly determined. If $\mathbf{Q}_i$ and $\mathbf{P}_i$ are diagonal matrices, then we have an m-layer decentralized cascade-saturation controller.

The control logic described by Eq. (7.80) is a generalized form of a saturation controller. A cascade-relay or poly-relay control algorithm similar to Eq. (7.80) has also been suggested in the literature using the signum function instead of the saturation function. The simplest form of a two-layer cascade-saturation control logic for a rigid spacecraft can be expressed as

$$\mathbf{u} = \mathbf{Q}_2 \, \underset{\sigma}{\text{sat}}[\mathbf{P}_2\boldsymbol{\omega} + \mathbf{Q}_1 \, \text{sat}(\mathbf{P}_1\mathbf{q})] \tag{7.81}$$

A typical rest-to-rest eigenaxis maneuver with slew rate constraint is now considered, which will be called a constrained rest-to-rest maneuver. The maneuver consists of the following three phases: 1) acceleration, 2) coast, and 3) deceleration. In the spin-up acceleration phase, the spacecraft will accelerate about the eigenaxis. In the coast phase, it rotates about the eigenaxis at a constant slew rate. In this coast phase, the control input and the body rates are kept in a quasi-steady mode. The following lemma guarantees the existence of such a quasi-steady coast phase.

Lemma 7.1

If a dynamic system described by Eqs. (7.76) is exponentially stabilized by the two-layer saturation control logic (7.81) in a constrained rest-to-rest maneuver problem, and if there exists a sufficiently large time instant t^* such that for $t \le t^*$, $|(\mathbf{P}_1\mathbf{q})_i| \ge 1$ $\forall i$, then there exists a time interval $[t, t^*]$ in which the angular velocity vector $\boldsymbol{\omega}$ is in a quasi-steady mode.

Proof: As t approaches to t^*, the closed-loop system becomes

$$\dot{\mathbf{q}} = \mathbf{f}(\mathbf{q}, \boldsymbol{\omega}) \tag{7.82a}$$

$$\dot{\boldsymbol{\omega}} = \mathbf{g}\left(\boldsymbol{\omega}, \mathbf{Q}_2 \, \underset{\sigma}{\text{sat}}[\mathbf{P}_2\boldsymbol{\omega} + \mathbf{Q}_1 \, \text{sgn}(\mathbf{P}_1\mathbf{q})]\right) \tag{7.82b}$$

For $t \in (t^*, \infty)$ the closed-loop system becomes

$$\dot{\mathbf{q}} = \mathbf{f}(\mathbf{q}, \boldsymbol{\omega}) \tag{7.83a}$$

$$\dot{\boldsymbol{\omega}} = \mathbf{g}\left(\boldsymbol{\omega}, \mathbf{Q}_2 \, \underset{\sigma}{\text{sat}}[\mathbf{P}_2\boldsymbol{\omega} + \mathbf{Q}_1 \, \text{sat}(\mathbf{P}_1\mathbf{q})]\right) \tag{7.83b}$$

Because the closed-loop system in a constrained rest-to-rest maneuver problem is assumed to be exponentially stabilized, both Eqs. (7.82) and (7.83) should be exponentially stable. As a result, along the trajectory of Eqs. (7.82) the slew rate $\|\boldsymbol{\omega}(t)\|$ should increase. If t^* is sufficiently large, $\|\boldsymbol{\omega}(t)\|$ will become a constant slew rate $\|\boldsymbol{\omega}^*\|$ whereas $\boldsymbol{\omega}^*$ satisfies

$$\mathbf{g}\left(\boldsymbol{\omega}^*, \mathbf{Q}_2 \operatorname*{sat}_{\sigma}[\mathbf{P}_2\boldsymbol{\omega}^* + \mathbf{Q}_1 \operatorname{sgn}(\mathbf{P}_1\mathbf{q})]\right) = 0 \qquad (7.84)$$

which is in fact an algebraic equation for $\boldsymbol{\omega}^*$. If the settling time t_s is smaller than t^*, then during the time interval $[t_s, t^*]$, $\boldsymbol{\omega}$ is very close to $\boldsymbol{\omega}^*$, i.e., $\boldsymbol{\omega}$ is in a quasi-steady mode; i.e., $\dot{\boldsymbol{\omega}} \approx 0$.

Lemma 7.2

If a dynamic system described by Eqs. (7.76) is exponentially stabilized by Eq. (7.81) in a constrained rest-to-rest maneuver problem, and if the following conditions are satisfied:

$$\boldsymbol{\omega}^T \mathbf{g}\left(\boldsymbol{\omega}, \mathbf{Q}_2 \operatorname*{sat}_{\sigma}[\mathbf{P}_2\boldsymbol{\omega} + \mathbf{Q}_1 \operatorname{sgn}(\mathbf{P}_1\mathbf{q})]\right) \geq 0, \qquad \text{if } \sigma(\mathbf{P}_1\mathbf{q}) \geq 1$$

$$\boldsymbol{\omega}^T \mathbf{g}\left(\boldsymbol{\omega}, \mathbf{Q}_2 \operatorname*{sat}_{\sigma}(\mathbf{P}_2\boldsymbol{\omega} + \mathbf{Q}_1\mathbf{P}_1\mathbf{q})\right) < 0, \qquad \text{if } \sigma(\mathbf{P}_1\mathbf{q}) < 1$$

then the slew rate $\|\boldsymbol{\omega}(t)\|$ will never exceed its upper bound $\|\boldsymbol{\omega}^*\|$.

Proof: By defining $V = \frac{1}{2}\boldsymbol{\omega}^T\boldsymbol{\omega}$ as a Lyapunov function, and considering the time derivative along the closed-loop trajectory and Lemma 7.1, we can obtain the proof of this lemma.

7.3.4 *Eigenaxis Rotation Under Slew Rate Constraint*

Consider a rigid spacecraft that is required to maneuver about an inertially fixed axis as fast as possible, but not exceeding the specified maximum slew rate about that eigenaxis. It will be shown that the following saturation control logic provides such a rest-to-rest eigenaxis rotation under slew rate constraint:

$$\mathbf{u} = -\mathbf{K} \operatorname{sat}(\mathbf{Pq}) - \mathbf{C}\boldsymbol{\omega} + \boldsymbol{\omega} \times \mathbf{J}\boldsymbol{\omega} \qquad (7.85)$$

where

$$\mathbf{K} = \operatorname{diag}(k_1, k_2, k_3)\, \mathbf{J}$$
$$\mathbf{P} = \operatorname{diag}(p_1, p_2, p_3)$$
$$\mathbf{C} = c\mathbf{J}$$

and k_i, p_i, and c are all positive scalar constants that are the control design parameters to be properly determined. Notice the similarity between this saturation control logic and the eigenaxis slew control logic (7.74).

The closed-loop attitude dynamics of a rigid spacecraft employing the saturation control logic of Eq. (7.85) are then described by

$$\mathbf{J}\dot{\boldsymbol{\omega}} = -\mathbf{K}\,\text{sat}(\mathbf{Pq}) - \mathbf{C}\boldsymbol{\omega} \tag{7.86a}$$

$$2\,\dot{\mathbf{q}} = q_4\boldsymbol{\omega} - \boldsymbol{\omega} \times \mathbf{q} \tag{7.86b}$$

$$2\dot{q}_4 = -\boldsymbol{\omega}^T\mathbf{q} \tag{7.86c}$$

The following lemma and theorem characterize the rotational motion of a rigid spacecraft described by Eqs. (7.86).

Lemma 7.3

Let $\dot{\theta}_{max}$ be the maximum slew rate about an eigenaxis allowed by saturating rate gyros; i.e., $|\dot{\theta}(t)| \le \dot{\theta}_{max}$, and nonzero $q_i(0)$ are specified with $\boldsymbol{\omega}(0) = 0$ for a rest-to-rest maneuver. It is assumed that t^* is a time instant at which there exists at least one axis such that $|p_iq_i(t^*)| = 1$, and $q_i(0)q_i(t) > 0$ for all i and for all $t \in [0, t^*]$. If we choose

$$k_i = c\frac{|q_i(0)|}{\|\mathbf{q}(0)\|}\,\dot{\theta}_{max} \tag{7.87a}$$

$$\mathbf{KP} = k\mathbf{J} \tag{7.87b}$$

where $k \equiv k_ip_i$ is a positive scalar constant, then we have the following results for all $t \in [0, t^*]$:

1) The rotational motion described by Eq. (7.86) is an eigenaxis rotation about $\mathbf{q}(0)$.

2) The actual slew rate about the eigenaxis is bounded as

$$\|\boldsymbol{\omega}(t)\| \le \dot{\theta}_{max}$$

and it increases monotonically.

3) The attitude error $\|\mathbf{q}(t)\|$ decreases monotonically.

4) At time t^*, we have

$$|p_iq_i(t^*)| = 1 \qquad \text{for all } i = 1, 2, 3$$

Proof: 1) Substituting Eqs. (7.87) into Eqs. (7.86), we obtain

$$\dot{\boldsymbol{\omega}} = -\frac{c\dot{\theta}_{max}}{\|\mathbf{q}(0)\|}\mathbf{q}(0) - c\boldsymbol{\omega} \tag{7.88}$$

The solution of Eq. (7.88) can be expressed as

$$\boldsymbol{\omega}(t) = e^{-ct}\boldsymbol{\omega}(0) - \int_0^t e^{-c(t-\tau)}\frac{c\dot{\theta}_{max}}{\|\mathbf{q}(0)\|}\mathbf{q}(0)\,d\tau \tag{7.89}$$

Because we are concerned with a rest-to-rest maneuver, i.e., $\boldsymbol{\omega}(0) = 0$, Eq. (7.89) is rewritten as

$$\boldsymbol{\omega}(t) = -f(t)\mathbf{q}(0) \tag{7.90}$$

where

$$f(t) = \frac{(1 - e^{-ct})\dot{\theta}_{max}}{\|\mathbf{q}(0)\|} > 0 \tag{7.91}$$

Because $\boldsymbol{\omega}(t)$ lies along $\mathbf{q}(0)$, Theorem 7.2 implies that $\mathbf{q}(t)$ also lies entirely along $\mathbf{q}(0)$, i.e., there exists a scalar function $g(t)$ such that

$$\mathbf{q}(t) = g(t)\mathbf{q}(0) \tag{7.92}$$

for all $t > 0$. This means that the resulting motion is an eigenaxis rotation.

2) From Eqs. (7.90) and (7.91), we obtain

$$\|\boldsymbol{\omega}(t)\| = \sqrt{\boldsymbol{\omega}^T\boldsymbol{\omega}} = (1 - e^{-ct})\dot{\theta}_{max}$$

which is obviously less than $\dot{\theta}_{max}$ and $\|\boldsymbol{\omega}(t)\|$ increases monotonically.

3) Substituting Eqs. (7.90) and (7.92) into the quaternion kinematic differential equations, we obtain

$$\dot{g}(t) = -\tfrac{1}{2}q_4 f(t) \tag{7.93}$$

The solution of Eq. (7.93) can be expressed as

$$g(t) = 1 - \frac{1}{2}\int_0^t q_4(\tau)f(\tau)\,d\tau \tag{7.94}$$

From Eqs. (7.91) and (7.93), we obtain

$$\dot{g}(t) < 0$$

and $\|\mathbf{q}(t)\|$ decreases monotonically.

4) Because $k_i p_i = k$ for $i = 1, 2, 3$, we have

$$p_i = \frac{k\|\mathbf{q}(0)\|}{c|q_i(0)|\dot{\theta}_{max}}$$

and

$$|p_i q_i(t)| = \frac{k\|\mathbf{q}(0)\|}{c\dot{\theta}_{max}}|g(t)|$$

Furthermore, there exists a time instant t^* satisfying

$$|g(t^*)| = \frac{c\dot{\theta}_{max}}{k\|\mathbf{q}(0)\|}$$

such that $|p_i q_i(t^*)| = 1$ for of all i, i.e., all of the elements of $\mathbf{q}(t)$ depart the coasting phase at the same time.

The preceding lemma characterizes the properties of the closed-loop system described by Eqs. (7.86) for $t \in [0, t^*]$. The following theorem characterizes the closed-loop system described by Eq. (7.86) in the entire time interval $0 \le t < \infty$.

Theorem 7.3

The closed-loop system described by Eqs. (7.86) has the following properties:

1) The entire time interval $[0, \infty]$ consists of the three motion phases, called the acceleration, coast, and deceleration phases, with the time intervals $[0, t_s]$, $[t_s, t^*]$ and $[t^*, \infty)$, respectively, whereas t_s and t^* can be approximated as

$$t_s \approx \frac{4}{c}$$

$$t^* \approx t_s + \frac{2}{\dot{\theta}_{max}} \tan^{-1} \left\{ \frac{\|q(0)\|}{q_4(t_s)} \right\}$$

2) The resulting rotational motion is an eigenaxis rotation for all $t \in [0, \infty)$.

3) The quaternion vector $q(t)$ and the angular velocity vector $\omega(t)$ become zero as t approaches ∞.

4) The slew rate is bounded by $\dot{\theta}_{max}$ for a properly chosen c.

Proof: 1) From part 2 of Lemma 7.3, we have the three motion phases: the acceleration, coast and deceleration phases. Because the angular velocity has the form (7.90) and the slew rate approaches $\dot{\theta}_{max}$ with the time constant $1/c$, the settling time for this acceleration phase can be approximated as t_s. For $t \le t_s$, $\|\omega(t)\|$ increases quickly and the acceleration phase is $[0, t_s]$.

Because $1 - e^{-ct} \approx 1$ for $t \ge t_s$, ω is in a quasi-steady mode for $t \ge t_s$. Hence, the coast phase starts from $t = t_s$.

Assuming that the coast phase ends at $t = t^*$, we can estimate t^* as follows. During the coast phase, ω is very slowly changing and we can approximate it as

$$\omega(t) \approx -\frac{\dot{\theta}_{max}}{\|q(0)\|} q(0) = \omega^*$$

Substituting ω^* into the quaternion kinematic differential equations of the following form:

$$\dot{\hat{q}} = f(\hat{q}, \omega) \tag{7.95}$$

where $\hat{q} = (q_1, q_2, q_3, q_4)$ and

$$f(\hat{q}, \omega) = F(\omega)\hat{q}$$

$$F(\omega) = \frac{1}{2} \begin{bmatrix} 0 & \omega_3 & -\omega_2 & \omega_1 \\ -\omega_3 & 0 & \omega_1 & \omega_2 \\ \omega_2 & -\omega_1 & 0 & \omega_3 \\ -\omega_1 & -\omega_2 & -\omega_3 & 0 \end{bmatrix}$$

and simplifying the resulting equations, we obtain the following equation:

$$\dot{\hat{q}} = F(\omega^*)\dot{\hat{q}}$$

$$= F(\omega^*)F(\omega^*)\hat{q} = -[\dot{\theta}_{max}/2]^2 \hat{q}$$

Because the motion is an eigenaxis rotation at $t = t_s$, we have

$$\dot{\mathbf{q}}(t_s) = \frac{1}{2}q_4(t_s)\boldsymbol{\omega}^* = -\frac{\dot{\theta}_{max}q_4(t_s)}{2\|\mathbf{q}(0)\|}\mathbf{q}(0)$$

For $t_s < t \leq t^*$, we have

$$\mathbf{q}(t) = \sqrt{1 + \frac{q_4^2(t_s)}{\|\mathbf{q}(0)\|^2}} \; \sin[\dot{\theta}_{max}(t - t_s)/2 - \phi]\mathbf{q}(0) \tag{7.96}$$

where

$$\phi = \tan^{-1}\left[\frac{\|\mathbf{q}(0)\|}{q_4(t_s)}\right]$$

At the end of the coast phase, $\mathbf{q}(t)$ becomes nearly zero and we have

$$t^* \approx t_s + \frac{2\phi}{\dot{\theta}_{max}} \tag{7.97}$$

For $t \geq t^*$, we have the deceleration phase.

2) From Lemma 7.3, we know that all of the components of $\mathbf{q}(t)$ reach the linear range of the saturation function at the same time, i.e., the closed-loop equations for $t \geq t^*$ are the same as Eqs. (7.75). At $t = t^*$, $\mathbf{q}(t^*)$ and $\boldsymbol{\omega}(t^*)$ lie along the vector $\mathbf{q}(0)$. From Theorem 7.1, the rotational motion is an eigenaxis rotation. Also part 1 of Lemma 3 leads to the following conclusion: the closed-loop rotational motion is, in fact, an eigenaxis rotation for all $t \in [0, \infty)$.

3) For the deceleration phase, consider the closed-loop system described by Eqs. (7.75) and a positive definite function of the form

$$V = (1/k)V_\omega + V_q$$

where the *quaternion Lyapunov function* V_q and the *angular velocity Lyapunov function* V_ω are defined as follows:

$$V_q = q_1^2 + q_2^2 + q_3^2 + (1 - q_4)^2 \tag{7.98}$$

$$V_\omega = \frac{1}{2}(\omega_1^2 + \omega_2^2 + \omega_3^2) \tag{7.99}$$

The time derivative of V along the closed-loop trajectory described by Eqs. (7.75) becomes

$$\dot{V} = -(c/k)\boldsymbol{\omega}^T\boldsymbol{\omega} < 0$$

This means that the closed-loop system is asymptotically stable, and $\mathbf{q}(t)$ and $\boldsymbol{\omega}(t)$ will become zero.

For $t \geq t^*$, however, we have

$$\dot{V}_q = \mathbf{q}^T\boldsymbol{\omega}$$
$$\dot{V}_\omega = -c\boldsymbol{\omega}^T\boldsymbol{\omega} - k\mathbf{q}^T\boldsymbol{\omega}$$

If $\dot{V}_q < 0$, then it is possible to have $\dot{V}_\omega > 0$, i.e., it is possible for the rate limitation to be violated. In the following statement, it will be shown that this situation can be avoided by properly choosing c.

4) The closed-loop stability is guaranteed by part 3 of this theorem during the deceleration phase, but the rate constraint is not necessarily guaranteed as discussed earlier. Because the maneuver is an eigenaxis rotation, Eq. (7.75) becomes

$$\ddot{\theta} + c\dot{\theta} + k\sin(\theta/2) = 0 \qquad \text{for } t \geq t^* \qquad (7.100)$$

with the following initial condition

$$\theta(t^*) = 2\sin^{-1}[g(t^*)]$$

where θ is the rotational angle about the eigenaxis.

At the end of the coast phase, it is reasonable to assume that θ is small. Consequently, we have

$$\ddot{\theta} + c\dot{\theta} + k\frac{\theta}{2} = 0 \qquad \text{for } t \geq t^*$$

Then we can properly choose k and c as follows:

$$k = 2\omega_n^2$$
$$c = 2\zeta\omega_n$$

where ζ and ω_n are the desired damping ratio and the natural frequency, which characterize the second-order dynamics of the desired slew rate during the deceleration phase. It is clear that the slew rate will not exceed $\dot{\theta}_{max}$ for all $t \in [0, \infty)$ if we do not choose a small ζ.

7.3.5 Slew Rate and Control Constraints

We now consider a rigid spacecraft that is required to maneuver about an inertially fixed axis as fast as possible, but within the saturation limits of reaction wheels as well as rate gyros.

Let τ_i denote the torque generated by the ith reaction wheel, and also assume that

$$|\tau_i| \leq \bar{\tau}_i, \qquad i = 1, \ldots, \ell \qquad (7.101)$$

where ℓ is the number of the reaction wheels and $\bar{\tau}_i$ is the maximum torque of the ith reaction wheel. Usually, $\ell \geq 3$ to allow the failure of at most $\ell - 3$ reaction wheels.

The control torque inputs u_1, u_2, and u_3 along the body-fixed control axes are generated by reaction wheels, and in general the control input vector $\mathbf{u}$ can be expressed as

$$\mathbf{u} = \mathbf{a}_1\tau_1 + \mathbf{a}_2\tau_2 + \cdots + \mathbf{a}_\ell\tau_\ell$$

where $\mathbf{a}_i \in \mathcal{R}^3$ is the torque distribution vector of the ith reaction wheel and $\mathbf{a}_i^T \mathbf{a}_i = 1$ for all i. The torque distribution matrix is defined as

$$\mathbf{A} = [\mathbf{a}_1 \, \mathbf{a}_2 \, \cdots \, \mathbf{a}_\ell]$$

and $\mathbf{u} = \mathbf{A}\tau$. At least three column vectors of $\mathbf{A}$ must be linearly independent or $(\mathbf{A}\mathbf{A}^T)^{-1}$ must exist for independent three-axis control.

For the commanded control input vector $\mathbf{u}_c$, the reaction wheel torque command is determined as

$$\tau_c = \mathbf{A}^+ \mathbf{u}_c$$

where $\mathbf{A}^+ = \mathbf{A}^T (\mathbf{A}\mathbf{A}^T)^{-1}$ is the pseudoinverse transformation matrix.* For an ideal case without actuator dynamics and saturation, the reaction wheel torque vector $\tau = \tau_c$ is then physically redistributed as $\mathbf{u} = \mathbf{A}\tau$, and the spacecraft will be acted on by the control input vector $\mathbf{u}$, which is the same as the the commanded control input vector $\mathbf{u}_c$. If a torque saturation occurs in one of reaction wheels, however, then $\mathbf{u} \neq \mathbf{u}_c$ and $\tau \neq \tau_c$.

Now we consider a control logic that accommodates possible torque saturation of the reaction wheels, but that still provides an eigenaxis rotation under slew rate constraint.

The commanded control input vector $\mathbf{u}_c$ that accommodates the slew rate constraint is given as

$$\mathbf{u}_c = -\mathbf{K}\,\mathrm{sat}(\mathbf{Pq}) - \mathbf{C}\boldsymbol{\omega}$$

where

$$\mathbf{P} = \mathrm{diag}(p_1, p_2, p_3)$$
$$\mathbf{K} = \mathrm{diag}(k_1, k_2, k_3)\,\mathbf{J}$$
$$\mathbf{C} = c\mathbf{J}$$

and, furthermore, we choose

$$k_i = c\frac{|q_i(0)|}{\|\mathbf{q}(0)\|}\dot{\theta}_{\max} \qquad (7.102a)$$

$$\mathbf{KP} = k\mathbf{J} \qquad (7.102b)$$

where k is a positive scalar and it is assumed that $q_i(0) \neq 0$ for all i.

The reaction wheel torque command τ_c for the commanded control torque $\mathbf{u}_c$ is then determined as

$$\tau_c = \mathbf{A}^+ \mathbf{u}_c$$

---

*$\mathbf{A}^+$ is also called the *Moore–Penrose inverse* of $\mathbf{A}$, or a generalized inverse of $\mathbf{A}$.

To keep the reaction wheel torque vector $\boldsymbol{\tau}$ even in the presence of saturation lie in the same direction as $\boldsymbol{\tau}_c$, we use the maximum value of the components of $\boldsymbol{\tau}_c$ to normalize it. Thus, we choose the following criterion:

$$\sigma(\mathbf{q}, \boldsymbol{\omega}) = \|\mathbf{T}\boldsymbol{\tau}_c\|_\infty = \max_i |(\mathbf{T}\boldsymbol{\tau}_c)_i| \tag{7.103}$$

where

$$\mathbf{T} = \text{diag}(1/\bar{\tau}_1, 1/\bar{\tau}_2, \ldots, 1/\bar{\tau}_\ell)$$

The actual reaction wheel torque vector acting on the spacecraft then becomes

$$\boldsymbol{\tau} = \underset{\sigma}{\text{sat}}(\boldsymbol{\tau}_c) = \begin{cases} \boldsymbol{\tau}_c & \text{if } \sigma(\mathbf{q}, \boldsymbol{\omega}) \le 1 \\ \boldsymbol{\tau}_c/\sigma(\mathbf{q}, \boldsymbol{\omega}) & \text{if } \sigma(\mathbf{q}, \boldsymbol{\omega}) > 1 \end{cases}$$

The control input vector $\mathbf{u}$ acting on the spacecraft, which is generated by the saturated torque vector $\boldsymbol{\tau}$, becomes

$$\mathbf{u} = \mathbf{A}\boldsymbol{\tau} = \mathbf{A} \underset{\sigma}{\text{sat}}(\boldsymbol{\tau}_c) = \mathbf{A} \underset{\sigma}{\text{sat}}(\mathbf{A}^+\mathbf{u}_c)$$

and, thus, we have

$$\mathbf{u} = \underset{\sigma}{\text{sat}}(\mathbf{u}_c)$$

Finally, we have the following equivalent expression of the saturation control logic:

$$\mathbf{u} = -\underset{\sigma}{\text{sat}}[\mathbf{K} \, \text{sat}(\mathbf{Pq}) + \mathbf{C}\boldsymbol{\omega}] \tag{7.104}$$

7.3.5.1 Case of Negligible Gyroscopic Coupling
The gyroscopic term is not significant during most practical rotational maneuvers, and the term can be neglected without much impact on performance and stability. For such a case, let

$$\mu(\mathbf{q}, \boldsymbol{\omega}) = \begin{cases} 1 & \text{if } \sigma(\mathbf{q}, \boldsymbol{\omega}) \le 1 \\ \sigma(\mathbf{q}, \boldsymbol{\omega}) & \text{if } \sigma(\mathbf{q}, \boldsymbol{\omega}) > 1 \end{cases}$$

Then the closed-loop dynamics of a rigid spacecraft during the acceleration and coast phases can be described by

$$\mu(\mathbf{q}, \boldsymbol{\omega})\dot{\boldsymbol{\omega}} = -\frac{c\dot{\theta}_{max}}{\|\mathbf{q}(0)\|}\mathbf{q}(0) - c\boldsymbol{\omega} \tag{7.105a}$$

$$2\,\dot{\mathbf{q}} = q_4\boldsymbol{\omega} - \boldsymbol{\omega} \times \mathbf{q} \tag{7.105b}$$

$$2\dot{q}_4 = -\boldsymbol{\omega}^T\mathbf{q} \tag{7.105c}$$

From the definition of $\sigma(\cdot)$, we have

$$\mu(\mathbf{q}, \boldsymbol{\omega}) = a + \mathbf{b}^T\boldsymbol{\omega}$$

where a is a scalar constant and $\mathbf{b}$ is a 3×1 column vector. If $\sigma(\mathbf{q}, \boldsymbol{\omega}) \leq 1$, then $a = 1$ and $\mathbf{b} = (0, 0, 0)$; otherwise there are no limitations on a and $\mathbf{b}$.

The solution $\bar{\boldsymbol{\omega}}(t)$ of the homogeneous differential equation of Eq. (7.105a) during the acceleration phase satisfies

$$(a + \mathbf{b}^T \bar{\boldsymbol{\omega}}) \, \dot{\bar{\boldsymbol{\omega}}} = -c\bar{\boldsymbol{\omega}}$$

or it satisfies the following iteration form:

$$\bar{\boldsymbol{\omega}}(t) = \exp\left[-\frac{c}{a + \mathbf{b}^T \bar{\boldsymbol{\omega}}(t)} t \right] \bar{\boldsymbol{\omega}}(0) = \bar{f}(t)\bar{\boldsymbol{\omega}}(0)$$

Hence, the solution of Eqs. (7.105) can be expressed as

$$\boldsymbol{\omega}(t) = \bar{f}(t)\boldsymbol{\omega}(0) - \int_0^t \bar{f}(t - \tau) \frac{c\dot{\theta}_{\max}}{\|\mathbf{q}(0)\|} \mathbf{q}(0) \, d\tau$$

If $\boldsymbol{\omega}(0) = 0$, we obtain

$$\boldsymbol{\omega}(t) = \hat{f}(t)\mathbf{q}(0)$$

From Theorem 7.2 we have

$$\mathbf{q}(t) = \hat{g}(t)\mathbf{q}(0)$$

i.e., the maneuver during the acceleration and coast phases is still an eigenaxis rotation even in the presence of control torque saturation.

During the deceleration phase, the closed-loop system is described by

$$\mu(\mathbf{q}, \boldsymbol{\omega})\dot{\boldsymbol{\omega}} = -k\mathbf{q} - c\boldsymbol{\omega} \tag{7.106a}$$

$$2\dot{\mathbf{q}} = q_4\boldsymbol{\omega} - \boldsymbol{\omega} \times \mathbf{q} \tag{7.106b}$$

$$2\dot{q}_4 = -\boldsymbol{\omega}^T \mathbf{q} \tag{7.106c}$$

and the resulting motion is still an eigenaxis rotation. Furthermore, we have

$$\ddot{\theta} + \frac{c}{\mu}\dot{\theta} + \frac{k}{\mu}\sin\left(\frac{\theta}{2}\right) = 0$$

This equation is the same as Eq. (7.100) except the positive factor μ. Hence, we have a similar result to Theorem 7.3, i.e., the maneuver in the deceleration phase is an eigenaxis rotation, $\mathbf{q}$ and $\boldsymbol{\omega}$ will be regulated to zero, and the magnitude of the angular velocity will never exceed $\dot{\theta}_{\max}$ for a properly chosen c.

7.3.5.2 Case of Significant Gyroscopic Coupling The closed-loop system including the gyroscopic coupling term is described by

$$\mathbf{J}\dot{\boldsymbol{\omega}} = -\boldsymbol{\omega} \times \mathbf{J}\boldsymbol{\omega} - \underset{\sigma}{\text{sat}}[\mathbf{K}\,\text{sat}(\mathbf{Pq}) + \mathbf{C}\boldsymbol{\omega}] \qquad (7.107a)$$

$$2\,\dot{\mathbf{q}} = q_4\boldsymbol{\omega} - \boldsymbol{\omega} \times \mathbf{q} \qquad (7.107b)$$

$$2\dot{q}_4 = -\boldsymbol{\omega}^T\mathbf{q} \qquad (7.107c)$$

During the acceleration and coast phases, we have

$$\mu(\mathbf{q}, \boldsymbol{\omega})\dot{\boldsymbol{\omega}} = -\mu(\mathbf{q}, \boldsymbol{\omega})\mathbf{J}^{-1}\boldsymbol{\omega} \times \mathbf{J}\boldsymbol{\omega} - \frac{c\dot{\theta}_{\max}}{\|\mathbf{q}(0)\|}\mathbf{q}(0) - c\boldsymbol{\omega}$$

and

$$|\,\text{sat}(\mathbf{Pq})_i| = 1 \qquad (7.108)$$

$$\left|\mu(\mathbf{q}, \boldsymbol{\omega})(\mathbf{J}^{-1}\boldsymbol{\omega} \times \mathbf{J}\boldsymbol{\omega})_i\right| < \left|\frac{c\dot{\theta}_{\max}}{\|\mathbf{q}(0)\|}q_i(0)\right| \qquad (7.109)$$

where subscript i denotes the ith component of a vector $(\cdot)$. This implies that

$$\text{sgn}(\dot{\omega}_i) = -\text{sgn}(q_i), \qquad i = 1, 2, 3$$

The time derivatives of V_q and V_ω along the closed-loop trajectory described by Eqs. (7.107) satisfy

$$\dot{V}_q < 0 \qquad \text{and} \qquad \dot{V}_\omega > 0$$

and during the acceleration phase, the slew rate increases and the quaternion decreases.

During the coast phase, the angular velocity vector becomes $\boldsymbol{\omega}^*$ satisfying

$$\boldsymbol{\omega}^* \times \mathbf{J}\boldsymbol{\omega}^* + \underset{\sigma}{\text{sat}}[\mathbf{K}\,\text{sgn}(\mathbf{Pq}) + \mathbf{C}\boldsymbol{\omega}^*] = 0 \qquad (7.110)$$

These are algebraic equations independent of $\mathbf{q}$. During this period, the control $\mathbf{u}$ is near constant and we have

$$\mathbf{u}^* = -\boldsymbol{\omega}^* \times \mathbf{J}\boldsymbol{\omega}^*$$

Problem

7.17 Consider the near-minimum-time eigenaxis reorientation problem of the XTE spacecraft subject to slew rate and control torque constraints. The XTE spacecraft with the inertia matrix

$$\mathbf{J} = \begin{bmatrix} 6292 & 0 & 0 \\ 0 & 5477 & 0 \\ 0 & 0 & 2687 \end{bmatrix} \text{kg} \cdot \text{m}^2$$

is equipped with four skewed reaction wheels. The torque distribution matrix from the reaction wheel torque vector $\boldsymbol{\tau}$ to the spacecraft control input vector $\mathbf{u}$ is given as

$$\mathbf{A} = \begin{bmatrix} \cos\beta & -\cos\beta & \cos\beta & -\cos\beta \\ \sin\beta & 0 & -\sin\beta & 0 \\ 0 & -\sin\beta & 0 & \sin\beta \end{bmatrix}$$

where the skew angle β is chosen as 45 deg. The maximum torque level of each reaction wheel is given as $\bar{\tau}_i = 0.3$ N·m, and the maximum slew rate is given as $\dot{\theta}_{max} = 0.2$ deg/s, which is about 90% of the low-rate gyro measurement capability.

The initial quaternions for a specific reorientation maneuver are given as

$$\hat{\mathbf{q}}(0) = (0.2652, 0.2652, -0.6930, 0.6157)$$

which corresponds to a 104-deg slew angle about the eigenaxis $\mathbf{q}(0) = (0.2652, 0.2652, -0.6930)$. The slew should ideally be completed in 8.7 min.

(a) Synthesize a saturation control logic of the form

$$\mathbf{u}_c = -\mathbf{K}\,\text{sat}(\mathbf{Pq}) - \mathbf{C}\boldsymbol{\omega}$$

$$\boldsymbol{\tau}_c = \mathbf{A}^+ \mathbf{u}_c \qquad \text{where } \mathbf{A}^+ = \mathbf{A}^T(\mathbf{A}\mathbf{A}^T)^{-1}$$

$$\boldsymbol{\tau} = \underset{\sigma}{\text{sat}}(\boldsymbol{\tau}_c)$$

$$\mathbf{u} = \mathbf{A}\boldsymbol{\tau}$$

Hint: If we choose $\zeta = 0.707$ and $\omega_n = 0.1$ rad/s, then we have

$$\mathbf{C} = \text{diag}(889, 774, 380)$$

$$\mathbf{KP} = \text{diag}(126, 110, 54)$$

$$\mathbf{K} = \text{diag}(k_1, k_2, k_3)\mathbf{J} = \text{diag}(1.0452, 0.9098, 1.1667)$$

$$\mathbf{P} = k\mathbf{K}^{-1}\mathbf{J} = \text{diag}(120, 120, 46)$$

(b) Perform computer simulation of the closed-loop system, and verify that a near bang-off-bang, eigenaxis maneuver is, in fact, achieved under the slew rate and control torque constraints.

Note: See Ref. 17 for additional information pertaining to this problem.

7.4 Attitude Control and Momentum Management

This section, based on Refs. 18–20, is concerned with the attitude control and momentum management problem of a large space vehicle in low Earth orbit, such as the International Space Station, as shown in Fig. 7.4.

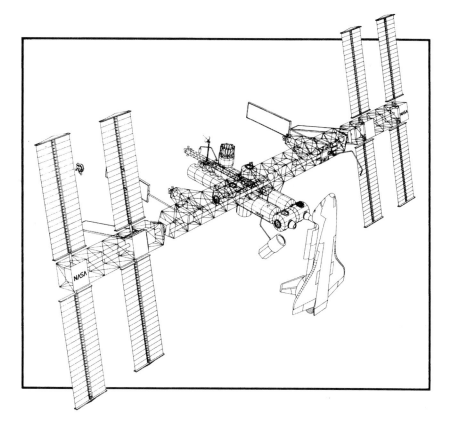

Fig. 7.4   Large space vehicle in low Earth orbit.

7.4.1 Introduction

Large space vehicles, such as Skylab of the 1970s, the Mir space station, and the International Space Station, employ control moment gyros (CMGs) as primary actuating devices during normal flight mode operation. Because the CMGs are momentum exchange devices, external control torques must be used to desaturate the CMGs, that is, bring the momentum back to its nominal value. Some methods for unloading CMG momentum include the use of magnetic torques, reaction jets, and gravity-gradient torque. For a large spacecraft in low Earth orbit, such as the International Space Station, the gravity-gradient torque approach is preferred because it requires no consumables or additional hardware. One approach to CMG momentum management is to integrate the momentum management and attitude control design. In this continuous, closed-loop control of both the CMG momentum and vehicle attitude, the design objective is to establish a proper tradeoff between spacecraft pointing and CMG momentum management, while satisfying the specific mission requirements.

The International Space Station will be controlled by four parallel mounted double-gimbal CMGs. A CMG of the International Space Station consists of a large wheel rotating at a constant speed (6600 rpm) and producing an angular momentum of 3500 ft-lb-s about its spin axis. This rotating wheel is mounted in a two-degree-of-freedom gimbal system that can point the spin axis (momentum vector) of the wheel in any direction, i.e., the tip of the angular momentum vector can be placed anywhere on a sphere of radius 3500 ft-lb-s. The CMG generates an output reaction torque that is applied to the Space Station by inertially changing the direction of its wheel momentum (spin axis). The CMG output torque has two components, one proportional to the rate of change of the CMG gimbals and a second proportional to the inertial body rate of the Space Station as sensed at the CMG base. Because the momentum along the direction of the spin axis is fixed (constant wheel speed), the output torque is constrained to lie in the plane of the wheel. As a result, one CMG is insufficient to provide the three-axis torque needed to control the attitude of the Space Station. To provide attitude control at least two CMGs are required.

The Space Station CMG assembly consists of four parallel mounted double-gimbal CMGs with two of the four CMGs mounted antiparallel with the other two. The four CMGs have a spherical momentum storage capability of 14,000 ft-lb-s, the scalar sum of the individual CMG wheel momentum. The momentum stored in the CMG system at any given time equals the vector sum of the individual CMG momentum vectors. To maintain the desired attitude, the CMG system must cancel, or absorb, the momentum generated by the disturbance torques acting on the Space Station. If the average disturbance torque is nonzero, the resulting CMG output torque will also be nonzero and momentum will build up in the CMG system. Once the CMG system saturates, i.e., all of the CMGs' momentum vectors have become parallel and controllability about the parallel line is lost, it is unable to generate the torque required to cancel the disturbance torque and loss of attitude control results. To prevent the CMG system from saturating, a continuous gravity-gradient momentum management system will be employed for the Space Station.

In this section the CMGs are assumed as ideal torquers, i.e., the CMG gimbal dynamics and CMG steering logic are ignored. The attitude control and momentum management technique to be presented in this section is, however, applicable to any CMG-equipped spacecraft. A brief overview of the CMG steering logic design problem associated with different CMG configurations will be presented later in Sec. 7.5.

Most large spacecraft are, in fact, flexible multibody vehicles with time-varying inertias, as can be seen in Fig. 7.4; however, they can be considered single rigid bodies for the practical design of a low-bandwidth, integrated attitude/momentum controller. All states of the rigid vehicle are usually available for feedback control from the strapdown inertial reference system of the vehicle.

7.4.2 Mathematical Models for Control Design

Consider a rigid spacecraft in low-Earth circular orbit, which is expected to maintain local-vertical and local-horizontal (LVLH) orientation during normal mode operation. As discussed in the preceding chapter, the nonlinear equations of motion of a rigid body in circular orbit can be written, in terms of components along the body-fixed control axes, as follows.

Attitude kinematics for the $\mathbf{C}_1(\theta_1) \leftarrow \mathbf{C}_3(\theta_3) \leftarrow \mathbf{C}_2(\theta_2)$ sequence:

$$
\begin{bmatrix} \dot{\theta}_1 \\ \dot{\theta}_2 \\ \dot{\theta}_3 \end{bmatrix} = \frac{1}{\cos \theta_3} \begin{bmatrix} \cos \theta_3 & -\cos \theta_1 \sin \theta_3 & \sin \theta_1 \sin \theta_3 \\ 0 & \cos \theta_1 & -\sin \theta_1 \\ 0 & \sin \theta_1 \cos \theta_3 & \cos \theta_1 \cos \theta_3 \end{bmatrix} \begin{bmatrix} \omega_1 \\ \omega_2 \\ \omega_3 \end{bmatrix} + \begin{bmatrix} 0 \\ n \\ 0 \end{bmatrix}
$$
(7.111)

Rigid-body dynamics:

$$
\begin{bmatrix} J_{11} & J_{12} & J_{13} \\ J_{21} & J_{22} & J_{23} \\ J_{31} & J_{32} & J_{33} \end{bmatrix} \begin{bmatrix} \dot{\omega}_1 \\ \dot{\omega}_2 \\ \dot{\omega}_3 \end{bmatrix} = - \begin{bmatrix} 0 & -\omega_3 & \omega_2 \\ \omega_3 & 0 & -\omega_1 \\ -\omega_2 & \omega_1 & 0 \end{bmatrix} \begin{bmatrix} J_{11} & J_{12} & J_{13} \\ J_{21} & J_{22} & J_{23} \\ J_{31} & J_{32} & J_{33} \end{bmatrix} \begin{bmatrix} \omega_1 \\ \omega_2 \\ \omega_3 \end{bmatrix}
$$

$$
+ 3n^2 \begin{bmatrix} 0 & -C_{13} & C_{12} \\ C_{13} & 0 & -C_{11} \\ -C_{12} & C_{11} & 0 \end{bmatrix} \begin{bmatrix} J_{11} & J_{12} & J_{13} \\ J_{21} & J_{22} & J_{23} \\ J_{31} & J_{32} & J_{33} \end{bmatrix} \begin{bmatrix} C_{11} \\ C_{12} \\ C_{13} \end{bmatrix} + \begin{bmatrix} -u_1 + d_1 \\ -u_2 + d_2 \\ -u_3 + d_3 \end{bmatrix}
$$
(7.112)

where

$$
C_{11} = -\sin \theta_2 \cos \theta_3
$$
$$
C_{12} = \cos \theta_1 \sin \theta_2 \sin \theta_3 + \sin \theta_1 \cos \theta_2
$$
$$
C_{13} = -\sin \theta_1 \sin \theta_2 \sin \theta_3 + \cos \theta_1 \cos \theta_2
$$

CMG momentum dynamics:

$$
\begin{bmatrix} \dot{h}_1 \\ \dot{h}_2 \\ \dot{h}_3 \end{bmatrix} + \begin{bmatrix} 0 & -\omega_3 & \omega_2 \\ \omega_3 & 0 & -\omega_1 \\ -\omega_2 & \omega_1 & 0 \end{bmatrix} \begin{bmatrix} h_1 \\ h_2 \\ h_3 \end{bmatrix} = \begin{bmatrix} u_1 \\ u_2 \\ u_3 \end{bmatrix}
$$
(7.113)

where subscripts 1, 2, 3 denote the roll, pitch, and yaw control axes whose origin is fixed at the mass center, with the roll axis in the flight direction, the pitch axis perpendicular to the orbit plane, and the yaw axis toward the Earth; $(\theta_1, \theta_2, \theta_3)$ are the roll, pitch, and yaw Euler angles of the body-fixed control axes with respect to the LVLH axes that rotate with the orbital angular velocity n; $(\omega_1, \omega_2, \omega_3)$ are the body-axis components of the absolute angular velocity of the vehicle; (J_{11}, J_{22}, J_{33}) are the moments of inertia; J_{ij} $(i \neq j)$ are the products of inertia; (h_1, h_2, h_3) are the body-axis components of the CMG momentum; (u_1, u_2, u_3) are the body-axis components of the control torque caused by CMG momentum change; (d_1, d_2, d_3) are the body-axis components of the external disturbance torque; and n is the orbital rate.

Without loss of generality, we consider in this section a large space vehicle shown in Fig. 7.4 with the following inertia matrix:

$$\begin{bmatrix} J_{11} & J_{12} & J_{13} \\ J_{21} & J_{22} & J_{23} \\ J_{31} & J_{32} & J_{33} \end{bmatrix} = \begin{bmatrix} 50.28 & -0.39 & 0.16 \\ -0.39 & 10.80 & 0.16 \\ 0.16 & 0.16 & 58.57 \end{bmatrix} \times 10^6 \text{ slug-ft}^2$$

The uncontrolled vehicle is in an unstable equilibrium when $\theta_1 = \theta_2 = \theta_3 = 0$. The external disturbance torque mainly consists of aerodynamic drag torque that can be modeled as bias plus cyclic terms in the body-fixed control axes; i.e., d_i in foot-pound are modeled as

$$d_1 = 1 + \sin(nt) + 0.5 \sin(2nt)$$
$$d_2 = 4 + 2 \sin(nt) + 0.5 \sin(2nt)$$
$$d_3 = 1 + \sin(nt) + 0.5 \sin(2nt)$$

where $n = 0.0011$ rad/s, and magnitudes and phases are unknown for control design. The cyclic component at orbital rate is due to the effect of the Earth's diurnal bulge, whereas the cyclic torque at twice the orbital rate is caused by the rotating solar panels.

For small attitude deviations from LVLH orientation, the linearized equations of motion can be obtained as follows.

Attitude kinematics:

$$\dot{\theta}_1 - n\theta_3 = \omega_1 \tag{7.114a}$$
$$\dot{\theta}_2 - n = \omega_2 \tag{7.114b}$$
$$\dot{\theta}_3 + n\theta_1 = \omega_3 \tag{7.114c}$$

Rigid-body dynamics:

$$\begin{bmatrix} J_{11} & J_{12} & J_{13} \\ J_{21} & J_{22} & J_{23} \\ J_{31} & J_{32} & J_{33} \end{bmatrix} \begin{bmatrix} \dot{\omega}_1 \\ \dot{\omega}_2 \\ \dot{\omega}_3 \end{bmatrix} = n \begin{bmatrix} J_{31} & 2J_{32} & J_{33} - J_{22} \\ -J_{32} & 0 & J_{12} \\ J_{22} - J_{11} & -2J_{12} & -J_{13} \end{bmatrix} \begin{bmatrix} \omega_1 \\ \omega_2 \\ \omega_3 \end{bmatrix}$$
$$+ 3n^2 \begin{bmatrix} J_{33} - J_{22} & J_{21} & 0 \\ J_{12} & J_{33} - J_{11} & 0 \\ -J_{13} & -J_{23} & 0 \end{bmatrix} \begin{bmatrix} \theta_1 \\ \theta_2 \\ \theta_3 \end{bmatrix} + n^2 \begin{bmatrix} -2J_{23} \\ 3J_{13} \\ -J_{12} \end{bmatrix} + \begin{bmatrix} -u_1 + d_1 \\ -u_2 + d_2 \\ -u_3 + d_3 \end{bmatrix}$$
$$\tag{7.115}$$

CMG momentum dynamics:

$$\dot{h}_1 - nh_3 = u_1 \tag{7.116a}$$
$$\dot{h}_2 = u_2 \tag{7.116b}$$
$$\dot{h}_3 + nh_1 = u_3 \tag{7.116c}$$

Combining Eqs. (7.114) and (7.115), we obtain the linearized equations of motion as

$$
\begin{bmatrix} J_{11} & J_{12} & J_{13} \\ J_{21} & J_{22} & J_{23} \\ J_{31} & J_{32} & J_{33} \end{bmatrix} \begin{bmatrix} \ddot{\theta}_1 \\ \ddot{\theta}_2 \\ \ddot{\theta}_3 \end{bmatrix}
$$

$$
= n \begin{bmatrix} 0 & 2J_{32} & J_{11} - J_{22} + J_{33} \\ -2J_{32} & 0 & 2J_{12} \\ -J_{11} + J_{22} - J_{33} & -2J_{12} & 0 \end{bmatrix} \begin{bmatrix} \dot{\theta}_1 \\ \dot{\theta}_2 \\ \dot{\theta}_3 \end{bmatrix}
$$

$$
+ n^2 \begin{bmatrix} 4(J_{33} - J_{22}) & 3J_{21} & -J_{31} \\ 4J_{12} & 3(J_{33} - J_{11}) & J_{32} \\ -4J_{13} & -3J_{23} & J_{11} - J_{22} \end{bmatrix} \begin{bmatrix} \theta_1 \\ \theta_2 \\ \theta_3 \end{bmatrix}
$$

$$
+ n^2 \begin{bmatrix} -4J_{23} \\ 3J_{13} \\ J_{12} \end{bmatrix} + \begin{bmatrix} -u_1 + d_1 \\ -u_2 + d_2 \\ -u_3 + d_3 \end{bmatrix} \tag{7.117}
$$

Note that the products of inertia cause three-axis coupling as well as a bias torque in each axis. Fortunately, most practical situations of interest with small products of inertia permit further simplification in such a way that pitch motion is uncoupled from roll/yaw motion. Otherwise, Eq. (7.117) should be used for three-axis coupled stability analysis and control design.

For most practical cases in which the control axes are nearly aligned with the principal axes, i.e., $J_1 = J_{11}$, $J_2 = J_{22}$, and $J_3 = J_{33}$, and attitude deviations from the desired LVLH orientation are small, we have the following set of linearized equations of motion for control design:

$$
J_1 \dot{\omega}_1 + n(J_2 - J_3)\omega_3 + 3n^2(J_2 - J_3)\theta_1 = -u_1 + d_1 \tag{7.118a}
$$

$$
J_2 \dot{\omega}_2 + 3n^2(J_1 - J_3)\theta_2 = -u_2 + d_2 \tag{7.118b}
$$

$$
J_3 \dot{\omega}_3 - n(J_2 - J_1)\omega_1 = -u_3 + d_3 \tag{7.118c}
$$

$$
\dot{\theta}_1 - n\theta_3 = \omega_1 \tag{7.118d}
$$

$$
\dot{\theta}_2 - n = \omega_2 \tag{7.118e}
$$

$$
\dot{\theta}_3 + n\theta_1 = \omega_3 \tag{7.118f}
$$

$$
\dot{h}_1 - nh_3 = u_1 \tag{7.118g}
$$

$$
\dot{h}_2 = u_2 \tag{7.118h}
$$

$$
\dot{h}_3 + nh_1 = u_3 \tag{7.118i}
$$

A set of linearized equations of motion in terms of θ_i and h_i can also be obtained as

$$
J_1 \ddot{\theta}_1 + 4n^2(J_2 - J_3)\theta_1 - n(J_1 - J_2 + J_3)\dot{\theta}_3 = -u_1 + d_1 \tag{7.119a}
$$

$$
J_2 \ddot{\theta}_2 + 3n^2(J_1 - J_3)\theta_2 = -u_2 + d_2 \tag{7.119b}
$$

$$J_3 \ddot{\theta}_3 + n^2(J_2 - J_1)\theta_3 + n(J_1 - J_2 + J_3)\dot{\theta}_1 = -u_3 + d_3 \tag{7.119c}$$

$$\dot{h}_1 - nh_3 = u_1 \tag{7.119d}$$

$$\dot{h}_2 = u_2 \tag{7.119e}$$

$$\dot{h}_3 + nh_1 = u_3 \tag{7.119f}$$

Certain assembly configurations of the large space vehicle shown in Fig. 7.4 may need a large torque equilibrium attitude (TEA) in the pitch axis because of the small gravity-gradient torque available in the pitch axis. In such cases, Eqs. (7.111) and (7.112) with small roll/yaw attitude errors and small products of inertia become

$$J_1 \ddot{\theta}_1 + \left(1 + 3\cos^2 \theta_2\right)n^2(J_2 - J_3)\theta_1 - n(J_1 - J_2 + J_3)\dot{\theta}_3$$
$$+ 3(J_2 - J_3)n^2(\sin \theta_2 \cos \theta_2)\theta_3 = -u_1 + d_1 \tag{7.120a}$$

$$J_2 \ddot{\theta}_2 + 3n^2(J_1 - J_3)\sin \theta_2 \cos \theta_2 = -u_2 + d_2 \tag{7.120b}$$

$$J_3 \ddot{\theta}_3 + \left(1 + 3\sin^2 \theta_2\right)n^2(J_2 - J_1)\theta_3 + n(J_1 - J_2 + J_3)\dot{\theta}_1$$
$$+ 3(J_2 - J_1)n^2(\sin \theta_2 \cos \theta_2)\theta_1 = -u_3 + d_3 \tag{7.120c}$$

It is evident from this set of equations that roll/yaw motion is affected by pitch motion. If pitch attitude is held constant with respect to the pitch TEA, the roll/yaw equations can be considered time invariant. These equations are also useful for roll/yaw controller design for the large space vehicle shown in Fig. 7.4 with a large pitch TEA.

7.4.3 Pitch Control Design

Equation (7.119b), which is uncoupled from the roll/yaw equations, is used as the basis for pitch control analysis and design. A block diagram of the pitch-axis momentum/attitude control system is shown in Fig. 7.5. First we consider a control design that does not include the indicated cyclic- or periodic-disturbance rejection filter $R_2(s)$ followed by a control design that includes the filter.

7.4.3.1 Pitch Control without Periodic-disturbance Rejection The pitch-axis controller consists of a single control input u_2 and four states: θ_2, $\dot{\theta}_2$, h_2, and $\int h_2$. The pitch control logic is then given by

$$u_2 = K_{2P}\theta_2 + K_{2D}\dot{\theta}_2 + K_{2H} h_2 + K_{2I} \int h_2 \tag{7.121}$$

where the pitch-axis CMG momentum and its integral are included to prevent momentum buildup.

Various methods may be employed for the selection of the four gains of Eq. (7.121). A pole placement technique may be used to place the closed-loop eigenvalues at any desired location. The practical problem with this approach is that it is not always clear where to place the eigenvalues for satisfactory performance and robustness.

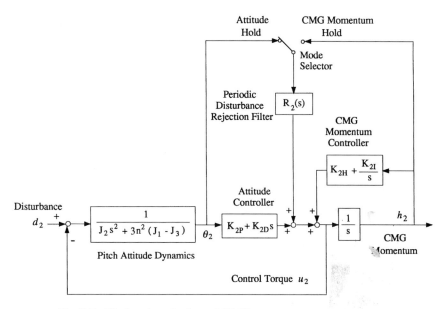

Fig. 7.5   Pitch-axis attitude and CMG momentum control system.

Pitch control design by iterative successive loop closures is possible, but quite tedious. The pitch controller can also be synthesized using an equivalent compensator. This approach provides physical insight into the proper tradeoff between the attitude control and the momentum management. Combining Eqs. (7.118h) and (7.121) to eliminate h_2 gives an equivalent compensator of the form

$$u_2 = \frac{s^2(K_{2P} + sK_{2D})}{s^2 - K_{2H}s - K_{2I}} \, \theta_2 \tag{7.122}$$

The integrated momentum/attitude controller can be interpreted as a second-order compensator with four parameters. The momentum controller consists of double zeros at the origin and complex poles. The attitude controller consists of one zero on the real axis, which is a conventional proportional-derivative controller. An unstable compensator is needed to stabilize the unstable system. It can be easily shown that $h_2(s)/d_2(s)$ of the closed-loop system has a zero at $s = 0$. Thus, $h_2(t)$ has zero steady-state value for a constant disturbance.

LQR synthesis of the pitch control with full-state feedback can be quickly accomplished using a computer code. The LQR technique makes use of a quadratic performance index to synthesize state-feedback gains that minimize the index, as was discussed in Chapter 2. It is especially useful for multivariable systems; all loops are closed simultaneously instead of successively as in classical frequency-domain methods. The problem with this approach, however, is that the proper selection of the weighting matrices is not obvious. It is not always possible to predict the effects of given weighting matrices on the closed-loop behavior.

A practical approach to the pitch-axis controller design would be to find the control gains and closed-loop poles that result from a wide range of weighting matrices, and to simulate the corresponding closed-loop system. The gain matrix that produces the closed-loop responses satisfying the various requirements would become the final selection.

7.4.3.2 Pitch Control with Periodic-disturbance Rejection

Depending on the circumstances, either pitch attitude or CMG momentum oscillation, caused by the aerodynamic disturbance torque, may be undesirable. In such cases, a periodic-disturbance rejection filter can be employed as was illustrated in Fig. 7.5.

The periodic-disturbance rejection filter is represented as

$$R_2(s) = \frac{N_2(s)}{[s^2 + n^2][s^2 + (2n)^2]} \qquad (7.123)$$

The filter poles will appear in the numerator of the closed-loop transfer function $\theta_2(s)/d_2(s)$ or $h_2(s)/d_2(s)$, depending on mode selection. This results in a disturbance rejection at frequencies of n and $2n$ for either θ_2 or h_2. The filter numerator $N_2(s)$ should be properly designed to stabilize the overall control system.

The disturbance rejection filter for θ_2 can also be represented as

$$\ddot{\alpha}_2 + (n)^2 \alpha_2 = \theta_2 \qquad (7.124a)$$
$$\ddot{\beta}_2 + (2n)^2 \beta_2 = \theta_2 \qquad (7.124b)$$

where initial conditions for α_2, β_2, $\dot{\alpha}_2$, and $\dot{\beta}_2$ can be arbitrarily selected (usually zero initial conditions). These filter equations are integrated in the onboard computer. Each filter equation passes the filter states that carry undesired frequency components of θ_2. For example, α_2 and $\dot{\alpha}_2$ carry the frequency component of θ_2 at the orbital frequency n. The undesired component is rejected by feeding back the filter states.

The pitch control logic, with additional disturbance rejection filter states, can then be expressed as

$$u_2 = K_{2P}\theta_2 + K_{2D}\dot{\theta}_2 + K_{2H}h_2 + K_{2I}\int h_2$$
$$+ K_{2\alpha}\alpha_2 + K_{2\dot{\alpha}}\dot{\alpha}_2 + K_{2\beta}\beta_2 + K_{2\dot{\beta}}\dot{\beta}_2 \qquad (7.125)$$

It is also possible to provide periodic-disturbance rejection for the pitch-axis CMG momentum. For this case, θ_2 of Eqs. (7.124) is replaced by h_2. Proper gains of Eq. (7.125) are then selected. Depending on the specific mission requirements, either pitch attitude or CMG momentum can be held constant by employing the pitch controller illustrated in Fig. 7.5.

Problems

7.18 Consider the pitch-axis control design of the large space vehicle shown in Fig. 7.4.

Table 7.1   Closed-loop eigenvalues in units of orbital rate, $n = 0.0011$ rad/s

	Momentum/attitude		Disturbance filter	
Pitch	$-1.0, -1.5$	$-1.5 \pm 1.5j$	$-0.3 \pm 1.0j$	$-0.3 \pm 2.0j$
Roll/yaw	$-0.23, -0.68$	$-0.66 \pm 1.51j$	$-0.23 \pm 0.92j$	$-0.20 \pm 2.02j$
	$-1.02 \pm 0.29j$	$-1.50 \pm 0.84j$	$-0.26 \pm 1.04j$	$-0.62 \pm 2.29j$

(a) Given the desired closed-loop eigenvalues $-1.5n$, $-1.0n$, and $(-1.5 \pm 1.5j)n$, determine a set of four gains of the pitch-axis control logic (7.121).

   Note: If an LQR code, instead of the pole-placement technique, is to be employed, then determine a set of four gains for a control bandwidth of approximately $1.5n$.

(b) Obtain the closed-loop, frequency-response magnitude plots of

$$\frac{\theta_2(s)}{d_2(s)} \quad \text{and} \quad \frac{h_2(s)}{d_2(s)}$$

(c) Perform computer simulation of the closed-loop system (only pitch axis) subject to the aerodynamic disturbance d_2 and initial conditions $\theta_2(0) = 1$ deg and $\dot{\theta}_2(0) = 0.001$ deg/s.

   Note: The periodic aerodynamic torque will cause the periodic responses of both pitch attitude and pitch-axis CMG momentum. The CMG momentum h_2 will be bounded with zero mean value, whereas θ_2 is oscillating with respect to a -7.5- deg pitch TEA.

7.19   Consider the periodic-disturbance rejection control design for the pitch-axis of the large space vehicle shown in Fig. 7.4.

(a) Determine a set of eight gains of the pitch-axis control logic (7.125) to achieve the desired closed-loop eigenvalues listed in Table 7.1.

   Note: If an LQR code, instead of the pole-placement technique, is to be employed, then design a controller for a control bandwidth of approximately $1.5n$ and for an asymptotic disturbance rejection within three orbits.

(b) Obtain the closed-loop, frequency-response magnitude plots of

$$\frac{\theta_2(s)}{d_2(s)} \quad \text{and} \quad \frac{h_2(s)}{d_2(s)}$$

(c) Perform computer simulation of the closed-loop system with the pitch-axis aerodynamic disturbance and initial conditions of $\theta_2(0) = 1$ deg and $\dot{\theta}_2(0) = 0.001$ deg/s.

   Note: Because a disturbance rejection filter is used here for pitch attitude, an asymptotic disturbance rejection of a periodic disturbance of frequencies n and $2n$ will be achieved for the pitch attitude. The CMG momentum will be bounded with zero mean value, while the pitch attitude is held constant at a -7.5-deg pitch TEA after approximately two orbits.

7.4.4 Roll/Yaw Control Design

Similar to the pitch-axis control design, we consider here roll/yaw control design. Two cases are considered: the first without periodic-disturbance rejection filters and the second with the filters.

7.4.4.1 Roll/yaw Control without Periodic-disturbance Rejection
The first case of roll/yaw control design consists of two inputs u_1 and u_3, and eight states, including two integral states for the CMG momentum. The eight states are the roll-axis θ_1, ω_1, h_1, and $\int h_1$; and the yaw-axis θ_3, ω_3, h_3, and $\int h_3$. Note that body rates ω_1 and ω_3, instead of $\dot{\theta}_1$ and $\dot{\theta}_3$, are used as state variables.

The roll/yaw full-state feedback controller can be designed, iteratively, using the linear quadratic regulator (LQR) or pole-placement techniques. The multi-input characteristics of the roll/yaw axes, however, provide for the calculation of various gain matrices that yield the same closed-loop eigenvalues. Although the gain matrix can be completely specified by assigning not only the closed-loop eigenvalues but also an allowable set of closed-loop eigenvectors, the conventional LQR technique can be employed to find a closed-loop system satisfying various requirements.

7.4.4.2 Roll/yaw Control with Periodic-disturbance Rejection
For the reduction of the steady-state oscillation of roll/yaw attitude and CMG momentum, periodic-disturbance rejection filtering for a multivariable system is described here.

Combining Eqs. (7.119a), (7.119c), (7.119d), and (7.119f), we obtain the transfer function matrix description of the coupled roll/yaw dynamics as

$$\begin{bmatrix} \theta_1(s) \\ \theta_3(s) \end{bmatrix} = \frac{1}{\Delta} \begin{bmatrix} G_{11} & G_{13} \\ G_{31} & G_{33} \end{bmatrix} \begin{bmatrix} u_1(s) - d_1(s) \\ u_3(s) - d_3(s) \end{bmatrix} \tag{7.126}$$

$$\begin{bmatrix} h_1(s) \\ h_3(s) \end{bmatrix} = \frac{1}{s^2 + n^2} \begin{bmatrix} s, & n \\ -n, & s \end{bmatrix} \begin{bmatrix} u_1(s) \\ u_3(s) \end{bmatrix} \tag{7.127}$$

where

$$\Delta = J_1 J_3 (s^2 + n^2)\left[s^4 + (1 + 3k_1 + k_1 k_3)n^2 s^2 + 4k_1 k_3 n^4 \right]$$

$$G_{11} = -\left[J_3 s^2 + (J_2 - J_1)n^2 \right](s^2 + n^2)$$

$$G_{13} = -(J_1 - J_2 + J_3)ns(s^2 + n^2) = -G_{31}$$

$$G_{33} = -\left[J_1 s^2 + 4(J_2 - J_3)n^2 \right](s^2 + n^2)$$

$$k_1 = (J_2 - J_3)/J_1 \qquad \text{and} \qquad k_3 = (J_2 - J_1)/J_3$$

It is apparent that $\pm nj$ are transmission zeros of Eq. (7.126). In other words, a periodic disturbance at orbital rate cannot be rejected for both roll and yaw attitude. In addition, it can be seen in Eq. (7.127) that resonance of CMG momentum can happen for sinusoidal control inputs of frequency n. For this case, where the CMG momentum dynamics are described by Eq. (7.116), it is not evident why a periodic disturbance at orbital rate can be rejected for the yaw attitude and not for the roll attitude.

To investigate such inherent characteristics of the coupled roll/yaw dynamics, the following CMG momentum dynamics, with a proper modification of the equations of motion of the vehicle, are considered:

$$\dot{h}_1 = u_1 \tag{7.128a}$$

$$\dot{h}_3 = u_3 \tag{7.128b}$$

For this case, the following transfer function matrix description can be obtained as:

$$\begin{bmatrix} \theta_1(s) \\ \theta_3(s) \end{bmatrix} = \frac{1}{\Delta} \begin{bmatrix} G_{11} & G_{13} \\ G_{31} & G_{33} \end{bmatrix} \begin{bmatrix} u_1(s) \\ u_3(s) \end{bmatrix}$$

$$+ \frac{s^2}{\Delta} \begin{bmatrix} J_3 s^2 + (J_2 - J_1)n^2 & (J_1 - J_2 + J_3)ns \\ -(J_1 - J_2 + J_3)ns & J_1 s^2 + 4(J_2 - J_3)n^2 \end{bmatrix} \begin{bmatrix} d_1(s) \\ d_3(s) \end{bmatrix} \tag{7.129}$$

$$\begin{bmatrix} h_1(s) \\ h_3(s) \end{bmatrix} = \frac{1}{s} \begin{bmatrix} 1 & 0 \\ 0 & 1 \end{bmatrix} \begin{bmatrix} u_1(s) \\ u_3(s) \end{bmatrix} \tag{7.130}$$

where

$$\Delta = J_1 J_3 s^2 \left[s^4 + (1 + 3k_1 + k_1 k_3)n^2 s^2 + 4k_1 k_3 n^4 \right]$$

$$G_{11} = -J_3 s^2 (s^2 + n^2)$$

$$G_{13} = (J_2 - J_1)ns(s^2 + n^2)$$

$$G_{31} = (J_3 - J_2)ns[s^2 + (2n)^2]$$

$$G_{33} = -s^2 \left[J_1 s^2 + (J_1 + 3J_2 - 3J_3)n^2 \right]$$

$$k_1 = (J_2 - J_3)/J_1 \quad \text{and} \quad k_3 = (J_2 - J_1)/J_3$$

It can be shown that $\pm nj$ are transmission zeros of the transfer function matrix from (u_1, u_3) to (θ_1, θ_3). The zero at $s = \pm nj$ appears in both the $\theta_1(s)/u_1(s)$ and $\theta_1(s)/u_3(s)$ transfer functions, but not in yaw attitude channels. Therefore, a periodic-disturbance rejection at the orbital rate is not possible for roll attitude, whereas it is possible for yaw attitude. Furthermore, it can be shown that $\pm nj$ are not transmission zeros of the transfer function matrix from (u_1, u_3) to (h_1, θ_3). Consequently, a periodic-disturbance rejection for the roll-axis CMG momentum and yaw attitude can be achieved by employing roll/yaw periodic-disturbance rejection filters of the form

$$\ddot{\alpha}_1 + (n)^2\alpha_1 = h_1 \qquad (7.131a)$$

$$\ddot{\beta}_1 + (2n)^2\beta_1 = h_1 \qquad (7.131b)$$

$$\ddot{\alpha}_3 + (n)^2\alpha_3 = \theta_3 \qquad (7.131c)$$

$$\ddot{\beta}_3 + (2n)^2\beta_3 = \theta_3 \qquad (7.131d)$$

Problems

7.20 Consider the large space vehicle shown in Fig. 7.4, which is unstable in roll/yaw and has open-loop poles of $(\pm 1.05 \pm 0.7j)n$, $\pm nj$. The roll/yaw control logic is described by

$$\begin{bmatrix} u_1 \\ u_3 \end{bmatrix} = \begin{bmatrix} \mathbf{K}_{11} & \mathbf{K}_{13} \\ \mathbf{K}_{31} & \mathbf{K}_{33} \end{bmatrix} \begin{bmatrix} \mathbf{x}_1 \\ \mathbf{x}_3 \end{bmatrix}$$

where $\mathbf{K}_{ij}$ are 1×4 gain matrices and

$$\mathbf{x}_1 = \begin{bmatrix} \theta_1 & \omega_1 & h_1 & \int h_1 \end{bmatrix}^T$$

$$\mathbf{x}_3 = \begin{bmatrix} \theta_3 & \omega_3 & h_3 & \int h_3 \end{bmatrix}^T$$

(a) Determine a 2×8 gain matrix of the roll/yaw control logic for the following desired closed-loop poles: $(-1.05 \pm 0.68j)n$, $(-1.04 \pm 0.72j)n$, $(-1.42 \pm 1.38j)n$, and $(-1.42 \pm 1.38j)n$.
   Note: If an LQR code, instead of the eigenstructure assignment technique, is to be employed, then design a controller for a control bandwidth of approximately $1.5n$.

(b) Obtain the closed-loop, frequency-response magnitude plots of

$$\frac{\theta_1(s)}{d_1(s)}, \qquad \frac{h_1(s)}{d_1(s)}, \qquad \frac{\theta_3(s)}{d_3(s)}, \qquad \text{and} \qquad \frac{h_3(s)}{d_3(s)}$$

(c) Perform computer simulation of the closed-loop system subject to roll/yaw aerodynamic disturbances and initial conditions $\theta_1(0) = \theta_3(0) = 1$ deg and $\omega_1(0) = \omega_3(0) = 0.001$ deg/s.

7.21 Consider the large space vehicle of the preceding problem but with the roll/yaw control logic, employing the periodic-disturbance filters described by Eqs. (7.131), of the form

$$\begin{bmatrix} u_1 \\ u_3 \end{bmatrix} = \begin{bmatrix} \mathbf{K}_{11} & \mathbf{K}_{13} \\ \mathbf{K}_{31} & \mathbf{K}_{33} \end{bmatrix} \begin{bmatrix} \mathbf{x}_1 \\ \mathbf{x}_3 \end{bmatrix}$$

where $\mathbf{K}_{ij}$ are 1×8 gain matrices and

$$\mathbf{x}_1 = \begin{bmatrix} \theta_1 & \omega_1 & h_1 & \int h_1 & \alpha_1 & \dot{\alpha}_1 & \beta_1 & \dot{\beta}_1 \end{bmatrix}^T$$

$$\mathbf{x}_3 = \begin{bmatrix} \theta_3 & \omega_3 & h_3 & \int h_3 & \alpha_3 & \dot{\alpha}_3 & \beta_3 & \dot{\beta}_3 \end{bmatrix}^T$$

(a) Determine a 2×16 gain matrix for the desired closed-loop eigenvalues as listed in Table 7.1.

Note: If an LQR code, instead of the eigenstructure assignment technique, is to be employed, then design a controller for a control bandwidth of approximately $1.5n$ and for an asymptotic disturbance rejection within three orbits.

(b) Obtain the closed-loop, frequency-response magnitude plots of

$$\frac{\theta_1(s)}{d_1(s)}, \quad \frac{h_1(s)}{d_1(s)}, \quad \frac{\theta_3(s)}{d_3(s)}, \quad \text{and} \quad \frac{h_3(s)}{d_3(s)}$$

(c) Show that $\pm nj$ are "blocking zeros" of the closed-loop transfer function matrix from (d_1, d_3) to (h_1, θ_3).

(d) Perform computer simulation of the closed-loop system subject to roll/yaw aerodynamic disturbances and initial conditions $\theta_1(0) = \theta_3(0) = 1$ deg and $\omega_1(0) = \omega_3(0) = 0.001$ deg/s.

Note: The roll-axis CMG momentum will approach zero steady-state value, whereas roll attitude oscillates at the orbital rate. The yaw attitude will approach a constant steady-state value, whereas the yaw-axis CMG momentum oscillates at the orbital rate. As a result, the overall attitude and CMG momentum oscillations are minimized.

7.22 Consider the large space vehicle of the preceding problems but with the roll/pitch/yaw control logic employing the periodic-disturbance rejection filters of the form

$$\ddot{\alpha}_1 + (n)^2 \alpha_1 = h_1$$
$$\ddot{\beta}_1 + (2n)^2 \beta_1 = h_1$$
$$\ddot{\alpha}_2 + (n)^2 \alpha_2 = h_2$$
$$\ddot{\beta}_2 + (2n)^2 \beta_2 = h_2$$
$$\ddot{\alpha}_3 + (n)^2 \alpha_3 = h_3$$
$$\ddot{\beta}_3 + (2n)^2 \beta_3 = h_3$$

(a) Design such a controller that will provide an asymptotic disturbance rejection for CMGs.

(b) Perform computer simulation of the closed-loop system subject to roll/pitch/yaw aerodynamic disturbances and initial conditions $\theta_1(0) = \theta_2(0) = \theta_3(0) = 1$ deg, $\omega_1(0) = \omega_3(0) = 0.001$ deg/s, and $\dot{\theta}_2(0) = 0.001$ deg/s.

7.4.5 *Robust Control Design*

We now consider a robust control design problem of the large space vehicle shown in Fig. 7.4. The objective is to design a constant-gain controller that may yield the largest stable hypercube in uncertain parameter space, subject to the nominal performance requirements. However, the robust control design does not deal with the problem of controlling the large space vehicle in the presence of significant changes of inertias during the assembly sequence and space shuttle docking.

7.4.5.1 Robust H_∞ Control Design

As discussed in Chapter 2, structured uncertainty modeling of dynamic systems with uncertain inertia matrices results in an uncertain plant model with $\mathbf{D}_{11} \neq 0$, where $\mathbf{D}_{11}$ is the matrix that relates the disturbance input and the output. In Ref. 19, the robust stabilization of the large space vehicle in the face of inertia matrix uncertainty is formulated as a robust H_∞ full-state feedback control problem with $\mathbf{D}_{11} \neq 0$, and the significance of employing an uncertain model with the nonzero $\mathbf{D}_{11}$ term is demonstrated.

The nominal closed-loop eigenvalues of the pitch axis with this new control design are listed in Table 7.2. These closed-loop poles are comparable to those of previous designs, as can be seen in Table 7.3 in which the various previous designs [18–22] are compared in terms of their fastest closed-loop poles. Note that a typical H_∞ control design often achieves the desired robustness by having a high bandwidth controller. The new design with the consideration of the nonzero $\mathbf{D}_{11}$ term, however, has a remarkable stability robustness margin with nearly the same bandwidth as the conventional LQR design, as will be discussed next.

Table 7.2   Closed-loop eigenvalues in units of orbital rate

	Momentum and attitude control	Disturbance rejection
Pitch	$-1.27, -2.43$	$-0.30 \pm 1.40j$
	$-0.23 \pm 0.25j$	$-1.40 \pm 2.65j$
Roll/yaw	$-0.65, -1.77$	$-0.10 \pm 1.00j$
	$-0.44 \pm 0.05j$	$-0.30 \pm 1.09j$
	$-0.10 \pm 1.10j$	$-0.17 \pm 2.00j$
	$-1.26 \pm 0.97j$	$-0.72 \pm 2.38j$

Table 7.3   Comparison of fastest closed-loop poles

Control designs	Fastest poles
No.1 Wie et al. [18]	-1.50
No.2 Byun et al. [20]	-8.29
No.3 Rhee and Speyer [21]	-4.77
No.4 Balas et al. [22]	-5.43
No.5 Wie et al. [19]	-2.43

7.4.5.2 Stability Robustness Analysis Consider the effects of the moments-of-inertia variations on the closed-loop stability of the large space vehicle.

From the definition of the moments of inertia, we have the following physical constraints for possible inertia variations in the three-dimensional parameter space (J_1, J_2, J_3):

$$J_1 + J_2 > J_3, \qquad J_1 + J_3 > J_2, \qquad J_2 + J_3 > J_1 \qquad (7.132)$$

A control designer may unknowingly consider inertia variations that result in inertia values that violate these physical constraints. In such a case, stability of the closed-loop control system is being tested for physically impossible inertia values.

When gravity-gradient torque is used in the control of an orbiting spacecraft, additional inertia constraints are also required as follows:

$$J_1 \neq J_2, \qquad J_1 \neq J_3, \qquad J_2 \neq J_3 \qquad (7.133)$$

Roll-axis gravity-gradient and gyroscopic coupling torques become zero if $J_2 = J_3$, pitch-axis gravity-gradient torque becomes zero if $J_1 = J_3$, and yaw-axis gyroscopic coupling torque becomes zero if $J_1 = J_2$. For the large space vehicle considered in this section, the physical constraints given by Eqs. (7.132) and (7.133) can be combined into the following constraints:

$$J_1 + J_2 > J_3, \qquad J_1 \neq J_2, \qquad J_1 \neq J_3 \qquad (7.134)$$

Table 7.4 summarizes the physical inertia bounds along the various directions of inertia variation for the particular configuration with $J_1 = 50.28 \times 10^6$, $J_2 = 10.80 \times 10^6$, and $J_3 = 58.57 \times 10^6$ slug-ft [2]. In Table 7.4, δ represents the amounts of directional parameter variations with respect to the nominal inertias J_1°, J_2°, and J_3°. It is evident in Table 7.4 that there exist physical bounds for δ due to the inherent physical properties of the gravity-gradient stabilization and the moments of inertia itself. In particular, the Δ_1-inertia variation is physically caused by the translational motion of a large payload along the pitch axis.

Table 7.4   Physical bounds for inertia variations

Variation type $\Delta_i = [\Delta J_1 \ \Delta J_2 \ \Delta J_3]$	Lower bound $\underline{\delta}, \%$	Upper bound $\bar{\delta}, \%$
$\Delta_1 = \delta[J_1^\circ \ 0 \ J_1^\circ]$	-78.5^a	∞
$\Delta_2 = \delta[J_1^\circ \ J_2^\circ \ J_3^\circ]$	-100.0^b	∞
$\Delta_3 = \delta[J_1^\circ \ 0 \ -J_3^\circ]$	-2.3^c	$+7.6^a$
$\Delta_4 = \delta[J_1^\circ \ -J_2^\circ \ 0]$	-6.4^c	$+16.4^a$
$\Delta_5 = \delta[J_1^\circ \ J_2^\circ \ -J_3^\circ]$	-2.1^c	$+7.6^a$

[a] Due to roll/yaw open-loop characteristic.
[b] Due to pitch open-loop characteristic.
[c] Due to triangle inequalities for the moments of inertia.

In Tables 7.5 and 7.6, stability margins of a new design (No. 5) presented in Ref. 19 are compared to those of the previous designs (Nos. 1 and 2), and one can notice that the new design has better stability margins for all Δ_i-inertia variations. A significant margin of 77% for the Δ_2-inertia variation was achieved for the pitch axis, compared to the 34% margin of the LQR design (No. 1) in Ref. 18. A significant margin of 77% for the Δ_2-inertia variation was also achieved for the roll/yaw axes, compared to the 43% margin of the standard LQR design.

Compared to the LQR design in Ref. 18, the overall stability robustness with respect to inertia variations has been significantly improved while meeting the nominal performance requirements. That is, the method with the consideration of the nonzero $\mathbf{D}_{11}$ term has resulted in a remarkable stability robustness margin with nearly the same bandwidth as the conventional LQR design.

A hypercube in the space of the plant parameters, centered at a nominal point, is often used as a stability robustness measure in the presence of parametric uncertainty, as discussed in Chapter 2. To determine the largest hypercube that will fit within the existing, but unknown, region of closed-loop stability in the plant's parameter space, consider the open-loop characteristic equations for the pitch axis:

$$J_2 s^2 + 3(J_1 - J_3) = 0 \qquad (7.135)$$

Table 7.5 Pitch-axis stability robustness comparison

Percent	No. 1		No. 2		No. 5	
	$\underline{\delta}$	$\bar{\delta}$	$\underline{\delta}$	$\bar{\delta}$	$\underline{\delta}$	$\bar{\delta}$
Δ_1	−99	∞	−99	∞	−99	∞
Δ_2	−89	34	−99	70	−99	77
Δ_3	−17	7.6	−27	7.6	−40	7.6
Δ_4	−19	16	−40	16	−45	16
Δ_5	−30	7.6	−31	7.6	−44	7.6

Table 7.6 Roll/yaw stability robustness comparison

Percent	No. 1		No. 2		No. 5	
	$\underline{\delta}$	$\bar{\delta}$	$\underline{\delta}$	$\bar{\delta}$	$\underline{\delta}$	$\bar{\delta}$
Δ_1	−78	44	−78	73	−78	76
Δ_2	−99	43	−99	71	−98	77
Δ_3	−61	80	−58	77	−79	79
Δ_4	−64	64	−64	99	−64	99
Δ_5	−51	68	−49	66	−74	67

and for the roll/yaw axes:

$$J_1 J_3 s^4 + \left(-J_1 J_2 + 2J_1 J_3 + J_2^2 + 2J_2 J_3 - 3J_3^2\right)s^2$$
$$+ 4\left(-J_1 J_2 + J_1 J_3 + J_2^2 - J_2 J_3\right) = 0 \qquad (7.136)$$

Equation (7.135) represents a characteristic equation of a conservative plant with multilinearly uncertain parameters. For such a system, the ∞-norm real parameter margin of the closed-loop system can be found simply by checking for instability in the corner directions of the parameter space hypercube, at a finite number of critical frequencies. The ∞-norm parameter margin for the pitch-axis controller can be found as 0.076 at a critical corner with

$$(\delta_1, \delta_2, \delta_3) = (0.076, 0.076, -0.076)$$

This corresponds to a critical corner with zero pitch-axis gravity-gradient control torque. Notice that this largest stable hypercube includes physically impossible inertia variations, however, and that it is too conservative because the controller has a 77% margin for the most physically possible Δ_2-inertia variation. The important point is that the control designer should consider only inertia variations that do not violate the physical constraints.

The roll/yaw characteristic equation (7.136) represents a plant that is conservative, but not multilinear with respect to the uncertain parameters (J_1, J_2, J_3). Because there is no guarantee that the roll/yaw closed-loop instability with respect to inertia variations occurs at one of the corners of the parameter space hypercube, the real parameter margin computation is not as simple as the case of pitch axis. The inertia boundaries may be computed by iteratively varying the inertia values until a physical bound is reached or the closed-loop system becomes unstable.

7.5   Steering Logic for Control Moment Gyros

In the preceding section, CMGs, also called control moment gyroscopes, were considered as ideal torque-generating actuators; i.e., CMG gimbal torquer dynamics and CMG steering logic were not considered in developing an attitude control and momentum management system of a large space vehicle, such as the International Space Station. Because CMG steering logic is, in fact, one of the most critical components of any CMG-based attitude control system of space vehicles, an overview of the CMG steering logic design problem is presented in this section.

Any reader who wishes to pursue further research in the development of advanced CMG control systems for space vehicles is referred to Refs. 23–51. No attempt is made to provide a complete bibliography for the CMGs; however, these references cover a wide variety of dynamics and control problems of CMG-equipped space vehicles, including agile spacecraft, the MIR space station and the International Space Station. A more comprehensive treatment of the CMG singularity problem is presented in Chapter 11.

7.5.1   Introduction

During the past three decades, CMGs as applied to spacecraft attitude control and momentum management have been studied extensively, and they have been successfully employed for a wide variety of space missions. However, there still exist various practical as well as theoretical issues inherent to CMGs. They include: system-level tradeoffs, e.g., reaction wheels vs CMGs, single-gimbal vs double-gimbal, etc.; optimal arrangements of CMG arrays, e.g., parallel vs skewed or orthogonal mounting, etc.; optimal CMG steering logic design, e.g., local vs global methods for singularity avoidance; and computational issues for real-time implementation.

7.5.1.1 Reaction Wheels vs Control Moment Gyros
A reaction wheel consists of a spinning rotor whose spin rate is nominally zero. Its spin axis is fixed to the spacecraft and its speed is increased or decreased to generate reaction torque about the spin axis. Reaction wheels are conventionally used to control three-axis stabilized spacecraft and smaller satellites. They are the simplest and least expensive of all momentum-exchange actuators; however, they have much smaller control torque capability than CMGs.

A CMG contains a spinning rotor with large, constant angular momentum, but whose angular momentum vector (direction) can be changed with respect to the spacecraft by gimbaling the spinning rotor. The spinning rotor is mounted on a gimbal (or a set of gimbals), and torquing the gimbal results in a precessional, gyroscopic reaction torque orthogonal to both the rotor spin and gimbal axes. The CMG is a torque amplification device because a small gimbal torque input produces a large control torque output on the spacecraft. Because the CMGs are capable of generating large control torques and storing large angular momentum over long periods of time, they are often favored for precision pointing and tracking control of agile spacecraft in low Earth orbit and momentum management of large spacecraft.

7.5.1.2 Single-gimbal vs Double-gimbal CMGs
In general, CMGs are characterized by their gimbaling arrangements and their configurations for the redundancy management and failure accommodation.

There are two basic types of control moment gyros: 1) single-gimbal control moment gyros (SGCMGs) and 2) double-gimbal control moment gyros (DGCMGs). For SGCMGs, the spinning rotor is constrained to rotate on a circle in a plane normal to the gimbal axis. For DGCMGs, the rotor is suspended inside two gimbals, and consequently the rotor momentum can be oriented on a sphere along any direction provided there are no restrictive gimbal stops. The SGCMGs are considerably simpler than DGCMGs from the hardware viewpoint. They offer significant cost, power, weight, and reliability advantages over DGCMGs. However, the gimbal steering problem is much simpler for DGCMGs because of the extra degree of freedom per device.

For the purposes of optimal redundancy management and failure accommodation, many different arrangements of CMGs have been developed in the past. They include: four SGCMGs of pyramid configuration, six parallel mounted SGCMGs, three orthogonally mounted DGCMGs, and four parallel mounted DGCMGs. The three orthogonally mounted DGCMGs were used in NASA's Skylab [24,29] and six parallel mounted SGCMGs have been successfully installed on the MIR space station. The International Space Station will be controlled by four parallel mounted DGCMGs with two of the four CMGs mounted antiparallel with the other two [48]. Control moment gyros have never been used in commercial satellites, although SGCMGs have been employed in agile spacecraft with classified missions. Recently, smaller SGCMGs have been developed by Space Systems, Honeywell, Inc., for small space vehicles where more agility is required than reaction wheels can provide. For such mini-CMGs, a high-speed rotor design is employed to achieve low weight and substantial output torque.

The use of CMGs necessitates the development of CMG steering logic, which generates the CMG gimbal rate commands for the commanded spacecraft control torques. The optimal steering logic is one for which the CMG-generated torques are equal to the commanded spacecraft control torques. One of the principal difficulties in using CMGs for spacecraft attitude control is the geometric singularity problem in which no control torque is generated for the commanded gimbal rates. Therefore, the development of an optimal CMG steering logic should also consider the avoidance of the singularities. SGCMG systems are more prone to lock up in singular configurations because of the reduced degrees of freedom; however, reaction wheel systems do not have such a geometric singularity problem inherent to any CMG system.

7.5.2 Mathematical Modeling of Spacecraft with Control Moment Gyros

The fundamental principles of control moment gyros are briefly described here, as applied to the attitude control of a rigid spacecraft. The goal here is to present a simple mathematical model for developing CMG steering and attitude control logic.

The rotational equation of motion of a rigid spacecraft equipped with momentum-exchange actuators such as control moment gyros is, in general, given by

$$\dot{\mathbf{H}}_s + \boldsymbol{\omega} \times \mathbf{H}_s = \mathbf{T}_{\text{ext}} \tag{7.137}$$

where $\mathbf{H}_s$ is the angular momentum vector of the total system expressed in the spacecraft body-fixed control axes and $\mathbf{T}_{\text{ext}}$ is the external torque vector, including the gravity-gradient, solar pressure and aerodynamic torques, expressed in the same body axes. The total angular momentum vector consists of the spacecraft main body angular momentum and the CMG angular momentum; i.e., we have

$$\mathbf{H}_s = \mathbf{J}\boldsymbol{\omega} + \mathbf{h} \tag{7.138}$$

where $\mathbf{J}$ is the inertia matrix of the spacecraft including CMGs, $\boldsymbol{\omega} = (\omega_1, \omega_2, \omega_3)$ is the spacecraft angular velocity vector, and $\mathbf{h} = (h_1, h_2, h_3)$ is the total CMG momentum vector, all expressed in the spacecraft body-fixed control axes.

Combining Eq. (7.137) and Eq. (7.138), we simply obtain

$$(\mathbf{J}\dot{\boldsymbol{\omega}} + \dot{\mathbf{h}}) + \boldsymbol{\omega} \times (\mathbf{J}\boldsymbol{\omega} + \mathbf{h}) = \mathbf{T}_{\text{ext}} \tag{7.139}$$

Furthermore, by introducing the internal control torque vector generated by CMGs, denoted as $\mathbf{u} = (u_1, u_2, u_3)$, we rewrite Eq. (7.139) as

$$\mathbf{J}\dot{\boldsymbol{\omega}} + \boldsymbol{\omega} \times \mathbf{J}\boldsymbol{\omega} = \mathbf{u} + \mathbf{T}_{\text{ext}} \qquad (7.140a)$$

$$\dot{\mathbf{h}} + \boldsymbol{\omega} \times \mathbf{h} = -\mathbf{u} \qquad (7.140b)$$

Using this set of equations of motion, with an additional set of kinematic differential equations of spacecraft attitude variables such as quaternions or Euler angles, one can design an attitude control and CMG momentum management system, as was described in the preceding section. Consequently, the spacecraft control torque input $\mathbf{u}$ can be assumed to be known for the subsequent steering logic design and the desired CMG momentum rate is often selected as

$$\dot{\mathbf{h}} = -\mathbf{u} - \boldsymbol{\omega} \times \mathbf{h} \qquad (7.141)$$

The CMG angular momentum vector $\mathbf{h}$ is in general a function of CMG gimbal angles $\boldsymbol{\delta} = (\delta_1, \ldots, \delta_n)$; i.e., we have

$$\mathbf{h} = \mathbf{h}(\boldsymbol{\delta}) \qquad (7.142)$$

One approach to the CMG steering logic design is simply to find an inversion of Eq. (7.142). In this inverse kinematic problem, the task is to determine optimal gimbal angle trajectories that generate the commanded $\mathbf{h}$ trajectory while meeting the various hardware constraints, such as the gimbal rate limits and gimbal stops, and also avoiding singularities.

The second approach involves the differential relationship between gimbal angles and the CMG momentum vector. For such local inversion or tangent methods, the time derivative of $\mathbf{h}$ is obtained as

$$\dot{\mathbf{h}} = \mathbf{A}(\boldsymbol{\delta})\,\dot{\boldsymbol{\delta}} \qquad (7.143)$$

where

$$\mathbf{A} = \frac{\partial \mathbf{h}}{\partial \boldsymbol{\delta}} \equiv \left[\frac{\partial h_i}{\partial \delta_j} \right]$$

is the $3 \times n$ Jacobian matrix. Equation (7.143) can also be written as

$$\dot{\mathbf{h}} = \mathbf{A}\dot{\boldsymbol{\delta}} = [\mathbf{a}_1 \quad \cdots \quad \mathbf{a}_n] \begin{bmatrix} \dot{\delta}_1 \\ \vdots \\ \dot{\delta}_n \end{bmatrix}$$

$$= \sum_{i=1}^{n} \mathbf{a}_i \dot{\delta}_i \qquad (7.144)$$

where $\mathbf{a}_i$ denotes the ith column of the Jacobian matrix $\mathbf{A}$. The CMG steering logic design is then to find an inversion of $\mathbf{A}\dot{\boldsymbol{\delta}} = \dot{\mathbf{h}}$, i.e., to determine the gimbal rates that can generate the commanded $\dot{\mathbf{h}}$ while meeting the various hardware constraints, such as the gimbal rate limits and gimbal stops, and also avoiding singularities.

Note that in this CMG steering problem formulation, the gimbal torquer dynamics has been ignored because the gimbal torque input is often much smaller than the control torque output generated by CMGs.

7.5.3 Steering Logic for Single-Gimbal Control Moment Gyros

7.5.3.1 SGCMG Arrays
Consider a typical pyramid mounting arrangement of four SGCMGs as shown in Fig. 7.6, in which four CMGs are constrained to gimbal on the faces of a pyramid and the gimbal axes are orthogonal to the pyramid faces. Each face is inclined with a skew angle of β from the horizontal, resulting in gimbal axes with a $(90 - \beta)$-deg inclination from the horizontal. When each CMG has the same angular momentum about its spin-rotor axis and the skew angle is chosen as $\beta = 54.73$ deg, the momentum envelope becomes nearly spherical. This minimally redundant, four CMG configuration with $\beta = 54.73$ deg has been extensively studied in the literature (see Refs. 38–41, 43–47, 49, and 50), and it presents a significant challenge for developing singularity-robust steering laws.

From a momentum storage point of view, however, the optimal skew angle has been found to be $\beta = 90$ deg, which results in a box configuration of four CMGs [51]. Mounting arrangements of SGCMGs, other than the pyramid mount, are also possible. For example, the six parallel mounted SGCMGs have been successfully used to control the MIR space station.

7.5.3.2 Pseudoinverse Steering Logic
For the conventional pyramid mount of four SGCMGs, the total CMG angular momentum vector $\mathbf{h} = (h_1, h_2, h_3)$

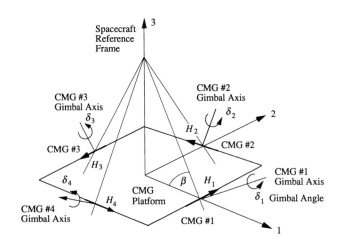

Fig. 7.6   Pyramid mounting arrangement of four SGCMGs.

is expressed in spacecraft reference frame as

$$\mathbf{h} = \sum_{i=1}^{4} \mathbf{H}_i(\delta_i)$$

$$= \begin{bmatrix} -c\beta \sin \delta_1 \\ \cos \delta_1 \\ s\beta \sin \delta_1 \end{bmatrix} + \begin{bmatrix} -\cos \delta_2 \\ -c\beta \sin \delta_2 \\ s\beta \sin \delta_2 \end{bmatrix} + \begin{bmatrix} c\beta \sin \delta_3 \\ -\cos \delta_3 \\ s\beta \sin \delta_3 \end{bmatrix} + \begin{bmatrix} \cos \delta_4 \\ c\beta \sin \delta_4 \\ s\beta \sin \delta_4 \end{bmatrix} \quad (7.145)$$

where $\mathbf{H}_i$ is the angular momentum vector of the ith CMG expressed in spacecraft reference frame, β is the pyramid skew angle, $c\beta \equiv \cos \beta$, $s\beta \equiv \sin \beta$, δ_i are the gimbal angles, and constant unit momentum magnitudes are assumed without loss of generality.

One way of determining gimbal angle trajectories that generate the commanded $\mathbf{h}$ trajectory is to solve a constrained optimization problem by minimizing a suitable performance index $J(\boldsymbol{\delta})$ subject to the nonlinear constraints (7.145); however, this approach is not suitable for real-time implementation.

A more practical approach to solving the inverse kinematic problem is to utilize the differential relationship between gimbal angles and CMG momentum vector. For such local inversion or tangent methods, the time derivative of the CMG angular momentum vector, Eq. (7.145), can be obtained as

$$\dot{\mathbf{h}} = \sum_{i=1}^{4} \dot{\mathbf{H}}_i = \sum_{i=1}^{4} \mathbf{a}_i(\delta_i) \dot{\delta}_i = \mathbf{A}\dot{\boldsymbol{\delta}} \quad (7.146)$$

where $\boldsymbol{\delta} = (\delta_1, \delta_2, \delta_3, \delta_4)$ is the gimbal angle vector, $\mathbf{a}_i$ is the ith column of $\mathbf{A}$, and

$$\mathbf{A} = \begin{bmatrix} -c\beta \cos \delta_1 & \sin \delta_2 & c\beta \cos \delta_3 & -\sin \delta_4 \\ -\sin \delta_1 & -c\beta \cos \delta_2 & \sin \delta_3 & c\beta \cos \delta_4 \\ s\beta \cos \delta_1 & s\beta \cos \delta_2 & s\beta \cos \delta_3 & s\beta \cos \delta_4 \end{bmatrix}$$

For the commanded control torque input $\mathbf{u}$, the CMG momentum rate command $\dot{\mathbf{h}}$ is chosen as

$$\dot{\mathbf{h}} = -\mathbf{u} - \boldsymbol{\omega} \times \mathbf{h}$$

and the gimbal rate command $\dot{\boldsymbol{\delta}}$ is then obtained as

$$\dot{\boldsymbol{\delta}} = \mathbf{A}^+ \dot{\mathbf{h}} = \mathbf{A}^T (\mathbf{A}\mathbf{A}^T)^{-1} \dot{\mathbf{h}} \quad (7.147)$$

which is often referred to as the *pseudoinverse steering logic*. Most CMG steering laws determine the gimbal rate commands with some variant of pseudoinverse.

7.5.3.3 Singular States

If rank($\mathbf{A}$) < 3 for certain sets of gimbal angles, or, equivalently, rank($\mathbf{A}\mathbf{A}^T$) < 3, the pseudoinverse does not exist and the pseudoinverse steering logic encounters singular states. This singular situation occurs when all individual CMG torque output vectors $\mathbf{a}_i$ are perpendicular to the commanded torque direction. Equivalently, the singular situation occurs when all individual CMG momentum vectors have extremal projections onto the commanded torque vector direction.

In general, the singularity condition

$$\det(\mathbf{A}\mathbf{A}^T) = 0$$

defines a set of surfaces in δ-space, or, equivalently, in $\mathbf{h}$-space. The simplest singular state is the momentum saturation singularity characterized by the so-called momentum envelope, which is a three-dimensional surface representing the maximum available angular momentum of CMGs along any given direction. Any singular state for which the total CMG momentum vector is inside the momentum envelope is called internal. For any system of n CMGs, there exist 2^n sets of gimbal angles for which no control torque can be generated along any arbitrary direction. Consequently, for the set of all directions, all of the internal singular states form closed surfaces both in $\mathbf{h}$-space and in δ-space, and the number of singular surfaces is equal to 2^n.

There are two types of internal singular states: "hyperbolic" states and "elliptic" states [32,40]. Because the pseudoinverse $\mathbf{A}^+ = \mathbf{A}^T(\mathbf{A}\mathbf{A}^T)^{-1}$ is the minimum two-norm solution of gimbal rates subject to Eq. (7.146), the pseudoinverse steering logic tends to leave inefficiently positioned CMGs alone, causing the gimbal angles to eventually hang-up in singular antiparallel arrangements. That is, it tends to steer the gimbals toward singular states. An approach to avoiding singular states is to introduce null motion into the CMG steering logic. Null motion is a motion of the gimbals that produces no net control torque on the spacecraft; however, the elliptic singular states cannot be escaped through null motion, whereas the hyperbolic singular states can be escaped through null motion [32,40]. For the pyramid-type system of four CMGs, for example, $\delta = (-90, 0, 90, 0)$ deg is an elliptic singularity with the singular direction along the first-axis and $\delta = (90, 180, -90, 0)$ deg is a hyperbolic singularity with the singular direction also along the first-axis.

The "impassable" elliptic singular states pose a major difficulty with SGCMG systems because they are not escapable without torquing the spacecraft. The impassability is defined locally in δ-space. An impassable surface in $\mathbf{h}$-space is not always impassable because it depends on δ, as discussed in Ref. 46. The size of spherical momentum space without impassable singular states is reduced to about half of the maximum momentum space of the standard pyramid type system. To fully utilize the available momentum space in the presence of impassable singular states, an intelligent steering algorithm needs to be developed that avoids the impassable singular states whenever possible, or rapidly transits unavoidable singularities, while minimizing their effects on the spacecraft attitude control.

7.5.3.4 Singularity-avoidance Steering Logic
Equation (7.147) can be considered as a particular solution to Eq. (7.146). The corresponding homogeneous solution is then obtained through null motion such that

$$\mathbf{A}\mathbf{n} = 0$$

where $\mathbf{n}$ denotes the null vector spanning the null space of $\mathbf{A}$. The general solution to Eq. (7.146) is then given by

$$\dot{\boldsymbol{\delta}} = \mathbf{A}^T (\mathbf{A}\mathbf{A}^T)^{-1} \, \dot{\mathbf{h}} + \gamma \mathbf{n} \qquad (7.148)$$

where γ represents the amount of null motion to be properly added.

The amount of null motion may be chosen as [39]:

$$\gamma = \begin{cases} m^6 & \text{for } m \geq 1 \\ m^{-6} & \text{for } m < 1 \end{cases}$$

where

m	=	singularity measure, also called CMG gain, $\sqrt{\det(\mathbf{A}\mathbf{A}^T)}$
$\mathbf{n}$	=	Jacobian null vector, (C_1, C_2, C_3, C_4)
C_i	=	order 3 Jacobian cofactor, $(-1)^{i+1} M_i$
M_i	=	order 3 Jacobian minor, $\det(\mathbf{A}_i)$
$\mathbf{A}_i$	=	$\mathbf{A}$ with ith column removed

This choice of scaling factor γ arises from the representation of m as a measure of distance from singularity, as well as the fact that

$$\det(\mathbf{A}\mathbf{A}^T) = \sum_{i=1}^{4} M_i^2 = \mathbf{n}^T \mathbf{n} \qquad (7.149)$$

This nondirectional null-motion approach introduces substantial null motion even when the system is far from being singular and tries to prevent the gimbal angles from settling into locally optimal configurations, which may eventually result in a singularity. This approach does not guarantee singularity avoidance, however, and has few potential shortcomings, as discussed in Ref. 39.

Although the null vector can be obtained through a variety of ways, e.g., using singular value decomposition, a projection operator, or the generalized cross or wedge product, it is often expressed as

$$\mathbf{n} = [\mathbf{I} - \mathbf{A}^+ \mathbf{A}]\mathbf{d} \qquad (7.150)$$

where $\mathbf{A}^+ = \mathbf{A}^T (\mathbf{A}\mathbf{A}^T)^{-1}$, $\mathbf{I}$ is an identity matrix, and $[\mathbf{I} - \mathbf{A}^+ \mathbf{A}]$ is a projection matrix* and $\mathbf{d}$ is an arbitrary n-dimensional nonzero vector.

A variety of analytic and heuristic approaches have been developed in the past to determine a proper null motion for singularity avoidance, i.e., to properly select the scalar γ and the n-dimensional vector $\mathbf{d}$. In a gradient-based method, a scalar function $f(\boldsymbol{\delta})$ is defined such that its gradient vector points toward the singular directions. The vector $\mathbf{d}$ is then chosen as

$$\mathbf{d} = \left(\frac{\partial f}{\partial \delta_1}, \dots, \frac{\partial f}{\partial \delta_n} \right)$$

---

*A symmetric matrix $\mathbf{P}$ is called a projection matrix if $\mathbf{P}^2 = \mathbf{P}$.

and the scalar function is often selected as the inverse of the square of the singularity measure; i.e.,

$$f(\delta) = \frac{1}{\det(\mathbf{A}\mathbf{A}^T)}$$

and the scalar γ is determined by minimizing $f(\delta)$ in the null vector direction.

This gradient-based method does not always work in directly avoiding internal singularities. Thus, an indirect singularity-avoidance steering law of feedback control form, which adds null motion to steer toward a set of desired gimbal angles, may be employed as [33]:

$$\dot{\delta} = \mathbf{A}^+ \dot{\mathbf{h}} + \gamma [\mathbf{I} - \mathbf{A}^+\mathbf{A}](\delta^* - \delta) \qquad (7.151)$$

where δ^* denotes a set of desired gimbal angles and γ is a positive scale factor.

A further heuristic modification of the pseudoinverse-based steering logic is to employ a singularity robust inverse algorithm of the form [52]:

$$\mathbf{A}^{\#} = \mathbf{A}^T (\mathbf{A}\mathbf{A}^T + \lambda \mathbf{I})^{-1} \qquad (7.152)$$

where $\mathbf{I}$ is an identity matrix and λ is a positive scale factor that may be automatically adjusted as

$$\lambda = \begin{cases} \lambda_0(1 - m/m_0)^2 & \text{for } m < m_0 \\ 0 & \text{for } m \geq m_0 \end{cases}$$

where $m = \sqrt{\det(\mathbf{A}\mathbf{A}^T)}$ and λ_0 and m_0 are small positive constants to be properly selected. However, a small positive constant of the order of 0.01 may be simply selected for λ.

Most existing pseudoinverse-based, local inversion or tangent methods do not always guarantee singularity avoidance and often unnecessarily constrain the operational momentum envelope of SGCMG systems. To guarantee successful singularity avoidance throughout the operational envelope of SGCMGs, the optimal steering algorithm needs to be "global" in nature, as suggested by Paradiso [44]. Any global method requires extensive computations, however, which may not be practical for real-time implementation.

7.5.3.5   Attitude Control and CMG Steering Logic

For large-angle (possibly, near-minimum-time eigenaxis) reorientation maneuvers of agile spacecraft in the presence of slew rate and gimbal rate limits, we may integrate the steering logic based on the singularity robust inverse $\mathbf{A}^{\#}$ and the attitude control logic based on quaternion feedback, presented in Sec. 7.3, as follows:

$$\mathbf{u} = -\operatorname*{sat}_{\sigma}[\mathbf{K} \operatorname{sat}(\mathbf{P}\mathbf{q}_e) + \mathbf{C}\boldsymbol{\omega}] \qquad (7.153a)$$

$$\mathbf{A}^{\#} = \mathbf{A}^T (\mathbf{A}\mathbf{A}^T + \lambda \mathbf{I})^{-1} \qquad (7.153b)$$

$$\dot{\delta}_c = \mathbf{A}^{\#}(-\mathbf{u} - \boldsymbol{\omega} \times \mathbf{h}) + \gamma [\mathbf{I} - \mathbf{A}^{\#}\mathbf{A}](\delta^* - \delta) \qquad (7.153c)$$

$$\dot{\delta} = \operatorname*{sat}_{\sigma}(\dot{\delta}_c) \qquad (7.153d)$$

where $\mathbf{q}_e = (q_{1e}, q_{2e}, q_{3e})$ is the attitude error quaternion vector. Furthermore, we have

$$
\begin{bmatrix} q_{1e} \\ q_{2e} \\ q_{3e} \\ q_{4e} \end{bmatrix} = \begin{bmatrix} q_{4c} & q_{3c} & -q_{2c} & -q_{1c} \\ -q_{3c} & q_{4c} & q_{1c} & -q_{2c} \\ q_{2c} & -q_{1c} & q_{4c} & -q_{3c} \\ q_{1c} & q_{2c} & q_{3c} & q_{4c} \end{bmatrix} \begin{bmatrix} q_1 \\ q_2 \\ q_3 \\ q_4 \end{bmatrix}
\tag{7.154}
$$

$$
\begin{bmatrix} \dot{q}_1 \\ \dot{q}_2 \\ \dot{q}_3 \\ \dot{q}_4 \end{bmatrix} = \frac{1}{2} \begin{bmatrix} 0 & \omega_3 & -\omega_2 & \omega_1 \\ -\omega_3 & 0 & \omega_1 & \omega_2 \\ \omega_2 & -\omega_1 & 0 & \omega_3 \\ -\omega_1 & -\omega_2 & -\omega_3 & 0 \end{bmatrix} \begin{bmatrix} q_1 \\ q_2 \\ q_3 \\ q_4 \end{bmatrix}
\tag{7.155}
$$

where $(q_{1c}, q_{2c}, q_{3c}, q_{4c})$ are the desired or commanded attitude quaternions and (q_1, q_2, q_3, q_4) the current attitude quaternions. (Further details of the quaternion-feedback control logic can be found in Sec. 7.3.)

7.5.4   Steering Logic for Double-Gimbal Control Moment Gyros

The singularity problem for DGCMGs is less severe than for SGCMGs because of the extra degree of freedom per device.

For DGCMGs, the rotor is suspended inside two gimbals and, consequently, the rotor momentum can be oriented on a sphere along any direction provided there are no restrictive gimbal stops. For the different purposes of redundancy management and failure accommodation, several different arrangements of DGCMGs have been developed, such as three orthogonally mounted DGCMGs used in the Skylab [24,29] and four parallel mounted DGCMGs to be employed for the International Space Station [48].

As shown by Kennel [34], the mounting of DGCMGs of unlimited outer gimbal angle freedom, with all their outer gimbal axes parallel, allows drastic simplification of the CMG steering law development in the redundancy management, failure accommodation, and in the mounting hardware.

Consider such a parallel mounting arrangement of n DGCMGs with the inner and outer gimbal angles α_i and β_i of the ith CMG as defined in Fig. 7.7. The total CMG angular momentum vector $\mathbf{h} = (h_1, h_2, h_3)$ is expressed in the CMG axes as

$$
\mathbf{h} = \sum_{i=1}^{n} \mathbf{H}_i = \begin{bmatrix} \sum H_i \sin \alpha_i \\ \sum H_i \cos \alpha_i \cos \beta_i \\ \sum H_i \cos \alpha_i \sin \beta_i \end{bmatrix}
\tag{7.156}
$$

where $\mathbf{H}_i$ is the angular momentum vector of the ith CMG and $H_i = \|\mathbf{H}_i\|$.

The time derivative of $\mathbf{h}$ becomes

$$
\dot{\mathbf{h}} = \begin{bmatrix} \dot{h}_1 \\ \dot{h}_2 \\ \dot{h}_3 \end{bmatrix} = \begin{bmatrix} \sum H_i \cos \alpha_i \dot{\alpha}_i \\ \sum (-H_i \sin \alpha_i \cos \beta_i \dot{\alpha}_i - H_i \cos \alpha_i \sin \beta_i \dot{\beta}_i) \\ \sum (-H_i \sin \alpha_i \sin \beta_i \dot{\alpha}_i + H_i \cos \alpha_i \cos \beta_i \dot{\beta}_i) \end{bmatrix}
\tag{7.157}
$$

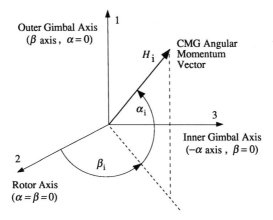

Fig. 7.7   Inner gimbal angle α_i and outer gimbal angle β_i of the ith DGCMG.

where all H_i are assumed to be constant without loss of generality. The CMG steering logic design is then to find a set of gimbal rate commands $\dot{\alpha}_i$ and $\dot{\beta}_i$, which deliver the commanded $\dot{\mathbf{h}}$ while utilizing the excess degrees of freedom to distribute the gimbal angles, such that certain hardware constraints are met and singular states are avoided. The n DGCMGs have $2n$ degrees of freedom. Three are needed for the independent, three-axis attitude control of spacecraft. The excess $(2n - 3)$ degrees of freedom are utilized to achieve a desirable gimbal angle distribution.

For the parallel mounting arrangement with gimbal angles as defined in Fig. 7.7, it can be seen that $\dot{h}_1$ is not a function of the outer gimbal motions. In particular, Kennel's CMG steering law distributes the CMG angular momentum vectors such that all inner gimbal angles are equal, which reduces the rate requirements on the outer gimbals; that is,

$$\dot{\alpha}_i = \frac{\dot{h}_1}{\sum H_i \cos \alpha_i} + K(\alpha^* - \alpha_i) \qquad (7.158)$$

where K is called the inner gimbal angle distribution gain and α^* is the desired inner gimbal angle for all CMGs chosen as

$$\alpha^* = \frac{\sum (H_i \cos \alpha_i)\alpha_i}{\sum H_i \cos \alpha_i}$$

For DGCMGs, a singular condition inside the total CMG momentum envelope can occur when some of the CMG momentum vectors are antiparallel and the rest parallel to their resultant total CMG momentum vector. Therefore, maintaining more or less equal spacing between the CMG momentum vectors will eliminate the possibility of singular states. Consequently, Kennel's steering law spreads the outer gimbals, which ensures avoidance of singularities internal to the CMG momentum envelope. For further details of Kennel's steering law and its implementation to the International Space Station, the reader is referred to Refs. 34, 35, and 48.

7.6 Optimal Jet Selection Logic

In this section, a brief introduction to the Space Shuttle on-orbit reaction control system is followed by a description of the optimal jet selection problem of advanced space vehicles with strongly three-axis coupled, redundant sets of thrusters for both translation and rotation control.

7.6.1 Space Shuttle On-Orbit Reaction Control System

The Space Shuttle has two orbital maneuvering system (OMS) engines, each producing about 26,700 N (6000 lb) of thrust. They are bipropellant rockets with a specific impulse of 313 s, and used mainly for large orbit-change ΔV maneuvers. Each OMS engine has redundant two-axis gimbal capability.

The Space Shuttle on-orbit reaction control system (RCS) consists of 38 primary jets, each producing about 3870 N (870 lb) of thrust, and 6 vernier jets, each producing about 107 N (24 lb) of thrust. All thrusters using the same bipropellant as the OMS have 80-ms minimum pulse granularity. The thrusters are located in three pods: forward, left aft, and right aft. The primary jets are divided into 14 clusters around the vehicle with both translational and rotational control capabilities and with multiple failure tolerance. The vernier jets, however, each fire in a different direction, only control rotation, and have no failure tolerance if any one of four of the thrusters malfunctions. Figure 7.8 illustrates thruster locations and plume directions with an explanation of the thruster identification nomenclature [53–55].

The Space Shuttle RCS processor with a 12.5-Hz sampling rate consists of phase-plane and jet selection logic. The phase-plane logic in general performs on–off modulation of jet torque commands as required to maintain vehicle attitude and rate errors within the specified deadbands. In addition, desired angular rate change is also computed to facilitate computation of jet burn times. The phase plane of angular rate vs attitude angle, per each axis, consists of a set of switch lines for specified attitude deadbands, angular rate limits, and expected RCS accelerations. As illustrated in Fig. 7.9, the phase plane is divided into nine regions defined by numbered boundaries. At any time, for each axis, the rigid vehicle state is defined by an attitude and rate error point that must lie in one of the defined regions because the regions cover the entire plane. The decision concerning whether to send a rotation command is made on the basis of logic unique to each region. Regions 1 and 5 always command jets. For primary jets, regions 2, 3, 6, and 7 always permit coast with no jet commands. Region 9 never causes commands to be generated, but a preference for vernier jet selection is computed. Regions 4 and 8 have hysteresis. If the phase point is in either region 4 or 8 and a jet firing is taking place, then the firing will continue until the phase point crosses the S13 switch curve [53].

The jet selection logic identifies specific jets to be used. For primary jets it uses a table lookup scheme with selection details dependent on flight phase, selected modes, and recognized jet failure status. For vernier jets it utilizes a dot-product algorithm. The principle of the dot-product algorithm is to take the dot product of each jet's rotational velocity increment vector with a vector from the rotation

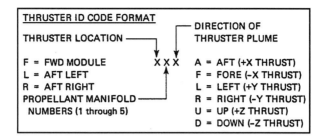

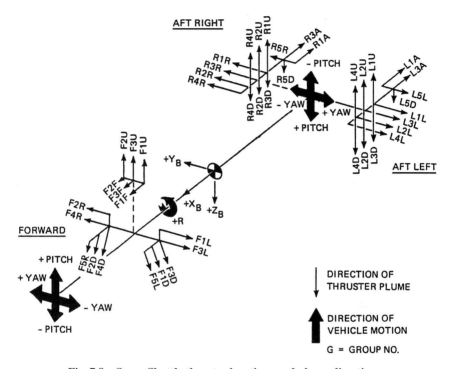

Fig. 7.8   Space Shuttle thruster locations and plume directions.

commands and then select the jets with the biggest dot-product values. The baseline RCS processor excludes mixed operations of primary and vernier jets.

During the early Shuttle program, a new autopilot concept was developed by Bergmann et al. [56] at the Charles Stark Draper Laboratory, which promised certain advantages over the conventional phase-plane logic based on a table lookup scheme for jet selection. The new autopilot concept incorporated a six-dimensional "phase space" control logic and a linear programming algorithm for jet selection. The unique features of this new autopilot scheme include: fuel-optimal jet selection, a high degree of adaptability to configuration changes and jet failures, combined primary and vernier jet usage, and closed-loop translation control.

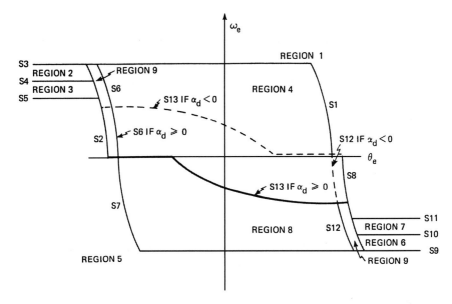

Fig. 7.9   **Phase-plane logic for the Space Shuttle RCS.**

Flight tests of this advanced autopilot system were successfully performed on STS missions 51G (June 1985) and 61B (November 1985).

7.6.2   Optimal Jet Selection Logic

The rotational equation of motion of a rigid spacecraft equipped with n jets can be approximated, neglecting the gyroscopic coupling terms, as

$$\mathbf{J}\dot{\boldsymbol{\omega}} = \mathbf{u} \Rightarrow \mathbf{J}\Delta\boldsymbol{\omega} = \sum_{j=1}^{n} \mathbf{a}_j \Delta t_j \qquad (7.159)$$

where $\mathbf{J}$ is the inertia matrix of the spacecraft, $\boldsymbol{\omega} = (\omega_1, \omega_2, \omega_3)$ is the spacecraft angular velocity vector, $\Delta\boldsymbol{\omega} = (\Delta\omega_1, \Delta\omega_2, \Delta\omega_3)$, $\mathbf{u} = (u_1, u_2, u_3)$ is the control torque vector, and $\mathbf{a}_j$ is the torque vector of the jth jet with on-time Δt_j, all expressed in the spacecraft body-fixed control axes.

Equation (7.159) can be rewritten as

$$\mathbf{b} = \mathbf{A}\mathbf{x} \qquad (7.160)$$

where $\mathbf{b} = \mathbf{J}\Delta\boldsymbol{\omega}$, $\mathbf{x} = (\Delta t_1, \dots, \Delta t_n)$, and $\mathbf{A} = [a_{ij}] = [\mathbf{a}_1, \dots, \mathbf{a}_n]$. That is, a_{ij} is the ith-axis torque caused by the jth jet firing with on-time x_j. It is assumed that $n > 3$ and rank$(\mathbf{A}) = 3$.

The translational equations of motion of a rigid spacecraft can be easily augmented to Eq. (7.160), resulting in an $m \times n$ torque/force distribution matrix $\mathbf{A}$ with $m \le 6$ and rank $(\mathbf{A}) = m < n$. The optimal jet selection problem is then to determine, in a fuel-optimal manner, a proper set of jets for the commanded angular/translational rate change in each axis. In general, a set of three jets are needed for the independent three-axis attitude control.

Consider an optimal jet selection logic concept that utilizes a linear programming approach for the minimum fuel problem formulated as

$$\min_{x_j} \sum_{j=1}^{n} c_j x_j = \mathbf{c}^T \mathbf{x} \qquad (7.161)$$

where c_j is the jth jet flow rate, subject to the constraints

$$\mathbf{Ax} = \mathbf{b} \qquad (7.162)$$

$$\mathbf{x} \geq 0 \qquad (7.163)$$

Several important definitions and theorems often encountered in a linear programming problem with an $m \times n$ constant matrix $\mathbf{A}$ are summarized as follows [57].

Definition 7.4

1) A *feasible solution* to the linear programming problem is a vector $\mathbf{x} = (x_1, \ldots, x_n)$ that satisfies the constraints (7.162) and (7.163).

2) A *basis matrix* is an $m \times m$ nonsingular matrix formed from some m columns of the constraint matrix $\mathbf{A}$.

3) A *basic solution* to the linear programming problem is the unique vector $\mathbf{x}$ determined by choosing a basis matrix, letting the $n - m$ variables associated with columns of $\mathbf{A}$ not in the basis matrix equal zero.

4) A *basic feasible solution* is a basic solution in which all variables have nonnegative values.

5) A *nondegenerate basic feasible solution* is a basic feasible solution with exactly m positive x_j.

6) An *optimal solution* is a feasible solution that also minimizes the performance index (objective function) $J = \mathbf{c}^T \mathbf{x}$ in Eq. (7.161).

Theorem 7.4

The objective function $J = \mathbf{c}^T \mathbf{x}$ assumes its minimum at an extreme point of the constraint set (7.162) and (7.163). If it assumes its minimum at more than one extreme point, then it takes on the same value at every point of the line segment joining any two optimal extreme points.

Theorem 7.5

A vector $\mathbf{x}$ is an extreme point of the constraint set of a linear programming problem if and only if $\mathbf{x}$ is a basic feasible solution of the constraints (7.162) and (7.163).

These theorems imply that, in searching for an optimal solution, one needs only consider extreme points, i.e., only basic feasible solutions. An upper bound to the number of basic feasible solutions is given by

$$\frac{n!}{(n-m)!m!} = \frac{n(n-1)(n-2)\cdots(n-m+1)}{m!}$$

Consider an optimal jet selection problem of minimizing the objective function $J = \mathbf{c}^T \mathbf{x}$ with the constraint set (7.162) and (7.163). Although such an optimal jet selection problem has been studied previously in terms of an iterative linear programming approach [56,57], a new approach developed by Glandorf [58] and Kubiak and Johnson [59] appears to be more analytic and relatively concise. The algorithm developed in Refs. 58 and 59 is briefly described herein to introduce the reader to the essential feature of an optimal jet selection logic that can be executed recursively.

Let the constraint equation (7.162) be rewritten as

$$[\hat{\mathbf{A}} \quad \tilde{\mathbf{A}}]\begin{bmatrix}\hat{\mathbf{x}}\\\tilde{\mathbf{x}}\end{bmatrix} = \mathbf{b} \tag{7.164}$$

and the objective function as

$$J = [\hat{\mathbf{c}}^T \quad \tilde{\mathbf{c}}^T]\begin{bmatrix}\hat{\mathbf{x}}\\\tilde{\mathbf{x}}\end{bmatrix} = \hat{\mathbf{c}}^T\hat{\mathbf{x}} + \tilde{\mathbf{c}}^T\tilde{\mathbf{x}} \tag{7.165}$$

where $\hat{\mathbf{x}} \geq 0$, $\tilde{\mathbf{x}} = 0$, and $\hat{\mathbf{A}}$ is a nonsingular $m \times m$ matrix. Then a basic feasible solution is simply obtained as

$$\hat{\mathbf{x}} = \hat{\mathbf{A}}^{-1}\mathbf{b} \tag{7.166}$$

At this point, $\hat{\mathbf{A}}$ is not unique, and it represents a nonsingular $m \times m$ acceleration matrix corresponding to any combination of jets taken m at a time as long as they are linearly independent. To determine which of these combinations of jets are potentially optimal, one can assume there is a more efficient solution by allowing alternate jets to be turned on for a differential time, $d\mathbf{x}$. That is, let

$$\hat{\mathbf{x}}' = \hat{\mathbf{x}} + d\hat{\mathbf{x}}$$
$$\tilde{\mathbf{x}}' = \tilde{\mathbf{x}} + d\tilde{\mathbf{x}} = d\tilde{\mathbf{x}} \geq 0$$
$$J' = J + dJ$$

the constraint equation (7.162) then becomes

$$\hat{\mathbf{A}}\,d\hat{\mathbf{x}} + \tilde{\mathbf{A}}\,d\tilde{\mathbf{x}} = 0 \quad \text{or} \quad d\hat{\mathbf{x}} = -\hat{\mathbf{A}}^{-1}\tilde{\mathbf{A}}\,d\tilde{\mathbf{x}} \tag{7.167}$$

Using the objective function $J' = J + dJ$, one can obtain

$$dJ = \hat{\mathbf{c}}^T\,d\hat{\mathbf{x}} + \tilde{\mathbf{c}}^T\,d\tilde{\mathbf{x}} \tag{7.168}$$

Combining Eqs. (7.167) and (7.168), one can obtain the cost differential caused by $d\tilde{x}$ as

$$dJ = \hat{c}^T[-\hat{A}^{-1}\tilde{A}\,d\tilde{x}] + \tilde{c}^T\,d\tilde{x}$$
$$= [\tilde{c}^T - \hat{c}^T\hat{A}^{-1}\tilde{A}]\,d\tilde{x} \qquad (7.169)$$

Defining

$$\lambda^T = \hat{c}^T\hat{A}^{-1}$$

where λ is called the optimality vector associated with $\hat{x}$, and also defining $\tilde{a}_i = $ the ith column of $\tilde{A}$, i.e., $\tilde{a}_i$ are the acceleration vectors for the jets not in the original selection, one can rewrite Eq. (7.169) as

$$dJ = \tilde{c}^T\,d\tilde{x} - \lambda^T\tilde{A}\,d\tilde{x}$$
$$= \sum_{i=1}^{n-m}\tilde{c}_i\,d\tilde{x}_i - \sum_{i=1}^{n-m}\lambda^T\tilde{a}_i\,d\tilde{x}_i$$
$$= \sum_{i=1}^{n-m}\left[\tilde{c}_i - \lambda^T\tilde{a}_i\right]d\tilde{x}_i$$

Assuming that all of the elements of c are unity without loss of generality, we obtain

$$dJ = \sum_{i=1}^{n-m}\left[1 - \lambda^T\tilde{a}_i\right]d\tilde{x}_i \qquad (7.170)$$

This equation is the key to determining whether the original selection of jets, i.e., $\hat{A}$, is optimal or not. Therefore, if $\lambda^T\tilde{a}_i > 1$ for any jet not in the original selection, the original selection is not optimal and should be discarded. On the other hand, if $\lambda^T\tilde{a}_i \leq 1$ for all of the jets not in the original selection, then the original selection is potentially optimal and should be saved. These remaining k combinations or groups are referred to as candidate optimal groups (COGs) in Refs. 58 and 59.

The key properties of the optimality vector λ and COGs can be summarized as follows [59]:

1) For each COG, λ is perpendicular to the hyperplane in m-dimensional space formed by the end points of the m vectors $\{\hat{a}_1, \ldots, \hat{a}_m\}$ where $\hat{a}_i$ is defined as the ith column of $\hat{A}$.

2) The tips of the $\hat{a}_i$ for a given COG define a surface segment of a hyperplane and all other surfaces or vertices for other COGs lie at or below this surface (closer to the origin than the surface).

3) The optimal COG is the COG with a maximum dot product of b and λ.

4) After finding the optimal COGs, the jet on times are computed as $\hat{x} = \hat{A}^{-1}b$ and $\tilde{x} = 0$.

A more detailed description of the optimal jet select algorithm originally developed by Glandorf and Kubiak can be found in Refs. 58 and 59.

7.7 Pulse-Modulated Attitude Control

In the preceding section, the Space Shuttle on-orbit reaction control system was briefly described and the optimal jet selection problem of advanced space vehicles were introduced. The phase-plane logic described in the preceding section was used successfully for the Apollo and the Space Shuttle and may well continue as a basic architecture for future space vehicles. The phase-plane logic can have a great level of sophistication when implemented on a digital flight computer; however, it can be easily replaced by a simple Schmitt trigger for certain applications. In this section, conventional pulse modulation techniques, including the Schmitt trigger, as applied to the single-axis, on–off attitude control systems are described.

7.7.1 Introduction

Pulse modulation represents the common control logic behind most reaction-jet control systems of various spacecraft. Unlike other actuators, such as reaction wheels or control moment gyros, thruster output consists of two values: on or off. Proportional thrusters, whose fuel valves open a distance proportional to the commanded thrust level, are not employed much in practice. Mechanical considerations prohibit proportional valve operation largely because of dirt particles that prevent complete closure for small valve openings; fuel leakage through the valves consequently produces opposing thruster firings. Pulse modulation techniques have been developed that fully open and close the fuel valves, while producing a nearly linear duty cycle. In general, pulse modulators produce a pulse command sequence to the thruster valves by adjusting the pulse width and/or pulse frequency.

Several commonly used, flight-proven pulse modulators are briefly described in this section. Static characteristics are summarized for each modulator. The pulse frequency of these modulators are usually fast compared to the spacecraft attitude control bandwidth, and the static characteristics are often used for the modulator design. We consider here a simplified single-axis reaction control system for a rigid spacecraft as illustrated in Fig. 7.10.

7.7.2 Schmitt Trigger

Strictly speaking, this device, which is often called a relay with dead-band/hysteresis, shown in Fig. 7.11, is not a pulse modulator. The advantage of this device, as opposed to other pulse modulators, is its simplicity. A disadvantage of the Schmitt trigger is the dependence of its static characteristics on the spacecraft inertia.

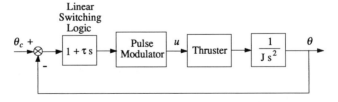

Fig. 7.10　Single-axis reaction control system.

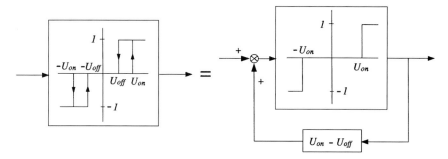

Fig. 7.11   Schmitt trigger.

If the pulse modulator in Fig. 7.10 is replaced by the Schmitt trigger, then the spacecraft dynamics can be described as

$$J\ddot{\theta} = T_c u \tag{7.171}$$

where

$$J = \text{spacecraft inertia}$$
$$\theta = \text{spacecraft attitude}$$
$$u = \text{Schmitt trigger output (0 or} \pm 1)$$
$$T_c = \text{control torque level}$$

and the input to the Schmitt trigger is determined by the linear switching logic $(1 + \tau s)(\theta_c - \theta)$.

The rigid-body limit cycling characteristics of this closed-loop system can be predicted, as follows:

$$\text{minimum pulse width} = \frac{Jh}{\tau T_c} \tag{7.172a}$$

$$\text{limit cycle amplitude} = \frac{T_c (U_{\text{on}} + U_{\text{off}})}{2} + \frac{J^2 h^2}{8\tau} \tag{7.172b}$$

$$\text{limit cycle period} = 4\tau \left(\frac{U_{\text{on}} + U_{\text{off}}}{h} + \frac{Jh}{2\tau^2} \right) \tag{7.172c}$$

where $h \equiv U_{\text{on}} - U_{\text{off}}$ and $\tau = $ the linear switching line slope.

Note that the minimum pulse width is a function of spacecraft parameters: the spacecraft inertia and thrust level. These parameters tend to change over time; as a result, the minimum pulse width will vary as well. Knowledge of the spacecraft properties is therefore required to estimate the thruster minimum pulse width.

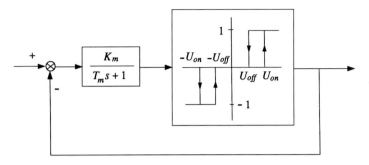

Fig. 7.12   PWPF modulator.

7.7.3   Pulse-Width Pulse-Frequency Modulator

The pulse-width pulse-frequency (PWPF) modulator has been used in the control systems of such spacecraft as the Agena satellite, INTELSAT V, INSAT, and ARABSAT. The device, as illustrated in Fig. 7.12, mainly comprises two components: a first-order lag filter and a Schmitt trigger inside a feedback loop. With a constant input, the PWPF modulator drives the thruster valve with an on–off pulse sequence having a nearly linear duty cycle with the input amplitude. The *duty cycle* or *modulation factor* is defined as the average output of the modulator.

The static characteristics of the PWPF modulator for a constant input E are summarized as follows for the thruster pulse width

$$T_{\text{on}} = -T_m \ln \left\{ \frac{(1-h)E_d - (E-1)}{E_d - (E-1)} \right\} \qquad (7.173\text{a})$$

the thruster off time

$$T_{\text{off}} = -T_m \ln \left\{ \frac{E_d - E}{(1-h)E_d - E} \right\} \qquad (7.173\text{b})$$

the pulse frequency

$$f = 1/(T_{\text{on}} + T_{\text{off}}) \qquad (7.173\text{c})$$

and the minimum pulse width

$$\Delta = -T_m \ln \left(1 - \frac{h}{K_m} \right) \approx \frac{h T_m}{K_m} \qquad (7.173\text{d})$$

where E = constant input magnitude, $E_d = U_{\text{on}}/K_m$ = equivalent internal deadband, and $h = U_{\text{on}} - U_{\text{off}}$. The duty cycle is given by $f T_{\text{on}}$ and it will be further discussed in Chapter 9.

In contrast to the Schmitt trigger, the static characteristics of the PWPF modulator are independent of the spacecraft inertia because of the feedback loop within the device. The presence of the filter and the feedback loop, however, inhibits linear analysis of the device's dynamic characteristics. The problem of designing a reaction jet control system employing the PWPF modulator will be further studied in Chapter 9.

7.7.4 Derived-Rate Modulator

The derived-rate and PWPF modulators are similar in format, as seen in Figs. 7.13 and 7.14, except that the first-order filter now compensates the Schmitt trigger output in the feedback path. The device is used much in the same way that the PWPF modulator is used, except that the derived-rate modulator introduces phase lead into a system as opposed to the PWPF modulator, which is a phase-lag device.

The static characteristics are similar to the PWPF modulator for the thruster pulse width

$$T_{\text{on}} = -T_m \ln \left\{ 1 - \frac{h}{K_m - (E - U_{\text{on}})} \right\} \tag{7.174a}$$

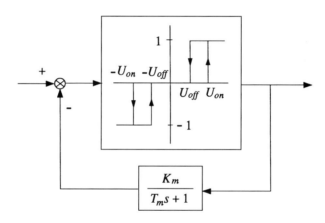

Fig. 7.13 Derived-rate modulator.

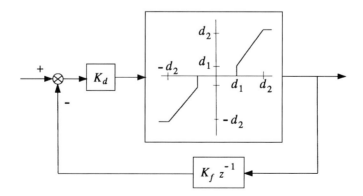

Fig. 7.14 Pulse-width modulator.

the thruster off time

$$T_{\text{off}} = T_m \, \ell n \left\{ 1 + \frac{h}{E - U_{\text{on}}} \right\} \qquad (7.174\text{b})$$

the pulse frequency

$$f = \frac{1}{T_{\text{on}} + T_{\text{off}}} \qquad (7.174\text{c})$$

and the minimum pulse width

$$\Delta \approx \frac{h \, T_m}{K_m} \qquad (7.174\text{d})$$

where E is the static input magnitude and the duty cycle is given by $f \, T_{\text{on}}$.

The derived-rate modulator is as difficult as the PWPF modulator to analyze in a limit cycling situation. In that case, the describing function method may be employed to characterize the modulator in terms of its gain and phase. This subject will be studied in detail in Chapter 9.

7.7.5 Pulse-Width Modulator

The pulse-width modulator (PWM) differs from the modulators discussed earlier in that it is essentially a discrete-time device; the PWPF and derived-rate modulators can be digitally implemented but are often analyzed as continuous-time systems.

The output of this particular device is not a thruster firing state as in the aforementioned devices; instead, the PWM output is thruster pulse width. A zero-order-hold (ZOH) device transmits that signal to the thrusters. The value d_1 represents the minimum pulse width of the system; this deadzone is directly proportional to the attitude deadband. The value d_2 represents the maximum pulse width of the RCS; it is equivalent to the microprocessor sampling period. As a result, the pulse frequency and the minimum pulse width of the modulator are fixed; a thruster firing command is given during every microprocessor sampling period even if the pulse is of zero width.

The delay in the feedback loop introduces damping to the system; maximum damping occurs when the feedback signal is smaller than the PWM input. If the input signal is not greater than the feedback signal, the modulator may limit the cycle itself. This criterion enables the designer to determine the feedback gain K_f. The feedforward gain K_d is selected as result of the minimum pulse criterion.

For more details of pulse modulation techniques as applied to spacecraft attitude control systems design, the reader is referred to Chapter 9, or Refs. 60–64.

References

[1] Rahn, C. D., and Barba, P. M., "Reorientation Maneuver for Spinning Spacecraft," *Journal of Guidance, Control, and Dynamics*, Vol. 14, No. 4, 1991, pp. 724–728.

[2] Bryson, A. E., Jr., *Control of Spacecraft and Aircraft*, Princeton Univ. Press, Princeton, NJ, 1994.

[3] Kaplan, M. H., *Modern Spacecraft Dynamics & Control*, Wiley, New York, 1976, pp. 367–379.

[4] Barba, P. M., and Aubrun, J. N., "Satellite Attitude Acquisition by Momentum Transfer," *AIAA Journal*, Vol. 14, No. 10, 1976, pp. 1382–1386.

[5] Scrivene, S. L., and Thompson, R. C., "Survey of Time-Optimal Attitude Maneuvers," *Journal of Guidance, Control, and Dynamics*, Vol. 17, No. 2, 1994, pp. 225–233.

[6] Bilimoria, K., and Wie, B., "Minimum-Time Large-Angle Reorientation of a Rigid Spacecraft," AIAA Paper 90-3486, Aug. 1990.

[7] Bilimoria, K., and Wie, B., "Time-Optimal Three-Axis Reorientation of a Rigid Spacecraft," *Journal of Guidance, Control, and Dynamics*, Vol. 16, No. 3, 1993, pp. 446–452.

[8] Bryson, A. E., Jr., and Ho, Y.-C., *Applied Optimal Control*, Hemisphere, Washington, DC, 1975.

[9] Athans, M., and Falb, P., *Optimal Control*, McGraw-Hill, New York, 1966.

[10] Li, F., and Bainum, P. M., "Numerical Approach for Solving Rigid Spacecraft Minimum Time Attitude Maneuvers," *Journal of Guidance, Control, and Dynamics*, Vol. 13, No. 1, 1990, pp. 38–45.

[11] Bilimoria, K., and Wie, B., "Time-Optimal Reorientation of a Rigid Axisymmetric Spacecraft," *Proceedings of the AIAA Guidance, Navigation, and Control Conference*, AIAA, Washington, DC, 1991, pp. 422–431.

[12] Byers, R. M., and Vadali, S. R., "Quasi-Closed-Form Solution to the Time-Optimal Rigid Spacecraft Reorientation Problem," *Journal of Guidance, Control, and Dynamics*, Vol. 16, No. 3, 1993, pp. 453–461.

[13] Seywald, H., and Kumar, R. R., "Singular Control in Minimum Time Spacecraft Reorientation," *Journal of Guidance, Control, and Dynamics*, Vol. 16, No. 4, 1993, pp. 686–694.

[14] Bauer, F. H., Femiano, M. D., and Mosier, G. E., "Attitude Control System Conceptual Design for the X-ray Timing Explorer," *Proceedings of the AIAA Guidance, Navigation, and Control Conference*, AIAA, Washington, DC, 1992, pp. 236–249.

[15] Wie, B., and Barba, P. M., "Quaternion Feedback for Spacecraft Large Angle Maneuvers," *Journal of Guidance, Control, and Dynamics*, Vol. 8, No. 3, 1985, pp. 360–365.

[16] Wie, B., Weiss, H., and Arapostathis, A., "Quaternion Feedback Regulator for Spacecraft Eigenaxis Rotations," *Journal of Guidance, Control, and Dynamics*, Vol. 12, No. 3, 1989, pp. 375–380.

[17] Wie, B. and Lu, J., "Feedback Control Logic for Spacecraft Eigenaxis Rotations Under Slew Rate and Control Constraints," *Journal of Guidance, Control, and Dynamics*, Vol. 18, No. 6, 1995, pp. 1372–1379.

[18] Wie, B., Byun, K.-W., Warren, W., Geller, D., Long, D., and Sunkel, J., "New Approach to Momentum/Attitude Control for the Space Station," *Journal of Guidance, Control, and Dynamics*, Vol. 12, No. 5, 1989, pp. 714–722.

[19] Wie, B., Liu, Q., and Sunkel, J., "Robust Stabilization of the Space Station in the Presence of Inertia Matrix Uncertainty," *Journal of Guidance, Control, and Dynamics*, Vol. 18, No. 3, 1995, pp. 611–617.

[20] Byun, K.-W., Wie, B., Geller, D., and Sunkel, J., "Robust H_∞ Control Design for the Space Station with Structured Parameter Uncertainty," *Journal of Guidance, Control, and Dynamics*, Vol. 14, No. 6, 1991, pp. 1115–1122.

[21] Rhee, I., and Speyer, J. L., "Robust Momentum Management and Attitude Control System for the Space Station," *Journal of Guidance, Control, and Dynamics*, Vol. 15, No. 2, 1992, pp. 342–351.

[22] Balas, G. J., Packard, A. K., and Harduvel, J. T., "Application of μ-Synthesis Technique to Momentum Management and Attitude Control of the Space Station," *Proceedings of the AIAA Guidance, Navigation, and Control Conference*, AIAA, Washington, DC, 1991, pp. 565–575.

[23] Jacot, A. D., and Liska, D. J., "Control Moment Gyros in Attitude Control," *Journal of Spacecraft and Rockets*, Vol. 3, No. 9, 1966, pp. 1313–1320.

[24] Chubb, W. B., and Epstein, M., "Application of Control Moment Gyros to the Attitude Control of the Apollo Telescope Unit," *Proceedings of AIAA Guidance, Control, and Flight Mechanics Conference*, AIAA, New York, 1968.

[25] Kennel, H. F., "A Control Law for Double-Gimballed Control Moment Gyros Used for Space Vehicle Attitude Control," NASA TM-64536, Aug. 1970.

[26] Austin, F., Liden, S., and Berman, H., "Study of Control Moment Gyroscope Applications to Space Base Wobble Damping and Attitude Control Systems," Grumman Aerospace Corp., CR NAS 9-10427, National Technical Information Service, N71-11401, Springfield, VA, Sept. 1970.

[27] Ranzenhofer, H. D., "Spacecraft Attitude Control with Control Moment Gyros," *Proceedings of AIAA Guidance, Control, and Flight Mechanics Conference*, AIAA, New York, 1971.

[28] Powell, B. K., Lang, G. E., Lieberman, S. I., and Rybak, S. C., "Synthesis of Double Gimbal Control Moment Gyro Systems for Spacecraft Attitude Control," *Proceedings of AIAA Guidance, Control, and Flight Mechanics Conference*, AIAA, New York, 1971.

[29] Ross, C. H., and Worley, E., "Optimized Momentum and Attitude Control System for Skylab," *Proceedings of AIAA Guidance, Control, and Flight Mechanics Conference*, AIAA, New York, 1971.

[30] Yoshikawa, T., "A Steering Law for Three Double-Gimballed Control Moment Gyro System," NASA TMX-64926, March 1975.

[31] Yoshikawa, T., "Steering Law for Roof-Type Configuration Control Moment Gyro System," *Automatica*, Vol. 13, No. 4, 1977, pp. 359–368.

[32] Margulies, G., and Aubrun, J.-N., "Geometric Theory of Single-Gimbal Control Moment Gyro Systems," *Journal of the Astronautical Sciences,* Vol. 26, No. 2, 1978, pp. 159–191.

[33] Cornick, D. E., "Singularity Avoidance Control Laws for Single Gimbal Control Moment Gyros," *Proceedings of the AIAA Guidance and Control Conference*, AIAA, New York, 1979, pp. 20–33.

[34] Kennel, H. F., "Steering Law for Parallel Mounted Double-Gimballed Control Moment Gyros—Revision A," NASA TM-82390, Jan. 1981.

[35] Kennel, H. F., "Double-Gimballed Control Moment Gyro Steering Law Update," NASA Internal Memo ED12-86-52, May, 1986.

[36] Kurokawa, H., Yajima, S., and Usui, S., "A New Steering Law of a Single-Gimbal CMG System of Pyramid Configuration," *Proceedings of the 10th IFAC Symposium on Automatic Control in Space*, Pergamon, Oxford, England, UK, 1985, pp. 249–255.

[37] Paradiso, J. A., "A Highly Adaptable Steering/Selection Procedure for Combined CMG/RCS Spacecraft Control," *Advances in Astronautical Sciences*, Vol. 61, Feb. 1986, pp. 263–280.

[38] Branets, V. N., et al., "Development Experience of the Attitude Control System Using Single-Axis Control Moment Gyros for Long-Term Orbiting Space Stations," *38th Congress of the International Astronautical Federation*, IAF-87-04, Oct. 1987.

[39] Bedrossian, N. S., Paradiso, J., Bergmann, E. V., and Rowell, D., "Steering Law Design for Redundant Single-Gimbal Control Moment Gyroscopes," *Journal of Guidance, Control, and Dynamics*, Vol. 13, No. 6, 1990, pp. 1083–1089.

[40] Bedrossian, N. S., Paradiso, J., Bergmann, E. V., and Rowell, D., "Redundant Single-Gimbal Control Moment Gyroscope Singularity Analysis," *Journal of Guidance, Control, and Dynamics*, Vol. 13, No. 6, 1990, pp. 1096–1101.

[41] Vadali, S. R., Oh, H., and Walker, S., "Preferred Gimbal Angles for Single Gimbal Control Moment Gyroscopes," *Journal of Guidance, Control, and Dynamics*, Vol. 13, No. 6, 1990, pp. 1090–1095.

[42] Dzielski, J., Bergmann, E., Paradiso, J. A., Rowell, D., and Wormley, D., "Approach to Control Moment Gyroscope Steering Using Feedback Linearization," *Journal of Guidance, Control, and Dynamics*, Vol. 14, No. 1, 1991, pp. 96–106.

[43] Oh, H.-S., and Vadali, S. R., "Feedback Control and Steering Laws for Spacecraft Using Single Gimbal Control Moment Gyroscopes," *Journal of the Astronautical Sciences*, Vol. 39, No. 2, 1991, pp. 183–203.

[44] Paradiso, J. A., "Global Steering of Single Gimballed Control Moment Gyroscopes Using a Direct Search," *Journal of Guidance, Control, and Dynamics*, Vol. 15, No. 5, 1992, pp. 1236–1244.

[45] Hoelscher, B. R., and Vadali, S. R., "Optimal Open-Loop and Feedback Control Using Single Gimbal Control Moment Gyroscopes," *Journal of the Astronautical Sciences*, Vol. 42, No. 2, 1994, pp. 189–206.

[46] Kurokawa, H., "Exact Singularity Avoidance Control of the Pyramid Type CMG System," *Proceedings of the AIAA Guidance, Navigation, and Control Conference*, AIAA, Washington, DC, 1994, pp. 170–180.

[47] Vadali, S. R., and Krishnan, S., "Suboptimal Command Generation for Control Moment Gyroscopes and Feedback Control of Spacecraft," *Journal of Guidance, Control, and Dynamics*, Vol. 18, No. 6, 1995, pp. 1350–1354.

[48] "Technical Description Document for the PG-1 Guidance, Navigation and Control System," Space Station Div., MDC 95H0223, McDonnell Douglas Aerospace, Houston, TX, March 1996.

[49] Kurokawa, H., "Constrained Steering Law of Pyramid-Type Control Moment Gyros and Ground Tests," *Journal of Guidance, Control, and Dynamics*, Vol. 20, No. 3, 1997, pp. 445–449.

[50] Heiberg, C. J., Bailey, D., and Wie, B., "Precision Pointing Control of Agile Spacecraft Using Single-Gimbal Control Moment Gyros," *Journal of Guidance, Control, and Dynamics*, Vol. 23, No. 1, 2000, pp. 77–85.

[51] Meffe, M., and Stocking, G., "Momentum Envelope Topology of Single-Gimbal CMG Arrays for Space Vehicle Control," AAS Paper No. 87-002, Jan. 31–Feb. 4, 1987.

[52] Nakamura, Y., and Hanafusa, H., "Inverse Kinematic Solutions with Singularity Robustness for Robot Manipulator Control," *Journal of Dynamic Systems, Measurement and Control*, Vol. 108, Sept. 1986, pp. 163–171.

[53] "Shuttle On-Orbit Flight Control Characterization (Simplified Digital Autopilot Model)," C. S. Draper Lab., NASA, JSC 18511, Aug. 1982.

[54] Hattis, P. D., "Qualitative Differences Between Shuttle On-Orbit and Transition Control," *Journal of Guidance, Control, and Dynamics*, Vol. 7, No. 1, 1984, pp. 4–8.

[55] Cox, K. J., and Hattis, P. D., "Shuttle Orbit Flight Control Design Lessons: Direction for Space Station," *Proceedings of the IEEE*, Vol. 75, No. 3, 1987, pp. 336–355.

[56] Bergmann, E. V., Croopnick, S. R., Turkovich, J. J., and Works, C. C., "An Advanced Spacecraft Autopilot Concept," *Journal of Guidance and Control*, Vol. 2, No. 3, 1979, pp. 161–168.

[57] Luenberger, D. G., *Linear and Non-linear Programming*, 2nd ed., Addison–Wesley, Reading, MA, 1989.

[58] Glandorf, D. R., "An Innovative Approach to the Solution of a Class of Linear Programming Problems," Lockheed Engineering and Management Services Co., Inc., LEMSCO-23244, Houston, TX, Nov. 1986.

[59] Kubiak, E. T., and Johnson, D. A., "Optimized Jet Selection Logic," NASA Internal Memo EH2-87L-038, March 1987.

[60] Wie, B., and Plescia, C. T., "Attitude Stabilization of a Flexible Spacecraft During Stationkeeping Maneuvers," *Journal of Guidance, Control, and Dynamics*, Vol. 7, No. 4, 1984, pp. 430–436.

[61] Anthony, T., Wie, B., and Carroll, S., "Pulse-Modulated Controller Synthesis for a Flexible Spacecraft," *Journal of Guidance, Control, and Dynamics*, Vol. 13, No. 6, 1990, pp. 1014–1022.

[62] Song, G., Buck, N., and Agrawal, B., "Spacecraft Vibration Reduction Using Pulse-Width Pulse-Frequency Modulated Input Shaper," *Proceedings of the 1997 AIAA Guidance, Navigation, and Control Conference*, AIAA, Reston, VA, 1997, pp. 1535–1549.

[63] Ieko, T., Ochi, Y., and Kanai, K., "A New Digital Redesign Method for Pulse-Width Modulation Control Systems," *Proceedings of the 1997 AIAA Guidance, Navigation, and Control Conference*, AIAA, Reston, VA, 1997, pp. 1730–1737.

[64] Bernelli-Zazzera, F., Mantegazza, P., and Nurzia, V., "Multi-Pulse-Width Modulated Control of Linear Systems," *Journal of Guidance, Control, and Dynamics*, Vol. 21, No. 1, 1998, pp. 64–70.

Part 4
Structural Dynamics and Control

8
Structural Dynamics

This chapter, which is based on Refs. 1–6, introduces physical concepts and mathematical tools that are useful for structural dynamic modeling and control of flexible spacecraft. Particular emphasis is placed on developing mathematical models of flexible structures for control analysis and design. Various generic models of flexible structures, which are simple enough to treat analytically, yet complicated enough to demonstrate the practicality of pole–zero modeling of structural dynamic systems, are treated in this chapter.

8.1 Introduction

Simple structural elements such as bars, beams, and plates are often treated as *distributed parameter systems* that are described by partial differential equations. Most spacecraft, including the INTELSAT V spacecraft shown in Fig. 8.1 and the International Space Station illustrated in Fig. 8.2, often consist of many lumped and beam- or trusslike subsystems with fairly complex interconnections. In practice, such complex structures are usually modeled as *lumped parameter systems* using finite element methods. In certain cases, distributed parameter modeling of beam- and trusslike structures is more effective than conventional lumped- or discrete-parameter modeling using finite element methods. Consequently, the modeling and control of hybrid or flexible multibody systems with complex interconnections of lumped and distributed parameter subsystems is of practical interest.

In this chapter, we consider generic models of flexible structures from the discrete-spectrum viewpoint of distributed parameter systems. The models are simple enough to treat analytically, yet complicated enough to demonstrate the practical usefulness of transcendental-transfer-function modeling for the purposes of control analysis and design [1–6]. Transfer functions of the various generic models are derived analytically, and their pole–zero patterns are studied. The generic models to be studied in this chapter will also provide physical insights into the dynamics and control of more complex structures. Although controlling flexible structures using many actuators and sensors is of much current research interest, we focus on the fundamental issue of controlling a flexible spacecraft using a single pair of actuator and sensor in each axis.

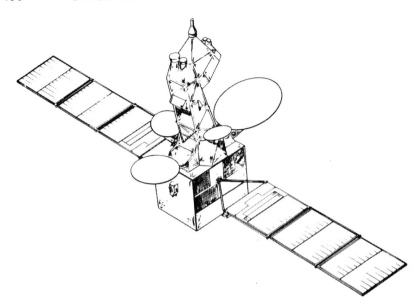

Fig. 8.1   INTELSAT V spacecraft. (Courtesy of Ford Aerospace and Communications Corp., Palo Alto, CA.)

Fig. 8.2   International Space Station. (Courtesy of John Frassanito and Associates.)

8.2   Uniform Bars

In this section we consider a uniform free-free bar as a generic model of bar-like structures with longitudinal vibrations. Basic concepts of modal analysis and modal truncation for distributed parameter systems are discussed using such a simple model. Transcendental transfer functions of a uniform free-free bar are derived, and the effects of actuator and sensor locations on the transfer function zeros are also discussed in this section.

8.2.1   Modal Analysis

Consider a uniform free-free bar of length ℓ with a longitudinal control force $u(t)$ applied at the end $x = 0$, as illustrated in Fig. 8.3. The equation of motion is given by the wave equation in dimensionless form

$$y''(x,t) - \ddot{y}(x,t) = 0 \tag{8.1}$$

where $y'' \equiv \partial^2 y/\partial x^2$, $\ddot{y} \equiv \partial^2 y/\partial t^2$, $y(x,t)$ is the longitudinal displacement at the location x and at time t, x and y are in units of ℓ, time is in units of $\sqrt{\sigma \ell^2/EA}$, EA denotes the axial stiffness, and σ denotes the mass per unit length.

The boundary conditions are given by

$$y'(1,t) = 0 \tag{8.2a}$$
$$-y'(0,t) = u(t) \tag{8.2b}$$

where $u(t)$ is in units of EA.

Let the displacement $y(x,t)$ be expressed as

$$y(x,t) = \sum_{i=0}^{\infty} \phi_i(x) q_i(t) \tag{8.3}$$

where $\phi_i(x)$ is the ith normal mode shape and $q_i(t)$ is the ith modal coordinate. Substituting Eq. (8.3) into Eq. (8.1) and using the boundary conditions (8.2), we obtain the modal equations

$$\ddot{q}_i(t) + \omega_i^2 q_i(t) = f_i(t), \qquad i = 0, \ldots, \infty \tag{8.4}$$

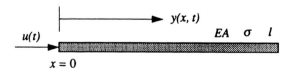

Fig. 8.3   Uniform free-free bar with longitudinal vibration.

where ω_i is the ith modal frequency in units of $\sqrt{\sigma \ell^2/EA}$ and $f_i(t)$ is the ith generalized force in units of EA. We also have

$$\omega_i = i\pi, \qquad\qquad i = 0,\ldots,\infty \qquad (8.5a)$$

$$\phi_i(x) = \begin{cases} 1, & i = 0 \\ \sqrt{2}\cos i\pi x, & i = 1,\ldots,\infty \end{cases} \qquad (8.5b)$$

$$f_i(t) = \phi_i(0)u(t) = \begin{cases} u(t), & i = 0 \\ \sqrt{2}u(t), & i = 1,\ldots,\infty \end{cases} \qquad (8.5c)$$

The transfer function from control force to structural displacement at the location x can then be written as

$$\frac{y(x,s)}{u(s)} = \sum_{i=0}^{\infty} \frac{a_i(x)}{s^2 + \omega_i^2} = \frac{a_0}{s^2} + \sum_{i=1}^{\infty} \frac{a_i(x)}{s^2 + \omega_i^2} \qquad (8.6)$$

where $a_i(x) \equiv \phi_i(0)\phi_i(x)$ is the residue of the ith mode, $a_0 = 1$, and $u(s)$ and $y(x,s)$ are the Laplace transforms of $u(t)$ and $y(x,t)$, respectively. The first term in Eq. (8.6) is called the rigid-body mode.

Because all structures dissipate energy during elastic deformations, we introduce the modal damping ratio ζ_i so that Eq. (8.6) is modified to

$$\frac{y(x,s)}{u(s)} = \frac{a_0}{s^2} + \sum_{i=1}^{\infty} \frac{a_i(x)}{s^2 + 2\zeta_i\omega_i s + \omega_i^2} \qquad (8.7)$$

The transfer functions (8.6) and (8.7) are said to be expressed as *partial fraction expansion* with exact poles and residues.

8.2.1.1 Controllability and Observability

The residue a_i of the ith mode can be considered the product of the controllability coefficient $\mathcal{C}_i$ and the observability coefficient $\mathcal{O}_i$ such that

$$a_i \equiv \mathcal{C}_i\mathcal{O}_i \qquad (8.8)$$

If either coefficient is zero, then the ith mode does not appear in the transfer function, which corresponds to the pole–zero cancellation in the transfer function expressed in product expansion.

For a uniform free-free bar with control force at the end $x = 0$, we have

$$\mathcal{C}_i = \phi_i(0) = \begin{cases} 1, & i = 0 \\ \sqrt{2}, & i = 1,\ldots,\infty \end{cases} \qquad (8.9a)$$

$$\mathcal{O}_i = \phi_i(x) = \begin{cases} 1, & i = 0 \\ \sqrt{2}\cos i\pi x, & i = 1,\ldots,\infty \end{cases} \qquad (8.9b)$$

Thus, all of the vibration modes, plus the rigid-body mode, are controllable by the control force applied at the boundary. This is intuitively trivial because the boundary cannot be a nodal point of any mode shape. The rigid-body mode is always observable for any sensor location, but observability of a vibration mode depends on the sensor location. This is also obvious because, if the sensor is located at the nodal point of the ith mode, we cannot observe the ith mode with that sensor. (See Chapter 2 for the more rigorous definitions of controllability and observability of linear dynamic systems.)

8.2.1.2 Modal Truncation The transfer functions (8.6) and (8.7) include an infinite number of modes. In practice, they are truncated to a finite-dimensional model. One of the simplest methods of reducing the order of dynamic systems is *modal truncation*. Modal truncation criteria can be obtained by considering the impulse response of Eq. (8.7) with zero initial conditions, given by

$$y(x,t) = a_0 t + \sum_{i=1}^{\infty} \frac{a_i(x)}{\omega_i \sqrt{1 - \zeta_i^2}} e^{-\zeta_i \omega_i t} \sin \sqrt{1 - \zeta_i^2} \omega_i t \qquad (8.10)$$

The modal gain $g_i(x,t)$ of the ith flexible mode at the location x and time t is defined as

$$g_i(x,t) = \frac{a_i(x)}{\omega_i \sqrt{1 - \zeta_i^2}} e^{-\zeta_i \omega_i t} \qquad (8.11)$$

Note that the modal gain g_i as defined here is, in fact, the amplitude of the ith mode, and that it is inversely proportional to the modal frequency ω_i. Also note that the exponential term in Eq. (8.11) determines the so-called slow and fast modes. Consequently, the high-frequency modes are often neglected in designing control logic for flexible structures if they have adequate natural damping. The residue a_i, however, is the most important factor to be considered in selecting the dominant modes for the given actuator and sensor locations. The residue a_i might be zero or almost zero for a certain actuator and sensor location, which corresponds to the exact pole–zero cancellation or near pole–zero cancellation, respectively.

From the foregoing discussion, we can conclude that infinite-dimensional models can be well approximated by finite-dimensional models by neglecting modes with frequencies that are much higher than the control system bandwidth, as well as modes with small residues. This simplifies control design by eliminating the modes that are nearly uncontrollable and/or unobservable for the given actuator and sensor pair. Also, order reduction by modal truncation may be regarded as minimizing the mean square error between the exact and an approximate impulse response of the system.

8.2.1.3 Collocated Actuator and Sensor If a displacement sensor is located at the end $x = 0$, then Eq. (8.6) becomes

$$\frac{y(0,s)}{u(s)} = \frac{1}{s^2} + \sum_{i=1}^{\infty} \frac{2}{s^2 + \omega_i^2} \qquad (8.12)$$

Note that all residues in Eq. (8.12) have the same sign. In general, if an actuator and a compatible output sensor are placed at the same location on an undamped elastic body, the resulting transfer function will have alternating poles and zeros on the imaginary axis. This can be proved by using the fact that the residues a_i all have the same sign when the actuator and sensor are collocated. Thus, the transfer function represented by Eq. (8.12) will have alternating poles and zeros on the imaginary axis because all of the residues in Eq. (8.12) have the same sign.

A reduced-order model of Eq. (8.12) is obtained by simply taking a finite number of modes, as follows:

$$\frac{y(0, s)}{u(s)} = \frac{1}{s^2} + \sum_{i=1}^{N} \frac{2}{s^2 + \omega_i^2} \tag{8.13}$$

For $N = 1$, we have

$$\frac{y(0, s)}{u(s)} = \frac{3[s^2 + (1.813)^2]}{s^2(s^2 + \omega_1^2)} \tag{8.14}$$

For $N = 2$, we obtain

$$\frac{y(0, s)}{u(s)} = \frac{5[s^2 + (1.708)^2][s^2 + (5.166)^2]}{s^2(s^2 + \omega_1^2)(s^2 + \omega_2^2)} \tag{8.15}$$

where $\omega_1 = \pi$ and $\omega_2 = 2\pi$. These transfer functions are finite dimensional, reduced-order models with exact residues of the model given by Eq. (8.6). The zeros are not identical to the exact zeros, which will be found later, and they depend on the number of modes retained. These approximate zeros are still on the imaginary axis, however, and they approach the exact zeros as the number of modes increases.

8.2.1.4 Separated Actuator and Sensor
When actuator and sensor are separated on flexible structures, there are no simple properties of the pole–zero patterns. To investigate the pole–zero patterns for the case of a separated actuator and sensor pair, consider a bar with actuator and sensor located at opposite ends ("boundary control and boundary observation").

If a displacement sensor is located at the right end $x = 1$, Eq. (8.6) with a finite number of modes becomes

$$\frac{y(1, s)}{u(s)} = \frac{1}{s^2} + \sum_{i=1}^{N} \frac{2(-1)^i}{s^2 + \omega_i^2} \tag{8.16}$$

The residues in this transfer function do not have the same sign for all of the modes. Thus, the transfer function with a separated actuator and sensor pair does not have alternating poles and zeros along the imaginary axis. For example, when $N = 1$, we have

$$\frac{y(1, s)}{u(s)} = \frac{-(s + \omega_1)(s - \omega_1)}{s^2(s^2 + \omega_1^2)} \tag{8.17}$$

which has real zeros. For $N = 2$, we have

$$\frac{y(1,s)}{u(s)} = \frac{(s \pm 4.7123)^2 \pm (0.2054)^2}{s^2(s^2 + \omega_1^2)(s^2 + \omega_2^2)} \tag{8.18}$$

which has complex zeros. These transfer functions (8.17) and (8.18) have nonminimum-phase zeros (right-half s-plane zeros). It will be shown later that the transfer function $y(1,s)/u(s)$ with an infinite number of modes has no zeros (except at infinity). Thus, the zeros in Eq. (8.17) and Eq. (8.18) are clearly due to modal truncation.

8.2.2 Finite Element Models

One of the methods of determining the vibration modes of a flexible structure is the finite element method. It is the most practical method for analyzing complex structures. The first step in the finite element modeling of an elastic structure is to divide it into several elements. Points where the elements are connected are called the "nodal" points. Unfortunately, the same terminology is also used in the discussion of mode shapes, but the difference is usually clear from the context. The displacements of these nodal points then become the *generalized coordinates* of the structure.

By evaluating the physical properties, such as stiffness and mass matrices, of the individual finite elements and combining them appropriately, we can find the equations of motion of the complete structure. The stiffness matrix of a single finite element can be derived by defining the so-called interpolation function and using the principle of virtual displacement.

The stiffness matrix of the complete structure can then be determined by adding the element stiffness coefficients, which is known as the direct stiffness method. Before the element stiffness coefficients are added, they must be expressed in global coordinates. The stiffness matrix of the complete structure is all that is needed for static analysis. For dynamic analysis the mass matrix of the structure must also be found. Two types of mass matrices are used in the finite element method: a lumped-mass matrix and a consistent-mass matrix.

8.2.2.1 Lumped-Mass Matrix The simplest way of defining the mass properties of a finite element model is to lump the element masses at the nodal points. The lumped-mass matrix is then a diagonal matrix. For example, the equations of motion of a uniform free-free bar consisting of two finite elements, as shown in Fig. 8.4a, is given by

$$\frac{1}{4}\begin{bmatrix} 1 & 0 & 0 \\ 0 & 2 & 0 \\ 0 & 0 & 1 \end{bmatrix}\begin{bmatrix} \ddot{y}_1 \\ \ddot{y}_2 \\ \ddot{y}_3 \end{bmatrix} + 2\begin{bmatrix} 1 & -1 & 0 \\ -1 & 2 & -1 \\ 0 & -1 & 1 \end{bmatrix}\begin{bmatrix} y_1 \\ y_2 \\ y_3 \end{bmatrix} = \begin{bmatrix} u \\ 0 \\ 0 \end{bmatrix} \tag{8.19}$$

where y_1, y_2, and y_3 are the nodal displacements in units of ℓ, u is the control force in units of EA applied at the left end, and time is in units of $\sqrt{\sigma \ell^2 / EA}$.

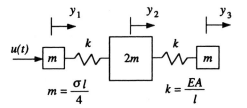

a) Lumped-mass matrix

b) Consistent-mass matrix

Fig. 8.4   Finite element models of a uniform free-free bar with two elements.

Taking the Laplace transforms of Eq. (8.19), we obtain

$$
\begin{bmatrix} y_1(s) \\ y_2(s) \\ y_3(s) \end{bmatrix} = \frac{1}{\Delta(s)} \begin{bmatrix} 4[s^2 + (1.530)^2][s^2 + (3.695)^2] \\ 16[s^2 + (2.828)^2] \\ 128 \end{bmatrix} u(s) \qquad (8.20)
$$

where

$$
\Delta(s) = s^2[s^2 + (2.828)^2][s^2 + (4.0)^2]
$$

The poles and zeros of these transfer functions are not the same as exact poles and zeros. Obviously, the accuracy increases as the number of elements increases. The exact pole–zero cancellation of the first mode in $y_2(s)/u(s)$ means that the first mode is unobservable with a sensor located at the middle of the bar.

8.2.2.2   Consistent-Mass Matrix

The consistent-mass matrix is found using the same interpolation functions that are used for evaluating the stiffness matrix of a finite element. It has an off-diagonal term leading to mass coupling, so that dynamic analysis with a consistent-mass matrix generally requires more computational time than with a lumped-mass matrix. However, a consistent-mass model produces more accurate natural frequencies and mode shapes than a lumped-mass model.

For example, the equations of motion of a uniform free-free bar consisting of two finite elements, as shown in Fig. 8.4b, is given by

$$
\frac{1}{12} \begin{bmatrix} 2 & 1 & 0 \\ 1 & 4 & 1 \\ 0 & 1 & 2 \end{bmatrix} \begin{bmatrix} \ddot{y}_1 \\ \ddot{y}_2 \\ \ddot{y}_3 \end{bmatrix} + 2 \begin{bmatrix} 1 & -1 & 0 \\ -1 & 2 & -1 \\ 0 & -1 & 1 \end{bmatrix} \begin{bmatrix} y_1 \\ y_2 \\ y_3 \end{bmatrix} = \begin{bmatrix} u \\ 0 \\ 0 \end{bmatrix} \qquad (8.21)
$$

where y_1, y_2, and y_3 are the nodal displacements in units of ℓ, u is the control force in units of EA applied at the left end, and time is in units of $\sqrt{\sigma \ell^2 / EA}$.

Taking the Laplace transforms of Eq. (8.21), we obtain

$$
\begin{bmatrix} y_1(s) \\ y_2(s) \\ y_3(s) \end{bmatrix} = \frac{1}{\Delta(s)} \begin{bmatrix} 7[s^2 + (1.611)^2][s^2 + (5.629)^2] \\ -2[s^2 + (3.464)^2][s \pm 4.898] \\ (s \pm 4.898)^2 \end{bmatrix} u(s) \qquad (8.22)
$$

where

$$
\Delta(s) = s^2[s^2 + (3.464)^2][s^2 + (6.928)^2]
$$

It is clear that the consistent-mass model produces more accurate poles than the lumped-mass model; however, the consistent-mass model produces a nonminimum-phase transfer function for the case of separated actuator and sensor. Such approximate zeros come from minimizing the error between the exact and approximate impulse responses of the system.

For more details of the finite element method as applied to structural dynamics, the reader is referred to Refs. 7–10.

8.2.3 Transcendental Transfer Functions

In the preceding sections we have studied pole–zero patterns of reduced-order models by modal truncation or by the finite element method. For a simple structure such as the uniform bar, we can derive exact transcendental transfer functions with exact poles and zeros.

Taking the Laplace transforms of Eqs. (8.1) and (8.2), we obtain

$$
y''(x, s) - s^2 y(x, s) = 0 \qquad (8.23)
$$
$$
y'(1, s) = 0 \qquad (8.24)
$$
$$
- y'(0, s) = u(s) \qquad (8.25)
$$

and we can find the transcendental transfer function as

$$
\frac{y(x, s)}{u(s)} = \frac{\cosh s(1 - x)}{s \sinh s} \qquad (8.26)
$$

which can be expanded into infinite product form

$$
\frac{y(x, s)}{u(s)} = \frac{1}{s^2} \prod_{i=1}^{\infty} \frac{(s/z_i)^2 + 1}{(s/\omega_i)^2 + 1} \qquad (8.27)
$$

where

$$
\omega_i = i\pi
$$
$$
z_i = \frac{i - 0.5}{1 - x}\pi, \qquad (x \neq 1)
$$

Thus, when the actuator and sensor are collocated at the end $x = 0$, the transfer function (8.27) has alternating poles $\omega_i = i\pi$ and zeros $z_i = (i - 0.5)\pi$ along the imaginary axis. When the sensor is located at the middle of the bar ($x = 0.5$), all of the zeros $z_i = (2i - 1)\pi$ are canceled by poles; these poles correspond to the symmetric vibration modes, which are unobservable with a sensor at $x = 0.5$. When the sensor is located at the opposite end ($x = 1$), there are no zeros in Eq. (8.27).

We can obtain reduced-order models with exact poles and zeros by truncating the infinite product expansion Eq. (8.27). Where to truncate depends on the bandwidth of the control system and on other factors such as near pole–zero cancellations and damping in the system. For control design purposes, it is not evident which finite-dimensional model (exact residues or exact zeros) is better. It is probably better to use the model with exact residues for low-gain control and the model with exact zeros for high-gain control because the closed-loop poles with high gain are close to the open-loop zeros.

When the actuator and sensor are collocated at $x = 0$, Eq. (8.26) becomes

$$\frac{y(0, s)}{u(s)} = \frac{\cosh s}{s \sinh s} \tag{8.28}$$

which can be written as

$$\frac{y(0, s)}{u(s)} = \frac{e^s + e^{-s}}{s(e^s - e^{-s})} = \frac{1}{s(1 - e^{-2s})} + \frac{e^{-2s}}{s(1 - e^{-2s})} \tag{8.29}$$

For the case of a sensor located at the right end $x = 1$, we have

$$\frac{y(1, s)}{u(s)} = \frac{1}{s \sinh s} \tag{8.30}$$

which becomes

$$\frac{y(1, s)}{u(s)} = \frac{2}{s(e^s - e^{-s})} = \frac{2e^{-s}}{s(1 - e^{-2s})} \tag{8.31}$$

Problems

8.1 For a uniform free-free bar with both actuator and sensor collocated at $x = 0$, sketch the exact impulse response $y(0, t)$ to a unit impulse control force, i.e., $u(s) = 1$. Then compare the exact impulse response and the impulse responses of reduced-order models given by Eqs. (8.14) and (8.15).

8.2 For a uniform free-free bar with actuator and sensor located at opposite ends (actuator at $x = 0$ and sensor at $x = 1$), sketch the exact impulse response $y(1, t)$ to a unit impulse control force, i.e., $u(s) = 1$. Then compare the exact impulse response and the impulse responses of reduced-order models given by Eqs. (8.17) and (8.18).

8.3 Uniform Beams

In this section we consider a uniform free-free beam as a generic model of a beamlike space structure with transverse bending vibrations. Simple beams have been often used as generic models of flexible launch vehicles in connection with preliminary control analysis and design.

We shall discuss the characteristics of beam vibrations in terms of transfer function poles and zeros. We shall review modal analysis of a uniform free-free beam and study the effects of actuator and sensor locations and modal truncation on transfer function zeros. Transfer functions from finite element models are also compared with transcendental transfer functions. In particular, nonminimum-phase characteristics of beamlike structures are discussed.

8.3.1 Modal Analysis

For a uniform free-free beam as shown in Fig. 8.5, the Bernoulli–Euler model, neglecting shear distortion and rotary inertia, is given in dimensionless form as

$$y''''(x,t) + \ddot{y}(x,t) = 0 \qquad (8.32)$$

where $y'''' \equiv \partial^4 y/\partial x^4$; $\ddot{y} \equiv \partial^2 y/\partial t^2$; $y(x,t)$ is the transverse displacement at the location x and at time t; x and y are in units of ℓ; time is in units of $\sqrt{\sigma \ell^4/EI}$; and EI, σ, and ℓ denote the bending stiffness, the mass density per unit length, and the total length of the beam, respectively.

For the case with a control force at one end, e.g., a flexible rocket, the boundary conditions are given by

$$y''(0,t) = y''(1,t) = 0 \qquad (8.33a)$$

$$y'''(0,t) = u(t) \qquad (8.33b)$$

$$y'''(1,t) = 0 \qquad (8.33c)$$

where $u(t)$ denotes transverse control force (in units of EI/ℓ^2) applied at the left end $x = 0$.

Let the displacement $y(x,t)$ be expressed as

$$y(x,t) = \sum_{i=0}^{\infty} \phi_i(x) q_i(t) \qquad (8.34)$$

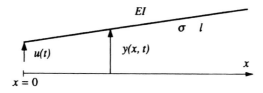

Fig. 8.5   Uniform free-free beam with transverse vibration.

where $\phi_i(x)$ is the ith normal mode shape and $q_i(t)$ is the ith modal coordinate. Substituting Eq. (8.34) into Eq. (8.32) and using the boundary conditions (8.33), we obtain the modal equations

$$\ddot{q}_i(t) + \omega_i^2 q_i(t) = f_i(t), \qquad i = 0, \ldots, \infty \qquad (8.35)$$

where ω_i is the ith modal frequency in units of $\sqrt{EI/\sigma\ell^4}$ and $f_i(t)$ is the ith generalized force in units of EI/ℓ^2. We also have

$$\omega_i^2 = \lambda_i^4, \qquad i = 0, \ldots, \infty \qquad (8.36a)$$

$$\phi_0(x) = 2 - 3x \qquad (8.36b)$$

$$\phi_i(x) = \cos \lambda_i x + \cosh \lambda_i x - \frac{\cos \lambda_i - \cosh \lambda_i}{\sin \lambda_i - \sinh \lambda_i}(\sin \lambda_i x + \sinh \lambda_i x) \qquad (8.36c)$$

$$f_i(t) = \phi_i(0)u(t) = 2u(t), \qquad i = 0, \ldots, \infty \qquad (8.36d)$$

where λ_i can be obtained from the characteristic equation

$$\cos \lambda_i \cosh \lambda_i - 1 = 0$$

For example, we have $\lambda_0 = 0$, $\lambda_1 = 1.506\pi$, and $\lambda_2 = 2.5\pi$.

The transfer function from control force u to structural displacement at the location x can then be written as

$$\frac{y(x, s)}{u(s)} = \sum_{i=0}^{\infty} \frac{a_i(x)}{s^2 + \omega_i^2} = \frac{a_0}{s^2} + \sum_{i=1}^{\infty} \frac{a_i(x)}{s^2 + \omega_i^2} \qquad (8.37)$$

where $a_0 = 4$; $a_i(x) \equiv \phi_i(0)\phi_i(x)$ is the residue of the ith mode; and $\omega_0 = 0$, $\omega_1 = 22.373$, $\omega_2 = 61.672$, and $\omega_3 = 120.903$, etc.

Because all structures dissipate energy during elastic deformations, we introduce the modal damping ratio ζ_i so that Eq. (8.37) is modified to

$$\frac{y(x, s)}{u(s)} = \frac{a_0}{s^2} + \sum_{i=1}^{\infty} \frac{a_i(x)}{s^2 + 2\zeta_i \omega_i s + \omega_i^2} \qquad (8.38)$$

8.3.1.1 Controllability and Observability

The residue a_i of the ith mode is the product of the controllability coefficient $\mathcal{C}_i$ and the observability coefficient $\mathcal{O}_i$ such that

$$a_i \equiv \mathcal{C}_i \mathcal{O}_i \qquad (8.39)$$

where

$$\mathcal{C}_i = \phi_i(0) = 2, \qquad i = 0, \ldots, \infty \qquad (8.40a)$$

$$\mathcal{O}_i = \begin{cases} 2, & i = 0 \\ \phi_i(x), & i = 1, \ldots, \infty \end{cases} \qquad (8.40b)$$

Hence, all of the vibration modes plus the rigid-body mode are controllable by the control force applied at one end. The rigid-body mode is always observable for any sensor location, but the observability of vibration modes depends on the sensor location. There are two rigid-body modes: a translational mode and a rotational mode. By using a single control force at one end, we can control only linear combination of the two rigid-body modes. Similarly, we can observe only a linear combination of those rigid-body modes by using a single sensor. Thus, strictly speaking, a beam with transverse vibration is not completely controllable and observable using a single control force and a single sensor. This is a special case of the repeated-eigenvalue problem encountered in controlling symmetric structures. At least two controls and two sensors are needed to control and observe these two rigid-body modes independently; however, all of the vibration modes are controllable and observable with single control force and single sensor at one end.

8.3.1.2 Collocated Actuator and Sensor

If a displacement sensor is located at the end $x = 0$, then Eq. (8.37) becomes

$$\frac{y(0,s)}{u(s)} = \frac{4}{s^2} + \sum_{i=1}^{\infty} \frac{4}{s^2 + \omega_i^2} \tag{8.41}$$

It is seen that the residues are all equal for all of the modes. Consequently, Eq. (8.41) will have alternating poles and zeros on the imaginary axis. A reduced-order model of Eq. (8.41) can be obtained by simply taking a finite number of modes, as was done for a uniform bar in the preceding section. A reduced-order model with exact residues will also have alternating poles and zeros on the imaginary axis. Hence, there are no serious modeling errors in the location of zeros for the case with a collocated actuator and sensor; this is not the case with a separated actuator and sensor.

8.3.1.3 Separated Actuator and Sensor

When actuator and sensor are separated on flexible structures, there are no simple properties of the pole–zero patterns. For the special case with a displacement sensor at the right end $x = 1$, Eq. (8.37) becomes

$$\frac{y(1,s)}{u(s)} = \frac{2}{s^2} + \sum_{i=1}^{\infty} \frac{4(-1)^{i+1}}{s^2 + \omega_i^2} \tag{8.42}$$

The residues in this transfer function do not have the same sign for all of the modes. Thus, this transfer function will not have alternating poles and zeros along the imaginary axis.

A reduced-order model with exact residues is then obtained by taking a finite number of modes. For example, we have

$$\frac{y(1,s)}{u(s)} = \frac{-2(s + \omega_1)(s - \omega_1)}{s^2(s^2 + \omega_1^2)} \tag{8.43}$$

or

$$\frac{y(1,s)}{u(s)} = \frac{-2[(s \pm 35.572)^2 \pm (10.698)^2]}{s^2(s^2 + \omega_1^2)(s^2 + \omega_2^2)} \qquad (8.44)$$

These transfer functions (8.43) and (8.44) have nonminimum-phase zeros. Clearly the zero locations depend on the number of modes chosen in modal truncation. It will be shown later that the exact transfer function $y(1,s)/u(s)$ with an infinite number of modes has right-half plane real zeros.

8.3.2 Finite Element Models

Before we derive an exact transfer function, consider here transfer functions obtained with finite element modeling. The equations of motion of a uniform free-free beam (see Fig. 8.6) using a one-element model with a consistent-mass matrix are given by

$$\frac{1}{420}\begin{bmatrix} 156 & 54 & 22 & -13 \\ 54 & 156 & 13 & -22 \\ 22 & 13 & 4 & -3 \\ -13 & -22 & -3 & 4 \end{bmatrix}\begin{bmatrix} \ddot{y}_1 \\ \ddot{y}_2 \\ \ddot{y}_3 \\ \ddot{y}_4 \end{bmatrix}$$

$$+ 2\begin{bmatrix} 6 & -6 & 3 & 3 \\ -6 & 6 & -3 & -3 \\ 3 & -3 & 2 & 1 \\ 3 & -3 & 1 & 2 \end{bmatrix}\begin{bmatrix} y_1 \\ y_2 \\ y_3 \\ y_4 \end{bmatrix} = \begin{bmatrix} u \\ 0 \\ 0 \\ 0 \end{bmatrix} \qquad (8.45)$$

where y_1 and y_2 are the nodal transverse displacements in units of ℓ, y_3 and y_4 are the nodal bending slopes, u is the control force in units of EI/ℓ^2 applied at the left end, and time is units of $\sqrt{\sigma \ell^4/EI}$.

Taking the Laplace transforms of Eq. (8.45), we obtain

$$\frac{y_1(s)}{u(s)} = \frac{16}{\Delta(s)}[s^2 + (17.544)^2][s^2 + (70.087)^2] \qquad (8.46)$$

$$\frac{y_2(s)}{u(s)} = \frac{-4}{\Delta(s)}[s \pm 62.496][s \pm 27.825] \qquad (8.47)$$

where

$$\Delta(s) = s^2[s^2 + (26.832)^2][s^2 + (91.651)^2]$$

The poles and zeros of these transfer functions are not the same as exact poles and zeros. Obviously, the accuracy increases as the number of element increases.

Fig. 8.6   Finite element model of a uniform free-free beam with one element.

8.3.3 Transcendental Transfer Function

For a uniform free-free beam we can find the exact zeros by using a transcendental transfer function. Taking the Laplace transforms of Eqs. (8.32) and (8.33), we obtain

$$y''''(x, s) - \lambda^4 y(x, s) = 0 \tag{8.48}$$

where $\lambda^4 \equiv -s^2$, the Laplace transform variable s is in units of $\sqrt{EI/\sigma \ell^4}$, and

$$y''(0, s) = y''(1, s) = 0 \tag{8.49a}$$
$$y'''(0, s) = u(s) \tag{8.49b}$$
$$y'''(1, s) = 0 \tag{8.49c}$$

The solution of Eq. (8.48) is given by

$$y(x, s) = A_1 \sin \lambda x + A_2 \cos \lambda x + A_3 \sinh \lambda x + A_4 \cosh \lambda x \tag{8.50}$$

Using the boundary conditions (8.49), we obtain the transcendental transfer function from the control force $u(s)$ to transverse displacement $y(x, s)$ at the location x, as follows:

$$\frac{y(x, s)}{u(s)} = \frac{1}{\Delta(s)}[(\sinh \lambda \cos \lambda - \cosh \lambda \cos \lambda)(\cosh \lambda x + \cos \lambda x)$$
$$+ \sinh \lambda \sin \lambda(\sinh \lambda x + \sin \lambda x) + (1 - \cosh \lambda)(\sinh \lambda x - \sin \lambda x)] \tag{8.51}$$

where

$$\Delta(s) = 2\lambda^3(1 - \cos \lambda \cosh \lambda)$$

Similarly, we can obtain the transfer function from control force $u(s)$ to slope $\theta(x, s) \equiv y'(x, s)$ at the location x, as follows:

$$\frac{\theta(x, s)}{u(s)} = \frac{1}{\Delta(s)}[(\sinh \lambda \cos \lambda - \cosh \lambda \cos \lambda)(\sinh \lambda x + \sin \lambda x)$$
$$+ \sinh \lambda \sin \lambda(\cosh \lambda x + \cos \lambda x) + (1 - \cosh \lambda)(\cosh \lambda x - \cos \lambda x)] \tag{8.52}$$

In particular, for control and observation (measurement) at the ends of the beam, we have

$$\frac{y(0, s)}{u(s)} = \frac{\sinh \lambda \cos \lambda - \cosh \lambda \sin \lambda}{\lambda^3(1 - \cos \lambda \cosh \lambda)} \tag{8.53a}$$

$$\frac{\theta(0, s)}{u(s)} = \frac{\sinh \lambda \sin \lambda}{\lambda^2(1 - \cos \lambda \cosh \lambda)} \tag{8.53b}$$

$$\frac{y(1, s)}{u(s)} = \frac{\sinh \lambda - \sin \lambda}{\lambda^3(1 - \cos \lambda \cosh \lambda)} \tag{8.53c}$$

$$\frac{\theta(1, s)}{u(s)} = \frac{\cosh \lambda - \cos \lambda}{\lambda^2(1 - \cos \lambda \cosh \lambda)} \tag{8.53d}$$

These transfer functions may be expressed as infinite products:

$$\frac{y(0,s)}{u(s)} = \frac{-4}{\lambda^4} \prod_{i=1}^{\infty} \frac{1-(\lambda/\gamma_i)^4}{1-(\lambda/\beta_i)^4} \tag{8.54a}$$

$$\frac{\theta(0,s)}{u(s)} = \frac{6}{\lambda^4} \prod_{i=1}^{\infty} \frac{1-(\lambda/i\pi)^4}{1-(\lambda/\beta_i)^4} \tag{8.54b}$$

$$\frac{y(1,s)}{u(s)} = \frac{6}{\lambda^4} \prod_{i=1}^{\infty} \frac{1-(\lambda/\sqrt{2}\alpha_i)^4}{1-(\lambda/\beta_i)^4} \tag{8.54c}$$

$$\frac{\theta(1,s)}{u(s)} = \frac{6}{\lambda^4} \prod_{i=1}^{\infty} \frac{1-(\lambda/\sqrt{2}i\pi)^4}{1-(\lambda/\beta_i)^4} \tag{8.54d}$$

where $\tan\alpha_i = \tanh\alpha_i$; $\cos\beta_i \cosh\beta_i = 1$; $\cos\gamma_i \sinh\gamma_i = \sin\gamma_i \cosh\gamma_i$; and α_i, β_i, and γ_i are all positive real numbers.

Because $\lambda^4 \equiv -s^2$ (see Fig. 8.7), the transfer functions $y(0,s)/u(s)$ and $\theta(0,s)/u(s)$ have alternating poles and zeros on the imaginary axis of the s plane. Their

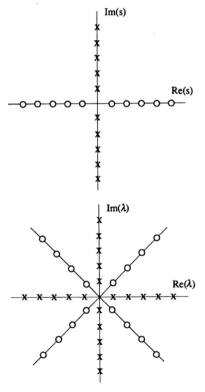

Fig. 8.7   Complex mapping between s plane and λ plane: $\lambda^4 = -s^2$.

lowest poles and zeros are shown in Fig. 8.8. The transfer functions $y(1,s)/u(s)$ and $\theta(1,s)/u(s)$ have an infinite number of zeros on the real axis of the s plane. Their lowest poles and zeros are shown in Fig. 8.9. Thus, the exact transfer function for a beam with separated actuator and sensor can be nonminimum phase with real zeros (no complex zeros), whereas the exact transfer function for a bar with separated actuator and sensor is minimum phase.

When the sensor is located at an arbitrary point (other than one of the boundaries), then, in general, the transfer function will have some sequence of nonalternating poles and zeros on the imaginary axis, along with zeros on the real axis. Also, some of the poles may be canceled by zeros, which means that those modes are unobservable by the sensor.

8.3.4 Physical Interpretation of Transfer Function Zeros

Transfer function poles are the natural frequencies of the structure with the actuator boundary condition equal to zero. Transfer function zeros are the eigenvalues of the structures with the sensor boundary condition equal to zero. For example, the numerator of $y(0,s)/u(s)$ in Eq. (8.53a) is the characteristic equation of a uniform beam with boundary conditions

$$y(0,t) = y''(0,t) = 0 \tag{8.55a}$$

$$y''(1,t) = y'''(1,t) = 0 \tag{8.55b}$$

Thus, the zeros of $y(0,s)/u(s)$ are the natural frequencies of a uniform pinned-free beam. The numerator of $y(1,s)/u(s)$ in Eq. (8.53c) is the characteristic equation of a uniform beam with boundary conditions

$$y''(0,t) = 0 \tag{8.56a}$$

$$y(1,t) = y''(1,t) = y'''(1,t) = 0 \tag{8.56b}$$

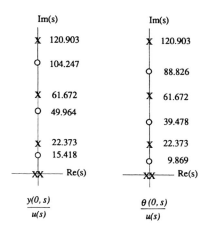

Fig. 8.8   Exact poles and zeros of $y(0,s)/u(s)$ and $\theta(0,s)/u(s)$.

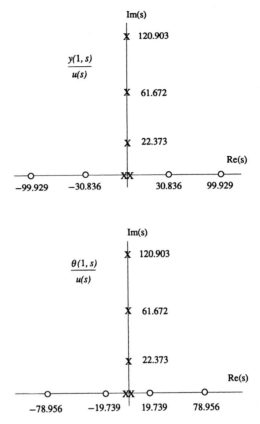

Fig. 8.9 Exact poles and zeros of $y(1,s)/u(s)$ and $\theta(1,s)/u(s)$.

The preceding boundary conditions indicate that the sensor boundary condition is zero whereas the actuator boundary condition is not zero.

Using this simple concept, we can make direct calculations of the exact zeros of the transfer function for a uniform free-free beam. This concept can be extended to the direct calculation of accurate transfer function zeros of complicated structures using the finite element method.

8.3.5 Natural Damping Models

In our earlier discussions we ignored energy dissipation mechanisms during elastic deformation. Structural damping mechanisms are still poorly understood, and appear to be quite complicated. Thus, there are many empirical damping models proposed in the literature. Here we discuss several of these models as they relate to reduced-order beam vibration models.

The equation of motion of a uniform beam with several rate-dependent damping terms may be written in dimensionless form, as follows:

$$y''''(x,t) + D_1\dot{y}'''' + D_2\dot{y}'' + D_3\dot{y}' + D_4\dot{y} + \ddot{y} = f(x,t) \qquad (8.57)$$

where $D_1 \dot{y}''''$ is called structural damping, $D_2 \dot{y}''$ is called square-root damping, $D_3 \dot{y}'$ is called strain-rate damping, $D_4 \dot{y}$ is called external viscous damping, and $f(x, t)$ represents external forces applied to the beam. All the damping coefficients are positive constants.

8.3.5.1 Structural Damping
Structural damping can be motivated by assuming a viscous resistance to straining of the beam material, as follows:

$$(y'' + D\dot{y}'')'' + \ddot{y} = f(x, t) \qquad (8.58)$$

where $D\dot{y}''$ is a bending moment proportional to the normal strain rate.

The modal equations associated with Eq. (8.58) can be found as

$$\ddot{q}_i(t) + D\omega_i^2 \dot{q}_i(t) + \omega_i^2 q_i(t) = f_i(t), \qquad i = 0, \ldots, \infty \qquad (8.59)$$

which indicate that the modal damping ratio is proportional to the modal frequency. This implies that the energy dissipated per cycle of oscillation is proportional to the square of the amplitude and to the first power of excitation frequency. It has been experimentally observed, however, that the energy dissipated per cycle of oscillation is proportional to the square of the amplitude and independent of the frequency for a large variety of materials.

Equation (8.58) can be empirically modified to fit the experiment data (for nearly simple harmonic motion) as follows:

$$\left(y'' + \frac{D}{\omega} \dot{y}'' \right)'' + \ddot{y} = F(x) \, e^{j\omega t} \qquad (8.60)$$

where $j \equiv \sqrt{-1}$ and ω is the excitation frequency. Then the modal equations can be written as

$$\ddot{q}_i(t) + D\omega_i \dot{q}_i(t) + \omega_i^2 q_i(t) = f_i e^{j\omega t} \qquad (8.61)$$

which has the same damping ratio for each vibration mode. This type of damping mechanism is often called structural hysteric damping, wherein the higher frequency modes damp out more quickly than the lower frequency modes. It is this behavior that makes it possible to use truncated models for control design.

Equations (8.60) and (8.61) are often expressed as

$$(1 + jD)y'''' + \ddot{y} = F(x) \, e^{j\omega t} \qquad (8.62)$$

$$\ddot{q}_i + (1 + jD)\omega_i^2 q_i = f_i e^{j\omega t} \qquad (8.63)$$

where D is called the structural damping factor. These equations are also known as the "complex stiffness" model.

8.3.5.2 External Viscous Damping
This damping model represents an external viscous resistance to transverse displacement of the beam. The modal equations are then simply given by

$$\ddot{q}_i(t) + D\dot{q}_i(t) + \omega_i^2 q_i(t) = f_i(t) \qquad (8.64)$$

which indicate the same damping coefficient for each mode, i.e., the same real part. This model is often used for the analysis of a damped beam; however, it should be noted that it does not represent an internal energy dissipation mechanism.

8.3.5.3   Strain-Rate Damping

One may interpret the strain-damping term $D\dot{y}'$ as an external viscous resistance to the time rate of change of the slope (strain-rate) of the beam. This term must be a transverse force per unit length, however, and it does not seem physically logical to have a force proportional to the time rate of change of the slope.

8.3.5.4   Square-Root Damping

This type of damping mechanism has often been used for analysis of flexible structures. Because the damping term is proportional to the square root of structural stiffness, we call this model square-root damping.

8.4   Rigid Body with Beamlike Appendages

In this section we consider a rigid body connected with uniform Bernoulli–Euler beams, as shown in Fig. 8.10, as a generic model of a spacecraft with symmetric flexible appendages. A planar single-axis rotation with small elastic deformation, but with possible large-angle rigid-body rotation, is considered.

8.4.1   Equations of Motion

In formulating the equations of motion of a flexible spacecraft with large rigid-body rotation, a "floating reference frame" is often employed. There are many different choices of a floating reference frame; however, we choose here a reference frame attached to the central rigid body as shown in Fig. 8.10.

From the theory of linear mechanical vibration, it is known that an orthogonality relationship exists between the rigid-body mode and the elastic vibration modes. Physically, such a relationship means zero net angular (or translational) momentum of the elastic modes. Such orthogonality relationship is mathematically identical to the constraints of the linearized Tisserand reference frame or Buckens frame. Without going into details of floating reference frames, we simply derive the exact

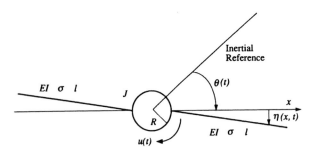

Fig. 8.10   Rigid body with beamlike appendages.

equations of motion of the simple model with large rigid-body rotation using Hamilton's principle, which is one of the systematic ways of deriving the equation of motion of elastic bodies,

$$\delta \int_{t_1}^{t_2} (T - V)\, dt + \int_{t_1}^{t_2} \delta W\, dt = 0 \qquad (8.65)$$

where

T = total kinetic energy
V = potential energy
W = work done by nonconservative forces
δ = variation taken during indicated time interval

For simplicity's sake we assume that the appendages are simple uniform beams (symmetric) and that the central rigid body has a spherical shape. Then, for the idealized model shown in Fig. 8.10, we have

$$T = \frac{1}{2}J\dot{\theta}^2 + \int_0^\ell \sigma[(R + x)\dot{\theta} + \dot{\eta}]^2\, dx \qquad (8.66a)$$

$$V = \frac{1}{2}\int_0^\ell EI[\eta''(x,t)]^2\, dx \qquad (8.66b)$$

$$\delta W = u(t)\delta\theta \qquad (8.66c)$$

where $\theta(t)$ is the arbitrarily large rotational angle of the central rigid body with respect to an inertial reference frame, $\eta(x,t)$ is the small elastic deformation of the appendages with respect to the reference frame attached to the central rigid body, $u(t)$ is the control torque applied to the central rigid body, EI is the bending stiffness of the appendage, σ is the mass density per unit length of the appendage, ℓ is the length of the single appendage, $J = 2MR^2/5$ is the rotational inertia of the central spherical rigid body, and R is the radius of the central rigid body.

Using Hamilton's principle, we obtain the rotational equation of motion, as follows:

$$[EI\eta''(x,t)]'' + \sigma[(R + x)\ddot{\theta}(t) + \ddot{\eta}(x,t)] = 0 \qquad (8.67)$$

and the boundary conditions are given by

$$J\ddot{\theta}(t) = u(t) - 2REI\eta'''(0,t) + 2EI\eta''(0,t) \qquad (8.68a)$$

$$\eta(0,t) = \eta'(0,t) = 0 \qquad (8.68b)$$

$$\eta''(\ell,t) = \eta'''(\ell,t) = 0 \qquad (8.68c)$$

where the prime and dot denote partial differentiation with respect to x and t, respectively.

Equation (8.67) has a variable coefficient, but if we define a new coordinate $y(x,t)$ as

$$y(x,t) \equiv (R + x)\theta(t) + \eta(x,t) \qquad (8.69)$$

then we have a simple uniform beam equation (with constant coefficient)

$$EIy''''(x,t) + \sigma\ddot{y}(x,t) = 0 \tag{8.70}$$

and the boundary conditions in terms of the new coordinate $y(x,t)$ become

$$J\ddot{\theta}(t) = u(t) - 2REIy'''(0,t) + 2EIy''(0,t) \tag{8.71a}$$
$$\theta(t) = y'(0,t) \tag{8.71b}$$
$$y(0,t) = Ry'(0,t) \tag{8.71c}$$
$$y''(\ell,t) = y'''(\ell,t) = 0 \tag{8.71d}$$

Note that Eq. (8.69) can also be expressed as

$$y(x,t) = (R+x)\theta(t) + \sum_{i=1}^{\infty}\phi_i(x)q_i(t) \tag{8.72}$$

where $\phi_i(x)$ is called the ith "cantilever" or "appendage" mode shape and $q_i(t)$ is the ith appendage modal coordinate. In this expression $\theta(t)$ is not the rigid-body modal coordinate but the rotational angle of the central rigid body. The slope of the appendage mode shape $\phi_i(x)$ at $x = 0$ is always zero for the cantilever modes.

8.4.2 Transcendental Transfer Function

Taking the Laplace transforms of Eqs. (8.70) and (8.71), we obtain the transcendental transfer function from control torque $u(s)$ to attitude angle $\theta(s)$ of the central rigid body, as follows:

$$\frac{\theta(s)}{u(s)} = \frac{-(1 + \cos\lambda\cosh\lambda)}{\Delta(s)} \tag{8.73}$$

where

$$\Delta(s) = 2\lambda\big[(1 + P_1\lambda^2)\cosh\lambda\sin\lambda + (P_1\lambda^2 - 1)\sinh\lambda\cos\lambda$$
$$+ 2P_1\lambda\sinh\lambda\sin\lambda + \tfrac{2}{5}P_1^2 P_2\lambda^3(1 + \cos\lambda\cosh\lambda)\big]$$

and $\lambda^4 \equiv -s^2$, with s in units of $\sqrt{EI/\sigma\ell^4}$ and $u(s)$ in units of EI/ℓ. The dimensionless structural parameters P_1 and P_2 are defined as

$$P_1 = \frac{R}{\ell} = \frac{\text{radius of spherical central body}}{\text{length of single appendage}}$$

$$P_2 = \frac{M}{2\sigma\ell} = \frac{\text{mass of central rigid body}}{\text{total mass of two appendages}}$$

The vanishing of the numerator polynomial of Eq. (8.73) is identical to the characteristic equation of a cantilevered beam of length ℓ. Thus, the zeros of the

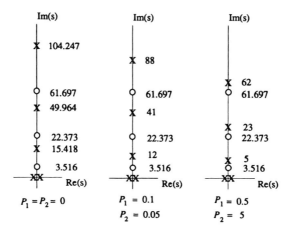

Fig. 8.11   Exact poles and zeros of $\theta(s)/u(s)$ for different values of structural parameters P_1 and P_2.

transfer function (8.73) are identical to the natural frequencies of a cantilevered beam of length ℓ.

For the special case without the central body, Eq. (8.73) becomes

$$\frac{\theta(s)}{u(s)} = \frac{1 + \cos\lambda\cosh\lambda}{2\lambda(\sinh\lambda\cos\lambda - \cosh\lambda\sin\lambda)} \tag{8.74}$$

which corresponds to the transfer function of a free-free beam of length 2ℓ with control torquer and rotational angle sensor at the center of the beam.

The pole–zero patterns of Eq. (8.73) for different values of P_1 and P_2 are shown in Fig. 8.11. As expected, the poles and zeros alternate along the imaginary axis, and the pole–zero pairs of each vibration mode depend on the structural parameters P_1 and P_2. For a free-free beam without the central rigid body each pole has an associated zero of higher frequency than the pole, and the lowest zeros are very close to the origin. As we increase the inertia of the central rigid body, the poles become associated with a zero of lower frequency, and the vibration modes become nearly uncontrollable (undisturbable) and unobservable by the control torquer and the attitude angle sensor at the central rigid body.

8.5   Rigid/Flexible Frame with a Pretensioned Membrane

In this section we consider a simplified model of a membrane-type solar array. We shall treat a flexible frame as well as a rigid frame with a pretensioned membrane to study the effect of membrane tension on the pole–zero pairs of vibration modes.

8.5.1   Rigid Frame with a Pretensioned Membrane

Consider a simplified model of a rigid frame with a membrane-type solar array shown in Fig. 8.12. The solar array (or blanket) is pretensioned in one direction

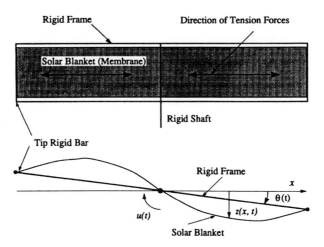

Fig. 8.12 Rigid frame with a pretensioned membrane.

and may be described by the wave equation

$$z''(x,t) - (\rho/T)\ddot{z}(x,t) = 0 \qquad (8.75)$$

where $z(x,t)$ is transverse displacement of the membrane, ρ is mass density per unit area of the membrane, and T is tension per unit length.

The boundary conditions for pure rotational motion are

$$z(\ell, t) = \ell\theta(t) \qquad (8.76a)$$

$$z(0, t) = 0 \qquad (8.76b)$$

where $\theta(t)$ is the rotational angle of the rigid frame and 2ℓ is total length of the frame.

The equation of motion of the rigid frame is

$$J\ddot{\theta} = 2aT[z'(\ell, t) + \theta(t)]\ell + u(t) \qquad (8.77)$$

where J is rotational inertia of the rigid frame, $u(t)$ is the control torque applied to the frame through the rigid shaft, and a is width of the frame. The rotational inertia of the rigid shaft is omitted for simplicity.

Taking the Laplace transform of Eq. (8.75), we obtain

$$z''(x, s) + \lambda^2 z(x, s) = 0 \qquad (8.78)$$

where $\lambda^2 \equiv -s^2$, with s in units of $\sqrt{T/\rho\ell^2}$ and z and x in units of ℓ. The solution of this Laplace transformed equation of motion is given by

$$z(x, s) = A_1 \sin \lambda x + A_2 \cos \lambda x \qquad (8.79)$$

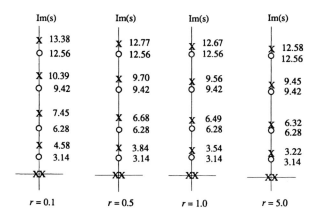

Fig. 8.13 Exact poles and zeros of $\theta(s)/u(s)$ for different values of r.

Using the boundary conditions (8.76), we obtain

$$z(x, s) = \frac{\theta(s)}{\sin \lambda} \sin \lambda x \qquad (8.80)$$

Combining Eqs. (8.77) and (8.80), we obtain the transcendental transfer function from control torque $u(s)$ to rotational angle $\theta(s)$ as follows:

$$\frac{\theta(s)}{u(s)} = \frac{\sin \lambda}{aT\ell[(3/2)r\lambda^2 \sin \lambda - 2(\sin \lambda + \lambda \cos \lambda)]} \qquad (8.81)$$

where

$$r = \frac{J}{\frac{2}{3}a\rho\ell^3} = \frac{\text{moment of inertia of rigid frame}}{\text{moment of inertia of solar blanket}}$$

As $\lambda \to 0$, Eq. (8.81) becomes

$$\frac{\theta(s)}{u(s)} = \frac{1}{\left(J + \frac{2}{3}a\rho\ell^3\right)s^2} \qquad (8.82)$$

which is identical to the transfer function of a rigid frame with a rigid solar blanket.

The numerator of Eq. (8.81) is identical to the characteristic equation of a vibrating string of length ℓ. The pole–zero patterns of Eq. (8.81) are shown in Fig. 8.13 for different values of inertia ratio r. As $r \to \infty$, we have near pole–zero cancellations of vibration modes, which means that the membrane vibration becomes negligible and the entire system behaves as a rigid body.

8.5.2 Flexible Frame with a Pretensioned Membrane

Consider a simple model of a flexible frame with a pretensioned membrane, as shown in Fig. 8.14. The equation of motion for the solar blanket is

$$z''(x,t) - (\rho/T)\ddot{z}(x,t) = 0 \tag{8.83}$$

with boundary conditions for pure rotational motion

$$z(0,t) = 0 \tag{8.84a}$$

$$z(\ell,t) = y(\ell,t) \tag{8.84b}$$

where $z(x,t)$ is the transverse displacement of solar blanket, $y(x,t)$ is the transverse displacement of the support boom, ρ is mass density per unit area of the membrane, and T is tension per unit length.

Because the support booms are compressed due to the tension in the solar blanket, they are described by the beam-column equation

$$EIy''''(x,t) + (aT/2)y''(x,t) + \sigma\ddot{y}(x,t) = 0 \tag{8.85}$$

with boundary conditions

$$4EIy''(0,t) = u(t) \tag{8.86a}$$

$$y(0,t) = y''(\ell,t) = 0 \tag{8.86b}$$

where EI is the bending stiffness of support booms, σ is the mass per unit length of booms, and $u(t)$ is the control torque applied through the rigid shaft.

For the tip rigid bar, we have

$$m\ddot{y}(\ell,t) - 2EIy''(\ell,t) - aTy'(\ell,t) + aTz'(\ell,t) = 0 \tag{8.87}$$

where m is mass of the tip rigid bar.

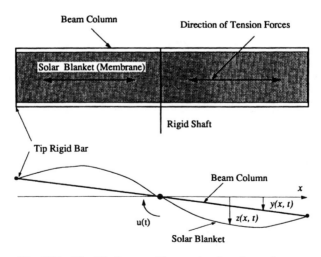

Fig. 8.14 Flexible frame with a pretensioned membrane.

The dimensionless structural parameters $\bar{m}$, $\bar{\sigma}$, and $\bar{T}$ are defined as

$$\bar{m} = \frac{m}{a\rho\ell} = \frac{\text{mass of tip rigid bar}}{\text{mass of solar blanket (one side)}} \tag{8.88a}$$

$$\bar{\sigma} = \frac{2\sigma\ell}{a\rho\ell} = \frac{\text{mass of support booms (one side)}}{\text{mass of solar blanket (one side)}} \tag{8.88b}$$

$$\bar{T} = \frac{aT\ell^2}{2EI} = \pi^2 \frac{\text{compressive load } (aT/2)}{\text{buckling load } (EI/\pi^2\ell^2)} \tag{8.88c}$$

The Laplace transformed equations of motion can then be written in dimensionless form as follows.
Solar blanket:

$$z''(x,s) + \lambda^2 z(x,s) = 0 \tag{8.89}$$

Beam column:

$$y''''(x,s) + \bar{T}y''(x,s) - \bar{\sigma}\bar{T}\lambda^2 y(x,s) = 0 \tag{8.90}$$

Tip rigid bar:

$$-\bar{m}\lambda^2 y(1,s) - \bar{T}^{-1}y'''(1,s) - y'(1,s) + z'(1,s) = 0 \tag{8.91}$$

with the boundary conditions

$$z(0,s) = 0 \tag{8.92}$$
$$z(1,s) = y(1,s) \tag{8.93}$$
$$y(0,s) = 0 \tag{8.94}$$
$$y''(1,s) = 0 \tag{8.95}$$
$$4y''(0,s) = u(s) \tag{8.96}$$

where $\lambda^2 \equiv -s^2$ with s in units of $\sqrt{T/\rho\ell^2}$; x, y, and z are in units of ℓ; and $u(s)$ is in units of EI/ℓ.
The solution of Eq. (8.89) is given by

$$z(x,s) = A_1 \sin \lambda x + A_2 \cos \lambda x \tag{8.97}$$

Substituting this into the boundary conditions (8.92) and (8.93), we obtain

$$z(x,s) = \frac{y(1,s)}{\sin \lambda} \sin \lambda x \tag{8.98}$$

The solution of Eq. (8.90) is given by

$$y(x,s) = B_1 \sinh \alpha x + B_2 \cosh \alpha x + B_3 \sin \beta x + B_4 \cos \beta x \tag{8.99}$$

where

$$\alpha = \sqrt{\dfrac{-\bar{T} + \sqrt{\bar{T}^2 + 4\bar{\sigma}\,\bar{T}\lambda^2}}{2}}$$

$$\beta = \sqrt{\dfrac{\bar{T} + \sqrt{\bar{T}^2 + 4\bar{\sigma}\,\bar{T}\lambda^2}}{2}}$$

Combining Eqs. (8.94) and (8.95), we have

$$B_2 = -B_4 \tag{8.100}$$

and

$$B_1\alpha^2 \sinh \alpha + B_2(\alpha^2 \cosh \alpha + \beta^2 \cos \beta) - B_3\beta^2 \sin \beta = 0 \tag{8.101}$$

From Eq. (8.91), we have

$$
\begin{aligned}
0 = {} & B_1\left[\left(-\bar{m}\lambda^2 + \dfrac{\lambda \cos \lambda}{\sin \lambda}\right)\sinh \alpha - \bar{T}^{-1}\alpha^3 \cosh \alpha - \alpha \cosh \alpha\right] \\
& + B_2\left[\left(-\bar{m}\lambda^2 + \dfrac{\lambda \cos \lambda}{\sin \lambda}\right)(\cosh \alpha - \cos \beta)\right. \\
& \left. - \bar{T}^{-1}(\alpha^3 \sinh \alpha - \beta^3 \sin \beta) - (\alpha \sinh \alpha + \beta \sin \beta)\right] \\
& + B_3\left[\left(-\bar{m}\lambda^2 + \dfrac{\lambda \cos \lambda}{\sin \lambda}\right)\sin \beta + \bar{T}^{-1}\beta^3 \cos \beta - \beta \cos \beta\right] \tag{8.102}
\end{aligned}
$$

From Eq. (8.96), we have

$$B_2 = \dfrac{u(s)}{4(\alpha^2 + \beta^2)} \tag{8.103}$$

Using Eq. (8.103), we can combine Eqs. (8.101) and (8.102) as

$$\begin{bmatrix} a_{11} & a_{12} \\ a_{21} & a_{22} \end{bmatrix}\begin{bmatrix} B_1 \\ B_3 \end{bmatrix} = \begin{bmatrix} b_1 \\ b_2 \end{bmatrix}u(s) \tag{8.104}$$

where

$$a_{11} = \alpha^2(\alpha^2 + \beta^2)\sinh \alpha$$

$$a_{12} = -\beta^2(\alpha^2 + \beta^2)\sin \beta$$

$$a_{21} = (\alpha^2 + \beta^2)[(-\bar{m}\lambda^2 \sin \lambda + \lambda \cos \lambda)\sinh \alpha$$
$$- (\bar{T}^{-1}\alpha^3 \cosh \alpha - \alpha \cosh \alpha)\sin \lambda]$$

$$a_{22} = (\alpha^2 + \beta^2)[(-\bar{m}\lambda^2 \sin \lambda + \lambda \cos \lambda)\sin \beta + (\bar{T}^{-1}\beta^3 \cos \beta - \beta \cos \beta)\sin \lambda]$$

$$b_1 = -\tfrac{1}{4}(\alpha^2 \cosh\alpha + \beta^2 \cos\beta)$$

$$b_2 = \tfrac{1}{4}[-(-\bar{m}\lambda^2 \sin\lambda + \lambda\cos\lambda)(\cosh\alpha - \cos\beta)$$

$$+ \bar{T}^{-1}(\alpha^3 \sinh\alpha - \beta^3 \sin\beta)\sin\lambda + (\alpha\sinh\alpha + \beta\sin\beta)\sin\lambda]$$

Finally the transcendental transfer function from control torque $u(s)$ to angle $\theta(s) \equiv y'(0, s)$ can then be found as

$$\frac{\theta(s)}{u(s)} = \frac{\alpha(b_1 a_{22} - b_2 a_{12}) + \beta(b_2 a_{11} - b_1 a_{21})}{a_{11}a_{22} - a_{12}a_{21}} \qquad (8.105)$$

The exact pole–zero patterns of Eq. (8.105) are shown in Fig. 8.15 for different values of $\bar{T}$. The nominal values are assumed as $\bar{m} = 0.2$, $\bar{\sigma} = 0.2$, and $\bar{T} = 2.0$. The lowest poles and zeros approach the origin as $\bar{T} \to \pi^2$; i.e., $aT/2 \to$ buckling load $(EI/\pi^2\ell^2)$. For the nominal value of $\bar{T} = 2.0$, each pole has an associated zero of higher frequency than the pole with near cancellation. The lowest zeros are quite close to the origin. Thus the reduced-order models obtained from the product expansion of transcendental transfer function will have the same number of poles and zeros, whereas the reduced-order models obtained by modal analysis always have more poles than zeros. Most physical systems are "strictly proper" systems, which have more poles than zeros in the transfer functions; however, some dynamic systems, such as the generic examples considered in this chapter, behave as "proper" systems with direct transmission.

To further illustrate such a direct transmission property of a certain class of structural dynamic systems, consider again a two-mass–spring system. The equations of motion of this system are

$$m_1\ddot{x}_1 + k(x_1 - x_2) = u \qquad (8.106a)$$

$$m_2\ddot{x}_2 + k(x_2 - x_1) = 0 \qquad (8.106b)$$

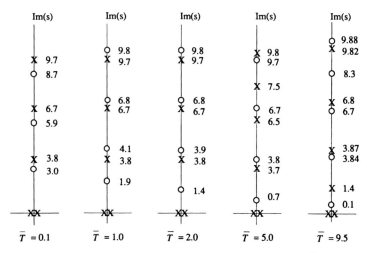

Fig. 8.15   Exact poles and zeros of $\theta(s)/u(s)$ for different values of $\bar{T}$ ($\bar{m} = \bar{\sigma} = 0.2$).

For the case of $m_1 \ll m_2$, we obtain the transfer function from $u(s)$ to $x_1(s)$ as follows:

$$\frac{x_1(s)}{u(s)} = \frac{m_2 s^2 + k}{m_2 k s^2} = \frac{1}{k} + \frac{1}{m_2 s^2} \tag{8.107}$$

which has the same number of poles and zeros and the direct transmission term $1/k$. The zero of this transfer function is the natural frequency of the system when x_1 is constrained equal to zero. The pole–zero pattern of this transfer function is similar to the reduced-order transfer function of the flexible frame when we neglect all the near pole–zero cancellations as well as high-frequency poles and zeros. The flexible frame then behaves as the simple two-mass–spring system with $m_1 \ll m_2$ for the low-frequency range.

8.6 Flexible Toroidal Structures

In this section we are concerned with the modeling of a flexible toroidal structure for attitude and structural control studies. We shall consider a two-dimensional model that can be analyzed somewhat easily and still be realistic. The models are simple enough to treat analytically, yet complicated enough to demonstrate the dynamic characteristics of future space structures with toroidal configurations. It will be shown that for such structures, the coupling between two elastic systems results in a transfer function expressed as a combination of an infinite product expansion and an infinite partial fraction expansion.

Because the coupling of bending and torsion is a basic property of the flexible toroid, we briefly discuss the attitude motion of a flexible toroid itself. We then consider the modeling of a flexible toroid, as well as a rigid toroid, with a pre-tensioned membrane in terms of the transfer functions. These models may be of academic interest without any practical relevance to future space structures, which will consist of many lumped and trusslike subsystems with fairly complex inter-connections. It is, however, emphasized that these models will provide physical insights into the dynamic characteristics of complex space structures.

8.6.1 Flexible Toroid

Consider the modeling and control of the flexible toroid shown in Fig. 8.16. Roll attitude control could be obtained by two identical torquers, e.g., reaction wheels, located at points A and B. Pitch control could be obtained by two identical torquers located at points C and D. Obviously, pitch and roll controls are decoupled and are identical, so we consider only roll attitude control.

The equations of motion of a flexible inextensional toroid for out-of-plane vibration can be written as follows.

Toroid bending:

$$\frac{EI}{R^3}\left[\frac{1}{R}y'''' + \theta''\right] + \frac{GJ}{R^3}\left[\theta'' - \frac{1}{R}y''\right] + \sigma\ddot{y} = 0 \tag{8.108}$$

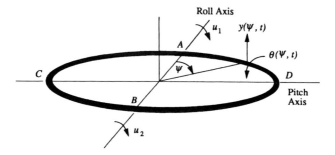

Fig. 8.16   Flexible toroid with bending-torsion coupling.

Toroid torsion:

$$\frac{GJ}{R^2}\left[\theta'' - \frac{1}{R}y''\right] - \frac{EI}{R^2}\left[\frac{1}{R}y'' + \theta\right] - \bar{J}\ddot{\theta} = 0 \qquad (8.109)$$

where

$$()' \equiv \frac{\partial()}{\partial\psi}, \quad (\dot{}) \equiv \frac{\partial()}{\partial t}$$

and where

$$
\begin{aligned}
y(\psi,t) &= \text{out-of-plane bending displacement}\\
\theta(\psi,t) &= \text{torsional displacement}\\
EI &= \text{uniform bending stiffness}\\
GJ &= \text{uniform torsional stiffness}\\
\sigma &= \text{mass per unit length}\\
\bar{J} &= \text{polar moment of inertia per unit length}\\
R &= \text{radius of toroid}
\end{aligned}
$$

The boundary conditions for pure roll motion at $\psi = 0$ and π are

$$y(\psi,t) = 0 \qquad (8.110a)$$

$$\theta(\psi,t) = 0 \qquad (8.110b)$$

$$u(t) = \frac{4EI}{R}\left[\frac{1}{R}y''(\psi,t) + \theta(\psi,t)\right] \qquad (8.110c)$$

where roll control torque $u(t)$ is defined as $u/2 = u_1 = u_2$.

The transfer function from roll control torque to elastic deformations at various points on the flexible toroid can be obtained using the Laplace transform ($t \to s$) and the finite sine transform ($\psi \to n$) because of the periodicity in ψ. The finite sine transform is defined as

$$y(n,t) = \int_0^\pi y(\psi,t) \sin n\psi \, d\psi \qquad (8.111a)$$

$$y(\psi,t) = \frac{2}{\pi}\sum_{n=1}^\infty y(n,t) \sin n\psi \qquad (8.111b)$$

The transformed equations of motion with the boundary conditions incorporated can be written as

$$\left[n^4 + \frac{GJ}{EI}n^2 + \frac{\sigma R^4}{EI}s^2 \right] \frac{1}{R} y(n,s) - \left(1 + \frac{GJ}{EI} \right) n^2 \theta(n,s)$$

$$= \frac{nR}{4EI}[(-1)^n - 1]u(s) \tag{8.112}$$

and

$$\left(1 + \frac{GJ}{EI} \right) n^2 \frac{1}{R} y(n,s) - \left(1 + \frac{GJ}{EI}n^2 \right) \theta(n,s) = 0 \tag{8.113}$$

where the torsional inertia of the thin toroid is neglected.

Finally, the transfer functions from $u(s)$ to $y(n,s)$ and $\theta(n,s)$ can be found as

$$\frac{y(n,s)}{u(s)} = \frac{R^2}{EI} \frac{[1 + (GJ/EI)n^2]n/2}{\Delta(n,s)} \tag{8.114}$$

$$\frac{\theta(n,s)}{u(s)} = \frac{R}{EI} \frac{[1 + (GJ/EI)]n^3/2}{\Delta(n,s)} \tag{8.115}$$

where

$$\Delta(n,s) = \left(1 + \frac{GJ}{EI} \right)^2 n^4 - \left(n^4 + \frac{GJ}{EI}n^2 + \frac{\sigma R^4}{EI}s^2 \right) \left(1 + \frac{GJ}{EI}n^2 \right)$$

By the definition of the finite sine transform, we have

$$\frac{y(\psi,s)}{u(s)} = \frac{2R^2}{\pi EI} \sum_{n=1,3}^{\infty} \frac{[1 + (GJ/EI)n^2]n/2}{\Delta(n,s)} \sin n\psi \tag{8.116}$$

$$\frac{\theta(\psi,s)}{u(s)} = \frac{2R}{\pi EI} \sum_{n=1,3}^{\infty} \frac{[1 + (GJ/EI)]n^3/2}{\Delta(n,s)} \sin n\psi \tag{8.117}$$

The transfer function from roll control torque $u(s)$ to roll angle $\phi(s) \equiv -y'$ $(0,s)/R$ can be written as

$$\frac{\phi(s)}{u(s)} = \frac{1}{J} \sum_{n=1,3}^{\infty} \frac{n^2}{s^2 + \omega_n^2} \tag{8.118}$$

where $J \equiv \sigma \pi R^3$ is roll moment of inertia of the rigid toroid. The natural frequencies ω_n are defined as

$$\omega_n = \left[n^4 + \frac{GJ}{EI}n^2 - \frac{(1 + GJ/EI)^2 n^4}{1 + (GJ/EI)n^2} \right]^{\frac{1}{2}} \tag{8.119}$$

where $\omega_1 = 0$ represents the rigid-body mode.

It can be seen that the residues or modal gains in Eq. (8.118) have the same sign. Thus, Eq. (8.118) will have alternating poles and zeros along the imaginary axis, which is a direct consequence of collocated actuator and sensor. For this case, attitude stabilization can be simply achieved using angle and rate feedback or attitude angle feedback with lead compensation.

8.6.2 Rigid Toroid with a Pretensioned Membrane

A simple model for roll attitude control of a rigid toroid with a pretensioned membrane is shown in Fig. 8.17. This can be considered as an approximate model for a space reflector (mirror) with a reflector surface stretched across a rigid toroidal frame.

The equation of motion for a circular membrane in polar coordinates (r, ψ) is

$$z_{rr} + (1/r)z_r + (1/r^2)z_{\psi\psi} - (1/c^2)\ddot{z}(r, \psi, t) = 0 \qquad (8.120)$$

where $(\ )_r \equiv \partial(\ )/\partial r$, $(\ )_\psi \equiv \partial(\ )/\partial\psi$, $z(r, \psi, t)$ is the transverse displacement of the uniform membrane, $c \equiv \sqrt{T/\rho}$, T is the tension per unit length, and ρ is the mass per unit area of the membrane.

The boundary conditions for pure roll motion are

$$z(r, 0, t) = 0 \qquad (8.121)$$

$$z(r, \psi, t) = -(R\sin\psi)\phi(t) \qquad (8.122)$$

$$z_\psi(r, \pi/2, t) = 0 \qquad (8.123)$$

$$u(t) = J\ddot{\phi} - 4\int_0^{\pi/2} TR[z_r(r, \psi, t) + \phi\sin\psi]R\sin\psi\,d\psi \qquad (8.124)$$

where $u(t)$ is roll control torque applied to the rigid toroid ($u/2 = u_1 = u_2$), $\phi(t)$ is roll attitude angle of the rigid toroid, and $J = \sigma\pi R^3$ is moment of inertia of the rigid toroid.

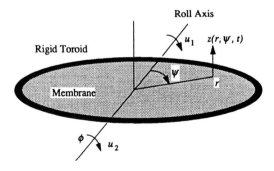

Fig. 8.17 Rigid toroid with a pretensioned membrane.

For a steady-state sinusoidal input, let

$$u(t) = u(\omega) \sin \omega t \tag{8.125a}$$
$$\phi(t) = \phi(\omega) \sin \omega t \tag{8.125b}$$

Using the boundary conditions (8.121) and (8.123), we obtain

$$z(r, \psi, t) = z(r, \omega) \sin \psi \sin \omega t \tag{8.126}$$

Substituting Eq. (8.126) into Eq. (8.120), we obtain

$$z_{rr} + \frac{1}{r} z_r + \left(\frac{\omega^2}{c^2} - \frac{1}{r^2} \right) z(r, \omega) = 0 \tag{8.127}$$

The solution of Eq. (8.127) is given by

$$z(r, \omega) = A J_1 \left(\frac{\omega r}{c} \right) + B Y_1 \left(\frac{\omega r}{c} \right) \tag{8.128}$$

where $J_1(\omega r/c)$ denotes the Bessel function of the first kind of order one and $Y_1(\omega r/c)$ denotes the Bessel function of the second kind of order one.

Because $z(0, \omega)$ is finite, we have $B = 0$. From the boundary condition (8.121), we have

$$z(r, \omega) = -R\phi(\omega) \frac{J_1(\omega r/c)}{J_1(R/c)} \tag{8.129}$$

Substituting Eq. (8.129) into Eq. (8.124), we obtain the transfer function from roll control torque $u(s)$ to roll attitude angle $\phi(s)$ as follows:

$$\frac{\phi(s)}{u(s)} = \frac{J_1(\lambda)}{-\pi T R^2 \left[\mu \lambda^2 J_1(\lambda)/4 + J_1(\lambda) - \lambda J_1'(\lambda) \right]} \tag{8.130}$$

where

$$s = j\omega = j \frac{c}{R} \lambda$$

$$\mu = \frac{J}{\pi \rho R^4 / 4} = \frac{\text{moment of inertia of rigid toroid}}{\text{moment of inertia of rigid membrane}}$$

$$J_1'(\lambda) = \frac{dJ_1(\lambda)}{d\lambda}$$

As $\lambda \to 0$, Eq. (8.130) becomes

$$\frac{\phi(s)}{u(s)} = \frac{1}{[J + \pi \rho R^4 / 4] s^2} \tag{8.131}$$

which is the transfer function of the rigid toroid with rigid membrane.

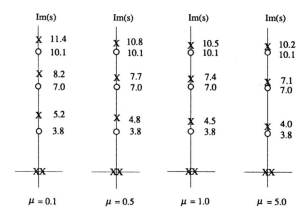

Fig. 8.18   Exact poles and zeros of a rigid toroid with a pretensioned membrane.

Pole–zero patterns for different values of μ are shown in Fig. 8.18. As $\lambda \rightarrow \infty$, we have near pole–zero cancellations of all the vibration modes that correspond to the case of a rigid toroid with negligible effects of membrane vibration.

The numerator of Eq. (8.130) is identical to the characteristic equation of a circular membrane with a fixed boundary and nodal line at the roll control axis. Thus, the zeros in Eq. (8.130) are independent of μ.

8.6.3   Flexible Toroid with Pretensioned Membrane

In this section we derive the equations of motion of the flexible toroid with a pretensioned membrane. The tension for out-of-plane buckling of the toroid and membrane will also be determined. Finally, the roll transfer function will be derived for the case of a collocated actuator and sensor.

8.6.3.1   Equations of Motion   Consider a flexible toroid with a pretensioned membrane as shown in Fig. 8.19.

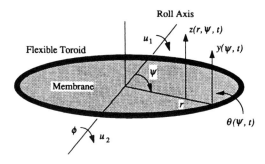

Fig. 8.19   Flexible toroid with a pretensioned membrane.

The membrane equation is the same as Eq. (8.120),

$$z_{rr} + (1/r)z_r + (1/r^2)z_{\psi\psi} - (1/c^2)\ddot{z}(r,\psi,t) = 0$$

The equations of motion of a flexible inextensional toroid with a pretensioned membrane for out-of-plane vibration can be found as follows.

Toroid bending:

$$\frac{EI}{R^3}\left[\frac{1}{R}y'''' + \theta''\right] + \frac{GJ}{R^3}\left[\theta'' - \frac{1}{R}y''\right] + \frac{T}{R}y'' + \sigma\ddot{y} = -Tz_r(R,\psi,t) \quad (8.132)$$

Toroid torsion:

$$\frac{GJ}{R^2}\left[\theta'' - \frac{1}{R}y''\right] - \frac{EI}{R^2}\left[\frac{1}{R}y'' + \theta\right] - aT\theta - \bar{J}\ddot{\theta} = -aTz_r(R,\psi,t) \quad (8.133)$$

where $2a$ is thickness of the toroid and other variables have been defined in the preceding sections.

The boundary conditions for pure roll motion are

$$z(r,0,t) = z'(r,\pi/2,t) = 0 \qquad (8.134a)$$
$$y(0,t) = y'(\pi/2,t) = 0 \qquad (8.134b)$$
$$\theta(0,t) = \theta'(\pi/2,t) = 0 \qquad (8.134c)$$
$$z(R,\psi,t) = y(\psi,t) - a\theta(\psi,t) \qquad (8.134d)$$
$$u(t) = \frac{4EI}{R}\left[\frac{1}{R}y''(0,t)\right] \qquad (8.134e)$$

where $u(t)$ is the roll control torque defined as $u/2 = u_1 = u_2$.

The equations of motion in dimensionless form can be written as follows.

Toroid bending:

$$y'''' + \theta'' + k_1[\theta'' - y''] + k_2 y'' + \ddot{y} = -k_2 z_r(1,\psi,t) \qquad (8.135)$$

Toroid torsion:

$$k_1[\theta'' - y''] - [y'' + \theta] - k_2 k_5\theta - k_4\ddot{\theta} = -k_2 k_5 z_r(1,\psi,t) \qquad (8.136)$$

Membrane:

$$z_{rr} + \frac{1}{r}z_r + \frac{1}{r^2}z'' - \frac{k_3}{k_2}\ddot{z} = 0 \qquad (8.137)$$

where

$$k_1 = \frac{GJ}{EI}, \qquad k_2 = \frac{TR^3}{EI}, \qquad k_3 = \frac{\rho R}{\sigma}, \qquad k_4 = \frac{\bar{J}}{\sigma R^2}, \qquad k_5 = \frac{a}{R}$$

y, z, and r are in units of R, and time in units of $\sqrt{\sigma R^4/EI}$.

The boundary conditions in dimensionless forms are

$$z(r, 0, t) = z'(r, \pi/2, t) = 0 \tag{8.138a}$$

$$y(0, t) = y'(\pi/2, t) = 0 \tag{8.138b}$$

$$\theta(0, t) = \theta'(\pi/2, t) = 0 \tag{8.138c}$$

$$z(1, \psi, t) = y(\psi, t) - k_5\theta(\psi, t) \tag{8.138d}$$

$$u(t) = 4y''(0, t) \tag{8.138e}$$

where $u(t)$ is in units of EI/R.

The equations of motion for a thin toroid ($k_4 \ll 1$ and $k_5 \ll 1$) are as follows. Toroid bending:

$$y'''' + (k_2 - k_1)y'' + (1 + k_1)\theta'' + \ddot{y} = -k_2 z_r(1, \psi, t) \tag{8.139}$$

Toroid torsion:

$$-(1 + k_1)y'' + k_1\theta'' - \theta = 0 \tag{8.140}$$

The membrane equation is the same as Eq. (8.137), and the boundary conditions are the same as Eq. (8.138) except that

$$z(1, \psi, t) = y(\psi, t) \tag{8.141}$$

8.6.3.2 Static Buckling Analysis The critical tension (static buckling load) can be determined assuming the buckled shape as

$$y(\psi) = A \sin 2\psi \tag{8.142a}$$

$$\theta(\psi) = B \sin 2\psi \tag{8.142b}$$

$$z(r, \psi) = z(r) \sin 2\psi \tag{8.142c}$$

Substituting Eqs. (8.142) into the static membrane equation, we find

$$z(r) = Ar^2 \tag{8.143}$$

which represents parabolic deflection of the membrane when the toroid buckles. Using Eqs. (8.142) and (8.143), and from the toroid bending and torsion equations of a thin toroid (static case), we can obtain the condition for the static out-of-plane buckling as

$$k_2 = 18k_1/(4k_1 + 1) \tag{8.144}$$

Because $k_1 = GJ/EI$ and $k_2 = TR^3/EI$, the critical tension is expressed as

$$T_c = \frac{3EI}{R^3} \frac{6}{4 + EI/GJ} \tag{8.145}$$

It is interesting to compare this result to other known values of critical loads for the different external loading conditions: 1) flexible ring (toroid) with externally applied constant uniform pressure (always in the same direction) and 2) flexible ring with externally applied load (uniformly distributed), which always points to the center of the ring.

For case 1, Eqs. (8.139) and (8.140) can be modified to

$$y'''' + (k_2 - k_1)y'' + (1 + k_1)\theta'' = 0 \qquad (8.146a)$$

$$- (1 + k_1)y'' + k_1\theta'' - \theta = 0 \qquad (8.146b)$$

where there is no transverse component of external pressure in this case. By assuming the buckled shape, $y = A \sin 2\psi$ and $\theta = B \sin 2\psi$, we obtain the following critical value of external pressure:

$$T_c = \frac{3EI}{R^3} \frac{3}{4 + EI/GJ} \qquad (8.147)$$

which is exactly the same as the critical pressure obtained by Timoshenko and Gere [11] in 1923.

For case 2, Eqs. (8.139) and (8.140) can be modified to

$$y'''' + (k_2 - k_1)y'' + (1 + k_1)\theta'' = -k_2 y \qquad (8.148a)$$

$$- (1 + k_1)y'' + k_1\theta'' - \theta = 0 \qquad (8.148b)$$

Similarly, we obtain the critical load for this case as follows:

$$T_c = \frac{3EI}{R^3} \frac{4}{4 + EI/GJ} \qquad (8.149)$$

which is identical to the critical load obtained by Hencky (see Ref. [11]) in 1921.

8.6.3.3 Transfer Functions

The transfer functions from roll control torque to structural displacements, at various points on the flexible thin toroid with a pretensioned membrane, can be obtained by taking the Fourier transform ($t \rightarrow \omega$) and the finite sine transform ($\psi \rightarrow n$) of the coupled partial differential equations and the boundary conditions [12].

The transformed equation of motion of the membrane becomes

$$z_{rr} + \frac{1}{r}z_r + \left[-\frac{n^2}{r^2} + \frac{k_3}{k_2}\omega^2 \right] z(r, n, \omega) = 0 \qquad (8.150)$$

Using the boundary conditions, we obtain the solution of Eq. (8.150) as

$$z(r, n, \omega) = \frac{y(n, \omega)J_n(\lambda r)}{J_n(\lambda)} \qquad (8.151)$$

where $\lambda = \omega\sqrt{k_3/k_2}$ and $J_n(\lambda)$ is the Bessel function of the first kind of order n.

The toroid bending equation (8.139) after transformations becomes

$$n^4 y(n,\omega) + n[(-1)^{n+1} y''(\pi,t) + y''(0,t)] - n^3[(-1)^{n+1} y(\pi,t) + y(0,t)]$$
$$+ (k_2 - k_1)\{-n^2 y(n,\omega) + n[(-1)^{n+1} y(\pi,t) + y(0,t)]\}$$
$$+ (1 + k_1)\{-n^2 \theta(n,\omega) + n[(-1)^{n+1} \theta(\pi,t) + \theta(0,t)]\}$$
$$- \omega^2 y(n,\omega)$$
$$= -k_2 \frac{y(n,\omega)\lambda J_n'(\lambda)}{J_n(\lambda)} \tag{8.152}$$

where

$$J_n'(\lambda) = \frac{dJ_n(\lambda)}{d\lambda}$$

Rearranging Eq. (8.152) and using $y(\pi,t) = y(0,t) = \theta(\pi,t) = \theta(0,t) = 0$ and $y''(0,\omega) = y''(\pi,\omega) = u(\omega)/4$, we get

$$\left[n^4 + (k_1 - k_2)n^2 - \omega^2 + \frac{k_2\lambda J_n'(\lambda)}{J_n(\lambda)}\right] y(n,\omega) - (1 + k_1)n^2 \theta(n,\omega)$$
$$= \frac{n}{4}[(-1)^n - 1]u(\omega) \tag{8.153}$$

Similarly, the toroid torsion equation (8.140) becomes

$$(1 + k_1)n^2 y(n,\omega) - \left(1 + k_1 n^2\right)\theta(n,\omega) = 0 \tag{8.154}$$

Combining Eqs. (8.153), (8.154), and (8.151), we obtain

$$\frac{y(n,\omega)}{u(\omega)} = \frac{\left\{\left(1 + k_1 n^2\right)n/2\right\} J_n(\lambda)}{\Delta(n,\omega)} \tag{8.155}$$

$$\frac{\theta(n,\omega)}{u(\omega)} = \frac{\left\{\left(1 + k_1\right)n^3/2\right\} J_n(\lambda)}{\Delta(n,\omega)} \tag{8.156}$$

$$\frac{z(r,n,\omega)}{u(\omega)} = \frac{\left\{\left(1 + k_1 n^2\right)n/2\right\} J_n(\lambda r)}{\Delta(n,\omega)} \tag{8.157}$$

where

$$\Delta(n,\omega) = -\left[n^4 J_n(\lambda) + (k_1 - k_2)n^2 J_n(\lambda) - \omega^2 J_n(\lambda)\right.$$
$$\left. + k_2\lambda J_n'(\lambda)\right]\left(1 + k_1 n^2\right) + (1 + k_1)^2 n^4 J_n(\lambda)$$

By using the definition of the finite sine transform, we obtain

$$\frac{y(\psi,\omega)}{u(\omega)} = \frac{2}{\pi} \sum_{n=1,3}^{\infty} \frac{\{(1+k_1n^2)n/2\}J_n(\lambda)}{\Delta(n,\omega)} \sin n\psi \qquad (8.158a)$$

$$\frac{\theta(\psi,\omega)}{u(\omega)} = \frac{2}{\pi} \sum_{n=1,3}^{\infty} \frac{\{(1+k_1)n/2\}J_n(\lambda)}{\Delta(n,\omega)} \sin n\psi \qquad (8.158b)$$

$$\frac{z(r,\psi,\omega)}{u(\omega)} = \frac{2}{\pi} \sum_{n=1,3}^{\infty} \frac{\{(1+k_1n^2)n/2\}J_n(\lambda r)}{\Delta(n,\omega)} \sin n\psi \qquad (8.158c)$$

which are the transfer functions from the roll control torque to the displacements at various points on the toroid and membrane.

In particular, the transfer function from the roll control torque to the collocated roll angle sensor, which measures the bending slope $\phi(\omega) = y'(0,\omega)$, can be written as

$$\frac{\phi(\omega)}{u(\omega)} = \frac{1}{\pi} \sum_{n=1,3}^{\infty} \frac{n^2(1+k_1n^2)J_n(\lambda)}{\Delta(n,\omega)} \qquad (8.159)$$

Using the Laplace transform variable s, the roll transfer function (8.159) can be rewritten as (in dimensional form)

$$\frac{\phi(s)}{u(s)} = \frac{1}{Js^2} + \sum_{n=1,3}^{\infty} a_n \prod_{m=1,2}^{\infty} \frac{1+(s/z_{nm})^2}{1+(s/\omega_{nm})^2} \qquad (8.160)$$

where J denotes total moment of inertia of rigid toroid and rigid membrane, ω_{nm} is the mth root of the characteristic equation $\Delta(n,\omega)$, z_{nm} is the mth root of $J_n(\lambda) = 0$, and a_n is defined as

$$a_1 = (1+k_1)/J \qquad (8.161a)$$

$$a_n = \frac{n^2(1+k_1n^2)}{J\{-[n^4+(k_1-k_2)n^2+k_2](1+k_1n^2)+(1+k_1)^2n^4\}} \qquad (8.161b)$$

for $n = 3, 5, \ldots, \infty$.

The transfer function (8.160) is not as simple as other transfer functions because of the combination of infinite product and infinite partial fraction expansions. This additional complexity is due to the coupling between two elastic systems, a flexible toroid and a membrane.

Because Eq. (8.160) is the transfer function between the collocated actuator and sensor, it has alternating poles and zeros along the imaginary axis. Unfortunately, we did not find the exact zeros, but Eq. (8.160) can be used to determine the reduced-order transfer function for the finite dimensional controller design.

8.7 Summary

In this chapter we have considered the pole–zero modeling of some generic models of flexible structures. In particular, we determined the exact transfer functions from applied torques to attitude angles at the points where the control torquers are located. Although analytical frequency-domain modeling of some hybrid systems was possible, the derivation of exact transfer functions involved a fair amount of effort. Perhaps a more logical next step will be the development of a computer-aided approach to the frequency-domain modeling of hybrid systems with complex interconnections of lumped and beam- or trusslike lattice substructures. Consequently, various algorithms and techniques for numerical or symbolic manipulation of frequency-domain continuum models are under development. The various hybrid models considered in this chapter will, therefore, be useful for checking or validating the computer-aided techniques so that the practical use of these techniques can be made with confidence where future large space structures are concerned.

References

[1] Wie, B., "On the Modeling and Control of Flexible Space Structures," Ph.D. Dissertation, Dept. of Aeronautics and Astronautics, SUDAAR 525, Stanford Univ., Stanford, CA, June 1981.

[2] Wie, B., and Bryson, A. E., Jr., "Attitude Control of a Triangular Truss in Space," International Federation of Automatic Control 8th World Congress, Paper 77-2, 1981.

[3] Wie, B., and Bryson, A. E., Jr., "Modeling and Control of Flexible Space Structures," *Proceedings of the 3rd VPI&SU/AIAA Symposium on the Dynamics and Control of Large Flexible Spacecraft*, Virginia Polytechnic Inst. and State Univ., Blacksburg, VA, 1981, pp. 153–174.

[4] Wie, B., and Bryson, A. E., Jr., "Pole–zero Modeling of Flexible Space Structures," *Journal of Guidance, Control, and Dynamics*, Vol. 11, No. 6, 1988, pp. 554–561.

[5] Wie, B., and Bryson, A. E., Jr., "Transfer Function Analysis of a Flexible Toroidal Structure," *Journal of Guidance, Control, and Dynamics*, Vol. 13, No. 5, 1990, pp. 881–886.

[6] Bryson, A. E., Jr., *Control of Spacecraft and Aircraft*, Princeton Univ. Press, Princeton, NJ, 1994, Chap. 9 and Appendix F.

[7] Przemieniecki, J. S., *Theory of Matrix Structural Analysis,* McGraw–Hill, New York, 1968.

[8] Bathe, K.-J., and Wilson, E. L., *Numerical Methods in Finite Element Analysis,* Prentice–Hall, Englewood Cliffs, NJ, 1976.

[9] Meirovitch, L., *Computational Methods in Structural Dynamics*, Sijhoff and Noordhoff, Alphen aan den Rijn, The Netherlands, 1980.

[10] Craig, R. R., *Structural Dynamics: An Introduction to Computer Methods*, Wiley, New York, 1981.

[11] Timoshenko, S. P., and Gere, J. M., *Theory of Elastic Stability*, McGraw–Hill, New York, 1961, pp. 317–318.

[12] Meirovitch, L., *Analytical Methods in Vibrations*, Macmillan, New York, 1967.

9
Attitude and Structural Control

This chapter is primarily concerned with the analysis and design of attitude control systems for space vehicles in the presence of propellant sloshing and/or structural flexibility. Active structural vibration control problems are also treated; however, many theoretical aspects of flexible structure controls are not elaborated upon. This chapter is mainly intended to provide the reader with various practical examples of attitude and structural control designs for space vehicles having flexible solar arrays and/or flexible appendages.

9.1 Thrust Vector Control Design for a Spacecraft with Propellant Sloshing

In this section a preliminary thrust vector control design for a spacecraft in the presence of propellant sloshing is presented. The intent here is not to present the final control system design in detail, but rather to describe mission requirements, dynamic modeling of propellant slosh, and thrust vector control design during the early phases of developing a liquid upper stage [1].

9.1.1 Introduction

Future spacecraft missions will require more efficient, low-cost orbital transfer from low Earth orbit to geosynchronous orbit. This requirement has resulted in the development of various new upper stages, including transfer orbit stage (TOS). The TOS is a three-axis stabilized perigee stage with the same solid propellant, first-stage motor that is used in the inertial upper stage (IUS); however, an integrated liquid upper stage that performs both the transfer and geosynchronous orbit injections may be more cost effective for some missions.

An integrated liquid-propellant stage for use as a reusable orbital transfer vehicle for a future geostationary platform has been extensively studied by NASA in the past. A feasibility study for the development of a new upper stage to minimize the cost of launching communications satellites into geosynchronous orbit via the Space Shuttle has also been studied [1]. One of the several systems studied is the integrated liquid upper stage that employs a nonspinning deployment from the Space Shuttle in conjunction with three-axis stabilization of the upper stage during perigee and apogee maneuvers. Figure 9.1 illustrates a mission scenario for deploying a geosynchronous communications satellite using this upper stage.

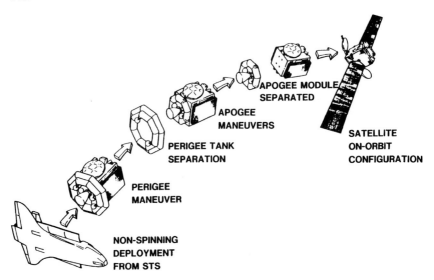

Fig. 9.1   Mission sequence for deploying a geosynchronous communications satellite. (Courtesy of Ford Aerospace and Communications Corp., Palo Alto, CA.)

As can be seen in this figure, this liquid-bipropellant (MMH/NTO) upper stage consists of separate perigee and apogee tanks with a single gimbaled main engine.

During the early phases of this new upper stage development, the large amount of liquid propellant has posed a major concern regarding propellant sloshing and thrust vector control design in the presence of significant center-of-mass uncertainties. The dynamic interactions of propellant sloshing with attitude control systems have been one of the major concerns in control systems design for various aerospace vehicles; see, e.g., Greensite [2] and Bryson [3]. There has also been a considerable amount of analytical and experimental work done in the past to account for the effects of propellant sloshing [4].

In this section we consider the problem of designing a thrust vector control (TVC) system for a liquid-bipropellant upper stage spacecraft. The objective here is not to present the final control system design in detail, but rather to describe an analytical design procedure, as well as a systematic way of specifying the overall system requirements, during the early phases of the new upper stage development. Mission requirements, dynamic modeling of propellant slosh, and a preliminary TVC design are described in this section. A simple model with a single slosh pendulum is introduced and an implicit "gravity turn" guidance scheme, often called *tangential steering*, is considered for TVC design. Overall system design requirements such as ΔV pointing error, spacecraft gain factor, center-of-mass uncertainty, gimbal angle and rate limit, gyro saturation limit, and slosh characteristics are discussed.

9.1.2   Mission Requirements

For a satellite mass range of 750–1400 kg, a 3750-lb thrust engine with a specific impulse of 328 s is selected for the preliminary mission analysis.

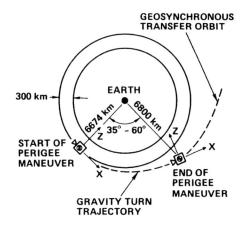

Fig. 9.2   Perigee maneuver with a 35–60-deg burn arc.

Approximately 8–14 min perigee burn times are required to achieve a ΔV of 2450 m/s. The long burn time is due to the relatively low-thrust level of 3750 lb, compared to the PAM-D thrust level of 14,000 lb, the Centaur thrust level of 30,000 lb, or the IUS thrust level of 50,000 lb. Because of the long burn time, a powered flight with a 35–60 deg burn arc is required, as illustrated in Fig. 9.2. Although it is most efficient to perform some orbital plane change at perigee, we consider here only a planar perigee injection maneuver. The apogee burn requires 2–4 min to achieve a ΔV of 178 m/s.

The total burn time may be completed in one burn or divided into several short burns. A single perigee firing and two apogee firings are selected as a baseline. The duration of each burn determines the efficiency of the ΔV maneuver. The difference between the actual ΔV and the equivalent impulsive ΔV is termed *gravity loss*; it is a measure of the burn efficiency. The efficiency of the firing determines the amount of propellant mass needed to achieve the desired transfer orbit. The efficiency of the ΔV maneuver also depends on the thrust-vector steering logic. A summary of the mission requirements is given in Table 9.1, which lists the ΔV requirements, burn time, and burn arc for each mission. The spacecraft parameters for each mission are also summarized in Table 9.2.

The implicit gravity turn guidance, often called tangential steering, is considered assuming that the actual flight trajectory is very close to the expected nominal trajectory. The gravity turn is defined as the trajectory that results from simply keeping the thrust vector always parallel to the velocity vector. It is known that an optimal thrust vector steering other than the simple tangential steering increases the fuel efficiency of any finite thrust maneuver; however, the burn efficiency of the tangential steering even with a 60-deg burn arc for the mission under consideration is still better than 98.5% because of the relatively higher acceleration level than the g level considered in most low-thrust maneuver studies [5]. Note that the 1.5% efficiency loss will only require approximately 50 kg of additional propellant. An explicit cross-product steering guidance [6] could also be employed to improve the fuel efficiency. Tradeoffs between fuel efficiency and implementation simplicity are needed in practice.

Table 9.1   Summary of the mission requirements

Mission description		Minimum	Maximum
Satellite mass (BOL[a]), kg		750	1400
PF[b]	ΔV, m/s	2450	2458
	Burn time, s	480	840
	Burn arc, deg	35	60
	ΔV pointing error,[e] deg	2	2
AF1[c]	ΔV, m/s	1700 (900)[f]	1700 (900)
	Burn time, s	150 (80)	250 (130)
	Burn arc, deg	0.5 (0.3)	1 (0.8)
	ΔV pointing error, deg	0.5	0.5
AF2[d]	ΔV, m/s	100 (900)	100 (900)
	Burn time, s	5 (80)	18 (130)
	Burn arc, deg	0.01 (0.3)	0.6 (0.8)
	ΔV pointing error, deg	0.5	0.5

[a]Beginning of life. [b]Perigee firing. [c]First apogee firing. [d]Second apogee firing. [e]Pointing error budget for TVC due to center-of-mass uncertainty; total pointing error requirements: 3 deg (perigee) and 1 deg (apogee). [f]95–5% (50–50%) split for AF1 and AF2.

Table 9.2   Summary of spacecraft (upper stage and satellite) parameters

Description	Minimum mission			Maximum mission		
	PF[a]	AF1[b]	AF2[c]	PF	AF1	AF2
Total mass, kg	4900	1950	1100	7800	3150	1800
Acceleration, $g = 9.8$ m/s^2	0.34	0.87	1.55	0.22	0.54	0.94
Inertia, kg·m^2						
Roll	10000	2143	1000	29000	5600	2000
Pitch	10000	2676	1000	24000	6100	2200
Yaw	10000	1576	1000	20000	500	2000
Center-of-mass distance, m	0.2	0.5	1.0	0.3	0.6	1.3
Gain factor, s^{-2}						
Pitch	0.3	5.3	16	0.25	2.2	10
Yaw	0.3	3.1	16	0.25	1.7	10
Propellant, kg	2500	790 (400)	40 (400)	4200	1270 (660)	65 (660)

[a]Perigee firing. [b]First apogee firing. [c]Second apogee firing.

9.1.3 Spacecraft Dynamics with Propellant Sloshing

Maintaining the thrust vector pointing of the spacecraft in the presence of significant propellant sloshing is of practical interest. If the propellant tanks are always 100% full, then there will be no sloshing problems. In practice, however, tanks sized for the maximum propellant requirement will also be used for some missions with a lower propellant requirement. As summarized in Table 9.1, the minimum payload mission needs only 60% of the propellant required for the maximum mission. This is the case in which special considerations of the propellant sloshing in liquid upper stage designs are required.

The propellant slosh motions are generally characterized as lateral, vertical, rotational (swirl motion), vortex formation, surface spray, dome impact, and low-gravity phenomena. Of these, lateral sloshing has the most significant effect on spacecraft attitude dynamics during thrusting maneuvers. Exact formulation and solution of propellant sloshing problems in various tanks with/without compartments and baffles are, in general, very complex. Thus, dynamically equivalent mechanical models have been used to represent the gross effects of propellant sloshing in various tanks.

A flat doughnut-shaped tank, as shown in Fig. 9.1, has been selected because of its high volume efficiency (shorter length); however, there are no theoretical analyses available for predicting the behavior of liquids in compartmented toroidal tanks. This is because liquid motions in such toroidal tanks are more nonlinear than those in cylindrical or spherical tanks. For a preliminary thrust vector control design, an ideal model with a single slosh pendulum, as illustrated in Fig. 9.3, can be used. This simple model makes it possible to perform a preliminary TVC design and to generate various system level requirements during the early phases of the new upper stage development. The tank fill ratio determines the parameters of an equivalent mechanical model. For example, in a spherical tank with a 50% fill ratio, the effective slosh mass is about 35% of the total propellant and the pendulum length is about 60% of the tank radius (pivoted at the center of tank).

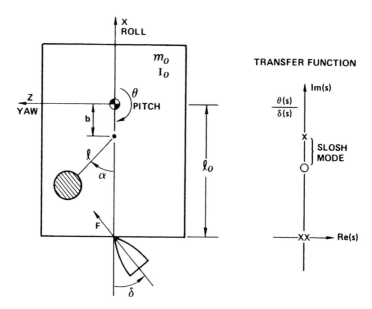

Fig. 9.3   Simple model with a single slosh pendulum.

Because it is a common practice to consider a decoupled single-axis control design, we consider here only the pitch-axis TVC design. For such a single-axis control design, the nonlinear equations of motion of the simple planar model with a single slosh pendulum, as shown in Fig. 9.3 can be derived as

$$m_o a_x = F \cos \delta - m[a_x + \ell(\ddot{\theta} + \ddot{\alpha}) \sin \alpha + b\dot{\theta}^2 + \ell(\dot{\theta} + \dot{\alpha})^2 \cos \alpha] \quad (9.1a)$$

$$m_o a_z = F \sin \delta - m[a_z + b\ddot{\theta} + \ell(\ddot{\theta} + \ddot{\alpha}) \cos \alpha - \ell(\dot{\theta} + \dot{\alpha})^2 \sin \alpha] \quad (9.1b)$$

$$J_o \ddot{\theta} = F\ell_o \sin \delta - mb[a_z + b\ddot{\theta} + \ell(\ddot{\theta} + \ddot{\alpha}) \cos \alpha - \ell(\dot{\theta} + \dot{\alpha})^2 \sin \alpha] \quad (9.1c)$$

$$m[a_z \cos \alpha + a_x \sin \alpha + b\ddot{\theta} \cos \alpha + \ell(\ddot{\theta} + \ddot{\alpha}) + b\dot{\theta}^2 \sin \alpha] + c\ell\dot{\alpha} = 0 \quad (9.1d)$$

where a_x and a_z are respectively the longitudinal and lateral accelerations of the main body center of mass; θ and α are the pitch angle and pendulum deflection angle, respectively; m_o and J_o are the mass and pitch inertia of the main body excluding the slosh pendulum, respectively; F is the thrust magnitude; ℓ_o is the distance from the gimbal pivot point to the main body center of mass; m and ℓ are the mass and length of the slosh pendulum, respectively; c is the slosh damping coefficient; and b is the distance between the center of mass of the main body and the pendulum pivot point.

The engine inertia effect that results in a so-called tail-wags-dog zero is not included in Eqs. (9.1) because of the relatively small engine inertia. Also notice that the preceding equations of motion have been written with respect to the center of mass of the main body, not to the composite center of mass. For a relatively small sloshing mass, it can be assumed that the composite center of mass is coincident with the main body center of mass. For a relatively large sloshing mass, however, the effects of a moving composite center of mass on the TVC design is likely to mislead the TVC designer. For example, the effects of initial propellant asymmetry may be considered additional thrust vector misalignment if the equations of motion are written with respect to the composite center of mass, which is, in fact, moving relative to the main body center of mass as the propellant sloshes. However, the initial propellant asymmetry must be distinguished from the thrust vector misalignment arising from the constant center-of-mass offset. Thus, the effects of large slosh initial conditions become clear if the equations are written by separating the main body and the slosh pendulum. The gyros sense not the motion of the composite center of mass, but the motion of the main body to which the gyros are attached. The thrust vector control system uses the main body attitude changes measured by the gyros for feedback controls.

For a preliminary TVC system design, Eqs. (9.1) can be linearized for small θ and α, and the following transfer function from the gimbal deflection angle δ to the pitch attitude θ can be used:

$$\frac{\theta}{\delta} = \frac{a_o(s^2 + \omega_z^2)}{s^2(s^2 + \omega_p^2)} \quad (9.2)$$

where $a_o = F\ell_o/J_o$ is often called the *gain factor* of the vehicle excluding propellant sloshing mass. The engine inertia effect, known as a tail-wags-dog zero, is not shown here because of the relatively small engine inertia. The pole and zero of the slosh mode are given by

$$\omega_p^2 = \omega_s^2 \left\{ 1 + \frac{m}{m_o} + \frac{mb(b + \ell)}{J_o} \right\} \tag{9.3a}$$

$$\omega_z^2 = \omega_p^2 - \omega_s^2 \left\{ \frac{mb}{m_o \ell_o} + \frac{mb(b + \ell)}{J_o} \right\} \tag{9.3b}$$

where ω_s denotes the slosh frequency defined as

$$\omega_s = \sqrt{F/(m_o + m)\ell} \tag{9.4}$$

The slosh mode pole–zero separation represents the magnitude of propellant sloshing. If the slosh pendulum is pivoted below the composite center of mass, then the slosh mode zero will always be lower than the pole. This results in a stable interaction of the slosh mode and the rigid-body mode. Thus, the problem of controlling a spacecraft with propellant sloshing is very similar to the problem of controlling a spacecraft with flexible solar arrays; however, the TVC designer has little direct control over propellant sloshing. His approach is to note its effect and, if objectionable, to specify a greater amount of mechanical passive damping.

Several passive methods can be employed to minimize the effects of propellant sloshing. Baffles of various configurations usually add passive damping ratios up to 5% and mainly affect the magnitude of the slosh forces and the amplitude of the slosh motion. The penalty of baffles is, obviously, more weight and, hence, less payload (or more launch cost). However, a significant weight saving can be realized without the loss of damping effectiveness by using perforated (usually 30%) baffles.

Compartmentation, or subdivision, of the tank has a very marked effect in increasing the fundamental slosh frequency. This is an ideal method of avoiding slosh coupling with rigid-body control by using spectral separation. Compartmentation also has the effect of lowering the second slosh frequency so that these two frequencies are not widely separated. Thus, integration of the tank and TVC design is needed during the early phases of the liquid upper stage development.

9.1.4 Thrust Vector Control Design

Consider a preliminary TVC design for the pitch-axis TVC system shown in Fig. 9.4. This pitch-axis thrust vector control system, which receives the gyro output and gimbal angle readout and feeds them to the gimbal servo, consists of an inner attitude control loop and an outer guidance loop. As mentioned earlier, a gravity turn is assumed so that the thrust vector of the upper stage spacecraft is nominally pointing along the velocity vector and only small deviations from the nominal trajectory are considered.

Although all of the spacecraft parameters are actually time varying (mass flow rate ≈ 5.19 kg/s), they are held fixed for analysis purposes because the variations are slow when compared to the attitude control dynamics. However, gain scheduling may be needed to compensate for the large variation that occurs in spacecraft mass and inertia properties as propellant are depleted and the stage is jettisoned.

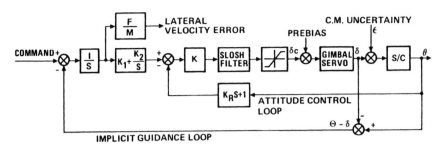

Fig. 9.4   Pitch-axis thrust vector control system.

Because the center-of-mass offset results in a constant disturbance torque, a steady-state attitude error is necessary to produce the gimbal deflection to counteract the disturbance. Thus, an attitude pointing error develops that must be either lessened by increasing the loop gain or eliminated by means of integral control. Integral trim is provided from the outer guidance loop as can be seen in Fig. 9.4. For guidance purposes, attitude or attitude rate is not the primary variable of interest in the TVC design. The guidance loop commands an attitude change to control the lateral velocity $\dot{z}$ of the spacecraft. Thus, control of the spacecraft velocity vector is a prime requirement in the TVC design. Control of the orbit injection velocity vector at main engine cutoff is most critical. Because the ultimate goal of thrust vector control is to control the lateral velocity error, the outer guidance loop generates an appropriate attitude angle command for the inner attitude control loop. Integral gain K_2 in the guidance loop eliminates the steady-state lateral velocity error due to the center-of-mass offset or thrust vector misalignment. The preprogrammed thrust vector steering command $(\theta - \delta)_c$ is approximately a "ramp" signal for a simple gravity turn maneuver. The TVC system shown in Fig. 9.4 will then have a steady-state lateral velocity error for a ramp command. This error is inversely proportional to the integral gain K_2.

The type of slosh filter (notch, phase lead, or rolloff) in the inner attitude control loop depends on the degree of passive control afforded by baffles and compartmentation. If the slosh mode frequency is widely separated from the control bandwidth, a gain stabilization of the slosh mode is preferred using a roll-off filter. If not, the slosh mode should be phase stabilized by carefully considering the additional phase lag from the actuator/sensor dynamics and control loop time delay.

If the outer guidance loop response time is kept an order of magnitude slower than the response time of the inner attitude control loop, then an equivalent control logic for the TVC system in terms of a single measurement of the pitch attitude can be represented as

$$\delta_c = -K(1 + K_R s)\left(1 + \frac{K_1}{s} + \frac{K_2}{s^2}\right)\theta \qquad (9.5)$$

where K is the attitude control loop gain, K_R the rate gain, and K_1 and K_2 are the guidance loop parameters. This equivalent single-loop control logic representation of the multiloop system allows a classical control design using the root locus method and/or frequency response methods.

Furthermore, neglecting the slosh mode, we can obtain the following approximate closed-loop relations for a parametric study of the TVC system:

ω_1 = attitude control bandwidth, $\sqrt{aK}$
ζ_1 = attitude control damping, $0.5K_R\sqrt{aK}$
ω_2 = guidance loop bandwidth, $\sqrt{K_2}$
ζ_2 = guidance loop damping, $0.5K_1\sqrt{K_2}$
$\dot{\theta}_{max}$ = attitude peak rate, $\epsilon\sqrt{a/K}\exp[-D^{-1}\tan^{-1}D]$
δ_{ss} = gimbal steady-state angle, $-\epsilon$
δ_{max} = gimbal peak rate, $-KK_Ra\epsilon$

where a is the rigidized spacecraft gain factor, ϵ is the center-of-mass offset angle, and $D = \sqrt{1 - \zeta_1^2/\zeta_1}$.

These approximate relationships among the TVC system parameters have been very useful in the early phases of the upper stage development, because they have provided high confidence in specifying requirements on the overall system without the need for extensive digital simulations [1].

During the early phases of the upper stage development, the TVC designer is often asked to specify the capabilities of the gimbal servo for system level tradeoffs in selecting the type of gimbal servo. An electrohydraulic servo can be selected because of its performance superiority to any other type of servo, such as electromechanical or pneumatic servos. Large inertia and torque loads can be handled with high accuracy and very rapid response. The electrohydraulic servo used to gimbal large rocket engines is a highly nonlinear device. For preliminary analysis purposes, however, the servo transfer function from the gimbal angle command δ_c to gimbal angle output δ can be simply modeled as

$$\frac{\delta}{\delta_c} = \frac{\omega_n^2}{(Ts + 1)(s^2 + 2\zeta\omega_n s + \omega_n^2)} \tag{9.6}$$

where T is the servo time constant and ω_n and ζ are the hydraulic natural frequency and damping ratio, respectively. The servo bandwidth $(1/T)$ is usually limited to 20–40% of the hydraulic natural frequency. Thus, a fast servo requires higher hydraulic natural frequency and large damping ratio. Both of these quantities are fixed once the power element is selected. Significant nonlinearities may occur in the servo valve and in the gimbal bearing friction. There is a saturation limit for hydraulic flow rate, which limits the slew rate for the gimbal actuator.

A preliminary TVC design for the specific mission requirements given in Table 9.1 has been conducted [1]. Such a preliminary design can best be performed using a root locus plot, illustrated in Fig. 9.5. This figure shows a root locus plot vs the attitude control loop gain K for a minimum payload mission at first apogee: $a_o = 5.0, K_1 = 0.1, K_2 = 0.01, K_R = 2.0$, and nominal $K = 0.7$. A servo time constant of 0.1 s and a slosh mode with 5% damping, $\omega_z = 4$ rad/s, and $\omega_p = 5$ rad/s have been assumed.

One of the difficult tasks in the preliminary TVC design and analysis is to specify requirements on slosh mode frequency and passive damping to ensure satisfactory TVC performance and stability. If the slosh mode characteristics are given to the TVC designer, then it may be straightforward to include the slosh mode in the

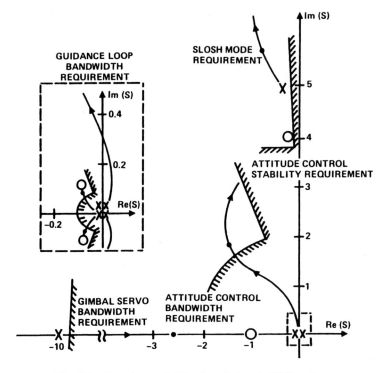

Fig. 9.5   Root locus plot for the pitch-axis TVC system.

TVC design. In general, tradeoffs between passive and active controls in terms of cost, complexity, and reliability are needed in a system level study.

A classical control design of the TVC system using the root locus method has been verified by computer simulations. Figure 9.6 shows the pitch-axis responses for a maximum mission at perigee. The center-of-mass offset of 4 deg and thrust vector steering command of 0.1 deg/s are assumed for the perigee maneuver. The peak attitude rate and peak lateral velocity error are relatively small because of the large inertia of vehicle at perigee. The peak lateral velocity error during the initial transient is well within the ΔV pointing requirement. The steady-state lateral velocity error of 0.4 m/s comes from the thrust vector steering ramp command of 0.1 deg/s. The lateral velocity error can be further reduced, if necessary, by increasing the guidance loop bandwidth or adding another integral control term. Note that the 0.4-m/s steady-state error was negligible; however, navigation error analysis and simulation in terms of actual orbital parameters should be performed as the system design becomes mature.

Also, the effect of large slosh initial conditions (up to 45 deg) has been studied using the nonlinear equations. Not surprisingly, the overall performance and stability are not changed much from the simulations using the linearized equations of motion. For a 45-deg initial slosh angle, both linear and nonlinear simulations closely predicted a spacecraft peak rate of about 8 deg/s at the first apogee firing with a 2-deg center-of-mass offset. Therefore, it can be concluded that the linear

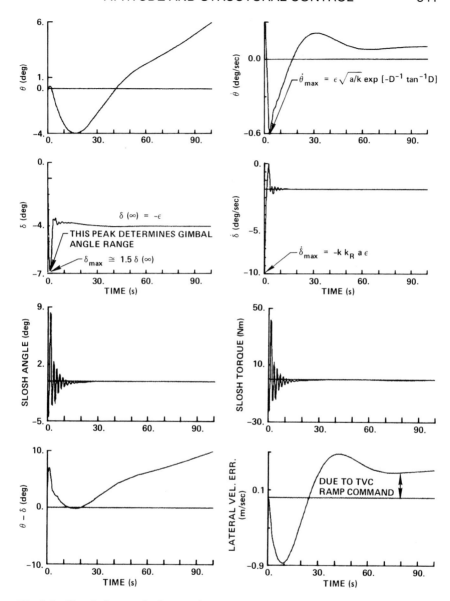

Fig. 9.6   Simulation results for a perigee maneuver with a 4-deg center-of-mass offset and 0.1-deg/s TVC command.

transfer function analysis is still valid in the presence of a time-varying composite center of mass. Because a large initial propellant asymmetry can result in an attitude peak rate of 8 deg/s, the propellant settling burns, often called "ullage" burns, may be needed to provide propellant settling prior to the main engine firings if the gyro has a saturation limit of about 10 deg/s.

9.2 Attitude Control Design for a Bias-Momentum Stabilized Spacecraft

In this section, which is based on [7], we consider an attitude control design problem of a bias-momentum stabilized geosynchronous communications satellite having flexible solar arrays.

9.2.1 Introduction

Three-axis bias-momentum stabilization has been employed for many geosynchronous communications satellites [7–10]. The unique feature of the bias-momentum stabilization for satellites in geosynchronous orbit is the ability to control yaw-axis pointing error passively without a direct yaw measurement. The bias momentum provides gyroscopic stiffness to the environmental disturbances, primarily to the solar radiation pressure torque.

The increased demand of electrical power for communications and/or direct TV broadcasting leads to large flexible solar panel arrays for three-axis stabilized spacecraft. Consequently, the structural flexibility of the solar arrays has been one of the primary concerns in the design of attitude control systems for a certain class of three-axis stabilized spacecraft. For most cases of practical concern, however, the structural flexibility of the solar arrays does not strongly interact with attitude control systems, and thus all of the structural modes are often gain stabilized by the steep rolloff at a frequency well below the first structural frequency. In this section we consider such a case in which the structural modes do not strongly interact with an attitude control system but need to be considered in the control design process to avoid a possible closed-loop instability of the structural modes.

Figure 9.7 shows a three-axis stabilized, geosynchronous communications satellite with large, flexible solar arrays. The antenna reflectors are rigidly mounted (after deployment) to the main body.

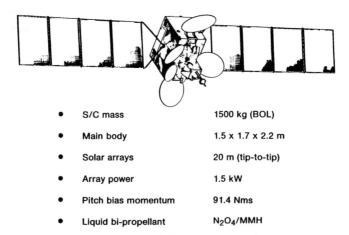

- • S/C mass   1500 kg (BOL)
- • Main body   1.5 x 1.7 x 2.2 m
- • Solar arrays   20 m (tip-to-tip)
- • Array power   1.5 kW
- • Pitch bias momentum   91.4 Nms
- • Liquid bi-propellant N_2O_4/MMH

Fig. 9.7   Three-axis stabilized, geosynchronous communications satellite.

9.2.2 Attitude Control System Description

Figure 9.8 shows an attitude control system configuration that consists of two momentum wheels skewed with respect to the pitch axis, a smaller yaw reaction wheel for backup mode, redundant two-axis Earth sensors to measure roll and pitch attitude references, and thrusters to provide wheel momentum desaturation torques. Other thrusters for stationkeeping maneuvers are not shown here. The x axis is nominally in the flight direction, the y axis is normal to the orbit plane, and the z axis is directed toward the Earth. Such roll, pitch, and yaw control axes are nearly coincident with the principal axes of the spacecraft.

For the spacecraft shown in Fig. 9.7, the principal moments of inertia are given by

$$(J_x, J_y, J_z) = (3026, 440, 3164) \text{ kg} \cdot \text{m}^2$$

and two skewed momentum wheels with a 2.5-deg skew angle provide pitch bias momentum of 91.4 N·m·s with ±2.5-N·m·s momentum modulation capability. The yaw reaction wheel has momentum modulation capability of ±0.65 N·m·s.

A functional block diagram representation of the on-orbit normal mode control system employing the skewed bias-momentum wheel configuration is also illustrated in Fig. 9.9. As can be seen in this figure, the on-orbit normal mode control system consists of the attitude control loops and wheel momentum desaturation loops.

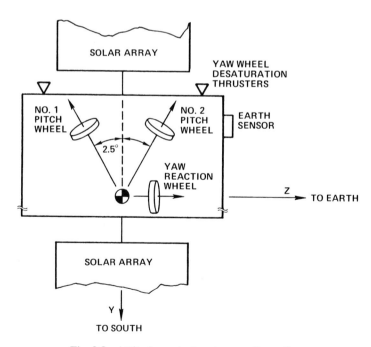

Fig. 9.8 Attitude control system configuration.

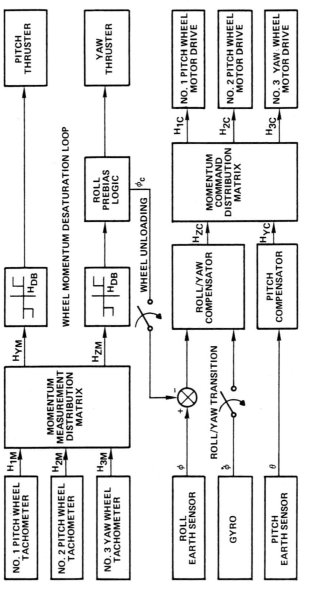

Fig. 9.9   On-orbit normal control system.

During the on-orbit normal mode operations of this control system, the yaw attitude error is neither measured nor estimated. The yaw error in a limit cycle at orbit rate is controlled indirectly by the yaw momentum desaturation loop with the measurement of wheel yaw momentum. The yaw momentum control loop provides active roll but passive yaw control with secularly increasing yaw error due to external disturbances. The roll/yaw secular momentum resulting from external disturbances is stored either as yaw attitude error or in wheel momentum, until a desaturation torque is applied to the spacecraft. The gyros provide yaw attitude reference and three-axis body rates during the stationkeeping mode operations, and are turned off after the transition mode operations. (The attitude stabilization problem of the spacecraft during stationkeeping maneuvers will be treated in Sec. 9.3.)

This control system has three redundant modes: 1) the primary mode (MW 1 and 2), 2) backup mode 1 (MW 1 and RW), and 3) backup mode 2 (MW 2 and RW). For the primary mode, the pitch axis is controlled by modulating the pitch bias momentum H_y of the two momentum wheels about their nominal bias momentum H_o in response to the error signal θ from the pitch channel of the Earth sensor. The roll/yaw axes are controlled by differentially modulating the yaw angular momentum H_z of the two momentum wheels in response to the error signal ϕ from the roll channel of the Earth sensor, as illustrated in Fig. 9.9.

The momentum command distribution matrix is used to convert the pitch and yaw momentum commands H_{yc} and H_{zc} to the wheel momentum commands H_{1c}, H_{2c}, and H_{3c}. Similarly, the momentum measurement distribution matrix is used to convert the angular momentum of each wheel, as measured by its tachometer (H_{1m}, H_{2m}, and H_{3m}), to the angular momentum in the control axes (H_{ym} and H_{zm}).

9.2.3 Flexible Spacecraft Model

9.2.3.1 Hybrid-Coordinate Model
The hybrid-coordinate modeling is very useful for the control analysis and design of a spacecraft having a main rigid body and flexible solar arrays [11]. A single solar array flexibility model referenced to the spacecraft center of mass is given in Table 9.3. The orientation of the solar arrays with respect to the spacecraft main body depends on orbital position and, thus, on orbital time. Solar array orientation at 6 a.m. will be considered as a nominal configuration for the subsequent analysis and design. Orbit time of 6 a.m. or 6 p.m. yields out-of-plane bending modes in the yaw axis and in-plane bending modes in the roll axis. (Low-frequency characteristics of the first in-plane bending mode is caused by array yoke deformation.)

During on-orbit normal mode operations, both solar arrays are always pointing toward the sun, whereas the main body is pointing toward the Earth. This results in very slowly changing modal frequencies and mode shapes. For control design purposes, however, the spacecraft model will be treated as a time-invariant system with a known range of modal characteristics.

Including a single dominant cantilever mode in each axis, the linearized equations of motion of the bias-momentum stabilized spacecraft in circular orbit can be written in terms of the hybrid coordinates as follows.

Table 9.3   Single solar array flexibility model at 6 a.m.

Cantilever mode description[a]	Cantilever frequency σ, rad/s	Coupling scalars, $\sqrt{kg \cdot m^2}$		
		Roll δ_x	Pitch δ_y	Yaw δ_z
OP-1	0.885	0	0	35.372
OP-2	6.852	0	0	4.772
OP-3	16.658	0	0	2.347
OP-4	33.326	0	0	0.548
T-1	5.534	0	2.532	0
T-2	17.668	0	0.864	0
T-3	33.805	0	0.381	0
IP-1	1.112	35.865	0	0
IP-2	36.362	2.768	0	0

[a]OP is out-of-plane, T is torsion, and IP is in-plane.

Rigid main body:

$$J_x \ddot{\phi} + (a + nH_o)\phi + (b + H_o)\dot{\psi} - nH_z + \sqrt{2}\delta_x \ddot{q}_x = M_x \qquad (9.7a)$$

$$J_y \ddot{\theta} + d\theta + \dot{H}_y + \sqrt{2}\delta_y \ddot{q}_y = M_y \qquad (9.7b)$$

$$J_z \ddot{\psi} + (c + nH_o)\psi - (b + H_o)\dot{\phi} + \dot{H}_z + \sqrt{2}\delta_z \ddot{q}_z = M_z \qquad (9.7c)$$

Two solar arrays:

$$\ddot{q}_x + \sigma_x^2 q_x + \sqrt{2}\delta_x \ddot{\phi} = 0 \qquad (9.8a)$$

$$\ddot{q}_y + \sigma_y^2 q_y + \sqrt{2}\delta_y \ddot{\theta} = 0 \qquad (9.8b)$$

$$\ddot{q}_z + \sigma_z^2 q_z + \sqrt{2}\delta_z \ddot{\psi} = 0 \qquad (9.8c)$$

Wheel/motor dynamics:

$$T_m \dot{H}_y + H_y = H_{yc} \qquad (9.9a)$$

$$T_m \dot{H}_z + H_z = H_{zc} \qquad (9.9b)$$

where

$$a = 4n^2(J_y - J_z)$$
$$b = -n(J_x - J_y + J_z)$$
$$c = n^2(J_y - J_x)$$
$$d = 3n^2(J_x - J_z)$$

and (ϕ, θ, ψ) are small roll, pitch, and yaw attitude errors of the spacecraft with respect to the local vertical and local horizontal (LVLH) reference frame, n is the orbital rate; H_o is the nominal pitch bias momentum; (J_x, J_y, J_z) are the spacecraft principal moments of inertia; (q_x, q_y, q_z) are the cantilever modal coordinates, $(\sigma_x, \sigma_y, \sigma_z)$ are the cantilever modal frequencies; $(\delta_x, \delta_y, \delta_z)$ are the rigid-elastic coupling scalars of a single solar array; (H_y, H_z) are the angular momentum components of the wheels; (H_{yc}, H_{zc}) are the wheel angular momentum commands; (M_x, M_y, M_z) are the external torques; and T_m the motor time constant, which is chosen as 4 s for the spacecraft under consideration.

It is apparent that the pitch-axis dynamics are decoupled from the roll/yaw dynamics. Furthermore, for the spacecraft in geosynchronous orbit the constants $(a, b, c, \text{and } d)$ can be neglected compared to the relatively large value of pitch bias momentum.

9.2.3.2 Nutation Dynamics
The open-loop characteristics of the spacecraft nutation dynamics can be easily examined using the following roll/yaw responses to the initial conditions, neglecting the effects of orbit rate and solar array flexibility:

$$\phi(t) = \phi(0) + \frac{\dot{\phi}(0)}{\lambda} \sin \lambda t - \frac{J_z \dot{\psi}(0)}{H_o}(1 - \cos \lambda t) \qquad (9.10)$$

$$\psi(t) = \psi(0) + \frac{\dot{\psi}(0)}{\lambda} \sin \lambda t + \frac{J_x \dot{\phi}(0)}{H_o}(1 - \cos \lambda t) \qquad (9.11)$$

where λ is the nutation frequency defined as

$$\lambda = H_o / \sqrt{J_x J_z}$$

These equations indicate that a positive initial roll rate produces a positive yaw response with a steady-state offset, and that the positive initial yaw rate produces a negative roll response with a steady-state offset. From these equations, the maximum roll/yaw errors due to a single roll/yaw thruster firing with a small pulse width can be easily estimated.

9.2.3.3 Transfer Function
The roll/yaw control analysis and design can be accomplished using the classical transfer function approach. The cross-axis transfer function from the wheel yaw momentum command $u = H_{zc}$ to the roll attitude error output ϕ can be obtained as

$$\frac{\phi(s)}{u(s)} = \frac{(s^2/z_1^2 + 1)(s^2/z_2^2 + 1)}{H_o(s^2/\lambda^2 + 1)(s^2/p_1^2 + 1)(s^2/p_2^2 + 1)(T_m s + 1)} \qquad (9.12)$$

where

$$z_1 = \sigma_z = 0.885 \text{ rad/s}$$

$$z_2 = \sigma_x = 1.112 \text{ rad/s}$$

$$p_1 \approx \frac{\sigma_z}{\sqrt{1 - 2\delta_z^2/J_z}} = 1.93 \text{ rad/s}$$

$$p_2 \approx \frac{\sigma_x}{\sqrt{1 - 2\delta_x^2/J_x}} = 2.87 \text{ rad/s}$$

$$T_m = 4 \text{ s}$$

and (σ_x, δ_x) and (σ_z, δ_z) are, respectively, the cantilever frequencies and coupling scalars of the dominant in-plane and out-of-plane bending cantilever modes of a single array given in Table 9.3. The approximate values of the poles as given here agree very closely with the values obtained for the flexible spacecraft model including all of the cantilever modes given in Table 9.3.

The pole and zero associated with the orbit rate mode are not included in Eq. (9.12) because they are of the same order of magnitude as the orbit rate n. However, their exact values can be determined from the equations of motion including the constants a, b, and c. The relative location determines the closed-loop stability of the orbit rate mode, and it depends on the pitch bias momentum direction. For a spacecraft with $J_z > J_y$, the pitch bias momentum vector is chosen along the negative pitch axis for a stable interaction between the nutation and orbit rate modes.

It is interesting to note that the structural mode poles and zeros of the cross-axis transfer function (9.12) are not alternating along the imaginary axis. Consequently, a control logic designed to stabilize the nutation mode neglecting the structural modes could destabilize the structural modes, because they are not stably interacting.

9.2.4 Roll/Yaw Control Design

The roll/yaw attitude control design problem with $u \equiv H_{zc}$ as a control input and ϕ as a sensor output is discussed here.

9.2.4.1 Low-Bandwidth Controller
Consider a nonminimum-phase control logic of the form

$$u(s) = -\frac{K(1 - T_z s)}{s(Ts + 1)}\phi(s) \tag{9.13}$$

where K is the positive loop gain, T_z the time constant of the nonminimum-phase zero, and T the time constant of the first-order low-pass filter pole. Referring to the conventional PID controller, we have $K_P = -KT_z$, $K_I = K$, and $K_D = 0$.

It is emphasized that this somewhat unconventional, nonminimum-phase control logic, which was originally proposed by Terasaki [12] in 1967, has been actually implemented for many geosynchronous communications satellites, including INSAT and ARABSAT spacecraft.

Neglecting the solar array flexible modes, the closed-loop roll transfer function from the spacecraft initial conditions can be found as

$$\phi(s) = \frac{\left[(s^2 + \lambda^2)\phi(0) + s\dot{\phi}(0) - \dot{\psi}(0)H_o/J_x\right](s + 1/T)}{s^4 + s^3/T + \lambda^2 s^2 + s(H_o - KT_z)\lambda^2/H_oT + K\lambda^2/H_oT} \tag{9.14}$$

in which the wheel motor dynamics with a time constant of 4 s has also been neglected. Without a feedback control, Eq. (9.14) simply becomes the open-loop roll transfer function from the initial conditions

$$\phi(s) = \frac{\phi(0)}{s} + \frac{\dot{\phi}(0)}{s^2 + \lambda^2} - \frac{H_o\dot{\psi}(0)/J_x}{s(s^2 + \lambda^2)} \tag{9.15}$$

which corresponds to Eq. (9.10) in time domain.

The practical significance of Eq. (9.14) is that it shows individual, as well as combined effects of the initial conditions on the closed-loop roll transient response. For example, one can notice that the roll responses to the initial roll and yaw rates have a 90-deg phase difference. Such property is independent of the control logic used. However, the yaw responses to the initial conditions do not have such a property.

The denominator of Eq. (9.14), which is in fact the closed-loop characteristic polynomial, can be factored into two pairs of roots with natural frequencies and damping ratios given approximately by

$$\omega_1 \approx \lambda, \qquad \zeta_1 = \frac{\omega_2^2 T_z}{2\omega_1} \qquad (9.16a)$$

$$\omega_2 \approx \sqrt{\frac{K}{H_o T}}, \qquad \zeta_2 = \frac{H_o - KT_z}{2\omega_2 H_o T} \qquad (9.16b)$$

Using the practical design rules given by Dougherty et al. [13] as

$$\zeta_1\omega_1 \approx \zeta_2\omega_2, \qquad \zeta_1 \approx 0.175, \qquad \zeta_2 \approx 0.707 \qquad (9.17)$$

we obtain the following approximate gain formula:

$$T = \frac{1}{0.7\lambda}, \qquad T_z = \frac{1}{0.175\lambda}, \qquad K = 0.0875 H_o\lambda \qquad (9.18)$$

or

$$K_P = -0.5H_o, \qquad K_I = 0.0875H_o\lambda, \qquad K_D = 0 \qquad (9.19)$$

A root locus plot vs the overall loop gain K is shown in Fig. 9.10. Note that the nutation mode has a small active damping ratio of 0.175 for the selected nominal gain. Only the lower frequency region is shown because the poles of the structural modes do not move significantly. Such an insignificant structural mode interaction can be seen in a Bode plot shown in Fig. 9.11, in which a passive structural damping ratio of 0.002 is assumed for all of the structural modes. All of the structural modes are gain stabilized with an 80-dB gain margin by the passive damping ratio of 0.002 and the steep rolloff of the control loop at a frequency well below the first structural mode frequency. As indicated in Fig. 9.11, the low-frequency control mode has a 5-dB gain margin and a 50-deg phase margin. Such a relatively small gain margin is due to the use of the nonminimum-phase control logic. However, the gain margin for the low-frequency control mode is independent of the spacecraft inertia, but dependent on the pitch bias momentum, which is measured quite accurately by the wheel tachometers; thus, the 5-dB gain margin of the control loop is acceptable in practice.

During on-orbit normal mode operations, the environment is very quiet except for the solar radiation pressure torques. Thus, the roll/yaw normal mode controller even with such a small active nutation damping ratio of 0.175 provides acceptable performance and stability margins. Figure 9.12 shows the results of digital simulation, where the initial conditions $\phi(0) = \psi(0) = -0.05$ deg are chosen

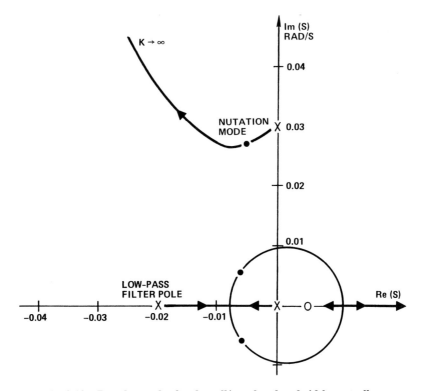

Fig. 9.10   Root locus plot for the roll/yaw low-bandwidth controller.

from a steady-state limit cycling at the end of a stationkeeping maneuver. The wheel torque demand is well below the saturation limit and the structural modes are almost unexcited. However, the nonminimum-phase nature of the controller has resulted in a large transient peak for the roll attitude error, which may not be acceptable for certain cases.

9.2.4.2 High-Bandwidth Controller Consider the design of a high-bandwidth transition controller for improving the transient performance during the transition period from a stationkeeping maneuver to the on-orbit normal mode operations.

The roll/yaw transient performance can be improved by employing a minimum-phase, PID-type nutation controller [14,15]. Figure 9.13 shows a roll/yaw transition controller, which utilizes the direct roll rate measurement from the gyros and the roll attitude error measurement from the Earth sensor.

For the control design purpose, the roll/yaw control logic, shown in Fig. 9.13, can be described as

$$u(s) = -\frac{K_D s^2 + K_P s + K_I}{s(Ts + 1)} \phi(s) \qquad (9.20)$$

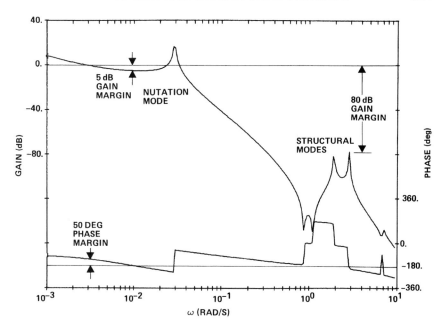

Fig. 9.11   Bode plot for the roll/yaw low-bandwidth controller.

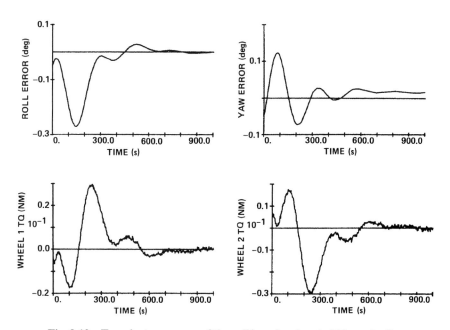

Fig. 9.12   Transient responses of the roll/yaw low-bandwidth controller.

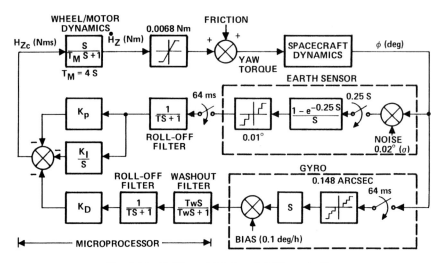

Fig. 9.13　Roll/yaw high-bandwidth controller.

Similar to Eq. (9.14), the closed-loop roll transfer function from the spacecraft's initial conditions neglecting the solar array flexibility can be obtained as

$$\phi(s) = \frac{[(s^2 + \lambda^2)\phi(0) + s\dot{\phi}(0) - \dot{\psi}(0)H_o/J_x](s + 1/T)}{s^4 + s^3/T + s^2(1 + K_D/H_oT)\lambda^2 + s(H_o + K_P)\lambda^2/H_oT + K_I\lambda^2/H_oT}$$

(9.21)

For the chosen closed-loop poles, the following approximate gain formula can be used for the selection of controller parameters:

$$T = (s_1 + s_2 + 2\zeta\omega)^{-1} \tag{9.22a}$$

$$K_P = \omega^2(s_1 + s_2)H_oT/\lambda^2 - H_o \tag{9.22b}$$

$$K_I = s_1s_2\omega^2 H_oT/\lambda^2 \tag{9.22c}$$

$$K_D = [\omega^2 + 2\zeta\omega(s_1 + s_2) - \lambda^2]H_oT/\lambda^2 \tag{9.22d}$$

where s_1 and s_2 are negative real roots and ω and ζ are closed-loop nutation frequency and damping ratio, respectively.

Figure 9.14 shows a root locus plot vs overall loop gain of the preceding PID controller. Figure 9.15 also shows a Bode plot, which indicates a 10-dB gain margin for the structural modes, and a 20-dB gain margin and a 50-deg phase margin for the nutation mode. Figure 9.16 shows digital simulation results with the same initial conditions as Fig. 9.12. It can be seen that the roll transient performance has been improved, but at the expense of initial torque saturation due to the high-bandwidth control. (The different positive and negative torque saturation limits shown in Fig. 9.16 are the result of the different effects of drag on the wheel with a biased speed: less drag effect for slowing down.)

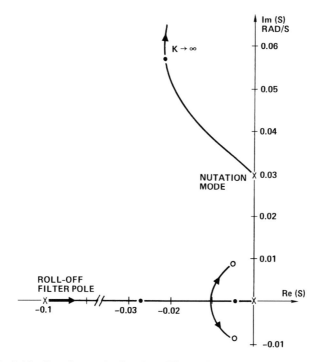

Fig. 9.14   Root locus plot for the roll/yaw high-bandwidth controller.

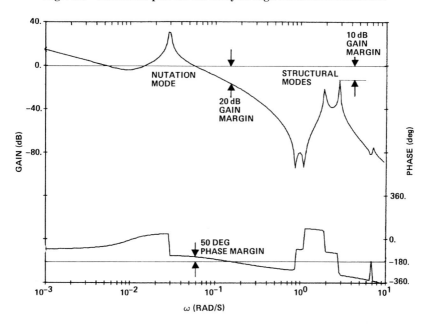

Fig. 9.15   Bode plot for the roll/yaw high-bandwidth controller.

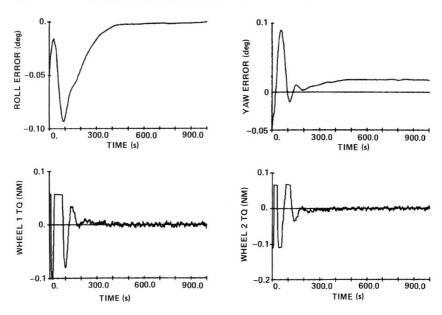

Fig. 9.16   Transient responses of the roll/yaw high-bandwidth controller.

A possible structural mode instability can be easily demonstrated without the use of the rolloff filter. The in-plane bending mode ($p_2 = 2.87$ rad/s) in the roll axis is destabilized in this case, which also validates the analytic prediction of the 10-dB gain margin for the structural modes. However, a passive damping ratio of 0.01 could make it stable even without the rolloff filter. Thus, it is emphasized that the so-called control and observation spillover problem should not be exaggerated without considering the degree of instability, which might be naturally avoided by reasonable passive damping or by a simple first-order roll-off filter.

In the next section, we shall consider a spacecraft attitude control problem in the presence of significant control–structure interactions.

9.3   Stationkeeping Attitude Control of a Flexible Spacecraft

In this section, as discussed in [16], we describe the problem of designing a reaction jet attitude control system for a spacecraft having flexible solar arrays during translational thrusting (stationkeeping) maneuvers.

9.3.1   Introduction

Many three-axis stabilized spacecraft have large solar panel arrays with significant structural flexibility. For most cases of practical concern, however, the structural flexibility of the solar arrays does not strongly interact with attitude control systems, and thus all of the structural modes are often simply gain stabilized by the steep rolloff at a frequency well below the first structural frequency. This is the case considered in the preceding section. However, in this section we examine a different case in which the structural modes strongly interact with an attitude control system.

In the presence of large disturbance torques during translational thrusting (stationkeeping) maneuvers, a high-bandwidth attitude controller is often needed to maintain accurate body pointing. During stationkeeping mode operations of three-axis stabilized spacecraft, reaction jets (thrusters) are used to stabilize the attitude motion of spacecraft. Unlike other actuators, such as reaction wheels, thruster output consists of two values: on or off. Furthermore, most on–off reaction jet control systems are, in practice, pulse modulated, as was discussed in Chapter 7. The solar array flexibility interacts strongly with such nonlinear control systems, and structural mode instability in the high-bandwidth nonlinear control system often manifests as a limit cycle.

In this section we consider the problem of designing a stationkeeping attitude control system for the same spacecraft model considered in the preceding section.

9.3.2 Stationkeeping Attitude Control System

To maintain the spacecraft position in geosynchronous orbit to within ± 0.1 deg, the spacecraft shown in Fig. 9.7, requires about six north/south stationkeeping maneuvers per year, each with a duration of maximum 2 min. East/west orbital corrections are made at approximately 12-day intervals with a duration of only a few seconds. The reaction jet attitude control system with 22-N thrusters must counteract disturbance torques due to thruster misalignment, plume impingement, and mismatch between pairs of thrusters used to impart velocity changes during these stationkeeping maneuvers.

It is assumed that the disturbance torques of ± 2.0 N $\cdot$ m can be prebiased leaving a residual uncertainty of ± 0.5 N $\cdot$ m to be controlled by the feedback control system. Attitude pointing requirements are to maintain short-term attitude errors within ± 0.06 deg in roll/pitch and ± 0.1 deg in yaw.

Figure 9.17 shows a block diagram representation of the stationkeeping attitude control system. The Earth sensor provides roll/pitch attitude references, whereas the gyros provide yaw attitude reference as well as three-axis body rates during the stationkeeping mode operations. The control logic for direct attitude and rate feedback, structural filtering, and pulse-width pulse-frequency (PWPF) modulation is implemented in the microprocessor with a 64-ms sampling period. There is no cross-axis feedback control as can be seen in Fig. 9.17. The sun sensor in the yaw channel provides yaw attitude reference in case of gyro failure. The rate loop has a high-pass filter to wash out any residual drift rate bias (up to 1 deg/h) of the gyros.

The signal after structural filtering is then passed through the loop deadband, the loop gain, and the PWPF modulator, which in turn activates the thruster valves. The PWPF modulator causes a pulse command on the average by adjusting the pulse width and pulse frequency. Because the roll thrusters are utilized in pairs for north/south stationkeeping maneuvers, they will be off-modulated to be used for roll attitude control while pitch and yaw thrusters are on-modulated. Similarly, when the pitch and yaw thrusters are used for east/west maneuvers, they provide the spacecraft velocity changes and are off-modulated for attitude control.

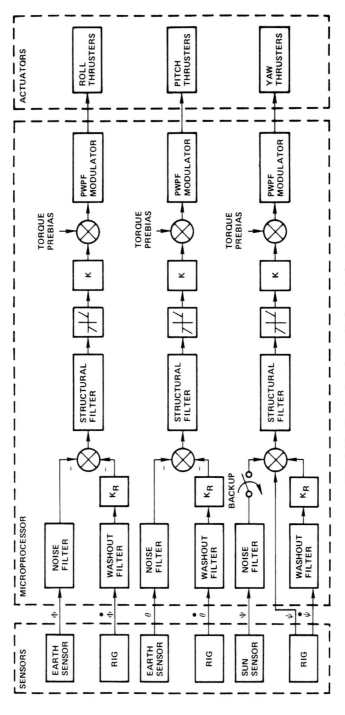

Fig. 9.17 Stationkeeping attitude control system.

9.3.3 Flexible Spacecraft Model

In the preceding section, the hybrid-coordinate modeling has been shown to be very useful for the control analysis and design of a spacecraft having a main rigid body and flexible solar arrays. Consider the same spacecraft model with the solar array configuration at 6 a.m. During the north/south or east/west stationkeeping maneuvers at 6 a.m., both arrays are pointing to the sun. Because the nutation frequency $\lambda = H_o/\sqrt{J_x J_z} = 0.02$ rad/s, is well below the controller bandwidth of about 1.0 rad/s, we can neglect both orbital and nutation dynamics in deriving the equations of motion for the stationkeeping attitude control design.

Including a single dominant cantilever mode in each axis, the linearized equations of motion of the spacecraft can be written in terms of the hybrid coordinates as follows.

Rigid main body:

$$J_x \ddot{\phi} + \sqrt{2}\delta_x \ddot{q}_x = u_x \tag{9.23a}$$

$$J_y \ddot{\theta} + \sqrt{2}\delta_y \ddot{q}_y = u_y \tag{9.23b}$$

$$J_z \ddot{\psi} + \sqrt{2}\delta_z \ddot{q}_z = u_z \tag{9.23c}$$

Two solar arrays:

$$\ddot{q}_x + \sigma_x^2 q_x + \sqrt{2}\delta_x \ddot{\phi} = 0 \tag{9.24a}$$

$$\ddot{q}_y + \sigma_y^2 q_y + \sqrt{2}\delta_y \ddot{\theta} = 0 \tag{9.24b}$$

$$\ddot{q}_z + \sigma_z^2 q_z + \sqrt{2}\delta_z \ddot{\psi} = 0 \tag{9.24c}$$

where (ϕ, θ, ψ) are small roll, pitch, yaw attitude errors of the spacecraft with respect to the LVLH reference frame; (J_x, J_y, J_z) are the spacecraft principal moments of inertia; (q_x, q_y, q_z) are the cantilever modal coordinates; $(\sigma_x, \sigma_y, \sigma_z)$ are the cantilever modal frequencies; $(\delta_x, \delta_y, \delta_z)$ are the rigid–elastic coupling scalars of a single solar array; and (u_x, u_y, u_z) are the control torques.

Because the roll, pitch, and yaw axes are decoupled, the following single-axis transfer function can be used for control design

$$G(s) = \frac{s^2 + \sigma^2}{Js^2[(1 - 2\delta^2/J)s^2 + \sigma^2]} \tag{9.25}$$

where σ and δ are the cantilever modal frequency and the rigid–elastic coupling scalar of the dominant mode of a single array given in Table 9.3. Table 9.4 summarizes the spacecraft nominal parameters for control design. The alternating pole–zero pattern of the structural modes clearly indicates that we have a collocated actuator and sensor problem here.

During the north/south stationkeeping maneuvers, the south array may need to be oriented to minimize the thermal effects of thruster firings on the solar cell. Because there are usually no thruster firings from the spacecraft north face, the north array is always pointing to the sun. If two arrays are perpendicular to each other during the north/south stationkeeping maneuvers at noon, the following transfer function with two structural modes can be used for control analysis and

Table 9.4   Spacecraft nominal parameters at 6 a.m.

Parameters	Roll	Pitch	Yaw
Spacecraft inertia, $kg \cdot m^2$	3026	440	3164
Control torque, $N \cdot m$	10	10	10
Residual disturbance, $N \cdot m$	0.5	0.5	0.5
Pointing requirement, deg	0.06	0.06	0.1
Structural mode zero, rad/s	1.11	5.53	0.88
Structural mode pole, rad/s	2.87	5.61	1.93

design:

$$G(s) = \frac{(s^2 + \sigma_1^2)(s^2 + \sigma_2^2)}{Js^2\{[1 - (\delta_1^2 + \delta_2^2)/J]s^4 + (\sigma_1^2 + \sigma_2^2 - \delta_1^2\sigma_2^2/J - \delta_2^2\sigma_1^2/J)s^2 + \sigma_1^2\sigma_2^2\}} \tag{9.26}$$

where (σ_1, δ_1) and (σ_2, δ_2) correspond to the dominant out-of-plane and in-plane bending modes of a single array given in Table 9.3.

9.3.4 Pulse-Width Pulse-Frequency Modulator Analysis and Design

A unique feature of the control system shown in Fig. 9.17 is the PWPF modulator, which activates the thruster valves by adjusting the pulse width and pulse frequency. Several different on–off pulse modulation techniques for reaction jet control systems have been briefly introduced in Chapter 7. The PWPF modulator shown in Fig. 9.18 is one of such pulse-modulation techniques, and it mainly comprises two components: a first-order lag filter and a Schmitt trigger inside a feedback loop.

The PWPF modulator produces a pulse command sequence to the thruster valves by adjusting the pulse width and pulse frequency. In the linear range, the average output produced equals the commanded input. With a constant input, the PWPF modulator drives the thruster valve with an on–off pulse sequence having a nearly linear duty cycle with the input amplitude. The *duty cycle* or *modulation factor* is defined as the average output of the modulator.

The static characteristics of the PWPF modulator for a constant input E can be summarized as follows.

Thruster on time:

$$T_{on} = -T_m \ln\left\{\frac{(1-h)E_d - (E-1)}{E_d - (E-1)}\right\} \tag{9.27a}$$

Thruster off time:

$$T_{off} = -T_m \ln\left\{\frac{E_d - E}{(1-h)E_d - E}\right\} \tag{9.27b}$$

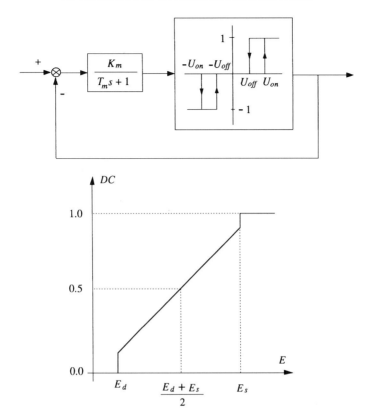

Fig. 9.18   PWPF modulator and its duty cycle vs constant input E.

Duty cycle:

$$DC = \frac{T_{\text{on}}}{T_{\text{on}} + T_{\text{off}}} \qquad (9.27c)$$

Minimum pulse width:

$$\Delta = -T_m \, \ell n \left(1 - \frac{h}{K_m}\right) \approx \frac{h\,T_m}{K_m} \qquad (9.27d)$$

where the equivalent internal deadband is

$$E_d = \frac{U_{\text{on}}}{K_m}$$

and the hysteresis width is

$$h = U_{\text{on}} - U_{\text{off}}$$

The duty cycle can be obtained as

$$DC = \left\{ \frac{\ell n \,[1 + a/(1 - x)]}{1 + \ell n[1 + a/x]} \right\}^{-1} \tag{9.28}$$

where the normalized hysteresis width is

$$a = \frac{h}{K_m(E_s - E_d)}$$

the normalized input is

$$x = \frac{E - E_d}{E_s - E_d}$$

and the saturation level is

$$E_s = 1 + \frac{U_{\text{off}}}{K_m}$$

A typical plot of the duty cycle, which is nearly linear over the range above deadband E_d and below saturation E_s is also shown in Fig. 9.18. A linearized duty cycle can be obtained by Taylor series approximation of Eq. (9.28) about $x = 0.5$ as follows:

$$DC \approx 0.5 + \frac{2a(x - 0.5)}{(1 + 2a)\ell n(1 + 2a)} \tag{9.29}$$

which is not valid near $x = 0$ and $x = 1$. From this linearized duty cycle expression, an effective deadband of about $E_d/2$ for $E_s = 1.0$ can be obtained.

9.3.5   Control Analysis and Design

The roll-axis control loop of the spacecraft under consideration is shown in Fig. 9.19. Other axes employ the same control architecture. As can be seen in Fig. 9.19, the PWPF modulator implemented in a microprocessor receives an input signal every 64 ms and causes pulse command updates every 16 ms with pulse-width quantization of 16 ms. The microprocessor sampling period of 64 ms is relatively fast compared to the control bandwidth of 0.5 rad/s and also to the dominant flexible mode frequency of about 3.0 rad/s. This makes it possible to employ the classical s-domain control design approach to digital control systems design, which was discussed in Chapter 2.

The combined effect of the microprocessor sampling period of 64 ms and pure computation delay of 48 ms can be approximated as a 80-ms pure time delay for control design here. The loop gain K determines the steady-state pointing accuracy for a constant disturbance torque during the stationkeeping maneuvers. The rate-to-position gain ratio K_R determines the closed-loop damping. The amplitude and rate of the rigid-body limit cycle are determined by the deadband and the PWPF modulator. The controller parameters obtained using the classical control design technique are also given in Fig. 9.19. Further details of this control system design and digital simulation results can be found in Wie and Plescia [16].

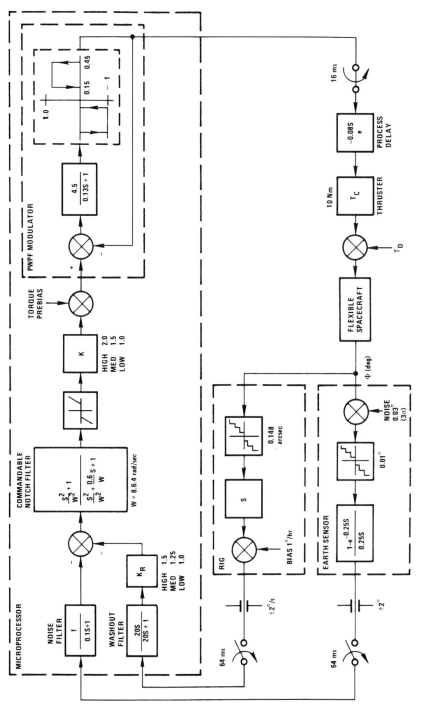

Fig. 9.19   Roll-axis stationkeeping control loop.

As mentioned earlier, this control loop has a collocated actuator and sensor pair, and thus all of the flexible modes are stably interacting with the rigid-body mode control. However, a structural filter that provides additional gain and phase margins for the structural mode is needed to compensate for the dynamic effect of the modulator. This subject will be treated further in the next section.

9.4 Nonlinear Pulse-Modulated Control Analysis and Design

In this section, which is based on [17], we examine the describing function method for the nonlinear control analysis and design of a flexible spacecraft equipped with pulse-modulated reaction jets. The method provides a means of characterizing the pulse modulator in terms of its gain and phase for the analysis of the structural mode limit cycle. Although the describing function method is inherently inexact, a new way of utilizing the method for practical control design problems is described. The approximations inherent in the method is accounted for as a modeling uncertainty for the nonlinear control robustness analysis.

9.4.1 Introduction

The nonlinear control instability in reaction jet control systems appears in the form of a structural mode limit cycle caused by thruster chatterings at the structural mode frequency. Although considerable attention has been directed toward the control-structure interaction problems of flexible structures in space, the nonlinear control interaction with flexible modes has not been studied much.

In this section, a frequency-domain approach is described for the design of a pulse-modulated nonlinear control system of a flexible spacecraft. The approach involves the use of the well-known describing function method, which is basically a frequency-response technique primarily used to predict the limit cycling of nonlinear systems. Many researchers have attempted, within the last several decades, to improve the accuracy of the describing function method by establishing new criteria to approximate the output of a nonlinear device; despite their efforts, the results obtained by researchers are not used much in practice. The reason is partly that the describing function method is inherently inexact and represents only a first-order approximation of the gain and phase characteristics of a nonlinear device.

This section, however, introduces a new way of utilizing the describing function method for designing nonlinear control systems. Analogs to the classical concept of gain/phase margins are applied to nonlinear systems to designate nonlinear stability margins; these margins give the designer a means of measuring the limit cycling tendency of a control system. The approximations inherent in the method are accounted for as the modeling uncertainty for designating the nonlinear stability margins. The describing function method is not employed here as a means of accurately predicting limit cycling information; instead, it is used as a method for characterizing a nonlinear component in terms of its gain and phase. In most cases, the control designer is more interested in determining whether or not the system will limit cycle as opposed to accurately identifying the oscillation characteristics, e.g., limit cycle frequency and amplitude.

Once the pulse modulator is adequately characterized by the describing function, one can iteratively synthesize a linear compensator and/or a pulse modulator such that the margins with respect to limit cycling condition are maximized. The methodology shall be applied to the design example of the INTELSAT V spacecraft controlled by a reaction jet attitude control system. The interaction of the spacecraft's flexible solar arrays with the pulse modulator was a major concern during the development of the spacecraft in the mid-1970s. Such interaction appears in the form of structural mode limit cycling. Control design example of the INTELSAT V spacecraft shall be studied to illustrate the concept and methodology in avoiding the structural mode limit cycling.

9.4.2 Describing Function Analysis

A proper selection of the modulator parameters, based on its static characteristics, does not guarantee the closed-loop stability of the control system. The pulse modulator, being a nonlinear device, cannot be adequately analyzed through application of linear analysis techniques. Still, the designer must be concerned about linear compensator design, linear/nonlinear stability margins, and structural mode limit cycle. These problems can be approached by using the describing function method to characterize the pulse modulator in terms of its gain and phase.

In the describing function analysis, it is assumed that the fundamental harmonic component of the output is significant when the input to a nonlinear element is sinusoidal. The describing function of a nonlinear element is then defined to be the complex ratio of the fundamental harmonic component of the output to the input; that is, we have

$$N(X, \omega) = (Y_1/X)e^{j\phi} \tag{9.30}$$

where

N = describing function
X = sinusoidal input amplitude
ω = frequency of input sinusoid
Y_1 = amplitude of the fundamental harmonic
Y_1/X = describing function gain
ϕ = describing function phase

Calculation of the describing function involves a conventional Fourier series analysis to obtain the fundamental component of the output, and Y_1 and ϕ may then be expressed as

$$Y_1 = \sqrt{A_1^2 + B_1^2} \tag{9.31}$$

$$\phi = \tan^{-1}(A_1/B_1) \tag{9.32}$$

where

$$A_1 = \frac{1}{\pi} \int_0^{2\pi} y(t) \cos \omega t \, d\omega t \tag{9.33}$$

$$B_1 = \frac{1}{\pi} \int_0^{2\pi} y(t) \sin \omega t \, d\omega t \tag{9.34}$$

and $y(t)$ is the output of the nonlinear device.

If the nonlinear element can be adequately characterized by the describing function $N(X, \omega)$, the loop transfer function of a negative feedback control system is simply given by $N(X, \omega) G(j\omega)$, where $G(j\omega)$ is the frequency-response function of the linear components of the loop. Thus, the study of limit cycle stability involves an equation of the form

$$1 + N(X, \omega)G(j\omega) = 0 \qquad (9.35)$$

or

$$G(j\omega) = -\frac{1}{N(X, \omega)} \qquad (9.36)$$

Equation (9.36) is often called the *harmonic balance equation*. The main advantage of the describing function analysis using Eq. (9.36) is its simplicity. For example, we simply plot the $-1/N$ locus and the $G(j\omega)$ locus. If two loci intersect, then the system exhibits a limit cycle, which is characterized by the amplitude X and frequency ω. Note that the amplitude and frequency of the limit cycle indicated by the intersection of the $-1/N$ locus and the $G(j\omega)$ locus are approximate values. Furthermore, if they intersect tangentially, then the system may not actually exhibit a limit cycle.

In this section, the term "DF (describing function) plot" designates a gain-phase plot of $-1/N(X, \omega)$. Figure 9.20 shows a DF plot for a PWPF modulator in the gain-phase plane. Because the PWPF modulator's describing function is dependent

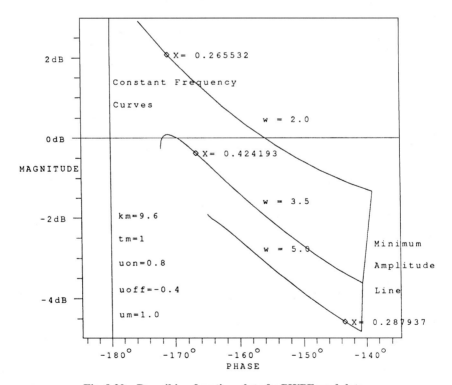

Fig. 9.20   Describing function plot of a PWPF modulator.

on ω, a limit cycle will only occur if the frequencies of the two curves are equivalent at the intersection point. The DF plot in Fig. 9.20 was generated by using the INCA computer program developed by Bauer and Downing [18]. The INCA program has been enhanced to allow the calculation of describing functions as well as inclusion of the DF plots in the frequency domain. To calculate describing functions for devices such as the PWPF modulator, for which the describing function cannot be expressed analytically, the INCA program solves Eqs. (9.33) and (9.34) for steady-state output. All of the describing function plots in this section were constructed using the enhanced INCA software.

The modulator DF plot is made up of two components, the minimum-amplitude line, that is, the vertical line on the right side of the curve set, and the roughly horizontal lines that comprise the DF curves for different values of frequency. The constant-frequency lines are a function of amplitude X. In the INCA program, these curves are only calculated as long as the modulator output contains just two pulses per cycle with amplitudes of 1 and -1; increasing the input amplitude causes the modulator to produce multiple pulses per cycle. It is difficult to accurately approximate such an irregular signal with a sinusoid. The sinusoidal approximation of the modulator output depends on the criteria used to equate the two functions. The approximation of such a discontinuous function with a single harmonic curve does not provide meaningful expression of the modulator gain and phase; as a result, all discussions of PWPF modulator describing functions refer to single-pulse output.

The challenge presented by the describing function method is to obtain the most realistic sinusoidal model of the pulse output to determine gain and phase change from the input signal. It is difficult to determine a criterion that best represents the pulse output in sinusoidal form. A popular representation is the fundamental harmonic of a Fourier series; however, the fundamental Fourier harmonic is only the first term of an infinite-order series. Despite the increased accuracy of the approximation, the extra computational effort necessitates exclusion of the extra harmonics from the describing function analysis. A common misperception is that the fundamental harmonic is, therefore, the best first-order sinusoidal approximation of that function. It is difficult to consider the fundamental Fourier harmonic the best approximation without acknowledging the contribution of the higher harmonics.

The gain and phase characteristics of a describing function depend on the method defined to approximate the periodic nonlinear output with a sinusoid of the same frequency. Several other types of describing function methods have been proposed in the literature. A method that is treated in this section involves approximating the modulator pulse signal with a sine curve of equal area. The first integral of torque with respect to time is total angular momentum; therefore, two torque profiles of equal area will impart an identical impulse to a linear plant. In an ideally limit cycling system, substitution of the nonlinear output with the sinusoidal approximation should exhibit no change in the system output.

Calculation of the area-matching describing function for the PWPF modulator is computationally simple. The area of a modulator pulse, integrated from 0 to π/ω, is simply the modulator output pulse width Δ. The area of a sine curve over the same period is $2Y_s/\omega$ where Y_s is the area-matching sinusoid amplitude and ω is the frequency of both the sinusoid and the pulse signal. Equating these two relations yields

$$Y_s = \omega\Delta/2 \qquad (9.37)$$

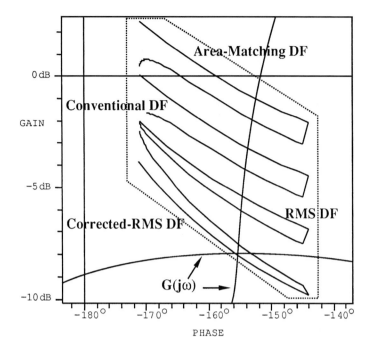

Fig. 9.21   Region of uncertainty for the PWPF modulator.

The amplitude of the Fourier fundamental harmonic Y_1 can be found by solving Eqs. (9.33) and (9.34) and substituting into Eq. (9.31)

$$Y_1 = (4/\pi) \sin(\omega\Delta/2) = (4/\pi) \sin Y_s \qquad (9.38)$$

Equation (9.38) states that for small values of Y_s, Y_1 differs from Y_s by a factor of 2.1 dB as shown in Fig. 9.21. This value can be interpreted as the uncertainty of the actual modulator gain. The modulator gain obviously depends on the criteria used to treat the output signal. As a result, the designer can utilize a region of gain uncertainty, instead of a single DF curve, to evaluate limit cycling tendency.

Another approximation method also involves sinusoidal input to the nonlinear device. Similar to the area-matching technique, the root-mean-square (rms) describing function method expresses the gain of a pulse-modulating device as a function of the pulse area. It has been used in evaluation of systems that contain odd-valued nonlinear devices.

In this procedure, the curve-approximating criterion specifies the sinusoidal output to have the same rms value as the actual modulator output. This is similar to Eqs. (9.31–9.34) in that the describing function gain $|N(X,\omega)_{rms}|$ is defined as

$$|N(X,\omega)_{rms}| = \sqrt{\frac{\int_0^{2\pi} [y(t)]^2 \, d\omega t}{\int_0^{2\pi} [X \sin(\omega t)]^2 \, d\omega t}} \qquad (9.39)$$

For a pulsed output, the amplitude of the DF sinusoid is

$$Y_{\text{rms}} = \sqrt{2\omega\Delta/\pi} \tag{9.40}$$

Note that the rms output sinusoidal amplitude is greater than that of the area matching and the fundamental harmonic DF evaluations; the reason is that $Y_{\text{rms}} > Y_1$ for $\omega\Delta < 1$. In fact, for small values of $\omega\Delta$, such that $\sin(\omega\Delta) \approx \omega\Delta$, $Y_1 = Y_{\text{rms}}^2$. In the DF plots, however, we are plotting $-1/N(X,\omega)$; as a result, the rms describing function gain is less than the two earlier types of DFs.

For nonpulse-modulated systems, it can be difficult to evaluate the numerator of the right-hand side of Eq. (9.39). If $y(t)$ is assumed to be the superposition of an infinite number of harmonics, the evaluation of the equation is formidable. For example, Eq. (9.39) would be evaluated as

$$|N(X,\omega)_{\text{rms}}| = \frac{\sqrt{Y_1^2 + Y_3^2 + Y_5^2 + \cdots + Y_{2n-1}^2 + \cdots}}{X} \tag{9.41}$$

or

$$Y_{\text{rms}} = \sqrt{Y_1^2 + Y_3^2 + Y_5^2 + \cdots + Y_{2n-1}^2 + \cdots} \tag{9.42}$$

where Y_{2n-1} represents amplitude of the odd Fourier harmonics. As a result, some authors have employed a corrected rms describing function method, which is based on the rms DF theory. In this technique, only the first two odd harmonics are used in evaluation of describing function. As a result, the corrected rms DF is expressed as

$$|N(X,\omega)_{c-\text{rms}}| = \sqrt{Y_1^2 + Y_3^2}/X \tag{9.43}$$

where

$$Y_1 = (4/\pi)\sin(\omega\Delta/2)$$
$$Y_3 = (4/3\pi)\sin(3\omega\Delta/2)$$

From inspection of Eq. (9.43), it is obvious that $Y_{c-\text{rms}} < Y_{\text{rms}}$. As a result, the corrected rms method yields the greater describing function gain when viewed in a gain-phase plot.

As mentioned earlier, evaluation of the pulse-modulator gain is not accomplished accurately using the describing function method. Table 9.5 summarizes the different gain approximations of a pulse-modulator limit cycle output. There are valid physical justifications for use of each of the different modeling methods. The fundamental flaw in each procedure, however, is the necessity of approximating a discontinuous pulse signal with a single-harmonic sine wave. This is the feature of describing function theory that inhibits its usefulness in accurately predicting limit cycle behavior.

Figure 9.21 represents an evaluation of the preceding modeling methods for a PWPF modulator. As displayed in the figure, the disparity in location of the curves is evidence of the disagreement of the methods outlined in Table 9.5. What is important to realize is that the methods are in relative agreement; they all lie within a few decibels of one another. By assuming the validity of the phase information of the describing functions, the different values of gain describe a region

Table 9.5   Describing function amplitude comparison

Describing function method	Sinusoidal amplitude
Conventional	$Y_1 = \dfrac{4}{\pi} \sin\left(\dfrac{\omega\Delta}{2}\right)$
Area matching	$Y_s = \dfrac{\omega\Delta}{2}$
Root mean square (rms)	$Y_{\text{rms}} = \sqrt{\dfrac{2\omega\Delta}{\pi}}$
Corrected rms	$Y_{c-\text{rms}} = \dfrac{4}{\pi}\sqrt{\sin^2\left(\dfrac{\omega\Delta}{2}\right) + \dfrac{1}{9}\sin^2\left(\dfrac{3\omega\Delta}{2}\right)}$

on the Nichols plot, which is shaded in Fig. 9.21. By acknowledging that the approximation methods in Table 9.5 have some physical justification, the shaded area can be referred to as a region of describing function uncertainty. This region can be defined on the Nichols plot, which loosely designates the modulator's DF plot, $-1/N(X, \omega)$. Instead of using this information to predict limit cycle behavior, the designer can use this plot to measure a control system's tendency to limit cycle. In practical cases, avoidance of the limit cycle is more important than predicting its oscillation characteristics.

A system whose $G(j\omega)$ locus passes through this region has a greater chance of exhibiting a sustained oscillation. It is apparent that the actual boundaries of the region depend on the parameters of the pulse modulator. The DF uncertainty region is only an approximation in itself; uncertainties in modeling the pulse modulator prevent any confidence in the gain boundaries. Therefore, a system whose frequency response is horizontal in the vicinity of the DF region has a smaller chance of actually passing through the region; these systems tend not to experience limit cycles. Conversely, frequency-response curves that are nearly vertical in the vicinity of the gain uncertainty region tend to experience limit cycle oscillation. This case is shown in Fig. 9.21.

As a result, the describing function uncertainty region can be used as a useful tool in robust compensation design for pulse-modulated control systems. A designer can synthesize linear compensation such that the $G(j\omega)$ locus avoids the uncertainty region of $-1/N$ locus as much as possible.

9.4.3 Nonlinear Control Design Methodology

The design methodology described here employs the describing function method to characterize a nonlinear element in terms of its gain and phase. The conventional describing function analysis emphasizes an accurate prediction of the limit cycle amplitude and frequency. The control system designer, however, is more concerned with the avoidance of the limit cycle rather than the accurate analysis of the limit cycle. A methodology or guideline is described here to aid the control designer in avoiding repetitive computer simulation while designing a pulse-modulated control system.

The following six steps can be used as a guideline for on–off control system design for flexible spacecraft:

1) The first step is rigid-body mode stabilization. The type of compensation for rigid-body mode control depends on the control bandwidth or settling time requirement. In general, a phase-lead filter is used when direct rate feedback is not possible.

2) The second step is pulse-modulator synthesis. Usually, the type of pulse modulator is fixed as a result of previous spacecraft design. Certain factors that must be considered in the modulator design include ease of hardware/software implementation, stabilization requirements of flexible modes, and the modulator static characteristics, i.e., minimum pulse width, duty cycle, etc.

3) The third step is linear analysis and structural mode stabilization. By neglecting the nonlinearity of the modulator, linear analysis methods such as root locus and frequency response will enable the control designer to investigate the stability of the flexible modes. Higher-order filters may be necessary, in addition to the rigid-body mode controller, to compensate for unstable flexible modes.

4) The fourth step is describing function analysis. Gain and phase characteristics of the nonlinear controller, as approximated in a linear sense, can be obtained through the describing function method. Investigate the $-1/N$ locus and the $G(j\omega)$ locus in the frequency domain.

5) The fifth step is nonlinear simulation. The results in step 4 can be verified by a numerical simulation. The existence a structural mode limit cycle can be verified in this step, as can the approximate nonlinear stability margins.

6) The sixth step is linear analysis iteration. If the designer is not satisfied with nonlinear stability margins or the possibility of limit cycling, the design process can be repeated from step 3 by altering or adding to the linear compensation. If needed, the pulse modulator itself can be modified to further increase the nonlinear stability margins.

9.4.4   Example: INTELSAT V Spacecraft

A generic example of a nonlinear control system of flexible spacecraft is examined here. The control system consists of three elements: the flexible spacecraft, the attitude sensor, and the controller, as illustrated in Fig. 9.22. The sensor is modeled as a sample and ZOH device, which introduces phase lag into the control loop. The controller comprises linear compensation and a pulse-modulation logic. The linear compensation consists of a phase-lead filter for rigid-body mode stabilization and a structural filter for flexible mode compensation.

Except for the structural filter and digital control implementation, the control system parameters are derived from the yaw-axis control system of the INTELSAT V spacecraft, which has a collocated actuator and sensor; however, most spacecraft with digital control systems can be represented in this format. A nonlinear instability that can arise from the interaction of the pulse modulation with the flexible structural modes was of primary concern for the INTELSAT V control system design; however, it will be shown that a robust nonlinear control system can be synthesized through use of classical linear compensation techniques.

Four different compensation schemes, consisting of different combinations of the phase-lead filters and structural filters are given in Table 9.6. In case 1, the

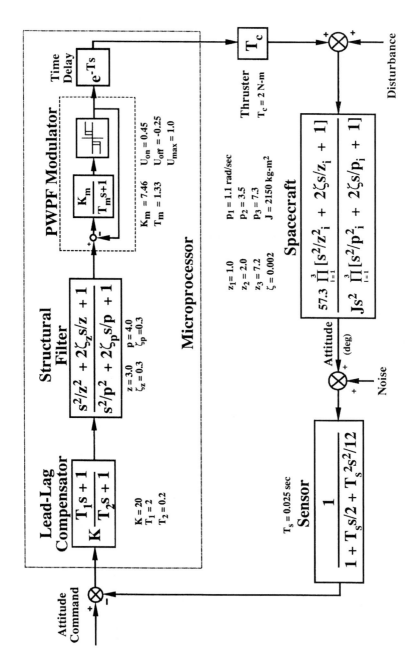

Fig. 9.22 Reaction jet control system for a flexible spacecraft ($K = 2, T_1 = 2, T_2 = 0.2$).

Table 9.6   Controller parameters

	Rigid-body controller	Structural filter	
Case 1	$T_1 = 2$		
	$T_2 = 0.2$		
Case 2	$T_1 = 2$	$z = 3.0$	$\zeta_z = 0.3$
	$T_2 = .2$	$p = 4.0$	$\zeta_p = 0.3$
Case 3	$T_1 = 2$	$z = 3.4$	$\zeta_z = 0.002$
	$T_2 = 0.2$	$p = 3.4$	$\zeta_p = 1$
Case 4	$T_1 = 2$	$z = 3.6$	$\zeta_z = 0.002$
	$T_2 = 0.2$	$p = 3.6$	$\zeta_p = 1$

linear compensation consists of a phase-lead filter without any structural filtering. The filter has the effect of phase stabilizing the rigid-body mode and the structural modes. Figure 9.23 contains several gain-phase (Nichols) plots; the leftmost plot shows the $G(j\omega)$ locus and the $-1/N(X, \omega)$ locus. The spacecraft's three flexible modes are visible with portions of the $G(j\omega)$ locus in the vicinity of the $-1/N$ locus. The frequencies of the upper and lower curves on the $-1/N$ locus are 3 and 4 rad/s, respectively, which correspond to the frequency range of the dominant mode. The phase separation between the second mode and the minimum amplitude line of the $-1/N$ locus is less than 15 deg.

A time delay is used to evaluate the phase margin of the control loop. By combining the time-delay model with the compensation, the $G(j\omega)$ locus can be

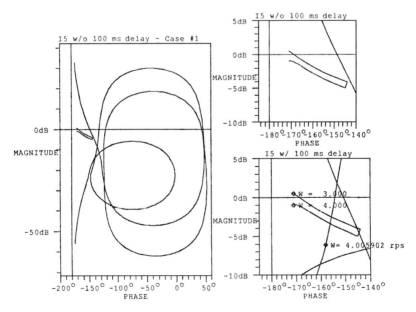

Fig. 9.23   Case 1 describing function analysis.

shifted to the left, creating an intersection with the $-1/N$ locus of the PWPF modulator. A second-order Padé approximation of a pure delay is used, where the delay time T is found by the relation

$$T = \phi_m/57.3\omega \qquad\qquad (9.44)$$

where ω (rad/s) is the frequency of the point of the $G(j\omega)$ locus close to the $-1/N$ locus and ϕ_m is the phase margin in degrees. The smallest value of ϕ_m that causes an intersection is defined as the *nonlinear phase margin*. A similar procedure can be followed for the nonlinear gain margin. The lower right-hand plot depicts the introduction of a 100-ms delay at 3.5 rad/s; the corresponding phase lag, according to Eq. (9.44), is 20.1 deg. Note that the second mode now crosses the $-1/N$ locus. The frequency of the $G(j\omega)$ locus near the describing function plot is close to 4 rad/s making a limit cycle at that frequency possible. Note the nearly orthogonal crossing of the $G(j\omega)$ locus with the $-1/N$ locus, which, as mentioned earlier, increases the accuracy of the describing function analysis.

Figure 9.24 shows the results of a nonlinear simulation of the case 1 compensation with the additional 100-ms delay. As predicted, a limit cycle is evident in all four states displayed in Fig. 9.24. The rigid-body mode is sufficiently damped but the second flexible mode is limit cycling.

The nonlinear phase margin of the system with case 1 compensation is less than 15 deg, which may not prove sufficient; increasing the nonlinear phase margin will further decrease the chance of structural limit cycling. This can be accomplished

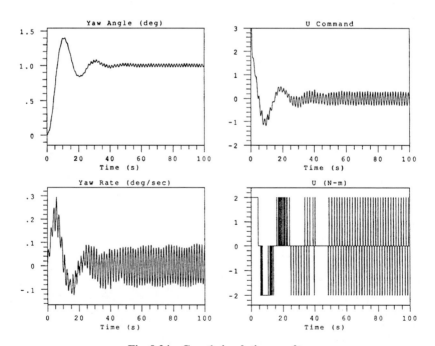

Fig. 9.24   Case 1 simulation results.

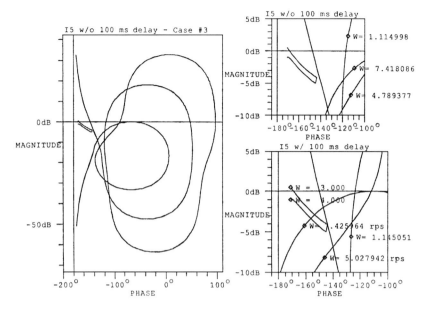

Fig. 9.25   Case 2 describing function analysis.

by introducing phase lead into the system. As a result, the case 2 compensation contains a phase-lead structural filter in addition to the lead filter for the rigid-body mode stabilization. A filter that would introduce significant phase lead between the frequencies of 3 and 4 rad/s is desired; the generalized structural filtering approach, described in Chapter 2, can be employed to select the minimum-phase lead filter. Figure 9.25 details the $G(j\omega)$ and $-1/N$ locus for case 2. It can be seen that the nonlinear phase margin has roughly doubled from case 1 to 30 deg. The addition of the 100-ms delay does not result in an intersection between the curves; there is still a 10-deg phase difference between the $-1/N$ locus and the $G(j\omega)$ locus. Therefore, no limit cycle is predicted from describing function analysis.

From Fig. 9.26, it can be seen that there is indeed no limit cycle; the yaw angle output contains gradually decreasing flexible mode excitation. No high-frequency thruster firings are evident; the low-frequency steady-state firings that are visible are rigid-body mode limit cycle firings. The case 2 compensation represents a method of phase stabilization of the nonlinear system; an alternative method of stabilization is gain stabilization.

Along with the rigid-body controller from case 1, case 3 contains a notch filter. This notch filter provides a sharp gain attenuation at 3.4 rad/s. As in the earlier cases, it can be shown that, for case 3, there is a wide separation between the $-1/N$ locus and the 3.5 rad/s point on the $G(j\omega)$ locus; as a result, the system has a large nonlinear phase margin. The utility of the case 3 compensation is, however, misleading because of the wide phase margin apparent in the system. The disparaging phase characteristics of the notch filter undermine its suitability as a gain-stabilizing device. The case 4 compensation resembles the filters used in case 3 except that the notch filter pole and zero lie on the opposite side of the

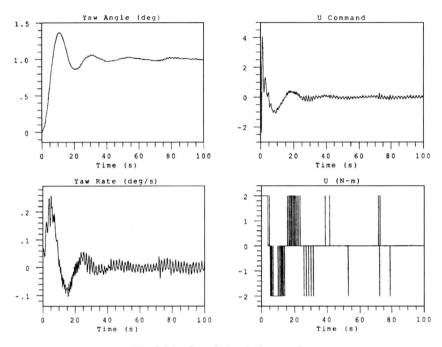

Fig. 9.26   Case 2 simulation results.

3.5 rad/s mode. This demonstrates the effect of structural mode uncertainty on a control system. The notch filter is not an example of robust compensation because of the linear instability that could arise from structural pole or zero uncertainty; case 4 could represent an analogous situation in the nonlinear sense. As a result of the notch frequency mismatching, the $-1/N$ and $G(j\omega)$ loci intersect.

9.4.5 Summary

A computer-aided design methodology based on the describing function method for a pulse-modulated control system has been described in this section. A brief discussion of the describing function method and the inherent modeling uncertainty of the method was also given. It was shown that if $-1/N$ locus and $G(j\omega)$ locus intersect almost perpendicularly, the modeling uncertainty for the pulse modulators does not significantly affect the accuracy of the describing function analysis. The INTELSAT V spacecraft was used as an example.

9.5   Attitude Control Redesign for the Hubble Space Telescope

In this section, as discussed in [19], we consider a control redesign problem of the Hubble Space Telescope (HST) for reducing the effects of solar array vibrations on telescope pointing jitter. The HST is a real example of a flexible spacecraft with noncollocated actuator and sensor pairs.

9.5.1 Introduction

The Hubble Space Telescope (HST) shown in Fig. 9.27, is a 13-ton free-flying telescope in space with a precision pointing stability requirement of 0.007 arcsec over a 24-h period, which is the most stringent pointing requirement imposed on any spacecraft to date. Following the successful deployment of the HST from the Space Shuttle Orbiter in April 1990, the HST experienced a pointing jitter problem caused by unexpected, solar-array-induced disturbances. (This pointing jitter problem was later solved by replacing the problematic arrays with new, improved solar arrays in an HST repair mission.) Flight results associated with the pointing jitter problem indicate that there appear to be two types of thermal flutter of the 20-ft-long solar arrays: 1) an end-to-end bending oscillation at 0.12 Hz when the spacecraft passes between sunlight and shadow and 2) a sideways bending oscillation at 0.66 Hz that occurs on the day side of the Earth. Under the worst conditions, the tips of the two 20-ft arrays could deflect as much as 3 ft. The effect of such solar-array oscillations is that the telescope moves 0.00022 in., enough to make the planned long observations of as much as 25 min not worthwhile.

In this section we consider the control redesign problem for reducing the effects of both 0.12- and 0.66-Hz solar array oscillations on jitter. Both classical and

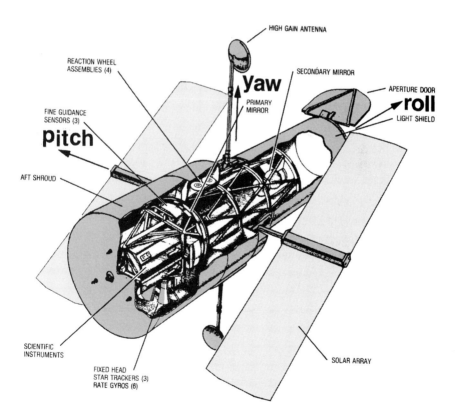

Fig. 9.27   Hubble Space Telescope.

modern control design methodologies presented in Chapter 2 will be employed for precision line-of-sight pointing control of the HST in the presence of significant structural vibrations. It will be shown that two disturbance rejection dipoles effectively reduce the effects of the solar-array-induced disturbances at 0.12 and 0.66 Hz on pointing jitter.

9.5.2 Problem Statement for Control Redesign

The pointing control system of the HST consists of fine-guidance sensors, star trackers, rate gyros, reaction wheels, magnetic torquers, and a digital computer [20,21]. The rate gyro assembly comprises six rate-integrating gyros and provides rate and attitude information that is supplemented by attitude data from star trackers and fine-guidance sensors. Magnetic torquers are used for momentum management. Control torques are provided by the four skewed reaction wheels, which are not collocated with the rate gyro assembly. The rate gyros are located with the star tracker on an equipment shelf on the back side of the optical telescope assembly, and the reaction wheels are located at the midsection of the main body, as can be seen in Fig. 9.27. As a result, the primary bending modes of the optical telescope assembly have large negative modal gains (see Table 9.7) and they are interacting unstably with the rigid-body pointing control system. The pointing control problem of the HST is thus an excellent practical example of the noncollocated control problem of flexible structures.

We consider here only the pitch-axis control design problem, which has most significant interaction with the solar arrays. The roll, pitch, and yaw control axes of the HST, illustrated in Fig. 9.27, are nearly aligned with the principal axes of the vehicle. The pitch-axis transfer function of the HST is given by

$$\frac{\theta}{u} = \frac{1}{Js^2} + \sum_{i=1} \frac{K_i/J}{s^2 + 2\zeta\omega_i s + \omega_i^2} \tag{9.45}$$

where θ is the pitch-axis pointing error output, u the pitch axis reaction wheel control torque input, s the Laplace transform variable, $J = 77{,}076 \, \text{kg} \cdot \text{m}^2$ the spacecraft pitch-axis moment of inertia, K_i the ith flexible modal gain in the pitch

Table 9.7   HST structural modes for the pitch axis

Mode number	ω_i, Hz	Modal gain K_i	Description
1	0.110	0.018	Solar array
2	0.432	0.012	High-gain antenna
3	0.912	0.057	Aperture door
4	10.834	0.024	Telescope structure
5	12.133	0.155	Telescope structure
6	13.201	−1.341	Telescope structure
7	14.068	−1.387	Telescope structure
8	14.285	−0.806	Telescope structure
9	15.264	−0.134	Telescope structure

axis, ω_i the ith flexible mode frequency in rad/s, and ζ the passive damping ratio assumed to be 0.005.

Table 9.7 lists the pitch-axis modal data of the HST, and the corresponding transfer function zeros in rad/s are

$$
\begin{array}{ll}
-0.0034 \pm 0.6850j, & -0.0134 \pm 2.6983j \\
-0.0272 \pm 5.5805j, & -0.3397 \pm 67.945j \\
-0.3790 \pm 75.818j, & -0.4255 \pm 85.079j \\
-0.4464 \pm 89.286j, & -0.4780 \pm 95.589j \\
-58.678, & +59.069
\end{array}
$$

The transfer function with nonalternating poles and zeros has a nonminimum-phase zero at $s = 59.069$ because the rate gyros are not collocated with the reaction wheels.

Figure 9.28 shows a simplified pitch-axis block diagram of the HST pointing control system. The outer loop with the fine-guidance sensor and a command

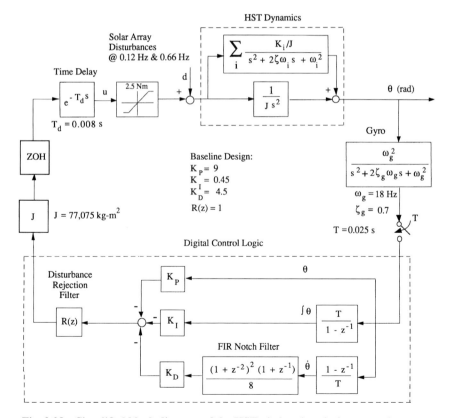

Fig. 9.28   Simplified block diagram of the HST pitch-axis pointing control system.

generator for the feedforward path are not shown here. Other axes of the HST employ the same control architecture.

The original controller on board the HST, which is basically a digital PID controller without the disturbance rejection filter, utilizes a finite impulse response (FIR) filter in the rate path to attenuate the high-frequency, main-body (optical telescope assembly) bending modes.

The solar-array-induced disturbances are modeled as

$$d(t) = A_1 \sin(p_1 t + \phi_1) + A_2 \sin(p_2 t + \phi_2) \tag{9.46}$$

where the frequencies are known as

$$\begin{aligned} p_1 &= 2\pi(0.12) \text{ rad/s} \\ p_2 &= 2\pi(0.66) \text{ rad/s} \end{aligned} \tag{9.47}$$

The magnitudes A_i and phases ϕ_i are unknown for control design, whereas the nominal magnitudes can be estimated as $A_1 = A_2 = 0.2 \text{ N} \cdot \text{m}$, from flight results.

The Bode magnitude plot of the loop transfer function of the original controller on board the HST is shown in Fig. 9.29. As can be seen in Fig. 9.29, the pitch-axis pointing control system with the original controller has a 1.03-Hz gain crossover frequency. The FIR filter in the rate path provides 2.3-dB gain suppression of the 13.2-Hz bending mode. The solar-array-induced disturbances for this controller lead to pointing jitter of 0.1-arcsec peak, which significantly exceeds the 0.007-arcsec pointing accuracy requirement.

Thus, a new digital control logic is to be designed to most effectively attenuate the effects of the solar array oscillations at 0.12 and 0.66 Hz. An integral compensation is also needed to attenuate a low-frequency gravity gradient and aerodynamic disturbances.

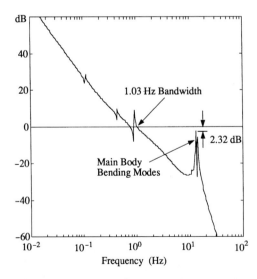

Fig. 9.29 Bode magnitude plot of the loop transfer function of the original HST control design.

The control redesign requirements and/or goals can be stated as follows:

1) Maintain at least 5-dB gain margin and 20-deg phase margin.

2) Provide at least 6-dB gain suppression (rolloff) of the high-frequency telescope structural modes at 14 Hz.

3) Provide at least 20-dB additional disturbance attenuation at both 0.12 and 0.66 Hz with respect to the original design.

4) Maintain the bandwidth (the open-loop gain crossover frequency) close to 1.5 Hz.

Next we consider a classical control redesign for the HST, followed by a modern state-space control redesign, to reduce the effects of solar array vibrations on telescope pointing jitter. Both designs will be compared to the original controller onboard the HST, and some practical usefulness of a modern design technique that deals with both structured and unstructured uncertainties will also be demonstrated. Theoretical aspects of robust control design will not be elaborated, although some detailed discussion on both the classical and robust H_∞ design techniques can be found in Chapter 2 or elsewhere.

9.5.3 Classical Control Design

For the classical s-domain synthesis of digital control logic, the computational delay and the sample/ZOH delay, shown in Fig. 9.28, are first approximated as

$$e^{-0.008s} \approx \frac{(0.008s)^2/12 - 0.008s/2 + 1}{(0.008s)^2/12 + 0.008s/2 + 1} \tag{9.48}$$

$$\frac{1 - e^{-0.025s}}{0.025s} \approx \frac{1}{(0.025s)^2/12 + 0.025s/2 + 1} \tag{9.49}$$

Following a classical design guideline described in Chapter 2, a new controller can be designed to accommodate the disturbances at 0.12 and 0.66 Hz. After a trial-and-error iteration, a PID controller for the control logic architecture shown in Fig. 9.28 can be found as $K_P = 7.720, K_I = 3.798$, and $K_D = 7.318$. This controller utilizes the same FIR filter as in the original controller and employs a periodic-disturbance rejection filter of the form

$$R(s) = \frac{(s/z_1)^2 + 2\zeta_{z_1}s/z_1 + 1}{(s/p_1)^2 + 1} \cdot \frac{(s/z_2)^2 + 2\zeta_{z_2}s/z_2 + 1}{(s/p_2)^2 + 1} \tag{9.50}$$

where

$$z_1 = 2\pi(0.124), \qquad \zeta_{z_1} = 0.364, \qquad p_1 = 2\pi(0.120)$$

$$z_2 = 2\pi(0.612), \qquad \zeta_{z_2} = 0.127, \qquad p_2 = 2\pi(0.660)$$

This new PID controller with two dipoles satisfies the 1.5-Hz gain crossover frequency requirement, as well as the gain and phase margin requirements. However, the FIR filter in the rate path does not provide enough gain attenuation for the high-frequency bending modes. The high-frequency structural modes near 14 Hz are, in fact, phase stabilized by the phase lag effects of the control system. Consequently, the second requirement of gain stabilizing the primary structural bending modes at 14 Hz has not been met, whereas all other design requirements and goals have been satisfied.

A new PID controller with notch filtering of the significantly interacting modes at 14 Hz can also be synthesized as

$$K(s) = (7.72 + 3.798/s + 7.318s)\, R(s)\, H(s) \tag{9.51}$$

where $R(s)$ is the same periodic-disturbance rejection filter as given in Eq. (9.50) and $H(s)$, which denotes the structural notch filter, is given by

$$H(s) = \frac{(s/87)^2 + 2(0.001)s/87 + 1}{(s/45)^2 + 2(0.70)s/45 + 1} \tag{9.52}$$

As can be seen in the Bode magnitude plot of the loop transfer function shown in Fig. 9.30, the controller given by Eq. (9.51) has a 1.5-Hz gain crossover frequency. The closed-loop system with this controller has a phase margin of 38 deg and a gain margin of 7.1 dB. The notch filter gain stabilizes the primary bending modes at 14 Hz with about 12-dB gain suppression. This new controller with two dipoles provides 40 dB more gain attenuation at both 0.12 and 0.66 Hz, which can be verified from the closed-loop frequency magnitude response from the solar array disturbance input d to the pitch attitude output θ.

As described here for the HST control redesign problem, a simple classical approach based on a dipole concept is a very effective means of achieving precision line-of-sight pointing control in the presence of significant structural vibrations or persistent external disturbances. The classical approach, however, requires a significant amount of trial-and-error iterations to satisfy the 1.5-Hz bandwidth

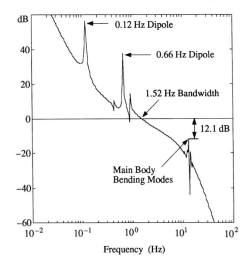

Fig. 9.30   Bode magnitude plot of the loop transfer function of a classical control redesign.

requirement and the stability requirements of the two dipoles and the rigid-body mode. Next we will employ a modern state-space design technique based on H_∞ control theory to overcome some drawbacks of a classical approach. It is, however, emphasized that even for such a modern systematic approach, trial-and-error iterations for selecting proper weightings are necessary.

9.5.4   Robust Control Design

A robust H_∞ control design methodology, which is essentially based on state-space solutions of standard H_∞ control problems, was briefly described in Chapter 2, with a special emphasis on modeling of both structured parametric uncertainty and unmodeled dynamics. Many theoretical aspects of robust control are not elaborated here. Instead, we emphasize a proper formulation of the HST pointing control redesign problem for the robust H_∞ control design in terms of structured and unstructured uncertainties.

The pole–zero pattern of the HST pitch-axis transfer function indicates that the pitch-axis dynamics can be approximated by a rigid-body mode and several dominant bending modes at 14 Hz. Because the control redesign requires gain stabilization of those bending modes with at least 6-dB gain suppression, the high-frequency bending modes are considered as unmodeled dynamics.

The pitch-axis dynamics of HST with only a rigid-body mode can then be described as

$$\frac{d}{dt}\begin{bmatrix}\theta\\\dot{\theta}\end{bmatrix}=\begin{bmatrix}0&1\\0&0\end{bmatrix}\begin{bmatrix}\theta\\\dot{\theta}\end{bmatrix}+\begin{bmatrix}0\\1+\delta\end{bmatrix}u+\begin{bmatrix}0\\1\end{bmatrix}d$$
$$y=\theta+v$$

where θ is the pitch attitude error, d is the external disturbance torque induced by solar arrays, v is the sensor noise, u is the pitch-axis control torque generated by reaction wheels, and δ represents a percentage variation of the overall loop gain perturbed mainly by vehicle inertia uncertainty, i.e., δ is a *structured uncertain parameter*.

As was discussed in Chapter 2, a fictitious input $\tilde{d}$ and a fictitious output $\tilde{z}$ can be introduced as follows:

$$\tilde{d}=\mathcal{E}u$$
$$\tilde{z}=-u$$
$$\mathcal{E}=\delta$$

where $\mathcal{E}$ is called the gain of a fictitious internal feedback loop.

Disturbance rejection filters are represented as

$$\dot{\alpha}=y \tag{9.53a}$$
$$\ddot{\beta}_1+p_1^2\beta_1=y \tag{9.53b}$$
$$\ddot{\beta}_2+p_2^2\beta_2=y \tag{9.53c}$$

where $p_1=2\pi(0.12)$ rad/s and $p_2=2\pi(0.66)$ rad/s are the frequencies of solar-array-induced disturbances. Note that a constant-disturbance rejection filter is also included, because an integral compensation is needed to attenuate low-frequency disturbances.

The disturbance filters can then be represented in state-space form as

$$\dot{\mathbf{x}}_d = \mathbf{A}_d \mathbf{x}_d + \mathbf{B}_d y \tag{9.54}$$

where

$$\mathbf{A}_d = \begin{bmatrix} 0 & 0 & 0 & 0 & 0 \\ 0 & 0 & 1 & 0 & 0 \\ 0 & -p_1^2 & 0 & 0 & 0 \\ 0 & 0 & 0 & 0 & 1 \\ 0 & 0 & 0 & -p_2^2 & 0 \end{bmatrix}$$

$$\mathbf{B}_d = \begin{bmatrix} 1 \\ 0 \\ 1 \\ 0 \\ 1 \end{bmatrix}, \qquad \mathbf{x}_d = \begin{bmatrix} \alpha \\ \beta_1 \\ \dot{\beta}_1 \\ \beta_2 \\ \dot{\beta}_2 \end{bmatrix}$$

Finally, the state-space plant model augmented by the disturbance filters and the fictitious internal feedback loop can be described as

$$\dot{\mathbf{x}}(t) = \mathbf{A}\mathbf{x}(t) + \mathbf{B}_1 \hat{\mathbf{d}}(t) + \mathbf{B}_2 u(t)$$
$$\hat{\mathbf{z}}(t) = \mathbf{C}_1 \mathbf{x}(t) + \mathbf{D}_{11} \hat{\mathbf{d}}(t) + \mathbf{D}_{12} u(t) \tag{9.55}$$
$$y(t) = \mathbf{C}_2 \mathbf{x}(t) + \mathbf{D}_{21} \hat{\mathbf{d}}(t) + \mathbf{D}_{22} u(t)$$

where

$$\mathbf{x} = [\theta \quad \dot{\theta} \quad \alpha \quad \beta_1 \quad \dot{\beta}_1 \quad \beta_2 \quad \dot{\beta}_2]^T$$
$$\hat{\mathbf{d}} = [\tilde{d} \quad d \quad v]^T, \qquad \hat{\mathbf{z}} = [\tilde{z} \quad z]^T$$

$$\mathbf{A} = \begin{bmatrix} 0 & 1 & 0 & 0 & 0 & 0 & 0 \\ 0 & 0 & 0 & 0 & 0 & 0 & 0 \\ 1 & 0 & 0 & 0 & 0 & 0 & 0 \\ 0 & 0 & 0 & 0 & 1 & 0 & 0 \\ 1 & 0 & 0 & -p_1^2 & 0 & 0 & 0 \\ 0 & 0 & 0 & 0 & 0 & 0 & 1 \\ 1 & 0 & 0 & 0 & 0 & -p_2^2 & 0 \end{bmatrix}$$

$$\mathbf{B}_1 = \begin{bmatrix} 0 & 0 & 0 \\ 1 & 1 & 0 \\ 0 & 0 & 1 \\ 0 & 0 & 0 \\ 0 & 0 & 1 \\ 0 & 0 & 0 \\ 0 & 0 & 1 \end{bmatrix} \mathbf{W}_{\hat{d}}, \qquad \mathbf{B}_2 = \begin{bmatrix} 0 \\ 1 \\ 0 \\ 0 \\ 0 \\ 0 \\ 0 \end{bmatrix}$$

$$\mathbf{C}_1 = \begin{bmatrix} 1 & 0 & w_\alpha & w_{\beta_1} & w_{\dot\beta_1} & w_{\beta_2} & w_{\dot\beta_2} \\ 0 & 0 & 0 & 0 & 0 & 0 & 0 \end{bmatrix}$$

$$\mathbf{C}_2 = [1 \quad 0 \quad 0 \quad 0 \quad 0 \quad 0 \quad 0 \,]$$

$$\mathbf{D}_{12} = \begin{bmatrix} 0 \\ 1 \end{bmatrix}, \qquad \mathbf{D}_{21} = [0 \quad 0 \quad 1]\mathbf{W}_{\hat{d}}$$

and $\mathbf{D}_{11} = 0$ and $\mathbf{D}_{22} = 0$; $\mathbf{W}_{\hat{d}}$ is the diagonal weighting matrix for $\hat{d}$; and w_α, w_{β_1}, $w_{\dot\beta_1}$, w_{β_2}, and $w_{\dot\beta_2}$ are weighting factors for α, $\beta_1, \dot\beta_1$ β_2, and $\dot\beta_2$, respectively.

The selection of a design bound γ and various weighting factors requires a trial-and-error iteration for proper tradeoffs between performance and robustness. The weighting factors on the disturbance rejection filter states are related to the separation between corresponding pole and zero of a dipole. Such separation represents the strength of the dipole, which in turn affects the settling time of the closed-loop system.

Because all of the structural modes are treated as unmodeled dynamics in the design process, we select the following weighting function:

$$W(s) = \frac{0.532(s/30 + 1)^2}{[s/(2\pi \cdot 13.8)]^2 + 2(0.004)s/(2\pi \cdot 13.8) + 1} \tag{9.56}$$

to meet the frequency domain robustness requirement with respect to unstructured uncertainty. After a certain amount of trial and error using a computer code, one may select the following

$$\gamma = 4.39$$

$$\mathbf{W}_{\hat{d}} = \text{diag}(0.05,\ 3.10,\ 0.002)$$

$$w_\alpha = 0.3, \qquad w_{\beta_1} = 2.0, \qquad w_{\dot\beta_1} = 1.8$$

$$w_{\beta_2} = 0.5, \qquad w_{\dot\beta_2} = 20$$

and find a compensator of the form

$$K(s) = (5.945 + 0.374/s + 10s)\, R(s)\, H(s) \tag{9.57}$$

where

$$R(s) = \frac{(s/0.946)^2 + 2(0.283)s/0.946 + 1}{(s/0.754)^2 + 1} \cdot \frac{(s/3.746)^2 + 2(0.021)s/3.746 + 1}{(s/4.147)^2 + 1}$$

$$H(s) = \frac{(s/86.71)^2 + 2(0.004)s/86.71 + 1}{(s/58.40)^2 + 2(0.862)s/58.40 + 1} \cdot \frac{1}{(s/42.76)^2 + 2(0.454)s/42.76 + 1}$$

For an effective rejection of the sinusoidal disturbances, the disturbance rejection filter $R(s)$ has poles at $\pm 2\pi(0.12)j$ and $\pm 2\pi(0.66)j$ with the associated zeros at $-0.267 \pm 0.907j$ and $-0.079 \pm 3.745j$, respectively.

The Bode magnitude plot of the loop transfer function with this new controller is shown in Fig. 9.31. As can be seen in this figure, this new controller has met the

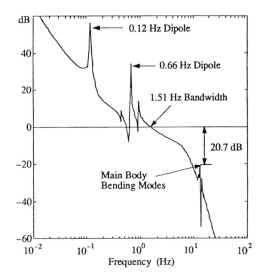

Fig. 9.31   Bode magnitude plot of the loop transfer function of a robust H_∞ control redesign.

bandwidth requirement of 1.5-Hz gain crossover frequency. It has a 5.1-dB gain margin and a 37-deg phase margin. The notch zeros at 13.8 Hz provide effective gain stabilization of the primary bending modes at 14 Hz with about 21-dB gain suppression. As can be verified from the closed-loop frequency magnitude response of this new controller from the solar array disturbance input d to the pitch attitude output θ, this new controller provides additional 40-dB disturbance attenuation at both 0.12 and 0.66 Hz with respect to the original controller.

9.5.5   Summary

Both classical and robust H_∞ control synthesis techniques were applied to the control redesign problem of the HST in this section. It was demonstrated that the proposed controllers with two disturbance rejection dipoles effectively accommodate the solar-array-induced disturbances at 0.12 and 0.66 Hz, resulting in a significant pointing performance improvement over the original controller on-board the HST. Some practical usefulness of a modern, but somewhat esoteric, robust H_∞ control design methodology has been also demonstrated, although trial-and-error iterations for selecting proper weightings were necessary.

Problem

9.1   Compare the Bode magnitude plots of the closed-loop transfer function $\theta(s)/d(s)$ of three different controllers:
   (a) The baseline controller.
   (b) The classical redesign.
   (c) The H_∞ control redesign.

9.6 Active Structural Vibration Control

9.6.1 Introduction

Thus far in this chapter, we have considered attitude control problems of various spacecraft in the presence of propellant sloshing and/or structural flexibility, with emphasis on practical aspects of attitude control systems design. As discussed in the preceding sections, most attitude control problems of practical concern do not require an active damping control of the structural or sloshing modes. The flexible modes in such problems are often naturally phase stabilized as a result of the collocated control, or gain stabilized in most cases by simply employing rolloff or notch filters.

In this section, however, we consider active structural vibration control problems in which the flexible modes need to be actively damped; however, many theoretical aspects of flexible structure controls will not be elaborated. Some fundamental concepts, as well as advanced control theory for the active structural control problem, were introduced in Chapter 2. This section is intended to provide the reader with some practical examples of structural control designs and ground experimental results.

9.6.2 Active Vibration Control of the Mast Flight System

The active structural vibration control problem of the control of flexible structures (COFS) Mast flight system is discussed here, and is based on [22].

9.6.2.1 Mast Flight System Description
The COFS Mast flight system illustrated in Fig. 9.32 was once envisioned by NASA in the mid-1980s as a means of experimenting with active structural control technologies in space. The basic element of this system is a 60-m-long, triangular-cross-sectional truss beam with a 275-kg distributed mass, including the 93-kg distributed mass of control hardware. A 180-kg tip mass includes primary actuators and sensors. A summary of the distributed actuator/sensor locations and the material properties of the truss beam are also provided in Fig. 9.32.

The proof-mass actuator was selected as the primary actuating device for the COFS Mast flight system. The proof-mass actuator can be considered the translational equivalent to the reaction wheel that is one of the primary actuating devices for the attitude control of spacecraft. The proof-mass actuator applies force, whereas the reaction wheel applies torque. Although the proof-mass actuator dynamics are inherently nonlinear and complex, the low-frequency characteristics of the proof-mass actuator can be modeled as a high-pass filter, similar to those of the reaction wheel. The transfer function from the actuator input command to the actual force output can be approximated as[*]

$$\frac{u}{u_c} = -\frac{ms}{Ts+1} \qquad (9.58)$$

---

[*]The proof-mass force actuator was developed mainly through the active control of space structures (ACOSS) project in the late 1970s. For example, see "ACOSS Five Phase 1A," Final Technical Report RADC-TR-82-21, Lockheed Missiles & Space Co., March 1982.

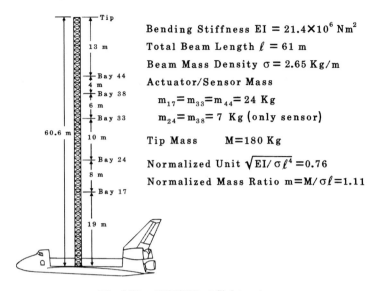

Fig. 9.32   COFS Mast flight system.

where m is the mass of the proof-mass, $T = m/K$, K is the velocity-controlled actuator servo gain, and $1/T$ becomes the actuator high-pass break frequency.

The proof-mass actuator selected for the COFS experiment has a high-pass break frequency of 1 Hz. This frequency must be distinguished from the 110-Hz actuator servo bandwidth. The first bending mode of 0.18 Hz is within the high-pass break frequency and the second bending mode of 1.98 Hz is near the high-pass break frequency. Therefore, special care should be taken with the high-pass filtering characteristics of the proof-mass actuator dynamics. Furthermore, the actuator stroke and force limits should be included in the detailed control design and simulation.

Because this truss beam with significant tip mass is cantilevered to the Orbiter, the overall transient responses are dominated by the first flexible mode in each axis. Because of the symmetric nature of the system, we consider here only the in-plane transverse bending vibration of the truss beam. The 1.98-Hz second bending mode is widely separated from the 0.18-Hz first bending mode and has an order of magnitude less modal contribution than the first mode. Consequently, the COFS Mast flight system does not have the closely spaced modal characteristics expected for flexible structures in space. On the other hand, this system represents a typical control-structure interaction problem in which controlling the primary modes always requires consideration of the control interactions with the secondary modes.

As discussed earlier in this book, when using a collocated actuator/sensor pair, a stable interaction exists; however, special consideration must be given to the unstable interaction when the actuator and sensor are not collocated. For the COFS Mast flight system, the first bending mode can be considered as a primary mode to be actively damped. The second and higher bending modes are considered the secondary modes that are not to be destabilized by the primary-mode controller.

9.6.2.2 *Pole–zero Modeling* Consider the problem of controlling the
COFS Mast flight system using the tip-mounted proof-mass actuator. As mentioned
earlier, we consider here only the in-plane transverse bending vibration of the truss
beam. An analytical transfer function from the control input force acting at the tip
to the deflections at various points on the truss beam can be determined using the
analytical approach introduced in Chapter 8. The additional lumped mass at bays
17, 24, 33, and 44 is neglected for the preliminary analytical modeling herein.
This analytical modeling provides a qualitative representation of the actual truss
beam. It also provides physical insights to the modeling and control problems of
the COFS Mast flight system.

For a uniform cantilevered beam with a tip mass M, the equation of motion is
given by

$$EI\,y''''(x,t) + \sigma\ddot{y}(x,t) = 0 \qquad (9.59)$$

where

$$y'''' \equiv \frac{\partial^4 y}{\partial x^4}, \qquad \ddot{y} \equiv \frac{\partial^2 y}{\partial t^2}$$

and $y(x,t)$ denotes the transverse deflection at the location x and time t; EI, σ, and
ℓ are the bending stiffness, mass density, and total length of the beam, respectively.
The boundary conditions are

$$y(0,t) = y'(0,t) = 0 \qquad (9.60a)$$
$$y''(\ell,t) = 0 \qquad (9.60b)$$
$$M\ddot{y}(\ell,t) - EI\,y''''(\ell,t) = u(t) \qquad (9.60c)$$

where M is the tip mass and u the proof-mass actuator control force.

Taking the Laplace transforms of Eqs. (9.59) and (9.60), we obtain

$$y''''(x,s) - \lambda^4 y(x,s) = 0 \qquad (9.61)$$

where $\lambda^4 = -s^2$, with s in units of $\sqrt{EI/\sigma\ell^4}$, and

$$y(0,s) = y'(0,s) = 0 \qquad (9.62a)$$
$$y''(1,s) = 0 \qquad (9.62b)$$
$$-\rho\lambda^4 y(1,s) - y'''(1,s) = u(s) \qquad (9.62c)$$

where $\rho = M/\sigma\ell$, and x and y are in units of ℓ, and $u(s)$ is in units of EI/ℓ^2.
The solution of Eq. (9.61) is given by

$$y(x,s) = A_1\sin\lambda x + A_2\cos\lambda x + A_3\sinh\lambda x + A_4\cosh\lambda x \qquad (9.63)$$

Combining Eqs. (9.62) and (9.63) we obtain the transcendental transfer function
from the control input force $u(s)$ to the beam deflection $y(x,s)$ and to the beam

slope $y'(x, s)$ at location x as follows:

$$\frac{y(x,s)}{u(s)} = \frac{(c\lambda + ch\lambda)(s\lambda x - sh\lambda x) - (s\lambda + sh\lambda)(c\lambda x - ch\lambda x)}{2\lambda^3(1 + c\lambda ch\lambda) + 2\rho\lambda^4(c\lambda sh\lambda - s\lambda ch\lambda)} \quad (9.64)$$

$$\frac{y'(x,s)}{u(s)} = \frac{\lambda(c\lambda + ch\lambda)(c\lambda x - ch\lambda x) + \lambda(s\lambda + sh\lambda)(s\lambda x + sh\lambda x)}{2\lambda^3(1 + c\lambda ch\lambda) + 2\rho\lambda^4(c\lambda sh\lambda - s\lambda ch\lambda)} \quad (9.65)$$

where $s(\ ) = \sin(\ )$, $sh(\ ) = \sinh(\ )$, etc., for brevity.

In particular, the transfer functions from the tip-mounted primary actuator to the tip deflection and to the tip slope become

$$\frac{y(1,s)}{u(s)} = \frac{s\lambda ch\lambda - c\lambda sh\lambda}{\lambda^3(1 + c\lambda ch\lambda) + \rho\lambda^4(c\lambda sh\lambda - s\lambda ch\lambda)} \quad (9.66)$$

$$\frac{y'(1,s)}{u(s)} = \frac{\lambda s\lambda sh\lambda}{\lambda^3(1 + c\lambda ch\lambda) + \rho\lambda^4(c\lambda sh\lambda - s\lambda ch\lambda)} \quad (9.67)$$

By determining the roots of the numerator and denominator of these transfer functions for given tip-mass ratio ρ, these transcendental transfer functions can be represented as products of poles and zeros.

Because the control of tip deflection is the primary consideration, Eq. (9.66) is further discussed here. It is observed that the transfer function zeros are independent of the tip-mass ratio ρ, whereas the poles (natural frequencies) are dependent on the mass ratio. The zeros of Eq. (9.66) are the natural frequencies of a clamped-hinged beam. The poles and zeros of Eq. (9.66) for different values of ρ are shown in Fig. 9.33. As can be observed in this figure, when ρ increases, the poles and zeros associated with the second and higher bending modes become closely spaced and the first mode becomes a single dominant mode of the cantilevered beam with a tip mass.

The frequencies in Fig. 9.33 are in units of $\sqrt{EI/\sigma\ell^4}$ and they should be multiplied by 0.76 to become frequencies in units of radians per second. For the case of

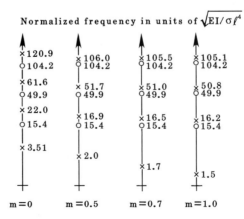

Fig. 9.33 Poles and zeros of the collocated transfer function, $y(1,s)/u(s)$, for different values of the tip-mass ratio $\rho = M/\sigma\ell$.

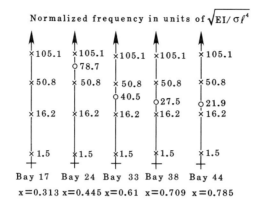

Fig. 9.34 **Poles and zeros of the transfer functions, $y(x,s)/u(s)$, at various sensor locations.**

$\rho = 1$, the lowest four natural frequencies become 0.18, 1.96, 6.14, and 12.7 Hz, which are in close agreement with the finite element modal data.

Figure 9.34 also shows the poles and zeros of transfer functions from the tip-mounted actuator force to the beam deflection at various locations. Because the actuator and the sensor are not collocated in this case, the poles and zeros are not alternating along the imaginary axis. The zero locations are sensitive to the actual mode shapes. The transfer function to the deflection at the bay 17 ($x = 0.313$) has no zeros associated with the lowest four modes. A case of controlling the beam vibration using the tip-mounted actuator and the sensor at bay 33 has been selected because the second mode has a maximum deflection at bay 33.

9.6.2.3 Collocated and Noncollocated Controls

A simplified diagram of a control system proposed for the COFS experiment is shown in Fig. 9.35. The proposed control system consists of the collocated control loop and the noncollocated control loop. A detailed discussion of the control design for this system, which can be found in [22], is omitted here because a similar control design problem will be considered in detail in the next section.

9.6.3 Active Structural Control Experiment for the Mini-Mast System

The COFS project, which was originally envisioned for the Mast flight system experiment in space, was later replaced by a ground experiment project for the Mini-Mast system located at the NASA Langley Research Center. Detailed control system designs and ground experiments have been performed by many control researchers for the Mini-Mast system, under the NASA Control Structure Interaction (CSI) Guest Investigator Program in the late 1980s.

In particular, the simplicity and practicality of a classical control approach, enhanced by the concepts of nonminimum-phase filtering and active rejection of persistent disturbances, have been demonstrated for the Mini-Mast system by Wie

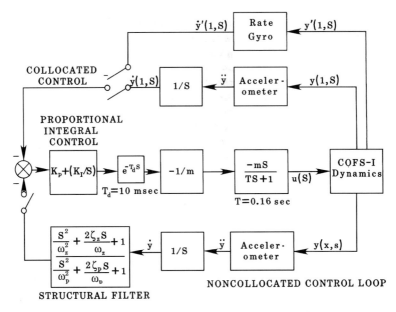

Fig. 9.35   Control system proposed for the COFS Mast flight system.

et al. [23] The intent here is to briefly describe such experimental results, with emphasis on the practical significance of the classical control concepts introduced in Chapter 2.

9.6.3.1 Mini-Mast System Description

The Mini-Mast system consists primarily of a 20-m-long, triangular-cross-sectional, joint-dominated truss structure, which is deployed vertically inside a high-bay tower, cantilevered from its base on a rigid-ground foundation. It represents future deployable trusses to be used in space. The overall system configuration and the actuator/sensor locations are illustrated in Fig. 9.36.

An approximately 160-kg tip mass includes three torque wheel actuators (TWAs) mounted on a tip plate at bay 18; the total mass of the system is about 300 kg. The three TWAs generate orthogonal control torques along the global X, Y, and Z axes. The origin of the global control axes is located at the base of the truss and at the centroid of the triangular cross section, as illustrated in Fig. 9.37. For disturbance input to the structure, three shakers at bay 9 can be utilized. These shakers are oriented normal to the faces of the truss at each of the three vertices. Displacement sensors used for control experiments are located on the platforms at the tip (bay 18) and near the midpoint of the truss (bay 10). These sensors are positioned to measure deflections normal to the face of the sensor, which are mounted parallel to the flat face on the corner joints on the truss, as shown in Fig. 9.37. Because the sensor axes are not aligned with the torque wheel axes, i.e., the global X, Y, and Z axes, a coordinate transformation of the sensor outputs is needed for

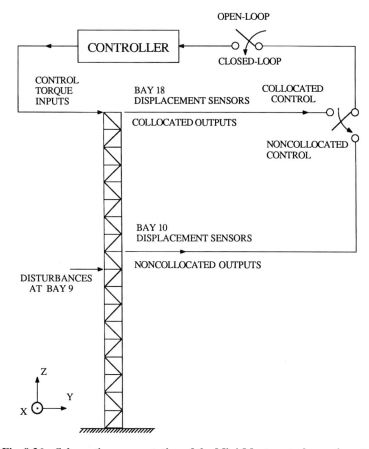

Fig. 9.36   Schematic representation of the Mini-Mast control experiments.

the classical decoupled control design. Such a coordinate transformation or sensor output decoupling matrix is given in Fig. 9.37.

Experimental, open-loop responses of the Mini-Mast excited by the shaker A with a pulse of 100 ms duration and 50 N magnitude are shown in Fig. 9.38 for the displacement sensor outputs (A, B, and C), and in Fig. 9.39 for the decoupled outputs $(X, Y,$ and $\theta)$. It is evident from Fig. 9.38 that the overall transient responses at the tip are dominated by the first flexible mode in each axis. The 6-Hz second bending mode, which can be seen at bay 10 responses, is widely separated from the 0.8-Hz first bending mode, and has an order-of-magnitude less modal contribution than the first mode. Although there are a cluster of local modes and additional global bending, torsion, and axial modes up to 100 Hz, they are not visible in Fig. 9.38. The unsymmetric behavior of the Mini-Mast truss structure is also evident in this figure. This effect is probably due to the joint nonlinearity of the truss structure. Despite such a joint nonlinearity problem, the decoupled outputs shown in Fig. 9.39 clearly indicate the effectiveness of the sensor decoupling concept illustrated in Fig. 9.37.

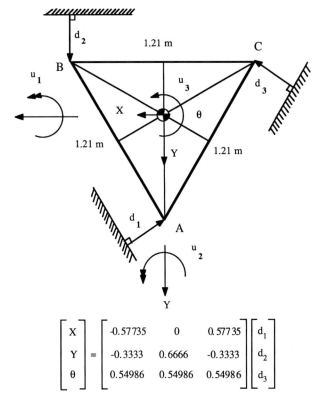

$$\begin{bmatrix} X \\ Y \\ \theta \end{bmatrix} = \begin{bmatrix} -0.57735 & 0 & 0.57735 \\ -0.3333 & 0.6666 & -0.3333 \\ 0.54986 & 0.54986 & 0.54986 \end{bmatrix} \begin{bmatrix} d_1 \\ d_2 \\ d_3 \end{bmatrix}$$

Fig. 9.37   Schematic diagram of the cross section of the Mini-Mast truss beam.

A simplified transfer function representation of the actuator dynamics used for control design is given as

$$\frac{u(s)}{u_c(s)} = \frac{s}{(s/p_1 + 1)(s/p_2 + 1)} \tag{9.68}$$

where $u(s)$ is the control torque in units of N·m, $u_c(s)$ is the torque wheel angular momentum command in units of N·m·s, and p_1 and p_2 are the actuator poles. The following numerical values are used for control design: $p_1 = 25\,\text{s}^{-1}$ and $p_2 = 300\,\text{s}^{-1}$.

9.6.3.2 Mathematical Modeling for Control Design
A finite element model of the Mini-Mast truss structure consists of 355 joints and 490 beam elements. Rigid plates are positioned at bays 18 and 10, with appropriate masses for the control hardware.

A summary of the modal frequencies and modal damping ratios for the first five modes identified by modal tests is provided in Table 9.8. Modal damping ratios for the higher frequency modes are assumed to be 1%. There are five modes below

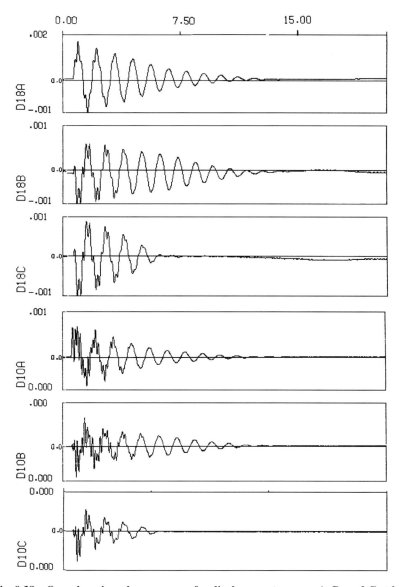

Fig. 9.38   Open-loop impulse responses for displacement sensors A, B, and C at bay 10 and bay 18.

10 Hz. The first two, at approximately 0.8 Hz, are first bending modes oriented nearly in the global X and Y axes. Next is the first torsion mode at 4.4 Hz. It is slightly coupled with bending. The fourth and fifth modes are second bending modes. Unlike the first bending modes, however, the direction of motion for second bending has rotated 45 deg from the global X and Y axes. This phenomenon results in the coupled second bending modes. Following the second bending modes, there

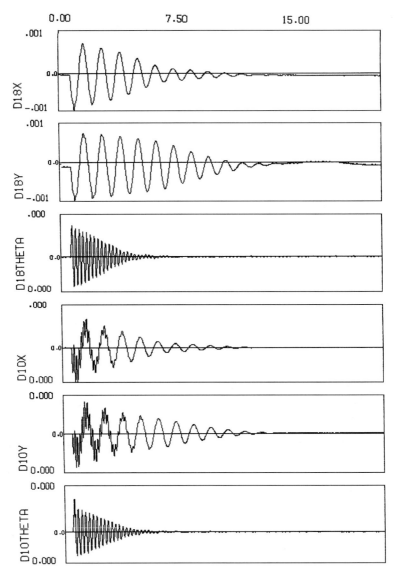

Fig. 9.39   Open-loop impulse responses for the decoupled outputs X, Y, and θ at bay 10 and bay 18.

are a cluster of 108 "local" modes. These modes involve mainly the bending of the diagonal truss members. Following this cluster of local modes, there are additional global bending, torsion, and axial modes up to 100 Hz. In the control design, the first five modes are considered the primary modes to be actively damped. The sixth and higher modes are considered the secondary modes, which are not to be destabilized by the primary-mode controller.

Table 9.8   Mini-Mast modal data

Mode	Frequency, Hz	Damping ratio, %
First X bending	0.801	1.964
First Y bending	0.801	1.964
First torsion	4.364	1.660
Second XY bending	6.104	1.194
Second YX bending	6.156	1.194

9.6.3.3 Control Experiment Objectives   Because this beam structure with significant tip mass is cantilevered to the ground, the overall transient responses are dominated by the first flexible mode in each axis. The 6-Hz second bending mode is widely separated from the 0.8-Hz first bending mode and has an order-of-magnitude less modal contribution than the first mode. Consequently, the Mini-Mast represents a typical control–structure interaction problem, where controlling the primary modes always requires consideration of the control interactions with the secondary modes. Even for a collocated actuator and sensor pair, special consideration must be given to the effect of the actuator and sensor dynamics and the phase lag caused by computational delay and prefiltering. In the preliminary control design, the first and second modes in each axis (except the second torsion mode) are considered as the primary modes to be actively damped. The third and higher modes in each axis are considered as the secondary modes, which are not to be destabilized by the primary-mode controller.

State-space methods for control design of flexible space structures have been emphasized in the literature and more widely explored than classical methods. There has also been a growing research interest in robust H_∞ control and robustness analysis applied to the control problems of flexible space structures. The classical control approach was employed in [23] for Mini-Mast control experiments, however, and the objective was to investigate the fundamental nature of control–structure interaction problems and to further understand the practical effects of many simplifying assumptions on a realistic space truss structure such as the Mini-Mast.

In particular, a new concept of generalized structural filtering for flexible-mode stabilization and a periodic disturbance rejection concept were to be validated by Mini-Mast control experiments. The simplicity and practicality of the classical control approach were to be demonstrated for the Mini-Mast. The practicality of a sensor decoupling approach was also to be demonstrated for the inherent multivariable control problem of the Mini-Mast.

The control objectives of Mini-Mast experiments described in Wie et al. [23] are: 1) a rapid transient vibration control of the tip deflection to an impulsive disturbance and 2) a steady-state vibration suppression to periodic disturbances. Control design and tests were performed for two different generic cases: 1) a collocated control using the tip-mounted actuator and sensor pairs at bay 18 and 2) a noncollocated control using the tip-mounted actuators at bay 18 and noncollocated sensors at bay 10. The external disturbances were generated by shakers at bay 9. In fact, there exists a multivariable control issue for both cases, because the sensor axes are not

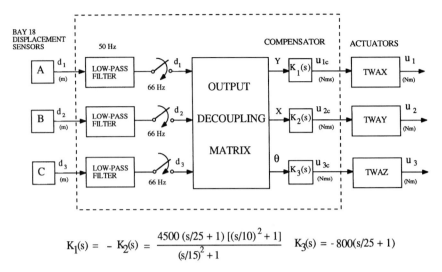

$$K_1(s) = -K_2(s) = \frac{4500\,(s/25 + 1)\,[(s/10)^2 + 1]}{(s/15)^2 + 1} \qquad K_3(s) = -800(s/25 + 1)$$

Fig. 9.40   Collocated control system with periodic disturbance rejection for X and Y at bay 18.

aligned with the control input axes. A sensor decoupling approach was, however, used to simplify the control design, which allows the use of a classical single-input single-output (SISO) control method.

9.6.3.4 Collocated Control Experiment
A simplified control system block diagram is shown in Fig. 9.40 for collocated control experiments using the tip-mounted actuators and displacement sensors A, B, and C at bay 18. The sensor outputs are low-pass filtered by an analog, three-pole Bessel filter with a corner frequency of 50 Hz in each axis. (Although a 50-Hz filter was used in the control design, it was found during the final stages of testing that a 20-Hz filter was mistakenly selected. The resulting effect was significant and will be discussed later.)

The global X- and Y-axes bending and Z-axis torsional displacements at bay 18 are generated by using the output decoupling matrix given in Fig. 9.37. The Y/u_1 transfer function has alternating poles and zeros along the imaginary axis. The X/u_2 and θ/u_3 transfer functions also have such a collocated actuator–sensor property. As a result, a SISO collocated control design is possible, where the TWAs at the tip provide the control inputs and the global displacements of the tip (bay 18) are fed back with proportional gains. Thus, direct feedback of the decoupled displacement outputs (X, Y, θ) is the basic element of the collocated control experiments for the Mini-Mast using the torque wheel actuators.

A proportional-and-derivative (PD) control was needed, however, to effectively compensate the actuator dynamics at near its high-pass break frequency. The PD compensator zero can be chosen near the actuator pole at $s = -25$. A mismatching of the actuator pole by the PD compensation zero is not significant. An effective control loop delay of 15 ms from digital control implementation with a sampling

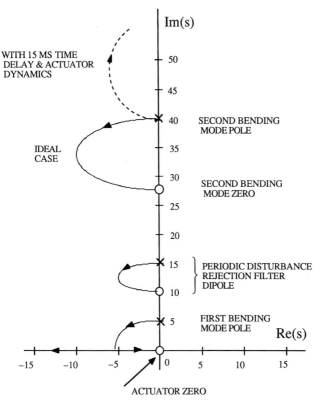

Fig. 9.41 **Root locus vs overall loop gain of the collocated *X*-axis controller.**

rate of 66.6 Hz was assumed. Such a time delay causes approximately 33-deg phase lag at the second bending mode frequency of 38 rad/s (6.1 Hz).

To validate the concept of active disturbance rejection for flexible space structures, a case with a sinusoidal disturbance with known frequency of 15 rad/s and unknown magnitude of 5 N was considered. Such a sinusoidal disturbance was generated by shaker A at bay 9, and disturbance rejection for the *X*- and *Y*-axes bending at the tip was considered. The compensator for this case, given in Fig. 9.40, includes a dipole for periodic disturbance rejection with the pole at $s = \pm j15$ and the zero at $s = \pm j10$, so that the zero lies between two consecutive poles of the loop transfer function. Figure 9.41 shows the locus of closed-loop poles vs the overall loop gain for the *X*-axis control loop with the periodic disturbance rejection filter dipole.

The first test of this collocated controller was to investigate the closed-loop behavior to an impulsive disturbance. Figure 9.42 shows the closed-loop tip response of this controller for the same impulse disturbance as for Figs. 9.38 and 9.39. A 20% active damping ratio for the first mode in each axis has been achieved, as predicted by design. An undesirable phenomenon is evident in Fig. 9.42, however, that is, the second bending modes become less stable. This anomaly, which

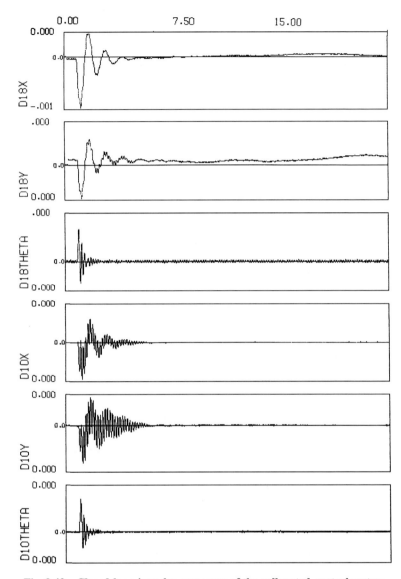

Fig. 9.42   Closed-loop impulse responses of the collocated control system.

was not predicted by design, was found to be caused probably by the mistakenly selected analog prefilter of 20-Hz (instead of 50-Hz) corner frequency. Additional test results with a 50-Hz filter indicate less destabilization of the second bending modes, but the second bending modes were still gain stabilized. The late discovery of this problem prevented a redesign of the collocated controller.

The test cycle proceeded with the sinusoidal disturbance. It was demonstrated that a periodic disturbance rejection can be achieved for the outputs X and Y at

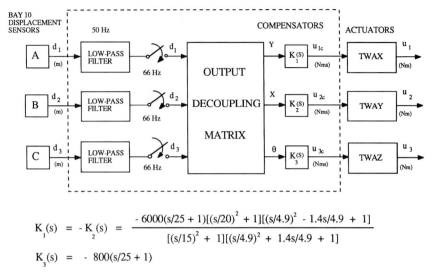

$$K_1(s) = -K_2(s) = \frac{-6000(s/25 + 1)[(s/20)^2 + 1][(s/4.9)^2 - 1.4s/4.9 + 1]}{[(s/15)^2 + 1][(s/4.9)^2 + 1.4s/4.9 + 1]}$$

$$K_3(s) = -800(s/25 + 1)$$

Fig. 9.43   Noncollocated control system with periodic disturbance rejection for X and Y at bay 10.

bay 18. In this case, the control inputs u_1 and u_2 present a steady-state oscillation caused by the dipole, which counteracts the periodic disturbance applied on the system.

9.6.3.5 Noncollocated Control Design and Experiments   A case of further interest is the case where the controller uses displacement sensors that are not located at the tip but at the midsection (bay 10), resulting in a noncollocated control configuration. Figure 9.43 shows a control system block diagram for this noncollocated case, with nonminimum-phase compensation and with periodic-disturbance rejection filtering for X and Y at bay 10.

Like the collocated case, the sensor outputs are low-pass filtered by an analog, three-pole Bessel filter with a corner frequency of 50 Hz in each axis. Then, the global X- and Y-axes bending and Z-axis torsional displacements at bay 10 are generated by using the output decoupling matrix. The Y/u_1 transfer function has nonalternating poles and zeros along the imaginary axis. The X/u_2 and θ/u_3 transfer functions also have such a noncollocated actuator–sensor property. A SISO control design, however, was still possible, where the TWAs at the tip provide the control inputs and the global displacements at bay 10 are fed back with proper compensation.

The test results, for a case where the Mini-Mast is excited by the shaker A with a pulse of 100-ms duration and 50-N magnitude, show that the noncollocated control system, shown in Fig. 9.43, without the disturbance rejection filter dipole, has a 15% damping ratio for the first mode in each axis.

To validate the concept of periodic disturbance rejection for flexible space structures even for the noncollocated case, a case with a sinusoidal disturbance with a

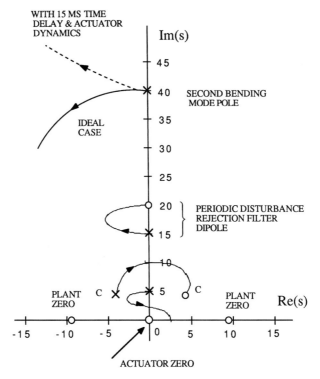

Fig. 9.44   Root locus vs overall loop gain of the noncollocated X-axis controller.

known frequency of 15 rad/s and disturbance rejection for the X- and Y-axes bending at bay 10 was considered. The compensator for this case, given in Fig. 9.43, includes a dipole for periodic disturbance rejection with the pole at $s = \pm j15$. The zero corresponding to the dipole is placed at $s = \pm j20$, between the consecutive poles of the dipole and the second bending mode.

Figure 9.44 shows the locus of closed-loop poles vs overall loop gain for the X-axis control loop with both nonminimum-phase compensation and the periodic-disturbance rejection filter dipole. Test results shown in Fig. 9.45 demonstrate that a periodic-disturbance rejection can be achieved for the outputs X and Y at bay 10. The shaker A is turned on at $t = 5$ s and turned off at $t = 20$ s. The closed-loop system is stable and the control inputs u_1 and u_2 (at bay 18) present a steady-state oscillation caused by the dipole, which counteracts the periodic disturbance applied on the system (at bay 9).

In the preliminary control design for the Mini-Mast, the first and second modes in each axis (except the second torsion mode) were considered as the primary modes to be actively damped. The third and higher modes in each axis were considered as the secondary modes, which are not to be destabilized by the primary-mode controller. During the test cycle it became clear, however, that active control, i.e.,

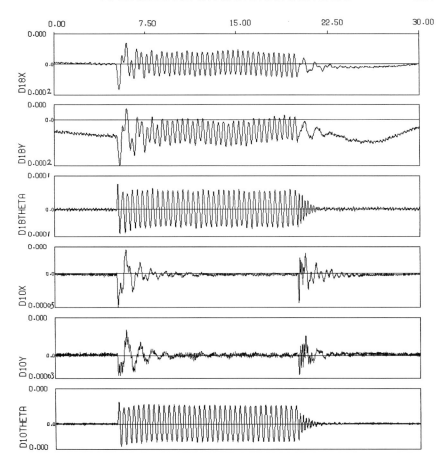

Fig. 9.45   Closed-loop sinusoidal responses of the noncollocated control system with disturbance rejection for X and Y at bay 10.

phase stabilization, of the second bending modes was not practical in the presence of significant phase lag caused by the analog prefilter and computational time delay. Consequently, the second bending mode in each axis were gain stabilized, which undoubtedly caused some transient performance degradation at bay 10, as can be seen in Fig. 9.42. An approach to actively control the second bending modes is to employ a complex phase lead filter; however, test constraints and unmodeled high-frequency dynamics prevented further refinement of the simple classical collocated controller shown in Fig. 9.40.

In summary, despite the second bending mode anomaly, the concept of nonminimum-phase compensation for the noncollocated control problem and the periodic-disturbance rejection filter concept were successfully demonstrated for the Mini-Mast. The effectiveness of the sensor decoupling concept was also successfully validated in [23].

9.6.4   Pointing and Vibration Control Experiment for the Advanced Control Evaluation for Structures Testbed

The effectiveness of a disturbance accommodating control concept for line-of-sight control of a pointing system mounted on a flexible structure has been demonstrated by Wie [24] for the advanced control evaluation for structures (ACES) testbed at the NASA Marshall Space Flight Center, under the NASA CSI Guest Investigator Program. Such experimental results, which demonstrate the simplicity and practicality of a classical "single-loop-at-a-time" approach to the active structural control design for a complex structure such as the ACES testbed, are described here.

9.6.4.1   ACES Testbed Description   The basic configuration of the ACES test facility is shown in Fig. 9.46. The basic component is a 2.27-kg, 13-m deployable Astromast, which served as the flight backup magnetometer boom for the Voyager spacecraft. It is a symmetric beam that is triangular in cross section. The Astromast has the equivalent continuous beam parameter of $EI = 2.3 \times 10^8$ N·m^2. Appendages are attached to the Astromast to emulate the closely spaced modal frequencies characteristic of large space structures. The appendages consist of an antenna and two counterbalance legs. The overall system has 43 structural modes under 8 Hz. Table 9.9 lists some of the dominant modes of the ACES structure.

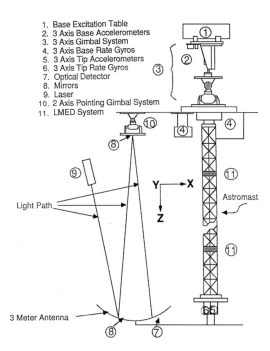

1. Base Excitation Table
2. 3 Axis Base Accelerometers
3. 3 Axis Gimbal System
4. 3 Axis Base Rate Gyros
5. 3 Axis Tip Accelerometers
6. 3 Axis Tip Rate Gyros
7. Optical Detector
8. Mirrors
9. Laser
10. 2 Axis Pointing Gimbal System
11. LMED System

Light Path

3 Meter Antenna

Astromast

Fig. 9.46   ACES ground test facility.

Table 9.9 ACES structural mode description

Mode, Hz	Description
0.14	First Astromast X bending
0.15	First Astromast Y bending
0.63	Second Astromast Y bending
0.75	Second Astromast X bending
0.82	Third Astromast Y bending
1.04	Third Astromast X bending
1.41	Antenna torsion
1.51	IMC[a] gimbals
2.35	Antenna torsion + astromast bending
2.49	Antenna torsion + astromast bending
2.73	Astromast bending

[a]Image motion compensation.

The base excitation table (BET) is hydraulically driven to provide two-axis translational disturbances at the point where the Astromast is attached to the overhead structure of the building. Several disturbances, representative of an actual space environment, can be provided by the BET. Two of these disturbances utilized in the experiment will be referred to as 1) the BET pulse (thruster firings) and 2) the BET step (crew motion disturbance).

The advanced gimbal system (AGS) is a precision, two-axis gimbal system designed for high-accuracy pointing applications, augmented by a third gimbal with a torquer and air-bearing system for azimuth control. The AGS provides torque actuation at the base mounting plate of the Astromast in response to voltage command over the range of ±10 V. The AGS torquers operate over ±30 deg, saturate at 50 N·m of torque, and have bandwidths in excess of 50 Hz. Rate gyros are provided at the base and tip of the Astromast, and they measure three-axis angular rates at each location. The AGS and the rate gyros at the base become collocated actuator/sensor pairs.

The linear momentum exchange devices (LMED) are proof-mass actuators, which produce translational forces in two axes at each location, as shown in Fig. 9.46. Each LMED has a collocated accelerometer and a proof-mass position transducer. The moving mass of 0.75 kg contains permanent magnets that can move ±3 cm over a voice coil actuator driven by a constant current source amplifier. Each LMED can deliver a peak force of 90 N; however, over the ±10-V range of the control input, a maximum continuous force of 18 N is available. The force applied to the proof mass appears as a reaction force to the structure.

An optical system is provided to measure two-axis angular displacement of the antenna frame and thus monitor the line-of-sight pointing performance. The system consists of a laser source, two mirrors, and a two-axis optical detector. One of the mirrors is mounted on a two-axis pointing gimbal so that the system can be used as a closed-loop image motion compensation (IMC) controller in addition to an optical performance sensor. The objective of this setup is to test an IMC controller that will minimize the laser-beam pointing error; this setup is representative of a secondary-mirror pointing control system of a large telescope.

A digital computer with a sampling rate of 50 Hz is provided to implement digital controllers and to store and postprocess test results.

9.6.4.2 ACES Control Objectives

State-space methods for control design of flexible space structures have been emphasized in the literature and have been more widely explored than a classical method, as evidenced by all of the prior control experiments conducted for the ACES testbed [25–27].

The primary goals of control design for the ACES testbed were: 1) to reduce the IMC line-of-sight error, i.e., to point the laser beam in the center of the detector, in the presence of two representative disturbances; 2) to ensure that the controller has a practical size; and 3) to ensure that the controller is tolerant of model uncertainties.

The performance measures employed to evaluate the controller effectiveness include the detector response, the base rate gyro response, and the controller complexity. The primary performance criterion was the IMC line-of-sight pointing accuracy.

During the previous experiments conducted by other researchers for the ACES, several problems were encountered. One of such problems was described as follows [27]:

> An unmodeled mode appeared as the dominant mode at the detector. Because the detector error was intended to be the evaluation parameter, the appearance of the unmodeled mode was disastrous to the original evaluation plan. The mode (0.15 Hz) did not destabilize any of the controllers; on the other hand, no controller attenuated the mode.

This 0.15-Hz mode problem, however, was not encountered by other CSI guest investigators because their controllers were not tested for the so-called BET step disturbance, which is one of three representative disturbances available for the ACES.

One of the major contributions of [24] was to show that such a control problem encountered during the previous experiments was actually caused by the "nearly uncontrollable" but "significantly disturbable" 0.15-Hz mode. The mode is nearly uncontrollable by the AGS torque input and completely uncontrollable by the IMC gimbals; however, it can be excited or disturbed significantly by the BET step disturbance and can be observed by the IMC detector.

It will be shown later in this section that a simple IMC controller utilizing the dipole concept can easily rectify such an uncontrollable but disturbable mode problem.

9.6.4.3 Classical Control Design

The first step in any control system design is to obtain a mathematical model of the physical system to be controlled. The math model of the ACES provided by the NASA Marshall Space Flight Center includes the modal frequencies and mode shapes of the lowest 43 modes, and transfer function models of the ACES were used for classical control design.

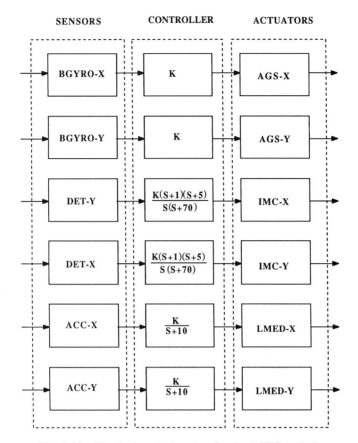

Fig. 9.47 Classical control system for the ACES testbed.

Figure 9.47 illustrates a classical control system architecture selected for the ACES. Basically, the control system consists of six actuators and six sensors without cross feedback. The control inputs are the X- and Y-axes torques of the AGS gimbals, the X- and Y-axes forces of LMED located at the lower section of the Astromast, and the X- and Y-axes torques of the IMC gimbals. The sensor measurements consist of the X- and Y-axes base rate gyros (BGYRO), the X- and Y-axes accelerometer (ACC) outputs of the LMED, and the X- and Y-axes detector (DET) position outputs.

The classical single-loop-at-a-time control designs for the ACES are briefly described as:

1) The AGS-X–BGYRO-X (also the AGS-Y–BGYRO-Y) loop basically consists of a collocated torque actuator and angular rate sensor pair. Although the effect of phase lag at high frequencies, caused by actuator/sensor dynamics and control loop time delay, must be considered in practical control design, a collocated direct rate feedback controller, as shown in Fig. 9.47, was chosen for testing. Because of relatively weak cross-axis coupling, it was not necessary to cross feedback the BGYRO outputs to the AGS gimbals.

2) The IMC-*X*–DET-*Y* (also the IMC-*Y*–DET-*X*) transfer function plot includes a few flexible modes near 1.5 Hz. As a result, a classical PID controller, as shown in Fig. 2, was chosen for each IMC loop. Similar to the AGS loops, it was not necessary to cross feedback the DET outputs to the IMC gimbals.

3) As can be seen in Fig. 9.47, a direct acceleration feedback with first-order rolloff filtering was chosen for each collocated LMED and AGS control loop.

9.6.4.4 Experimental Results   Some of the experimental results are presented here, which support the effectiveness of a simple classical controller, shown in Fig. 9.47, for achieving an excellent closed-loop performance.

Figure 9.48 shows a direct comparison of the open-loop and AGS closed-loop responses of BGYRO-*X* to a BET-*Y* pulse disturbance. Figure 9.48a shows the BGYRO-*X* open-loop response dominated by a 2.3-Hz mode. It can be seen in Fig. 9.48b that the 2.3-Hz mode is actively damped by the AGS controller (with nominal gain). However, an undesirable phenomenon of high-frequency instability is evident in Fig. 9.48c when the AGS rate gain is increased by a factor of two for more active damping of the 2.3-Hz mode. This experimental result confirms that there is no such case as a perfect collocated rate feedback control with an infinity

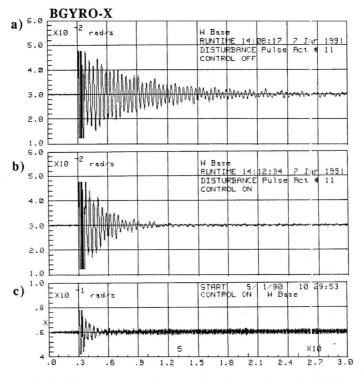

Fig. 9.48   BGYRO-*X* responses to BET-*Y* pulse: a) open-loop, b) AGS closed-loop with nominal gain, and c) AGS closed-loop with high gain.

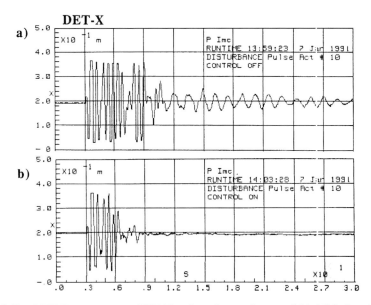

Fig. 9.49   DET-*X* responses to BET-*X* pulse: a) open-loop and b) AGS closed-loop with nominal gain.

gain margin. Consequently, the AGS controller with nominal gain has a 6-dB gain margin.

Figure 9.49 shows a direct comparison of the open-loop and AGS closed-loop responses of DET-*X* to a BET-*X* pulse disturbance. Figure 9.49a shows the open-loop response of the detector output, dominated by two structural modes at 0.75 Hz and 2.3 Hz. As can be seen in Fig. 9.49b, significant performance improvement is achieved by the AGS controller (with nominal gain). Both the 0.75-Hz and 2.3-Hz modes are effectively damped out by the collocated rate feedback controller in the detector response.

Figure 9.50 shows the closed-loop responses of DET-*X* to a BET-*X* pulse disturbance. Figure 9.50a demonstrates the effectiveness of controlling the 2.3-Hz mode (but not the 0.75-Hz mode) by the LMED. As can be seen in Fig. 9.50b, very significant performance improvement in both the line-of-sight error and vibration suppression is achieved by the integrated AGS and IMC controller, as compared to the open-loop response shown in Fig. 9.49a.

The complete controller, i.e., AGS + IMC + LMED, has also been tested, resulting in responses similar to those of the integrated AGS + IMC controller. Because the AGS controller does provide sufficient active damping to structural modes, it is unnecessary, from the practical viewpoint, to use both the LMED and AGS controllers simultaneously.

Based on the experimental results summarized as in Figs. 9.48–9.50, it can be said that an excellent closed-loop performance has been achieved by a rather simple classical controller shown in Fig. 9.47.

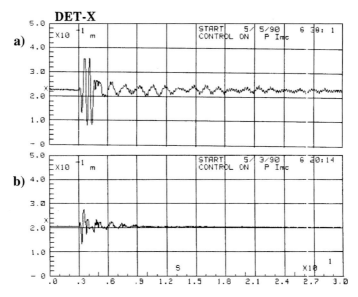

Fig. 9.50   DET-X responses to BET-X pulse: a) LMED closed-loop and b) AGS + IMC closed-loop with nominal gain.

A case of further interest, however, is the case with the BET step disturbance. Figure 9.51 compares the open-loop and closed-loop responses of DET-X to a BET-X step disturbance. Figure 9.51a shows the open-loop response, dominated by a 0.15-Hz mode and other lower frequency modes. Figure 9.51b shows the baseline integrated IMC + AGS closed-loop response, which is undoubtedly unacceptable. The reason for such unacceptable (also unpredicted) closed-loop performance is due to the presence of the 0.15-Hz mode, which can be seen in both Figs. 9.51a and 9.51b.

Several prior ACES experiments had been hampered by the uncontrolled 0.15-Hz mode. It is important to note that such a low-frequency mode is nearly uncontrollable by the AGS torque input, nearly unobservable by the base rate gyros, and completely uncontrollable by the IMC gimbals. However, it can be excited or disturbed significantly by a BET step disturbance and can be observed by the IMC detector, which is not collocated with the AGS and the rate gyros at the base.

For the purpose of IMC controller redesign, the 0.15-Hz mode excitation can be simply considered as a persistent external disturbance. To isolate such an undesirable disturbance, a new IMC controller shown in Fig. 9.52 was designed. The new controller simply includes a dipole for disturbance rejection with the pole at $s = \pm j0.9$. The zero corresponding to the dipole is placed at $s = \pm j0.1$. A detailed treatment of this dipole concept can be found in Chapter 2.

The closed-loop test result shown in Fig. 9.51c clearly demonstrates the effectiveness of the redesigned IMC controller for rejecting the 0.15-Hz mode in the detector output.

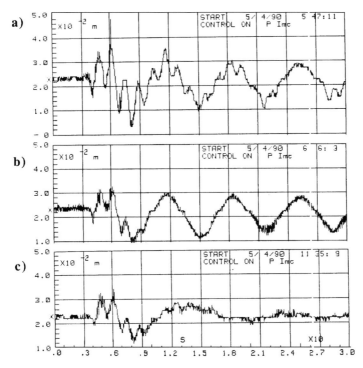

Fig. 9.51 DET-X responses to BET-X step: a) open-loop, b) AGS + IMC closed-loop, and c) AGS + IMC with periodic disturbance rejection.

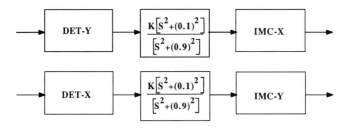

Fig. 9.52 IMC controller with periodic disturbance accommodating filters.

9.7 Summary

In this chapter we have considered various practical examples of attitude and structural control designs for space vehicles having flexible solar arrays and/or flexible appendages.

State-space methods for control design of flexible space vehicles have been emphasized by many control researchers and more widely explored than classical methods. This arises from the convenience of obtaining a compensator for the whole system given one set of weighting parameters. The fundamental question remains, however, of how to choose these parameters and what choice provides the best optimal design. The designer must find an acceptable set of parameters for a good optimal design. The use of state-space methods for control design usually results in a compensator of the same order as the system to be controlled. This means that for systems having several flexible modes, the compensator adds compensation even to modes that are stable and need no compensation. This may result in a high-order compensator design.

In classical design, on the other hand, a compensator must be constructed piece by piece, or mode by mode. As shown in this chapter, the classical design is particularly convenient for the control of flexible space vehicles with properly selected collocated actuator/sensor pairs. The concept of nonminimum-phase compensation also provides an extremely convenient way of stabilizing unstably interacting flexible modes for the case in which actuators and sensors are not collocated. The resulting compensator is usually of a lower order than the system to be controlled because not all flexible modes in a structure tend to be destabilized by a reduced-order controller.

A helpful characteristic of most flexible space structures is their inherent passive damping. This gives the designer the opportunity to phase stabilize significant modes and to gain stabilize all other higher frequency modes, which have less influence on the structure, as discussed in this chapter.

References

[1] Wie, B., "Thrust Vector Control Design for a Liquid Upper Stage Spacecraft," *Journal of Guidance, Control, and Dynamics*, Vol. 8, No. 5, 1985, pp. 566–572.

[2] Greensite, A. L., *Analysis and Design of Space Vehicle Flight Control Systems*, Spartan, New York, 1970.

[3] Bryson, A. E., Jr., *Control of Spacecraft and Aircraft*, Princeton Univ. Press, Princeton, NJ, 1994, Chap. 9 and Appendix F.

[4] Abramson, H. N. (ed.), *The Dynamic Behavior of Liquids in Moving Containers with Applications to Space Vehicle Technology*, NASA SP-106, 1966.

[5] Redding, D., "Highly Efficient Very Low-Thrust Transfer to Geosynchronous Orbit: Exact and Approximate Solutions," *Journal of Guidance, Control, and Dynamics*, Vol. 7, No. 2, 1984, pp. 141–147.

[6] Penchuk, A., and Croopnick, S., "The Digital Autopilot for Thrust Vector Control of the Shuttle Orbital Maneuvering System," *Journal of Guidance, Control, and Dynamics*, Vol. 6, No. 6, 1983, pp. 436–441.

[7] Wie, B., Lehner, J. A., and Plescia, C. T., "Roll/Yaw Control of a Flexible Spacecraft Using Skewed Bias Momentum Wheels," *Journal of Guidance, Control, and Dynamics*, Vol. 8, No. 4, 1985, pp. 447–453.

[8] Roberson, R. E., "Two Decades of Spacecraft Attitude Control," *Journal of Guidance and Control*, Vol. 2, No. 1, 1979, pp. 3–8.

[9] Muhlfelder, L., "Evolution of an Attitude Control System for Body-Stabilized Communication Spacecraft," *Journal of Guidance, Control, and Dynamics*, Vol. 9, No. 1, 1986, pp. 108–112.

[10] Agrawal, B. N., *Design of Geosynchronous Spacecraft*, Prentice–Hall, Englewood Cliffs, NJ, 1986.

[11] Likins, P. W., Ohkami, Y., and Wong, C., "Appendage Modal Coordinate Truncation Criteria in Hybrid Coordinate Dynamic Analysis," *Journal of Spacecraft and Rockets*, Vol. 13, No. 10, 1976, pp. 611–617.

[12] Terasaki, R. M., "Dual Reaction Wheel Control of Spacecraft Pointing," Symposium of Attitude Stabilization and Control of Dual Spin Spacecraft, The Aerospace Corp., Los Angeles, CA, Aug. 1967.

[13] Dougherty, H. J., Scott, E. D., and Rodden, J. J., "Attitude Stabilization of Synchronous Communications Satellites Employing Narrow-Beam Antennas," *Journal of Spacecraft and Rockets,* Vol. 8, No. 8, 1971, pp. 834–841.

[14] Manabe, S., Tsuchiya, K., and Inoue, M., "Zero PID Control for Bias Momentum Satellites," International Federation of Automatic Control 8th World Congress, Paper 76-4, 1981.

[15] Laskin, R. A., and Kopf, E. H., "High Precision Active Nutation Control of a Momentum Biased Spacecraft with Flexible Appendages," *Journal of Guidance, Control, and Dynamics*, Vol. 8, No. 1, 1985, pp. 78–85.

[16] Wie, B., and Plescia, C. T., "Attitude Stabilization of a Flexible Spacecraft During Stationkeeping Maneuvers," *Journal of Guidance, Control, and Dynamics*, Vol. 7, No. 4, 1984, pp. 430–436.

[17] Anthony, T., Wie, B., and Carroll, S., "Pulse-Modulated Controller Synthesis for a Flexible Spacecraft," *Journal of Guidance, Control, and Dynamics*, Vol. 13, No. 6, 1990, pp. 1014–1022.

[18] Bauer, F., and Downing, J. P., "Control System Design and Analysis Using Interactive Control Analysis (INCA) Program," *Proceedings of AIAA Guidance, Navigation and Control Conference*, AIAA, Washington, DC, 1987, pp. 1138–1148.

[19] Wie, B., Liu, Q., and Bauer, F., "Classical and Robust H_∞ Control Redesign for the Hubble Space Telescope," *Journal of Guidance, Control, and Dynamics*, Vol. 16, No. 6, 1993, pp. 1069–1077.

[20] Dougherty, H., Tompetrini, K., Levinthal, J., and Nurre, G., "Space Telescope Pointing Control System," *Journal of Guidance, Control, and Dynamics*, Vol. 5, No. 4, 1982, pp. 403–409.

[21] Beals, G. A., Crum, R. C., Dougherty, H. J., Hegel, D. K., Kelley, J. L., and Rodden, J. J., "Hubble Space Telescope Precision Pointing Control System," *Journal of Guidance, Control, and Dynamics*, Vol. 11, No. 2, 1988, pp. 119–123.

[22] Wie, B., "Active Vibration Control Synthesis for the COFS (Control of Flexible Structures) Mast Flight System," *Journal of Guidance, Control, and Dynamics*, Vol. 11, No. 3, 1988, pp. 271–276.

[23] Wie, B., Horta, L., and Sulla, J., "Active Vibration Control Synthesis and Experiment for the Mini-Mast," *Journal of Guidance, Control, and Dynamics*, Vol. 14, No. 4, 1991, pp. 778–784.

[24] Wie, B., "Experimental Demonstration of a Classical Approach to Flexible Structure Control," *Journal of Guidance, Control, and Dynamics*, Vol. 15., No. 6, 1992, pp. 1327–1333.

[25] Waites, H., "Active Control Technique Evaluation for Spacecraft (ACES)," Final Technical Rept. F33615-86-C-3225, NASA/Marshall Space Flight Center and Control Dynamics Co., Huntsville, AL, March 1988.

to other direct numerical optimization approaches for the time-optimal control problem, is that robustness constraints with respect to plant parameter uncertainty can be easily augmented.

10.1.1 Time-Optimal Rest-to-Rest Maneuver

10.1.1.1 Problem Formulation Consider a structural dynamic system described by

$$\mathbf{M\ddot{q} + Kq = Gu} \tag{10.1}$$

where $\mathbf{q}$ is a generalized displacement vector, $\mathbf{M}$ a mass matrix, $\mathbf{K}$ a stiffness matrix, $\mathbf{G}$ a control input distribution matrix, and $\mathbf{u}$ a control input vector.

In this section, we consider a case with a scalar control input $u(t)$ bounded as

$$-1 \leq u \leq 1 \tag{10.2}$$

This simple case, however, represents a typical single-axis control problem of most flexible spacecraft controlled by reaction jets.

Equation (10.1) is transformed into the decoupled modal equations

$$
\begin{aligned}
\ddot{y}_1 + \omega_1^2 y_1 &= \phi_1 u \\
\ddot{y}_2 + \omega_2^2 y_2 &= \phi_2 u \\
&\vdots \\
\ddot{y}_n + \omega_n^2 y_n &= \phi_n u
\end{aligned}
\tag{10.3}
$$

where y_i is the ith modal coordinate, ω_i is the ith modal frequency, ϕ_i is ith modal gain, and n is the number of modes considered in control design.

The problem is to find the control input that minimizes the performance index

$$J = \int_0^{t_f} dt = t_f$$

subject to Eqs. (10.2) and (10.3), and the given boundary conditions.

For an undamped dynamic system of n degrees of freedom described by Eq. (10.3), the time-optimal solution for a rest-to-rest maneuver has, in most cases, $(2n - 1)$ switches, and the solution is symmetric about $t_f/2$. That is, for a case with $(2n - 1)$ switches, we have

$$t_j = t_{2n} - t_{2n-j}, \qquad j = 1, \ldots, n$$

where t_{2n} is the maneuver time t_f and t_j for $j = 1, \ldots, n$ represents the switching time.

A bang-bang input with $(2n - 1)$ switches can then be represented as

$$u(t) = \sum_{j=0}^{2n} B_j u_s(t - t_j) \tag{10.4}$$

where B_j is the magnitude of a unit step function $u_s(t)$ at t_j. This function can be characterized by its switch pattern as

$$B = \{B_0, \quad B_1, \quad B_2, \ldots, B_{2n}\} \tag{10.5a}$$

$$T = \{t_0, \quad t_1, \quad t_2, \ldots, t_{2n}\} \tag{10.5b}$$

where B represents a set of B_j with $B_0 = B_{2n} = \pm 1$ and $B_j = \pm 2$ for $j = 1, \ldots,$ $2n - 1$; T represents a set of switching times $(t_1, \ldots, t_{2n-1})$ and the initial and final times $(t_0 = 0$ and $t_f = t_{2n})$.

10.1.1.2 Rest-to-Rest Maneuver Constraints

Consider the rigid-body mode equation with $\omega_1 = 0$

$$\ddot{y}_1 = \phi_1 u \tag{10.6}$$

with the rest-to-rest maneuvering boundary conditions

$$\begin{aligned} y_1(0) = 0, \quad & y_1(t_f) \neq 0 \\ \dot{y}_1(0) = 0, \quad & \dot{y}_1(t_f) = 0 \end{aligned} \tag{10.7}$$

Substituting Eq. (10.4) into Eq. (10.6) and solving for the time response of the rigid-body mode, we obtain

$$y_1(t \geq t_f) = \frac{\phi_1}{2} \sum_{j=0}^{2n} (t_f - t_j)^2 B_j \tag{10.8}$$

The rest-to-rest maneuvering constraint for the rigid-body mode can then be written as

$$\frac{\phi_1}{2} \sum_{j=0}^{2n} (t_f - t_j)^2 B_j - y_1(t_f) = 0 \tag{10.9}$$

Next consider the structural modes described by

$$\ddot{y}_i + \omega_i^2 y_i = \phi_i u, \qquad i = 2, \ldots, n \tag{10.10}$$

with the corresponding boundary conditions for the rest-to-rest maneuver

$$\begin{aligned} y_i(0) = 0, \quad & y_i(t_f) = 0 \\ \dot{y}_i(0) = 0, \quad & \dot{y}_i(t_f) = 0 \end{aligned} \tag{10.11}$$

for each flexible mode.

Substituting Eq. (10.4) into the ith structural mode equation and solving for the time response for $t \geq t_f$, we obtain

$$
y_i(t) = -\frac{\phi_i}{\omega_i^2} \sum_{j=0}^{2n} B_j \cos \omega_i(t - t_j)
$$

$$
= -\frac{\phi_i}{\omega_i^2} \left[\cos \omega_i(t - t_n) \sum_{j=0}^{2n} B_j \cos \omega_i(t_j - t_n) \right.
$$

$$
\left. + \sin \omega_i(t - t_n) \sum_{j=0}^{2n} B_j \sin \omega_i(t_j - t_n) \right] \qquad (10.12)
$$

It can be shown that the following constraint equation for each mode

$$
\sum_{j=0}^{2n} B_j \sin \omega_i(t_j - t_n) = 0 \qquad (10.13)
$$

is always satisfied for any bang-bang input that is symmetric about the midmaneuver time t_n. Consequently, we have the following flexible mode constraints for no residual structural vibration, i.e., $y_i(t) = 0$ for $t \geq t_f$,

$$
\sum_{j=0}^{2n} B_j \cos \omega_i(t_j - t_n) = 0 \qquad (10.14)
$$

for each flexible mode.

10.1.1.3 Parameter Optimization Problem
For an undamped structural dynamic system of n degrees of freedom, the time-optimal solution represented by Eq. (10.4) has the $(2n - 1)$ unknown switching times and the final time t_f to be determined. The time-optimal control problem can now be formulated as a constrained parameter optimization problem as follows.

Determine a control input of the form given by Eq. (10.4), which minimizes the performance index

$$
J = t_f \qquad (10.15)
$$

subject to

$$
\frac{\phi_1}{2} \sum_{j=0}^{2n} (t_f - t_j)^2 B_j - y_1(t_f) = 0 \qquad (10.16a)
$$

$$
\sum_{j=0}^{2n} B_j \cos \omega_i(t_j - t_n) = 0, \qquad i = 2, \ldots, n \qquad (10.16b)
$$

$$
t_j > 0, \qquad j = 1, \ldots, 2n \qquad (10.16c)
$$

where $t_f = t_{2n}$.

Standard numerical optimization packages, e.g., International Mathematical and Statistical Library (IMSL) subroutines or MATLAB optimization toolbox, can be used to obtain the solution of the preceding optimization problem. The major advantage of this approach, compared to other direct numerical optimization approaches for the time-optimal control problem, is that robustness constraints with respect to plant parameter uncertainty can be easily augmented. This subject will be discussed in detail later in this chapter.

10.1.2 Sufficient Condition for Optimality

Equations (10.16) are only necessary conditions for the time-optimal control problem, and sufficient conditions for optimality are derived here.

Equation (10.3) with $\omega_1 = 0$ can be expressed in state-space form as

$$\dot{\mathbf{x}} = \mathbf{A}\mathbf{x} + \mathbf{B}u \tag{10.17}$$

where

$$\mathbf{x} = (y_1, \ \dot{y}_1, \ \omega_2 y_2, \ \dot{y}_2, \dots, \ \omega_n y_n, \ \dot{y}_n)$$

$$\mathbf{A} = \begin{bmatrix} 0 & 1 & & & & & \\ 0 & 0 & & & & & \\ & & 0 & \omega_2 & & & \\ & & -\omega_2 & 0 & & & \\ & & & \vdots & \vdots & & \\ & & & & & 0 & \omega_n \\ & & & & & -\omega_n & 0 \end{bmatrix}$$

$$\mathbf{B} = (0, \ \phi_1, \ 0, \ \phi_2, \dots, \ 0, \ \phi_n)$$

Because the system described by Eq. (10.17) is linear, time invariant, and controllable, the optimal control input does not contain any singular or undetermined intervals. Using Pontryagin's minimum principle to characterize the optimal solution, we define the Hamiltonian as

$$H = 1 + \boldsymbol{\lambda}^T(\mathbf{A}\mathbf{x} + \mathbf{B}u) \tag{10.18}$$

where

$$\boldsymbol{\lambda} = (p_1, \ q_1, \dots, \ p_n, \ q_n) \tag{10.19}$$

is the costate vector.

For a rest-to-rest maneuver, the necessary and sufficient conditions for optimal solution are

$$\dot{\boldsymbol{\lambda}} = -\mathbf{A}^T \boldsymbol{\lambda} \tag{10.20a}$$

$$u = -\operatorname{sgn}\{S(t)\} \tag{10.20b}$$

$$H(t_f) = 0 \tag{10.20c}$$

where $t \in [0, t_f]$, the signum function $\mathrm{sgn}(S)$ and switching function $S(t)$ are defined as

$$\mathrm{sgn}(S) = \begin{cases} +1, & S \geq 0 \\ -1, & S < 0 \end{cases} \qquad (10.21a)$$

$$S(t) = \mathbf{B}^T \boldsymbol{\lambda}(t) \qquad (10.21b)$$

The costate equations corresponding to Eq. (10.17) can be obtained from Eq. (10.20) as

$$\dot{p}_1(t) = 0 \qquad (10.22a)$$

$$\dot{q}_1(t) = -p_1(t) \qquad (10.22b)$$

$$\dot{p}_i(t) = \omega_i q_i(t) \qquad (10.22c)$$

$$\dot{q}_i(t) = -\omega_i p_i(t) \qquad (10.22d)$$

for $i = 2, 3, \ldots, n$.

It can be shown that at midmaneuver, we have

$$\boldsymbol{\lambda}(t_n) = [p_1(t_n), \ 0, \ p_2(t_n), \ 0, \ldots, \ p_n(t_n), \ 0]^T \qquad (10.23)$$

Thus, the costate vector can be solved from Eq. (10.22) as

$$p_1(t) = p_1(t_n) \qquad (10.24a)$$

$$q_1(t) = -(t - t_n)p_1(t_n) \qquad (10.24b)$$

$$p_i(t) = p_i(t_n) \cos \omega_i(t - t_n) \qquad (10.24c)$$

$$q_i(t) = -p_i(t_n) \sin \omega_i(t - t_n) \qquad (10.24d)$$

for each flexible mode.

Assume that a solution obtained from the constrained parameter optimization problem satisfies

$$0 < t_1 < t_2 < \cdots < t_{2n-1} < t_f \qquad (10.25)$$

Then the costate vector at midmaneuver, as described by Eq. (10.23), can be found from the following n linear equations:

$$1 + \phi_1 p_1(t_n)t_n + \sum_{i=2}^{n} \phi_i p_i(t_n) \sin \omega_i t_n = 0 \qquad (10.26)$$

$$\phi_1 p_1(t_n)(t_j - t_n) + \sum_{i=2}^{n} \phi_i p_i(t_n) \sin \omega_i(t_j - t_n) = 0 \qquad (10.27)$$

where $j = n + 1, \ldots, 2n - 1$. Note that Eqs. (10.26) and (10.27) correspond to $H(t_f) = 0$ and $S(t_i) = 0$, respectively.

Once the costates are obtained as in Eqs. (10.24), we can determine a control input

$$u(t) = -\text{sgn}\{\mathbf{B}^T\boldsymbol{\lambda}\} \tag{10.28}$$

which brings the system to the desired target set.

If the switching function

$$S(t) = \mathbf{B}^T\boldsymbol{\lambda}, \qquad t \in [0, t_f] \tag{10.29}$$

only vanishes at t_i in Eq. (10.25), then $\boldsymbol{\lambda}$ and u satisfy all of the necessary and sufficient conditions in Eq. (10.20) and are, therefore, the optimal costate vector and control input.

Consequently, we have the following theorem.

Theorem 10.1   Sufficient Condition

The solution obtained by minimizing t_f subject to Eq. (10.16) becomes time optimal provided that

$$S(t) = -\phi_1 p_1(t_n)(t - t_n) - \sum_{i=2}^{n} \phi_i p_i(t_n) \sin \omega_i(t - t_n) \neq 0 \tag{10.30}$$

for $t \in (t_n, t_{2n})$ and $t \neq t_j, j = n + 1, \ldots, 2n$.

10.1.3   Properties of Time-Optimal Solution

Some important properties of the time-optimal solution are briefly summarized here.

The time-optimal bang-bang solution for a rest-to-rest maneuver has an odd number of switches and is symmetric about $t_f/2$. For the case of a rigid-body mode and one flexible mode, the time-optimal bang-bang solution has at most three switches. If modal frequencies are integer multiples of the fundamental frequency ω_2 and

$$\omega_2 = 2\ell\pi\sqrt{\frac{\phi_1}{y_1(t_f)}}, \qquad \ell = 1, 2, \ldots \tag{10.31}$$

then the time-optimal solution has only one switch and is equivalent to the solution of a rigidized case.

Consider a system with one rigid-body mode and one flexible-body mode described by

$$\ddot{y}_1 = \phi_1 u \tag{10.32a}$$

$$\ddot{y}_2 + \omega_2^2 y_2 = \phi_2 u \tag{10.32b}$$

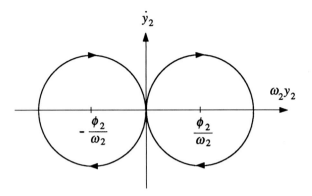

Fig. 10.1   Phase-plane trajectory.

The time-optimal solution for the rigidized model can be obtained as

$$t_f = 2\sqrt{\frac{y_1(t_f)}{\phi_1}} \tag{10.33a}$$

$$u(t) = u_s(t) - 2u_s[t - (t_f/2)] + u_s(t - t_f) \tag{10.33b}$$

where the optimal control input has only one switch.
    For the flexible-body mode, we can obtain

$$\left(\omega_2 y_2 - \frac{\phi_2}{\omega_2}\right)^2 + \dot{y}_2 = \left(\frac{\phi_2}{\omega_2}\right)^2 \qquad \text{if } u = 1 \tag{10.34a}$$

$$\left(\omega_2 y_2 + \frac{\phi_2}{\omega_2}\right)^2 + \dot{y}_2 = \left(\frac{\phi_2}{\omega_2}\right)^2 \qquad \text{if } u = -1 \tag{10.34b}$$

which correspond to two circles on the $(\omega_2 y_2, \dot{y}_2)$ plane, as shown in Fig. 10.1. The travel time along a full circle is $2\pi/\omega_2$. Therefore, if the control input has only one switch, we must have

$$t_f = \frac{4\ell\pi}{\omega_2}, \qquad \ell = 1, 2, \ldots \tag{10.35}$$

for a rest-to-rest maneuver.
    Combining Eqs. (10.33a) and (10.35), we obtain Eq. (10.31). In this case, the solution given by Eq. (10.33) is also the optimal solution for the system described by Eqs. (10.32).
    For the case of more than one flexible mode, if Eq. (10.31) is satisfied and

$$\omega_i = N_i \omega_2, \qquad i = 3, 4, \ldots \tag{10.36}$$

where $N_i > 1$ is any integer, then the optimal phase-plane trajectory for each flexible mode is similar to the one shown in Fig. 10.1. Therefore, the time-optimal solution is equivalent to the solution of rigidized case as given in Eq. (10.33).

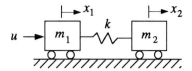

Fig. 10.2   Two-mass–spring example.

The rigid-body optimal, phase-plane trajectory in the $(y_1, \dot{y}_1)$ plane is symmetric about the line $y_1 = y_1(t_f)/2$, and the ith flexible-body mode phase-plane trajectory in the $(\omega_i y_i, \dot{y}_i)$ plane is symmetric about the line $\omega_i y_i = 0$. It comes directly from the symmetry property of time-optimal bang-bang solution for a rest-to-rest maneuver. The time-optimal control input is independent of actuator location for a system described by Eq. (10.3) with a scalar input.

Note that ϕ_i $(i = 2, \ldots, n)$ do not appear in Eq. (10.16) due to the zero boundary conditions for the flexible modes; that is, the optimal solution to this problem is independent of the flexible mode shapes.

10.1.4   Example: Two-Mass–Spring System

Consider a simple example, shown in Fig. 10.2, which is a generic representation of a flexible spacecraft with a rigid-body mode and one flexible mode. It is assumed that the nominal parameters are $m_1 = m_2 = k = 1$ with appropriate units, and time is in seconds. A control input force u, which is bounded as $|u| \leq 1$, is applied on body 1.

10.1.4.1   Rigid-Body Time-Optimal Control   For a rigidized model of the nominal system shown in Fig. 10.2, the equation of motion is simply

$$(m_1 + m_2)\ddot{y} = u \qquad\qquad (10.37)$$

The rest-to-rest, time-optimal solution for $y(0) = 0$ and $y(t_f) = 1$ can be found as

$$u^*(t) = u_s(t) - 2u_s[t - (t_f/2)] + u_s(t - t_f) \qquad\qquad (10.38)$$

where $t_f = 2\sqrt{(m_1 + m_2)y(t_f)} = 2.828$ s.

If this time-optimal input force is exerted on the nominal system with a flexible mode, a significant residual structural vibration occurs, as shown in Fig. 10.3.

10.1.4.2   Flexible-Body Time-Optimal Control   Consider a time-optimal control problem for the flexible-body model shown in Fig. 10.2. The equations of motion are

$$m_1\ddot{x}_1 + k(x_1 - x_2) = u \qquad\qquad (10.39a)$$
$$m_2\ddot{x}_2 - k(x_1 - x_2) = 0 \qquad\qquad (10.39b)$$

where x_1 and x_2 are the positions of body 1 and body 2, respectively.

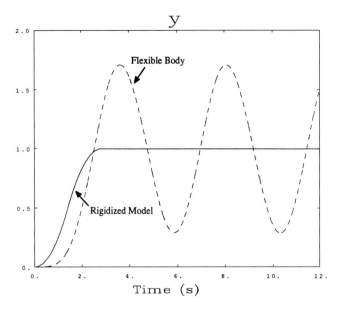

Fig. 10.3   Responses to rigid-body, time-optimal control input.

The boundary conditions for a rest-to-rest maneuver are given as

$$x_1(0) = x_2(0) = 0, \qquad x_1(t_f) = x_2(t_f) = 1$$
$$\dot{x}_1(0) = \dot{x}_2(0) = 0, \qquad \dot{x}_1(t_f) = \dot{x}_2(t_f) = 0 \tag{10.40}$$

The modal equations are

$$\ddot{y}_1 = u/2 \tag{10.41a}$$
$$\ddot{y}_2 + \omega^2 y_2 = u/2 \tag{10.41b}$$

where $\omega = \sqrt{2}$ rad/s is the nominal flexible mode frequency. The corresponding boundary conditions for modal coordinates are

$$y_1(0) = y_2(0) = 0, \qquad y_1(t_f) = 1, \qquad y_2(t_f) = 0$$
$$\dot{y}_1(0) = \dot{y}_2(0) = 0, \qquad \dot{y}_1(t_f) = \dot{y}_2(t_f) = 0 \tag{10.42}$$

Because there are three switches, the time-optimal switch pattern for the given boundary conditions is represented as

$$B = \{B_0, \ B_1, \ B_2, \ B_3, \ B_4\}$$
$$= \{1, \ -2, \ 2, \ -2, \ 1\}$$
$$T = \{t_0, \ t_1, \ t_2, \ t_3, \ t_4\}$$

with the symmetry conditions

$$t_4 = 2t_2$$
$$t_3 = 2t_2 - t_1$$

The time-optimal control problem is then formulated as the following constrained minimization problem:

$$\min J = 2t_2 \qquad (10.43)$$

subject to

$$2 + 2t_1^2 + t_2^2 - 4t_1t_2 = 0 \qquad (10.44a)$$
$$1 - 2\cos\omega(t_2 - t_1) + \cos\omega t_2 = 0 \qquad (10.44b)$$
$$t_1, \ t_2 > 0 \qquad (10.44c)$$

One can obtain a solution as $t_1 = 1.003$ and $t_2 = 2.109$. Substituting this solution into Eqs. (10.26) and (10.27), the costate vector at midmaneuver can be solved as

$$\boldsymbol{\lambda}(t_2) = [p_1(t_2), q_1(t_2), p_2(t_2), q_2(t_2)]^T$$
$$= [-1.0342, 0, 1.1438, 0]^T \qquad (10.45)$$

The switching function is then obtained from Eq. (10.30) as

$$S(t) = -0.5171(t - t_2) - 0.5719\sin\omega(t - 2.109) \qquad (10.46)$$

which is shown in Fig. 10.4.

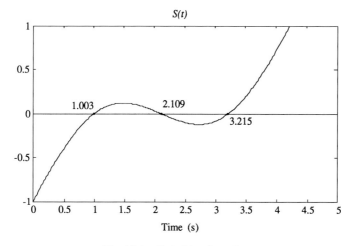

Fig. 10.4   Switching function.

The computed solution satisfies the optimality conditions (10.30); i.e., the switching function vanishes only at $t = t_1, t_2$ and t_3, as shown in Fig. 10.4. Thus, the solution is indeed time-optimal, and it can be expressed as

$$u(t) = u_s(t) - 2u_s(t - 1.003) + 2u_s(t - 2.109)$$
$$- 2u_s(t - 3.215) + u_s(t - 4.218) \tag{10.47}$$

It is noted that, for a different initial guess of the solution, it is also possible to find other solutions (for example, $t_1 = 1.737$, $t_2 = 5.481$), which also satisfy the constraints given in Eq. (10.44), but fail to meet the sufficient condition described by Theorem 10.1.

The maneuver time and switch pattern are plotted as functions of spring stiffness k and maneuver distance $y_1(t_f)$ in Fig. 10.5. In general, the maneuver time

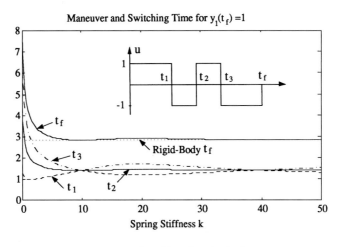

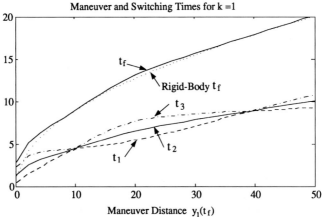

Fig. 10.5   Maneuver time and switch pattern vs spring stiffness and maneuver distance.

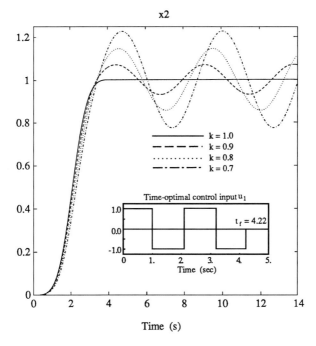

Fig. 10.6   Responses to flexible-body, time-optimal control input.

decreases as k increases and eventually converges to the rigid-body solution; however, it does not decrease monotonically. It is also observed that, at certain points satisfying Eq. (10.31), the flexible-body optimal solution coincides with the rigid-body solution.

The time responses of x_2 to the time-optimal control input (10.47) are shown in Fig. 10.6 for four different values of k. It can be seen that the resulting responses are very sensitive to variations in model parameter k.

In this section an approach has been described to transform the standard, time-optimal control problem for flexible systems into a constrained parameter optimization problem. It has also been shown that the ideal time-optimal solutions are very sensitive to model parameter variations. In the next section, robustness constraints are augmented into the problem to generate robust time-optimal control inputs for flexible system subject to model parameter uncertainties.

10.2   Robust Time-Optimal Control

As shown in the preceding section, the standard, time-optimal control problem of flexible spacecraft requires an accurate mathematical model, and thus the resulting solution is often sensitive to variations in model parameters. Consequently, the development of a "robustified" open-loop approach is of practical interest; however, most open-loop approaches attempt to find a smooth, continuous forcing function, e.g., a versine function, that begins and ends with zero slope. The basic

idea behind such approaches is that a smooth control input without sharp tran-
sitions is less likely to excite structural modes during maneuvers. Such an input
function, however, does not fully utilize the available maximum maneuvering force
and results in a slower response time as well as residual structural vibrations even
for a nominal system.

In this section, a new approach, expanding on the approach introduced in the
preceding section, is described for generating robust time-optimal control inputs
for the single-axis, rest-to-rest maneuvering problem of flexible spacecraft in the
presence of structural frequency uncertainty [2]. A parameter optimization problem, where the objective function to be minimized is the maneuvering time, is
formulated with additional constraints for robustness with respect to structural
frequency uncertainty. The resulting robustified, time-optimal solution is a multiswitch bang-bang control that can be implemented for spacecraft equipped with
on–off reaction jets.

10.2.1 Robustness Constraints

Constraint equations (10.14) were derived earlier for each flexible mode under
consideration to guarantee that no residual structural vibration occurs at the end
of rest-to-rest maneuvers. Such constraints can be represented as a vector function
as follows:

$$\mathbf{f}(T, \mathbf{p}) = 0 \qquad (10.48)$$

where T represents a set of switching times as described by Eq. (10.5b) and $\mathbf{p}$
represents an uncertain parameter vector. For our problem, $\mathbf{p}$ consists of flexible-
mode frequencies that are considered uncertain.

Expanding $\mathbf{f}(T, \mathbf{p})$ about its nominal value $\mathbf{p}^\circ$, we obtain

$$\mathbf{f}(T, \mathbf{p}) = \mathbf{f}(T, \mathbf{p}^\circ) + \left. \frac{\partial \mathbf{f}}{\partial \mathbf{p}} \right|_{(T,\mathbf{p}^\circ)} (\mathbf{p} - \mathbf{p}^\circ) + \cdots \qquad (10.49)$$

and T can then be redesigned to satisfy the following two sets of constraints:

$$\mathbf{f}(T, \mathbf{p}^\circ) = 0 \qquad (10.50a)$$

$$\left. \frac{\partial \mathbf{f}}{\partial \mathbf{p}} \right|_{(T,\mathbf{p}^\circ)} = 0 \qquad (10.50b)$$

where the second restraint is called the first-order robustness constraint, because
it limits the amplitude of residual structural vibrations caused by mismodeling of
$\mathbf{p}$ to be second-order or higher in the modeling error $(\mathbf{p} - \mathbf{p}^\circ)$.

By taking the derivative of Eq. (10.12) with respect to ω_i, we obtain

$$\frac{\mathrm{d}y_i(t)}{\mathrm{d}\omega_i} = \frac{\phi_i}{\omega_i^2} \cos \omega_i \left(t - \frac{t_f}{2} \right) \sum_{j=0}^{2n} \left(t_j - \frac{t_f}{2} \right) B_j \sin \omega_i \left(t_j - \frac{t_f}{2} \right) \qquad (10.51)$$

for each flexible mode. Letting $dy_i(t)/d\omega_i = 0$ for all $t \geq t_f$, we have

$$\sum_{j=0}^{2n} \left(t_j - \frac{t_f}{2} \right) B_j \sin \omega_i \left(t_j - \frac{t_f}{2} \right) = 0, \qquad i = 2, \ldots, n \qquad (10.52)$$

which is called the first-order robustness constraint.

Similarly, taking the derivative of Eq. (10.12) r_i times with respect to ω_i results in the r_ith-order robustness constraints for each flexible mode, as follows:

$$\sum_{j=0}^{2n} \left(t_j - \frac{t_f}{2} \right)^m B_j \sin \omega_i \left(t_j - \frac{t_f}{2} \right) = 0 \qquad \text{for } m = 1, 3, \ldots \leq r_i$$

$$(10.53a)$$

$$\sum_{j=0}^{2n} \left(t_j - \frac{t_f}{2} \right)^m B_j \cos \omega_i \left(t_j - \frac{t_f}{2} \right) = 0 \qquad \text{for } m = 2, 4, \ldots \leq r_i$$

$$(10.53b)$$

There are totally r robustness constraints for $(n - 1)$ flexible modes, where

$$r = \sum_{i=2}^{n} r_i \qquad (10.54)$$

If these robustness constraints are included in the constrained minimization problem formulation described by Eq. (10.16), the number of switches in the bang-bang control input, in most cases, must be increased to match the number of the constraint equations. Because of the symmetric nature of the rest-to-rest maneuvering problem, adding one robustness constraint will require, at the very least, two more switches.

10.2.2 Robust Time-Optimal Control

If r robustness constraints are considered for a flexible system of n modes, the corresponding robust bang-bang control input becomes

$$u(t) = \sum_{j=0}^{2(n+r)} B_j u_s(t - t_j) \qquad (10.55)$$

which has $2(n + r)$ unknown switching times. Because of the symmetry property of the optimal solution for the rest-to-rest maneuvering problem, we have

$$t_j = t_{2(n+r)} - t_{2(n+r)-j}, \qquad j = 1, \ldots, n + r \qquad (10.56)$$

Therefore, there are only $(n + r)$ unknowns to be determined in Eq. (10.55). These unknowns can be determined by minimizing t_f subject to the $(n + r)$ constraint equations: one positioning constraint for the rigid-body mode, $(n - 1)$ no-vibration constraints, and r robustness constraints.

Although many theoretical issues, e.g., the uniqueness of the optimal solution, need to be discussed, a solution can be obtained by solving the following constrained parameter optimization problem:

$$\min J = t_f = t_{2(n+r)} \qquad (10.57)$$

subject to

$$\frac{\phi_1}{2} \sum_{j=0}^{2(n+r)} (t_{2(n+r)} - t_j)^2 B_j - y_1(t_f) = 0$$

$$\sum_{j=0}^{2(n+r)} B_j \cos \omega_i (t_j - t_{n+r}) = 0$$

$$\sum_{j=0}^{2(n+r)} (t_j - t_{n+r})^m B_j \sin \omega_i (t_j - t_{n+r}) = 0, \qquad \text{for } m = 1, 3, \ldots \leq r_i$$

$$\sum_{j=0}^{2(n+r)} (t_j - t_{n+r})^m B_j \cos \omega_i (t_j - t_{n+r}) = 0, \qquad \text{for } m = 2, 4, \ldots \leq r_i$$

$$t_j > 0, \qquad\qquad j = 1, \ldots, 2(n+r)$$

for each flexible mode. The resulting bang-bang control input, which has $2r$ more switches than the time-optimal bang-bang solution of the preceding section, will be called a robust (or robustified) time-optimal solution throughout this chapter.

10.2.3 Example: Two-Mass–Spring System

For the generic model of a flexible spacecraft shown in Fig. 10.2, the time-optimal control is a three-switch bang-bang function, but the resulting response was shown to be very sensitive to variations in model parameter k. A robustified, time-optimal solution of the same problem is now developed as follows. The switching pattern for a case with the first-order robustness constraint is assumed as

$$B = \{B_0, B_1, B_2, B_3, B_4, B_5, B_6\}$$
$$= \{1, -2, 2, -2, 2, -2, 1\} \qquad (10.58a)$$
$$T = \{t_0, t_1, t_2, t_3, t_4, t_5, t_6\} \qquad (10.58b)$$

with the symmetry conditions

$$t_4 = 2t_3 - t_2$$
$$t_5 = 2t_3 - t_1 \qquad (10.59)$$
$$t_6 = 2t_3$$

The constrained optimization problem with the first-order robustness constraint can be formulated as

$$\min J = 2t_3 \qquad (10.60)$$

subject to

$$2 + 2t_1^2 - 2t_2^2 - t_3^2 - 4t_1t_3 + 4t_2t_3 = 0 \qquad (10.61\text{a})$$

$$\cos \omega t_3 - 2\cos \omega(t_3 - t_1) + 2\cos \omega(t_3 - t_2) - 1 = 0 \qquad (10.61\text{b})$$

$$t_3 \sin \omega t_3 - 2(t_3 - t_1)\sin \omega(t_3 - t_1) + 2(t_3 - t_2)\sin \omega(t_3 - t_2) = 0 \quad (10.61\text{c})$$

$$t_1, \; t_2, \; t_3 > 0$$

A robust time-optimal solution with five switches can be found as

$$t_1 = 0.7124, \qquad t_2 = 1.6563$$
$$t_3 = 2.9330, \qquad t_4 = 4.2097$$
$$t_5 = 5.1536, \qquad t_6 = 5.8660$$

The time responses of x_2 to this robust, time-optimal control input are shown in Fig. 10.7 for four different values of k. It can be seen that the resulting response is less sensitive to parameter variations compared to the responses to the ideal, time-optimal control input as shown in Fig. 10.6. The performance robustness has been increased at the expense of the increased maneuvering time of 5.866 s, compared to the ideal minimum time of 4.218 s. It is, however, emphasized that

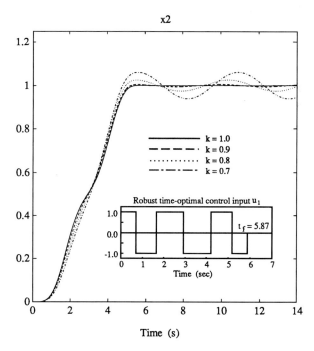

Fig. 10.7   **Responses to robust time-optimal control input with first-order robustness constraint.**

simply prolonging the maneuver time does not help to reduce residual structural vibrations caused by modeling uncertainty.

If the second-order robustness constraint is included in the parameter optimization problem, a solution for robust bang-bang control can be obtained as

$$t_1 = 0.5383, \quad t_2 = 1.4019$$
$$t_3 = 2.6459, \quad t_4 = 3.8010$$
$$t_5 = 4.9561, \quad t_6 = 6.2000$$
$$t_7 = 7.0636, \quad t_8 = 7.6019$$

Parameter robustness with respect to spring constant variations is compared in Fig. 10.8 for ideal time-optimal and robustified time-optimal control schemes. It is evident that the robustness has been improved by the robustness constraints, resulting in a zero slope for the dashed and dotted curves at a nominal value of k. Comparing with other robustified feedforward approaches, the approach described here provides a faster and more robust maneuver in the presence of model parameter uncertainty. Also, unlike other approaches, the resulting solution of this approach is a multiswitch bang-bang control, which can be implemented for spacecraft with on–off reaction jets.

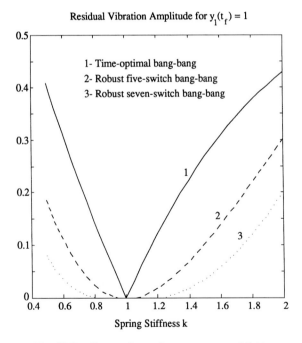

Fig. 10.8 Comparison of parameter sensitivity.

10.3   Robust Time-Optimal Control Using One-Sided Control Inputs

In this section we consider a case, illustrated in Fig. 10.9, with two one-sided control inputs bounded as

$$0 \le u_1 \le +1 \tag{10.62a}$$

$$-1 \le u_2 \le 0 \tag{10.62b}$$

10.3.1   Problem Formulation

Because the control inputs are one sided, each control input for the time-optimal solution need not be an odd function about the midmaneuver time. Thus, the problem with one-sided control inputs becomes much more difficult to solve than the standard problem with two-sided control inputs, and many theoretical issues, e.g., the uniqueness and structure of time-optimal solutions, need further consideration.

The modal equations of the system with nominal parameter values are obtained as

$$\ddot{y}_1 = \tfrac{1}{2}(u_1 + u_2) \tag{10.63a}$$

$$\ddot{y}_2 + \omega^2 y_2 = \tfrac{1}{2}(u_1 - u_2) \tag{10.63b}$$

where $\omega = \sqrt{2}$ rad/s is the nominal flexible mode frequency.

For the control input constraint given by Eq. (10.62), the control inputs can be expressed as

$$u_1 = \sum_{j=0,2,4,\dots}^{N-1} [u_s(t - t_j) - u_s(t - t_j - \Delta_j)] \tag{10.64a}$$

$$u_2 = - \sum_{j=1,3,5,\dots}^{N} [u_s(t - t_j) - u_s(t - t_j - \Delta_j)] \tag{10.64b}$$

which is in the form of one-sided pulse sequences as shown in Fig. 10.10. The jth pulse starts at t_j and ends at $(t_j + \Delta_j)$. Because of the symmetric nature of the rest-to-rest maneuvering problem, we assume that u_1 and u_2 have the same number of pulses, $(N + 1)/2$, where N is defined as in Fig. 10.10.

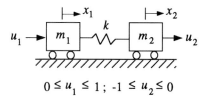

Fig. 10.9   Two-mass–spring example with two, one-sided control inputs.

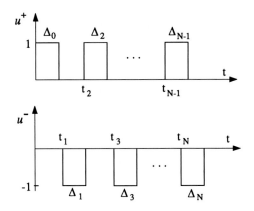

Fig. 10.10   Pulse sequences.

Substituting Eq. (10.64) into Eq. (10.63) and solving for the time response of the rigid-body mode, we obtain

$$y_1(t \geq t_f) = \frac{1}{4} \sum_{j=0}^{N} (-1)^j \left[2t\Delta_j - 2t_j\Delta_j - \Delta_j^2 \right] \qquad (10.65)$$

For the desired boundary condition, $y_1(t \geq t_f) = 1$, the following constraint must hold:

$$\sum_{j=0}^{N} (-1)^j \Delta_j = 0 \qquad (10.66)$$

The positioning constraint for the rigid-body mode then becomes

$$\sum_{j=0}^{N} (-1)^j \left[2t_j\Delta_j + \Delta_j^2 \right] + 4 = 0 \qquad (10.67)$$

Substituting Eq. (10.64) into Eq. (10.63b) and solving for the time response of the flexible mode, we obtain

$$y_2(t) = -\frac{1}{4} \cos \omega t \sum_{j=0}^{N} \left[\cos \omega t_j - \cos \omega (t_j + \Delta_j) \right]$$

$$-\frac{1}{4} \sin \omega t \sum_{j=0}^{N} \left[\sin \omega t_j - \sin \omega (t_j + \Delta_j) \right] \qquad (10.68)$$

for $t \geq t_f$.

Also, rest-to-rest maneuvering requires that $y_2(t) = 0$ for $t \geq t_f$; i.e., we have

$$\sum_{j=0}^{N} [\cos \omega t_j - \cos \omega(t_j + \Delta_j)] = 0 \tag{10.69a}$$

$$\sum_{j=0}^{N} [\sin \omega t_j - \sin \omega(t_j + \Delta_j)] = 0 \tag{10.69b}$$

which become the no-vibration constraints for the rest-to-rest maneuvering problem.

10.3.2 Time-Optimal Control

Let the time-optimal control inputs for the rest-to-rest maneuver problem be of the form

$$u_1 = u_s(t) - u_s(t - \Delta)$$
$$u_2 = -u_s(t - t_1) + u_s(t - t_1 - \Delta)$$

where each input has a single pulse with the same pulse width of Δ, t_1 is defined as shown in Fig. 10.10, and the maneuver time $t_f = t_1 + \Delta$.

The rest-to-rest maneuver constraints can be obtained from Eqs. (10.67) and (10.69) as

$$t_f - (2/\Delta) - \Delta = 0 \tag{10.70a}$$

$$\sin \frac{\omega t_f}{2} + \sin \omega \left(\Delta - \frac{t_f}{2} \right) = 0 \tag{10.70b}$$

which can be combined as

$$\sin \frac{\omega \Delta}{2} \cos \frac{\omega}{2\Delta} = 0 \tag{10.71}$$

The time-optimal solution can then be obtained by solving the constrained minimization problem

$$\min J = t_f = (2/\Delta) + \Delta \tag{10.72}$$

subject to the constraint given by Eq. (10.71). The solution of this problem can be found as $\Delta = 0.9003$ and $t_f = 3.1218$.

The time responses of x_2 to the time-optimal control inputs are shown in Fig. 10.11 for four different values of k. The maneuver time and control on-time are, respectively, 3.12 and 1.8 s. As expected, the resulting solution is sensitive to parameter variations.

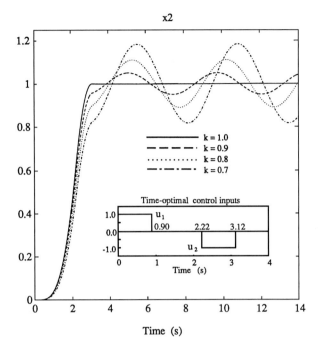

Fig. 10.11   Responses to time-optimal, one-sided control inputs.

10.3.3 Robust Time-Optimal Control

We now consider the robustification of the time-optimal solution obtained earlier.

Letting the derivative of Eq. (10.68), with respect to ω, be zero, we have

$$\frac{dy_2}{d\omega} = -\frac{1}{4} \sin \omega t \sum_{j=0}^{N} [t_j \cos \omega t_j - (t_j + \Delta_j) \cos \omega (t_j + \Delta_j)]$$

$$+ \frac{1}{4} \cos \omega t \sum_{j=0}^{N} [t_j \sin \omega t_j - (t_j + \Delta_j) \sin \omega (t_j + \Delta_j)] = 0 \quad (10.73)$$

For this derivative to be zero for arbitrary $t \geq t_f$, we must have

$$\sum_{j=0}^{N} [t_j \cos \omega t_j - (t_j + \Delta_j) \cos \omega (t_j + \Delta_j)] = 0 \qquad (10.74a)$$

$$\sum_{j=0}^{N} [t_j \sin \omega t_j - (t_j + \Delta_j) \sin \omega (t_j + \Delta_j)] = 0 \qquad (10.74b)$$

which are called the first-order robustness constraints.

Taking the derivative of Eq. (10.68) r times with respect to ω, we obtain the rth-order robustness constraint equations for input pulse sequences, as follows:

$$\sum_{j=0}^{N} \left[(t_j)^m \cos \omega t_j - (t_j + \Delta_j)^m \cos \omega (t_j + \Delta_j) \right] = 0 \qquad (10.75)$$

$$\sum_{j=0}^{N} \left[(t_j)^m \sin \omega t_j - (t_j + \Delta_j)^m \sin \omega (t_j + \Delta_j) \right] = 0 \qquad (10.76)$$

for $m = 1, 2, \ldots, r$.

As an example, we consider the first-order robustness constraint, incorporated with the rest-to-rest maneuver constraints, to construct robust time-optimal pulse sequences. Assuming that each input has two pulses, we can represent the control inputs as follows:

$$u_1 = u_s(t) - u_s(t - \Delta_0) + u_s(t - t_2) - u_s(t - t_2 - \Delta_2) \qquad (10.77a)$$

$$u_2 = -u_s(t - t_1) + u_s(t - t_1 - \Delta_1) - u_s(t - t_3) + u_s(t - t_3 - \Delta_3) \qquad (10.77b)$$

in which we have seven unknowns to be determined, and where t_j and Δ_j are defined as shown in Fig. 10.10.

The robust time-optimal solution can then be obtained by solving the constrained parameter optimization problem:

$$\min J = t_3 + \Delta_3 \qquad (10.78)$$

subject to

$$\Delta_0 - \Delta_1 + \Delta_2 - \Delta_3 + \Delta_4 = 0$$

$$\sum_{j=0}^{3} (-1)^j \left[2t_j \Delta_j + \Delta_j^2 \right] + 4 = 0$$

$$\sum_{j=0}^{3} \left[\cos \omega t_j - \cos \omega (t_j + \Delta_j) \right] = 0$$

$$\sum_{j=0}^{3} \left[\sin \omega t_j - \sin \omega (t_j + \Delta_j) \right] = 0 \qquad (10.79)$$

$$\sum_{j=0}^{3} \left[t_j \cos \omega t_j - (t_j + \Delta_j) \cos \omega (t_j + \Delta_j) \right] = 0$$

$$\sum_{j=0}^{3} \left[t_j \sin \omega t_j - (t_j + \Delta_j) \sin \omega (t_j + \Delta_j) \right] = 0$$

$$\Delta_j \geq 0; \qquad j = 0, 1, 2, 3$$

$$t_1, \ t_2, \ t_3 > 0$$

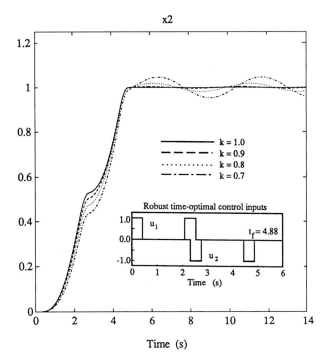

Fig. 10.12　Responses to robust time-optimal one-sided control inputs.

The solution to this problem can be obtained as

$$
\begin{array}{ll}
t_0 = 0.0, & \Delta_0 = 0.4274 \\
t_1 = 2.3357, & \Delta_1 = 0.4329 \\
t_2 = 2.1132, & \Delta_2 = 0.4329 \\
t_3 = 4.4544, & \Delta_3 = 0.4274
\end{array}
\tag{10.80}
$$

The time responses of x_2 to the robustified, time-optimal control inputs are shown in Fig. 10.12 for four different values of k. It can be seen that the robustness has been increased at the expense of the increased maneuvering time of 4.882 s, compared with the ideal minimum time of 3.122 s. Note, however, that the control on-time is only 1.721 s, compared to the control on-time of 1.8 s of the ideal, time-optimal solution.

A most interesting feature of the optimal solutions for noncollocated jets is that the overall input shape shown in Figs. 10.11 and 10.12 is of a "bang-off-bang" type, resulting in the control on-time being significantly smaller than the maneuver time. For the case of collocated reaction jets, the maneuver time and jet on-time are the same, which is clearly undesirable from the viewpoint of fuel consumption (jet on-time). Therefore, the actuator configuration with noncollocated jets is considered to be optimal in the sense that it minimizes both the maneuver time and fuel consumption [3,4]. Furthermore, the results also indicate that properly coordinated, on–off pulse sequences can achieve a fast maneuvering time with a minimum of structural residual vibrations, even in the face of plant modeling uncertainty.

10.4 Robust Fuel- and Time-Optimal Control

In the preceding sections, the problem of computing robustified, open-loop, time-optimal control inputs for uncertain flexible spacecraft has been considered. The primary control objective in such a robust time-optimal control problem is to achieve a fast maneuvering time with minimum structural vibrations during and/or after a maneuver in the face of modeling uncertainty. Two different cases, as summarized in Figs. 10.13 and 10.14, were considered to address such a challenging control problem. For case 1 (Fig. 10.13), both positive and negative jets are placed on body 1, which represents a typical situation in which two opposing jets are collocated. For case 2 (Fig. 10.14), a positive jet is placed on body 1 and a negative jet on body 2; that is, two opposing jets are not collocated. This two-mass–spring model represents a generic model of spacecraft with one rigid-body mode and one dominant structural mode for the single-axis attitude control design.

For a certain case of noncollocated positive and negative jets, as shown in Fig. 10.14, the robustified time-optimal control inputs are of a bang-off-bang type, resulting in a control on-time that is much smaller than the total maneuver time. On the other hand, the robustified time-optimal control inputs for a case of collocated positive and negative jets are of a bang-bang type, resulting in a control on-time that is the same as the total maneuver time. Although a significant desensitization with respect to model uncertainty is possible, as demonstrated in the preceding sections, such bang-bang type solutions may not be acceptable from the practical viewpoint of fuel consumption.

In this section, as was covered in [5], we further study the feasibility of finding open-loop, on–off pulse control sequences for a robust time-optimal maneuver with less use of fuel. The ideal, fuel- and time-optimal control problem of flexible spacecraft without modeling uncertainty is first reviewed.

10.4.1 Fuel- and Time-Optimal Control

10.4.1.1 Problem Formulation Consider a flexible space structure described by the modal equation

$$\ddot{y} + \Omega^2 y = \Phi u \tag{10.81}$$

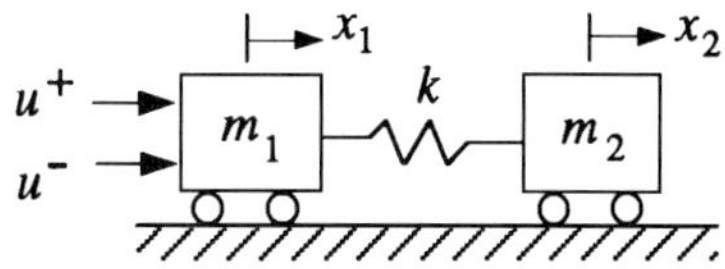

$$0 \le u^+ \le 1 \quad \text{and} \quad -1 \le u^- \le 0$$

Fig. 10.13   Case 1.

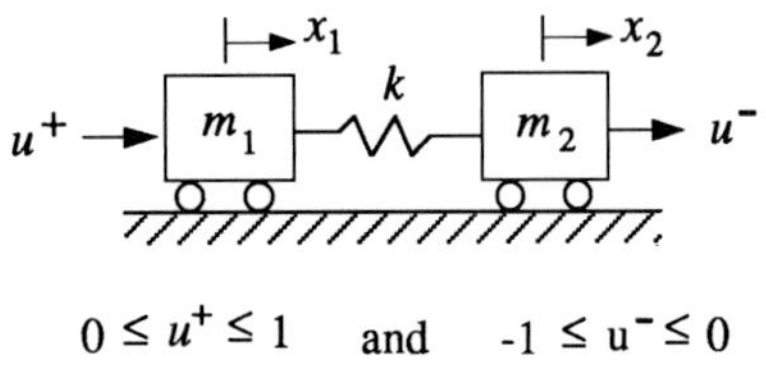

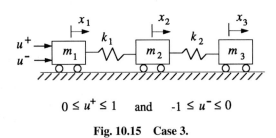

$$0 \le u^+ \le 1 \quad \text{and} \quad -1 \le u^- \le 0$$

Fig. 10.14   Case 2.

where $\mathbf{y} = (y_1, \ldots, y_n)$ is the modal coordinate vector, $\boldsymbol{\Omega}^2 = \mathrm{diag}(\omega_i^2)$, ω_i is the ith modal frequency, $\boldsymbol{\Phi}$ is the modal input distribution matrix, and $\mathbf{u}$ is the control input vector.

Because we are interested in a typical, single-axis, rest-to-rest maneuvering control problem of flexible spacecraft, we consider a flexible spacecraft model with one rigid-body mode and a few dominant flexible modes in the corresponding axis. The boundary conditions of such a rest-to-rest maneuver problem are assumed to be of the form

$$
\begin{aligned}
y_1(0) &= 0, & y_1(t_f) &= 1 \\
\dot{y}_1(0) &= 0, & \dot{y}_1(t_f) &= 0 \\
y_i(0) &= 0, & y_i(t_f) &= 0 \\
\dot{y}_i(0) &= 0, & \dot{y}_i(t_f) &= 0
\end{aligned}
\tag{10.82}
$$

for $i = 2, \ldots, n$, and n is the number of modes.

It is also assumed that we are given a positive jet u^+ and a negative jet u^-

$$0 \le u^+ \le +1 \tag{10.83a}$$

$$-1 \le u^- \le 0 \tag{10.83b}$$

For case 1, shown in Fig. 10.13, both u^+ and u^- are acting on body 1, resulting in a typical case with a two-sided control input u with $|u| \le 1$. For case 2, shown in Fig. 10.14, u^+ is placed on body 1 and u^- is placed on body 2. Case 3, shown in Fig. 10.15 is similar to case 1, in which both the positive and negative jets are acting on body 1.

$$0 \le u^+ \le 1 \quad \text{and} \quad -1 \le u^- \le 0$$

Fig. 10.15   Case 3.

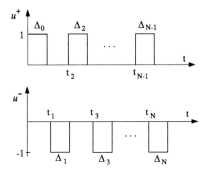

Fig. 10.16   Pulse sequences.

Pulse sequences for the control inputs u^+ and u^-, shown in Fig. 10.16 can be expressed as a combination of unit-step function $u_s(t)$ as

$$u^+(t) = \sum_{j=0,2}^{N-1} [u_s(t - t_j) - u_s(t - t_j - \Delta_j)] \qquad (10.84a)$$

$$u^-(t) = \sum_{j=1,3}^{N} [-u_s(t - t_j) + u_s(t - t_j - \Delta_j)] \qquad (10.84b)$$

The jth pulse starts at t_j and ends at $(t_j + \Delta_j)$. The jth pulse duration is denoted by Δ_j. Because of the symmetric nature of the rest-to-rest maneuvering problem, we assume that u^+ and u^- have the same number of pulses, $(N + 1)/2$, where N is defined as in Fig. 10.16.

For a fuel- and time-optimal control problem, the objective function to be minimized is, in general, a weighted sum of the maneuvering time and the consumed fuel. We assume that the mass of the consumed fuel is small compared with the mass of the spacecraft, and that the consumed fuel is proportional to the jet on-time. We further assume that the structural flexibility and mass distribution of the vehicle are more uncertain than the total mass of the system. Consequently, we focus on the robust control problem of flexible spacecraft in the face of modal frequency uncertainty.

Consider the following objective function, which is simply a sum of the maneuvering time and the product of the weighting parameter α and the total jet on-time:

$$J = \int_0^{t_f} \{1 + \alpha(|u^+| + |u^-|)\} \, dt \qquad (10.85)$$

The problem is then to find the control inputs that minimize the performance index J subject to the equations of motion. In the next section we study the preceding problem, but with additional robustness constraints with respect to plant modeling uncertainty.

10.4.1.2   Rigid-Body Fuel- and Time-Optimal Control   First consider a rigidized model of case 1 and case 2. The nominal parameters are assumed as $m_1 = m_2 = 1$ with appropriate units, and time is in seconds. The equation of motion is simply given by

$$2\ddot{x} = u \qquad (10.86)$$

where x is the rigid-body displacement and the control input is bounded as $|u| \leq 1$. Given the fuel- and time-optimal control performance index

$$J = \int_0^{t_f} \{1 + \alpha|u|\} \, dt$$

the bang-off-bang type solution for a rest-to-rest maneuver with $x(t_f) = 1$ can be found as

$$\Delta = \sqrt{2/(2\alpha + 1)}$$
$$t_f = 2\Delta(\alpha + 1)$$

where Δ is the pulse width.

When a bang-off-bang solution with $\Delta = 0.816$ s and $t_f = 3.27$ s (for $\alpha = 1$) is applied to the actual flexible model of case 1, significant excitation of the flexible mode can be noticed. Even for a case with a shorter pulse width of $\Delta = 0.308$ s and a prolonged maneuver time of $t_f = 7.35$ s for $\alpha = 10$, significant excitation of the flexible mode can be seen as shown in Fig. 10.17. Thus, simply prolonging the maneuver time and using short pulses does not help to reduce structural vibrations; a flexible-body model must be used in determining fuel- and time-optimal control inputs.

10.4.1.3   Flexible-Body Fuel- and Time-Optimal Control   Substituting Eq. (10.84) into Eq. (10.81) and incorporating the boundary conditions (10.82), we obtain the time response of the rigid-body mode as

$$y_1(t \geq t_f) = \frac{1}{2} \left[\phi_{11} \sum_{j=0,2}^{N-1} \left(2t\Delta_j - 2t_j\Delta_j - \Delta_j^2\right) \right.$$
$$\left. - \phi_{12} \sum_{j=1,3}^{N} \left(2t\Delta_j - 2t_j\Delta_j - \Delta_j^2\right) \right] \qquad (10.87)$$

where ϕ_{i1} and ϕ_{i2} denote the elements of the modal input distribution matrix Φ for the positive jet u^+ and the negative jet u^-, respectively. For a given boundary condition, $y_1(t \geq t_f) = 1$, we have the following constraint:

$$\phi_{11} \sum_{j=0,2}^{N-1} \Delta_j - \phi_{12} \sum_{j=1,3}^{N} \Delta_j = 0 \qquad (10.88)$$

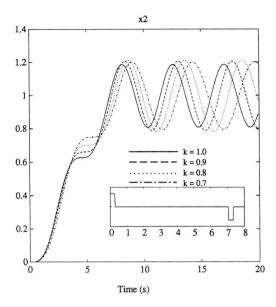

Fig. 10.17 **Flexible body responses to the fuel- and time-optimal control inputs of rigidized case 1 with $\alpha = 10$.**

The rest-to-rest constraint for the rigid-body mode then becomes

$$2 - \phi_{11} \sum_{\substack{j=0,2}}^{N-1} \left[2t_j\Delta_j - \Delta_j^2\right] + \phi_{12} \sum_{\substack{j=1,3}}^{N} \left[2t_j\Delta_j - \Delta_j^2\right] = 0 \qquad (10.89)$$

Substituting Eq. (10.84) into Eq. (10.81) and solving for the time response of the ith flexible mode, we obtain

$$
\begin{aligned}
y_i(t) &= \frac{1}{\omega_i^2} \cos(\omega_i t) \left[-\phi_{i1} \sum_{\substack{j=0,2}}^{N-1} c_{ij} + \phi_{i2} \sum_{\substack{j=1,3}}^{N} c_{ij} \right] \\
&+ \frac{1}{\omega_i^2} \sin(\omega_i t) \left[\phi_{i1} \sum_{\substack{j=0,2}}^{N-1} s_{ij} - \phi_{i2} \sum_{\substack{j=1,3}}^{N} s_{ij} \right]
\end{aligned} \qquad (10.90)
$$

where

$$c_{ij} = \cos(\omega_i t_j) - \cos[\omega_i(t_j + \Delta_j)]$$
$$s_{ij} = \sin(\omega_i t_j) - \sin[\omega_i(t_j + \Delta_j)]$$

for $t \geq t_f$.

Also, rest-to-rest maneuvering requires $y_i(t) = 0$ for $t \geq t_f$; i.e., we have the following flexible mode constraints for no-residual structural vibration:

$$-\phi_{i1} \sum_{j=0,2}^{N-1} c_{ij} + \phi_{i2} \sum_{j=1,3}^{N} c_{ij} = 0 \qquad (10.91a)$$

$$\phi_{i1} \sum_{j=0,2}^{N-1} s_{ij} - \phi_{i2} \sum_{j=1,3}^{N} s_{ij} = 0 \qquad (10.91b)$$

for each flexible mode.

The fuel- and time-optimal control problem can then be formulated as a constrained parameter optimization problem as follows:

$$\min J = t_f + \alpha \sum_{j=0,1,2}^{N} \Delta_j \qquad (10.92)$$

subject to the constraints given by Eqs. (10.88), (10.89), and (10.91).

10.4.1.4 Case 1 Consider case 1, in which two control inputs u^+ and u^- are both acting on body 1. The nominal parameters are $m_1 = m_2 = k = 1$ with appropriate units, and time is in seconds. The corresponding matrices Ω^2 and Φ in Eq. (10.81) are

$$\Omega^2 = \begin{bmatrix} 0 & 0 \\ 0 & 2 \end{bmatrix}, \qquad \Phi = \frac{1}{2} \begin{bmatrix} 1 & 1 \\ 1 & 1 \end{bmatrix}$$

The control inputs are assumed to have one pulse per jet with the same pulse width Δ. The maneuver time is $t_f = t_1 + \Delta$, where t_1 is defined as in Fig. 10.16. The constraint equations for a rest-to-rest maneuver with $y_1(t_f) = 1$ are

$$t_f - \Delta - 2/\Delta = 0 \qquad (10.93a)$$

$$\cos(\omega t_f/2) - \cos[\omega(t_f/2 - \Delta)] = 0 \qquad (10.93b)$$

where $\omega = \sqrt{2}$ rad/s (period $= 4.44288$ s).

The fuel- and time-optimal solution for $\alpha = 1$ can be obtained by solving the constrained minimization problem

$$\min J = t_f + 2\Delta$$

subject to the constraints. A solution to this problem can be obtained as

$$\Delta = 0.45016$$

$$t_1 = 4.44288$$

$$t_f = 4.89304$$

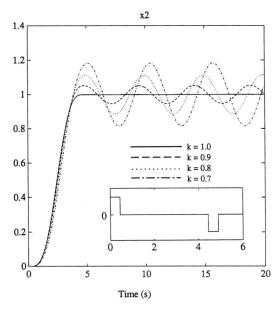

Fig. 10.18   Fuel- and time-optimal control of case 1 with $\alpha = 1$.

The time responses of x_2 (the position of body 2) to the bang-off-bang-type, fuel- and time-optimal control inputs are shown in Fig. 10.18 for four different values of spring stiffness k. For the nominal case with $k = 1$, the flexible mode is perfectly controlled and the jet on-time is now significantly reduced by 79%, whereas the maneuvering time is increased by only 18%, compared to the time-optimal control solution of this case. (The time-optimal solution for case 1 has both the maneuvering time and control on-time of 4.128 s.) It can be seen in Fig. 10.18, however, that the responses are quite sensitive to variations in the model parameter k. Significant residual structural vibrations caused by model uncertainty can be seen in this figure.

In an attempt to achieve a maneuver with less excitation of the structural mode, consider a case with a longer maneuver time by increasing the weighting parameter α in the performance index

$$\min J = t_f + 2\alpha \Delta$$

The solution of this problem for $\alpha = 10$ can be obtained as

$$\Delta = 0.225$$
$$t_1 = 8.880$$
$$t_f = 9.105$$

The time responses of x_2 (the position of body 2) to the fuel- and time-optimal control inputs of a longer maneuver time of 9.11 s and a short pulse of 0.225 s are

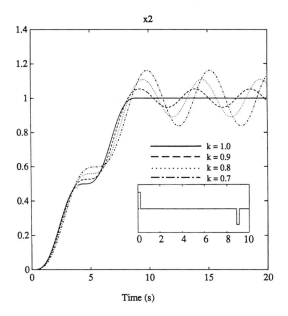

Fig. 10.19   Fuel- and time-optimal control of case 1 with $\alpha = 10$.

shown in Fig. 10.19 for four different values of spring stiffness k. The residual vibrations due to uncertain parameter variations are still significant; that is, simply prolonging the maneuver time and employing short pulses does not help to reduce residual structural mode vibration.

10.4.1.5   Case 2   To explore the effects of using a pair of noncollocated, one-sided jets on the maneuver time and the fuel consumption, consider case 2 shown in Fig. 10.14 with two control inputs u^+ and u^- acting on body 1 and body 2, respectively. The control inputs are assumed to have one pulse per jet and the maneuver time $t_f = t_1 + \Delta$. The corresponding matrices $\mathbf{\Omega}^2$ and $\mathbf{\Phi}$ for this case are

$$\mathbf{\Omega}^2 = \begin{bmatrix} 0 & 0 \\ 0 & 2 \end{bmatrix}, \qquad \mathbf{\Phi} = \frac{1}{2}\begin{bmatrix} 1 & 1 \\ 1 & -1 \end{bmatrix}$$

The constraint equations are

$$t_f - 2/\Delta - \Delta = 0$$
$$\sin(\omega t_f/2) + \sin[\omega(\Delta - t_f/2)] = 0$$

The fuel- and time-optimal solution for $\alpha = 1$ is obtained by solving the constrained minimization problem

$$\min J = t_f + 2\Delta$$

subject to the preceding constraints.

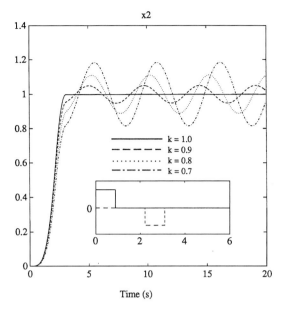

Fig. 10.20   Fuel- and time-optimal control of case 2 with $\alpha = 1$.

A solution is obtained as

$$\Delta = 0.9003$$

$$t_1 = 2.2215$$

$$t_f = 3.1218$$

The time responses of x_2 to this solution are shown in Fig. 10.20 for four different values of spring stiffness k. Compared to the time-optimal solution of case 1, the jet on-time is now significantly reduced by 56%, whereas the maneuvering time is also reduced by 24%. This result indicates some important effect of jet placement in the fuel- and time-optimal control problem. As expected, however, significant residual structural vibrations can be still seen in Fig. 10.20 due to variations in the uncertain parameter k.

10.4.2   Robust Fuel- and Time-Optimal Control

10.4.2.1   Problem Formulation   As illustrated in Figs. 10.19 and 10.20, a standard optimal control approach requires an accurate plant model, and, thus, the resulting solution is not robust to plant modeling uncertainty. To obtain robust, open-loop, fuel- and time-optimal control inputs for flexible spacecraft, we formulate a parameter optimization problem, where the objective function to be minimized is a weighted sum of the consumed fuel and the maneuvering time, with additional constraints for robustness with respect to structural frequency uncertainty.

To attenuate residual vibrations of the flexible modes, the energy of the residual vibrations should be minimized, where the residual vibration energy is proportional to the square of the amplitude. Thus, Eq. (10.90) can be written as

$$y_i(t \geq t_f) = A \sin(\omega_i t) + B \cos(\omega_i t)$$
$$= C \sin(\omega_i t + \text{phase}) \tag{10.94}$$

where A and B are functions of ω_i and ϕ_{ij} and

$$C^2 = A^2 + B^2 \tag{10.95a}$$

$$\frac{dC^2}{dp_i} = \sum_k \frac{dC^2}{d\omega_k} \frac{d\omega_k}{dp_i} \tag{10.95b}$$

where p_i is the ith uncertain parameter. For these derivatives to be zero $(t \geq t_f)$ when ω_i themselves are uncertain parameters, we obtain

$$-\phi_{i1} \sum_{\substack{j=0,2}}^{N-1} g_{ij} + \phi_{i2} \sum_{\substack{j=1,3}}^{N} g_{ij} = 0 \tag{10.96a}$$

$$\phi_{i1} \sum_{\substack{j=0,2}}^{N-1} h_{ij} - \phi_{i2} \sum_{\substack{j=1,3}}^{N} h_{ij} = 0 \tag{10.96b}$$

where

$$g_{ij} = t_j \cos(\omega_i t_j) - (t_j + \Delta_j) \cos[\omega_i(t_j + \Delta_j)] \tag{10.97a}$$
$$h_{ij} = t_j \sin(\omega_i t_j) - (t_j + \Delta_j) \sin[\omega_i(t_j + \Delta_j)] \tag{10.97b}$$

Equations (10.96) are called the first-order robustness constraints. Similarly, the rth order robustness constraints can be expressed as

$$g_{ij} = (t_j)^m \cos(\omega_i t_j) - (t_j + \Delta_j)^m \cos[\omega_i(t_j + \Delta_j)] \tag{10.98a}$$
$$h_{ij} = (t_j)^m \sin(\omega_i t_j) - (t_j + \Delta_j)^m \sin[\omega_i(t_j + \Delta_j)] \tag{10.98b}$$

for $m = 1, 2, \ldots, r$.

These robustness constraints are incorporated in the constrained parameter optimization problem formulation. Consequently, the number of pulses for each control input is changed to match the increased number of constraints. As an example, we consider the first-order robustness constraint, incorporated with the rest-to-rest maneuver constraints, to determine robust fuel- and time-optimal pulse sequences.

10.4.2.2 Case 1

The control inputs are assumed to have two pulses per jet, as defined in Fig. 10.16. The robust fuel- and time-optimal solution can then be obtained by solving the constrained parameter optimization problem

$$\min J = t_3 + \Delta_3 + \alpha(\Delta_0 + \Delta_1 + \Delta_2 + \Delta_3) \tag{10.99}$$

subject to

$$\Delta_0 - \Delta_1 + \Delta_2 - \Delta_3 = 0$$

$$\sum_{j=0}^{3} (-1)^j \left[2t_j \Delta_j + \Delta_j^2 \right] + 4 = 0$$

$$\sum_{j=0}^{3} (-1)^j [\cos \omega t_j - \cos \omega (t_j + \Delta_j)] = 0$$

$$\sum_{j=0}^{3} (-1)^j [\sin \omega t_j - \sin \omega (t_j + \Delta_j)] = 0 \qquad (10.100)$$

$$\sum_{j=0}^{3} (-1)^j [t_j \cos \omega t_j - (t_j + \Delta_j) \cos \omega (t_j + \Delta_j)] = 0$$

$$\sum_{j=0}^{3} (-1)^j [t_j \sin \omega t_j - (t_j + \Delta_j) \sin \omega (t_j + \Delta_j)] = 0$$

$$\Delta_j \geq 0; \qquad j = 0, 1, 2, 3$$

$$t_1, \ t_2, \ t_3 > 0$$

A solution to this problem for $\alpha = 1$ is obtained as

$$
\begin{array}{ll}
t_0 = 0.0, & \Delta_0 = 0.2379 \\
t_1 = 4.1575, & \Delta_1 = 0.2489 \\
t_2 = 2.3732, & \Delta_2 = 0.2489 \\
t_3 = 6.5418, & \Delta_3 = 0.2379
\end{array}
$$

The time responses of x_2 to this robust fuel- and time-optimal control solution are shown in Fig. 10.21 for four different values of k. It can be seen that the robustness with respect to the uncertain parameter variations has been significantly increased and the jet on-time is only 0.97 s compared to the maneuver time of 6.7797 s.

10.4.2.3 *Case 2*   For case 2, we also assume that the control inputs have two pulses per jet. The problem is to minimize the following objective function:

$$\min J = t_3 + \Delta_3 + \alpha(\Delta_0 + \Delta_1 + \Delta_2 + \Delta_3) \qquad (10.101)$$

A solution to this problem for $\alpha = 1$ is obtained as

$$
\begin{array}{ll}
t_0 = 0.0, & \Delta_0 = 0.3762 \\
t_1 = 2.5918, & \Delta_1 = 0.4238 \\
t_2 = 1.9151, & \Delta_2 = 0.4238 \\
t_3 = 4.5545, & \Delta_3 = 0.3762
\end{array}
\qquad (10.102)
$$

The time responses of x_2 to this robust fuel- and time-optimal control solution are shown in Fig. 10.22 for four different values of k. From

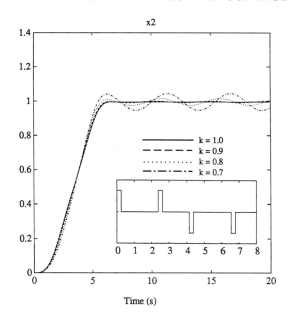

Fig. 10.21   Robust fuel- and time-optimal control of case 1 with $\alpha = 1$.

Fig. 10.22, we notice that a robust fuel- and time-optimal performance has been achieved, compared with Fig. 10.20. Because of the properly coordinated pulse sequences, the flexible modes are not significantly excited during maneuvers and the residual responses after the maneuvers are well desensitized.

10.4.3 Summary

The robust fuel- and time-optimal control problem of flexible spacecraft in the face of modeling uncertainty has been presented in this section. It was shown that it is possible to generate a bang-off-bang type pulse sequence for the case of collocated jets, resulting in less use of fuel with robust time-optimal performance. Contrary to a common notion, the results further confirm the possible existence of on–off pulse sequences for robust fuel- and time-optimal maneuvering of flexible spacecraft. It is emphasized that simply prolonging the maneuver time and employing short pulses does not help to achieve robust performance; a proper coordination of pulse sequences satisfying the robustness constraints is necessary as demonstrated in this section.

Problem

10.1   Consider the three-mass–spring model illustrated in Fig. 10.15 (case 3). The nominal parameter values are assumed as $m_1 = m_2 = m_3 = k_1 = k_2 = 1$ with appropriate units, and time is in seconds. The modal equations for

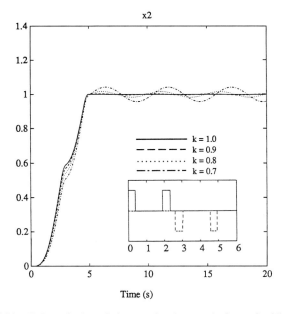

Fig. 10.22 Robust fuel- and time-optimal control of case 2 with $\alpha = 1$.

this case are

$$\ddot{y}_1 = 0.3333(u^+ + u^-)$$

$$\ddot{y}_2 + \omega_2^2 y_2 = 0.5(u^+ + u^-)$$

$$\ddot{y}_3 + \omega_3^2 y_3 = 0.1667(u^+ + u^-)$$

where $\omega_2 = 1$ and $\omega_3 = \sqrt{3}$.

(a) Verify the time responses shown in Fig. 10.23 of the fuel- and time-optimal control solution for a rigidized case with $\alpha = 1$.

(b) Find a fuel- and time-optimal control solution of the flexible model for $\alpha = 1$ as

$$t_1 = 4.536, \qquad \Delta_0 = 0.400$$
$$t_2 = 2.582, \qquad \Delta_1 = 0.135$$
$$t_3 = 6.854, \qquad \Delta_2 = 0.135$$
$$t_f = 7.251, \qquad \Delta_3 = 0.400$$

Also verify the time responses of x_3 shown in Fig. 10.24 for four different values of $k = k_1 = k_2$.

(c) In an attempt to achieve a maneuver with less excitation of structural modes, consider a case with a longer maneuver time by increasing the weight α on the fuel in the cost function. Find a solution of this

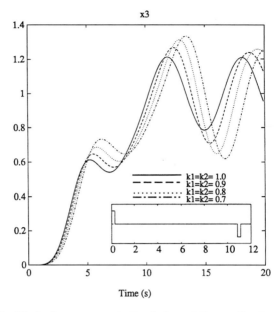

Fig. 10.23　Flexible body responses to the fuel- and time-optimal control inputs of rigidized case 3 with $\alpha = 1$.

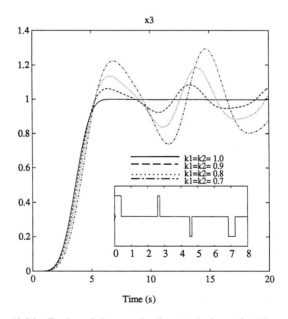

Fig. 10.24　Fuel- and time-optimal control of case 3 with $\alpha = 1$.

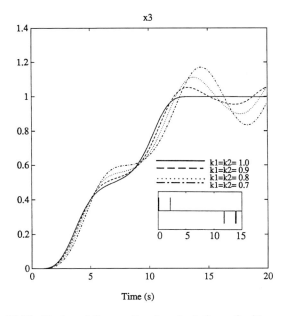

Fig. 10.25   Fuel- and time-optimal control of case 3 with $\alpha = 14$.

problem for $\alpha = 14$ as

$$t_1 = 11.762, \qquad \Delta_0 = 0.153$$
$$t_2 = 2.152, \qquad \Delta_1 = 0.092$$
$$t_3 = 13.854, \qquad \Delta_2 = 0.092$$
$$t_f = 14.007, \qquad \Delta_3 = 0.153$$

with the total jet on-time of 0.49 s.

Also verify the time responses of x_3 shown in Fig. 10.25 to the fuel and time-optimal control inputs. Note that the residual vibrations are still significant and are not reduced by simply increasing the maneuver time.

(d) Find a robust fuel- and time-optimal solution for $\alpha = 1$ as

$$t_1 = 6.6158, \qquad \Delta_0 = 0.1209$$
$$t_2 = 2.3022, \qquad \Delta_1 = 0.1411$$
$$t_3 = 9.0502, \qquad \Delta_2 = 0.2018$$
$$t_4 = 4.7973, \qquad \Delta_3 = 0.2018$$
$$t_5 = 11.433, \qquad \Delta_4 = 0.1411$$
$$t_f = 11.554, \qquad \Delta_5 = 0.1209$$

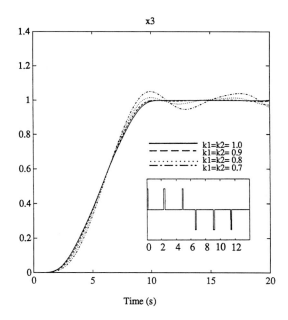

Fig. 10.26   Robust fuel- and time-optimal control of case 3 with $\alpha = 1$.

Note that the jet on-time is only 0.9276 s, compared to the maneuvering time of 11.554 s.

Also verify the time responses of x_3 shown in Fig. 10.26 to the robust fuel- and time-optimal control inputs for four different values of $k = k_1 = k_2$.

Note: Because of the properly coordinated pulse sequences, the flexible modes are not significantly excited during maneuvers and the residual responses after the maneuvers are well desensitized. It should be noted that the preceding solution to case 3 with the switching times truncated to two decimal places also provides robust fuel- and time-optimal responses very similar to those of Fig. 10.26. Responses similar to those of Fig. 10.26 can also be observed for arbitrarily combined variations of k_i and m_i, but with the total mass kept constant ($m_1 + m_2 + m_3 = 3$).

10.5   Robustified Feedforward and Feedback Control

One of the difficult control design objectives of flexible spacecraft is to achieve a fast settling time with minimum structural mode vibrations in the presence of limited actuator capability and plant modeling uncertainty. In the preceding sections, a method was described to precompute the robustified time-optimal, bang-bang control logic for flexible spacecraft subject to plant modeling uncertainties. It is, however, emphasized that one of the primary motivations for the use of closed-loop rather than open-loop control systems in practice is to cope with unexpected disturbances, which an open-loop controller cannot do.

In this section we study a preshaped feedforward command generator and a robustified feedback controller with nonzero set-point command for a reference-input tracking problem. We then consider an alternative approach of control design, which attempts to combine the advantages of the robust H_∞ feedback control technique presented in Chapter 2 and the robust time-optimal, feedforward control approach presented in the preceding sections. The approach taken here is to design a controller that uses robustified open-loop control for fast maneuvering and vibration suppression, followed by robustified feedback control for damping and disturbance rejection. The simple two-mass–spring model shown in Fig. 10.2 is again used to illustrate the control concepts and methodologies developed in Ref. 6.

10.5.1 Preshaped Time-Optimal Control

The so-called impulse-sequence preshaping technique developed by Singer and Seering [7] is briefly described here. It will be shown that the preshaping technique simply utilizes a tapped-delay filter with proper weightings and time delays.

Consider a sequence of m impulses expressed in time domain as

$$f(t) = \sum_{i=1}^{m} A_i \delta(t - t_i) \tag{10.103}$$

with the following normalization:

$$\sum_{i=1}^{m} A_i = 1 \tag{10.104}$$

where A_i is the magnitude of the ith impulse at $t = t_i$ and the last impulse occurs at $t = t_m$.

A bang-bang function with $(n - 2)$ switches can be represented as

$$u(t) = \sum_{j=1}^{n} B_j u_s(t - t_j) \tag{10.105}$$

where B_j is the magnitude of a step function at $t = t_j$. This bang-bang function ends at $t = t_n$.

The convolution of $u(t)$ and $f(t)$ will result in a new multiswitch, multilevel, bang-bang function

$$\tilde{u}(t) = \sum_{i=1}^{n} \sum_{j=1}^{m} A_i B_j u_s(t - t_i - t_j) \tag{10.106}$$

This function has $(mn - 2)$ switching times and ends at $t = (t_m + t_n)$.

A proper sequence of impulses, whose power spectrum has a notch at a structural resonant frequency, can be found as follows. If a sequence of m impulses in

Eq. (10.103) is applied to an undamped second-order system with natural frequency of ω, the system response for $t > t_m$ can be represented as

$$\sum_{i=1}^{m} A_i \omega \sin \omega(t - t_i) = A \sin (\omega t - \phi) \tag{10.107}$$

where

$$A = \sqrt{\left(\sum_{i=1}^{m} A_i \omega \cos \omega t_i\right)^2 + \left(\sum_{i=1}^{m} A_i \omega \sin \omega t_i\right)^2}$$

$$\phi = \tan^{-1}\left\{\frac{\sum_{i=1}^{m} A_i \sin \omega t_i}{\sum_{i=1}^{m} A_i \cos \omega t_i}\right\}$$

If A_i and t_i are chosen such that $A = 0$, i.e.,

$$A_1 \cos \omega t_1 + A_2 \cos \omega t_2 + \cdots + A_m \cos \omega t_m = 0 \tag{10.108a}$$
$$A_1 \sin \omega t_1 + A_2 \sin \omega t_2 + \cdots + A_m \sin \omega t_m = 0 \tag{10.108b}$$

then the residual vibration will not occur after $t = t_m$.

Taking derivatives of the preceding two equations for $(m - 2)$ times with respect to ω, we obtain the following $2(m - 2)$ robustness constraint equations:

$$A_1(t_1)^j \sin \omega t_1 + A_2(t_2)^j \sin \omega t_2 + \cdots + A_m(t_m)^j \sin \omega t_m = 0 \tag{10.109a}$$
$$A_1(t_1)^j \cos \omega t_1 + A_2(t_2)^j \cos \omega t_2 + \cdots + A_m(t_m)^j \cos \omega t_m = 0 \tag{10.109b}$$

where $j = 1, \ldots, m - 2$. For an m-impulse sequence with $t_1 = 0$, we now have $(2m - 1)$ equations for $(2m - 1)$ unknowns.

Figure 10.27 illustrates three different impulse-sequences with proper A_i and the time-delay interval of $\Delta T = \pi/\omega$, where ω is the natural frequency of the flexible mode under consideration. Note that the magnitudes of A_i are independent of ω.

The frequency response characteristics of this impulse-sequence shaping technique can be analyzed simply by taking the Laplace transform of an m-impulse sequence as follows:

$$\mathcal{L}[f(t)] = \sum_{i=1}^{m} A_i e^{-t_i s}$$

$$= \sum_{i=1}^{m} A_i e^{-\Delta T(i-1)s} \tag{10.110}$$

which can be interpreted as a tapped-delay filter as illustrated in Fig. 10.28. The frequency responses of this tapped-delay filter for $m = 2, 3, 4$ and $\omega = \sqrt{2}$ rad/s are shown in Fig. 10.29. It can be seen that the frequency component around the resonant frequency is notched out. The wider notch width indicates more robustness to frequency uncertainty, but a longer response time.

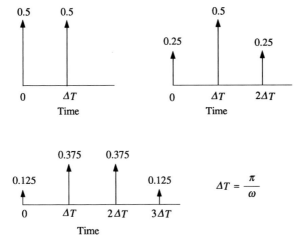

Fig. 10.27   Impulse-sequence shaping.

For example, the flexible-body time-optimal input of the two-mass–spring model (Fig. 10.2) is preshaped using a tapped-delay filter with $m = 3$, resulting in the preshaped control input command

$$\begin{aligned}
\tilde{u}(t) = {} & 0.25u_s(t) - 0.5u_s(t - 1.003) + 0.5u_s(t - 2.109) \\
& + 0.5u_s(t - 2.221) - 0.5u_s(t - 3.215) \\
& - u_s(t - 3.224) + 0.25u_s(t - 4.218) \\
& + u_s(t - 4.330) + 0.25u_s(t - 4.442) \\
& - u_s(t - 5.436) - 0.5u_s(t - 5.445) \\
& + 0.5u_s(t - 6.439) + 0.5u_s(t - 6.551) \\
& - 0.5u_s(t - 7.657) + 0.25u_s(t - 8.660)
\end{aligned} \tag{10.111}$$

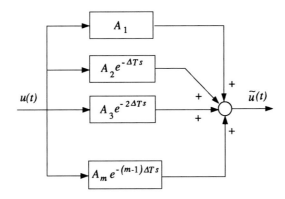

Fig. 10.28   Tapped-delay filter.

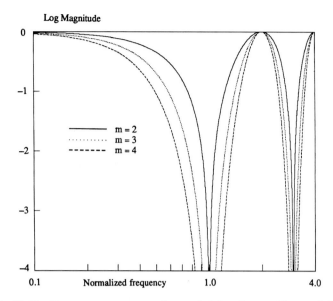

Fig. 10.29   Frequency responses of tapped-delay filters with $m = 2, 3, 4$.

This preshaped input takes values of $\pm 0.25, \pm 0.5$, and ± 0.75, as illustrated in Fig. 10.30. The time responses of the system to this preshaped input are shown in Fig. 10.31 for four different values of k. It is evident that the performance robustness with respect to flexible mode frequency variations has been significantly increased, but at the expense of increased maneuver time of 8.66 s, compared with the ideal minimum time of 4.218 s.

The performance robustness with respect to model parameter variations for the preshaped inputs are also very comparable to that of robustified, time-optimal inputs. However, the preshaped inputs take some intermediate values and switch more times. For more recent advances in this subject [8–15].

10.5.2   Robust Nonzero Set-Point Control

As discussed in Chapter 2, a nonminimum-phase compensation is particularly useful for practical tradeoffs between performance and robustness for a certain class of noncollocated structural control problems. It is, however, often criticized because of its sluggish response and its loop gain limitation. In this section, a robust H_∞ feedback compensator design is discussed with special emphasis on a proper implementation of a nonminimum-phase compensator, incorporating a nonzero set-point control scheme. It is shown that a properly designed feedback controller with a nonzero set-point command performs well compared with a time-optimal, open-loop controller with special preshaping for robustness.

Consider a single-input single-output (SISO) control system as illustrated in Fig. 10.32, which is the most commonly used configuration for a two-degree-of-freedom controller. The plant and compensator transfer functions are

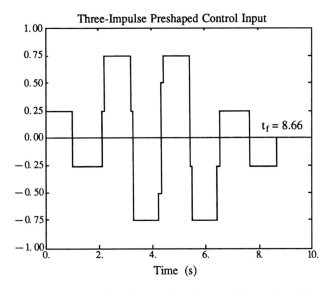

Fig. 10.30   Three-impulse preshaped control input ($m = 3$).

represented as

$$K(s) = \frac{N_c(s)}{D_c(s)} \qquad (10.112a)$$

$$G(s) = \frac{N(s)}{D(s)} \qquad (10.112b)$$

where $N_c(s), D_c(s), N(s)$, and $D(s)$ are polynomials of the Laplace transform variable s. The closed-loop transfer function from the desired output command y^* to the actual output y is then

$$\begin{aligned}
\frac{y(s)}{y^*(s)} &= \frac{K(s)G(s)}{1 + K(s)G(s)} F(s) \\
&= \frac{N_c N}{D_c D + N_c N} F(s) \qquad (10.113)
\end{aligned}$$

Thus, for the conventional feedback control system of Fig. 10.32, the zeros of the closed-loop transfer function are identical with the zeros of the loop transfer function $K(s)G(s)$. These zeros sometimes cause an excessive, transient peak overshoot even when the closed-loop poles are properly selected. In this case, a prefilter $F(s)$ is often used for the cancellation of the undesirable zeros of the closed-loop transfer function. (Of course, the nonminimum-phase zeros cannot be canceled.)

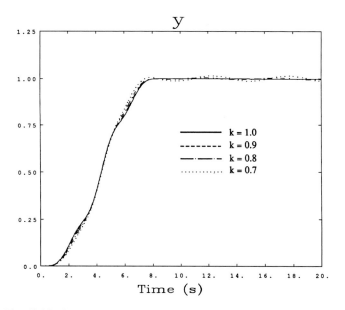

Fig. 10.31   Responses to the three-impulse preshaped control input.

If the compensator is placed in the feedback path, the closed-loop transfer function becomes

$$\frac{y(s)}{y^*(s)} = \frac{G(s)}{1 + K(s)G(s)} F(s)$$

$$= \frac{D_c N}{D_c D + N_c N} F(s) \tag{10.114}$$

where the compensator zeros do not appear as zeros of the closed-loop transfer function and a prefilter $F(s)$ must be properly designed for the generation of a control input command.

Next, we show a proper way of implementing a compensator to minimize such excessive, transient peak overshoot caused by the compensator zeros. A nonzero set-point control scheme for an H_∞-based controller will be presented, followed by an example design.

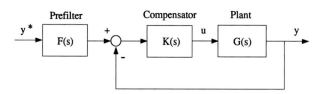

Fig. 10.32   Conventional feedforward/feedback control system configuration.

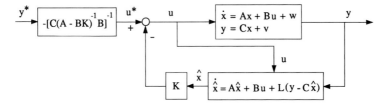

Fig. 10.33 Control system configuration with LQG type controller.

The separation principle of the conventional linear-quadratic-Gaussian (LQG) technique does not hold for an H_∞ controller. Consequently, the nonzero set-point control scheme, which has been well established for LQG control synthesis, needs some minor modification, as is discussed next.

A block diagram representation of a SISO closed-loop system for a conventional LQG type controller with a nonzero set-point command is illustrated in Fig. 10.33. For this configuration, the control input command u^* corresponding to the desired (constant) output command y^* is simply given as

$$u^* = -[C(A - BK)^{-1}B]^{-1}y^* \qquad (10.115)$$

which is independent of the estimator gain matrix L. In this case, it can be shown that for dynamic systems having a rigid-body mode, u^* depends only on the regulator parameters (not on the plant parameters such as m_1, m_2, and k of the example model shown in Fig. 10.2). Hence, the nonzero set-point control scheme is inherently robust to plant parameter uncertainty for a certain class of dynamic systems with at least one pole at the origin, i.e., a type one system.

Now consider a closed-loop control system with an H_∞ controller as shown in Fig. 10.34, which is described in state-space form as

$$\begin{bmatrix} \dot{x} \\ \dot{x}_c \end{bmatrix} = \begin{bmatrix} A & -B_2K \\ LC_2 & A_c \end{bmatrix} \begin{bmatrix} x \\ x_c \end{bmatrix} + \begin{bmatrix} B_2 \\ B_2 \end{bmatrix} u^*$$

$$y = [C_2 \quad 0] \begin{bmatrix} x \\ x_c \end{bmatrix} \qquad (10.116)$$

An H_∞ suboptimal controller that satisfies $\|T_{zw}\|_\infty < \gamma$, where γ is a design variable specifying an upper bound of the perturbed closed-loop performance T_{zw}, can be obtained as

$$u(s) = -K[sI - A_c]^{-1}Ly(s) \qquad (10.117)$$

or

$$\dot{x}_c = A_c x_c + Ly \qquad (10.118a)$$

$$u = -Kx_c \qquad (10.118b)$$

where

$$K = B_2^T X \qquad (10.119a)$$

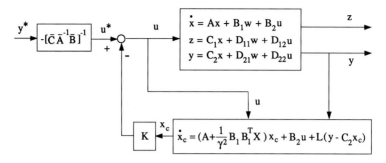

Fig. 10.34 Control system configuration with H_∞ type controller.

$$L = \left(I - \frac{1}{\gamma^2}YX\right)^{-1} YC_2^T \qquad (10.119b)$$

$$A_c = A + \frac{1}{\gamma^2}B_1B_1^TX - B_2K - LC_2 \qquad (10.119c)$$

The input command u^* corresponding to the desired output y^* can be simply found as

$$u^* = -[\bar{C}\bar{A}^{-1}\bar{B}]^{-1}y^* \qquad (10.120)$$

where

$$\bar{A} = \begin{bmatrix} A & -B_2K \\ LC_2 & A_c \end{bmatrix}, \qquad \bar{B} = \begin{bmatrix} B_2 \\ B_2 \end{bmatrix}$$

$$\bar{C} = [C_2 \quad 0] \qquad (10.121)$$

Similar to the LQG case, it can be shown that for dynamic systems having a rigid-body mode, u^* for an H_∞ controller depends only on the controller parameters (not on the plant parameters such as m_1, m_2, and k of the example model shown in Fig. 10.2). However, u^* now depends on both the gain matrices K and L, not just on the regulator gain matrix K as it does for the LQG case.

We now consider the two-mass–spring model shown in Fig. 10.2 with nominal values of $m_1 = m_2 = k = 1$. A control input force u acts on body 1 and the position of body 2 is measured as y. The output command is $y^* = 1$ and the control input is bounded as $|u| \leq 1$.

A robust H_∞ controller with the following gain matrices K and L are considered here

$$K = [1.506 \quad -0.494 \quad 1.738 \quad 0.932] \qquad (10.122a)$$

$$L = [0.720 \quad 2.973 \quad -3.370 \quad 4.419]^T \qquad (10.122b)$$

which result in a nonminimum-phase compensator.

The time response of the closed-loop system implemented as in Fig. 10.32, for $F(s) = 1$ and $y^* = 1$, is shown in Fig. 10.35. A nonminimum-phase behavior of

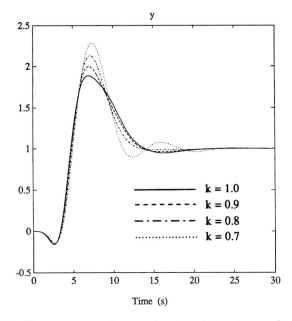

Fig. 10.35   Time responses of conventional control system configuration.

the closed-loop system is evident and the nominal system has a peak overshoot of about 80% and a settling time of 15 s. When compared with the response of a feedforward controller, shown in Fig. 10.31, the overall response is not acceptable. It can be shown that the excessive overshoot is due to the compensator zero at $s = -0.145$. This zero may be canceled by a prefilter $F(s)$, but the resulting slower settling time may not be desirable.

Figure 10.36 shows the closed-loop responses for four different values of k to an output command of $y^* = 1$ (consequently, $u^* = 0.9959$), when the same controller is implemented as in Fig. 10.34. Clearly, the responses no longer have excessive overshoot and the settling time is quite short, as compared with the response in Fig. 10.35. The overall responses are also comparable with those of a three-impulse preshaped feedforward controller, shown in Fig. 10.31. The control input $u(t)$ is always within the saturation limit of one.

It may be concluded that a feedback controller, when implemented properly, could achieve good performance and robustness, for both command following as well as disturbance rejection problems. The feedforward/feedback control approach described in this section is robust for a certain class of uncertain dynamic systems, because the control input command computed for a given desired output does not depend on the plant parameters.

10.5.3 Summary

In this section, both feedforward and feedback control approaches for rapid maneuvering control of uncertain flexible spacecraft have been presented and compared. It was shown that a time-optimal control input, preshaped using

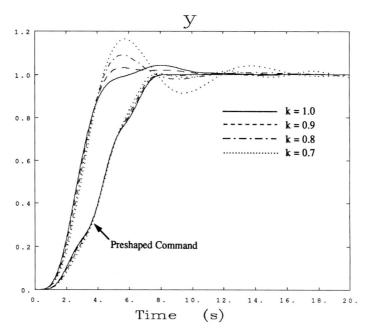

Fig. 10.36   Time responses of nonzero set-point control configuration.

a tapped-delay filter, provides a rapid maneuver and robust suppression of residual structural vibrations. A proper implementation of a nonminimum-phase compensator with a nonzero set-point control command was discussed. It was demonstrated that a properly implemented feedback controller performs well, when compared with a time-optimal, open-loop controller with special preshaping for performance robustness. To achieve a fast settling time for a constant output command and robust performance with respect to plant modeling uncertainty, a robustified feedforward/feedback control approach has been presented.

References

[1] Liu, Q., "Robust Feedforward/Feedback Control Synthesis for Uncertain Dynamical Systems," Ph.D. Thesis, Dept. of Mechanical and Aerospace Engineering, Arizona State Univ., Tempe, AZ, Dec. 1992.

[2] Liu, Q., and Wie, B., "Robust Time-Optimal Control of Uncertain Flexible Spacecraft," *Journal of Guidance, Control, and Dynamics*, Vol. 15, No. 3, 1992, pp. 597–604.

[3] Wie, B., Sinha, R., and Liu, Q., "On Actuator Placement for Robust Time-Optimal Control of Uncertain Flexible Spacecraft," *Proceedings of the 1992 AIAA Guidance, Navigation, and Control Conference*, AIAA, Washington, DC, 1992, pp. 1352–1360.

[4] Wie, B., Sinha, R., and Liu, Q., "Robust Time–Optimal Control of Uncertain Structural Dynamic Systems," *Journal of Guidance, Control, and Dynamics*, Vol. 16, No. 5, 1993, pp. 980–982.

[5] Wie, B., Sinha, R., Sunkel, J., and Cox, K., "Robust Fuel- and Time-Optimal Control of Uncertain Flexible Space Structures," *Proceedings of the 1993 AIAA Guidance, Navigation, and Control Conference*, AIAA, Washington, DC, 1993, pp. 939–948.

[6] Wie, B., and Liu, Q., "Comparison Between Robustified Feedforward and Feedback for Achieving Parameter Robustness," *Journal of Guidance, Control, and Dynamics*, Vol. 15, No. 4, 1992, pp. 935–943.

[7] Singer, N. C., and Seering, W. P., "Preshaping Command Inputs to Reduce System Vibration," *Journal of Dynamic Systems, Measurements and Control*, Vol. 112, March 1990, pp. 76–82.

[8] Singh, T., and Vadali, S. R., "Input-Shaped Control of Three-Dimensional Maneuvers of Flexible Spacecraft," *Journal of Guidance, Control, and Dynamics*, Vol. 16, No. 6, 1993, pp. 1061–1068.

[9] Singh, T., and Vadali, S. R., "Robust Time-Optimal Control: A Frequency-Domain Approach," *Journal of Guidance, Control, and Dynamics*, Vol. 17, No. 2, 1994, pp. 346–353.

[10] Singh, T., "Fuel/Time Optimal Control of the Benchmark Problem," *Journal of Guidance, Control, and Dynamics*, Vol. 18, No. 6, 1995, pp. 1225–1231.

[11] Singh, T., and Ali, H., "Exact Time-Optimal Control of the Wave Equation," *Journal of Guidance, Control, and Dynamics*, Vol. 19, No. 1, 1996, pp. 130–134.

[12] Singhose, W., Derezinski, S., and Singer, N., "Extra-Insensitive Input Shapers for Controlling Flexible Spacecraft," *Journal of Guidance, Control, and Dynamics*, Vol. 19, No. 2, 1996, pp. 385–391.

[13] Singhose, W., Bohlke, K., and Seering, W., "Fuel-Efficient Pulse Command Profiles for Flexible Spacecraft," *Journal of Guidance, Control, and Dynamics*, Vol. 19, No. 4, 1996, pp. 954–960.

[14] Singhose, W. E., Banerjee, A. K., and Seering, W. P., "Slewing Flexible Spacecraft with Deflection-Limiting Input Shaping," *Journal of Guidance, Control, and Dynamics*, Vol. 20, No. 2, 1997, pp. 291–298.

[15] Song, G., Buck, N., and Agrawal, B., "Spacecraft Vibration Reduction Using Pulse-Width Pulse-Frequency Modulated Input Shaper," *Proceedings of the 1997 AIAA Guidance, Navigation, and Control Conference*, AIAA Reston, VA, 1997, pp. 1535–1549.

Part 5
Advanced Spacecraft Dynamics and Control

11
Control Moment Gyros for Agile Imaging Satellites

The fundamental principles of control moment gyros (CMGs) applied to space-craft attitude control were briefly treated in Sec. 7.5. A CMG is a powerful torque amplification actuator; however, the redundant CMG systems have an inherent geometric singularity problem. This chapter, which is based on [1–7], presents a comprehensive treatment of the CMG singularity problem. It also presents practical CMG steering algorithms that provide a simple means of avoiding or escaping troublesome, internal elliptic singularities commonly encountered by most pseudo-inverse-based steering logic. This chapter also introduces practical feedback control logic for large-angle, rapid multitarget acquisition and pointing control of agile imaging satellites in the presence of CMG hardware constraints.

11.1   Introduction

The next-generation Earth imaging satellites will require rapid rotational agility as well as precision steady-state pointing accuracy for high-resolution images. Rather than sweep a gimbaled imaging system from side to side, the spacecraft body will turn rapidly. Pointing the entire spacecraft body allows the body-fixed imaging system to achieve a higher definition and improves the resolution for its images. Because the overall cost and effectiveness of agile imaging satellites are greatly affected by the average retargeting time, the development of an agile attitude control system employing CMGs is of current practical importance.

For example, the Pleiades High-Resolution (HR) imaging satellite has a slew rate requirement of 4 deg/s. The solar arrays are mounted directly on the bus without any solar-array drive mechanism to ensure a maximum structural rigidity. Their stiffness is further increased by the use of Carpentier joints when deployed, resulting in a lowest structural frequency of about 120 Hz. As described in [8–10], the Pleiades-HR imaging satellite is equipped with four small single-gimbal CMGs of a pyramid arrangement with a skew angle of 30 deg. Each CMG has an angular momentum of 15 N·m·s, a maximum gimbal rate of 3 rad/s, a peak output torque of 45 N·m, and an average maximum torque of 20 N·m. The 1000-kg Pleiades-HR imaging satellite with a roll/pitch inertia of approximately 800 kg·m^2 will be able to perform a 60-deg roll (cross-track) slew maneuver in less than 25 s. It will provide a precision steady-state pointing accuracy of ±0.03 deg during its planned imaging phase after completing the slew maneuver. However,

667

Fig. 11.1   BCP 5000 (WorldView 1) equipped with single-gimbal CMGs. (Image courtesy of Digital Globe.)

during large-angle cross-track slew maneuvers, precision pointing of the spacecraft body is not required.

Ball Aerospace and Technologies Corporation, teamed with Digital Globe, has been developing an agile imaging satellite bus, called BCP 5000, to accommodate the next-generation optical and synthetic aperture radar (SAR) remote-sensing payloads. For increased spacecraft agility, the BCP 5000 is equipped with single-gimbal CMGs to provide excellent re-targeting capability. Successfully launched in 2007, the 2500-kg BSP 5000 (WorldView 1) imaging satellite, illustrated in Fig. 11.1, will become the first commercial high-resolution imaging satellite equipped with the state-of-the-art single-gimbal CMGs. It will be capable of collecting images with 0.5-m panchromatic resolution and 2-m multispectral resolution.*

The BCP 5000 (WorldView 2) satellite, illustrated in Fig. 11.2, is scheduled to be launched in 2009. It will incorporate the industry standard four multispectral bands (red, blue, green, and near-infrared) and will also include four new bands (coastal, yellow, red edge, and near-infrared 2). The 2800-kg WorldView 2 with a 110-cm aperture telescope, flying at a higher altitude of 770 km, will provide the same panchromatic 0.5-m resolution imagery as WorldView 1, in addition to 1.8-m multispectral resolution imagery.

For future scientific missions such as near-Earth object rendezvous and high-resolution Earth observation using agile small satellites (<300 kg) with a slew rate requirement of 6 deg/s, a single-gimbal CMG (smaller than Astrium's CMG 15-45S) has been being developed at Surrey Space Centre [7,11].

A simple block diagram representation of a CMG-based attitude control system of agile imaging satellites is illustrated in Fig. 11.3. This chapter is intended to

---

*http://www.skyrocket.de/space/doc_sat/ball_bcp-5000.htm.

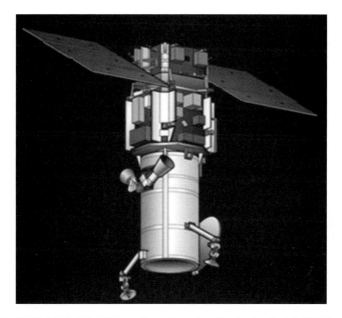

Fig. 11.2 BCP 5000 (WorldView 2) equipped with single-gimbal CMGs. (Image courtesy of Digital Globe.)

provide the reader with a comprehensive treatment of the analysis and design problem of such a CMG-based attitude control system for agile imaging satellites.

11.2 Single-Gimbal CMG Systems

This section provides a summary of a few representative single-gimbal CMG systems. These CMG systems will be used as illustrative examples throughout this chapter to illustrate the various concepts and approaches useful for characterizing and analyzing the CMG singularities.

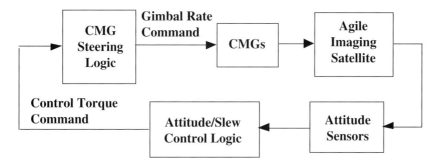

Fig. 11.3 A CMG-based attitude control system of agile imaging satellites.

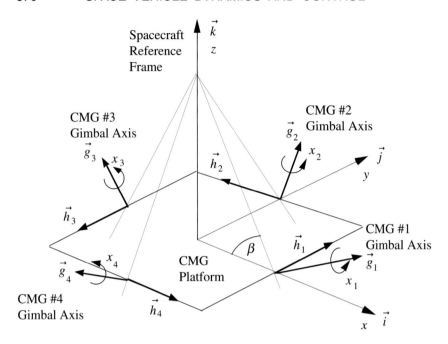

Fig. 11.4   Pyramid mounting arrangement of four single-gimbal CMGs.

11.2.1   Pyramid Array of Four Single-Gimbal CMGs

A typical pyramid array of four single-gimbal CMGs is illustrated in Fig. 11.4. The total angular momentum vector $\vec{H}$ of four single-gimbal CMGs is simply given by

$$\vec{H} = \vec{h}_1 + \vec{h}_2 + \vec{h}_3 + \vec{h}_4 \tag{11.1}$$

where $\vec{h}_i$ is the angular momentum vector of the ith CMG and it is assumed that $|\vec{h}_i| = 1$ without loss of generality. The ith gimbal-angle x_i describes rotation of $\vec{h}_i$ about the normalized gimbal-axis vector $\vec{g}_i$ ($|\vec{g}_i| = 1$); that is, $\vec{g}_i \cdot \vec{h}_i = 0$, and $\vec{h}_i = \vec{h}_i(x_i)$.

For the pyramid mount of four single-gimbal CMGs with skew angle of β as illustrated in Fig. 11.4, the gimbal-axis vectors can be simply represented as $\vec{g}_1 = \sin \beta \vec{i} + \cos \beta \vec{k}$, $\vec{g}_2 = \sin \beta \vec{j} + \cos \beta \vec{k}$, $\vec{g}_3 = -\sin \beta \vec{i} + \cos \beta \vec{k}$, and $\vec{g}_4 = -\sin \beta \vec{j} + \cos \beta \vec{k}$.

The total CMG momentum vector $\vec{H}$ is often expressed in a spacecraft reference frame (x, y, z) with a set of orthogonal unit vectors $\{\vec{i}, \vec{j}, \vec{k}\}$, as follows:

$$\vec{H} = H_x i + H_y j + H_z k$$

$$= \begin{bmatrix} i & j & k \end{bmatrix} \begin{bmatrix} H_x \\ H_y \\ H_z \end{bmatrix} = \begin{bmatrix} i & j & k \end{bmatrix} \mathbf{H} \tag{11.2}$$

where $\mathbf{H} = (H_x, H_y, H_z) \equiv [H_x \ H_y \ H_z]^T$ is the representation of the vector $\vec{H}$ with respect to the basis vectors $\{\vec{i}, \vec{j}, \vec{k}\}$. Although the column vector $\mathbf{H}$ should be distinguished from the vector $\vec{H}$ itself, it is often called a vector. However, the meaning should be clear from the context, and in general we must be clear on what is meant by a vector.

For the pyramid mount of four single-gimbal CMGs with skew angle of β, the total CMG momentum vector can be represented in matrix form as

$$\mathbf{H} = \mathbf{h}_1(x_1) + \mathbf{h}_2(x_2) + \mathbf{h}_3(x_3) + \mathbf{h}_4(x_4)$$

$$= \begin{bmatrix} -c\beta \sin x_1 \\ \cos x_1 \\ s\beta \sin x_1 \end{bmatrix} + \begin{bmatrix} -\cos x_2 \\ -c\beta \sin x_2 \\ s\beta \sin x_2 \end{bmatrix} + \begin{bmatrix} c\beta \sin x_3 \\ -\cos x_3 \\ s\beta \sin x_3 \end{bmatrix} + \begin{bmatrix} \cos x_4 \\ c\beta \sin x_4 \\ s\beta \sin x_4 \end{bmatrix} \quad (11.3)$$

where $c\beta \equiv \cos \beta$, $s\beta \equiv \sin \beta$. Note that $\mathbf{h}_i$ are periodic with a period of 2π and that

$$\frac{d^2 \mathbf{h}_i}{dx_i^2} = -\mathbf{h}_i \quad \text{or} \quad \frac{d^2 \vec{h}_i}{dx_i^2} = -\vec{h}_i \quad (11.4)$$

The differential of $\mathbf{H}$ becomes

$$dH = \mathbf{f}_1 dx_1 + \mathbf{f}_2 dx_2 + \mathbf{f}_3 dx_3 + \mathbf{f}_4 dx_4$$

$$= \mathbf{A} d\mathbf{x} \quad (11.5)$$

where $d\mathbf{x} = (dx_1, dx_2, dx_3, dx_4)$, and $\mathbf{A}$ is the Jacobian matrix defined as

$$\mathbf{A} = \begin{bmatrix} \mathbf{f}_1 & \mathbf{f}_2 & \mathbf{f}_3 & \mathbf{f}_4 \end{bmatrix}$$

$$= \begin{bmatrix} -c\beta \cos x_1 & \sin x_2 & c\beta \cos x_3 & -\sin x_4 \\ -\sin x_1 & -c\beta \cos x_2 & \sin x_3 & c\beta \cos x_4 \\ s\beta \cos x_1 & s\beta \cos x_2 & s\beta \cos x_3 & s\beta \cos x_4 \end{bmatrix} \quad (11.6)$$

Equation (11.5) represents a linear mapping from $d\mathbf{x} = (dx_1, dx_2, dx_3, dx_4)$ to $d\mathbf{H} = (dH_x, dH_y, dH_z)$. Consequently, we obtain

$$\dot{\mathbf{H}} = \mathbf{A}\dot{\mathbf{x}} \quad (11.7)$$

where $\dot{\mathbf{H}} = d\mathbf{H}/dt$ and $\dot{\mathbf{x}} = d\mathbf{x}/dt$.

11.2.2 Two-Scissored-Pair-Ensemble, Explicit-Distribution Single-Gimbal CMG System

A special case with $\beta = \pi/2$ was employed by Crenshaw [12] for the so-called 2-SPEED (two-scissored-pair-ensemble, explicit-distribution) single-gimbal CMG control system. For this configuration, we simply have two orthogonal pairs of two parallel single-gimbal CMGs with a Jacobian matrix of the form

$$\mathbf{A} = \begin{bmatrix} 0 & \sin x_2 & 0 & -\sin x_4 \\ -\sin x_1 & 0 & \sin x_3 & 0 \\ \cos x_1 & \cos x_2 & \cos x_3 & \cos x_4 \end{bmatrix} \quad (11.8)$$

This special configuration is also of practical importance. Many other CMG configurations are in fact some variants of this basic arrangement of two orthogonal pairs of two parallel CMGs, known as the 2-SPEED CMG system in the literature.

11.2.3 Two and Three Parallel Single-Gimbal CMG Configurations

Two and three single-gimbal CMG configurations with parallel gimbal axes have also been studied for two-axis control applications in [13,14]. Consider first a case with only two CMGs without redundancy. The momentum vectors $\vec{H}_1$ and $\vec{H}_2$ move in the (x, y) plane normal to the gimbal axis, as shown in Fig. 11.5. For such scissored single-gimbal CMGs, the total CMG momentum vector can be

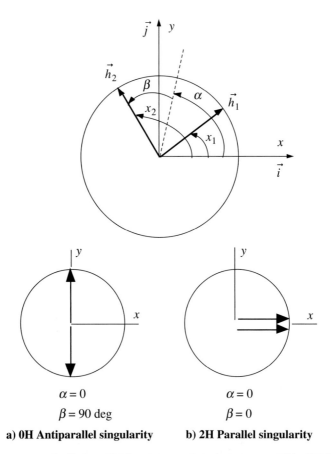

a) 0H Antiparallel singularity   b) 2H Parallel singularity

Fig. 11.5 Two single-gimbal CMGs with parallel gimbal axes. This CMG system remains singular for any α motion when $\beta = 0$ or 90 deg.

represented in matrix form as

$$\mathbf{H} = \begin{bmatrix} \cos x_1 + \cos x_2 \\ \sin x_1 + \sin x_2 \end{bmatrix} \tag{11.9}$$

where a constant unit momentum for each CMG is assumed.

Defining a new set of gimbal-angles (α, β) as

$$\alpha = \frac{x_1 + x_2}{2}, \quad \beta = \frac{x_2 - x_1}{2} \tag{11.10}$$

where α is called the "rotation" angle and β the "scissor" angle [12], we can express the CMG momentum vector as

$$\mathbf{H} = 2 \begin{bmatrix} \cos \alpha \; \cos \beta \\ \sin \alpha \; \cos \beta \end{bmatrix} \tag{11.11}$$

and we have

$$\dot{\mathbf{H}} = \mathbf{A}\dot{\mathbf{x}} \tag{11.12}$$

where $\dot{\mathbf{H}} = (\dot{H}_x, \dot{H}_y)$, $\dot{\mathbf{x}} = (\dot{\alpha}, \dot{\beta})$, and $\mathbf{A}$ is the Jacobian matrix defined as

$$\mathbf{A} = 2 \begin{bmatrix} -\sin \alpha \cos \beta & -\cos \alpha \sin \beta \\ \cos \alpha \cos \beta & -\sin \alpha \sin \beta \end{bmatrix} \tag{11.13}$$

For a system of three single-gimbal CMGs with parallel gimbal axes, the total CMG angular momentum vector is given by

$$\mathbf{H} = \begin{bmatrix} \cos x_1 + \cos x_2 + \cos x_3 \\ \sin x_1 + \sin x_2 + \sin x_3 \end{bmatrix} \tag{11.14}$$

and the Jacobian matrix for $\mathbf{x} = (x_1, x_2, x_3)$ is

$$\mathbf{A} = \begin{bmatrix} -\sin x_1 & -\sin x_2 & -\sin x_3 \\ \cos x_1 & \cos x_2 & \cos x_3 \end{bmatrix} \tag{11.15}$$

11.3   CMG Singularities and Singular Surfaces

This section briefly introduces a method developed by Margulies and Aubrun [13] for analyzing and visualizing the singular momentum surfaces. Illustrative examples are presented with some significant new results.

Similar to $\vec{H}$ expressed as in Eq. (11.2), an arbitrary vector $\vec{u}$ can also be represented as

$$\vec{u} = u_x \vec{i} + u_y \vec{j} + u_z \vec{k}$$

$$= \begin{bmatrix} \vec{i} & \vec{j} & \vec{k} \end{bmatrix} \begin{bmatrix} u_x \\ u_y \\ u_z \end{bmatrix} = \begin{bmatrix} \vec{i} & \vec{j} & \vec{k} \end{bmatrix} \mathbf{u} \tag{11.16}$$

where $\mathbf{u} = (u_x, u_y, u_z) \equiv [u_x \; u_y \; u_z]^T$.

As introduced by Margulies and Aubrun [13], let $\vec{u}$ be a unit vector of the punctured unit sphere defined as

$$S = \{\vec{u} : |\vec{u}| = 1, \vec{u} \neq \pm\vec{g}_i, i = 1, \ldots, n\}$$

where $\vec{g}_i$ is the gimbal-axis vector ($|\vec{g}_i| = 1$). Such a unit vector along all possible directions in three-dimensional space (except along the gimbal-axis directions) can be parameterized as

$$\vec{u} = u_x\vec{i} + u_y\vec{j} + u_z\vec{k}$$
$$= \sin\theta_2\vec{i} - \sin\theta_1\cos\theta_2\vec{j} + \cos\theta_1\cos\theta_2\vec{k} \qquad (11.17)$$

where θ_1 and θ_2 are the rotation angles of two successive rotations about the x and y axes. The longitude and latitude angles of spherical coordinates, commonly used for a unit vector description along all possible directions in three-dimensional space, were found to be numerically ill suited for visualization of the singular surfaces. (Note that such a specific representation of $\vec{u}$ as Eq. (11.17) was not provided in [13].)

For a system of n single-gimbal CMGs, the differential of the total CMG momentum vector becomes

$$d\vec{H} = \sum_{i=1}^{n} d\vec{h}_i = \sum_{i=1}^{n} \frac{d\vec{h}_i}{dx_i}dx_i = \sum_{i=1}^{n} \vec{f}_i dx_i \qquad (11.18)$$

where $\vec{f}_i$ are unit tangent vectors defined as

$$\vec{f}_i = \vec{f}_i(x_i) = \frac{d\vec{h}_i}{dx_i} = \vec{g}_i \times \vec{h}_i \qquad (11.19)$$

Note that $d\vec{H}$ and $\vec{f}_i$ are the equivalent vector representations of $d\mathbf{H}$ and $\mathbf{f}_i$, respectively, used in Eq. (11.5). The three vectors $\{\vec{f}_i, \vec{g}_i, \vec{h}_i\}$ form a set of orthogonal unit vectors rotating about each gimbal axis $\vec{g}_i$. In Margulies and Aubrun [13], this set of orthonormal vectors plays a major role in developing the geometric theory of redundant single-gimbal CMGs.

The $3 \times n$ Jacobian matrix $\mathbf{A}$, introduced in the preceding section, has maximum rank of 3. When the gimbal axes are not arranged to be coplanar, the minimum rank of $\mathbf{A}$ is two. However, when rank($\mathbf{A}$) = 2, all various $\vec{f}_i$ become coplanar, and there exists a unit vector $\vec{u}$ normal to that plane, that is,

$$\vec{f}_i(x_i) \cdot \vec{u} \equiv \mathbf{f}_i^T\mathbf{u} = 0 \qquad i = 1, \ldots, n \qquad (11.20)$$

Consequently, for such a case, we have

$$d\vec{H} \cdot \vec{u} = 0 \qquad (11.21)$$

and the CMG array cannot produce any momentum vector change (or torque) along the direction of $\vec{u}$ regardless of the gimbal rates. Such a unit vector $\vec{u}$ is called the *singular vector*, and a set of gimbal-angles when $\text{rank}(\mathbf{A}) = 2$ is called the *singular gimbal-angles*.

Because $\vec{f_i}$ is also orthogonal to $\vec{g_i}$ and $|\vec{f_i}| = 1$, the singularity condition (11.20) can be rewritten as

$$\vec{f_i} = \pm \frac{\vec{g_i} \times \vec{u}}{|\vec{g_i} \times \vec{u}|} \tag{11.22}$$

Because $\vec{h_i} = \vec{f_i} \times \vec{g_i}$, the singularity condition for $\vec{h_i}(x_i)$ can also be written as

$$\vec{h_i} = \pm \frac{(\vec{g_i} \times \vec{u}) \times \vec{g_i}}{|\vec{g_i} \times \vec{u}|} \tag{11.23}$$

which leads to the following inner product of $\vec{h_i}$ and $\vec{u}$:

$$e_i = \vec{h_i} \cdot \vec{u} = \pm |\vec{g_i} \times \vec{u}| \equiv \pm \sqrt{1 - (\vec{g_i} \cdot \vec{u})^2} \tag{11.24}$$

The *singular momentum vector*, corresponding to the singular vector $\vec{u}$ and the singular gimbal-angles $\mathbf{x}$, is expressed as

$$\vec{H} = \vec{H}\left[\mathbf{x}(\vec{u})\right] = \vec{H}(\vec{u})$$

$$= \sum_i \frac{1}{e_i}(\vec{g_i} \times \vec{u}) \times \vec{g_i}$$

$$= \sum_i \frac{\epsilon_i[\vec{u} - \vec{g_i}(\vec{g_i} \cdot \vec{u})]}{\sqrt{1 - (\vec{g_i} \cdot \vec{u})^2}} \tag{11.25}$$

where $\epsilon_i = \text{sign}(e_i) = \text{sign}(\vec{h_i} \cdot \vec{u}) = \pm 1$. Note that there are 2^n combinations of ϵ_i for a cluster of n CMGs. See Margulies and Aubrun [13] for a detailed discussion of these 2^n combinations.

Because $\vec{u} = \vec{u}(\theta_1, \theta_2)$, we have $\vec{H} = \vec{H}(\theta_1, \theta_2)$; that is, $\vec{H}$ is also parameterized by θ_1 and θ_2. Singular momentum surfaces can be directly obtained using Eq. (11.25) without recourse to singular gimbal angles.

Example 11.1 Two Single-Gimbal CMGs

The singular momentum vector of the two single-gimbal CMGs with parallel gimbal axes (see Fig. 11.5) is given by

$$\vec{H} = \sum_{i=1}^{2} \frac{1}{e_i}(\vec{g_i} \times \vec{u}) \times \vec{g_i}$$

$$= \epsilon_1 \vec{u} + \epsilon_2 \vec{u} = 0 \quad \text{or} \quad \pm 2\vec{u} \tag{11.26}$$

The singular momentum vector is then described as

$$\vec{H} = H_x \vec{i} + H_y \vec{j}$$

$$= \pm 2\vec{u} = \pm 2(\cos\theta \vec{i} + \sin\theta \vec{j}); \quad 0 \le \theta \le 2\pi$$

or

$$H_x^2 + H_y^2 = 4 \qquad (11.27)$$

which represents a circle on the (H_x, H_y) plane.

Example 11.2   Two-SPEED CMG System

Consider a special case ($\beta = 90$ deg) of the pyramid configuration of four single-gimbal CMGs, called a 2-SPEED system by Crenshaw [12]. For such two orthogonal pairs of two parallel single-gimbal CMGs, the gimbal-axis vectors are simply represented by $\vec{g}_1 = \vec{i}$, $\vec{g}_2 = \vec{j}$, $\vec{g}_3 = -\vec{i}$, and $\vec{g}_4 = -\vec{j}$.

We then obtain the singular momentum vector, for all positive ϵ_i, as

$$\vec{H} = \vec{H}(\theta_1, \theta_2) = \frac{2(u_y \vec{j} + u_z \vec{k})}{\sqrt{u_y^2 + u_z^2}} + \frac{2(u_x \vec{i} + u_z \vec{k})}{\sqrt{u_x^2 + u_z^2}} \qquad (11.28)$$

where $u_x = \sin\theta_2$, $u_y = -\sin\theta_1 \cos\theta_2$, and $u_z = \cos\theta_1 \cos\theta_2$. The so-called 4H momentum saturation singularity surface for this case when all four ϵ_i are positive can then be obtained as shown in Fig. 11.6. Note that the 4H saturation singularity surface itself does not cover the complete momentum envelope and that it has four circular holes caused by the four conditions $\vec{u} \neq \vec{g}_i$ ($i = 1, \ldots, 4$).

A special case when two e_i are zero and the other two e_i are $+1$ produces a circular flat "window" matched with a circular hole shown in Fig. 11.6. There are four circular flat windows, which are in fact the 2H singularity surfaces. The 4H saturation surface and the four circular flat windows provide the complete momentum envelope (workspace). There are also 2H singular curves, inside the momentum envelope, consisting of two perpendicular circles described by

$$H_x^2 + H_z^2 = 4 \quad (H_y = 0)$$
$$H_y^2 + H_z^2 = 4 \quad (H_x = 0)$$

Similarly, we can also obtain the 0H singularity surface as shown in Fig. 11.7. The 0H surface represents a singularity condition where two CMGs are aligned along the desired momentum vector direction and the other two CMGs in the opposite direction. The 0H surface does not necessarily mean a zero total momentum. Only if they are antiparallel does the total momentum become zero. In a composite singularity surface plot, consisting of both the 0H and 4H surfaces, the four trumpet-like funnels of the 0H surface are smoothly patched to the 4H saturation surface along the edges of the circular windows (2H surfaces).

Example 11.3   Pyramid Array with a Skew Angle

For a typical pyramid configuration of four single-gimbal CMGs with skew angle of β, we have

$$e_1 = \pm\sqrt{1 - (s\beta\, u_x + c\beta\, u_z)^2}$$

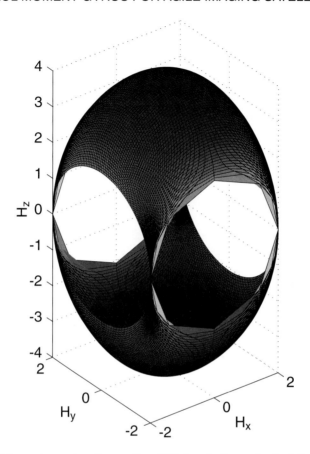

Fig. 11.6 **Momentum saturation surface (4H singularity surface) of the 2-SPEED system of four single-gimbal CMGs ($\beta = 90$ deg). The complete momentum envelope consists of the 4H singularity surface and the four circular flat windows (2H singularity surface).**

$$e_2 = \pm\sqrt{1 - (s\beta\ u_y + c\beta\ u_z)^2}$$

$$e_3 = \pm\sqrt{1 - (-s\beta\ u_x + c\beta\ u_z)^2}$$

$$e_4 = \pm\sqrt{1 - (-s\beta\ u_y + c\beta\ u_z)^2}$$

Furthermore, analytic expressions for the singular momentum surfaces (H_x, H_y, H_z) can be obtained as

$$H_x = \frac{c\beta(-s\beta\ u_z + c\beta\ u_x)}{e_1} + \frac{u_x}{e_2} + \frac{c\beta(s\beta\ u_z + c\beta\ u_x)}{e_3} + \frac{u_x}{e_4}$$

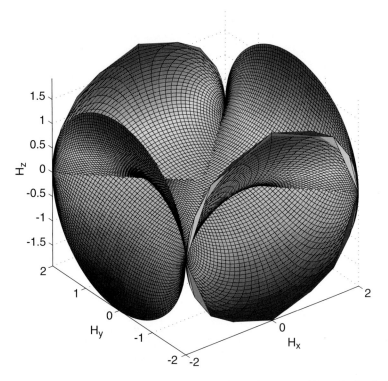

Fig. 11.7   Internal 0H singularity surface of the 2-SPEED system of four single-gimbal CMGs ($\beta = 90$ deg).

$$H_y = \frac{u_y}{e_1} - \frac{c\beta(s\beta\,u_z - c\beta\,u_y)}{e_2} + \frac{u_y}{e_3} + \frac{c\beta(s\beta\,u_z + c\beta\,u_y)}{e_4}$$

$$H_z = \frac{s\beta(-c\beta\,u_x + s\beta\,u_z)}{e_1} + \frac{s\beta(s\beta\,u_z - c\beta\,u_y)}{e_2} + \frac{s\beta(s\beta\,u_z + c\beta\,u_x)}{e_3}$$

$$+ \frac{s\beta(s\beta\,u_z + c\beta\,u_y)}{e_4}$$

where $u_x = \sin\theta_2$, $u_y = -\sin\theta_1\cos\theta_2$, and $u_z = \cos\theta_1\cos\theta_2$.

Using these expressions, one can obtain 4H, 2H, and 0H singularity surfaces. In a composite singularity surface plot, consisting of the 2H and 4H singularity surfaces, the eight trumpet-like funnels of the 2H surface are smoothly patched to the 4H saturation surface along the edges of the circular holes, resulting in the momentum envelope but with eight nonflat windows.

11.4   Singularity Analysis Using the Binet–Cauchy Identity

In the preceding section, a method developed by Margulies and Auburn [13] was used to analyze and visualize the singular momentum surfaces. A different approach, called the cutting-plane method, for determining the singular momentum surfaces can also be found in Stocking and Meffe [15]. These two techniques do not explicitly require the singular gimbal-angle information to determine the singular momentum surfaces.

However, it is often preferred to determine all possible singular gimbal-angles 1) to understand and characterize all possible singularities of CMG systems, 2) to be used for a direct singularity-avoidance logic, or 3) to determine singular momentum surfaces indirectly.

The singularity condition (11.20) can be rewritten in matrix form as

$$\mathbf{A}^T \mathbf{u} = 0 \qquad (11.29)$$

In other words, the singular vector $\mathbf{u}$ of the Jacobian matrix $\mathbf{A}$ is the null space vector of $\mathbf{A}^T$; that is, $\mathbf{u} = \text{null}(\mathbf{A}^T)$. Nontrivial solutions, $\mathbf{u} \neq 0$, exist for $(\mathbf{A}\mathbf{A}^T)\mathbf{u} = 0$ if and only if $\mathbf{A}\mathbf{A}^T$ is singular. That is, we have the singularity condition of the form

$$\det(\mathbf{A}\mathbf{A}^T) = 0 \qquad (11.30)$$

In general, the singularity condition defines a set of surfaces in $\mathbf{x}$ space, or equivalently, in $\mathbf{H}$ space. The simplest singular state is the momentum saturation singularity characterized by the momentum envelope, which is a three-dimensional surface representing the maximum available angular momentum of CMGs along any given direction. Any singular state for which the total CMG momentum vector is inside the momentum envelope is called "internal."

The singularity condition can also be written by using the Binet–Cauchy identity [16], as follows:

$$\det(\mathbf{A}\mathbf{A}^T) \equiv \sum_{i=1}^{n} M_i^2 = 0 \qquad (11.31)$$

where $M_i = \det(\mathbf{A}_i)$ are the Jacobian minors of order 3 and $\mathbf{A}_i = \mathbf{A}$ with the ith column removed.

The singularity conditions, Eqs. (11.31), for the pyramid array of four single-gimbal CMGs become

$$M_1 = s\beta[(s_2 s_3 c_4 + c_2 s_3 s_4) + c\beta(c_2 c_3 s_4 - s_2 c_3 c_4) + 2(c\beta)^2 c_2 c_3 c_4] = 0 \qquad (11.32a)$$

$$M_2 = s\beta[(s_3 s_4 c_1 + c_3 s_4 s_1) + c\beta(c_3 c_4 s_1 - s_3 c_4 c_1) + 2(c\beta)^2 c_3 c_4 c_1] = 0 \qquad (11.32b)$$

$$M_3 = s\beta[(s_4 s_1 c_2 + c_4 s_1 s_2) + c\beta(c_4 c_1 s_2 - s_4 c_1 c_2) + 2(c\beta)^2 c_4 c_1 c_2] = 0 \qquad (11.32c)$$

$$M_4 = s\beta[(s_1 s_2 c_3 + c_1 s_2 s_3) + c\beta(c_1 c_2 s_3 - s_1 c_2 c_3) + 2(c\beta)^2 c_1 c_2 c_3] = 0 \qquad (11.32d)$$

where $s_i \equiv \sin x_i$ and $c_i \equiv \cos x_i$. Although these singularity conditions based on the Binet–Cauchy identity have been discussed in the literature, a new approach to computing the singularity momentum surfaces using the singularity conditions, $M_i = 0$, is presented here.

Because the minimum rank of the Jacobian matrix $\mathbf{A}$ is two, the four conditions, $M_1 = M_2 = M_3 = M_4 = 0$, are not independent of one another, and only two of them are independent. Consequently, any two of these four conditions can be used to find the singular gimbal-angles, as follows.

When $c_i \neq 0$, the singularity conditions can be simplified as

$$\tan x_3(\tan x_2 + \tan x_4) + c\beta(\tan x_4 - \tan x_2) = -2(c\beta)^2 \quad (11.33)$$

$$\tan x_4(\tan x_3 + \tan x_1) + c\beta(\tan x_1 - \tan x_3) = -2(c\beta)^2 \quad (11.34)$$

$$\tan x_1(\tan x_4 + \tan x_2) + c\beta(\tan x_2 - \tan x_4) = -2(c\beta)^2 \quad (11.35)$$

$$\tan x_2(\tan x_1 + \tan x_3) + c\beta(\tan x_3 - \tan x_1) = -2(c\beta)^2 \quad (11.36)$$

Because the minimum rank of the Jacobian matrix $\mathbf{A}$ is two, only two of these four conditions are independent. Therefore, these four equations, (11.33–11.36), yield six singular gimbal-angle combinations.

For example, we have the following:
Case 1: For all (x_1, x_3), determine (x_2, x_4) using Eqs. (11.34) and (11.36),

$$x_2 = \tan^{-1}\left(\frac{-2(c\beta)^2 - c\beta \tan x_3 + c\beta \tan x_1}{\tan x_1 + \tan x_3}\right) \quad (11.37)$$

$$x_4 = \tan^{-1}\left(\frac{-2(c\beta)^2 - c\beta \tan x_1 + c\beta \tan x_3}{\tan x_1 + \tan x_3}\right) \quad (11.38)$$

The other five cases can also be found in Dominguez and Wie [4].

There are additional six cases when $c_i = 0$ (i.e., $\tan x_i = \pm\infty$). For example, we have the following:
Case 7:

$$\sin(x_2 + x_4) = 0 \quad \text{when} \quad \cos x_1 = \cos x_3 = 0 \quad (11.39)$$

Case 8:

$$\sin(x_1 + x_3) = 0 \quad \text{when} \quad \cos x_2 = \cos x_4 = 0 \quad (11.40)$$

Case 9:

$$\tan x_2 = -\tan x_4 = c\beta \quad \text{when} \quad \cos x_1 = 0 \text{ (for all } x_3) \quad (11.41)$$

Case 10:

$$\tan x_3 = -\tan x_1 = c\beta \quad \text{when} \quad \cos x_2 = 0 \text{ (for all } x_4) \quad (11.42)$$

Case 11:

$$\tan x_4 = -\tan x_2 = c\beta \quad \text{when} \quad \cos x_3 = 0 \text{ (for all } x_1) \quad (11.43)$$

Case 12:

$$\tan x_1 = -\tan x_3 = c\beta \qquad \text{when} \qquad \cos x_4 = 0 \text{ (for all } x_2) \quad (11.44)$$

Singular surfaces in the three-dimensional vector momentum space, $\mathbf{H} = (H_x, H_y, H_z)$, are then defined as surfaces mapped by

$$H_x = -c\beta \sin x_1 - \cos x_2 + c\beta \sin x_3 + \cos x_4 \qquad (11.45a)$$

$$H_y = \cos x_1 - c\beta \sin x_2 - \cos x_3 + c\beta \sin x_4 \qquad (11.45b)$$

$$H_z = s\beta \sin x_1 + s\beta \sin x_2 + s\beta \sin x_3 + s\beta \sin x_4 \qquad (11.45c)$$

As an illustrative example, the singular momentum projection on the (H_x, H_z) plane for case 7 (with $\beta = 54.7$ deg) is shown in Fig. 11.8. For this case, the singularity conditions become

$$(x_1, x_3) = (\pm\pi/2, \pm\pi/2)$$

$$x_2 + x_4 = 0 \qquad \text{or} \qquad x_2 + x_4 = \pi$$

For $x_2 + x_4 = 0$, there are four line singularities passing through the four points denoted by $*$ as shown in Fig. 11.8. For $x_2 + x_4 = \pi$, the four ellipses on the

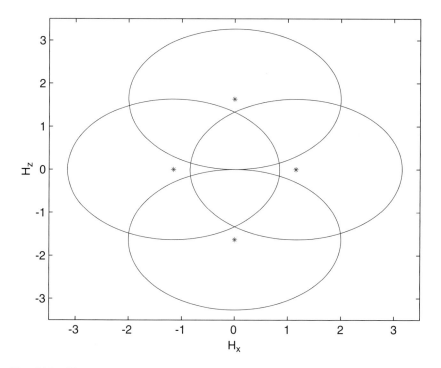

Fig. 11.8 Singular momentum projection on the (H_x, H_z) plane for case 7 with $\beta = 54.7$ deg.

(H_x, H_z) plane with $H_y \equiv 0$ can be found as

$$H_x^2 + \frac{(H_z \pm 2 \sin \beta)^2}{\sin^2 \beta} = 4$$

$$(H_x \pm 2 \cos \beta)^2 + \frac{H_z^2}{\sin^2 \beta} = 4$$

Detailed results for computing all possible singular gimbal-angles and also visualizing the singular momentum surfaces using this new approach can be found in Dominguez and Wie [4]. The proposed method provides detailed information about singular gimbal-angles and their direct relationships to the resulting singular momentum surfaces. It also provides a systematic way of determining all possible singular gimbal-angles, whereas the method by Margulies and Aubrun [13] does not.

11.5 CMG Null Motions

The so-called "null motions" of CMGs are defined as CMG gimbal motions that generate no net torque from the CMGs. One of the principal difficulties in using CMGs for spacecraft attitude control and momentum management is the geometric singularity problem in which no control torque is generated for the commanded control torque along a particular direction. At a singularity, CMG torque is available in all but one direction. In practice, the null motions are often employed to avoid and/or to escape such a singularity situation.

There are two types of singular states: hyperbolic states and elliptic states. The hyperbolic singular states can sometimes be escaped through null motion, whereas the elliptic singular states cannot be escaped through any null motion. Although null motion can be generated at the hyperbolic singularity, the mere existence of null motion does not guarantee escape from the hyperbolic singularity. This so-called degenerate null motion problem was briefly mentioned in Bedrossian et al. [17,18] without any specific example.

In this section, a mathematical framework for analyzing such null motions as well as for determining the singularity types is established by expanding upon the previous work by Bedrossian et al. [17,18]. This section is also intended to provide the reader with some new physical insights into the null motions (in particular, the degenerate null motions). Several illustrative examples with significant new results are presented.

The virtual momentum vector $\delta \vec{h}_i$, associated with the ith CMG, is defined such that

$$\delta \vec{H} = \sum_{i=1}^{n} \delta \vec{h}_i = 0 \qquad (11.46)$$

which is in fact the condition for null motion, also called virtual or zero-torque motion. The virtual momentum vector $\delta \vec{h}_i$ is different from the actual momentum vector change $\vec{\mathrm{d}h_i}$, but it must be compatible with the null motion constraint, Eq. (11.46), irrespective of time.

The virtual (null motion) momentum vector $\delta\vec{h}_i$ can be expanded in a Taylor series, as follows:

$$\delta\vec{H} = \sum_{i=1}^{n} \delta\vec{h}_i$$

$$= \sum_{i=1}^{n} \left[\frac{\mathrm{d}\vec{h}_i}{\mathrm{d}x_i}\delta x_i + \frac{1}{2!}\frac{\mathrm{d}^2\vec{h}_i}{\mathrm{d}x_i^2}\delta x_i^2 + \frac{1}{3!}\frac{\mathrm{d}^3\vec{h}_i}{\mathrm{d}x_i^3}\delta x_i^3 + \cdots \right]$$

$$= \sum_{i=1}^{n} \left[\vec{f}_i\delta x_i - \frac{1}{2!}\vec{h}_i\delta x_i^2 - \frac{1}{3!}\vec{f}_i\delta x_i^3 + \cdots \right] = 0 \qquad (11.47)$$

where δx_i are the virtual (null motion) gimbal-angle displacements from arbitrary gimbal-angles x_i, and $\vec{f}_i = \mathrm{d}\vec{h}_i/\mathrm{d}x_i$.

The first-order necessary condition for null motion is then given by

$$\sum_{i=1}^{n} \vec{f}_i\delta x_i = 0 \qquad (11.48)$$

which can be rewritten in matrix form as

$$\sum_{i=1}^{n} \mathbf{f}_i\delta x_i = \mathbf{A}\delta\mathbf{x} = 0 \qquad (11.49)$$

where $\delta\mathbf{x} = (\delta x_1, \ldots, \delta x_n)$ is called the null-motion displacement vector and $\mathbf{A}$ is the Jacobian matrix defined as in Eq. (11.6). In other words, the null vector $\delta\mathbf{x}$ is the null-space vector of $\mathbf{A}$; that is, $\delta\mathbf{x} = \mathbf{n} = \text{null}(\mathbf{A})$ and $\mathbf{An} = 0$. A null-space vector can be obtained as

$$\mathbf{n} = (C_1, C_2, C_3, C_4) = \text{Jacobian null vector}$$
$$C_i = (-1)^{i+1}M_i = \text{order 3 Jacobian cofactor}$$
$$M_i = \det(\mathbf{A}_i) = \text{order 3 Jacobian minor}$$
$$\mathbf{A}_i = \mathbf{A} \text{ with } i\text{th column removed}$$

However, the mere existence of the local null vectors (the first-order necessary condition for null motion) is not sufficient for an escape by null motion from singularity. The second-order necessary condition needs to be checked.

To test whether null motion is possible at a given singularity or to determine its type of singularity (i.e., elliptic or hyperbolic), consider the null motion constraint expressed in matrix form as

$$\delta\mathbf{H} = \mathbf{H}(\mathbf{x} + \delta\mathbf{x}) - \mathbf{H}(\mathbf{x})$$

$$= \sum_{i=1}^{n} \left[\mathbf{f}_i \delta x_i - \frac{1}{2!}\mathbf{h}_i \delta x_i^2 - \frac{1}{3!}\mathbf{f}_i \delta x_i^3 + \cdots \right]$$

$$= 0 \qquad\qquad (11.50)$$

which is the equivalent matrix representation of Eq. (11.47).

Taking the inner product of $\delta\mathbf{H}$ with an arbitrary vector $\mathbf{u}$, we obtain

$$\mathbf{u}^T \delta\mathbf{H} = \mathbf{u}^T \left\{ \sum_{i=1}^{n} \left[\mathbf{f}_i \delta x_i - \frac{1}{2!}\mathbf{h}_i(x_i)\delta x_i^2 - \frac{1}{3!}\mathbf{f}_i \delta x_i^3 + \cdots \right] \right\}$$

$$= 0 \qquad\qquad (11.51)$$

Because $\mathbf{u}^T \mathbf{f}_i = 0$ ($i = 1,\dots,4$) when $\mathbf{u}$ is along the singular vector direction, we obtain the constraint equation for null motion, as follows:

$$0 = \mathbf{u}^T \left\{ \sum_{i=1}^{n} \left[-\frac{1}{2!}\mathbf{h}_i \delta x_i^2 + \frac{1}{4!}\mathbf{h}_i \delta x_i^n + \cdots \right] \right\}$$

$$= \sum_{i=1}^{n} (\mathbf{u}^T \mathbf{h}_i) \left(-\frac{1}{2!}\delta x_i^2 + \frac{1}{4!}\delta x_i^n - \cdots \right)$$

$$= \sum_{i=1}^{n} e_i (\cos \delta x_i - 1) \qquad\qquad (11.52)$$

where $e_i = \mathbf{u}^T \mathbf{h}_i = \vec{u} \cdot \vec{h}_i$.

Considering only the second-order terms, we obtain the second-order necessary condition for null motion, as follows:

$$\sum_{i=1}^{n} e_i \delta x_i^2 = 0 \qquad\qquad (11.53)$$

which can be rewritten as

$$\delta\mathbf{x}^T \mathbf{E} \delta\mathbf{x} = 0 \qquad\qquad (11.54)$$

where $\delta\mathbf{x} = (\delta x_1, \dots, \delta x_n)$ and $\mathbf{E}$ is a diagonal matrix defined as $\mathbf{E} = \text{diag}(e_i)$. If $\mathbf{E}$ is a sign-definite matrix, the only solution to Eq. (11.54) is $\delta\mathbf{x} = 0$, and null motion is not possible. However, the sign-definiteness of $\mathbf{E}$ is only the sufficient but not necessary condition for the trivial solution, $\delta\mathbf{x} = 0$, to Eq. (11.54).

The virtual null motion of gimbal-angles can be expressed using the first-order necessary condition, as follows:

$$\delta \mathbf{x} = \sum_{i=1}^{n-2} c_i \mathbf{n}_i = \mathbf{N}\mathbf{c} \tag{11.55}$$

where c_i is the ith weighting coefficient, $\mathbf{c} = (c_1, \ldots, c_{n-2})$ and $\mathbf{n}_i$ are the null space basis vectors of the Jacobian matrix $\mathbf{A}$ such that $\mathbf{A}\mathbf{n}_i = 0$ or $\mathbf{N} = \text{null}(\mathbf{A})$. Note that at a singularity, $\text{rank}(\mathbf{A}) = 2$, and $\text{nullity}(\mathbf{A}) = n - \text{rank}(\mathbf{A})$.

Substituting Eq. (11.55) into Eq. (11.54), we obtain the second-order necessary condition of the form

$$\mathbf{c}^T \mathbf{M} \mathbf{c} = 0 \tag{11.56}$$

where $\mathbf{M} = \mathbf{N}^T \mathbf{E} \mathbf{N}$.

If $\mathbf{M}$ is a sign-definite matrix, the only solution to Eq. (11.56) is $\mathbf{c} = 0$, and null-motion is not possible. This type of singularity is referred to as an elliptic singularity, and consequently it cannot be escaped by null-motion. The other possibility for $\mathbf{M}$ is to be sign-indefinite (or singular). This type of singularity is referred to as a hyperbolic singularity. As to be illustrated for a system of two single-gimbal CMGs, however, the mere possibility of null-motion does not guarantee escape from singularity. Degenerate null-motion solutions that do not affect the rank of the Jacobian matrix must be excluded.

Example 11.4   Two Parallel Single-Gimbal CMGs

A system of two parallel single-gimbal CMGs has a Jacobian matrix $\mathbf{A}$ of the form

$$\mathbf{A} = 2 \begin{bmatrix} -\sin\alpha\cos\beta, & -\cos\alpha\sin\beta \\ \cos\alpha\cos\beta, & -\sin\alpha\sin\beta \end{bmatrix} \tag{11.57}$$

where α is the "rotation" angle and β the "scissor" angle as shown in Fig. 11.5. The singularity condition becomes

$$|\mathbf{A}| = \sin\beta\cos\beta = 0 \tag{11.58a}$$

$$\Rightarrow \beta = 0, \pm\pi/2, \pm\pi \qquad \forall \alpha \tag{11.58b}$$

The singular momentum surface for $\beta = 0$ is simply a circle of the form: $H_x^2 + H_y^2 = 4$. This also confirms the singular momentum surface expression (11.27), previously found directly without recourse to singular gimbal-angles. An internal antiparallel 0H singularity for $\beta = \pm\pi/2$ is located at the origin: $(H_x, H_y) = (0, 0)$.

For the 0H singularity at $\mathbf{x} = (\alpha, \beta) = (0, \pi/2)$, illustrated in Fig. 11.5, the Jacobian matrix becomes

$$\mathbf{A} = \begin{bmatrix} 0 & -2 \\ 0 & 0 \end{bmatrix}$$

with its singular vector of $\mathbf{u} = (0, 1)$, that is, $\mathbf{A}^T\mathbf{u} = 0$. A null space vector of $\mathbf{A}$ such that $\mathbf{A}\dot{\mathbf{x}} = 0$ is $\dot{\mathbf{x}} = \text{null}(\mathbf{A}) = (1, 0)$. This 0H (antiparallel) singularity has the following null motion constraint:

$$0 = \sum_{i=1}^{2} e_i(\cos \delta x_i - 1) = \cos \delta x_1 - \cos \delta x_2 \qquad (11.59)$$

Its null-motion solution is $\delta x_1 = \delta x_2 =$ arbitrary, which corresponds to $\beta = \pi/2$ for any α. This type of singularity with possible null motion is referred to as a hyperbolic singularity. However, this hyperbolic singularity cannot be escaped by null motion because the singular configuration ($\beta = \pi/2$) remains undisturbed during null motion along the null manifold (or a degenerate null trajectory), as illustrated in Fig. 11.9. This example demonstrates that the mere existence of null motion does not guarantee escape from a hyperbolic singularity.

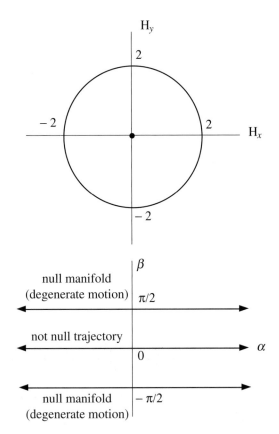

Fig. 11.9 The (α, β) trajectories for two parallel single-gimbal CMGs.

For the 2H saturation singularity with $\mathbf{x} = (\alpha, \beta) = (0, 0)$, the Jacobian matrix becomes

$$\mathbf{A} = \begin{bmatrix} 0 & 0 \\ 2 & 0 \end{bmatrix}$$

with its singular vector of $\mathbf{u} = (-1, 0)$, that is, $\mathbf{A}^T \mathbf{u} = 0$. Its null space vector is $\dot{\mathbf{x}} = \text{null}(\mathbf{A}) = (0, 1)$. This saturation singularity has the null motion constraint as

$$0 = \sum_{i=1}^{2} e_i(\cos \delta x_i - 1) = \cos \delta x_1 + \cos \delta x_2 - 2 \qquad (11.60)$$

The only solution to this equation is $\delta x_1 = \delta x_2 = 0$, that is, null motion does not exist. This type of singularity with no possible null motion is referred to as an elliptic singularity. All saturation (external) singularities are elliptic that is, they cannot be escaped by null motion.

Example 11.5   Pyramid Array of Four CMGs

Consider a pyramid array of four CMGs with $\beta = 53.13$ deg. It can be shown that a set of gimbal-angles, $\mathbf{x} = (-\pi/2, 0, \pi/2, 0)$, shown in Fig. 11.10a, is an internal elliptic singularity with its singular vector of $\mathbf{u} = (1, 0, 0)$, as follows:

$$\mathbf{A} = \begin{bmatrix} 0 & 0 & 0 & 0 \\ 1 & -0.6 & 1 & 0.6 \\ 0 & 0.8 & 0 & 0.8 \end{bmatrix}$$

$$\text{null}(\mathbf{A}) = \mathbf{N} = \begin{bmatrix} -0.7952 & 0 \\ -0.2774 & -0.5392 \\ 0.4623 & -0.6470 \\ 0.2774 & 0.5392 \end{bmatrix}$$

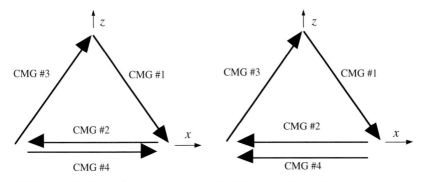

a) 2H internal elliptic singularity   b) 0H internal hyperbolic singularity

Fig. 11.10   Singularity illustration for a pyramid mounting arrangement of four single-gimbal CMGs.

$$\text{null}(\mathbf{A}^T) = \mathbf{u} = \begin{bmatrix} 1 & 0 & 0 \end{bmatrix}^T$$

$$\mathbf{E} = \text{diag}(\mathbf{u}^T \mathbf{h}_i) = \text{diag}(0.6, -1.0, 0.6, 1.0)$$

$$\mathbf{M} = \mathbf{N}^T \mathbf{EN} = \begin{bmatrix} 0.5077 & -0.1795 \\ -0.1795 & 0.2512 \end{bmatrix}$$

$$\text{eig}(\mathbf{M}) = 0.1588, \quad 0.6$$

However, a set of gimbal-angles, $\mathbf{x} = (-\pi/2, 0, \pi/2, \pi)$, shown in Fig. 11.10b, can be shown to be a hyperbolic singularity with its singular vector of $\mathbf{u} = (1,0,0)$. It can be further shown that this hyperbolic singularity can be escaped by null motion because its null motion is not a degenerate solution.

Example 11.6   Three Parallel Single-Gimbal CMGs

For a system of three single-gimbal CMGs with parallel gimbal axes, described by Eqs. (11.14) and (11.15), the singularity condition becomes

$$\det\left(\mathbf{A}\mathbf{A}^T\right) = M_1^2 + M_2^2 + M_3^2$$

$$= \sin^2(x_3 - x_2) + \sin^2(x_3 - x_1) + \sin^2(x_2 - x_1)$$

$$= 0 \tag{11.61}$$

and singular gimbal-angles can be found as follows:
External 3H singularity:

$$x_1 = x_2 = x_3$$

Internal 1H singularity:

$$x_i = x_j, \quad x_k = x_i \pm \pi$$

The singular momentum surfaces are then simply described by two circles, as shown in Fig. 11.11. The internal antiparallel (1H) singularity and the external parallel (3H) singularity are also illustrated in Fig. 11.11.

The 1H singularity has the null-motion constraint as

$$0 = \sum_{i=1}^{3} e_i(\cos \delta x_i - 1)$$

$$= (\cos \delta x_1 - 1) + (\cos \delta x_2 - 1) - (\cos \delta x_3 - 1)$$

$$= \cos \delta x_1 + \cos \delta x_2 - \cos \delta x_3 - 1 \tag{11.62}$$

Its nondegenerate null-motion solution can be found as $\delta x_2 = \delta x_3$ and $\delta x_1 = 0$, and thus it is a hyperbolic singularity that can be escaped by null motion.

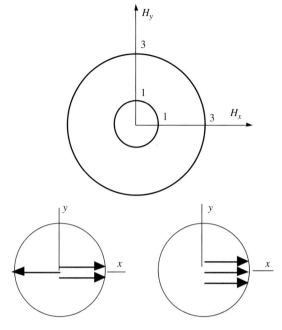

a) 1H antiparallel singularity   b) 3H parallel singularity

Fig. 11.11   Singularity illustration for three single-gimbal CMGs with parallel gimbal axes.

A new set of orthogonal coordinates (x, y, z) for gimbal-angles, described by Margulies and Aubrun (but not explicitly provided in [13]), can be found as

$$
\begin{bmatrix} x \\ y \\ z \end{bmatrix} = \begin{bmatrix} 1/\sqrt{2} & -1/\sqrt{2} & 0 \\ 1/\sqrt{6} & 1/\sqrt{6} & -2/\sqrt{6} \\ 1/\sqrt{3} & 1/\sqrt{3} & 1/\sqrt{3} \end{bmatrix} \begin{bmatrix} x_1 \\ x_2 \\ x_3 \end{bmatrix}
\tag{11.63}
$$

or

$$
\begin{bmatrix} x_1 \\ x_2 \\ x_3 \end{bmatrix} = \begin{bmatrix} 1/\sqrt{2} & 1/\sqrt{6} & 1/\sqrt{3} \\ -1/\sqrt{2} & 1/\sqrt{6} & 1/\sqrt{3} \\ 0 & -2/\sqrt{6} & 1/\sqrt{3} \end{bmatrix} \begin{bmatrix} x \\ y \\ z \end{bmatrix}
\tag{11.64}
$$

Using Eq. (11.64), we can then express $\det\left(\mathbf{AA}^T\right)$ and the total angular momentum H as [13]

$$
\det\left(\mathbf{AA}^T\right) = 2 - \cos^2 \frac{2x}{\sqrt{2}} - \cos \frac{2x}{\sqrt{2}} \cos \frac{6y}{\sqrt{6}}
\tag{11.65a}
$$

$$
H^2 = 1 + 4\cos^2 \frac{x}{\sqrt{2}} + 4\cos \frac{x}{\sqrt{2}} \cos \frac{3y}{\sqrt{6}}
\tag{11.65b}
$$

which are independent of z.

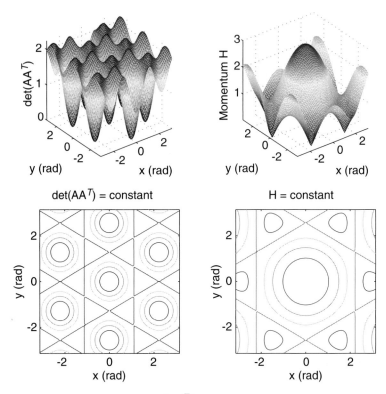

Fig. 11.12   Contour plots of $\det\left(\mathbf{AA}^T\right)$ **and** H **for a system of three parallel CMGs. Note that** (x,y) **are the new transformed gimbal-angle coordinates defined by Eq. (11.63).**

Contour plots of $\det\left(\mathbf{AA}^T\right)$ and H vs (x,y) are shown in Fig. 11.12. In the contour plot of H, one can easily identify an elliptic singularity at the center and hyperbolic singularities at the six saddle-like points. The constant-H lines of $H = 1$, connecting the internal hyperbolic singularities, are null-motion trajectories. The zero-momentum points are not singular points.

Instead of employing (x, y, z) introduced by Margulies and Aubrun, we can introduce two angles (α, β) as

$$\alpha = x_2 - x_1 \tag{11.66a}$$

$$\beta = x_3 - x_1 \tag{11.66b}$$

Then, we simply obtain

$$\det\left(\mathbf{AA}^T\right) = \sin^2\alpha + \sin^2(\alpha - \beta) + \sin^2\beta \tag{11.67}$$

$$H^2 = (1 + \cos\alpha + \cos\beta)^2 + (\sin\alpha + \sin\beta)^2 \tag{11.68}$$

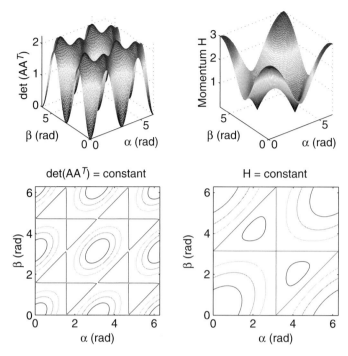

Fig. 11.13 **Contour plots of det $\left(\mathbf{AA}^T\right)$ and H vs (α, β) for a system of three parallel CMGs.**

Contour plots of det $\left(\mathbf{AA}^T\right)$ and H vs (α, β) are also shown in Fig. 11.13. The external elliptic singularities, internal hyperbolic singularities, and null-motion manifolds can also be easily identified in the contour plot of H in Fig. 11.13.

11.6 Surface Theory of Differential Geometry

In this section, the singularity surface problem is further examined by applying the surface theory of differential geometry [19,20]. This subject is certainly mathematical in nature; however, it is also of practical importance for a better understanding of CMG systems and also even for developing CMG steering laws, as attempted by many previous researchers, including Margulies and Aubrun [13].

One can analyze, compute, and visualize the singular surfaces without employing the differential surface theory, as demonstrated in the preceding sections. However, this section is intended to provide the reader with the basic mathematical foundation of the theory of differential geometry as being applied to the geometric theory of CMG systems. A further study is necessary to fully exploit the differential surface geometry theory, and this section will certainly help the reader further advance this subject.

Let a unit vector along all possible directions in three-dimensional space (except along the gimbal-axis directions) be parameterized as

$$\vec{u} = u_x\vec{i} + u_y\vec{j} + u_z\vec{k}$$
$$= \sin\theta_2\vec{i} - \sin\theta_1\cos\theta_2\vec{j} + \cos\theta_1\cos\theta_2\vec{k} \qquad (11.69)$$

which was previously introduced in Eq. (11.17). Then, we obtain the differential of $\vec{u}$ as

$$d\vec{u} = \vec{u}_1\,d\theta_1 + \vec{u}_2\,d\theta_2 \qquad (11.70)$$

where

$$\vec{u}_1 \equiv \frac{\partial\vec{u}}{\partial\theta_1} = -\cos\theta_1\cos\theta_2\vec{j} - \sin\theta_1\cos\theta_2\vec{k}$$

$$\vec{u}_2 \equiv \frac{\partial\vec{u}}{\partial\theta_2} = \cos\theta_2\vec{i} + \sin\theta_1\sin\theta_2\vec{j} - \cos\theta_1\sin\theta_2\vec{k}$$

Note that $|\vec{u}_1| = \cos^2\theta_2$, $|\vec{u}_2| = 1$, and $\vec{u}_1 \cdot \vec{u}_2 = 0$.

Similarly, the differential of the singular momentum vector or singularity surface vector $\vec{H}(\theta_1,\theta_2)$ can also be represented as

$$d\vec{H} = \frac{\partial\vec{H}}{\partial\theta_1}d\theta_1 + \frac{\partial\vec{H}}{\partial\theta_2}d\theta_2 = \vec{H}_1 d\theta_1 + \vec{H}_2 d\theta_2 \qquad (11.71)$$

where

$$\vec{H}_i \equiv \frac{\partial\vec{H}}{\partial\theta_i} = \frac{\partial H_x}{\partial\theta_i}\vec{i} + \frac{\partial H_y}{\partial\theta_i}\vec{j} + \frac{\partial H_z}{\partial\theta_i}\vec{k}$$

The *first fundamental form* of the singularity surface vector $\vec{H}$ is given by

$$I \equiv d\vec{H} \cdot d\vec{H} = (\vec{H}_1 d\theta_1 + \vec{H}_2 d\theta_2) \cdot (\vec{H}_1 d\theta_1 + \vec{H}_2 d\theta_2)$$

$$= \sum_{i,j=1}^{2} G_{ij}d\theta_i d\theta_j = d\boldsymbol{\theta}^T \mathbf{G} d\boldsymbol{\theta}$$

$$= [d\theta_1 \quad d\theta_2]\begin{bmatrix} G_{11} & G_{12} \\ G_{12} & G_{22} \end{bmatrix}\begin{bmatrix} d\theta_1 \\ d\theta_2 \end{bmatrix} \qquad (11.72)$$

where $d\boldsymbol{\theta} = (d\theta_1, d\theta_2)$, $\mathbf{G} = [G_{ij}] = \mathbf{G}^T > 0$, and

$$G_{ij} \equiv \vec{H}_i \cdot \vec{H}_j = \frac{\partial H_x}{\partial\theta_i}\frac{\partial H_x}{\partial\theta_j} + \frac{\partial H_y}{\partial\theta_i}\frac{\partial H_y}{\partial\theta_j} + \frac{\partial H_z}{\partial\theta_i}\frac{\partial H_z}{\partial\theta_j}$$

In this section, we will employ the following vector/matrix dot product operation:

$$\begin{bmatrix} \vec{H}_1 \\ \vec{H}_2 \end{bmatrix} \cdot [\vec{H}_1 \quad \vec{H}_2] \equiv \begin{bmatrix} \vec{H}_1 \cdot \vec{H}_1 & \vec{H}_1 \cdot \vec{H}_2 \\ \vec{H}_2 \cdot \vec{H}_1 & \vec{H}_2 \cdot \vec{H}_2 \end{bmatrix} = \mathbf{G} \qquad (11.73)$$

The square of an element of arc length ds^2 is also often used to denote the first fundamental form of the surface as

$$ds^2 = G_{11}(d\theta_1)^2 + 2G_{12}d\theta_1 d\theta_2 + G_{22}(d\theta_2)^2 \tag{11.74}$$

and G_{ij} are the components of the so-called *metric tensor* or *fundamental tensor* of the two-dimensional Riemannian geometry. The first fundamental form enables us to measure lengths, angles, and areas in a surface.

The *second fundamental form* of the singularity surface vector $\vec{H}$ is given by

$$\mathrm{II} \equiv -d\vec{H} \cdot d\vec{u} = -(\vec{H}_1 d\theta_1 + \vec{H}_2 d\theta_2) \cdot (\vec{u}_1 d\theta_1 + \vec{u}_2 d\theta_2)$$

$$= \sum_{i,j=1}^{2} B_{ij}d\theta_i d\theta_j = d\boldsymbol{\theta}^T \mathbf{B} d\boldsymbol{\theta}$$

$$= [d\theta_1 \quad d\theta_2] \begin{bmatrix} B_{11} & B_{12} \\ B_{21} & B_{22} \end{bmatrix} \begin{bmatrix} d\theta_1 \\ d\theta_2 \end{bmatrix} \tag{11.75}$$

where $B_{ij} \equiv -\vec{H}_i \cdot \vec{u}_j$.

Differentiating the orthogonality relation $\vec{H}_i \cdot \vec{u} = 0$, we obtain

$$\vec{H}_{ij} \cdot \vec{u} + \vec{H}_i \cdot \vec{u}_j = 0 \tag{11.76}$$

where $\vec{H}_{ij} \equiv \partial^2 \vec{H}/\partial\theta_i \partial\theta_j$. Consequently, we have

$$B_{ij} \equiv -\vec{H}_i \cdot \vec{u}_j = \vec{H}_{ij} \cdot \vec{u} \tag{11.77}$$

and $\mathbf{B} = \mathbf{B}^T$ because $\vec{H}_{ij} = \vec{H}_{ji}$.

The *third fundamental form* of the surface is given by

$$\mathrm{III} \equiv d\vec{u} \cdot d\vec{u} = (\vec{u}_1 d\theta_1 + \vec{u}_2 d\theta_2) \cdot (\vec{u}_1 d\theta_1 + \vec{u}_2 d\theta_2)$$

$$= \sum_{i,j=1}^{2} C_{ij}d\theta_i d\theta_j = d\boldsymbol{\theta}^T \mathbf{C} d\boldsymbol{\theta}$$

$$= [d\theta_1 \quad d\theta_2] \begin{bmatrix} C_{11} & C_{12} \\ C_{12} & C_{22} \end{bmatrix} \begin{bmatrix} d\theta_1 \\ d\theta_2 \end{bmatrix} \tag{11.78}$$

where $C_{ij} \equiv \vec{u}_i \cdot \vec{u}_j$ and $\mathbf{C} = \mathbf{C}^T > 0$.

Defining the so-called Weingarten transformation as

$$\begin{bmatrix} \vec{u}_1 \\ \vec{u}_2 \end{bmatrix} = - \begin{bmatrix} D_{11} & D_{12} \\ D_{21} & D_{22} \end{bmatrix} \begin{bmatrix} \vec{H}_1 \\ \vec{H}_2 \end{bmatrix} = -\mathbf{D} \begin{bmatrix} \vec{H}_1 \\ \vec{H}_2 \end{bmatrix} \tag{11.79}$$

we simply obtain

$$\mathbf{B} \equiv - \begin{bmatrix} \vec{H}_1 \\ \vec{H}_2 \end{bmatrix} \cdot [\vec{u}_1 \quad \vec{u}_2]$$

$$= \begin{bmatrix} \vec{H}_1 \\ \vec{H}_2 \end{bmatrix} \cdot [\vec{H}_1 \quad \vec{H}_2] \mathbf{D}^T = \mathbf{G} \mathbf{D}^T \tag{11.80}$$

The Weingarten transformation matrix is then obtained as

$$\mathbf{D} = \mathbf{G}^{-1}\mathbf{B} = \mathbf{B}\mathbf{G}^{-1} \tag{11.81}$$

and $\mathbf{D} = \mathbf{D}^T$. Also we have

$$\mathbf{C} \equiv \begin{bmatrix} \vec{u}_1 \\ \vec{u}_2 \end{bmatrix} \cdot [\, \vec{u}_1 \quad \vec{u}_2 \,] = \mathbf{D} \begin{bmatrix} \vec{H}_1 \\ \vec{H}_2 \end{bmatrix} \cdot [\, \vec{H}_1 \quad \vec{H}_2 \,] \mathbf{D}^T$$
$$= \mathbf{D}\mathbf{G}\mathbf{D}^T = \mathbf{D}\mathbf{B} = \mathbf{G}^{-1}\mathbf{B}^2 = \mathbf{B}^2\mathbf{G}^{-1} \tag{11.82}$$

Similar to the Weingarten transformation, consider a *congruence transformation* of the form

$$\begin{bmatrix} d\theta_1 \\ d\theta_2 \end{bmatrix} = \begin{bmatrix} E_{11} & E_{12} \\ E_{21} & E_{22} \end{bmatrix} \begin{bmatrix} \omega_1 \\ \omega_2 \end{bmatrix} \tag{11.83}$$

or $d\theta = \mathbf{E}\omega$, where $\mathbf{E} = [E_{ij}]$ is a nonsingular matrix, which diagonalizes the first fundamental form as

$$I \equiv d\vec{H} \cdot d\vec{H} = d\theta^T \mathbf{G} d\theta = \omega^T \mathbf{E}^T \mathbf{G} \mathbf{E} \omega$$
$$= \omega^T \mathbf{I}\omega = \omega_1^2 + \omega_2^2 \tag{11.84}$$

where $\mathbf{I}$ is an identity matrix. The differential form $d\vec{H}$ can then be simply written in terms of the linear differential forms, ω_1 and ω_2, as

$$d\vec{H} = \omega_1 \vec{e}_1 + \omega_2 \vec{e}_2 \tag{11.85}$$

where $\vec{e}_1$ and $\vec{e}_2$ are orthogonal unit vectors, that is, $\vec{e}_i \cdot \vec{e}_j = \delta_{ij}$.

By applying the congruence transformation $d\theta = \mathbf{E}\omega$, we rewrite the second fundamental form as

$$II \equiv -d\vec{H} \cdot d\vec{u} = d\theta^T \mathbf{B} d\theta = \omega^T \mathbf{E}^T \mathbf{B} \mathbf{E} \omega = \omega^T \mathbf{L}\omega$$
$$= [\omega_1 \quad \omega_2] \begin{bmatrix} L_{11} & L_{12} \\ L_{12} & L_{22} \end{bmatrix} \begin{bmatrix} \omega_1 \\ \omega_2 \end{bmatrix} \tag{11.86}$$

where $\mathbf{L} \equiv \mathbf{E}^T \mathbf{B} \mathbf{E}$ is known as the *Weingarten mapping matrix* in the theory of differential geometry. According to *Sylvester's Law of Inertia*, the signs of the eigenvalues of $\mathbf{B}$ are preserved by a congruence transformation because $\det \mathbf{L} = \det \mathbf{B}(\det \mathbf{E})^2$. Note that the symmetry of $\mathbf{B}$ is also preserved by a congruence transformation that is, $\mathbf{L} = \mathbf{L}^T$.

Furthermore, the second fundamental form can be diagonalized as

$$II \equiv -d\vec{H} \cdot d\vec{u} = [\omega_1 \quad \omega_2] \begin{bmatrix} L_{11} & L_{12} \\ L_{12} & L_{22} \end{bmatrix} \begin{bmatrix} \omega_1 \\ \omega_2 \end{bmatrix}$$
$$= [\, \omega_1' \quad \omega_2' \,] \boldsymbol{\Phi}^T \mathbf{L} \boldsymbol{\Phi} \begin{bmatrix} \omega_1' \\ \omega_2' \end{bmatrix}$$
$$= [\, \omega_1' \quad \omega_2' \,] \begin{bmatrix} k_1 & 0 \\ 0 & k_2 \end{bmatrix} \begin{bmatrix} \omega_1' \\ \omega_2' \end{bmatrix} \tag{11.87}$$

where k_1 and k_2 are the eigenvalues of $\mathbf{L}$ and $\boldsymbol{\Phi}$ is a similarity transformation matrix defined as

$$d\vec{H} = \omega_1 \vec{e}_1 + \omega_2 \vec{e}_2 = \omega'_1 \vec{e}_{1'} + \omega'_2 \vec{e}_{2'} \qquad (11.88)$$

where

$$\begin{bmatrix} \vec{e}_1 \\ \vec{e}_2 \end{bmatrix} = \begin{bmatrix} \cos\phi & -\sin\phi \\ \sin\phi & \cos\phi \end{bmatrix} \begin{bmatrix} \vec{e}_{1'} \\ \vec{e}_{2'} \end{bmatrix} = \boldsymbol{\Phi} \begin{bmatrix} \vec{e}_{1'} \\ \vec{e}_{2'} \end{bmatrix}$$

$$\begin{bmatrix} \omega_1 \\ \omega_2 \end{bmatrix} = \begin{bmatrix} \cos\phi & -\sin\phi \\ \sin\phi & \cos\phi \end{bmatrix} \begin{bmatrix} \omega'_1 \\ \omega'_2 \end{bmatrix} = \boldsymbol{\Phi} \begin{bmatrix} \omega'_1 \\ \omega'_2 \end{bmatrix}$$

$$\tan 2\phi = \frac{2L_{12}}{L_{11} - L_{22}} \qquad (11.89)$$

Note that $\boldsymbol{\Phi}$ is an orthonormal matrix with $\boldsymbol{\Phi}^T = \boldsymbol{\Phi}^{-1}$ and $\det \boldsymbol{\Phi} = 1$, that is, $\boldsymbol{\Phi}$ is a rotation matrix for $\boldsymbol{\Phi}^T \mathbf{L} \boldsymbol{\Phi} = \mathrm{diag}\{k_1, k_2\}$, and $\det \mathbf{L} = L_{11}L_{22} - L_{12}^2 = k_1 k_2$.

The normal curvature k_n, Gauss curvature k, and mean curvature $\tilde{k}$ are then defined as

$$k_n = k_1 \cos^2\phi + k_2 \sin^2\phi \equiv \frac{\mathrm{II}}{\mathrm{I}} \qquad (11.90a)$$

$$k = k_1 k_2 = \det \mathbf{L} = L_{11}L_{22} - L_{12}^2 \qquad (11.90b)$$

$$\tilde{k} = \frac{1}{2}(k_1 + k_2) = \frac{1}{2}\,\mathrm{tr}\,\mathbf{L} = \frac{1}{2}(L_{11} + L_{22}) \qquad (11.90c)$$

The principal curvatures k_1 and k_2 are the roots of the following characteristic equation:

$$\det(\mathbf{L} - \lambda \mathbf{I}) = 0 \qquad \text{or} \qquad \det(\mathbf{B} - \lambda \mathbf{G}) = 0 \qquad (11.91)$$

Furthermore, we have the following relationship from [19,20]:

$$k\mathrm{I} - 2\tilde{k}\mathrm{II} + \mathrm{III} = 0 \qquad (11.92)$$

where $\mathrm{I} = \omega^T \mathbf{I}\omega$, $\mathrm{II} = \omega^T \mathbf{L}\omega$, and $\mathrm{III} = \omega^T \mathbf{L}^2 \omega$. Equivalently, we have

$$k\mathbf{G} - 2\tilde{k}\mathbf{B} + \mathbf{C} = 0 \qquad (11.93)$$

For the known surfaces, the angle ψ between the parameter lines, the unit normal vector $\vec{u}$, and B_{ij} can be obtained from the following relationships:

$$\cos\psi = \frac{\vec{H}_1 \cdot \vec{H}_2}{|\vec{H}_1||\vec{H}_2|} = \frac{G_{12}}{\sqrt{G_{11}G_{22}}} \qquad (11.94a)$$

$$\vec{u} = \frac{\vec{H}_1 \times \vec{H}_2}{|\vec{H}_1 \times \vec{H}_2|} = \frac{\vec{H}_1 \times \vec{H}_2}{\sqrt{\det \mathbf{G}}} \qquad (11.94b)$$

$$B_{ij} \equiv -\vec{H}_i \cdot \vec{u}_j = \vec{H}_{ij} \cdot \vec{u} = \frac{\vec{H}_{ij} \cdot \vec{H}_1 \times \vec{H}_2}{\sqrt{\det \mathbf{G}}} \qquad (11.94c)$$

Note that $\det \mathbf{G} > 0$ because $\mathbf{G}$ is a positive-definite matrix.

Taking the differential of the orthogonality condition $d\vec{H} \cdot \vec{u} = 0$, we obtain

$$d^2\vec{H} \cdot \vec{u} + d\vec{H} \cdot d\vec{u} = 0 \qquad (11.95)$$

and the second fundamental form can then be written as

$$II \equiv d^2\vec{H} \cdot \vec{u} = \frac{d^2\vec{H} \cdot \vec{H}_1 \times \vec{H}_2}{\sqrt{\det \mathbf{G}}} \qquad (11.96)$$

where

$$d^2\vec{H} \equiv \frac{\partial^2\vec{H}}{\partial\theta_1^2}(d\theta_1)^2 + \frac{\partial^2\vec{H}}{\partial\theta_1\partial\theta_2}d\theta_1 d\theta_2 + \frac{\partial^2\vec{H}}{\partial\theta_2^2}(d\theta_2)^2 \qquad (11.97)$$

Using dyadic notation, we can rewrite the singular momentum vector Eq. (11.25) as

$$\vec{H} = \vec{H}(\vec{u}) = \sum_{i=1}^n \frac{1}{e_i}(\hat{I} - \vec{g}_i\vec{g}_i) \cdot \vec{u} \qquad (11.98)$$

where $\hat{I}$ is a unit dyadic and $e_i = \vec{h}_i \cdot \vec{u} = \pm|\vec{g}_i \times \vec{u}|$. The differential of the singular momentum vector is then written as [13]

$$d\vec{H} = \hat{J} \cdot d\vec{u} \qquad (11.99)$$

where

$$\hat{J} = \sum_{i=1}^n \frac{1}{e_i}\vec{f}_i\vec{f}_i \qquad (11.100)$$

Because $\vec{f}_i \cdot \vec{u} = 0$ at singularity, we have $\hat{J} \cdot \vec{u} = 0$. Thus, it is said that $\hat{J}$ is singular of rank 2 and is a symmetric planar dyadic in the tangent plane orthogonal to $\vec{u}$.

Representing the symmetric planar dyadic $\hat{J}$ as

$$\hat{J} = \begin{bmatrix} \vec{u}_1 & \vec{u}_2 \end{bmatrix} \begin{bmatrix} J_{11} & J_{12} \\ J_{12} & J_{22} \end{bmatrix} \begin{bmatrix} \vec{u}_1 \\ \vec{u}_2 \end{bmatrix} = \sum_{i,j=1}^2 J_{ij}\vec{u}_i\vec{u}_j \qquad (11.101)$$

where $\vec{u}_i \equiv \partial\vec{u}/\partial\theta_i$ and $J_{ij} = \vec{u}_i \cdot \hat{J} \cdot \vec{u}_j$, we also obtain the two fundamental forms of the singularity surface as

$$I \equiv d\vec{H} \cdot d\vec{H} = d\vec{u} \cdot \hat{J}^2 \cdot d\vec{u}$$
$$= d\theta^T \mathbf{CJCJC}d\theta = d\theta^T \mathbf{G}d\theta \qquad (11.102)$$
$$II \equiv -d\vec{H} \cdot d\vec{u} = -d\vec{u} \cdot \hat{J} \cdot d\vec{u}$$
$$= -d\theta^T \mathbf{CJC}d\theta = d\theta^T \mathbf{B}d\theta \qquad (11.103)$$

and we have

$$\mathbf{B} = -\mathbf{CJC} \qquad (11.104a)$$

$$\mathbf{G} = \mathbf{BC}^{-1}\mathbf{B} \qquad (11.104b)$$

Note that the matrix $\mathbf{C}$, which is not an identity matrix in general, is missing in Eq. (13) of [13].

In summary, we have $\mathbf{E}^T\mathbf{GE} = \mathbf{I}$, $\mathbf{E}^T\mathbf{BE} = \mathbf{L}$, $\mathbf{E}^T\mathbf{CE} = \mathbf{L}^2$, $\mathbf{C} = \mathbf{B}^2\mathbf{G}^{-1}$, and $\mathbf{D} = \mathbf{BG}^{-1}$. Consequently, we obtain

$$\det \mathbf{C} = \frac{\det \mathbf{B}^2}{\det \mathbf{G}} \qquad (11.105a)$$

$$\det \mathbf{D} = \frac{\det \mathbf{B}}{\det \mathbf{G}} \qquad (11.105b)$$

and

$$k \equiv k_1 k_2 = \det \mathbf{L} = L_{11}L_{22} - L_{12}^2$$

$$= \det \mathbf{D} = \frac{\det \mathbf{B}}{\det \mathbf{G}} = \frac{B_{11}B_{22} - B_{12}^2}{G_{11}G_{22} - G_{12}^2}$$

$$= \frac{1}{\det \mathbf{J}}\frac{1}{\det \mathbf{C}} = \frac{1}{J_{11}J_{22} - J_{12}^2}\frac{1}{\det \mathbf{C}}$$

$$= \frac{1}{\lambda_1\lambda_2}\frac{1}{\det \mathbf{C}} \qquad (11.106)$$

where λ_1 and λ_2 are the eigenvalues of $\mathbf{J}$ and $\det \mathbf{C} > 0$. Finally, we have

$$\text{sign}(k) = \text{sign}(\det \mathbf{L}) = \text{sign}(\det \mathbf{D})$$

$$= \text{sign}(\det \mathbf{B}) = \text{sign}(\det \mathbf{J}) \qquad (11.107)$$

The sign of k provides important information about the shape of the surface in the neighborhood of a given point $\vec{H}^*$. If $k(\vec{H}^*) > 0$, then $\vec{H}^*$ is an elliptic point of the surface. If $k(\vec{H}^*) = 0$, then $\vec{H}^*$ is a parabolic point or a planar umbilic of the surface. If $k(\vec{H}^*) < 0$, then $\vec{H}^*$ is a hyperbolic point of the surface. The two local surfaces are said to be *isometric* if and only if they have identical Gauss curvatures. A closed surface always has elliptic points. A line for which the second fundamental form becomes diagonal is a principal direction. A curve all of whose tangents are in principal directions is a curvature line. The normal curvature of a curvature line is a principal curvature of the surface.

The inverse mapping problem is to determine a direction $d\vec{u}$ such that $\hat{J} \cdot d\vec{u} = d\vec{H}$ when $d\vec{H}$ is given. If the tangent direction of the singularity surface is given by

$$d\vec{H} = \hat{J} \cdot d\vec{u} \qquad (11.108)$$

where $\vec{u}$ is an arbitrary vector defined over the singularity surface, then the inverse mapping theorem [13] gives

$$d\vec{u} = k\hat{J} \cdot d\vec{H} \qquad (11.109)$$

where $k = k(\vec{u})$ is the Gauss curvature given by

$$k = 2 \sum_{i,j=1}^{n} \frac{e_i e_j}{(\det[\mathbf{f}_i, \mathbf{f}_j, \mathbf{u}])^2} \qquad (11.110)$$

$e_i \equiv \vec{h}_i \cdot \vec{u}$ and $\det[\mathbf{f}_i, \mathbf{f}_j, \mathbf{u}] \equiv \vec{f}_i \times \vec{f}_j \cdot \vec{u}$.

Example 11.7   Four Single-Gimbal CMGs

For the two orthogonal pairs of two parallel single-gimbal CMGs, with its momentum saturation surface shown in Fig. 11.6, the singular momentum at $(H_x, H_y, H_z) = (0, 0, 4)$ has $\vec{u}(0, 0) = \vec{k}$ for $\vec{u} = \sin\theta_2 \vec{i} - \sin\theta_1 \cos\theta_2 \vec{j} + \cos\theta_1 \cos\theta_2 \vec{k}$, and

$$\vec{u}_1 = -\vec{j}, \quad \vec{u}_2 = \vec{i}, \quad \vec{H}_1 = -2\vec{j}, \quad \vec{H}_2 = 2\vec{i}$$
$$d\vec{u} = \vec{u}_1 d\theta_1 + \vec{u}_2 d\theta_2 = -\vec{j} d\theta_1 + \vec{i} d\theta_2$$
$$d\vec{H} = \vec{H}_1 d\theta_1 + \vec{H}_2 d\theta_2 = -2\vec{j} d\theta_1 + 2\vec{i} d\theta_2$$

For this somewhat trivial case, we have

$$\mathbf{G} = \begin{bmatrix} 4 & 0 \\ 0 & 4 \end{bmatrix}, \quad \mathbf{B} = \begin{bmatrix} -2 & 0 \\ 0 & -2 \end{bmatrix}, \quad \mathbf{C} = \begin{bmatrix} 1 & 0 \\ 0 & 1 \end{bmatrix}$$

$$\mathbf{D} = \begin{bmatrix} -1/2 & 0 \\ 0 & -1/2 \end{bmatrix}, \quad \mathbf{E} = \begin{bmatrix} 1/2 & 0 \\ 0 & 1/2 \end{bmatrix}$$

$$\mathbf{L} = \begin{bmatrix} -1/2 & 0 \\ 0 & -1/2 \end{bmatrix}, \quad \mathbf{J} = \begin{bmatrix} 2 & 0 \\ 0 & 2 \end{bmatrix}$$

$$k_1 = k_2 = -\frac{1}{2}, \quad k = k_1 k_2 = \frac{1}{4}, \quad \tilde{k} = -\frac{1}{2}$$

The singularity at $(H_x, H_y, H_z) = (0, 0, 4)$ is thus an elliptic point because $k > 0$, as can also be confirmed by the convexity of the momentum saturation surface shown in Fig. 11.6.

This example is mainly intended to verify some of the mathematical formulas derived in this section. It is not intended to demonstrate any significant practicality of this approach to the CMG singularity problem. A further study is necessary to fully exploit the differential surface geometry theory, as applied to the CMG singularity problem.

11.7 Singularity-Robust Steering Logic

In the preceding sections, the singularity problem inherent to the redundant CMG systems was examined by characterizing and visualizing the CMG singularities. This section presents the generalized singularity-robust steering logic (U.S. Patent No. 6,039,290, 21 March, 2000) by Wie et al. [2,3], which is a simple yet effective way of passing through, and also escaping from, any internal singularities.

11.7.1 Standard Pseudoinverse

Consider a CMG torque model simply given by

$$\mathbf{A}\dot{\mathbf{x}} = \boldsymbol{\tau} \tag{11.111}$$

where $\mathbf{A}$ is a $3 \times n$ Jacobian matrix, $\dot{\mathbf{x}} = (\dot{x}_1, \ldots, \dot{x}_n)$, and x_i is the ith gimbal-angle. For the given control torque command $\boldsymbol{\tau}$, a gimbal rate command $\dot{\mathbf{x}}$, often referred to as the *pseudoinverse steering logic*, is then obtained as

$$\dot{\mathbf{x}} = \mathbf{A}^+ \boldsymbol{\tau} \tag{11.112}$$

where

$$\mathbf{A}^+ = \mathbf{A}^T \left(\mathbf{A}\mathbf{A}^T\right)^{-1} \tag{11.113}$$

This pseudoinverse is the minimum two-norm solution of the following constrained minimization problem:

$$\min_{\mathbf{x}} \|\dot{\mathbf{x}}\|^2 \quad \text{subject to} \quad \mathbf{A}\dot{\mathbf{x}} = \boldsymbol{\tau} \tag{11.114}$$

where $\|\dot{\mathbf{x}}\|^2 = \dot{\mathbf{x}}^T \dot{\mathbf{x}}$. Most CMG steering laws determine the gimbal rate commands with some variant of the pseudoinverse.

The pseudoinverse is a special case of the weighted minimum two-norm solution

$$\dot{\mathbf{x}} = \mathbf{A}^+ \boldsymbol{\tau} \quad \text{where} \quad \mathbf{A}^+ = \mathbf{Q}^{-1}\mathbf{A}^T[\mathbf{A}\mathbf{Q}^{-1}\mathbf{A}^T]^{-1} \tag{11.115}$$

of the following constrained minimization problem:

$$\min_{\dot{\mathbf{x}}} \|\dot{\mathbf{x}}\|_{\mathbf{Q}}^2 \quad \text{subject to} \quad \mathbf{A}\dot{\mathbf{x}} = \boldsymbol{\tau} \tag{11.116}$$

where $\|\dot{\mathbf{x}}\|_{Q}^2 = \dot{\mathbf{x}}^T \mathbf{Q}\dot{\mathbf{x}}$ and $\mathbf{Q} = \mathbf{Q}^T > 0$.

If $\text{rank}(\mathbf{A}) < m$ for certain sets of gimbal-angles, or equivalently $\text{rank}(\mathbf{A}\mathbf{A}^T) < m$, when $\mathbf{A}$ is an $m \times n$ matrix, the pseudoinverse does not exist, and it is said that the pseudoinverse steering logic encounters singular states. This singular situation occurs when all individual CMG torque output vectors are perpendicular to the commanded torque direction. Equivalently, the singular situation occurs when all individual CMG momentum vectors have extremal projections onto the commanded torque $\boldsymbol{\tau}$.

Because the pseudoinverse, $\mathbf{A}^+ = \mathbf{A}^T(\mathbf{A}\mathbf{A}^T)^{-1}$, is the minimum two-norm solution of gimbal rates subject to the constraint $\mathbf{A}\dot{\mathbf{x}} = \boldsymbol{\tau}$, the pseudoinverse steering logic and all other pseudoinverse-based steering logic tend to leave inefficiently positioned CMGs alone causing the gimbal-angles to eventually "hang-up" in antiparallel singular arrangements. That is, they tend to steer the gimbals toward antiparallel singular states. Despite this deficiency, the pseudoinverse steering logic, or some variant of pseudoinverse, is commonly employed for most CMG systems because of its simplicity for onboard, real-time implementation.

11.7.2 Generalized Singularity-Robust Inverse

Most existing pseudoinverse-based, local inversion or tangent methods, including the singularity-robust (SR) inverse logic of the form [21]

$$\dot{\boldsymbol{\delta}} = \mathbf{A}^{\#}\mathbf{u} \qquad \text{where} \qquad \mathbf{A}^{\#} = \mathbf{A}^T(\mathbf{A}\mathbf{A}^T + \lambda\mathbf{I})^{-1} \tag{11.117}$$

do not always guarantee singularity avoidance and often unnecessarily constrain the operational momentum envelope of single-gimbal CMG systems. A single-gimbal CMG system with the SR inverse logic can become singular even in the presence of sensor noise. Furthermore, if it does become singular and a control torque is commanded along the singular direction, the system becomes trapped in the singular state because the SR inverse is unable to command nonzero gimbal rates.

A simple yet effective way of passing through, and also escaping from, any internal singularities has been developed by Wie et al. [2,3]. The proposed CMG steering logic of [2,3] is mainly intended for typical reorientation maneuvers in which precision pointing or tracking is not required during reorientation maneuvers, and it fully utilizes the available CMG momentum space in the presence of any singularities. Although there are special missions in which prescribed attitude trajectories are to be "exactly" tracked in the presence of internal singularities, most practical cases, however, will require a tradeoff between robust singularity transit/escape and the resulting, transient pointing errors.

The generalized SR inverse proposed in [2,3] is described by

$$\begin{aligned}
\mathbf{A}^{\#} &= [\mathbf{A}^T\mathbf{P}\mathbf{A} + \lambda\mathbf{I}_4]^{-1}\mathbf{A}^T\mathbf{P} \\
&= \mathbf{A}^T[\mathbf{P}\mathbf{A}\mathbf{A}^T + \lambda\mathbf{I}_3]^{-1}\mathbf{P} \\
&= \mathbf{A}^T[\mathbf{A}\mathbf{A}^T + \lambda\mathbf{P}^{-1}]^{-1} \\
&= \mathbf{A}^T[\mathbf{A}\mathbf{A}^T + \lambda\mathbf{E}]^{-1}
\end{aligned} \tag{11.118}$$

where

$$\mathbf{P}^{-1} \equiv \mathbf{E} = \begin{bmatrix} 1 & \epsilon_3 & \epsilon_2 \\ \epsilon_3 & 1 & \epsilon_1 \\ \epsilon_2 & \epsilon_1 & 1 \end{bmatrix} > 0 \tag{11.119}$$

The scalar λ and the off-diagonal elements ϵ_i are to be properly selected such that $\mathbf{A}^{\#}\mathbf{u} \neq 0$ for any nonzero constant $\mathbf{u}$.

Note that there exists always a null vector of $A^\#$ because rank($A^\#$) < 3 for any λ and ϵ_i when the Jacobian matrix A is singular. Consequently, a simple way of guaranteeing that $A^\# u \neq 0$ for any nonzero constant u command is to continuously modulate ϵ_i, for example, as follows:

$$\epsilon_i = \epsilon_0 \sin(\omega t + \phi_i) \tag{11.120}$$

where the amplitude ϵ_0, the modulation frequency ω, and the phases ϕ_i need to be appropriately selected [2,3].

The generalized SR inverse is based on the mixed, two-norm, and weighted least-squares minimization although the resulting effect is somewhat similar to that of artificially misaligning the commanded control torque vector from the singular vector directions. Because the proposed steering logic is based on the minimum two-norm, pseudoinverse solution, it does not explicitly avoid singularity encounters, but it rather approaches and rapidly transits unavoidable singularities whenever needed. The proposed logic effectively generates deterministic dither signals when the system becomes near singular. Any internal singularities can be escaped for any nonzero constant torque commands using the proposed steering logic.

Next, the singular-value-decomposition technique will be applied to the various pseudoinverse solutions, including the SR inverse. The Moore–Penrose generalized pseudoinverse will also be described for a possible application to the CMG steering problem.

11.7.3 Singular Value Decomposition of A, A⁺, and A#

Consider a 3 × 4 Jacobian matrix A of rank 3 without loss of generality. For such $A \in R^{3 \times 4}$ of rank 3, there exist orthonormal matrices $U \in R^{3 \times 3}$ and $V \in R^{4 \times 4}$ such that $U^T U = I_3$ and $V^T V = I_4$, and

$$A = U \Sigma V^T \tag{11.121}$$

where

$$\Sigma = \begin{bmatrix} \sigma_1 & 0 & 0 & 0 \\ 0 & \sigma_2 & 0 & 0 \\ 0 & 0 & \sigma_3 & 0 \end{bmatrix} \tag{11.122}$$

The positive numbers σ_1, σ_2, and σ_3 are called the *singular values* of A.

From Eq. (11.121), we have

$$(AA^T)U = U(\Sigma\Sigma^T) \quad \text{or} \quad (AA^T)u_i = \sigma_i^2 u_i \tag{11.123a}$$

$$(A^T A)V = V(\Sigma^T \Sigma) \quad \text{or} \quad (A^T A)v_i = \sigma_i^2 v_i \tag{11.123b}$$

where

$$U = \begin{bmatrix} u_1 & u_2 & u_3 \end{bmatrix} \tag{11.124a}$$

$$V = \begin{bmatrix} v_1 & v_2 & v_3 & v_4 \end{bmatrix} \tag{11.124b}$$

and $\mathbf{U}$ is the orthonormal modal matrix of $\mathbf{AA}^T$, whereas $\mathbf{V}$ is the orthonormal modal matrix of $\mathbf{A}^T\mathbf{A}$. The modal form of $\mathbf{AA}^T$ is $\mathbf{\Sigma\Sigma}^T$, whereas the modal form of $\mathbf{A}^T\mathbf{A}$ is $\mathbf{\Sigma}^T\mathbf{\Sigma}$. The columns of $\mathbf{U}$ are called the left singular vectors of $\mathbf{A}$ or the orthonormal eigenvectors of $\mathbf{AA}^T$. Similarly, the columns of $\mathbf{V}$ are called the right singular vectors of $\mathbf{A}$ or the orthonormal eigenvectors of $\mathbf{A}^T\mathbf{A}$.

The singular values of $\mathbf{A}$ are thus defined to be the positive square roots of the eigenvalues of $\mathbf{AA}^T$, that is,

$$\sigma_i(\mathbf{A}) = \sqrt{\lambda_i(\mathbf{AA}^T)} \qquad (11.125)$$

where $\lambda_i(\mathbf{AA}^T)$ denotes the ith eigenvalue of $\mathbf{AA}^T$ and all $\lambda_i(\mathbf{AA}^T) \geq 0$. Furthermore, we have

$$\det(\mathbf{AA}^T) = (\sigma_1\sigma_2\sigma_3)^2 \qquad (11.126)$$

The largest and smallest singular values of $\mathbf{A}$, denoted by $\overline{\sigma}(\mathbf{A})$ and $\underline{\sigma}(\mathbf{A})$, respectively, are given by

$$\overline{\sigma}(\mathbf{A}) = \sqrt{\lambda_{\max}(\mathbf{AA}^T)} \qquad (11.127a)$$

$$\underline{\sigma}(\mathbf{A}) = \sqrt{\lambda_{\min}(\mathbf{AA}^T)} \qquad (11.127b)$$

The choice of $\mathbf{AA}^T$ rather than $\mathbf{A}^T\mathbf{A}$ in the definition of singular values is arbitrary. Only the nonzero singular values are usually of real interest, and their number is the rank of the matrix. The matrix $\mathbf{A}^T\mathbf{A}$ is a square matrix of order four and is a positive semidefinite symmetric matrix.

From Eq. (11.121), $\mathbf{A}$ can be expanded in terms of the singular vectors $\mathbf{u}_i$ and $\mathbf{v}_i$ as follows:

$$\mathbf{A} = \begin{bmatrix} \mathbf{u}_1 & \mathbf{u}_2 & \mathbf{u}_3 \end{bmatrix} \begin{bmatrix} \sigma_1 & 0 & 0 & 0 \\ 0 & \sigma_2 & 0 & 0 \\ 0 & 0 & \sigma_3 & 0 \end{bmatrix} \begin{bmatrix} \mathbf{v}_1^T \\ \mathbf{v}_2^T \\ \mathbf{v}_3^T \\ \mathbf{v}_4^T \end{bmatrix}$$

$$= \sum_{i=1}^{3} \sigma_i \mathbf{u}_i \mathbf{v}_i^T \qquad (11.128)$$

Using Eq. (11.121), it can be shown that the pseudoinverse $\mathbf{A}^+ = \mathbf{A}^T(\mathbf{AA}^T)^{-1}$ can also be expanded in terms of the singular vectors $\mathbf{u}_i$ and $\mathbf{v}_i$ as follows:

$$\mathbf{A}^+ = \mathbf{V}\mathbf{\Sigma}^+\mathbf{U}^T \qquad (11.129)$$

where

$$\mathbf{\Sigma}^+ = \begin{bmatrix} 1/\sigma_1 & 0 & 0 \\ 0 & 1/\sigma_2 & 0 \\ 0 & 0 & 1/\sigma_3 \\ 0 & 0 & 0 \end{bmatrix} \qquad (11.130)$$

Consequently, $\mathbf{A}^+$ can be expanded in terms of the singular vectors $\mathbf{u}_i$ and $\mathbf{v}_i$ as follows:

$$\mathbf{A}^+ = \sum_{i=1}^{3} \left(\frac{1}{\sigma_i}\right) \mathbf{v}_i \mathbf{u}_i^T \qquad (11.131)$$

and we have

$$\dot{\delta} = \mathbf{A}^+ \mathbf{u} = \sum_{i=1}^{3} \left(\frac{1}{\sigma_i}\right) \mathbf{v}_i \mathbf{u}_i^T \mathbf{u} \qquad (11.132)$$

If the commanded torque vector $\mathbf{u}$ lies along one of the column vectors of $\mathbf{U}$, that is, $\mathbf{u}_i$, the gimbal rates computed by the pseudoinverse become

$$\dot{\delta} = \mathbf{A}^+ \mathbf{u} = \left(\frac{1}{\sigma_i}\right) \mathbf{v}_i \qquad (11.133)$$

where $\mathbf{v}_i$ is the ith column of $\mathbf{V}$. Furthermore, if σ_i is zero, then the gimbal rate command for the commanded torque vector along $\mathbf{u}_i$ becomes infinity.

If $\mathrm{rank}(\mathbf{A}) < 3$ with nonzero singular values σ_1 and σ_2, we can directly use the Moore–Penrose generalized pseudoinverse defined as

$$\mathbf{A}^+ = \begin{bmatrix} \mathbf{v}_1 & \mathbf{v}_2 & \mathbf{v}_3 & \mathbf{v}_4 \end{bmatrix} \begin{bmatrix} 1/\sigma_1 & 0 & 0 \\ 0 & 1/\sigma_2 & 0 \\ 0 & 0 & 0 \\ 0 & 0 & 0 \end{bmatrix} \begin{bmatrix} \mathbf{u}_1^T \\ \mathbf{u}_2^T \\ \mathbf{u}_3^T \end{bmatrix}$$

$$= \sum_{i=1}^{2} \left(\frac{1}{\sigma_i}\right) \mathbf{v}_i \mathbf{u}_i^T \qquad (11.134)$$

Note that the Moore–Penrose generalized pseudoinverse is defined for any rank-deficient matrix $\mathbf{A}$, that is, when neither $(\mathbf{A}\mathbf{A}^T)^{-1}$ nor $(\mathbf{A}^T\mathbf{A})^{-1}$ exists. The MATLAB® command for computing the Moore–Penrose generalized pseudoinverse is pinv($\mathbf{A}$).

The SR inverse $\mathbf{A}^{\#} = \mathbf{A}^T[\mathbf{A}\mathbf{A}^T + \lambda\mathbf{I}]^{-1}$ can also be expanded as

$$\mathbf{A}^{\#} = \mathbf{V}\Sigma^{\#}\mathbf{U}^T \qquad (11.135)$$

where

$$\Sigma^{\#} = \begin{bmatrix} \sigma_1/(\sigma_1^2 + \lambda) & 0 & 0 \\ 0 & \sigma_2/(\sigma_2^2 + \lambda) & 0 \\ 0 & 0 & \sigma_3/(\sigma_3^2 + \lambda) \\ 0 & 0 & 0 \end{bmatrix} \qquad (11.136)$$

Consequently, $\mathbf{A}^{\#}$ can be expanded in terms of the singular vectors $\mathbf{u}_i$ and $\mathbf{v}_i$ as follows:

$$\mathbf{A}^{\#} = \sum_{i=1}^{3} \left(\frac{\sigma_i}{\sigma_i^2 + \lambda} \right) \mathbf{v}_i \mathbf{u}_i^T \qquad (11.137)$$

and we have

$$\dot{\boldsymbol{\delta}} = \mathbf{A}^{\#}\mathbf{u} = \sum_{i=1}^{3} \left(\frac{\sigma_i}{\sigma_i^2 + \lambda} \right) \mathbf{v}_i \mathbf{u}_i^T \mathbf{u} \qquad (11.138)$$

If the commanded torque vector $\mathbf{u}$ lies along one of the column vectors of $\mathbf{U}$, that is, $\mathbf{u}_i$, the gimbal rates computed by the SR inverse become

$$\dot{\boldsymbol{\delta}} = \mathbf{A}^{\#}\mathbf{u} = \frac{\sigma_i}{\sigma_i^2 + \lambda} \mathbf{v}_i \qquad (11.139)$$

where $\mathbf{v}_i$ is the ith column of $\mathbf{V}$. Furthermore, if σ_i is zero, then the gimbal rate command for the commanded torque vector along $\mathbf{u}_i$ becomes zero. At singular points, the gimbal rate command for the commanded $\mathbf{u}$ along the singular direction is zero. Consequently, singular points cannot be escaped simply using the standard SR inverse.

Now consider the generalized singularity-robust steering logic of the form

$$\dot{\boldsymbol{\delta}} = \mathbf{A}^{\#}\mathbf{u} = \mathbf{A}^T[\mathbf{A}\mathbf{A}^T + \lambda\mathbf{E}]^{-1}\mathbf{u} \qquad (11.140)$$

where

$$\mathbf{E} = \begin{bmatrix} 1 & \epsilon_3 & \epsilon_2 \\ \epsilon_3 & 1 & \epsilon_1 \\ \epsilon_2 & \epsilon_1 & 1 \end{bmatrix} > 0 \qquad (11.141)$$

Because λ and ϵ_i are to be properly selected such that $\mathbf{A}^{\#}\mathbf{u} \neq 0$ for any nonzero constant $\mathbf{u}$, the gimbal rate command will never become zero for any nonzero constant $\mathbf{u}$, and consequently, any internal singularities can be escaped using the the generalized singularity-robust steering logic.

Example 11.8   Four Single-Gimbal CMGs

Consider the typical pyramid mounting arrangement of four single-gimbal CMGs. A skew angle β of 53.13 deg (i.e., $\cos\beta = 0.6$, $\sin\beta = 0.8$) and constant unit momentum magnitude for each CMG are assumed. Initial gimbal-angles are $\boldsymbol{\delta} = (90, 0, -90, 0)$ deg, and the commanded torque vector is given by $\mathbf{u} = (-1, 0, 0)$. That is, the system is at an internal elliptic singularity, and the commanded torque is along the singular, vector direction.

For this case, the Jacobian matrix is simply given by

$$\mathbf{A} = \begin{bmatrix} 0 & 0 & 0 & 0 \\ -1 & -0.6 & -1 & 0.6 \\ 0 & 0.8 & 0 & 0.8 \end{bmatrix} \qquad (11.142)$$

Because $\mathrm{rank}(\mathbf{A}) = 2 < 3$, the standard pseudoinverse $\mathbf{A}^+ = \mathbf{A}^T(\mathbf{A}\mathbf{A}^T)^{-1}$ does not exist.

However, the singular value decomposition of $\mathbf{A}$ can be obtained using the MATLAB® as

$$\mathbf{A} = \mathbf{U}\mathbf{\Sigma}\mathbf{V} \tag{11.143}$$

where

$$\mathbf{U} = \begin{bmatrix} 0 & 0 & -1 \\ -1 & 0 & 0 \\ 0 & -1 & 0 \end{bmatrix} \tag{11.144}$$

$$\mathbf{\Sigma} = \begin{bmatrix} 1.6492 & 0 & 0 & 0 \\ 0 & 1.1314 & 0 & 0 \\ 0 & 0 & 0 & 0 \end{bmatrix} \tag{11.145}$$

$$\mathbf{V} = \begin{bmatrix} 0.6063 & 0 & -0.7952 & 0 \\ 0.3638 & -0.7071 & 0.2774 & -0.5392 \\ 0.6063 & 0 & 0.4623 & 0.6470 \\ -0.3638 & -0.7071 & -0.2774 & 0.5392 \end{bmatrix} \tag{11.146}$$

Furthermore, the Moore–Penrose generalized pseudoinverse can be obtained as

$$\mathrm{pinv}(\mathbf{A}) = \begin{bmatrix} 0 & -0.3676 & 0 \\ 0 & -0.2206 & 0.6250 \\ 0 & -0.3676 & 0 \\ 0 & 0.2206 & 0.6250 \end{bmatrix} \tag{11.147}$$

$$\Rightarrow \dot{\boldsymbol{\delta}} = \mathrm{pinv}(\mathbf{A})\begin{bmatrix} -1 \\ 0 \\ 0 \end{bmatrix} = \begin{bmatrix} 0 \\ 0 \\ 0 \\ 0 \end{bmatrix} \tag{11.148}$$

Because the commanded gimbal rates are zero, this singularity cannot be escaped via the Moore–Penrose generalized pseudoinverse.

The SR inverse with $\mathbf{P} = \mathbf{I}_3$ and $\mathbf{Q} = 0.01\mathbf{I}_4$ can be obtained as

$$\mathbf{A}^{\#} = [\mathbf{A}^T\mathbf{A} + 0.01\mathbf{I}_4]^{-1}\mathbf{A}^T = \mathbf{A}^T[\mathbf{A}\mathbf{A}^T + 0.01\mathbf{I}_3]^{-1}$$

$$= \begin{bmatrix} 0 & -0.3663 & 0 \\ 0 & -0.2198 & 0.6202 \\ 0 & -0.3663 & 0 \\ 0 & 0.2198 & 0.6202 \end{bmatrix} \tag{11.149}$$

$$\Rightarrow \dot{\boldsymbol{\delta}} = \mathbf{A}^{\#}\begin{bmatrix} -1 \\ 0 \\ 0 \end{bmatrix} = \begin{bmatrix} 0 \\ 0 \\ 0 \\ 0 \end{bmatrix} \tag{11.150}$$

Again, the commanded gimbal rates are zero. It can be further verified that this singularity cannot be escaped via the singularity robust inverse with any diagonal weighting matrices $\mathbf{P}$ and $\mathbf{Q}$.

Now consider the generalized SR inverse with $\lambda = 0.01$ and $\epsilon_i = 0.01$, as follows:

$$\mathbf{A}^{\#} = \mathbf{A}^T[\mathbf{A}\mathbf{A}^T + 0.01\mathbf{E}]^{-1} \tag{11.151a}$$

$$= \begin{bmatrix} 3.6627\text{e-}3 & -3.6630\text{e-}1 & 2.8111\text{e-}5 \\ -4.0037\text{e-}3 & -2.1980\text{e-}1 & 6.2017\text{e-}1 \\ 3.6627\text{e-}3 & -3.6630\text{e-}1 & 2.8111\text{e-}5 \\ -8.3990\text{e-}3 & 2.1976\text{e-}1 & 6.2014\text{e-}1 \end{bmatrix} \tag{11.151b}$$

Consequently, the commanded gimbal rates become

$$\dot{\boldsymbol{\delta}} = \mathbf{A}^{\#}\begin{bmatrix} -1 \\ 0 \\ 0 \end{bmatrix} = \begin{bmatrix} -3.6627\text{e-}3 \\ 4.0037\text{e-}3 \\ -3.6627\text{e-}3 \\ 8.3990\text{e-}3 \end{bmatrix} \tag{11.152}$$

which is a nonzero vector, and the actual torque vector output becomes

$$\mathbf{u} = \mathbf{A}\mathbf{A}^{\#}\begin{bmatrix} -1 \\ 0 \\ 0 \end{bmatrix}$$

$$= \begin{bmatrix} 0 & 0 & 0 \\ -9.9626\text{e-}3 & 9.9634\text{e-}1 & -7.6463\text{e-}5 \\ -9.9221\text{e-}3 & -3.5983\text{e-}5 & -9.9225\text{e-}1 \end{bmatrix}\begin{bmatrix} -1 \\ 0 \\ 0 \end{bmatrix}$$

$$= \begin{bmatrix} 0 \\ 9.9626\text{e-}3 \\ 9.9221\text{e-}3 \end{bmatrix} \tag{11.153}$$

11.8 Singularity Avoidance/Escape Logic

The singularity-robust CMG steering logic proposed by Wie et al. [2,3] is a simple, yet effective way of passing through, and also escaping from, any internal singularities. However, it is unable to escape the external saturation singularities of certain special CMG configurations, even when momentum desaturation is requested. Consequently, a new steering logic (U.S. Patent No. 6,917,862, 12 July 2005), based on a mixed weighted two-norm and least-squares optimization solution, has been developed in [6] to overcome this deficiency of being trapped in the momentum saturation singularities. The new steering logic also provides a simple means of avoiding troublesome, internal elliptic singularities that are commonly encountered by most pseudoinverse-based steering logic, including the singularity-robust steering logic of [2,3].

In this section, the simplicity and effectiveness of the new steering logic will be demonstrated using various CMG systems, such as two or three parallel single-gimbal CMG configuration, a pyramid array of four single-gimbal CMGs, and four parallel double-gimbal CMGs of the International Space Station. As to be noticed in various simulation results of this section, transient CMG torque errors are inevitable while escaping or passing through elliptic singularities. However, the resulting attitude transient dynamics are often acceptable because precision pointing is not required during large-angle slew maneuvers or CMG momentum desaturation maneuvers of most imaging satellites. Such slew maneuvers are usually performed by a closed-loop attitude control system.

11.8.1 Mixed Two-Norm and Least-Squares Minimization

Consider

$$\dot{x} = A^\# \tau \tag{11.154}$$

where

$$
\begin{aligned}
A^\# &= [A^T P A + Q]^{-1} A^T P \\
&\equiv Q^{-1} A^T [A Q^{-1} A^T + P^{-1}]^{-1} \\
&\equiv W A^T [A W A^T + V]^{-1} \tag{11.155}
\end{aligned}
$$

where $W \equiv Q^{-1}$ and $V \equiv P^{-1}$. For a $3 \times n$ Jacobian matrix A, $[A^T P A + Q]$ is an $n \times n$ matrix and $[A W A^T + V]$ is a 3×3 matrix.

The singularity-robust inverse of the form (11.155) is the solution of the well-known, mixed two-norm and least-squares minimization problem:

$$\min_{\dot{x}}(e^T P e + \dot{x}^T Q \dot{x}) \tag{11.156}$$

where $e = A\dot{x} - \tau$ is the torque error vector. Because P and Q are positive definite matrices, $[A^T P A + Q]$ and $[A W A^T + V]$ are always nonsingular.

The weighting matrices P and Q (equivalently, W and V) must be properly chosen 1) to obtain acceptable levels of torque errors and gimbal rates, 2) to escape any internal as well as external singularities, and 3) to pass through singularities or avoid singularity encounters.

For example, P is chosen such that $A^T P \tau \neq 0$ for any torque command τ, as follows:

$$
P^{-1} \equiv V = \lambda \begin{bmatrix} 1 & \epsilon_3 & \epsilon_2 \\ \epsilon_3 & 1 & \epsilon_1 \\ \epsilon_2 & \epsilon_1 & 1 \end{bmatrix} > 0 \tag{11.157}
$$

and λ and ϵ_i are appropriately selected as discussed earlier.

Because the condition $A^T P \tau \neq 0$, which is a necessary condition, is not sufficient for escaping and/or avoiding singularities, the matrix Q also needs to be properly chosen as

$$
Q^{-1} \equiv W = \begin{bmatrix} W_1 & \lambda & \lambda & \lambda \\ \lambda & W_2 & \lambda & \lambda \\ \lambda & \lambda & W_3 & \lambda \\ \lambda & \lambda & \lambda & W_4 \end{bmatrix} > 0 \tag{11.158}
$$

for a case of $n = 4$. Different values of W_i and/or nonzero off-diagonal elements are required to escape all types of singularities, including the external saturation singularities when CMG momentum desaturation is requested. Furthermore, a proper choice of $\mathbf{W} \neq \mathbf{I}$ provides an effective means for "explicitly" avoiding singularity encounters.

Several examples of demonstrating the significance of $\mathbf{W} \neq \mathbf{I}$ for escaping a certain type of external saturation singularities for which the singularity-robust steering logic of [2,3] fails will be presented in this section. Examples of avoiding an internal elliptic singularity that the singularity-robust steering logic of [2,3] had to pass through will also be presented in this section.

11.8.2 Two Parallel Single-Gimbal CMGs

Most agile imaging satellites, such as the Earth-observing satellite, require large control torques only along the roll and pitch axes, not along the Earth-pointing yaw axis. For such a case, two or three single-gimbal CMGs with parallel gimbal axes can be employed for the roll/pitch control of an Earth-observing spacecraft, while the yaw axis is controlled by a smaller reaction wheel.

For a system of two CMGs without redundancy, the momentum vectors $\vec{h}_1$ and $\vec{h}_2$ move in the (x, y) plane normal to the gimbal axis. For such scissored single-gimbal CMGs, the total CMG momentum vector is simply represented as

$$\mathbf{H} = \begin{bmatrix} \cos x_1 + \cos x_2 \\ \sin x_1 + \sin x_2 \end{bmatrix} \tag{11.159}$$

where a constant unit momentum for each CMG is assumed.

Defining a new set of gimbal-angles α and β as follows

$$\alpha = \frac{x_1 + x_2}{2}, \quad \beta = \frac{x_2 - x_1}{2} \tag{11.160}$$

where α is called the rotation angle and β the scissor angle, we obtain

$$\mathbf{H} = 2 \begin{bmatrix} \cos \alpha \cos \beta \\ \sin \alpha \cos \beta \end{bmatrix} \tag{11.161}$$

and

$$\dot{\mathbf{H}} = \mathbf{A}\dot{\mathbf{x}} \tag{11.162}$$

where $\dot{\mathbf{H}} = (\dot{H}_x, \dot{H}_y)$, $\dot{\mathbf{x}} = (\dot{\alpha}, \dot{\beta})$, and $\mathbf{A}$ is the Jacobian matrix defined as

$$\mathbf{A} = 2 \begin{bmatrix} -\sin \alpha \cos \beta & -\cos \alpha \sin \beta \\ \cos \alpha \cos \beta & -\sin \alpha \sin \beta \end{bmatrix} \tag{11.163}$$

For the internal antiparallel singularity at $\mathbf{x} = (\alpha, \beta) = (0, \pi/2)$, the Jacobian matrix becomes

$$\mathbf{A} = \begin{bmatrix} 0 & -2 \\ 0 & 0 \end{bmatrix}$$

with its singular vector $\mathbf{u} = (0, 1)$; that is, $\mathbf{A}^T \mathbf{u} = 0$. A null space vector of $\mathbf{A}$ such that $\mathbf{A}\dot{\mathbf{x}} = 0$ is $\dot{\mathbf{x}} = \text{null}(\mathbf{A}) = (1, 0)$.

Although null motion does exist from this singularity, this hyperbolic singularity cannot be escaped by null motion because the singular configuration ($\beta = \pi/2$) remains undisturbed during this null motion along the null manifold (or a degenerate trajectory). Although null motion can be generated at this hyperbolic singularity, this example demonstrates that the mere existence of null motion does not guarantee escape from a singularity. However, this type of internal singularity can be easily escaped by the singularity-robust steering logic of [2,3].

For the saturation singularity at $\mathbf{x} = (\alpha, \beta) = (0, 0)$, the Jacobian matrix becomes

$$\mathbf{A} = \begin{bmatrix} 0 & 0 \\ 2 & 0 \end{bmatrix}$$

with its singular vector $\mathbf{u} = (-1, 0)$ and its null space vector $\dot{\mathbf{x}} = \text{null}(\mathbf{A}) = (0, 1)$. This external elliptic singularity cannot be escaped by null motion because null motion does not exist in the vicinity of this momentum saturation singularity.

The generalized SR inverse with $\lambda = 0.1$ and $\epsilon = 0.1$ becomes

$$\mathbf{A}^{\#} = \mathbf{A}^T [\mathbf{A}\mathbf{A}^T + 0.1\mathbf{E}]^{-1} = \begin{bmatrix} -0.0488 & 0.4879 \\ 0 & 0 \end{bmatrix}$$

$$\Rightarrow \dot{\mathbf{x}} = \mathbf{A}^{\#} \begin{bmatrix} -1 \\ 0 \end{bmatrix} = \begin{bmatrix} 0.0488 \\ 0 \end{bmatrix}$$

Note that $\dot{\mathbf{x}} = (\dot{\alpha}, \dot{\beta}) \neq 0$, but $\dot{\beta} = 0$, that is, the system remains singular while α changes. In fact, $\dot{\beta} = 0$ for any torque command vector $\boldsymbol{\tau}$ and any ϵ_i. Consequently, the singularity-robust steering logic of [2,3] is unable to command a nonzero $\dot{\beta}$ for this special case.

However, this external singularity can be escaped by the new singularity escape/avoidance steering logic of the form $\dot{\mathbf{x}} = \mathbf{A}^{\#} \boldsymbol{\tau}$, where

$$\mathbf{A}^{\#} = \mathbf{W}\mathbf{A}^T [\mathbf{A}\mathbf{W}\mathbf{A}^T + \mathbf{V}]^{-1}$$

As can be seen in Fig. 11.14, the momentum saturation singularity is escaped using the new steering logic with

$$\mathbf{V} = \lambda \begin{bmatrix} 1 & \epsilon \\ \epsilon & 1 \end{bmatrix}, \quad \mathbf{W} = \begin{bmatrix} 1 & \lambda \\ \lambda & 1 \end{bmatrix}$$

where $\epsilon = 0.1 \cos t$ and $\lambda = 0.01 \exp[-10 \det (\mathbf{A}\mathbf{A}^T)]$.

In Fig. 11.14, (u_x, u_y) represents the actual torques generated by the CMG system for the commanded torques of $(\tau_x, \tau_y) = (-1, 0)$ for $0 \leq t \leq 3$ s. The CMG momentum is completely desaturated at $t = 2$ s, but the desaturation torque is commanded until $t = 3$ s to further explore any problem of encountering an internal singularity. (For all of the simulation results presented in this section, the commanded torque level as well as the CMG momentum is normalized as one

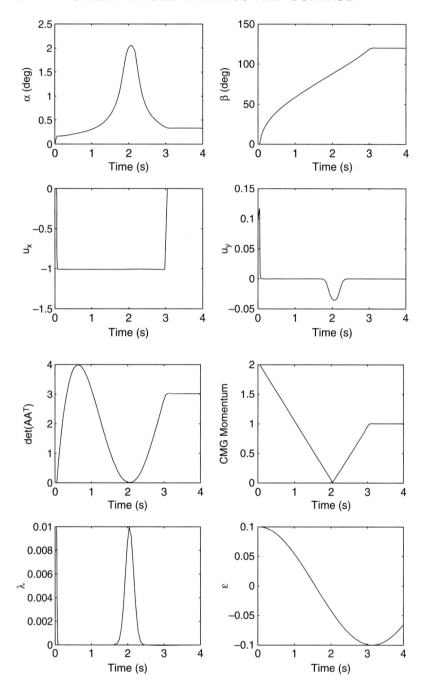

Fig. 11.14   Singularity escape (momentum desaturation) simulation results for two parallel single-gimbal CMGs with the proposed new steering logic.

without loss of generality. Thus it should be appropriately adjusted depending on the actual values of gimbal rate limit and CMG momentum magnitude.) As can be seen in Fig. 11.14, the new steering logic also provides a simple yet effective way of passing through the 0H internal hyperbolic singularity, but with a small transient torque error in u_y, which is inevitable while passing through the hyperbolic singularity.

For this special case of two parallel single-gimbal CMGs, there exists a direct inverse solution. A steering logic based on the direct inverse of $\mathbf{A}$ can be obtained for this simple 2×2 problem, as follows:

$$\dot{\mathbf{x}} = \mathbf{A}^{-1} \boldsymbol{\tau} \tag{11.164}$$

where

$$
\begin{aligned}
\mathbf{A}^{-1} &= \frac{1}{2|\mathbf{A}|} \begin{bmatrix} -\sin \alpha \cos \beta & -\cos \alpha \sin \beta \\ \cos \alpha \cos \beta & -\sin \alpha \sin \beta \end{bmatrix} \\
&= \frac{1}{2} \begin{bmatrix} -\sin \alpha / \cos \beta & \cos \alpha / \cos \beta \\ -\cos \alpha / \sin \beta & -\sin \alpha / \sin \beta \end{bmatrix}
\end{aligned} \tag{11.165}
$$

Then, we obtain

$$\dot{\alpha} = \frac{-\sin \alpha \, \tau_x + \cos \alpha \, \tau_y}{2(\cos \beta + \lambda_1)} \tag{11.166}$$

$$\dot{\beta} = \frac{-\cos \alpha \, \tau_x - \sin \alpha \, \tau_y}{2(\sin \beta + \lambda_2)} \tag{11.167}$$

where

$$\lambda_1 = \lambda_0 \exp\left[-\mu \det\left(\mathbf{A}\mathbf{A}^T\right)\right] \sin \beta \tag{11.168}$$

$$\lambda_2 = \lambda_0 \exp\left[-\mu \det\left(\mathbf{A}\mathbf{A}^T\right)\right] \cos \beta \tag{11.169}$$

which are included in the steering law to avoid dividing by zero when the system becomes singular.

As shown in Fig. 11.15, the momentum saturation singularity is easily escaped using this direct-inverse steering logic for $(\tau_x, \tau_y) = (-1, 0)$ for $0 \le t \le 3$ s with $\lambda_0 = 0.01$ and $\mu = 10$. In Fig. 11.15, (u_x, u_y) represents the actual torques generated by the CMG system for the commanded torques: $(\tau_x, \tau_y) = (-1, 0)$ for $0 \le t \le 3$ s. This direct inverse steering logic provides a very smooth transit through the 0H internal singularity, without any torque transient error.

However, such an exact direct-inverse steering logic does not exist for the practical cases with redundant CMGs, as to be discussed in the remainder of this section.

11.8.3 Three Parallel Single-Gimbal CMGs

For a system of three single-gimbal CMGs with parallel gimbal axes, $\mathbf{W}$ should be selected as $W_1 \ne W_2 \ne W_3$ to escape the 3H saturation singularity at $(x_1, x_2, x_3) = (0, 0, 0)$ with $(\tau_x, \tau_y) = (-1, 0)$, whereas $\mathbf{V} = \lambda \mathbf{E}$ is simply

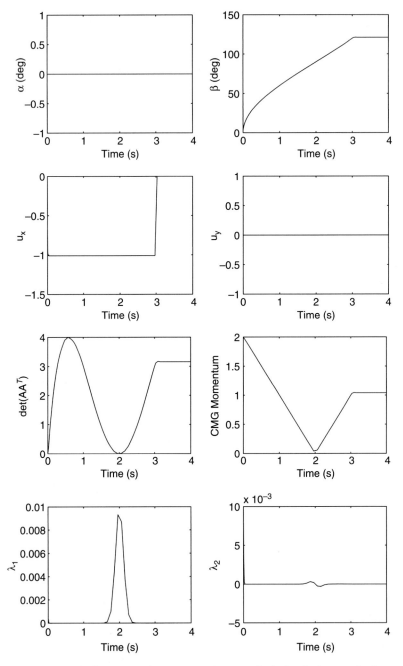

Fig. 11.15   Singularity escape (momentum desaturation) simulation results for two parallel single-gimbal CMGs with the direct-inverse steering logic.

selected as $\epsilon_i = 0.1 \cos t$ and $\lambda = 0.01 \exp[-10 \det (\mathbf{AA}^T)]$. Simulation results of $(\tau_x, \tau_y) = (-1, 0)$ for $0 \leq t \leq 3$ s with $\mathbf{W} = \text{diag}\{1, 2, 3\}$ are shown in Fig. 11.16.

11.8.4 *Pyramid Array of Four Single-Gimbal CMGs*

For a typical pyramid mount of four single-gimbal CMGs with skew angle of β, we have the Jacobian matrix as

$$\mathbf{A} = \begin{bmatrix} -c\beta \cos x_1 & \sin x_2 & c\beta \cos x_3 & -\sin x_4 \\ -\sin x_1 & -c\beta \cos x_2 & \sin x_3 & c\beta \cos x_4 \\ s\beta \cos x_1 & s\beta \cos x_2 & s\beta \cos x_3 & s\beta \cos x_4 \end{bmatrix} \qquad (11.170)$$

where x_i is the ith gimbal-angle, $c\beta \equiv \cos \beta$, and $s\beta \equiv \sin \beta$.

The Pleiades-HR imaging satellite described in Sec. 11.1 will be controlled by a pyramid cluster of four small single-gimbal CMGs. The z axis of the pyramid cluster is aligned along the line of sight of the Earth-pointing optical imaging system of Pleiades-HR imaging satellite, and a 30-deg skew angle is selected to have a larger angular momentum capability along the x and y axes.

When $\beta = 90$ deg, we have a four-CMG configuration with two orthogonal pairs of scissored CMGs. This case is also of practical importance because of its simple arrangement of four single-gimbal CMGs for three-axis control applications. Similar to the case of three parallel single-gimbal CMGs, $\mathbf{W} \neq \mathbf{I}$ is also required to escape the saturation singularity, as shown in Fig. 11.17. For this simulation, $\mathbf{W} = \text{diag}\{1, 2, 3, 4\}$ and $\mathbf{V} = \lambda \mathbf{E}$ with $\epsilon_i = 0.1 \cos t$ and $\lambda = 0.01 \exp[-10 \det (\mathbf{AA}^T)]$ were used to escape the saturation singularity at $(\pi/2, \pi/2, \pi/2, \pi/2)$ with $(\tau_x, \tau_y, \tau_z) = (0, 0, -1)$ for $0 \leq t \leq 4$ s. It can be seen in Fig. 11.17 that an internal singularity is encountered at $t = 2$ s, but it is rapidly passed through.

For all other simulation results presented in the remainder of this section, $\mathbf{V} = \lambda \mathbf{E}$ with $\epsilon_i = 0.1 \cos t$ and $\lambda = 0.01 \exp\left[-10 \det (\mathbf{AA}^T)\right]$ were used.

In Fig. 11.18, simulation results are shown for a typical pyramid array of four single-gimbal CMGs ($\beta = 53.13$ deg) with initial gimbal-angles of $(0, 0, 0, 0)$ and a torque command of $(\tau_x, \tau_y, \tau_z) = (1, 0, 0)$. For this typical roll torque command simulation with zero initial gimbal-angles, $\mathbf{W} = \text{diag}\{1, 1, 1, 1\}$ was used, that is, the singularity-robust steering logic of [2,3] was used for this simulation. It can be seen that the well-known internal elliptic singularity at $(-\pi/2, 0, \pi/2, 0)$ is encountered, but it is successfully passed through. The inevitable transient torque errors during the singularity transit can be seen in Fig. 11.18.

In Fig. 11.19, simulation results with $\mathbf{W} = \text{diag}\{10^{-4}, 1, 1, 1\}$ are shown for the preceding case of a typical pyramid array of four single-gimbal CMGs ($\beta = 53.13$ deg) with initial gimbal-angles of $(0, 0, 0, 0)$ and a torque command of $(\tau_x, \tau_y, \tau_z) = (1, 0, 0)$. It can be seen that the well-known internal elliptic singularity at $(-\pi/2, 0, \pi/2, 0)$ is not "directly" encountered. However, the inevitable transient torque errors, caused by skirting such a troublesome, impassable elliptic singularity, can also be seen in Fig. 11.19. A different weighting matrix of $\mathbf{W} = \text{diag}\{1, 1, 10^{-4}, 1\}$ can also be employed for this case.

For the simulations shown in Figs. 11.18 and 11.19, the maximum gimbal rate of each CMG was explicitly limited to ± 2 rad/s. Such a gimbal rate saturation

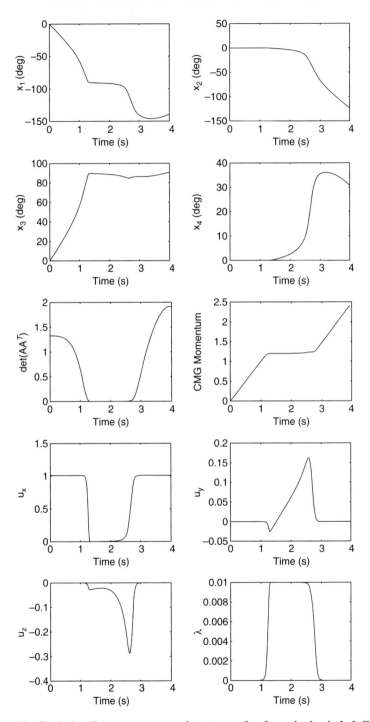

Fig. 11.18 Typical roll torque command test case for four single-gimbal CMGs ($\beta = 53.13$ deg), starting from nonsingular, zero-initial gimbal-angles and passing through the well-known internal elliptic singularity, with the steering logic of [2,3] with $W = \text{diag}\{1, 1, 1, 1\}$.

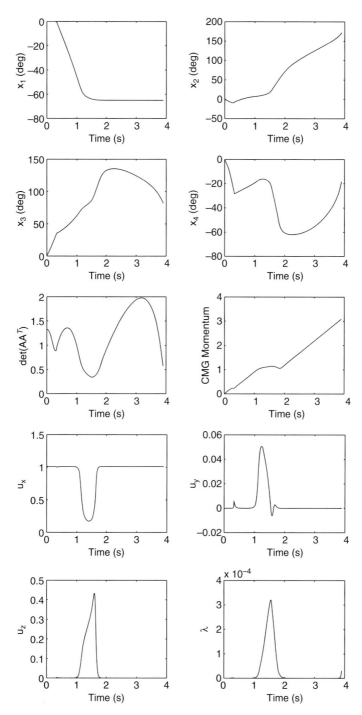

Fig. 11.19 Typical roll torque command test case for four single-gimbal CMGs ($\beta = 53.13$ deg), starting from nonsingular, zero-initial gimbal-angles, but with the new steering logic of $W = \mathrm{diag}\{10^{-4}, 1, 1, 1\}$

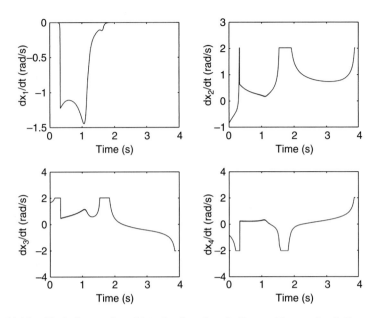

Fig. 11.20   Gimbal rate time histories for singularity avoidance simulation with a gimbal rate limit of ±2 rad/s.

limit did not affect the singularity-avoidance performance of the proposed steering logic. The gimbal rate time histories associated with the simulation shown in Fig. 11.19 are provided in Fig. 11.20. Although the gimbal rate command is limited to ±2 rad/s, the gimbal-angle and CMG torque time histories shown in Fig. 11.19 are all acceptable.

11.8.5   Four Parallel Double-Gimbal CMGs

The simplicity and effectiveness of the proposed steering logic can be demonstrated also for a system of double-gimbal control moment gyros (DGCMG). For a DGCMG, the rotor is suspended inside two gimbals, and consequently the rotor momentum can be oriented on a sphere along any direction provided no restrictive gimbal stops. For the different purposes of redundancy management and failure accommodation, several different arrangements of DGCMGs have been developed, such as three orthogonally mounted DGCMGs used in the Skylab and four parallel mounted DGCMGs employed for the International Space Station.

As shown by Kennel [22], mounting of DGCMGs of unlimited outer gimbal-angle freedom with all of their outer gimbal axes parallel allows drastic simplification of the CMG steering law development in the redundancy management and failure accommodation and in the mounting hardware.

Consider such a parallel mounting arrangement of four double-gimbal CMGs with the inner and outer gimbal-angles, α_i and β_i of the ith CMG. The total CMG momentum vector $\mathbf{H} = (H_x, H_y, H_z)$ is expressed in the (x, y, z) axes as

$$\mathbf{H} = \begin{bmatrix} \sum \sin \alpha_i \\ \sum \cos \alpha_i \cos \beta_i \\ \sum \cos \alpha_i \sin \beta_i \end{bmatrix} \tag{11.171}$$

where a constant unit momentum is assumed for each CMG. The time derivative of $\mathbf{H}$ becomes

$$\dot{\mathbf{H}} = \begin{bmatrix} \sum \cos \alpha_i \dot{\alpha}_i \\ \sum (-\sin \alpha_i \cos \beta_i \dot{\alpha}_i - \cos \alpha_i \sin \beta_i \dot{\beta}_i) \\ \sum (-\sin \alpha_i \sin \beta_i \dot{\alpha}_i + \cos \alpha_i \cos \beta_i \dot{\beta}_i) \end{bmatrix} \tag{11.172}$$

Note that the x-axis torque component is not a function of the outer gimbal β_i motions. Consequently, in Kennel's CMG steering law implemented on the International Space Station, the inner gimbal rate commands $\dot{\alpha}_i$ are determined first for the commanded x-axis torque, and then the outer gimbal rate commands $\dot{\beta}_i$ for the commanded y- and z-axis torques.

Typical singularities of a system of four parallel double-gimbal CMGs are illustrated in Fig. 11.21. The 4H saturation singularity is an elliptic singularity

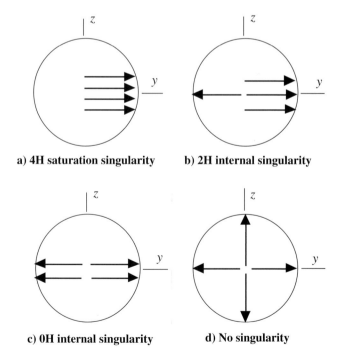

a) 4H saturation singularity   b) 2H internal singularity

c) 0H internal singularity   d) No singularity

Fig. 11.21   Singularities of a system of four parallel double-gimbal CMGs.

that cannot be escaped by null motion. The 2H and 0H singularities, shown in Figs. 11.21b and 11.21e, respectively, are hyperbolic singularities that can be escaped by null motion. A nonsingular configuration but with a zero momentum is also shown in Fig. 11.21d.

The new steering logic of the form (11.155) can also be employed for a system of four parallel double-gimbal CMGs described by $A\dot{x} = \tau$, where $x = (\alpha_1, \alpha_2, \alpha_3, \alpha_4, \beta_1, \beta_2, \beta_3, \beta_4)$, $\tau = (\tau_x, \tau_y, \tau_z)$, and A is a 3×8 Jacobian matrix that can be easily constructed from Eq. (11.172).

The saturation singularity escape capability of the new steering logic is demonstrated in Fig. 11.22. The simulation conditions for this case are as follows: initial gimbal-angles are $\alpha_i = \beta_i = 0$ for all i; commanded torques is $(\tau_x, \tau_y, \tau_z) = (0, 1, 0)$; $W = \text{diag}\{1, 1, 2, 2, 1, 1, 2, 2\}$; and $V = \lambda E$ with $\lambda = 0.01 \exp[-\det(AA^T)]$ and $\epsilon_i = 0.1 \cos t$.

In Fig. 11.22, we have $\alpha_1 = \alpha_2 > 0$, $\alpha_3 = \alpha_4 < 0$ and $\beta_1 = \beta_2 > 0$, $\beta_3 = \beta_4 < 0$. For this case, it escapes the saturation singularity, but it passes through the 0H singularity, as can be seen in Fig. 11.22. However, for $W = I$ with any V, this system remains singular at its singular momentum envelope, that is, the singularity-robust steering logic of [2,3] is unable to escape the saturation singularity of this system of four double-gimbal CMGs even when CMG momentum desaturation is requested.

Furthermore, when we choose $W = \text{diag}\{1, 2, 3, 4, 1, 2, 3, 4\}$ and the same V as before, Fig. 11.23 shows that it escapes the saturation singularity and that the 0H singularity is completely avoided with no transient torque errors. The CMG momentum is completely desaturated at $t = 4\,s$, but the desaturation torque is continuously commanded even after $t = 4\,s$ to further explore any problem of encountering an internal singularity.

This example of a system of four double-gimbal CMGs demonstrates that it is feasible to completely avoid singularity encounters (with no transient torque errors) with a proper selection of W, in conjunction with the deterministic dither signals generated in V.

Simulation results also show that the new steering logic rapidly transits through the 2H singularity shown in Fig. 11.21b.

11.8.6 Mixed SGCMG/DGCMG System

Four parallel mounted double-gimbal CMGs with a total weight of about 2400 lb and with a design life of 10 years are employed on the International Space Station (ISS). However, one of the ISS's four CMGs failed on 8 June 2002 after a little more than a year through its planned 10-year service life. Consequently, there is a possibility of augmenting the existing three DGCMGs of the ISS with an array of much smaller SGCMGs manufactured by Honeywell. The proposed steering logic is directly applicable to such a mixed SGCMG/DGCMG system. Simulation results indicate that both W and V provide a simple yet effective way of combining DGCMGs with much smaller SGCMGs for a possible application to the ISS.

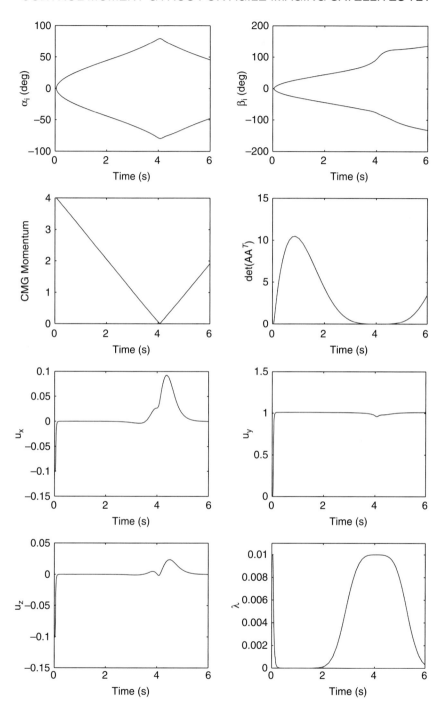

Fig. 11.22 Momentum desaturation simulation results for escaping the 4H saturation singularity but passing through the 0H singularity of four parallel double-gimbal CMGs with $W = \text{diag}\{1, 1, 2, 2, 1, 1, 2, 2\}$.

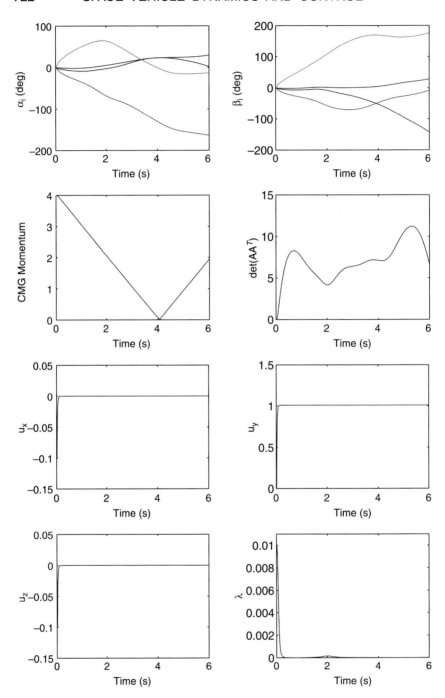

Fig. 11.23   Momentum desaturation simulation results for escaping the saturation singularity and also avoiding the 0H singularity of four parallel double-gimbal CMGs with W = diag{1, 2, 3, 4, 1, 2, 3, 4}.

11.9 Rapid Multitarget Acquisition and Pointing Control of Agile Spacecraft

An attitude control design problem for agile spacecraft that require rapid retargeting and fast transient settling is studied in this section. In particular, a practical feedback control logic is presented for large-angle, rapid multitarget acquisition and pointing maneuvers in the presence of CMG hardware constraints. The rapid multitarget acquisition and pointing capability of the proposed attitude control system is demonstrated for an agile spacecraft equipped with redundant single-gimbal CMGs. A realistic case of pointing the line of sight of an imaging satellite in low-Earth orbit toward multiple targets on the ground is also briefly discussed. A heuristic modification of the singularity-robust CMG steering logic of [2,3] to steer CMG gimbal-angles away from a mechanical gimbal-angle constraint is also discussed. This section is based on [3] and [7].

11.9.1 Single-Axis, Rapid Multitarget Slew Maneuver

For the purpose of illustrating the proposed practical feedback control logic, consider the single-axis attitude control problem of a rigid spacecraft described by

$$J\ddot{\theta} = u \qquad\qquad |u(t)| \leq U \qquad (11.173)$$

where J is the spacecraft moment of inertia, θ the attitude angle, and u the control torque input with the saturation limit of $\pm U$.

The time-optimal feedback control logic for the commanded constant attitude angle of θ_c is given by

$$u = -U \, \text{sgn}\left(e + \frac{1}{2a}\dot{\theta}|\dot{\theta}|\right) \qquad (11.174)$$

where $e = \theta - \theta_c$ and $a = U/J$ is the maximum control acceleration. The signum function is defined as $\text{sgn}(x) = 1$ if $x > 0$ and $\text{sgn}(x) = -1$ if $x < 0$.

In practice, a direct implementation of such an ideal, time-optimal switching control logic results in a chattering problem. Consequently, there exist various ways of avoiding such chattering problem inherent to the ideal, time-optimal switching control logic. Consider a feedback control logic of the form

$$u = - \underset{U}{\text{sat}}\left\{K \underset{L}{\text{sat}}(e) + C\dot{\theta}\right\} \qquad (11.175)$$

where $e = \theta - \theta_c$, and K and C are respectively the attitude and attitude rate gains to be properly determined. The saturation function is defined as

$$\underset{L}{\text{sat}}(e) = \begin{cases} L & \text{if} \quad e \geq L \\ e & \text{if} \quad |e| < L \\ -L & \text{if} \quad e \leq -L \end{cases} \qquad (11.176)$$

and it can also be represented as

$$\underset{L}{\text{sat}}(e) = \text{sgn}(e)\min\{|e|,\, L\} \qquad (11.177)$$

Because of the presence of a limiter in the attitude-error feedback loop, the attitude rate becomes constrained as

$$-|\dot{\theta}|_{max} \leq \dot{\theta} \leq |\dot{\theta}|_{max} \qquad (11.178)$$

where $|\dot{\theta}|_{max} = LK/C$. For most practical cases, a proper use of the feedback control logic (11.175) will result in a typical "bang-off-bang" control.

For the nominal range of attitude-error signals that do not saturate the actuator, the controller gains $K = kJ$ and $C = cJ$ can be determined such that

$$k = \omega_n^2 \qquad \text{and} \qquad c = 2\zeta\omega_n \qquad (11.179)$$

where ω_n and ζ are respectively the desired or specified linear control bandwidth and damping ratio. Furthermore, if the maximum slew rate is specified as $|\dot{\theta}|_{max}$, then the limiter in the attitude-error feedback loop can be selected simply as

$$L = \frac{C}{K}|\dot{\theta}|_{max} = \frac{c}{k}|\dot{\theta}|_{max} \qquad (11.180)$$

However, as the attitude-error signal gets larger and also as the slew rate limit becomes larger for rapid maneuvers, the overall response becomes sluggish with increased transient overshoot because of the actuator saturation. To achieve rapid transient settlings even for large commanded attitude angles, the slew rate limit needs to be adjusted as

$$|\dot{\theta}|_{max} = \min\left\{\sqrt{2a|e|},\ |\omega|_{max}\right\} \qquad (11.181)$$

where $e = \theta - \theta_c$ is the attitude error, $|\omega|_{max}$ is the specified maximum slew rate, and $a = U/J$ is the maximum control acceleration. A smaller value than the nominal a is to be used to accommodate various uncertainties in the spacecraft inertia and actuator dynamics.

Such a variable limiter in the attitude-error feedback loop has the self-adjusting saturation limit:

$$L = \frac{C}{K}|\dot{\theta}|_{max} = \frac{c}{k}|\dot{\theta}|_{max} = \frac{c}{k}\min\left\{\sqrt{2a|e|},\ |\omega|_{max}\right\} \qquad (11.182)$$

and we obtain a nonlinear control logic of the form

$$u = -\operatorname*{sat}_{U}\left\{K \operatorname*{sat}_{L}(e) + C\dot{\theta}\right\}$$

$$= -\operatorname*{sat}_{U}\left\{kJ \operatorname{sgn}(e) \min\left(|e|, \frac{c}{k}\sqrt{2a|e|}, \frac{c}{k}|\omega|_{max}\right) + cJ\dot{\theta}\right\} \qquad (11.183)$$

If an integral control is necessary to eliminate a steady-state pointing error caused by any constant external disturbance, the feedback control logic (11.183) can be further modified into the following proportional-integral-derivative (PID) saturation control logic:

$$u = -\operatorname*{sat}_{U}\left\{K \operatorname*{sat}_{L}\left(e + \frac{1}{T}\int e\right) + C\dot{\theta}\right\} \qquad (11.184)$$

where T is the time constant of integral control and L is given by Eq. (11.182). In terms of the standard notation for PID controller gains, K_P, K_I, and K_D, we have

$$K_P = K, \quad K_I = \frac{K}{T}, \quad K_D = C \quad (11.185)$$

The PID controller gains can be determined as

$$K_P = J\left(\omega_n^2 + \frac{2\zeta\omega_n}{T}\right) \quad (11.186a)$$

$$K_I = J\left(\frac{\omega_n^2}{T}\right) \quad (11.186b)$$

$$K_D = J\left(2\zeta\omega_n + \frac{1}{T}\right) \quad (11.186c)$$

and the time constant T of integral control is often selected as $T \approx 10/(\zeta\omega_n)$.

If the attitude reference input to be tracked is a smooth function, instead of a multistep input, we can employ a PID saturation control logic of the following form:

$$u = -\operatorname*{sat}_{U}\left\{kJ\operatorname*{sat}_{L}\left(e + \frac{1}{T}\int e\right) + cJ\dot{e}\right\} \quad (11.187)$$

where $e = \theta - \theta_c$.

Also note that the equivalent single-axis representation of the quaternion-error feedback control logic to be described in the next section is given by

$$u = -\operatorname*{sat}_{U}\left\{2kJ\operatorname*{sat}_{L}\left(e + \frac{1}{T}\int e\right) + cJ\dot{\theta}\right\} \quad (11.188)$$

where $e = \sin\frac{1}{2}(\theta - \theta_c)$ and

$$L = \frac{c}{2k}|\dot{\theta}|_{\max} = \frac{c}{2k}\min\left\{\sqrt{4a|e|}, |\omega|_{\max}\right\} \quad (11.189)$$

For a PID-type saturation control logic of the form (11.184), (11.187), or (11.188), the so-called "integrator antiwindup" or "integrator synchronization" is necessary to avoid the phenomenon known as "integrator windup," inherent to all PID-type controllers with actuator saturation. Such integrator windup results in substantial transient overshoot and control effort. If the controller is implemented on a digital computer, integrator antiwindup can be simply achieved by turning off the integral action as soon as the actuator or any other limiter in the control loop saturates.

Example 11.9

Consider the single-axis attitude control problem of a flexible spacecraft with colocated actuator and sensor, described by the transfer function

$$\frac{\theta(s)}{u(s)} = \frac{1}{Js^2} \frac{s^2/\omega_z^2 + 1}{s^2/\omega_p^2 + 1} \qquad (11.190)$$

The nominal spacecraft parameters are assumed as $J = 21,400$ kg·m^2, $\omega_z = 5$ rad/s, and $\omega_p = 6$ rad/s. For simplicity, only one dominant flexible mode is included. Actuator dynamics is assumed as

$$\frac{u}{u_c} = \frac{(50)^2}{s^2 + 2(0.7)(50)s + (50)^2} \qquad (11.191)$$

where u is the actual control torque acting on the spacecraft and u_c is the control torque command with the saturation limit of ± 1000 N·m. A PID saturation feedback control logic of the form (11.184) is considered.

Assuming $\omega_n = 3$ rad/s, $\zeta = 0.9$, and $T = 10$ s, we can determine the controller gains $K = kJ$ and $C = cJ$ as

$$k = \omega_n^2 + \frac{2\zeta\omega_n}{T} = 9.54$$

$$c = 2\zeta\omega_n + \frac{1}{T} = 5.5$$

Commanded attitude angles for two successive rest-to-rest maneuvers are assumed as $\theta_c = 10$ deg for $0 \leq t < 20$ s and $\theta_c = 50$ deg for $t \geq 20$ s.

Simulation results for the following two controllers are compared in Figs. 11.24 and 11.25.

Controller 1:

$$u_c = -\operatorname*{sat}_{U}\left\{ kJ\left(e + \frac{1}{T}\int e\right) + cJ\dot\theta \right\} \qquad (11.192a)$$

Controller 2:

$$u_c = -\operatorname*{sat}_{U}\left\{ kJ \operatorname*{sat}_{L}\left(e + \frac{1}{T}\int e\right) + cJ\dot\theta \right\} \qquad (11.192b)$$

$$L = \frac{c}{k} \min\left\{ \sqrt{2a|e|}, \ |\omega|_{\max} \right\} \qquad (11.192c)$$

where $e = \theta - \theta_c$. The maximum slew rate is assumed as $|\omega|_{\max} = 10$ deg/s for controller 2. The maximum control acceleration a is chosen as 70% of U/J to accommodate the actuator dynamics as well as the control acceleration uncertainty.

Figure 11.24 shows the simulation results for controller 1; the sluggish response with excessive transient overshoot caused by the actuator saturation is evident for the large commanded attitude angle of 50 deg. As shown in Fig. 11.25, however, the rapid retargeting capability (without excessive transient overshoot) of the proposed constant-gain, PID-type control system with the variable limiter L is demonstrated.

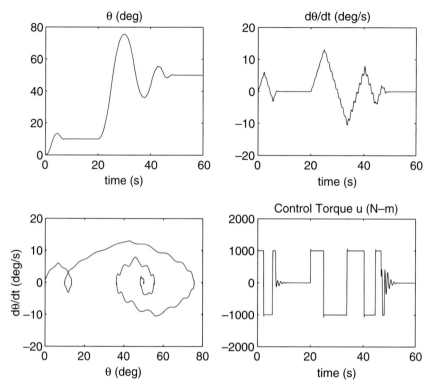

Fig. 11.24 **Simulation results of the standard PID saturation control logic without variable limiter L (controller 1).**

11.9.2 Three-Axis, Rapid Multitarget Slew Maneuver

Consider the rotational equations of motion of a rigid spacecraft described by

$$\mathbf{J}\dot{\boldsymbol{\omega}} + \boldsymbol{\omega} \times \mathbf{J}\boldsymbol{\omega} = \mathbf{u} \tag{11.193}$$

where $\mathbf{J}$ is the inertia matrix, $\boldsymbol{\omega} = (\omega_1, \omega_2, \omega_3)$ the angular velocity vector, and $\mathbf{u} = (u_1, u_2, u_3)$ the control torque input vector. It is assumed that the angular velocity vector components ω_i along the body-fixed control axes are measured by rate gyros.

The quaternion kinematic differential equations are given by

$$\begin{bmatrix} \dot{q}_1 \\ \dot{q}_2 \\ \dot{q}_3 \\ \dot{q}_4 \end{bmatrix} = \frac{1}{2} \begin{bmatrix} 0 & \omega_3 & -\omega_2 & \omega_1 \\ -\omega_3 & 0 & \omega_1 & \omega_2 \\ \omega_2 & -\omega_1 & 0 & \omega_3 \\ -\omega_1 & -\omega_2 & -\omega_3 & 0 \end{bmatrix} \begin{bmatrix} q_1 \\ q_2 \\ q_3 \\ q_4 \end{bmatrix} \tag{11.194}$$

The commanded attitude quaternions $(q_{1c}, q_{2c}, q_{3c}, q_{4c})$ and the current attitude quaternions (q_1, q_2, q_3, q_4) are related to the attitude-error quaternions

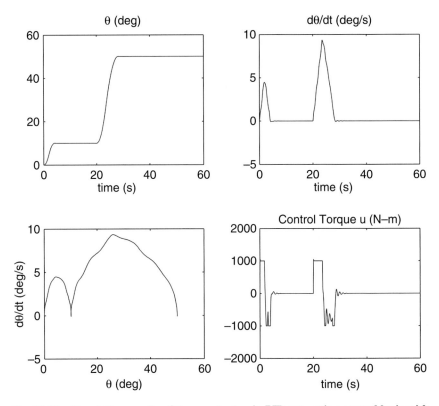

Fig. 11.25 Simulation results of the constant-gain PID saturation control logic with variable limiter L (controller 2).

(e_1, e_2, e_3, e_4) as follows:

$$\begin{bmatrix} e_1 \\ e_2 \\ e_3 \\ e_4 \end{bmatrix} = \begin{bmatrix} q_{4c} & q_{3c} & -q_{2c} & -q_{1c} \\ -q_{3c} & q_{4c} & q_{1c} & -q_{2c} \\ q_{2c} & -q_{1c} & q_{4c} & -q_{3c} \\ q_{1c} & q_{2c} & q_{3c} & q_{4c} \end{bmatrix} \begin{bmatrix} q_1 \\ q_2 \\ q_3 \\ q_4 \end{bmatrix} \tag{11.195}$$

Consider a rigid spacecraft that is required to maneuver about an inertially fixed axis as fast as possible, but not exceeding the specified maximum slew rate about that eigenaxis. The following saturation control logic provides such a rest-to-rest eigenaxis rotation under slew rate constraint:

$$\begin{aligned} \mathbf{u} &= -\mathbf{K} \operatorname*{sat}_{L_i} \left(\mathbf{e} + \frac{1}{T} \int \mathbf{e} \right) - \mathbf{C}\boldsymbol{\omega} \\ &= -\mathbf{J} \left\{ 2k \operatorname*{sat}_{L_i} \left(\mathbf{e} + \frac{1}{T} \int \mathbf{e} \right) + c\boldsymbol{\omega} \right\} \end{aligned} \tag{11.196}$$

where $\mathbf{e} = (e_1, e_2, e_3)$ is the quaternion-error vector, $\mathbf{K} = 2k\mathbf{J}$, $\mathbf{C} = c\mathbf{J}$, and the saturation limits L_i are determined as

$$L_i = \frac{c}{2k}|\omega_i|_{\max} \tag{11.197}$$

where $|\omega_i|_{\max}$ is the specified maximum angular rate about each axis.

Assume that the control torque input for each axis is constrained as

$$-U \le u_i(t) \le +U \qquad\qquad i = 1, 2, 3 \tag{11.198}$$

where U is the saturation limit of each control input. Then, a control logic that accommodates possible control torque input saturation but that still provides an eigenaxis rotation under slew rate constraint can be expressed as

$$\boldsymbol{\tau} = -\mathbf{J}\left\{2k \operatorname*{sat}_{L_i}\left(\mathbf{e} + \frac{1}{T}\int\mathbf{e}\right) + c\boldsymbol{\omega}\right\} \tag{11.199}$$

$$\mathbf{u} = \operatorname*{sat}_U(\boldsymbol{\tau}) = \begin{cases} \boldsymbol{\tau} & \text{if } \|\boldsymbol{\tau}\|_\infty < U \\ U\dfrac{\boldsymbol{\tau}}{\|\boldsymbol{\tau}\|_\infty} & \text{if } \|\boldsymbol{\tau}\|_\infty \ge U \end{cases} \tag{11.200}$$

where $\|\boldsymbol{\tau}\|_\infty = \max\{|\tau_1|, |\tau_2|, |\tau_3|\}$.

Similar to the case of single-axis attitude control for achieving rapid transient settlings for large attitude-error signals discussed earlier, the slew rate limit is adjusted as

$$L_i = \frac{c}{2k}\min\left\{\sqrt{4a_i|e_i|},\ |\omega_i|_{\max}\right\} \tag{11.201}$$

where $a_i = U/J_{ii}$ is the maximum control acceleration about the ith control axis and $|\omega_i|_{\max}$ is the specified maximum angular rate about each axis.

Example 11.10

Consider the three-axis attitude control problem of a rigid spacecraft with the following nominal inertia matrix in units of kg·m^2:

$$\mathbf{J} = \begin{bmatrix} 21{,}400 & 2{,}100 & 1{,}800 \\ 2{,}100 & 20{,}100 & 500 \\ 1{,}800 & 500 & 5{,}000 \end{bmatrix} \tag{11.202}$$

Actuator dynamics in each axis is assumed as

$$\frac{u_i}{u_{ic}} = \frac{(50)^2}{s^2 + 2(0.7)(50)s + (50)^2} \tag{11.203}$$

where u_i is the actual control torque acting along the ith control axis of the spacecraft and u_{ic} is the control torque command with the saturation limit of ± 1000 N·m.

The proposed three-axis control logic is then given by

$$\boldsymbol{\tau} = -\mathbf{J}\left\{2k \underset{L_i}{\text{sat}}\left(\mathbf{e} + \frac{1}{T}\int\mathbf{e}\right) + c\boldsymbol{\omega}\right\} \tag{11.204}$$

where $\mathbf{e}$ is the quaternion-error vector and

$$L_i = \frac{c}{2k}\min\left\{\sqrt{4a_i|e_i|},\ |\omega_i|_{\max}\right\} \tag{11.205}$$

$$\mathbf{u} = \underset{U}{\text{sat}}(\boldsymbol{\tau}) = \begin{cases} \boldsymbol{\tau} & \text{if} \quad \|\boldsymbol{\tau}\|_\infty < U \\ U\dfrac{\boldsymbol{\tau}}{\|\boldsymbol{\tau}\|_\infty} & \text{if} \quad \|\boldsymbol{\tau}\|_\infty \geq U \end{cases} \tag{11.206}$$

Assuming $\omega_n = 3$ rad/s, $\zeta = 0.9$, and $T = 10$ s, we can determine the controller gains k and c as $k = \omega_n^2 + 2\zeta\omega_n/T = 9.54$ and $c = 2\zeta\omega_n + 1/T = 5.5$. The maximum angular rate is assumed as $|\omega_i|_{\max} = 10$ deg/s. The maximum control acceleration a_i is chosen as 40% of U/J_{ii} to accommodate the actuator

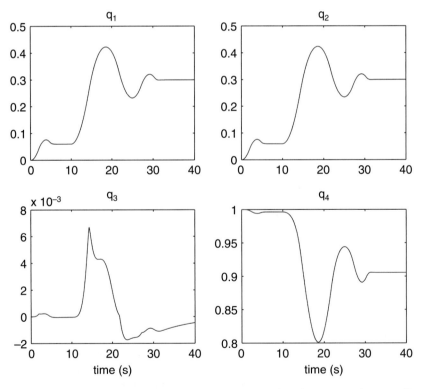

Fig. 11.26   Time histories of spacecraft attitude quaternions (constant-gain controller without variable limiter L).

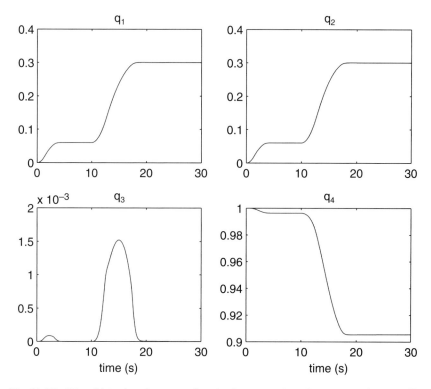

Fig. 11.27 Time histories of spacecraft attitude quaternions (constant-gain controller with variable limiter L).

dynamics, nonlinear nature of quaternion-based, phase-plane dynamics, and control acceleration uncertainty.

Figure 11.26 shows the time histories of attitude quaternions for the proposed three-axis control system but without employing the variable limiter. The sluggish response with excessive transient overshoot is evident for two successive, large-angle roll/pitch maneuvers. As shown in Fig. 11.27, the rapid multitarget acquisition and pointing capability of the proposed constant-gain control system with the variable limiter L is demonstrated. Figure 11.28 shows the time histories of control input torques required to achieve such large-angle, rapid retargeting maneuvers.

11.9.3 Agile Spacecraft Control Using CMGs

The proposed practical feedback control logic is now applied to a more realistic problem of controlling an agile spacecraft using redundant single-gimbal CMGs. The control objective is to maneuver the spacecraft as fast as possible in the presence of the CMG internal singularities, momentum saturation, slew rate limit, and gimbal rate limits.

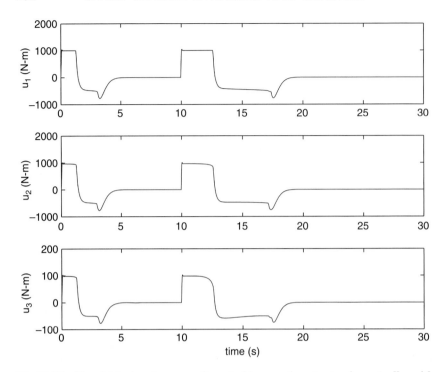

Fig. 11.28   Time histories of spacecraft control torques (constant-gain controller with variable limiter L).

Example 11.11   Spacecraft Attitude Control Using CMGs

A spacecraft with the following inertia matrix is considered:

$$\mathbf{J} = \begin{bmatrix} 21{,}400 & 2{,}100 & 1{,}800 \\ 2{,}100 & 20{,}100 & 500 \\ 1{,}800 & 500 & 5{,}000 \end{bmatrix} \text{kg} \cdot \text{m}^2 \qquad (11.207)$$

The yaw axis with the smallest moment of inertia is pointing toward a target.

Consider a typical pyramid mounting arrangement of four single-gimbal CMGs. A skew angle β of 53.13 deg (i.e., $\cos\beta = 0.6$, $\sin\beta = 0.8$) and 1000-N·m·s momentum magnitude for each CMG are assumed. The total maximum CMG momentum available then becomes 3200 N·m·s (not 4000 N·m·s) because of the pyramid mounting arrangement of four CMGs with $\beta = 53.13$ deg. The gimbal rate command limit of each CMG is assumed as 1 rad/s. It is also assumed that the attitude control bandwidth needs to be lower than 5 rad/s and the maximum slew rate less than 10 deg/s.

The proposed nonlinear attitude control system, consisting of the quaternion-error feedback control logic with the variable limiter and the CMG steering logic based on the generalized SR inverse $\mathbf{A}^{\#}$, is then described by

$$\boldsymbol{\tau} = -\mathbf{J} \left\{ 2k \operatorname*{sat}_{L_i} \left(\mathbf{e} + \frac{1}{T} \int \mathbf{e} \right) + c\boldsymbol{\omega} \right\} \tag{11.208a}$$

$$L_i = \frac{c}{2k} \min \left\{ \sqrt{4a_i |e_i|}, \quad |\omega_i|_{max} \right\} \tag{11.208b}$$

$$\mathbf{A}^{\#} = \mathbf{A}^T [\mathbf{A}\mathbf{A}^T + \lambda \mathbf{E}]^{-1} \tag{11.208c}$$

$$\mathbf{u} = -\boldsymbol{\tau} - \boldsymbol{\omega} \times \mathbf{h} \tag{11.208d}$$

$$\dot{\boldsymbol{\delta}}_c = \operatorname*{sat}_{\pm \dot{\delta}_{max}} \left\{ \mathbf{A}^{\#} \mathbf{u} \right\} \tag{11.208e}$$

$$\dot{\boldsymbol{\delta}} = \frac{(50)^2}{s^2 + 2(0.7)(50)s + (50)^2} \dot{\boldsymbol{\delta}}_c \tag{11.208f}$$

where $\dot{\delta}_{max} = 1$ rad/s and $|\omega_i|_{max} = 10$ deg/s. Similar to Example 11.10, assuming $\omega_n = 3$ rad/s, $\zeta = 0.9$, and $T = 10$ s, the controller gains k and c are chosen as $k = \omega_n^2 + 2\zeta\omega_n/T = 9.54$ and $c = 2\zeta\omega_n + 1/T = 5.5$. The maximum control acceleration a_i is chosen as 40% of U/J_{ii} to accommodate the actuator dynamics, nonlinear nature of quaternion kinematics, and control acceleration uncertainty.

For the normalized Jacobian matrix $\mathbf{A}$, the scale factors λ and $\mathbf{E}$ are chosen as

$$\lambda = 0.01 \exp[-10 \det \left(\mathbf{A}\mathbf{A}^T \right)] \tag{11.209a}$$

$$\mathbf{E} = \begin{bmatrix} 1 & \epsilon_3 & \epsilon_2 \\ \epsilon_3 & 1 & \epsilon_1 \\ \epsilon_2 & \epsilon_1 & 1 \end{bmatrix} \tag{11.209b}$$

where $\epsilon_i = 0.01 \sin(0.5\pi t + \phi_i)$ with $\phi_1 = 0$, $\phi_2 = \pi/2$, and $\phi_3 = \pi$.

Simulation results for two successive, large-angle roll/pitch maneuvers are presented in Figs. 11.29–11.32. The rapid retargeting capability of the proposed nonlinear control system is demonstrated, as shown in Fig. 11.29. The plots of the total CMG momentum and singularity measures, shown in Fig. 11.32, indicate that the CMG system with the proposed control logic successfully passed through the internal singularities and also utilizes the maximum CMG momentum of 3200 N·m·s to achieve a rapid reorientation of the spacecraft. As can be seen in Fig. 11.32, the new CMG steering logic does not explicitly avoid singularity encounters, but rather it approaches and rapidly transits the internal singularities whenever necessary.

In the next section, we will briefly examine a more realistic case of pointing the line of sight of an imaging satellite in low-Earth orbit toward multiple targets on the ground.

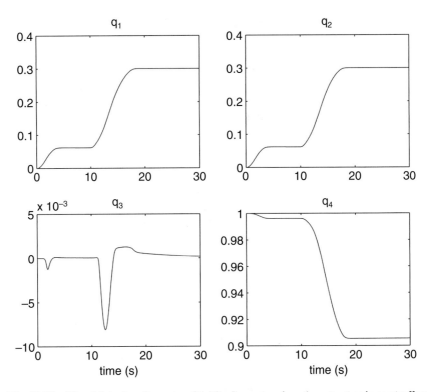

Fig. 11.29   Time histories of spacecraft attitude quaternions (constant-gain controller with variablelimiter L).

11.9.4   Multiple Ground Target Tracking

A simulation study of the proposed attitude control logic described by Eq. (11.208) was performed in [3] for a realistic mission with many closely placed ground targets. In [3], a spacecraft with an inertia matrix of $\mathbf{J} =$ diag (3,400, 3,400, 1,200) $kg \cdot m^2$ and with a six-sided pyramid arrangement of 100 N·m·s CMGs was considered. The CMG gimbal axes have a skew angle of 65 deg from the roll-pitch plane. This produces an angular momentum ellipsoid with two equal major axes on the roll-pitch plane and a minor axis in the yaw direction. The lower angular momentum in the yaw direction still gives yaw approximately 20% more angular rate capability because of the small yaw inertia. This means that the mission planner can plan based on line-of-sight (LOS) angle changes not accounting for the yaw component of the maneuver.

The orbit used in a simulation was 250 by 500 n miles. It is inclined about 98 deg. Perigee occurs about one-third the way through the simulation run. The distribution of sensor operations was a fixed distribution of the satellite maneuver angle. The location on the ground was such that when including the time of maneuver, the orbital motion of the satellite and the rotation of the Earth would be from a

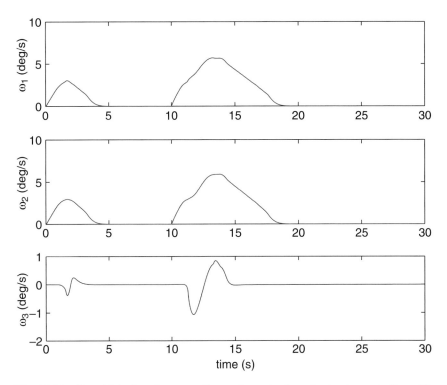

Fig. 11.30 Time histories of spacecraft angular rates (constant-gain controller with variable limiter L).

distribution with more small angles, but still having some large-angle maneuvers. The location on the ground was not optimized in a minimum time order but a desired maneuvering load.

The measure of an agile satellite attitude control system is its ability to collect the maximum data from an area on the Earth that is rich in data-collection opportunities. The planner selects the orientation of the satellite such that all data are scanned parallel to a latitude or longitude line. The scan was done in pure roll with respect to the Earth and in either direction. Each scan direction was selected such that the minimum yaw was needed for the maneuver and the scan direction continued in the direction of slew maneuver.

The schedule data transmitted to the satellite were latitude, longitude, altitude, and time for the beginning of scan and the end of scan. The control trajectory generator used the two points and the times to generate a constant rate trajectory that would pass through the two points. At the end of a scan (data collection), the commanded trajectory was stepped to the next scan trajectory. A particular simulation result is shown in Fig. 11.33.

The time to make a maneuver was not repeatable for a given change in LOS angle, but depended on the stored angular momentum in the CMG array, the direction of the maneuver, and the amount of yaw angle (not accounted for by the

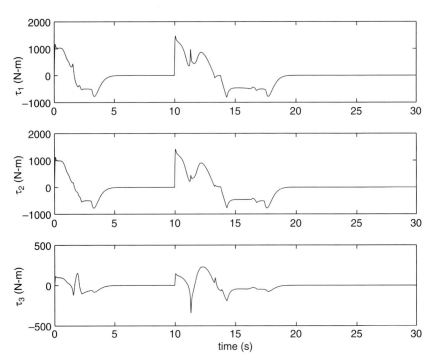

Fig. 11.31   Time histories of spacecraft control torques generated by CMGs (constant-gain controller with variable limiter L).

planner). The stored angular momentum was maintained at less than 10% of the total angular momentum, and consequently it was not a significant factor. Angular momentum determines the maneuver velocity. Saturation of angular momentum determined the velocity for large maneuvers, and that surface is not an ellipsoidal surface.

This complex problem of pointing the LOS of an imaging satellite in low-Earth orbit toward multiple targets on the ground was the main motivation for developing the practical feedback control logic as described by Eq. (11.208).

11.9.5   CMG Steering Logic with Gimbal-Angle Constraint

A miniature CMG smaller than Astrium's CMG 15-45S, which has been being developed at Surrey Space Centre, is a low-cost approach to providing small imaging satellites (<300 kg) with an agile slew maneuver capability (a slew rate of 6 deg/s). However, the miniature CMG has been designed with a mechanical stop that does not allow a full 360-deg rotation but constrains the gimbal-angle to ±180 deg. A mechanical gimbal-angle constraint significantly simplifies the hardware design of CMGs as discussed in [7] and [11].

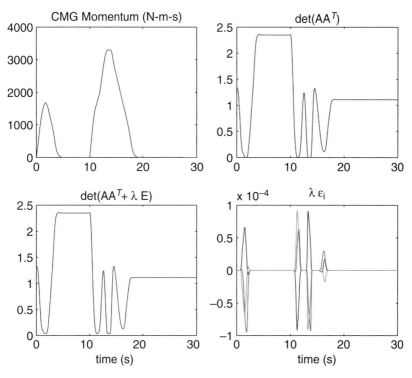

Fig. 11.32 Time histories of CMG momentum and singularity measures (constant-gain controller with variable limiter L).

For such a miniature CMG system with a gimbal-angle constraint, Lappas and Wie [7] have proposed a CMG steering logic of the form:

$$\tau = -\mathbf{J}\left\{2k \ \underset{L_i}{\text{sat}}\left(\mathbf{e} + \frac{1}{T}\int \mathbf{e}\right) + c\boldsymbol{\omega}\right\} \tag{11.210a}$$

$$L_i = \frac{c}{2k}\min\left\{\sqrt{4a_i|e_i|}, \quad |\omega_i|_{\max}\right\} \tag{11.210b}$$

$$\mathbf{A}^{\#} = \mathbf{A}^T[\mathbf{A}\mathbf{A}^T + \lambda\mathbf{E}]^{-1} \tag{11.210c}$$

$$\mathbf{u} = -\tau - \boldsymbol{\omega} \times \mathbf{h} \tag{11.210d}$$

$$\dot{\boldsymbol{\delta}}_c = \underset{\pm\dot{\delta}_{\max}}{\text{sat}}\left\{\mathbf{A}^{\#}\mathbf{u} + \gamma[\mathbf{I} - \mathbf{A}^{\#}\mathbf{A}](\boldsymbol{\delta}^* - \boldsymbol{\delta})\right\} \tag{11.210e}$$

$$\dot{\delta}_{c_i} = 0 \quad \text{if} \quad |\delta_i| \geq \delta_{\max} \quad \text{and} \quad \text{sgn}(\delta_i)\dot{\delta}_i > 0 \tag{11.210f}$$

where $\delta_{\max}$ is the gimbal-angle limit, $\dot{\delta}_{\max}$ is the gimbal rate limit, $\boldsymbol{\delta}^*$ is a preferred set of steady-state gimbal-angles, and γ is a positive weighting scalar to be properly chosen for the null-motion command term.

As demonstrated in [7], such a heuristic modification of the singularity-robust CMG steering logic of [2,3] steers CMG gimbal-angles away from the ± 180-deg constraint, even in the presence of CMG singularities, without degradating the spacecraft slew-maneuver performance. It is emphasized that both the null-motion term in Eq. (11.210) and the explicit gimbal-angle constraint term in Eq. (11.210) are required to steer CMG gimbal-angles away from the gimbal-angle constraint.

11.10   Variable-Speed Control Moment Gyros

Recently, the variable-speed control moment gyros (VSCMGs) have been extensively studied by CMG researchers, as evidenced in [23–29], as an alternative solution to the singularity problem inherent to standard CMGs with constant wheel speed. The wheel speed of a VSCMG is allowed to change continuously.

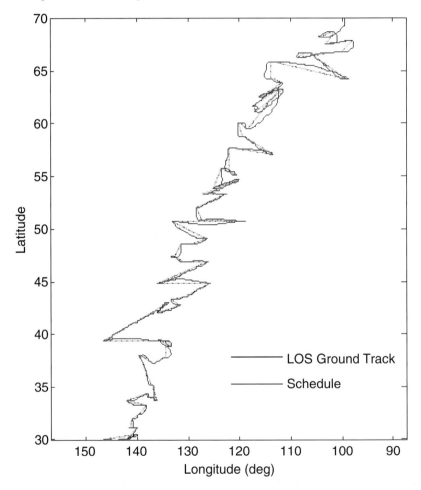

Fig. 11.33   Multiple ground target tracking simulation results [3].

Consequently, a VSCMG can generate a control torque along any direction that lies in the plane normal to its gimbal axis. An array of VSCMGs can thus generate a control torque along an arbitrary direction in the three-dimensional space. Furthermore, the extra degree of freedom of a VSCMG over a standard CMG can be used to achieve additional objectives of combined attitude and power tracking control or singularity avoidance, as described in [23–29].

The recent advent of miniature single-gimbal CMGs has further spawned interest in the VSCMGs for energy storage and attitude control systems on small satellites [28]. A recent study in [29] also examined a feasibility of linking various steering algorithms of conventional CMGs and the recent VSCMGs for the more general case of simultaneous attitude and power tracking using a cluster of VSCMGs.

Detailed discussions of the VSCMGs can be found in [23–29].

References

[1] Heiberg, C., Bailey, D., and Wie, B., "Precision Pointing Control of Agile Spacecraft Using Single Gimbal Control Moment Gyros," *Journal of Guidance, Control, and Dynamics*, Vol. 23, No. 1, 2000, pp. 77–85.

[2] Wie, B., Heiberg, C., and Bailey, D., "Singularity Robust Steering Logic for Redundant Single-Gimbal Control Moment Gyros," *Journal of Guidance, Control, and Dynamics*, Vol. 24, No. 5, 2001, pp. 865–872.

[3] Wie, B., Heiberg, C., and Bailey, D., "Rapid Multi-Target Acquisition and Pointing Control of Agile Spacecraft," *Journal of Guidance, Control, and Dynamics*, Vol. 25, No. 1, 2002, pp. 96–104.

[4] Dominguez, J., and Wie, B., "Computation and Visualization of Control Moment Gyroscope Singularities," AIAA Paper 2002-4570, Aug. 2002.

[5] Wie, B., "Singularity Analysis and Visualization for Single-Gimbal Control Moment Gyro Systems," *Journal of Guidance, Control, and Dynamics*, Vol. 27, No. 2, 2004, pp. 271–282.

[6] Wie, B., "Singularity Escape/Avoidance Steering Logic for Control Moment Gyro Systems," *Journal of Guidance, Control, and Dynamics*, Vol. 28, No. 5, 2005, pp. 948–956.

[7] Lappas, V., and Wie, B., "Robust CMG Steering Logic with Gimbal Constraints," AIAA Paper 2006-6651, Aug. 2006.

[8] Damilano, P., "Pleiades High Resolution Satellite: a Solution for Military and Civilian Needs in Metric-Class Optical Observation," AIAA/USU, Paper SSC01-I-5, Aug. 2001.

[9] Girouart, B., Sebbag, I., and Lachiver, J.-M., "Performances of the Pleiades-HR Agile Attitude Control System," *Proceedings of the 5th International ESA Conference on Spacecraft Guidance, Navigation and Control Systems*, 2002.

[10] Defendini, A., Faucheux, P., Guay, P., Morand, J., and Heimel, H., "A Compact CMG Products for Agile Satellites," *Proceedings of the 5th International ESA Conference on Spacecraft Guidance, Navigation and Control Systems*, 2002.

[11] Lappas, V. J., Steyn, W. H., and Underwood, C. I., "Attitude Control Systems for Agile Small Satellites Using Control Moment Gyros," *Acta Astronautica*, Vol. 51, No. 1, 2002, pp. 101–111.

[12] Crenshaw, J., "2-SPEED, A Single-Gimbal Control Moment Gyro Attitude Control System," AIAA Paper 73-895, Aug. 1973.

[13] Margulies, G., and Aubrun, J. N., "Geometric Theory of Single-Gimbal Control Moment Gyro Systems," *Journal of the Astronautical Sciences,* Vol. 26, No. 2, 1978, pp. 159–191.

[14] Cornick, D. E., "Singularity Avoidance Control Laws for Single Gimbal Control Moment Gyros," AIAA Paper 79-1698, *Proceedings of the AIAA Guidance and Control Conference,* AIAA, New York, 1979, pp. 20–33.

[15] Stocking, G., and Meffe, M., "Momentum Envelope Topology of Single-Gimbal CMG Arrays for Space Vehicle Control," American Astronomical Society, Paper 87-002, Jan. – Feb. 1987.

[16] Gantmacher, F. R., *The Theory of Matrices,* Vol. 1, Chelsea Publishing, New York, 1959, pp. 9–10.

[17] Bedrossian, N. S., Paradiso, J., Bergmann, E. V., and Rowell, D., "Redundant Single-Gimbal Control Moment Gyroscope Singularity Analysis," *Journal of Guidance, Control, and Dynamics,* Vol. 13, No. 6, 1990, pp. 1096–1101.

[18] Bedrossian, N. S., Paradiso, J., Bergmann, E. V., and Rowell, D., "Steering Law Design for Redundant Single-Gimbal Control Moment Gyroscopes," *Journal of Guidance, Control, and Dynamics,* Vol. 13, No. 6, 1990, pp. 1083–1089; also Bedrossian, N. S., M.S. thesis, Massachusetts Inst. of Technology, Cambridge, MA, Aug. 1987.

[19] Guggenheimer, H. W., *Differential Geometry,* Dover, New York, 1963.

[20] Kreyszig, E., *Differential Geometry,* Dover, New York, 1991.

[21] Nakamura, Y., and Hanafusa, H., "Inverse Kinematic Solutions with Singularity Robustness for Robot Manipulator Control," *Journal of Dynamic Systems, Measurement, and Control,* Vol. 108, Sept. 1986, pp. 163–171.

[22] Kennel, H. F., "Steering Law for Parallel Mounted Double-Gimballed Control Moment Gyros: Revision A," NASA TM-82390, Jan. 1981.

[23] Junkins, J. L., and Schaub, H., "Feedback Control Law for Variable Speed Control Moment Gyroscopes," *Journal of the Astronautical Sciences,* Vol. 46, No. 3, 1998, pp. 307–328.

[24] Schaub, H., and Junkins, J. L., "Singularity Avoidance Using Null Motion and Variable-Speed Control Moment Gyros," *Journal of Guidance, Control, and Dynamics,* Vol. 23, No. 1, 2000, pp. 11–16.

[25] Yoon, H., and Tsiotras, P., "Spacecraft Adaptive Attitude and Power Tracking with Variable Speed Control Moment Gyroscopes," *Journal of Guidance, Control, and Dynamics,* Vol. 25, No. 6, 2002, pp. 1081–1090.

[26] Yoon, H., and Tsiotras, P., "Singularity Analysis of Variable-Speed Control Moment Gyros," *Journal of Guidance, Control, and Dynamics,* Vol. 27, No. 3, 2004, pp. 374–386.

[27] Yoon, H., and Tsiotras, P., "Spacecraft Line-of-Sight Control Using a Single Variable-Speed Control Moment Gyro," *Journal of Guidance, Control, and Dynamics,* Vol. 29, No. 6, 2006, pp. 1295–1308.

[28] Richie, D., Lappas, V., and Palmer, P., "Sizing/Optimization of a Small Satellite Energy Storage and Attitude Control System," American Astronomical Society, Paper 07-177, Jan.–Feb. 2007.

[29] Lappas, V., Asghar, S., Richie, D., Palmer, P., and Fertin, D., "Combined Singularity Avoidance for Variable-Speed Control Moment Gyroscope Clusters," American Astronomical Society, Paper 07-181, Jan.–Feb. 2007.

12
Solar-Sail Dynamics and Control

Solar sails are large, lightweight reflectors in space that are propelled by sunlight. This chapter presents a comprehensive treatment of various dynamical modeling and control problems of solar-sail spacecraft. It also presents the analysis and design of solar-sail attitude control systems for interplanetary solar-sailing missions as well as a solar-sail flight validation mission in a sun-synchronous orbit. In Chapter 13, we will study solar-sail mission applications to a complex astrodynamical problem of changing the trajectory of near-Earth objects to mitigate their impact threat to the Earth.

12.1 Introduction

A rendezvous mission for Halley's comet, employing a large solar-sail as illustrated in Fig. 12.1, was proposed by Jet Propulsion Laboratory (JPL) in 1977 [1]. Although it became an ill-fated mission concept of 1970s, it introduced the propellantless solar-sailing concept for achieving a large orbital inclination change (>90 deg) to reverse the orbital flight direction. Detailed historical as well as technical discussions of various solar-sailing missions and the associated technologies can be found in [1–3].

A renewed interest in solar-sailing as a result of its potential for propellantless space propulsion is spurring recent developments of near-term sail missions and the associated solar-sail technologies [4–8]. Non-Keplerian orbits, high-velocity missions to the outer planets, and high-velocity interstellar precursor missions (all based on solar-sailing technology) are envisioned by NASA and European Space Agency (ESA). Near-term applications of solar-sailing technology also include high-performance science missions to the inner solar system. Such near-term solar-sail missions will most likely require solar-sails smaller than 100 m due to recent advances in ultralightweight sail films, lightweight deployable booms, and the miniaturization of spacecraft hardware.

In support of solar-sail road-map missions, NASA's In-Space Propulsion program has been focusing on the quantitative demonstration of scalability of current solar-sail system architectures to future mission requirements through ground testing of key hardware systems [7,8]. In April 2005, NASA and ATK Space Systems (formerly, AEC-ABLE Engineering) successfully deployed a 20-m solar-sail, as shown in Fig. 12.2, in the 30-m thermal vacuum chamber at the Plum Brook Space Power Facility of the NASA Glenn Research Center.

Fig. 12.1   An 800 × 800-m solar sail proposed by JPL in 1977 for a rendezvous mission with Halley's comet for the 1986 passage [1].

A 20-m solar-sail of L'Garde has also been successfully deployed in July 2005, as can be seen in Fig. 12.3.

A spaceflight experiment of the 30-m, 105-kg Cosmos 1 solar-sail spacecraft was attempted by The Planetary Society on 21 June, 2005. However, because of a boost rocket failure, the Cosmos 1 solar-sail project did not achieve its mission goal

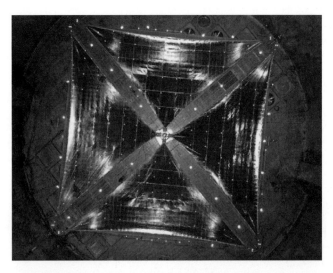

Fig. 12.2   ATK's 20-m solar sail successfully deployed in the 30-m thermal vacuum chamber at the Plum Brook Space Power Facility of the NASA Glenn Research Center in 2005 [7]. (Image courtesy of ATK Space Systems.)

Fig. 12.3 L'Garde's 20-m solar sail successfully deployed in the 30-m thermal vacuum chamber at the Plum Brook Space Power Facility of the NASA Glenn Research Center in 2005 [8]. (Image courtesy of L'Garde.)

of demonstrating the first controlled solar-sail flight as the spacecraft is propelled by photons from sunlight.

Recent advances in solar-sail technologies and near-term mission concepts were presented at the 1st International Symposium on Solar Sailing (ISSS 2007), Herrsching, Germany, 27–29 June, 2007 (http://www.isss.spacesailing.net/).

12.2   Solar-Sail Attitude Control Issues

Following Garwin's publication on solar-sailing in 1958 [9], attitude stabilization of a space vehicle by means of solar radiation pressure was first proposed by Sohn in 1959 [10]. Since then, the concept of using solar radiation pressure for attitude stabilization as well as stationkeeping control of various satellites has been studied extensively by many researchers during the last three decades [11–15].

In fact, such a solar-pressure attitude control concept has been successfully implemented on a certain type of geostationary satellite as well as on several interplanetary spacecraft. For example, the large solar radiation disturbance torque caused by an asymmetrical solar-array configuration of INSAT and GOES satellites with only one solar-array wing on the south side is countered by a conical-shaped solar-sail on a long boom mounted on the north side, as shown in Fig. 1.12.

The roll/yaw control systems of geosynchronous communications satellites such as OTS, TELECOM 1, and INMARSAT 2 successfully utilized the solar pressure attitude control concept. An asymmetrical offsetting of the solar-array wings from their nominal sun-pointing orientation generates the so-called windmill torque, as illustrated in Fig. 12.4a. The nominal sun-pointing orientation of the north and south solar-array wings is shown in Fig. 12.4b. In Fig. 12.4, the Earth-pointing main body of the spacecraft is not shown. For typical geosynchronous communications satellites, the body-fixed pitch axis is perpendicular to the orbital plane, the roll axis along the flight direction, and the yaw axis toward

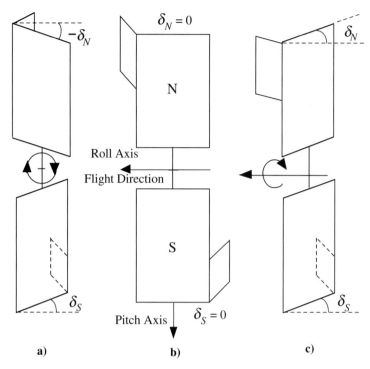

Fig. 12.4   Illustration of a solar-pressure attitude control concept successfully imple-mented on geosynchronous communications satellites, such as OTS, TELECOM 1, and INMARSAT 2 [16–19].

the Earth. If the north and south solar-array wings are not rotated asymmetrically (i.e., $\delta_N \neq -\delta_S$), then an additional roll/yaw torque, perpendicular to the windmill torque, is also generated as illustrated in Fig. 12.4c. Additional flaps mounted to the outermost solar panels substantially increase the roll/yaw control torques. Note that such flaps produce the windmill torque even when $\delta_N = \delta_S = 0$. The maximum offsetting of each array is constrained by the acceptable electrical power loss (nominally 1%). Detailed descriptions of such a flight-proven, solar-pressure attitude control technique can be found in [16–19].

The idea of asymmetrically twisting solar panels to create the windmill torque about the roll axis was also successfully applied to the Mariner 10 spacecraft during its flight to Mercury. However, its roll attitude control was done by commands from the mission controllers because there was no onboard controller for that task.

Even though the effectiveness of the solar-pressure attitude control has been in-flight validated as discussed earlier in this section, the solar radiation pressure is often considered as an external disturbance for most satellites. For example, the solar-pressure effect on formation flying of small satellites was investigated in [20], and the long-term attitude drift problem of spinning spacecraft caused by solar-pressure disturbance torques was recently examined in [21]. The significant effects of solar radiation pressure on attitude and orbit control of a very large platform in geostationary orbit were studied in [22].

During the last several decades, a variety of advanced dynamic modeling and spacecraft control techniques has been developed. Detailed descriptions of such advanced technologies associated with spacecraft dynamics and control problems can be found in [19, 23–25].

However, there exist various practical implementation issues to be resolved in applying these advanced control techniques to active three-axis attitude control of near-term sailcraft as well as to future advanced sailcraft. All practical spacecraft control designs are often subjected to the physical limits of actuators, sensors, spacecraft structural rigidity, and other mission constraints. In particular, when a gimbaled control boom and/or sail control vanes (instead of conventional thrusters, reaction wheels, and magnetic torquers) are to be employed as primary actuators for active three-axis attitude control of solar-sail spacecraft, there exist a variety of practical issues to be resolved. Therefore, solar-sail attitude and flight control technology needs to be rapidly advanced so that a sail spaceflight experiment for validating sail attitude stability and thrust vector pointing performance can be conducted in the near future.

Three basic types of near-term solar-sails are shown in Fig. 12.5. These configurations do have their own advantages and disadvantages in terms of control

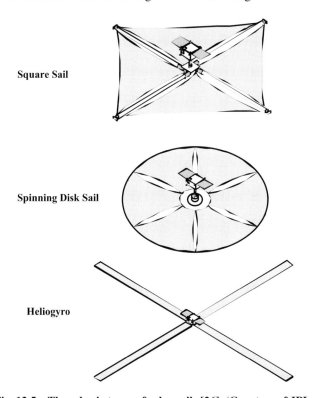

Square Sail

Spinning Disk Sail

Heliogyro

Fig. 12.5   Three basic types of solar sails [26]. (Courtesy of JPL.)

authority, controllability, packaging, deployment, and other system-level tradeoff issues (i.e., mass, cost, etc.). Selecting a particular sail configuration for a specific mission is a complex problem, requiring detailed system-level tradeoffs. We focus on a square sail configuration, which is most likely to be chosen for various near-term sail missions.

As discussed in [10] and [11], an interplanetary spacecraft is often said to be *statically stable* when its center of mass lies between the sun and its center of pressure. Although any point along the resultant solar pressure force direction can be considered as the center of pressure, the location along a spacecraft reference line through which the resultant force is acting is often defined as the center of pressure. Whenever a statically stable sailcraft rotates away from its neutral sun-pointing orientation, a restoring (stabilizing) torque is generated. The dynamical behavior of such a statically stable sailcraft is analogous to that of a gravity-gradient stabilized satellite. That is, if disturbed, the sailcraft will oscillate indefinitely. If the center of pressure lies between the sun and the center of mass, a destabilizing torque is generated whenever the sailcraft rotates away from its null or trim orientation.

A spin-stabilized, 76×76-m square sailcraft, shown in Fig. 12.6, has been proposed for the New Millennium Program Space Technology 5 (ST5) Geostorm warning mission which would provide real-time monitoring of solar activity [4]. It would operate inside the L_1 point of the sun-Earth system toward the sun and increase the warning time for geomagnetic storms compared to a vantage point closer to the Earth. For such a large sailcraft with moments of inertia of $(44{,}000; 22{,}000; 24{,}000)$ kg·m^2, an uncertain c.m./c.p. offset of approximately 1 m was assumed by the Geostorm sail study team, and a spin rate of 0.45 deg/s was selected to keep the angular momentum vector within 1 deg of the sunline.

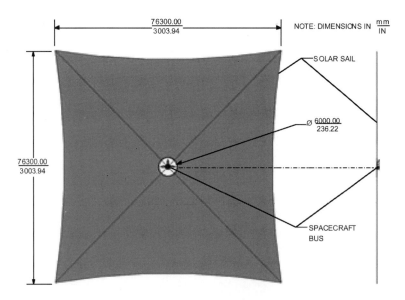

Fig. 12.6   ST5 Geostorm solar sail [4]. (Courtesy of JPL.)

Thrusters are used for precession/nutation control as well as spin rate control of this sailcraft.

A 76 × 76-m sailcraft was also once considered for the Team Encounter mission [5]. It was planned to be launched as a secondary payload on an Ariane 5. A combination of passive and active attitude control techniques was considered for this sailcraft. The Team Encounter sailcraft of a total mass of 18 kg was required to achieve solar-system escape within three or five years. Its sun-pointing orientation was passively stabilized. A constant pitch angle of 25 deg with respect to the sun was required during the first 300 days after separation from a carrier spacecraft. The 25-deg pitch trim angle was passively maintained by an intentional c.m./c.p. offset caused by a 3-kg payload tied to the side with a burn wire. The rotational motion about the sun vector is actively controlled. An onboard star camera measures the sailcraft orientation with respect to a fixed star field, and the control vanes provide the necessary control torque to counteract a windmill disturbance torque of 0.1 mN · m. After 300 days, an onboard timer will power the burn wire to release the payload restrained by a suspension wire. Consequently, the center of mass will move to the sailcraft center, and the sailcraft will be passively stabilized for a zero pitch trim angle. Detailed preliminary design results of the Team Encounter sailcraft can be found in [5].

The ST5 Geostorm solar sail as well as the Team Encounter solar sail were designed by L'Garde. A square sail with four triangular control vanes designed by L'Garde is illustrated in Fig. 12.7. Although a passive or spin-stabilization technique can be very cost effective for certain missions, an active three-axis attitude control will be necessary for most sailcraft requiring continuous thrust vector steering maneuvers.

One method of actively controlling the attitude of a three-axis stabilized sailcraft is to employ small reflective control vanes mounted at the spar tips, as can be seen in Figs. 12.1 and 12.7. Another method is to change its center-of-mass location

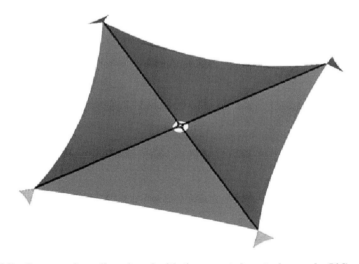

Fig. 12.7   Square solar sail equipped with tip-mounted control vanes by L'Garde [8].

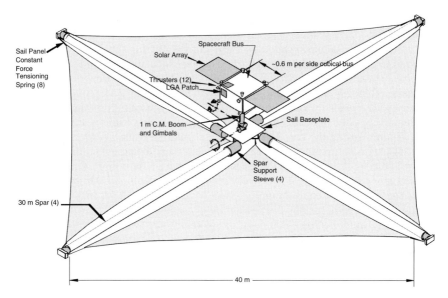

Fig. 12.8   A 40 × 40-m, 160-kg sailcraft (not to scale), proposed by JPL and AEC-ABLE Engineering for the NMP ST7 sail flight experiment [26].

relative to its center-of-pressure location. This can be achieved by articulating a control boom with a tip-mounted payload/bus. Such a concept of articulating a two-axis gimbaled control boom was investigated for a 40 × 40-m square sailcraft (Fig. 12.8), and it was once proposed for the New Millennium Program Space Technology 7 (ST7) sail flight validation experiment [26].

Similar to the problem inherent to the various sail configurations, these different attitude control methods do also have their own advantages and disadvantages in terms of control authority, controllability, and other system-level tradeoff issues.

Although the essential idea behind all of these c.m./c.p. methods appears simple, there are challenging hardware implementation problems to be solved. Some technical issues inherent to the development of an attitude control subsystem for a square sailcraft are briefly discussed next.

Attitude control and stabilization of a sailcraft might also be possible by use of a typical attitude control subsystem, which is often a necessary part of the sailcraft bus. However, small reaction wheels and/or a propulsion subsystem with a limited amount of propellant to be employed for a typical 100-kg bus might be inefficient or ineffective for a fully deployed sailcraft because of its large moments of inertia, its large solar-pressure disturbance torque, and its extended sailing voyages. For example, a 40 × 40-m, 160-kg sailcraft with a nominal solar-pressure force of 0.01 N and a c.m./c.p. offset of ±0.1 m has a solar-pressure disturbance torque of ±0.001 N·m, which is about 100 times larger than that of typical geosynchronous communications satellites. A conventional three-axis attitude control system will require large reaction wheels and also a prohibitively large amount of propellant to counter such a major disturbance torque acting on a sailcraft.

Consequently, the use of a gimbaled control boom, control vanes, sail panel translation/rotation, control-mass translation, or possibly reflectivity modulation is necessary for three-axis attitude control of sailcraft. In addition to these propellantless sail control actuators, three-axis attitude information is crucial for active three-axis attitude control and thrust vector pointing/steering. It is assumed that three-axis inertial attitude information will be available from an attitude determination subsystem, consisting of sun sensors, star cameras, and rate gyros.

One of the critical parameters of sailcraft is the sun angle, often denoted by α, between the sail surface normal and the sun. Its significant effects on the overall performance, stability, and control of a sailcraft are similar to the effects of the aircraft angle of attack α on aircraft performance, stability, and control. Similar to the so-called high-α control problem of high-performance aircraft, a high-performance sailcraft might also have a similar high-α control problem because of its thrust vector pointing requirement of typically $\alpha \approx 35$ deg.

The basic principle behind various aerodynamic control surfaces of aircraft, such as ailerons, elevator, rudder, flaps, trim tabs, and spoilers, should be exploited in developing a sail attitude and flight control subsystem. Furthermore, uncertainties inherent to solar radiation pressure modeling of nonperfect and nonflat sails should be taken into account in designing a sail attitude control subsystem.

12.3 Solar-Radiation-Pressure Models

In this section, solar-radiation-pressure (SRP) models are described for the purpose of sail attitude control analysis and design. These models do not include the effects of sail film wrinkles, thermal deformation, and structural vibration. Throughout this chapter, solar-sails are assumed to be rigid (although they are in fact large, flexible membrane structures) because attitude control and thrust vector steering are to be performed very slowly not to excite structural mode vibrations.

The SRP forces are caused by photons impinging on a surface in space. If a fraction ρ_a of the impinging photons is absorbed, a fraction ρ_s is specularly reflected, and a fraction ρ_d is diffusely reflected by a surface, and then we have

$$\rho_a + \rho_s + \rho_d = 1 \tag{12.1}$$

The SRP force acting on such a flat, Lambertian surface located at 1 astronomical unit (AU) from the sun is modeled as

$$\begin{aligned} \vec{F} &= PA \left[\rho_a(\vec{S} \cdot \vec{n})\vec{S} + 2\rho_s(\vec{S} \cdot \vec{n})^2 \vec{n} + \rho_d(\vec{S} \cdot \vec{n}) \left(\vec{S} + \frac{2}{3}\vec{n} \right) \right] \\ &= PA(\vec{S} \cdot \vec{n}) \left\{ (\rho_a + \rho_d)\vec{S} + \left[2\rho_s(\vec{S} \cdot \vec{n}) + \frac{2}{3}\rho_d \right]\vec{n} \right\} \\ &= PA(\vec{S} \cdot \vec{n}) \left\{ (1 - \rho_s)\vec{S} + \left[2\rho_s(\vec{S} \cdot \vec{n}) + \frac{2}{3}\rho_d \right]\vec{n} \right\} \end{aligned} \tag{12.2}$$

where $P = 4.563 \times 10^{-6}$ N/m^2 is the nominal solar-radiation-pressure constant at 1 AU from the sun, A is the surface area, $\vec{n}$ is a unit vector normal to the surface,

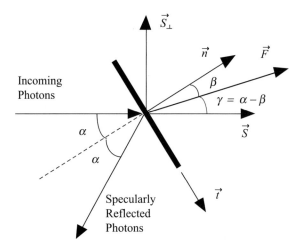

Fig. 12.9   Solar-radiation-pressure force model of a nonperfect flat surface.

and $\vec{S}$ is a unit vector pointing from the sun to the surface as shown in Fig. 12.9. The solar radiation pressure varies inversely with the square of the distance from the sun.

Because $\vec{S} = \cos\alpha\vec{n} + \sin\alpha\vec{t}$, where α is the sun angle between the surface normal and the sunline and $\vec{t}$ is the transverse unit vector, as shown in Fig. 12.9, the SRP force can also be expressed as

$$\vec{F} = F_n\vec{n} + F_t\vec{t} \tag{12.3}$$

where

$$F_n = PA\left\{(1+\rho_s)\cos^2\alpha + \frac{2}{3}\rho_d\cos\alpha\right\}$$

$$F_t = PA(1-\rho_s)\cos\alpha\,\sin\alpha$$

For a case with $\rho_d \approx 0$, we can also express the SRP force as

$$\begin{aligned}
\vec{F} &= PA\,\cos\alpha\left\{(1-\rho_s)\vec{S} + 2\rho_s\cos\alpha\vec{n}\right\} \\
&= PA\,\cos\alpha\left\{(1-\rho_s)\vec{S} + 2\rho_s\cos\alpha(\cos\alpha\vec{S} + \sin\alpha\vec{S}_\perp)\right\} \\
&= PA\,\cos\alpha\left\{(1-\rho_s + 2\rho_s\cos^2\alpha)\vec{S} + 2\rho_s\cos\alpha\,\sin\alpha\vec{S}_\perp\right\} \\
&= PA\,\cos\alpha\left\{(1+\rho_s\cos 2\alpha)\vec{S} + \rho_s\sin 2\alpha\vec{S}_\perp\right\} \\
&= F_s\vec{S} + F_\perp\vec{S}_\perp \tag{12.4}
\end{aligned}$$

where $\vec{S}_\perp$ is a unit vector perpendicular to $\vec{S}$ and is in the same plane as $\vec{S}$ and $\vec{n}$.

For sails in sun-centered orbits, the components along the sun line and perpendicular to the sun line are sometimes called the "drag" and "lift" components, respectively. The transverse component $F_\perp$ provides an effective thrust control for orbital maneuvering of sailcraft in sun-centered orbits. For example, the sun angle that maximizes the transverse component can be found as

$$\frac{d}{d\alpha}F_\perp = 0 \implies \frac{d}{d\alpha}\cos^2\alpha \sin\alpha = 0 \tag{12.5}$$

which gives $\alpha = \tan^{-1}(1/\sqrt{2}) = 35.26$ deg. This optimal sun angle of 35.26 deg for maximizing the transverse component is often selected as the desired pitch orientation of an interplanetary sailcraft in a sun-centered orbit.

The normal and transverse components of the SRP force acting on a flat sail surface with its more detailed optical and thermal properties are also described by (Appendix A of [1] and Chapter 2 of [3])

$$\frac{F_n}{PA} = (1 + rs)\cos^2\alpha + B_f r(1 - s)\cos\alpha + \frac{e_f B_f - e_b B_b}{e_f + e_b}(1 - r)\cos\alpha \tag{12.6}$$

$$\frac{F_t}{PA} = (1 - rs)\cos\alpha \sin\alpha \tag{12.7}$$

$$F = \sqrt{F_n^2 + F_t^2} \tag{12.8}$$

$$\tan\beta = \frac{F_t}{F_n} \tag{12.9}$$

where

B_f, B_b = non-Lambertian coefficients for front and back surfaces
e_f, e_b = front and back surface emission coefficients
r = reflectivity of front surface
s = specular reflection coefficient
β = angle of SRP force vector from surface normal

For a square sailcraft similar to the one illustrated in Fig. 12.1, we have the following optical properties (Appendix A of [1]):

$$B_f = 0.79; \qquad B_b = 0.55$$
$$e_f = 0.05; \qquad e_b = 0.55$$
$$r = 0.88; \qquad s = 0.94$$

and the resulting normal and transverse components of the SRP force become

$$\frac{F_n}{PA} = 1.8272\cos^2\alpha + 0.0417\cos\alpha - 0.0526\cos\alpha$$

$$= 1.8272\cos^2\alpha - 0.0109\cos\alpha \tag{12.10a}$$

$$\frac{F_t}{PA} = 0.1728\cos\alpha \sin\alpha \tag{12.10b}$$

In practice, the pressure distribution is not uniform across the surface of a sail because of curvature (billow). A numerical integration of the F_n and F_t equations across the curved surface of the sail is needed to determine the resulting pressure distribution. This requires an iterative process because the pressure distribution is a function of the sail shape, and, vice versa, the shape is a function of the pressure distribution. For a sailcraft shown in Fig. 12.1, such an iterative process was used by JPL to find a parameterized model of the following form (Appendix B of [1]):

$$F = \eta PA(0.349 + 0.662 \cos 2\gamma - 0.011 \cos 4\gamma) \qquad (12.11)$$

where $\eta = 1.816$ and $\gamma = \alpha - \beta$.

The SRP force acting on a sail surface with an area A is also often approximated as

$$F \approx \eta PA \cos^2 \alpha \qquad (12.12)$$

where η is called the overall sail thrust coefficient, typically around 1.8 for a real sailcraft with sail wrinkles and billowing, with an ideal maximum value of $\eta_{max} = 2$.

12.4   Spin Stabilization of Sailcraft

A simple solution to the problem of maintaining a desired orientation of a sailcraft in the presence of a c.m./c.p. offset is to spin the sailcraft. A thrust vector misalignment with the center of mass will cause the sailcraft to tumble in the absence of spinning or active three-axis control. However, a spinning sailcraft possesses a gyroscopic stiffness to external disturbances, and its motion under the influence of external disturbances is characterized by the precession and nutation of the spin axis. The orientation of a spinning sailcraft can be changed by precession of the sailcraft using thrusters. Tilting and/or translating sail panels can also provide an effective precession control torque to a spinning sailcraft with a large angular momentum.

For example, a spin-stabilization approach was chosen for a 76×76-m square sailcraft of the NMP ST5 Geostorm warning mission [4]. For such a large sailcraft with moments of inertia of $(44,000; 22,000; 24,000)$ kg·m^2, an uncertain c.m./c.p. offset of approximately 1 m was assumed by the Geostorm sail study team, and a spin rate of 0.45 deg/s was then selected to keep the angular momentum vector within 1 deg of the sun line.

In this section, an analytic approach often employed for the dynamic analysis of a spinning body with thrust vector misalignment, discussed in Sec. 6.8, is applied to a spinning solar-sail with a c.m./c.p. offset.

12.4.1   Spinning Sailcraft with a CM/CP Offset

Consider a sailcraft possessing a body-fixed reference frame B with basis vectors $\{\vec{b}_1, \vec{b}_2, \vec{b}_3\}$ and with its origin at the center of mass. The reference frame B coincides with principal axes. It is assumed that the first axis is the roll (spin) axis perpendicular to the sail surface and the second and third axes are the pitch/yaw (transverse) axes. The solar-pressure force vector is nominally aligned along $\vec{b}_1$ through the center of pressure of the sailcraft.

Euler's rotational equations of motion of a rigid sailcraft are simply given by

$$J_1\dot{\omega}_1 - (J_2 - J_3)\omega_2\omega_3 = T_1 \tag{12.13a}$$

$$J_2\dot{\omega}_2 - (J_3 - J_1)\omega_3\omega_1 = T_2 \tag{12.13b}$$

$$J_3\dot{\omega}_3 - (J_1 - J_2)\omega_1\omega_2 = T_3 \tag{12.13c}$$

where $\omega_i \equiv \vec{b}_i \cdot \vec{\omega}$ are the body-axis components of the angular velocity of the sailcraft and T_i are the external torque vector components along the body axes.

For a square (or an axisymmetric circular) sailcraft with $J_2 = J_3 = J$, the rotational equations of motion become

$$J_1\dot{\omega}_1 = 0 \tag{12.14a}$$

$$J\dot{\omega}_2 - (J - J_1)\omega_3\omega_1 = T_2 \tag{12.14b}$$

$$J\dot{\omega}_3 - (J_1 - J)\omega_1\omega_2 = T_3 \tag{12.14c}$$

where T_2 and T_3 are the solar-pressure torque vector components caused by a c.m./c.p. offset. The windmill torque about the spin (roll) axis is ignored here, that is, it is assumed that $T_1 \approx 0$.

From Eq. (12.14a), we have

$$\omega_1 = \text{constant} = \Omega \tag{12.15}$$

where the constant Ω is called the spin rate of the sailcraft about its roll axis $\vec{b}_1$. For simplicity, it is assumed that the pitch/yaw transverse axes are chosen such that $T_2 = 0$ and $T_3 = \epsilon F$, where ϵ is a c.m./c.p. offset distance and F is the solar-pressure force. It is further assumed that the solar-pressure force is nearly constant regardless of a coning motion of the roll axis, although it is a function of the spin-axis orientation relative to the sun.

Equations (12.14b) and (12.14c) then become

$$\dot{\omega}_2 = -\lambda\omega_3 \tag{12.16a}$$

$$\dot{\omega}_3 = \lambda\omega_2 + a \tag{12.16b}$$

where $\lambda = \Omega(J_1 - J)/J$ and $a \equiv \epsilon F/J$ denote the disturbance acceleration resulting from a c.m./c.p. offset. Note that a is assumed to be constant.

To describe the rotational motion of the spinning sailcraft as seen from an inertial reference frame, we consider the roll $\leftarrow$ pitch $\leftarrow$ yaw rotational sequence: $C_1(\theta_1) \leftarrow C_2(\theta_2) \leftarrow C_3(\theta_3)$. For this rotational sequence, we have the following kinematic differential equations:

$$\dot{\theta}_1 = \omega_1 + (\omega_2 \sin\theta_1 + \omega_3 \cos\theta_1)\tan\theta_2 \tag{12.17a}$$

$$\dot{\theta}_2 = \omega_2 \cos\theta_1 - \omega_3 \sin\theta_1 \tag{12.17b}$$

$$\dot{\theta}_3 = \frac{\omega_2 \sin\theta_1 + \omega_3 \cos\theta_1}{\cos\theta_2} \tag{12.17c}$$

For small θ_2, these kinematic differential equations become

$$\dot{\theta}_1 = \omega_1 + \dot{\theta}_3 \theta_2 \tag{12.18a}$$

$$\dot{\theta}_2 = \omega_2 \cos \theta_1 - \omega_3 \sin \theta_1 \tag{12.18b}$$

$$\dot{\theta}_3 = \omega_2 \sin \theta_1 + \omega_3 \cos \theta_1 \tag{12.18c}$$

Assuming $\theta_2 \dot{\theta}_3 << \omega_1$, we can further approximate $\dot{\theta}_1$ as

$$\dot{\theta}_1 \approx \omega_1 = \Omega = \text{constant} \tag{12.19}$$

and $\theta_1 \approx \Omega t$.

Finally, we obtain a set of linearized equations of motion as follows:

$$\dot{\omega}_2 = -\lambda \omega_3 \tag{12.20a}$$

$$\dot{\omega}_3 = \lambda \omega_2 + a \tag{12.20b}$$

$$\dot{\theta}_2 = \omega_2 \cos \Omega t - \omega_3 \sin \Omega t \tag{12.20c}$$

$$\dot{\theta}_3 = \omega_2 \sin \Omega t + \omega_3 \cos \Omega t \tag{12.20d}$$

The solutions of Eqs. (12.20a) and (12.20b) for a constant disturbance acceleration a can be found as

$$\omega_2(t) = \omega_2(0) \sin \lambda t - \omega_3(0) \cos \lambda t - \frac{a}{\lambda}(1 - \cos \lambda t)$$

$$\omega_3(t) = -\omega_2(0) \cos \lambda t - \omega_3(0) \sin \lambda t + \frac{a}{\lambda} \sin \lambda t$$

For a case with $\omega_2(0) = \omega_3(0) = 0$, Eqs. (12.20c) and (12.20d) become

$$\dot{\theta}_2 = \frac{a}{\lambda} \left\{ \cos \frac{J_1}{J} \Omega t - \cos \Omega t \right\} \tag{12.21a}$$

$$\dot{\theta}_3 = \frac{a}{\lambda} \left\{ \sin \frac{J_1}{J} \Omega t - \sin \Omega t \right\} \tag{12.21b}$$

Integrating these equations with respect to time for the initial conditions of $\theta_2(0) = \theta_3(0) = 0$, we obtain

$$\theta_2 = A_p \sin \omega_p t - A_n \sin \omega_n t \tag{12.22a}$$

$$\theta_3 = A_p(1 - \cos \omega_p t) - A_n(1 - \cos \omega_n t) \tag{12.22b}$$

where

$$A_p = \frac{a}{\lambda \Omega} \frac{J}{J_1} = \text{precessional amplitude}$$

$$A_n = \frac{a}{\lambda \Omega} = \text{nutational amplitude}$$

$$\omega_p = \frac{J_1}{J} \Omega = \text{precessional frequency}$$

$$\omega_n = \Omega = \text{nutational frequency}$$

These equations can be used for preliminary dynamic analyses and/or tradeoffs for designing a spin-stabilized solar-sail.

Example 12.1

The feasibility of spin stabilizing a baseline $40 \times 40\text{-}m$ sailcraft shown in Fig. 12.8, without employing a two-axis gimbaled control boom, is examined here. For this sailcraft with an estimated c.m./c.p. offset uncertainty of ± 0.1 m and a 0.01-N solar-pressure force, a spin rate of 0.5 deg/s is considered. The spin dynamics of this sailcraft are then characterized as follows:

$$(J_1, J_2, J_3) = (6000, 3000, 3000) \text{ kg-m}^2$$

$$\frac{J_1}{J} = 2; \qquad \Omega = 0.5 \text{ deg/s}$$

$$\lambda = \frac{J_1 - J}{J}\Omega = 0.5 \text{ deg/s}$$

$$F = 0.01 \text{ N}; \qquad \epsilon = 0.1 \text{ m}$$

$$a = \frac{\epsilon F}{J} = 3.3 \times 10^{-7} \text{ rad/s}^2$$

For these parameters, we obtain the precessional and nutation amplitudes as $A_p = 0.12$ deg and $A_n = 0.25$ deg. Therefore, it can be concluded that a low spin rate of 0.1–0.5 deg/s can keep the thrust vector pointing error well within ± 1 deg for this sailcraft with a 0.1-m c.m./c.p. offset and a 0.01-N solar-pressure force. The simulation results are shown in Fig. 12.10. The plot of θ_3 vs θ_2 shows the path of the tip of the roll axis in space. However, high-fidelity dynamic modeling of a spinning sail will be needed to validate the simplified model used here, if such a spin-stabilization approach is to be actually employed for near-term solar-sails.

12.5 Sailcraft in an Earth-Centered Elliptic Orbit

Although most future solar-sails will be flying in sun-centered orbits, a variety of solar-sail application missions in Earth-centered orbits is also being studied by NASA and ESA. This section provides a brief formulation of attitude dynamics of solar-sail in an Earth-centered elliptic orbit for the purpose of attitude control analysis and design. Various orientations of a sailcraft in an Earth-centered orbit are illustrated in Fig. 12.11.

12.5.1 Dynamic Equations of Motion

Consider a sailcraft in an Earth-centered elliptic orbit. A local vertical and local horizontal (LVLH) reference frame A with its origin at the center of mass of an orbiting sailcraft has a set of unit vectors $\{\vec{a}_1, \vec{a}_2, \vec{a}_3\}$ with $\vec{a}_3$ locally vertical toward the Earth, $\vec{a}_1$ along the locally horizontal (transverse) direction, and $\vec{a}_2$ perpendicular to the orbit plane.

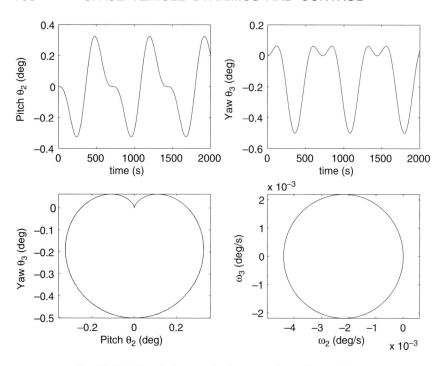

Fig. 12.10   Simulation results for a c.m./c.p. offset of 0.1 m.

The angular velocity of such an LVLH reference frame A with respect to the Earth-centered-inertial (ECI) reference frame is given by

$$\vec{\omega}^{A/E} = -\dot{\theta}\vec{a}_2 \tag{12.23}$$

where $\dot{\theta}$ is the orbital rate and θ is the true anomaly (satellite angular position) as measured from the perigee. The angular velocity of the body-fixed reference frame B with basis vectors $\{\vec{b}_1, \vec{b}_2, \vec{b}_3\}$ is then given by

$$\vec{\omega}^{B/E} = \vec{\omega}^{B/A} + \vec{\omega}^{A/E} = \vec{\omega}^{B/A} - \dot{\theta}\vec{a}_2 \tag{12.24}$$

where $\vec{\omega}^{B/A}$ is the angular velocity of B relative to A.

To describe the orientation of the body-fixed reference frame B with respect to the LVLH reference frame A in terms of three Euler angles θ_i $(i = 1, 2, 3)$, consider the sequence of $\mathbf{C}_1(\theta_1) \leftarrow \mathbf{C}_3(\theta_3) \leftarrow \mathbf{C}_2(\theta_2)$ from the LVLH reference frame A to a body-fixed reference frame B. For this rotational sequence we have

$$
\begin{bmatrix} \vec{b}_1 \\ \vec{b}_2 \\ \vec{b}_3 \end{bmatrix} =
\begin{bmatrix} C_{11} & C_{12} & C_{13} \\ C_{21} & C_{22} & C_{23} \\ C_{31} & C_{32} & C_{33} \end{bmatrix}
\begin{bmatrix} \vec{a}_1 \\ \vec{a}_2 \\ \vec{a}_3 \end{bmatrix} \tag{12.25}
$$

where C_{ij} are the direction cosine elements.

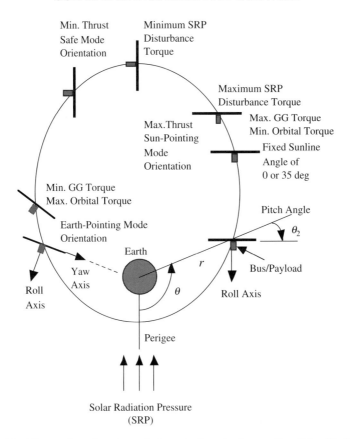

Min. Thrust
Safe Mode
Orientation

Minimum SRP
Disturbance
Torque

Maximum SRP
Disturbance Torque

Max. Thrust
Sun-Pointing
Mode
Orientation

Max. GG Torque
Min. Orbital Torque

Fixed Sunline
Angle of
0 or 35 deg

Min. GG Torque
Max. Orbital Torque

Earth-Pointing Mode
Orientation

Earth

Pitch Angle

θ_2

Yaw
Axis

Roll
Axis

r

θ

Bus/Payload

Roll Axis

Perigee

Solar Radiation Pressure
(SRP)

Fig. 12.11   Illustration of various pitch control modes for a sailcraft in elliptic orbit.

The rotational equation of motion of a rigid body with an angular momentum $\vec{H}$ is given by

$$\left\{\frac{d\vec{H}}{dt}\right\}_E \equiv \left\{\frac{d\vec{H}}{dt}\right\}_B + \vec{\omega}^{B/E} \times \vec{H} = \vec{M} \qquad (12.26)$$

where $\{d/dt\}_E$ indicates differentiation with respect to time in reference frame E and $\{d/dt\}_B$ indicates differentiation with respect to time in reference frame B. The external torque vector $\vec{M}$ is expressed as

$$\vec{M} = \frac{3\mu}{r^3}\vec{a}_3 \times \hat{J} \cdot \vec{a}_3 + \vec{u} \qquad (12.27)$$

where $\vec{a}_3 \equiv -\vec{r}/r$, r is the radial distance of the sailcraft from the center of the Earth, $\hat{J}$ is the inertia dyadic of the sailcraft with respect to its mass center, and $\vec{u}$ is the control torque vector.

Because $\vec{H} = \hat{J} \cdot \vec{\omega}^{B/E}$, the attitude dynamical equations of motion can be rewritten as

$$\hat{J} \cdot \dot{\vec{\omega}} + \vec{\omega} \times \hat{J} \cdot \vec{\omega} = \frac{3\mu}{r^3} \vec{a}_3 \times \hat{J} \cdot \vec{a}_3 + \vec{u} \tag{12.28}$$

where $\vec{\omega} \equiv \vec{\omega}^{B/E}$ and $\dot{\vec{\omega}} = \{d\vec{\omega}/dt\}_E \equiv \{d\vec{\omega}/dt\}_B$. Expressing $\vec{\omega}$, $\vec{a}_3$, and $\hat{J}$ in terms of basis vectors of the body-fixed reference frame B as

$$\vec{\omega} = \omega_1 \vec{b}_1 + \omega_2 \vec{b}_2 + \omega_3 \vec{b}_3 \tag{12.29a}$$

$$\vec{a}_3 = C_{13} \vec{b}_1 + C_{23} \vec{b}_2 + C_{33} \vec{b}_3 \tag{12.29b}$$

$$\hat{J} = \sum_{i=1}^{3} \sum_{j=1}^{3} J_{ij} \vec{b}_i \vec{b}_j \tag{12.29c}$$

we obtain the dynamical equations of motion about the body-fixed principal axes as

$$J_1 \dot{\omega}_1 - (J_2 - J_3)\omega_2\omega_3 = -\frac{3\mu}{r^3}(J_2 - J_3)C_{23}C_{33} + u_1 \tag{12.30a}$$

$$J_2 \dot{\omega}_2 - (J_3 - J_1)\omega_3\omega_1 = -\frac{3\mu}{r^3}(J_3 - J_1)C_{33}C_{13} + u_2 \tag{12.30b}$$

$$J_3 \dot{\omega}_3 - (J_1 - J_2)\omega_1\omega_2 = -\frac{3\mu}{r^3}(J_1 - J_2)C_{13}C_{23} + u_3 \tag{12.30c}$$

where (u_1, u_2, u_3) are the control torque components along the body-fixed reference frame B and

$$C_{13} = -\sin\theta_2 \cos\theta_3 \tag{12.31a}$$
$$C_{23} = \cos\theta_1 \sin\theta_2 \sin\theta_3 + \sin\theta_1 \cos\theta_2 \tag{12.31b}$$
$$C_{33} = -\sin\theta_1 \sin\theta_2 \sin\theta_3 + \cos\theta_1 \cos\theta_2 \tag{12.31c}$$

for the sequence of $C_1(\theta_1) \leftarrow C_3(\theta_3) \leftarrow C_2(\theta_2)$ under consideration. For this rotational sequence we have the following kinematic differential equations:

$$\begin{bmatrix} \dot{\theta}_1 \\ \dot{\theta}_2 \\ \dot{\theta}_3 \end{bmatrix} = \frac{1}{c\theta_3} \begin{bmatrix} c\theta_3 & -c\theta_1 s\theta_3 & s\theta_1 s\theta_3 \\ 0 & c\theta_1 & -s\theta_1 \\ 0 & s\theta_1 c\theta_3 & c\theta_1 c\theta_3 \end{bmatrix} \begin{bmatrix} \omega_1 \\ \omega_2 \\ \omega_3 \end{bmatrix} + \begin{bmatrix} 0 \\ \dot{\theta} \\ 0 \end{bmatrix} \tag{12.32}$$

where $\dot{\theta}$ is the orbital rate, $c\theta_i \equiv \cos\theta_i$, and $s\theta_i \equiv \sin\theta_i$.

12.5.2 Earth-Pointing Sailcraft in an Earth-Centered Elliptic Orbit

In this section, the attitude equations of motion of a sailcraft that might need to be continuously pointing its yaw axis toward the Earth (i.e., edgewise toward the earth) are derived. Such an Earth-pointing mode for minimizing the effect of the gravity-gradient disturbance torque is illustrated in Fig. 12.11.

Assuming that θ_1 and θ_3 are small, θ_2 can be arbitrarily large, ω_1 and ω_3 are also small, and that

$$\omega_1 \approx \dot{\theta}_1 - \dot{\theta}\theta_3 \tag{12.33a}$$

$$\omega_2 \approx \dot{\theta}_2 - \dot{\theta} \tag{12.33b}$$

$$\omega_3 \approx \dot{\theta}_3 + \dot{\theta}\theta_1 \tag{12.33c}$$

we obtain the attitude equations of motion as

$$J_1\ddot{\theta}_1 + (\dot{\theta}^2 + \frac{3\mu}{r^3}\cos^2\theta_2)(J_2 - J_3)\theta_1 - \dot{\theta}(J_1 - J_2 + J_3)\dot{\theta}_3$$

$$+ \frac{3\mu}{r^3}(J_2 - J_3)(\sin\theta_2 \cos\theta_2)\theta_3 = u_1 \tag{12.34a}$$

$$J_2\ddot{\theta}_2 + \frac{3\mu}{r^3}(J_1 - J_3)\sin\theta_2 \cos\theta_2 = J_2\ddot{\theta} + u_2 \tag{12.34b}$$

$$J_3\ddot{\theta}_3 + (\dot{\theta}^2 + \frac{3\mu}{r^3}\sin^2\theta_2)(J_2 - J_1)\theta_3 + \dot{\theta}(J_1 - J_2 + J_3)\dot{\theta}_1$$

$$+ \frac{3\mu}{r^3}(J_2 - J_1)(\sin\theta_2 \cos\theta_2)\theta_1 = u_3 \tag{12.34c}$$

As can be seen in Eq. (12.34b), the pitch attitude equation has an orbital disturbance torque $J_2\ddot{\theta}$, caused by the time-varying orbital rate $\dot{\theta}$.

Consider a sailcraft in an Earth-centered elliptic orbit with an eccentricity of e and a semimajor axis of a. Its orbital motion with slowly changing orbital elements is then described by

$$r = \frac{p}{1 + e\,\cos\theta} \tag{12.35}$$

where $p = a(1 - e^2)$, r is the radial distance of the sailcraft from the center of the Earth, and θ is the true anomaly. Furthermore, we have

$$\dot{r} = \sqrt{\frac{\mu}{p}}\, e\,\sin\theta \tag{12.36a}$$

$$\dot{\theta} = \sqrt{\frac{\mu}{p^3}}\,(1 + e\,\cos\theta)^2 \tag{12.36b}$$

$$\ddot{\theta} = \frac{2\mu}{p^3}\,(1 + e\,\cos\theta)^3 e\,\sin\theta \tag{12.36c}$$

where μ is the Earth's gravitational parameter. The orbital mean motion is defined as $n = \sqrt{\mu/a^3}$.

For small roll/pitch/yaw angles with respect to the LVLH frame, we obtain

$$J_1\ddot{\theta}_1 + (\dot{\theta}^2 + \frac{3\mu}{r^3})(J_2 - J_3)\theta_1 - \dot{\theta}(J_1 - J_2 + J_3)\dot{\theta}_3 = u_1$$

$$J_2\ddot{\theta}_2 + \frac{3\mu}{r^3}(J_1 - J_3)\theta_2 = J_2\ddot{\theta} + u_2$$

$$J_3\ddot{\theta}_3 + \dot{\theta}^2(J_2 - J_1)\theta_3 + \dot{\theta}(J_1 - J_2 + J_3)\dot{\theta}_1 = u_3 \qquad (12.37)$$

For a circular orbit with the constant orbital rate of $\dot{\theta} = n = \sqrt{\mu/a^3}$, we have

$$J_1\ddot{\theta}_1 + 4n^2(J_2 - J_3)\theta_1 - n(J_1 - J_2 + J_3)\dot{\theta}_3 = u_1$$

$$J_2\ddot{\theta}_2 + 3n^2(J_1 - J_3)\theta_2 = u_2$$

$$J_3\ddot{\theta}_3 + n^2(J_2 - J_1)\theta_3 + n(J_1 - J_2 + J_3)\dot{\theta}_1 = u_3 \qquad (12.38)$$

which are the well-known linearized equations of motion of an Earth-pointing spacecraft in a circular orbit.

12.5.3 Sun-Pointing Sailcraft in an Earth-Centered Elliptic Orbit

The pitch axis of a sun-pointing sailcraft considered in this section is assumed to be perpendicular to the orbital plane (not to the ecliptic plane), as illustrated in Fig. 12.11. For such a sun-pointing sailcraft in an Earth-centered elliptic orbit with small body rates, ω_i $(i = 1, 2, 3)$, and small roll/yaw angles, θ_1 and θ_3, the kinematic differential equations, Eq. (12.32), can be approximated as

$$\omega_1 \approx \dot{\theta}_1 \qquad (12.39a)$$

$$\omega_2 \approx \dot{\theta}_2 - \dot{\theta} \qquad (12.39b)$$

$$\omega_3 \approx \dot{\theta}_3 \qquad (12.39c)$$

The attitude equations of motion with small roll and yaw angles in an Earth-centered elliptic orbit can then be obtained as

$$J_1\ddot{\theta}_1 + (\dot{\theta}^2 + \frac{3\mu}{r^3}\cos^2\theta_2)(J_2 - J_3)\theta_1 + \frac{3\mu}{r^3}(J_2 - J_3)(\sin\theta_2 \cos\theta_2)\theta_3 = u_1$$
$$(12.40a)$$

$$J_2\ddot{\theta}_2 + \frac{3\mu}{r^3}(J_1 - J_3)\sin\theta_2 \cos\theta_2 = J_2\ddot{\theta} + u_2 \qquad (12.40b)$$

$$J_3\ddot{\theta}_3 + (\dot{\theta}^2 + \frac{3\mu}{r^3}\sin^2\theta_2)(J_2 - J_1)\theta_3 + \frac{3\mu}{r^3}(J_2 - J_1)(\sin\theta_2 \cos\theta_2)\theta_1 = u_3$$
$$(12.40c)$$

The pitch angle relative to the LVLH frame θ_2 can be expressed as

$$\theta_2 = \theta - \frac{\pi}{2} + \alpha \qquad (12.41)$$

where α is the sun angle between the surface normal and the sunline. In Fig. 12.11, the pitch angle θ_2 and the true anomaly θ are shown for an ideal case of $\alpha = 0$.

The pitch equation of motion of a sun-pointing sailcraft in terms of its sun angle α then becomes

$$J_2 \ddot{\alpha} - \frac{3\mu}{r^3}(J_1 - J_3) \sin(\alpha - \theta) \cos(\alpha - \theta) = u_2 \qquad (12.42)$$

12.5.4 Attitude Control of a Sailcraft Using Reaction Wheels

The feasibility, as well as limitation, of employing a typical attitude control subsystem that is often a necessary part of the sailcraft bus is discussed here. Various pitch control modes for an experimental sailcraft in an elliptic orbit, investigated for the NMP ST7 mission study, are illustrated in Fig. 12.11. Preliminary pitch control analyses and simulation results are discussed here to emphasize the significant effects of the coupled attitude and orbital dynamics and the solar-pressure disturbance torque on sail attitude control using reaction wheels.

The Earth-pointing pitch model of a sailcraft in an Earth-centered elliptic orbit is given by

$$J_2 \ddot{\theta}_2 + \frac{3\mu}{r^3}(J_1 - J_3) \sin \theta_2 \cos \theta_2 = J_2 \ddot{\theta} + u_2 + d_2 \qquad (12.43)$$

where u_2 is the pitch control torque and d_2 is the solar-pressure disturbance torque. The pitch equation, Eq. (12.43), is often transformed to a dynamic model of the form

$$(1 + e \cos \theta)\theta_2'' - (2e \sin \theta)\theta_2' + \frac{3(J_1 - J_3)}{J_2} \sin \theta_2 \cos \theta_2$$
$$= 2e \sin \theta + \frac{(1 + e)^3}{J_2(1 + e \cos \theta)^3}(u_2 + d_2) \qquad (12.44)$$

where $(\ )' \equiv d(\ )/d\theta$. This model has been investigated extensively in the literature for analyzing the effect of the periodic pitching excitation, $2e \sin \theta$, on the pitch attitude motion of a rigid spacecraft in an elliptic orbit.

The sun-pointing pitch model of a sailcraft in an Earth-centered elliptic orbit is also given by

$$J_2 \ddot{\alpha} - \frac{3\mu}{r^3}(J_1 - J_3) \sin(\alpha - \theta) \cos(\alpha - \theta) = u_2 + d_2 \qquad (12.45)$$

where α is the sun angle related to θ and θ_2 as $\alpha = \theta_2 - \theta + \pi/2$.

For the purpose of studying the feasibility of stabilizing a sailcraft using a conventional attitude control system employing reaction wheels, we consider a simple model of reaction-wheel dynamics of the form

$$\dot{h}_2 = -u_2 \qquad (12.46)$$

where h_2 is the pitch-wheel angular momentum. The solar-pressure disturbance torque caused by a c.m./c.p. offset is also assumed as

$$d_2 = \epsilon F \cos^2\alpha \, \cos^2 i \qquad (12.47)$$

where F is the nominal solar-pressure force, ϵ is the c.m./c.p. offset, and i is the orbital inclination angle from the ecliptic plane. Note that the nominal pitch axis is assumed to be perpendicular to the orbital plane and that small roll/yaw attitude angles are assumed.

For a 40×40 m sailcraft, we assume that $F = 0.01$ N, $\epsilon = \pm 0.1$ m, and $(J_1, J_2, J_3) = (6000, 3000, 3000)$ kg·m^2. An Earth-centered elliptic orbit, called the super-synchronous transfer orbit (SSTO), proposed for a sail validation mission is characterized as follows:

$$r_p = 6374 + 2000 = 8{,}374 \, \text{km}$$

$$r_a = 6374 + 78{,}108 = 84{,}482 \, \text{km}$$

$$a = \frac{(r_p + r_a)}{2} = 46{,}428 \, \text{km}$$

$$e = \frac{r_a - r_p}{r_a + r_p} = 0.8196$$

$$i = 12 \, \text{deg (from the ecliptic plane)}$$

$$p = a(1 - e^2) = 15{,}238 \, \text{km}$$

$$n = \sqrt{\frac{\mu}{a^3}} = 6.311 \times 10^{-5} \, \text{rad/s}$$

and the orbital period of 27.65 h.

Pitch control logic of the Earth-pointing mode is assumed as

$$u_2 = -K_P\theta_2 - K_D\dot{\theta}_2 \qquad (12.48)$$

and pitch control logic of the sun-pointing mode as

$$u_2 = -K_P(\alpha - \alpha_c) - K_D\dot{\alpha} \qquad (12.49)$$

where α_c is the commanded sun angle. Controller gains of $K_P = 0.0865$ N·m/rad and $K_D = 22.78$ N·m/rad/s were selected for the closed-loop eigenvalues of $-0.005 \pm 0.005j$ rad/s.

Simulation results of flight validating a sailcraft during its continuous Earth-pointing mode operation are shown in Fig. 12.12 for two consecutive orbits. The significant effect of the orbital disturbance torque $J_2\ddot{\theta}$ on the pitch-wheel momentum requirement of a peak value of about 3.3 N·m·s can be seen in Fig. 12.12. The c.m./c.p. offset was not included in this simulation.

Simulation results of the continuous Earth-pointing mode operation, including a c.m./c.p. offset of 0.1 m, indicate that the sailcraft in an Earth-pointing mode might

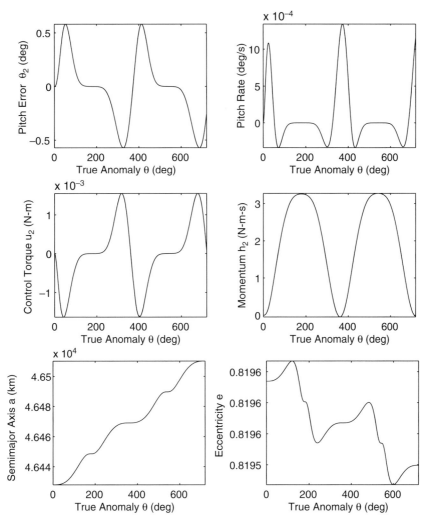

Fig. 12.12 **Simulation results for the Earth-pointing mode, not including a c.m./c.p. offset of 0.1 m.**

require a pitch wheel of ± 15 N·m·s momentum storage capability to counter both the solar pressure and orbital disturbance torques.

However, a continuous sun-pointing mode operation of the 40-m sailcraft with a desired fixed sun angle of $\alpha = 35$ deg causes the reaction-wheel momentum growth rate of about 100 N·m·s per orbit for countering a c.m./c.p. offset of 0.1 m. Such a momentum growth rate is too excessive. Consequently, either a spin stabilization or a c.m./c.p. method needs to be employed for a sun-pointing sailcraft. The net change of the semimajor axis during each orbit is zero for such a continuous sun-pointing mode operation with a fixed sun angle. With such a fixed sun angle, the semimajor axis increases when the sail is moving away from the sun, but it decreases when the sail is moving sunward. To increase the orbital energy or the semimajor axis, the sail needs to be oriented edgewise toward the sun when the sailcraft is moving sunward. Such a simple orbit-raising sail steering profile requires two rapid 90-deg pitch maneuvers twice per orbit.

Simulation results for the baseline 40-m sailcraft in a zero-thrust mode are shown in Fig. 12.13. Although the effect of the solar-pressure disturbance torque

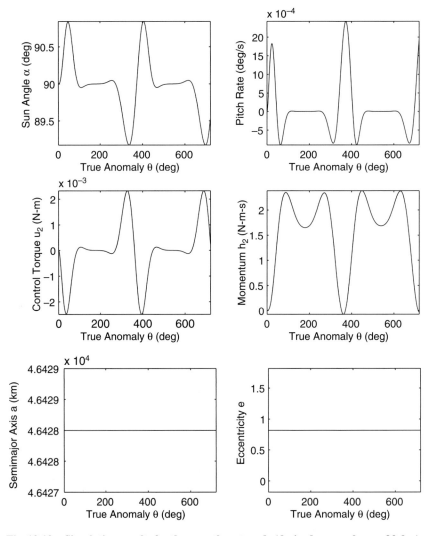

Fig. 12.13 Simulation results for the zero-thrust mode (desired sun angle $\alpha = 90$ deg), including a c.m./c.p. offset of 0.1 m.

is not significant because the edge of the sail is oriented toward the sun, the effect of the gravity-gradient torque on the pitch-wheel momentum storage requirement of about 2.5 N·m·s is also evident for this case.

In summary, a solar-pressure disturbance torque of 1 mN·m of the 40-m sail-craft when it is nominally sun pointing is about 100 times larger than that of typical geosynchronous communications satellites. Consequently, a nominally sun-pointing sailcraft in an Earth-centered orbit will need to be either spin stabilized or three-axis stabilized using control vanes or a gimbaled control boom.

12.5.5 Spin Stabilization of a Sailcraft in Earth-Centered Elliptic Orbit

A spin-stabilized, sun-pointing mode of a solar-sail in an Earth-centered elliptic orbit is illustrated in Fig. 12.14. As derived earlier, the attitude dynamical equations of motion of such a spinning sailcraft in an Earth-centered elliptic orbit can be summarized as

$$J_1\dot{\omega}_1 - (J_2 - J_3)\omega_2\omega_3 = -\frac{3\mu}{r^3}(J_2 - J_3)C_{23}C_{33} \qquad (12.50a)$$

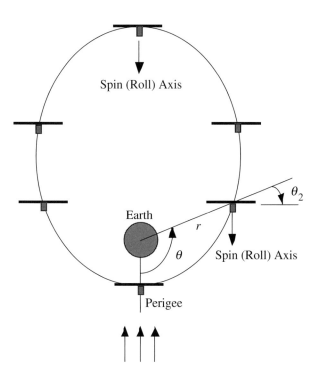

Fig. 12.14   Illustration of a spin-stabilized, sun-pointing mode for the sailcraft in elliptic orbit.

$$J_2\dot{\omega}_2 - (J_3 - J_1)\omega_3\omega_1 = -\frac{3\mu}{r^3}(J_3 - J_1)C_{33}C_{13} + \epsilon F \qquad (12.50b)$$

$$J_3\dot{\omega}_3 - (J_1 - J_2)\omega_1\omega_2 = -\frac{3\mu}{r^3}(J_1 - J_2)C_{13}C_{23} + \epsilon F \qquad (12.50c)$$

$$\begin{bmatrix} \dot{\theta}_1 \\ \dot{\theta}_2 \\ \dot{\theta}_3 \end{bmatrix} = \frac{1}{c\,\theta_3}\begin{bmatrix} c\,\theta_3 & -c\,\theta_1\,s\,\theta_3 & s\,\theta_1\,s\,\theta_3 \\ 0 & c\,\theta_1 & -s\,\theta_1 \\ 0 & s\,\theta_1\,c\,\theta_3 & c\,\theta_1\,c\,\theta_3 \end{bmatrix}\begin{bmatrix} \omega_1 \\ \omega_2 \\ \omega_3 \end{bmatrix} + \begin{bmatrix} 0 \\ \dot{\theta} \\ 0 \end{bmatrix} \qquad (12.51)$$

where $\dot{\theta}$ is the time-varying orbital rate, $c\,\theta_i \equiv \cos\theta_i$, $s\,\theta_i \equiv \sin\theta_i$, and the direction cosine elements C_{ij} are given by Eq. (12.31). The solar-pressure disturbance torque is simply assumed as ϵF for both the pitch and yaw axes.

For the baseline 40 × 40-m sailcraft in a supersynchronous transfer orbit, nominal spin mode conditions are assumed as follows: a spin rate of $\omega_1 = 0.5$ deg/s about the roll axis, a solar-pressure force of 0.01 N, a c.m./c.p. offset of 0.1 m, and a nominal sun angle of 0 deg. Figure 12.15 shows two consecutive orbit simulation results of this nominal sun-pointing spin mode. In this figure, the sun angle, $\alpha = \theta_2 - \theta + \pi/2$, is used as a pitch pointing error, and the gravity-gradient torques are expressed along the pitch/yaw axes fixed to the spinning sailcraft. The plot of θ_3 vs α shows the path of the tip of the roll axis in space. The gravity-gradient

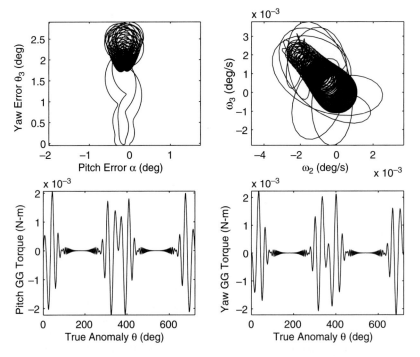

Fig. 12.15 Spin mode simulation with a c.m./c.p. offset of 0.1 m.

torques have caused a sun-pointing error of about 3 deg. The complex, but cyclic, nature of the gravity-gradient torques on a spinning sail in an Earth-centered elliptic orbit is evident here.

High-fidelity dynamic modeling of a spinning sail will be needed to validate the simplified model used here, if such a spin-stabilization approach is to be actually employed for near-term solar-sails.

12.6   Solar-Sail Attitude Control Using a Gimbaled Control Boom and Control Vanes

12.6.1   Introduction

A 40 × 40-m sailcraft configuration and its system architecture, as illustrated in Fig. 12.8, was proposed by JPL and AEC-ABLE Engineering (now, ATK Space Systems) for the NMP ST7 flight validation mission [26]. A significant feature of this baseline ST7 sailcraft is the use of a two-axis gimbaled control boom, instead of control vanes, for propellantless sail attitude control. Although the proposed solar-sail mission was not selected as an actual flight validation mission of the NMP ST7, 20-m scaled solar-sails were further developed by NASA, ATK Space Systems, and L'Garde for ground validation experiments in 2005 [7,8].

In this section, a 40 × 40-m, 160-kg sailcraft with a nominal solar-pressure force of 0.01 N, an uncertain c.m./c.p. offset of ±0.1 m, and moments of inertia of $(6000, 3000, 3000)$ kg·m^2 are further studied to illustrate the various concepts and principles involved in dynamic modeling and attitude control design [27]. Particular emphasis is placed on various control design options for countering the significant solar-pressure disturbance torque caused by an uncertain c.m./c.p. offset.

As discussed in the preceding sections, one method of controlling the attitude of a three-axis stabilized sailcraft is to change its center-of-mass location relative to its center-of-pressure location. This can be achieved by articulating a control boom with a tip-mounted mass. Another method is to employ small reflective control vanes mounted at the spar tips. A dynamic model of a generic three-axis stabilized sailcraft with such tip-mounted vanes and a control boom, as illustrated in Fig. 12.16, is developed here. The complexity of the modeling and control problem inherent to even such a simple rigid sail, but with a moving mass, will be discussed.

The problem of a rigid spacecraft with internal moving mass was first investigated in the early 1960s. For spacecraft dynamical problems with internal moving mass, one can choose the composite center of mass of the total system as a reference point for the equations of motion. This formulation leads to a time-varying inertia matrix of the main rigid body because the reference point is not fixed at the main body as the internal mass moves relative to the main body. On the other hand, one can choose the center of mass of the main body as the reference point, which leads to a constant inertia matrix of the main body relative to the reference point, but resulting in complex equations of motion. In this section, the second approach of choosing the center of mass of the main body as the reference point is employed.

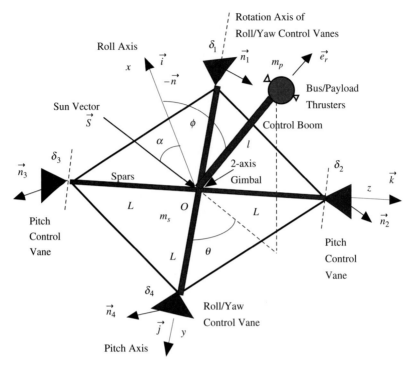

Fig. 12.16   Three-axis stabilized square sailcraft with a two-axis gimbaled control boom and four tip-mounted control vanes.

12.6.2   Dynamic Equations of Motion

Consider an ideal sailcraft model consisting of a rigid sail subsystem of mass m_s and a payload/bus of mass m_p located at the end of a massless control boom of length ℓ, as shown in Fig. 12.16. The origin of the body-fixed reference frame (x, y, z) is located at point O, which is assumed to be the center of mass of a rigid sail subsystem of mass m_s.

The position vector of the payload/bus mass from the reference point O is expressed as

$$\vec{r} = x\vec{i} + y\vec{j} + z\vec{k} = \ell\vec{e}_r = \ell(\cos\phi\,\vec{i} + \sin\phi\,\cos\theta\,\vec{j} + \sin\phi\,\sin\theta\,\vec{k}) \quad (12.52)$$

where ϕ is the boom tilt angle and θ is the boom azimuth angle relative to the sailcraft body axes (x, y, z). These two gimbal angles can be considered as control inputs; however, the gimbal dynamics needs to be included later for detailed control design.

There are four different ways of modeling the attitude dynamics of a space-craft with a moving mass. In general, the four different angular momentum equations are given by

$$\dot{\vec{H}}_o + \dot{\vec{R}}_o \times m\dot{\vec{r}}_c = \vec{M}_o \qquad (12.53a)$$

$$\dot{\vec{h}}_o + \vec{r}_c \times m\vec{a}_o = \vec{M}_o \qquad (12.53b)$$

$$\dot{\vec{H}}_c + \vec{r}_c \times m\vec{a}_c = \vec{M}_o \qquad (12.53c)$$

$$\dot{\vec{H}}_c = \vec{M}_c \qquad (12.53d)$$

where m is the total system mass, $\vec{H}_o$ and $\vec{h}_o$ are, respectively, the absolute and relative angular momentum about an arbitrary reference point O, $\vec{a}_o$ is the absolute acceleration of the point O, $\vec{a}_c$ is the absolute acceleration of the composite center of mass, $\vec{H}_c$ is the absolute angular momentum about the composite center of mass, $\vec{M}_o$ is the moment of the external forces about the point O, and $\vec{M}_c$ is the moment of the external forces about the composite center of mass.

An inertially fixed point or the moving center of mass is often selected as a reference point O. However, there are many cases in which a reference point is selected to be neither inertially fixed nor the center of mass of the system.

For a sailcraft with a gimbaled control boom, one can choose the angular momentum equation of the form

$$\dot{\vec{h}}_o + m\vec{r}_c \times \vec{a}_o = \vec{M}_o \qquad (12.54)$$

where $\vec{h}_o$ is the relative angular momentum of the total system about point O, $m = m_s + m_p$ is the total mass, $\vec{r}_c$ is the position vector of the composite center of mass from the reference point O, and $\vec{a}_o$ is the inertial acceleration of the point O, and $\vec{M}_o$ is the external torque vector about the point O. The position vector of the composite center of mass is simply given by

$$\vec{r}_c = \frac{m_p}{m_s + m_p}\vec{r} = \frac{m_p}{m}\ell\vec{e}_r \qquad (12.55)$$

The relative angular momentum of the total system about point O is given by

$$\vec{h}_o = \hat{J} \cdot \vec{\omega} + m_p\vec{r} \times \dot{\vec{r}} \qquad (12.56)$$

where $\hat{J}$ is inertia dyadic of the system and $\vec{\omega}$ is the angular velocity vector of the main body (i.e., sail subsystem) expressed as

$$\vec{\omega} = \omega_x\vec{i} + \omega_y\vec{j} + \omega_z\vec{k} \qquad (12.57)$$

The inertia dyadic of the system, including the payload mass, about the reference point O is given by

$$\hat{J} = \begin{bmatrix} \vec{i} & \vec{j} & \vec{k} \end{bmatrix} \begin{bmatrix} J_{11} & J_{12} & J_{13} \\ J_{21} & J_{22} & J_{23} \\ J_{31} & J_{32} & J_{33} \end{bmatrix} \begin{bmatrix} \vec{i} \\ \vec{j} \\ \vec{k} \end{bmatrix}$$

where

$$J_{11} = J_x + m_p(y^2 + z^2), \qquad J_{22} = J_y + m_p(x^2 + z^2)$$

$$J_{33} = J_z + m_p(x^2 + y^2), \qquad J_{12} = J_{21} = -m_p xy$$

$$J_{13} = J_{31} = -m_p xz, \qquad J_{23} = J_{32} = -m_p yz(J_x, J_y, J_z)$$

are the principal moments of inertia of the sail subsystem, not including the tip mass m_p.

As discussed earlier, the resultant SRP force can be simply modeled as

$$\vec{F} = \eta PA(\vec{S} \cdot \vec{n})^2 \, \vec{n} = F_s \cos^2\alpha \, \vec{n} \tag{12.58}$$

where η is the overall sail thrust coefficient ($\eta_{max} = 2$), $P = 4.563 \times 10^{-6} \text{ N/m}^2$, A is the total sail area, $F_s \equiv \eta PA$ is the maximum sail thrust, $\vec{S}$ is a unit vector from the sun, and $\vec{n} \equiv -\vec{i}$ is the unit vector normal to the sail surface. The resultant SRP force, nominally acting on the reference point O but with a possible uncertain offset, generates a control torque about the composite center of mass.

In addition to the gimbaled control boom, two or four vanes at the spar tips can be used, each vane with one or two degrees of freedom. Control vanes, each with two degrees of freedom, provide more control redundancy and also more control authority. A variety of control vane configurations is possible for generating proper three-axis control torques. An arrangement of four triangular vanes is considered here without loss of generality. The SRP control force acting on a triangular control vane acts at a point two-thirds of the distance out of the spar axis, called the vane center of pressure.

The control torque vector about the point O generated by deflecting the four control vanes is expressed as

$$\vec{M}_o = \sum_{i=1}^{4} \vec{\ell}_i \times F_c(\vec{S} \cdot \vec{n}_i)^2 \vec{n}_i \tag{12.59}$$

where $F_c = \eta PA_c$, A_c is the control vane area, with all vanes assumed identical, $\vec{\ell}_i$ is the position vector of the ith vane center of pressure from the point O, and

$$\vec{n}_1 = -\cos\delta_1 \vec{i} - \sin\delta_1 \vec{k}$$

$$\vec{n}_2 = -\cos\delta_2 \vec{i} + \sin\delta_2 \vec{k}$$

$$\vec{n}_3 = -\cos\delta_3 \vec{i} + \sin\delta_3 \vec{k}$$

$$\vec{n}_4 = -\cos\delta_4 \vec{i} - \sin\delta_4 \vec{k}$$

The inertial acceleration of the reference point O is related to the inertial acceleration of the composite center of mass as

$$\vec{a}_o = \vec{a}_c - \ddot{\vec{r}}_c$$

$$= \frac{1}{m} \left[F_s \cos^2\alpha \, \vec{n} + \sum_{i=1}^{4} F_c(\vec{S} \cdot \vec{n}_i)^2 \, \vec{n}_i \right] - \frac{m_p}{m} \ddot{\vec{r}} \tag{12.60}$$

The attitude equation of motion, Eq. (12.54), can then be rewritten as

$$\frac{d}{dt}\left(\bar{J}\cdot\vec{\omega}+m_p\vec{r}\times\dot{\vec{r}}\right)+\frac{m_p}{m}\vec{r}\times\left\{\left[F_s\cos^2\alpha\,\vec{n}+\sum_{i=1}^{4}F_c(\vec{S}\cdot\vec{n}_i)^2\,\vec{n}_i\right]-m_p\ddot{\vec{r}}\right\}$$

$$=\sum_{i=1}^{4}\vec{\ell}_i\times F_c(\vec{S}\cdot\vec{n}_i)^2\,\vec{n}_i \qquad (12.61)$$

where $\vec{r}=x\vec{i}+y\vec{j}+z\vec{k}$. The complexity of the equations of motion, caused by the time-varying center-of-mass location, can be seen in Eq. (12.61).

Considering the rotational equations of motion about the composite center of mass, instead of the body-fixed reference point O, we obtain the following equations of motion:

$$J_x\dot{\omega}_x+\cdots=F_c\cos^2\alpha(\ell_1\cos^2\delta_1\sin\delta_1-\ell_4\cos^2\delta_4\sin\delta_4) \qquad (12.62)$$

$$J_y\dot{\omega}_y+\cdots=\frac{m_p}{m}F_s\ell\cos^2\alpha\,\sin\phi\,\sin\theta+F_c[-\ell_2\cos^2(\alpha-\delta_2)\cos\delta_2$$

$$+\ell_3\cos^2(\alpha-\delta_3)\cos\delta_3] \qquad (12.63)$$

$$J_z\dot{\omega}_z+\cdots=-\frac{m_p}{m}F_s\ell\cos^2\alpha\,\sin\phi\,\cos\theta-F_c\cos^2\alpha(\ell_1\cos^3\delta_1-\ell_4\cos^3\delta_4\,)$$

$$(12.64)$$

where $(\omega_x,\omega_y,\omega_z)$ are the angular velocity components and (J_x,J_y,J_z) are the moments of inertia of the total system about the composite center of mass, which are functions of gimbal angles ϕ and θ. Furthermore, the distance from the composite center of mass to each vane ℓ_i is also a function of ϕ and θ. There are also additional terms caused by the time-varying inertias and gyroscopic couplings in the left-hand side of these equations. The complexity of a dynamic model caused by the time-varying center-of-mass location is evident here again.

12.6.3 Sailcraft with Control Vanes

For a sailcraft with control vanes (but without a gimbaled control boom), we have the following equations of motion:

$$J_x\dot{\omega}_x=F_cL\cos^2\alpha(\cos^2\delta_1\sin\delta_1-\cos^2\delta_4\sin\delta_4) \qquad (12.65)$$

$$J_y\dot{\omega}_y=F_cL[-\cos^2(\alpha-\delta_2)\cos\delta_2+\cos^2(\alpha-\delta_3)\cos\delta_3] \qquad (12.66)$$

$$J_z\dot{\omega}_z=-F_cL\cos^2\alpha(\cos^3\delta_1-\cos^3\delta_4) \qquad (12.67)$$

where L is the distance from the center of mass to the center of pressure of each control vane ($\approx$ the spar length).

Given the desired control torques (T_x, T_y, T_z) from an attitude control system, we have

$$T_x = F_c L \cos^2\alpha (\cos^2\delta_1 \sin\delta_1 - \cos^2\delta_4 \sin\delta_4) \approx F_c L \cos^2\alpha (\delta_1 - \delta_4) \quad (12.68a)$$

$$T_y = F_c L[-\cos^2(\alpha - \delta_2)\cos\delta_2 + \cos^2(\alpha - \delta_3)\cos\delta_3] \quad (12.68b)$$

$$T_z = -F_c L \cos^2\alpha (\cos^3\delta_1 - \cos^3\delta_4) \approx F_c L \cos^2\alpha (\delta_1^2 - \delta_4^2)$$

$$\approx F_c L \cos^2\alpha (\delta_1 - \delta_4)(\delta_1 + \delta_4) \quad (12.68c)$$

Defining $\Delta = \delta_1 - \delta_4$ and $\Theta = \delta_1 + \delta_4$, we obtain the commanded roll/yaw control vane angles as

$$\delta_{1c} = \frac{(\Delta_c + \Theta_c)}{2} \quad (12.69a)$$

$$\delta_{4c} = \frac{(\Delta_c - \Theta_c)}{2} \quad (12.69b)$$

where Δ_c and Θ_c are determined from Eq. (12.68) as

$$\Delta_c = \frac{T_x}{F_c L \cos^2\alpha} \quad (12.70a)$$

$$\Theta_c = \frac{T_z}{T_x} \quad \text{if } T_x \neq 0 \quad (12.70b)$$

When $T_x = 0$, the preceding vane steering logic has a singularity problem. That is, the yaw torque cannot be generated when $\delta_1 = \delta_4$ for $T_x = 0$. This is the main reason for requiring additional flaps, mounted to the outermost solar panels, for the OTS, TELECOM 1, and INMARSAT 2 satellites.

Because there exists a solution $\delta_1 \neq \delta_4$ of the following nonlinear equations

$$\cos^2\delta_1 \sin\delta_1 - \cos^2\delta_4 \sin\delta_4 = 0 \quad (12.71a)$$

$$\cos^3\delta_1 - \cos^3\delta_4 = \frac{-T_z}{F_c L \cos^2\alpha} \quad (12.71b)$$

it is also possible to generate an "unbalance" yaw torque without inducing a roll-axis windmill torque. However, an actual implementation of this method will require a further detailed study.

12.6.4 Statically Stable Sailcraft

An interplanetary spacecraft is often said to be *statically stable* when its center of mass lies between the sun and its center of pressure, as illustrated in Fig. 12.17. Whenever a statically stable sailcraft rotates away from its neutral sun-pointing orientation, a restoring (stabilizing) torque is generated. The dynamical behavior of such a statically stable sailcraft is analogous to that of a gravity-gradient stabilized satellite. That is, if disturbed, the sailcraft will oscillate indefinitely. If the center

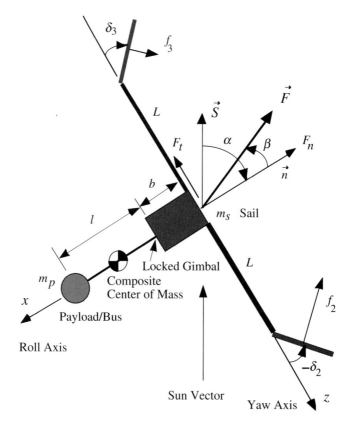

Fig. 12.17   Pitch-axis model of a statically stable sailcraft.

of pressure lies between the sun and the center of mass, a destabilizing torque is generated whenever the sailcraft rotates away from its null or trim orientation.

Consider a simple pitch-axis model of a sailcraft with control vanes, as illustrated in Fig. 12.17. The pitch-axis equation of motion becomes

$$J_y\ddot{\alpha} = F_c L[-\cos^2(\alpha - \delta_2)\cos\delta_2 + \cos^2(\alpha - \delta_3)\cos\delta_3] - \frac{m_s}{m_s + m_p}(b + \ell)F_t$$

$$\approx F_c L[-\cos^3\delta_2 - 2(\cos^2\delta_2 \sin\delta_2) + \cos^3\delta_3$$

$$+ 2(\cos^2\delta_3 \sin\delta_3)\,\alpha] - \frac{m_s}{m_s + m_p}(b + \ell)F_t \qquad (12.72)$$

where J_y is the pitch moment of inertia of the complete system.

For fixed control vane angles of $\delta = -\delta_2 = \delta_3 > 0$, we have a pitch-axis dynamical model of the form

$$\ddot{\alpha} + \omega_n^2 \alpha = 0 \qquad (12.73)$$

where

$$\omega_n^2 = \frac{1}{J_y} \left[4F_c L \cos^2\delta \sin\delta + \frac{m_p(b+\ell)}{m_s + m_p} PA(1 - \rho_s) \right]$$

12.7   Gimbaled Thrust Vector Control Design for a Sailcraft

A simplified pitch-axis model of the baseline ST7 sailcraft is illustrated in Fig. 12.18. The sailcraft consists of a sail subsystem and a payload/bus system. The sail subsystem is treated as a gimbaled engine, and a gimbaled thrust vector control (TVC) design problem is formulated here as discussed in [27].

12.7.1   Pitch-Axis Dynamic Model

The equations of motion of the gimbaled two-body system shown in Fig. 12.18 can be obtained as

$$m_s a_x = G_x - F_n \qquad\qquad (12.74a)$$

$$m_s a_z = G_z - F_t \qquad\qquad (12.74b)$$

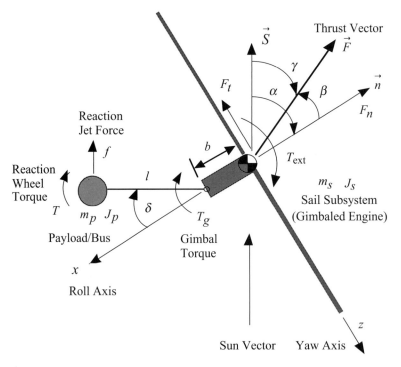

Fig. 12.18   Simplified pitch-axis model of the ST7 sailcraft controlled by a gimbaled TVC system, reaction wheels, and/or reaction jets.

$$m_p[a_x - b\dot{\alpha}^2 - \ell(\ddot{\alpha} + \ddot{\delta})\sin\delta - \ell(\dot{\alpha} + \dot{\delta})^2\cos\delta] = -G_x - f\sin\delta \quad (12.74c)$$

$$m_p[a_z - b\ddot{\alpha} - \ell(\ddot{\alpha} + \ddot{\delta})\cos\delta + \ell(\dot{\alpha} + \dot{\delta})^2\sin\delta] = -G_z - f\cos\delta \quad (12.74d)$$

$$J_s\ddot{\alpha} = -bG_z - T_g + T_{\text{ext}} \quad (12.74e)$$

$$J_p(\ddot{\alpha} + \ddot{\delta}) = T + T_g - G_x\ell\sin\delta - G_z\ell\cos\delta \quad (12.74f)$$

where (a_x, a_z) are the body-axis components of the sail subsystem center-of-mass acceleration; (G_x, G_z) the body-axis components of the gimbal joint reaction force; δ the gimbal angle; α the sail pitch (sun) angle; m_s and J_s the mass and pitch inertia of the sail subsystem, respectively; (F_n, F_t) the normal and transverse components of total solar-pressure force; m_p and J_p the payload/bus system mass and inertia, respectively; ℓ the control boom length; b the distance between the sail subsystem center-of-mass and the gimbal joint; T_g the gimbal (internal) control torque; T the control torque from a pitch reaction wheel located at the payload/bus system; and f the reaction-jet force.

The external disturbance torque T_{ext} in Eq. (12.74e) consists of the solar-pressure disturbance torque and the gravity-gradient torque as follows:

$$T_{\text{ext}} = \epsilon F_n + \frac{3\mu}{r^3}(J_x - J_z)\sin(\alpha - \theta)\cos(\alpha - \theta) \quad (12.75)$$

where ϵ is the c.m./c.p. offset of the sail subsystem along the yaw axis, μ the Earth's gravitational parameter, (J_x, J_z) the roll and yaw moments of inertia of the sail subsystem, r the radial distance of the sailcraft from the center of the Earth, and θ the true anomaly.

Eliminating G_x and G_z from Eqs. (12.74), we obtain

$$m_s a_x = -m_p[a_x - b\dot{\alpha}^2 - \ell(\ddot{\alpha} + \ddot{\delta})\sin\delta - \ell(\dot{\alpha} + \dot{\delta})^2\cos\delta] - f\sin\delta - F_n \quad (12.76a)$$

$$m_s a_z = -m_p[a_z - b\ddot{\alpha} - \ell(\ddot{\alpha} + \ddot{\delta})\cos\delta + \ell(\dot{\alpha} + \dot{\delta})^2\sin\delta] - f\cos\delta - F_t \quad (12.76b)$$

$$J_s\ddot{\alpha} = -b(m_s a_z + F_t) - T_g + T_{\text{ext}} \quad (12.76c)$$

$$J_p(\ddot{\alpha} + \ddot{\delta}) = -(m_s a_x + F_n)\ell\sin\delta - (m_s a_z + F_t)\ell\cos\delta + T + T_g \quad (12.76d)$$

For small angles and rates, we obtain the following set of linearized equations of motion:

$$(m_s + m_p)a_x = -f\delta - F_n \quad (12.77a)$$

$$(m_s + m_p)a_z = m_p[b\ddot{\alpha} + \ell(\ddot{\alpha} + \ddot{\delta})] - f - F_t \quad (12.77b)$$

$$J_s\ddot{\alpha} = -b(m_s a_z + F_t) - T_g + T_{\text{ext}} \quad (12.77c)$$

$$J_p(\ddot{\alpha} + \ddot{\delta}) = -(m_s a_x + F_n)\ell\delta - (m_s a_z + F_t)\ell + T + T_g \quad (12.77d)$$

Eliminating a_x and a_z gives

$$\left[J_s + \frac{m_s m_p}{m} b(b + \ell)\right] \ddot{\alpha} + \frac{m_s m_p}{m} b\ell\ddot{\delta} = -\frac{m_p}{m} bF_t + \frac{m_s}{m} bf - T_g + T_{ext}$$

(12.78a)

$$\left[J_p + \frac{m_s m_p}{m} \ell(b + \ell)\right] \ddot{\alpha} + \left[J_p + \frac{m_s m_p}{m} \ell^2\right] \ddot{\delta} = -\frac{m_p}{m} \ell F_t$$

$$-\frac{m_p}{m} \ell F_n \delta + \frac{m_s}{m} \ell f + T_g + T$$

(12.78b)

where $m = m_s + m_p$. The solar radiation pressure force components (F_n, F_t) can be expressed either as

$$F_n = PA \left\{ (1 + \rho_s)\cos^2\alpha + \frac{2}{3}\rho_d \cos\alpha \right\}$$

(12.79a)

$$F_t = PA(1 - \rho_s) \cos\alpha \sin\alpha$$

(12.79b)

or

$$F_n = PA[(1 + rs) \cos^2\alpha + B_f r(1 - s) \cos\alpha + \frac{e_f B_f - e_b B_b}{e_f + e_b}(1 - r) \cos\alpha]$$

(12.80a)

$$F_t = PA(1 - rs) \cos\alpha \sin\alpha$$

(12.80b)

12.7.2 Preliminary TVC Design and Simulation Results

A baseline configuration is assumed as $m_s = 40$ kg, $m_p = 120$ kg, $b = 0.5$ m, $\epsilon = \pm 0.1$ m, $\ell = 2$ m, $J_s = 3000$ kg·m^2, $J_x = 6000$ kg·m^2, $J_z = 3000$ kg·m^2, $J_p = 20$ kg·m^2, $P = 4.563 \times 10^{-6}$ N/m^2, and $A = 1400$ m^2. More detailed mass properties and basic characteristics of this 40×40 m sailcraft are summarized in Table 12.1. The following optical properties were also assumed: $B_f = 0.79$, $B_b = 0.55$, $e_f = 0.05$, $e_b = 0.55$, $r = 0.88$, and $s = 0.94$.

An elliptic orbit, called the supersynchronous transfer orbit, proposed for the ST7 sail validation mission is characterized as follows:

$$r_p = 6374 + 2000 = 8{,}374 \text{ km}$$

$$r_a = 6374 + 78,108 = 84{,}482 \text{ km}$$

$$a = \frac{(r_p + r_a)}{2} = 46{,}428 \text{ km}$$

$$e = \frac{r_a - r_p}{r_a + r_p} = 0.8196$$

$$i = 12 \text{ deg (from the ecliptic plane)}$$

$$p = a(1 - e^2) = 15{,}238 \text{ km}$$

Table 12.1   ST7 sailcraft characteristics

Characteristic	Value
Sail film	$m_f = 6.1\,\text{kg}$
Booms (4x)	$m_b = 4 \times 3.575 = 14.3\,\text{kg}$
	$EI \approx 1000\text{--}2000\,\text{N} \cdot \text{m}^2$
	$L = 30\,\text{m}$
	$\rho = 0.1191\,\text{kg/m}$
Hub platform	$m_h = 19.6\,\text{kg}$
Solar sail	$m_s = m_f + m_b + m_h = 40\,\text{kg}$
Payload/bus	$m_p = 116\,\text{kg}$
Total mass	$m = m_s + m_p = 156\,\text{kg}$
Sail area	$A = 1400\,\text{m}^2$
Solar pressure	$P = 4.563\,\text{e-6 N/m}^2$
Thrust coefficient	$\eta = 1.816$ (ideal $\eta_{max} = 2$)
Maximum thrust	$F_{max} = \eta PA = 0.0116\,\text{N}$
Area-to-mass ratio	$A/m = 8.97\,\text{m}^2/\text{kg}$
Areal density	$\sigma = m/A = 0.111\,\text{kg/m}^2$
Acceleration	$a_c = F_{max}/m = \eta PA/m$
	$= \eta P/\sigma = 73.7\,\text{e-6 m/s}^2$

$$n = \sqrt{\frac{\mu}{a^3}} = 6.311 \times 10^{-5}\ \text{rad/s}$$

and the orbital period of 27.65 h.

A linear state-space model for control design can be obtained as

$$\frac{d}{dt}\begin{bmatrix} \alpha \\ \dot{\alpha} \\ \delta \\ \dot{\delta} \end{bmatrix} = \begin{bmatrix} 0 & 1 & 0 & 0 \\ -1.9705 \times 10^{-8} & 0 & 1.2316 \times 10^{-6} & 0 \\ 0 & 0 & 0 & 1 \\ -1.1803 \times 10^{-5} & 0 & -1.2470 \times 10^{-4} & 0 \end{bmatrix}\begin{bmatrix} \alpha \\ \dot{\alpha} \\ \delta \\ \dot{\delta} \end{bmatrix}$$

$$+ \begin{bmatrix} 0 \\ -4.0462 \times 10^{-4} \\ 0 \\ 7.6342 \times 10^{-3} \end{bmatrix} T_g$$

with its open-loop eigenvalues of $\pm 0.0112j$ and $\pm 0.0004j$. We can then employ a gimbal control logic of the form

$$T_g = -K_\alpha(\alpha - \alpha_c) - K_i \int (\alpha - \alpha_c) - K_{\dot{\alpha}}\dot{\alpha} - K_\delta\delta - K_{\dot{\delta}}\dot{\delta} \qquad (12.81)$$

where $(K_\alpha, K_i, K_{\dot{\alpha}}, K_\delta, K_{\dot{\delta}})$ are feedback control gains and α_c is the commanded pitch (sun) angle. A linear-quadratic-regulator (LQR) design technique was used to determine a set of control gains $(K_\alpha, K_i, K_{\dot{\alpha}}, K_\delta, K_{\dot{\delta}})$, resulting in the following closed-loop eigenvalues: $-0.23 \pm j0.23$, $-0.0009 \pm j0.0012$, and -0.0002. The feasibility of employing a gimbaled control system for a 35-deg pitch maneuver with an initial pitch rate of -0.05 deg/s in the presence of a disturbance torque caused by a c.m./c.p. offset of ± 0.1 m is demonstrated in Fig. 12.19. Certain variables such as the pitch rate, gimbal rate, and gimbal torque are shown in Fig. 12.19 for the initial 55-s period to indicate their peak values.

Further detailed control design tradeoffs are needed for selecting a proper control bandwidth in the presence of mission/hardware constraints (e.g., pointing accuracy, sail turning rate, maximum gimbal torque, maximum gimbal angle, maximum gimbal rate, gimbal friction, sail structural flexibility, etc.). Furthermore, a complete three-axis simulation validation of a two-axis TVC design needs to be performed in a further detailed study.

12.8 Attitude Control System Architecture for Solar Sails

This section presents a solar-sail attitude control system (ACS) architecture developed by Murphy and Wie in [28] and [29]. The proposed solar-sail ACS architecture consists of a propellantless primary ACS and a microthruster-based secondary ACS. The primary ACS employs two ballast masses running along mast lanyards for pitch/yaw trim control and roll stabilizer bars at the mast tips for roll control. The secondary ACS utilizes lightweight pulsed-plasma-thruster (PPT) modules mounted at the mast tips. The proposed solar-sail ACS will be applicable with minimal modifications to a wide range of future solar-sailing missions with varying requirements and mission complexity.

12.8.1 Introduction

Some potential near-term solar-sail missions envisioned by NASA are summarized in Table 12.2. The first three cases represent likely options for a near-Earth flight validation experiment. The first case is a low-Earth orbit (LEO) sufficiently high to minimize the influence of atmospheric drag and atomic oxygen on solar-sails. Such an orbit is attainable with several affordable dedicated launch options; however, the gravity-gradient disturbance torque can dominate the ACS design unless a particular low-Earth sun-synchronous (LSS) orbit and a proper sail orientation are selected as discussed later. The second is a geosynchronous transfer orbit (GTO), as this was the basis for a solar-sail study proposal to the New Millennium Program Space Technology 7. This orbit is attractive because the spacecraft would spend 80% of its time above 30,000 km, where the disturbance from gravity-gradient torques is negligible.

For these near-term missions, solar-sail sizes of 40, 80, and 160 m were chosen to allow the most straightforward scaling from the 80-m class hardware currently under development and test by ATK Space Systems [30]. A 20-m scalable solar-sail by ATK Space Systems, which was successfully deployed in the 30-m thermal vacuum chamber at the Plum Brook Space Power Facility of the NASA Glenn Research Center, is illustrated in Fig. 12.20.

The mass properties and the solar disturbance torques of such a scalable sailcraft are summarized in Table 12.3. An uncertainty of 0.25% of sail size is assumed for a

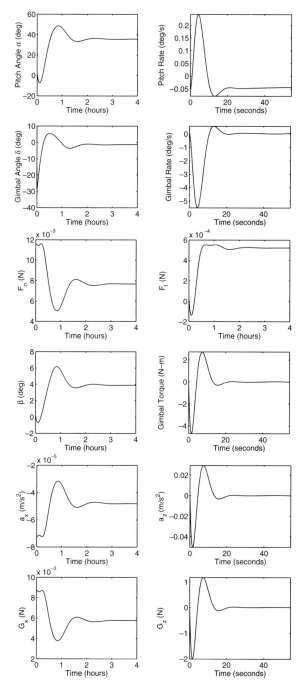

Fig. 12.19 Pitch-axis control simulation results for the ST7 sailcraft controlled by a gimbaled TVC system ($\alpha_c = 35$ deg, $\epsilon = 0.1$ m, initial pitch rate $= -0.05$ deg/s).

Table 12.2   Near-term solar-sail missions [28]

Case	Mission description	Mission life	Size, m
LSS	1,600 km sunsynchronous	90 days	40
GTO	2,000 × 40,000 km elliptic	90 days	40
GEO	36,000-km circular orbit	90 days	40
EMX	Earth Magnetotail Explorer	5 years	40
L1S	Lagrangian point, 0.95 AU	5 years	80
SPI	Solar Polar Imager, 0.5 AU	5 years	160

preliminary estimation of the worst-case c.m./c.p. offset. The pinwheel (windmill) disturbance torque about the roll axis is assumed to be about 50% of the pitch/yaw solar torque. Note that the solar disturbance torque is proportional to the cube of the sail size. Assumptions for masses of various components are based on hardware currently under development and in test. Variations are related to the scale of the sail and the timeline for launch. That is, the larger sails require larger structural geometry, but evolution of the engineering will allow some components to be more mass efficient. For example, the sail thickness will be reduced over time, as will the overall mass of integrated subsystems. Much uncertainty exists in the near-term capability of industry to build bus and payload hardware light enough to allow some future missions to be accomplished with reasonably sized sails. However, the masses and inertia of typical spacecraft bus and payload do not affect the overall sailcraft inertia, which are a primary input to the sail ACS sizing and design.

12.8.2   Solar-Sail ACS Architecture

A solar-sail ACS architecture developed by Murphy and Wie [28,29] consists of a propellantless primary ACS and a microthruster-based secondary ACS. The primary ACS employs trim ballasts running along mast lanyards for pitch/yaw

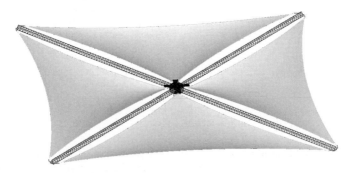

Fig. 12.20   A 20-m scalable solar sail by ATK Space Systems (formerly, AEC-ABLE Engineering). (Image Courtesy of ATK Space Systems.)

Table 12.3 Mass properties and solar disturbance torques of ATK's scalable sailcraft [28–30]

Sail size	40	80	160
Geometry			
Mast length, m	28	56	113
Mast diameter, m	0.4	0.4	0.6
Bending EI, $N \cdot m^2$	82,441	82,441	438,847
Torsional GJ, $N \cdot m^2$	197	197	453
Scallop factor, %	75	75	75
Sail area, m^2	1,200	4,800	19,200
Thrust (max), N	0.01	0.04	0.16
c.m./c.p. offset[a], m	0.1	0.2	0.4
Mass			
Sails, kg	6	19	67
Masts, kg	7	14	60
Tip mass (each), kg	1	2	2
Central assembly, kg	8	10	15
Sail ACS (primary), kg	3	5	7
Payload, kg	7	19	43
Bus (microsat), kg	50	75	100
Bus (standard), kg	150	200	250
Total (w/microsat bus), kg	85	150	300
Total (w/standard bus), kg	185	275	450
Acceleration (w/micro bus), mm/s^2	0.11	0.26	0.53
Inertia			
I_x (roll), $kg \cdot m^2$	4,340	40,262	642,876
I_y (pitch), $kg \cdot m^2$	2,171	20,136	321,490
I_z (yaw), $kg \cdot m^2$	2,171	20,136	321,490
Solar disturbance torque[b]			
Pitch/yaw, $mN \cdot m$	1	8	64
Roll[c], $mN \cdot m$	0.5	4	32

[a]0.25% of the overall sail size is assumed for a nominally worst case.
[b]A nominally worst, maximum disturbance torque for untrimmed sailcraft.
[c]50% of the pitch/yaw solar torque is assumed for the windmill disturbance torque.

control together with roll stabilizer bars (also called spreader bars) at the mast tips for quadrant tilt control (Fig. 12.21). The stability and robustness of such a propellantless primary ACS are further enhanced by a secondary ACS utilizing tip-mounted, lightweight pulsed plasma thrusters. Such a micro-PPT-based ACS can be employed for attitude recovery maneuvers from off-nominal conditions as well as for a spin-stabilized safe mode. It can also be employed as a backup to a conventional ACS (e.g., on a validation flight) prior to sail deployment and also during preflight sail checkout operation.

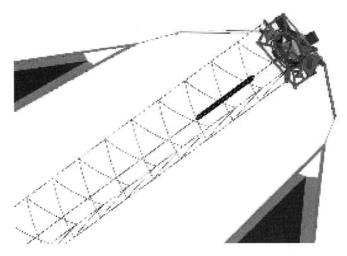

Fig. 12.21   ATK's solar-sail mast with a translating ballast mass (running along a lanyard tape), tip-mounted roll stabilizer bars attached to sail panels, and tip-mounted micro-PPT module.

As illustrated in Fig. 12.22, a solar-sail ACS architecture proposed for near-term solar-sailing missions consists of 1) the propellantless primary ACS employing trim ballasts for pitch/yaw control and roll stabilizer bars, 2) the secondary ACS utilizing tip-mounted, lightweight pulsed plasma thrusters, and 3) the spacecraft bus ACS.

The trim actuator positions for balancing at the nominal orientation might not provide precise trim for substantially different sailcraft orientations. Trim balance can be enhanced by integrating the mechanical balance system with thruster firing. For example, an excessive firing rate of one thruster will command the balance system to adjust on the related control axis. The autotrimming control system will result in low-frequency, two-sided limit-cycle thruster firings within the attitude error deadband required by navigation and payload pointing requirements. The proposed sailcraft ACS architecture possesses large stability margins, multiple redundancies, and the agility to enable the thrust vector and thrust magnitude to be adjusted independently to optimize low-thrust trajectories. It also provides a full three-axis, robust control authority for any sail orientation including an edge-on flight orientation toward the sun. The proposed sail flight control system architecture will be applicable with minimal modifications to a wide range of future solar-sail flight missions with varying requirements and mission complexity.

The attitude determination system (ADS) is a critical subsystem of the spacecraft attitude determination and control control system (ADCS). An ADS of particular interest for solar-sail applications is the Inertial Stellar Compass (ISC) developed by Draper Laboratory for the New Millennium Program ST6 flight validation experiment [31]. The ISC is a miniature, low-power ADS developed for use with low-cost microsatellites. It is suitable for a wide range of future solar-sail

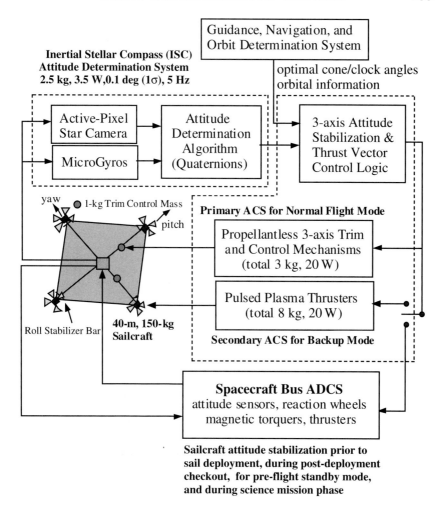

Fig. 12.22   Integrated attitude and orbit control system (AOCS) architecture for solar sails.

missions because of its low-mass, low-power, and low-volume design and its self-initializing, autonomous operational capability. The ISC is composed of a wide field-of-view active-pixel star camera and microgyros, with associated data processing and power electronics. It has a total mass of 2.5 kg, a power requirement of 3.5 W, and an accuracy of 0.1 deg (1σ).

The proposed solar-sail ADCS architecture integrating various ADCS options also includes a conventional ADCS of sailcraft bus, as illustrated in Fig. 12.22. For most solar-sailing missions, a conventional spacecraft bus will be required, but it might be jettisoned after sail deployment depending on mission requirements.

As part of an AOCS of solar sails, a TVC system maintains the proper orientation of solar sails to provide the desired thrust vector pointing/steering. A guidance, navigation, and orbit determination system, also included in Fig. 12.22, will also be required for most solar-sailing missions. Because of the inherent inability to precisely model the solar radiation pressure and to accurately point the true thrust vector direction, most solar-sail TVC systems will require frequent updates of both orbital parameters and TVC steering commands for frequent trajectory corrections. The guidance, navigation, and orbit determination system provides such required updates, and the update frequency is determined by many factors, such as the trajectory dispersions, the target-body ephemeris uncertainty, the calibration of a solar radiation pressure model, and the operational constraints.

The recent advances in microsatellite bus technologies also need to be exploited to develop such an integrated low-cost, low-risk, low-mass, and low-power ADCS of solar-sail spacecraft.

12.9   Solar-Sail Flight Validation (SSFV) Mission

This section presents the preliminary orbit analysis and simulation results of a solar sail in a dawn-dusk sun-synchronous orbit proposed in [28] and [29] for the NMP ST9 solar-sail flight-validation mission.

12.9.1   Dawn-Dusk Sun-Synchronous (DDSS) Orbit

A circular sun-synchronous orbit is characterized by its nodal regression rate $\dot{\Omega}$ as follows:

$$\dot{\Omega} \approx -\frac{3J_2 R_\oplus^2}{2a^2} n \cos i \qquad (12.82)$$

where $R_\oplus$ is the Earth's radius, $J_2 = 0.001082$ is the Earth's oblateness, $n = \sqrt{\mu/a^3}$, μ is the Earth's gravitational parameter, $a = R_\oplus + h$, h is the orbital altitude, and i is the inclination.

The Earth revolves around the sun in a nearly circular orbit ($e = 0.016726$) with a period of 365.24 solar days. The Earth also rotates about its own axis with a period of one siderial day (23 h 56 min 4.09 s). The sun-synchronous orbit has a nodal regression rate $\dot{\Omega}$ equal to the Earth's mean rate of revolution around the sun (i.e., 360 deg in 365.24 solar days or 0.985 deg/day). This regression must be in the direction of the Earth's rotation because the Earth rotates about its axis in the same direction that it revolves around the sun. Therefore, a sun-synchronous satellite must have a retrograde orbit so that its nodal regression can be prograde. Also, the satellite must have a combination of altitude and inclination that produces 0.985 deg/day regression. The sun-synchronous orbits maintain their initial orientation relative to sun. Such retrograde orbits lie between inclination angles of 95.7 deg and 180 deg at altitudes up to 5970 km. Sun-synchronous orbits below 1400 km can experience some shadowing, and the orbital debris density remains significant from 1400 to 1550 km. As these environments pose some hazard to a sailcraft, a preferable orbit might be in the range of 1600 km. It is desirable to avoid shadowing, as the penumbra transits will cause rapid thermal shocks that will dynamically excite the structure, stress the sail, and swiftly exercise the negator mechanisms used to maintain constant sail tension. Eclipse will also cause large charging swings, and the effects on a large sail can be detrimental. These issues should be explored further, but at present the aggregate benefits of a 1600-km orbit with $i = 102.5$ deg, $n = 8.86e\text{-}4$ rad/s, and an orbital period of 118.2 min are enticing.

If the launch is scheduled to place the sailcraft in a dawn-dusk orientation, the line of nodes will be (always) perpendicular to the sun line, and on two occasions per year the orbit will be normal to the sun. For a short-duration sail validation experiment, the opportunity exists to conduct operations with the orbit plane nearly normal to the sun. There are two windows per year, beginning near the vernal equinox (for a dusk launch) or the autumn equinox (for a dawn launch), where the so-called sun angle β will remain within 10 deg for nearly six months.

In contrast to the GTO, a dawn-dusk sun-synchronous orbit would mean 1) the sailcraft would not need a propulsion stage to lift the perigee to 2000 km; 2) the nominal orientation of the sail could be kept nearly in plane with the orbit plane, thus minimizing the gravity-gradient disturbance torque; 3) the sailcraft would not pass through the Van Allen belts, the environmental effects of high radiation on the sails are not yet quantified and additional shielding mass will not be required on all electronics; 4) the thermal environment for the bus would have reduced variations; and 5) operational ranging and communication can be simplified.

12.9.2 Zero-Thrust Mode (Sail Deployment Mode)

After orbit insertion into the DDSS orbit, the first major task is completion of a successful solar-sail deployment. Large sails take a considerable time to deploy, and if management of unfolding sails becomes unsynchronized, solar pressure will place a large torque on the sailcraft and would quickly accelerate any free section. To minimize the consequences of such events, it would be prudent to orient edge on to the sun initially, as illustrated in Fig. 12.23. If the sailcraft were continuously aligned with the nadir line as illustrated in Fig. 12.23, the sailcraft would be gravity-gradient stabilized in pitch with negligible gravity-gradient torque in roll. In fact, the roll, pitch, and yaw axes of the sailcraft in this zero-thrust mode are to be aligned with the LVLH reference frame. Consequently, a conventional bus ACS can be employed for this fully deployed sailcraft only during the zero-thrust mode operation without the presence of significant solar-pressure disturbance torques. During this zero-thrust "safe" mode operation, various initial testing and checkout of the fully deployed sailcraft can be safely conducted. A proper orientation/articulation of the solar arrays of sail carrier spacecraft would be required for long-duration zero-thrust mode operation.

12.9.3 Yaw Reorientation Mode

After successful sail deployment and completion of checkout activities in the zero-thrust mode, a −90-deg yaw maneuver will then be needed to achieve the proper sail orientation of the full-thrust standby mode illustrated in Fig. 12.24. The orbit would be nearly normal to the sun line if the launch were timed for the minimal sun angle β, which varies with the season.

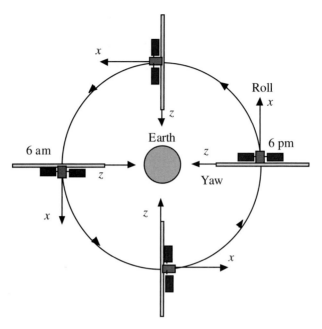

Fig. 12.23 Gravity-gradient stable sail flight orientation of the zero-thrust mode (sail deployment mode) for sail deployment and initial checkout. The sun vector is perpendicular to the orbit plane with an offset sun angle β.

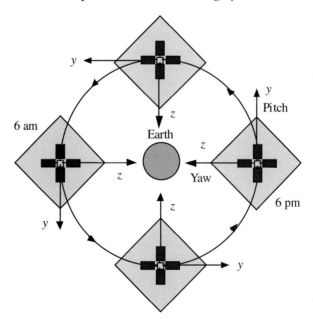

Fig. 12.24 Sail flight orientation for the full-thrust standby mode (sun-pointing mode): no significant orbital effects of the solar thrust.

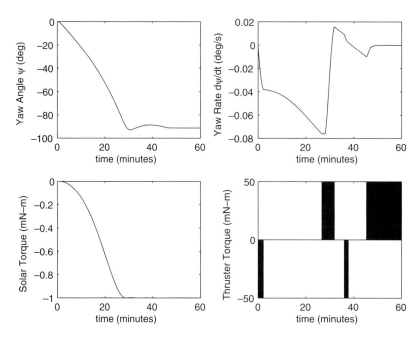

Fig. 12.25   A −90 deg yaw maneuver of an untrimmed 40-m sailcraft (with c.m./c.p. offset of 0.1 m) using reaction jets for transition to the standby full-thrust mode.

The conventional bus ACS (or the tip-mounted microthruster system) is best suited to perform this reorientation about the yaw axis, as the ballast masses have negligible control authority when the sail is oriented near edge-on to the sun. As indicated in Figs. 12.25 and 12.26, a transition to the propellantless sail ACS will be required, immediately after achieving the desired yaw orientation to avoid excessive propellant usage or excessive wheel momentum growth for counteracting the solar disturbance torque. For simulations in Fig. 12.25, a 0.1-N thruster with a 0.5-m moment arm was assumed. As can be noticed in Fig. 12.26, a small reaction wheel cannot be employed for such a maneuver because a typical 2-kg wheel with a peak torque of 0.01 N · m has less than 2 N·m·s momentum storage capability. However, small magnetic torquers can be employed for momentum dumping, and such a combination of reaction wheels and magnetic torquers will be examined later in Sec. 12.17 although such a conventional bus-based ACS cannot be employed for most interplanetary solar-sailing missions. While the sailcraft is held on-sun by a thruster-based ACS of spacecraft bus, the trim ballast system and roll stabilizer bars would be activated to trim the transverse solar torque and also null any pinwheel motion.

12.9.4   Full-Thrust Standby Mode (Sun-Pointing Mode)

After successful sail deployment, further key validations of sail technology can begin: the demonstration of effective propulsion and robust steering control of the solar thrust vector. At a minimum, it is proposed that attitude maneuvers and

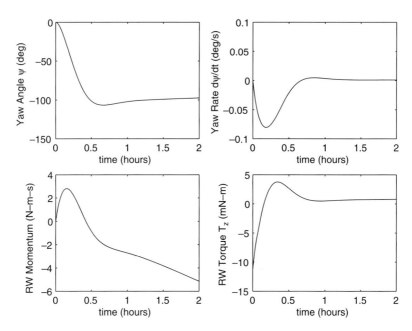

Fig. 12.26   A −90 deg yaw maneuver of an untrimmed 40-m sailcraft (with c.m./c.p. offset of 0.1 m) using a reaction wheel for transition to the standby full-thrust mode.

orbit-changing propulsion be validated. The sun line would be nearly normal to the orbital plane if the launch were timed to the seasonal variations in sun angle β. Although the sailcraft is at full thrust, the orbital motion is not much affected by the solar-pressure force when it is nearly perpendicular to the orbital plane, as can be seen in Fig. 12.27. Note that the nodal regression rate of 1 deg/day is still maintained.

12.9.5   TVC Mode

A constant yaw angle of −55 deg (equivalently, a 35-deg tilt angle of the roll axis relative to the sun line) will add energy to the orbit. Therefore, a reorientation maneuver back to a yaw angle of −55 deg will be needed to continuously increase the orbital energy. It might be possible to alternate the yaw angle to maintain the orbit as operational confidence is developed. Over time the ability to orient for viewing from the ground can be tested. The validation can conclude with an intentional deorbit, unless the thrust-to-mass ratio allows an Earth escape. Simulation results of the TVC mode indicate that approximately 8-km/day orbit raising will be possible as shown in Fig. 12.28.

12.10   Attitude Dynamics of a Solar Sail with Translating Control Masses

The problem of a spacecraft with internal moving mass was first investigated in the early 1960s. For spacecraft dynamical problems with internal moving mass,

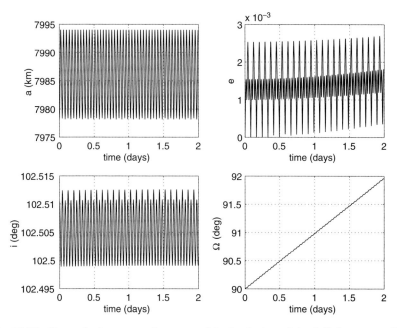

Fig. 12.27   Dawn-dusk sun-synchronous orbit simulation of the full-thrust standby mode.

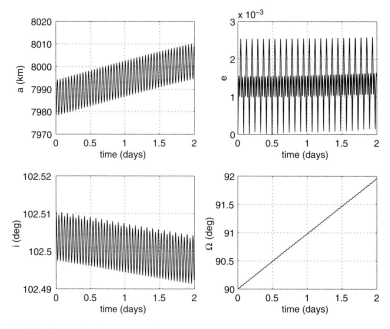

Fig. 12.28   Dawn-dusk sun-synchronous orbit simulation of the TVC mode: an 8-km/day orbit raise.

one can choose the composite center of mass of the total system as a reference point for the equations of motion. This formulation leads to a time-varying inertia matrix because the reference point moves as the internal mass moves relative to the main body. On the other hand, one can choose the center of mass of the main body as the reference point, which leads to a constant inertia matrix of the main body relative to the reference point, but results in complex equations of motion. Because of the convenience of dealing with a constant inertia matrix of the main body, we employ the second approach of choosing the center of mass of the main body as the reference point for a sailcraft with moving trim ballasts.

12.10.1 Formulation of a Dynamic Model

Consider a solar-sail spacecraft in an Earth-centered orbit as illustrated in Fig. 12.29. The spacecraft has the body-fixed principal axes as indicated in this figure. The roll axis is perpendicular to the sail plane, and it often nominally points toward the sun (but not necessarily). The pitch and yaw axes are the transverse axes along the masts. The dynamical model consists of a rigid sailcraft of mass M and two trim ballasts of each mass of m, as shown in Fig. 12.29. The origin of the body-fixed reference frame with a set of basis vectors $\{\vec{i}, \vec{j}, \vec{k}\}$ is located at point O, which is assumed to be the center of mass of a sailcraft of mass M excluding the ballast masses. It is also assumed that ballasts are running along mast lanyards (i.e., along the pitch and yaw axes).

For a spacecraft with translating control masses, one can choose the angular momentum equation as

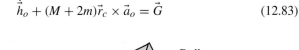

$$\dot{\vec{h}}_o + (M + 2m)\vec{r}_c \times \vec{a}_o = \vec{G} \qquad (12.83)$$

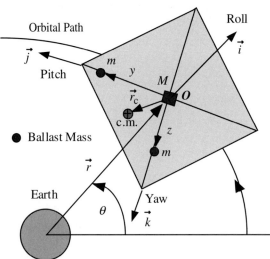

Fig. 12.29   Solar-sail spacecraft with translating trim ballasts in an Earth-centered orbit.

where $\vec{h}_o$ is the relative angular momentum of the total system about point O, $(M + 2m)$ is the total mass, $\vec{a}_o$ is the inertial acceleration of the point O, and $\vec{G}$ is the gravity-gradient torque vector about the point O, and $\vec{r}_c$ is the position vector of the composite center of mass from the reference point O given by

$$\vec{r}_c = \frac{m}{M + 2m}(y\vec{j} + z\vec{k}) \qquad (12.84)$$

The relative angular momentum of the total system about point O is given by

$$\vec{h}_o = \hat{I} \cdot \vec{\omega} + my\vec{j} \times \frac{d}{dt}(y\vec{j}) + mz\vec{k} \times \frac{d}{dt}(z\vec{k}) \qquad (12.85)$$

where $\hat{I}$ is inertia dyadic of the total system when the ballast is not moving and $\vec{\omega}$ is the angular velocity vector of the main body (i.e., the sail platform) expressed as

$$\vec{\omega} = \omega_x \vec{i} + \omega_y \vec{j} + \omega_z \vec{k} \qquad (12.86)$$

The inertia dyadic of the total system, including the ballast mass, about the reference point O is given by

$$\hat{I} = \begin{bmatrix} \vec{i} & \vec{j} & \vec{k} \end{bmatrix} \begin{bmatrix} \tilde{I}_x & 0 & 0 \\ 0 & \tilde{I}_y & 0 \\ 0 & 0 & \tilde{I}_z \end{bmatrix} \begin{bmatrix} \vec{i} \\ \vec{j} \\ \vec{k} \end{bmatrix}$$

where $\tilde{I}_x = I_x + m(y^2 + z^2)$, $\tilde{I}_y = I_y + mz^2$, $\tilde{I}_z = I_z + my^2$, and (I_x, I_y, I_z) are the principal moments of inertia of the sailcraft, not including the ballast mass m.

The inertial acceleration of the reference point O (caused by the solar-radiation-pressure force) is related to the inertial acceleration of the composite center of mass as

$$\vec{a}_o = \vec{a}_c - \ddot{\vec{r}}_c = \frac{\vec{F}}{M + 2m} - \frac{m}{M + 2m}\frac{d^2}{dt^2}(y\vec{j} + z\vec{k}) \qquad (12.87)$$

where $\vec{F}$ is the solar-radiation-pressure force vector acting on the sailcraft.
The attitude equation of motion, Eq. (12.83), can then be rewritten as

$$\frac{d}{dt}\left[\hat{I} \cdot \vec{\omega} + my\vec{j} \times \frac{d}{dt}(y\vec{j}) + mz\vec{k} \times \frac{d}{dt}(z\vec{k}) \right]$$

$$+ \frac{m}{M + 2m}(y\vec{j} + z\vec{k}) \times \left[\vec{F} - m\frac{d^2}{dt^2}(y\vec{j} + z\vec{k}) \right] = \vec{G} \qquad (12.88)$$

Ignoring the various nonlinear terms associated with $\dot{y}$, $\dot{z}$, etc., and by expressing $\vec{F}$ and $\vec{G}$ as

$$\vec{F} = F_x \vec{i} + F_y \vec{j} + F_z \vec{k}$$
$$\vec{G} = G_x \vec{i} + G_y \vec{j} + G_z \vec{k}$$

we obtain the attitude equations of motion as

$$J_x \dot{\omega}_x = (J_y - J_z)\omega_y\omega_z - \frac{m}{M+2m}(yF_z - zF_y) + G_x \qquad (12.89)$$

$$J_y \dot{\omega}_y = (J_z - J_x)\omega_z\omega_x - \frac{m}{M+2m}zF_x + G_y \qquad (12.90)$$

$$J_z \dot{\omega}_z = (J_x - J_y)\omega_x\omega_y + \frac{m}{M+2m}yF_x + G_z \qquad (12.91)$$

and

$$J_x = I_x + m_r(y^2 + z^2)$$
$$J_y = I_y + m_r z^2$$
$$J_z = I_z + m_r y^2$$

where m_r is the so-called reduced mass defined as

$$m_r = \frac{m(M+m)}{M+2m} \qquad (12.92)$$

For solar-sail applications, we have $m_r \approx m$ because $M \gg m$.

12.10.2 Attitude Dynamics in a DDSS Orbit

For the description of angular orientations of sailcraft in an Earth-centered circular orbit, the LVLH reference frame is considered here. This LVLH frame with its origin at the center of mass of an orbiting sailcraft has a set of unit vectors $\{\vec{a}_1, \vec{a}_2, \vec{a}_3\}$ with $\vec{a}_3$ locally vertical toward the Earth, $\vec{a}_1$ along the locally horizontal (transverse) direction, and $\vec{a}_2$ perpendicular to the orbit plane. The body-fixed reference frame B has a set of basis vectors $\{\vec{i}, \vec{j}, \vec{k}\}$. The relative orientation of the sailcraft with respect to the LVLH frame is described by three Euler angles (ϕ, θ, ψ) of the rotational seq uence of $C_1(\phi) \leftarrow C_2(\theta) \leftarrow C_3(\psi)$ from the LVLH to B.

The angular velocity components of the sailcraft are then given by

$$\begin{bmatrix} \omega_x \\ \omega_y \\ \omega_z \end{bmatrix} = \begin{bmatrix} 1 & 0 & -s\theta \\ 0 & c\phi & s\phi c\theta \\ 0 & -s\phi & c\phi c\theta \end{bmatrix} \begin{bmatrix} \dot{\phi} \\ \dot{\theta} \\ \dot{\psi} \end{bmatrix} - n \begin{bmatrix} c\theta s\psi \\ s\phi s\theta s\psi + c\phi c\psi \\ c\phi s\theta s\psi - s\phi c\psi \end{bmatrix} \qquad (12.93)$$

where $n = \sqrt{\mu/a^3}$ is the orbital rate, (ϕ, θ, ψ) are called the roll, pitch, and yaw angles of a spacecraft relative to the LVLH orientation; and $c\phi = \cos\phi$, etc. The kinematic differential equations can then be found as

$$\begin{bmatrix} \dot{\phi} \\ \dot{\theta} \\ \dot{\psi} \end{bmatrix} = \frac{1}{c\theta} \begin{bmatrix} c\theta & s\phi s\theta & c\phi s\theta \\ 0 & c\phi c\theta & -s\phi c\theta \\ 0 & s\phi & c\phi \end{bmatrix} \begin{bmatrix} \omega_x \\ \omega_y \\ \omega_z \end{bmatrix} + \frac{n}{c\theta} \begin{bmatrix} s\psi \\ c\theta c\psi \\ s\theta s\psi \end{bmatrix} \qquad (12.94)$$

The gravity-gradient torque components are given by

$$G_x = -3n^2(J_y - J_z)R_yR_z \tag{12.95a}$$

$$G_y = -3n^2(J_z - J_x)R_zR_x \tag{12.95b}$$

$$G_z = -3n^2(J_x - J_y)R_xR_y \tag{12.95c}$$

where $R_x = -\sin\theta$, $R_y = \sin\phi\cos\theta$, and $R_z = \cos\phi\cos\theta$. For small roll/pitch angles, the gravity-gradient disturbance torques become

$$G_x = -3n^2(J_y - J_z)\phi \tag{12.96a}$$

$$G_y = -3n^2(J_x - J_z)\theta \tag{12.96b}$$

$$G_z = 0 \tag{12.96c}$$

Note that there is no gravity-gradient torque about the yaw axis and that the gravity-gradient torque about the roll axis becomes zero when $J_y = J_z$.

Finally, we obtain the attitude equations of motion as follows:

$$J_x\dot{\omega}_x = (J_y - J_z)\omega_y\omega_z - 3n^2(J_y - J_z)\phi + 0.5\epsilon F + T_x \tag{12.97a}$$

$$J_y\dot{\omega}_y = (J_z - J_x)\omega_z\omega_x - 3n^2(J_x - J_z)\theta + \frac{m}{M + 2m}zF + \epsilon F + T_y \tag{12.97b}$$

$$J_z\dot{\omega}_z = (J_x - J_y)\omega_x\omega_y - \frac{m}{M + 2m}yF + \epsilon F + T_z \tag{12.97c}$$

where (T_x, T_y, T_z) are the control torques generated by reaction wheels, thrusters, or roll stabilizer bars. The solar-radiation-pressure force is assumed here as $\vec{F} = F_x\vec{i} = -F\vec{i}$ and $F = F_s\cos^2\alpha$, where $F_s \equiv \eta PA$ is the maximum sail thrust, η is the overall sail thrust coefficient ($\eta_{max} = 2$), $P = 4.563 \times 10^{-6}$ N/m^2, A is the total sail area, and α is the sun angle between the sun line and the roll axis. It is also assumed that the c.m./c.p. distance has an uncertainty of ϵ. The roll-axis solar disturbance torque is assumed as 50% of the pitch/yaw solar disturbance torque ϵF.

For large-angle reorientations about the yaw axis (during a transition from the zero-thrust mode to the full-thrust mode), but with small roll and pitch angles, we have

$$\omega_x \approx \dot{\phi} - n\sin\psi \tag{12.98a}$$

$$\omega_y \approx \dot{\theta} - n\cos\psi \tag{12.98b}$$

$$\omega_z \approx \dot{\psi} - n(\theta\sin\psi - \phi\cos\psi) \tag{12.98c}$$

The final set of attitude equations of motion for the design and simulation of the various sail flight modes in a DDSS orbit, as illustrated in Figs. 12.23 and 12.24, can then be obtained as

$$J_x\ddot{\phi} + n^2(J_y - J_z)(3 + \cos^2\psi)\phi - n^2(J_y - J_z)(\cos\psi\,\sin\psi)\theta$$
$$- n(J_x - J_y + J_z)(\cos\psi)\dot{\psi} = 0.5\epsilon F + T_x \tag{12.99}$$
$$J_y\ddot{\theta} + n^2(J_x - J_z)(3 + \sin^2\psi)\theta - n^2(J_x - J_z)(\cos\psi\,\sin\psi)\phi$$
$$- n(J_x - J_y - J_z)(\sin\psi)\dot{\psi} = \frac{m}{M + 2m}zF + \epsilon F + T_y \tag{12.100}$$
$$J_z\ddot{\psi} + n^2(J_y - J_x)\sin\psi\,\cos\psi + n(J_x - J_y + J_z)(\cos\psi)\dot{\phi}$$
$$+ n(J_x - J_y - J_z)(\sin\psi)\dot{\theta} = -\frac{m}{M + 2m}yF + \epsilon F + T_z \tag{12.101}$$

where $F = F_s\cos^2\alpha$. It is further assumed that the sun angle $\beta \approx 0$ and $\alpha = \pi/2 + \psi$. Note that the zero-thrust mode has a zero-yaw angle and that the full-thrust mode has a yaw angle of -90 deg.

For a special case of standard spacecraft with small roll/pitch/yaw angles relative to the LVLH reference frame, we obtain the well-known attitude equations of motion of the following form:

$$J_x\ddot{\phi} + 4n^2(J_y - J_z)\phi - n(J_x - J_y + J_z)\dot{\psi} = T_x \tag{12.102}$$
$$J_y\ddot{\theta} + 3n^2(J_x - J_z)\theta = T_y \tag{12.103}$$
$$J_z\ddot{\psi} + n^2(J_y - J_x)\psi + n(J_x - J_y + J_z)\dot{\phi} = T_z \tag{12.104}$$

12.11   Preliminary Design of the Propellantless Primary ACS

Ignoring the coupling terms in Eqs. (12.99–12.101), one can design sail attitude control logic for each axis as described in this section. A 40-m, 150-kg sailcraft is assumed here.

12.11.1   Yaw-Axis Control Design

A yaw-axis control design model, as illustrated in Fig. 12.30, is assumed as

$$J_z\ddot{\psi} + \frac{1}{2}n^2(J_y - J_x)\sin 2\psi = -\frac{F_s\cos^2\alpha m}{M + 2m}y + \epsilon F_s\cos^2\alpha \tag{12.105}$$

where $J_x = I_x + m_r(y^2 + z^2)$, $J_y = I_y + m_rz^2$, $J_z = I_z + m_ry^2$, ψ is the yaw angle defined as $\psi = -\pi/2 + \alpha$, and m_r is the reduced mass defined as $m_r = m(M + m)/(M + 2m) \approx m$. The steady-state trim position of the ballast for countering the effect of the c.m./c.p. offset ϵ can be estimated as

$$y_{ss} = \frac{M + 2m}{m}\epsilon \tag{12.106}$$

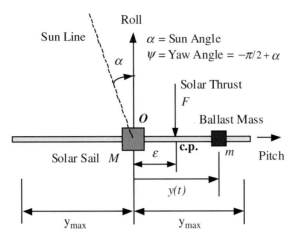

Fig. 12.30 **Yaw-axis dynamic model of sailcraft with a translating ballast mass.**

The actuator dynamics of the translating ballast mass is assumed as

$$T\dot{y} + y = y_c; \qquad |y_c(t)| \le y_{max} \qquad (12.107)$$

where T is the actuator time constant, y is the actual position, and y_c is the commanded position (the control input) with a maximum value of y_{max}. The maximum speed $\dot{y}_{max}$ can be estimated as

$$\dot{y}_{max} = \frac{y_{max}}{T} \qquad (12.108)$$

As an example, we have $y_{ss} = \pm 14.9$ m and $\dot{y}_{max} = \pm 0.05$ m/s for a 40-m, 150-kg sailcraft with $M = 148$ kg, $m = 1$ kg, $\epsilon = \pm 0.1$ m, $y_{max} = \pm 28$ m, and $T = 560$ s.
Consider a feedback control logic of the form

$$y_c = \underset{y_{max}}{\text{sat}} \left\{ K_p \underset{L}{\text{sat}}(e) + K_d \dot{\psi} \right\} \qquad (12.109)$$

where $e = \psi - \psi_c$, and K_p and K_d are, respectively, the attitude and attitude rate gains to be properly determined. The saturation function is defined as

$$\underset{L}{\text{sat}}(e) = \begin{cases} L & \text{if } e \ge L \\ e & \text{if } |e| < L \\ -L & \text{if } e \le -L \end{cases} \qquad (12.110)$$

and it can also be represented as

$$\underset{L}{\text{sat}}(e) = \text{sgn}(e) \min\{|e|, L\} \qquad (12.111)$$

Because of the presence of a limiter in the attitude-error feedback loop, the attitude rate becomes constrained as

$$-\dot{\psi}_{max} \le \dot{\psi} \le \dot{\psi}_{max} \qquad (12.112)$$

where $\dot{\psi}_{max} = LK_p/K_d$.

Furthermore, if the maximum slew rate is specified as $|\dot{\psi}|_{max}$, then the limiter in the attitude-error feedback loop can be selected simply as

$$L = \frac{K_d}{K_p} \dot{\psi}_{max} \tag{12.113}$$

However, as the attitude-error signal gets larger and also as the slew rate limit becomes larger for rapid maneuvers, the overall response becomes sluggish with increased transient overshoot because of the actuator saturation. To achieve rapid transient settlings even for large commanded attitude angles, the slew rate limit needs to be adjusted as follows:

$$\dot{\psi}_{max} = \min \left\{ \sqrt{2a|e|}, \; \omega_{max} \right\} \tag{12.114}$$

where ω_{max} is the specified maximum slew rate and a is the maximum angular acceleration described by

$$a = \frac{F_s m y_{max}}{J_z (M + 2m)} \tag{12.115}$$

A smaller value than the nominal a can be used to accommodate various uncertainties in the spacecraft inertia and actuator dynamics. Such a variable limiter in the attitude-error feedback loop has the self-adjusting saturation limit as

$$L = \frac{K_d}{K_p} \min \left\{ \sqrt{2a|e|}, \; \omega_{max} \right\} \tag{12.116}$$

If an integral control is necessary to eliminate a steady-state pointing error caused by the constant external disturbance caused by a c.m./c.p. offset, the feedback control logic (12.109) can be further modified into the nonlinear (proportional-integral-derivative (PID)) control logic of the form

$$y_c = \underset{y_{max}}{\text{sat}} \left\{ K_p \underset{L}{\text{sat}} \left[(\psi - \psi_c) + \frac{1}{\tau} \int (\psi - \psi_c) \right] + K_d \dot{\psi} \right\} \tag{12.117}$$

where τ is the time constant of integral control and L is given by Eq. (12.116).

12.11.2 Pitch-Axis Control Design

Similar to the yaw-axis control design, the pitch-axis control logic for a translating control mass along the z axis is assumed as

$$z_c = - \underset{z_{max}}{\text{sat}} \left\{ K_p \underset{L}{\text{sat}} \left[(\theta - \theta_c) + \frac{1}{\tau} \int (\theta - \theta_c) \right] + K_d \dot{\theta} \right\} \tag{12.118}$$

where z_c is the commanded position of the control mass.

12.11.3 Roll-Axis Control Design

The roll-axis control logic of the roll stabilizer bar (RSB) is also assumed as

$$T_x = - \operatorname*{sat}_{T_{\max}} \left\{ K_p \operatorname{sat}_L \left[(\phi - \phi_c) + \frac{1}{\tau} \int (\phi - \phi_c) \right] + K_d \dot{\phi} \right\} \tag{12.119}$$

where T_x is the roll control torque command with a saturation limit of $T_{\max}$.

The roll control torque T_x (in units of N·m) is related to the tilt angle Θ of the RSB of an assumed length of 1 m for a baseline 40-m sailcraft, as follows:

$$T_x = \left(\frac{0.5}{20}\right)(13.3)F_s \sin \Theta \tag{12.120}$$

where it is assumed that all four RSBs are rotated simultaneously. Consequently, we have $T_{\max}$ of ± 2.3 mN·m for a maximum tilt angle of ± 45 deg.

The RSB actuator dynamics is also assumed as

$$T\dot{\Theta} + \Theta = \Theta_c; \qquad |\Theta_c(t)| \le \Theta_{\max} \tag{12.121}$$

where T is the actuator time constant, Θ is the actual tilt angle, and Θ_c is the commanded tilt angle (the control input) with a maximum value of $\Theta_{\max}$. The commanded tilt angle of the RSB is then given by

$$\Theta_c = \arcsin \left\{ \frac{20T_x}{(0.5)(13.3)F_s} \right\} \tag{12.122}$$

12.11.4 Three-Axis Coupled Dynamic Simulation

The effectiveness of the proposed sail attitude control logic was validated using three-axis coupled dynamic simulations. Simulation results of the TVC mode with $\psi_c = -55$ deg are shown in Fig. 12.31. A 35-deg yaw maneuver, while maintaining small roll/pitch attitude errors, is performed within the saturation limits of propellantless control actuators. This case confirms the feasibility of demonstrating an orbit raising maneuver in a 1600-km DDSS orbit by employing the proposed propellantless primary ACS.

More detailed discussions of the proposed propellantless ACS can be found in [32–35].

12.12 Micro-PPT-Based Secondary ACS for Solar Sails

A secondary ACS, utilizing tip-mounted, lightweight PPT, developed in [36–38], is described in this section. Such a microthruster-based ACS provides reliable capability for recovery of attitude given any off-nominal conditions, including tumbling, which cannot be handled by either the propellantless primary ACS or by conventional ACS within the sail spacecraft bus. Micro-PPT-based ACS is also useful for three-axis stabilization of the sailcraft after release from the launch vehicle, (most critically) during deployment, and also during preflight

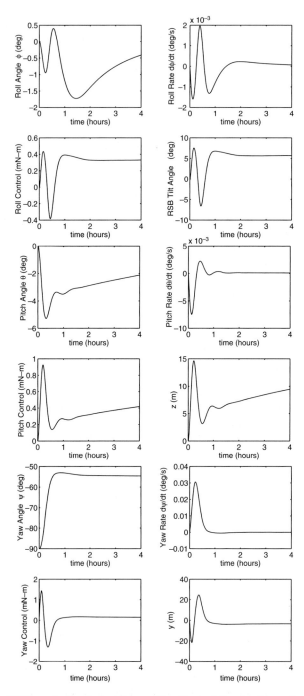

Fig. 12.31   TVC mode simulation for an 8-km/day orbit raise ($\phi_c = \theta_c = 0$ deg and $\psi_c = -55$ deg).

sail checkout operations as well. Alternately, these secondary functions can in general be performed by the conventional ACS of a spacecraft bus; however, such systems would be significantly more massive. But the control offered by standard-size conventional ACS during sail deployment would not be robust to variations in symmetry during deployment. Micro-PPT technology enables a tremendous gain in sailcraft performance, most critically in the areas of agility, mass, and redundancy.

An overview of the state-of-the-art PPT technology, PPT performance requirements for solar-sails, and the pulse-modulated PPT control design and simulation is presented in this section.

12.12.1 Recent Advances in Micro-PPT Technology

A brief overview of the recent advances in micro-PPT technology applicable to solar-sail attitude control is presented here.

A variety of microthrusters for stationkeeping and attitude control of microsatellites has been developed by NASA Glenn Research Center, Air Force Research Laboratory (AFRL), Jet Propulsion Laboratory, and industry [39–43]. They include the vaporizing liquid microthruster, the micro-Newton-sized cold-gas thruster, the micro Hall thruster, the micro ion engine, the micro pulsed plasma thruster, the free molecule microresistojet, and the digital microthruster array. Most of these microthrusters have an inherent problem of low specific impulse and/or low efficiency. However, such a low-I_{sp} and/or low-efficiency drawback of microthrusters might not be of significant importance for solar-sail attitude control applications, whereas other factors such as low thrust, low power, low mass, low volume, low voltage, and low impulse bit are more important. Among these microthrusters, a micro PPT is judged most suitable for solar-sail attitude control applications because of its inherent simplicity (not requiring propellant tanks, micromachined valves, and complex feed systems) and its use of a solid Teflon® propellant. However, the spacecraft contamination issues involved with using Teflon® propellant need to be further examined, and the selection of a particular type of attitude-control propulsion system for a given mission is in general a complex problem, and it is strongly mission dependent.

The PPT uses electric power to ionize and electromagnetically accelerate a plasma to high exhaust velocities. As illustrated schematically in Fig. 12.32, the PPT consists of a Teflon® fuel bar, a negator spring, a power processing unit (PPU), capacitors, electrodes, a spark plug, and a trigger circuit. The main discharge, ignited by the spark plug, ablates and ionizes a small amount of Teflon® from the face of the fuel bar into a plasma slug. The interaction of the current and the self-imposed magnetic field generates the $\vec{j} \times \vec{B}$ Lorentz force, which accelerates the plasma to high exhaust velocities.

The PPT technology has a long history of reliable spaceflight operation, for example, Russian Zond-2 Mars Probe in 1964, LES-6 spacecraft (by MIT Lincoln Laboratory) in 1968, LES-8/9 satellites, and NOVA satellites in early 1980s [39]. Recently, several miniaturized PPTs have been developed for a variety of satellite applications. For example, a 5-kg PPT module by Primex Aerospace (now Aerojet-Redmond Operation) was flight validated on the EO-1 New Millennium Program

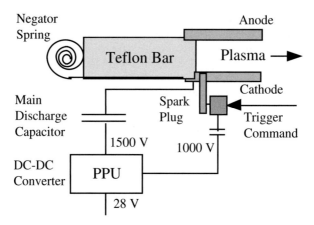

Fig. 12.32   Schematic of a pulsed plasma thruster.

mission [40]. A precision pitch attitude pointing capability of a PPT-based ACS was demonstrated for the EO-1 spacecraft while meeting stringent electromagnetic and contamination constraints. A 1-kg PPT module was also recently developed for the Dawgstar microsatellite [41,42]. Although the Dawgstar satellite did not become part of an actual flight program, a flight-qualified unit of such a 1-kg PPT module has been developed for use in microsatellites [42]. A much smaller 0.1-kg PPT unit was developed by the AFRL [43]. The 10-μN micro-PPTs by the AFRL were originally planned to fly on the TechSat21 mission as a technology demonstration, but they were never flown because of program cancellation of the TechSat21 mission. Such a significant miniaturization was made possible by using a self-igniting discharge, thus eliminating the separate igniter circuit from a standard PPT. Busek Corporation has taken over commercial development of the micro-PPT design from the AFRL and is providing micro PPTs for the FalconSat3 program at the U.S. Air Force Academy. FalconSat3 is a 50-kg, 0.46-m cube microsatellite. Among its other mission objectives, FalconSat3 will demonstrate a micropropulsion attitude control system (MPACS) technology. A micro-PPT module is also installed at the end of a gravity-gradient boom. Each module of a total mass of 1.6 kg includes three thrusters, a shared main capacitor, and individual triggering units for each thruster with an impulse bit of 100 μN·s. The overall characteristics of these recently developed microPPTs are compared in Table 12.4.

12.12.2   PPT Requirements for Solar-Sail Attitude Control

As illustrated in Table 12.5, a thrust level of 150 μN is selected for use with a 40-m untrimmed sailcraft with large torque margins (and also for a larger solar sail up to 80 m but with less torque margins). A pair of thrusters can also be employed to further increase control torque margins. The EO-1 PPT module with a total mass of 4.9 kg is too heavy, and the thrust level of AFRL's micro PPT is too low to be employed for 150-kg class sailcraft. The Dawgstar PPT module with a total mass of 1 kg has a thrust level of 120 μN (at 2-Hz pulsing frequency), but its current

Table 12.4 Comparison of recently developed micro-PPT modules

Parameter	EO-1	Dawgstar	AFRL	FalconSat3
Thrust,[a] μN	860	120	10	100[b]
Thrusters per module	2	2	1	3
Pulsing freq., Hz	1	2	1	2
Impulse bit, $\mu N \cdot s$	90–860	60	10	50[b]
Pulse energy, J	8.5–56	5	6.6	2
Total mass, kg	4.9	1	0.1	1.6
Power, W	70	15	1	8
Isp, s	650–1350	242	——	——
Efficiency,[c] %	9.8	1.8	——	——
Total impulse,[d] $N \cdot s$	925	140	——	——
Propellant,[d] kg	0.07	0.03	——	——

[a]Maximum steady thrust at maximum pulsing frequency.
[c]Thruster efficiency, not including the PPU efficiency.
[d]Per thruster.
[b]Beginning of life. Possibly, 50 at End of life.

Table 12.5 Design requirements for a 150-μN PPT

Sail size	40	80
Mast length, m	28	56
c.m./c.p. offset[a], m	0.1	0.2
Solar disturbance torque, mN·m[a]	1	8
Control torque[b], mN·m	4.2	8.4[c]
Total impulse (required)[d], $\mu N \cdot s$	1126	4505
Total pulses (required)[d], million	7.5	30

[a]The normally worst case for untrimmed sailcraft.
[b]Using one thruster.
[c]This control torque can be doubled using a pair of thrusters.
[d]Per year.

design has a total impulse of only 140 N·s (per thruster), which is too small for solar-sail attitude control applications.

Consequently, a prototype lightweight PPT module, named PPT150, has been developed for use in solar-sail attitude control by employing the design methodology of the flight-proven EO-1 PPT module. As shown in Fig. 12.33, the PPT150 module has four thruster units. Preliminary design characteristics of the PPT150 module are summarized in Table 12.6. Similar to mounting control vanes at the mast tips, four PPT modules will be mounted at the mast tips, as illustrated in Fig. 12.34, to utilize the largest moment arm length. A yaw-axis dynamic model with tip-mounted micro PPTs and a translating control mass are also illustrated in Fig. 12.35. A mast tip-mounted PPT150 of a maximum steady thrust level

Fig. 12.33   A 150-μN, 15-W, 2-kg micro-PPT module currently under preliminary development and testing for solar-sail attitude control applications (prearranged for the purpose of illustrating its overall cross-axis configuration and components).

Table 12.6   Design parameters of the PPT150 module for solar-sail attitude control

Parameter	Value
Total mass	2 kg
Thrusters	4
Impulse bit	150 μN·s
Pulse frequency (max)	1 Hz
Steady thrust (max)[a]	150 μN
Pulse energy	15 J
Peak power[a]	20 W
Average power[a]	15 W
I_{sp}	500 s
Efficiency	5%
Total impulse[a]	1500 N·s
Propellant[a]	0.2 kg

[a]Per a single thruster unit.

of 150 μN (at 1-Hz pulsing frequency) provides a maximum control torque of 4.2 mN·m for a 40-m sailcraft (using a single thruster).

12.12.3   Pulse-Modulated On–Off Control Using Micro PPTs

Pulse modulation represents the common control logic behind most reaction-jet control systems of spacecraft. Unlike other actuators, such as reaction wheels,

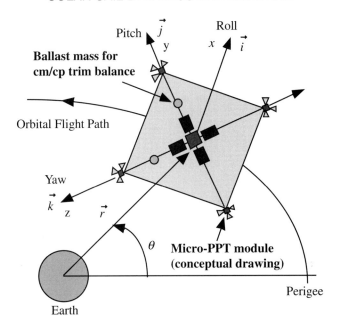

Fig. 12.34 Sailcraft equipped with a propellantless primary ACS and a micro-PPT-based secondary ACS.

thruster output consists of two values: on or off. Proportional thrusters, whose fuel valves open a distance proportional to the commanded thrust level, are not often employed in practice. Mechanical considerations prohibit proportional valve operation largely because of dirt particles that prevent complete closure for small

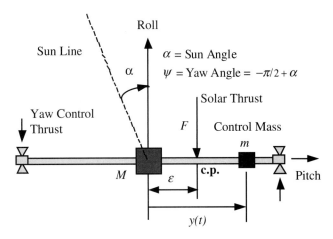

Fig. 12.35 Yaw-axis dynamic model with tip-mounted micro PPTs and a moving control mass.

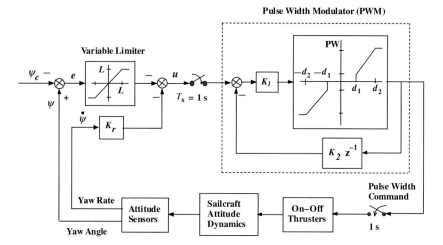

Fig. 12.36 Single-axis control-loop design model of a microthruster-based ACS.

valve openings; fuel leakage through the valves consequently produces opposing thruster firings. In general, pulse modulators produce a pulse command sequence to the thruster valves by adjusting the pulse width and/or pulse frequency.

The pulse width modulator (PWM) shown in Fig. 12.36 differs from other modulators, such as a pulse width and pulse frequency (PWPF) modulator, that it is essentially a discrete-time device. The PWPF modulator can be digitally implemented but is often analyzed as a continuous-time system. The output of a PWM is not a thruster firing state; instead, the PWM output is thruster pulse width, as illustrated in Fig. 12.37.

A single-axis control-loop design model of a reaction-jet control system with a PWM is illustrated in Fig. 12.36. The value d_1 represents the minimum pulse width

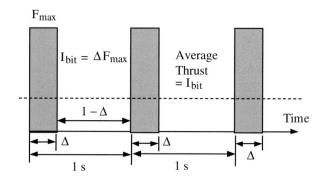

Fig. 12.37 Steady-state pulse firings of a PPT at 1-Hz pulse frequency.

of the system; this dead zone is directly proportional to the attitude deadband. The value d_2 represents the maximum pulse width of a thruster; it is often chosen to be the digital control sampling period. The delay in the feedback loop introduces damping to the system; maximum damping occurs when the feedback signal is smaller than the PWM input. If the input signal is not greater than the feedback signal, the modulator can limit cycle itself. This criterion enables the designer to determine the feedback gain K_2. The feedforward gain K_1 is selected as result of the minimum pulse width and the attitude deadband.

As illustrated in Fig. 12.36 for the yaw axis, the pulse control logic is given by

$$u = -\left\{ \underset{L}{\text{sat}}(e) + K_r \dot{\psi} \right\} \qquad (12.123)$$

where

$$e = \psi - \psi_c \qquad (12.124)$$

$$L = K_r \min \left\{ \sqrt{2a|e|},\ \omega_{max} \right\} \qquad (12.125)$$

For a PPT with a fixed impulse bit, I_{bit} and a maximum pulsing frequency of 1 Hz, as illustrated in Fig. 12.37, we simply choose $K_1 = 1$, $K_2 = 0$, and $d_1 = d_2$. A single pulse firing is commanded if $|u| > d_1$, where d_1 is the desired attitude deadband. Preliminary design values of the PPT-based ACS are provided in Table 12.7.

The feasibility of controlling a 40-m untrimmed sailcraft in a 1600-km DDSS orbit using the proposed micro-PPT-based ACS is demonstrated in Fig. 12.38, using three-axis coupled dynamic simulation in the presence of orbital as well as solar-pressure disturbances. In Fig. 12.38, an attitude recovery maneuver

Table 12.7   Preliminary design of a micro-PPT-based ACS for a 40-m sailcraft

Parameter	Value
Sampling time T_s	1 s
Pulsing freq. (max)	1 Hz
I_{bit}	150 μN·s
Δ	15 μs
F_{max}	10 N
Average thrust (max)	150 μN
Average torque (max)	4.2 mN·m
d_1	0.002 rad
d_2	0.002 rad
Attitude deadband	±0.1 deg
K_1	1
K_2	0
K_r	300 rad^{-1}
ω_{max}	0.05 deg/s

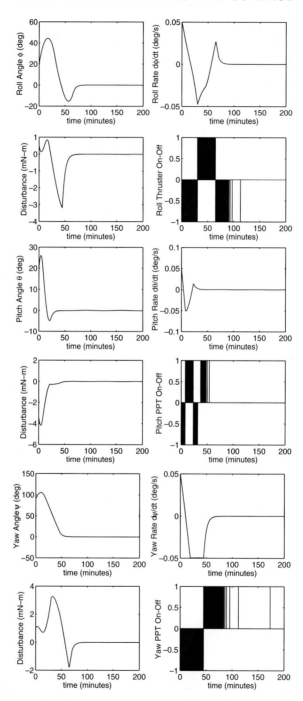

Fig. 12.38 Attitude recovery maneuver using the PPT-based ACS for an untrimmed sailcraft in a 1600-km DDSS orbit ($\phi_c = \theta_c = \psi_c = 0$ deg).

for achieving the LVLH orientation of the zero-thrust mode was simulated. The initial tumbling conditions were $\phi(0) = \theta(0) = 20$ deg, $\psi(0) = 90$ deg, and $\dot{\phi}(0) = \dot{\theta}(0) = \dot{\psi}(0) = 0.05$ deg/s. As demonstrated in this simulation, the proposed micro-PPT-based ACS provides a backup control capability for achieving the zero-thrust safe mode with $\psi(0) = 0$ deg.

12.13 Orbital Dynamics of Solar Sails

In this section we examine various basic forms of interplanetary orbital equations for solar-sail trajectory design and simulation. In general, the orbital equation of a sailcraft in a heliocentric orbit is simply described by

$$\ddot{\vec{r}} + \frac{\mu}{r^3}\vec{r} = \vec{F} + \vec{G} \qquad (12.126)$$

where $\vec{r}$ is the position vector of the sailcraft from the center of the sun, $\mu \approx \mu_\odot = 132{,}715\mathrm{E}6\,\mathrm{km}^3/\mathrm{s}^2$, $\vec{F}$ is the solar radiation pressure force vector (per unit mass) acting on the sailcraft, and $\vec{G}$ is the sum of all the perturbing gravitational forces (per unit mass) acting on the sailcraft. In this section, all of the perturbing gravitational forces are ignored without loss of generality.

12.13.1 Introduction

In support of interplanetary solar-sailing missions, including the Comet Halley Rendezvous Mission [1] and the Solar Polar Imager (SPI) mission [44], the solar-sail trajectory optimization problem has been extensively studied in the past [46–51]. For typical solar-sail trajectory optimization problems, trajectory models that are decoupled from attitude dynamics were often used in the past. However, the effect of attitude motion of large solar sails on the solar-sailing trajectory is of practical concern for solar-sail mission designs. Consequently, a six-degree-of-freedom, orbit-attitude coupled dynamical model of solar-sail spacecraft is also considered for the solar-sail trajectory optimization and simulation in [52].

As part of an AOCS of solar sails, a TVC system maintains the proper orientation of solar sails to provide the desired thrust vector pointing/steering. Most solar-sail TVC systems, with the inherent inability to precisely model the solar radiation pressure and to accurately point the true thrust vector direction, will require frequent updates of both orbital parameters and TVC steering commands for frequent trajectory corrections [53]. The frequency of such orbit determination and TVC command updates is determined by many factors, such as the trajectory dispersions, the target-body ephemeris uncertainty, the calibration of a solar radiation pressure model, and the operational constraints.

The orientation of the solar-sail thrust vector, ideally normal to the sail plane, is often described in terms of the cone and clock angles. These two angles are the typical trajectory control inputs used in solar-sail trajectory optimization. There are at least two different sets of cone/clock angles used in the literature. In this section, the fundamentals of orbital equations of motion in various coordinates employing such two different sets of cone/clock angles are described for the purpose of trajectory design, TVC design, and simulation.

12.13.2 Cone and Clock Angles

Let $\{\hat{I}, \hat{J}, \hat{K}\}$ and $\{\hat{r}, \hat{\psi}, \hat{\phi}\}$ be respectively a set of right-handed, orthonormal vectors of the heliocentric ecliptic rectangular and spherical coordinate reference frames, as illustrated in Fig. 12.39. These two sets of basis vectors are related as

$$\begin{bmatrix} \hat{r} \\ \hat{\psi} \\ \hat{\phi} \end{bmatrix} = \begin{bmatrix} \cos\phi & 0 & \sin\phi \\ 0 & 1 & 0 \\ -\sin\phi & 0 & \cos\phi \end{bmatrix} \begin{bmatrix} \cos\psi & \sin\psi & 0 \\ -\sin\psi & \cos\psi & 0 \\ 0 & 0 & 1 \end{bmatrix} \begin{bmatrix} \hat{I} \\ \hat{J} \\ \hat{K} \end{bmatrix} \qquad (12.127)$$

where ψ and ϕ are called the ecliptic longitude and latitude of the sailcraft position, respectively; $0 \le \psi \le 360\,\deg$ and $-90\,\deg \le \phi \le +90\,\deg$.

The sailcraft position vector is then expressed as

$$\begin{aligned} \vec{r} &= r\hat{r} \\ &= (r\cos\phi\cos\psi)\hat{I} + (r\cos\phi\sin\psi)\hat{J} + (r\sin\phi)\hat{K} \\ &= X\hat{I} + Y\hat{J} + Z\hat{K} \end{aligned} \qquad (12.128)$$

where $r = |\vec{r}|$ is the distance from the sun to the sailcraft.

The orientation of a unit vector normal to the sail plane $\hat{n}$ is described in terms of the cone angle α and the clock angle β, illustrated in Fig. 12.40, as follows:

$$\hat{n} = (\cos\alpha)\hat{r} + (\sin\alpha\sin\beta)\hat{\psi} + (\sin\alpha\cos\beta)\hat{\phi} \qquad (12.129)$$

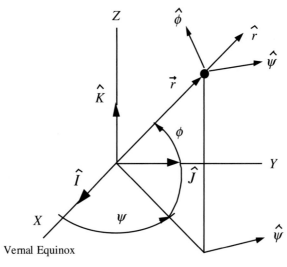

Fig. 12.39   Heliocentric ecliptic coordinates (X, Y, Z) and spherical coordinates (r, ψ, ϕ).

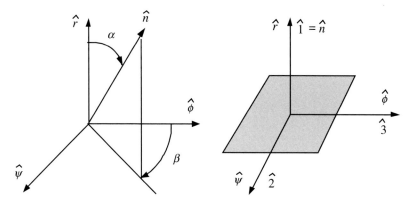

Fig. 12.40   Cone angle α, clock angle β, and sailcraft orientation when $\alpha = \beta = 0$.

where

$$\cos \alpha = \hat{r} \cdot \hat{n}$$

$$\cos \beta = \frac{\hat{r} \times (\hat{n} \times \hat{r})}{|\hat{r} \times (\hat{n} \times \hat{r})|} \cdot \hat{\phi}$$

$$0 \le \alpha \le 90 \deg$$

$$0 \le \beta \le 360 \deg$$

As also illustrated in Fig. 12.40, the sailcraft body-fixed basis vectors $\{\hat{1}, \hat{2}, \hat{3}\}$ are assumed to be aligned with $\{\hat{r}, \hat{\psi}, \hat{\phi}\}$ when $\alpha = \beta = 0$, and the sailcraft roll axis is defined to be perpendicular to the sail surface; that is, $\hat{1} \equiv \hat{n}$. The sailcraft body-fixed basis vectors $\{\hat{1}, \hat{2}, \hat{3}\}$ are then related to $\{\hat{r}, \hat{\psi}, \hat{\phi}\}$ as follows:

$$\begin{bmatrix} \hat{1} \\ \hat{2} \\ \hat{3} \end{bmatrix} = \begin{bmatrix} \cos \alpha & 0 & \sin \alpha \\ 0 & 1 & 0 \\ -\sin \alpha & 0 & \cos \alpha \end{bmatrix} \begin{bmatrix} 1 & 0 & 0 \\ 0 & \cos \beta & -\sin \beta \\ 0 & \sin \beta & \cos \beta \end{bmatrix} \begin{bmatrix} \hat{r} \\ \hat{\psi} \\ \hat{\phi} \end{bmatrix} \qquad (12.130)$$

Let $\{\hat{r}, \hat{\theta}, \hat{k}\}$ be a set of basis vectors of an osculating orbital plane, as illustrated in Fig. 12.41. A different set of the cone and clock angles (α, δ) can then be defined as shown in Fig. 12.41. A body-fixed rotational sequence to $\{\hat{1}, \hat{2}, \hat{3}\}$ from $\{\hat{I}, \hat{J}, \hat{K}\}$ is then described by successive coordinate transformations of the form

$$\mathbf{C}_2(-\alpha) \leftarrow \mathbf{C}_1(-\delta) \leftarrow \mathbf{C}_3(\theta) \leftarrow \mathbf{C}_3(\omega) \leftarrow \mathbf{C}_1(i) \leftarrow \mathbf{C}_3(\Omega)$$

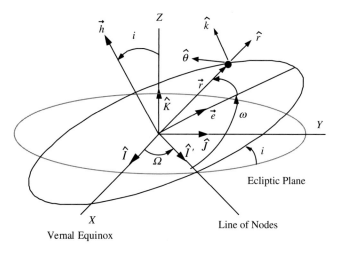

Fig. 12.41   Orbital geometry (illustrated for a near-circular orbit).

which becomes

$$
\begin{bmatrix} \hat{r} \\ \hat{\theta} \\ \hat{k} \end{bmatrix} = \begin{bmatrix} \cos(\omega+\theta) & \sin(\omega+\theta) & 0 \\ -\sin(\omega+\theta) & \cos(\omega+\theta) & 0 \\ 0 & 0 & 1 \end{bmatrix} \begin{bmatrix} 1 & 0 & 0 \\ 0 & \cos i & \sin i \\ 0 & -\sin i & \cos i \end{bmatrix}
$$

$$
\times \begin{bmatrix} \cos\Omega & \sin\Omega & 0 \\ -\sin\Omega & \cos\Omega & 0 \\ 0 & 0 & 1 \end{bmatrix} \begin{bmatrix} \hat{I} \\ \hat{J} \\ \hat{K} \end{bmatrix} \tag{12.131}
$$

$$
\begin{bmatrix} \hat{1} \\ \hat{2} \\ \hat{3} \end{bmatrix} = \begin{bmatrix} \cos\alpha & 0 & \sin\alpha \\ 0 & 1 & 0 \\ -\sin\alpha & 0 & \cos\alpha \end{bmatrix} \begin{bmatrix} 1 & 0 & 0 \\ 0 & \cos\delta & -\sin\delta \\ 0 & \sin\delta & \cos\delta \end{bmatrix} \begin{bmatrix} \hat{r} \\ \hat{\theta} \\ \hat{k} \end{bmatrix} \tag{12.132}
$$

The orientation of a unit vector normal to the sail plane $\hat{n}$ is then described in terms of α and δ, as illustrated in Fig. 12.42, as follows:

$$
\hat{n} = (\cos\alpha)\hat{r} + (\sin\alpha\sin\delta)\hat{\theta} + (\sin\alpha\cos\delta)\hat{k} \tag{12.133}
$$

and

$$
\cos\alpha = \hat{r}\cdot\hat{n}
$$

$$
\cos\delta = \frac{\hat{r}\times(\hat{n}\times\hat{r})}{|\hat{r}\times(\hat{n}\times\hat{r})|}\cdot\hat{k}
$$

$$
0 \le \alpha \le 90 \text{ deg}
$$

$$
0 \le \delta \le 360 \text{ deg}
$$

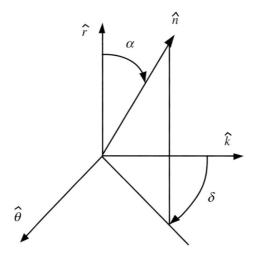

Fig. 12.42 Cone angle α and clock angle δ.

12.13.3 Solar Radiation Pressure

An ideal model of the SRP is used here. The SRP force vector (per unit mass) acting on the sailcraft is described in various coordinates as follows:

$$\vec{F} = F_0(\hat{r} \cdot \hat{n})^2 \hat{n}$$
$$= F_r \hat{r} + F_\psi \hat{\psi} + F_\phi \hat{\phi}$$
$$= R\hat{r} + T\hat{\theta} + N\hat{k}$$
$$= F_X \hat{I} + F_Y \hat{J} + F_Z \hat{K} \tag{12.134}$$

where

$$F_0 = \left(\frac{r_\oplus}{r}\right)^2 a_c \tag{12.135}$$

where $r_\oplus = 1\ \mathrm{AU} = 149,597,870.691$ km is the distance from the sun to the Earth and a_c is the so-called characteristic acceleration of the sailcraft at 1 AU.

Furthermore, we have the following relationships:

$$\begin{bmatrix} F_r \\ F_\psi \\ F_\phi \end{bmatrix} = F_0 \cos^2 \alpha \begin{bmatrix} \cos \alpha \\ \sin \alpha \sin \beta \\ \sin \alpha \cos \beta \end{bmatrix} \tag{12.136}$$

$$\begin{bmatrix} R \\ T \\ N \end{bmatrix} = F_0 \cos^2 \alpha \begin{bmatrix} \cos \alpha \\ \sin \alpha \sin \delta \\ \sin \alpha \cos \delta \end{bmatrix} \tag{12.137}$$

$$\begin{bmatrix} F_r \\ F_\psi \\ F_\phi \end{bmatrix} = \begin{bmatrix} \cos \phi & 0 & \sin \phi \\ 0 & 1 & 0 \\ -\sin \phi & 0 & \cos \phi \end{bmatrix} \begin{bmatrix} \cos \psi & \sin \psi & 0 \\ -\sin \psi & \cos \psi & 0 \\ 0 & 0 & 1 \end{bmatrix} \begin{bmatrix} F_X \\ F_Y \\ F_Z \end{bmatrix} \tag{12.138}$$

$$\begin{bmatrix} R \\ T \\ N \end{bmatrix} = \begin{bmatrix} \cos(\omega+\theta) & \sin(\omega+\theta) & 0 \\ -\sin(\omega+\theta) & \cos(\omega+\theta) & 0 \\ 0 & 0 & 1 \end{bmatrix} \begin{bmatrix} 1 & 0 & 0 \\ 0 & \cos i & \sin i \\ 0 & -\sin i & \cos i \end{bmatrix}$$

$$\times \begin{bmatrix} \cos\Omega & \sin\Omega & 0 \\ -\sin\Omega & \cos\Omega & 0 \\ 0 & 0 & 1 \end{bmatrix} \begin{bmatrix} F_X \\ F_Y \\ F_Z \end{bmatrix} \tag{12.139}$$

12.13.4 Orbital Equations in Rectangular Coordinates

The orbital equation of motion in vector form, Eq. (12.126), can be expressed in the rectangular coordinates as follows:

$$\ddot{X} = -\frac{\mu X}{r^3} + F_X \tag{12.140a}$$

$$\ddot{Y} = -\frac{\mu Y}{r^3} + F_Y \tag{12.140b}$$

$$\ddot{Z} = -\frac{\mu Z}{r^3} + F_Z \tag{12.140c}$$

where $r = \sqrt{X^2 + Y^2 + Z^2}$ and

$$\begin{bmatrix} F_X \\ F_Y \\ F_Z \end{bmatrix} = \begin{bmatrix} \cos\psi & -\sin\psi & 0 \\ \sin\psi & \cos\psi & 0 \\ 0 & 0 & 1 \end{bmatrix} \begin{bmatrix} \cos\phi & 0 & -\sin\phi \\ 0 & 1 & 0 \\ \sin\phi & 0 & \cos\phi \end{bmatrix} \begin{bmatrix} F_r \\ F_\psi \\ F_\phi \end{bmatrix} \tag{12.141}$$

$$\sin\psi = \frac{Y}{\sqrt{X^2+Y^2}}; \quad \cos\psi = \frac{X}{\sqrt{X^2+Y^2}}$$

$$\sin\phi = \frac{Z}{r}; \quad \cos\phi = \frac{\sqrt{X^2+Y^2}}{r}$$

When the orientation of $\hat{n}$ is described in terms of α and δ, as shown in Fig. 12.41, we use the following coordinate transformation:

$$\begin{bmatrix} F_X \\ F_Y \\ F_Z \end{bmatrix} = \begin{bmatrix} \cos\Omega & -\sin\Omega & 0 \\ \sin\Omega & \cos\Omega & 0 \\ 0 & 0 & 1 \end{bmatrix} \begin{bmatrix} 1 & 0 & 0 \\ 0 & \cos i & -\sin i \\ 0 & \sin i & \cos i \end{bmatrix}$$

$$\times \begin{bmatrix} \cos(\omega+\theta) & -\sin(\omega+\theta) & 0 \\ \sin(\omega+\theta) & \cos(\omega+\theta) & 0 \\ 0 & 0 & 1 \end{bmatrix} \begin{bmatrix} R \\ T \\ N \end{bmatrix}$$

For this case of employing (R, T, N), we need to determine $(\Omega, i, \omega, \theta)$ for given $(X, Y, Z, \dot{X}, \dot{Y}, \dot{Z})$.

The six classical orbital elements $(a, e, i, \Omega, \omega, M)$ can be determined for given $\vec{r} = X\vec{I} + Y\vec{J} + Z\vec{K}$ and $\vec{v} = \dot{X}\vec{I} + \dot{Y}\vec{J} + \dot{Z}\vec{K}$.

12.13.5   Orbital Equations in Spherical Coordinates

The orbital equations of motion in the spherical coordinates (r, ψ, ϕ) are given by

$$\ddot{r} - r\dot{\phi}^2 - r\dot{\psi}^2 \cos^2 \phi = -\frac{\mu}{r^2} + F_r \tag{12.142a}$$

$$r\ddot{\psi} \cos \phi + 2\dot{r}\dot{\psi} \cos \phi - 2r\dot{\psi}\dot{\phi} \sin \phi = F_\psi \tag{12.142b}$$

$$r\ddot{\phi} + 2\dot{r}\dot{\phi} + r\dot{\psi}^2 \sin \phi \cos \phi = F_\phi \tag{12.142c}$$

Let $v_r = \dot{r}$, $v_\psi = r\dot{\psi} \cos \phi$, and $v_\phi = r\dot{\phi}$, and then we obtain

$$\dot{r} = v_r \tag{12.143a}$$

$$\dot{\psi} = \frac{1}{r \cos \phi} v_\psi \tag{12.143b}$$

$$\dot{\phi} = \frac{1}{r} v_\phi \tag{12.143c}$$

$$\dot{v}_r = \frac{1}{r}(v_\psi^2 + v_\phi^2) - \frac{\mu}{r^2} + F_0 \cos^3 \alpha \tag{12.143d}$$

$$\dot{v}_\psi = \frac{1}{r}(v_\psi v_\phi \tan \phi - v_r v_\psi) + F_0 \cos^2 \alpha \sin \alpha \sin \beta \tag{12.143e}$$

$$\dot{v}_\phi = -\frac{1}{r}(v_\psi^2 \tan \phi + v_r v_\phi) + F_0 \cos^2 \alpha \sin \alpha \cos \beta \tag{12.143f}$$

where

$$F_0 = \left(\frac{r_\oplus}{r}\right)^2 a_c$$

This set of six trajectory equations is often employed to find the time histories of optimal control inputs (α, β). By numerically integrating this set of trajectory equations for known time histories of (α, β), we can obtain $(X, Y, Z, \dot{X}, \dot{Y}, \dot{Z})$ as follows:

$$X = r \cos \phi \cos \psi \tag{12.144a}$$

$$Y = r \cos \phi \sin \psi \tag{12.144b}$$

$$Z = r \sin \phi \tag{12.144c}$$

$$\begin{bmatrix} \dot{X} \\ \dot{Y} \\ \dot{Z} \end{bmatrix} = \begin{bmatrix} \cos \phi \cos \psi & -r \cos \phi \sin \psi & -r \sin \phi \cos \psi \\ \cos \phi \sin \psi & r \cos \phi \cos \psi & -r \sin \phi \sin \psi \\ \sin \phi & 0 & r \cos \phi \end{bmatrix} \begin{bmatrix} \dot{r} \\ \dot{\psi} \\ \dot{\phi} \end{bmatrix} \tag{12.145}$$

An interesting case of applying the orbital equations of motion expressed in spherical coordinates is the logarithmic spiral trajectory problem [54]. Although such logarithmic spiral trajectories are not practically useful for an interplanetary transfer between circular orbits, a simple steering law with a fixed sun angle is required.

For a simple planar case with $\phi = 0$ and $\beta = 0$, we have

$$\ddot{r} - r\dot{\psi}^2 = -\frac{\mu}{r^2} + F_0 \cos^3 \alpha \qquad (12.146a)$$

$$r\ddot{\psi} + 2\dot{r}\dot{\psi} = F_0 \cos^2 \alpha \, \sin \alpha \qquad (12.146b)$$

where

$$F_0 = \left(\frac{r_\oplus}{r}\right)^2 a_c$$

For this planar case, α is often called a pitch angle with -90 deg $\leq \alpha \leq 90$ deg.

By defining a sailcraft lightness number λ as

$$\lambda = \frac{F_0}{\mu/r^2} = \frac{r_\oplus^2 a_c}{\mu} = \frac{(149,597,870E3)^2}{132,715E15} a_c = 168.6284 a_c \qquad (12.147)$$

we rewrite the orbital equations of motion as

$$\ddot{r} - r\dot{\psi}^2 = -(1-\lambda)\frac{\mu}{r^2} \cos^3 \alpha \qquad (12.148a)$$

$$r\ddot{\psi} + 2\dot{r}\dot{\psi} = \lambda \frac{\mu}{r^2} \cos^2 \alpha \, \sin \alpha \qquad (12.148b)$$

12.13.6 Gauss's Form of the Variational Equations (Osculating Orbital Elements)

A set of six first-order differential equations, called Gauss's form of the variational equations [55], in terms of osculating orbital elements, is given by

$$\dot{a} = \frac{2a^2}{h}[eR \sin \theta + T(1 + e \cos \theta)] \equiv \frac{2a^2}{h}\left[eR \sin \theta + \frac{pT}{r}\right] \qquad (12.149a)$$

$$\dot{e} = \sqrt{\frac{p}{\mu}} \, [R \sin \theta + T(\cos \theta + \cos E)]$$

$$= \frac{1}{h}\{pR \sin \theta + [(p+r)\cos \theta + re]T\} \qquad (12.149b)$$

$$\dot{i} = \frac{r \cos(\omega + \theta)}{h} N \qquad (12.149c)$$

$$\dot{\Omega} = \frac{r \sin(\omega + \theta)}{h \sin i} N \qquad (12.149d)$$

$$\dot{\omega} = -\frac{r \sin(\omega + \theta)}{h \tan i} N + \frac{1}{eh}\left[-pR \cos \theta + (p+r)T \sin \theta\right] \qquad (12.149e)$$

$$\dot{\theta} = \frac{h}{r^2} + \frac{1}{eh}\left[pR \cos \theta - (p+r)T \sin \theta\right] \qquad (12.149f)$$

where

$$p = a(1 - e^2)$$

$$r = \frac{p}{1 + e \cos \theta} \equiv a(1 - e \cos E)$$

$$h = \sqrt{\mu p} = na^2 \sqrt{1 - e^2}$$

$$n = \sqrt{\mu / a^3}$$

$$\begin{bmatrix} R \\ T \\ N \end{bmatrix} = F_0 \cos^2 \alpha \begin{bmatrix} \cos \alpha \\ \sin \alpha \sin \delta \\ \sin \alpha \cos \delta \end{bmatrix}$$

$$F_0 = \left(\frac{r_\oplus}{r}\right)^2 a_c$$

In particular, the inclination equation becomes

$$\dot{i} = \frac{r \cos(\omega + \theta)}{h} N = \frac{\lambda}{r} \sqrt{\frac{\mu}{p}} \cos^2 \alpha \, \sin \alpha \, \cos \delta \, \cos(\omega + \theta) \qquad (12.150)$$

We then obtain a simple sail-steering law for maximizing the rate of change of inclination as

$$\alpha = \tan^{-1}(1/\sqrt{2}) = 35.26 \text{ deg}$$

and

$$\delta = \begin{cases} 0 \text{ deg} & \text{for} \quad \cos(\omega + \theta) \geq 0 \\ 180 \text{ deg} & \text{for} \quad \cos(\omega + \theta) < 0 \end{cases}$$

Although this simple steering law indicates that the clock angle has to change ±180 deg instantaneously, an equivalent attitude motion is a ±70-deg single-axis slew maneuver every half-orbit (i.e., at ascending and descending nodes). Furthermore, this simple steering law demonstrates an advantage of employing (α, δ) instead of (α, β) because (α, δ) is more conveniently tied to the classical orbital elements

12.14   Examples of Solar-Sail Trajectory Design

12.14.1   Solar Polar Imager Mission

Solar sails are envisioned as a propellantless, high-energy propulsion system for future space exploration missions. NASA's future missions enabled by solar-sail propulsion include the Solar Polar Imager (SPI), L1-Diamond, Particle Acceleration Solar Orbiter (PASO), and Interstellar Probe, which are the sun–Earth connections (SEC) solar-sail road-map missions. Our current understanding of the sun is limited by a lack of observations of its polar regions. The SPI mission utilizes a large solar-sail to place a spacecraft in a 0.48-AU heliocentric circular orbit with an inclination of 75 deg. Viewing of the polar regions of the sun provides a unique opportunity to more fully investigate the structure and dynamics of its interior, the generation of solar magnetic fields, the origin of the solar cycle, the causes of solar activity, and the structure and dynamics of the corona.

The SPI mission consists of the initial cruise phase to a 0.48-AU circular orbit, the cranking orbit phase, and the science mission phase [44]. A 160-m, 450-kg solar-sail spacecraft is considered as a reference model for such a solar-sailing mission. A Delta II launch vehicle is able to inject the 450-kg SPI spacecraft into an Earth escaping orbit with $C_3 = 0.25$ km^2/s^2. The solar sail is to be deployed at the beginning of the interplanetary cruise phase. The SPI sailcraft first spirals inwards from 1 AU to a heliocentric circular orbit at 0.48 AU, and then the cranking orbit phase begins to achieve a 75-deg inclination. The solar sail will be jettisoned after achieving the science mission orbit, and the total sailing time is approximately 6.6 yr.

Figure 12.43 shows an optimization-based trajectory design by Carl Sauer at NASA/JPL for achieving a circular orbit at 0.48 AU with a 75-deg inclination. A set of the cone and clock angles (α, β) was used by Carl Sauer for such a baseline SPI mission trajectory design. The monotonically decreasing semimajor axis and the corresponding variation of eccentricity can be seen in this figure during the initial cruise phase to 0.48 AU. The eccentricity remains constant during the orbit cranking phase. The eccentricity is finally nulled after cranking is complete. A somewhat complicated nature of the desired, optimal clock angle β command can be noticed in Fig. 12.43, although the cone angle is nearly kept constant except for the final orbit correction phase to null the eccentricity.

A possibility of employing simple sail-steering laws (a combination of constant cone/clock angles), based on Gauss's form of the variational equations with the cone and clock angles (α, δ), can be explored for the SPI mission. Figure 12.44 shows the result of applying such simple sail-steering laws with (α, δ) to the SPI mission. The switch from orbit radius reduction to the cranking phase occurs once the sail reaches the target semimajor axis of 0.48 AU. However, an actual transition to the cranking orbit phase was executed at a proper orbital location such that the large eccentricity variation of the initial cruise phase can be removed as can be seen in Fig. 12.44. The simulation used a nominal characteristic acceleration of 0.3 mm/s^2 (at 1 AU), an 8-deg initial inclination, and $C_3 = 0.25$ km^2/s^2. The sailcraft achieves the desired semimajor axis and inclination, but the final orbit is slightly eccentric. The corresponding three-dimensional orbital trajectory is illustrated in Fig. 12.45.

A comparison of the desired clock angle commands β and δ, shown in Figs. 12.43 and 12.44 respectively, suggests that the clock angle δ is a better choice for an actual TVC steering command implementation because of its simplicity. More detailed trajectory optimization study results for the SPI mission can be found in [56].

12.14.2   Solar-Sailing Kinetic Energy Impactor Mission

A fictional asteroid mitigation problem of AIAA is briefly described as follows. On 4 July, 2004, NASA/JPL's Near-Earth Asteroid Tracking (NEAT) camera at

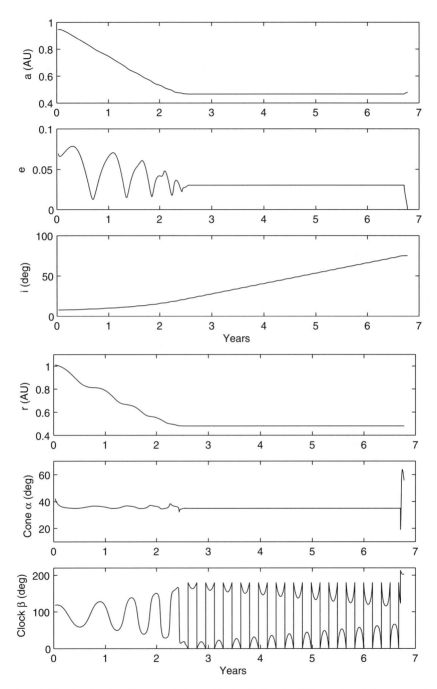

Fig. 12.43   Optimal trajectory design with (α, β) for the SPI mission (plots created using data from Carl Sauer at NASA/JPL; image appeard in Wie, *Journal of Spacecraft and Rockets***, Vol. 4, No. 3, 2007).**

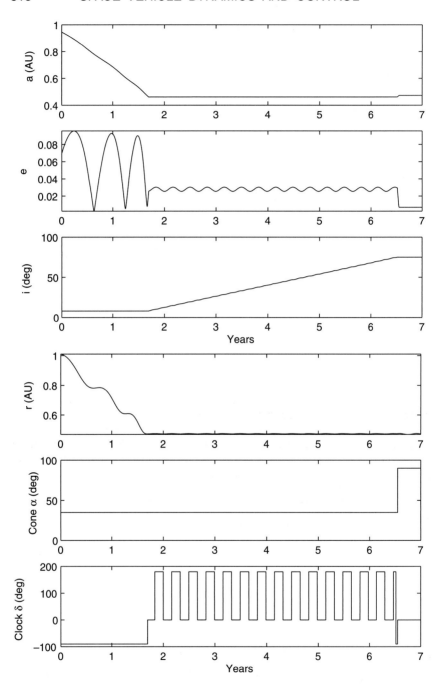

Fig. 12.44   SPI mission trajectory design using simple sail-steering laws with (α, δ).

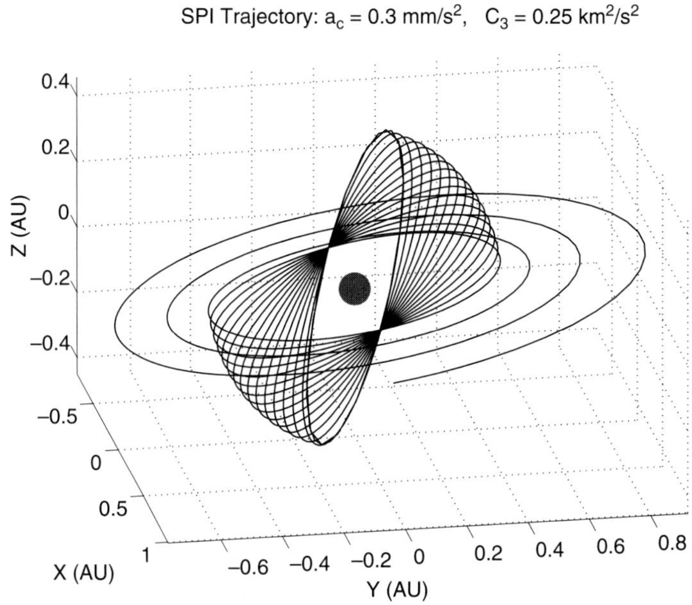

SPI Trajectory: $a_c = 0.3$ mm/s^2, $C_3 = 0.25$ km^2/s^2

Fig. 12.45   SPI mission trajectory design result using (α, δ).

the Maui Space Surveillance Site discovered a 0.205-km-dia Apollo asteroid designated 2004WR. This asteroid has been assigned a Torino Impact Scale rating of 9.0 on the basis of subsequent observations that indicate there is a 95% probability that 2004WR will impact the Earth. The expected impact will occur in the Southern Hemisphere on 14, January 2015, causing catastrophic damage throughout the Pacific region. The mission is to design a space system that can rendezvous with 2004WR in a timely manner, inspect it, and remove the hazard to Earth by changing its orbit and/or destroying it. The classical orbital elements of 2004WR are given in the J2000 heliocentric ecliptic reference frame as follows:

$$\text{Epoch} = 53200 \text{ TDB (14, July 2004)}$$

$$a = 2.15374076 \text{ AU}$$

$$e = 0.649820926$$

$$i = 11.6660258 \text{ deg}$$

$$\omega = 66.2021796 \text{ deg}$$

$$\Omega = 114.4749665 \text{ deg}$$

$$M = 229.8987151 \text{ deg}$$

A solar-sailing kinetic energy impactor (KEI) mission concept applied to this fictional asteroid mitigation problem is illustrated in Fig. 12.46. The proposed

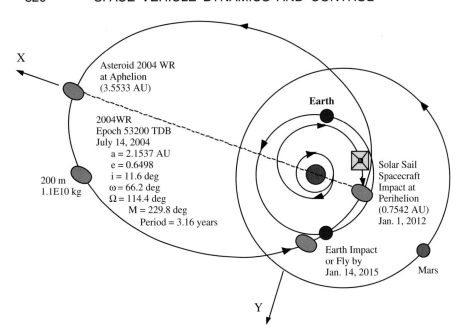

Fig. 12.46   Illustration of the solar-sailing KEI mission for impacting and deflecting a near-Earth asteroid. The final, retrograde heliocentric orbit phase (starting from 0.25 AU) results in a head-on collision with the target asteroid at its perihelion (0.75 AU) with an impact speed larger than 70 km/s.

mission requires at least ten 160-m, 300-kg solar-sail spacecraft with a characteristic acceleration of 0.5 mm/s^2, as proposed in [57,58] as a viable near-term option for mitigating the threat posed by near-Earth asteroids (NEAs). The solar-sailing phase of the proposed KEI mission, which is very similar to that of the SPI mission, is composed of the initial cruise phase from 1 to 0.25 AU, the cranking orbit phase (for a 168-deg inclination change), and the final retrograde orbit phase prior to impacting the target asteroid at its perihelion with an impact speed larger than 70 km/s.

Simple sail-steering laws based on Gauss's form of the variational equations can also be employed for the preliminary KEI mission design. Figures 12.47–12.49 show the result of applying the simple sail-steering laws to the KEI mission [57]. A solar-sail KEI trajectory design for optimally intercepting, impacting, and deflecting asteroid 2004WR is also provided in Fig. 12.50. More detailed discussions on the optimal KEI mission trajectory design can be found in [58,59]. In practice, because of the inherent difficulty to precisely point the true thrust vector direction, frequent updates of the sail-steering commands will be required.

Such a technically challenging, asteroid-impact mitigation problem and its viable engineering solution, which utilizes the recent advances in solar-sail technology, will be further studied in Chapter 13.

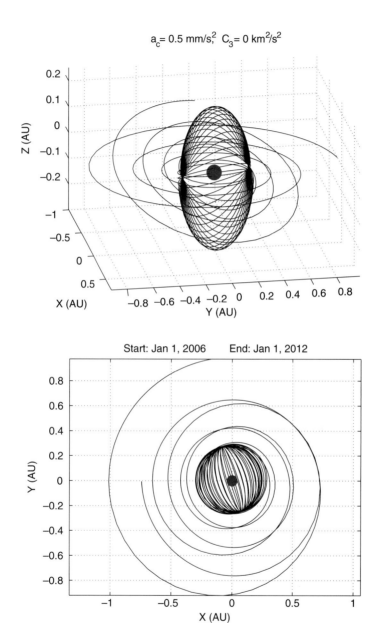

$a_c = 0.5 \text{ mm/s}^2, \quad C_3 = 0 \text{ km}^2/\text{s}^2$

Start: Jan 1, 2006   End: Jan 1, 2012

Fig. 12.47 Solar-sail KEI mission trajectory resulting in a head-on collision with the target asteroid at its perihelion.

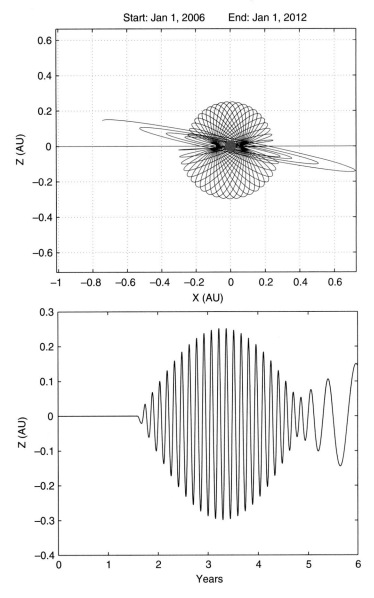

Fig. 12.48   Solar-sail KEI mission trajectory resulting in a head-on collision with the target asteroid at its perihelion.

12.15   Solar-Sail Thrust Vector Control Design

As described in the preceding sections, various forms of orbital trajectory equations employing two sets of the cone and clock angles are available for the solar-sail trajectory design and simulation. It was shown that a preliminary trajectory design

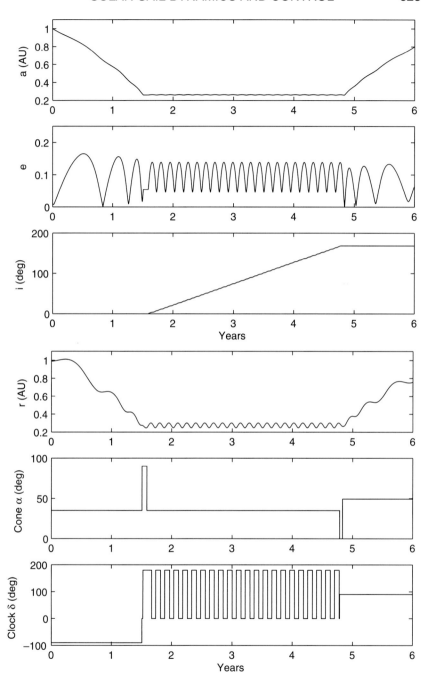

Fig. 12.49   Solar-sail KEI trajectory design using simple sail-steering laws [57].

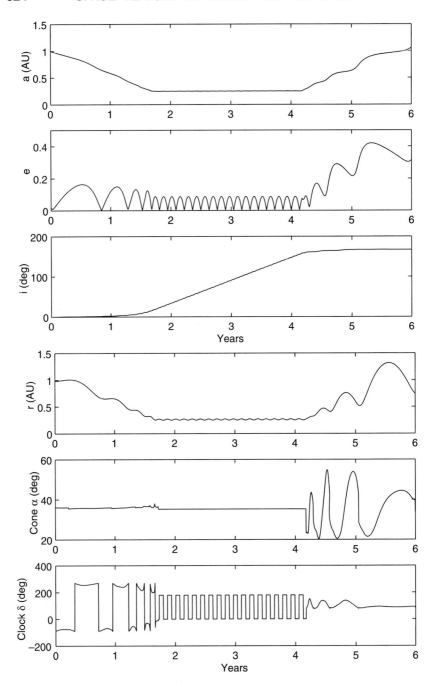

Fig. 12.50   Solar-sail KEI trajectory design using an optimal sail-steering law [58].

can be easily performed by employing simple sail-steering laws as applied to a set of orbital equations, called Gauss's form of the variational equations with (α, δ). In this section a quaternion-feedback attitude control scheme is proposed for solar-sail TVC applications because of its simple interface with the typical trajectory control inputs (the desired cone and clock angles of the solar thrust vector).

12.15.1 Quaternion-Feedback Attitude Control

Euler's attitude dynamical equations of motion of a rigid sailcraft are given by

$$I_1\dot{\omega}_1 - (I_2 - I_3)\omega_2\omega_3 = u_1 + d_1 \qquad (12.151a)$$

$$I_2\dot{\omega}_2 - (I_3 - I_1)\omega_3\omega_1 = u_2 + d_2 \qquad (12.151b)$$

$$I_3\dot{\omega}_3 - (I_1 - I_2)\omega_1\omega_2 = u_3 + d_3 \qquad (12.151c)$$

where $(\omega_1, \omega_2, \omega_3)$ are the angular velocity components, (I_1, I_2, I_3) the principal moments of inertia, (u_1, u_2, u_3) the attitude control torques, and (d_1, d_2, d_3) the disturbance torques.

The kinematic differential equations in terms of attitude quaternions are given by

$$\begin{bmatrix} \dot{q}_1 \\ \dot{q}_2 \\ \dot{q}_3 \\ \dot{q}_4 \end{bmatrix} = \frac{1}{2} \begin{bmatrix} 0 & \omega_3 & -\omega_2 & \omega_1 \\ -\omega_3 & 0 & \omega_1 & \omega_2 \\ \omega_2 & -\omega_1 & 0 & \omega_3 \\ -\omega_1 & -\omega_2 & -\omega_3 & 0 \end{bmatrix} \begin{bmatrix} q_1 \\ q_2 \\ q_3 \\ q_4 \end{bmatrix} \qquad (12.152)$$

where (q_1, q_2, q_3, q_4) are the inertial attitude quaternions.

The quaternion-feedback attitude control logic, proposed for solar-sailing applications, is simply a PID control logic of the form

$$u_1 = -k_1 \left(e_1 + \frac{1}{\tau} \int e_1 \, dt \right) - c_1\omega_1 \qquad (12.153a)$$

$$u_2 = -k_2 \left(e_2 + \frac{1}{\tau} \int e_2 \, dt \right) - c_2\omega_2 \qquad (12.153b)$$

$$u_3 = -k_3 \left(e_3 + \frac{1}{\tau} \int e_3 \, dt \right) - c_3\omega_3 \qquad (12.153c)$$

where (e_1, e_2, e_3) are the roll, pitch, and yaw components of attitude-error quaternions (e_1, e_2, e_3, e_4) and (k_i, τ, c_i) are control gains to be properly determined. The attitude-error quaternions are computed using the desired or commanded attitude quaternions $(q_{1c}, q_{2c}, q_{3c}, q_{4c})$ and the actual attitude quaternions (q_1, q_2, q_3, q_4), as follows:

$$\begin{bmatrix} e_1 \\ e_2 \\ e_3 \\ e_4 \end{bmatrix} = \begin{bmatrix} q_{4c} & q_{3c} & -q_{2c} & -q_{1c} \\ -q_{3c} & q_{4c} & q_{1c} & -q_{2c} \\ q_{2c} & -q_{1c} & q_{4c} & -q_{3c} \\ q_{1c} & q_{2c} & q_{3c} & q_{4c} \end{bmatrix} \begin{bmatrix} q_1 \\ q_2 \\ q_3 \\ q_4 \end{bmatrix} \qquad (12.154)$$

A saturation control logic accommodating the actuator torque and slew rate constraints is given by

$$u_i = -\operatorname*{sat}_{U_{max}} \left\{ k_i \operatorname{sat} \left[e_i + \frac{1}{L_i} \int e_i \right] + c_i \omega_i \right\}; \quad i = 1, 2, 3 \tag{12.155}$$

and the variable limiter L_i is self-adjusted as

$$L_i = \frac{c_i}{k_i} \min \left\{ \sqrt{2a_i |e_i|}, \omega_{max} \right\} \tag{12.156}$$

where ω_{max} is the maximum slew rate (if required) and a_i is the maximum angular acceleration.

12.15.2 Trajectory Control Inputs (α, β)

Consider the body-fixed rotational sequence of the form: $C_2(-\alpha) \leftarrow C_1(-\beta) \leftarrow C_2(-\phi) \leftarrow C_3(\psi)$. For this case of employing (α, β) as the trajectory control inputs or the TVC steering commands, we have

$$\begin{bmatrix} \hat{r} \\ \hat{\psi} \\ \hat{\phi} \end{bmatrix} = \begin{bmatrix} \cos\phi & 0 & \sin\phi \\ 0 & 1 & 0 \\ -\sin\phi & 0 & \cos\phi \end{bmatrix} \begin{bmatrix} \cos\psi & \sin\psi & 0 \\ -\sin\psi & \cos\psi & 0 \\ 0 & 0 & 1 \end{bmatrix} \begin{bmatrix} \hat{I} \\ \hat{J} \\ \hat{K} \end{bmatrix} \tag{12.157}$$

$$\begin{bmatrix} \hat{1} \\ \hat{2} \\ \hat{3} \end{bmatrix} = \begin{bmatrix} \cos\alpha & 0 & \sin\alpha \\ 0 & 1 & 0 \\ -\sin\alpha & 0 & \cos\alpha \end{bmatrix} \begin{bmatrix} 1 & 0 & 0 \\ 0 & \cos\beta & -\sin\beta \\ 0 & \sin\beta & \cos\beta \end{bmatrix} \begin{bmatrix} \hat{r} \\ \hat{\psi} \\ \hat{\phi} \end{bmatrix} \tag{12.158}$$

The equivalent coordinate transformation in terms of quaternions is expressed as

$$\begin{bmatrix} 0 \\ -\sin(\alpha/2) \\ 0 \\ \cos(\alpha/2) \end{bmatrix} \leftarrow \begin{bmatrix} -\sin(\beta/2) \\ 0 \\ 0 \\ \cos(\beta/2) \end{bmatrix} \leftarrow \begin{bmatrix} 0 \\ -\sin(\phi/2) \\ 0 \\ \cos(\phi/2) \end{bmatrix} \leftarrow \begin{bmatrix} 0 \\ 0 \\ \sin(\psi/2) \\ \cos(\psi/2) \end{bmatrix} \tag{12.159}$$

which results in

$$\begin{bmatrix} q_{1c} \\ q_{2c} \\ q_{3c} \\ q_{4c} \end{bmatrix} = \begin{bmatrix} \cos(\alpha/2) & 0 & \sin(\alpha/2) & 0 \\ 0 & \cos(\alpha/2) & 0 & -\sin(\alpha/2) \\ -\sin(\alpha/2) & 0 & \cos(\alpha/2) & 0 \\ 0 & \sin(\alpha/2) & 0 & \cos(\alpha/2) \end{bmatrix}$$

$$\times \begin{bmatrix} \cos(\beta/2) & 0 & 0 & -\sin(\beta/2) \\ 0 & \cos(\beta/2) & -\sin(\beta/2) & 0 \\ 0 & \sin(\beta/2) & \cos(\beta/2) & 0 \\ \sin(\beta/2) & 0 & 0 & \cos(\beta/2) \end{bmatrix}$$

$$\times \begin{bmatrix} \cos(\phi/2) & 0 & \sin(\phi/2) & 0 \\ 0 & \cos(\phi/2) & 0 & -\sin(\phi/2) \\ -\sin(\phi/2) & 0 & \cos(\phi/2) & 0 \\ 0 & \sin(\phi/2) & 0 & \cos(\phi/2) \end{bmatrix} \begin{bmatrix} 0 \\ 0 \\ \sin(\psi/2) \\ \cos(\psi/2) \end{bmatrix}$$

$$\tag{12.160}$$

where $(q_{1c}, q_{2c}, q_{3c}, q_{4c})$ are the desired attitude quaternions of a sailcraft whose orientation provides the desired thrust vector direction as commanded by (α, β). The preceding result indicates that a TVC system will require frequent updates of both the orbital information (ϕ, ψ) and TVC steering commands (α, β) caused by the inherent difficulty to precisely point the true thrust vector of a large solar-sail.

Given the actual attitude quaternions (q_1, q_2, q_3, q_4), the actual cone and clock angles (α, β) can be determined from the following relationship:

$$
\begin{bmatrix} 1 - 2(q_2^2 + q_3^2) & 2(q_1q_2 + q_3q_4) & 2(q_1q_3 - q_2q_4) \\ 2(q_2q_1 - q_3q_4) & 1 - 2(q_1^2 + q_3^2) & 2(q_2q_3 + q_1q_4) \\ 2(q_3q_1 + q_2q_4) & 2(q_3q_2 - q_1q_4) & 1 - 2(q_1^2 + q_2^2) \end{bmatrix} = \begin{bmatrix} C_{11} & C_{12} & C_{13} \\ C_{21} & C_{22} & C_{23} \\ C_{31} & C_{32} & C_{33} \end{bmatrix}
$$

$$
= \begin{bmatrix} \cos\alpha & 0 & \sin\alpha \\ 0 & 1 & 0 \\ -\sin\alpha & 0 & \cos\alpha \end{bmatrix} \begin{bmatrix} 1 & 0 & 0 \\ 0 & \cos\beta & -\sin\beta \\ 0 & \sin\beta & \cos\beta \end{bmatrix}
$$

$$
\times \begin{bmatrix} \cos\phi & 0 & \sin\phi \\ 0 & 1 & 0 \\ -\sin\phi & 0 & \cos\phi \end{bmatrix} \begin{bmatrix} \cos\psi & \sin\psi & 0 \\ -\sin\psi & \cos\psi & 0 \\ 0 & 0 & 1 \end{bmatrix}
$$

Given a direction cosine matrix, quaternions can also be determined as

$$
q_4 = \frac{1}{2}\sqrt{1 + C_{11} + C_{22} + C_{33}}
$$

$$
\begin{bmatrix} q_1 \\ q_2 \\ q_3 \end{bmatrix} = \frac{1}{4q_4} \begin{bmatrix} C_{23} - C_{32} \\ C_{31} - C_{13} \\ C_{12} - C_{21} \end{bmatrix}
$$

when $q_4 \neq 0$

However, this approach has a singularity problem when $q_4 = 0$.

12.15.3 Trajectory Control Inputs (α, δ)

For a case of employing (α, δ) as the trajectory control inputs or the TVC steering commands, consider a body-fixed rotational sequence of the form: $\mathbf{C}_2(-\alpha) \leftarrow \mathbf{C}_1(-\delta) \leftarrow \mathbf{C}_3(\theta) \leftarrow \mathbf{C}_3(\omega) \leftarrow \mathbf{C}_1(i) \leftarrow \mathbf{C}_3(\Omega)$. For this case, we have

$$
\begin{bmatrix} \hat{r} \\ \hat{\theta} \\ \hat{k} \end{bmatrix} = \begin{bmatrix} \cos(\omega + \theta) & \sin(\omega + \theta) & 0 \\ -\sin(\omega + \theta) & \cos(\omega + \theta) & 0 \\ 0 & 0 & 1 \end{bmatrix} \begin{bmatrix} 1 & 0 & 0 \\ 0 & \cos i & \sin i \\ 0 & -\sin i & \cos i \end{bmatrix}
$$

$$
\times \begin{bmatrix} \cos\Omega & \sin\Omega & 0 \\ -\sin\Omega & \cos\Omega & 0 \\ 0 & 0 & 1 \end{bmatrix} \begin{bmatrix} \hat{I} \\ \hat{J} \\ \hat{K} \end{bmatrix} \qquad (12.161)
$$

$$\begin{bmatrix} \hat{1} \\ \hat{2} \\ \hat{3} \end{bmatrix} = \begin{bmatrix} \cos\alpha & 0 & \sin\alpha \\ 0 & 1 & 0 \\ -\sin\alpha & 0 & \cos\alpha \end{bmatrix} \begin{bmatrix} 1 & 0 & 0 \\ 0 & \cos\delta & -\sin\delta \\ 0 & \sin\delta & \cos\delta \end{bmatrix} \begin{bmatrix} \hat{r} \\ \hat{\theta} \\ \hat{k} \end{bmatrix} \qquad (12.162)$$

$$\begin{bmatrix} 0 \\ -\sin(\alpha/2) \\ 0 \\ \cos(\alpha/2) \end{bmatrix} \leftarrow \begin{bmatrix} -\sin(\delta/2) \\ 0 \\ 0 \\ \cos(\delta/2) \end{bmatrix} \leftarrow \begin{bmatrix} 0 \\ 0 \\ \sin\left(\dfrac{\omega+\theta}{2}\right) \\ \cos\left(\dfrac{\omega+\theta}{2}\right) \end{bmatrix}$$

$$\leftarrow \begin{bmatrix} \sin(i/2) \\ 0 \\ 0 \\ \cos(i/2) \end{bmatrix} \leftarrow \begin{bmatrix} 0 \\ 0 \\ \sin(\Omega/2) \\ \cos(\Omega/2) \end{bmatrix} \qquad (12.163)$$

$$\begin{bmatrix} q_{1c} \\ q_{2c} \\ q_{3c} \\ q_{4c} \end{bmatrix} = \begin{bmatrix} \cos(\alpha/2) & 0 & \sin(\alpha/2) & 0 \\ 0 & \cos(\alpha/2) & 0 & -\sin(\alpha/2) \\ -\sin(\alpha/2) & 0 & \cos(\alpha/2) & 0 \\ 0 & \sin(\alpha/2) & 0 & \cos(\alpha/2) \end{bmatrix}$$

$$\times \begin{bmatrix} \cos(\delta/2) & 0 & 0 & -\sin(\delta/2) \\ 0 & \cos(\delta/2) & -\sin(\delta/2) & 0 \\ 0 & \sin(\delta/2) & \cos(\delta/2) & 0 \\ \sin(\delta/2) & 0 & 0 & \cos(\delta/2) \end{bmatrix}$$

$$\times \begin{bmatrix} \cos\left(\dfrac{\omega+\theta}{2}\right) & \sin\left(\dfrac{\omega+\theta}{2}\right) & 0 & 0 \\ -\sin\left(\dfrac{\omega+\theta}{2}\right) & \cos\left(\dfrac{\omega+\theta}{2}\right) & 0 & 0 \\ 0 & 0 & \cos\left(\dfrac{\omega+\theta}{2}\right) & \sin\left(\dfrac{\omega+\theta}{2}\right) \\ 0 & 0 & -\sin\left(\dfrac{\omega+\theta}{2}\right) & \cos\left(\dfrac{\omega+\theta}{2}\right) \end{bmatrix}$$

$$\times \begin{bmatrix} \cos(i/2) & 0 & 0 & \sin(i/2) \\ 0 & \cos(i/2) & \sin(i/2) & 0 \\ 0 & -\sin(i/2) & \cos(i/2) & 0 \\ -\sin(i/2) & 0 & 0 & \cos(i/2) \end{bmatrix}$$

$$\times \begin{bmatrix} 0 \\ 0 \\ \sin(\Omega/2) \\ \cos(\Omega/2) \end{bmatrix} \qquad (12.164)$$

The preceding result indicates that a TVC system employing (α, δ) will also require frequent updates of both the orbital parameters $(\Omega, i, \omega, \theta)$ and TVC steering commands (α, δ) because of the inherent inability to point the true thrust vector direction of a large solar-sail.

Equation (12.164) provides a simple computational algorithm for determining the desired attitude quaternions from the commanded cone/clock angles (α, δ).

Given the actual attitude quaternions (q_1, q_2, q_3, q_4), the actual cone and clock angles (α, δ) can be determined from the following relationship:

$$
\begin{bmatrix}
1 - 2(q_2^2 + q_3^2) & 2(q_1q_2 + q_3q_4) & 2(q_1q_3 - q_2q_4) \\
2(q_2q_1 - q_3q_4) & 1 - 2(q_1^2 + q_3^2) & 2(q_2q_3 + q_1q_4) \\
2(q_3q_1 + q_2q_4) & 2(q_3q_2 - q_1q_4) & 1 - 2(q_1^2 + q_2^2)
\end{bmatrix}
$$

$$
= \begin{bmatrix} \cos\alpha & 0 & \sin\alpha \\ 0 & 1 & 0 \\ -\sin\alpha & 0 & \cos\alpha \end{bmatrix}
\begin{bmatrix} 1 & 0 & 0 \\ 0 & \cos\delta & -\sin\delta \\ 0 & \sin\delta & \cos\delta \end{bmatrix}
$$

$$
\times \begin{bmatrix} \cos(\omega+\theta) & \sin(\omega+\theta) & 0 \\ -\sin(\omega+\theta) & \cos(\omega+\theta) & 0 \\ 0 & 0 & 1 \end{bmatrix}
\begin{bmatrix} 1 & 0 & 0 \\ 0 & \cos i & \sin i \\ 0 & -\sin i & \cos i \end{bmatrix}
$$

$$
\times \begin{bmatrix} \cos\Omega & \sin\Omega & 0 \\ -\sin\Omega & \cos\Omega & 0 \\ 0 & 0 & 1 \end{bmatrix}
\tag{12.165}
$$

12.15.4 Relationship of β and δ

The two different clock angles β and δ are related as

$$
\begin{bmatrix} 1 & 0 & 0 \\ 0 & \cos\beta & -\sin\beta \\ 0 & \sin\beta & \cos\beta \end{bmatrix}
\begin{bmatrix} \cos\phi & 0 & \sin\phi \\ 0 & 1 & 0 \\ -\sin\phi & 0 & \cos\phi \end{bmatrix}
\begin{bmatrix} \cos\psi & \sin\psi & 0 \\ -\sin\psi & \cos\psi & 0 \\ 0 & 0 & 1 \end{bmatrix}
$$

$$
= \begin{bmatrix} 1 & 0 & 0 \\ 0 & \cos\delta & -\sin\delta \\ 0 & \sin\delta & \cos\delta \end{bmatrix}
\begin{bmatrix} \cos(\omega+\theta) & \sin(\omega+\theta) & 0 \\ -\sin(\omega+\theta) & \cos(\omega+\theta) & 0 \\ 0 & 0 & 1 \end{bmatrix}
$$

$$
\times \begin{bmatrix} 1 & 0 & 0 \\ 0 & \cos i & \sin i \\ 0 & -\sin i & \cos i \end{bmatrix}
\begin{bmatrix} \cos\Omega & \sin\Omega & 0 \\ -\sin\Omega & \cos\Omega & 0 \\ 0 & 0 & 1 \end{bmatrix}
$$

12.16 TVC Design and Simulation for the SPI Mission

In this section the preliminary TVC design and simulation results for the SPI mission are presented.

12.16.1 Simplified Control Design Model

Although various forms of orbital trajectory equations employing two different sets of the cone and clock angles are available for the solar-sail trajectory design

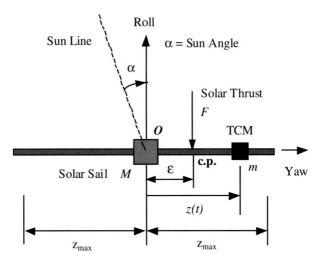

Fig. 12.51 Dynamic model of sailcraft with a trim control mass moving along the yaw axis to generate a pitch control torque.

and simulation, we choose here a set of orbital equations, called Gauss's form of the variational equations with (α, δ), for the preliminary TVC design and simulation.

A dynamic model of sailcraft with a trim control mass (TCM) translating along the yaw axis to generate a pitch control torque is illustrated in Fig. 12.51. A set of simplified attitude equations of motion is considered for control design, as follows:

$$J_1 \dot{\omega}_1 = u_1 + d_1 \tag{12.166}$$

$$J_2 \dot{\omega}_2 = u_2 + d_2 \tag{12.167}$$

$$J_3 \dot{\omega}_3 = u_3 + d_3 \tag{12.168}$$

$$u_1 = C \sin \Theta$$

$$u_2 = -\frac{m_r F}{M + m} z$$

$$u_3 = \frac{m_r F}{M + m} y$$

$$d_1 = \pm 0.5 \epsilon F, \quad d_2 = d_3 = \pm \epsilon F$$

$$J_1 = I_1 + m_r (y^2 + z^2)$$

$$J_2 = I_2 + m_r z^2$$

$$J_3 = I_3 + m_r y^2$$

where Θ is the RSB tilt angle, C is the RSB control scale factor, y and z are the TCM positions along the pitch and yaw axes, respectively, F is the solar radiation pressure force defined as

$$F \approx \left(\frac{r_\oplus}{r}\right)^2 F_s \cos^2 \alpha = 2\eta PA \left(\frac{r_\oplus}{r}\right)^2 \cos^2 \alpha \qquad (12.169)$$

and m_r is the so-called reduced mass defined as

$$m_r = \frac{m(M+m)}{M+2m} \qquad (12.170)$$

Note that $m_r \approx m$ because $M \gg m$.

A reference 160-m SPI sailcraft model for TVC/AOCS design is assumed as follows:

Sail size = 160 m × 160 m
Scallop factor = 75%
Sail area $A = 19,200$ m^2 with $\eta = 0.84$
Solar thrust force $F_s = 0.16$ N (at 1 AU), 0.69 N (at 0.48 AU)
Sailcraft total mass = 450 kg (150-kg sail, 250-kg bus, and 50-kg payload)
Characteristic acceleration $a_c = 0.35$ mm/s^2 (at 1 AU)
Moments of inertia $(I_1, I_2, I_3) = (642,876; 321,490; 321,490)$ kg·m^2
Cm-cp offset $\epsilon = 0.4$ m (0.25% of 160 m)
Disturbance torque = 0.064 N·m (1 AU), 0.256 N·m (at 0.48 AU)
Trim control mass $m = 5$ kg (each axis)
Main-body mass $M = 440$ kg
TCM speed limit = 5 cm/s
TCM $y_{max} = z_{max} = \pm100$ m
RSB maximum deflection angle $\Theta_{max} = \pm45$ deg
RSB moment arm length = 1.7 m
$(a, e, i, \Omega, \omega, \theta) = (0.48$ AU, 1E-5, 45 deg, 0, 0, 0) at $t = 0$

12.16.2 Roll-Axis Control Design

The roll-axis control logic is assumed as

$$u_1 = -\sup_{U_{max}} \text{sat} \left\{ K_1 \text{ sat} \left[e_1 + \frac{1}{\tau_i} \int e_1 \, dt \right] + D_1 \omega_1 \right\} \qquad (12.171)$$

where u_1 is the roll control torque command with a saturation limit of $U_{max} = C \sin \Theta_{max}$ and the variable limiter L_1 is self-adjusted as

$$L_1 = \frac{C_1}{D_1} \min \left\{ \sqrt{2a_1|e_1|}, \omega_{max} \right\} \qquad (12.172)$$

where $a_1 = U_{max}/J_1$ and ω_{max} is assumed to be constrained as 0.1 deg/s.

The roll control torque u_1 (in units of N·m) is related to the tilt angle Θ of the RSB of an assumed moment arm length of 1.7 m for a 160-m sailcraft, as follows:

$$u_1 = C \sin \Theta \approx \frac{(1.7)(53.2)}{80} F \sin \Theta \qquad (12.173)$$

It is assumed that all four RSBs are rotated simultaneously. Consequently, we have U_{max} of ±0.15 N·m (at 1 AU) for a maximum tilt angle of ±45 deg.

The RSB actuator dynamics is also assumed as

$$T\dot{\Theta} + \Theta = \Theta_c; \qquad |\Theta_c(t)| \le \Theta_{max} \tag{12.174}$$

where T is the actuator time constant, Θ is the actual tilt angle, and Θ_c is the commanded tilt angle (the control input) with a maximum value of Θ_{max}. The commanded tilt angle of the RSB is then given by

$$\Theta_c = \arcsin\left\{ \frac{80\, u_1}{(1.7)(53.2)F} \right\} \approx \frac{80\, u_1}{(1.7)(53.2)F} \tag{12.175}$$

where

$$F \approx 0.16 \left(\frac{r_\oplus}{r}\right)^2 \cos^2\alpha \tag{12.176}$$

and u_1 is given by Eq. (12.171).

12.16.3 Pitch/Yaw Control Design

The steady-state trim position of the pitch TCM for countering the effect of the c.m./c.p. offset ϵ can be estimated as

$$z_{ss} = \frac{M+m}{m_r}\epsilon \approx \frac{M+m}{m}\epsilon = \pm35 \text{ m} \tag{12.177}$$

The actuator dynamics of the pitch TCM is assumed as

$$T\dot{z} + z = z_c; \qquad |z_c(t)| \le z_{max} \tag{12.178}$$

where T is the actuator time constant, z is the actual position, and z_c is the commanded position (the control input) with a maximum value of z_{max}. For a reference control design, it is assumed that $z_{max} = \pm100$ m, $\dot{z}_{max} = \pm0.05$ m/s, and $T = 200$ sec for a 160-m sailcraft with $M = 440$ kg, $m = 5$ kg, and $\epsilon = \pm0.4$ m.

The pitch control logic is assumed as

$$z_c = \operatorname*{sat}_{z_{max}}\left\{ K_2 \operatorname*{sat}_{L_2}\left[e_2 + \frac{1}{\tau_2}\int e_2\, dt \right] + D_2\omega_2 \right\} \tag{12.179}$$

where τ is the time constant of integral control and variable limiter L_2 is self-adjusted as

$$L_2 = \frac{D_2}{K_2}\min\left\{ \sqrt{2a_2|e_2|},\ \omega_{max} \right\} \tag{12.180}$$

Similarly, we have the yaw control logic as

$$y_c = -\operatorname*{sat}_{y_{max}}\left\{ K_3 \operatorname*{sat}_{L_3}\left[e_3 + \frac{1}{\tau_3}\int e_3\, dt \right] + D_3\omega_3 \right\} \tag{12.181}$$

where $y_{max} = \pm100$ m.

12.16.4   *Simulation Example: Cranking Orbit Phase at 0.48 AU*

An attitude maneuver for achieving a desired 180-deg δ clock-angle change within 3 h (with a fixed cone-angle command of 35 deg) is illustrated in Figs. 12.52–12.54. An equivalent, direct 70-deg pitch maneuver within 3 hr was also validated but not included here.

12.17   Attitude Control of a Solar Sail in Sun-Synchronous Orbit Using Reaction Wheels and Magnetic Torquers

As discussed in Sec. 12.9, a dawn-dusk sun-synchronous orbit was proposed in [28] and [29] for the NMP ST9 solar-sail flight validation mission. In Sec. 12.9.3, it was shown that a small reaction wheel might not be suitable for the attitude control of a 40-m solar sail of such a flight validation mission because a typical 2-kg wheel with a peak torque of 0.01 N·m has less than 2 N·m·s momentum storage capability. However, magnetic torquers can be employed for momentum unloading of reaction wheels without affecting the orbital motion, as investigated in [60]. Such a conventional combination of reaction wheels (RWs) and magnetic torquers (MTs) of small satellites is examined in this section, whereas such a conventional RW/MT-based ACS will not be suitable for interplanetary solar-sailing missions.

For a large-angle yaw reorientation maneuver (but with small roll/pitch attitude errors) considered in Sec. 12.10, the attitude equations of motion of a solar-sail spacecraft employing RWs and MTs in a DDSS orbit can be written as

$$J_x\ddot{\phi} + n^2(J_y - J_z)(3 + \cos^2\psi)\phi - n^2(J_y - J_z)(\cos\psi\,\sin\psi)\theta$$
$$- n(J_x - J_y + J_z)(\cos\psi)\dot{\psi} = T_x + N_x + D_x \qquad (12.182)$$
$$J_y\ddot{\theta} + n^2(J_x - J_z)(3 + \sin^2\psi)\theta - n^2(J_x - J_z)(\cos\psi\,\sin\psi)\phi$$
$$- n(J_x - J_y - J_z)(\sin\psi)\dot{\psi} = T_y + N_y + D_y \qquad (12.183)$$
$$J_z\ddot{\psi} + n^2(J_y - J_x)\sin\psi\,\cos\psi + n(J_x - J_y + J_z)(\cos\psi)\dot{\phi}$$
$$+ n(J_x - J_y - J_z)(\sin\psi)\dot{\theta} = T_z + N_z + D_z \qquad (12.184)$$

where (ϕ, θ, ψ) are the roll, pitch, and yaw attitude angles with respect to the LVLH orbital reference frame, (T_x, T_y, T_z) are the reaction-wheel torque components, (N_x, N_y, N_z) are the magnetic control torque components, and (D_x, D_y, D_z) are the solar disturbance torque components. It is assumed that $D_x = 0.5\epsilon F$ and $D_y = D_z = \epsilon F$, where $F = F_s\cos^2\alpha$ and ϵ is the c.m.-c.p. offset uncertainty. It is also assumed that the sun angle $\beta \approx 0$ and $\alpha = \pi/2 + \psi$. Note that the zero-thrust mode has a zero yaw angle and that the standby full-thrust mode has a yaw angle of -90 deg, as described in Sec. 12.9.

The three-axis reaction-wheel dynamics are simply modeled as

$$\dot{H}_x = -T_x; \quad \dot{H}_y = -T_y; \quad \dot{H}_z = -T_z \qquad (12.185)$$

where (H_x, H_y, H_z) are the reaction-wheel momentum components in the body axes.

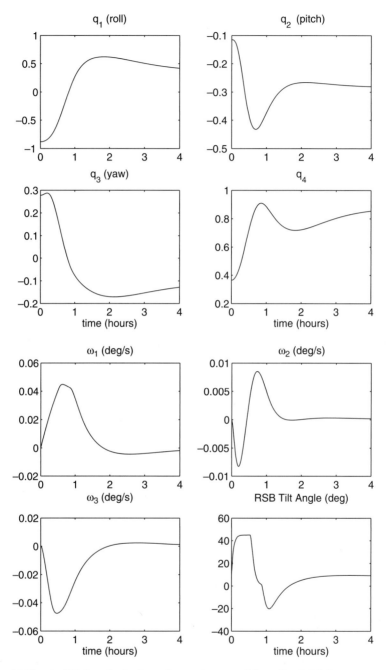

Fig. 12.52   A 180-deg δ clock-angle maneuver with a fixed 35-deg cone-angle command.

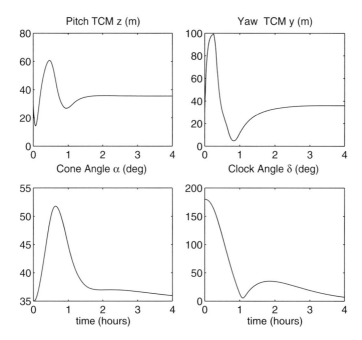

Fig. 12.53   A 180-deg δ clock-angle maneuver with a fixed 35-deg cone-angle command.

Attitude control logic employing RWs is chosen as

$$T_x = -\operatorname*{sat}_{T_{\max}} \left\{ K_x \operatorname*{sat}_L \phi + C_x \dot{\phi} \right\} \tag{12.186a}$$

$$T_y = -\operatorname*{sat}_{T_{\max}} \left\{ K_y \operatorname*{sat}_L \theta + C_y \dot{\theta} \right\} \tag{12.186b}$$

$$T_z = -\operatorname*{sat}_{T_{\max}} \left\{ K_z \operatorname*{sat}_L (\psi - \psi_c) + C_z \dot{\psi} \right\} \tag{12.186c}$$

where $T_{\max}$ is the peak torque available from a RW and it is assumed as $T_{\max} = 0.02$ N·m for a 3-kg RW with a peak momentum storage capability of $H_{\max} = 5$ N·m·s.

Magnetic torquers, also known as magnetorquers or torque bars, have been employed extensively in the attitude control of spacecraft as well as for the momentum unloading of reaction wheels [61–67]. When the magnetic torquer is electrically energized, it generates a magnetic dipole moment. Interaction between a magnetic dipole moment, $\vec{M}$ in units of A·m^2, with the Earth's magnetic field vector, $\vec{B}$ in units of Wb/m^2, produces a control torque vector, $\vec{N}$ in units of N·m,

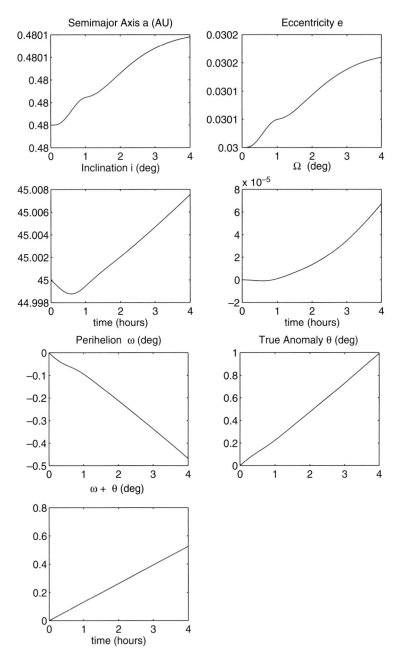

Fig. 12.54 A 180-deg δ clock-angle maneuver with a fixed 35-deg cone-angle command.

acting on the spacecraft as

$$\vec{N} = \vec{M} \times \vec{B} \tag{12.187}$$

which can be rewritten in component form as

$$N_x = M_y B_z - M_z B_y \tag{12.188a}$$
$$N_y = M_z B_x - M_x B_z \tag{12.188b}$$
$$N_z = M_x B_y - M_y B_x \tag{12.188c}$$

The Earth's magnetic field vector components (B_x, B_y, B_z) measured along the spacecraft body axes (x, y, z) are related to the components of $\vec{B}$ represented along the LVLH orbital reference frame as follows:

$$
\begin{bmatrix} B_x \\ B_y \\ B_z \end{bmatrix} =
\begin{bmatrix} 1 & 0 & 0 \\ 0 & \cos\phi & \sin\phi \\ 0 & -\sin\phi & \cos\phi \end{bmatrix}
\begin{bmatrix} \cos\theta & 0 & -\sin\theta \\ 0 & 1 & 0 \\ \sin\theta & 0 & \cos\theta \end{bmatrix}
$$
$$
\times
\begin{bmatrix} \cos\psi & \sin\psi & 0 \\ -\sin\psi & \cos\psi & 0 \\ 0 & 0 & 1 \end{bmatrix}
\begin{bmatrix} B_1 \\ B_2 \\ B_3 \end{bmatrix}
\tag{12.189}
$$

where (B_1, B_2, B_3) are the components of $\vec{B}$ represented along the LVLH reference frame. For small ϕ and θ, we have

$$B_x = \cos\psi\, B_1 + \sin\psi\, B_2 \tag{12.190a}$$
$$B_y = -\sin\psi\, B_1 + \cos\psi\, B_2 \tag{12.190b}$$
$$B_z = B_3 \tag{12.190c}$$

A simplified model of the Earth's magnetic field is expressed as [64–66]

$$B_1 = \frac{\mu_m}{r^3} \sin i \cos nt \tag{12.191a}$$

$$B_2 = -\frac{\mu_m}{r^3} \cos i \tag{12.191b}$$

$$B_3 = 2\frac{\mu_m}{r^3} \sin i \sin nt \tag{12.191c}$$

where $\mu_m = 7.96 \times 10^{15}$ (in units of Wb·m) is the Earth's magnetic dipole strength, r is the geocentric orbital position, and i is the orbital inclination angle with respect to the magnetic equator. It is assumed that $r = R_\oplus + h = 7{,}978 \times 10^3$ m, $i = 102$ deg, and $n = 8.8589 \times 10^{-4}$ rad/s for the proposed 1600-km DDSS orbit.

A simple, heuristic MT control logic for momentum unloading is given by ([19], pp. 191,192)

$$N_{xc} = -k\Delta H_x = -k(H_x - H_{x0}) \tag{12.192a}$$
$$N_{yc} = -k\Delta H_y = -k(H_y - H_{y0}) \tag{12.192b}$$
$$N_{zc} = -k\Delta H_z = -k(H_z - H_{z0}) \tag{12.192c}$$

where (N_{xc}, N_{yc}, N_{zc}) are the magnetic control torque command, (H_x, H_y, H_z) are the reaction-wheel momentum components in the body axes, and $(k, H_{x0}, H_{y0}, H_{z0})$ are the MT control parameters to be properly determined mainly by a typical trial-and-error approach using simulations.

Utilizing a cross-product MT control logic of the form ([19], pp. 191, 192)

$$\vec{M} = \frac{\vec{B} \times \vec{N}_c}{B^2} \tag{12.193}$$

where $\vec{N}_c$ is the magnetic control torque command vector and $B = \sqrt{B_x^2 + B_y^2 + B_z^2}$ in units of Wb/m^2, we obtain the following MT control logic:

$$M_x = \underset{M_{max}}{\text{sat}} \left\{ \frac{(B_y N_{zc} - B_z N_{yc})}{B^2} \right\} \tag{12.194a}$$

$$M_y = \underset{M_{max}}{\text{sat}} \left\{ \frac{(B_z N_{xc} - B_x N_{zc})}{B^2} \right\} \tag{12.194b}$$

$$M_z = \underset{M_{max}}{\text{sat}} \left\{ \frac{(B_x N_{yc} - B_y N_{xc})}{B^2} \right\} \tag{12.194c}$$

where M_{max} is the peak dipole moment chosen as $M_{max} = 100$ A·m^2 for a medium-sized, 4-kg magnetic torquer. Note that the Earth's magnetic field vector components (B_x, B_y, B_z) are directly measured by three-axis magnetometers.

For an ideal case without the MT dipole moment saturation limit, the actual magnetic control torque acting on the spacecraft then becomes

$$\vec{N} = \vec{M} \times \vec{B} = \frac{(\vec{B} \times \vec{N}_c)}{B^2} \times \vec{B} = \vec{N}_c - \frac{(\vec{N}_c \cdot \vec{B})}{B^2} \vec{B} \tag{12.195}$$

which indicates that $\vec{N} \neq \vec{N}_c$ unless $\vec{N}_c \cdot \vec{B} = 0$. Thus, the overall effectiveness of this simple MT control system relies on the orbit-averaged removal of the wheel momentum utilizing the cyclic nature of the Earth's magnetic field $\vec{B}$.

Simulation results are shown in Figs. 12.55–12.58 for a -90-deg yaw maneuver of an untrimmed 40-m solar sail (with $\epsilon = 0.1$ m and $F_s = 0.01$ N·m) for a transition from the zero-thrust mode to the standby full-thrust mode. The roll and pitch attitude errors are maintained within ± 1 deg. The MT control parameters were assumed as $k = 0.01$, $H_{x0} = H_{y0} = 0$, and $H_{z0} = -5$ N·m·s. The angular momentum of reaction wheels is properly managed by the simple MT control system in the presence of the three-axis, secular solar-pressure disturbance torque during the standby full-thrust mode operation ($\psi = -90$ deg).

It is again emphasized that a conventional combination of small reaction wheels and medium-sized magnetic torquers of small satellites can be employed for the proposed solar-sail flight validation mission in a DDSS orbit, whereas such a conventional RW/MT-based ACS will not be suitable for interplanetary solar-sailing missions.

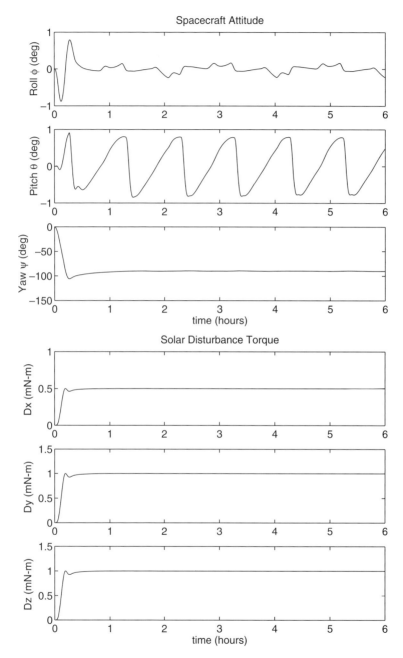

Fig. 12.55   A −90-deg yaw maneuver of an untrimmed 40-m solar sail (with $\epsilon =$ 0.1 m) using RWs and MTs for a transition from the zero-thrust mode to the standby full-thrust mode.

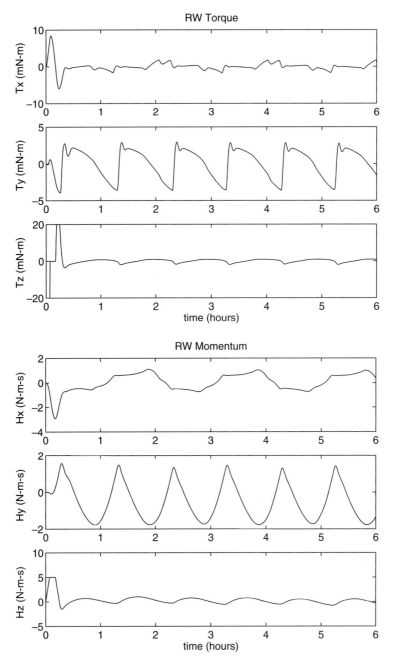

Fig. 12.56   A −90-deg yaw maneuver of an untrimmed 40-m solar sail (with $\epsilon =$ 0.1 m) using RWs and MTs for a transition from the zero-thrust mode to the standby full-thrust mode.

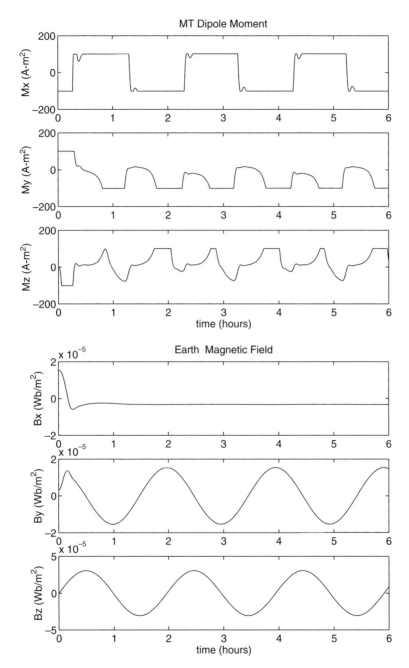

Fig. 12.57 A −90-deg yaw maneuver of an untrimmed 40-m solar sail (with $\epsilon =$ 0.1 m) using RWs and MTs for a transition from the zero-thrust mode to the standby full-thrust mode.

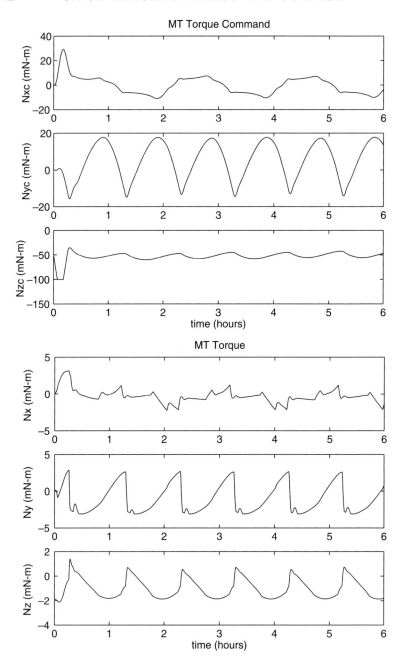

Fig. 12.58 A −90-deg yaw maneuver of an untrimmed 40-m solar sail (with $\epsilon =$ 0.1 m) using RWs and MTs for a transition from the zero-thrust mode to the standby full-thrust mode.

References

[1] Wright, J. L., *Space Sailing*, Gordon and Breach, New York, 1992.

[2] Friedman, L., *Star Sailing: Solar Sails and Interstellar Travel*, Wiley, New York, 1988.

[3] McInnes, C. R., *Solar Sailing Technology, Dynamics and Mission Applications*, Springer-Verlag, Chichester, England, U.K., 1999.

[4] West, J. L., and Derbes, B., "Solar-Sail Vehicle System Design for the Geostorm Warning Mission," AIAA Paper 2000-5326, Sept. 2000.

[5] Cohen, D., Gloyer, P., and Rogan, J., "Preliminary Design of a High Performance Solar Sailing Mission," 16th Annual AIAA/USU Conference on Small Satellites, SSC02-II-5, Aug. 2002.

[6] Garbe, G., and Montgomery, E., "An Overview of NASA's Solar-Sail Propulsion Project," AIAA Paper 2003-4662, July 2003.

[7] Murphy, D. M., McEachen, M. E., Macy, B. D., and Gaspar, J. L., "Demonstration of a 20-m Solar-Sail System," AIAA Paper 2005-2126, April 2005.

[8] Lichodziejewski, D., Derbès, B., Slade, K., and Mann, T., "Vacuum Deployment and Testing of a 4-Quadrant Scalable Inflatable Rigidizable Solar-Sail System," AIAA Paper 2005-3927, July 2005.

[9] Garwin, R. L., "Solar Sailing–A Practical Method of Propulsion Within the Solar System," *Jet Propulsion*, Vol. 28, No. 3, March 1958, pp. 188–190.

[10] Sohn, R. L., "Attitude Stabilization by Means of Solar Radiation Pressure," *ARS Journal*, Vol. 29, May 1959, pp. 371–373.

[11] Acord, J. D., and Nicklas, J. C., "Theoretical and Practical Aspects of Solar Pressure Attitude Control for Interplanetary Spacecraft," *Guidance and Control II*, Progress in Astronautics and Aeronautics, Vol. 13, Academic Press, New York, 1964, pp. 73–101.

[12] Modi, V. J., and Kumar, K. "Attitude Control of Satellites Using the Solar Radiation Pressure," *Journal of Spacecraft and Rockets*, Vol. 9, No. 9, 1972, pp. 711–713.

[13] Joshi, V. K., and Kumar, K., "New Solar Attitude Control Approach for Satellites in Elliptic Orbits," *Journal of Guidance and Control*, Vol. 3, No. 1, 1980, pp. 42–47.

[14] Stuck, B. W., "Solar Pressure Three-Axis Attitude Control," *Journal of Guidance and Control*, Vol. 3, No. 2, 1980, pp. 132–139.

[15] Angrilli, F., and Bortolami, S., "Attitude and Orbital Modelling of Solar-Sail Spacecraft," *European Space Agency Journal*, Vol. 14, No. 4, 1990, pp. 431–446.

[16] Renner, U., "Attitude Control by Solar Sailing: A Promising Experiment with OTS-2," *European Space Agency Journal*, Vol. 3, No. 1, 1979, pp. 35–40.

[17] Lievre, J., "Solar Sailing Attitude Control of Large Geostationary Satellite," *IFAC Automatic Control in Space*, Pergamon, Oxford, England, U.K., 1985, pp. 29–33.

[18] Azor, R., "Solar Attitude Control Including Active Nutation Damping in a Fixed-Momentum Wheel Satellite," *Proceedings of AIAA Guidance, Navigation, and Control Conference*, AIAA, Washington, D.C., 1992, pp. 226–235.

[19] Sidi, M. J., *Spacecraft Dynamics and Control: A Practical Engineering Approach*, Cambridge Univ. Press, Cambridge, England, U.K., 1997, pp. 191–192.

[20] Burns, R., Gabor, M., McLaughlin, C., Luu, K., and Sabol, C., "Solar Radiation Pressure Effects on Formation Flying of Satellites with Different Area-to-Mass Ratios," AIAA Paper 2000-4132, Aug. 2000.

[21] Van der Ha, J. C., and Lappas, V., "Long-Term Attitude Drift of Spinning Spacecraft Under Solar Radiation Torques," *Journal of Guidance, Control, and Dynamics*, Vol. 30, No. 5, 2007, pp. 1470–1479.

[22] Wie, B., and Roithmayr, C., "Attitude and Orbit Control of a Very Large Geostationary Solar Power Satellite," *Journal of Guidance, Control, and Dynamics*, Vol. 28, No. 3, 2005, pp. 439–451.

[23] Kaplan, M., *Modern Spacecraft Dynamics and Control*, Wiley, New York, 1976.

[24] Bryson, A. E., Jr., *Control of Spacecraft and Aircraft*, Princeton Univ. Press, Princeton, NJ, 1994.

[25] Wie, B., *Space Vehicle Dynamics and Control*, AIAA Education Series, AIAA, Washington, DC, 1998.

[26] Price, H., et al.,"Design for a Solar-Sail Demonstration Mission," Space Technology and Applications International Forum (STAIF 2000), Feb. 2001.

[27] Wie, B., "Solar-Sail Attitude Control and Dynamics: Parts 1 and 2," *Journal of Guidance, Control, and Dynamics*, Vol. 27, No. 4, 2004, pp. 526–544.

[28] Murphy, D., and Wie, B.,"Robust Thrust Control Authority for a Scalable Sailcraft," American Astronomical Society, AAS 04-285, Feb. 2004.

[29] Wie, B., and Murphy, D., "Solar-Sail Attitude Control System Design for a Flight Validation Mission in Sun-Synchronous Orbit," *Journal of Spacecraft and Rockets*, Vol. 44, No. 4, 2007, pp. 809–821.

[30] Murphy, D. M., Murphey, T. W., and Gierow, P. A., "Scalable Solar-Sail Subsystem Design Concept," *Journal of Spacecraft and Rockets*, Vol. 40, No. 4, 2003, pp. 539–547.

[31] Brady, T., Tillier, C., Brown, R., Jimenez, A., and Kourepenis, A., "The Inertial Stellar Compass: A New Direction in Spacecraft Attitude Determination," 16th Annual AIAA/USU Conference on Small Satellites, SSC02-II-1, Aug. 2002.

[32] Wie, B., Murphy, D., Thomas, S., and Paluszek, M., "Robust Attitude Control Systems Design for Solar-Sail Spacecraft (Part One): Propellantless Primary ACS," AIAA Paper 2004-5010, Aug. 2004.

[33] Thomas, S., Paluszek, M., Wie, B., and Murphy, D., "Design and Simulation of Sailcraft Attitude Control Systems Using Solar-Sail Control Toolbox," AIAA Paper 2004-4890, Aug. 2004.

[34] Wie, B., Thomas, S., Paluszek, M., and Murphy, D., "Propellantless AOCS Design for a 160-m, 450-kg Solar-Sail Spacecraft of the Solar Polar Imager Mission," AIAA Paper 2005-3928, July 2005.

[35] Wie, B., "Thrust Vector Control Analysis and Design for Solar-Sail Spacecraft," *Journal of Spacecraft and Rockets*, Vol. 44, No. 3, 2007, pp. 545–547.

[36] Pryor, K., Wie, B., and Mikellides, P., "Development of a Lightweight Pulsed Plasma Thruster Module for Solar-Sail Attitude Control," 18th Annual AIAA/USU Conference on Small Satellites, Paper SSC04-XI-4, Aug. 2004.

[37] Wie, B., Murphy, D., Thomas, S., and Paluszek, M., "Robust Attitude Control Systems Design for Solar-Sail Spacecraft (Part Two): microPPT-based Backup ACS," AIAA Paper 2004-5011, Aug. 2004.

[38] Wie, B., and Murphy, D., "MicroPPT-Based Secondary/Backup ACS for a 160-m, 450-kg Solar-Sail Spacecraft," AIAA Paper 2005-3724, July 2005.

[39] Micci, M., and Ketsdever, A. (ed.), *Micropropulsion for Small Spacecraft*, Progress in Astronautics and Aeronautics, Vol. 187, AIAA, Washington, DC, 2000, Chap. 3.

[40] Benson, S., Arrington, L., Hoskins, A., and Meckel, N., "Development of a PPT for the EO-1 Spacecraft," AIAA Paper 99-2276, June 1999.

[41] Cassady, R., Hoskins, A., Campbell, M., and Rayburn, C., "A Micro Pulsed Plasma Thruster (PPT) for the Dawgstar Spacecraft," *Proceedings of the 2000 IEEE Aerospace Conference*, Vol. 4, IEEE Press, Piscataway, NJ, 2000, pp. 7–14.

[42] Rayburn, C., Campbell, M., and Mattick, A., "Pulsed Plasma Thruster System for Microsatellites," *Journal of Spacecraft and Rockets*, Vol. 42, No. 1, 2005, pp. 161–170.

[43] Spanjers, G., Bromaghim, D., Lake, J., Dulligan, M., White, D., Schilling J., Bushman, S., Antonsen, E., Burton, R., Keidar, M., and Boyd, I., "AFRL MicroPPT Development for Small Spacecraft Propulsion," AIAA Paper 2002-3974, July 2002.

[44] Murphy, N., "Solar Polar Imager Vision Mission Overview," http://lws.gsfc.nasa.gov/solar_sails_conf/NMurphy.pdf [retrieved 22 Jan. 2005].

[45] Sauer, C., "A Comparison of Solar-Sail and Ion Drive Trajectories for a Halley's Comet Rendezvous Mission," American Astronautical Society, AAS 77-104, Sept. 1977.

[46] Hur, S.-H., and Bryson, A. E., "Minimum Time Solar Sailing from Geosynchronous Orbit to the Sun-Earth L_2 Point," *Proceedings of 1992 AIAA/AAS Astrodynamics Conference*, AIAA, Washington, DC, 1992, pp. 538–543.

[47] Sauer, C., "Solar-Sail Trajectories for Solar Polar and Interstellar Probe Mission," *Advances in the Astronautical Sciences*, Vol. 103, Univelt, Inc., San Diego, CA, 2000, pp. 547–562.

[48] Coverstone-Carroll, V. L., and Prussing, J. E., "Technique for Escape from Geosynchronous Transfer Orbit Using a Solar-Sail," *Journal of Guidance, Control, and Dynamics*, Vol. 26, No. 4, 2003, pp. 628–634.

[49] Dachwald, B., "Optimal Solar-Sail Trajectories for Missions to the Outer Solar System," *Journal of Guidance, Control, and Dynamics*, Vol. 28, No. 6, 2005, pp. 1187–1193.

[50] Leipold, M., et al., "Heliopause Explorer: A Sailcraft Mission to the Outer Boundaries of the Solar System," *Acta Astronautica*, Vol. 59, No. 8–11, Oct.–Dec. 2006, pp. 785–796.

[51] Mengali, G., Quarta, A. A., Circi, C., and Dachwlad, B., "Refined Solar-Sail Force Model with Mission Application," *Journal of Guidance, Control, and Dynamics*, Vol. 30, No. 2, 2007, pp. 512–520.

[52] Lisano, M. E., "A Practical Six-Degree-of-Freedom Solar-Sail Dynamics Model for Optimizing Solar-Sail Trajectories with Torque Constraints," AIAA Paper 2004-4891, Aug. 2004.

[53] Jacobson, R., and Thornton, C., "Elements of Solar-Sail Navigation with Application to a Halley's Comet Rendezvous," *Journal of Guidance and Control*, Vol. 1, No. 5, 1978, pp. 365–371.

[54] Bacon, R. H., "Logarithmic Spiral—an Ideal Trajectory for an Interplanetary Vehicle with Engines of Low Sustained Thrust," *American Journal of Physics*, Vol. 27, No. 3, 1959, pp. 12–18.

[55] Battin, R. H., *An Introduction to the Mathematics and Methods of Astrodynamics*, AIAA Education Series, AIAA, New York, 1987, pp. 389, 447–450.

[56] Dachwald, B., Ohndorf, A., and Wie, B., "Solar-Sail Trajectory Optimization for a Solar Polar Imager (SPI) Mission," AIAA Paper 2006-6177, Aug. 2006.

[57] Wie, B., "Solar Sailing Kinetic Energy Interceptor (KEI) Mission for Impacting and Deflecting Near-Earth Asteroids," AIAA Paper 2005-6175, Aug. 2005.

[58] Dachwald, B., and Wie, B., "Solar-Sail Kinetic Energy Impactor Trajectory Optimization for an Asteroid Deflection," *Journal of Spacecraft and Rockets*, Vol. 44, No. 4, 2007, pp. 755–764.

[59] Dachwald, B., Kahle, R., and Wie, B., "Solar Sailing KEI Mission Design Trade-offs for Impacting and Deflecting Asteroid 99942 Apophis," AIAA Paper 2006-6178, Aug. 2006.

[60] Lappas, V., and Wie, B., "Advanced Small Satellite Bus Technology for Near-Term Solar-Sail Missions," AIAA Paper 2006-6179, Aug. 2006.

[61] Stickler, A. C., and Alfriend, K., "Elementary Magnetic Attitude Control System," *Journal of Spacecraft and Rockets*, Vol. 13, No. 5, 1976, pp. 282–287.

[62] Camillo, P. J., and Markley, F. L., "Orbit-Averaged Behavior of Magnetic Control Laws for Momentum Unloading," *Journal of Guidance and Control*, Vol. 3, No. 6, 1980, pp. 563–568.

[63] Pittelkau, M. E., "Optimal Periodic Control for Spacecraft Pointing and Attitude Determination," *Journal of Guidance, Control*, and Dynamics, Vol. 16, No. 6, 1993, pp. 1078–1084.

[64] Hablani, H. B., "Pole-Placement Technique for Magnetic Momentum Removal of Earth-Pointing Spacecraft," *Journal of Guidance, Control, and Dynamics*, Vol. 20, No. 2, 1997, pp. 268–275.

[65] Chen, X., Steyn, W., Hodgart, S., and Hashida, Y., "Optimal Combined Reaction-Wheel Momentum Management for Earth-Pointing Satellites," *Journal of Guidance, Control, and Dynamics*, Vol. 22, No. 4, 1999, pp. 543–550.

[66] Psiaki, M. L., "Magnetic Torquer Attitude Control via Asymptotic Periodic Linear Quadratic Regulation," *Journal of Guidance, Control, and Dynamics*, Vol. 24, No. 2, 2001, pp. 386–394.

[67] Giulietti, F., Quarta, A. A., and Tortora, P., "Optimal Control Laws for Momentum-Wheel Desaturation Using Magetorquers," *Journal of Guidance, Control, and Dynamics*, Vol. 29, No. 6, 2006, pp. 1464–1467.

13

Solar-Sail Missions for Asteroid Deflection

In Chapter 12 we studied the dynamical modeling and control problems of solar-sail spacecraft. Solar sails are large, lightweight reflectors in space that are propelled by sunlight. The near-term solar-sail missions, including a solar-sail flight validation mission in a 1600-km sun-synchronous orbit, were studied in Chapter 12. In this chapter we study solar-sail mission applications to a complex astrodynamical problem of changing the trajectory of near-Earth objects (NEOs) to mitigate their impact threat to the Earth. This chapter is intended to briefly introduce such a technically challenging astrodynamical problem of deflecting NEOs and also to describe the nonnuclear deflection options utilizing kinetic impactors, gravity tractors, and solar sails. The dynamical modeling and control aspects of such an emerging astrodynamical problem are emphasized throughout this chapter.

13.1 Introduction to the NEO Deflection Problem

13.1.1 Introduction

The spectacular collision of comet Shoemaker–Levy 9 with Jupiter in July 1994 was a clear evidence of the fact that the risk of impacts upon the Earth by NEOs is very real. In response, the U.S. Congress funded a 10-year survey to locate and track 90% of the NEOs with diameters of 1 km or greater, the impacts of which could threaten the extinction of civilization. In the course of this search, hundreds of thousands of smaller asteroids have been discovered, many similar in size to the 60-m object that exploded above Tunguska, Siberia, on 30 June, 1908 with an energy level of 10 megatons of TNT, destroying essentially everything within a 25-km radius. Air bursts with an energy level of 5 kilotons of TNT, such as that caused by a 10-m object that disintegrated over Tagish Lake, British Columbia, Canada, in 2000, are estimated to occur on an annual basis [1].

Scientists widely accept that an impact by a large asteroid of greater than 10 km in diameter caused the extinction of the dinosaurs 65 million years ago. A 2-km object is known to be capable of causing catastrophic alteration to the global ecosystem, which could lead to the end of civilization. Ocean impacts from even smaller objects are of some concern because the destructive potential caused by the resulting tsunamis could be higher than that from a same-size object's impact on land. The probability of a major impact causing the extinction of humanity is extremely low, but it is not zero. Unlike many other natural disasters, such as earthquakes, tsunamis, hurricanes, and tornadoes, which cannot be prevented, the impact threat posed by NEOs can be mitigated given adequate warning time. The impact of an object smaller than 30 m in diameter is often naturally mitigated by the Earth's atmosphere. As the typical small meteoroids enter the atmosphere, they often burn up or explode before they hit the ground. If they burn up, they are called meteors; if they explode, they are called bolides.

A near-Earth asteroid (NEA) refers to any asteroid with a perihelion of less than 1.3 astronomical units (AU). If comets are included, then we speak of NEOs. If an NEA's perihelion is less than that of Earth, and its aphelion is greater than that of Earth, it is referred to as an Earth-crossing asteroid (ECA). All asteroids with an Earth minimum orbit intersection distance (MOID) of 0.05 AU or less and an absolute magnitude of 22.0 or less are considered potentially hazardous asteroids (PHAs). Asteroids that cannot get any closer to the Earth than 0.05 AU ($\approx 117R_\oplus$) or are smaller than about 150 m in diameter are not considered PHAs. A comet sometimes experiences net thrust caused by evaporating ices; this thrust varies significantly as a function of radial distance from the sun, the comet's rotational axis and period, and the distribution of ices within the comet's structure. The precise trajectories of comets are thus less predictable, and an accurate intercept correspondingly is more complex. Fortunately, the threat posed by comets appears to be small compared to the risks of impacts by NEAs, and thus NEO researchers currently focus on mitigating the threat posed mostly by NEAs. However, a further study should continue to address the difficult task of detecting and deflecting comets.

13.1.2 NEO Detection, Characterization, and Deflection Issues

To establish a national plan for providing adequate warning and mitigation of the potential NEO hazard, the following text became law as part of the NASA Authorization Act of 2005 [2] (now known as the George E. Brown, Jr. Near-Earth Object Survey Act), passed by the Congress on 22 December, 2005, and subsequently signed by the President:

> The U.S. Congress has declared that the general welfare and security of the United States require that the unique competence of NASA be directed to detecting, tracking, cataloguing, and characterizing near-Earth asteroids and comets in order to provide warning and mitigation of the potential hazard of such near-Earth objects to the Earth. The NASA Administrator shall plan, develop, and implement a Near-Earth Object Survey program to detect, track, catalogue, and characterize the physical characteristics of near-Earth objects equal to or greater than 140 meters in diameter in order to assess the threat of such near-Earth objects to the Earth. It shall be the goal of the Survey program to achieve 90% completion of its near-Earth object catalogue (based on statistically predicted populations of near-Earth objects) within 15 years after the date of enactment of this Act. The NASA Administrator shall transmit to Congress not later than 1 year after the date of enactment of this Act an initial report that provides the following: (A) An analysis of possible alternatives that NASA may employ to carry out the Survey program, including ground-based

and space-based alternatives with technical descriptions. (B) A recommended option and proposed budget to carry out the Survey program pursuant to the recommended option. (C) Analysis of possible alternatives that NASA could employ to divert an object on a likely collision course with Earth.

In response to this congressional direction, a NASA Workshop on NEO Detection, Characterization and Threat Mitigation was held in Vail, Colorodo, 26–29 June 2006. The purpose of this workshop was to engage experts from the NEO scientific and technical communities to identify the fullest possible set of alternatives for meeting congressional direction in the following three areas: 1) detection, tracking and cataloging NEOs; 2) characterization of NEOs; and 3) deflection or other forms of NEO threat mitigation.

In March 2007, NASA's NEO Report to Congress, entitled "Near-Earth Object Survey and Deflection Analysis of Alternatives," was released to the public. (Data available online at http://www.nasa.gov/pdf/171331main_NEO_report_march07.pdf.) The report described a range of possible options from public and private sources and then analyzed their capabilities and levels of performance including development schedules and technical risks. The report assessed a variety of issues associated with the asteroid survey program, and it also assessed a series of approaches that could be used to divert a NEO potentially on a collision course with Earth. The report summarized key findings as follows.

Key Findings for the Survey Program:

- The goal of the Survey Program should be modified to detect, track, catalogue, and characterize, by the end of 2020, 90 percent of all Potentially Hazardous Objects (PHOs) greater than 140 meters whose orbits pass within 0.05 AU of the Earths orbit (as opposed to surveying for all NEOs).
- The Agency could achieve the specified goal of surveying for 90 percent of the potentially hazardous NEOs by the end of 2020 by partnering with other government agencies on potential future optical ground-based observatories and building a dedicated NEO survey asset assuming the partners potential ground assets come online by 2010 and 2014, and a dedicated asset by 2015.
- Together, the two observatories potentially to be developed by other government agencies could complete 83 percent of the survey by 2020 if observing time at these observatories is shared with NASA's NEO Survey Program.
- New space-based infrared systems, combined with shared ground-based assets, could reduce the overall time to reach the 90 percent goal by at least three years. Space systems have additional benefits as well as costs and risks compared to ground-based alternatives.
- Radar systems cannot contribute to the search for potentially hazardous objects, but may be used to rapidly refine tracking and to determine object sizes for a few NEOs of potentially high interest. Existing radar systems are currently oversubscribed by other missions.
- Determining a NEO's mass and orbit is required to determine whether it represents a potential threat and to provide required information for most

alternatives to mitigate such a threat. Beyond these parameters, characterization requirements and capabilities are tied directly to the mitigation strategy selected.

Key Findings for Diverting a Potentially Hazardous Object (PHO):

- Nuclear standoff explosions are assessed to be 10–100 times more effective than the non-nuclear alternatives analyzed in this study. Other techniques involving the surface or subsurface use of nuclear explosives may be more efficient, but they run an increased risk of fracturing the target NEO. They also carry higher development and operations risks.
- Non-nuclear kinetic impactors are the most mature approach and could be used in some deflection/mitigation scenarios, especially for NEOs that consist of a single small, solid body.
- "Slow push" mitigation techniques are the most expensive, have the lowest level of technical readiness, and their ability to both travel to and divert a threatening NEO would be limited unless mission durations of many years to decades are possible.
- 30–80 percent of potentially hazardous NEOs are in orbits that are beyond the capability of current or planned launch systems. Therefore, planetary gravity assist swingby trajectories or on-orbit assembly of modular propulsion systems may be needed to augment launch vehicle performance, if these objects need to be deflected.

(*Note*: The Final Report on the 2006 Near-Earth Object Survey and Deflection Study is also available at http://www.hq.nasa.gov/office/pao/FOIA/NEO_Analysis_Doc.pdf. This 272-page report contains detailed technical information and supporting material that were not included in NASA's 27-page NEO Report to Congress. Also interesting ongoing technical discussions on nuclear vs nonnuclear alternatives can be found in the B612 Foundation's Web site at http://www.b612foundation.org/press/press.html.)

At the 2007 Planetary Defense Conference: Protecting Earth from Asteroids, held in Washington, D.C., 5–8 March 2007, a variety of concepts and issues associated with detecting, tracking, characterizing, and deflecting NEOs was discussed. The objectives of the meeting were 1) to highlight the current state of the art in NEO detection, characterization and mitigation; 2) to understand the threat posed by asteroids and comets and possible responses to an asteroid impact; and 3) to determine political, policy, and legal issues that would affect our ability to mount an effective defense. A white paper, which described the major findings and recommendations of this meeting, was published on 25 April 2007, and it has been posted at the conference Web site (http://www.aero.org/conferences/planetarydefense/2007papers.html).

The primary findings of the 2007 Planetary Defense Conference, as summarized in its white paper, are as follows:

- While our search and discovery efforts have successfully found most of the large, "civilization-killer" 1-km and larger objects, we are just beginning to

find the much more prevalent and, for that reason, more frequently danger-
ous objects in the 140- to 300-meter size range. An impact by an object in
this size range could occur with little or no warning and could cause serious
loss of life and property over a broad area.

- Earth-based resources such as the Arecibo radar are critical for refining
  a PHO's orbit and providing basic information required for deflection.
  Arecibo has an essential role in refining the threat posed by PHOs such
  as Apophis.

- Deflection of a threatening object is currently in the conceptual phase. We
  are just beginning to identify the options available to deflect an object and
  have yet to design or test techniques that might be used. Further, we have yet
  to design complete missions to deliver one or more deflection devices, and
  have not considered what is required to assure a high probability of success
  for an overall deflection campaign.

- There are serious technical, political, policy, legal and societal issues
  involved in deciding whether and how to respond to a threat of a NEO
  impact. NEO impacts have the potential to cause disasters that would equal
  or exceed anything ever faced by recent civilizations. Moreover, this type of
  threat has never been seriously considered by any agencies that would have
  responsibility for responding. In addition, it is uncertain where responsibility for coordination of all aspects of the NEO threat lies, from detection to
  deflection to impact aftermath.

- Understanding, analyzing, and dealing with a potential NEO threat is an
  international problem demanding international cooperation. Considerable
  work is required to develop a foundation for international cooperation
  and action in all areas related to planetary defense. This foundation may
  extend beyond defense and include benefits from international manned and
  unmanned space exploration.

The white paper also describes detailed recommendations for future research
and development activities associated with detecting, tracking, characterizing, and
deflecting NEOs, including the following specific recommendations associated
with deflecting NEOs:

- Research, characterize and demonstrate technologies associated with the
  most promising impulsive and slow-push techniques. Except for some technologies that might be used for impulsive missions, very little work to
  characterize deflection techniques has been done. Research is required to
  move these techniques from concepts to viable options for NEO deflection.
  Research should identify technologies critical to each method. Research
  should also consider approaches that might be synergetic and improve the
  overall certainty of a deflection mission. Included should be microgravity
  experiments to illustrate the response of NEO materials to impacts or to
  methods that might be used to attach or couple to the surface of such objects
  in microgravity conditions (i.e., attaching transponders or other instrument
  packages).

- Identify and pursue opportunities to demonstrate potential deflection technologies during characterization missions that are in formulation or early development. At present, designers of deflection missions must allow for large uncertainties in the response of a target NEO to a deflection attempt, and additional research is required to increase confidence in our ability to predict and control the effectiveness of a deflection attempt. Compatible opportunities during characterization missions should be identified to demonstrate potential deflection technologies (e.g., attaching transponders, testing kinetic impactors, using low impulse ion engines and slow-push techniques, etc.). The European Space Agency's Don Quijote mission is an example of the type of mission that might be used to characterize a NEO and to test deflection technologies.
- Develop and document complete designs of a deflection campaign, including launch vehicle and payload requirements, ground support requirements, overall mission reliability, mission timelines and milestones, and costs. Our ability to deliver a deflection option to a threatening NEO with a high probability of success must also be considered in detail. Results of these studies would feed into an overall NEO deflection plan and help develop a roadmap for the architecture of a deflection campaign using current and near-term technology and capabilities. This plan should be updated on a periodic basis.

As evidenced in the preceding descriptions of the current state of the art in NEO detection, characterization, and deflection, there still exists a wide variety of technical (e.g., ground-based vs space-based detection, nuclear vs nonnuclear deflection, etc.) and political/sociological issues, which need to be further examined and resolved. Because of the limited scope of this chapter, no attempt is made for a complete discussion of such technical and sociological issues; however, some important technical issues are discussed briefly at appropriate places throughout this chapter.

13.1.3 NEA Deflection Technologies

Early detection, accurate tracking, reliable precision orbit calculation, and characterization of physical properties of NEAs are prerequisites to any space mission of deflecting NEAs. The early discovery of NEAs prior to impact using current ground-based optical sensors is not assured, and detection/tracking of small (1 km or less) NEAs is a difficult task given their low albedo and small size. Various concepts and approaches for advanced ground-based as well as space-based detection systems are being developed to allow for adequate warning time [3].

Assuming that NEAs on a collision course can be detected prior to impact with a mission lead time of at least 10 years, however, the challenge becomes eliminating their threat, either by destroying the asteroid, or by altering its trajectory so that it will miss Earth. A variety of schemes, including a nuclear standoff detonation, mass drivers, kinetic-energy projectiles, laser beaming, and low-thrust deflection via electric propulsion or solar sails, has already been extensively investigated in the past for such a technically challenging, asteroid deflection problem [4–8]. The feasibility of each approach to deflect an incoming hazardous object depends on its

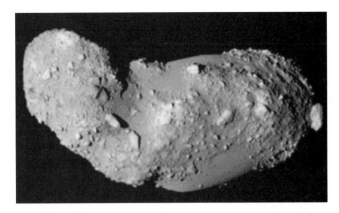

Fig. 13.1   Image of near-Earth asteroid 25143 Itokawa taken by Japan's Hayabusa spacecraft in 2005. [Image courtesy of JAXA (Japan Aerospace Exploration Agency).]

size, spin rate, composition, the mission lead time, and many other factors. Nuclear standoff explosions are sometimes assessed to be much more effective than the nonnuclear alternatives, especially for larger asteroids with a short mission lead time. Other techniques involving the surface or subsurface use of nuclear explosives are also assessed to be more efficient, although they can run an increased risk of fracturing the target asteroid. The NEAR Shoemaker study of asteroid Mathilde [3] and the Japanese Hayabusa mission for exploring the near-Earth asteroid Itokawa suggest that many asteroids are essentially "rubble piles." An image of asteroid Itokawa, shown in Fig. 13.1, reveals a rugged surface of collected dust and debris. In comparison, an image of asteroid Ida, which is a solid monolithic body, is shown in Fig. 13.2. Computational studies show that a thermonuclear detonation within or near a body of rubble piles would not effectively disperse the constituent fragments, which would continue following the same trajectory toward Earth [9].

Another option would be an impulsive velocity change of the NEA, accomplished either in a single event, or in multiple events over an extended period. Applied correctly without causing fragmentation of a large asteroid into smaller pieces, the effect of such a ΔV would magnify over decades (or even centuries), eliminating the risk of collision with Earth. A gradual impulsive change might be accomplished by taking advantage of the Yarkovsky effect, in which a rotating asteroid experiences a minute nonradial thrust as a result of the absorption of sunlight and subsequent reemission of heat. By varying the reflective and thermal characteristics of one area of an asteroid's surface, thrust could be created in the desired direction. Unfortunately, the requisite technologies for such an operation will not be readily available in the near future. Many of the previously proposed deflection schemes utilizing such a low-thrust push/pull idea appear to be impractical. These include attaching large solar sails, mass drivers, or high-efficiency electric propulsion systems to a tumbling or spinning asteroid, painting an asteroid to change its albedo to utilize the Yarkovsky effect, and laser beaming to ablate small amounts of material from the surface of a tumbling asteroid. Some of these schemes might also require an extremely large number of heavy launch vehicles.

Fig. 13.2   Asteroid 243 Ida (a main-belt asteroid, not a NEA) imaged by the Galileo spacecraft's solid-state imaging system at ranges of 3057 to 3821 km on 28 August 1993. Galileo flew within 2400 km of the 52-km Ida at a relative velocity of 12.4 km/s. (Image courtesy of NASA.)

A technology does currently exist for an impulsive velocity change, caused by the targeted kinetic impact of a spacecraft on the asteroid's surface. Again, the immediate effect would be small, but if applied long enough prior to a projected Earth impact, the deflection could be sufficient to cause a miss [10–15]. To be most effective, the impacting spacecraft would either have to be massive or be moving very fast relative to the asteroid. Because current launch technology limits the mass (including propellant) that can be lifted into an interplanetary trajectory, we are therefore led to consider designs that would maximize impact velocity and that would not require large amounts of fuel.

Propellantless solar-sail propulsion, therefore, emerges as a realistic near-term option to such a technically challenging problem of deflecting NEAs. A previously proposed concept of using solar sails to tow or tug an asteroid requires an unrealistically large solar sail, which is not technically feasible to assemble in space. Furthermore, attaching such an extremely large solar sail to a tumbling asteroid will not be a trivial task. However, solar sails have the potential to provide cost effective, propellantless propulsion that enables longer mission lifetimes, increased payload mass fraction, and access to previously inaccessible orbits (e.g., high solar latitude, retrograde heliocentric, and non-Keplerian).

In the past, various solar-sailing rendezvous missions with a comet or an asteroid have been studied (e.g., see [16–20]), as illustrated in Figs. 13.3 and 13.4. An innovative solar-sailing mission concept was studied by the Jet Propulsion Laboratory (JPL) in 1977 for a rendezvous mission with Halley's comet for the 1986 passage [16, 17]. Although it became an ill-fated mission concept of 1970s, which required a very large, 800-m solar sail to be deployed in space, it introduced a propellantless solar-sailing concept to achieve a large, 145-deg orbital inclination change at 0.25 AU in order to rendezvous with Halley's comet in a retrograde orbit.

Fig. 13.3   Large solar sail proposed by JPL in 1977 for a rendezvous mission with Halley's comet for the 1986 passage [16, 17]. (Image courtesy of NASA/JPL.)

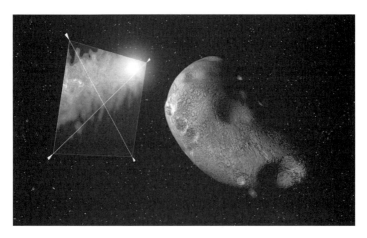

Fig. 13.4   Solar-sail rendezvous mission with an asteroid. (Image courtesy of Olivier Boisard/U3P/2006—www.u3p.net.)

The recent advances in lightweight deployable booms, ultralightweight sail films, and small satellite technologies are spurring a renewed interest in solar sailing and the missions it enables. Consequently, various near-term solar-sailing missions and the associated technologies are being developed for future space exploration missions [21–28].

One of such near-term solar-sailing missions is the Solar Polar Imager (SPI) mission. Our current understanding of the sun is limited by a lack of observations of its polar regions. The SPI mission utilizes a large solar sail to place a spacecraft in a 0.48-AU heliocentric circular orbit with an inclination of 75 deg. Viewing of the polar regions of the sun provides a unique opportunity to more fully investigate the structure and dynamics of its interior, the generation of solar magnetic fields, the origin of the solar cycle, the causes of solar activity, and the structure and dynamics of the corona. The SPI mission consists of the initial cruise phase to a 0.48-AU circular orbit, the cranking orbit phase, and the science mission phase. A 160-m, 450-kg solar-sail spacecraft was once studied by NASA for the SPI mission. A Delta II launch vehicle is able to inject the 450-kg SPI spacecraft into an Earth escaping orbit with $C_3 = 0.25$ km^2/s^2. The solar-sail is to be deployed at the beginning of the interplanetary cruise phase. The SPI sailcraft first spirals inwards from 1 AU to a heliocentric circular orbit at 0.48 AU, and then the cranking orbit phase begins to achieve a 75-deg inclination. The solar sail will be jettisoned after achieving the science mission orbit, and the total sailing time is approximately 6.6 yr. The corresponding three-dimensional orbital trajectory of the solar-sailing phase of the SPI mission is illustrated in Fig. 13.5.

As stated in NASA's NEO Report to Congress (March 2007), 30–80% of potentially hazardous NEOs are in orbits that are beyond the capability of current or planned launch systems. Therefore, solar sails with a large propellantless ΔV capability might be needed to augment conventional launch-vehicle performance, if these objects need to be deflected.

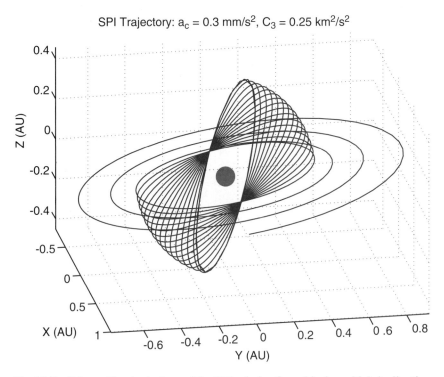

Fig. 13.5   Solar-sailing trajectory of the SPI mission for achieving a high-inclination heliocentric orbit.

A solar-sailing mission to be described in the next section utilizes a solar sail to deliver a kinetic energy impactor into a heliocentric retrograde orbit, which will result in a head-on collision with a target asteroid at its perihelion, thus increasing its impact velocity to at least 70 km/s. McInnes [29] has recently investigated the feasibility of such an innovative solar-sailing concept as applied to the asteroid deflection problem, expanding upon the work of Wright [17]. A solar-sailing mission architecture, which employs 160-m, 300-kg solar-sail spacecraft with a characteristic acceleration of 0.5 mm/s^2, will be presented in the next section as a realistic near-term option for mitigating the threat posed by NEAs. Its mission feasibility will be demonstrated for a fictional asteroid deflection problem of AIAA.

A practical concern of any kinetic-impact approach of mitigating the threat of asteroids is the risk that the impact could result in the fragmentation of the asteroid, which could substantially increase the damage upon Earth impact. In Secs. 13.3 and 13.4, we will also study an asteroid deflection concept utilizing the mutual gravitational force between a hovering spacecraft and a target asteroid as a towline.

13.2 Solar-Sail Mission Concept for Asteroid Deflection

13.2.1 Fictional Asteroid Deflection Problem

A fictional asteroid deflection problem was created by AIAA for the 2004/2005 AIAA Foundation Undergraduate Team Space Design Competition. A similar fictional asteroid deflection problem, called the defined threat (DEFT) scenarios, has been also created for the 2004 Planetary Defense Conference. One of the four DEFT scenarios is about mitigating a fictional 200-m Athos asteroid with the predicted impact date of 29 February 2016.

The fictional asteroid mitigation problem of AIAA is briefly described as follows. On 4 July 2004, NASA/JPL's Near Earth Asteroid Tracking (NEAT) camera at the Maui Space Surveillance Site discovered a 0.205-km-diam Apollo asteroid designated 2004WR. This asteroid has been assigned a Torino Impact Scale rating of 9.0 on the basis of subsequent observations that indicate there is a 95% probability that 2004WR will impact the Earth. The expected impact will occur in the Southern Hemisphere on 14 January 2015 causing catastrophic damage throughout the Pacific region. The mission is to design a space system that can rendezvous with 2004WR in a timely manner, inspect it, and remove the hazard to Earth by changing its orbit and/or destroying it. The classical orbital elements of 2004WR are given in the J2000 heliocentric ecliptic reference frame as follows:

$$\text{Epoch} = 53200 \text{ TDB (14 July 2004)}$$
$$a = 2.15374076 \text{ AU}$$
$$e = 0.649820926$$
$$i = 11.6660258 \text{ deg}$$
$$\omega = 66.2021796 \text{ deg}$$
$$\Omega = 114.4749665 \text{ deg}$$
$$M = 229.8987151 \text{ deg (at epoch)}$$

The STK 5.0.4 software package, with a ninth-order Runge–Kutta integrator with variable step size and the planetary positions from JPL's DE405, was used by AIAA to create this set of orbital parameters of 2004WR.

It is further assumed that 2004WR is an S-class (stony-silicate) asteroid with a density of $2720 \, \text{kg/m}^3$ and that its estimated mass is 1.1×10^{10} kg. If 2004WR is an M-class (nickel-iron) asteroid, then its estimated mass would be 2.2×10^{10} kg.

The initial position and velocity components in the heliocentric ecliptic coordinates are then obtained as

$$(X, \, Y, \, Z) = (3.17670340, \quad 0.84877205, \quad -0.66956611) \text{ AU}$$

$$(\dot{X}, \, \dot{Y}, \, \dot{Z}) = (-0.0038834223, \quad 0.0048780152, \quad 0.00031250049) \text{ AU/day}$$

where 1 AU = 149,597,870.691 km and 1 day = 24 h = 86,400 s.

Other orbital parameters of 2004WR in an ideal Keplerian orbit can be found as follows:

$$r_p = 0.7542 \text{ AU (perihelion)}$$

$$r_a = 3.5533 \text{ AU (aphelion)}$$

$$v_p = 44 \text{ km/s (perihelion speed)}$$

$$v_a = 9.3 \text{ km/s (aphelion speed)}$$

$$P = 3.16 \text{ year (orbital period)}$$

An ideal Keplerian orbit simulation of 2004WR was performed first. The result indicated that its closest approach to Earth is about 0.035 AU, which is less than the MOID of 0.05 AU of a PHA. It also had a close encounter with Mars by 0.1 AU. After checking the ideal orbital characteristics of 2004WR, three different n-body software packages were used to confirm 2004WR's collision with Earth on 14 January 2015. These software packages were JPL's Horizons [30], CODES (data available online at http://home.earthlink.net/jimbaer1/), and SSCT [31], all utilizing JPL's DE405 ephemeris data for the planetary positions. Orbit simulation results of using these packages indicate that 2004WR misses Earth by $1.6R_\oplus$ ($\approx$10,000 km from the Earth center). This Earth miss-distance prediction of approximately 10,000 km is in fact caused by the inherent uncertainty (not the numerical integration error) associated with the complex n-body orbital simulation problem.

Although AIAA's asteroid problem statement claims that the expected impact of 2004WR will occur in the Southern Hemisphere on 14 January 2015, it is important to point out some inherent uncertainties associated with the practical orbit determination problem. In practice, ground-based optical observations of an asteroid during the first several days after its discovery are known to result in an orbit determination uncertainty of 70,000 km in position and 50 m/s in velocity. Further continuous optical tracking and observations, probably for over one year, can reduce the orbit determination uncertainty to 100 km and 5 cm/s. Additional radar observations can further reduce the orbit determination uncertainty to 10 km and 0.5 cm/s. A velocity uncertainty of 0.5 cm/s for an asteroid at epoch results in an Earth miss-distance prediction uncertainty of 15,000 km after 10-year orbit propagation. Therefore, a future asteroid detection/tracking system will require the orbit determination accuracy better than 10 km and 0.5 cm/s to avoid serious false alarms, as well as to increase the mitigation mission reliability.

13.2.2   Kinetic-Impact ΔV Estimation

The simplest deflection approach is to impact the target NEA with a massive projectile at a high relative speed. However, a successful asteroid deflection mission will require accurate modeling and prediction of the change in velocity caused by the interceptor's impact. The effective impulse imparted to the asteroid will be the sum of the pure kinetic impulse (linear momentum) of the interceptor, plus the impulse caused by the thrust of material being ejected from the impact crater. This last term can be very significant (even dominant), but its magnitude depends strongly upon the density and yield strength of the material of which the asteroid is composed, as well as the mass and relative velocity of the interceptor.

Using the conservation principle of linear momentum, we can estimate the resulting impact ΔV (the instantaneous velocity change of the target asteroid) as

$$\Delta V \approx \beta \frac{m}{M + m} U \approx \beta \frac{m}{M} U \qquad (13.1)$$

where β is the impact efficiency factor ($\beta = 1$ for an ideal inelastic impact), m the impactor mass, M the target asteroid mass, and U the relative impact velocity.

For example, a head-on impact (at a relative velocity of 70 km/s) of a 150-kg impactor on a 200-m, S-class asteroid (with a density of 2720 kg/m^3 and a mass of 1.1×10^{10} kg) results in a pure kinetic-impact ΔV of approximately 0.1 cm/s. If the asteroid is composed of hard rock, then the modeling of crater ejecta impulse from previous studies [5, 6] would predict an additional ΔV of 0.2 cm/s, which is double the pure kinetic-impact ΔV, resulting in $\beta = 3$. However, if the asteroid were composed of soft rock, the previous studies would instead predict an additional ΔV of 0.55 cm/s, which is more than five times the pure kinetic-impact ΔV, resulting in $\beta = 6.5$. Thus, an accurate modeling and prediction of ejecta impulse for various asteroid compositions is a critical part of the most kinetic-impact approaches. Recent empirical research [32] would not only allow us to substantially improve the accuracy of these predictions, especially for high velocity impacts, but also to extend them to cases where the impact area is composed of ice or lunar-type regolith, as well as to cases where the asteroid is a porous rubble pile. Furthermore, the cratering efficiency could be improved through the use of a small conventional explosive payload, an option that would likely require tradeoffs in the impactor design and mission architecture.

For an impulsive ΔV change along the orbital direction, the resulting deflection Δx after a coasting time of t_c can be estimated as

$$\Delta x \approx 3t_c \Delta V \qquad (13.2)$$

for an asteroid in a near circular orbit. For example, an impulsive ΔV of 1 cm/s with a coasting time of 10 years will result in a deflection distance of 9460 km. This deflection formula with an orbital amplification factor of three will be derived and further discussed in Sec. 13.6.

A practical concern of any kinetic-impact approach of mitigating the threat of asteroids is the risk that the impact could result in the fragmentation of the asteroid, which could substantially increase the damage upon Earth impact. The energy required to fragment an asteroid depends critically upon the asteroid's size, composition, and structure.

In astrophysics, the energy required to disassemble a celestial body consisting of loose material, which is held together by gravity alone, into space debris such as dust and gas is called the *gravitational binding energy*. The gravitational binding energy of a spherical body of mass M, uniform density ρ, and radius R is given by [33, 34]

$$E = \frac{3GM^2}{5R} = \frac{3G}{5R}\left(\frac{4\pi\rho R^3}{3}\right)^2 = \frac{\pi^2\rho^2 G}{30}D^5 \qquad (13.3)$$

where $G = 6.67259 \times 10^{-11}$ N·m^2/kg^2 is the universal gravitational constant and $D = 2R$ is the diameter of a spherical body. The escape speed from its surface is given by

$$V_e = \sqrt{\frac{2GM}{R}} \qquad (13.4)$$

For example, consider a 200-m (diameter) asteroid with a uniform density of $\rho = 2720$ kg/m^3 and a mass of $M = 1.1 \times 10^{10}$ kg. Its gravitational binding energy is 4.8×10^7 J. Because the kinetic energy of a 150-kg impactor at an impact velocity of 70 km/s is 3.7×10^{11} J, one might expect that a gravity-dominated, 200-m asteroid would be disrupted and dispersed by such a high-energy impactor. However, its escape velocity of 12 cm/s is about 120 times the impact ΔV of 0.1 cm/s. This large ratio of the escape velocity to the impact ΔV might suggest that if the asteroid disperses, the resulting fragments might scatter around their deflected center of mass [34]. In [4] and [5] (pp. 135, 136), the disruption energy per unit asteroid mass is predicted to be 150 J/kg for strength-dominated asteroids. This indicates that a strength-dominated, 200-m asteroid would not be disrupted by a 150-kg impactor at a high impact velocity of 70 km/s. Also in [5] (pp. 135, 136), the energy (per unit asteroid mass) required for both disruption and dispersion of a 1-km asteroid is predicted to be 5 kJ/kg. Thus, the feasibility of the most kinetic-impact approaches for either disrupting or deflecting an incoming NEO depends on its size and composition (e.g., solid body, porous rubble pile, etc.), as well as the time available to change its orbit. An accurate determination of the composition of the target asteroid is a critical part of the kinetic-impact approaches, which might require a separate inspection mission.

A further study is also needed to optimize impactor size, relative impact velocity, and the total number of impactors as functions of asteroid size and composition to ensure a deflection attempt does not cause fragmentation/dispersal.

13.2.3 Solar-Sailing Kinetic Energy Impactor Mission Concept

The proposed solar-sailing mission of mitigating the threat posed by NEAs is illustrated here using AIAA's asteroid deflection problem described in Sec. 13.2.1.

The proposed solar-sailing mission is basically composed of the initial cruise phase from 1 AU to a heliocentric orbit at 0.25 AU (1.5 years), the cranking orbit phase of a 168-deg inclination change (3.5 years), and the final retrograde-orbit phase (1 year) prior to impacting the asteroid at its perihelion. The proposed mission of intercepting, impacting, and deflecting NEAs is basically exploiting the unique, propellantless nature of a solar-sail propulsion system capable of achieving a retrograde heliocentric orbit in order to increase the relative speed of a kinetic impactor. The solar-sailing phase of the proposed mission architecture is similar to that of a rendezvous mission with Halley's comet, extensively studied by NASA/JPL in the mid-1970s [16–19]. Although the rendezvous mission with Halley's comet became an ill-fated mission concept requiring an 800-m solar-sail for an 850-kg payload/bus, the mission concept demonstrated the unique capability of a solar-sail for achieving a 145-deg orbital inclination change at 0.25 AU to rendezvous with Halley's comet for the 1986 passage. However, the mission proposed here requires a moderate size, 160-m solar sail. Such a moderate-size solar sail is also being considered for the SPI mission to achieve a heliocentric mission orbit with an inclination of 75 deg at 0.48 AU from the sun. The solar-sailing trajectory design for achieving a 75-deg inclination at 0.48 AU was discussed in Sec. 12.14.1 for the SPI mission. The corresponding three-dimensional orbital trajectory of the solar-sailing phase of the SPI mission is illustrated in Fig. 13.5. The mission of intercepting an asteroid will continue such an orbit cranking maneuver (but at

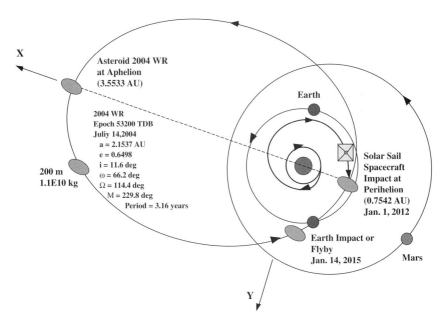

Fig. 13.6   Illustration of the solar-sailing kinetic-energy-impactor (KEI) mission for intercepting/impacting/deflecting a near-Earth asteroid. The final, retrograde heliocentric orbit phase (starting from 0.25 AU) results in a head-on collision with the target asteroid at its perihelion (0.75 AU).

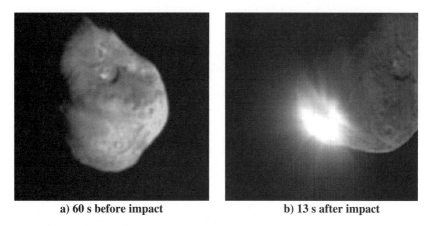

a) 60 s before impact   **b) 13 s after impact**

Fig. 13.7   A 370-kg impactor spacecraft released from the 650-kg flyby spacecraft has successfully collided with the 5-km target comet on 4 July 2005. (Image courtesy of NASA/JPL-Caltech/UMD.)

0.25 AU) to achieve a 168-deg inclination change. The final retrograde-orbit phase prior to impacting the asteroid at its perihelion (0.75 AU) is illustrated in Fig. 13.6.

The recent success of NASA's Deep Impact mission in 2005 has significantly enhanced the kinetic-impact approaches, including the proposed solar-sailing KEI

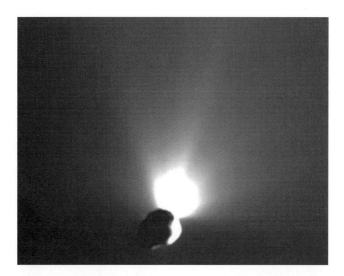

Fig. 13.8   Image of Tempel 1, which was taken by the high-resolution camera of the flyby spacecraft, 50 minutes after impact. Large plumes of ejected material can be seen here streaming away from the back side of the comet. The resulting impact ΔV was extremely small and was not measurable, however. (Image courtesy of NASA/JPL-Caltech/UMD. http://deepimpact.jpl.nasa.gov/gallery/index.html.)

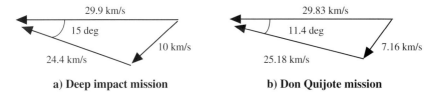

a) Deep impact mission   b) Don Quijote mission

Fig. 13.9   Velocity vector diagrams of the Deep Impact mission and the Don Quijote mission.

mission concept. Its mission goals were to explore the internal structure and composition of the nucleus of comet Tempel 1 before, during, and after impacts, and to return the observations to Earth, as shown in Figs. 13.7 and 13.8. The Deep Impact spacecraft was launched by a Delta II launch vehicle on 12 January, 2005 and released a 370-kg impactor spacecraft, which collided with Tempel 1 on 4 July, 2005 to create a large crater on the surface of the 5-km target comet. The crater is estimated to be 20 m deep and 100 m wide. In fact, the 5-km comet with a heliocentric speed of 29.9 km/s crashed into the 370-kg impactor, which was moving at a slower heliocentric speed of 22.4 km/s. This resulted in a rear-end collision of the impactor spacecraft at a 10-km/s impact speed but with an impact approach angle of 15 deg, as illustrated in Fig. 13.9a. The kinetic energy of the impactor was 1.9×10^{10} J, and the resulting impact ΔV was practically zero. The Deep Impact mission was not intended to deflect the orbit of such a large 5-km comet. The attitude/position of the impactor spacecraft after being released from the flyby spacecraft was precisely controlled by the autonomous optical navigation (AutoNav) system to achieve a 300-m targeting accuracy [35]. The target

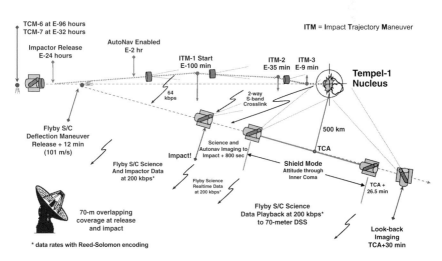

Fig. 13.10   Illustration of the target encounter phase for the Deep Impact mission. (Image courtesy of Ball Aerospace and Technologies Corporation and JPL.)

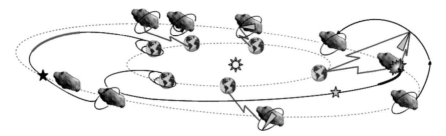

Fig. 13.11   Conceptual illustration of the Don Quijote mission [36]. (Image courtesy of ESA.)

encounter phase of the Deep Impact mission is illustrated in Fig. 13.10. The proposed solar-sail KEI mission will also require such a terminal targeting phase (but with an accuracy better than 100 m at a closing velocity of 70 km/s).

The European Space Agency has been developing a small kinetic-energy impactor precursor mission (named the Don Quijote mission) targeted for a 500-m asteroid [36]. The Don Quijote mission consists of two nearly identical spacecraft: an orbiter (Sancho) and an impactor (Hidalgo). Its overall mission objective is to measure the actual translational/rotational momentum transfer of the impact, impact crater/ejecta, and surface/internal properties before/after impact. Its conceptual mission scenario is illustrated in Fig. 13.11. The orbiter Sancho will be launched first. It will rendezvous with the asteroid to conduct an orbit determination of the target asteroid and also to measure its mass, size, gravity, and surface properties. After accomplishing a successful mission of the orbiter at the asteroid, the impactor Hidalgo will be launched from Earth. Similar to the Deep Impact mission, a target asteroid with a 29.83-km/s heliocentric speed will crash into an impactor spacecraft moving at a 25.18-km/s heliocentric speed. This will result in a rear-end collision of the impactor spacecraft at an impact velocity of 7.16 km/s with an impact approach angle of 11.4 deg, as illustrated in Fig. 13.9b. The resulting impact ΔV will be very small (less than 0.01 cm/s), but it is expected to be sufficient to achieve a 100-m change in the semimajor axis that can be measured with 10% accuracy [36]. To mitigate the real threat of NEAs in the future, a separate inspection mission similar to Deep Impact or Don Quijote might be required as an integral part of any large-scale space mission of deflecting NEAs.

The proposed solar-sailing KEI mission will provide a relatively high impact velocity of at least 70 km/s, compared to a typical impact velocity of 10 to 15 km/s of conventional missions with the gravity-assist flyby maneuvers. However, the proposed mission will require at least 10 KEI sailcraft. Each KEI sailcraft consists of a 160-m, 150-kg solar-sail and a 150-kg microspacecraft impactor. A characteristic acceleration of at least 0.5 mm/s^2 is required to intercept the target asteroid in five to six years. The KEI sailcraft with a fully deployed solar sail first spirals inwards from 1 AU to a heliocentric orbit at 0.25 AU, followed by a cranking orbit phase for a 168-deg inclination change. After completing the cranking orbit phase, the KEI sailcraft then spirals outwards in a retrograde orbit to intercept the target asteroid 2004WR, as illustrated in Fig. 13.6. A total flight time of six years will be required prior to impacting 2004WR at its perihelion on 1 January, 2012. The large solar sail will be jettisoned from the KEI spacecraft prior to the target encounter phase. Each impactor, with a relative impact velocity of at least 70 km/s, will cause a conservatively estimated ΔV of 0.3 cm/s in the trajectory of the 200-m target asteroid, caused largely by the impulsive effect of material ejected from the newly formed crater. The deflection caused by a single impactor will increase the Earth miss distance by $0.45 R_\oplus$, where $R_\oplus$ denotes the Earth radius of 6378 km. Therefore, at least 10 KEI sailcraft will be required to increase the Earth miss distance by $4.5 R_\oplus$. Because of a possible launch failure, physical modeling uncertainties, and mission reliability, we might have to launch more than 10 KEI sailcraft to mitigate the threat posed by a 200-m asteroid.

A conventional Delta II 2925 launch vehicle is capable of injecting at least two KEI sailcraft into an Earth escaping orbit at $C_3 = 0.25 \text{ km}^2/\text{s}^2$. Although the proposed mission concept requires a mission lead time of at least 10 years, it can be applicable to asteroids larger than 200 m by simply increasing the total number of 160-m, 300-kg sailcraft. Note that a Delta IV-Heavy (4250H-19) launch vehicle is capable of injecting a 9300-kg payload into an Earth escaping orbit at $C_3 = 0 \text{ km}^2/\text{s}^2$.

13.2.4 Solar-Sailing KEI Mission Trajectory Design and Simulation

The preliminary trajectory design and simulation results were discussed in Sec. 12.14.2, and further discussions of the solar-sailing KEI mission design can be found in [37–41]. An optimal impact velocity of 81 km/s was achieved by Dachwald and Wie [40], resulting in a conservatively estimated ΔV of 0.35 cm/s for the target asteroid 2004WR. Figure 13.12 shows the resulting trajectory and the orbital elements that define the shape and inclination of the final retrograde orbit prior to impacting the target asteroid at its perihelion. Numerical integration of the asteroid orbit after the impact, including all planetary disturbances, yielded a deflection distance of 4656 km (0.73 Earth radii) at 14 January 2015, the date of Earth impact. A detailed tradeoff study between different mission parameters (e.g., characteristic acceleration, sail temperature limit, hyperbolic excess energy for interplanetary insertion, and optical solar-sail degradation) was also performed in [40].

Dachwald et al. [41] have also applied the solar-sailing KEI approach to near-Earth asteroid 99942 Apophis, which will have a close Earth encounter in 2029 with potential very close subsequent Earth encounters (or even an impact) in 2036 or later. In Sec. 13.3.1 we will briefly study the orbital characteristics of asteroid Apophis and its impact threat concern.

13.2.5 Technology Advances Required for the Solar Sailing KEI Mission

A variety of technical issues inherent to the proposed solar-sailing KEI concept needs to be further examined to make the proposed concept to become a viable option for mitigating the threat of NEAs. Tradeoffs will be required to design a baseline mission architecture by considering technology readiness levels, cost, system complexity, feasibility, reliability, etc.

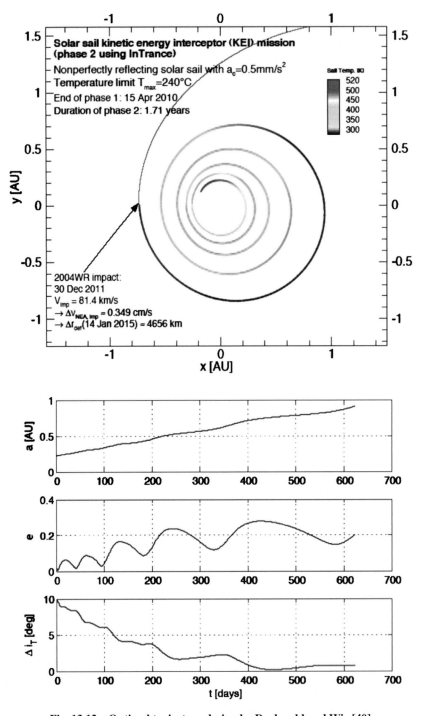

Fig. 13.12   Optimal trajectory design by Dachwald and Wie [40].

Table 13.1   Mass properties of scalable sailcraft of ATK Space Systems [24, 26, 27]

Sail size, m	40	80	160	240	320
Geometry					
Mast length, m	28	56	113	170	226
Mast diameter, m	0.4	0.4	0.6	0.6	0.6
Bending EI, N·m^2	82,441	82,441	438,847	438,847	438,847
Torsional GJ, N·m^2	197	197	453	453	453
Scallop factor, %	75	75	75	75	75
Sail area, m^2	1,200	4,800	19,200	43,200	76,800
Solar thrust (max), N	0.01	0.04	0.16	0.36	0.64
c.m./c.p. offset, m	0.1	0.2	0.4	0.6	0.8
Mass, kg					
Sails	6	19	67	100	200
Masts	7	14	60	90	120
Tip mass (each)	1	2	2	3	5
Central assembly	8	10	15	20	30
Sailcraft bus/impactor	100	125	150	400	500
Total mass	125	176	300	622	870
Characteristic acc., mm/s^2	0.08	0.23	0.53	0.58	0.73
Inertia, kg·m^2					
I_x (roll)	4,340	40,262	642,876	3.0E6	6E6
I_y (pitch)	2,171	20,136	321,490	1.5E6	3E6
I_z (yaw)	2,171	20,136	321,490	1.5E6	3E6

For example, mass properties of solar sails of various sizes are provided in Table 13.1. A 240-m solar sail can deliver a 400-kg impactor spacecraft in five years to a larger target asteroid; however, it will be more difficult to deploy and control such a 240-m solar sail, and thus system-level tradeoffs will be needed to optimize the overall mission architecture. Furthermore, for larger asteroids, the impactor spacecraft might not have to be separated from the solar sail; the complete solar-sail spacecraft could be designed to impact with a larger asteroid and increase the resulting ΔV. The differing sizes and compositions of asteroids could require a family of impactors, some containing small conventional explosive payloads to increase cratering efficiency. Concerns as to fragmentation might require the use of smaller impactors or even different orbits that allow deflection at lower relative velocities.

The critical, enabling technologies required for the proposed solar-sailing KEI mission include 1) deployment and control of a 160-m solar sail; 2) development of microspacecraft bus able to withstand the space environment only 0.25 AU from the sun; 3) precision solar-sailing navigation and terminal guidance and targeting (accuracy better than 100 m at an impactor speed of 70 km/s); and 4) impact-crater ejecta modeling and accurate ΔV prediction. A 160-m solar sail is not currently available, and the deployment and control of such a large solar sail in space will not be a trivial task. However, a variety of near-term solar sailing missions requiring

160-m solar sails and the associated solar-sail technologies is being developed [21–28]. In particular, a 160-m solar sail will be required for the SPI mission, which is one of the sun–Earth connections solar-sail road-map missions currently envisioned by NASA [22, 23]. Furthermore, the Deep Impact mission by NASA, with the spectacular collision with comet Tempel 1 on 4 July, 2005, has already provided significant technological advances required for the proposed solar-sailing KEI mission of mitigating the threat of NEAs.

13.3   Asteroid Deflection Using a Gravity Tractor

A practical concern of the kinetic-impact approach for mitigating the threat of NEAs is the risk that the impact could result in the fragmentation of NEAs, which could substantially increase the damage upon Earth impact. Consequently, Lu and Love [42] have recently proposed an asteroid deflection concept utilizing the mutual gravitational force between a hovering spacecraft and a target asteroid as a towline. For an apparently more fuel-efficient way of towing asteroids, McInnes [43] further discussed the use of a displaced, non-Keplerian orbit rather than a static hovering, which requires canted thrusters to avoid plume impingement on the NEA surface. Utilizing the same physical principle of gravitationally "anchoring" the spacecraft to the asteroid, without physical contact between the spacecraft and the asteroid, we can employ solar sails rather than electric-propulsion systems to produce the required continuous low-thrust force. Such a solar-sail gravity tractor, to be described in Sec. 13.5, exploits the "propellantless" nature of solar-sails for towing asteroids. The concept of gravitational coupling/towing using solar radiation pressure has been explored previously for large-scale astronomical problems by Shkadov [44] in 1987 and also by McInnes [45] in 2002.

In the remainder of this chapter, we use asteroid 99942 Apophis as an example target asteroid to illustrate the asteroid deflection concept utilizing the mutual gravitational force between a spacecraft and a target asteroid as a towline.

13.3.1   Asteroid 99942 Apophis

Asteroid 99942 Apophis, previously known by its provisional designation 2004 MN4, was discovered on 19 June, 2004. It is a 320-m NEA that is currently predicted to swing by the Earth in 2029 with a possible resonant return to impact the Earth in 2036. Apophis is an Aten-class asteroid with an orbital semimajor axis less than 1 AU, and its mass is estimated to be 4.6×10^{10} kg. It has an orbital period of 323 days about the sun. After its close flyby of the Earth in 2029, it will become an Apollo-class asteroid. It was previously predicted that Apophis will pass about 36,350 km from the Earth's surface on 13 April, 2029, slightly higher than the 35,786-km altitude of geosynchronous satellites. Recent observations using Doppler radar at the giant Arecibo radio telescope in Puerto Rico have further confirmed that Apophis will in fact swing by at around 32,000 km from the Earth's surface in 2029, but with a very slim chance of resonant return in 2036 (with an impact probability of 1 in 45,000, equivalently, with a no-impact probability of 99.998%).

Its orbital elements in the J2000 heliocentric ecliptic reference frame are as follows:

$$Epoch = JD\ 2454200.5\ TDB\ (10\ April,\ 2007)$$
$$a = 0.9222614\ AU$$
$$e = 0.191059$$
$$i = 3.331\ deg$$
$$\omega = 126.365\ deg$$
$$\Omega = 204.462\ deg$$
$$M = 222.273\ deg\ (at\ epoch)$$

Its other orbital properties are $r_p = 0.746\ AU$, $r_a = 1.0986\ AU$, $v_p = 37.6\ km/s$, $v_a = 25.5\ km/s$, the orbital period $= 323.574$ days, the mean orbital rate $n = 2.2515 \times 10^{-7}$ rad/s, and the mean orbital speed $= 30.73$ km/s. Further accurate observations of its orbit are expected when it makes fairly close flybys at 0.1 AU from Earth in 2013 and 2021.

As discussed in [34, 41, 46, and 47], an extremely small amount of impact ΔV (approximately 0.04 mm/s) in 2026 is required to move Apophis out of a 600-m keyhole area by approximately 10 km in 2029, in case it is going to pass through a keyhole, to completely eliminate any possibility of its resonant return impact with the Earth in 2036. Keyholes are very small regions of the first encounter b-plane such that if an asteroid passes through them, it will have a resonant return impact with the Earth [48–51].

Dachwald et al. [41] have applied the solar-sailing KEI approach to deflecting Apophis assuming that it is going to pass through a 600-m keyhole in 2029. About 20,000 potential Apophis orbits were generated in [47] by random variation of the orbital elements within the 3-σ accuracy. Two of them (here termed Ap1 and Ap2) have been found to collide with the Earth, both during a 7:6 resonant return on 13 April, 2036. They are used in [41] as potential impact trajectories of Apophis. Their orbital elements, before and after the 2029 encounter, are listed in Table 13.2, and their collision orbits with the Earth are shown in Fig. 13.13.

Table 13.2   Orbital elements of the Earth-impacting Apophis orbit variations [41]

Elements	Before 2029 encounter		After 2029 encounter	
	Ap1	Ap2	Ap1	Ap2
MJD	53459.0	53459.0	64699.0	64699.0
a, AU	0.9223913	0.9223912	1.1082428	1.1082581
e	0.191038	0.191038	0.190763	0.190753
i, deg	3.331	3.331	2.166	2.169
ω, deg	126.384	126.383	70.230	70.227
Ω, deg	204.472	204.472	203.523	203.523
M, deg	203.974	203.974	227.857	227.854

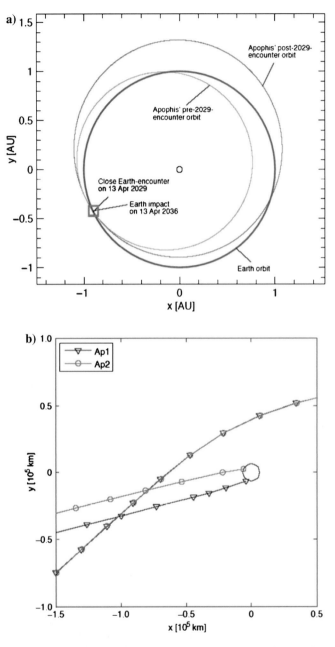

Fig. 13.13   Earth-impacting Apophis orbit variations [41]: a) Comparison of Ap1's and Ap2's pre- and post-2029-encounter orbit and b) closeup of the 2029 encounter and the impact (geocentric reference frame).

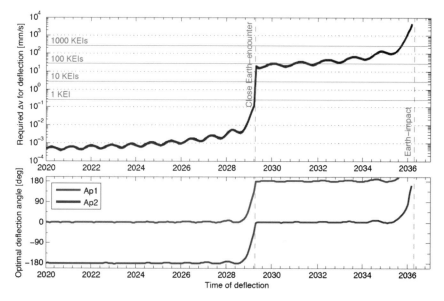

Fig. 13.14   Required velocity change and optimal deflection angle for solar-sail KEI missions for Apophis [41].

Figure 13.14 shows ΔV required for a successful deflection of Ap1 and Ap2 to a safe distance of two Earth radii, as well as the optimal deflection angles. One can see that after the 2029 encounter a successful deflection of Apophis requires in the best case about 100 KEIs, assuming that every consecutive KEI impact has the optimal deflection angle and provides the same ΔV of 0.27 mm/s. Of course, a pre-encounter impact is clearly the better option for deflecting Apophis, but this option might not be available for other NEOs that do not have a close encounter before they impact the Earth.

13.3.2 Gravity Tractor Concept

The gravity tractor (GT) concept by Lu and Love [42] utilizes the mutual gravitational force between a hovering spacecraft and a target asteroid as a towline as illustrated in Fig. 13.15. Although a 20-ton spacecraft propelled by a nuclear-electric propulsion system is considered in [42], we consider here a 1000-kg spacecraft as an illustrative example applied to asteroid Apophis. To avoid exhaust plume impingement on the asteroid surface, two ion engines are properly tilted outward, and the hovering distance is accordingly selected as $d = 1.5r$ and $\phi = 20$ deg. This illustrative combination yields an engine cant angle of 60 deg, and the two tilted thrusters (each with a thrust T) then produce a total towing thrust T as illustrated in Fig. 13.15. (For an apparently more fuel-efficient way of towing asteroids, McInnes [43] proposed the use of a displaced, non-Keplerian orbit rather than a static hovering, which requires such canted thrusters to avoid plume impingement on the NEA surface.)

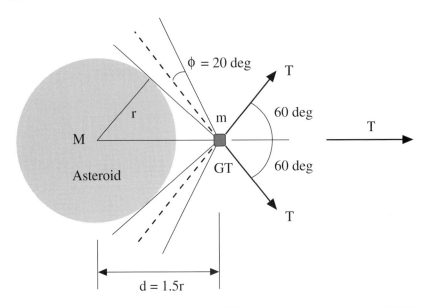

Fig. 13.15   Geometrical illustration of the GT concept for towing an asteroid [42].

A simplified dynamical model for the target asteroid Apophis (ignoring its orbital motion) is

$$M \frac{\Delta V}{\Delta t} = \frac{GMm}{d^2} = T \qquad (13.5)$$

or

$$\frac{\Delta V}{\Delta t} = \frac{Gm}{d^2} = \frac{T}{M} = A \qquad (13.6)$$

where $G = 6.6695 \times 10^{-11}$ N·m^2/kg^2, $M = 4.6 \times 10^{10}$ kg, $m = 1000$ kg, $r = 160$ m. Here $d = 240$ m, $T = 0.05326$ N, $A = 1.1579 \times 10^{-9}$ mm/s^2 is the characteristic acceleration, and

$$\Delta V = A \Delta t \qquad (13.7a)$$

$$\Delta X = \frac{1}{2} A (\Delta t)^2 \qquad (13.7b)$$

where ΔV and ΔX are, respectively, the resulting velocity and position changes for the total towing period of Δt. For example, we have $\Delta V = 0.036$ mm/s and $\Delta X = 575$ m for $\Delta t =$ one year.

Including the orbital "amplification" effect (to be further discussed in Sec. 3.6), we have

$$\Delta V = 3A \Delta t \qquad (13.8a)$$

$$\Delta X = \frac{3}{2} A (\Delta t)^2 \tag{13.8b}$$

Consequently, we have $\Delta V = 0.1$ mm/s and $\Delta X = 1.7$ km for one-year towing.

Including an additional coasting time of t_c, we have the total position change given by

$$\Delta X = \frac{3}{2} A \Delta t (\Delta t + 2 t_c) \tag{13.9}$$

This deflection formula will be derived in Sec. 3.6. One-year towing in 2026 with an additional coasting time of three years will cause a total position change of approximately 12 km in 2029, which is more than sufficient to move Apophis out of its 600-m keyhole in 2029.

The propellant amount required for maintaining a desired hovering altitude of 80 m can be estimated as

$$\Delta m_f = \frac{2 T \Delta t}{g_o I_{\text{sp}}} \approx 0.3 \text{ kg per day} \approx 114 \text{ kg per year}$$

where $T = 0.053$ N, $g_o = 9.8$ m/s^2, and $I_{\text{sp}} = 3000$ s (assumed for typical ion engines).

Therefore, a 1000-kg GT spacecraft equipped with ion engines can be considered as a viable option for a pre-2029 deflection mission for Apophis. However, it is emphasized that a 1000-kg spacecraft, colliding with Apophis at a modest impact velocity of 10 km/s in 2026, will cause a much larger, instantaneous velocity change of at least 0.22 mm/s for Apophis, resulting in an orbital deflection of 62 km in 2029. Such a high-energy kinetic impactor approach might not be applicable to highly porous, rubble-pile asteroids, and a GT spacecraft mission might need an additional large ΔV to rendezvous with a target asteroid. Consequently, a further study on various issues, such as the total mission ΔV requirement, low-thrust gravity towing vs high-energy kinetic impact, asteroid dispersal/fragmentation concern, etc., is needed.

13.4 Multiple Gravity Tractors in Halo Orbits

13.4.1 Introduction

Because a hovering GT in a static equilibrium standoff position requires canted thrusters to avoid plume impingement on the NEA surface, McInnes [43] has recently investigated a GT flying in a displaced non-Keplerian orbit (also often called a halo orbit) for a possible fuel-efficient way of towing asteroids. From a practical viewpoint, such a GT in a displaced orbit does not have an overall system-level advantage over a single hovering GT with two canted thrusters. A gravity tractor in a displaced orbit will require a much heavier spacecraft (about 2.8 times heavier than a single hovering GT) if its orbital displacement is the same as the standoff distance of a hovering GT. Or it will need to be placed much closer to the

of 16 h and a center-of-mass/center-of-gravity offset of 10 m. An ideal exhaust plume model of a uniform cone with a half-angle of 20 deg [42] is also assumed.

As can be seen in Figs. 13.18–13.20, an ideal circular halo orbit is achieved for case 1. Simulation results shown in Figs. 13.21–13.23 demonstrate that near-circular halo orbits can also be achieved for a realistic case 2. The 1000-kg GT spacecraft can be replaced by two 500-kg small satellites by properly placing them in a halo orbit. The resulting asteroid ΔV (Fig. 13.23) can be increased, if needed, by simply adding more GTs in halo orbits. However, a further system-level design tradeoff study is needed for overall mission design tradeoffs, propulsion system selection tradeoffs, halo orbit insertion/maintenance, high-fidelity dynamical modeling, orbit control design, and simulations.

13.5   Hovering Solar-Sail Gravity Tractor

Utilizing the same physical principle of employing the mutual gravitational force between a hovering spacecraft and a target asteroid as a towline, we can choose solar sails rather than electric propulsion systems to produce the required continuous low-thrust towing force. This section presents a preliminary conceptual development of such a solar-sail gravity tractor for asteroid deflection as proposed in [52].

13.5.1   Solar-Sail Gravity Tractor Options

The basic physical principle of the solar-sail gravity tractor (SSGT) spacecraft hovering over the NEA surface is illustrated in Fig. 13.24. The concept of gravitational coupling/towing by the use of solar radiation pressure has been proposed previously for somewhat science-fictional, astronomical problems by Shkadov [44] in 1987 and also by McInnes [45] in 2002. Solar sails are large, lightweight reflectors in space that are propelled by sunlight. The SSGT spacecraft discussed in this section exploits the "propellantless" nature of solar sails for towing asteroids; consequently, its probable advantage over a GT spacecraft propelled by ion engines is its longer mission lifetimes (>10 years) with a larger propellantless ΔV capability. Furthermore, it has no concern of rocket plume impingement on the asteroid surface.

A NEA deflection system architecture consisting of GT spacecraft options, SSGT spacecraft options, a solar-sail KEI, and a NEO orbiting surveyor spacecraft is illustrated in Fig. 13.25. For a solar-sail KEI mission, its solar sail will be deployed at the beginning of an interplanetary solar-sailing phase toward a target asteroid, and the KEI spacecraft will be separated from the solar sail prior to impacting a target asteroid. For the SSGT spacecraft mission, its solar sail will be deployed after completing a rendezvous with a target asteroid.

As illustrated in Fig. 13.25, three different options of GT/SSGT spacecraft are possible as follows:

1) *Baseline*: A GT with two canted ion engines hovers at $(x, y) = (240, 0)$ m.

2) *Option 1*: A GT with orthogonally mounted ion engines hovers at $(x, y) = (200, 200)$ m.

3) *Option 2*: An SSGT with an ion engine and a solar-sail hovers at $(x, y) = (200, 200)$ m.

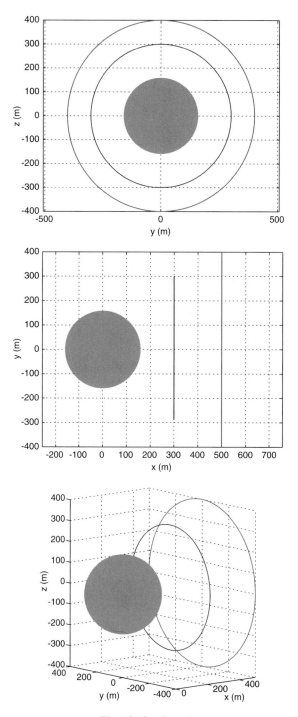

Fig. 13.18   Case 1.

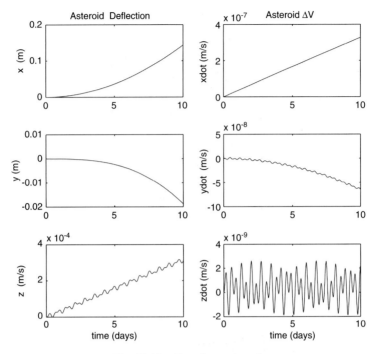

Fig. 13.19   Case 1, continued.

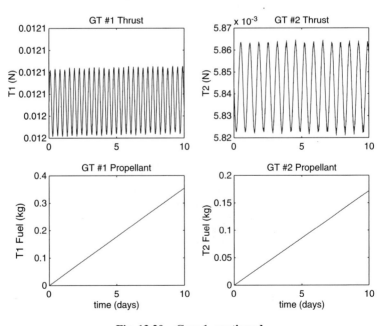

Fig. 13.20   Case 1, continued.

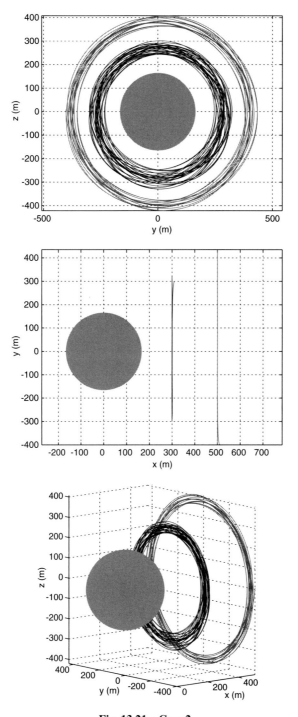

Fig. 13.21   Case 2.

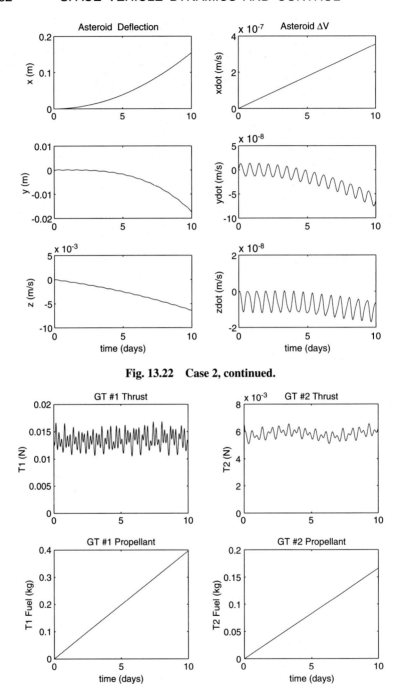

Fig. 13.22 Case 2, continued.

Fig. 13.23 Case 2, continued.

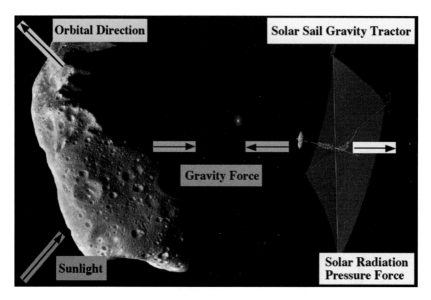

Fig. 13.24   Illustration of the SSGT concept for asteroid deflection.

4) *Option 3*: An SSGT with a solar-sail (35-deg sun angle) hovers at $(x, y) =$ $(286, 409)$ m.

As illustrated in Fig. 13.25, it is feasible to employ a GT with two orthogonally mounted ion engines (option 1). Because only the x-axis thrust force provides an effective ΔV of the target asteroid, the y-axis ion engine of the option 1 GT can be replaced by a 70-m solar-sail resulting in an option 2 SSGT spacecraft. An option 3 SSGT spacecraft propelled using only a solar sail (35-deg sun angle) is also proposed for a case that might require much longer mission lifetimes (>5 years) with a much larger propellantless ΔV capability. However, the option 3 SSGT spacecraft requires a heavier, 2500-kg spacecraft with a 90-m solar sail to be able to hover at a slightly higher altitude of 340 m, compared to the other spacecraft hovering at an altitude of 120 m. In practice, multiple ion engines and redundant position/attitude control thrusters will be required for the GT/SSGT spacecraft, and thus a detailed system-level tradeoff study (e.g., ion engines vs solar sails) is needed for the GT/SSGT spacecraft design.

As an example, for Apophis, we might need a 2500-kg SSGT spacecraft, equipped with a 90×90-m solar sail of a 0.03-N solar thrust with a 35-deg sun angle as illustrated in Fig. 13.26. Such a particular hovering position with an offset angle of $\theta = 55$ deg from an asteroid's flight direction is necessary because of the 35-deg sun angle requirement of a typical solar-sail/collector to produce a maximum solar-pressure thrust. A larger (2500-kg) SSGT, compared to a 1000-kg GT equipped with ion engines, is to be placed at a higher altitude of 350 m because of its large solar-sail/collector. This 2500-kg SSGT produces an along-track acceleration of $A_x = 3.8 \times 10^{-10}$ mm/s^2 and a radial acceleration of $A_y = 5.4 \times 10^{-10}$ mm/s^2 of the target asteroid Apophis. However, its radial acceleration component

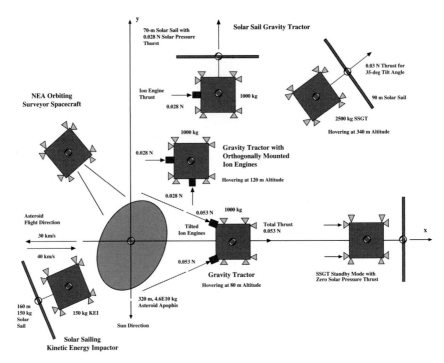

Fig. 13.25   NEA deflection system architecture consisting of a gravity tractor (without solar sail), an SSGT, a solar-sail KEI, and an NEA orbiting surveyor spacecraft.

A_y has a negligible effect on the asteroid deflection. This fact can be considered as an inherent drawback of the SSGT although a solar sail is a propellantless propulsion system. A five-year towing of Apophis using this 2500-kg SSGT spacecraft and a three-year coasting time will result in an orbital deflection of 30 km in 2029, which is more than sufficient to move Apophis out of its 600-m keyhole. However, the SSGT spacecraft is intended to be employed for a much longer mission lifetime (>5 years) to fully exploit its propellantless ΔV capability, which is its only advantage over a GT equipped with ion engines.

In the next section, simplified dynamical modeling and hovering control of the SSGT spacecraft will be presented to validate the technical feasibility of the SSGT concept for towing asteroids.

13.5.2   Dynamic Modeling and Hovering Control

A simple planar model of the hovering dynamics of an SSGT spacecraft towing a target asteroid is illustrated in Fig. 13.26. Utilizing the Clohessy–Wiltshire–Hill equations of motion in astrodynamics, we can derive the equations of motion of the asteroid-spacecraft system orbiting around the sun as follows:

$$\ddot{x}_1 = 2n\dot{y}_1 + Gm_2\frac{x_2 - x_1}{r^3} \tag{13.23}$$

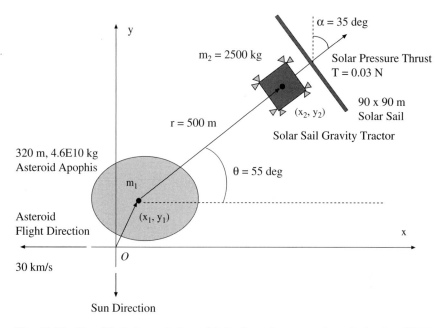

Fig. 13.26   Simplified dynamical model for hovering control analysis of a SSGT spacecraft.

$$\ddot{y}_1 = -2n\dot{x}_1 + 3n^2 y_1 + Gm_2 \frac{y_2 - y_1}{r^3} \qquad (13.24)$$

$$\ddot{x}_2 = 2n\dot{y}_2 - Gm_1 \frac{x_2 - x_1}{r^3} + \frac{1}{m_2}(T_x + F_x) \qquad (13.25)$$

$$\ddot{y}_2 = -2n\dot{x}_2 + 3n^2 y_2 - Gm_1 \frac{y_2 - y_1}{r^3} + \frac{1}{m_2}(T_y + F_y) \qquad (13.26)$$

where (x_1, y_1) are the coordinates of the target asteroid with respect to an orbiting reference frame, (x_2, y_2) the coordinates of the SSGT spacecraft, (T_x, T_y) solar-pressure thrust components, (F_x, F_y) control thrust components, $r = \sqrt{(x_2 - x_1)^2 + (y_2 - y_1)^2}$, $G = 6.6695 \times 10^{-11}$ N·m^2/kg^2, m_1 the asteroid mass, m_2 the SSGT spacecraft mass, and n the orbital rate of the reference frame (x, y). For simplicity, a circular orbital motion of the reference frame is considered here. The eccentric orbital effect of a target asteroid will be discussed in Sec. 13.6. The gravitational perturbation effect caused by a spinning motion of an irregularly shaped asteroid will be included in simulations.

Preliminary hovering control logic is considered as follows:

$$\tan \theta = \frac{y_2 - y_1}{x_2 - x_1}$$

$$\alpha = \frac{\pi}{2 - \theta}$$

$$T_x = T_o \cos^2 \alpha \sin \alpha$$

$$T_y = T_o \cos^2 \alpha \cos \alpha$$

$$x = x_2 - x_1 = r \cos \theta$$

$$y = y_2 - y_1 = r \sin \theta$$

$$F_x = -K_P(x - x_c) - K_d \dot{x}$$

$$F_y = -K_P(y - y_c) - K_d \dot{y}$$

$$\text{if } |F_x| > F_{max}, \ F_x = \text{sgn}(F_x)F_{max}$$

$$\text{if } |F_y| > F_{max}, \ F_y = \text{sgn}(F_y)F_{max}$$

$$\text{if } |x - x_c| < \epsilon_x, \ F_x = 0$$

$$\text{if } |y - y_c| < \epsilon_y, \ F_y = 0$$

where $T_o = 0.045$ N for a 90-m solar sail at 1 AU, $F_{max} = 0.1$ N, $\epsilon_x = \epsilon_y = 10$ m, (x_c, y_c) is the desired hovering position command, $K_p = 0.00001m_2$, and $K_d = 0.03m_2$. For simplicity, electric thrusters with a maximum thrust of 0.1 N and a specific impulse of 3000 s are considered for the hovering control of SSGT spacecraft. An attitude control problem of the SSGT spacecraft is not considered here.

Hovering control simulation results are shown in Figs. 13.27–13.31 for a 2500-kg SSGT spacecraft (option 3). For these simulations, a spin period of 16 h and a center-of-mass/center-of-gravity offset of 10 m are assumed. As can be estimated from Fig. 13.31, less than 7-kg propellant per year (worst case) might be needed for a 2500-kg SSGT spacecraft to hover above a target asteroid with its quite uncertain gravitational environment. Any undesirable cyclic thruster firings can be eliminated by employing a cyclic-disturbance rejection control scheme to reduce the hovering control propellant consumption to less than 2 kg per year. However, a more rigorous dynamical model of the gravitational field of a slowly rotating irregularly shaped asteroid must be used for the detailed hovering control design and high-fidelity simulations, as discussed in [53]. The Yarkovsky effect must also be included in such a rigorous dynamical model.

13.5.3 Gravity-Gradient Torque Estimation

Consider the pitch-axis attitude dynamical model of the SSGT described by

$$I_2 \ddot{\theta}_2 - \frac{3\mu}{R^3}(I_3 - I_1)\theta_2 = 0 \tag{13.27}$$

where $\mu = GM = 3$ N·m²/kg ($G = 6.6695 \times 10^{-11}$ N·m²/kg², $M = 4.6 \times 10^{10}$ kg); $I_1 = I_2 = 300{,}000$ kg·m²; $I_3 = 600{,}000$ kg·m²; and $R = 500$ m. For this case, we have

$$I_2 \ddot{\theta}_2 - 0.02\theta_2 = 0 \tag{13.28}$$

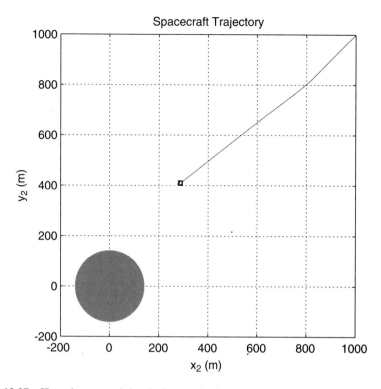

Fig. 13.27 **Hovering control simulation results for the SSGT (option 3). Starting point at $(x, y) = (1000, 1000)$ m and a desired hovering point $(x, y) = (286, 409)$ m.**

The peak gravity-gradient torque can be estimated as 0.01 N·m for an assumed peak pitch attitude error of 30 deg. Although such a small, peak gravity-gradient disturbance torque is not of practical concern, a detailed attitude control study is needed as discussed in [53].

13.6 Asteroid Deflection Dynamics

In this section, we employ Clohessy–Wiltshire–Hill equations to discuss the fundamentals of asteroid deflection formulas [54–56]. The long-term effect of target asteroid's orbital eccentricity on asteroid deflection is also discussed using simulation results.

13.6.1 Orbital Ampification Effect on the Miss Distance

Consider the Clohessy–Wiltshire–Hill equations of motion of a target asteroid (in an assumed heliocentric circular orbit) described by

$$\ddot{x} = 2n\dot{y} + A_x \tag{13.29}$$

$$\ddot{y} = -2n\dot{x} + 3n^2 y + A_y \tag{13.30}$$

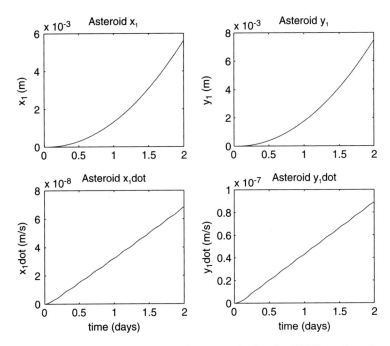

Fig. 13.28   Hovering control simulation results for the SSGT, continued.

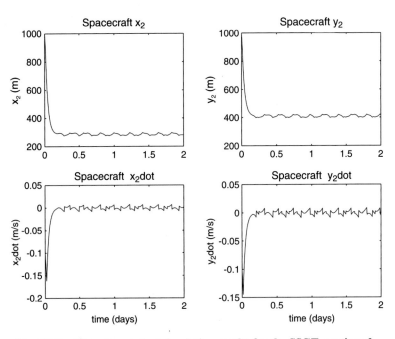

Fig. 13.29   Hovering control simulation results for the SSGT, continued.

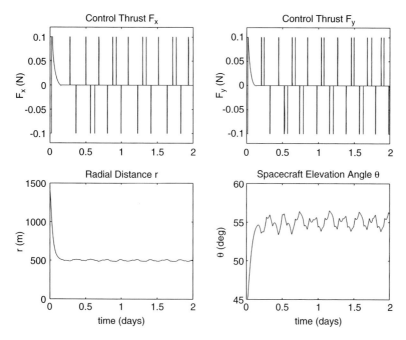

Fig. 13.30   Hovering control simulation results for the SSGT, continued.

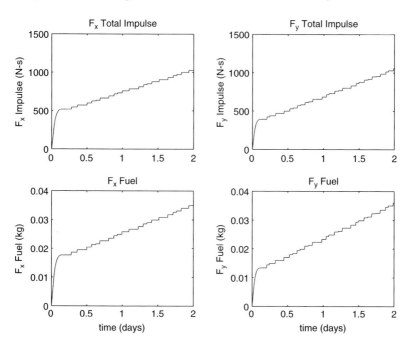

Fig. 13.31   Hovering control simulation results for the SSGT, continued.

where (x, y) are the coordinates of an asteroid with respect to a circular orbit reference frame shown in Fig. 13.26 and (A_x, A_y) are the "gravitational towing" acceleration components acting on the asteroid. The out-of-plane motion is not considered here. A simple case with $A_x = A = $ constant and $A_y = 0$ is further assumed here without loss of generality because the asteroid deflection effect of a nonzero A_y is practically negligible. For an asteroid towed by a single hovering GT, we have $A_y \equiv 0$.

Integrating the x-axis equation, we obtain

$$\dot{x} = \dot{x}(0) + 2ny + At \tag{13.31}$$

where $\dot{x}(0)$ denotes the along-track velocity at $t = 0^-$. All other initial conditions will be ignored here. For a kinetic-energy impactor problem, the initial impact ΔV along the x-axis direction becomes $\dot{x}(0)$.

Substituting Eq. (13.31) into the y-axis equation, we obtain

$$\ddot{y} + n^2 y = -2n\dot{x}(0) - 2nAt \tag{13.32}$$

Its solution can be found as

$$y(t) = -\frac{2}{n}\dot{x}(0)(1 - \cos nt) - \frac{2}{n}A\left(t - \frac{1}{n}\sin nt\right) \tag{13.33}$$

and

$$\dot{y}(t) = -2\dot{x}(0)\sin nt - \frac{2}{n}A(1 - \cos nt) \tag{13.34}$$

We then obtain

$$\dot{x}(t) = -\dot{x}(0)(3 - 4\cos nt) - 3At + \frac{4}{n}A\sin nt \tag{13.35}$$

which can be integrated as

$$x(t) = -\dot{x}(0)\left(3t - \frac{4}{n}\sin nt\right) - \frac{3}{2}At^2 + \frac{4}{n^2}A(1 - \cos nt) \tag{13.36a}$$

$$\approx -3\dot{x}(0)t - \frac{3}{2}At^2 \qquad\qquad \text{for large t} \tag{13.36b}$$

The orbital "amplification" factor of three can be seen from the preceding equation. Note that the positive values of $\dot{x}(0)$ and A slow down the asteroid and reduce its orbital energy. Consequently, its along-track position becomes negative (i.e., ahead of its unperturbed virtual position in a circular reference orbit).

Consider an asteroid with the accelerated towing time of t_a and the additional coasting time of t_c. It is assumed that $\dot{x}(0) = 0$ here. A new set of initial conditions at the end of towing period becomes:

$$x_0 = -\frac{3}{2}At_a^2 + \frac{4}{n^2}A(1 - \cos nt_a)$$

$$\dot{x}_0 = -3At_a + \frac{4}{n}A\sin nt_a$$

$$y_0 = -\frac{2}{n}A\left(t_a - \frac{1}{n}\sin nt_a\right)$$

$$\dot{y}_0 = -\frac{2}{n}A(1 - \cos nt_a)$$

The final position changes at the end of the coasting phase can then be found as

$$\Delta x = x_0 + (6ny_0 - 3\dot{x}_0)t_c + \frac{2\dot{y}_0}{n}(1 - \cos nt_c) + \left(\frac{4\dot{x}_0}{n} - 6y_0\right)\sin nt_c \quad (13.37)$$

$$\Delta y = 4y_0 - \frac{2\dot{x}_0}{n} + \left(\frac{2\dot{x}_0}{n} - 3y_0\right)\cos nt_c + \frac{\dot{y}_0}{n}\sin nt_c \quad (13.38)$$

Substituting the initial conditions into Eq. (13.37), we obtain

$$\Delta x \approx -\frac{3}{2}At_a(t_a + 2t_c) \quad (13.39)$$

which is the low-thrust deflection formula derived in [54–56] using different approaches. Note that $\Delta y \approx 0$ compared to Δx.

Equation (13.39) can be rewritten as

$$\Delta x = -\left(\frac{3}{2}At_a^2 + \Delta Vt_c\right) \quad \text{where} \quad \Delta V = 3At_a \quad (13.40)$$

Note that Δx is caused by various initial conditions including $\dot{x}_0$ and y_0, as can be seen in Eq. (13.37). Such a combined effect of $\dot{x}_0$ and y_0 results in the term ΔVt_c (not $3\Delta Vt_c$ as one might expect) in Eq. (13.40).

For an asteroid in an eccentric orbit colliding with the Earth, we have

$$V = V_\oplus\sqrt{2 - \frac{r_\oplus}{a}} \quad (13.41)$$

$$e^2 = (\lambda - 1)^2\cos^2\gamma + \sin^2\gamma \quad (13.42)$$

where V is its heliocentric velocity at an impact point, a its semimajor axis, e its eccentricity, $r_\oplus = 1$ AU $= 1.496 \times 10^8$ km, $V_\oplus = 29.784$ km/s, γ the intersection angle between $\vec{V}$ and $\vec{V}_\oplus$, and $\lambda = (V/V_\oplus)^2$. The heliocentric elevation angle γ is also called the flight-path angle.

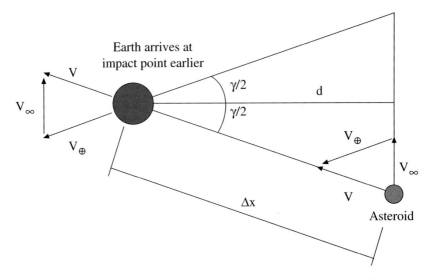

Fig. 13.32   Asteroid deflection geometry for a case with $a \approx 1$ AU.

As illustrated in Fig. 13.32 for a case with $a \approx r_\oplus$, we have

$$V \approx V_\oplus$$

$$e \approx \sin \gamma$$

$$d = \Delta x \cos \left(\frac{\gamma}{2}\right)$$

where d is the approach distance (also called the b-plane miss distance). The impact parameter b, which defines the radius of a collision cross section, is given by

$$b = R_\oplus \sqrt{1 + \frac{V_e^2}{V_\infty^2}} \qquad (13.43)$$

where $R_\oplus$ is the radius of the Earth ($= 6378$ km), $V_e = \sqrt{2\mu_\oplus/R_\oplus}$ the escape velocity from the surface of the Earth ($= 11.18$ km/s), and V_∞ the hyperbolic approach velocity. To avoid an impact, we simply need $d > b$.

For a special case of a near head-on collision of an NEO with the Earth, we have

$$d = \Delta x \sin(\gamma/2) \approx -\frac{3e}{4} A t_a (t_a + 2t_c)$$

This formula was previously presented in [56] without mentioning that this formula is only applicable to a retrograde (head-on) collision case. In practice, it will be extremely difficult to deflect an NEO, which is in a head-on collision path toward the Earth, also evidenced in this formula.

For an impulsive ΔV along the x-axis direction, the resulting deflection Δx after a coasting time of t_c is simply given by

$$\Delta x = -3\Delta V t_c \tag{13.44}$$

For a kinetic impactor approach, ΔV can be estimated as

$$\Delta V \approx \beta \frac{m}{M+m} U \approx \beta \frac{m}{M} U \tag{13.45}$$

where β is the impact efficiency factor, m the impactor mass, M the target asteroid mass, and U the relative impact velocity.

Consider a numerical test case (option 3 SSGT) for an asteroid with a circular orbital period of 323 days ($n = 2.2515 \times 10^{-7}$ rad/s), an assumed low-thrust acceleration of $A_x = 3.8284 \times 10^{-10}$ mm/s^2, $A_y = 5.4667 \times 10^{-10}$ mm/s^2, $t_a = 5$ years, and $t_c = 3$ years. The final position change Δx can be estimated as -30 km, as can also be noticed in Fig. 13.33.

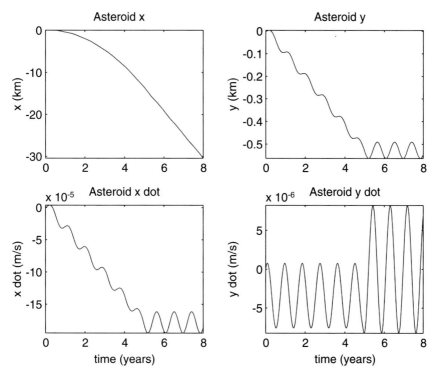

Fig. 13.33   Long-term simulation of the SSGT for Apophis in an assumed circular obit.

13.6.2 Eccentricity Effect on the Miss Distance

Consider the Clohessy–Wiltshire–Hill equations of motion of a target asteroid in an elliptical reference orbit as described by

$$\ddot{x} = 2\dot{\theta}\dot{y} + \ddot{\theta}y + \dot{\theta}^2 x - \frac{\mu}{r^3}x + A_x \tag{13.46}$$

$$\ddot{y} = -2\dot{\theta}\dot{x} - \ddot{\theta}x + \dot{\theta}^2 y + \frac{2\mu}{r^3}y + A_y \tag{13.47}$$

$$\ddot{r} = r\dot{\theta}^2 - \frac{\mu}{r^2} \tag{13.48}$$

$$\ddot{\theta} = -\frac{2\dot{r}\dot{\theta}}{r} \tag{13.49}$$

where (x, y) are the relative coordinates of the target asteroid with respect to a reference point of its nominal elliptical orbit, r is the radial distance of the reference orbit from the sun, θ is the true anomaly, and μ is the gravitational parameter of the sun. Furthermore, we have

$$r = \frac{p}{1 + e\cos\theta} \tag{13.50a}$$

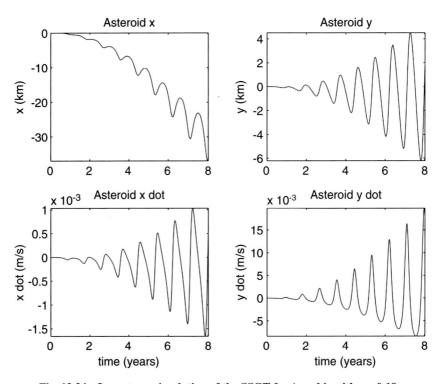

Fig. 13.34   Long-term simulation of the SSGT for Apophis with $e = 0.19$.

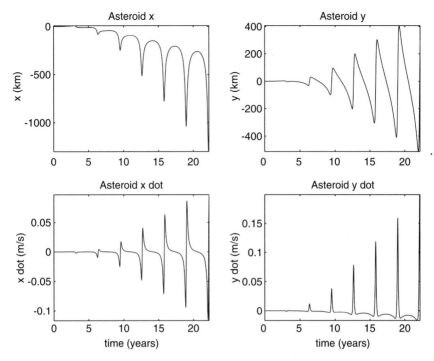

Fig. 13.35   Long-term simulation of the SSGT for a 200-m asteroid with $M = 1.1 \times 10^{10}$ kg, $a = 2.1537\,\text{AU}, e = 0.6498, A_x = 1.76 \times 10^{-9}$ mm/s^2, $A_y = 2.51 \times 10^{-9}$ mm/s^2, $t_a = 10$ years, and $t_c = 12$ years.

$$\dot{r} = \sqrt{\mu/p}\,(e\sin\theta) \tag{13.50b}$$

$$\dot{\theta} = \sqrt{\mu/p^3}\,(1 + e\cos\theta)^2 \tag{13.50c}$$

where $p = a(1 - e^2)$.

Consider the option 3 SSGT case for Apophis with $a = 0.92239\,\text{AU}$, $e = 0.19$, an assumed low-thrust acceleration of $A_x = 3.8284 \times 10^{-10}$ mm/s^2, $A_y = 5.4667 \times 10^{-10}$ mm/s^2, $t_a = 5$ years, and $t_c = 3$ years. The eccentricity effect on the nonsecular terms is evident in Fig. 13.34, as compared to Fig. 13.33 for the same case but with $e = 0$.

To further examine the eccentricity effect, consider the option 3 SSGT case for a 200-m asteroid with $M = 1.1 \times 10^{10}$ kg, $a = 2.1537\,\text{AU}$, $e = 0.6498$, $A_x = 1.76 \times 10^{-9}$ mm/s^2, and $A_y = 2.51 \times 10^{-9}$ mm/s^2. Simulation results for $t_a = 10$ years and $t_c = 12$ years are shown in Figure 13.35. The significant effect of a large eccentricity ($e = 0.6498$) is evident in Fig. 13.35. Consequently, further studies are needed to explore 1) recent research results of the eccentric C-W-H problem described in [57–60] and 2) a more realistic hovering control problem near asteroids examined in [61–64].

References

[1] Brown, P., Spalding, R. E., ReVelle, D. O., Tagliaferri, E., and Worden, S. P. "The Flux of Small near-Earth Objects Colliding with the Earth," *Nature*, Vol. 420, 21 Nov., 2002, pp. 294–296.

[2] NASA Authorization Act of 2005 (now known as the George E. Brown, Jr., Near-Earth Object Survey Act), U.S. Congress, Washington, D.C., 22 Dec. 2005.

[3] Cheng, A., "Near Earth Asteroid Rendezvous: Mission Summary," *Asteroids III*, Univ. of Arizona Press, Tucson, AZ, 2002, pp. 351–366.

[4] Ahrens, T. J., and Harris, A. W., "Deflection and Fragmentation of near-Earth Asteroids," *Hazards due to Comets and Asteroids*, edited by T. Gehrels, Univ. of Arizona Press, Tucson, AZ, 1994, pp. 897–927.

[5] Holsapple, K., "About Deflecting Asteroids and Comets," *Mitigation of Hazardous Comets and Asteroids*, Cambridge Univ. Press, New York, 2005, pp. 113–140.

[6] Gold, R. E., "SHIELD: A Comprehensive Earth Protection System," A Phase 1 Report, NASA Inst. for Advanced Concepts, NIAC, 28 May, 1999, http://www.niac.usra.edu/.

[7] Schweickart, R., Lu, E., Hut, P., and Chapman, C., "The Asteroid Tugboat," *Scientific American*, Nov. 2003, pp. 54–61.

[8] Adams, R. B., Alexander, R., Bonometti, J., Chapman, J., Fincher, S., Hopkins, R., Kalkstein, M., Polsgrove, T., Statham, G., and White, S., "Survey of Technologies Relevant to Defense from near-Earth Objects," NASA-TP-2004-213089, NASA-MSFC, July 2004.

[9] Richardson, D. C., Leinhardt, Z. M., Melosh, H. J., Bottke, W. F., Jr., and Asphaug, E., "Gravitational Aggregates: Evidence and Evolution," *Asteroids III*, Univ. of Arizona Press, Tucson, AZ, 2002, pp. 501–515.

[10] Solem, J. C., "Interception of Comets and Asteroids on Collision Course with Earth," *Journal of Spacecraft and Rockets*, Vol. 30, No. 2, 1993, pp. 222–228.

[11] Hall, C. D., and Ross, I. M., "Dynamics and Control Problems in the Deflection of Near-Earth Objects," American Astronautical Society, AAS-97-640, Aug. 1997.

[12] Park, S.-Y., and Ross, I. M., "Two-Body Optimization for Deflecting Earth-Crossing Asteroids," *Journal of Guidance, Control, and Dynamics*, Vol. 22, No. 3, 1999, pp. 415–420.

[13] Conway, B. A., "Near-Optimal Deflection of Earth Approaching Asteroids," *Journal of Guidance, Control, and Dynamics*, Vol. 24, No. 5, 2001, pp. 1035–1037.

[14] Park, S.-Y., and Mazanek, D. D., "Mission Functionality for Deflecting Earth-Crossing Asteroids/Comets," *Journal of Guidance, Control, and Dynamics*, Vol. 26, No. 5, 2003, pp. 734–742.

[15] Izzo, D., Negueruela, C., Ongaro, F., and Walker, R., "Strategies for near Earth Object Impact Hazard Mitigation," American Astronautical Society, AAS-05-147, Jan. 2005.

[16] Sauer, C., "A Comparison of Solar Sail and Ion Drive Trajectories for a Halley's Comet Rendezvous Mission," American Astronautical Society, AAS-77-104, Sept. 1977.

[17] Wright, J. L., *Space Sailing,* Gordon and Breach, Philadelphia, 1992.

[18] Friedman, L., *Star Sailing: Solar Sails and Interstellar Travel*, Wiley, New York, 1988.

[19] McInnes, C. R., *Solar Sailing: Technology, Dynamics and Mission Applications,* Springer-Praxis, New York, 1999.

[20] Dachwald, B., "Optimal Solar Sail Trajectories for Missions to the Outer Solar System," *Journal of Guidance, Control, and Dynamics*, Vol. 28, No. 6, 2005, pp. 1187–1193.

[21] Cosmos 1 Solar Sail, http://www.planetary.org/solarsail/update_20050209.html.

[22] Murphy, N., "Solar Polar Images," http://lws.gsfc.nasa.gov/solar_sails_conf/NMurphy.pdf.

[23] Garbe, G., and Montgomery, E., "An Overview of NASA's Solar Sail Propulsion Project," AIAA Paper 2003-4662, July 2003.

[24] Murphy, D. M., Murphey, T. W., and Gierow, P. A., "Scalable Solar Sail Subsystem Design Concept," *Journal of Spacecraft and Rockets*, Vol. 40, No. 4, 2003, pp. 539–547.

[25] Wie, B., "Solar Sail Attitude Control and Dynamics, Parts 1 and 2," *Journal of Guidance, Control, and Dynamics*, Vol. 27, No. 4, 2004, pp. 526–544.

[26] Murphy, D., and Wie, B., "Robust Thrust Control Authority for a Scalable Sailcraft," American Astronomical Society, AAS-04-285, Feb. 2004.

[27] Wie, B., Murphy, D., Thomas, S., and Paluszek, M., "Robust Attitude Control Systems Design for Solar Sail Spacecraft: Parts One and Two," AIAA Paper 2004-5010, AIAA Paper 2004-5011, Aug. 2004.

[28] Wie, B., Thomas, S., Paluszek, M., and Murphy, D., "Propellantless AOCS Design for a 160-m, 450-kg Solar Sail Spacecraft of the Solar Polar Imager Mission," AIAA Paper 2005-3928, July 2005.

[29] McInnes, C. R., "Deflection of near-Earth Asteroids by Kinetic Energy Impacts from Retrograde Orbits," *Planetary and Space Science*, Vol. 52, June 2004, pp. 587–590.

[30] The JPL HORIZONS On-Line Solar System Data and Ephemeris Computation Service, http://ssd.jpl.nasa.gov/horizons.html.

[31] Thomas, S., "Simulating the AIAA Asteroid Problem in SSCT (Solar Sail Control Toolbox)," Internal Memo, Princeton Satellite Systems, Princeton, NJ, 14 Dec., 2004.

[32] Holsapple, K., Giblin, I., Housen, K., Nakamura, A., and Ryan, E., "Asteroid Impacts—Laboratory Experiments and Scaling Laws," *Asteroids III*, Univ. of Arizona Press, Tucson, AZ, 2002, pp. 443–462.

[33] Lang, K. R., *Astrophysical Formulae*, Springer-Verlag, Berlin, 1980, p. 272.

[34] Gennery, D. B., "Scenarios for Dealing with Apophis," presented at 2007 Planetary Defense Conference, March 2007.

[35] Kubitschek, D. G., "Impactor Spacecraft Targeting for the Deep Impact Mission to Comet Tempel 1," American Astronautical Society, AAS 03-615, Aug. 2003.

[36] Carnelli, I., "Industrial Assessment for the Don Quijote Mission," presented at 2007 Planetary Defense Conference, March 2007.

[37] Wie, B., "Solar Sailing Kinetic Energy Interceptor Mission for Impacting and Deflecting Near-Earth Asteroids," AIAA Paper 2005-3725, July 2005.

[38] Wie, B., "Thrust Vector Control Analysis and Design for Solar-Sail Spacecraft," *Journal of Spacecraft and Rockets*, Vol. 44, No. 3, 2007, pp. 545–557.

[39] Wie, B., "Solar Sailing Kinetic Energy Impactor Mission Design for Impacting and Deflecting near-Earth Asteroids," NASA Workshop on NEO Detection, Characterization, and Threat Mitigation, White Paper No. 009, June 2006.

[40] Dachwald, B., and Wie, B., "Solar Sail Kinetic Energy Impactor Trajectory Optimization for an Asteroid-Deflection Mission," *Journal of Spacecraft and Rockets*, Vol. 44, No. 4, 2007, pp. 755–764.

[41] Dachwald, B., Kahle, R., and Wie, B., "Solar Sailing KEI Mission Design Trade-offs for Impacting and Deflecting Asteroid 99942 Apophis," AIAA Paper 2006-6178, Aug. 2006.

[42] Lu, E., and Love, S., "Gravitational Tractor for Towing Asteroids," *Nature*, Vol. 438, 10 Nov. 2005, pp. 177–178.

[43] McInnes, C. R., "Near Earth Object Orbit Modification Using Gravitational Coupling," *Journal of Guidance, Control, and Dynamics*, Vol. 30, No. 3, 2007, pp. 870–872.

[44] Shkadov, L. M., "Possibility of Controlling Solar System Motion in the Galaxy," International Astronautical Federation, Paper IAA-87-613, Oct. 1987.

[45] McInnes, C. R., "Astronomical Engineering Revisited: Planetary Orbit Modification Using Solar Radiation Pressure," *Astrophysics and Space Science*, Vol. 282, No. 4, 2002, pp. 765–772.

[46] Gennery, D. B., "What Should Be Done About Asteroid Apophis (2004 MN4)?," 7 Aug. 2005, http://www.spaceref.com/news/viewsr.html?pid=17666.

[47] Kahle, R., Hahn, G., and K'uhrt, E., "Optimal Deflection of NEOs in Route of Collision with the Earth," *Icarus*, Vol. 182, No. 2, June 2006, pp. 482–488.

[48] Chodas, P., and Yeomans, D., "Predicting Close Approaches and Estimating Impact Probabilities for near-Earth Objects," American Astronomical Society, AAS 99-462, Aug. 1999.

[49] Valsecchi, G., Milani, A., Rossi, A., and Tommei, G., "2004 MN4 Keyholes," IAU Symposium No. 229: Asteroids, Comets, Meteors, Aug. 2005.

[50] Chesley, S. R., "Potential Impact Detection for near-Earth Asteroids: The Case of 99942 Apophis (2004 MN4)," IAU Symposium No. 229: Asteroids, Comets, Meteors, Aug. 2005.

[51] Junkins, J., Singla, P., and Davis, J., "Impact Keyholes and Collision Probability Analysis for Resonant Encounter Asteroids," NASA Workshop on NEO Detection, Characterization, and Threat Mitigation, White Paper No. 070, June 2006.

[52] Wie, B., "Hovering Control of a Solar Sail Gravity Tractor Spacecraft for Asteroid Deflection," American Astronautical Society, AAS 07-145, Jan–Feb. 2007; also 2007 Planetary Defense Conference, March 2007.

[53] Fahnestock, E. G., and Scheeres, D. J., "Dynamical Characterization and Stabilization of Large Gravity Tractor Designs," 2007 Planetary Defense Conference, March 2007; also *Journal of Guidance, Control, and Dynamics* (to be published).

[54] Scheeres, D. J., and Schweickart, R. L., "The Mechanics of Moving Asteroids," AIAA Paper 2004-1440, Feb. 2004.

[55] Izzo, D., Bourdoux, A., Walker, R., and Ongaro, F., "Optimal Trajectories for the Impulsive Deflection of near-Earth Objects," *Acta Astronautica*, Vol. 59, No. 1–5, April 2006, pp. 294–300.

[56] Izzo, D., "Optimization of Interplanetary Trajectories for Impulsive and Continuous Asteroid Deflection," *Journal of Guidance, Control, and Dynamics*, Vol. 30, No. 2, 2007, pp. 401–408.

[57] Inalhan, G., Tillerson, M., and How, J.,"Relative Dynamics and Control of Spacecraft Formations in Eccentric Orbits," *Journal of Guidance, Control, and Dynamics*, Vol. 25, No. 1, 2002, pp. 48–59.

[58] Vaddi, S. S., Vadali, S. R., and Alfriend, K. T., "Formation Flying: Accommodating Nonlinearity and Eccentricity Perturbations," *Journal of Guidance, Control, and Dynamics*, Vol. 26, No. 2, 2003, pp. 214–223.

[59] Alfriend, K. T., and Yan, H., "Evaluation and Comparison of Relative Motion Theories," *Journal of Guidance, Control, and Dynamics*, Vol. 28, No. 2, 2005, pp. 254–261.

[60] Palmer, P., and Imre, E., "Relative Motion Between Satellites on Neighboring Keplerian Orbits," *Journal of Guidance, Control, and Dynamics*, Vol. 30, No. 2, 2007, pp. 521–528.

[61] Broschart, S. B., and Scheeres, D. J., "Control of Hovering Spacecraft near Small Bodies: Application to Asteroid 25142 Itokawa," *Journal of Guidance, Control, and Dynamics*, Vol. 28, No. 2, 2005, pp. 343–354.

[62] Broschart, S. B., and Scheeres, D. J., "Boundness of Spacecraft Hovering Under Dead-Band Control in Time-Invariant Systems," *Journal of Guidance, Control, and Dynamics*, Vol. 30, No. 2, 2007, pp. 601–608.

[63] Kawaguchi, J., "Hayabusa: Summary of Guidance, Navigation and Control Achievement in Its Proximity Phase," AIAA Paper 2006-6533, Aug. 2006.

[64] Gabern, F., Koon, W. S., and Marsden, J. E., "Parking a Spacecraft near an Asteroid Pair," *Journal of Guidance, Control, and Dynamics*, Vol. 29, No. 3, 2006, pp. 544–553.

14
Attitude and Orbit Control of Space Solar Power Satellites

This chapter presents a preliminary conceptual design of an attitude and orbit control system (AOCS) architecture for a very large, 3.2×3.2 km space solar power (SSP) satellite in geostationary orbit. (In this chapter, the term "geostationary orbit" means a geosynchronous orbit with a small inclination.) The proposed AOCS architecture utilizes electric thrusters for integrated attitude and orbit eccentricity control by counteracting, simultaneously, attitude disturbance torques and a large orbital perturbing force. Significant control-structure interaction, possible for such a very large flexible structure, is avoided by employing a low-bandwidth attitude control system. However, a cyclic-disturbance accommodating control concept is utilized to provide proper attitude stabilization in the presence of dynamic modeling uncertainties and cyclic external disturbances. Such a very large SSP satellite concept might not be realizable in the near future; however, the proposed AOCS architecture will be applicable with minimal modifications to future large geostationary platforms of practical sizes. This chapter is an adaptation of [1] and [2].

14.1 Introduction

A renewed interest in space solar power has been spurring a reexamination of the prospects for generating large amounts of electricity from large-scale, space-based solar power systems.* In [3] and [4], Peter Glaser first proposed the satellite solar-power station (SSPS) concept in 1968 and received a U.S. patent on a conceptual design for such a satellite in 1973. As a result of a series of technical and economic feasibility studies by NASA and Department of Energy in the 1970s, an SSPS reference system was developed in the late 1970s. The 1979 SSPS reference system, as it is called, featured a very large solar-array platform (5.3×10.7 km) and a double-gimbaled microwave beam transmitting antenna (1-km diameter), as illustrated in Fig. 14.1. The total mass was estimated to be 50×10^6 kg. A ground- or ocean-based rectenna (rectifying antenna) measuring 10×13 km would receive the microwave beam on the Earth and deliver up to 5 GW of electricity. Unfortunately, such a large-scale space solar-power development plan ceased in the early 1980s.

---

*"Space-Based Solar Power as an Opportunity for Strategic Security," *Phase O Architecture Feasibility Study Report*, National Security Space Office, 10 Oct. 2007.

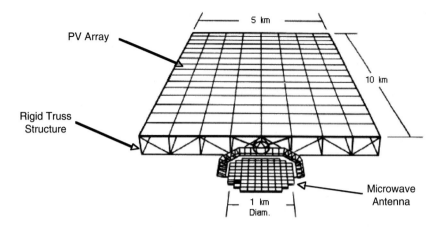

Fig. 14.1   The 1979 SSPS reference system. (Image courtesy of NASA).

In 1995, NASA revisited the SSP concept to assess whether SSP-related technologies had advanced enough to alter significantly the outlook on the economic and technical feasibility of space solar power. The "Fresh Look" study, conducted by NASA during 1995–1997, found that in fact a great deal had changed and that multimegawatt SSP satellites appear viable, with important spacecraft-to-spacecraft applications [5]. The study also found that ambitious research, technology development, and validation over a period of perhaps 15 to 20 years are required to enable SSP concepts to be considered "ready" for commercial development. Figure 14.2 provides a conceptual illustration of a "sun tower" solar-power satellite, which emerged from NASA's 1995–1997 "Fresh Look" study.

Fig. 14.2   "Sun tower" SSP satellite configuration [5].

Further studies during 1998–2001 by NASA as part of the SSP Exploratory Research and Technology (SERT) program have produced a variety of new configurations of SSP satellites, as illustrated in Fig. 14.3. Some of these configurations are based on the passive gravity-gradient stabilization concept. However, most other configurations require active three-axis attitude control to maintain continuous sun tracking of the solar arrays in the presence of external disturbances including the gravity-gradient torque. As illustrated in Fig. 14.4, a cylindrical configuration, which is not affected by the troublesome pitch gravity-gradient torque, has also been considered by NASA.

This chapter, which is based on [1] and [2], focuses on the 1.2-GW geostationary "Abacus" satellite configuration shown in Fig. 14.5, characterized by its configuration consisting of an inertially oriented, 3.2 × 3.2-km solar-array platform, a 500 m-diam microwave beam transmitting antenna fixed to the inertially oriented platform, and a 500 × 700-m rotating reflector that tracks the Earth. The pitch axis is nominally perpendicular to the Earth's equatorial plane (the plane of the orbit). The solar array must in fact make one rotation per year about an axis perpendicular to the orbit plane, making its attitude *quasi* inertial, but this detail can be neglected in the preliminary analysis and design. Various structural concepts for providing the required stiffness and rigidity of such a large space platform are illustrated in Fig. 14.6. An assembly sequence of such a large space platform in geostationary orbit is illustrated in Fig. 14.7. A preliminary conceptual design of the microwave transmitting antenna and the lightweight rotating reflector is illustrated in Figs. 14.8 and 14.9, respectively.

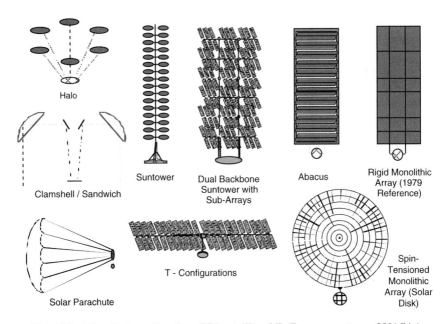

Fig. 14.3 Morphology of various SSP satellites [6]. (Image courtesy of NASA.)

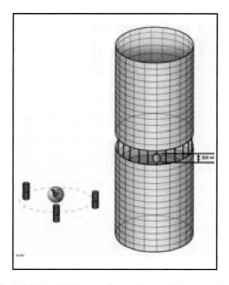

Fig. 14.4   Cylindrical SSP satellite configuration with zero-pitch gravity-gradient torque. (Image courtesy of NASA.)

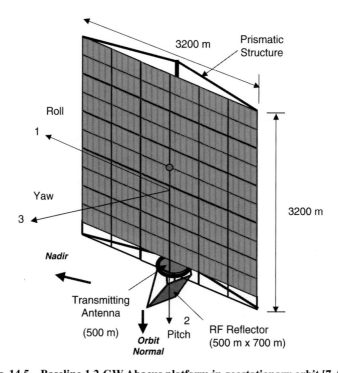

Fig. 14.5   Baseline 1.2-GW Abacus platform in geostationary orbit [7–9].

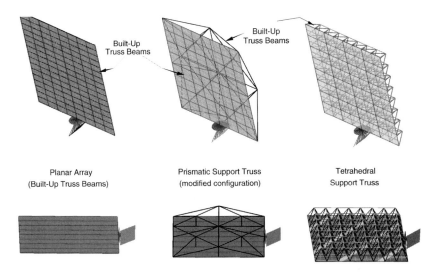

Fig. 14.6   Various structural support configurations [7].

Some unique features of the Abacus configuration relative to the 1979 SSPS reference system are as follows:

1) The massive transmitting antenna is not gimbaled.

2) The lightweight rotating reflector design thus eliminates massive rotary joint and slip rings of the 1979 SSPS reference system.

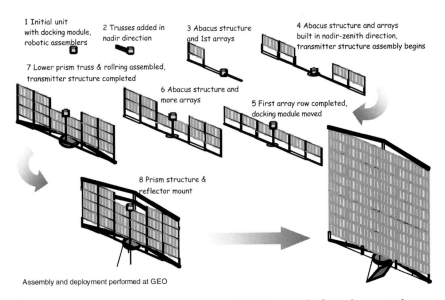

Fig. 14.7   Assembly sequence of the 1.2-GW Abacus platform in geostationary orbit [8–9]. (Image courtesy of NASA.)

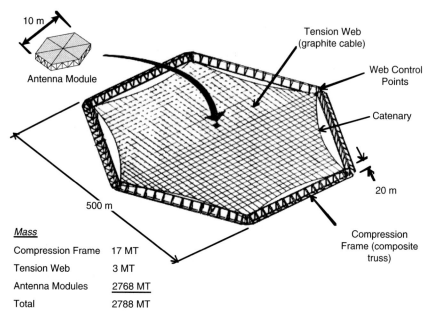

Fig. 14.8   Microwave transmitting antenna [7].

3) Links activated by ball-screw mechanisms can also tilt the reflector to point to ground stations at various latitudes.

The study objectives in the area of dynamics and control of large geostationary SSP satellites are the following: 1) to develop preliminary concepts for orbit, attitude, and structural control of very large SSP satellites using a variety of

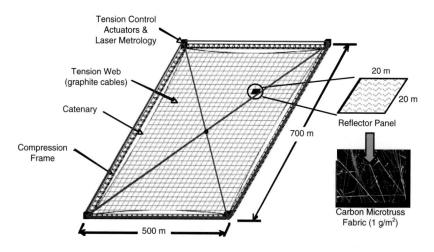

Fig. 14.9   Microwave beam reflector [7].

actuators such as control moment gyros, momentum wheels, and electric-propulsion thrusters; 2) to develop mathematical models, define a top-level control system architecture, and perform control system design and analysis for a baseline Abacus platform configuration in geostationary orbit; and 3) to determine the required number, size, placement, mass, and power for the actuators to control the orbit, attitude, and structural motions of the baseline Abacus platform.

14.2 System Description

14.2.1 Geometry and Mass Distribution

The baseline Abacus platform consists of three major parts whose dimensions are indicated in Fig. 14.5: an inertially oriented solar array, a transmitting antenna, and an Earth-pointing reflector that moves relative to the solar array. The massive transmitting antenna is *fixed* to the solar array, whereas the reflector is fastened to the solar array with two rotational joints that allow the reflected microwave beam to be pointed at a particular point on the Earth's surface. The first of these is an azimuth roll-ring that permits rotation once per orbit about the solar-array pitch axis, nominally perpendicular to the Earth's equatorial plane, and the second is a set of ball-screw activated links that change the tilt of the reflector to a constant offset so that the beam can be aimed at different latitudes. This design, with a lightweight rotating reflector, eliminates the massive rotary joint and slip rings that were required for the 1979 SSPS reference system.

The total mass and area of the spacecraft are given in Table 14.1, together with the mass of each of the major parts. The mass of the reflector is less than 4% of the total mass; therefore, the reflector's mass and inertia can be neglected in the analysis of attitude motion, simplifying the task in two important respects. First, the Abacus platform can be treated as a single body rather than a multibody spacecraft. When the Abacus platform is regarded as rigid, the spacecraft's moments and products

Table 14.1 Geometric and mass properties of the 1.2-GW Abacus platform

Parameters	Values
Solar-array mass	21×10^6 kg
Transmitting antenna mass	3×10^6 kg
Reflector mass	0.8×10^6 kg
Total mass	$m_B = 25 \times 10^6$ kg
Platform area	$A = 3200\,\text{m} \times 3200$ m
Area-to-mass ratio	$A/m_B = 0.4\ \text{m}^2/\text{kg}$
Roll inertia	$I_1 = 2.8 \times 10^{13}$ kg·m^2
Pitch inertia	$I_2 = 1.8 \times 10^{13}$ kg·m^2
Yaw inertia	$I_3 = 4.6 \times 10^{13}$ kg·m^2
c.m.-c.p. offset	200 m (along pitch axis)
c.m.-c.p. offset (uncertainty)	± 20 m (along roll axis)

of inertia for a set of axes fixed in the solar array do not vary with time. Second, when the asymmetrical mass distribution of the reflector is left out of account, the principal axes of inertia of the spacecraft with respect to the spacecraft's mass center are parallel to the roll, pitch, and yaw axes illustrated in Fig. 14.5. The moments of inertia for these axes, henceforth considered to be principal moments of inertia, are given in Table 14.1. The center of pressure is located approximately 100 m below the geometric center of the square platform, the center of mass is located approximately 300 m below the geometric center along the pitch axis, and ±20% overall uncertainty in the mass properties is assumed in control design.

14.2.2 Orbital Parameters and Control Requirements

The attitude control requirement specifies that the pitch axis (see Fig. 14.5) is to remain perpendicular to the Earth's equatorial plane, and the yaw axis, normal to the array, is to remain parallel to the equatorial plane. This orientation of the pitch axis is the standard, nominal attitude for all geostationary satellites with solar arrays. At the times of equinox, approximately March 21 and September 21, the angle between the sun's rays and the orbit plane, known as the solar beta angle, will vanish; therefore, there is no power loss because the rays arrive perpendicular to the solar array. At the times of solstice, approximately June 21 and December 21, the solar beta angle is equal to the angle between the equatorial and ecliptic planes, referred to as the obliquity of the ecliptic, with a value of 23.44 deg. On these dates power loss reaches a maximum value of only 8%, which can be remedied by oversizing the array slightly, if necessary. In practice, this is how geostationary satellites accommodate power loss; the attitude of the spacecraft bus is not adjusted to compensate for variation in the solar beta angle.

On a conventional geostationary spacecraft the attitude of the solar array, sized to accommodate the seasonal power loss, is normally controlled to 0.5 deg in roll and pitch in order for solar energy to be collected. On the Abacus platform such a requirement is superseded because the orientation of the 500 × 700 m reflector must be controlled to 5 min of arc (arcmin) so that the microwave beam is aimed precisely at a point fixed to the Earth. The precise performance characteristics of the rotational joints that attach the reflector to the array were unknown during the preliminary study phase; therefore, in the absence of this information, we consider the precision pointing requirements to be imposed on the array itself and pursue suitable attitude control. The design of the 500-m-diam microwave beam transmitting antenna, and the steerable reflector, and characterization of their precision pointing performance is of course a challenging engineering problem.

Basic orbital characteristics and control requirements for the Abacus platform in geostationary orbit are summarized in Table 14.2.

14.3 Orbital Dynamics

The orbital motion of two bodies is in general governed by the second-order, ordinary vector differential equation

$$\frac{d^2 \vec{r}}{dt^2} + \frac{\mu}{r^3} \vec{r} = \vec{f} = \vec{f}_B - \vec{f}_P \qquad (14.1)$$

Table 14.2   Orbit parameters and control requirements

Parameters	Values
Earth's gravitational parameter	$\mu_\oplus = 398,601 \text{ km}^3/\text{s}^2$
Geostationary orbit	$a = 42,164$ km
Orbit period	23 h 56 min 4 s
Orbit rate	$n = 7.292 \times 10^{-5}$ rad/s
Stationkeeping accuracy	± 0.1 deg
Eccentricity	$e < 0.0005$
Inclination	$i < 0.005$ deg
Solar-array pointing accuracy	± 0.5 deg for roll/pitch
Microwave beam pointing accuracy	± 5 arcmin (± 0.083 deg)

where $\vec{r}$ is the position vector from the mass center $P^\star$ of a planet P to the mass center $B^\star$ of a body B, r is the magnitude of $\vec{r}$, $d^2\vec{r}/dt^2$ indicates the second derivative of $\vec{r}$ with respect to time t in an inertial or Newtonian reference frame N, and $\mu \triangleq G(m_P + m_B)$, where G is the universal gravitational constant, m_B is the mass of P, and m_B is the mass of B.

If P were a sphere with uniform mass distribution, or a particle, and if B were a particle, then the gravitational force exerted by P on B would be given by $\vec{g} = -G m_P m_B \vec{r}/r^3$. The force exerted by B on P would be simply $-\vec{g}$. The vector $\vec{f}_B$ represents the resultant force per unit mass acting on B, *other than* $\vec{g}/m_B$; $\vec{f}_P$ represents the resultant force per unit mass acting on P, *other than* $-\vec{g}/m_P$.

In the case of geostationary satellites, the perturbing force per unit mass $\vec{f}$ receives significant contributions from the gravitational attraction of the sun and moon, Earth's tesseral gravitational harmonics of degree 2 and orders 0 and 2, and solar radiation pressure. The following material gives a brief development of expressions for each of these contributions, denoted respectively as $\vec{f}_s, \vec{f}_m$, $\vec{f}_{2,0}, \vec{f}_{2,2}$, and $\vec{f}_p$, such that

$$\vec{f} = \vec{f}_s + \vec{f}_m + \vec{f}_{2,0} + \vec{f}_{2,2} + \vec{f}_p \tag{14.2}$$

14.3.1   Solar and Lunar Gravitational Attraction

The gravitational force per unit mass exerted by the sun on P is given by $\mu_s \vec{r}_s/r_s^3$, where μ_s is the product of G and the sun's mass, $\vec{r}_s$ is the position vector from $P^\star$ to the sun's mass center, and r_s is the magnitude of $\vec{r}_s$. Likewise, the gravitational force per unit mass exerted by the sun on B is given by $\mu_s(\vec{r}_s - \vec{r})/|\vec{r}_s - \vec{r}|^3$. Therefore,

$$\vec{f}_s = \frac{\mu_s(\vec{r}_s - \vec{r})}{|\vec{r}_s - \vec{r}|^3} - \frac{\mu_s \vec{r}_s}{r_s^3} \tag{14.3}$$

When $\vec{r}$ is small in comparison to $\vec{r}_s$, numerical difficulties can be encountered in the evaluation of the right-hand member of Eq. (14.3); therefore, an alternate form of $\vec{f}_s$ is used, as suggested in Eq. (8.61) of [10]:

$$\vec{f}_s = -\frac{\mu_s}{|\vec{r}_s - \vec{r}|^3}[\vec{r} + f(q_s)\vec{r}_s] \tag{14.4}$$

where

$$q_s \overset{\Delta}{=} \frac{\vec{r} \cdot (\vec{r} - 2\vec{r}_s)}{\vec{r}_s \cdot \vec{r}_s} \tag{14.5}$$

and the function f of q is given by

$$f(q) = q\frac{3 + 3q + q^2}{1 + (1 + q)^{\frac{3}{2}}} \tag{14.6}$$

The contribution of lunar gravitational attraction to $\vec{f}$ is given by an expression similar to Eq. (14.3), and numerical difficulties are avoided with the aid of

$$\vec{f}_m = -\frac{\mu_m}{|\vec{r}_m - \vec{r}|^3}[\vec{r} + f(q_m)\vec{r}_m] \tag{14.7}$$

where μ_m is the product of G and the moon's mass and $\vec{r}_m$ is the position vector from $P^\star$ to the moon's mass center.

14.3.2 Tesseral Harmonics

Equation (12) of [11] can be used to account for the gravitational harmonics of P, for any degree n and order m; in this study, n and m are limited to 2. Numerical values of the gravitational coefficients, gravitational parameter of Earth, and mean equatorial radius are those of the Goddard Earth Model T1 as presented in [12].

Earth's oblateness is represented by a zonal harmonic of degree 2 and order 0 and is responsible for precessions in a satellite's orbit plane and argument of perigee. The contribution of this harmonic to the force per unit mass exerted by P on B is given in Eq. (45) of [11] [also Prob. 3.7(b) in [13]] as

$$\vec{f}_{2,0} = -3\mu_\oplus J_2\frac{R_\oplus{}^2}{r^4}\left[\sin\phi\,\hat{e}_3 + \frac{1}{2}(1 - 5\sin^2\phi)\hat{r}\right] \tag{14.8}$$

where $\mu_\oplus$ is the gravitational parameter of the Earth, the product of G and the Earth's mass; $R_\oplus$ is the mean equatorial radius of the Earth (6378.137 km); r is the magnitude of $\vec{r}$; $\hat{r}$ is a unit vector in the direction of $\vec{r}$; and ϕ is the geocentric latitude of B. Unit vector $\hat{e}_3$ is fixed in the Earth in the direction of the north polar axis. The zonal harmonic coefficient of degree 2 is represented by the familiar symbol J_2.

The contribution of oblateness to the force per unit mass exerted by B on P is given by $-m_B\vec{f}_{2,0}/m_P$, and the contribution of oblateness to $\vec{f}$ is thus $[1 + (m_B/m_P)]\vec{f}_{2,0}$. In the case of the Abacus platform orbiting Earth, $m_B = 25 \times 10^6$ kg, and $m_P = 5.98 \times 10^{24}$ kg, so that $m_B/m_P = 4 \times 10^{-18}$, which can be neglected in comparison to 1; therefore, the entire contribution of oblateness to $\vec{f}$ is essentially equal to $\vec{f}_{2,0}$.

The contribution $\vec{f}_{2,1}$ of the tesseral harmonic of degree 2 and order 1 vanishes because the tesseral harmonic coefficients $S_{2,1}$ and $C_{2,1}$ are both zero. The sectoral harmonic of degree 2 and order 2 can cause the longitude of a geostationary spacecraft to drift; from Eq. (12) of [11], contribution to the force per unit mass exerted by P on B is given by

$$\vec{f}_{2,2} = \frac{\mu_\oplus R_\oplus{}^2}{r^5} \left\{ \frac{C_{2,2}C_2 + S_{2,2}S_2}{r} \left[A_{2,3}\hat{e}_3 - \left(\sin\phi A_{2,3} + 5A_{2,2} \right)\hat{r} \right] \right.$$

$$\left. + 2A_{2,2} \left[(C_{2,2}C_1 + S_{2,2}S_1)\hat{e}_1 + (S_{2,2}C_1 - C_{2,2}S_1)\hat{e}_2 \right] \right\} \quad (14.9)$$

where unit vectors $\hat{e}_1$ and $\hat{e}_2$ are fixed in the Earth: $\hat{e}_1$ lies in the equatorial plane parallel to a line intersecting Earth's geometric center and the Greenwich meridian, and $\hat{e}_2 = \hat{e}_3 \times \hat{e}_1$. Sectoral harmonic coefficients of degree 2 and order 2 are denoted by $C_{2,2}$ and $S_{2,2}$.

Equations (6) and (7) of [11] indicate that the required derived Legendre polynomials are $A_{2,2} = 3$ and $A_{2,3} = 0$. In addition, Eqs. (9) and (10) of [11] show that

$$\mathcal{S}_1 = \vec{r} \cdot \hat{e}_2 = r\cos\phi\sin\lambda, \quad \mathcal{C}_1 = \vec{r} \cdot \hat{e}_1 = r\cos\phi\cos\lambda \quad (14.10)$$

$$\mathcal{S}_2 = 2(r\cos\phi)^2\sin\lambda\cos\lambda, \quad \mathcal{C}_2 = (r\cos\phi)^2(\cos^2\lambda - \sin^2\lambda) \quad (14.11)$$

where λ is the geographic longitude of B measured eastward from the Greenwich meridian. Therefore,

$$\vec{f}_{2,2} = \frac{\mu_\oplus R_\oplus{}^2}{r^4} \left\{ -15\cos^2\phi[C_{2,2}(\cos^2\lambda - \sin^2\lambda) \right.$$

$$+ 2S_{2,2}\sin\lambda\,\cos\lambda]\hat{r} + 6\cos\phi\left[(C_{2,2}\cos\lambda + S_{2,2}\sin\lambda)\hat{e}_1 \right.$$

$$\left. + (S_{2,2}\cos\lambda - C_{2,2}\sin\lambda)\hat{e}_2 \right] \Big\} \quad (14.12)$$

As in the case of $\vec{f}_{2,0}$, m_B/m_P is neglected in comparison to 1, and $\vec{f}_{2,2}$ thus constitutes the entire contribution of the present harmonic to $\vec{f}$.

14.3.3 Solar Radiation Pressure

Many researchers have investigated the significant perturbation in the orbit of a large spacecraft with a large area-to-mass ratio, caused by solar radiation pressure; the orbit of the Abacus platform (and other large SSPS) is adversely affected by an area-to-mass ratio, which is very large compared to contemporary, higher-density spacecraft. A detailed physical description of the solar radiation pressure can be found in [14].

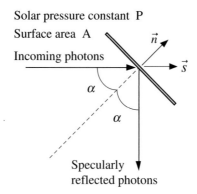

Solar pressure constant P

Surface area A

Incoming photons

$\vec{n}$

$\vec{s}$

α

α

Specularly reflected photons

Fig. 14.10   Solar radiation pressure force acting on an ideal flat surface.

The solar radiation pressure forces are caused by photons impinging on a surface in space, as illustrated in Fig. 14.10. Assuming that a fraction ρ_s of the impinging photons is specularly reflected, a fraction ρ_d is diffusely reflected, and a fraction ρ_a is absorbed by the surface, we have

$$\rho_s + \rho_d + \rho_a = 1 \qquad (14.13)$$

The solar radiation pressure force $\vec{F}_p$ acting on an ideal flat surface is then expressed as in Eq. (11) of [15],

$$\vec{F}_p = PA(\hat{n} \cdot \hat{s}) \left\{ (\rho_a + \rho_d)\hat{s} + \left[2\rho_s(\hat{n} \cdot \hat{s}) + \frac{2}{3}\rho_d \right] \hat{n} \right\} \qquad (14.14)$$

where $P = 4.5 \times 10^{-6}$ N/m^2 is the nominal solar radiation pressure constant, A is the surface area, $\hat{n}$ is a unit vector normal to the surface, and $\hat{s}$ is a unit vector having the same direction as $\vec{r} - \vec{r}_s$, the position vector from the sun's mass center to B^*. The angle between $\hat{n}$ and $\hat{s}$ is α.

For an ideal case of a perfect mirror with $\rho_d = \rho_a = 0$ and $\rho_s = 1$, we have $\vec{F}_p = 2PA(\cos\alpha)^2\hat{n}$. Also for an ideal case of a black body with $\rho_s = \rho_d = 0$ and $\rho_a = 1$, we have $\vec{F}_p = PA \cos\alpha \, \hat{s}$. For most practical cases of satellites with a small angle of incidence α, the radiation pressure force per unit mass exerted on B is given simply as

$$\vec{f}_p = \vec{F}_p/m_B = P(1 + \rho)(A/m_B)\hat{s} \triangleq k\hat{s} \qquad (14.15)$$

where ρ is the overall surface reflectance (0 for a blackbody and 1 for a mirror) and A/m_B is the area-to-mass ratio. The magnitude of $\vec{f}_p$ is a constant k. We neglect the solar radiation pressure exerted on the Earth and allow this expression to represent the total contribution of solar radiation pressure in Eq. (14.2).

An analytical estimate of the short-term orbital perturbations resulting from the significant in-plane component of solar radiation pressure force is obtained as follows. First, we express $\vec{f}_p$ as

$$\vec{f}_p = f_r \hat{e}_r + f_\theta \hat{e}_\theta + f_z \hat{e}_z \tag{14.16}$$

where unit vector $\hat{e}_r$ is identical to $\hat{r}$, unit vector $\hat{e}_z$ is normal to the orbit plane and has the same direction as the specific angular momentum of $B^\star$ in N, and $\hat{e}_r \times \hat{e}_\theta = \hat{e}_z$. Ignoring the effects of seasonal variations of the sun vector $\hat{s}$, we simply assume that $f_r \approx k \sin\theta$ and $f_\theta \approx k \cos\theta$, where θ is the true anomaly of the orbit. Although f_z is neglected for the moment, it is accounted for in detail in the numerical analysis discussed in Sec. 14.3.4. It is well known that there is no average change over one circular orbit to the inclination i or longitude of ascending node Ω when f_z is considered constant during the orbit, a result easily shown by analytical means.

From the orbit perturbation analysis discussed in [13] and [16], the time derivatives of semimajor axis a and eccentricity e are given by

$$\frac{da}{dt} = \frac{2}{n\sqrt{1-e^2}}[f_r e \sin\theta + f_\theta(1 + e\cos\theta)] \tag{14.17a}$$

$$\frac{de}{dt} = \frac{\sqrt{1-e^2}}{na}[f_r \sin\theta + f_\theta(\cos\theta + \cos E)] \tag{14.17b}$$

where E is the eccentric anomaly and n is the mean motion.

For geostationary satellites with a constant orbital rate of n and $e \approx 0$, we obtain variations caused by solar radiation pressure

$$\frac{da}{dt} = \frac{2}{n}f_\theta = \frac{2k}{n}\cos\theta = \frac{2k}{n}\cos nt$$

$$\Rightarrow a(t) = a(0) + \frac{2k}{n^2}\sin nt$$

$$\Rightarrow \Delta a = 0 \quad \text{per day} \tag{14.18}$$

and

$$\frac{de}{dt} = \frac{1}{na}(f_r \sin\theta + 2f_\theta \cos\theta)$$

$$= \frac{1}{na}(k\sin^2\theta + 2k\cos^2\theta)$$

$$= \frac{k}{na}\left(\frac{3}{2} + \frac{1}{2}\cos 2\theta\right)$$

$$\Rightarrow \quad e(t) = \frac{k}{na}\left(\frac{3}{2}t + \frac{1}{4n}\sin 2nt\right)$$

$$\Rightarrow \quad \Delta e \approx \frac{3\pi k}{n^2 a} \quad \text{per day} \tag{14.19}$$

Changes in the direction of $\hat{s}$ caused by Earth's heliocentric motion should be accounted for when making long-term predictions of eccentricity growth. One such approach is described in [17] and [18]. Another is presented in [19], where it is shown that eccentricity varies as $\sin^2$ over a year [Appendix A.2, Eq. (A.21)] and that the direction of the average change per orbit in the eccentricity vector is perpendicular to $\vec{r}_s$ [Eq. (3.120)].

The effect of solar radiation pressure on the change in longitude can be found by solving the Clohessy–Wiltshire–Hill equations for in-plane motion relative to a nominal circular orbit,

$$\ddot{x} - 2n\dot{y} - 3n^2x = f_r = k\sin nt \qquad (14.20a)$$

$$\ddot{y} + 2n\dot{x} = f_\theta = k\cos nt \qquad (14.20b)$$

where x is the relative displacement in the radial direction and y is the relative displacement in the direction of the velocity of the circular reference orbit. One can verify that

$$x = -\frac{3k}{2n}t\cos nt \qquad (14.21a)$$

$$y = \frac{k}{n}\left[3t\sin nt + \frac{2}{n}(\cos nt - 1)\right] \qquad (14.21b)$$

furnish a solution. For small angles, the longitude λ can be approximated as y/a; hence,

$$\lambda(t) \approx \frac{k}{na}\left[3t\sin nt + \frac{2}{n}(\cos nt - 1)\right] \qquad (14.22)$$

For the Abacus platform with an area-to-mass ratio $A/m_B \approx 0.4$ m^2/kg, assuming an overall surface reflectance $\rho = 0.3$, we have

$$F_p = (4.5 \times 10^{-6})(1.3)(10.24 \times 10^6) \approx 60 \text{ N}$$

$$k = F_p/m_B \approx 2.4 \times 10^{-6} \text{ m/s}^2$$

$$\Delta e = \frac{3\pi k}{n^2 a} \approx 1 \times 10^{-4} \quad \text{per day}$$

In view of Eq. (14.22), or Eq. (3.123) of [19], the longitude drift is given as $\Delta\lambda = 2\Delta e \approx 0.0115$ deg/day. Thus the stationkeeping requirement of 0.1 deg set forth in Table 14.2 is exceeded after less than 10 days of uncontrolled eccentricity growth. The sun-pointing-perigee technique of directing the eccentricity vector toward the sun, discussed in [18], is often used to control eccentricity growth of typical geosynchronous satellites; however, this approach might not be applicable to the Abacus platform with a stationkeeping requirement of 0.1 deg in the presence of a large area-to-mass ratio of 0.4 m^2/kg. Use of this strategy is also questionable in light of the need for continuous cancellation of gravitational moment for attitude control as discussed in the next section. Consequently, continuous orbit eccentricity control using high-specific-impulse (I_{sp}) ion engines becomes necessary.

14.3.4   Orbital Simulation Results

Orbital motion of the Abacus platform is simulated by solving Eqs. (14.1) with Encke's method, as described in Sec. 9.4 of [10] and Sec. 9.3 of [20]. The right-hand member of Eqs. (14.1) is given by Eq. (14.2) and evaluated by means of Eqs. (14.4), (14.7), (14.8), (14.12), and (14.15), which account in detail for all components of perturbing force per unit mass.

The initial values used in the simulations correspond to a circular, equatorial orbit of radius 42164.169 km; therefore, the initial orbital elements are $a = 42164.169$ km and $e = i = \Omega = \omega = 0$. The epoch used to calculate the solar and lunar positions, as well as the Earth's orientation in inertial space according to formulas given in [21], is 21 June 2000. On this date $\vec{r}_s$ is approximately 90 deg ahead of the vernal equinox direction in the ecliptic plane. To place the spacecraft at an initial terrestrial longitude of 75.07 deg (one of the stable longitudes), a true anomaly θ of -15.43 deg is used. These elements correspond to an initial position and velocity of

$$\vec{r} = X\hat{I} + Y\hat{J} + Z\hat{K} = 40644.757\,\hat{I} - 11216.990\,\hat{J} \quad (\text{km})$$

$$\vec{v} = 0.818\,\hat{I} + 2.964\,\hat{J} \quad (\text{km/s})$$

where $\hat{I}, \hat{J}$, and $\hat{K}$ make up a set of unit vectors fixed in an Earth-centered-inertial (ECI) reference frame.

Solution of the requisite six scalar, first-order, ordinary differential equations is obtained by numerical integration with a variable step, Runge–Kutta scheme, with relative and absolute error tolerances set to 1×10^{-8}. The true position and velocity are computed and used to determine values of classical orbital elements.

The results of a 30-day simulation of uncontrolled orbital motion, including the effects of Earth's oblateness and triaxiality, luni-solar perturbations, and a 60-N solar-pressure force, are shown in Fig. 14.11. The 30-day secular growth in eccentricity of 3×10^{-3} confirms the conclusion reached in Sec. 14.3.3 that solar radiation pressure must be offset if stationkeeping requirements are to be met. The average slope of inclination in Fig. 14.11 is about 0.0783 deg/month, corresponding to a yearly growth of 0.940 deg; this is caused by the solar and lunar gravitational perturbations expressed in Eqs. (14.4) and (14.7) and is within the expected range given on p. 82 of [16].

In the next section we develop a dynamic model of attitude motion of an inertially oriented spacecraft in geostationary orbit, in preparation for attitude control system architecture design.

14.4   Attitude Motion of an Inertially Oriented Spacecraft

Euler's equations for rotational motion of a rigid body B can be expressed in vector-dyadic form as

$$\bar{J} \cdot \dot{\vec{\omega}} + \vec{\omega} \times \bar{J} \cdot \vec{\omega} = \vec{\tau} = \vec{M} + \vec{u} + \vec{d} \tag{14.23}$$

where $\bar{J}$ is the inertia dyadic of B with respect to its mass center $B^\star$, $\vec{\omega}$ is the angular velocity of B in an inertial or Newtonian reference frame N, and $\dot{\vec{\omega}}$ is the

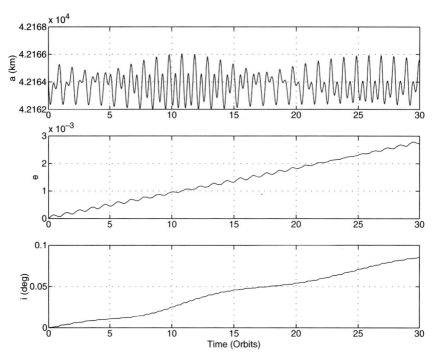

Fig. 14.11   Orbit simulation results of the Abacus platform with the effects of the Earth's oblateness and triaxiality, luni-solar perturbations, and 60-N solar radiation pressure force.

angular acceleration of B in N. The vector $\vec{\tau}$ represents the moment about $B^{\star}$ of external forces acting on B and is regarded as the sum of the gravitational moment $\vec{M}$ exerted by the primary P, the moment $\vec{u}$ exerted by control actuators, and the remaining contributions $\vec{d}$.

The gravitational moment exerted by P is often calculated by neglecting its gravitational harmonics and treating it as a particle or a sphere with uniform mass distribution, in which case $\vec{M}$, also known as gravity-gradient torque, can be expressed according to [13] and [22],

$$\vec{M} = \frac{3\mu_{\oplus}}{r^3} \hat{r} \times \bar{J} \cdot \hat{r} \tag{14.24}$$

where $\hat{r}$ is a unit vector having the same direction as the position vector from $P^{\star}$ to $B^{\star}$.

It is convenient to work with a set of right-handed, mutually perpendicular unit vectors $\vec{b}_1$, $\vec{b}_2$, and $\vec{b}_3$ fixed in B and introduce the following dot products:

$$\omega_i \triangleq \vec{\omega} \cdot \vec{b}_i, \quad r_i \triangleq \vec{r} \cdot \vec{b}_i, \quad u_i \triangleq \vec{u} \cdot \vec{b}_i, \quad d_i \triangleq \vec{d} \cdot \vec{b}_i \quad (i = 1, 2, 3) \tag{14.25}$$

If $\vec{b}_1$, $\vec{b}_2$, and $\vec{b}_3$ are taken to be parallel to central principal axes of inertia of B, then the central principal moments of inertia of B can be expressed as

$$I_i \triangleq \vec{b}_i \cdot \vec{J} \cdot \vec{b}_i \quad (i = 1, 2, 3) \tag{14.26}$$

If it is assumed that B travels in a circular orbit with constant orbital rate n given by $\sqrt{\mu_\oplus/r^3}$, the nonlinear equations (14.23) can be expressed in scalar form as

$$I_1\dot{\omega}_1 + (I_3 - I_2)\omega_2\omega_3 = 3n^2(I_3 - I_2)r_2r_3 + u_1 + d_1 \tag{14.27a}$$

$$I_2\dot{\omega}_2 + (I_1 - I_3)\omega_3\omega_1 = 3n^2(I_1 - I_3)r_3r_1 + u_2 + d_2 \tag{14.27b}$$

$$I_3\dot{\omega}_3 + (I_2 - I_1)\omega_1\omega_2 = 3n^2(I_2 - I_1)r_1r_2 + u_3 + d_3 \tag{14.27c}$$

Because Abacus platform is regarded here as a single body B whose nominal orientation is fixed in an inertial reference frame N, we introduce three unit vectors $\vec{n}_1$, $\vec{n}_2$, and $\vec{n}_3$ fixed in N such that Earth's equatorial plane contains $\vec{n}_1$ and $\vec{n}_3$, while $\vec{n}_2$ is perpendicular to the equatorial plane. To describe the orientation of B in N, we choose a pitch-yaw-roll sequence described as follows. With $\vec{b}_i$ having the same direction as $\vec{n}_i$ $(i = 1, 2, 3)$, proceed with a rotation first about $\vec{b}_2$ in the amount θ_2, followed by a rotation about $\vec{b}_3$ in the amount θ_3, and conclude with a rotation about $\vec{b}_1$ in the amount θ_1. For this rotational sequence, we have

$$\begin{bmatrix} \vec{b}_1 \\ \vec{b}_2 \\ \vec{b}_3 \end{bmatrix} = \begin{bmatrix} C_2C_3 & S_3 & -S_2C_3 \\ -C_2S_3C_1 + S_1S_2 & C_3C_1 & S_2S_3C_1 + S_1C_2 \\ C_2S_3S_1 + C_1S_2 & -C_3S_1 & -S_2S_3S_1 + C_1C_2 \end{bmatrix} \begin{bmatrix} \vec{n}_1 \\ \vec{n}_2 \\ \vec{n}_3 \end{bmatrix} \tag{14.28}$$

where $C_1 \triangleq \cos\theta_1$, $S_2 \triangleq \sin\theta_2$, etc.

The unit vector $\hat{r}$ can be expressed as

$$\hat{r} = \cos nt \, \vec{n}_3 - \sin nt \, \vec{n}_1 \tag{14.29}$$

and we have

$$r_1 = -S_2C_3 \cos nt - C_2C_3 \sin nt \tag{14.30a}$$

$$r_2 = (S_2S_3C_1 + S_1C_2) \cos nt - (-C_2S_3C_1 + S_1S_2) \sin nt \tag{14.30b}$$

$$r_3 = (-S_2S_3S_1 + C_1C_2) \cos nt - (C_2S_3S_1 + C_1S_2) \sin nt \tag{14.30c}$$

It is clear that Eqs. (14.27) and (14.30) are nonlinear in ω_i and θ_i, variables that can take on arbitrarily large values in the general case. However, ω_i and θ_i would all be zero if the attitude motion of the Abacus platform were controlled as intended. (The yearly rotation of the solar array about an axis perpendicular to the orbit plane is neglected.) Thus, we introduce quantities that are assumed to remain small, denoted with a tilde:

$$\omega_i = 0 + \tilde{\omega}_i, \qquad \theta_i = 0 + \tilde{\theta}_i \quad (i = 1, 2, 3) \tag{14.31}$$

It is easily shown that, expressed in terms of these quantities, the kinematical differential equations for the orientation angles are simply

$$\dot{\tilde{\theta}}_i \approx \tilde{\omega}_i, \qquad (i = 1, 2, 3) \tag{14.32}$$

After substituting from Eqs. (14.30) into (14.27), expanding trigonometric functions into power series, substituting from Eqs. (14.31) and (14.32), and omitting all terms of second or higher degree in the quantities assumed to be small, the dynamical equations linearized about the nominal motion are written as

$$I_1\ddot{\tilde{\theta}}_1 = 3n^2(I_3 - I_2)[(\cos^2 nt)\tilde{\theta}_1 + (\sin nt \cos nt)\tilde{\theta}_3] + u_1 + d_1 \tag{14.33a}$$

$$I_2\ddot{\tilde{\theta}}_2 = 3n^2(I_3 - I_1)[(\cos^2 nt - \sin^2 nt)\tilde{\theta}_2 + \sin nt \cos nt] + u_2 + d_2 \tag{14.33b}$$

$$I_3\ddot{\tilde{\theta}}_3 = 3n^2(I_1 - I_2)[(\sin^2 nt)\tilde{\theta}_3 + (\sin nt \cos nt)\tilde{\theta}_1] + u_3 + d_3 \tag{14.33c}$$

Equations (14.33c) are the attitude equations of motion of the Abacus platform for control design. It can be seen that the pitch motion is decoupled from the roll/yaw motion, but the pitching motion is significantly disturbed by the time-varying gravity-gradient torque. Various ways of dealing with this issue are discussed in Sec. 14.4.1. Furthermore, the roll/yaw motion is strongly coupled because of the time-varying roll/yaw gravity-gradient torques. The other external disturbances d_i will be discussed in Sec. 14.4.2.

14.4.1 Pitch Gravity-Gradient Torque

In the absence of any pitch attitude error $\tilde{\theta}_2 = 0$, the pitch equation of motion for a rigid inertially oriented spacecraft in circular orbit, Eq. (14.33b), can be rewritten as

$$I_2\ddot{\tilde{\theta}}_2 = \frac{3n^2}{2}(I_3 - I_1)\sin 2nt + u_2 + d_2 \tag{14.34}$$

Ignoring for the moment any other disturbance d_2, the pitch control torque u_2 required to counter the cyclic gravity-gradient torque simply becomes

$$u_2 = -\frac{3n^2}{2}(I_3 - I_1)\sin 2nt \tag{14.35}$$

with peak values of $\pm 143{,}000$ N·m.

If angular momentum exchange devices, such as momentum wheels (MWs) or control moment gyros (CMGs), are to be employed for pitch control, the peak angular momentum to be stored can then be estimated as

$$H_{\max} = \frac{3n}{2}(I_3 - I_1) = 2 \times 10^9 \text{ N·m·s} \tag{14.36}$$

This is about 100,000 times the angular momentum storage requirement of the International Space Station (ISS), which is controlled by four double-gimbaled

Table 14.3 Large single-gimbal CMG (Courtesy of Honeywell Space Systems, Glendale, Arizona)

Cost	$1M
Momentum	7000 N·m·s
Max torque	4000 N·m
Peak power	500 W
Mass	250 kg
Momentum/mass	28 N·m·s/kg

CMGs with a total momentum storage capability of about 20,000 N·m·s and a momentum density of 17.5 N·m·s/kg. Basic characteristics of a large single-gimbal CMG are summarized in Table 14.3; these devices have a momentum density of 28 N·m·s/kg. Future advanced flywheels can have a larger momentum density of 150 N·m·s/kg. In view of all this, it can be concluded that a traditional momentum management approach employing conventional CMGs (or even future advanced flywheels) is not a viable option for controlling a very large space platform.

To meet the momentum storage requirement of very large space platform, a concept of constructing large-diameter momentum wheels in space was proposed in the late 1970s in [23]. In an attempt to resolve the angular momentum storage problem of large sun-pointing spacecraft, a quasi-inertial sun-pointing, pitch control concept was developed by Elrod [24] in 1972 and was further investigated in [25]. However, such a "free-drift" concept is not a viable option for the Abacus platform because of the large pitch attitude peak error of 18.8 deg and the inherent sensitivity with respect to initial phasing and other orbital perturbations.

Because the pitch gravity-gradient torque becomes naturally zero for cylindrical, spherical, or beam-like satellites with $I_1 = I_3$, a cylindrical configuration (Fig. 14.4) was also studied by NASA as a simple way to avoid such a troublesome pitch gravity-gradient torque problem.

14.4.2 Microwave and Solar Radiation Pressure Torques

The remaining contributions to disturbance torque come from solar radiation pressure force $\vec{f}_p$ applied at the center of pressure, as discussed in Sec. 14.3.3, and a force applied at the center of the microwave reflector in reaction to the discharge of the microwave beam. These disturbances, with ±20% overall uncertainties, are summarized in Table 14.4. Disturbance torque in units of N·m, caused by solar pressure, microwave radiation, c.m.-c.p. offset, and c.m.-c.p. offset uncertainty, can be expressed along the platform-fixed control axes as

$$d_1 \approx 12,000 - 11,900 \cos nt \qquad \text{(Roll)} \qquad (14.37a)$$

$$d_2 \approx 1200 \qquad \text{(Pitch)} \qquad (14.37b)$$

$$d_3 \approx -11,900 \sin nt \qquad \text{(Yaw)} \qquad (14.37c)$$

The constant pitch disturbance torque of 1200 N·m is small in comparison to the sinusoidal gravity-gradient torque and is caused by the assumed c.m.-c.p. offset of

Table 14.4 Solar pressure and microwave radiation disturbances

Disturbances	Values
Solar-pressure force	(4.5E-6)(1.3)(A) = 60 N
Solar-pressure torque (roll)	60 N × 200 m
Solar-pressure torque (pitch)	60 N × 20 m
Reflector radiation force	7 N (rotating force)
Reflector radiation torque	7 N × 1700 m

20 m along the roll axis; ±20% uncertainty in this disturbance model is assumed in control design.

It is assumed that the electric currents circulate in the solar-array structure in such a way that magnetic fields cancel out, and the Abacus platform is not affected by the magnetic field of the Earth.

14.5 Structural Control Issues

A significant problem with control-structure interaction is quite naturally a major concern in connection with a very large platform such as the 3.2 × 3.2 km Abacus platform, whose lowest structural mode has a frequency of about 0.0018 Hz as can be seen in Fig. 14.12. Dynamics and control of such large flexible structures has been investigated by many researchers in the past (for example, see

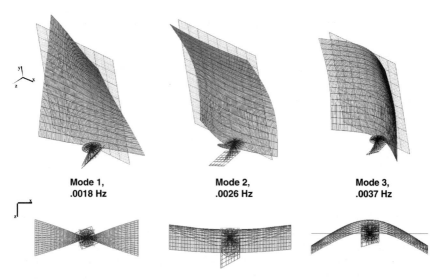

Mode 1,   Mode 2,   Mode 3,
.0018 Hz   .0026 Hz   .0037 Hz

Fig. 14.12   Flexible structural modes of the 3.2 × 3.2-km Abacus platform [7].

[26–28]), and active structural vibration control is a topic of continuing practical as well as theoretical interest.

The best way of dealing with such a problem is to avoid the conditions under which it occurs in the first place, an objective accomplished simply by employing a control bandwidth lower than 1×10^{-5} Hz in the systems that control the orbit and attitude. This is not to say that structural control of the Abacus platform is completely unnecessary, but the primary concern here is the preliminary design of a system for controlling the orbit and attitude without exciting structural vibrations. Although the problem of active structural control is not studied further in what follows, additional investigation of possible structural dynamic interaction between the transmitter and reflector in the Abacus configuration is certainly advised.

A low-bandwidth attitude control system is proposed in the following section; a concept of cyclic-disturbance accommodation control is utilized to provide the required ±5-arcmin pointing accuracy in the presence of large, but slowly varying, external disturbances and dynamic modeling uncertainties.

14.6   Attitude and Orbit Control

As demonstrated in Sec. 14.3.3, solar radiation pressure can, if left unopposed, cause a drift in the longitude of the Abacus platform of 0.3 deg per month, east or west. Current geosynchronous satellite control systems are perfectly capable of meeting a north–south and east–west stationkeeping requirement of ±0.1 deg and would suffice for the Abacus were it not for the unusually large perturbation from solar radiation pressure. To the extent that this perturbation can be eliminated through the use of thrusters, the major difference between the Abacus and current geosynchronous spacecraft vanishes, and the problem of stationkeeping is reduced to one that has already been solved. Hence, we concentrate in what follows on using ion thrusters with high specific impulse to control the secular growth in eccentricity and inclination exhibited in Fig. 14.11, rather than on design and performance of a traditional stationkeeping control system.

As discussed in Secs. 14.4.1 and 14.4.2, continuous sun tracking by the Abacus platform requires large control torques to counter various disturbance torques. Thus, the control system architecture proposed here utilizes properly distributed and oriented ion thrusters to negate the solar radiation pressure force; by applying this thrust in a cyclic fashion, the cyclic pitch gravity-gradient torque is opposed at the same time. Consequently, roll and pitch attitude control requires no additional expenditure of fuel over and above what is already required for orbit maintenance.

14.6.1   Electric Propulsion System

In principle, an electric propulsion system employs electrical energy to accelerate ionized particles to extremely high velocities, giving a large total impulse for a small consumption of propellant. In contrast to standard propulsion, in which the products of chemical combustion are expelled from a rocket engine, ion propulsion is accomplished by giving a gas, such as xenon (which is like neon or helium, but heavier), an electrical charge and electrically accelerating the ionized gas to a speed of about 30 km/s. When xenon ions are emitted at such high speed as exhaust from a spacecraft, they push the spacecraft in the opposite direction. Basic characteristics of an electric-propulsion system for the Abacus platform are summarized in Table 14.5.

The amount of propellant required for typical north–south and east–west stationkeeping maneuvers for the Abacus platform is estimated as

$$\Delta m = m_B \left[1 - \exp\left(-\frac{\Delta V}{g I_{sp}} \right) \right] \approx 30{,}000 \text{ kg/year}$$

where $m_B = 25 \times 10^6$ kg, $\Delta V = 50$ m/s per year ([16] Table 2.4 or [18] p. 106), $g = 9.8$ m/s^2, and $I_{sp} = 5000$ s. An estimate for the yearly propellant requirement to counter the solar-pressure force of approximately 60 N is given by

$$\Delta m = \frac{(60)(24 \times 3600 \times 365)}{5000 \times 9.8} \approx 40{,}000 \text{ kg/year}$$

and is associated with a ΔV of 76 m/s per year. Including 15,000 kg of reserve propellant, the estimated total yearly propellant consumption for orbit maintenance alone is 85,000 kg when using electric propulsion thrusters that have an I_{sp} of 5,000 s. Even though this amount of propellant is larger than the current annual worldwide production of xenon, about 40,000 kg, it must be remembered that the spacecraft has a total mass of 25×10^6 kg, so that the yearly stationkeeping propellant load is not at all unreasonable. If a way is found to construct a 3.2 × 3.2-km

Table 14.5 Electric-propulsion system for the 1.2-GW Abacus platform[a]

Parameters	Values
Thrust, T	≥ 1 N
Specific impulse, $I_{sp} = T/(\dot{m}g)$	$\geq 5{,}000$ s
Exhaust velocity, $V_e = I_{sp}g$	≥ 49 km/s
Total efficiency, $\eta = P_o/P_i$	$\geq 80\%$
Power/thrust ratio, P_i/T	≤ 30 kW/N
Mass/power ratio	≤ 5 kg/kW
Total peak thrust	200 N
Total peak power	6 MW
Total average thrust	80 N
Total average power	2.4 MW
Number of 1-N thrusters	≥ 500
Total dry mass	$\geq 75{,}000$ kg
Propellant consumption	85,000 kg/year

[a]Note: $T = \dot{m}V_e$, $P_o = \frac{1}{2}\dot{m}V_e^2 = \frac{1}{2}TV_e$, $P_o/T = \frac{1}{2}V_e =$ ideal power/thrust ratio, $P_i/T = \frac{1}{2\eta}V_e$, $I_{sp} = T/(\dot{m}g) = V_e/g$, $V_e = I_{sp}g$, where $g = 9.8$ m/s^2, $\dot{m}$ is the exhaust mass flow rate, P_i is the input power, and P_o is the output power.

platform in geostationary orbit, then supplying 85,000 kg of propellant each year and using 500 thrusters can certainly be considered feasible. The yearly propellant requirement is reduced to 21,000 kg if an I_{sp} of 20,000 s can be achieved (as was assumed for the 1979 SSPS reference system). As I_{sp} is increased, the propellant mass decreases, but the electric power requirement increases; consequently, the mass of solar arrays and power processing units increases. Based on a minimum of 500 1-N thrusters, a mass/power ratio of 5 kg/kW, and a power/thrust ratio of 30 kW/N, the total dry mass (power processing units, thrusters, tanks, feed systems, etc.) of an electric-propulsion system proposed for the Abacus platform is estimated as 75,000 kg.

The capability of present electric thrusters is orders of magnitude below that required for the Abacus platform. If the xenon fueled, 1-kW level, off-the-shelf ion engines available today are to be employed, the number of thrusters would be increased to 15,000. The actual total number of ion engines will further increase significantly when we consider the ion engine's lifetime, reliability, duty cycle, and redundancy.

For example, the 2.3-kW, 30-cm-diam ion engine of the Deep Space 1 spacecraft has a maximum thrust level of 92 mN. Throttling down is achieved by reducing the voltage and the amount of xenon propellant injected into the engine. Specific impulse ranges from 1900 s at the minimum throttle level to 3200 s.

14.6.2 Preliminary Control System Design

A preliminary control system architecture, shown in Fig. 14.13, utilizes properly distributed and oriented ion thrusters to counteract the solar radiation pressure force whose average value is about 60 N. The same thrusters are used to counter the secular roll torque caused by an offset of the center of mass and center of pressure; furthermore, the thrusters are employed in a cyclic fashion to oppose the cyclic pitch gravity-gradient and roll/yaw microwave radiation torques.

It is assumed that three-axis inertial attitude information in terms of quaternions will be available from an attitude determination subsystem, consisting of sun sensors, star camera/trackers, and inertial measurement units (IMUs). The IMUs contain rate gyros, and they are critical components of a spacecraft attitude determination subsystem.

The significant control-structure interaction problem, which is a major concern for such a very large Abacus platform with the lowest structural mode frequency of 0.0018 Hz, is simply avoided by designing an attitude control system with very low bandwidth (<orbit frequency). All structural modes are "gain-stabilized" by a large spectral separation between the rigid-body control and dominant flexible modes. A major drawback of such a low-bandwidth attitude control system is its low pointing performance in the presence of persistent disturbances.

However, the proposed low-bandwidth attitude control system utilizes a concept of cyclic-disturbance accommodating control to provide ±5 arcmin pointing in the presence of large, but slowly varying, external disturbances and dynamic modeling uncertainties. The concept of cyclic-disturbance accommodation control has been successfully applied to a variety of space-vehicle control problems, including the Hubble Space Telescope [29], the International Space Station [30, 31], and flexible space structures [32–34]. Detailed technical discussions of practical

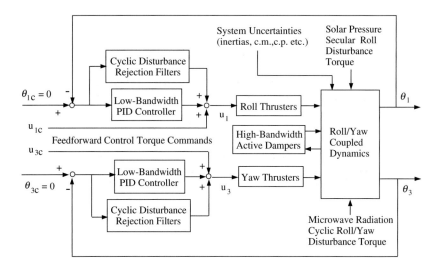

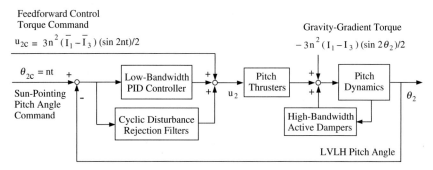

Fig. 14.13   AOCS architecture employing electric propulsion thrusters.

control system design for space vehicles in the presence structural flexibility as well as persistent external disturbances can be found in [13] and [29–34]. Thus, theoretical aspects of the control law design problem of the Abacus platform are not elaborated upon in this chapter. Even though the need for structural control is by and large eliminated by gain stabilization, it is of course prudent to anticipate the use of high-bandwidth, collocated direct velocity feedback, active dampers, properly distributed over the platform. Detailed design and analysis of damper performance has not been performed, although this should be undertaken in the future; therefore, Fig. 14.13 reflects the possible use of active dampers.

Although the roll/yaw motions are strongly coupled, a feedforward decoupling control approach provides a practical way of avoiding a time-varying, multivariable control design. A constant-gain, decoupled controller is designed for the following desired closed-loop poles (in units of the orbital rate, n) for each axis:

PID attitude controller:

$$-0.5 \pm 0.5j, \quad -0.2$$

Cyclic-disturbance rejection filters:

$$-0.2 \pm j, \quad -0.2 \pm 2j$$

For a standard PID control design, the integral control gain (or equivalently, the real eigenvalue associated with the integral control) is chosen as discussed on p. 125 of [13].

Placement of approximately 500 1-N electric propulsion thrusters at 12 different locations is illustrated in Fig. 14.14. In contrast to a typical placement of thrusters at the four corners, for example, employed for the 1979 SSPS reference system, the proposed placement shown in Fig. 14.14 minimizes roll/pitch thruster couplings as well as the excitation of platform out-of-plane structural modes. Thrusters #2 and #4 are 1600 m from the center of mass; therefore, approximately 90 N of thrust must

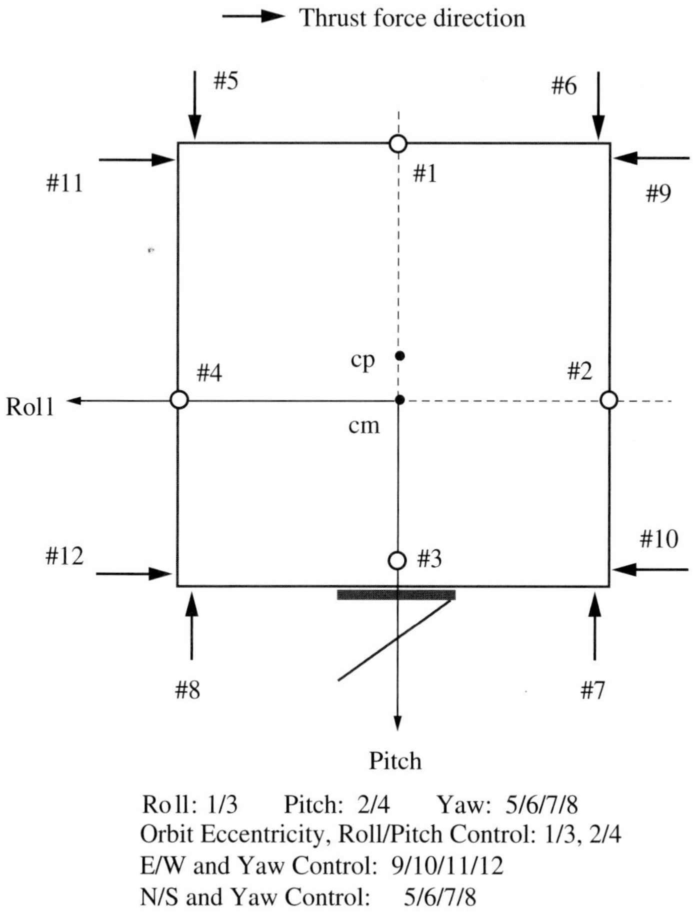
Thrust force direction
#5
#6
#11
#1
#9
cp
#4
#2
Roll
cm
#12
#3
#10
#8
#7
Pitch
Roll: 1/3 Pitch: 2/4 Yaw: 5/6/7/8
Orbit Eccentricity, Roll/Pitch Control: 1/3, 2/4
E/W and Yaw Control: 9/10/11/12
N/S and Yaw Control: 5/6/7/8

Fig. 14.14 Placement of a minimum of 500 1-N electric propulsion thrusters at 12 different locations, with 100 thrusters each at locations #2 and #4.

be applied at one of these locations in order to counteract a gravitational moment of 143,000 N·m as indicated in Eq. (14.35). A sinusoidal moment can be exerted if, for example, 1-N thrusters are used and 100 thrusters (to be conservative) are placed at each of locations #2 and #4. Likewise, approximately 15 1-N thrusters are required to offset the maximum roll torque from Eq. (14.37a), and a conservative number of 30 1-N thrusters is placed at each of the 10 locations #1, #3, and #5–12. Consequently, a minimum of 500 ion engines of 1-N thrust level are required for simultaneous attitude and orbit control. When reliability, lifetime, duty cycle, lower thrust level, and redundancy of ion engines are considered, this number will increase significantly.

Thrusters #1 and #3, which control roll attitude, and thrusters #2 and #4, which control pitch, all apply thrust in the positive yaw direction; that is, the thrust is directed toward the side of the array that faces the sun. With this arrangement, a cyclic pitch control torque is produced through the alternate use of thrusters #2 and #4; no matter which one is used, a force is produced perpendicular to the array, and thus the largest component of solar radiation pressure force is opposed at the same time that pitch attitude is controlled. Likewise, roll control also contributes to elimination of the solar radiation pressure force. During those times of the year when there is a component of solar radiation pressure force perpendicular to the equatorial (orbit) plane, thrusters #5–8 can be used to cancel it. Thus, *the propellant needed solely for orbit control is applied in such a way that attitude control is accomplished essentially at no cost.*

14.6.3　Attitude and Orbit Control Simulation Results

Attitude and orbit control system simulation results from an illustrative case involving a single rigid-body model, with 10-deg initial attitude errors in each axis, are shown in Figs. 14.15–14.18. Control gains are determined with the values given in Table 14.1 for the principal moments of inertia and c.m.-c.p. offset; however, the simulation is carried out with values that differ by ±20% in order to introduce intentional dynamic modeling errors. No active dampers were included for this simulation study. As can be seen in Fig. 14.16, the proposed low-bandwidth attitude control system, which effectively utilizes the concept of cyclic-disturbance accommodation control, maintains the required steady-state pointing of the Abacus platform in the presence of large, but slowly varying, external disturbances. In this example, the closed-loop system is shown to be insensitive to dynamic modeling uncertainties of ±20%.

Although the precision microwave beam pointing control by means of precision attitude control seems to be feasible as demonstrated in Fig. 14.16, there remain the challenging engineering problems of designing a very large microwave beam transmitting antenna and reflector system and characterizing its precision pointing performance. The rotating reflector concept of the Abacus platform eliminates massive rotary joint and slip rings of the 1979 SSPS reference concept; however, the lack of fine-pointing rotational joints attaching the reflector to the platform would result in very tight pointing requirements imposed on the platform itself. Consequently, further study needs to be performed for achieving the 5-arcmin microwave beam pointing accuracy in the presence of dynamic coupling between the transmitter and reflector, Abacus platform thermal distortion and vibrations, hardware constraints, and other short-term impulsive disturbances.

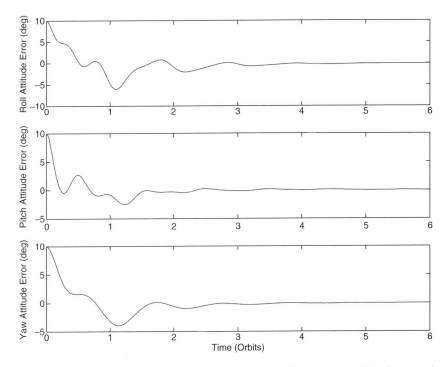

Fig. 14.15   Rigid-body control simulation results for the proposed attitude control system architecture.

The total thrusting force from the roll/pitch thrusters #1–4 nearly counters both the 60-N solar radiation pressure force and the cyclic pitch gravity-gradient torque; however, any residual ΔV caused by various dynamic modeling and control uncertainties should be corrected during standard north–south and east–west stationkeeping maneuvers.

As indicated in Eq. (14.35), the control torque needed to offset gravity-gradient torque in the pitch axis varies as $\sin 2nt$, and Fig. 14.17 shows that this is accomplished through alternate use of thrusters #2 and #4 to produce a force proportional to $|\sin 2nt|$. The total impulse over one orbit needed to counter the force of solar radiation pressure is $2\pi F_p/n$, which can be produced with a force whose magnitude is given by $(\pi F_p/2)|\sin 2nt|$ and whose direction is opposite to $\hat{s}$. In addition, north–south stationkeeping can be accomplished with thrusters #5–8 delivering a force of 0.1 N opposite to the spacecraft's out-of-plane displacement. All together, the thrusters contribute to $\vec{f}$ [see Eq. (14.2)] a force per unit mass

$$
\begin{aligned}
\vec{f}_p' &= -\left(\frac{\pi F_p}{2m_B}\right) |\sin 2nt|\,\hat{s} - \left(\frac{0.1}{m_B}\right)(\vec{r}\cdot\hat{e}_z)\hat{e}_z \\
&= -\left(\frac{\pi k}{2}\right)|\sin 2nt|\,\hat{s} - \left(\frac{0.1\vec{r}\cdot\hat{e}_z}{m_B}\right)\hat{e}_z
\end{aligned}
\tag{14.38}
$$

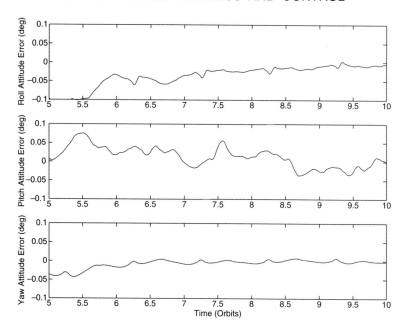

Fig. 14.16   Rigid-body control simulation results for demonstrating the steady-state pointing performance of meeting the 5 arcmin ($\approx$ 0.083 deg) pointing requirement.

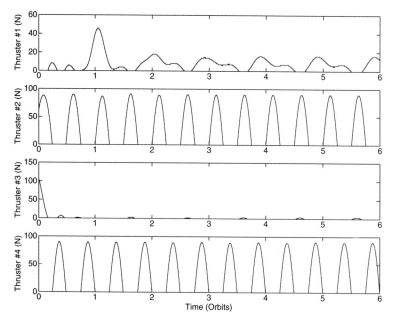

Fig. 14.17   Rigid-body control simulation results for the proposed attitude control system architecture.

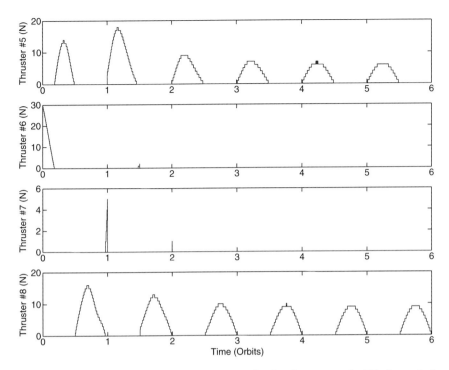

Fig. 14.18   Rigid-body control simulation results for the proposed attitude control system architecture.

The simulation described in Sec. 14.3.4 is performed again with this additional contribution, yielding the results displayed in Fig. 14.19, where it is evident that the secular growth in eccentricity and inclination have been eliminated. Compared to what is shown in Fig. 14.11, the remaining cyclic growth in eccentricity and inclination is smaller by factors of 20 and 25, respectively, thus returning the issue of stationkeeping to the realm of current solutions for geosynchronous spacecraft. As pointed out in Sec. 14.6.1, 85,000 kg of propellant per year are required for orbit maintenance.

The feasibility of using continuous (nonimpulsive) firings of ion thrusters for simultaneous attitude and orbit eccentricity control has been demonstrated for the Abacus platform; however, a further detailed orbit control study is needed. The precision orbit control problem of geosynchronous satellites is a topic of continuing practical interest, as discussed in [35–38].

14.7   Summary

This chapter, which is an adaptation of [1] and [2], has presented a preliminary AOCS architecture design for a very large, 3.2 × 3.2-km solar-array platform in geostationary orbit. It has shown that, in addition to the well-known significant

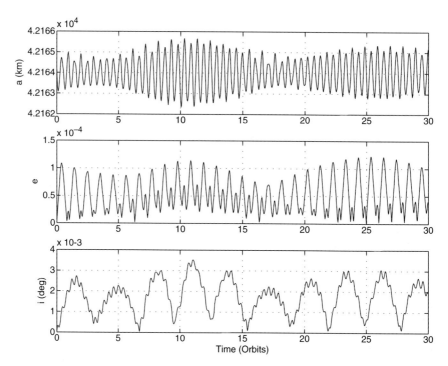

Fig. 14.19   Orbit simulation results with continuous (nonimpulsive) eccentricity and inclination control.

cyclic pitch gravity-gradient torque, a very large inertially oriented satellite in geo-stationary orbit is subject to a significant disturbance torque caused by solar radiation pressure, and the associated force produces considerable perturbation to the orbit. The Abacus platform has an area-to-mass ratio $A/m = 0.4$ m^2/kg that is very large compared to 0.02 m^2/kg, the value associated with contemporary, higher-density spacecraft. If left uncontrolled, 60 N of solar radiation pressure force can cause an eccentricity increase of 0.003 and a longitude drift of 0.3 deg per month.

Attitude dynamical equations of motion of an inertially oriented spacecraft, with constant moments of inertia but with time-varying external disturbances, have been developed and used for attitude control design and simulation.

Even though the Abacus platform is very large and its structural frequencies are low, a requirement for precision attitude control can be met with a control bandwidth that is much lower (<orbit frequency) than the lowest structural frequency of 0.002 Hz. Hence, the significant control-structure interaction problem, which can be a major concern for such a very large, flexible Abacus platform, appears to be a tractable one. The proposed low-bandwidth attitude control system effectively utilizes a concept of cyclic-disturbance accommodating control, successfully employed for attitude control and momentum management of the ISS, to provide precision pointing of the Abacus platform in the presence of dynamic modeling uncertainties and large, but slowly varying external disturbances.

Approximately 85,000 kg of propellant per year is required for simultaneous orbit and attitude control using a minimum of 500 1-N electric propulsion thrusters with $I_{sp} = 5000$ s. The total dry mass (power processing units, thrusters, tanks, feed systems, etc.) of an electric-propulsion system proposed for the Abacus platform is estimated as 75,000 kg.

The proposed AOCS architecture utilizes electric thrusters, properly distributed and oriented, to counteract the large solar radiation pressure force. Instead of using momentum exchange devices for attitude control, the same thrusters are used to counter the secular roll torque caused by an offset of the center of mass and center of pressure; furthermore, the thrusters are employed in a cyclic fashion to oppose the cyclic pitch gravity-gradient and roll/yaw microwave radiation torques. The resultant force and torque applied by the electric thrusters *simultaneously* counter the solar radiation pressure force and the cyclic external disturbance torques. This represents a fundamental and novel contribution of [1] and [2]: *the propellant needed solely for orbit control is applied in such a way that attitude control is accomplished essentially at no cost.*

The baseline control system architecture developed for the Abacus platform presupposes the availability of ion thrusters meeting certain specifications. Consequently, a 30-kW, 1-N level electric propulsion thruster with a specific impulse greater than 5000 s needs to be developed in order to avoid requiring an excessively large number of thrusters.

Several high-power electric-propulsion systems are currently under development. For example, the NASA T-220 10-kW Hall thruster provides 500 mN of thrust at a specific impulse of 2450 s and 59% total efficiency. Dual-mode Hall thrusters, which can operate in either high-thrust mode or high-I_{sp} mode for efficient propellant usage, are also being currently developed.

The exhaust gas from an electric propulsion system consists of large numbers of positive and negative ions that form an essentially neutral plasma beam extending for large distances in space. Because little is known yet about the long-term effect of an extensive plasma on geosynchronous satellites with regard to communications, solar-cell degradation, environmental contamination, etc., the use of lightweight, space-assembled large-diameter momentum wheels can also be considered as an option for the Abacus platform; therefore, these devices warrant further study.

The technology advances required for the Abacus platform are summarized here: electric thrusters–30 kW, 1 N, $I_{sp} > 5000$ s (500–1000 thrusters); CMGs—2000 N·m·s/kg, 500,000 N·m·s/unit; and space-assembled momentum wheels—66,000 N·m·s/kg, 4×10^8 N·m·s/unit (5–10 MWs). It is emphasized that both electric propulsion and momentum-wheel technologies require significant advancement to support the development of large SSPS.

Although the control-structure interaction problem appears to be tractable, further detailed study needs to be performed for achieving precision microwave beam pointing in the presence of structural dynamical coupling between the transmitter and reflector, Abacus platform thermal distortion and vibrations, hardware constraints, and other short-term impulsive disturbances.

The lack of fine-pointing rotational joints for attaching the reflector to the antenna would result in very tight pointing requirements imposed on such a very large platform. Further system-level tradeoffs will be required for design of the microwave-transmitting antenna and reflector, such as whether or not to gimbal the antenna with respect to the platform, use mechanical or electronic beam steering, or employ precision-pointing rotational joints for attaching the reflector to the antenna.

The very large SSP satellite concept might not be realizable in the near future; however, the AOCS architecture studied in this chapter will be applicable with minimal modifications to future space-based solar power satellites of practical sizes [39, 40].

References

[1] Wie, B., and Roithmayr, C. M., "Integrated Orbit, Attitude, and Structural Control Systems Design for Space Solar Power Satellites (SSPS)," NASA/TM-2001-210854, June 2001.

[2] Wie, B., and Roithmayr, C., "Attitude and Orbit Control of a Very Large Geostationary Solar Power Satellite," *Journal of Guidance, Control, and Dynamics*, Vol. 28, No. 3, 2005, pp. 439–451.

[3] Glaser, P. E., "Power from the Sun: Its Future," *Science*, Vol. 162, No. 3856, 22 Nov. 1968, pp. 857–861.

[4] Glaser, P. E., "The Potential of Satellite Solar Power," *Proceedings of the IEEE*, Vol. 65, No. 8, Aug. 1977, pp. 1162–1176.

[5] Mankins, J. C., "A Fresh Look at Space Solar Power: New Architecture, Concepts, and Technologies," International Astronautical Congress, IAF-97-R.2.03, Oct. 1997.

[6] Moore, C., "Structural Concepts for Space Solar Power Satellites," SSP Systems Workshop, Sept. 1999.

[7] Moore, C., "Structures, Materials, Controls, and Thermal Management," SSP Technical Interchange Meeting #3, June 2000.

[8] Carrington, C., Fikes, J., Gerry, M., Perkinson, D., Feingold, H., and Olds, J., "The Abacus/Reflector and Integrated Symmetrical Concentrator: Concepts for Space Solar Power Collection and Transmission," AIAA Paper 2000-3067, July 2000.

[9] Carrington, C., and Feingold, H., "Space Solar Power Concepts: Demonstrations to Pilot Plants," International Astronautical Congress, IAC-02-R.P.12, Oct. 2002.

[10] Battin, R. H., *An Introduction to The Mathematics and Methods of Astrodynamics*, AIAA, New York, 1987, pp. 389, 447–450.

[11] Roithmayr, C. M., "Contributions of Spherical Harmonics to Magnetic and Gravitational Fields," NASA, Paper EG2-96-02, Jan. 1996, URL: ftp://nssdcftp.gsfc.nasa.gov/models/geomagnetic/igrf/old_matlab_igrf/Contributions.pdf [retrieved 12 Jan. 2004].

[12] Marsh, J. G., et al., "A New Gravitational Model for the Earth from Satellite Tracking Data: GEM-T1," *Journal of Geophysical Research*, Vol. 93, No. B6, 10 June 1988, pp. 6169–6215.

[13] Wie, B., *Space Vehicle Dynamics and Control*, AIAA Education Series, AIAA, Washington, D.C., 1998, pp. 125, 233–239, 282–285, 290–292, 366–367.

[14] McInnes, C. R., *Solar Sailing: Technology, Dynamics and Mission Applications*, Springer-Praxis, Chichester, England, U.K., 1999, Chap. 2.

[15] Hughes, P. C., *Spacecraft Attitude Dynamics*, Wiley, New York, 1986, p. 262.

[16] Agrawal, B. N., *Design of Geosynchronous Spacecraft*, Prentice–Hall, Upper Saddle River, NJ, 1986, pp. 70–88.

[17] Chobotov, V. A. (ed.), *Orbital Mechanics*, 3rd ed., AIAA Education Series, AIAA, Reston, VA, 2002, pp. 223–226.

[18] Chao, C. C., and Baker, J. M., "On the Propagation and Control of Geosynchronous Orbits," *The Journal of the Astronautical Sciences*, Vol. XXXI, No. 1, 1983, pp. 99–115.

[19] Noton, M., *Spacecraft Navigation and Guidance*, Springer-Verlag, London, 1998, pp. 74, 170, 171.

[20] Bate, R. R., Mueller, D. D., and White, J. E., *Fundamentals of Astrodynamics*, Dover, New York, 1971, pp. 390–396.

[21] *The Astronomical Almanac for the Year 1999*, Nautical Almanac Office, United States Naval Observatory, U.S. Government Printing Office, Washinton, D.C., pp. D46, E4, E5.

[22] Roithmayr, C. M., "Gravitational Moment Exerted on a Small Body by an Oblate Body," *Journal of Guidance, Control, and Dynamics*, Vol. 12, No. 3, 1989, pp. 441–444.

[23] Oglevie, R. E., "Attitude Control of Large Solar Power Satellites," *Proceedings of the AIAA Guidance and Control Conference*, AIAA, New York, 1978, pp. 571–578.

[24] Elrod, B. D., "A Quasi-Inertial Attitude Mode for Orbiting Spacecraft," *Journal of Spacecraft and Rockets*, Vol. 9, No. 12, 1972, pp. 889–895.

[25] Juang, J.-N., and Wang, S.-J., "An Investigation of Quasi-Inertial Attitude Control for a Solar Power Satellite," *Space Solar Power Review*, Vol. 3, 1982, pp. 337–352.

[26] Reddy, A. S., Bainum, P. M., Krishna., R., and Hamer, H. A., "Control of a Large Flexible Platform in Orbit," *Journal of Guidance and Control*, Vol. 4, No. 6, 1981, pp. 642–649.

[27] Krishna, R., and Bainum, P. M., "Dynamics and Control of Orbiting Flexible Structures Exposed to Solar Radiation," *Journal of Guidance, Control, and Dynamics*, Vol. 8, No. 5, 1985, pp. 591–596.

[28] Rajasingh, C. K., and Shrivastava, S. K., "Orbit and Attitude Control of Geostationary Inertially Oriented Large Flexible Plate-Like Spacecraft," International Astronautical Congress, IAF Paper 86-245, Oct. 1986.

[29] Wie, B., Liu, Q., and Bauer, F., "Classical and Robust H_∞ Control Redesign for the Hubble Space Telescope," *Journal of Guidance, Control, and Dynamics*, Vol. 16, No. 6, 1993, pp. 1069–1077.

[30] Wie, B., Byun, K.W., Warren, W., Geller, D., Long, D., and Sunkel, J., "New Approach to Momentum/Attitude Control for the Space Station," *Journal of Guidance, Control, and Dynamics*, Vol. 12, No. 5, 1989, pp. 714–722.

[31] Wie, B., Liu, Q., and Sunkel, J., "Robust Stabilization of the Space Station in the Presence of Inertia Matrix Uncertainty," *Journal of Guidance, Control, and Dynamics*, Vol. 18, No. 3, 1995, pp. 611–617.

[32] Wie, B., "Active Vibration Control Synthesis for the COFS (Control of Flexible Structures) Mast Flight System," *Journal of Guidance, Control, and Dynamics*, Vol. 11, No. 3, 1988, pp. 271–276.

[33] Wie, B., Horta, L., and Sulla, J., "Active Vibration Control Synthesis and Experiment for the Mini-Mast," *Journal of Guidance, Control, and Dynamics*, Vol. 14, No. 4, 1991, pp. 778–784.

[34] Wie, B., "Experimental Demonstration of a Classical Approach to Flexible Structure Control," *Journal of Guidance, Control, and Dynamics*, Vol. 15, No. 6, 1992, pp. 1327–1333.

[35] Gartrell, C. F., "Simultaneous Eccentricity and Drift Rate Control," *Journal of Guidance and Control*, Vol. 4, No. 3, 1981, pp. 310–315.

[36] Kamel, A. A., and Wagner, C. A., "On the Orbital Eccentricity Control of Synchronous Satellites," *The Journal of the Astronautical Sciences*, Vol. XXX, No. 1, 1982, pp. 61–73.

[37] Kelly, T. J., White, L. K., and Gamble, D. W., "Stationkeeping of Geostationary Satellites with Simultaneous Eccentricity and Longitudinal Control," *Journal of Guidance, Control, and Dynamics*, Vol. 17, No. 4, 1994, pp. 769–777.

[38] Emma, B. P., and Pernicka, H. J., "Algorithm for Autonomous Longitude and Eccentricity Control for Geostationary Spacecraft," *Journal of Guidance, Control, and Dynamics*, Vol. 26, No. 3, 2003, pp. 483–490.

[39] "Space-Based Solar Power as an Opportunity for Strategic Security," Phase O Architecture Feasibility Study Report, National Security Space Office, 10 Oct. 2007; http://www.nss.org/settlement/ssp/library/nsso.htm.

[40] "Special Report on Space-Based Solar Power," *Ad Astra*, Vol. 20, No. 1, Spring 2008.

Index

Supporting Materials

Many of the topics introduced in this book are discussed in more detail in other AIAA publications. For a complete listing of titles in the Education Series, as well as other AIAA publications, please visit http://www.aiaa.org.